Design Hydrology and Sedimentology for Small Catchments

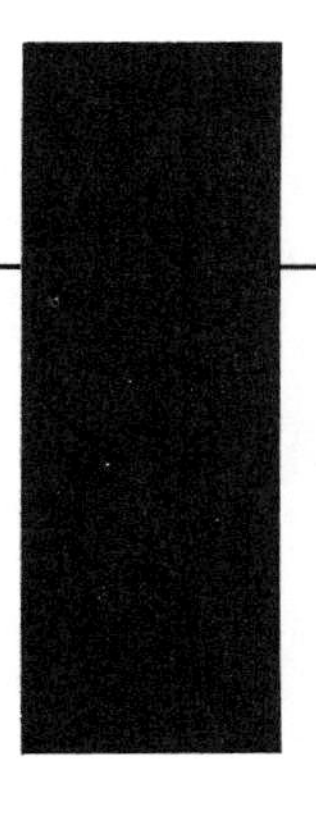

Design Hydrology and Sedimentology for Small Catchments

C. T. Haan
Biosystems and Agricultural Engineering Department
Oklahoma State University
Stillwater, Oklahoma

B. J. Barfield
Biosystems and Agricultural Engineering Department
Oklahoma State University
Stillwater, Oklahoma

J. C. Hayes
Agricultural and Biological Engineering Department
Clemson University
Clemson, South Carolina

Academic Press
A Division of Harcourt Brace & Company
San Diego New York Boston London Sydney Tokyo Toronto

This book is printed on acid-free paper. ♾

Academic Press, Inc.
525 B Street, Suite 1900, San Diego, California 92101-4495

United Kingdom Edition published by
Academic Press Limited
24–28 Oval Road, London NW1 7DX

Library of Congress Cataloging-in-Publication Data

Haan, C. T. (Charles Thomas), Date
Design hydrology and sedimentology for small catchments / C. T. Haan, B. J. Barfield, J. C. Hayes.
p. cm.
Includes bibliographical references and index.
ISBN 0-12-312340-2
1. Watershed management. 2. Sediment control. 3. Runoff. 4. Hydrology. I. Barfield, Billy J. II. Hayes, J. C. III. Title
TC409.H3 1993
627--dc20 93-11165
CIP

PRINTED IN THE UNITED STATES OF AMERICA
94 95 96 97 98 99 EB 9 8 7 6 5 4 3 2 1

Jan for 25 years of love and support;
Patti, Chris, and Pam for helping me grow;
Mom and Dad for getting me started.

CTH

Annette for being there for me for 25-plus years,
Michelle and Will for teaching me about myself,
Merrell and Ola for believing in me when it mattered.

BJB

Mary Anne for sharing my life and for showing love and faith,
Jay and Matthew for always making life exciting,
Mama and Daddy for providing strength and comfort.

JCH

To our Heavenly Father for His continual guidance.

Contents

8 Erosion and Sediment Yield

9 Sediment Control Structures

Fluvial Geomorphology: Fluvial Channel Analysis and Design

Ground Water

12

Monitoring Hydrologic Systems

13

Hydrologic Modeling

Chapter Appendixes

Preface

In recent years a number of excellent books on hydrology and related topics have been published. One might wonder about the need for yet another book on the topic. In the authors' experiences working with practicing engineers and hydrologists, we have found a definite need for a treatment of the design aspects of hydrology and sedimentology, especially for small catchments. Most practicing engineers and hydrologists work on relatively small watersheds, designing storm water control facilities, drainage facilities, erosion and sediment control practices, detention ponds, small channels and storm drains, and the like. This book attempts to provide a single source of design procedures for most aspects of runoff and sediment control in small catchments. Sections 208 and 319 of "The Clean Water Act" and their emphasis on storm water control in urban and rural areas have made the application of this technology imperative.

The approach used in the book is to present state-of-the-art design methodologies with enough explanation of basic principles to ensure understanding of the rationale behind the methodology. The mathematical and theoretical aspects are fully developed only when required for an understanding of the methodology. Adequate data are presented in tables and charts for many designs; however, the book does not attempt to replace design manuals currently being used by many local, state, and federal agencies. References to more extensive data tabulations are given where required.

The authors have taught basic hydrology and sedimentology courses to thousands of practicing engineers, as well as courses on the application of computer models to the analysis and design of hydrologic systems. It has been our experience that those who use computer models without an understanding of the theory and principles behind them do a poor job of applying them. Inappropriate designs are the frequent result. This book was written to provide a knowledge base for practitioners.

In practice, computer software would indeed be used to carry out many of the required computations. As microcomputers become more powerful, computer codes are being continually improved. User-friendly interfaces for computer programs are making it possible to use many hydrologic programs with little knowledge of the hydrology being simulated. This book provides the background required to understand most of the techniques used in current hydrology software and represents an excellent companion to program user manuals, which often contain almost no explanation of the hydrologic techniques being employed.

The book contains many solved example problems as well as numerous problems for solution at the end of each chapter. These problems will assist in developing a fuller understanding of design procedures. For use in classroom or continuing education settings, the problems can be easily adopted to local conditions by using local rainfall, soil, and other types of information.

This book has evolved over the past 15 years from a set of mimeographed notes used in continuing education courses for engineers interested in learning how to meet design requirements for permitting of areas to be surface mined. During its evolution it has appeared as two privately published books entitled "Hydrology and Sedimentology of Surface Mined Lands" by Haan and Barfield and "Applied Hydrology and Sedimentology for Disturbed Areas" by Barfield, Warner, and Haan. In the latter form it was widely used in the surface mining industry to design water and sediment control facilities. The current version is a complete rewrite of the previous texts in nearly all aspects. The material on erosion and sediment control presents extensive new technology that has evolved since the previous publication.

Chapter 1 provides an overview to the volume. Chapter 2 deals with hydrologic frequency analysis. Chapter 3 covers the estimation of runoff rates, volumes, and hydrographs. Chapters 4 and 5 deal with the hydraulics of open channel flow and hydraulic control structures. The design of channels in stable and erodible materials as well as the design of small hydraulic structures are covered here. Chapter 6 deals with flow routing in channels, ponds, and reservoirs. Chapters 7, 8, and 9 examine sediment properties and transport, the principles of erosion and sediment yield, and the design of practices to reduce erosion and control sediment. Chapter 10 discusses channel morphology and the natural equilibrium of erodible channels. Ground water is covered in Chapter 11, monitoring hydrologic systems in Chapter 12, and hydrologic modeling in Chapter 13. The appendices for all of the chapters are at the end of the book and contain design information too voluminous to include in the body of the text.

Every effort has been made to eliminate textual errors. In an undertaking of this magnitude, however, errors inevitably creep in. The authors would appreciate notification of any errors so that they may be corrected in future printings.

We must acknowledge the patience and support of our wives and families through the long process of bringing this book to fruition. Without this support, we would have been forced to abandon the project long ago. We are also grateful to a number of colleagues who have reviewed or discussed many aspects of the book. Among them are Bruce Wilson, Alex Fogle, Ron Elliott, Dan Storm, and Flint Holbrook. Graduate students in the Agricultural Engineering programs at Oklahoma State University, the University of Kentucky, and Clemson University have also contributed directly through their reading of parts of the book and indirectly through the stimulation they provided to the authors' research. A number of professionals who have commented on earlier versions of the work have also unknowingly made valuable contributions. Of course, we have benefitted from the intellectual atmosphere and the working conditions provided by our three universities. In particular, we are grateful to John Walker, who, as Chairman of the Agricultural Engineering Department at the University of Kentucky, encouraged and supported Tom Haan and Bill Barfield through the publication of the first volume. Finally, we would be remiss if we did not express appreciation to Charles Arthur of Academic Press for his patience with us as deadlines went unmet.

C. T. Haan
B. J. Barfield*
J. C. Hayes

* Formerly of the Agricultural Engineering Deparment, University of Kentucky, Lexington, Kentucky.

1 Introduction

THE PROBLEM

The human propensity to occupy areas subject to occasional flooding, to alter natural watercourses, to alter land forms, and to engage in other activities that impact natural hydrologic and sedimentologic processes creates a need to offset those impacts thought to be detrimental through designed flow and sediment control systems. Land clearing, agricultural activities, construction, mining, urban and industrial development, and similar activities can have a major impact on the quantity and rate of water runoff and on the rates of erosion and sediment transport that take place. Environmental concerns often make it desirable, if not necessary, to provide means of controlling runoff and erosion from altered land areas to the level that would be present if no alterations were made or to other legally specified levels.

Such an approach assumes that the hydrologic and sedimentologic responses of an area to climatic events can be quantified for both the unaltered and the altered state and that techniques for limiting the difference in responses to these two states are available. Further it assumes that quantitative methods for evaluating control techniques are available.

Figure 1.1, adapted from McBurnie *et al.* (1990), illustrates the impact of land use changes on peak runoff rates and sediment yields. This figure shows that as the land use changes from forest to a continuously disturbed condition, the sediment yield over a 2-year period goes from essentially nothing to about 1300 tons for this 20-acre catchment in Maryland having West Maryland silt loam soil on a 10% slope. No conservation practices were used except for the continuously disturbed condition, which, in some cases, used mulch and sediment detention ponds. The figure illustrates the effectiveness of sediment control practices in that the sediment yield was reduced on the continuously disturbed area from about 1300 tons with no control practices to about 350 tons with mulch and a sediment detention pond. Procedures presented in this book can be used to make comparative assessments like this for many different situations.

Environmental regulations present great challenges to engineers. They must design water and sediment control facilities that will enhance functional and aesthetic aspects of projects, will not hamper construction or operational activities, and yet will meet exacting state and federal regulations. Thus a detailed knowledge of the principles and practices of hydrology and sediment control is required.

SCOPE AND OBJECTIVES OF COVERAGE

This book has been written to acquaint engineers with hydrologic, hydraulic, and sedimentation principles that will be useful in designing water and sediment

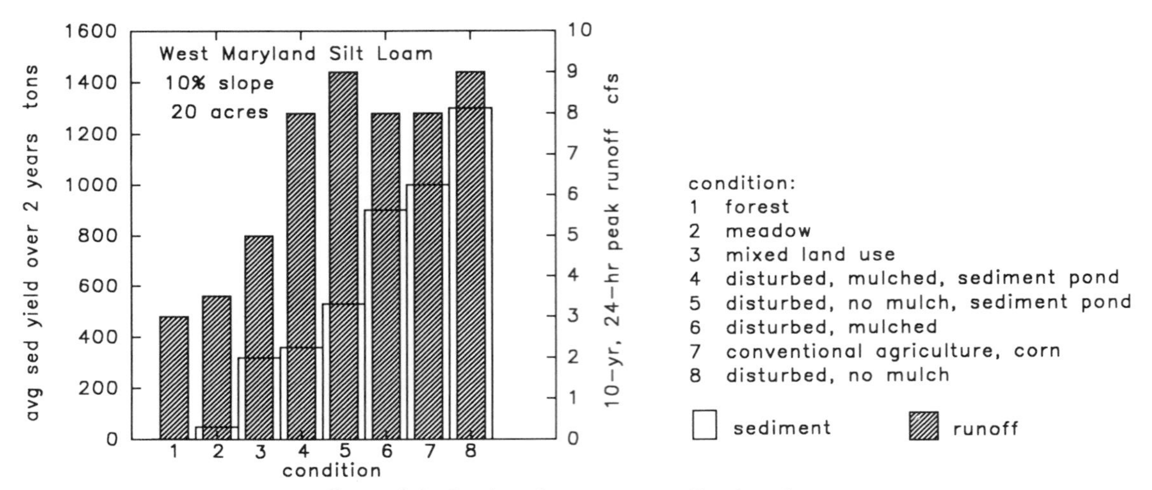

Figure 1.1 Land use impacts on runoff and erosion.

control facilities. Some of the material will be a review to some readers; however, nearly all engineers will find material in this treatment that supplements their current knowledge.

The treatment presented here is not theoretical; however, adequate theory is presented to develop a firm understanding of principles. The effort is directed toward practical design methodologies. The procedures are presented in sufficient detail so that the methods can be applied directly to actual situations. Numerous realistic, solved problems using the methods presented are included.

The book is not intended to replace or compete with federal and state publications regarding acceptable design procedures. It is not a design manual as such but presents design techniques that will apply in many situations. Often more than one solution technique may be possible and appropriate. The design engineer must select the best technique to use under a given set of circumstances.

To prevent the book from becoming excessively long, considerable material has been included by reference only. For instance, state manuals, U.S. Soil Conservation Service reports, U.S Corps of Engineers manuals, and U.S. Environmental Protection Agency publications contain much valuable information but are in themselves voluminous documents and thus not reproduced herein.

In view of the uncertain and dynamic nature of federal and state environmental regulations regarding water and sediment control from disturbed areas, it is not the purpose of this treatment to tell the engineer what must be done to comply with existing laws and regulations. Rather, the purpose of this treatment is to provide those in charge of sediment and water control with an understanding of:

1. hydrologic principles and techniques sufficient to estimate runoff rates, volumes, and hydrographs from a variety of scenarios;
2. open channel hydraulics in depth sufficient to design stable channels in erodible and nonerodible materials;
3. hydraulics in depth sufficient for the design of simple drop structures, pipe spillways, emergency spillways, and culverts;
4. flow routing in detail sufficient for the design of water retention and sediment detention basins;
5. the principles of sediment production and sediment control, including both structural and nonstructural methods;
6. principles for evaluating stable alluvial channels and for predicting the impact of changes in channel properties due to anthropogenic and natural changes;
7. basic definitions and principles of ground water hydrology;
8. requirements and techniques for monitoring hydrologic systems; and
9. the basis for and use of hydrologic models.

The selection of the notation to use in the book presented special problems since information from several disciplines including hydrology, hydraulics, sediment transport, erosion, geomorphology, and statistics is included. The decision was made to honor traditional notation in these various fields as much as possible to simplify supplementing the material from other reference sources. Thus the notation from chapter to chapter may differ depending on the topic under discussion. Within a chapter, a consistent notation was used if possible.

There are many worked problems throughout the text. Often the calculations for the problems were carried out using computer programs and spread sheets with the final results rounded to the number of significant figures shown in the problems after all calculations were completed. In some cases, this rounding at the end of the calculations produces slightly different results in the third significant figure than is obtained by hand calculations when rounding is done after each step in the calculation. This is especially noticeable when logarithmic or exponential relationships are involved.

GENERAL CONSIDERATIONS

Water and sediment management must be considered from the very beginning in developing plans for altering the physical setting of a drainage area. Certainly the final desired configuration of the area will play a dominate role in the design. The preoperation plan must consider such things as the installation of perimeter controls to prevent excessive water from entering the site and from leaving the site in an uncontrolled fashion while the site is undergoing change. The rate and extent of vegetation removal ahead of the operation, the placement of spoil and topsoil, and the amount of packing or sealing of the final graded fill must all be considered.

When developing plans, every opportunity should be taken to control water where it falls and sediment where it lies as this will generally be more effective, more permanent, and cheaper than control at some other point. Preventing erosion or accelerated runoff is preferred to trying to reduce runoff rates and sediment concentrations at later stages in the runoff process.

Consideration must be given to controlling sediment production on facilities constructed in conjunction with the site itself. For instance, haul and construction roads are major contributors of sediment and runoff. Likewise, access roads and construction activities around sediment detention basins, staging areas, and other high traffic areas often result in substantial sediment production.

Water control facilities such as channels, diversions and culverts must be properly designed and maintained. Improper design often results in failures caused by excessive flows or by erosion and sedimentation associated with normal flows. Inadequate consideration of the dissipation of energy at the outfall of a culvert can result in a scour hole and eventual failure of the culvert. Excessive flow velocities in channels and diversions can result in the formation of gullies. Inadequate cross-drainage on slopes, embankments, and haul roads may result in gullies. Delays in vegetating exposed slopes can result in substantial sheet and rill erosion, which, if unchecked, may lead to gullies.

It should be kept in mind that natural streams have developed over the centuries a state of dynamic equilibrium with the amount of sediment and water they carry. When this equilibrium is disturbed, the stream attempts to adjust to the new conditions. Thus, increased water and sediment loads may result in stream channel erosion in the form of bed and bank erosion. It may result in sediment deposition within the channel and thus a reduction in channel water-carrying capacity. It may result in a combination of these things depending on the local situation.

Effective water and sediment management is greatly aided by

1. preplanning of water and sediment control strategies;
2. installing diversions and vegetated waterways well ahead of the actual operation so that the vegetation can be established prior to disturbance;
3. keeping all reclamation activities current;
4. exposing and working as small an area at any one time as practical;
5. controlling water and sediment on the site;
6. using good engineering practice in designing water conveyances; and
7. having an aggressive maintenance program.

Complete preoperation investigation and planning are essential. Not only must the natural topography and drainage system be inventoried, but the desired final topography, drainage configuration, and land use anticipated. The amounts and types of spoil and soil material and where it is to be stockpiled or finally placed must be determined. The location of roadways and diversions must be specified. The nature and frequency of road culverts must be determined as well as the type of road ditches and erosion protection to be used in conjunction with these ditches.

The entire operation should be scheduled so that vegetated channels and diversions can be established before major land disturbances start and so that all reclamation activities can remain current. The amount of sediment production from an exposed site is somewhat proportional to the length of time the site is exposed. Thus, a slope exposed for 2 months will likely yield twice the sediment as one exposed for 1 month.

Water and sediment control practices should be installed at the problem location as much as possible. It is considerably cheaper and more permanent to do this as opposed to an attempt at a more downstream control. In general, water control is also an effective sediment control. Erosion tends to increase as the peak flow rate and the runoff volume increase.

Downstream sediment control measures, largely sediment basins, should be installed early before the operation begins. This gives time for a good job in constructing the facility and for stabilizing all slopes before the facility is asked to perform the duty for which it was designed. In some instances it may be necessary to introduce chemicals into the sediment–water mixture in order to cause the sediment particles to aggregate and thus be more easily removed by a sediment basin.

All water conveyances must be designed to have adequate capacity, to be stable over the range of flows under which they will be expected to function, and to have adequate energy dissipation.

It must be recognized that every hydrologic design is subject to the random vagaries of natural weather. Regardless of the design used, a certain level of probability exists that the design condition will be exceeded. Determining the acceptable risk of such a failure becomes a part of the design process.

Finally, importance of timely, effective, and routine maintenance cannot be overemphasized. The first sign of a developing gully or of scour around a structure is the sign for immediate and effective maintenance. The maintenance operations and procedures must go on for some time after the completion of the actual operation.

ACCEPTED DESIGN PRACTICE VERSUS STATE OF THE ART

Accepted design practices are those practices that have come into general usage because of their simplicity and relative accuracy. Accepted design practices are emphasized in this treatment. In new areas, sufficient time has not elapsed for accepted design practices to emerge. In such cases, state of the art procedures can be used. State of the art procedures are based on the latest research but are not yet in general practice. For example, for small detention structures accepted design practices are available for certain aspects such as flood retardation. However, accepted design procedures for the design of these basins based on detention time are not available. In this book, a state of the art procedure is proposed. Similarly, state of the art procedures for sediment yield and sediment and erosion problems are given.

Reference

McBurnie, J. C., Barfield, B. J., Clar, M. L., and Shaver, E. (1990). Maryland sediment detention pond design criteria and performance. *Appl. Eng. Agric.* **6**(2):167–173.

2

Hydrologic Frequency Analysis

In any discussion of hydrology one constantly hears such terms as the 100-year flood or the 50-year rainfall. Many times these terms are used rather loosely, and rarely are they understood by the layman. Frequently, the person using these terms does not fully appreciate their meaning, the implications associated with them, the difficulty of estimating the magnitude of events associated with the terms, and the uncertainty or variability of an estimate for the magnitude of an event associated with the terms.

Hydrologic literature is filled with discussions concerning flood frequency analysis. A review of this literature would require a book unto itself. A four-volume set of papers edited by Singh (1987a, b, c, d) provides a comprehensive treatment of many aspects of flood frequency analysis and provides references to hundreds of other works. What follows here is a basic treatment of frequency analysis and its application to flood flow estimation. A user of these techniques must keep in mind that the statistical techniques set forth are hydrologic tools and not hydrologic laws. The section "Discussion of Flood Frequency Determinations" appearing in this chapter should be read prior to the actual application of the techniques set forth. Haan (1977) can be consulted for a more detailed treatment of the application of statistics in hydrology.

Perhaps the most comprehensive study on flood flow estimation was conducted under the auspices of the Natural Environment Research Council (1975) of Great Britain. A five-volume set of reports details the study and the resulting recommendations. The procedures used and the general conclusions reached in that study are of general interest. The procedures and relationships will likely have to be adjusted for catchments outside the geographical region covered by the reports.

Throughout this chapter, a generalized notation is used to denote the events of interest. T-year event denotes an event with a return period of T years (return period is yet to be defined). Q_T denotes the magnitude of peak discharge of a T-year flood; Q_T is never known with certainty. One must always deal with an estimate for Q_T.

All of the statistical procedures, tables, and relationships that are used in this chapter are independent of the units employed. Thus any consistent set of units may be used. It does not matter if flows are in cubic feet per second, cubic meters per second, or acre-feet per day; the equations and tables in this chapter may be used without employing any conversion factors. For this reason, all of the example computations in this chapter are carried out using only one set of units.

RETURN PERIOD AND PROBABILITY

It is well known that maximum observed streamflow (the peak flow) observed on any stream over a period of 1 year varies from year to year in an apparently random fashion. This randomness has led to the use of probability and statistics in selecting the hydraulic capacity of storm water facilities. Reference should be made to Haan (1977) for a more complete treatment of this topic. The following is a generalized treatment of hydrologic frequency analysis.

A T-year event is formally defined as an event of such magnitude that over a long period of time (much much longer than T years), the average time between events having a magnitude greater than the T-year event is T years. Thus the expected number of occurrences of a T-year event in an N-year period is N/T. For example, Stillwater, Oklahoma, has a 25-year, 24-hr rainfall of 6.8 in. One would expect four occurrences of this 25-year event in a period of 100 years. In a 100-year record of annual maximum 24-hr rainfalls at Stillwater, the expectation is that in 4 of the years, the 24-hr maximum rainfall would exceed 6.8 in. This is another way of saying that on the average, one expects a T-year event to occur once every T years. It is to be emphasized that there is no regularity associated with a T-year event. It is not to be implied that a T-year event occurs once every T years, nor taken that in any T-year period there will always be one and only one occurrence of a T-year event, nor assumed that the T-year event will occur exactly N/T times in N years. These are the expectations in a statistical sense but are not certainties. In fact, later we show that there is a chance that in any T-year period, a T-year event can occur $0, 1, 2, \ldots, T$ times. Further, we show how to calculate the probabilities of these various possibilities.

The return period of a T-year event as defined above is T years. Often the actual time between occurrences of a T-year event is called the recurrence interval. The average value of recurrence interval is equal to the return period. Most discussions of return period and recurrence interval assume that the two terms are synonymous. In most instances, when one uses the term recurrence interval, the average recurrence interval is meant.

Since the average time between occurrences of a T-year event is T years, the probability of a T-year event in any given year is $1/T$. Thus we have the relationship

$$p_T = 1/T, \qquad (2.1)$$

where T is the return period associated with an event Q_T and p_T is the probability of Q_T in any given year. Probability is expressed as a number between 0 and 1 inclusively. For example, the probabilities associated with 10-, 25-, and 50-year events are 0.10, 0.04, and 0.02, respectively. A probability of 0 means that the event cannot happen, while a probability of 1 means the event will certainly happen. Sometimes probability is expressed as a percentage chance, in which case the true probability is multiplied by 100.

So far we have made several assumptions that must be emphasized. The assumptions involve the variable Q, the peak flow in any year. First, we have assumed that the peak flows from year-to-year are independent of each other. This means that the magnitude of a peak in any year is unaffected by the magnitude of a peak in any other year. Second, we have assumed that the statistical properties of the peak flows are not changing with time. This means that there are no changes going on within the watershed that result in changes in the peak flow characteristics of the watershed. It further means that the watershed characteristics have remained constant over the period of time producing the data we are using. In the language of statistics, we assume that the data are from a stationary time series.

RISK ANALYSIS

Under the assumptions set forth above, the occurrence of a T-year event is a random process meeting the requirements of a particular stochastic process known as a Bernoulli process. The probability of Q_T being exceeded in any year is p_T for all time and is unaffected by any prior history of occurrence of Q_T. Let us now denote any event exceeding Q_T as Q_T^*. We do not know the actual magnitude of Q_T^*; we know only that it exceeds Q_T ($Q_T^* > Q_T$). Q_T is a Bernoulli random variable. The probability of k occurrences of Q_T^* in n years can be evaluated from the binomial distribution[1]

$$f(k; p_T, n) = \frac{n!}{(n-k)!k!} p_T^k (1 - p_T)^{n-k}, \qquad (2.2)$$

where $f(k; p_T, n)$ is the probability of exactly k occurrences of Q_T^* in n years if the probability of Q_T^* in any single year is p_T. For example, the probability of two occurrences of a 20-year event ($p_T = 0.05$) in 30 years is

$$f(2; .05, 30) = \frac{30!}{28!2!} 0.05^2 0.95^{28} = 0.26.$$

The interpretation of this is that in a large number of 30-year records, we would expect 26% of the record to contain exactly 2 peaks that exceed Q_{20}. The other 74% of the 30-year records would contain $0, 1, 3, 4, \ldots,$ or 30 peaks that equal or exceed Q_{20}. The probabilities

[1] $n! = n(n-1)(n-2)\ldots 1;\ 0! = 1.$

of the latter number of exceedances can be evaluated from Eq. (2.2) also. If this is done, the summation of the probabilities of $0, 1, 2, 3, \ldots,$ 30 peaks in 30 years equal to or greater than Q_{20} must equal 1.00 since all possibilities have been exhausted.

Equation (2.2) can be used to calculate the probability that a T-year event will be exceeded at least once in an n-year period by noting that "at least once" means one or more. The probability of one or more exceedances plus the probability of no exceedances must equal 1.00. Therefore the probability of at least one exceedance is given by 1 minus the probability of no exceedances or

$$1 - f(0; p_T, n) = 1 - \frac{n!}{n!0!} p_T^0 (1 - p_T)^n.$$

Since $p_T = 1/T$ and $0! = 1$, this relationship reduces to

$$f(p_T, n) = 1 - (1 - 1/T)^n, \qquad (2.3)$$

where $f(p_T, n)$ is the probability that a T-year event will be exceeded at least once in an n-year period. If n is equal to T in Eq. (2.3), it can be shown that $f(p_T, T)$ approaches the constant 0.632 for large T (for $T = 10$, $f(p_T, T) = f(0.1, 10) = 0.65$). What this means is that if a structure having a design life of T years is designed on the basis of a T-year event, the probability is about 0.63 that the design capacity will be exceeded at least once during the design life.

By specifying the acceptable probability of the design capacity being exceeded during the design life of the structure, Eq. (2.3) can be used to calculate the required design return period. For example, if one wants to be 90% sure of not exceeding the design capacity of a structure in a 25-year period, $f(p_T, 25)$ would be $1 - 0.90 = 0.10$. Thus from Eq. (2.3),

$$0.10 = 1 - (1 - 1/T)^{25},$$

which can be solved to yield $T = 238$ years. To be 90% sure of not exceeding the design capacity in a 25-year period, the design capacity must be based on an event with a return period of 238 years. In this case, the acceptable risk was 10%, the degree of confidence was 90%, the design life was 25 years, and the required design return period was 238 years. Calculations like this can be carried out for various design lives, design return periods, and acceptable risks. Figure 2.1 is based on such calculations and can be used to quickly determine the required design return period based on the design life and acceptable risk or probability of having the design capacity exceeded.

In these discussions, it should be kept in mind that a high risk of having the design capacity exceeded may

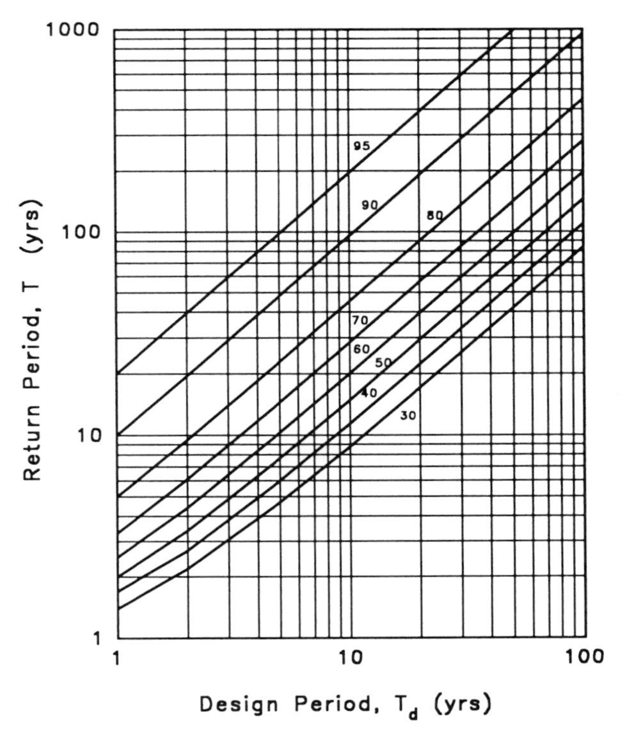

Figure 2.1 Design return period required as a function of design life to be given percentage confident (curve parameter) that the design condition is not exceeded.

be acceptable since what is meant by exceeded is failure of the structure to handle the resulting flow in the manner the structure was designed to operate. Failure in this sense does not necessarily mean that the structure will be destroyed. For example, the failure of a road culvert to pass a peak flow may result in only minor flooding of a roadway or adjacent area and may be acceptable on a fairly frequent basis. On the other hand, failure of a storm water detention basin may result in overtopping of the structure with considerable damage to property and high risk of loss of life downstream. Thus the selection of the acceptable risk and design return period depends on the consequences of the design capacity being exceeded. Building the structure large enough to protect against extremely rare events is quite expensive, while allowing the design capacity to be exceeded on a frequent basis may result in an accumulation of considerable economic loss. Thus, in addition to social and political considerations, the selection of the proper design return period is a problem in economic optimization.

Figure 2.2 illustrates the selection of a design return period based on economic optimization. The vertical scale contains average annual costs or benefits and the horizontal scale contains return period in years. Average annual costs should reflect all costs such as

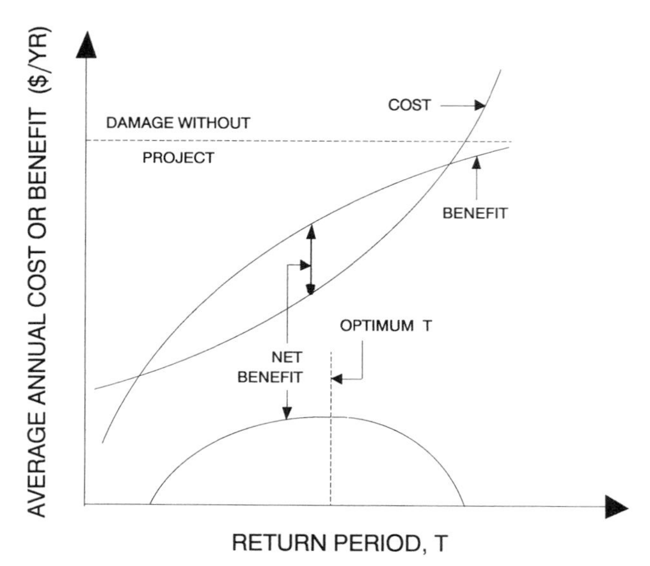

Figure 2.2 Determination of economically optimum design return period.

planning, right-of-way, construction, operation, and maintenance. The horizontal dashed line indicates the average annual damages that are occurring without the project. Obviously as the design return period increases, the design capacity would increase, and average annual costs would increase as well. Average annual benefits are taken as damages prevented on an average annual basis. As the capacity of the system increases, the level of protection against damage increases until presumably all damages could be eliminated. The net effect is that average annual benefits generally are low for low-return period designs and asymptotically approach a constant value as the design return period becomes very large.

The average annual net benefits is the difference in average annual costs and benefits. As shown in Fig. 2.2, a definite maximum average annual net benefit is apparent and represents the economically optimum design return period. These concepts are difficult to apply if damages include intangibles such as loss of life or destruction of nonreplaceable items.

One of the factors that inhibit the application of the economic approach to the selection of design return periods is the sensitivity of the approach to the interest rate used in the analysis. Many designers are reluctant to let the prevailing interest rate determine the capacity of a structure and thus its risk of failure.

Many governmental units have regulations governing the design period to be used. Often these return periods are based on the size of the structure and the consequences of the structural hydraulic capacity being exceeded. For example, in rural areas, road culverts might be based on a 10-year return period. Minor structures in urban areas might be based on the 25-year event, and major structures and flood plain delineations might be based on the 100-year event.

FREQUENCY DETERMINATIONS

Assigning a flood magnitude to a given return period requires knowledge of the flood flow characteristics of the basin of concern. The approach that is used to determine this relationship depends largely on the type, quantity, and quality of hydrologic data available and on the importance of the determination. If a minor culvert or channel is to be designed, one cannot justify a time-consuming, expensive flood frequency analysis. On the other hand, if a major component of a drainage system is under construction, the best possible flow estimates are desired.

In this treatment, five cases or situations a designer might be faced with are considered:

Case I: A reasonably long record of streamflow is available at or near the point of interest on the stream of interest.

Case II: A reasonably long record of streamflow is available on the stream of interest, but at a point somewhat removed from the location of interest.

Case III: A short streamflow record is available on the stream of interest.

Case IV: No records are available on the stream of interest, but records are available on nearby streams.

Case V: No streamflow records are available in the vicinity.

The cases are listed in the order they are considered. They are also listed in the order of increasing difficulty. Unfortunately, they are listed in the inverse order of their frequency of occurrence. That is, the designer is more likely to be faced with Case V than with Case I, especially for small watersheds. In spite of this, we devote a major part of our attention to the treatment of Case I, because it is essential that the Case I procedures and their limitations be understood before one can appreciate the problems associated with any of the other cases. The Case I analysis is basic to any flood frequency analysis. Case V is treated extensively in the next chapter.

Case I. Long Flow Record at Site

Several agencies of the United States Government sponsored a study to develop a uniform technique for flood frequency analysis. The result was a publication "Guidelines for Determining Flood Flow Frequency," Bulletin 17B (Interagency Advisory Committee on Water Data, 1981). Some of the material in this section comes from that report. This guide states, "Major problems are encountered when developing guides for flood flow frequency determinations. There is no procedure or set of procedures that can be adopted which, when rigidly applied to the available data, will accurately define the flood potential of any given watershed. Statistical analysis alone will not resolve all flood frequency problems."

If one is extremely fortunate, a relatively long record of peak flows may be available on the stream at the point where an estimate for a flood peak of a given frequency is desired. Such a listing might appear as in Table 2.1 for the Middle Fork of Beargrass Creek at Cannons Lane in Louisville, Kentucky. Any collection of data such as contained in Table 2.1 represents a sample of data from a population and under certain assumptions can be treated using probability and statistics. The population in this case would be the maximum annual flood peak for all time, both past and future. The data of Table 2.1 represents a sample from this population.

Quantities descriptive of a population are known as parameters. Population parameters are never known in a flood frequency study and must be estimated from the sample of data. Estimates of population parameters are known as sample statistics. Some parameters of interest are the mean, μ_X; the standard deviation, σ_X; the coefficient of variation, C_V; and the skewness, γ. Sample estimates for μ_X, σ_X, C_V, and γ are given by $\bar{X}$, S_X, $\hat{C}_V$, and C_S, respectively, and calculated from the equations

$$\bar{X} = \sum \frac{X_i}{n} \tag{2.4}$$

$$S_X = \sqrt{\frac{\sum X_i^2 - n\bar{X}^2}{n-1}} \tag{2.5}$$

$$\hat{C}_V = \frac{S_X}{\bar{X}} \tag{2.6}$$

$$C_S = \frac{n\sum(X_i - \bar{X})^3}{(n-1)(n-2)S_X^3} \tag{2.7a}$$

$$C_S = \frac{n^2\sum X_i^3 - 3n\sum X_i \sum X_i^2 + 2(\sum X_i)^3}{n(n-1)(n-2)S_X^3}, \tag{2.7b}$$

where X_i represents the ith data value, n is the sample size, and all summations are from 1 to n. Applying these equations to the Beargrass Creek data results in $\bar{X} = 1599$ cfs, $S_X = 1006$ cfs, $\hat{C}_V = 0.619$, and $C_S = 2.13$.

The mean is simply a measure of the central location of a group of data. The standard deviation is a measure of the spread of the data. The larger the standard deviation, the greater the spread in the data. The square of the standard deviation is known as the variance. The units on the standard deviation are the same as the units on the raw data. A dimensionless measure of the spread of a set of data is desirable so that comparisons of relative variability can be made among variables having widely differing means or among measures having different units. For example, the variance of peak flows in the Mississippi River is much greater than for a small tributary stream, yet the relative variability (relative to the mean) of flow in the tributary would be larger than that of the Mississippi. One such measure is the coefficient of variation which is the standard deviation divided by the mean. A compact data set will have a smaller coefficient of variation than will a wide ranging set of data.

The skewness is a measure of the symmetry of a distribution. The normal distribution has a skewness of zero. If the data tends to spread, or tail, to the right more than it does to the left with respect to its mean, the data are positively skewed and C_S will be positive. Data tailing to the left more than to the right is negatively skewed and C_S will be negative.

Equations (2.4) through (2.7) indicate that statistics are descriptive and not causal. Any statistical anlaysis is an attempt to describe, often in a probabilistic manner, the behavior of a set of data. Obviously this description must be updated as new information becomes available. As additional years of data are incor-

Table 2.1 Peak Discharge (cfs): Middle Fork, Beargrass Creek, Cannons Lane, Louisville, Kentucky

Year	Peak flow	Year	Peak flow	Year	Peak flow
1945	1810	1956	1060	1966	874
1946	791	1957	1490	1967	712
1947	839	1958	884	1968	1450
1948	1750	1959	1320	1969	707
1949	898	1960	3300	1970	5200
1950	2120	1961	2400	1971	2150
1951	1220	1962	976	1972	1170
1952	1290	1963	918	1973	2080
1953	768	1964	3920	1974	1250
1954	1570	1965	1150	1975	2270
1955	1240				

porated into an analysis, the statistics given by Eqs. (2.4) through (2.7) will change, and thus any flow estimates that depend on these statistics will change as well. In general, as the number of observations increases, the statistics become better estimates of the population parameters.

If data such as contained in Table 2.1 meet certain assumptions, we can consider them to be independent random variables and subject them to a frequency analysis. The main assumptions are that the data are independent of each other and are from a stationary time series. A stationary time series is a data series collected over time and having statistical properties that do not change over time.

In hydrologic terms, the statistical assumptions require:

1. There are no trends in the data.
2. The data represent independent hydrologic events.
3. There is one underlying meteorologic/hydrologic cause for the flows so that the flows can be assumed to be from a single population.
4. Measurement errors are random, unbiased, and have a relatively small variance.

Trends in data may be caused by climatic shifts, natural events, or human activities. Hydrologists generally consider the time scale of climatic change to be vary large in comparison to the period of concern in any analysis and thus do not consider possible climatic shifts. Major natural events such as earthquakes, land slides, and forest fires can cause changes in the hydrologic regime of a catchment and thus introduce nonhomogeneity into the flow record. If such is the case, this nonhomogeneity must be dealt with prior to any frequency analysis.

The most common cause of changes in the flow regime of a catchment is human activity. This may be in the form of land-use changes such as urbanization, deforestation, or surface mining activities. It may be in the form of reservoir construction, stream diversions, or channel work. Sometimes these changes are sudden and easily detected. Closure of a major reservoir can have immediate and obvious impacts on flood flow magnitude. Sometimes the changes are gradual. Urbanization may show up in the data as a trend toward higher peak flows. Generally trends of this type are difficult to detect over a short period of time due to the random nature of flood flows. If changes of this type are present, the data must be adjusted for the changes before they can be treated in a straightforward way using statistics. References can be made to Haan (1977) for possible ways to adjust for nonhomogeneity in the flow record.

In some locations, flood flows may be the result of two distinct meteorologic causes. For example, winter flows may be the result of frontal storms and summer flows may be the results of convective thunderstorms. If this type of nonhomogeneity is present and flows can be easily divided into two groups according to the storm type, it may be desirable to treat the two storm types separately and then combine the results probabilistically. The problems with this approach are the difficulty of actually dividing the flows along causative lines, and the length of record available in each part of the divided record may be too short to provide reliable estimates of the required statistical parameters. Haan (1977) can be consulted for more details on the use of mixed populations and mixed distributions in flow frequency analysis.

In any data analysis, measurement errors are of concern. In flood frequency determinations, it is generally assumed that the data are measured without error. If actual measurement errors are independent from one measurement to another, tend to overestimate flow as well as underestimate flow so as to have a mean error of zero, and are small in comparison to the flow itself, the assumption of no measurement errors is generally acceptable from a hydrologic standpoint. Obviously if measurements always produce low estimates or high estimates and/or are grossly in error, any analysis based on the data will be in error as well.

Two types of data series are commonly used in flow frequency analysis—the annual series and the partial duration series sometimes known as the "peaks over threshold" series. In the annual series, the data consist of the largest observed peak flow for each year of data. For the partial duration series, the data consists of all peak flows greater than some base or threshold value. The annual series produces one data value per year. The partial duration series may produce none, one, or more than one data value in any year depending on the flows for the year and the magnitude of the threshold value.

For return periods greater than about 10 years, the return period flow estimate for the two series are practically the same. For more frequent but smaller floods, the relationship between the estimates from the two series is somewhat dependent on the probability distribution selected. In this treatment, the annual series is used.

Probability Plotting

Summarizing the data in the form of a probability plot is often the first step in a frequency analysis. An intuitive estimate for the magnitude of frequent floods on Beargrass Creek can be made based on our under-

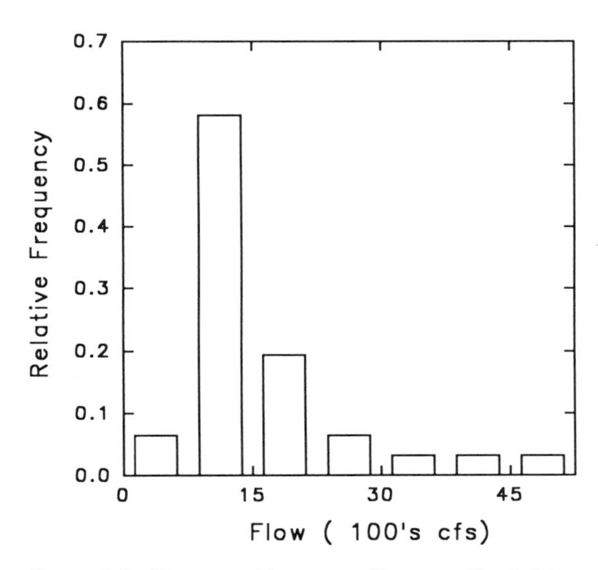

Figure 2.3 Frequency histogram—Beargrass Creek data.

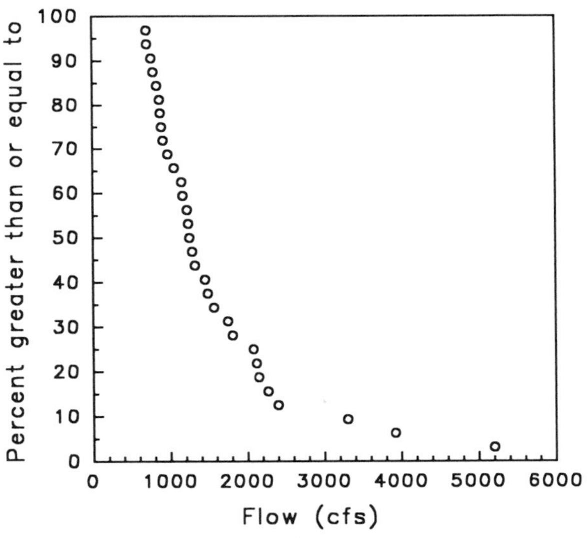

Figure 2.4 Empirical flood frequency—Beargrass Creek data.

standing of the concept of return period. For example, the 5-year flood is one that is equaled or exceeded on the average once every 5 years or about 20% of the time. Looking at Table 2.1, we see that about 20% (six events) of the peaks exceed 2120 cfs. Therefore, we might estimate the magnitude of the 5-year flood as 2120 cfs. Similarly 10% of the flows exceed 2400 cfs so we can estimate the 10-year event as 2400 cfs.

A difficulty with the intuitive approach to flood frequency analysis is that the magnitude of events having return periods longer than the length of the available record cannot be estimated. Also the magnitude of events having return periods close to the record length is dependent on very few observations and is thus somewhat uncertain. For example, the 10-year event in the above example depends on only three observations. What is needed is a procedure for using all of the data to describe the probabilistic nature of the peak flows. A start in this direction can be made by plotting the data in the form of a frequency histogram. This is merely a plot of the frequency of occurrence of peak flows in some class interval versus the class interval. Figure 2.3 is such a plot using a class interval of 750 cfs. Similarly a plot of the percentage of the values greater than or equal to a given value versus the magnitude of the value can be made. Figure 2.4 is a plot of this nature for the Beargrass Creek data. From Fig. 2.4, the magnitude of the 5-year flood ($p = 1/T = 1/5 = 0.20$ or 20% chance of occurrence) can be estimated as about 2150 cfs and the 10-year flood (10% chance of occurrence) is about 3250 cfs.

When considerable data are available, this is a reasonable procedure to use for estimating low-return periods floods. Inspection of Fig. 2.4 shows that the data exhibit some "roughness" and that perhaps a better estimate for low-return period floods could be obtained by drawing a smooth curve through the data and then using the curve to define the magnitude of floods with various return periods.

Unfortunately a plot such as Fig. 2.4 is generally not sufficient for estimating the magnitude of a longer return period flood. For example, the 25-year flood can be determined from Fig. 2.4 by reading the smooth curve at the 4% point. This is not a very reliable estimate, however, because it depends almost entirely on the magnitude of the two largest events in the record. If the largest flood event in the record had been 7000 or 4200 cfs or some other value, this would have greatly altered our estimate for the 25-year flood.

Furthermore, the estimation of a 100-year flood based on these data requires the smooth curve be extrapolated to the 1% point. This extrapolation, and indeed the entire smooth curve, would be extremely dependent on the whims of the individual doing the extrapolation. Different individuals would estimate different values for the 100-year flood, and the values could differ by 50% or more.

What is needed is an analytic method for placing a curve through the plotted points. This analytic curve could then be used to estimate the magnitude of floods with various return periods. Before discussing analytic techniques for flood frequency analysis, the matter of plotting random data (flood peaks) requires further attention.

The procedure arrived at in preparing Fig. 2.4 results in the point 707 cfs being plotted at the 100% point. This is equivalent to stating that 100% of all annual flood peaks on this stream will be greater than 707 cfs. Even though this is true for the particular 31-year record that is available, we do not know that it is true for all time and would suspect that there is a chance that in some future year an annual peak of less than 707 cfs might occur. Thus we would like to avoid assigning a 100% chance or probability of 1 to any event.

A second consideration in plotting flood peaks against probability is that when arithmetic graph paper is used as in Fig. 2.4, the points generally form an extremely curved pattern with the larger floods widely spaced. To overcome this inconvenience, special paper known as probability paper has been developed. Several kinds of probability paper are available. The most widely available are normal probability paper and lognormal probability paper. Lognormal probability paper is used in this treatment.

The steps to be followed in plotting random data on probability paper are to

1. rank the data from the largest to the smallest;
2. calculate the plotting position, p, based on the rank, m, and the number of years of data, n; and
3. plot the observation on probability paper with p along the probability scale and magnitude along the variable scale.

Several plotting position relationships are in use. A general relationship is

$$p = \frac{m - a}{n - a - b + 1},$$

where a and b are constants. The California (California State Department or Public Works, 1923) plotting position is $p = m/n$. The Hazen (1930) relation is $p = (2m - 1)/2n$. The Natural Environment Research Council (1975) of the United Kingdom used $p = (m - 0.44)/(n + 0.12)$. The most widely used relationship in the U.S. is the Weibull (1939) relationship given by

$$p = \frac{m}{n + 1}. \tag{2.8}$$

As an example of probability plotting, consider the Beargrass Creek data. These data are ranked and the plotting positions determined in Table 2.2. Figure 2.5 is a plot of the data on lognormal probability paper. Since the data were ranked from the largest to the smallest, the plotting position, p, represents the fraction of the values greater than or equal to the corresponding value of the data. The data do not plot as a straight line on lognormal paper, but the curvature is greatly reduced over that shown in Fig. 2.4.

Table 2.2 Plotting Position: Middle Fork, Beargrass Creek, Cannons Lane, Louisville, Kentucky

Year	Discharge	Rank	Plotting position
1945	1810	9	0.281
1946	791	28	0.875
1947	839	27	0.844
1948	1750	10	0.313
1949	898	24	0.750
1950	2120	7	0.219
1951	1220	18	0.563
1952	1290	15	0.469
1953	768	29	0.906
1954	1570	11	0.344
1955	1240	17	0.531
1956	1060	21	0.656
1957	1490	12	0.375
1958	884	25	0.781
1959	1320	14	0.438
1960	3300	3	0.094
1961	2400	4	0.125
1962	976	22	0.688
1963	918	23	0.719
1964	3920	2	0.063
1965	1150	20	0.625
1966	874	26	0.813
1967	712	30	0.938
1968	1450	13	0.406
1969	707	31	0.969
1970	5200	1	0.031
1971	2150	6	0.188
1972	1170	19	0.594
1973	2080	8	0.250
1974	1250	16	0.500
1975	2270	5	0.156

At this point a smooth curve can be sketched through the data or we can use analytical methods to "fit" a line through the points. In this latter approach, an equation having unknown parameters is used to describe the data much like the straight line $y = a + bX$ is fitted through plotted points on regular graph paper.

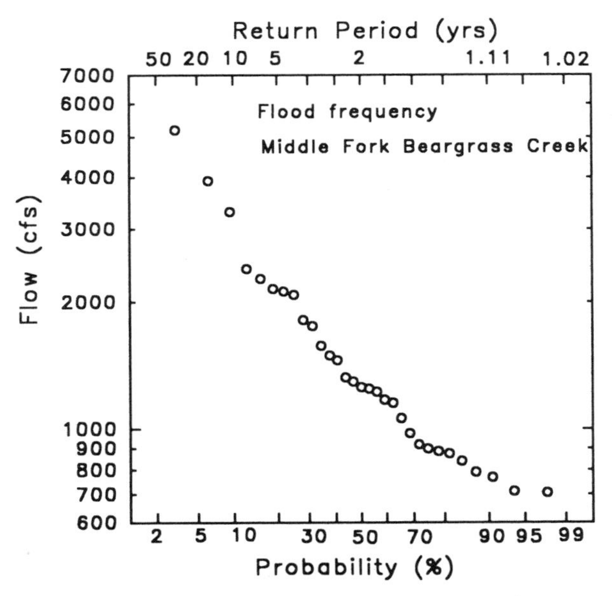

Figure 2.5 Probability plot—Beargrass Creek data.

The difficulty we now face is selecting the "equation" to use and in estimating the parameters of this equation.

Probability Distributions

Equations for describing the probability of occurrence of random events are known as probability density functions (pdf) and cumulative distribution functions (cdf). A pdf can be used to evaluate the probability of a random event in a specified interval. A cdf can be used to evaluate the probability of an event being equal to or less than a given value. We use the notation $p_X(x)$ and $P_X(x)$ to denote the pdf and cdf of the random variable X evaluated at $X = x$. These two are related by

$$P_X(x) = \int_{-\infty}^{x} p_X(x)\, dx. \tag{2.9}$$

There are a limitless number of functions that can be used for pdf's. Requirements for a function to be a pdf are

$$p_X(x) \geq 0 \qquad \text{for all } x$$

$$\int_{-\infty}^{\infty} p_X(x)\, dx = 1.$$

Pdf's may take on any number of shapes. The most familiar is the bell-shaped curve of the normal probability density function shown in Fig. 2.6. The normal

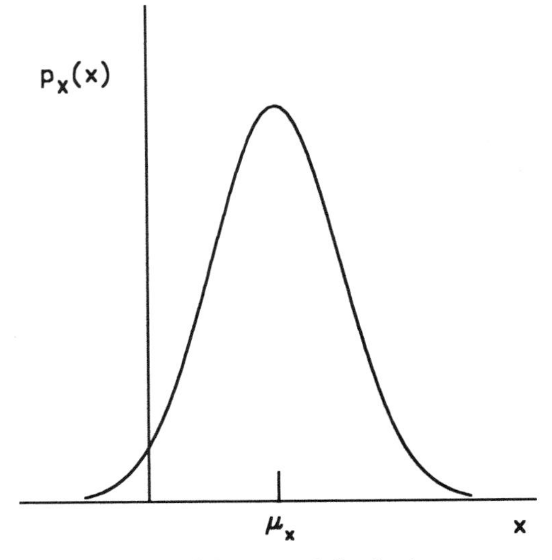

Figure 2.6 Normal distribution.

pdf is given by

$$p_X(x) = \frac{1}{\sigma_X\sqrt{2\pi}} \exp\left[\frac{-(x-\mu_X)^2}{2(\sigma_X)^2}\right]. \tag{2.10}$$

The normal distribution is symmetrical about the mean μ_X and ranges from $-\infty$ to ∞. The normal distribution is generally not used in flood frequency determinations because it permits negative values and because flood frequency distributions are generally not symmetrical. For example, the Beargrass Creek data in Fig. 2.3 exhibit a pronounced tailing off to the right, which is typical of flood peak data. Even though the normal distribution is generally not used in flood frequency analyses, we continue to consider it since an understanding of it is essential for statistical work.

The cdf of the normal distribution is

$$P_X(x) = \int_{-\infty}^{x} \frac{1}{\sigma_X\sqrt{2\pi}} \exp\left[\frac{-(x-\mu_X)^2}{2(\sigma_X)^2}\right] dx, \tag{2.11}$$

which gives the probability that $X \leq x$.

$$P_X(x) = \text{prob}\,(X \leq x).$$

The probability that X is between a and b can be evaluated from

$$\begin{aligned} \text{prob}(a \leq X \leq b) &= \text{prob}(X \leq b) - \text{prob}(X \leq a) \\ &= P_X(b) - P_X(a) \\ &= \int_a^b p_X(x)\, dx. \end{aligned} \tag{2.13}$$

The normal distribution is a two-parameter distribution with the parameters being μ_X, the mean of X and

σ_X, the standard deviation of X. For any application of the normal distribution, we must estimate the population mean and standard deviation, μ_X and σ_X, by their sample estimates, $\overline{X}$ and S_X.

Using Eqs. (2.4) and (2.5), the mean and standard deviation of the Beargrass Creek data are found to be 1599 and 1006 cfs, respectively. Now *if* the normal distribution was an adequate representation of the Beargrass Creek data, it could be used to make probabilistic statements concerning the data. For example, the probability of a peak less than or equal to 2500 cfs could be evaluated as

$$\text{prob}(Q \leq 2500) = P_Q(2500)$$

$$P_Q(q) = \int_{-\infty}^{2500} \frac{1}{1006\sqrt{2\pi}} \exp\left[\frac{-(x-1599)^2}{2(1006)^2}\right] dx. \tag{2.14}$$

Unfortunately this latter expression cannot be analytically evaluated, and numerical procedures must be used. To overcome the problem of requiring a separate numerical integration for the normal distribution for every possible combination of the parameters μ_X and σ_X, a transformation of variables is defined as

$$Z = \frac{(X - \mu_X)}{\sigma_X}. \tag{2.15}$$

Z is called a standardized normal variable and has the property that $\mu_Z = 0$ and $\sigma_Z = 1$. The expression

$$p_Z(z) = \frac{e^{-z^2/2}}{\sqrt{2\pi}} \tag{2.16}$$

is known as the standard normal distribution. Equation (2.14) can now be evaluated as

$$\text{prob}(Q \leq x) = \text{prob}(Z \leq (x - \mu_X)/\sigma_X) \tag{2.17}$$

or

$$\begin{aligned}\text{prob}(Q \leq 2500) &= \text{prob}(Z \leq (2500 - 1599)/1006)\\ &= \text{prob}(Z \leq 0.896)\\ &= \int_{-\infty}^{0.896} \frac{e^{-z^2/2}}{\sqrt{2\pi}}\, dz.\end{aligned}$$

The latter expression can be evaluated using tables of the standard normal distribution such as that contained in Appendix 2A. Appendix 2A gives the desired probability, prob(($Z \leq 0.896$), as 0.814. This corresponds to the prob($Q \leq 2500$).

The interpretation of this calculation is that *if* the flood peaks on Beargrass Creek can be described by a normal distribution with a mean of 1599 cfs and a standard deviation of 1006 cfs, then 81.4% of the annual peaks should be less than 2500 cfs and 18.6% of the annual peaks should be greater than 2500 cfs. Thus 2500 cfs is assigned a return period of 1/0.186 or 5.4 years under the normality assumption. Under this assumption, we expect 2500 cfs to be exceeded on the average once every 5.4 years. The data tabulation actually shows that 28 of the 31 values or 90.3% are less than or equal to 2500 cfs.

Looking back at Eq. (2.13), it is apparent that prob($a \leq X \leq b$) is the area under the pdf, $p_X(x)$, between $X = a$ and $X = b$. Thus, the probability of a random observation falling in the interval a to b is the area under the pdf between a and b. In a sense, the relative frequency histogram of Fig. 2.3 gives similar information. Based on the data in hand, we would estimate for example, that the probability that a random annual peak would fall in the interval 1500 to 2250 cfs is 0.190. There is apparently a relationship between relative frequency and probability. Denote by $f_X(x_i)$ the relative frequency of observations in an interval of width Δx centered on x_i. The probability of an observation falling in this interval is

$$\begin{aligned}&\text{prob}(x_i - \Delta x/2 \leq X \leq x_i + \Delta x/2)\\ &\quad= \int_{x_i - \Delta x/2}^{x_i + \Delta x/2} p_X(x)\, dx,\end{aligned} \tag{2.18}$$

which is the area under $p_X(x)$ between $x_i - \Delta x/2$ and $x_i + \Delta x/2$. This area can be approximated by $\Delta x p_X(x_i)$, which is the width of the interval times the height of $p_X(x)$ evaluated at x_i (Fig. 2.7).

Therefore, the relationship between the relative frequency of observations in an interval Δx and the pdf is

$$f_X(x_i) = \Delta x p_x(x_i). \tag{2.19}$$

As a side note, since probability is related to the area under the pdf, it is apparent that prob($X = x$) for a continuous random variable must be zero since

$$\text{prob}(X = x) = \int_x^x p_X(x)\, dx. \tag{2.20}$$

We can use Eq. (2.19) and Fig. 2.3 to visually judge the appropriateness of using the normal distribution to describe the Beargrass Creek data. Table 2.3 shows, under the assumption of a normal distribution, the observed and expected frequency of observations in several classes. The data are plotted in Fig. 2.8. Entries in the expected relative frequency column of Table 2.3 are based on Eq. (2.19) and the normal distribution.

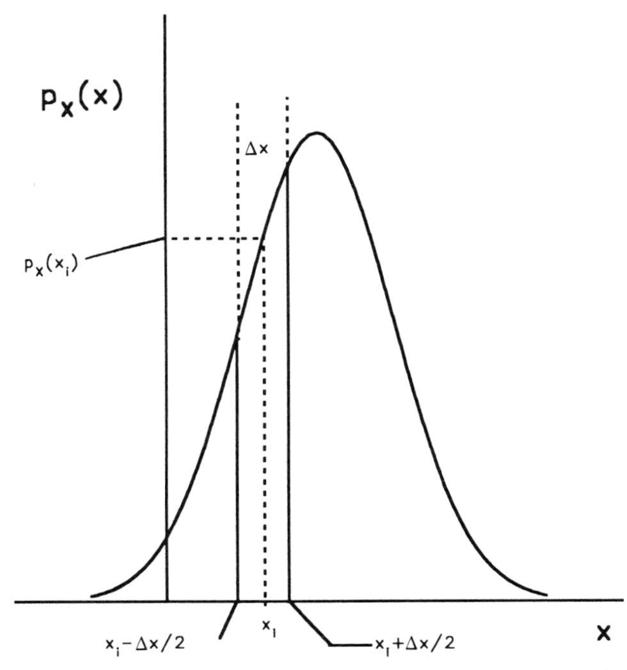

Figure 2.7 Calculation of prob($x_i - \Delta x/2 < x < x_i + \Delta x/2$).

For example, for the second class

$$f_X(1125) = 750 \frac{1}{1006\sqrt{2\pi}} \exp\left[\frac{-(1125 - 1599)^2}{2(1006)^2}\right]$$

$$= 0.267.$$

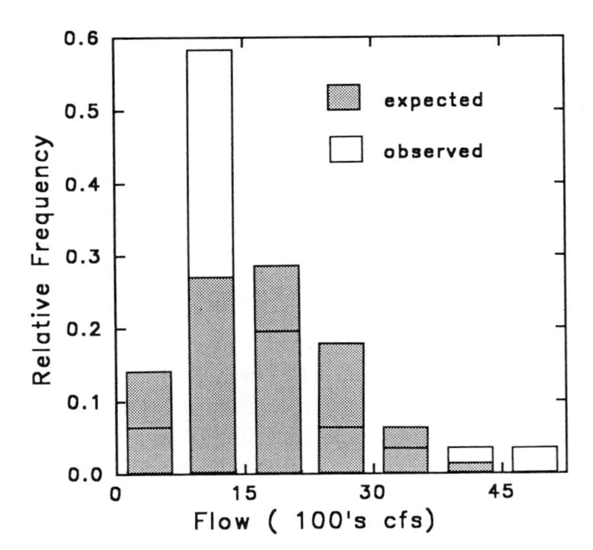

Figure 2.8 Comparison of observed and expected flow frequency (under assumption of normal distribution)—Beargrass Creek data.

Table 2.3 Observed and Expected Frequency: Beargrass Creek Data (Normal Distribution)

Class interval	Observed relative frequency	Expected relative frequency
0–750	0.064	0.141
750–1500	0.581	0.267
1500–2250	0.194	0.286
2250–3000	0.064	0.177
3000–3750	0.032	0.063
3750–4500	0.032	0.012
4500–5250	0 032	0.001
	0.999	0.947

A second visual comparison between the observed data and their assumed distribution (the normal distribution) can be made by using the normal distribution as the equation for the line describing the data in Fig. 2.4. Equation (2.11) can be used to draw a line through the points of Fig. 2.4, assuming the points are from a normal distribution. All that is required is to calculate the prob($Q > q$) for the various values of Q, and then to plot this probability versus q. The prob($Q > q$) is equal to 1 − prob($Q \leq q$) since prob($Q = q$) is zero and Q must either be $\leq q$ or $> q$. To obtain prob($Q > q$), we first evaluate prob($Q \leq q$). Equations (2.14) and (2.17) show such a calculation for Q = 2500 cfs. Table 2.4 shows the results of similar calculations for several values of Q. The prob($Q > q$) is plotted in Fig. 2.9.

Table 2.4 Comparison of Observed and Expected Cumulative Probabilities (Normal Distribution)

Q	Observed percentage $\geq Q$	Expected percentage $\leq Q$	Expected percentage $\geq Q$
700	100.0	18.7	81.3
1000	67.7	28.8	71.2
1500	35.5	46.0	54.0
2000	25.8	65.6	34.4
2500	9.6	81.6	18.4
3000	9.6	91.8	8.2
4000	3.2	99.2	0.8
5000	3.2	100.0	0.0
6000	0	100.0	0.0

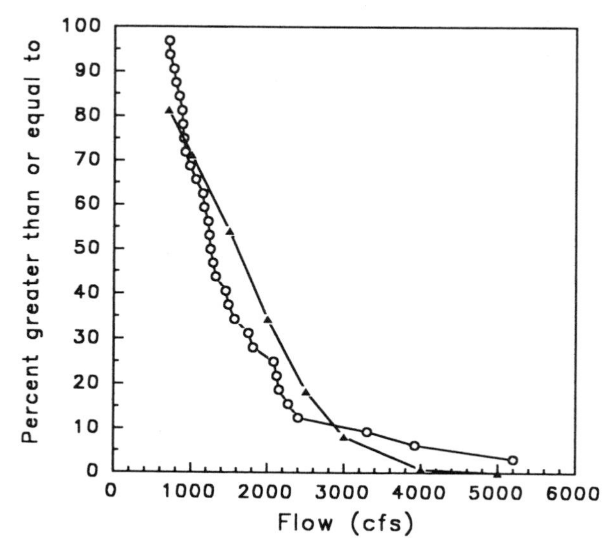

Figure 2.9 Flood frequency comparison for Beargrass Creek data —observed and normal distribution.

From either Fig. 2.8 or 2.9 it is apparent that the normal distribution is not a satisfactory approximation to the observed data of Beargrass Creek. Another probability distribution must be found to describe the data. This involves finding another mathematical function to use as a pdf and cdf in place of Eqs. (2.10) and (2.11) used for the normal distribution. A large number of such expressions are available. Again, these expressions are known as probability distributions.

The three probability distributions that receive the most attention for describing flood frequencies are the lognormal (LN), extreme value type I (EVI), and log Pearson type III (LP3). This treatment is restricted to these three distributions. Other distributions are discussed in Haan (1977).

The pdf for the lognormal distribution is

$$p_X(x) = \frac{1}{x\sigma_Y\sqrt{2\pi}} \exp\left[\frac{-(\ln x - \mu_Y)^2}{2(\sigma_Y)^2}\right],$$

where μ_Y and σ_Y are the mean and standard deviation of the natural logarithms of X.

The pdf for the extreme value type I distribution for maximums is

$$p_X(x) = \frac{1}{\alpha} \exp\left[\frac{-(x-\beta)}{\alpha} - \exp\left(\frac{-(x-\beta)}{\alpha}\right)\right],$$

where α and β can be estimated as

$$\hat{\alpha} = \frac{S_X\sqrt{6}}{\pi}, \qquad \hat{\beta} = 0.45s_X.$$

The pdf for the Pearson type III distribution is

$$p_X(x) = p_0 e^{-(x-\alpha)/\delta}(x/\alpha)^{\alpha/\delta}$$

with the mode at $X = \alpha$ and the lower bound at $X = 0$. The difference in the mode and the mean is δ and p_0 is $p_X(\alpha)$. Most applications of the Pearson type III distributions in hydrologic frequency analysis have been based on the log Pearson type III distribution obtained by converting all of the observations to their logarithms and then applying the Pearson type III to these logarithms. Resulting estimates must of course be transformed back to their original units.

As noted earlier, the Beargrass Creek data, when plotted in the form of a relative frequency histogram, tailed off to the right much more than to the left. This tailing off to the right results in a positive skewness. The normal distribution is symmetrical about the mean and, as such, has a skewness of zero. The LN, EVI, and LP3 distributions can all accommodate positively skewed data.

For the LN distribution, the skewness γ and the coefficient of variation C_v are related by

$$\gamma = 3C_V + C_V^3. \tag{2.21}$$

For the EVI distribution γ is a constant 1.139. There are no restrictions on γ for the LP3 distribution since it can take on any value the sample of data yields. The skewness is an important identifier of potential distributions that might be used to describe a set of data. For the distributions we have considered, the skewness is given by

Distribution	Skewness
Normal	0
Lognormal	$3C_v + C_v^3$
Extreme value I	1.139
Log Pearson III	Any value

Chow (1951) has shown that many types of frequency analyses can be reduced to

$$X_T = \bar{X}(1 + C_V K_T), \tag{2.22}$$

where X_T is the magnitude of the event with return period T, $\bar{X}$ is the mean of the original data, C_V is the coefficient of variation of the original data, and K_T is a frequency factor that is a function of the probability distribution selected and properties of the original data.

A comparison of Eqs. (2.15) and (2.22) shows that K_T for the normal distribution is equal to the standardized normal variate Z. Thus, Appendix 2A serves as a table of frequency factors for the normal distribution. The frequency factors for the LN distribution as a function of C_V are contained in Table 2.5. Table 2.6 contains the frequency factors for the EVI distribution. All that is required for selecting K_T for this distribution is knowledge of the sample size and the desired return period.

Table 2.5 Frequency Factors for Lognormal Distribution (Chow, 1964)

Return period					
1.01	2	5	20	100	Corresponding C_v
−2.33	0	0.84	1.64	2.33	0
−2.25	−0.02	0.84	1.67	2.40	0.033
−2.18	−0.04	0.83	1.70	2.47	0.067
−2.11	−0.06	0.82	1.72	2.55	0.100
−2.04	−0.07	0.81	1.75	2.62	0.136
−1.98	−0.09	0.80	1.77	2.70	0.166
−1.91	−0.10	0.79	1.79	2.77	0.197
−1.85	−0.11	0.78	1.81	2.84	0.230
−1.79	−0.13	0.77	1.82	2.90	0.262
−1.74	−0.14	0.76	1.84	2.97	0.292
−1.68	−0.15	0.75	1.85	3.03	0.324
−1.63	−0.16	0.73	1.86	3.09	0.351
−1.58	−0.17	0.72	1.87	3.15	0.381
−1.54	−0.18	0.71	1.88	3.21	0.409
−1.49	−0.19	0.69	1.88	3.26	0.436
−1.45	−0.20	0.68	1.89	3.31	0.462
−1.41	−0.21	0.67	1.89	3.36	0.490
−1.38	−0.22	0.65	1.89	3.40	0.517
−1.34	−0.22	0.64	1.89	3.44	0.544
−1.31	−0.23	0.63	1.89	3.48	0.570
−1.28	−0.24	0.61	1.89	3.52	0.596
−1.25	−0.24	0.60	1.89	3.55	0.620
−1.22	−0.25	0.59	1.89	3.59	0.643
−1.20	−0.25	0.58	1.88	3.62	0.667
−1.17	−0.26	0.57	1.88	3.65	0.691
−1.15	−0.26	0.56	1.88	3.67	0.713
−1.12	−0.26	0.55	1.87	3.70	0.734
−1.10	−0.27	0.54	1.87	3.72	0.755
−1.08	−0.27	0.53	1.86	3.74	0.776
−1.06	−0.27	0.52	1.86	3.76	0.796
−1.04	−0.28	0.51	1.85	3.78	0.818
−1.01	−0.28	0.49	1.84	3.81	0.857
−0.98	−0.29	0.47	1.83	3.84	0.895
−0.95	−0.29	0.46	1.81	3.87	0.930
−0.92	−0.29	0.44	1.80	3.89	0.966
−0.90	−0.29	0.42	1.78	3.91	1.000
−0.84	−0.30	9.39	1.75	3.93	1.081
−0.80	−0.30	0.37	1.71	3.95	1.155

The steps in using the LP3 distribution are:

1. transform the n original observations, X_i, to their logarithmic values, Y_i, by the relation
$$Y_i = \log X_i; \qquad (2.23)$$
2. compute the mean logarithm, $\bar{Y}$;
3. compute the standard deviation of the logarithm, S_Y;
4. compute the coefficient of skewness C_S from
$$C_S = \frac{n^2\Sigma Y_i^3 - 3n\Sigma Y_i \Sigma Y_i^2 + 2(\Sigma Y_i)^3}{n(n-1)(n-2)S_Y^3} \qquad (2.24)$$
or
$$C_S = \frac{n\Sigma(Y_i - \bar{Y})^3}{(n-1)(n-2)S_Y^3};$$
5. compute
$$Y_T = \bar{Y} + S_Y K_T, \qquad (2.25)$$
where K_T is from Table 2.7. This relationship is identical to Eq. (2.22) except it is based on logarithms; and
6. calculate
$$X_T = \text{antilog } Y_T. \qquad (2.26)$$

The skew coefficient is sensitive to extreme flood values and thus difficult to estimate from small samples typically available for many hydrologic studies. Figure 2.10 presents a map of generalized skew coefficients for the logs of peak flows taken from Bulletin 17B of the Interagency Committee. The station skew coefficient calculated from observed data and generalized skew coefficients can be combined to improve the overall estimate for the skew coefficient. Under the assumption that the generalized skew is unbiased and independent of the station skew, the mean square error (MSE) of the weighted estimate is minimized by weighting the station and generalized skew in inverse proportion to their individual mean square errors according to the equation (Tasker, 1978)

$$G_W = \frac{MSE_{\bar{G}}(G) + MSE_G(\bar{G})}{MSE_{\bar{G}} + MSE_G}, \qquad (2.27)$$

where G_w is the weighted skew coefficient, G is the station skew [from Eq. (2.24)], $\bar{G}$ is the generalized skew (from Fig. 2.10), $MSE_{\bar{G}}$ is the mean square error of the generalized skew, and MSE_G is the mean square

Table 2.6 Frequency Factors for Extreme Value Type I Distribution

Sample size n	5	10	15	20	25	50	75	100	1000
15	0.967	1.703	2.117	2.410	2.632	3.321	3.721	4.005	6.265
20	0.919	1.625	2.023	2.302	2.517	3.179	3.563	3.836	6.006
25	0.888	1.575	1.963	2.235	2.444	3.088	3.463	3.729	5.842
30	0.866	1.541	1.922	2.188	2.393	3.026	3.393	3.653	5.727
35	0.851	1.516	1.891	2.152	2.354	2.979	3.341	3.598	
40	0.838	1.495	1.866	2.126	2.326	2.943	3.301	3.554	5.576
50	0.820	1.466	1.831	2.086	2.283	2.889	3.241	3.491	5.478
60	0.807	1.446	1.806	2.059	2.253	2.852	3.200	3.446	
70	0.797	1.430	1.788	2.038	2.230	2.824	3.169	3.413	5.359
80	0.788	1.417	1.773	2.020	2.212	2.802	3.145	3.387	
90	0.782	1.409	1.762	2.007	2.198	2.785	3.125	3.367	
100	0.779	1.401	1.752	1.998	2.187	2.770	3.109	3.349	5.261
∞	0.719	1.305	1.635	1.866	2.044	2.592	2.911	3.137	4.936

error of the station skew. $\text{MSE}_{\overline{G}}$ is taken as a constant 0.302 when the generalized skew is estimated from Fig. 2.10. MSE_G can be estimated from (Wallis *et al.*, 1974):

$$\text{MSE}_G = \text{antilog}_{10}[A - B\log_{10}(N/10)], \quad (2.28)$$

where

$$A = -0.33 + 0.08|G| \quad \text{if } |G| \leq 0.90 \quad (2.29)$$

$$= -0.52 + 0.30|G| \quad \text{if } |G| > 0.90$$

$$B = 0.94 - 0.26|G| \quad \text{if } |G| \leq 1.50$$

$$= 0.55 \quad \text{if } |G| > 1.50$$

N = record length.

It is recommended that if the generalized and station skews differ by more than 0.5, the data and flood producing characteristics of the watershed should be examined and possibly greater weight given to the station skew.

Confidence Intervals

Any streamflow record is but a sample of all possible such records. How well the sample represents the population depends on the sample size and the underlying population probability distribution that is unknown. Both the form and parameters of the underlying distribution must be estimated. If a second sample of data were available, certainly different estimates would result for the parameters of the distribution even if the same distribution were selected. Different parameter estimates will obviously result in different return period flow estimates. If many samples were available, many estimates could be made of the distribution parameters and consequently many estimates could be made of return period flows—say Q_{100}. One could then examine the probabilistic behavior of these estimates of Q_{100}. The fraction of the Q_{100}'s that fell between certain limits could be determined.

In actuality we have just one sample of data from which to make estimates of Q_T. Statistical procedures are available for estimating confidence intervals about

Table 2.7 Frequency Factors for Pearson Type III Distribution (Interagency Advisory Committee on Water Data, 1981)

Skew coefficient c_S	Recurrence interval in years							
	1.0101	2	5	10	25	50	100	200
3.0	−0.667	−0.396	0.420	1.180	2.278	3.152	4.051	4.970
2.8	−0.714	−0.384	0.460	1.210	2.275	3.114	3.973	4.847
2.6	−0.769	−0.368	0.499	1.238	2.267	3.071	3.889	4.718
2.4	−0.832	−0.351	0.537	1.262	2.256	3.023	3.800	4.584
2.2	−0.905	−0.330	0.574	1.284	2.240	2.970	3.705	4.444
2.0	−0.990	−0.307	0.609	1.302	2.219	2.912	3.605	4.298
1.8	−1.087	−0.282	0.643	1.318	2.193	2.848	3.499	4.147
1.6	−1.197	−0.254	0.675	1.329	2.163	2.780	3.388	3.990
1.4	−1.318	−0.225	0.705	1.337	2.128	2.706	3.271	3.828
1.2	−1.449	−0.195	0.732	1.340	2.087	2.626	3.149	3.661
1.0	−1.588	−0.164	0.758	1.340	2.043	2.542	3.022	3.489
0.8	−1.733	−0.132	0.780	1.336	1.993	2.453	2.891	3.312
0.6	−1.880	−0.099	0.800	1.328	1.939	2.359	2.755	3.132
0.4	−2.029	−0.066	0.816	1.317	1.880	2.261	2.615	2.949
0.2	−2.178	−0.033	0.830	1.301	1.818	2.159	2.472	2.763
0	−2.326	0	0.842	1.282	1.751	2.054	2.326	2.576
−0.2	−2.472	0.033	0.850	1.258	1.680	1.945	2.178	2.388
−0.4	−2.615	0.066	0.855	1.231	1.606	1.834	2.029	2.201
−0.6	−2.755	0.099	0.857	1.200	1.528	1.720	1.880	2.016
−0.8	−2.891	0.132	0.856	1.166	1.448	1.606	1.733	1.837
−1.0	−3.022	0.164	0.852	1.128	1.366	1.492	1.588	1.664
−1.2	−3.149	0.195	0.844	1.086	1.282	1.379	1.449	1.501
−1.4	−3.271	0.225	0.832	1.041	1.198	1.270	1.318	1.351
−1.6	−3.388	0.254	0.817	0.994	1.116	1.166	1.197	1.216
−1.8	−3.499	0.282	0.799	0.945	1.035	1.069	1.087	1.097
−2.0	−3.605	0.307	0.777	0.895	0.959	0.980	0.990	0.995
−2.2	−3.705	0.330	0.752	0.844	0.888	0.900	0.905	0.907
−2.4	−3.800	0.351	0.725	0.795	0.823	0.830	0.832	0.833
−2.6	−3.889	0.368	0.696	0.747	0.764	0.768	0.769	0.769
−2.8	−3.973	0.384	0.666	0.702	0.712	0.714	0.714	0.714
−3.0	−4.051	0.396	0.636	0.660	0.666	0.666	0.667	0.667

estimated values of Q_T that will give a measure of uncertainty associated with Q_T. Confidence limits give a probability that the confidence limits contain the true value for Q_T. A 95% confidence limit indicates that 95% of the time intervals so calculated will contain the true estimate for Q_T.

Letting L_T and U_T be the lower and upper confidence intervals

$$\text{prob}(L_T \leq Q_T \leq U_T) = \alpha, \tag{2.30}$$

where α is the degree of confidence. Exact determination of L_T and U_T depend on the underlying parent population. Bulletin 17B of the Interagency Committee presents some approximate relationships for confidence intervals,

$$L_T = \overline{X} + S_X K_{T,L}$$
$$U_T = \overline{X} + S_X K_{T,U}, \tag{2.31}$$

where $\overline{X}$ and S_X are the sample means and standard deviations and $K_{T,L}$ and $K_{T,U}$ are the lower and

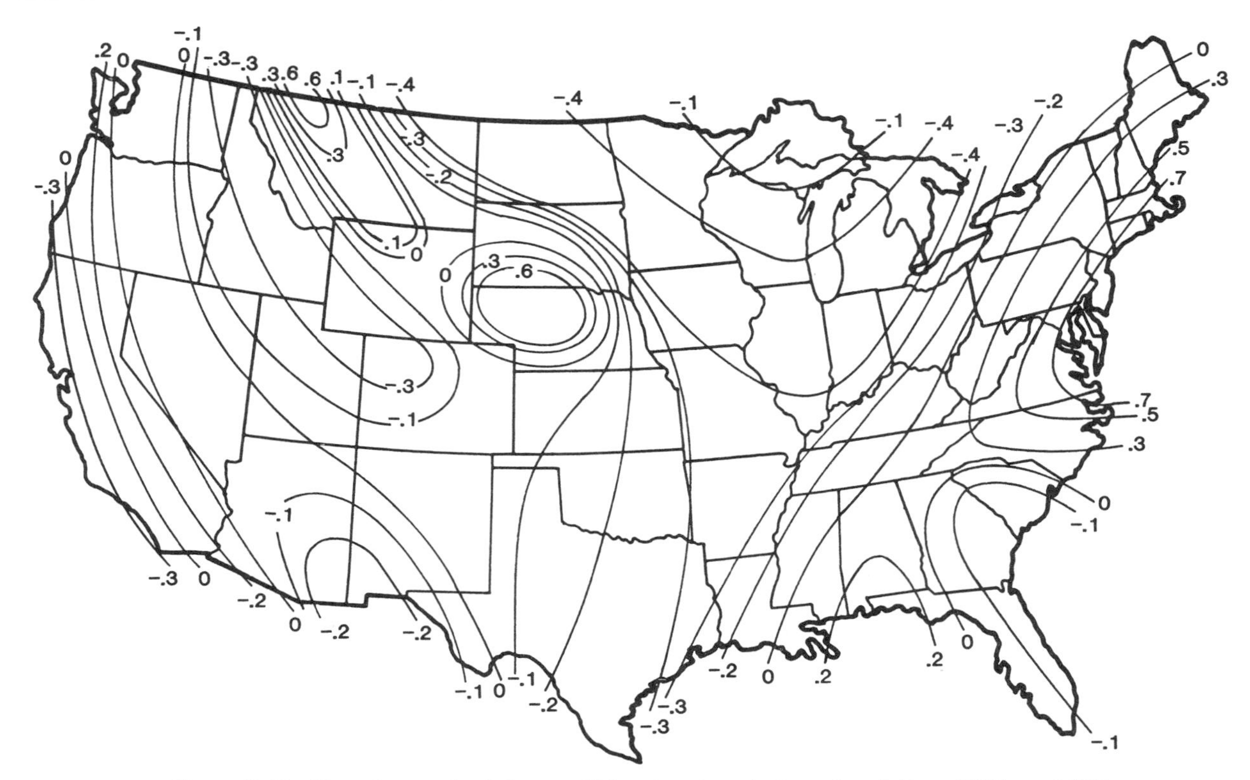

Figure 2.10 Map of generalized skew coefficient based on the logs of peak flows ($MSE_{\bar{G}} = 0.302$).

upper confidence coefficients. If a distribution like the LP3 distribution is used, $\bar{X}$ and S_X are based on the logarithms of the data and L_T and U_T are the logarithms of the confidence limits.

Approximations for $K_{T,L}$ and $K_{T,U}$ based on large samples and the noncentral t-distribution are

$$K_{T,L} = \frac{K_T - \sqrt{K_T^2 - ab}}{a} \tag{2.32}$$

$$K_{T,U} = \frac{K_T + \sqrt{K_T^2 - ab}}{a},$$

where

$$a = 1 - \frac{Z_\alpha^2}{2(n-1)} \tag{2.33a}$$

$$b = K_T^2 - \frac{Z_\alpha^2}{n}. \tag{2.33b}$$

In these relationships, K_T is the frequency factor of Eq. (2.22), Z_α is the standard normal deviate with cumulative probability α (Appendix 2A), and n is the sample size. Confidence limits can be placed on frequency curves plotted on probability paper by making calculations such as above for several values of T.

Example Analytical Frequency Analysis

As an example of applying these three distributions, again consider the data of Table 2.1. The mean and standard deviation of the original data were found to be 1599 and 1006, respectively. The C_V is 0.629. Values of K_T for various values of T for the lognormal distribution are selected from Table 2.5, and Eq. (2.22) gives the corresponding values of X_T. These results are shown in Table 2.8.

Values of K_T for various return periods for the extreme value type I distribution are selected from Table 2.6, and X_T again comes from Eq. (2.22). These results are shown in Table 2.8.

In applying the log Pearson type III distribution, $\bar{Y}$ based on the natural logarithms of the flow data is found to be 7.237, S_Y is 0.507, and C_S from Eq. (2.24) is 0.87. K_T values are then selected from Table 2.7, Y_T calculated from Eq. (2.25), and X_T from Eq. (2.26). The results of these calculations are in Table 2.8.

This example shows that the distribution that is selected can have a substantial affect on the estimated flood magnitude for a given frequency. This is especially apparent for the longer return periods. To make this point even more emphatic, if a normal distribution

Table 2.8 Flood Frequency Analysis for Middle Fork, Beargrass Creek, Cannons Lane, Louisville, Kentucky

Return period (years):	5	25	100
Lognormal distribution			
K_T	0.60	2.11	3.57
X_T	2202	3718	5190
Extreme value type I			
K_T	0.863	2.385	3.642
X_T	2467	3998	5262
Log Pearson type III			
K_T	0.772	2.011	2.937
Y_T	7.628	8.257	8.726
X_T	2056	3853	6161

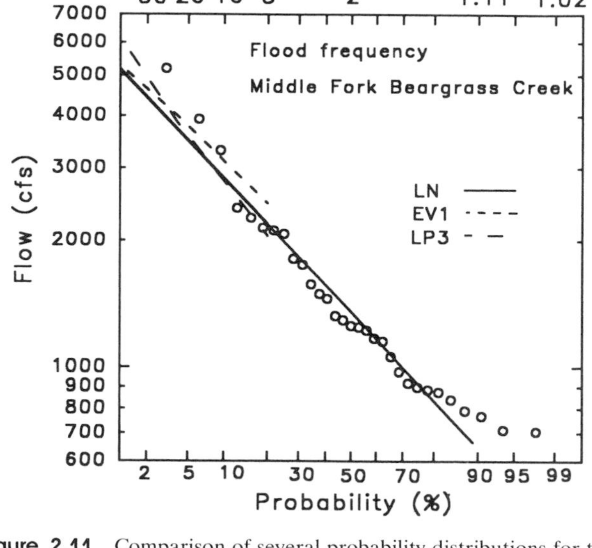

Figure 2.11 Comparison of several probability distributions for the Beargrass Creek data.

had been used, the 100-year estimate would have been 3943 cfs, clearly an inferior estimate. The best fitting lines according to three of the distributions used are shown in Fig. 2.11.

The plotting paper used to construct Fig. 2.11 is known as lognormal probability paper. It consists of a logarithmic scale and a normal probability scale. Any distribution may be plotted on lognormal probability paper, but only the lognormal distribution will plot as a straight line on this paper. Other types of probability paper are available, including normal and extreme value probability paper.

From Fig. 2.11 it is apparent that it is difficult to select which of the three distributions best describe these data. Considering the coefficient of skew, γ, the EVI might be discarded since γ for the EVI is 1.139 while it is 2.13 for these data. For the LN, γ and C_V are related through Eq. (2.21). The estimated value of γ is 2.14, which agrees quite well with the requirement of the LN. One discouraging factor concerning the LN is that γ of the logarithms should be zero for the LN distribution, while for these data it is 0.87. Apparently the LN is not a precise approximation for the data.

The skewness cannot be used to make decisions concerning the LP3 since the LP3 is a three-parameter distribution that uses γ to estimate these parameters. Looking at Fig. 2.11, it does appear that the LP3 is a better approximation to the data at the upper end of the frequency curve.

Confidence intervals on the LP3 can be calculated from Eqs. (2.30) through (2.33). The 95% confidence intervals are being calculated. $Z_\alpha = Z_{0.95}$ is found from Appendix 2A to be 1.645. From Eq. (2.33)

$$a = 1 - \frac{Z_\alpha^2}{2(n-1)} = 1 - \frac{1.645^2}{2(31-1)} = 0.955.$$

Table 2.9 contains the calculations of the confidence limits. These limits are plotted in Fig. 2.12. The interpretation of this plot is that, on the basis of the assumptions made, one can be 95% confident that the calculated confidence intervals will contain the true flood frequency relationship for Beargrass Creek.

As discussed later, one problem with these data is the possible nonstationarity of peak flows due to changing watershed conditions. The difficulties experienced with this set of data demonstrates why many have been led to the recommendation that single short records are not reliable, and regional frequency analysis procedures as discussed under Case IV should be used for single sites.

In summary, the Case I situation is that where a relatively long record of peak flows is available on the stream of interest at the point of interest. The method of analysis is to select a probability density function to describe the data, extract certain statistics from the data, and use Eq. (2.22) along with appropriate frequency factors to estimate flows within the desired return period.

Table 2.9 95% Confidence Intervals on LP3 for Beargrass Creek

T	K_T[a]	b[b]	$K_{T,L}$[c]	$K_{T,U}$[c]	L_T[d]	U_T[d]	Q_L[e]	Q_U[f]
1.01	−1.64	−2.60	−2.19	−1.24	6.13	6.61	458	740
2	−0.15	−0.06	−0.46	0.15	7.00	7.31	1100	1498
5	0.77	0.51	0.46	1.15	7.47	7.82	1754	2495
10	1.34	1.71	0.98	1.83	7.73	8.16	2283	3510
25	2.02	3.99	1.57	2.66	8.04	8.58	3087	5344
50	2.48	6.06	1.97	3.23	8.24	8.87	3771	7131
100	2.94	8.56	2.36	3.80	8.43	9.16	4596	9535

[a]From Table 2.7.
[b]From Eq. (2.33).
[c]From Eq. (2.32).
[d]From Eq. (2.31).
[e]$Q_L = \exp(L_T)$.
[f]$Q_U = \exp(U_T)$.

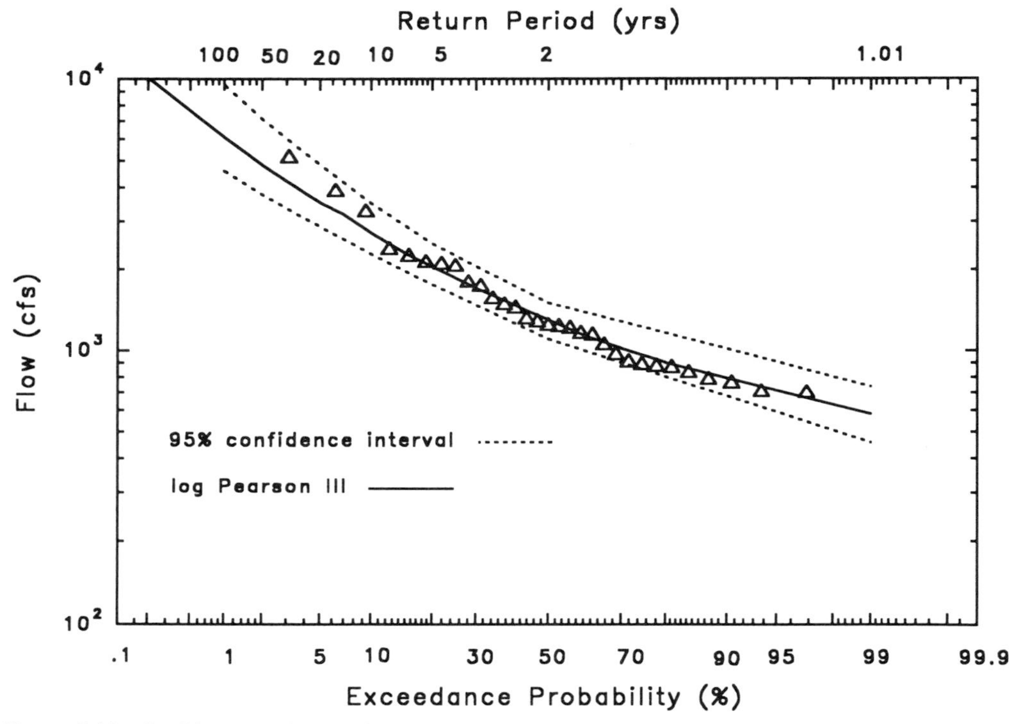

Figure 2.12 Confidence on the LP3 for Beargrass Creek.

Some of the assumptions made in the Case I situation are:

1. The data are sufficient in quantity and quality to produce reliable estimates for the parameters of the probability distribution selected.
2. The flow characteristics of the stream have not been changing over time (stationary data series).
3. The peak flow observations are statistically independent from year to year.
4. The data are representative of the flow behavior expected during the life of the project being considered.

Assumption 4 merely extends the stationarity assumption to future flows. In watersheds with changing land use, assumptions 2 and 4 are especially troublesome in that the changing land use alters streamflow characteristics. Many times this nonhomogeneity in streamflow data is very subtle and only becomes apparent over a long period of time or when sudden and large-scale changes occur. For example, Fig. 2.13 is a plot of the 31-year record for the Beargrass Creek data. The bulk of the peak flows are in the range 750 to 2000 cfs. It appears as though the frequency of occurrence of peaks in excess of 2000 cfs is increasing with time. The random nature of the data makes it difficult to make firm statements in this regard.

In closing the discussion on the Case I flood frequency analysis, a word of caution is offered concerning the extrapolation of frequency data to estimate the magnitude of an event with a return period much greater than the period of record. In looking at Fig. 2.12, it appears that the extrapolation of the frequency lines from the 31-year record to estimate the 100-year or even 500-year event is not much of an extrapolation. In the sense of the physical distance on the probability paper, the extrapolation is not very great; however, in the sense of extrapolating the data to 3 or possibly 15 times its original length, it is a very significant extrapolation. The nature of random data makes an extrapolation of this kind very speculative and produces estimates of low reliability or ones that possess considerable uncertainty. This is apparent from the width of the confidence limits at high return periods. Haan (1977) gives a procedure for evaluating the uncertainty that is present.

To illustrate this point, a simulation was made assuming a lognormal distribution with a mean of 1599 cfs and a standard deviation of 1006 cfs. The procedure was to randomly select 15 observations from this lognormal distribution and then use these 15 observations to estimate the magnitude of the 100-year event. The 100-year event was estimated on the basis of the lognormal distribution using the mean and standard deviation of the 15 generated values. This entire process was repeated 100 times, producing 100 estimates for the 100-year flood. The probabilistic behavior of these 100 estimates was then analyzed.

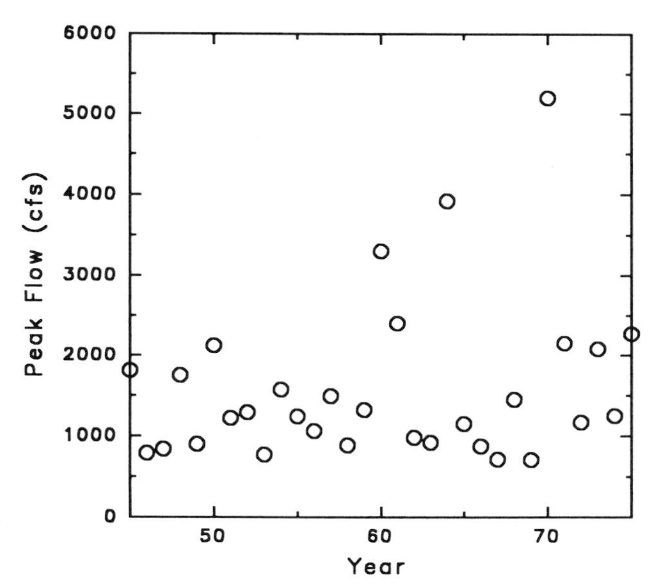

Figure 2.13 Changing flow condition resulting in nonhomogeneous time series.

The following tabulation shows the frequency of occurrence of the estimate for the 100-year flood. The true value is 5190 cfs. The mean estimated value was 4945 cfs. The following tabulation can be used to get some idea of the possible errors involved in using a short record to estimate a rare flood.

Flow (cfs)	No. of estimates
2500–3000	7
3000–3500	8
3500–4000	13
4000–4500	10
4500–5000	21
5000–5500	15
5500–6000	7
6000–6500	5
6500–7000	6
7000–7500	3
7500–8000	2
8000–8500	2
14000–∞	1
Total	100

Note that only 36 values or about one-third of the estimates are within 10% (500 cfs) of the actual 100-year flood. Nearly half (47 values) of the estimates are in error by more than 20%, 15% of the values are in error by more than 40%, and 1 estimate was actually high by a factor of nearly 3. These numbers illustrate what can happen when a short record (15 years in this case) is used to estimate a rare flood (100-year flood in this case). This is the reason that extrapolation is speculative. In any real situation, the variability would be even greater than shown here, because the true underlying probability distribution would not be known. We assumed a lognormal distribution in this example and based our calculations on this assumption as though it were the population distribution. It is understood that one must use procedures like those set forth here to make flood frequency estimates. The purpose of this latter illustration is to shed some light on the uncertainty that is inherent in dealing with random data.

Case II. Long Record near Site

In some instances, flow data are available on the stream of interest but at a location some distance either upstream or downstream from the point of interest. In this situation, there are several procedures that can be used to estimate the flood frequency relationship at the point of interest. One method is to perform a flood frequency analysis on the available record as described in Case I, and then adjust the record to the point of interest.

The adjustment of the flow record from one location on a stream to another can be done in a number of ways. If the flow record includes the entire flood hydrograph, this hydrograph could be routed to the point of interest making the proper adjustments for local inflows along the routing reach. This method is a very good one, but requires more data than are generally available and is quite time consuming.

A second method of flow adjustment is to correlate flood peaks with drainage basin characteristics and then use this correlation to adjust the flow rate. The most common characteristic used in a situation like this is the basin area. Quite frequently the peak discharge for a given frequency is related to the basin area by an equation of the form

$$Q_T = aA^b, \tag{2.34}$$

where Q_T is the T-year flood magnitude, A is the basin area, and a and b are constants. The coefficient b generally ranges from 0.5 to 1.0.

If data on a stream are available at two locations, the coefficients can be estimated from those data. For example, consider the data in Table 2.10. Here, information on the 2-, 5-, and 10-year floods is available at two locations. The exponents a and b of Eq. (2.34) can be estimated for each return period by substitution. Consider the 10-year data:

Table 2.10 Hypothetical Flood Frequency Data

T (years)	Area (mile^2) 0.7	2.3		
	Q_T (cfs):		a	b
2	750	2250	1041	0.923
5	950	2800	1318	0.909
10	1100	3300	1527	0.923

$$Q_{10} = aA^b$$

(i) $3300 = a\,(2.3)^b$ for the larger basin,
(ii) $1100 = a\,(0.7)^b$ for the smaller basin.

The ratio of (i) and (ii),

$$\frac{3300}{1100} = \left(\frac{2.3}{0.7}\right)^b,$$

can be solved for b using logarithms resulting in $b = 0.923$. Substituting this estimate for b into (i) results in

$$3300 = a\,(2.3)^{0.923},$$

which gives an estimate for a of 1527. Similar calculations for the 2- and 5-year floods result in the estimates shown in Table 2.10. From this, an average value of b is 0.92. The coefficient a is seen to be a function of return period T.

To estimate the 10-year flood at a point on the stream where the drainage area is 1.5 mile^2, Eq. (2.34) and the estimated 10-year coefficients are used:

$$Q_{10} = 1527\,(1.5)^{0.92} = 2217 \text{ cfs}.$$

If data at only one other location on the stream are available, one can still estimate the flow at a second location if an assumption as to the coefficient b is made. For small differences in area, a reasonable assumption for b is 1.0. For example, considering the 10-year data for the 0.7 mile^2 basin only and taking b as 1.0 results in an estimate for a of

$$a = 1100/0.7 = 1571.$$

The 10-year flood can then be estimated on the 1.5 mile^2 basin as

$$Q_{10} = 1571\,(1.5)^1 = 2357 \text{ cfs}.$$

When these approaches are used, several precautions must be exercised. First and most importantly, the basic flood-producing characteristics of all the basins must be the same. There cannot be a mix of drastically differing land uses unless some type of land-use variable is included in the prediction equation for a and b. Returning to the above example, it has been assumed that all three watersheds are similar. For many watersheds, this is a severe limitation and generally means that there cannot be a very large difference in the areas of watersheds considered or the land use on the watersheds.

If the watershed characteristics are changing along the steam, then calculations such as shown here can be used as an aid in estimating a flow of a given return period, but will not give the final estimate. If the available data represent a mixture of land uses and the point of interest is below only one land use within the basin, the flow estimated by a straightforward application of the techniques presented here will most likely have to be adjusted to reflect the fact that it is below a single land use while the coefficients a and b were estimated on the basis of a mixed land-use basin.

Case III. Short Stream Record

It is not uncommon to find that a streamflow record at the point of interest may be too short to use in a flood frequency analysis. This may be the result of a newly installed gage or a gaging program that was only recently changed so that much of the earlier portion of the streamflow record is no longer representative of the basin.

A short streamflow record can be a great aid in checking calculations and procedures used in flood frequency estimation. If a major drainage project is to be planned, the local governing body would be wise to install a streamgage early in the feasibility part of the project planning process. In this way, by the time the final design is made, some streamflow data would be available. For relatively large drainage projects, this short-term gaging approach is relatively inexpensive and can easily pay for itself through the resulting improvement in the design of the drainage system.

A short streamflow record is one of less than 10 years in length. A record such as this will contain a great deal of information, but will be insufficient for a Case I frequency analysis. Presumably, a record of the rainfall that produced the recorded runoff will be available or can be estimated from nearby raingages. These records on rainfall and runoff can now be used to estimate the empirical coefficients in an approximate model. The model might be a continuous simulation model, an event or hydrograph model, or a model for estimating peak flow only. The type of model selected will depend on the quantity and quality of available data and the purpose of the analysis.

Regardless of the type of model selected, the model will have empirical coefficients associated with it that must be estimated. In general, the number of empirical coefficients required is proportional to the complexity of the model, with the continuous simulation models having the most coefficients. The importance of actual data on the stream of interest in estimating the empirical coefficients of these models cannot be overemphasized.

Once the model coefficients have been estimated, a long-term rainfall record can be processed through the model to produce a long-term streamflow record. This simulated, long-term streamflow record can then be subjected to a frequency analysis as discussed under Case I if necessary.

In the event that a long-term rainfall record is not available for the site under study, one can use records from the nearest raingage. Fortunately, in many parts of the country, rainfall records can be transferred a few miles without introducing significant effects on the estimated peak runoff rates from major events. The long-term records from nearby raingages may not be usable for runoff parameter estimation since the recorded rainfall may have been considerably different from what actually fell on the watershed. The record can be used for simulation, however, because the long-term statistical properties may be the same as those of rains that actually fell on the watershed.

In the absence of any applicable long-term rainfall records, it may be possible to use a stochastic rainfall generation model to produce a synthetic rainfall record to use in simulations with the runoff model.

Another approach to using the information contained in a short record is to use the short record to estimate a low-return period index flood. This might be the 2-year flood, for example. Knowing the magnitude of this index flood, a regionalized relationship between the ratio of the index flood and floods of a greater return period can be used to estimate the magnitude of less frequent events. The determination of the regionalized relationship is covered under the Case IV situation discussed below.

A third option for using a short stream record is to correlate the annual peak flows from the short record with peak flows from another station in the vicinity with a longer record. The longer record and the correlation can then be used to extend the shorter record.

Let Y and X represent peak flows on two streams. Let n_1 be the number of observations in common on Y and X. Let n_2 be the number of additional observations on X not in common with Y. Assume the observations on each stream are independent from year to year (no serial correlation on either stream). Consider $\rho_{Y,X}$ as the correlation coefficient between Y and X for the n_1 common observations. The correlation coefficient $\rho_{Y,X}$ is estimated by $r_{Y,X}$ given by

$$r_{Y,X} = \frac{\sum_{i=1}^{n_1} (X_i - \bar{X})(Y_i - \bar{Y})}{\sqrt{\sum_{i=1}^{n_1} (X_i - \bar{X})^2 \sum_{i=1}^{n_1} (Y_i - \bar{Y})^2}}. \quad (2.35)$$

We can estimate n_2 values of Y from the n_2 values of X for which Y values are not available from

$$Y = \frac{r_{Y,X} S_Y (X - \bar{X})}{S_X} + \bar{Y}, \quad (2.36)$$

where $\bar{Y}$, $\bar{X}$, S_Y, and S_X are based on the n_1 common observations.

Let $\bar{Y}_1$ and $\bar{Y}_2$ represent the mean based on the original n_1 observations and the n_2 estimated observations, respectively. A new weighted mean for Y based on $n_1 + n_2$ observations is given by

$$\bar{Y} = \frac{n_1 \bar{Y}_1 + n_2 \bar{Y}_2}{n_1 + n_2}. \quad (2.37)$$

For the n_2 added observations to improve the estimate for $\bar{Y}$, it is necessary that $r_{Y,X}$ be greater than $1/(n_1 - 2)$ according to Matalas and Langbein (1962). If Y and X contain significant serial correlation, the situation is somewhat more complex (Haan, 1977).

Case IV. Regional Analysis

Often one finds that streamflow records are available at several nearby locations, while none or a very short record is available at the point of interest. Several methods for using the information on the nearby stations to augment whatever is known at the site of interest are available. These methods generally fall under the heading of "regional flood frequency analysis."

One widely used method of regional flood frequency analysis is discussed by Dalrymple (1960). The method consists of computing a base flood frequency relationship in terms of the return period and the ratio of the peak flow for a given return period to an index flood (usually the mean annual flood) for several streams in the region. The median value of this ratio is then plotted versus the return period. Figure 2.14 is such a plot for 18 stations in Alberta an Saskatchewan, Canada, as reported by Durant and Blackwell (1959).

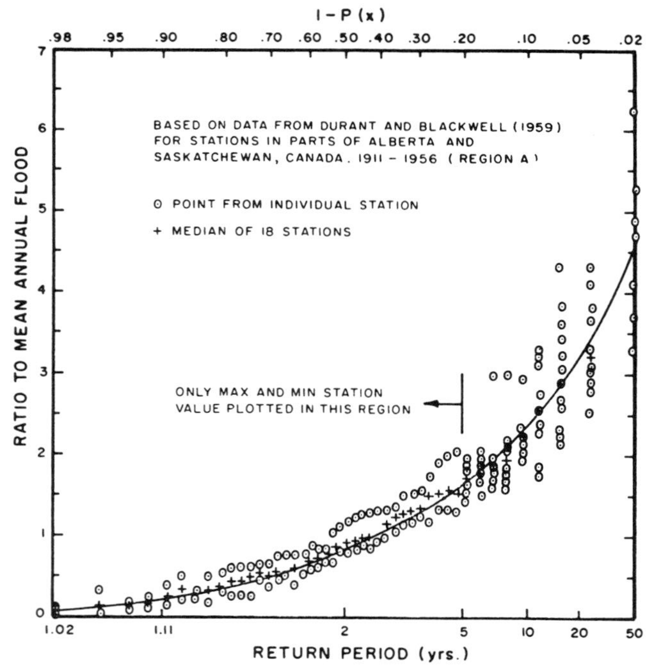

Figure 2.14 Regional flood frequency analysis.

The first step in this regional approach is to select several streamflow records from nearby locations that are "hydrologically similar" to the basin of interest. At each of these locations, a Case I flood frequency analysis is made. An index flood is then defined. This might be the 2-year flood. The ratio of the magnitude of the T-year flood to the index flood is computed for several values of T at each location. The ratio is then plotted versus T and a smooth curve drawn through the points (Fig. 2.14).

The next step is to relate the index flood to watershed characteristics. The area of the watershed is generally used along with other geomorphic, physical, and meteorological factors. This step is generally done through a regression analysis to produce an equation of the form

$$Q_I = a X_1^b X_2^c \dots X_n^q \quad (2.38a)$$

or

$$Q_I = a + bX_1 + cX_2 + \cdots + qX_n, \quad (2.38b)$$

where Q_I is the magnitude of the index flood; $a, b, \ldots, q$ are estimated coefficients; and $X_1, X_2, \ldots, X_n$ are watershed and climatic factors. Regressions of this type are discussed in more detail by Haan (1977).

The third step is to use Eq. (2.38) to estimate the index flood for the location of interest. Alternatively, if a short record is available at the location of interest, the index flood can be estimated from that record. The final step is to use the regional flood frequency curve and the estimated Q_{I} to calculate Q_T for the desired values of T.

A variation of the above technique for regional flood frequency analysis is to estimate Q_T for several values of T at each gaged location as explained above, and then relate Q_T to watershed factors and climatic data by regression to produce an equation like (2.38) with Q_{I} replaced by Q_T. A separate equation is needed for each value of T. These equations can then be used to estimate the desired value of Q_T at the study location. One disadvantage of this approach is the possibility of not retaining the proper relation among the Q_T's for different values of T. That is, one could conceivably estimate Q_{25} as being less than Q_{10}. Haan (1977) discussed the use of multivariate multiple regression to overcome this difficulty.

Case V. No Flow Records

When there are no streamflow records available on the stream of interest or on nearby streams or when available records are from basins whose characteristics are considerably different from those on the basin of interest, one is forced to use some type of empirical procedure for estimating the magnitude of runoff events of the desired frequency. We are now out of the realm of frequency analysis in the sense of determining the frequency of occurrence of events based on a probabilistic analysis of data.

For this situation, a hydrologic model of some type must be employed. The model may range in complexity from the Rational Equation to a complete, continuous simulation model. The type of model selected will depend on the data available for model fitting, the user's familiarity with various models, the purpose of the modeling effort, the time and money available for completing the modeling effort, and the importance of flow estimates.

If a continuous simulation model is selected and several years of streamflow are generated, these generated data can be subjected to a flood frequency analysis. If an event-based model (such as a unit hydrograph approach) is selected, the most severe rainfall events each year can be analyzed, with these data subsequently subjected to a flood frequency analysis.

If an approach like the Rational Equation is used, one is assuming that the frequency of the estimated flow peak is the same as the frequency of the rainfall used in the equation. This is not a bad assumption over the long run. For individual events, the return period of the rainfall and the resulting runoff are not necessarily the same because of the effect of such factors as antecedent soil water content and annual variation in land use. However, over the long run, the expected or average return period of the runoff will nearly equal the return period of the rainfall.

Since the Case V situation is really a modeling effort or requires the use of hydrologic techniques not generally thought of as being frequency analysis, its treatment is deferred to Chapter 3, which provides a detailed treatment of peak flow estimation. Chapter 13 contains a discussion of hydrologic models of various types.

SPECIAL CONSIDERATIONS

Historic Data

Occasionally flood information outside of the systematic flow record is available from historical sources such as newspaper reports or earlier flood investigations. Such data contain valuable information that should not be ignored in a frequency analysis. Bulletin 17B of the Interagency Advisory Committee on Water Data (1981) discusses this topic in detail. Basically, what one does is to compute the plotting position of the historical observations on the basis of the historical record length. Likewise the plotting position of the systematic data is computed on the basis of the historic record length except the rank used in the calculation is adjusted by a factor W depending on the historic record length, H, the number of historic flows, Z, and the length of the systematic record, N. These are related by

$$W = \frac{H - Z}{N}. \tag{2.39}$$

The adjusted rank for the systematic data is

$$m_{\mathrm{a}} = Wm - (W - 1)(Z + 0.5), \tag{2.40}$$

with m being the unadjusted rank of the total record (systematic plus historic).

Thus, if 20 years of systematic data and two historic observations are available from a 50-year period preceding the systematic record, the plotting position for the two largest values would be $1/71 = 0.014$ and $2/71 = 0.028$. The weighting factor would be

$$W = (70 - 2)/20 = 3.40.$$

The remaining plotting positions would be calculated from the adjusted rank given by

$$m_a = 3.40m - (3.40 - 1)(2 + 0.5) = 3.40m - 6.0.$$

The adjusted rank is then used in the plotting position relationship [Eq. (2.8)]. Thus for $m = 3$ (the largest systematic flow observation), the plotting position would be [3.40(3) − 6]/71 or 0.0592 and for $m = 22$ (the smallest value) the plotting position would be [3.40(22) − 6]/71 or 0.9690. This compares to plotting positions of 1/21 or 0.0476 and 20/21 or 0.9523, respectively, if the historic data had been ignored. If the historic data had simply been used to augment the systematic record without using the weighting factor, the plotting positions for these two events would have been 3/23 or 0.1304 and 22/23 or 0.9565, respectively. Clearly a plotting position of 0.1304 assigns too high a probability of occurrence to the largest systematic value. It is also apparent that the weighting procedure adjusts the plotting position toward a more frequent occurrence for the largest systematic value, thus taking into account the fact that two flows of magnitude greater than that of the largest systematic flow occurred.

Bulletin 17B also suggests the flow statistics be computed by weighting the contribution of the systematic record to the various statistics by the factor W. Thus, the adjusted mean is

$$\bar{X}_a = \frac{W\Sigma X + \Sigma X_Z}{H}, \tag{2.41}$$

where X represents the systematic record and X_Z the historic data. Similarly the variance and skew can be determined from

$$S_a^2 = \frac{W\Sigma(X - \bar{X}_a)^2 + \Sigma(X_Z - \bar{X}_a)^2}{H - 1} \tag{2.42}$$

and

$$C_{Sa} = \frac{H}{(H-1)(H-2)} \times \left[\frac{W\Sigma(X - \bar{X}_a)^3 + \Sigma(X_Z - \bar{X}_a)^3}{S_a^3}\right]. \tag{2.43}$$

If the LP3 distribution is being used, the X's and X_Z's would be based on logarithms.

Treatment of Zeros

The following is taken from Haan (1977). Most hydrologic variables are bounded on the left by zero. A zero in a set of data that is being logarithmically transformed requires special handling. One solution is to add a small constant to all of the observations. Another method is to analyze the nonzero values and then adjust the relation to the full period of record. This method biases the results as the zeros are essentially ignored. A third and theoretically more sound method is to use the theorem of total probability

$$\text{prob}(X \geq x) = \text{prob}(X \geq x | X = 0)\text{prob}(X = 0) + \text{prob}(X \geq x | X \neq 0)\text{prob}(X \neq 0).$$

Since $\text{prob}(X \geq x | X = 0)$ is zero, the relationship reduces to

$$\text{prob}(X \geq x) = \text{prob}(X \neq 0)\text{prob}(X \geq x | X \neq 0).$$

In this relationship $\text{prob}(X \neq 0)$ would be estimated by the fraction of nonzero values and the $\text{prob}(X \geq x | X \neq 0)$ would be estimated by a standard analysis of the nonzero values with the sample size taken as equal to the number of nonzero values. This relation can be written as a function of cumulative distributions,

$$1 - P_X(x) = k[1 - P_X^*(x)]$$

or

$$P_X(x) = 1 - k + kP_X^*(x), \tag{2.44}$$

where $P_X(x)$ is the cumulative probability distribution of all X ($\text{prob}(X \leq x | X \geq 0)$), k is the probability X is not zero, and $P_X^*(x)$ is the cdf of the nonzero values of X (i.e., $\text{prob}(X \leq x | X \neq 0)$. This type of mixed distribution with a finite probability that $X = 0$ and a continuous distribution of probability for $X > 0$ has been shown by Jennings and Benson (1969) to be applicable for flow frequencies with zeros present.

Equation (2.44) can be used to estimate the magnitude of an event with a return period T by solving first for $P_X^*(x)$ and then using the inverse transformation of $P_X^*(x)$ to obtain the value of X. For example the 10-year event ($P_X(x) = 0.90$) with $k = 0.95$ is found to be the value of X satisfying

$$P_X^*(x) = \frac{P_X(x) - 1 + k}{k} = \frac{0.90 - 1 + 0.95}{0.95} = 0.89.$$

To determine the corresponding value for X, this equation must be solved for X based on a probability plot or the assumed probability distribution. Note that it is possible to generate negative estimates for $P_X^*(x)$ from Eq. (2.44). For example, if $k = 0.50$ and $P_X(x) = 0.05$, the estimate for $P_X^*(x)$ is

$$P_X^*(x) = \frac{0.05 - 1 + 0.50}{0.5} = -0.9.$$

This merely means that the value of X corresponding to $P_X(x)$ is zero.

Outliers

Occasionally, a systematic record of peak flows may contain one or more observations that deviate greatly, either high or low, from the rest of the data. For example, it is entirely possible that a 100-year event is contained in 10 years of record. If this is the case, assigning a normal plotting position of 1/11 to this value would not be reflective of its true return period. Bulletin 17B suggests that outliers can be identified from

$$X_H = \overline{X} + K_n S_X$$
$$X_L = \overline{X} - K_n S_X, \quad (2.45)$$

where X_H and X_L are threshold values for high and low outliers and K_n can be approximated from

$$K_n \approx 1.055 + 0.981 \log_{10} n, \quad (2.46)$$

where n is the number of observations.

If a peak in the record exceeds X_H and historical information of the type discussed earlier is available regarding that peak, it should be removed from the systematic record and treated as historical observation as discussed under Historical Data. If historical information on the flow is not available, it should be retained as a part of the systematic data.

If a flow is less than X_L, that value should be deleted from the record and conditional probability procedures as explained in the section Treatment of Zeros should be employed.

More detail on the treatment of outliers is contained in Bulletin 17B. Equation (2.46) is an approximation for a table in Bulletin 17B. For $10 \leq n \leq 149$, the maximum error in the equation is less than 2.5% and averages about 1%.

DISCUSSION OF FLOOD FREQUENCY DETERMINATIONS

The foundation of any frequency analysis is the procedure described under Case I where a particular probability distribution is selected for describing a set of data. The parameters of this distribution are estimated, and the magnitude of events for various return periods are calculated. Methods for plotting the observed data on probability paper and for constructing the best fitting line according to the selected distribution have also been discussed.

At this point, it should be clear that there is nothing inherently hydrologic about frequency analysis procedures. They are simply statistical techniques that operate on numbers. The fact that the numbers being used are peak flows is of no concern to the technique. It should be of great concern to the analyst, however.

Statistical frequency analysis simply attempts to extract information about the probabilistic behavior of a set of numbers from the numbers themselves. In hydrologic frequency analysis, this probabilistic behavior is then generally extrapolated by the analyst to frequencies of occurrence well beyond that contained in the original set of numbers. From these extrapolations, the flows having return periods of 25, 50, 100, or even 500 years are determined. The straightforward application of hydrologic frequency analysis as generally employed uses no or very little hydrologic knowledge. In actuality, rare flows are determined by the hydrologic conditions that exist at the time of these flows and not by the statistical behavior of a sample of maximum peak flows that may have occurred some time in the past. Resolving the apparent conflict between these statements is what separates the hydrologist from the statistician.

Statistics are descriptive of a set of observed data. Statistics do not define a cause and effect relationship or a physical relationship. Any conclusion drawn on the basis of a statistical frequency analysis assumes that the sample of data on hand is representative of a wider range of data known as the population. In hydrologic terms, this means that if we have a sample of 15 years or so of observed annual maximum peak flows and use these data to estimate the 100-year flood, we are assuming that the hydrologic behavior of the basin during the 100-year flood is somehow imprinted in the 15 years of observed data and that the statistical technique being used can uncover this imprint and use it. To determine if this is truly the case, the hydrology of the basin must be examined. Some of the questions that must be answered are:

a. Is the type of storm that is likely to produce the 100-year flow represented in the observed sample?
b. Is the contributing area of the basin the same for extreme floods as it is for small ones?
c. Are there ponds and reservoirs that may discharge at high rates during rare floods and not during smaller flows? What is the possibility of a dam or levee breach and what would be the resulting flow?
d. Are the channel flow and storage characteristics the same for extreme flows as they are for smaller flows?
e. Are land-use and soil characteristics such that flows from rare storms may relate to precipitation in a manner different from more common storms?
f. Are there seasonal effects such that rare floods are more likely to occur in a different season than the more common floods?
g. Is the rare flood represented in the sample of data? If so, how is it treated? Is it assigned a return period of 15 years where in fact its return period may be much greater than that?
h. Are there changes going on within the basin that may cause change in the hydrologic response of the basin to rainstorms?

These last few paragraphs paint a discouraging picture for flood frequency analysis. That need not be the case as long as one does not discard hydrologic knowledge in the process. Often the questions posed can be answered in such a way as to make the statistical analysis valid. At other times, when problems with the statistical procedures are recognized, adjustments can be made in the resulting flow estimates to more accurately reflect the hydrology of the situation.

Hydrologic frequency analysis should be used as an aid in estimating rare floods. Sometimes the estimates made on the basis of the statistical frequency analysis can be taken as the final estimate. Sometimes the statistical estimate may need to be adjusted to better reflect the hydrology of the situation.

It should be kept in mind that other hydrologic estimation techniques suffer from some of the same difficulties as do the statistical techniques. For example if a hydrologic model is being employed, the parameters of the model must be estimated in some way. This is generally done on the basis of observed data from the basin in question, from observed data from a similar basin, or from so-called physical relationships such as Manning's equation, infiltration parameters, etc., and a set of accompanying tables. Regardless of how the parameters are estimated, the same type of questions regarding these estimates and the nature of the hydrologic model itself must be answered as outlined above for frequency analysis estimates. We cannot substitute mathematical and empirical relationships for hydrologic knowledge any more than we can substitute statistics for hydrologic knowledge.

Based on these last few paragraphs, one might conclude that the magnitude of rare events should not be estimated since the estimates may be so uncertain. Generally, however, this is not one of the options available. An estimate must be made. Hydrology must not be ignored in making this estimate. Statistical, modeling, and/or empirical flow estimates should be made and then adjusted, if required, to reflect the hydrologic situation. This is not to say a factor of safety is to be applied. Adjustments should be based on hydrology, not rules of thumb.

Example Problem 2.1. Flow probabilities

If we assume the data on peak discharge for Rose Creek (Table 2.11) can be described by a normal distribution, in any year what is the:

a. Prob($Q_p > 1200$ cfs),
b. prob($Q_p < 1200$ cfs),
c. return period for a flow of 1200 cfs,
d. magnitude of a 20-year flood, and
e. prob($800 < Q_p < 1000$)?

Table 2.11 Peak Discharge Data: Rose Creek at Nebo, Kentucky

Year	Discharge (cfs)	Year	Discharge (cfs)
1952	624	1962	730
1953	722	1963	680
1954	358	1964	800
1955	500	1965	622
1956	884	1966	571
1957	689	1967	350
1958	1230	1968	920
1959	1000	1969	1240
1960	900	1970	818
1961	860		

Solution:
Calculation of sums and means

$$\Sigma Q_p = 14498 \qquad \Sigma(\ln Q_p)^2 = 825.92$$

$$\Sigma Q_p^2 = 1.2131 \times 10^7 \qquad \Sigma(\ln Q_p)^3 = 5466.13$$

$$\Sigma \ln Q_p = 125.11 \qquad \bar{Q}_p = \Sigma Q_p/n = 763 \text{ cfs}$$

a. $$\text{prob}(Q_p > 1200 \text{ cfs}) = \text{prob}\left(Z > \frac{1200 - \bar{Q}_p}{S_Q}\right)$$

$$S_Q = \sqrt{\frac{\Sigma Q_p^2 - n\bar{Q}_p^2}{n-1}}$$

$$= \sqrt{\frac{1.2131 \times 10^7 - 19(763)^2}{18}} = 244$$

$$\text{prob}(Q_p > 1200) = \text{prob}\left(Z > \frac{1200 - 763}{244}\right)$$

$$= \text{prob}(Z > 1.79).$$

From the table of the standard normal distribution in Appendix 2A the prob($Z > 1.79$) is found to be 0.037. Therefore, the desired probability is

$$\text{prob}(Q_p > 1200) = 0.037$$

b. $$\text{prob}(Q_p < 1200) = 1 - \text{prob}(Q_p > 1200)$$

$$= 1 - 0.037 = 0.963$$

c. $$T = 1/p = 1/\text{prob}(Q_p > 1200)$$

$$= 1/0.037 = 27 \text{ years}$$

d. $$T = 20; \text{ therefore } p = 1/T = 0.05$$

$$\text{prob}(Q_p > Q_{20}) = 0.05$$

$$\text{prob}(Q_p < Q_{20}) = 0.95$$

$$\text{prob}(Q_p < Q_{20}) = \text{prob}(Z < Z_{20}) = 0.95$$

from the standard normal distribution (See Appendix 2A)

$$Z_{20} = 1.645$$

$$\frac{Q_{20} - \bar{Q}_p}{S_Q} = 1.645.$$

From part a, $S_Q = 244$ and $\bar{Q}_p = 763$.

$$Q_{20} = Z_{20}S_Q + \bar{Q}_p = 1.645(244) + 763 = 1164 \text{ cfs}.$$

e. $\text{prob}(800 < Q_p < 1000)$

$$= \text{prob}\left(\frac{800 - \bar{Q}_p}{S_Q} < Z < \frac{1000 - \bar{Q}_p}{S_Q}\right)$$

$$= \text{prob}(0.15 < Z < 0.97)$$

$$= \text{prob}(Z < 0.97) - \text{prob}(Z < 0.15)$$

$$= 0.8334 - 0.5596 = 0.274.$$

Example Problem 2.2. Frequency analysis

Using the Rose Creek data in Table 2.11,

a. Plot the data on lognormal probability paper.
b. Draw in the best fitting line according to the
  (i) lognormal distribution
  (ii) extreme value type I distribution
  (iii) log Pearson type III distribution.
c. Estimate the 100-year flood using the
  (i) lognormal distribution
  (ii) extreme value type I distribution
  (iii) log Pearson type III distribution.

Solution:

a. From Eq. (2.8), the plotting position is given by $m/(n+1)$, where m is rank and $n = 19$, the number of data values ($p = m/20$).

m	Q	p	m	Q	p
1	1240	0.05	11	722	0.55
2	1230	0.10	12	689	0.60
3	1000	0.15	13	680	0.65
4	920	0.20	14	624	0.70
5	900	0.25	15	622	0.75
6	884	0.30	16	571	0.80
7	860	0.35	17	500	0.85
8	818	0.40	18	358	0.90
9	800	0.45	19	350	0.95
10	730	0.50			

A plot of Q versus p on lognormal paper is shown in Fig. 2.15.

b.

$$\bar{Q}_p = \frac{\Sigma Q_p}{n} = \frac{14498}{19} = 763$$

$$S_Q = \sqrt{\frac{\Sigma Q_p^2 - n\bar{Q}_p^2}{n-1}}$$

$$= \sqrt{\frac{1.2131 \times 10^7 - 19(763)^2}{18}} = 244$$

$$C_V = \frac{S_Q}{\bar{Q}_p} = \frac{244}{763} = 0.320.$$

To draw the best fitting lines, calculate several points from

$$Q_T = \bar{Q}_p(1 + C_V K_T).$$

Lognormal (K_T from Table 2.5):

T	Return period 2	5	20
K_T	−0.15	0.75	1.85
Q_T	726	946	1215

Plot Q_T versus $1/T$. Since lognormal probability paper is used, the resulting plot should be a straight line.
Extreme value type I ($n = 20$ was used in determining K_T from Table 2.6):

T	5	10	15	20	25	50
K_T	0.919	1.625	2.023	2.302	2.517	3.179
Q_T	987	1159	1256	1325	1378	1539

Plot Q_T versus $1/T$. The plot is not necessarily a straight line since lognormal paper, not EVI paper, is being used. Note that the LN and EVI produce results that are close to each other. The EVI and the LN are nearly identical if the coefficient of skew of the data is 1.139. For these data,

$$C_S = \frac{n^2\Sigma Q_p^3 - 3n\Sigma Q_p \Sigma Q_p^2 + 2(\Sigma Q_p)^3}{n(n-1)(n-2)S_Q^3}$$

$$= \frac{(19)^2 1.0954 \times 10^{10} - 3(19)(14498)1.2131 \times 10^7 + 2(14498)^3}{19(18)(17)(244)^3}$$

$$= 0.29.$$

On the basis of the coefficient of skew, the two distributions are expected to differ. This difference will start to show up in the estimation of Q_{100}. The fact that the coefficient of skew for the extreme value type I is a constant 1.139 suggests the EVI may not be a good descriptor of this data.

Similarly for the LN, γ and C_V are related by

$$\gamma = 3C_V + C_V^3.$$

Substituting $C_V = 0.32$ into this relationship results in an estimate for γ of 0.99 compared to the calculated 0.29. This deviates from the requirements for the LN to some extent but not as severely as for the EVI.

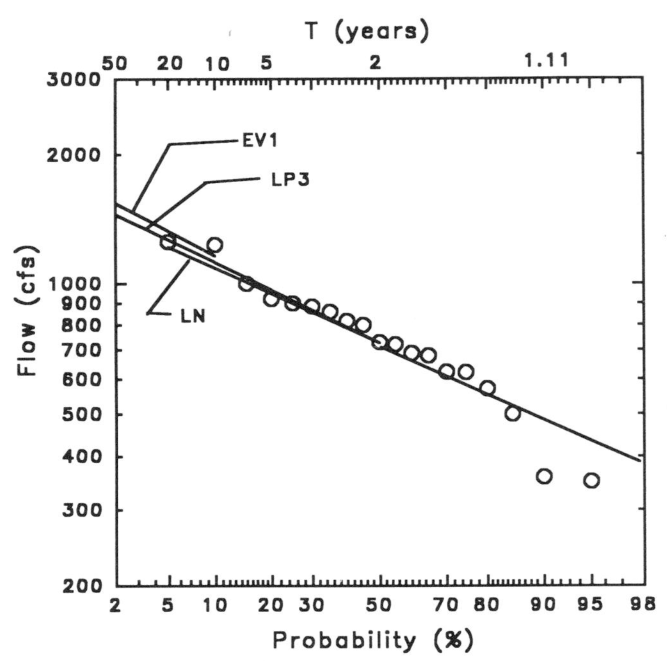

Figure 2.15 Frequency plot for Problem 2.2.

Log Pearson type III:

To draw the best fitting straight line, calculate

$$Q_T = \exp(Y_T),$$

where

$$Y_T = \overline{\ln Q_p} + S_{\ln Q_p} K_T.$$

Based on results in Example Problem 2.1,

$$\overline{\ln Q_p} = \frac{\Sigma \ln Q_p}{n} = \frac{125.11}{19} = 6.585$$

$$S_{\ln Q_p} = \sqrt{\frac{\Sigma(\ln Q_p)^2 - n(\overline{\ln Q_p})^2}{n-1}}$$

$$= \sqrt{\frac{825.92 - 19(6.585)^2}{18}} = 0.34.$$

The C_V of the logarithms is 0.34/6.585 or 0.052 indicating that the logarithms are nearly symmetrically distributed. K_T is a function of C_S of $\ln Q_p$. Let $Y = \ln Q_p$:

$$C_S = \frac{n^2 \Sigma Y^3 - 3n \Sigma Y \Sigma Y^2 + 2(\Sigma Y)^3}{n(n-1)(n-2)S_Y^3} = -0.058.$$

K_T is determined from Table 2.7.

T	2	5	10	25	50
K_T	0.008	0.844	1.276	1.730	2.027
Y_T	6.587	6.873	7.021	7.186	7.278
Q_T	726	966	1120	1310	1448

Q_T is plotted against $1/T$. The LP3 results and the LN results are very close to each other. Note that the skewness of the logarithms is -0.058 or close to zero. The skewness of the logarithms for a LN distribution is zero since the logarithms of the values are normally distributed and the normal distribution has zero skew.

c. Q_{100} is estimated in the same manner as the points on the straight line were calculated. The results are:

	LN	EVI	LP3
Q_{100}	1502	1699	1584

Note that the LN and the EVI differ by nearly 200 cfs, while the LN and LP3 differ by only 83 cfs.

A good choice for the best estimate of Q_{100} would be around 1550 cfs for this stream. If designing a facility on the basis of a 100-year return period and failure of the facility to properly handle the design flow were serious (high economic loss or loss of life), the more conservative estimate of around 1700 cfs could be used.

Example Problem 2.3. Regional analysis

It is desired to estimate the 50-year flood on a 4 mile^2 watershed. No streamflow data on the watershed are available. The investigator has collected flow data on six nearby streams. An analysis of the flood frequency relationship is shown in the following table. Use these data and a regional flood frequency approach to estimate the 50-year flows on the 4 mile^2 watershed.

	Stream					
	A	B	C	D	E	F
Q_{50}	52,000	26,000	6,400	1,800	8,300	4,000
Q_{20}	34,000	15,000	4,600	1,400	6,400	2,950
Q_{10}	24,000	9,200	3,400	1,150	5,100	2,200
Q_5	15,000	5,100	2,400	860	3,800	1,580
Q_2	6,300	1,640	1,200	520	2,600	810
$Q_{1.11}$	1,700	295	420	240	960	300
Area(mile^2)	20	5	3	1	7	2

Solution:

First develop a plot of Q_T/Q_1 versus T. Use Q_2 for Q_1.

	Watershed					
	A	B	C	D	E	F
Q_{50}/Q_2	8.25	15.85	5.33	3.46	3.19	4.94
Q_{20}/Q_2	5.40	9.15	3.83	2.69	2.46	3.64
Q_{10}/Q_2	3.81	5.61	2.83	2.21	1.96	2.72
Q_5/Q_2	2.38	3.11	2.00	1.65	1.46	1.95
Q_2/Q_2	1.00	1.00	1.00	1.00	1.00	1.00
$Q_{1.11}/Q_2$	0.27	0.18	0.35	0.46	0.37	0.37

Next plot Q_T/Q_2 versus T and draw a smooth line connecting the median of the data (Fig. 2.16). It appears as though watershed B does not fit the other watersheds, so it is not used in computing this median. Plot Q_2 versus watershed area (Fig. 2.17). Use this relationship to determine the index flood (Q_2). From this information for a 4 mile^2 watershed, Q_2 is estimated as 1500 cfs. The ratio Q_{50}/Q_2 is 5. Therefore,

$$Q_{50} = \frac{Q_{50}}{Q_2} \times Q_2 = 5 \times 1500 = 7500.$$

The estimated 50-year flow is 7500 cfs.

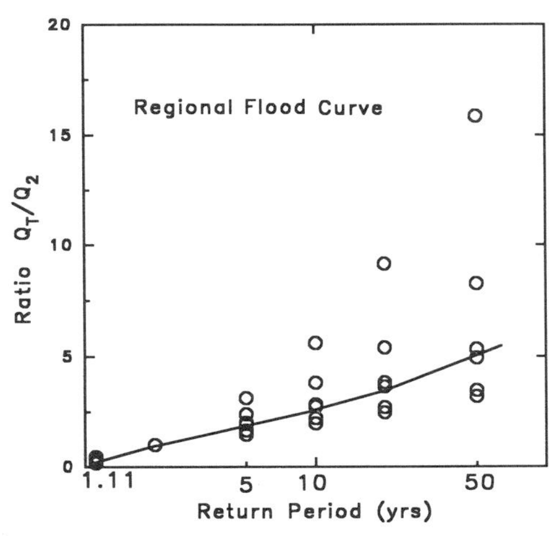

Figure 2.16 Regional flood frequency curve for Problem 2.3.

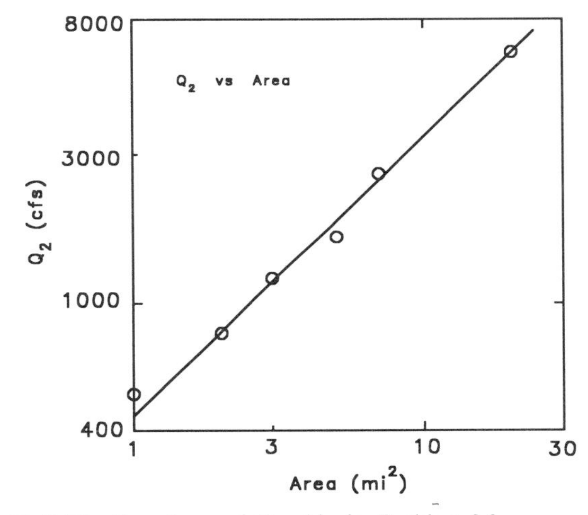

Figure 2.17 Q_2—Area relationship for Problem 2.3.

Example Problem 2.4. Weighted skew coefficient

Data from a stream near St. Louis, Missouri, yielded the following statistics based on the natural logarithms of the flow. Note that Y is equal to the log of the peak flow ($Y = \ln Q_p$).

$$\bar{Y} = 7.313, \qquad C_s = -0.108$$
$$S_Y = 0.250, \qquad N = 25.$$

Estimate the 100-year peak flow based on the log Pearson type III distribution using the weighting procedure on the generalized and data derived skews.

Solution:

A weighted skew coefficient based on Eq. (2.27) is calculated. From Fig. 2.10,

$$\bar{G} = -0.4.$$

The mean square error for Fig. 2.10 is

$$MSE_{\bar{G}} = 0.302.$$

The data derived estimate for the skew is

$$G = -0.108.$$

From Eq. (2.28) and (2.29),

$$MSE_G = \text{antilog}_{10}[A - B \log_{10}(N/10)]$$
$$A = -0.33 + 0.08|G| = -0.33 + 0.08(0.108) = -0.321$$
$$B = 0.94 - 0.26|G| = 0.94 - 0.26(0.108) = 0.912$$
$$MSE_G = \text{antilog}_{10}[-0.321 - 0.912 \log_{10}(25/10)] = 0.207$$

From eq. (2.27)

$$G_W = \frac{MSE_{\bar{G}}(G) + MSE_G(\bar{G})}{MSE_{\bar{G}} + MSE_G}$$
$$= \frac{0.302(-0.108) + 0.207(-0.4)}{0.302 + 0.207} = -0.227.$$

From Table 2.7 using a skew of -0.227, K_{100} is interpolated as

$$\frac{2.178 - K_{100}}{2.178 - 2.029} = \frac{-0.2 + 0.227}{-0.2 + 0.4}$$
$$K_{100} = 2.158$$
$$Y_{100} = \bar{Y} + S_Y K_T = 7.313 + 0.25(2.158) = 7.853$$
$$Q_{100} = e^{7.853} = 2573 \text{ cfs}.$$

Example Problem 2.5. Historical flow data

Assume that in 1923 a peak flow of 1800 cfs was estimated for Rose Creek at Nebo, Kentucky. Estimate the 100-year flow according to the log Pearson type III distribution.

Solution:

Equations (2.39) through (2.43) are used to adjust the observed flow data for the historical observation. $H = 1961-1923 = 48$ yrs.

$$W = \frac{H - Z}{N} = \frac{48 - 1}{19} = 2.47.$$

Based on data from Example Problem 2.1 and Eq. (2.41),

$$Y_a = \frac{W\Sigma Y + \Sigma Y_Z}{H} = \frac{2.47 \times 125.11 + \ln 1800}{48} = 6.594.$$

Equation (2.42) gives

$$S_a^2 = \frac{W\Sigma(Y - Y_a)^2 + \Sigma(Y_Z - Y_a)^2}{H - 1}$$

$$\Sigma(Y - Y_a)^2 = \Sigma(\ln Q_p - 6.594)^2 = 2.145$$

$$\Sigma(Y - Y_a)^3 = \Sigma(\ln Q_p - 6.594)^3 = -0.486$$

$$S_a^2 = \frac{2.47(2.145) + (\ln 1800 - 6.594)^2}{47} = 0.132$$

$$S_a = 0.363$$

$$C_{Sa} = \frac{H}{(H-1)(H-2)}\left[\frac{W\Sigma(Y - Y_a)^3 + \Sigma(Y_Z - Y_a)^3}{S_a^3}\right]$$

$$= \frac{48}{(47)(46)}\left[\frac{2.47(-0.486) + (\ln 1800 - 6.594)^3}{(0.363)^3}\right]$$

$$= -0.217.$$

From Eqs. (2.25) and (2.26) and Table 2.7 with $K_{100} \approx 2.165$

$$Y_{100} = Y_a + S_a K_{100} = 6.594 + 0.363(2.165) = 7.380$$

$$Q_{100} = e^{7.380} = 1603 \text{ cfs}.$$

Problems

(2.1) Using a hand calculator, determine the mean, standard deviation, coefficient of variation, and skewness of the following data:

20, 45, 13, 80, 12, 30, 18, 22, 17, 32, 22.

(2.2) Using a calculator and tables in the text, estimate the 20-year peak flow based on the lognormal and extreme value type I distribution for the data of Problem 1 assuming the data are annual peak flows.

(2.3) Using a calculator and tables in the text, estimate the 25-year peak flow based on the log Pearson type III distribution for the data of Problem 2.1 assuming the data are annual peak flows.

(2.4) The data given below are annual peak flows for the period 1936 through 1971 for the Chikaskia River near Blackwell, Oklahoma.

a. Plot the data on lognormal probability paper.

b. Plot the estimated flood frequency curve for return periods of 2 through 100 years based on the lognormal, extreme value type I and log Pearson type III distributions on the same graph as part a.

c. Which distribution gives the best estimate of the flood frequency relationship for this location?

d. What is the magnitude of the 100-year peak flow for this location?

Flow (cfs)			
10,800	35,800	3,120	15,500
12,900	6,200	39,300	8,460
26,800	31,000	14,600	64,000
8,340	23,100	55,000	1,200
6,040	13,300	9,050	5,650
8,820	8,070	20,000	31,000
85,000	53,000	48,000	26,000
12,200	8,130	36,500	45,200
82,000	7,280	27,000	5,350

(2.5) Should the value of 1200 cfs in the data of Problem 2.4 be considered an outlier?

(2.6) It was established through interviews of local residents that in 1923 a flood with an estimated peak flow of 100,000 cfs occurred on the stream of Problem 2.4. Based on the log Pearson type III distribution, estimate the magnitude of the 100-year peak flow incorporating this additional information in the estimate.

(2.7) The following data represent the gage height and annual peak discharge for the streamgaging station on the Arkansas River at Ralston, Oklahoma. These gage heights are in feet, and the discharges are in cfs. The period of record is 1923 through 1971. Based on this data:

a. plot discharge versus stage,

b. develop an equation that fits the plot of part a,

c. plot the discharge data on lognormal probability paper,

d. plot a log Pearson type III distribution for the discharge data on the plot of part c.

e. A local resident claims that in the early 1900's a flood that would correspond to a stage of 30 ft occurred at this location. Estimate the return period of the flow associated with this reported event.

Stage	Discharge	Stage	Discharge
23.0	200,000	18.50	114,000
11.8	42,000	14.93	70,200
6.4	11,300	15.30	70,700
10.4	32,400	17.60	92,800
18.7	108,000	21.45	135,000
15.0	73,000	10.48	25,800
15.3	76,500	8.80	17,500
12.1	47,800	9.07	18,700
9.5	28,200	12.71	36,300
10.6	33,700	14.64	49,200
9.3	25,700	21.41	120,000
6.4	11,700	14.86	56,800
16.0	77,800	14.65	54,800
9.9	26,600	21.62	158,000
13.0	47,500	21.22	165,000
16.44	75,600	17.83	103,000
8.48	19,200	8.76	19,700
10.26	27,800	9.00	21,100
13.59	51,000	22.60	171,000
18.54	94,000	6.74	10,400
18.12	97,200	12.54	42,000
22.82	179,000	14.10	52,800
19.55	124,000	16.42	77,000
19.48	110,000	18.33	101,000
		8.14	17,100

(2.8) Blackwell, Oklahoma, is in north central Oklahoma. Figure 2.10 indicates a generalized skew coefficient of -0.2 for this area. Determine a weighted skew coefficient on the basis of the generalized value and the value calculated for Problem 2.4. How much does this affect the estimate for 100-year peak flow? Would you prefer an estimate based on the skew coefficient determined only from the data, only from Fig. 2.10, or a combination? Why?

(2.9) Calculate the 90% confidence intervals for the log Pearson type III flood frequency curve of Problem 2.1. Plot the data, the log Pearson frequency curve, and the confidence intervals on lognormal probability paper.

(2.10) Repeat Problem 2.9 using the data of Problem 2.7.

(2.11) The following data are for Black Bear Creek at Pawnee, Oklahoma, for the period 1943 to 1971. This stream and the streams of Problems 2.4 and 2.7 are in the same general region of Oklahoma. The drainage areas are 576, 1859, and 46,850 mile^2 for Black Bear Creek, Chikaskia River, and the Arkansas River, respectively. Develop a regional flood frequency relationship relating Q_T/Q_2 to drainage area based on these three streams. Use T ranging up to 100 years.

17,800	4,280	3,880	2,520
17,500	4,790	6,890	4,100
4,900	2,610	30,200	1,780
9,390	2,810	15,400	3,040
4,890	8,720	8,880	4,440
4,790	5,430	3,620	3,490
3,830	12,200	2,250	3,050

(2.12) The following data are for the Salt Fork Arkansas River near Cherokee, Oklahoma, for the period of 1941 to 1950. The drainage area is 2439 mile^2. On the basis of these data and the regional relationship developed for Problem 2.11, estimate the 100-year flow at this location. (Note: In practice considerably more data should be used to develop a regional frequency curve.)

4,680	14,000	23,300
35,000	5,760	32,300
10,300	13,800	9,380
14,800		

(2.13) The following data are for the Cimarron River running through Oklahoma. On the basis of these data, estimate the 2-, 5-, 10-, 25-, 50-, and 100-year flows at points where the drainage area is 2050 mile^2.

Area (mile^2)	Q_2	Q_5	Q_{10}	Q_{25}	Q_{50}	Q_{100}
1,038	7,830	17,800	27,000	42,100	55,800	71,600
1,879	8,880	20,000	31,700	53,300		

(2.14) An analysis of 56 years of data indicated that the probability of a flood peak exceeding 1500 cfs was 0.02. During a 10-year period two such peaks occurred. If the original estimate of exceedance probability was correct, what is the probability of getting two such exceedances in 10 years?

(2.15) Forty-five years of peak flow data are available from an annual data series. Six of the values are zero. The remaining 39 values have a mean of 2150 cfs and a standard deviation of 1200 cfs and follow a lognormal distribution.

a. What is the probability of a peak flow exceeding 2750 cfs?

b. Estimate the 20-year peak flow.

(2.16) A project is designed on the basis of a 25-year return period peak flow.

a. What is the probability that the design will be exceeded during the first year?

b. What is the probability that the design will be exceeded (at least once) during the first 10 years?

c. What is the probability that the design will be exceeded (at least once) during the second 10 years?

d. What is the probability that the design will be exceeded (at least once) during the first 20 years?

e. What is the probability that the design will be exceeded exactly once during the first 10 years?

(2.17) What design return period should be used to ensure a 95% chance that a design will not be exceeded in (a) 10 years, (b) 25 years, (c) 50 years, and (d) 100 years?

(2.18) Repeat Problem 2.17 using a 50% chance.

(2.19) What design return period should be used to be 90% confident of no more than one exceedance in a 10-year period?

(2.20) Starting at 0 and using class intervals of 40,000 cfs, plot a frequency histogram of the flow data in Problem 2.7. Superimpose on the frequency histogram a lognormal distribution. Does the lognormal appear to be a good approximation for the histogram?

(2.21) A flood detention structure has its spillway designed on the basis of an estimated 1000-year flood. What is the probability that the design flow will be exceeded in a (a) 10-year period, (b) 50-year period, (c) 100-year period or (d) 1000-year period?

(2.22) Why do you think it is common to perform flood frequency analysis on the logarithm of peak flows rather than the peak flows themselves?

(2.23) What is the difference between an annual series of peak flows and a partial duration series?

(2.24) What is the reason for using regional information on the flood peak skewness coefficient rather than an estimate based strictly on observed data?

(2.25) What assumptions are made in applying flood frequency analysis techniques to estimate design flows?

(2.26) For a normal distribution with a mean of 25 and a variance of 400:

a. What is the probability of exceeding 35?

b. What is the probability of a value less than 30?

c. What is the probability of a negative value?

d. What value has a 1% chance of being exceeded?

(2.27) Discuss when a regional analysis would be beneficial to a frequency analysis?

(2.28) Define the 50-year flood.

(2.29) What is the difference between analytical and graphical flood frequency analysis? Describe the advantages and disadvantages of each.

References

California State Department of Public Works (1923). "Flow in California Streams," Bulletin 5, Chap. 5. [Original not seen, cited in Chow (1964)]

Chow, V. T. (1951). A generalized formula for hydrologic frequency analysis. *Trans. Am. Geophys. Union* **32**(2): 231–237.

Chow, V. T., ed. (1964). "Handbook of Applied Hydrology." McGraw-Hill, New York.

Dalrymple, T. (1960). Flood frequency analysis, U.S. Geological Survey Water Supply Paper 1543-A. *In* "Manual of Hydrology," Part 4, "Flood Flow Techniques." U.S. Government Printing Office, Washington, DC.

Durant, E. F., and Blackwell, S. R. (1959). The magnitude and frequency of floods on the Canadian prairies in spillway design floods. *In* "Proceedings, Hydrology Symposium 1, Ottawa, Canada, 1959."

Haan, C. T. (1977). "Statistical Methods in Hydrology." Iowa State Univ. Press, Ames, IA.

Hazen, A. (1930). "Flood Flows, A Study of Frequencies and Magnitudes." New York.

Interagency Advisory Committee on Water Data (1981). "Guidelines for Determining Flood Flow Frequency," Bulletin 17B of the Hydrology Subcommittee. U.S. Department of Interior, Geological Survey, Office of Water Data Coordination, Reston, VA.

Jennings, M. E., and Benson, M. A. (1969). Frequency curves for annual series with some zero events or incomplete data. *Water Resources Res.* **5**(1): 276–280.

Matalas, N. C., and Langbein, W. B. (1962). Information content of the mean. *J. Geophys. Res.* **67** (9): 3441–3448.

Natural Environment Research Council (1975). "Flood Studies Report." Natural Environment Research Council, London.

Singh, V. P. (1987a). Hydrologic frequency modeling. *In* "Proceedings, International Symposium on Flood Frequency and Risk Analysis, 14–17 May 1986, Baton Rouge, LA." Riedel, Dordrecht, Holland.

Singh, V. P. (1987b). Regional flood frequency analysis. *In* "Proceedings, International Symposium on Flood Frequency and Risk Analysis, 14–17 May 1986, Baton Rouge, LA." Riedel, Dordrecht, Holland.

Singh, V. P. (1987c). Flood hydrology. *In* "Proceedings, International Symposium on Flood Frequency and Risk Analysis, 14–17 May 1986, Baton Rouge, LA." Riedel, Dordrecht, Holland.

Singh, V. P. (1987d). Application of frequency and risk in water resources. *In* "Proceedings, International Symposium Flood Frequency and Risk Analysis, 14–17 May 1986, Baton Rouge, LA." Riedel, Dordrecht, Holland.

Tasker, G. D. (1978). Flood frequency analysis with a generalized skew coefficient. *Water Resources Res.* **14**(2): 373–376.

Wallis, J. R., Matalas, N. C., and Slack, J. R. (1974). Just a moment. *Water Resources Res.* **10**(2): 211–219.

Weibull, W. (1939). A statistical study of the strength of materials. *Ing. Vetenskaps Akad. Handl. (Stockholm).* [Original not seen, cited in Chow (1964)]

3

Rainfall-Runoff Estimation in Storm Water Computations

A hydrologist has an obligation to make the best hydrologic estimates possible, commensurate with the cost and scope of a particular water management problem. Hydrologic calculations are estimates, with the error in these estimates increasing as the degree of approximation increases or as the estimation procedure is applied beyond the range of conditions for which it is intended. The hydrologist must determine if the scope, cost, or importance of a particular project justifies collecting more data and using less approximate methods or whether less precise techniques can be applied. The hydrologist must make the best possible hydrologic estimate and then proceed on that basis. Hydrologists should constantly check back on the projects they have completed to determine the adequacy of the procedures they have used. Needed changes can then be incorporated in the estimation technique that is used.

The empirical and approximate nature of hydrologic estimation methods has led to the development and use of a great number of procedures for estimating runoff, whether it be peak flows, runoff volumes, or complete hydrographs. It is difficult to say that one method is absolutely better than another method or that there is a best method. One can talk of a simple method, but not necessarily of a best method. What is best in one location may produce poor estimates in another location. What is required to evaluate the adequacy of a hydrologic procedure is actual hydrologic data. There is no substitute for real, locally applicable data.

Methods for estimating hydrologic quantities such as streamflow range from the relatively simple and widely used Rational Equation to complex, computer simulation methods. The widespread availability of computers means that one can employ more detailed and tedious methods than were formerly feasible. Again, the method selected should be able to provide the information needed to design the system in question. If the design involves a small drain pipe or culvert, possibly all that is required is an estimate of the maximum flow the facility will be called upon to carry. If the design involves storage and delay of the runoff, a complete runoff hydrograph for the storm may be required. If the storage facility is large and will not empty to "permanent pool" elevation between storms, then possibly a continuous simulation of flow will be required.

Recent interest in water quality, especially ground water quality, has made it even more important to accurately model actual flow paths. The quality of water in various phases of the hydrologic cycle is dependent on the flow paths taken and materials the water has come in contact with. Simply modeling ground water as a lumped storage or even treating all ground water flow as Darcian in nature can be misleading in terms of water quality. Non-Darcian flow or large-pore (macropore) flow or flow in subsurface cracks and channels is known to exist and generally occurs at flow rates several orders of magnitude larger than Darcian flow. Thus, transported pollutants may move into and within ground water systems much quicker than anticipated based on a strictly Darcian analysis. This is just one example of why it is necessary to match procedures used in hydrologic analysis with the purpose of the analysis.

HYDROLOGIC CYCLE

The concepts of the hydrologic cycle are well known and covered by many excellent texts on the subject. The treatment of hydrology in this chapter is largely limited to those parts of the hydrologic cycle of major importance in storm water management. This means that primary emphasis will be placed on precipitation, abstractions from precipitation, and the runoff process. Such things as vegetal interception of precipitation, evaporation, transpiration, and soil water movement will be covered in less detail. Ground water is covered in a separate chapter. These latter parts of the hydrologic cycle are of extreme importance when one is attempting to continuously simulate streamflow. The emphasis in this chapter is on the analysis of single events.

Figure 3.1 depicts a portion of a watershed during a precipitation event. Shown in this figure are the processes of rainfall, interception, evaporation, transpiration, infiltration, percolation, ground water flow, overland flow, subsurface flow, surface storage, detention storage, and channel precipitation. All of these processes have a role to play in hydrology; however, precipitation, infiltration, overland flow, surface storage, detention storage, and of course streamflow are of major importance in storm water hydrology.

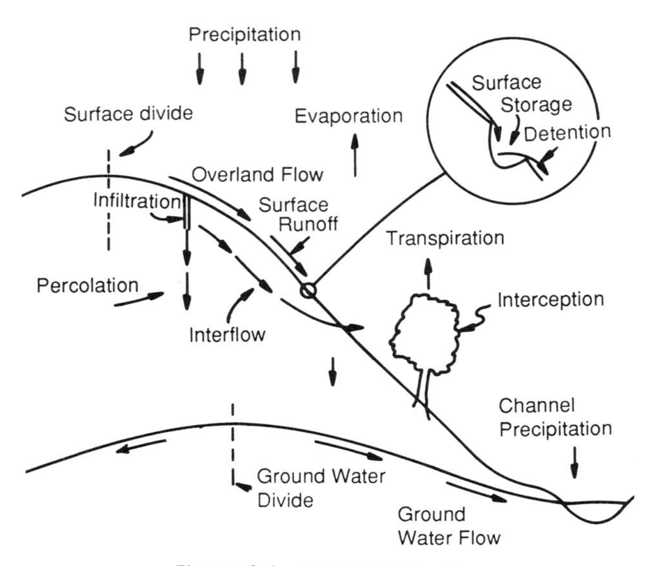

Figure 3.1 Hydrologic diagram.

The most basic equation in hydrology is the continuity equation, which states that over any time interval and for any hydrologic system the difference in the volume of water entering the system, I, and leaving the system, O, must equal the change in the volume of water stored in the system, S:

$$I - O = \Delta S. \tag{3.1}$$

If the hydrologic system is a small catchment, the inflow to the system would be precipitation. The outflow from the system would be streamflow, deep seepage, and evapotranspiration. Storage within the watershed would include soil water, ground water, ponds, lakes, reservoirs, channel storage, surface storage, detention storage, and interception.

For short time intervals (hours), evapotranspiration and deep seepage can generally be ignored. For long time intervals (weeks), surface storage, surface detention, and interception can often be ignored. In the absence of ponds, lakes, or reservoirs, the hydrologic equation is further simplified.

This chapter primarily addresses hydrologic problems that require a time scale of hours. Storm water computations are the major concern. Methods for estimating storm rainfall amounts, intensities, and time distributions are presented. Also presented are methods for estimating runoff peak flows, volumes, and hydrographs from storm rainfall. This type of hydrology is generally thought of as event-based hydrology. An overwhelming majority of storm water and erosion control facilities are designed on the basis of single events or single rainstorms of a given frequency. For this reason, event-based hydrology is emphasized in this chapter. Possibly the two most important factors determining the runoff hydrograph from a small catchment for a given volume of rainfall are the time distribution of the rainfall and the infiltration characteristics of the catchment. A major part of this chapter is devoted to these two considerations. The third important factor in hydrograph development, which also receives considerable attention, is the routing of a rainfall excess from the point of its generation to the catchment outlet.

Also covered in the last major section of the chapter but to a lesser degree are methods for computing hydrologic balances on ponds or subareas within watersheds. Such calculations are useful for sizing ponds, estimating the volume of waste water that can be disposed of via irrigation of vegetated plots, estimating the volume of water that might be required to maintain a preset water level in a pond, and other problems where long-term (weeks) hydrologic balances are of concern. Chapter 13 deals with continuous simulation models or models for simulating continuous records of runoff including storm water runoff as well as runoff that occurs between rainstorms.

PRECIPITATION

The starting point for most storm water management studies is a consideration of precipitation. Our concern is limited to precipitation in the form of rainfall. It will be further limited to storm rainfall. Rainfall data are much more widely available than streamflow data and much less affected by land-use changes. In the U.S., the National Weather Service maintains a system of raingages and has an extensive system of publications for disseminating the data collected. The network of gages consists of both recording and nonrecording gages. The nonrecording or standard gages are used primarily for 24-hr rainfall. The recording gages provide a complete time-intensity history of rainfall events. For small watershed storm water studies, relatively short duration but high-intensity rainfalls are of extreme importance. Thus, data derived from recording raingages are generally required.

Mean Area Precipitation

If one is attempting to reconstruct a historical rainfall on a watershed, interest may exist in methods for estimating the average rainfall on the watershed. The three most common methods for computing average watershed rainfall are the arithmetic mean, the Thiessen polygon method, and the isohyetal method. These methods are illustrated in Fig. 3.2. All of the methods compute the average watershed rainfall as a weighted average of nearby raingages. The equation used is

$$\bar{R} = \Sigma W_i R_i / \Sigma W_i, \tag{3.2}$$

where $\bar{R}$ is the average watershed rainfall, W_i is the weighting factor, and R_i is the ith rainfall amount.

The arithmetic mean, or station average method, is the easiest but least accurate of the three methods. The arithmetic average is the average of all applicable raingages. Thus, the weighting factors for Eq. (3.2) are all unity, and account of factors such as the placement of the gage with respect to the catchment boundaries is not taken. A gage central in the catchment gets the same weight as one on the boundary or outside the catchment.

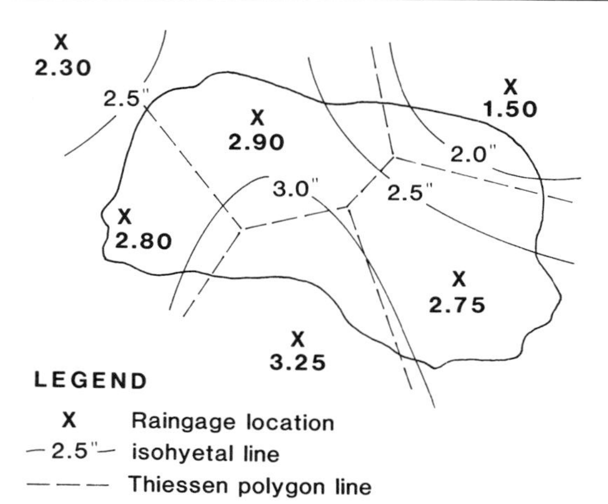

Figure 3.2 Estimating area depth of precipitation.

The Thiessen method is second in ease of application and accuracy. The Thiessen method weighting factor is the area of the watershed nearest to the particular gage. This area is determined by drawing polygons whose sides are the perpendicular bisectors of lines connecting the gages. The Thiessen method has the advantage that as long as the same raingages are used, the weighting factors remain the same for each rainstorm. Thus, many storms can be analyzed without recalculating the weighting factors.

The isohyetal method weighting factor is the area of the watershed enclosed between adjacent isohyetes (lines of constant rainfall). Isohyetal lines are drawn on a watershed in the same manner as topographic lines are drawn on a topographical map except rain depth rather than elevation is the controlling variable. With the isohyetal method, R_i in Eq. (3.2) is the average rainfall depth associated with the weighting factor W_i and is generally taken as the average of the two enclosing isohyetal lines. Exceptions to this are along watershed boundaries. The accuracy of this method is better than the previous two methods; however, the amount of work is greater. The isohyetal map, and consequently the weighting factors, must be determined for each rainstorm.

Example Problem 3.1 Mean Areal Rainfall

Compute the mean areal rainfall for the situation depicted in Fig. 3.2.

Solution: Table 3.1 shows the calculations that are involved. The weighting factors shown in the table are the areas enclosed by the isohyetal lines or Thiessen polygons. The total area is 175 mile^2.

Table 3.1 Calculation of Mean Annual Rainfall in Inches and Square Miles

Arithmetic average

$R_a = (2.30 + 2.90 + 2.80 + 3.25 + 2.75 + 1.50)/6 = 2.58$ in.

R_i	W_i	R_iW_i	
Thiessen polygon			
2.3	0	0	
2.9	48	139.2	
2.8	27	75.6	
3.25	21	68.3	$R_a = 479.1/175$
2.75	62	170.5	$= 2.74$ in.
1.5	17	25.5	
Total	175	479.1	
Isohyetal method			
3.1	34	105.4	
2.75	104	286.0	$R_a = 471.1/175$
2.25	29	65.3	$= 2.69$ in.
1.8	8	14.4	
Total	175	471.1	

Comment: Simple visual inspection indicates that either the Thiessen polygon or isohyetal estimate is superior to the arithmetic average. If one were to sketch the 2.70-in. isohyetal line representing approximately the average rainfall based on either the Theissen polygon or isohyetal method, it would traverse the basin in a manner approximating the center of the catchment. The same cannot be said of the arithmetic average 2.58-in. isohyetal line, which would only catch the upper right part of the catchment.

Point Precipitation Patterns

Rainfall Depth–Duration–Frequency

Often interest exists in rainfalls that can be expected in the future rather than what has happened in the past. Historical rainfalls certainly guide us in estimating future rainfalls, but only on a probabilistic basis. Many types of hydrologic analyses require estimates of rainfall depths (or intensities) for certain durations and frequencies of occurrence. Rainfall depth–duration–frequency (DDF) or intensity–duration–frequency (IDF) data are generally available in the form of tables, graphs, or maps on which isohyetal lines are drawn. U.S. Weather Bureau TP 40 and similar documents provide this information (Hershfield, 1961; Frederick *et al.*, 1977) in the form of maps for the U.S.

As discussed earlier, many hydrologic designs are done on the basis of a selected frequency. This means one must be able to determine rainfall depths for the selected frequency. In specifying a rainfall, it is necessary to specify the depth of the rainfall, the duration of the rainfall, and the frequency of occurrence of the rainfall. Alternatively, the intensity, duration, and frequency can be specified since the intensity, i, duration, T, and depth, D, are related by

$$D = iT. \tag{3.3}$$

DDF information thus indicates how much rain is expected, the duration or period of time over which the rain occurs, and how frequently the rain can be expected. The DDF relationship for point rainfall has been expressed by the general relationship

$$i = KF^x/(T + b)^n, \tag{3.4}$$

where F is frequency and K, b, x, and n are constants for a particular location (Bernard, 1942). Values for K, b, x, and n can be estimated by nonlinear regression if IDF or DDF data are available. These coefficients show considerable variation throughout the U.S. Bernard presented some typical values for these constants. Equation (3.4) is dimensional, and thus the coefficient K depends on the dimensions of i.

In the U.S., the use of Eq. (3.4) has been largely replaced by U.S. Weather Bureau TP 40 (Hershfield, 1961) and similar documents. TP 40 contains DDF data for the U.S. for durations of 30 min to 24 hr and frequencies of 1 to 100 years. A more recent publication (Frederick *et al.*, 1977), HYDRO-35, gives DDF data for the eastern U.S. for durations of 5 to 60 min and frequencies to 100 years. HYDRO-35 has largely superceded TP 40 for durations of 1 hr or less. Many local drainage authorities, highway departments, water resources agencies, etc., have prepared DDF data for use in specific localities. Rainfall depths for 24-hr durations and various frequencies for the U.S. can be found in Appendix 3A. For the western U.S., reference should be made to National Oceanic and Atmospheric Administration (NOAA) Atlases as listed in the references at the end of this chapter. These atlases give detailed rainfall information reflective of the rapid variability in the rainfall regime in the western U.S. due to topographic effects.

The procedure used to develop TP 40 was to prepare four key base maps showing the 2-year, 1-hr; 2-year, 24-hr; 100-year, 1-hr; and 100-year, 24-hr rainfalls. Annual series data consisting of the maximum 60-min and 24-hr rainfall depths converted to a partial duration series were used. The 2-year rainfall amounts were determined by plotting on log–log paper the return period versus rainfall depth, drawing a smooth curve through the data, and reading the 2-year value.

The 100-year depths were determined by using the Extreme Value type I distribution for selected stations with long rainfall records. The ratio of the 100- to 2-year depths was determined for these stations and a map prepared showing this ratio. The 100-year rainfall depths for the stations with short records were estimated by the 100- to 2-year ratio.

Rainfall depths for other return periods were determined by plotting the 2- and 100-year depths on special paper, connecting the points by a straight line, and reading off the desired depths for other return periods. The spacing along the return period scale of this special paper was empirical from 1 to 10 years and theoretical based on the Extreme Value type I distribution from 20 to 100 years.

Rainfall depths for durations other than 1 hr or 24 hr were obtained by plotting the 1- and 24-hr values on a second special paper and connecting the points with a straight line. This paper was obtained empirically from an analysis of records from 200 fist-order U.S. Weather Bureau stations. The depth for the 30-min duration was obtained by multiplying the 1-hr value by 0.79.

Example Problem 3.2 IDF curves

The data tabulated below were obtained from TP 40 for Stillwater, Oklahoma. Use these data to construct a smoothed set of IDF curves.

Solution: The curves in Fig. 3.3 were plotted directly from the TP 40 without any smoothing. The apparent roughness in the curves is due to the difficulty of reading values from TP 40. Equation (3.4) was used as a smoothing relationship. The coefficients K, b, x, and n were estimated by nonlinear regression. Nonlinear regression programs are widely available for microcomputers. A program by Wilkinson (1987) was used to estimate these coefficients for Stillwater, Oklahoma, with the result that $K = 1.75$ if units are in inches, $b = 0.12$, $x = 0.21$, and $n = 0.80$. Using these coefficients, the Stillwater data were smoothed and replotted in Fig. 3.4.

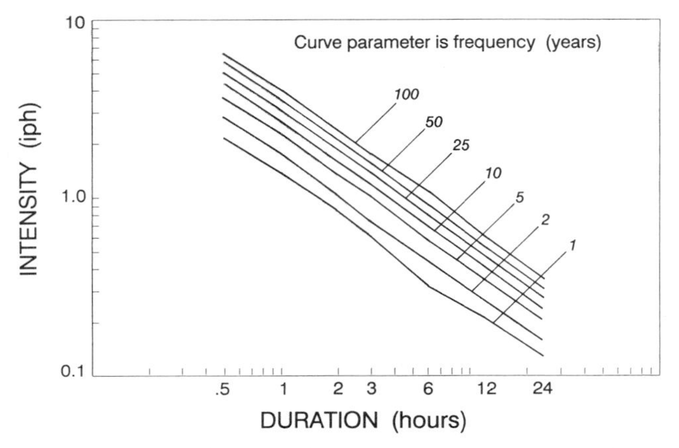

Figure 3.3 Raw intensity–duration–frequency curves, Stillwater, Oklahoma.

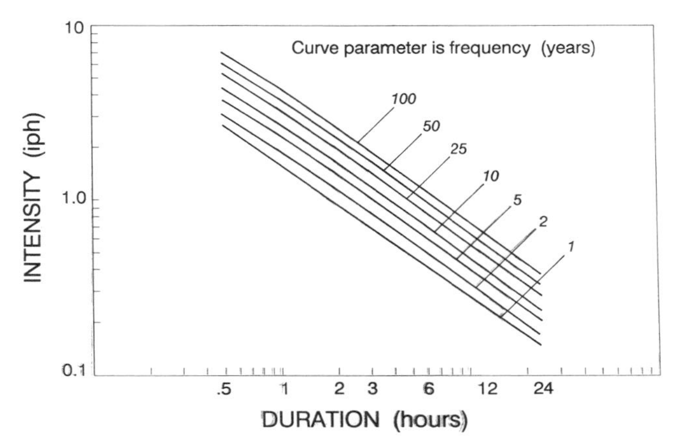

Figure 3.4 Smoothed intensity–duration–frequency curves, Stillwater, Oklahoma, based on Eq. (3.4) with $K = 1.75$, $x = 0.21$, $b = 0.12$, and $n = 0.80$.

Intensity–Duration–Frequency Data for Stillwater, Oklahoma

	T (hr)						
F (years)	0.5	1	2	3	6	12	24
				I (iph)			
1	2.20	1.45	0.85	0.62	0.33	0.22	0.13
2	2.90	1.80	1.07	0.75	0.45	0.27	0.16
5	3.70	2.35	1.37	1.05	0.60	0.36	0.21
10	4.40	2.75	1.63	1.20	0.71	0.42	0.24
25	5.10	3.20	1.90	1.42	0.83	0.49	0.28
50	5.80	3.60	2.15	1.58	0.92	0.55	0.32
100	6.50	4.10	2.40	1.77	1.10	0.62	0.36

Comment: These curves represent a consistant set of IDF information for Stillwater, Oklahoma, that can be used for a number of storm water design procedures. Figure 3.5 shows how well Eq. (3.4) predicts rainfall intensities for Stillwater. Such agreement makes it possible to replace the TP 40 maps with Eq. (3.4) in computer programs if the coefficients for the equation are known. Of course, the coefficients must be estimated for each particular location to which the equation is to be applied. One of the problems at the end of this chapter suggests a method for estimating the coefficients of Eq. (3.4) without using a nonlinear regression approach.

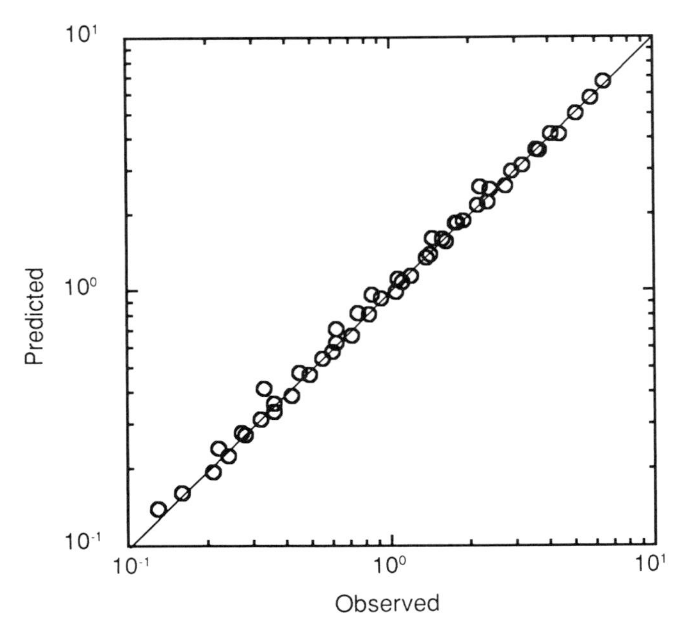

Figure 3.5 Predicted versus observed rainfall intensities, Stillwater, Oklahoma.

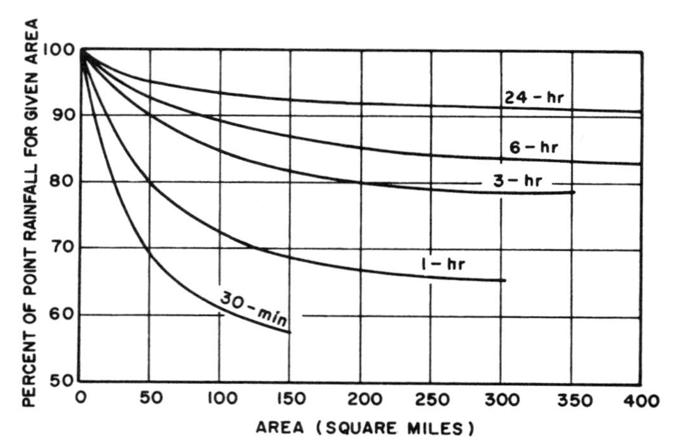

Figure 3.6 Reduction factor for areal precipitation (Hershfield, 1961).

Equation 3.4 can be used to extrapolate IDF curves toward shorter durations, although in the U.S. the use of HYDRO-35 is preferable where it applies. For example, the 10-min, 10-year rainfall intensity for Stillwater (Example Problem 3.2) can be estimated as

$$i = \frac{1.75(10)^{0.21}}{(10/60 + 0.12)^{0.80}} = 7.71 \text{ iph}.$$

IDF curves can be constructed for any locality in the U.S. (and many other places as well). From the curves one can easily determine the rainfall intensity (and depth) associated with any duration and frequency. Similar data for durations of up to 10 days are available (U.S. Department of Commerce, 1964), but these longer duration storms are seldom used in small watershed storm water management work.

The determination of DDF or IDF curves from local rainfall data requires several years of data. The frequency analysis techniques of Chapter 2 can be used. The probability distribution most often used is the Extreme Value type I distribution. The data would consist of the annual maximum rainfall depths for various durations over the period of record. The data for each duration would be analyzed separately and plotted in the form of an IDF curve. Equation (3.4) or a similar equation could then be used to smooth the data to ensure consistency at a given duration for various frequencies.

If only a short record of rainfall is available, it may be desirable to use a partial duration series rather than an annual series of data so that more than 1 data point per year may be included in the analysis. The results of a partial duration analysis can be converted to an annual series by using the multiplying factors of 0.88, 0.96, and 0.99 for return periods of 2, 5, and 10 years (Hershfield, 1961). For longer return periods, no adjustment is required.

From the discussion on the development of the data used to produce TP 40, it should be apparent that the data representing any particular frequency (any particular line in Fig. 3.4) are derived from not one, but a possible multitude of rain storms. This fact should be kept in mind when the development of temporal patterns for rainstorms is discussed.

Rainfall DDF data are derived from point rainfall information. When they are applied to an area rainfall situation, reduction factors as shown in Fig. 3.6 should be applied (Hershfield, 1961). This has the affect of reducing the rainfall depths. The correction is much more significant for short duration storms, which might typically be thunderstorms. For small watersheds, areal correction is not required.

The areal reduction factors shown in Fig. 3.6 are generalized curves presenting approximate relationships for the U.S. It is possible to develop such curves for a specific locality. For small catchments, the correction factor is small and generally ignored. Reasons the correction factor curves have the shape they do include: (1) High intensity, short duration storms tend to cover a relatively small area; (2) long duration storms generally are widespread in areal extent; and (3) data used to produce DDF curves are maximum observed values for a particular storm and as one moves from the center of maximum intensity, the recorded rainfall

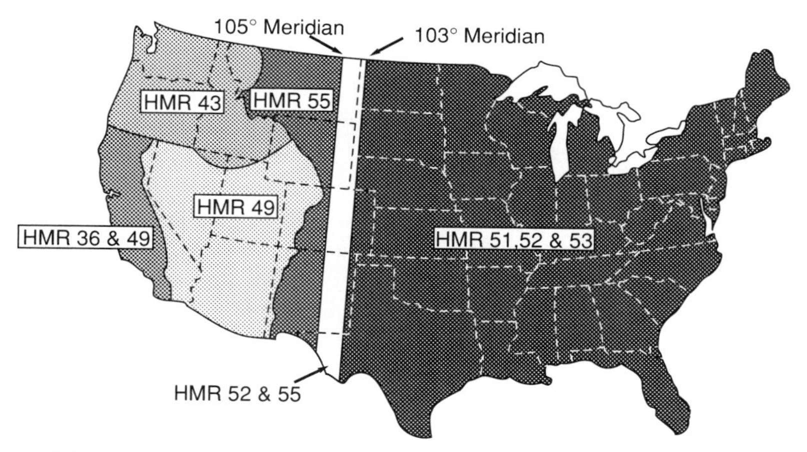

Figure 3.7 Regions of the conterminous U.S. for which PMP estimates are provided in the indicated Hydrometeorological Reports (U.S. National Weather Service, 1984).

depth decreases, thus lowering the average areal rainfall.

Probable Maximum Precipitation

The design of floodwater structures in situations where failure of the structure might endanger human life, disrupt vital community services, or cause unusual economic loss must be done with extreme care. Most engineers and agencies want to minimize the possibility of failure of structures under such conditions. They do not want to design with a known probability of failure no matter how small that probability might be. This has led to the adoption of design rainfalls based on the concept of the Probable Maximum Precipitation or PMP. The PMP is defined as the theoretically greatest depth of precipitation for a given duration that is physically possible over a given size storm area at a particular location at a certain time of the year (National Research Council, 1985).

The major steps in estimating the PMP are:

1. Study major rainstorms to determine maximum areal rainfalls and ascertain the meteorological factors important to the rainfall.
2. Transpose the major storms within topographically and meteorologically homogeneous regions.
3. Adjust the rainfall (for each transposed storm) by the ratio of maximum atmospheric moisture in the place of occurrence to that which existed during the storm.
4. Smoothly envelope the resulting rainfall values durationally, areally, and if generalized PMP is being developed, regionally. Explanations should be given for discontinuities.

PMP values are quite conservative. The ratio of PMP to the 100-year rainfall for 10 mile^2 (25.4 km^2) and 24-hr durations range from 2 to 6 in the U.S. with a majority of the ratios being 4 or 5. Many storms that exceed 50% of the PMP for several combinations of durations and area have been recorded (Riedel and Schreiner, 1980). PMP estimates for the U.S. have been published in a number of different Hydrometeorological Reports by the U.S. National Weather Service (1943–1984) as indicated in Fig. 3.7. Figure 3.8 shows one set of estimates for the 6-hr, 6-mile^2 (15.5 km^2) PMP.

In the U.S., federal and state agencies classify dams on the basis of hazard criteria. These criteria generally refer to the consequences of dam failure in terms of potential loss of life, amount of economic loss, disruption of vital services, damage to major transportation systems, etc. The more severe the potential losses in terms of these criteria, the greater the hazard classification. An example set of hazard classifications is given in Table 9.5. Generally the spillway of a structure receiving a high hazard classification is required to pass the flood resulting from the PMP. For some agencies, this capacity criteria is relaxed somewhat for less hazardous structures so that the spillways may be designed to pass a fraction of the flood resulting from the PMP. A report by the National Research Council (1985) summarizes these criteria for the U.S. Note that these criteria are generally spillway capacity criteria and not storage criteria.

The concepts of Probable Maximum Precipitation and Probable Maximum Flood have been criticized because they are neither probable nor maximum and their likelihood of occurrence cannot be stated (Yevjevich, 1968). Yet the concepts have found wide application and are embedded in many regulations.

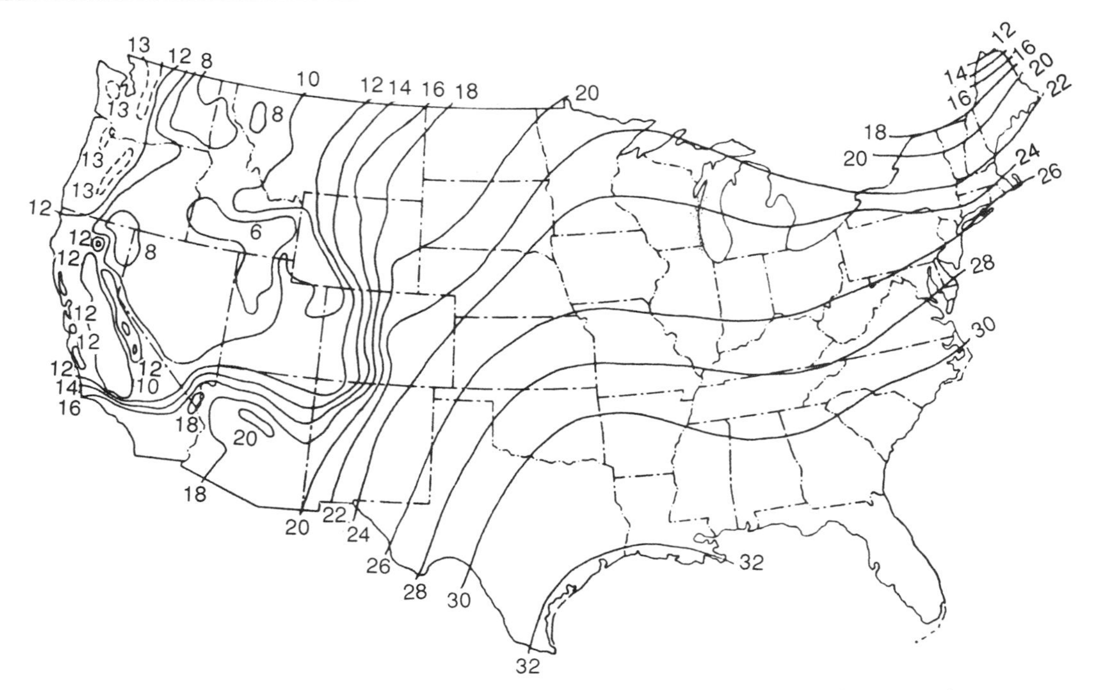

Figure 3.8 6-hr, 6-mile^2 (25.9 km^2) PMP. Multiply values by 25.4 to get mm (Chow, 1964).

Linsley *et al.* (1982) indicate that the world's greatest observed point rainfalls can be enveloped by the relationship

$$R_{max} = aD^{0.48}, \qquad (3.5)$$

where R_{max} is the maximum rainfall observed for duration, D, in hours. The value for the coefficient a is about 16.4 if rainfall is in inches and 417 if rainfall is in millimeters. Other references give slightly different estimates for the coefficient a (Chow *et al.*, 1988). This relationship apparently covers durations ranging from 1 min to 2 years. Values from Eq. (3.5) should not be confused with PMP values. Equation (3.5) is empirical and does not take into account any geographic or meteorologic factors. Values calculated from this equation generally exceed PMP values.

Rainfall Time Distribution

The analysis of modern storm water management systems often requires hydrographs of storm water flow —not just peak flow or runoff volume estimates. Hydrographs in turn require knowledge of the rainfall time-intensity pattern that produces the hydrograph. Thus, not only is it necessary to know the depth of a rainfall of a given duration and frequency, but the time distribution of the rainfall within its duration must be known as well. A plot of the time distribution of rainfall intensity is known as the rainfall hyetograph or simply the hyetograph. There are several possible methods of arriving at a design hyetograph. The two most common methods are either to adopt a historical storm that has occurred in the vicinity and has caused considerable damage or to develop a synthetic design rainstorm.

The historical rainfall approach has the advantage of being readily identified and explained to engineers and to the public. It has the distinct disadvantage of an unknown frequency of occurrence. As a matter of fact, it is difficult to associate a return period with a particular rainfall time-intensity pattern. For example, the rainfall depicted in Fig. 3.9 for Stillwater, Oklahoma,

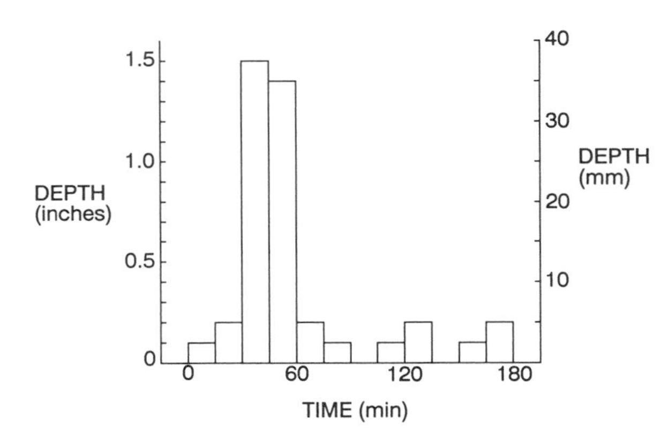

Figure 3.9 Historical rainfall.

Table 3.2 Frequency of Sample Historic Rainfall

Duration (min)	Depth (in.)	Frequency (years)
15	1.5	8
30	2.9	49
60	3.3	32
90	3.5	25
120	3.6	20
180	4.1	24

has different return periods depending on the duration being considered. Table 3.2 shows the maximum rainfall intensities from this storm and the associated return period as determined from Fig. 3.4 or Eq. (3.4). It is apparent that if this storm were adopted as a design storm, the resulting runoff would be assigned a different return period depending on the critical flow-time (rainfall duration) parameters for the watershed.

One approach to using historical rainfall data is to select severe storms that have occurred on or near the catchment of interest over a period of several years and use each of these storms in a rainfall-runoff model to estimate the resulting runoff, especially the peak flow. These estimated flows can then be analyzed by standard frequency analysis techniques and the runoff with the desired return period estimated. This procedure eliminates the problem of having to assign a return period to a rainstorm but does require data on severe rainstorms over a long period of time.

The difficulty of assigning a return period to a total rainstorm, or conversely of estimating a rainstorm pattern given a return period, has led to the development of synthetic storms to which return periods are assigned. These synthetic storms have the advantage of being a consistent basis for design, but have the disadvantage of having a very remote possibility of every happening exactly as specified. McPherson (1969) also comments on this problem. (It should be noted that any particular historical storm also has a very remote chance of being duplicated.)

Synthetic storms represent reasonable approximations to actual storm patterns that might be expected for a given return period. There are a number of ways to develop a synthetic storm pattern. Methods that produce storms that have the same return period regardless of the duration selected from within the storm are appealing from the standpoint of providing a consistant basis of design regardless of the size of the catchment. Storm patterns of this type are produced by critically "stacking" or arranging time increments of rainfall so that the largest possible depth is obtained for any duration and frequency. Such a storm pattern is sometimes referred to as a balanced storm. A balanced storm produces a rainfall depth or intensity whose frequency is independent of the duration of the storm. Thus a 60-min rainfall selected from a balanced storm would have the same return period as a 6-hr rainfall selected from the same balanced storm.

DDF Rainfall Pattern

One method of synthetic balanced storm development has been to read the IDF curve (for example, Fig. 3.4) for a given frequency at selected durations. From these intensities, incremental rainfall volumes and intensities are computed and then rearranged to form a storm pattern. Example Problem 3.3 illustrates this procedure.

Example Problem 3.3 Time distribution of rainfall—DDF method

Based on data in TP 40, develop a synthetic rainfall time distribution for a 3-hr, 25-year rainfall event at Stillwater, Oklahoma. Use a time increment of 15 min.

Solution: Table 3.3 contains the required calculations. The calculations will be illustrated by considering the 30-min line. The intensity in iph is determined from Eq. (3.4) using $K = 1.75$, $b = 0.12$, $x = 0.21$, and $n = 0.80$ as determined in Example Problem 3.2.

$$i = \frac{KF^x}{(T + b)^n} = \frac{1.75(25)^{0.21}}{(30/60 + 0.12)^{0.80}} = 5.04.$$

Intensities could also be obtained from IDF curves such as Fig. 3.4. The accumulated depth is from Eq. (3.3).

$$D = iT = 5.04(30/60) = 2.52.$$

The incremental depth is the difference in the current depth and the depth at the previous time step or

$$2.52 - 1.91 = 0.61.$$

The last column is obtained by rearranging the entries in column 4 while maintaining the concept of a balanced storm.

Comment: The actual pattern or sequence of occurrence of the blocks of rainfall depth shown in the last column of Table 3.3 is not well defined. Several patterns are possible. What must be done for a balanced storm pattern is to keep the blocks of rainfall depth grouped so that no matter what duration is being considered, the blocks of greatest depth are contained within that duration.

The advanced type storm that results without rearranging the values in Table 3.3 is a storm pattern meeting the

Table 3.3 Synthetic 3-hr, 25-Year Rainfall, Stillwater, Oklahoma, DDF Method

Duration (min)	Intensity[a] (iph)	Accumulated depth[b] (in.)	Incremental depth[c] (in.)	Depth increment[d] (in.)
15	7.62	1.91	1.91	0.10
30	5.04	2.52	0.61	0.12
45	3.85	2.88	0.36	0.17
60	3.14	3.14	0.26	0.26
75	2.67	3.34	0.20	0.61
90	2.34	3.51	0.17	1.91
105	2.08	3.65	0.14	0.36
120	1.86	3.77	0.12	0.20
135	1.72	3.88	0.11	0.14
150	1.59	3.98	0.10	0.11
165	1.48	4.07	0.09	0.09
180	1.38	4.15	0.08	0.08
		Total	4.15	4.15

[a]From Fig. 3.4 or Eq. (3.4).
[b]Column (2) × Column (1)/60.
[c]Based on differencing Column 3.
[d]Column 4 rearranged.

balanced storm criteria. A pattern of this type indicates that the highest intensity rainfall occurs at the beginning of the storm and then gradually decreases to the end of the storm. An alternative would be a delayed pattern obtained by arranging the rainfall in order of increasing depth so that the storm gradually builds in intensity and its highest intensity is at the end of the storm.

Obviously both the advanced and delayed rainfall patterns represent unusual storms in that in one case the rainfall begins in its most severe state and in the other it ends this way. A compromise that is widely used is to place the most intense increment near the center of the storm and pyramid the values by placing the next highest alternately at the beginning and end of the pattern as it develops both forward and backward in time from the central high intensity.

The last column of Table 3.3 shows the resulting time distribution of rainfall depths using this latter or central pattern. This method of arriving at a time distribution for rainfall can be used for any frequency and any total storm duration. In this book this method is referred to as the depth–duration–frequency (DDF) method.

SCS Rainfall Pattern

The Soil Conservation Service (SCS) (1973) has adopted a method similar to the DDF method. Several dimensionless rainfall temporal patterns called type curves as shown in Fig. 3.10 are used. Figure 3.11 shows the regions of the U.S. where the various type curves apply. For other locations from throughout the world, the best fitting type curve can be found by developing a storm time distribution using the DDF method and by using each of the type curves as explained below. The Type curve that best describes the DDF result can be adopted for that location. The type II curve is applicable to the majority of the U.S. and

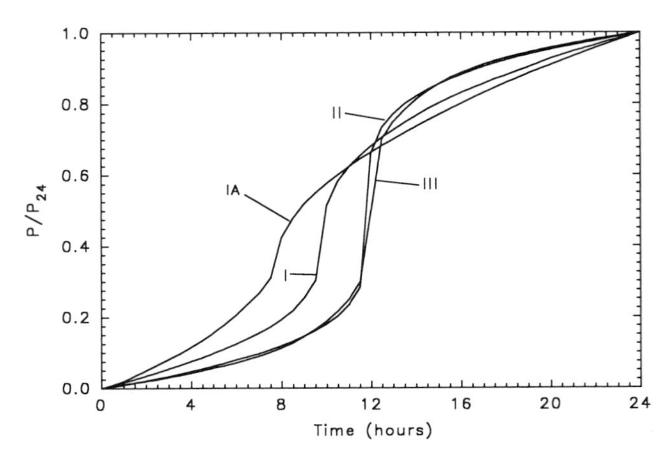

Figure 3.10 SCS type curves for distribution of 24-hr rainfalls.

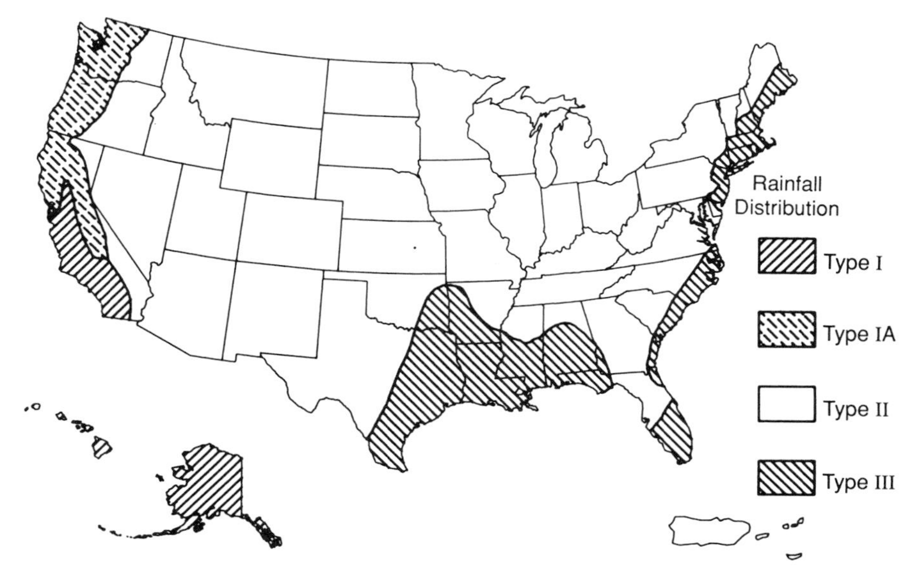

Figure 3.11 Applicable region for various SCS Type curves (Soil Conservation Service, 1986).

represents the most intense storm pattern. The type IA is the least intense pattern for short durations. Table 3.4 contains the coordinates for the various SCS type curves.

The SCS method is based on the 24-hr rainfall of the desired frequency. This rainfall is proportioned throughout the 24-hr period using the appropriate curve shown in Fig. 3.10 in the form of percentage mass curves. These mass curves were derived so that for the selected frequency, the depth–duration relationship based on the curves would be very close to the depth–duration curve developed from a frequency analysis of actual rainfall. Thus, the time–depth patterns based on the SCS method and the DDF method should be very similar. Example Problem 3.4 demonstrates the SCS method for storm pattern development. For storm durations less than 24 hr, the steepest part of the type curves is selected. For example, using the type II curve, a 3-hr storm uses the values extending from 10.5 to 13.5 hr in Fig. 3.10.

Example Problem 3.4 Time distribution of rainfall—SCS method

Develop a 3-hr, 25-year rainfall temporal distribution in 15-min time increments using the SCS dimensionless type curves.

Solution: Table 3.5 contains the required computations. The 25-year, 24-hr rainfall for Stillwater is estimated from Appendix 3A as 6.80 in. The values in column 2 are from Fig. 3.10 or Table 3.4. The most intense 3-hr part of the type II curve that is applicable to Stillwater (Fig. 3.11) is used. This corresponds to the time from 10.5 to 13.5 hr on the type II curve. The values in column 3 are obtained by multiplying the column 2 values by the 24-hr rainfall of 6.80 in. The values of the last column represent differences of successive values in column 3.

To facilitate computer analysis, it would be desirable to have an equation that approximates the actual shape of the type curves. The equation

$$\frac{P(t)}{P_{24}} = 0.5 + \frac{T}{24}\left[\frac{24.04}{2|T| + 0.04}\right]^{0.75}, \qquad (3.6)$$

where t is time and T is time $-$ 12 in hours fits the type II curve with a slight discrepancy on either side of 12 hr. The relationship is also a very good approximation of the type III curve. The discrepancy causes no noticeable difference in resulting runoff. This equation was furnished by Cronshey (1981) who credited it to Norman (1981). In some computer-generated runoff calculations contained later in this chapter, the above equation was used to describe the type II curve.

The SCS type curves can be used to estimate the depth of rainfall for any duration and frequency from the 24-hr rainfall for the same frequency. This is done by taking the difference in the ordinates of the type curve for the steepest part of the curve encompassing the desired duration. For example, the 6-hr, 25-year rainfall for Stillwater can be estimated by multiplying the 24-hr, 25-year rainfall of 6.5 in. (165 mm) by the largest difference in ordinates of the type II curve over a 6-hr period. This difference is found by using the ordinates at t = 15 hours and t = 9 hours to be

Table 3.4 Coordinates for SCS Type Curves: Table Entries Are P/P24

Time	Type I	Type Ia	Type II	Type III
0.0	0.000	0.000	0.000	0.000
0.5	0.008	0.010	0.005	0.005
1.0	0.017	0.020	0.011	0.010
1.5	0.026	0.035	0.016	0.015
2.0	0.035	0.050	0.022	0.020
2.5	0.045	0.066	0.028	0.025
3.0	0.055	0.082	0.034	0.031
3.5	0.065	0.098	0.041	0.037
4.0	0.076	0.116	0.048	0.043
4.5	0.087	0.135	0.055	0.050
5.0	0.099	0.156	0.063	0.057
5.5	0.111	0.180	0.071	0.064
6.0	0.125	0.206	0.080	0.072
6.5	0.140	0.237	0.089	0.081
7.0	0.156	0.268	0.098	0.091
7.5	0.173	0.310	0.109	0.102
8.0	0.194	0.425	0.120	0.114
8.5	0.219	0.480	0.133	0.128
9.0	0.254	0.520	0.147	0.146
9.5	0.303	0.550	0.163	0.166
10.0	0.515	0.577	0.181	0.189
10.5	0.583	0.601	0.204	0.217
11.0	0.624	0.624	0.235	0.250
11.5	0.654	0.645	0.283	0.298
12.0	0.682	0.664	0.663	0.500
12.5	0.706	0.683	0.735	0.702
13.0	0.727	0.701	0.772	0.750
13.5	0.748	0.719	0.799	0.783
14.0	0.767	0.736	0.820	0.811
14.5	0.785	0.753	0.838	0.834
15.0	0.801	0.769	0.854	0.854
15.5	0.816	0.785	0.868	0.872
16.0	0.830	0.800	0.880	0.886
16.5	0.843	0.815	0.892	0.898
17.0	0.856	0.830	0.903	0.910
17.5	0.868	0.844	0.913	0.919
18.0	0.879	0.858	0.922	0.928
18.5	0.891	0.871	0.930	0.936
19.0	0.902	0.884	0.938	0.943
19.5	0.914	0.896	0.945	0.950
20.0	0.926	0.908	0.952	0.957
20.5	0.936	0.920	0.958	0.963
21.0	0.946	0.932	0.964	0.969
21.5	0.955	0.944	0.970	0.975
22.0	0.964	0.956	0.976	0.981
22.5	0.974	0.967	0.982	0.986
23.0	0.982	0.978	0.988	0.991
23.5	0.991	0.989	0.994	0.996
24.0	1.000	1.000	1.000	1.000

$0.85 - 0.15 = 0.70$. The estimated 6-hr, 25-year rain is found to be $0.70 \times 6.5 = 4.55$ in. (116 mm). This 6 hr of rain can be distributed in time according to the type curves by applying the relationship

$$\frac{P'(t)}{PD} = \frac{P(12 + t - D/2) - P(12 - D/2)}{P(12 + D/2) - P(12 - D/2)}, \quad (3.7)$$

where t is the time within the storm in hours ($0 \leq t \leq D$); D is the storm duration in hours ($D \leq 24$); PD is the rainfall volume for the duration D and desired frequency; $P(t)$ is the value from the appropriate type curve, and $P'(t)$ is the accumulated volume of rainfall to time t.

Example Problem 3.5. Synthetic time distribution for a 6-hr storm

Compute the temporal pattern for a 6-hr, 25-year rain of 4.55 in. (116 mm) for Stillwater, Oklahoma, using the type II curve and a time increment of 1 hr.

Solution: Use Eq. (3.7).

$$P(12 - D/2) = P(12 - 6/2) = P(9) = 0.147$$

$$P(12 + D/2) = P(12 + 6/2) = P(15) = 0.854$$

$$PD = 4.55 \text{ in. } (116 \text{ mm})$$

$$P'(t) = 4.55\left[\frac{P(12 + t - 3) - 0.147}{0.854 - 0.147}\right]$$

$$= 6.44[P(9 + t) - 0.147]$$

t	$P(9+t)$	$P'(t)$
0	0.147	0.00
1	0.181	0.22
2	0.235	0.57
3	0.663	3.32
4	0.772	4.03
5	0.820	4.33
6	0.854	4.55

Reevaluation of the SCS type curves is ongoing. It is not unreasonable to expect that modifications for these curves will appear. Table 3.6 compares the DDF estimates and SCS type II estimates of the 25-year rainfall for several durations at a number of locations in the U.S. and Canada. Considering the wide range of topographic and climatic conditions, the type II curve does a reasonably good job of approximating depth–duration–frequency data found in TP 40.

Table 3.5 Synthetic 3-hr, 25-Year Rainfall, Stillwater, Oklahoma, SCS Method

Time[a] (hrs)	Ordinate[b] (P_t/P_{24})	Depth[c] (in.)	Increment Depth[d] (in.)
10.50	0.204	1.39	0.00
10.75	0.219	1.49	0.10
11.00	0.235	1.60	0.11
11.25	0.257	1.75	0.15
11.50	0.283	1.92	0.17
11.75	0.387	2.63	0.71
12.00	0.663	4.51	1.88
12.25	0.712	4.84	0.33
12.50	0.735	5.00	0.16
12.75	0.758	5.15	0.15
13.00	0.772	5.25	0.10
13.25	0.786	5.35	0.10
13.50	0.799	5.43	0.08
		Total	4.04

[a]Most intense 3-hr part of type II curve.
[b]Ordinates of type II curve.
[c]Column 2 × 25-year, 24-hr rain of 6.80 in.
[d]Differencing of Column 3.

Chicago Hyetograph

Keifer and Chu (1957) developed a balanced storm pattern known as the Chicago hyetograph by using an equation similar to Eq. (3.4). They partitioned the storm pattern so that at any intensity the time from the center of the storm back to the rising limb, t_a, divided by the time from the rising limb to the falling limb of the storm, $t_a + t_b$, was a constant r (see Fig. 3.12). From Eq. (3.3),

$$D = i_{\text{ave}} T$$

where i_{ave} is the average intensity over the duration T, and D is the total depth of rainfall. The instantaneous intensity, i, is given by the time rate of change of depth or

$$i = \frac{dD}{dT} = i_{\text{ave}} + T\frac{di_{\text{ave}}}{dT}.$$

From Eq. (3.4)

$$i_{\text{ave}} = \frac{KF^x}{(T+b)^n} = \frac{K'}{(T+b)^n},$$

where $K' = KF^x$ for a fixed frequency, F. Therefore

$$\frac{di_{\text{ave}}}{dT} = -nK'(T+b)^{-n-1}$$

Table 3.6 Comparison of TP 40 and SCS Type II Rainfall Depths for a 25-Year Rainfall of Various Durations: Type II Values Are First Row and TP 40 Values Are Second Row for Each Location

	Duration (hr)						
Location	0.5	1	2	3	6	12	24
Type II ordinate differences	0.37	0.46	0.54	0.59	0.70	0.84	1.00
Lexington, Kentucky	1.85	2.30	2.70	2.95	3.50	4.20	5.00
	1.85	2.30	2.75	3.15	3.60	4:20	5.00
St. Louis, Missouri	2.13	2.65	3.11	3.39	4.03	4.83	5.75
	2.10	2.65	3.22	3.50	4.30	4.90	5.75
Oklahoma City, Oklahoma	2.52	3.13	3.67	4.01	4.76	5.71	6.80
	2.55	3.25	3.90	4.20	5.10	6.10	6.80
Gillette, Wyoming	1.11	1.38	1.62	1.77	2.10	2.52	3.00
	1.20	1.40	1.80	1.90	2.30	2.70	3.00
Denver, Colorado	1.18	1.47	1.73	1.89	2.24	2.69	3.20
	1.32	1.70	1.90	2.20	2.50	2.75	3.20

and from an earlier relationship

$$i = i_{\text{ave}} - \frac{TnK'}{(T+b)^{n+1}} = \frac{K'[(1-n)T+b]}{(T+b)^{n+1}}. \quad (3.8)$$

From the definition of r

$$r = \frac{t_a}{t_a + t_b} = \frac{t_a}{T} \quad (3.9)$$

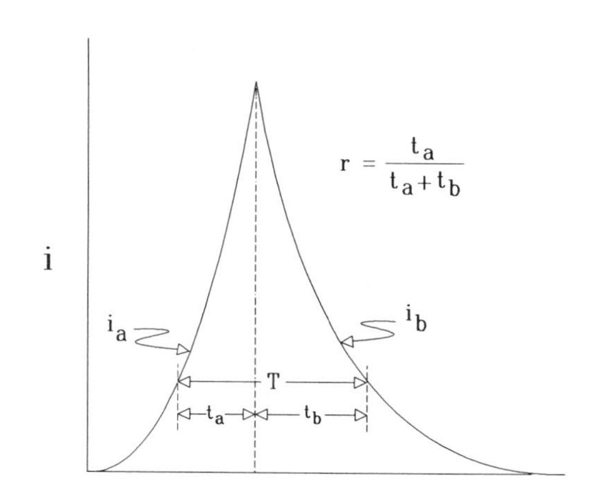

Figure 3.12 Chicago hyetograph.

or

$$T = \frac{t_a}{r} = \frac{t_b}{1-r}.$$

Using these relationships in Eq. (3.8) for T yields for the rising limb

$$i_a = \frac{K'[(1-n)(t_a/r)+b]}{(t_a/r+b)^{n+1}} \quad (3.10)$$

and for the falling limb

$$i_b = \frac{K'[(1-n)(t_b/(1-r))+b]}{(t_b/(1-r)+b)^{n+1}}. \quad (3.11)$$

Note that $r = 0.5$ produces a symmetrical pattern. A value of r between 0 and 0.5 produces an advanced pattern and a value between 0.5 and 1.0 produces a delayed pattern.

Example Problem 3.6. Time distribution of rainfall—Chicago method

Develop a 25-year, 3-hr storm at Stillwater, Oklahoma, using the Chicago hyetograph method. Use 15-min time increments.

Solution: Recall from Example Problem 3.2 that $K = 1.75$, $b = 0.12$, $x = 0.21$, and $n = 0.80$ for Stillwater. The value for

Table 3.7 Synthetic 3-hr, 25-yr rain, Stillwater, Oklahoma, Chicago Hyetograph Method

Time (min)	t_a or t_b (hr)	Intensity[a] (iph)	Depth (in.)	Modified depth (in.)
0	1.375	0.35	0.00	0.00
15	1.125	0.41	0.10	0.10
30	0.875	0.52	0.12	0.12
45	0.625	0.72	0.16	0.16
60	0.375	1.19	0.24	0.24
75	0.125	3.50	0.59	0.59
	0[b]	7.35		
90	0.125	3.50	1.36[c]	1.65[d]
105	0.375	1.19	0.59	0.59
120	0.625	0.72	0.24	0.24
135	0.875	0.52	0.16	0.16
150	1.125	0.41	0.12	0.12
165	1.375	0.35	0.10	0.10
180	1.625	0.30	0.08	0.08
		Total	3.86	4.15

[a]From Eq. (3.10) or (3.11) with $r = 0.5$, $K' = 3.44$, $b = 0.12$, $n = 0.8$.
[b]Time increment of 0.05 hr used to define storm center.
[c]Depth for time increment 75 to 90 min.
[d]See Example Problem 3.6 for explanation.

r is taken as 0.5. K' of Eqs. (3.10) and (3.11) is found to be

$$K' = KF^x = 1.75(25)^{0.21} = 3.44$$

for rainfall in inches. Values for i_a and i_b are calculated from Eqs. (3.10) and (3.11) and shown in Table 3.7. So that the storm pattern will be in agreement with the patterns developed using the SCS and DDF methods, the center of the storm is taken at the center of the time increment from 75 to 90 min or at 82.5 min. The calculations are illustrated for the time of 60 min. At 60 min, t_a is $82.5 - 60 = 22.5$ min or 0.375 hr. The value of i_a is determined from Eq. (3.10):

$$i_a = \frac{K'[(1-n)(t_a/r) + b]}{(t_a/r + b)^{n+1}}$$

$$= \frac{3.44[(1-0.8)(0.375/0.5) + 0.12]}{(0.375/0.5 + 0.12)^{0.8+1}} = 1.19.$$

The depth increment is the average intensity over the time interval times the duration of the interval or

$$\text{depth} = 0.25(1.19 + 0.72)/2 = 0.24.$$

It can be seen that at $t_a = t_b = 0$, $i_a = i_b = K'/b^n = 18.8$ iph. Since this is unreasonably high and represents an instantaneous value, a time of 0.05 hr was arbitrarily used to define the center of the pattern. The depth values are based on the average intensity for the interval and the duration of the interval.

If Eq. (3.4) is used to compute the 25-year, 3-hr rainfall for this location, the result is 4.15 in. The total indicated in the fourth column of Table 3.7 is only 3.86 in. The difference is largely due to the assumption used in calculating the intensity at $t_a = t_b = 0$. To bring the storm total in line with the estimate from Eq. (3.4), the depth at the center of the pattern was adjusted by $4.15 - 3.86 = 0.29$ inches to 1.65 in.

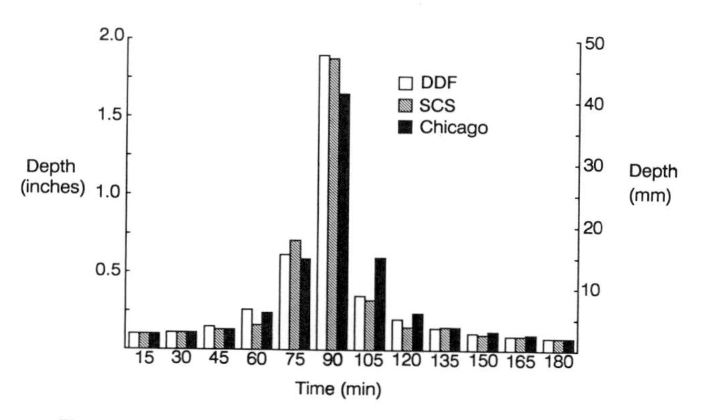

Figure 3.13 3-hr, 25-year pattern, Stillwater, Oklahoma.

Figure 3.13 compares the rainstorms synthesized for Stillwater, Oklahoma, in Example Problems 3.3, 3.4, and 3.6 based on the DDF method, the SCS method using the type II curve, and the Chicago hyetograph. The methods produce essentially identical results. The Chicago pattern could be brought more in line with the other two patterns by using a value of r less than 0.5. In view of the uncertainties that exist in the storm patterns and uncertainties in other factors that contribute to runoff, the methods can be considered to be essentially equal. Most examples in this text are based on the SCS type II distribution as approximated by Eq. (3.6).

There have been a few attempts at studying the actual time distribution of observed rainfalls. Major difficulties with studies of this type are obtaining long-term records from which short duration intensities can be extracted, the mass of data required, and the variability of natural rainfall. Huff (1967) has studied the time distribution of heavy rainfalls with durations of 3 to 48 hr in central Illinois. He divided the storms into four groups depending on the time quartile in which the majority of the rain occurred. The time distributions of the storm rainfalls within the various quartiles were computed as were the percentage of the storms having that particular time distribution or the one above it. The difficulty of assigning a return period to

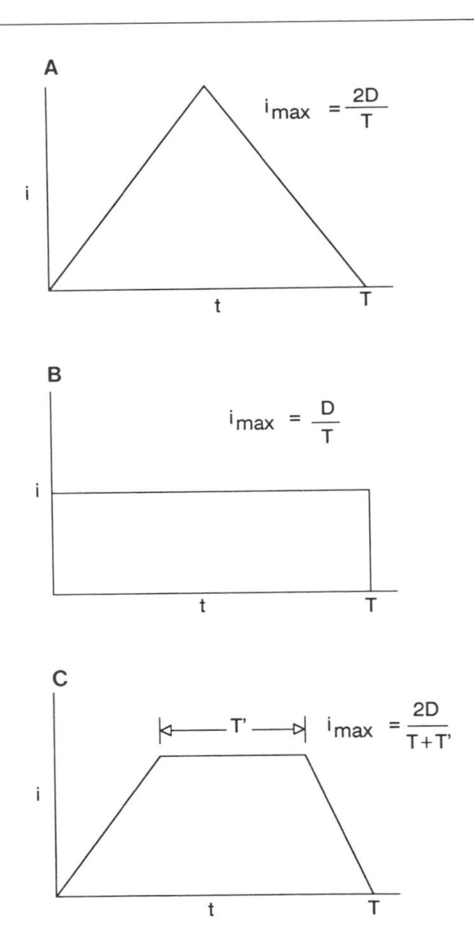

Figure 3.14 Hypothetical, unbalanced storm patterns.

the various Huff patterns and the selection of the proper quartile storm have precluded their widespread adoption. The Huff study definitely demonstrated the variability of time distributions that occur at particular location. This in turn means that any particular pattern that is adopted is unlikely to actually occur. Design patterns are simply possible, good estimates of reasonable time patterns.

For preliminary designs and rough calculations, storm patterns might be approximated as shown in Fig. 3.14. In this figure, i is the intensity and D represents the rainfall depth for the desired frequency and duration, T. The proper relation for determining i_{max} for each of the three assumed patterns is shown. It must be kept in mind that these patterns are *not balanced* patterns. In pattern (a) the time of the maximum intensity is arbitrary. In pattern (c) the duration, T', of the constant intensity portion and its location within T are arbitrary. Both patterns (a) and (b) are special cases of pattern (c).

ABSTRACTIONS FROM PRECIPITATION

Abstractions from precipitation are losses from precipitation that do not show up as storm water runoff. The volume of storm water runoff is the volume of precipitation minus the volume of abstractions. Precipitation minus abstractions is also known as rainfall excess or effective rainfall. Thus, the volume of storm water runoff is equal to the volume of effective rainfall. Abstractions include interception, evapotranspiration, surface storage and surface detention, bank storage and infiltration. The combination of interception and surface storage is sometimes denoted as surface storage.

Interception

Some rainfall is intercepted by vegetation before it reaches the ground. The amount of interception varies with the type, density and stage of growth of the vegetation, intensity of the rainfall, and wind speed. On an annual basis, interception may involve a significant percentage of the total rainfall. A dense forest canopy may result in interception values of 25% of the annual rainfall. This would be the case in climates with frequent, light rainfalls; low wind speeds; and evergreen vegetation.

On a per storm basis, interception storage may range up to 0.4 in. (10 mm) for dense forested vegetation but is generally considerably less than this for a mixed cover catchment. For significant rainfall events likely to produce large stormwater flows from small catchments, the amount of rainfall going to satisfy interception storage is generally a small percentage of total storm rainfall.

Evapotranspiration

Evapotranspiration or ET is the combination of evaporation and transpiration. Evaporation is the phase change of water from a liquid to a vapor. Evaporation from plant surfaces of water that has traversed from the soil through the plant is termed transpiration. All evaporation from a leaf surface is not transpiration since intercepted water is also evaporated. On an annual basis, ET generally involves a large fraction of the total precipitation. In arid climates most of the rainfall, 90% or more, may be lost through ET. In more humid climates, ET may account for 40 to 70% of the annual precipitation.

In spite of the high total fraction of rainfall involved in the ET process on an annual basis, for individual rainstorms, ET is generally a minor factor and not included in storm water computations. Infiltration, as we shall see, is a significant abstraction during storm events. Much of the water that is abstracted by infiltration is eventually lost from the catchment system via ET. This loss occurs at a relatively slow rate (0 to 0.4 in. or 10 mm per day) and is the most significant during the times between storm events, not during the storm events themselves. When rain falls on warm surfaces, some evaporation occurs. Sometimes an evaporative loss of 0.1 in. (2.5 mm) is used to reflect this loss if a large part of the basin is covered by concrete, asphalt, roofs, etc.

Bank Storage

Bank storage represents losses from streamflow during a period of rising stage in a stream and the subsequent seepage of streamflow into the banks of the stream. During the falling stage, this water generally seeps back into the stream. Thus bank storage is not actually a loss from runoff but a storage and delay in the runoff process. For small catchments, bank storage generally plays a small role in storm water runoff.

Surface Storage and Detention

Surface storage is the volume of water required to fill depressions and other storages before surface runoff begins. Detention storage is the buildup of small depths of water required to support the runoff process. Actual measurements of surface storage and detention are extremely difficult to make and consequently are practically nonexistent. Wright-McLaughlin Engineers (1969) in a special study of urban hydrology in the Denver, Colorado area, recommended the values shown in Table 3.8 for surface storage. Terstriep and Stall (1974) recommend a value of 0.20 in. (5 mm) for detention storage for bluegrass turf. Some investigators (Linsley *et al.*, 1949) recognized that a watershed surface is made up of depressions of various sizes and that as some of the smaller depressions were filled, surface runoff could begin even though the larger depressions were still filling. An exponential relationship

$$V_d = S_d[1 - e^{-K_d(P-F)}] \tag{3.12}$$

has been proposed where V_d is the volume of water in surface storage, S_d is the available surface storage, $P - F$ is the accumulated mass of surface storage supply (i.e., accumulated rainfall minus infiltration and other losses except surface storage), and K_d is a constant.

Table 3.8 Typical Values for Surface Storage (Wright-Mclaughlin Engineers, 1969)

Land cover	Surface storage (in.)	Recommended value (in.)
Impervious		
Large paved area	0.05–0.15	0.10
Roofs, flat	0.10–0.30	0.10
Roofs, sloping	0.05–0.10	0.05
Pervious		
Lawn grass	0.20–0.50	0.30
Wooded area	0.20–0.60	0.40
Open fields	0.20–0.60	0.40

The value of the constant K_d can be estimated by noting that when $P - F$ is near zero, all of the water goes to filling depressions so that $dV_d/(d(P - F))$ is essentially 1. Based on this reasoning, K_d is equal to $1/S_d$.

Neglecting interception losses, the rate that water becomes available for surface runoff, σ, is $i - f - \psi$, where i is the precipitation rate, f is the infiltration rate, and ψ is equal to dV_d/dt. Based on these assumptions, the surface runoff supply rate becomes

$$\sigma = (i - f)(1 - e^{-(P-F)/S_d}). \tag{3.13}$$

The ratio of surface runoff supply rate to the difference in the rainfall and infiltration rates becomes

$$\sigma/(i - f) = 1 - e^{-(P-F)/S_d} \tag{3.14}$$

and ranges from 0, at the beginning of the precipitation event ($P - F = 0$), to 1 when $P \gg F$.

Equation (3.14) is plotted in Fig. 3.15 based on a turf area with an average overall S_d of 0.25 in. or a pavement with S_d equal to 0.0625 in. The vertical dashed line in Fig. 3.15 represents the surface runoff supply ratio if it is assumed that the overall average surface storage must be filled before any runoff can begin. This would be the case if the abstractions indicated in Table 3.8 were subtracted directly from the beginning of a storm before any water was allowed to become available for surface runoff.

Tholin and Keifer (1960) surmised that the actual situation might be between that given by Table 3.8 and that given by Eq. (3.14). They found that the curve of the normal distribution, as shown in Fig. 3.15, fell within their desired range. This curve can be

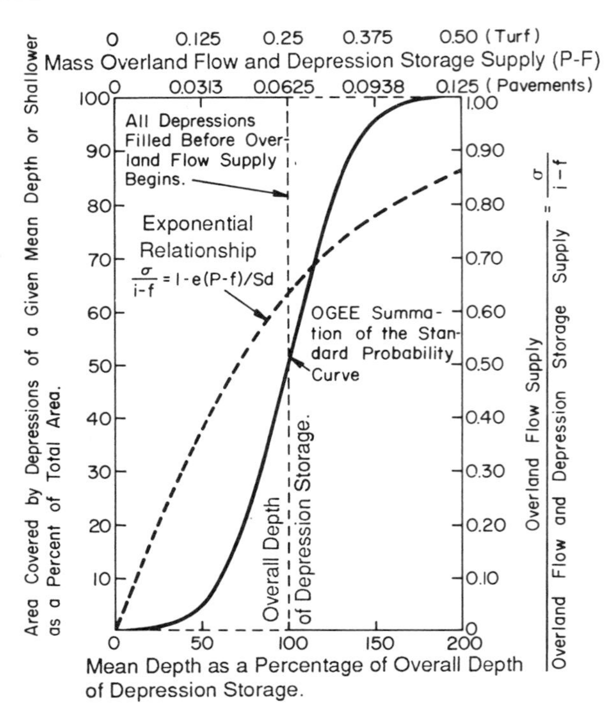

Figure 3.15 Depth distribution curve of depression storage. Enter graph from top, read down to selected curve, and project right or left as desired (modified from Tholin and Kiefer, 1960).

approximated using a normal distribution with a mean equal to S_d and a standard deviation of $S_d/3$ or

$$\frac{\sigma}{i-f} = \int_{-\infty}^{D/S_d} \frac{3}{S_d\sqrt{2\pi}} \exp\left[-\frac{1}{2}\left(\frac{x-S_d}{S_d/3}\right)^2\right] dx.$$

The value of S_d might be estimated from the data in Table 3.8.

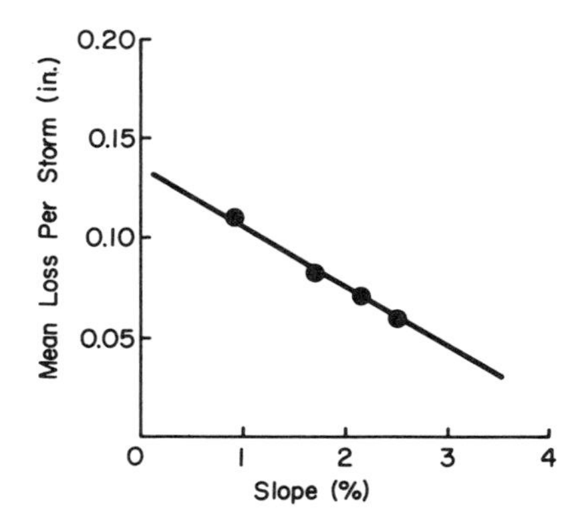

Figure 3.16 Depression-storage loss versus slope for four impervious drainage areas (Viessman, 1967).

As might be expected, surface storage is of greater importance on flat surfaces than on steep surfaces. Viessman (1967) found the relationship shown in Fig. 3.16 for four impervious drainage areas. The line in Fig. 3.16 should be extrapolated with care. More likely the surface storage would decrease exponentially with slope approaching zero at very steep slopes.

If long duration rainfalls are being studied, the values of surface storage will not appreciably affect estimated runoff rates since the early part of the storm would fill this storage prior to the occurrence of the major runoff producing part of the rainfall. Note that the values in Table 3.8 do not include built-in storage in the form of detention basins.

Infiltration

The major abstraction from rainfall during a significant runoff-producing storm is infiltration of water into pervious areas including soils, infiltration basins, and forest litter. The process of infiltration of water and subsequent water movement is an exceedingly complex process. In this discussion, soil is used in a general sense. The upper part of Fig. 3.17 shows the soil water content as a function of time and depth during a rainfall event. The curved lines represent water content at various times with time increasing as the wetting profile advances deeper in the soil profile. The soil was at a uniform initial water content and the soil properties are uniform with depth.

In general, the infiltration rate is dependent on soil physical properties, vegetative cover, antecedent soil water conditions, rainfall intensity, and the slope of the infiltrating surface. Referring again to Fig. 3.17, if the soil physical properties are not uniform with depth, the pattern of soil wetting may be greatly altered. It is not uncommon to find a soil layer that is less permeable than the surface layer. If this less permeable layer is located near the soil surface, it restricts the wetting front and reduces the infiltration rate. This restricting layer may be a shallow rock formation, a soil layer higher in clay content, a layer compacted by heavy equipment, a plow layer, or a fragipan. Final grading on construction, landscaping, or mining jobs may involve spreading stockpiled materials. In this way, a more permeable layer of soil may be placed over a compacted layer. This is especially common when sodding is done, as the sod is laid directly on relatively compact materials. Consequently, water can easily enter the surface layer, but is restricted by the compacted zone. Thus, light rainfalls are easily absorbed, but heavier rains soon saturate the surface soil layer. The saturated layer has a very low infiltration capacity,

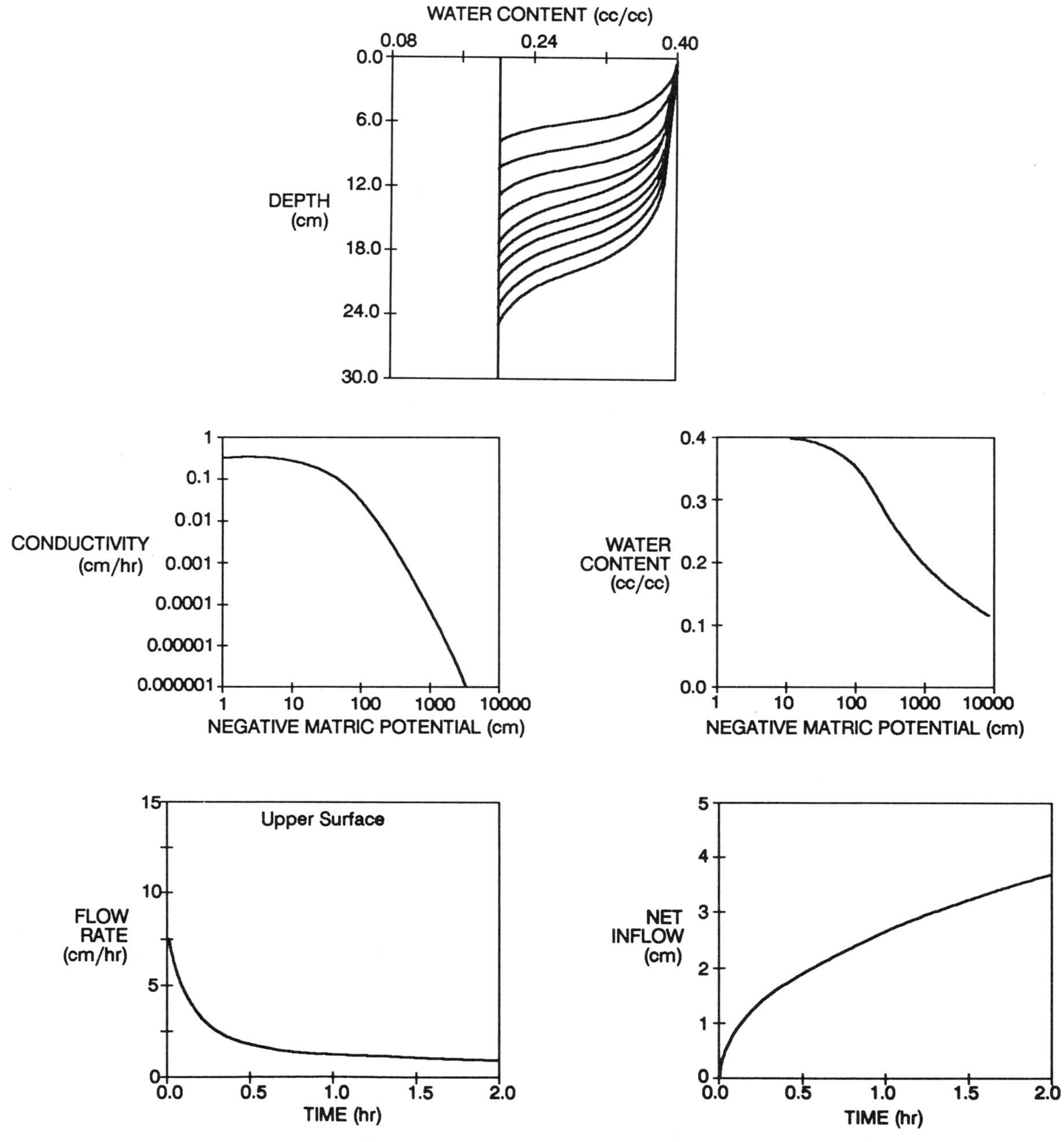

Figure 3.17 Water content profiles.

which results in runoff rates being near the rainfall rates.

Bare soils tend to have lower infiltration rates than soils protected by a vegetative cover. On bare soil, the impacting raindrops tend to puddle the soil. The energy of the falling rain breaks down soil aggregates and small particles are carried into the soil pores. The net result is a lowering of the infiltration rate.

The antecedent soil water content also alters the infiltration rate. Generally, a wet soil has a lower infiltration rate than a dry one. Thus a rain falling on a wet soil will produce more runoff at a higher rate than the same rain on a dry soil.

The infiltration opportunity time is a function of the slope of the infiltrating surface. On a steep slope, the water tends to run off rapidly and thus have less opportunity for infiltration than on a gentle slope. Also, the soil type found on steeper slopes is generally not the same as on flatter slopes. Often the more sloping soils, especially if the soils have been used for agriculture, have experienced more erosion than flatter soils. This in turn generally results in lower infiltration rates.

Rainfall intensity affects the infiltration rate in two ways. For high-intensity rains, the raindrops tend to be larger and have more energy when they strike the soil. Thus high-intensity rains are more effective in sealing the soil surface than are low-intensity rains. A good vegetative cover can minimize this effect. Obviously, the infiltration rate cannot exceed the rainfall rate for prolonged periods of time. In the absence of any ponded water or water flowing over the soil, the maximum possible infiltration rate is the lesser of the rainfall rate or the soils infiltration capacity. In the presence of ponded water or surface flowing water, the infiltration rate equals the infiltration capacity until this surface supply of water is exhausted.

The combination of all of the factors governing infiltration throughout a watershed interact in such a fashion as to result in a very complex spatial and temporal distribution of infiltration capacity. At some locations, the infiltration capacity may be so high as to practically never produce surface runoff, whereas other areas may have low infiltration capacities and produce surface flow from light rainfalls. Betson (1964) has termed these latter areas as source areas.

In recent years, an alternative theory of streamflow generation known as the variable source area concept has been proposed for hilly terrain (Hewlett and Hibbert, 1967). The concept particularly applies to highly pervious soils. At the beginning of a rainfall event, a water table exists with a capillary zone above. The water table near a stream would generally be closer to the soil surface than at some distance from the stream since water table slopes are generally flatter than surface topography slopes in hilly terrain. As the rain continues, the water table near the stream rises faster than that at higher elevations because the percolating water has less distance to travel. This rise near the stream may eventually cause the water table to reach the soil surface resulting in a saturated condition. Further rain on the saturated area becomes saturated return flow and quickly reaches the stream. Further rain causes the saturated area to grow in size and to move upstream. As the saturated area grows, a larger portion of the watershed is contributing to saturated return flow. When the rainfall rate diminishes or stops, the saturated areas drain, resulting in a shrinking of the saturated zone. This growing and shrinking saturated zone is known as a variable source area. Currently, it is difficult to incorporate this concept into hydrologic analyses of ungaged areas since parameters defining the extent and response of the variable source areas are not available.

Runoff estimation based on classical infiltration approaches are known as Hortonian approaches and currently dominate in the area of hydrologic analyses. A great deal of effort has been expended in developing the mathematical theory of the infiltration of water into soils and the subsequent movement of this water within the soil. The physical principles and mathematical relationships are well defined.

Richards Equation

One-dimensional, unsaturated flow is governed by Darcy's law and the continuity equation and is given by

$$v_z = -K(h)\frac{\partial h}{\partial z} \tag{3.15a}$$

and

$$\frac{\partial \theta}{\partial t} = \frac{\partial}{\partial z} - K(h)\frac{\partial h}{\partial z}, \tag{3.15b}$$

where v_z is flow velocity, z is the coordinate direction positive upward, h is the soil water potential, θ is the soil water content, and $K(h)$ is the hydraulic conductivity. The soil water potential is the sum of two components—the matric potential ψ and the gravitational potential z. The hydraulic conductivity is a strong function of soil water potential and may change by 7 orders of magnitude over the range of soil water potentials commonly encountered (Fig. 3.17). $K(h)$ is also dependent on the soil being considered and may change with depth. $K(h)$ is very small for dry soils and increases in value as the soil gets wetter. At saturation, $K(h)$ is the saturated hydraulic conductivity. The soil water content is also a function of the water potential of the soil. Equation (3.15) is based on the continuity equation and Darcy's Law applied to unsaturated flow. The book by Hillel (1971) can be consulted for more detail on the relationship. Chapter 11 of this book treats saturated ground water flow.

If one considers a uniform, deep soil that is initially at a constant water content and subjected to a constant rainfall rate, the change in infiltration rate with time can be deduced. Infiltration is simply equal to v_z at the soil surface computed from Eq. (3.15a). If the soil is initially dry, $K(h)$ will be small. When the first rain occurs, it will wet up the surface of the soil so that $\partial h/\partial z$ will be large. The product $K(h)\partial h/\partial z$ will thus be relatively high and will result in high infiltration rates. As time goes on, the soil will uniformly wet up and $\partial h/\partial z$ will become small. The second term of Eq. (3.15a) will then be small; however, $K(h)$ will now be higher than it was initially and will control the infiltration process. Thus, early in the rainfall event, the water potential gradient at the soil surface governs the infiltration rate, while later in the event, the hydraulic conductivity governs.

Figure 3.17 shows the results of applying Eq. (3.15) to a Bethany soil having an initial water content of 0.20 cc/cc and a rainfall rate of 15 cm/hr. Within the figure are shown the water potential–water content functions and water potential–hydraulic conductivity functions. Also shown are water content profiles with depth at time increments of 0.2 hr. Finally, the infiltration rate in cm/hr and the cumulative infiltration are shown up to 2 hr.

Equations (3.15) have found limited application in design hydrology. Difficulties experienced in using the equation are the nonuniformity of soils, both spatially and with depth; the great number of measurements needed to define the required parameters; and the difficulty of solving the relationships when the required data are available. A further difficulty is that of specifying the applicable boundary conditions for the equation. Finally, the relationship between water potential and soil properties is hysteretic, further complicating the solution of the equations.

Horton's Equation

Because of the difficulty of using theoretically based equations to describe the infiltration process, a great many empirical relationships have been proposed. Horton (1940) found that an equation of the form

$$f(t) = f_c + (f_0 - f_c)e^{-kt} \tag{3.16}$$

fit experimentally decreasing infiltration rates as a function of time. In this equation, $f(t)$ is the infiltration rate for any time, t; f_c and f_0 are the final and initial infiltration rates, respectively; and k is a measure of the rate of decrease in the infiltration rate. Horton's equation requires knowledge of three soil parameters, f_0, f_c, and k.

A difficulty with the Horton equation is that it makes infiltration rate a function of time and does not account for variations in rainfall intensity. The equation has no provision for a recovery of infiltration capacity during periods of low or no rainfall.

There are not general tables or guidelines for selecting values for the three parameters of Horton's equation. Occasionally, locally derived data are used. If infiltration data are available as from an infiltrometer study, the parameters of Eq. (3.16) that best describe the infiltration data can be determined. If none of the parameters are known, nonlinear regression techniques may be used. To illustrate the application of nonlinear

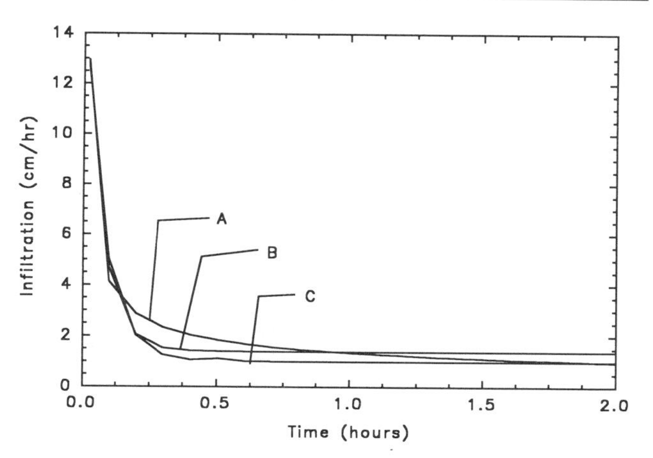

Figure 3.18 Horton infiltration curves.

regression, points every 0.2 hr were taken from the infiltration curve of Fig. 3.17. These time–infiltration pairs were subjected to a nonlinear least-squares estimation using SYSTAT (Wilkinson, 1987) with the result that $f_c = 1.391$ cm/hr, $f_0 = 17.093$ cm/hr, and $k = 15.68$ hr^{-1}. This estimate is plotted in Fig. 3.18 as curve B. Curve A is the infiltration curve of Fig. 3.17. Note that even though curve B is the best overall fit to curve A, the final infiltration rate is overestimated by about 0.4 cm/hr.

By constraining f_c to be 1.00 cm/hr and estimating f_0 and k, the results are 16.44 cm/hr and 13.54 hr^{-1}, respectively. This result is plotted as curve C in Fig. 3.18 and would be preferred for hydrologic design since it fits the data better for long times or for events of more than an hour or so in length.

If one can estimate f_c from data, the other parameters of Eq. (3.16) can be estimated by simple regression by transforming the equation to

$$f(t) - f_c = (f_0 - f_c)e^{-kt}$$

and taking natural logarithms to get

$$\ln[f(t) - f_c] = \ln(f_0 - f_c) - kt.$$

Now a linear regression of $\ln[f(t) - f_c]$ versus t yields $-k$ as its slope and f_0 as $\exp(a) + f_c$, where a is the intercept of the regression.

The Horton equation really suffers from the same difficulties as Eq. (3.15) in that spatially nonhomogeneous soils require spatial variability in the parameter values. Furthermore, if the soil is nonhomogeneous with depth, i.e., has a restricting layer at some shallow depth, the infiltration rate will not smoothly decrease, but will have a rather abrupt drop in infiltration rate as the wetting front reaches the restricting layer.

Table 3.9 Infiltration Constants for Bluegrass Turf (Terstriep and Stall, 1974)

HSG		A	B	C	D
f_c	iph	1	0.5	0.25	0.1
f_c	mm/hr	25.4	12.7	6.3	2.50
f_0	iph	10	8	5	3
f_0	mm/hr	254	203	127	76
k	hr^{-1}	2	2	2	2
F_p	in.	6	4	3	2
F_p	mm	152	102	76	51
n		1.4	1.4	1.4	1.4
a	in./hr	1.0	1.0	1.0	1.0
a	mm/hr	0.274	0.274	0.274	0.274

Table 3.10 Infiltration Calculations Based on Horton and Holtan Models

t (hr)	Horton $f(t)$(cm/hr)[a]	F_p (mm)[b]	Holtan f (cm/hr)[c]
0.0	12.7	76.0	12.4
0.2	8.7	51.2	7.4
0.4	6.0	36.4	4.8
0.6	4.3	26.7	3.4
0.8	3.0	20.0	2.4
1.0	2.3	15.2	1.9
1.2	1.7	11.4	1.5
1.4	1.4	8.5	1.2
1.6	1.1	6.1	1.0
1.8	1.0	4.1	0.8
2.0	0.9	1.7	0.7

[a]Based on Eq. (3.16) with $f_c = 0.63$ cm/hr, $f_o = 12.7$ cm/hr, $k = 2$ hr^{-1}.

[b]$F_p(t+1) = F_p(t) - f(t)\,dt$.

[c]Based on Eq. (3.17) with $f_c = 0.63$ cm/hr, $n = 1.4$, $a = 0.274$.

For bluegrass turf, Terstriep and Stall (1974) recommended values for the constants of Horton's equation as shown in Table 3.9. In this table, HSG denotes the SCS Hydrologic Soil Group designation explained later in this chapter. The values for f_c shown in the table exceed the water transmission rates for the HSGs. These rates are given in Table 3.15 for the various HSGs. For soils having less vegetative protection than provided by a bluegrass turf, values for f_c less than those shown in the table should be used.

Holtan's Equation

Holtan (1961) has advanced an empirical infiltration equation based on the concept that the infiltration rate is proportional to the unfilled capacity of the soil to hold water. The Holtan model for infiltration is

$$f = aF_p^n + f_c, \tag{3.17}$$

where f is the infiltration rate, f_c is the final infiltration rate, F_p is the unfilled capacity of the soil to store water, and a and n are constants. The exponent n has been found to be about 1.4 for many soils. The value of F_p ranges from a maximum of the available water capacity (AWC) to zero. The AWC is a measure of the ability of a soil to store water. Values for AWC are given for many soils in an Agricultural Research Service publication (U.S. Department of Agriculture, 1968).

The Holtan model for infiltration has the advantage over the Horton model in that it has a more physical basis and can describe infiltration and the recovery of infiltration capacity during periods of low or no rainfall.

Table 3.9 contains values for F_p as recommended by Terstriep and Stall (1974) for bluegrass turf. They also recommend that a be taken as 1.0 for f in iph (0.274 for f in millimeters per hour) and n be taken as 1.4. For conditions other than a bluegrass turf, the values of Table 3.9 would have to be adjusted. Note that F_p is the unfilled soil water storage capacity. The values in Table 3.9 are for an initially dry soil. As soon as infiltration begins, F_p starts decreasing since the infiltrated water fills some of the originally unfilled soil water storage capacity. F_p decreases continuously as long as infiltration exceeds drainage from the soil profile.

Example Problem 3.7. Holtan and Horton infiltration

Calculate the infiltration curve for a Bethany soil under bluegrass turf using both the Horton and Holtan equations.

Solution: From Appendix 3B the Bethany soil is found to be in HSG C. Based on parameters estimated from Table 3.9, infiltration can be computed based on either Eq. (3.16) or (3.17). Table 3.10 shows the calculations that are illustrated for a time of 1 hr. The parameters of the Horton equation are $f_c = 6.3$ mm/hr, $f_0 = 127$ mm/hr, and $k = 2$ hr^{-1}. The parameters of the Holtan equation are $f_c = 6.3$ mm/hr, $n = 1.4$, and $a = 0.274$. Horton's equation gives

$$f(t) = f_c + (f_0 - f_c)e^{-kt} = 6.3 + (127 - 6.3)e^{-2(1)}$$
$$= 23 \text{ mm/hr} = 2.3 \text{ cm/hr}.$$

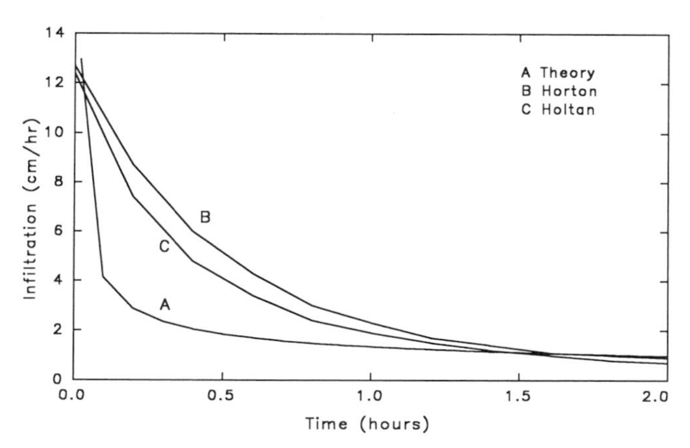

Figure 3.19 Infiltration comparison.

F_p for the Holtan equation is given as F_p for the previous time minus the infiltration during the previous time increment or 20.00 − 0.2(24) = 15.2 mm. Thus the Holtan equation gives

$$f = aF_p^n + f_c = 0.274(15.2)^{1.4} + 6.3$$
$$= 19 \text{ mm/hr} = 1.9 \text{ cm/hr}.$$

The results of the calculations are plotted in Fig. 3.19. Special note should be made of the calculations applying to Eqs. (3.17). The F_p term must be continually decreased by the volume of infiltration that has taken place. In Fig. 3.19, curve A is from Fig. 3.17, curve B is from the Horton model, and curve C is from the Holtan model.

Green–Ampt. Equation

In 1911, Green and Ampt (1911) developed an approximate infiltration model based on Darcy's law. They assumed vertical flow, a uniform water content, a sharp boundary between the wetted soil zone, and the soil zone unaffected by infiltration and that water movement occurs as "piston" flow or "slug" flow. The Green and Ampt model can be "derived" by applying Eq. (3.15a) to the situation depicted in Fig. 3.20:

$$f = K(\text{change in potential})/(\text{distance})$$

or

$$f = K(d + L_f + \psi)/L_f, \tag{3.18}$$

where f is the infiltration rate that varies with time as the wetting front advances into the soil, K is the hydraulic conductivity of the wetted soil part of the soil profile, L_f is the depth of the wetting front, ψ is the pressure head for wetting at the wetting front, and d is the depth of ponding of water on the soil surface.

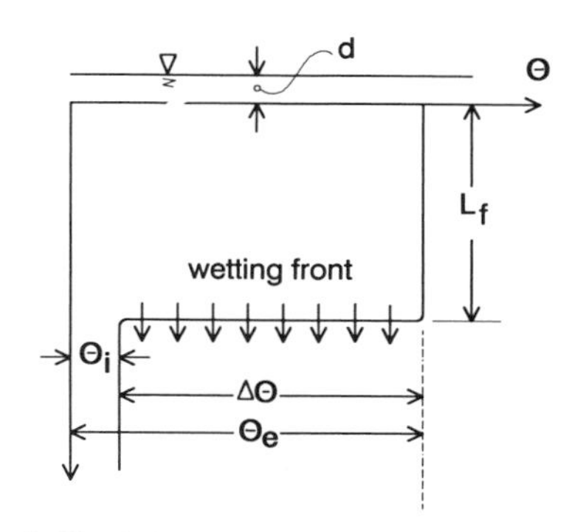

Figure 3.20 Schematic for Green–Ampt infiltration.

Skaggs and Kahleel (1982) discuss the determination of parameters of the Green and Ampt model.

Generally the ponded depth, d, is small and may be neglected in Eq. (3.18). The change in water content across the wetting front, $\Delta\theta$, depends on the initial water content, θ_i; the thoroughly drained or residual water content, θ_r; the effective saturation, s_e, and the total porosity, η. The relationship is

$$\Delta\theta = \theta_e - \theta_i = \theta_e - s_e\theta_e = (1 - s_e)\theta_e,$$

where θ_e is the effective porosity given by $\theta_e = \eta - \theta_r$ (Chow *et al.*, 1988). Rawls *et al.* (1983) present values for η, θ_e, ψ, and K as shown n Table 3.11.

By noting that $F(t) = L_f\Delta\theta$ or $L_f = F(t)/\Delta\theta$ and taking d as zero, Eq. (3.18) can be expressed as

$$f(t) = K\left[\frac{\psi\Delta\theta}{F(t)} + 1\right], \tag{3.19}$$

Table 3.11 Green and Ampt Infiltration Parameters (Rawls *et al.*, 1983)[a]

Soil texture	η	θ_e	ψ (cm)	K (cm/hr)
Sand	0.437	0.417	4.95	11.78
Loamy sand	0.437	0.401	6.13	2.99
Sandy loam	0.453	0.412	11.01	1.09
Loam	0.463	0.434	8.89	0.34
Silt loam	0.501	0.486	16.68	0.65
Sandy clay loam	0.398	0.330	21.85	0.15
Clay loam	0.464	0.309	20.88	0.10
Silty clay loam	0.471	0.432	27.30	0.10
Sandy clay	0.430	0.321	23.90	0.06
Silty clay	0.479	0.423	29.22	0.05
Clay	0.475	0.385	31.63	0.03

[a]Rawls *et al.* (1983) contains more information on these parameters including their standard deviation and values for various soil horizons.

where $F(t)$ is the cumulative infiltration at time t. The cumulative infiltration is found by integration of Eq. (3.19) as

$$F(t) = Kt + \psi\Delta\theta \ln\left(1 + \frac{F(t)}{\psi\Delta\theta}\right). \quad (3.20)$$

Equations (3.19) and (3.20) apply for the case where the ponded depth is negligible. If this is not the case, ψ should be replaced by $\psi + d$. Since Eq. (3.20) cannot be solved explicitly for $F(t)$, iteration is required. A trial value for $F(t)$ is substituted into the right-hand side of the equation, which is then compared to the left-hand side. This process is repeated until agreement between the two values is obtained. A good first estimate is Kt for $F(t)$. Possibly an easier way to calculate $f(t)$ from Eq. (3.19) is to solve Eq. (3.20) for t for various values of $F(t)$. $F(t)$ can then be used in Eq. (3.19) to determine $f(t)$ for the corresponding t.

Example Problem 3.8. Green–Ampt infiltration

Calculate the infiltration curve for a silty clay loam at 30% effective saturation.

Solution: Table 3.12 illustrates the approach. Values for θ_e, ψ, and K are taken from Table 3.11 as 0.432, 27.3 cm, and 0.1 cm/hr, respectively. $F(t)$ values are assumed. The calculations are illustrated for $F(t) = 1$ cm. $\Delta\theta$ is given by

$$\Delta\theta = (1 - s_e)\theta_e = (1 - 0.3)0.432 = 0.302.$$

t is computed by rearranging Eq. (3.20) to

$$t = \frac{F(t) - \psi\Delta\theta \ln(1 + F(t)/\psi\Delta\theta)}{K}$$

$$t = \frac{1 - 27.3(0.302)\ln(1 + 1/27.3(0.302))}{0.1} = 0.561 \text{ hr}.$$

From Eq. (3.19),

$$f(t) = K\left[\frac{\psi\Delta\theta}{F(t)} + 1\right] = 0.1\left[\frac{27.3(0.302)}{1} + 1\right] = 0.926 \text{ cm/hr}.$$

A plot of column 4 versus column 2 or 3 yields the infiltration curve.

Table 3.12 Green–Ampt Infiltration Calculations

$F(t)$ (cm)[a]	t (hr)[b]	t (min)[c]	f (cm/hr)[d]
0.0	0.000	0.000	0.000
0.1	0.006	0.360	8.356
0.2	0.024	1.431	4.228
0.3	0.053	3.193	2.852
0.4	0.094	5.633	2.164
0.5	0.146	8.734	1.751
0.6	0.208	12.481	1.476
0.7	0.281	16.860	1.279
0.8	0.364	21.856	1.132
0.9	0.458	27.456	1.017
1.0	0.561	33.648	0.926
1.1	0.674	40.418	0.851
1.2	0.796	47.753	0.788
1.3	0.927	55.643	0.735
1.4	1.068	64.075	0.690
1.5	1.217	73.039	0.650
1.6	1.375	82.523	0.616
1.7	1.542	92.517	0.586
1.8	1.717	103.010	0.559
1.9	1.900	113.994	0.535
2.0	2.091	125.458	0.513

[a] Assumed.
[b] From Eq. (3.20) with $S_e = 0.3$, $\theta_e = 0.432$, $\psi = 27.3$ cm, and $K = 0.1$ cm/hr.
[c] Column 2 × 60 min/hr.
[d] From Eq. (3.19).

Φ Index

Over the years many other empirical infiltration models have been proposed. Because of the general lack of values for the parameters for these various models and the nonhomogeneity of soils, these models have not been widely applied in storm water management. Instead, a steady infiltration loss rate from the rainfall rate has been defined to obtain the effective rainfall rate. Sometimes a two-stage constant loss rate is used. For example, for the Denver region, Wright-McLaughlin Engineers (1969) proposed that the following constant infiltration loss rates be used in the absence of measured data:

Storm frequency	First half hour	Remainder
2 to 5 years	1 iph	½ iph
	2.5 cm/hr	1.3 cm/hr
10 to 100 years	½ iph	½ iph
	1.3 cm/hr	1.3 cm/hr

Wright-McLaughlin Engineers (1969) went on to urge that each area being considered be field tested for infiltration rates and that the measured values be used in preference to those shown in the above table. Often the constant infiltration loss rate is termed the Φ index.

Effective Rainfall

In runoff calculations, effective rainfall is the rainfall that runs off. Effective rainfall represents the supply of water to the runoff process. The volume of effective rainfall is equal to the volume of storm water runoff. Effective rainfall is also known as rainfall excess since it represents the amount of rainfall in excess of abstractions or losses from runoff. Considering only the abstractions of surface storage and infiltration, the generation of effective rainfall can be envisioned as follows. All rainfall goes into infiltration until the rainfall rate exceeds the infiltration rate. At that point, surface storage begins to fill. As long as the rainfall rate exceeds the infiltration rate, surface storage will continue to fill. Runoff will begin on those parts of the catchment where the surface storage is filled. When the rainfall rate drops below the infiltration rate, water in surface storage will begin infiltrating. When the surface storage capacity at a location is no longer filled, no additional water becomes available for surface runoff; thus, the generation of effective rainfall ceases. This cyclic interplay between rainfall, surface storage, and effective rainfall may be repeated many times during a natural rainfall event with fluctuating rainfall intensities.

Example Problem 3.9. Effective rainfall from the Holtan equation

Estimate the effective rainfall rate based on a constant surface detention of 0.30 in. (7.5 mm) typical of a sod and an infiltration curve for a soil in HSG C based on the Holtan equation for the rainfall given in Table 3.3.

Solution: The results are tabulated in Table 3.13. The first two columns defining the rainfall are from Table 3.3. Several points based on Table 3.13 should be noted. First, the effective rainfall is zero unless surface storage is filled and the rainfall rate exceeds the infiltration rate. Second, the actual infiltration equals the rainfall rate until the rainfall rate exceeds the potential infiltration rate as determined from the Holtan relationship. Third, the actual infiltration may exceed the rainfall rate if surface storage is not zero. Fourth, the

Table 3.13 Calculation of Effective Rainfall Based on Holtan Equation

Time (min)[a]	Rain depth (in.)[b]	Surface detention (in.)[c]	F_p (in.)[d]	f (iph)[e]	Potential infiltration volume (in.)[f]	F_a (in.)[g]	Effective rain (in.)[h]
0							
15	0.10	0.00	3.00	4.91	1.23	0.10	0.00
30	0.12	0.00	2.90	4.69	1.17	0.12	0.00
45	0.17	0.00	2.78	4.43	1.11	0.17	0.00
60	0.26	0.00	2.61	4.08	1.02	0.26	0.00
75	0.61	0.00	2.35	3.56	0.89	0.61	0.00
90	1.91	0.30	1.74	2.42	0.61	0.61	1.00
105	0.36	0.30	1.13	1.44	0.36	0.36	0.00
120	0.20	0.26	0.77	0.94	0.24	0.24	0.00
135	0.14	0.23	0.53	0.66	0.17	0.17	0 00
150	0.11	0.22	0.36	0.49	0.12	0.12	0 00
165	0.09	0.21	0.24	0.39	0.10	0.10	0.00
180	0.08	0.21	0.14	0.31	0.08	0.08	0.00
Total	4.15	0.21				2.94	1.00

[a] From Table 3.3.
[b] From Table 3.3
[c] First 0.3 in. of rain in excess of infiltration goes to detention storage.
[d] $F_p(t) = F_p(t-1) - F_a(t-1)$.
[e] From Eq. (3.17).
[f] Column 5 × 0.25 hr.
[g] Actual incremental infiltration volume.
[h] Rainfall depth in excess of actual infiltration depth. Note: At end of storm, detention + total infiltration + total effective rainfall = total rainfall.

value of F_p is decreased by the actual, not the potential, infiltration volume. Consider the calculations for the time of 60 min. The rainfall depth is 0.26 in. during the 15-min time increment. There is no water in surface storage since the potential infiltration rate to this time has always exceeded the rainfall rate. F_p of column 4 is computed from $F_p(60) = F_p(45)$—actual infiltration during the preceding time interval given in column 7 = 2.78 − 0.17 = 2.61.

Column 5 represents the potential infiltration rate if rainfall is not limiting. It is determined from Eq. (3.17). The equation parameters are determined from Table 3.9 as f_c = 0.25 iph, F_p (initial) = 3.00 in., n = 1.4, and a = 1:

$$f = aF_p^n + f_c = 1.0(2.61)^{1.4} + 0.25 = 4.08 \text{ iph.}$$

The potential infiltration volume is given by the product of the potential rate times the time increment or 4.08 × 0.25 = 1.02 in. Since the rainfall volume during the time increment was only 0.26 in., the actual infiltration volume is 0.26 in. F_p for the next time increment is thus reduced by 0.26 in.

The time increment ending at 90 min is the first to produce any effective rainfall. For this time increment, the rainfall volume is 1.91 in. The potential infiltration volume is only 0.61 in. Thus 0.30 in. goes to surface storage, 0.61 in. goes to infiltration, and the remainder goes to effective rainfall.

The time increment ending at 135 min has a rainfall volume of 0.14 in. and a potential infiltration volume of 0.17 in. Thus, 0.03 in. are taken from surface storage and no effective rainfall is generated.

At the end of the storm, the sum of surface storage, actual infiltration, and effective rainfall must equal the total rainfall volume:

$$0.21 + 2.94 + 1.00 = 4.15.$$

Example Problem 3.10. Effective rainfall from the Green–Ampt equation

Calculate the effective rainfall pattern for the rain of Table 3.3 using the Green–Ampt model for a silt loam soil with 30% effective saturation. The maximum surface storage is 0.75 cm.

Solution: The parameters for the Green–Ampt equation [Eq. (3.19)] from Table 3.11 are $\theta_e = 0.486$, $\psi = 16.68$ cm, and $K = 0.65$ cm/hr. The change in water content as a result of the passing of the wetting front is calculated as

$$\Delta\theta = (1 - s_e)\theta_e = (1 - 0.3)0.486 = 0.34.$$

Table 3.14 Calculation of Effective Rainfall: Green–Ampt Approach

t (min)	ΔR (cm)[a]	f_p (cm/hr)[b]	ΔF_p (cm)[c]	ΔF_a (cm)[d]	F_p (cm)[e]	S (cm)[f]	ΔR_c (cm)[g]
0	0.00				0.00	0.00	0.00
15	0.25	[h]	[h]	0.25	0.25	0.00	0.00
30	0.30	15.40	3.85	0.30	0.55	0.00	0.00
45	0.43	7.36	1.84	0.43	0.98	0.00	0.00
60	0.66	4.42	1.10	0.66	1.64	0.00	0.00
75	1.55	2.90	0.73	0.73	2.37	0.75	0.07
90	4.85	2.21	0.55	0.55	2.92	0.75	4.30
105	0.91	1.91	0.48	0.48	3.40	0.75	0.43
120	0.51	1.74	0.43	0.43	3.83	0.75	0.08
135	0.36	1.61	0.40	0.40	4.23	0.71	0.00
150	0.28	1.52	0.38	0.38	4.61	0.61	0.00
165	0.23	1.45	0.36	0.36	4.97	0.48	0.00
180	0.20	1.39	0.35	0.35	5.32	0.33	0.00
Totals	10.53			5.32			4.88

[a]From Table 3.3 converted to centimeters.
[b]Potential infiltration rate from Green–Ampt equation.
[c]Potential infiltration volume $f_p\Delta t$.
[d]Actual infiltration volume.
[e]$F_p(t) = F_p(t\text{–}1) + \Delta F_a$.
[f]Surface storage.
[g]Effective rainfall.
[h]Very large.

The infiltration rate, f, and cumulative infiltration are related by Eq. (3.19) as

$$f = K\left[\frac{\Psi\Delta\theta}{F} + 1\right] = 0.65\left[\frac{16.68 \times 0.34}{F} + 1\right].$$

Table 3.14 shows the resulting calculations. The rainfall of Table 3.3 is converted to centimeters. The f_p of column 3 represents the potential infiltration rate calculated from the above equation using F_p from the previous time increment. ΔF_p of column 4 is the potential infiltration volume calculated as $f_p\Delta t$. ΔF_a represents the actual infiltration volume for the time increment. The actual infiltration rate may be less than the potential rate. Early in the storm, the potential infiltration rate exceeds the rainfall rate so the actual infiltration volume is limited to the rainfall volume. F_p at a particular time increment is equal to F_p from the previous time increment plus ΔF_a for the current time increment.

In the time increment from 60 to 75 min, the potential infiltration rate falls below the rainfall rate so some surface storage and rainfall excess is generated. Since at this time the surface storage is empty, the rainfall is divided with 0.73 cm going to satisfy infiltration, 0.75 cm going to surface storage, and 0.07 cm becoming rainfall excess since the surface storage is filled. The process is continued to the end of the storm. It can be seen that for the time interval from 120 to 135 min, ΔF is 0.40 cm while the rainfall is only 0.36 cm. Thus 0.04 cm of water is taken from surface storage. ΔF cannot exceed the rainfall for the time increment plus the depth of water in surface storage.

Table 3.15 Definition of SCS Hydrologic Soil Groups (Soil Conservation Service, 1986)

Group A soils have low runoff potential and high infiltration rates even when thoroughly wetted. They consist chiefly of deep, well to excessively drained sands or gravels and have a high rate of water transmission (greater than 0.30 in./hr).

Group B soils have moderate infiltration rates when thoroughly wetted and consist chiefly of moderately deep to deep, moderately well to well drained soils with moderately fine to moderately coarse textures. These soils have a moderate rate of water transmission (0.15–0.30 in./hr).

Group C soils have low infiltration rates when thoroughly wetted and consist chiefly of soils with a layer that impedes downward movement of water and soils with moderately fine to fine texture. These soils have a low rate of water transmission (0.05–0.15 in./hr).

Group D soils have high runoff potential. They have very low infiltration rates when thoroughly wetted and consist chiefly of clay soils with a high swelling potential, soils with a permanent high water table, soils with a claypan or clay layer at or near the surface, and shallow soils over nearly impervious material. These soils have a very low rate of water transmission (0–0.05 in./hr).

Some soils in the list are in group D because of a high water table that creates a drainage problem. Once these soils are effectively drained, they are placed in a different group. For example, Ackerman soil is classified as A/D. This indicates that the drained Ackerman soil is in group A and the undrained soil is in group D.

Curve Number Approach

The Soil Conservation Services (SCS) (1972, 1985) of the U.S. Department of Agriculture combines infiltration losses with initial abstractions and estimates rainfall excess or equivalently the runoff volume by the relationship

$$Q = \frac{(P - 0.2S)^2}{P + 0.8S}, \qquad P > 0.2S, \qquad (3.21)$$

where Q is the accumulated runoff volume or rainfall excess, P is the accumulated precipitation, and S is a parameter, sometimes called a maximum soil water retention parameter, given by

$$S = \frac{1000}{\text{CN}} - 10 \qquad (Q, P, S \text{ in.})$$
$$S = \frac{25400}{\text{CN}} - 254 \qquad (Q, P, S \text{ mm}), \qquad (3.22)$$

where CN is known as the curve number. Curve number tables are available from a number of sources. Table 3.16 and Appendix 3C contain representative values. Equation (3.21) indicates that P must exceed $0.2S$ before any runoff is generated. Thus a rainfall volume of $0.2S$ must fall before runoff is initiated. It should be noted that Eq. (3.21) is a runoff equation and not an infiltration equation. Using it as an infiltration equation can lead to errors.

The SCS has classified more than 4000 soils into four hydrologic soil groups (HSG) according to their minimum infiltration rate obtained for bare soil after prolonged wetting. Listings of soils and their hydrologic soil group may be found in a number of SCS publications (SCS, 1972, 1985). Local SCS offices can provide information on local soils. Appendix 3B contains the HSG classification for a large number of soils. The four hydrologic soil groups are denoted by the letters A, B, C, and D. Table 3.15 contains the definition of these four soil groups. In determining the hydrologic soil group, consideration should be given to compaction via heavy equipment, exposure of subsoil, etc. These things may cause a soil that would normally be in one soil group to be less pervious and thus behave as another soil group. For example, a normal B soil may behave as a C soil. When the soil is greatly disturbed but not significantly compacted, an estimate of the HSG can be made based on the texture of the exposed or surface soil as follows (Brakensiek and Rawls, 1983):

HSG	Soil texture
A	Sand, loamy sand, or sandy loam
B	Silt loam or loam
C	Sandy clay loam
D	Clay loam, silty clay loam, sandy clay, silty clay, or clay

The curve number of an area indicates the runoff potential of the area. Table 3.16 is a summary of CNs for various land-use and treatment combinations. Appendix 3C is a more extensive table of CNs. Impervious areas and water surfaces are assigned a CN of 98–100. Recognizing that abstractions from rainfall depend on the antecedent conditions that exist at the time a rainstorm occurs, three antecedent conditions have been defined. The curve numbers given in Table 3.16 are for antecedent condition II, which is based on median values for CN taken from sample rainfall and runoff data. Antecedent condition I is used when there has been little rainfall preceding the rainfall in question and condition III is used where there has been considerable rainfall prior to the rain in question. Curve numbers for antecedent conditions I or III can be estimated by (Chow *et al.*, 1988)

$$\mathrm{CN(I)} = \frac{4.2\ \mathrm{CN(II)}}{10 - 0.058\ \mathrm{CN(II)}}$$
$$\mathrm{CN(III)} = \frac{23\ \mathrm{CN(II)}}{10 + 0.13\ \mathrm{CN(II)}}, \tag{3.23}$$

where CN(I), CN(II), and CN(III) represent curve numbers for antecedent conditions I, II, and III, respectively.

Originally, the CN values were assigned by plotting observed runoff versus measured rainfall for a number of experimental plots scattered throughout the U.S. The CNs were then correlated with land use. The terms good condition or poor condition in CN tables refer to the relative runoff potential. An area in good hydrologic condition would have higher infiltration rates and lower runoff rates than an area in poor condition. Again, note that the CN approach is a runoff approach and not an infiltration approach. Certainly infiltration is a factor affecting runoff, but so is quick return flow and initial abstractions. Combining the CN approach with infiltration approaches such as minimum retention rates carries the CN concept beyond its original intent and beyond the data on which the CNs are based. Since the derivation of curve numbers includes factors in addition to infiltration, it is, in fact, a non-Hortonian approach to runoff estimation.

As with any determination of soil parameters, special attention must be given to situations where impervious areas such as streets, buildings, and parking lots are present. One must consider the extent of the impervious area and the manner in which flow from the impervious area reaches a drainage channel. The extent of the impervious area is generally expressed as a percentage of the total area. The path the flow uses to reach a channel is often specified by stating the impervious area is either "directly connected" or "indirectly connected." Directly connected impervious areas have flow that travels directly to the drainage system (channels, sewers, gutters, etc.) or occurs as concentrated flow over a pervious area. Indirectly connected impervious areas discharge flow in a diffuse manner as overland flow onto a pervious area presenting the runoff an opportunity for infiltration into the pervious area.

The CNs of Table 3.16 were developed for typical land-use relationships and specific assumed percentages of impervious area. The assumptions were that the pervious urban area corresponds to a pasture in good hydrologic condition and that the impervious area was directly connected with a CN of 98.

If all of the impervious area is directly connected but the pervious area percentage or the pervious land-use assumptions are not met, the following relationship based on SCS TR-55 (Soil Conservation Service, 1986) can be used to calculate a composite CN

$$\mathrm{CN_c} = \mathrm{CN_p} + (P_{\mathrm{imp}}/100)(98 - \mathrm{CN_p}), \tag{3.24a}$$

where $\mathrm{CN_c}$ is the composite CN, $\mathrm{CN_p}$ is the CN for the pervious area, and P_{imp} is the percentage impervious.

If runoff from the unconnected impervious area is spread out over a pervious area as sheet flow, the composite CN should be computed from

$$\mathrm{CN_c} = \mathrm{CN_p} + (P_{\mathrm{imp}}/100)(98 - \mathrm{CN_p})(1 - 0.5R) \tag{3.24b}$$

for P_{imp} less than 30% and from Eq. (3.24a) for P_{imp} greater than 30%. R represents the ratio of unconnected impervious area to total impervious area.

It is important to consider all impervious areas and other areas of high runoff potential such as soils in HSG D no matter how small since they produce high rates of runoff per unit of rainfall.

An area-weighted CN for mixed land uses and HSGs can be computed from

$$\mathrm{CN} = \frac{\Sigma_i A_i \mathrm{CN}_i}{\Sigma_i A_i}, \tag{3.25}$$

Table 3.16 Runoff Curve Numbers for Selected Land Uses (Soil Conservation Service, 1986)

Land use description		Hydrologic soil group			
		A	B	C	D
Cultivated land[a]					
Without conservation treatment		72	81	88	91
With conservation treatment		62	71	78	81
Pasture or range land					
Poor condition		68	79	86	89
Good condition		39	61	74	80
Meadow					
Good condition		30	58	71	78
Wood or forest land					
Thin stand, poor cover, no mulch		45	66	77	83
Good cover[b]		25	55	70	77
Open Spaces, lawns, parks, golf courses, cemeteries, etc.					
Good condition (grass cover on 75% or more of the area)		39	61	74	80
Fair condition (grass cover on 50 to 75% of the area)		49	69	79	84
Commercial and business areas (85% impervious)		89	92	94	95
Industrial districts (72% impervious)		81	88	91	93
Residential[c]					
Average lot size	Average percentage impervious[d]				
$\frac{1}{8}$ acre or less	65	77	85	90	92
$\frac{1}{4}$ acre	38	61	75	83	87
$\frac{1}{3}$ acre	30	57	72	81	86
$\frac{1}{2}$ acre	25	54	70	80	85
1 acre	20	51	68	79	84
Paved parking lots, roofs, driveways, etc.[e]		98	98	98	98
Streets and roads					
Paved with curbs and storm sewers[e]		98	98	98	98
Gravel		76	85	89	91
Dirt		72	82	87	89

[a]For a more detailed descrlption of agricultural and land use curve numbers refer to "National Engineering Handbook," Sect. 4, "Hydrology" Chap. 9, 1972.
[b]Good cover is protected from grazing, litter, and brush cover soil.
[c]Curve numbers are computed assuming the runoff from the house and driveway is directed toward the street with a minimum of roof water directed to lawns where additional infiltrations could occur.
[d]The remaining pervious areas (lawn) are considered to be in good pasture condition for these curve numbers.
[e]In some warmer climates of the country a curve number of 95 may be used.

where CN_i corresponds to the appropriate CN for the part of the catchment having area A_i.

Once the proper CN is obtained, Eq. (3.21) and (3.22) can be used to estimate the accumulated rainfall excess as a function of total accumulated rainfall. Figure 3.21 has been prepared to simplify the solution of Eq. (3.21).

In some cases, a straightforward weighting of infiltration indices, whether they are CNs or Φ indices or some other index, may not be appropriate. Such would be the case when there is a large difference in the indices and the areas with a high runoff potential are directly connected to the drainage system. In such cases, runoff from the nearly impervious area may

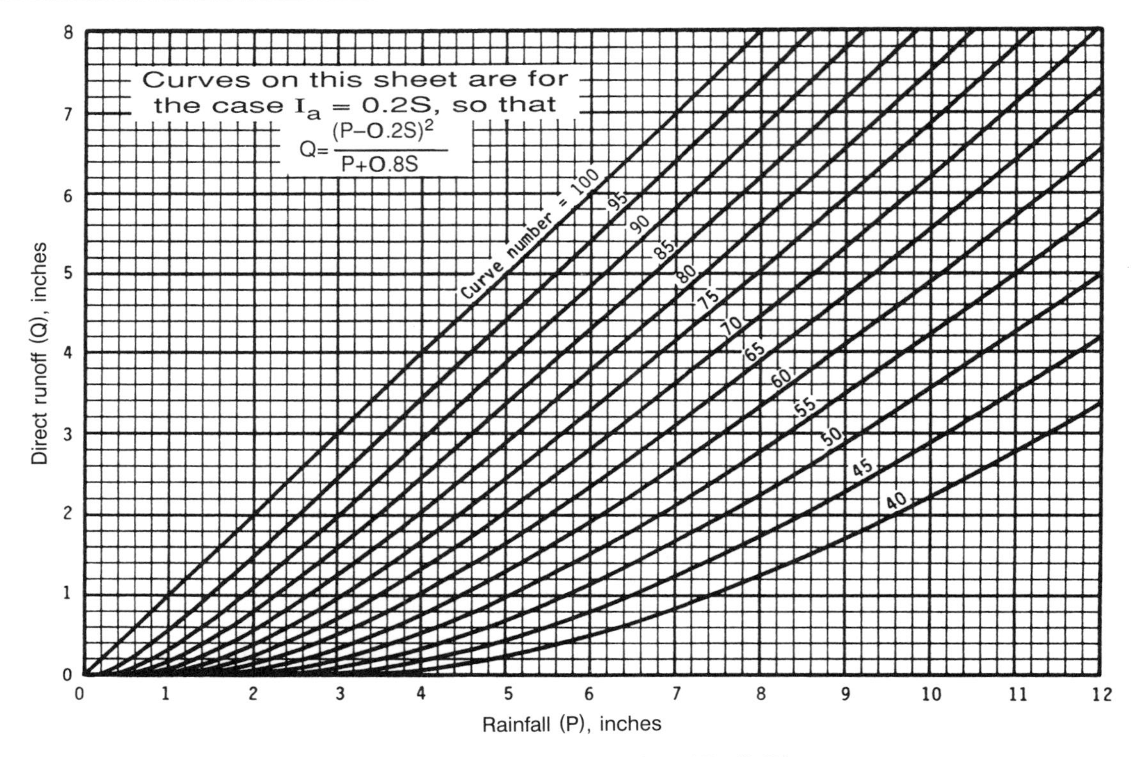

Figure 3.21 Graphical solution of Eq. (3.21).

represent a significant part of the total runoff and should not be diminished by averaging with a more pervious area. The nearly impervious areas may also respond very quickly to rainfall producing runoff well before the pervious areas. If the nearly impervious area drains across a pervious area, then some of the water from the nearly impervious area may infiltrate into the pervious area. In this case, the assignment of a large percentage runoff from the nearly impervious area could overestimate the actual runoff. The losses from the nearly impervious area runoff would depend on the infiltration rate of the intervening pervious area and the opportunity for infiltration.

Example Problem 3.11. Effective rainfall based on SCS curve number

Assume the rainstorm of Table 3.5 falls on a watershed that is 35% bare soil in hydrologic soil group D and has 30% of its soils in hydrologic soil group B under grass and 35% in hydrologic soil group C under forest. Determine the effective rainfall pattern.

Solution: The appropriate CN can be calculated by referring to Table 3.16 and noting that the bare soil area has a CN of 91, the B soil has a CN of 58, and the C soil has a CN of 77. Thus, the weighted CN is

$$\text{CN} = 0.35(91) + 0.35(77) + 0.30(58) = 77.$$

In this example, it is assumed that the various soils are randomly and somewhat uniformly scattered throughout the watershed and an unknown antecedent condition exists. The total runoff from the 4.04 in. (103 mm) of rain is computed from Eq. (3.21) and (3.22) as

$$S = \frac{1000}{\text{CN}} - 10 = \frac{1000}{77} - 10 = 2.99 \text{ in.}$$

or

$$S = \frac{25400}{\text{CN}} - 254 = \frac{25400}{77} - 254 = 75.9 \text{ mm}$$

and

$$Q = \frac{(P - 0.2S)^2}{P + 0.8S} = \frac{(4.04 - 0.2 \times 2.99)^2}{4.04 + 0.8 \times 2.99} = 1.84 \text{ in.}$$

or

$$Q = \frac{(P - 0.2S)^2}{P + 0.8S} = \frac{(102.6 - 0.2 \times 75.9)^2}{102.6 + 0.8 \times 75.9} = 46.8 \text{ mm.}$$

Table 3.17 shows the calculations required to arrive at the effective rainfall pattern. The calculations can be illustrated by considering the time interval from 11.75 to 12.00 hr. The accumulated precipitation to 12 hr is obtained by summing the entries in column 2 up to 12 hr as 3.12 in. Using $P = 3.12$

Table 3.17 Calculation of Effective Rainfall Using CN Approach

Time[a]	Incremental rainfall[a] (in.)	Accumulated rainfall[b] (in.)	Accumulated effective rainfall[c] (in.)	Incremental effective fainfall[d] (in.)
10.50	0.00	0.00	0.00	0.00
10.75	0.10	0.10	0.00	0.00
11.00	0.11	0.21	0.00	0.00
11.25	0.15	0.36	0.00	0.00
11.50	0.17	0.53	0.00	0.00
11.75	0.71	1.24	0.11	0.11
12.00	1.88	3.12	1.15	1.04
12.25	0.33	3.45	1.39	0.24
12.50	0.16	3.61	1.51	0.12
12.75	0.15	3.76	1.63	0.12
13.00	0.10	3.86	1.70	0.07
13.25	0.10	3.96	1.78	0.08
13.50	0.08	4.04	1.84	0.06
Total	4.04	4.04	1.84	1.84

[a]From Table 3.5.
[b]Summing Column 2.
[c]Based on Eq. (3.21) and a CN of 77 and P of Column 3.
[d]Differencing Column 4.

in. and $S = 2.99$ in. in Eq. (3.21) results in a Q of 1.15 in. The Q for the time increment is the difference in the accumulated runoff at 12.00 and 11.75 hr or $1.15 - 0.11 = 1.04$ in.

RUNOFF ESTIMATION

Runoff is the flow resulting from precipitation events. Some of this flow may occur during and immediately following the precipitation event and as such is known as storm water runoff or storm flow. Flow occurring between precipitation events is generally supported by seepage and ground water discharge and as such is known as baseflow. There is not a definable point or time at which storm flow ceases and base flow begins. One process gradually grades into the other. Generally on small catchments the magnitude of base flow is quite small and is neglected in the computation of storm water runoff.

A runoff hydrograph is a continuous record of streamflow over time. A complete runoff hydrograph contains information on runoff volume as the area under the hydrograph and peak flow rates as the maximum flow or peak of the runoff hydrograph as well as a complete time history of flow. Hydrographs of storm water runoff are often required in the design of storm water-retarding structures and sediment control ponds. If only an estimate of the peak flow or runoff volume is needed, it may not be necessary to develop the entire runoff hydrograph.

Storm Water Runoff Volume

In this section, reference to runoff means storm water runoff or the runoff occurring during and immediately following a major precipitation event. The volume of storm water runoff is equal to the volume of rainfall excess or effective precipitation. Thus runoff volume is rainfall minus abstractions. Any of the methods previously discussed for estimating abstractions or rainfall losses can be used in the computation of the volume of runoff.

The infiltration approach to estimating runoff volume consists of estimating initial abstractions and infiltration and deducting these losses from rainfall. Several methods for estimating these quantities have already been discussed and illustrated in the previous sections of this chapter.

When the SCS curve number approach is used, Eq. (3.21) and (3.22) define the runoff volume, Q, as a function of the total precipitation, P, and the curve number, CN, for storms with durations of 24 hr or less. For example, if 5 in. (127 mm) of rain falls in 24 hr on a watershed having a CN of 75, the estimated runoff volume is 2.45 in. (62 mm). Note that this is an estimate of storm water runoff and does not include base flow generated from ground water discharges that are dependent on infiltrated water.

Development of Runoff Hydrographs

To this point, the generation of rainfall excess volumes and time distributions and runoff volumes have been discussed. What is often of interest in storm water management is the hydrograph of runoff at the watershed outlet and possibly at selected points within the watershed. A hydrograph is simply a plot of flow rate versus time. The hydrograph is a result of a particular effective rainfall hyetograph as modified by basin flow characteristics. By definition, the volume of water under an effective rainfall hyetograph is equal to the volume of surface runoff.

Most of the terminology used in discussing runoff hydrographs is shown in Fig. 3.22. An effective rainfall hyetograph consisting of a single block of rainfall with

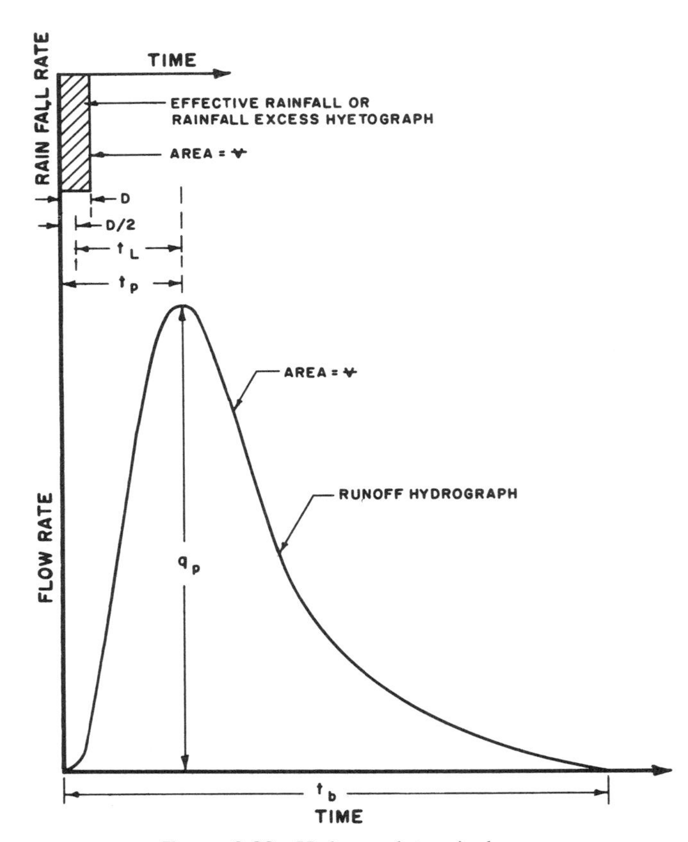

Figure 3.22 Hydrograph terminology.

duration D is shown in the upper left part of the figure. The runoff hydrograph constitutes the lower part of the figure. The areas enclosed by the hyetograph and by the hydrograph each represent the same volume, V, of water. Obviously, the scale for rainfall and flow are not the same in Fig. 3.22. The maximum flow rate on the hydrograph is the peak flow, q_p, while the time from the start of the hydrograph to q_p is the time to peak, t_p. The total time duration of the hydrograph is known as the base time, t_b.

The lag time, t_L is the time from the center of mass of the effective rainfall to the peak of the runoff hydrograph. It is apparent that

$$t_p = t_L + D/2, \tag{3.26}$$

using this definition. Some define lag time as the time from center of mass of effective rainfall to the center of mass of the runoff hydrograph. Methods for estimating these various hydrograph parameters are covered in this chapter.

A time parameter not shown in Fig. 3.22 is the time of concentration, t_c. The time of concentration is defined as the time it takes water to flow from the

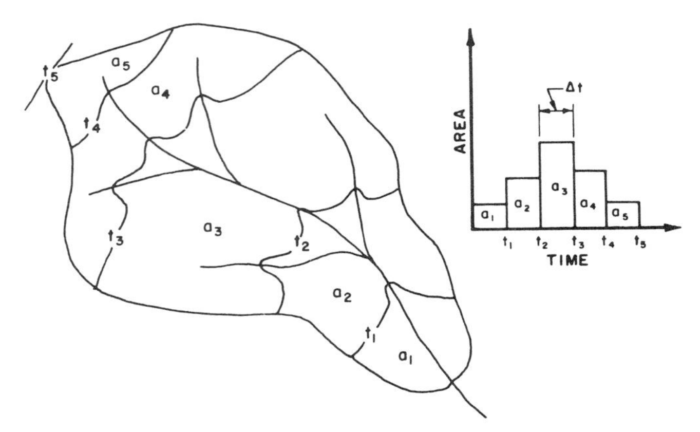

Figure 3.23 Watershed showing isochrones and time area diagram.

hydraulically most remote point in a basin to the basin outlet.

A Conceptual Model

A conceptual model of the runoff process can be developed starting with the time–area diagram of a watershed and considering only a translation of the rainfall excess hyetograph to the basin outlet. The time–area diagram is obtained by drawing isochrones (lines of constant time) on the watershed with the isochrones separated in time by Δt (Fig. 3.23). The value of time on each isochrone represents the travel time of water from the isochrone to the basin outlet. This travel time is the sum of the overland flow travel time and channel flow travel time. The time–area diagram is then a plot of a_i versus t_i, where a_i is the area enclosed between isochrone t_i and t_{i-1}.

Next visualize a block of effective rainfall of uniform intensity r_1 and duration Δt falling uniformly on the watershed. Also, envision that the runoff process is simply a translation of the rainfall from the point it strikes the basin to the watershed outlet with the travel time defined by the time–area diagram. In this event, the runoff from a_1 would be $q = 0$ at $t = 0$, increase linearly to $q = a_1 r_1$ at $t = t_1$, at which point the rainfall ceases. The runoff would then decrease linearly to $q = 0$ at $t = t_2$. Similarly, the runoff from a_2 would start at $q = 0$ at $t = t_1$ since it takes t_1 for the runoff to reach the outlet from a_2, increase linearly to $q = a_2 r_1$ at $t = t_2$, and decrease to $q = 0$ at $t = t_3$. Similar patterns result for the runoff from the remaining areas. The total runoff hydrograph from the first block of effective rainfall would then equal the sum of the individual triangular subhydrographs as shown in Fig. 3.24. The hydrograph peaks at $t = t_3$.

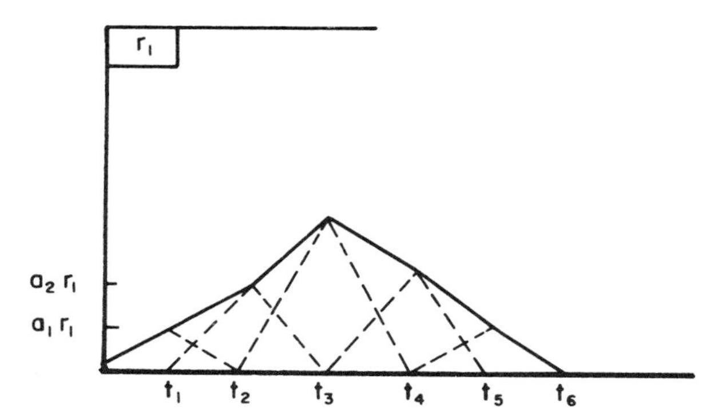

Figure 3.24 Illustration of runoff hydrograph from conceptual watershed due to a single block of short duration rainfall.

If a second block of effective rainfall of uniform intensity r_2 falls uniformly on the basin during the time interval t_1 to t_2, the resulting runoff hydrograph from this second block of rainfall would be obtained in a similar fashion with the hydrograph starting at $t = t_1$, peaking at $t = t_4$, and ending at $t = t_7$. The runoff hydrograph for the two blocks of rainfall r_1 and r_2 would be the sum of the runoff from the individual blocks of rainfall. Figure 3.25 shows the total runoff hydrograph from five blocks of rainfall on the watershed shown in Fig. 3.23.

The ordinate for runoff q_j at time t_j is in general given by

$$q_j = \sum_{i=1}^{j \leq m} r_i a_{j-i+1} \qquad \begin{aligned} r_i &= 0 \text{ for } i > m \\ a_i &= 0 \text{ for } i > n, \end{aligned} \tag{3.27}$$

where m is the number of rainfall blocks and n is the number of area blocks in the time–area diagram.

As can be seen from Fig. 3.25, the conceptual hydrograph is a very steeply rising and falling hydrograph with a time base only slightly longer than the duration of the rainfall excess. An actual hydrograph would have a lower peak and much slower flow recession. This is because in reality the flow system on a watershed contains considerable storage, which retards surface and channel flow. The process of simple translation of the water from its point of incidence on the watershed to the basin outlet neglects this storage aspect of the flow process.

To obtain a more realistic runoff hydrograph, the conceptual hydrograph shown in Fig. 3.25 might be thought of as the inflow into a hypothetical reservoir whose outflow would then be the runoff hydrograph. The effect of the reservoir would be to delay and lessen the peak of the runoff hydrograph. Such an approach is developed in the next section.

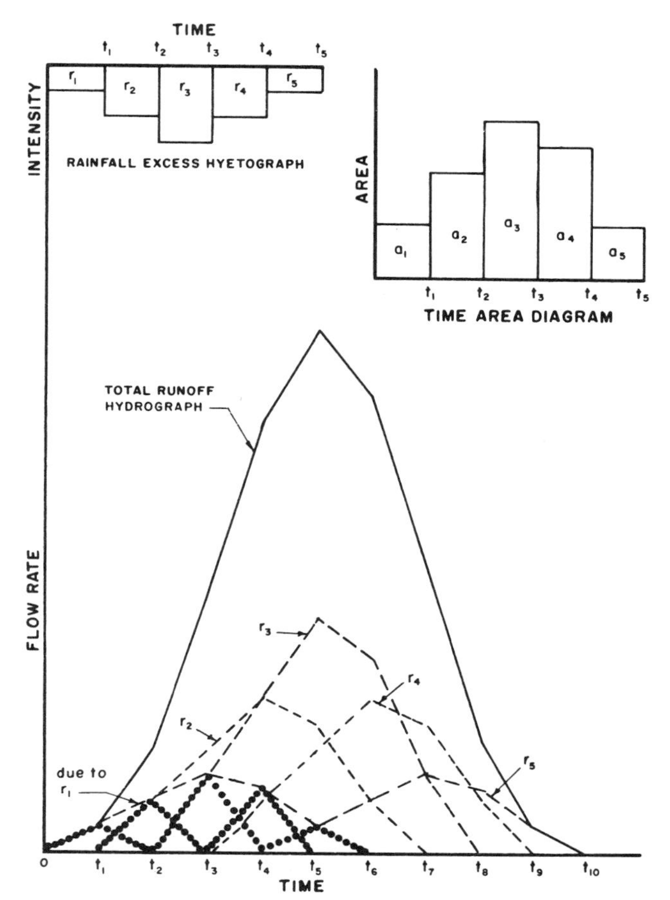

Figure 3.25 Hydrograph from rainstorm on conceptual watershed.

The Santa Barbara Urban Hydrograph Method

The Santa Barbara Urban Hydrograph (SBUH) method is an example of a procedure that produces a runoff hydrograph by routing a rainfall excess hyetograph through a conceptual reservoir (Stubchaer, 1975). The computations consist of applying the equation

$$Q_t = (1 - 2K)Q_{t-1} + K(I_t + I_{t-1}), \tag{3.28}$$

where Q_t is the runoff hydrograph ordinate at time t, I_t is the depth of rainfall excess in the time interval Δt, and K is a coefficient defined by

$$K = \Delta t/(2t_c + \Delta t), \tag{3.29}$$

where t_c is the watershed time of concentration. Any consistent set of units can be used. With t in hours and I in cfs-hours, I_t in cfs-hours can be determined from the rainfall excess hyetograph, R_t, in inches, from the relationship

$$I_t = 1.008\ AR_t/\Delta t, \tag{3.30}$$

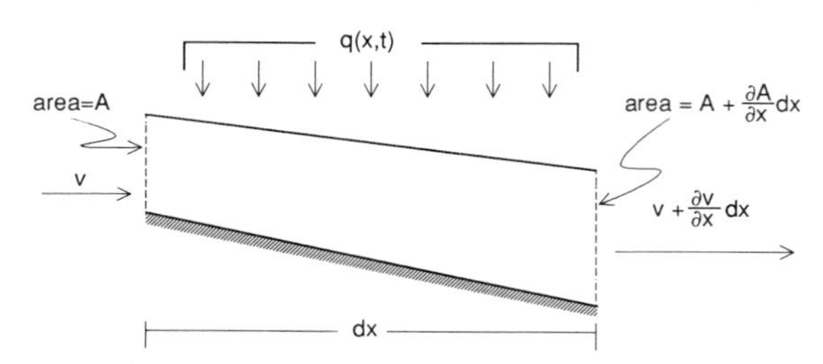

Figure 3.26 Control element for derivation of Eq. (3.36).

where A is the watershed area in acres. This method is equivalent to flow routing using the Muskingum method defined in Chapter 6 with $x = 0$ and $k = t_c$.

A Hydrodynamic Model

The method for obtaining runoff hydrographs that is the most theoretically elegant is to route the rainfall excess hyetograph as overland flow to established channels and as channel flow to the basin outlet. The procedure would rely on the continuity equation and a flow equation, as well as relationships between the various hydraulic elements of slope, roughness, channel shape, hydraulic radius, etc.

The continuity equation simply states that in a control element such as shown in Fig. 3.26, the difference in the rate of inflow and outflow equals the rate of change of water stored in the element. The volume of the element is equal to the average area times the length or

$$\text{Volume} = V = \left(A + \frac{\partial A}{\partial x}\frac{dx}{2}\right) dx. \tag{3.31}$$

The inflow to and outflow from the element are

$$\text{Inflow rate} = I = vA + q(x,t)\,dx \tag{3.32}$$

$$\text{Outflow rate} = O = \left(v + \frac{\partial v}{\partial x}dx\right)\left(A + \frac{\partial A}{\partial x}dx\right). \tag{3.33}$$

In these expressions, A is the cross-sectional area of the flow, x is the horizontal coordinate, v is the flow velocity (mean velocity) over the cross-sectional area, and $q(x,t)$ is the lateral inflow per unit length along the flow. The continuity equation states

$$I - O = \frac{\partial V}{\partial t}. \tag{3.34}$$

Substituting the correct quantities results in

$$vA + q(x,t)\,dx - \left(v + \frac{\partial v}{\partial x}dx\right)\left(A + \frac{\partial A}{\partial x}dx\right) = \frac{\partial[A + (\partial A/\partial x)(dx/2)\,dx]}{\partial t}, \tag{3.35}$$

neglecting higher-order differential terms (i.e., $(dx)^2$) and collecting terms result in

$$\frac{\partial A}{\partial t} + A\frac{\partial v}{\partial x} + v\frac{\partial A}{\partial x} = q(x,t). \tag{3.36}$$

Equation (3.36) has two unknowns, area and velocity. A second equation is required. The second equation is the momentum equation. Several authors (Overton and Meadows, 1976; Henderson, 1966; Chow *et al.*, 1988; Strelkoff, 1969) present the developments that lead to the momentum equation in the form

$$\frac{\partial v}{\partial t} + v\frac{\partial v}{\partial x} + g\frac{\partial y}{\partial x} + \frac{qv}{y} - g(S_o - S_f) = 0, \tag{3.37}$$

where $\partial v/\partial t + v(\partial v/\partial x)$ is due to dynamic waves, $g(\partial y/\partial x)$ is due to diffusion waves, qv/y is due to lateral inflow, and $g(S_o - S_f)$ is due to kinematic waves. Woolhiser and Liggett (1967) proposed a dimensionless parameter, k, given by

$$k = \frac{S_o L}{HF^2}, \tag{3.38}$$

where S_o is the slope of the flow plane, L is the length of the flow plane, H is the equilibrium depth at the bottom of the plane, and F is the equilibrium Froude number at the bottom of the plane. They found that for k greater than 10, neglecting all terms in Eq. (3.37) except the kinematic term had no appreciable effect on the solution. Since S_f represents the friction slope, the implication is that $S_o = S_f$ and a uniform flow equation can be used in place of Eq. (3.37). Such an equation is given by

$$Q = aA^n. \tag{3.39}$$

This is known as the kinematic approach where a and n are rating coefficients. Overton and Meadows (1976), Wooding (1965a, b, 1966), and Woolhiser and Liggett (1967) give more information on when the kinematic approach may be used. Essentially the kinematic approach can be used when the dynamic terms of the momentum equation are negligible, backwater effects are small, and a rating equation like Eq. (3.39) applies. In practice, the kinematic approach is widely used.

If one is considering overland flow over a plane surface, Eq. (3.36) can be written on a per unit of width basis as

$$\frac{\partial y}{\partial t} + y\frac{\partial v}{\partial x} + v\frac{\partial y}{\partial x} = q(x,t), \tag{3.40}$$

where $q(x,t)$ is the effective rainfall per unit area and y is the depth of flow. Several attempts have been made to use Eq. (3.39) and (3.40) as a hydrodynamic model of the overland flow process.

Equations (3.36) and (3.39) may also be used to route flow in channels. In this case, $q(x, t)$ would represent tributary or other local inflow to the channel per unit of channel length. Since $Q = vA$ and

$$\frac{\partial Q}{\partial x} = \frac{\partial vA}{\partial x} = v\frac{\partial A}{\partial x} + A\frac{\partial v}{\partial x}, \tag{3.41}$$

Eq. (3.36) can be written

$$\frac{\partial A}{\partial t} + \frac{\partial Q}{\partial x} = q(x, t). \tag{3.42}$$

Note that if a channel reach with no lateral inflow is being considered, $q(x, t) = 0$ for that reach. Brakensiek (1967), Overton and Meadows (1976), and Viessman *et al.*, (1977) discuss numerical solutions for Eqs. (3.39) and (3.42). The books by Eagleson (1970) and Viessman *et al.*, (1977) give more complete derivations of the routing equations.

A possible modeling procedure using the hydrodynamic equations is to divide the watershed into idealized planes of overland flow and a series of channels. Equation (3.39) and (3.40) could be used to route the effective rainfall to the channel system. Equation (3.39) and (3.42) could then be used to route the resulting overland flow through the channel system. The lateral inflow terms, $q(x, t)$ for the collector channels would be the routed overland flow hydrograph. For larger channels such as waterways and natural channels, $q(x, t)$ would represent tributary inflows such as flows from smaller channels. This modeling approach has been used by Brakensiek (1967), Wooding (1965a, b, 1966), Woolhiser (1969), and Woolhiser and Liggett (1967). The approach is also discussed by Eagleson (1970), Overton and Meadows (1976), and Viessman *et al.* (1977).

Advantages of the method include minimum reliance on any observed runoff data, theoretical appeal, and the directness with which the parameters of the equations can be related to land use with the consequent ease of evaluating land-use changes.

Disadvantages of the method include the difficulty of idealizing the overland flow planes and data requirements to define channel geometries. A further conceptual difficulty revolves around the fact that it is unlikely that overland flow as envisioned by this approach actually occurs except on very uniform impervious surfaces.

Figure 3.27 shows schematically some of the idealizations of plane surfaces that have been used by the previously referenced authors. Obviously, use of the kinematic approach to routing overland flow requires

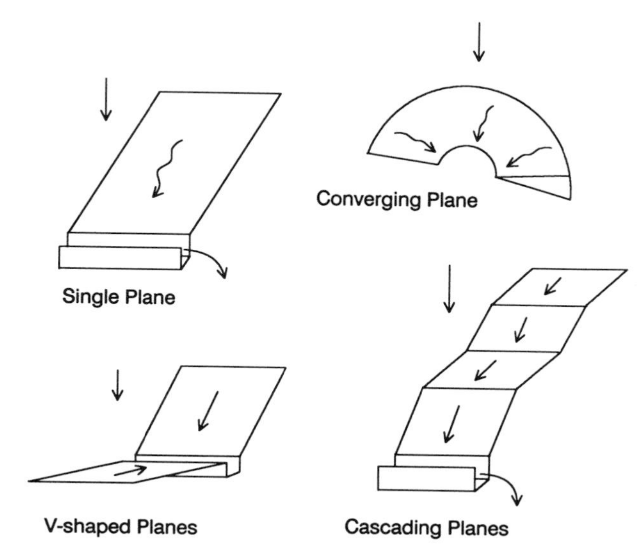

Figure 3.27 Idealized flow planes.

the use of a computer. Some hydrologic models use this approach or have it as an available option.

The Unit Hydrograph Approach

Sherman (1932) developed the concept of hydrograph estimation via a unit hydrograph. A unit hydrograph is a hydrograph of runoff resulting from a unit of rainfall excess occurring at a uniform rate, uniformly distributed over a watershed in a specified duration of time. The unit hydrograph approach to the development of runoff hydrographs is empirical and based on several assumptions. Some of these assumptions are contained in the definition. These are the assumptions of uniform distribution of rainfall excess over the watershed and uniform rate of rainfall excess. Possibly the most important and controversial assumption is that of superposition or linearity. This assumption states that the unit hydrograph reflects all basin characteristics to the degree that the runoff rate is simply proportional to the runoff volume for a rainfall excess of a given duration.

A unit hydrograph has attached to it the duration of the rainfall excess that generated the unit hydrograph. Thus, one might speak of a 20-min unit hydrograph or a 1-hr unit hydrograph. Conceptually, an infinite number of unit hydrographs, each of a different duration, can be developed for every watershed. Practically, a unit hydrograph is applied to rainfall excesses of durations as much as 25% different than the duration of the unit hydrograph. Thus, a 20-min unit hydrograph might be applied to rainfall excess of durations between 15 and 25 min. Prior to a discussion of the development of

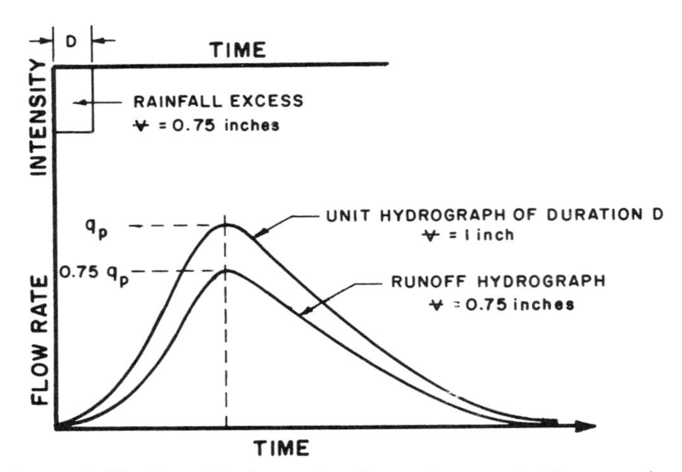

Figure 3.28 Runoff hydrograph ordinates from a rainfall excess of duration D are proportional to the ordinates of a D-min unit hydrograph. The constant of proportionality is the volume of rainfall excess.

unit hydrographs, their use in calculating runoff hydrographs is described.

Runoff Hydrographs from Unit Hydrographs

Figure 3.28 shows how a unit hydrograph is used in the development of a runoff hydrograph from a block of rainfall excess of duration equal to the duration of the unit hydrograph. Under the unit hydrograph assumptions, the runoff rate is proportional to the volume of rainfall excess. For a single block of rainfall excess, the runoff hydrograph ordinates are simply equal to the volume of rainfall excess times the unit hydrograph ordinates. Note that the intensity of the rainfall excess is 0.75 in./D units of time. Note also that the rainfall excess and not the rainfall itself is used.

Figure 3.29 illustrates the development of a runoff hydrograph from a complex rainfall excess pattern. The rainfall excess is divided into blocks of uniform intensity and of duration D. A component hydrograph for each block of rainfall excess is then obtained by multiplying the ordinates of D-min unit hydrograph by the volume of rainfall excess in each block. Hydrograph number 1 is a D-min unit hydrograph. The component hydrographs are plotted making sure that the initial point or beginning of each component hydrograph occurs at the start of the appropriate block of rainfall excess. Thus, hydrograph number 2, corresponding to the first block of effective rain of 0.2 in., is simply 0.2 times the unit hydrograph. Hydrograph 3, corresponding to the second block of effective rain of 0.15 in., is 0.15 times the unit hydrograph and shifted in time D-min to the right. Hydrograph 4 is 0.35 times the unit hydrograph and shifted $2D$ time units to the right. This process can be continued for as many blocks of rainfall excess as necessary. The resulting total runoff hydrograph, number 5 in Fig. 3.29, is then obtained by adding the ordinates of the component hydrographs at each point in time.

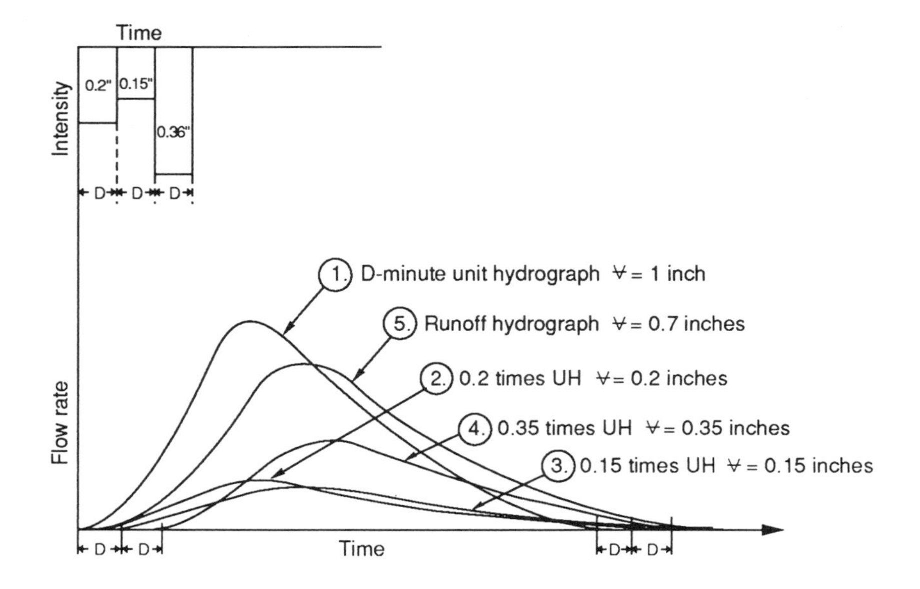

Figure 3.29 Runoff hydrograph from a complex storm is obtained by summing component hydrographs from D-min blocks of rainfall excess.

Mathematically, the ordinate of the total runoff hydrograph at any time $t = jD$ is given by

$$q_j = \sum_{i=1}^{j \le m} r_i u_{j-i+1} \qquad \begin{aligned} r_i &= 0, i > m \\ u_i &= 0, i > n, \end{aligned} \tag{3.43}$$

where q_j is the hydrograph ordinate at time $t = jD$, r_i is the volume of rainfall excess in the ith block of rainfall excess of duration D, u_i represents the ordinate of the D-min unit hydrograph at time iD, $n + 1$ is the number of unit hydrograph ordinates with u_0 and u_n both being zero, and m is the number of blocks of effective rainfall. Equation (3.43) is the same as Eq. (3.27), except the unit hydrograph has replaced the time–area diagram. The unit hydrograph has embedded within it the final routing that is lacking in the hydrograph generated by Eq. (3.27). Thus, basin flow time, lag time, and storage are incorporated into Eq. (3.43) and not (3.27). Figure 3.30 illustrates how unit hydrographs of a given duration can be used to develop runoff hydrographs for storms of any multiple of that duration.

In plots such as Fig. 3.29, it is conventional to plot the effective rainfall along the top of the diagram using the same time scale as used for the hydrograph, but a different rate scale. Many times the rate scale runs from zero at the top of the diagram with positive values in the downward direction. In a situation such as that shown in Fig. 3.29, the unit hydrograph duration must correspond to the duration of rainfall excess. That is, a D-min unit hydrograph times the volume of rainfall excess received in $3D$ min (for example) does not equal a $3D$-min (or a D-min) runoff hydrograph.

Converting Unit Hydrographs

Figure 3.30 illustrates how a unit hydrograph of a given duration can be used to develop a unit hydrograph for rainfall excess durations of any multiple of the duration of the original unit hydrograph. To obtain a unit hydrograph of duration nD, simply add n unit hydrographs of duration D each displaced by D time units and divide the sum by n. Keep in mind that the duration, D, is the duration of the blocks of rainfall excess—not the total time base of the unit hydrograph. Using this procedure a $2D$, $3D, \ldots, nD$ unit hydrograph can be derived. Only unit hydrographs of the same duration can be used in this summing process. The sum of unit hydrographs of durations D and $2D$ cannot be used to determine a unit hydrograph of duration $3D$.

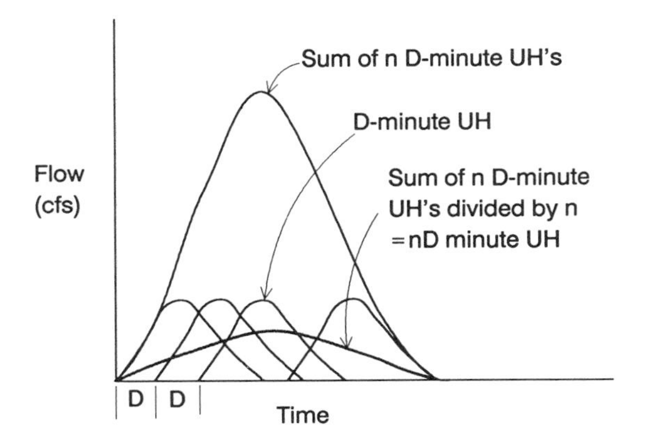

Figure 3.30 Development of a unit hydrograph of duration nD.

A procedure known as the S-curve technique can be used to derive a unit hydrograph of any duration, D', from a unit hydrograph of any other duration, D. An S-curve, like a unit hydrograph, has a duration attached to it. The duration of the S-curve is equal to the duration of the unit hydrograph from which the S-curve was derived. Conceptually, an S-curve represents the results of a steady, continuous excess rainfall rate of $1/D$ units of rain per unit of time. An S-curve can be computed by summing the ordinates of a large number of D unit hydrographs spaced D time units apart (Fig. 3.31). Computationally the S-curve can be more easily obtained from a single D unit hydrograph by computing the cumulative sum of the D unit hydrograph ordinates spaced D time units apart. Table 3.18 shows the calculation of a 30-min S-curve from a 30-min unit hydrograph when the ordinates of the unit hydrograph are tabulated in 15-min increments. The results are shown in Fig. 3.31.

Table 3.18 shows that if D is the duration of the unit hydrograph and S-curve, then the ordinates of the S-curve are given by

$$\begin{aligned} S(t) &= u(t) && \text{for } t = 0, D \\ S(t) &= S(t-D) + u(t) && \text{for } t > D. \end{aligned} \tag{3.44}$$

If $S(t)$ represents the ordinates for a D-min S-curve, the ordinates for a D'-min unit hydrograph can be computed from

$$u'(t) = (D/D')[S(t) - S(t-D')]. \tag{3.45}$$

The terms of this relationship are illustrated in Fig. 3.32.

Since the S-curve corresponds to a unit hydrograph of a specified duration, D, deriving a unit hydrograph of some other duration, D', requires a volume correction factor, which is the ratio of the old to the new unit hydrograph durations or D/D' as shown in Eq. (3.45). The original unit hydrograph corresponds to a uniform rainfall excess rate of one unit of rain per D time units,

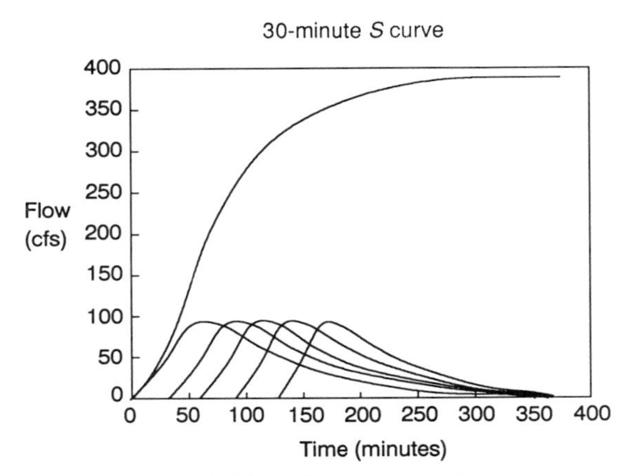

Figure 3.31 S-Curve computation.

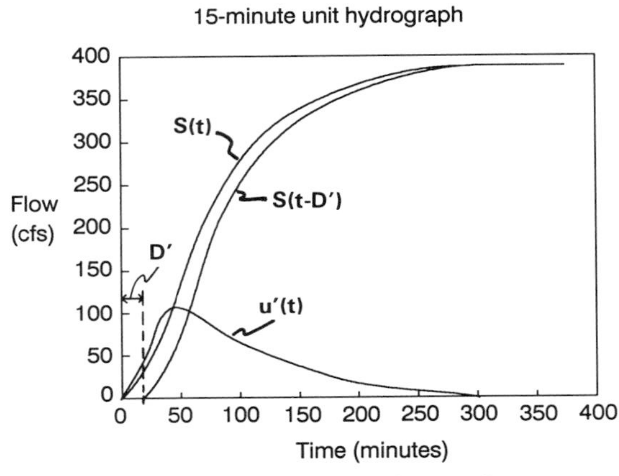

Figure 3.32 Unit hydrograph from an S-curve.

Table 3.18 S-Curve Computation from 30-min Unit Hydrograph

Time (hr)	Time (min)	UH (cfs)	Sum[a]	S-curve (cfs)
0.00	0	0	0	0
0.25	15	29	29	29
0.50	30	68	68	68
0.75	45	93	93+29	122
1.00	60	100	100+68	168
1.25	75	95	95+122	217
1.50	90	83	83+168	251
1.75	105	68	68+217	285
2.00	120	54	54+251	305
2.25	135	46	46+285	331
2.50	150	32	32+305	337
2.75	165	28	28+331	359
3.00	180	16	16+337	353
3.25	195	13	13+359	372
3.50	210	8	8+353	361
3.75	225	7	7+372	379
4.00	240	5	5+361	366
4.25	255	4	4+379	383
4.50	270	3	3+366	369
4.75	285	2	2+383	385
5.00	300	1	1+369	370
5.25	315	1	1+385	386
5.50	330	0	0+370	370
5.75	345	0	0+386	386
6.00	360	0	0+370	370

[a]Current UH ordinate plus S-curve ordinate 30 min earlier.

while the new one corresponds to a rate of one unit of rain per D' time units.

In computing S-curves, some manual smoothing is generally required. The unit hydrograph of Table 3.18 is from a basin with an area of 190 acres (77 hectares). Since the unit hydrograph has a duration of 30-min and a volume of 1 in., the equilibrium rate for the S-curve should correspond to a runoff rate of 1 in. per 30 min over 190 acres, 2 in./hr over 190 acres, or

$$2\frac{\text{inches}}{\text{hour}} \times 190 \text{ acres} \times 1.008\frac{\text{cfs}}{\text{acre-inch/hour}}$$
$$= 383 \text{ cfs}.$$

Table 3.19 shows the computation of a 15-min unit hydrograph based on the S-curve of Table 3.18. Note the required smoothing. The smoothing can be accomplished by using one of the unit hydrograph models discussed later in this chapter. The volume under the unit hydrograph should be checked to ensure that is in fact a unit hydrograph. If the time increment, Δt, is in minutes, the flow, q, in cfs, the catchment area, A, in acres, and the first and last flow values are zero, the volume, V, in inches under a hydrograph is given by

$$V = \frac{\Delta t \Sigma q_i}{60.5A}, \tag{3.46}$$

where Σq_i represents the sum of all of the ordinates of the hydrograph spaced Δt apart. Applying this equation to the last column of Table 3.18 results in a volume of 1.003 in. This sum could be made even closer to unity by a slight adjustment in the hydrograph ordinates; however, 1.003 is sufficiently close to 1.00 for most applications, making this further refinement unnecessary.

Unit hydrographs can be derived from observed records of streamflow and rainfall. Data requirements

Table 3.19 15-min Unit Hydrograph from S-Curve

Time (min)	S-curve (cfs)	Smoothed S-curve	Displaced S-curve[a]	UH[b] (cfs)	UH smoothed
0	0	0	0	0	0
15	29	29	0	58	58
30	68	68	29	78	78
45	122	122	68	108	112
60	168	168	122	92	100
75	217	217	168	98	96
90	251	251	217	68	85
105	285	285	251	68	64
120	305	305	285	40	44
135	331	331	305	52	36
150	337	342	331	22	28
165	359	355	342	26	20
180	353	360	355	10	14
195	372	368	360	16	12
210	361	375	368	14	10
225	379	377	375	4	6
240	366	378	377	2	4
255	383	379	378	2	2
270	369	382	379	6	0
285	385	383	382	2	0
300	370	383	383	0	0
315	386	383	383	0	0
330	370	383	383	0	0
345	386	383	383	0	0
360	370	383	383	0	0
				Sum	769

[a] $S(t-D')$
[b] $u(t) = [S(t) - S(t-D')]D/D' = [S(t) - S(t-15)]30/15$.

are extensive and generally prevent direct derivation of unit hydrographs for small catchments. Unit hydrographs represent direct stormwater runoff. Baseflow and/or ground water discharges to streams must be removed from the flow record before unit hydrographs can be defined from the record. Linsley *et al.* (1982) can be consulted or details. For small catchments, synthetic unit hydrographs are generally used. Synthetic unit hydrographs are discussed in detail in the following sections of this chapter. Several synthetic unit hydrograph models have been proposed. Generally they provide the ordinates of the unit hydrograph as a function of the time to peak, t_p, peak flow rate, q_p, and a mathematical or empirical shape description. After presenting procedures for estimating these attributes, several unit hydrograph models are presented.

Estimation of Time Parameters

This section deals with the estimation of the time parameters D, t_L, t_p, and t_b as shown in Fig. 3.22 and the time of concentration, t_c. Several methods for estimating these parameters are available. The method that produces results consistent with good engineering judgement should be selected for a particular study area.

The time of concentration is the time it takes for flow to reach the basin outlet from the hydraulically most remote point on the watershed. For some areas, this parameter can be estimated by summing the flow time for the various flow segments as the water travels toward the watershed outlet. These segments generally are overland flow, shallow channel flow toward larger channels, and flow in open channels, both natural and improved. The travel time in these various flow segments depends on the length of travel and the flow velocity.

Once the velocity in each flow segment is determined, the time of concentration is determined from

$$t_c = \sum_{i=1}^{n} \frac{L_i}{v_i}, \tag{3.47}$$

where n is the number of flow segments and L_i is the length and v_i the flow velocity for the ith segment.

Flow velocity of overland flow and shallow channel flow can be estimated using results such as those of Izzard (1946), Regan and Duru (1972), Overton and Meadows (1976), or from the relationship

$$v = aS^{1/2} \tag{3.48}$$

based on information in SCS (1975), where S is in ft/ft and v is in fps. The coefficient a is contained in Table 3.20.

Regan and Duru (1972) present a method for estimating travel time, t_t, over a plane surface based on the kinematic wave equation [Eq. (3.40)]. The equation is valid for turbulent flow or when the product of the rainfall excess intensity, i_e, in iph and the flow length, L, in feet is greater than 500. The equation is

$$t_t = \frac{0.0155\,(nL)^{0.6}}{i_e^{0.4} S^{0.3}}, \tag{3.49}$$

where t_t is in hours, n is Manning's n, L is in feet, i_e is in iph, and S is the slope in ft/ft. Table 3.21 presents some values for n for overland flow surfaces.

The Soil Conservation Service (1986) presents a relationship attributed to Overton and Meadows (1976) for

Table 3.20 Coefficient a for Eq. (3.48)[a]

Surface	a
Overland flow	
Forest with heavy ground litter	2.5
Hay; meadow	2.5
Trash fallow; minimum tillage	5.1
Contour; strip cropped	5.1
Woodland	5.1
Short grass	7.0
Straight row cultivation	8.6
Bare; untilled	10.1
Paved	20.3
Shallow concentrated flow	
Alluvial fans	10.1
Grassed waterways	16.1
Small upland gullies	20.3

[a]Results in fps; multiply by 0.305 to get m/sec.

Table 3.21 Manning's n for Travel Time Computations for Flow over Plane Surfaces (Soil Conservation Service, 1986)

Surface description	n[a]
Smooth surfaces (concrete, asphalt, gravel, or bare soil	0.011
Fallow (no residue)	0.05
Cultivated soils	
Residue cover ≤20%	0.06
Residue cover >20%	0.17
Grass	
Short grass prairie	0.15
Dense grasses[b]	0.24
Bermudagrass	0.41
Range (natural)	0.13
Woods[c]	
Light underbrush	0.40
Dense underbrush	0.80

[a]The n values are a composite of information compiled by Engman (1986).

[b]Includes species such as weeping lovegrass, bluegrass, buffalo grass, blue grama grass, and native grass mixtures.

[c]When selecting n, consider cover to a height of about 0.1 ft. This is the only part of the plant cover that will obstruct sheet flow.

travel time for sheet flow over plane surfaces based on Manning's equation and a kinematic approximation to the flow equations. The equation is for flow lengths of less than 300 ft. The friction value or Manning's n is an effective roughness coefficient that includes the effect of raindrop impact; drag over plane surfaces; obstacles such as litter, crop residue, ridges, and rocks; and the erosion and transport of sediment. These n values are for very shallow flow depths of about 0.1 ft or so. Table 3.21 gives Manning's n values for these conditions. The relationship for travel time is

$$T_t = \frac{0.007(nL)^{0.8}}{P_2^{0.5}S^{0.4}}, \quad (3.50)$$

where P_2 is the 2-year, 24-hr rainfall in inches and the other terms are as defined for Eq. (3.49). This relationship is based on shallow, steady, uniform flow; a constant rainfall excess intensity; and minor effects from infiltration.

In urban areas, the travel time may have to be based on a travel time to a storm drain inlet plus the travel time through the storm drain itself. Inlet travel time can generally be computed as the sum of overland flow and shallow channel flow travel times. Flow in storm drains would be considered as open channel flow with the storm drain pipe flowing full. Often large storms produce runoff rates that exceed the capacity of the storm drains and some of the runoff bypasses the drains in the form of concentrated surface flow as open channel flow. Such flow should be considered in computing the time of concentration.

Undersized culverts and bridge openings can cause ponding of flow and a reduction in the average flow velocity. For small ponds and situations where water is passing through the pond with little or no storage build up, the actual travel time through the pond may be very small. If significant storage results, the travel time is lengthened over that for normal channel flow, and flow routing as discussed in Chapter 6 must be used.

Flow velocity for open channels can be estimated from Manning's equation, which is treated in detail in Chapter 4.

Other methods are available in the form of empirical equations for estimating t_c. One such relationship that is widely used but based on limited data is expressed by Kirpich (1940)

$$t_c = 0.0078L^{0.77}(L/H)^{0.385}, \quad (3.51)$$

where t_c is in minutes, L is the maximum length of flow in feet, and H is the difference in elevation in feet between the outlet of the watershed and the hydraulically most remote point in the watershed. Obviously, Eq. (3.51) does not consider flow resistance in the form of overland flow and channel roughness.

Several methods for estimating the lag time of a watershed are available. One simple method for lag time estimation is (Soil Conservation Service, 1973)

$$t_{\mathrm{L}} = 0.6t_{\mathrm{c}}. \tag{3.52}$$

The SCS (1975) has developed a lag equation based on natural watersheds

$$t_{\mathrm{L}} = \frac{L^{0.8}(S+1)^{0.7}}{1900Y^{0.5}} \quad (50 \leq \mathrm{CN} \leq 95), \tag{3.53}$$

where t_{L} is the lag in hours, L is the hydraulic length of the watershed in feet, S is related to the curve number by Eq. (3.22), and Y is the average land slope in percentage. The S in Eq. (3.53) should be based on an antecedent condition II curve number since it is being used as a measure of surface roughness and not runoff potential.

Many local studies relating t_{L} or t_{p} or t_{c} to watershed physical characteristics have been conducted. For example, Putnam (1972) in a study of 34 watersheds in North Carolina, presented the relationship

$$t_{\mathrm{L}} = 0.49\left(\frac{L}{\sqrt{S}}\right)^{0.50} I^{-0.57}, \tag{3.54}$$

where t_{L} is the basin lag in hours, L is the length of the main water course in miles, S is the main stream slope in feet per mile, and I is fraction of impervious area. Here t_{L} was defined as the time from the center of mass of rainfall to the center of mass of runoff. Before an equation like (3.54) is used, care must be exercised to see that the conditions under which the equation was developed match the conditions of interest.

The duration, D, of the rainfall excess that is generally associated with a unit hydrograph should be one-fifth to one-third of the time to peak. The time to peak is given by Eq. (3.26) as

$$t_{\mathrm{p}} = t_{\mathrm{L}} + D/2.$$

Epsey *et al.* (1977) studied rainfall-runoff records from 41 watersheds located in several states (Texas, 16; North Carolina, 9; Kentucky, 6; Indiana, 4; Colorado, 2; Mississippi, 2; Tennessee, 1; and Pennsylvania, 1). The watersheds ranged in size from about 9 to 9600 acres (3.5 to 3900 hectares). They developed an estimation equation for the time to peak of 10-min unit hydrographs as

$$t_{\mathrm{p}} = 3.1L^{0.23}S^{-0.25}I^{-0.18}\Phi^{1.57}, \tag{3.55}$$

where t_{p} is the time to peak in minutes, L is the main channel length from the upper watershed boundary in feet, S is the slope in feet per foot of the lower 80% (in terms of length) of the main channel, I is the percentage impervious area with an assumed minimum value of 5% for an undeveloped area, and Φ is a conveyance factor that depends on the percentage impervious area and Manning's n for the main channel. Table 3.22 contains some representative values for Φ.

Table 3.22 Φ Values for Eq. (3.55) (Epsey *et al.*, 1977)

	Manning's n				
Percentage imp.	0.015	0.03	0.05	0.10	0.15
0	0.82	0.86	0.93	1.15	1.30
20	0.74	0.80	0.88	1.09	1.27
40	0.65	0.72	0.81	1.03	1.22
60	0.60	0.68	0.79	1.00	1.19

The base time of a unit hydrograph is somewhat arbitrary. Some hydrologists use a base time of five times the time to peak. Some unit hydrograph models have a recession limb that asymptotically approaches $q = 0$, so that the base time is theoretically infinity.

Estimation of Peak Flow Parameters

The peak flow rate of a unit hydrograph is often given by an equation of the form

$$q_{\mathrm{p}} = KA/t_{\mathrm{p}}. \tag{3.56}$$

Based on a triangular unit hydrograph with a base time of $2.67t_p$, the SCS (1972) estimates the peak flow of a unit hydrograph from the equation

$$q_{\mathrm{p}} = \frac{484A}{t_{\mathrm{p}}}, \tag{3.57}$$

where q_{p} is the peak flow in cfs, A is the basin area in square miles, and t_{p} is the time to peak in hours.

Epsey *et al.* (1977) recommend that for 10-min unit hydrographs, the relation

$$q_{\mathrm{p}} = 31620\left(A^{0.96}/t_{\mathrm{p}}^{1.07}\right) \tag{3.58}$$

be used where q_{p} is in cfs, A is the drainage area in square miles, and t_{p} is the time to peak in minutes.

As was the case for lag time, many studies have been conducted in an effort to relate q_{p} to watershed physical conditions. Before any of these empirically derived equations are used, their applicability should be carefully determined.

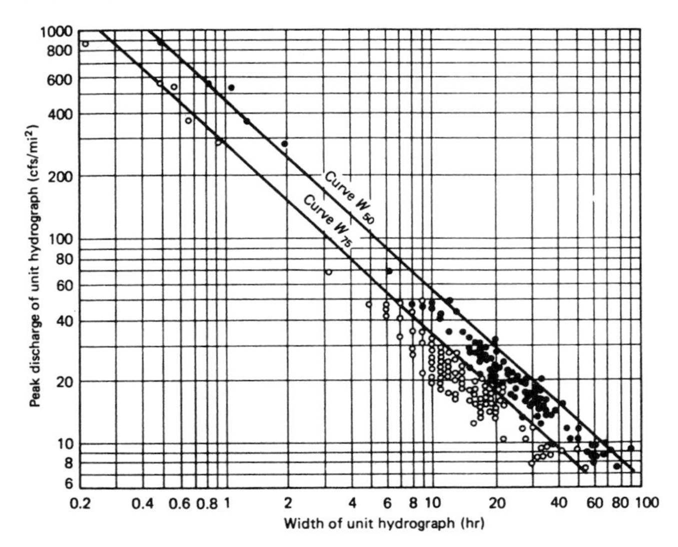

Figure 3.33 Unit hydrograph width at 50 and 75% of peak flow (U.S. Corps of Engineers, 1959).

Shape of Unit Hydrographs

If a short duration unit hydrograph is applied to a long rainstorm pattern, the actual shape of the unit hydrograph is not nearly as important as the time to peak and the peak flow rate. Some hydrograph procedures actually call for a simple sketching of the approximate shape of the unit hydrograph, making sure the volume of the hydrograph is 1 in.

Based on the work of Snyder (1938), the U.S. Army Corps of Engineers (1959) produced Fig. 3.33, which can be used as an aid in determining the shape of a unit hydrograph. The curves in Fig. 3.33 give the width of the unit hydrograph at flow rates equal to 0.75 q_p and 0.50 q_p. As a rule of thumb, the widths at these points can be proportioned so that one-third of the width occurs prior to the peak discharge.

Knowledge of q_p and t_p along with Fig. 3.33 make it possible to sketch a proposed unit hydrograph. Any time a sketching procedure is used, it is essential to ensure that the volume of the proposed unit hydrograph is 1 in. If it is not, the sketch should be altered until a volume of 1 in. is achieved.

The SCS (1972) uses a dimensionless unit hydrograph as shown in Fig. 3.34. They also employ a triangular unit hydrograph derived to have the same time to peak and peak flow rate as their dimensionless unit hydrograph. The SCS triangular unit hydrograph is also shown in Fig. 3.34. If either of the SCS dimensionless unit hydrographs is used, q_p and t_p must be related by Eq. (3.57).

Haan (1970), DeCoursey (1966), and others have proposed dimensionless hydrograph equations. Starting with the empirical equation

$$q(t) = at^b e^{-ct}$$

as a hydrograph equation and using the conditions

$$(1)\ \text{at } t = t_p,\ q = q_p$$

$$(2)\ \text{at } t = t_p,\ \frac{dq(t)}{dt} = 0$$

$$(3)\ \int_0^\infty q(t)\,dt = V.$$

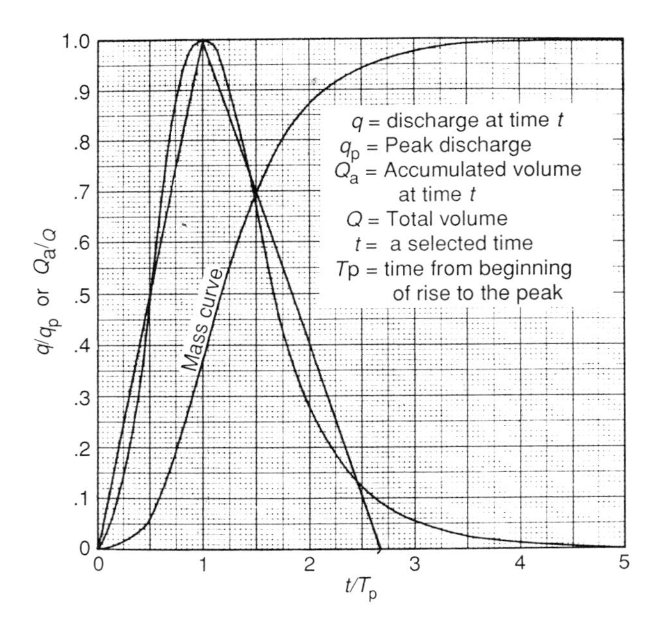

Figure 3.34 Dimensionless unit hydrograph and mass curve with triangular unit hydrograph superimposed.

Haan (1970) developed the relationship

$$\frac{q(t)}{q_p} = \left[\frac{t}{t_p} e^{1-t/t_p}\right]^K, \qquad (3.59)$$

where $q(t)$ is the hydrograph ordinate at any time t, q_p is the peak flow rate (iph), t_p is the time to peak (hours), and K is a parameter defined by the equation

$$V = q_p t_p [e/K]^K \Gamma(K), \qquad (3.60)$$

where V is the runoff volume (inches—1 in. for a unit hydrograph), e is the base of the natural logarithms, and Γ represents the gamma function. Figure 3.35 presents a quick solution of Eq. (3.60) for K. The relationship between K and $q_p t_p/V$ can be approximated by

$$K = 6.5\left(\frac{q_p t_p}{V}\right)^{1.92}. \qquad (3.61)$$

The SCS dimensionless unit hydrograph corresponds to a K value of about 3.77. This can be seen by using Eq. (3.57) and making the proper unit conversions.

Figure 3.36 shows how the shape of the hydrograph defined by Eq. (3.59) changes as K changes. Table 3.23 gives the ordinates for the dimensionless hydrograph of

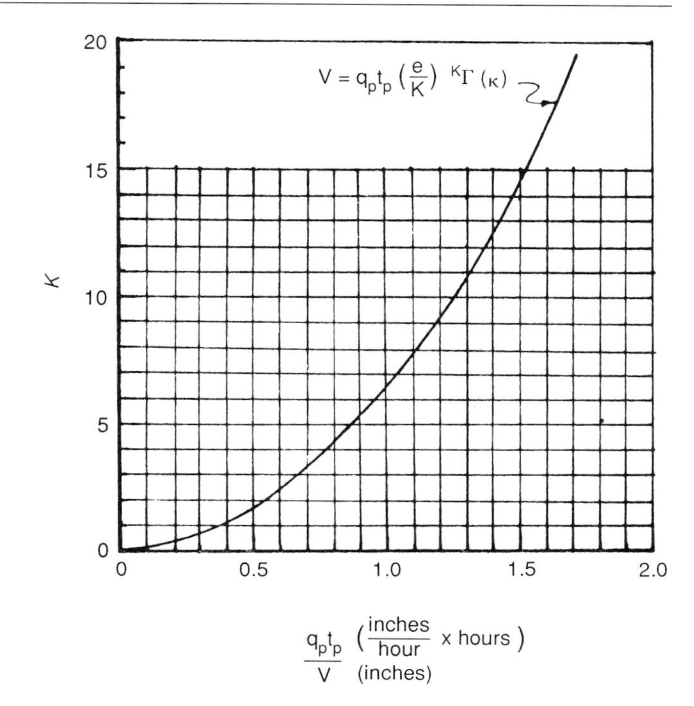

Figure 3.35 Relationship between $q_p t_p/V$ and K.

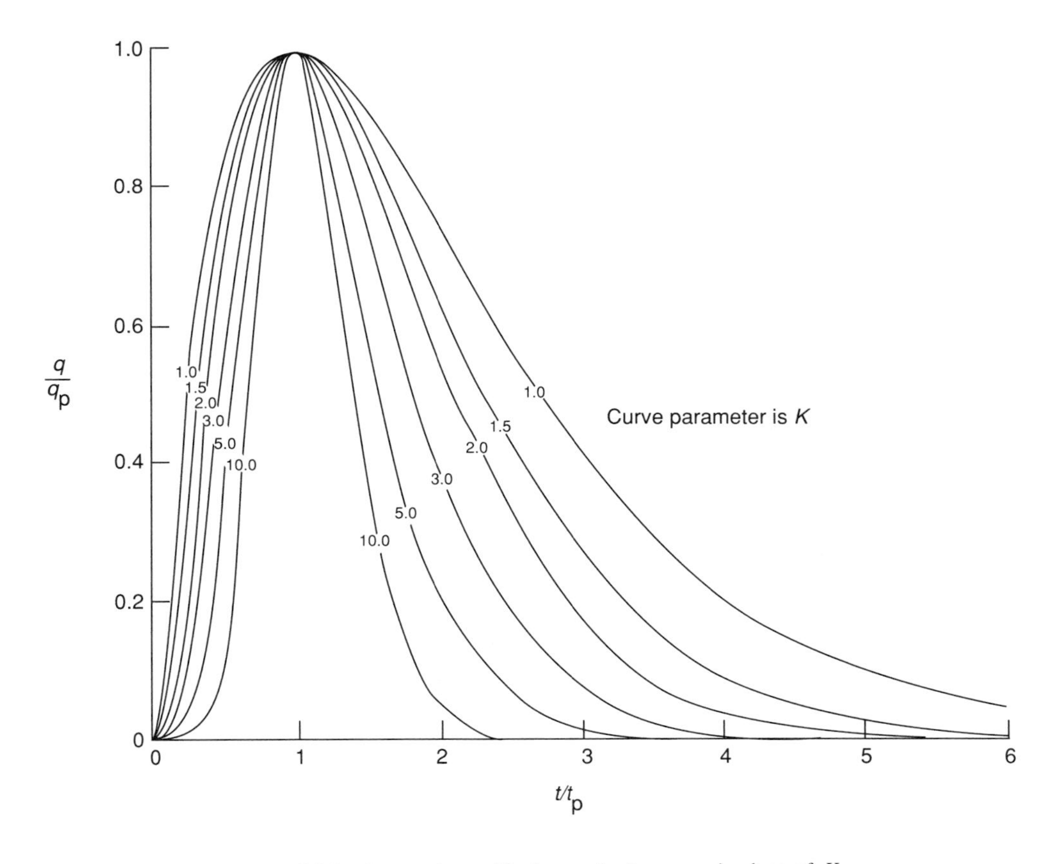

Figure 3.36 Comparison of hydrographs for several values of K.

Table 3.23 Coordinates for Dimensionless Unit Hydrographs

	K										
	1	1.5	2	3	4	5	6	7	8	9	10
t/t_p						q/q_p					
0.00	0.00	0.00	0.00	0.00	0.00	0.00	0.00	0.00	0.00	0.00	0.00
0.25	0.54	0.39	0.29	0.15	0.08	0.04	0.02	0.01	0.01	0.00	0.00
0.50	0.83	0.75	0.68	0.56	0.46	0.38	0.32	0.26	0.22	0.18	0.15
0.75	0.97	0.95	0.93	0.90	0.87	0.84	0.81	0.78	0.75	0.73	0.70
1.00	1.00	1.00	1.00	1.00	1.00	1.00	1.00	1.00	1.00	1.00	1.00
1.25	0.97	0.96	0.95	0.92	0.90	0.87	0.85	0.83	0.81	0.78	0.76
1.50	0.91	0.87	0.83	0.75	0.69	0.62	0.57	0.52	0.47	0.43	0.39
1.75	0.83	0.75	0.68	0.56	0.47	0.38	0.32	0.26	0.22	0.18	0.15
2.00	0.74	0.63	0.54	0.40	0.29	0.22	0.16	0.12	0.09	0.06	0.04
2.50	0.57	0.43	0.32	0.18	0.10	0.06	0.03	0.02	0.01	0.01	
3.00	0.41	0.26	0.16	0.07	0.03	0.01					
3.50	0.29	0.15	0.08	0.02	0.01						
4.00	0.20	0.09	0.04	0.01							
5.00	0.10	0.03	0.01								
6.00	0.04	0.01									

Eq. (3.59) for various values of K. The derivation of Eqs. (3.59) and (3.60) is in Haan (1970).

Ardis (1973) developed a double triangle unit hydrograph in an attempt to incorporate both a quick and delayed runoff response. He assumed that the delayed response peak would coincide with the time when the quick response ended and that both responses would start at the same time. Figure 3.37 depicts the components of and the resultant double triangle unit hydrograph.

Wilson *et al.* (1983) have adapted the double triangle unit hydrograph model for small watersheds. They plot the unit hydrograph in the form shown in Fig. 3.38. The units used are inches per hour for $q(t)$ and hours for time. The coordinates for the points labeled a, b, and c are

Land use	a	b	c
Forested	1.0, 0.268	4.105, 0.054	18.068, 0.0
Agricultural	1.0, 0.526	2.375, 0.113	6.982, 0.0
Urban	1.0, 0.756	2.000, 0.151	4.333, 0.0
Disturbed	1.0, 0.756	2.000, 0.151	4.333, 0.0

An effective rainfall pattern is used with a time interval of 3 min. If the time to peak is less than 6 min, no unit hydrograph is used and the runoff rate is taken as being equal to the effective rainfall rate using a time increment of 6 min.

Example Problem 3.12. Double triangle unit hydrograph

Develop a double triangle unit hydrograph for an urban area of 100 acres with a time to peak of 15 min.

Solution: The points defining the unit hydrograph are computed using the coordinates tabulated above.

Point a:

$$t_p q(t) = 0.756 \text{ in.}$$

$$q(t) = 0.756 \text{ in.}/0.25 \text{ hr} = 3.02 \text{ iph}$$

$$q_p = 3.02 \frac{\text{inches}}{\text{hour}} \times 100 \text{ acres} \times 1.008 \frac{\text{cfs}}{\text{acre-inch}} = 304 \text{ cfs}$$

$$t = t_p = 15 \text{ min.}$$

Point b:

$$q(t) = \frac{0.151}{0.25} \times 100 \times 1.008 = 61 \text{ cfs}$$

$$t = 2 \times 15 = 30 \text{ min.}$$

Point c:

$$q(t) = 0$$

$$t = 4.33 \times 15 = 65 \text{ min.}$$

The volume under the hydrograph can be computed by dividing the hydrograph into three parts at the three defining

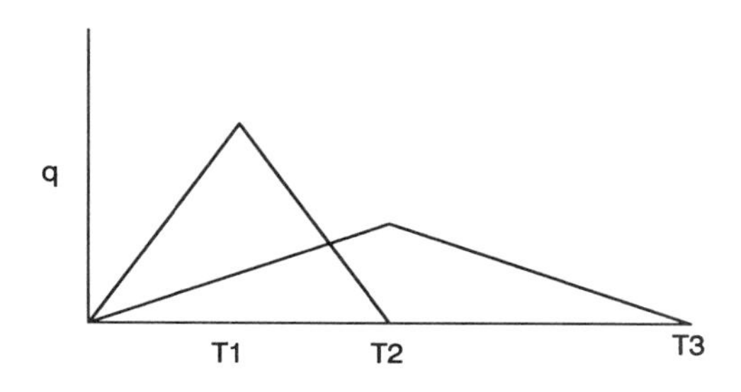

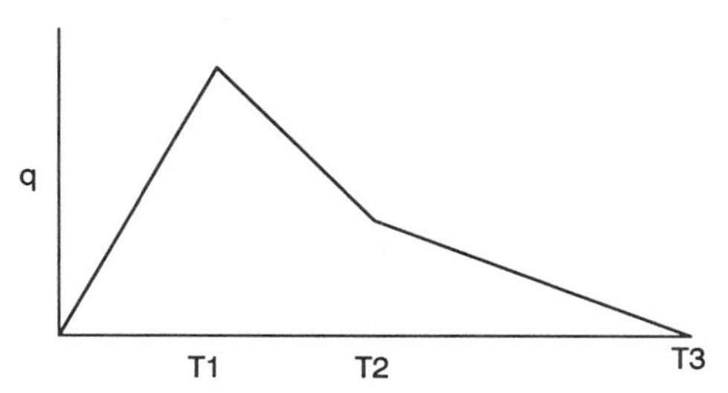

Figure 3.37 Components of double triangle unit hydrograph.

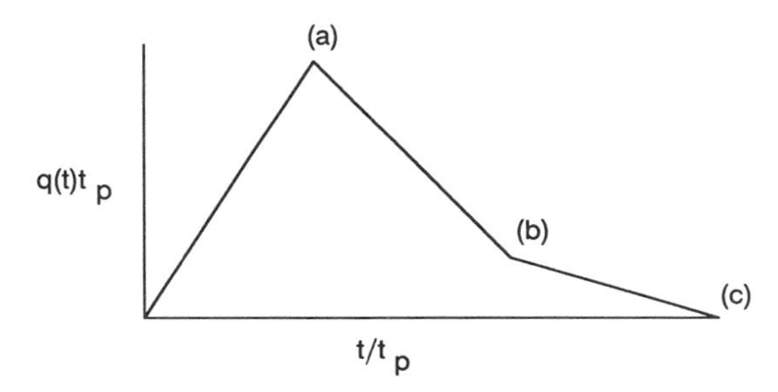

Figure 3.38 Double triangle unit hydrograph of Wilson *et al.* (1983).

time points:

$$V = V_{o-a} + V_{a-b} + V_{b-c}$$

$$V = \frac{q_a}{2} t_a + (t_b - t_a)\frac{q_a + q_b}{2} + \frac{q_b}{2}(t_c - t_b)$$

$$V = \frac{3.02}{2} \times 0.25 + (0.50 - 0.25)\frac{3.02 + 0.61}{2} + \frac{0.61}{2}\left(\frac{65}{60} - 0.5\right)$$

$$V = 1.009 \text{ in.}$$

which is acceptable for a unit hydrograph.

Ordinates at other time points can be easily determined from the straight line relationships defining the unit hydrograph.

The Natural Environmental Research Council (1975) of Great Britain conducted a comprehensive flood study involving a large number of catchments from throughout Great Britain. One aspect of the study involved developing a procedure for obtaining synthetic unit hydrographs. The prediction equations they developed were derived from 1631 events from 143 catchments ranging in size from 3.5 to 500 km^2 (864 to 123,550 acres). The result was a triangular unit hydrograph. The volume associated with their unit hydrograph was 10 mm:

$$V = 10 \text{ mm}.$$

The time to peak for the hydrograph is estimated from

$$t_p = 46.6 S^{-0.38} \text{URBT}^{-1.99} \text{RSMD}^{-0.4} L^{0.14}, \quad (3.62)$$

where t_p is in hours, S is the slope of the main channel between the points 10 and 85% of the length of the channel measured up from the outlet in meters per kilometer, URBT is 1 plus the fraction of the area that is urbanized, RSMD is the 5-year, 24-hr rainfall excess in millimeters, and L is the length of the main channel in kilometers. A statistical analysis presented based on the regressions used to arrive at the above equation indicates that a simpler version given by

$$t_p = 22.6 S^{-0.597} \text{URBT}^{-2.028} \quad (3.63)$$

is also an acceptable equation. The t_p given by these equations is for a unit hydrograph having a duration of 1 hr. The study recommends that the unit hydrograph t_p should be about five times the duration. To adjust the t_p to a new duration, D, in hours, they recommend the relationship

$$t'_p = t_p + (D - 1)/2, \quad (3.64)$$

where t'_p is the adjusted time to peak.

It is recommended that if any rainfall and runoff data are available, the lag time should be evaluated empirically and t_p computed from

$$t_p = 0.9 t_L, \quad (3.65)$$

where t_L is the lag time defined as the time from the center of mass of rainfall to the peak of the unit hydrograph.

Once the proper t_p is obtained, the q_p can be determined from

$$q_p = 2.20 \text{A}/t_p, \quad (3.66)$$

where q_p is in cubic meters per second and A is in square kilometers. The base time of the triangular unit hydrograph is given by

$$t_b = 2.52 t_p. \quad (3.67)$$

The Flood Studies Report gives an example for a catchment with $A = 63.2$ km^2, $L = 17.22$ km, $S =$

14.7 m/km, and RSMD = 55.2 mm. The catchment is 5% urbanized, so URBT is 1.05. Substituting these quantities into Eq. (3.62) results in a t_p of 4.6 hr. If Eq. (3.63) is used, t_p is found to be 4.1 hr.

Rainfall and runoff data indicate that for this catchment, a lag time of 2.3 hr is appropriate. Based on Eq. (3.65), the t_p is found to be 2.1 hr. The appropriate duration for the unit hydrograph is thus 2.1/5, which rounded to a convenient fraction of an hour is 0.5 hr. t_p for a 0.5 hr unit hydrograph is computed from Eq. (3.64) as

$$t'_p = 2.1 + (0.5 - 1)/2 = 1.9 \text{ hr..}$$

The peak flow of the unit hydrograph is found from Eq. (3.66) as

$$q_p = 2.20 \times 63.2/1.9 = 73.2 \text{ m}^3/\text{sec.}$$

The base time of the unit hydrograph is

$$t_b = 2.52 \times 1.9 = 4.8 \text{ hr.}$$

The volume of this unit hydrograph can be easily computed as the area under the triangular unit hydrograph giving 0.010 m or 10 mm as it should be under the conditions of the study.

Equations (3.62) through (3.67) should be applied only to conditions, especially meteorological conditions, similar to those existing in Great Britian. Rainfall intensities are generally lower than those prevalent in the U.S. for storms with similar return periods.

Edson (1951) and later Nash (1959) proposed that a hydrograph can be represented by an equation of the form

$$q(t) = \frac{V}{\Gamma(n)} k^{-n} e^{-t/k} t^{n-1}, \qquad (3.68)$$

where n and k are parameters and Γ represents the gamma function. Equation (3.68) is sometimes known as the Nash model or the gamma distribution. The equation was derived by assuming that an instantaneous rainfall routed through a series of n linear reservoirs represents the runoff process.

In applying Eq. (3.68), it is necessary to be able to estimate the parameters in the equation. Gray (1961) employed the method of maximum likelihood. Nash (1959) advocated the use of the method of moments. Both of these methods tend to produce hydrographs that are "best fits" over the entire range of the hydrographs and can have substantial deviations from observed hydrographs at the peak. Wu (1963) used the time to peak and a recession constant to estimate the parameters of his runoff equation. Bloomsburg (1960) fit the hydrograph parameters to the peak flow and the time to peak.

Gray (1961), Bleek (1975), Wittenburg (1975), and many others discuss relating the derived hydrograph parameters to watershed physical characteristics so that they may be applied to ungaged basins.

It is apparent that there are a number of relationships that can be used to estimate the parameters of unit hydrographs. The technique selected should be an accepted technique for the location under study. The conditions of the study area should be hydrologically similar to the conditions under which the technique was derived.

Consistency should be maintained in the application of a unit hydrograph procedure. It was previously stated that if the SCS unit hydrograph is used, Eq. (3.57) should be used to estimate q_p. One should not mix methods. That is, if t_p is based on a particular method, q_p should be based on the same method as should the unit hydrograph shape. The reason for this is that generally particular hydrograph methods are developed as a unified whole. The elements of the methods are not independent of each other. The coefficients in a particular estimating equation are somewhat dependent on the coefficients in the companion equations that make up the method.

Some hydrograph equations such as Eq. (3.59) are independent of the method used to estimate q_p, t_p, and V. However, even in this case, a consistent procedure should be used to estimate these three parameters. In general, one should not estimate t_p from one method and q_p from another.

Example Problem 3.13. Unit hydrograph

Develop a unit hydrograph for a 100-acre watershed with a main channel length of 5000 ft, Manning's n of 0.025, and a slope of 1% in the lower 80% of the channel. The watershed is undeveloped.

Solution: The method of Epsey *et al.* (1977) is used to estimate t_p and q_p. The choice is dictated by the information provided in the problem. If land-use and soils data were provided, the SCS method might be preferred. From Eq. (3.55),

$$t_p = 3.1 L^{0.23} S^{-0.25} I^{-0.18} \Phi^{1.57}.$$

For $n = 0.025$ and 0% impervious, Φ is found from Table 3.22 to be 0.83:

$$t_p = 3.1(5000)^{0.23}(0.01)^{-0.25}(5)^{-0.18}(0.83)^{1.57} = 39 \text{ min.}$$

q_p is determined from Eq. (3.58) as

$$q_p = 31620\left(A^{0.96}/t_p^{1.07}\right)$$
$$= 31620(100/640)^{0.96}(39)^{-1.07} = 106 \text{ cfs.}$$

Use Eq. (3.59) to define the unit hydrograph:

$$\frac{q(t)}{q_p} = \left[\frac{t}{t_p} e^{1-t/t_p}\right]^K$$

$$q_p = 106 \text{ cfs or about } 1.06 \text{ iph}$$

$$t_p = 39 \text{ min} = 0.65 \text{ hr}$$

$$V = 1 \text{ in.}$$

K is computed from Eq. (3.61) as

$$K = 6.5\left(\frac{1.06 \times 0.65}{1}\right)^{1.92} = 3.18.$$

The unit hydrograph is defined by

$$q(t) = 106\left[\frac{t}{0.65} e^{(1-t/0.65)}\right]^{3.18}$$

with q in cfs and t in hours. q at any time can be easily determined. For example, q at $t = 1$ hr is

$$q(1) = 106\left[\frac{1}{0.65} e^{(1-1/0.65)}\right]^{3.18} = 75 \text{ cfs.}$$

Note that this is a 10-min unit hydrograph since Eq. (3.55) and (3.58) were used. Thus the effective rainfall pattern should be developed in 10-min increments.

Runoff Hydrographs from Unit Hydrographs

Several procedures are currently in use for estimating runoff hydrographs using synthetic unit hydrographs. In general, these procedures consist of developing a design rainstorm, deducting abstractions from the rain to get the excess rainfall hyetograph, estimating the time to peak and the peak flow rate for a unit hydrograph, defining the remaining shape of the unit hydrograph, and applying the unit hydrograph to the derived rainfall excess hyetograph. The Santa Barbara Urban Hydrograph procedure and the routing of overland flow using a kinematic approach has been previously discussed.

The Soil Conservation Service (1972) hydrograph procedure consists of: (1) estimating rainfall from the 24-hr rainfall depth and the appropriate type curves of Fig. 3.10, (2) estimating rainfall excess using Eq. (3.21), (3) estimating unit hydrograph timing parameters from Eqs. (3.26) and (3.53) or Eqs. (3.47), (3.48), and (3.52), (4) estimating unit hydrograph peak flow rate from Eq. (3.57), and (5) using a hydrograph shape as shown in Fig. 3.34 (either the dimensionless hydrograph or the triangular hydrograph may be used). An incremental rainfall duration, D, of $t_p/3$ is recommended so that D is computed from

$$D = 0.4t_L. \tag{3.69}$$

For small catchments, the parameters of a 10-min unit hydrograph can be estimated from Eqs. (3.55) and (3.58) and the unit hydrograph model of Eq. (3.59) used.

ESTIMATION OF PEAK RUNOFF RATES

There are many occasions when all that is required to design a facility is the peak flow rate that might be experienced on a certain frequency. Of course, the hydrograph methods discussed in the previous sections can be used to estimate a peak flow. However, for small structures, it may be desirable to have quick and simple procedures to estimate the peak flow.

Rational Method

By far the most common method used for peak flow estimation is the Rational Method. A joint report by the American Society of Civil Engineers and the Water Pollution Control Federation is perhaps the most comprehensive treatment of the Rational Method (Water Pollution Control Federation, 1969). The Rational Method has many limitations and shortcomings. A report by McPherson (1969) discusses these problems in considerable detail. In spite of the recognized shortcomings of the Rational Method, it continues to be widely used because of its simplicity, entrenchment in practice, coverage in texts, and lack of a comparable alternative.

The Rational Equation is

$$q_p = CiA, \tag{3.70}$$

where q_p is the peak flow rate in cfs, C is a dimensionless coefficient, i is the rainfall intensity in iph with a duration equal to t_c, and A is the drainage area in acres. To be dimensionally correct, a conversion factor of 1.008 should be included to convert acre-inches per hour to cubic feet per second; however, this factor is generally neglected.

The "rationale" behind the Rational Equation is that if a steady rainfall occurs on a watershed, the runoff rate will increase until the entire watershed is contributing runoff. If a rainfall of duration less than t_c occurs, the entire basin would not be contributing so the resulting runoff rate would be less than from a rainfall with a duration equal to t_c. If a rainfall of duration greater than t_c occurs, the point relationship between average rainfall intensity and duration for a given frequency show that the average intensity would be less than if the duration were equal to t_c. Thus it is reasoned that a rainfall with an average peak intensity of duration t_c produces the maximum flow rate.

The Rational Equation is based on certain assumptions.

1. The rainfall occurs uniformly over the drainage area.
2. The peak rate of runoff can be reflected by the rainfall intensity averaged over a time period equal to the time of concentration of the drainage area.
3. The frequency of runoff is the same as the frequency of the rainfall used in the equation.

The coefficient C is called the runoff coefficient and is the most difficult factor to accurately determine. C must reflect factors such as interception, infiltration, surface detention, and antecedent conditions. Obviously, no single factor can quantify all of these things and their effect on peak runoff rates.

Several studies have shown that C is not a constant but varies with the frequency of the runoff event among other things (Haan, 1972; Horner, 1910; McPherson, 1969). The increase in C with return period is generally attributed to wetter conditions that would be expected to exist during more extreme events.

Many tables have been prepared for estimating C. When using these tables, care must be exercised to see that they reflect conditions that exist today. This is necessary because urbanized areas of years ago tended to have more temporary storage in the form of road ditches and many less impervious areas. Table 3.24 was taken from the Water Pollution Control Federation (1969) report for urban areas and from Schwab *et al.* (1971) for rural areas.

Average coefficients for composite areas may be calculated on an area weighted basis from

$$\overline{C} = \frac{\Sigma C_i A_i}{\Sigma A_i}, \qquad (3.71)$$

where C_i is the coefficient applicable to the area A_i. In areas where large parts are laid out in typical, repeating patterns such as subdivisions, the weighting factors and weighted C can be determined by considering a single, typical layout.

As with any estimation procedure, considerable care should be exercised when applying the Rational Equation to estimate peak flows. For instance, the location of relatively impervious areas with respect to the point of flow estimation must be carefully considered. If flow from an impervious area must cross an infiltrating area such as grass, the flows may be greatly reduced. If large impervious areas are present, they should be analyzed as separate units. The reason for this can be seen by considering the situation shown in Fig. 3.39. In case A,

Table 3.24 Runoff Coefficients

Urban areas The use of average coefficients for various surface types, which are assumed not to vary through the duration of the storm, is common. The range of coefficients, classified with respect to the general character of the tributary reported in use is:

Description of area	Runoff coefficients
Business	
Downtown areas	0.70 to 0.95
Neighborhood areas	0.50 to 0.70
Residential	
Single-family areas	0.30 to 0.50
Multiunits, detached	0.40 to 0.60
Multiunits, attached	0.60 to 0.75
Residential (suburban)	0.25 to 0.40
Apartment dwelling areas	0.50 to 0.70
Industrial	
Light areas	0.50 to 0.80
Heavy areas	0.60 to 0.90
Parks, cemeteries	0.10 to 0.25
Playgrounds	0.20 to 0.35
Railroad yard areas	0.20 to 0.35
Unimproved areas	0.10 to 0.30

Note: It is often desirable to develop a composite runoff coefficient based on the percentage of different types of surface in the drainage area. This procedure is often applied to typical 'sample' blocks as a guide to selection of reasonable values of the coefficient for an entire area. Coefficients with respect to surface type currently in use are:

Character of surface	Runoff coefficients
Streets	
Asphaltic and concrete	0.70 to 0.95
Brick	0.70 to 0.85
Roofs	0.75 to 0.95
Lawns; sandy soil	
Flat, 2%	0.05 to 0.10
Average, 2 to 7%	0.10 to 0.15
Steep, 7%	0.15 to 0.20
Lawns, heavy soil	
Flat, 2%	0.13 to 0.17
Average, 2 to 7%	0.18 to 0.22
Steep, 7%	0.25 to 0.35

Note: The coefficients in these two tabulations are applicable for storms of 5-year to 10-year frequencies. Less frequent higher intensity storms will require the use of higher coefficients because infiltration and other losses have a proportionally smaller effect on runoff. The coefficients are based on the assumption that the design storm does not occur when the ground surface is frozen.

Table 3.24—*Continued*

Rural areas

Topography and vegetation	Soil texture		
	Open sandy loam	Clay and silt loam	Tight clay
Woodland			
Flat 0–5% slope	0.10	0.30	0.40
Rolling 5–10% slope	0.25	0.35	0.50
Hilly 10–30% slope	0.30	0.50	0.60
Pasture			
Flat	0.10	0.30	0.40
Rolling	0.16	0.36	0.55
Hilly	0.22	0.42	0.60
Cultivated			
Flat	0.30	0.50	0.60
Rolling	0.40	0.60	0.70
Hilly	0.52	0.72	0.82

the impervious area is next to the outlet, while in case B the grass area is next to the outlet. Straightforward application of the Rational Equation would result in the same flow estimate for both cases since t_c and the weighted C would be the same.

A closer look at case A, however, shows that it is possible that the peak flow from the impervious area alone could exceed the peak estimated for the whole area based on a weighted C. Using Eq. (3.49), a flow time of 10 min is estimated for the 500 ft of grass. Using a similar procedure, a flow time of only about 5 min is estimated for the impervious area. Thus, the total flow time is 15 min. A weighted C is

$$\bar{C} = 0.5(0.9 + 0.2) = 0.55.$$

A 10-year, 15-min rain at Stillwater, Oklahoma, has an i of 6.29 iph. Thus,

$$q_{\text{total}} = 0.55(6.29)A = 3.15A.$$

Considering only the impervious area, C is 0.9 and t_c is 5 min. The corresponding i is 10.15 iph

$$q_{\text{imp}} = 0.9(10.15)A/2 = 4.56A,$$

or the 10-year flow peak from the lower impervious area would exceed the original estimate for the entire area. This illustrates the need to carefully consider the location of impervious or high runoff-producing areas. Of course, this same high flow rate would be estimated to occur at the lower end of the impervious part of case B. Here, however, the flow through the pervious area would at least partially attenuate this peak.

Finally, it must be kept in mind that this is a method of peak flow estimation only. The t_c bears no relation to the time from the beginning of a rainfall. One cannot predict from this method when the peak will occur. The t_c applies to the time from the beginning of an intense rainfall with a duration equal to t_c. This intense rainfall may occur anytime during the rainstorm. The use of the Rational Equation to develop runoff hydrographs carries the method well beyond its original intent. Tabulated values of the Rational C have generally been derived based on peak flows, not entire hydrographs, and thus may not be valid when used to develop hydrographs.

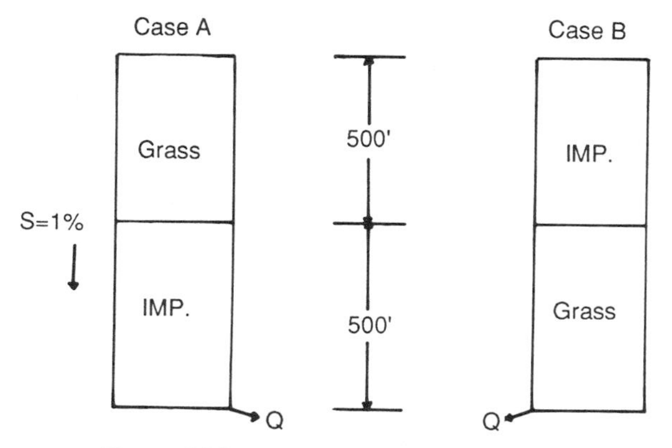

Figure 3.39 Hypothetical runoff situations.

SCS-TR55 Method

The Soil Conservation Service (1986) presents a method for estimating peak flows from small catchments based on an analysis of a large number of computer runs with their TR20 computer program (SCS, 1983). This program computes runoff hydrographs in a manner analogous to the procedure detailed earlier in this chapter and attributed to the SCS. The results of the peak flow analysis can be expressed as

$$q_p = q_u A Q F_p, \tag{3.72}$$

where q_p is the peak discharge in cfs, q_u is the unit peak discharge in cfs per inch of runoff per square mile, A is the drainage area in square miles, Q is the runoff in inches from a 24-hr storm of the desired

Table 3.25 Adjustment Factor, F_p, for Pond and Swamp Areas That Are Spread throughout the Watershed

Percentage of pond and swamp areas	F_p
0	1.00
0.2	0.97
1.0	0.87
3.0	0.75
5.0	0.72

Table 3.26 Coefficients for Eq. (3.73)

Rainfall type	I_a/P	C_0	C_1	C_2
I	0.10	2.30550	−0.51429	−0.11750
	0.20	2.23537	−0.50387	−0.08929
	0.25	2.18219	−0.48488	−0.06589
	0.30	2.10624	−0.45695	−0.02835
	0.35	2.00303	−0.40769	0.01983
	0.40	1.87733	−0.32274	0.05754
	0.45	1.76312	−0.15644	0.00453
	0.50	1.67889	−0.06930	0.0
IA	0.10	2.03250	−0.31583	−0.13748
	0.20	1.91978	−0.28215	−0.07020
	0.25	1.83842	−0.25543	−0.02597
	0.30	1.72657	−0.19826	0.02633
	0.50	1.63417	−0.09100	0.0
II	0.10	2.55323	−0.61512	−0.16403
	0.30	2.46532	−0.62257	−0.11657
	0.35	2.41896	−0.61594	−0.08820
	0.40	2.36409	−0.59857	−0.05621
	0.45	2.29238	−0.57005	−0.02281
	0.50	2.20282	−0.51599	−0.01259
III	0.10	2.47317	−0.51848	−0.17083
	0.30	2.39628	−0.51202	−0.13245
	0.35	2.35477	−0.49735	−0.11985
	0.40	2.30726	−0.46541	−0.11094
	0.45	2.24876	−0.41314	−0.11508
	0.50	2.17772	−0.36803	−0.09525

frequency, and F_p is a pond and swamp adjustment factor.

Q is computed directly from the curve number equation, Eq. (3.21), based on a 24-hr P with a return period equal to the desired return period of the peak flow. F_p is taken from Table 3.25 assuming the ponds and/or swampy areas are distributed throughout the watershed. The value of q_u is computed from

$$\log(q_u) = C_0 + C_1 \log t_c + C_2(\log t_c)^2, \quad (3.73)$$

where the C's come from Table 3.26, and logarithms to the base 10 are used. The t_c in hours has limits of 0.1 to 10 hr. In Table 3.26, the value of I_a is given by $I_a = 0.2S$. S can be computed from Eq. (3.22) using the appropriate curve number. Once I_a/P is computed, the line in Table 3.26 with the closest value of I_a/P to the computed value should be used. Linear interpolation may be used in Table 3.26; however, care must be exercised to see that all three C coefficients are consistent with each other and with other tabulated values. A graphical solution to Eq. (3.72) from the SCS (1986) is presented in Fig. 3.40. A computer program that implements TR55 methods has been developed (SCS, 1986).

Frequency Method

Hydrologic frequency methods are often combined with regression analyses or correlation studies to develop empirical peak flow prediction equations. The method requires observed streamflow records from which estimates are made of peak flows for various return periods. These peak flows are then related to watershed physical conditions and rainfall. A typical relationship might be

$$q_T = aA^bS^cI^dL^eR_T{}^g, \quad (3.74)$$

where q_T and R_T are the T-year peak flow and rainfall; A is the area, S the slope, I the impervious fraction, L the axial length of the watershed, and a, b, c, d, e, and g are coefficients.

Generally, it is difficult to find adequate streamflow records for estimating q_T for T over about 10 years. Various studies use different combinations of variables in relationships like Eq. (3.74). It is important to determine the conditions under which these relationships were developed before they are applied to a particular situation. Often they do not extrapolate well to conditions outside those under which they were developed. This approach was discussed in Chapter 2 under "Regional Analysis" where Eq. (2.38), which is similar in form to Eq. (3.74), was suggested.

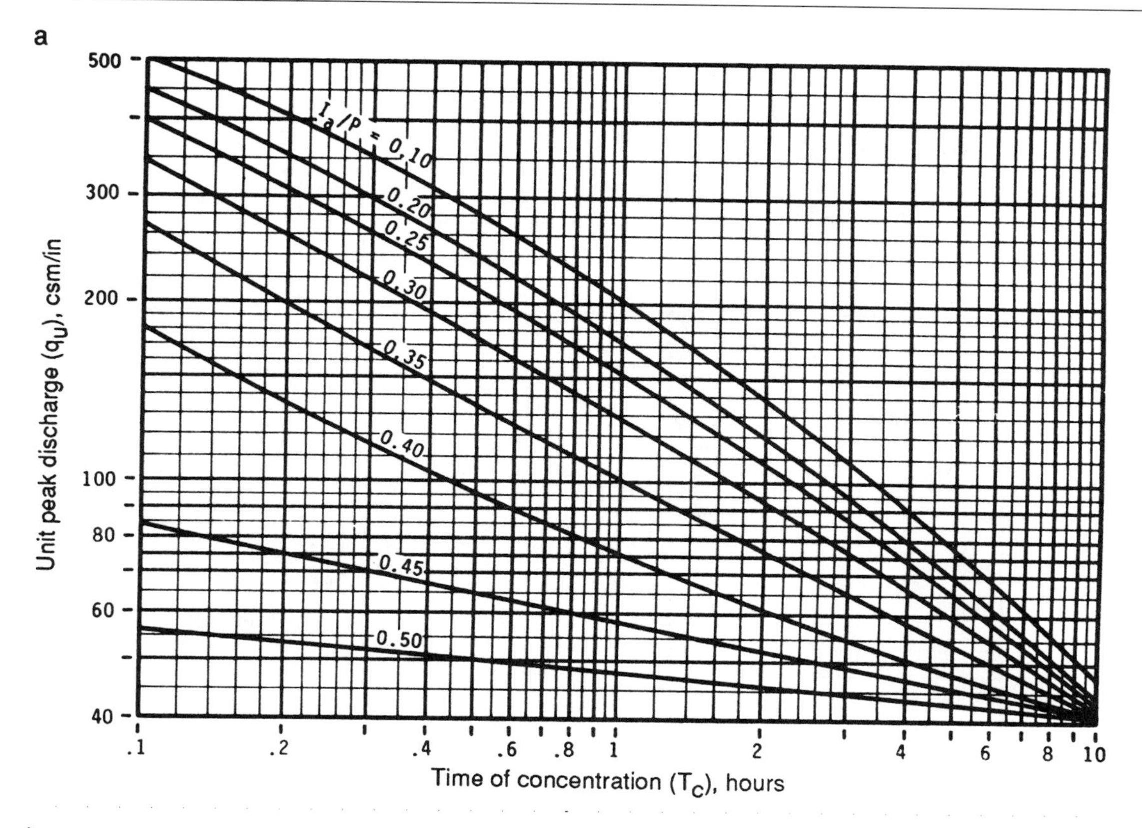

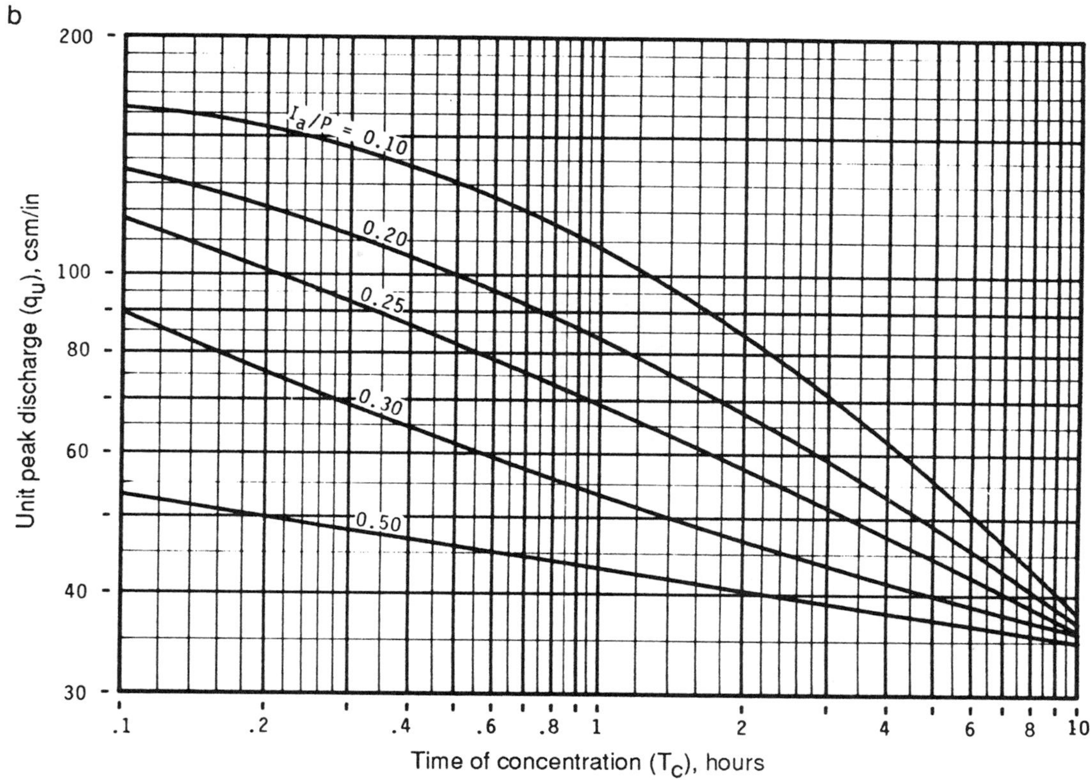

Figure 3.40 Graphical solution for Eq. (3.72). SCS rainfall distribution (a) type I, (b) type 1A, (c) type II, and (d) type III.

c

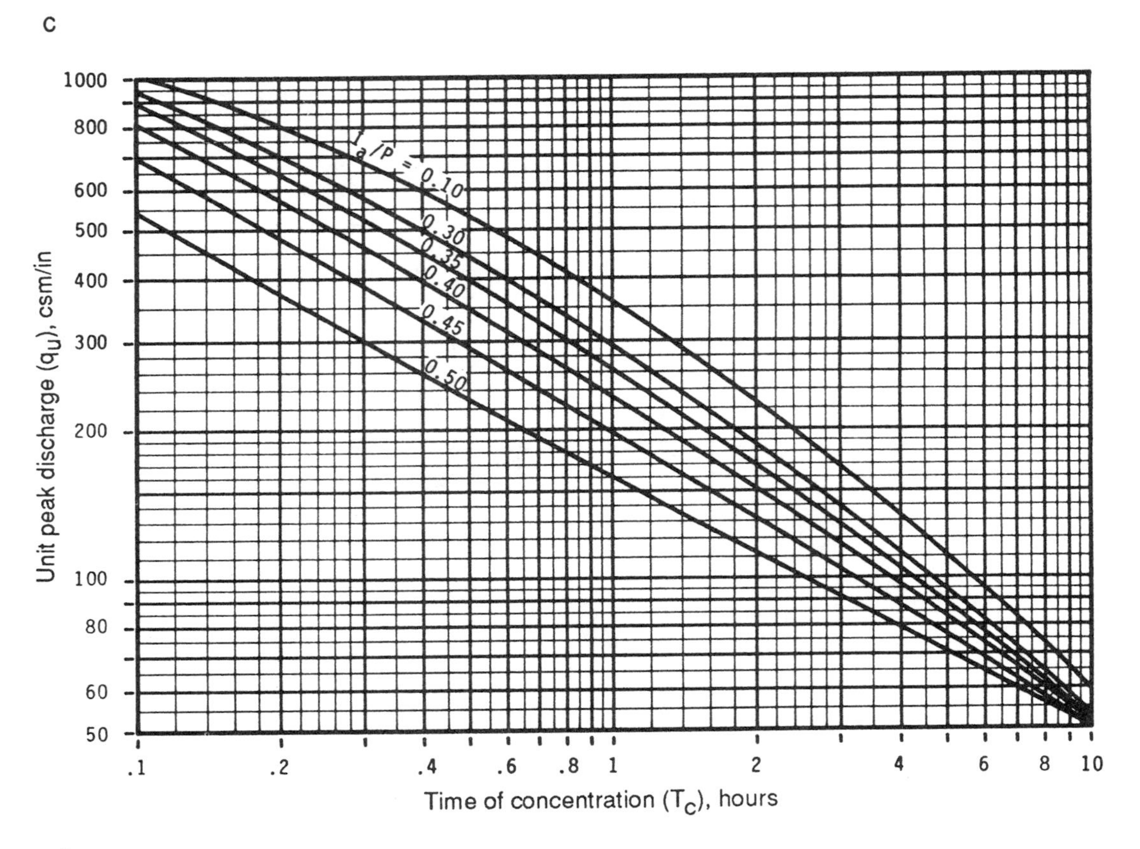

d

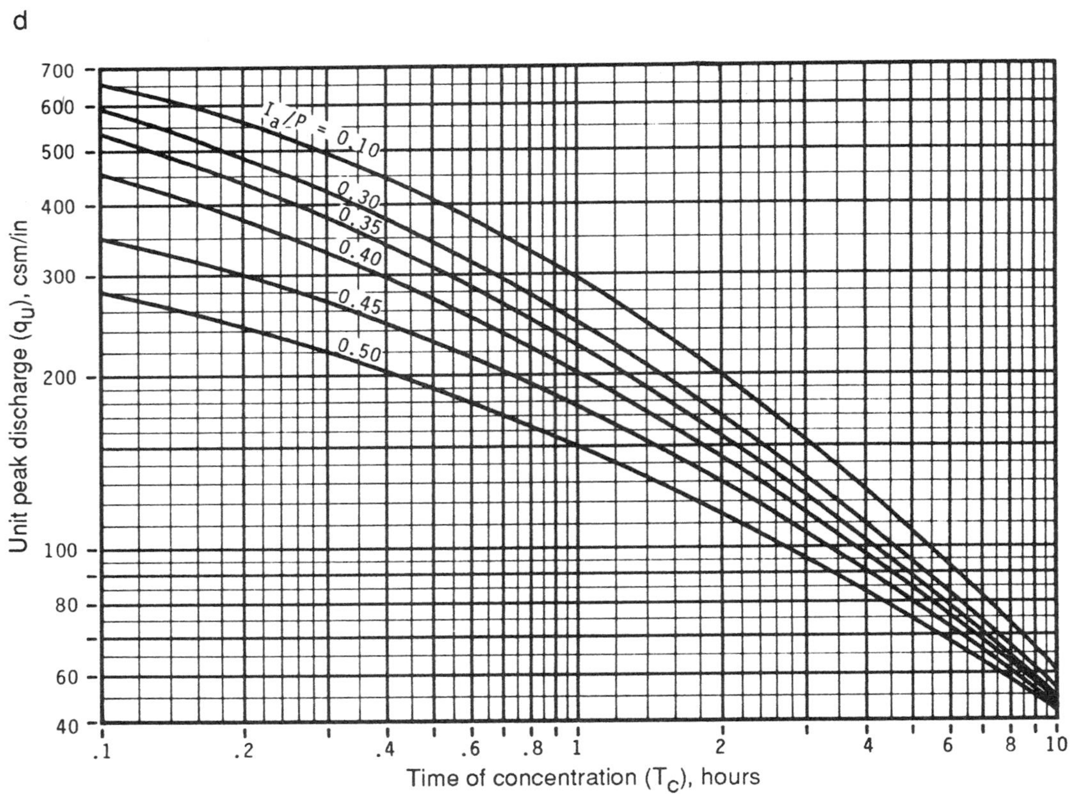

Figure 3.40 *Continued*

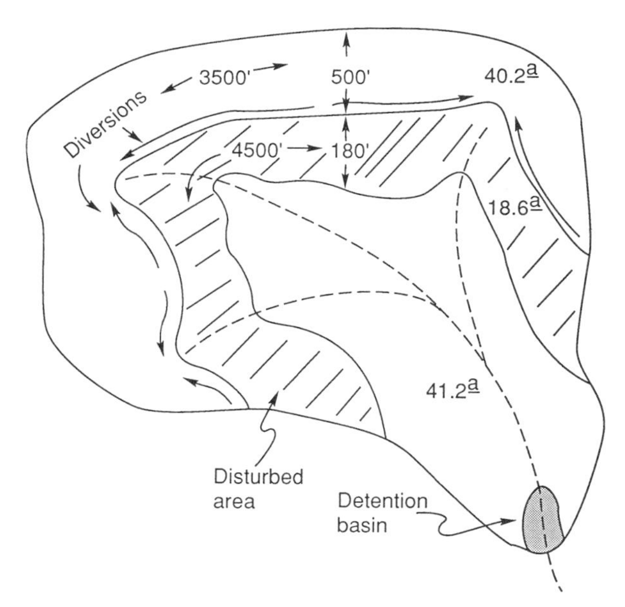

Figure 3.41 Watershed for Example Problem 3.14.

Example Problem 3.14. Impact of mining on runoff hydrograph

The watershed shown in Fig. 3.41 is located in Eastern Kentucky. One seam of coal is to be surface mined as indicated. The entire watershed is 100 acres in size. The average width of the area to be stripped is 180 ft. The stripped area is about 4500 ft long. The average width of the area above the stripped area is 500 ft, and it is about 3500 ft in length. The watershed is predominately forested. The soils are predominately Muskingum and Shelocta with about 50% of the watershed in each. Average land slopes are around 55% with stream slopes averaging 5%. The maximum elevation rise in the watershed is 450 ft, and the maximum flow length is around 5000 ft. A diversion channel is to be placed along the upper periphery of the area to be stripped.

a. Estimate the 10-year, 24-hr runoff volume before and immediately after mining.
b. Estimate the 10-year peak inflow into the diversion channel.
c. Estimate the 25-year runoff hydrograph both before and after mining.

Solution:

a. From Appendix 3A, the 10-year, 24-hr rain is found to be 4.25 in.

Appendix 3B indicates that the hydrologic soil groups are Muskingum (HSG C) and Shelocta (HSG B). The CNs before mining (Table 3.16) are Muskingum (HSG C, CN = 70) and Shelocta (HSG B, CN = 55):

$$\overline{CN} = [0.5(70) + 0.5(55)]/(0.5 + 0.5) = 62.5.$$

From Fig. 3.21 for $P = 4.25$ and CN = 62.5, read $Q = 1.03$ in. Q can also be computed from Eqs. (3.21) and (3.22). Therefore, the runoff volume from the 10-year, 24-hr rainfall before mining is 1.03 in. or $1.03/12 \times 100 = 8.6$ acre-feet.

After mining, the disturbed area is 180 ft × 4500 ft or 18.6 acres. The undisturbed area is 100 − 18.6 = 81.4 acres.

The CNs on the disturbed area (Table 3.16) are Muskingum (HSG C, CN = 88) and Shelocta (HSG B, CN = 81):

$$\overline{CN} = \frac{[(18.6/2)(88) + (18.6/2)(81) + (81.4/2)(70) + (81.4/2)(55)]}{100}$$
$$= 66.6.$$

From Fig. 3.21 for $P = 4.25$ and CN = 66.6, read

$$Q = 1.28 \text{ inches}$$
$$= 10.63 \text{ acre-feet}.$$

b. Equation (3.72) will be used to calculate the peak flow in the diversion. The maximum length of flow within the diversion depends on the direction the diversion is sloped. Here we assume the diversion slopes toward the nearest main draw, so the maximum flow length is about 600 ft if the draws are evenly spaced. We use 700 ft to reflect the uneven spacing and to be conservative. The maximum area draining into a diversion is thus

$$A = \frac{700 \times 500}{43560} = 8 \text{ acres}.$$

The time of concentration is the time it takes overland flow to travel 500 ft down the slope plus 700 ft to the

Table 3.27 Unit Hydrograph for Example Problem 3.14

t	t/t_p	q/q_p	q
0	0	0	0
5	0.25	0.15	34
10	0.50	0.50	114
15	0.75	0.87	197
20	1.00	1.00	227
25	1.25	0.90	204
30	1.50	0.67	152
35	1.75	0.43	98
40	2.00	0.28	64
45	2.25	0.19	43
50	2.50	0.13	30
55	2.75	0.08	18
60	3.00	0.05	11
65	3.25	0.04	9
70	3.50	0.03	7
75	3.75	0.02	5
80	4.00	0.01	2
		Sum	1215

diversion outlet. The overland flow velocity is estimated from Eq. (3.48). The coefficient a from Table 3.20 is 2.5. The overland flow velocity is

$$v = 2.5(0.55)^{1/2} = 1.85 \text{ or about 2 fps}.$$

A reasonable design velocity estimate for the diversion yet to be designed is 6 fps.

$$t_c = \frac{500 \text{ ft}}{2 \text{ fps}} + \frac{700 \text{ ft}}{6 \text{ fps}} = 250 + 116 \simeq 6 \text{ min}.$$

The CN for the area above the diversion is 62.5. Therefore the 10-year, 24-hr runoff volume is 1.03 in. as computed earlier. The pond factor, F_p, from Table 3.25 is 1.0. I_a is computed from $I_a = 0.2S$, where S is given by

$$S = \frac{1000}{CN} - 10 = \frac{1000}{62.5} - 10 = 6.$$

Therefore $I_a = 0.2(6) = 1.2$. I_a/P is 1.2/4.25 or 0.28. Figure 3.11 shows Kentucky is in the SCS type II rainfall region. Table 3.26 gives $C_0 = 2.47$, $C_1 = -0.62$, and $C_2 = -0.12$. Equation (3.73) gives

$$\log(q_u) = 2.47 - 0.62\log(0.1) - 0.12[\log(0.1)]^2 = 2.97$$

or

$$q_u = 10^{2.97} = 933 \text{ cfs/in./mile}^2.$$

A similar estimate can be read from Fig. 3.40. Equation

Table 3.28 Effective Rainfall for Example Problem 3.14

			Premining CN = 62.5		Postmining CN = 66.6	
Time	Ordinate[a]	Accumulated rain[b]	Accumulated effect rain[c]	Incremental effect rain[d]	Accumulated effect rain[c]	Incremental effect rain[d]
11.33	0.26	1.30	0.00	0.00	0.00	0.00
11.42	0.27	1.35	0.00	0.00	0.02	0.02
11.50	0.29	1.45	0.00	0.00	0.04	0.02
11.58	0.31	1.55	0.02	0.02	0.05	0.01
11.67	0.34	1.70	0.04	0.02	0.09	0.04
11.75	0.39	1.95	0.08	0.04	0.15	0.06
11.83	0.51	2.55	0.25	0.17	0.36	0.21
11.92	0.62	3.10	0.46	0.21	0.62	0.26
12.00	0.66	3.30	0.54	0.08	0.72	0.10
12.08	0.68	3.40	0.59	0.05	0.78	0.06
12.17	0.69	3.45	0.61	0.02	0.80	0.02
12.25	0.71	3.55	0.66	0.05	0.86	0.06
12.33	0.72	3.60	0.69	0.03	0.87	0.01
12.42	0.73	3.65	0.71	0.02	0.91	0.04
12.50	0.73	3.65	0.71	0.00	0.91	0.00
12.58	0.74	3.70	0.74	0.03	0.94	0.03
12.67	0.74	3.70	0.74	0.00	0.94	0.00
12.75	0.75	3.75	0.76	0.02	0.97	0.03
12.83	0.75	3.75	0.76	0.00	0.97	0.00
12.92	0.76	3.80	0.79	0.03	1.00	0.03
13.00	0.77	3.85	0.81	0.02	1.03	0.03
				0.81		1.03

[a] From Fig. 2.19.
[b] 25-year, 24-hr rainfall of 5.00 inches × Column 2.
[c] From Fig. 2.26 using rain in Column 3 and appropriate CN.
[d] Difference in successive values in previous column.

(3.72) gives q_p as

$$q_p = q_u AQF_p = 933 \times (8/640) \times 1.03 \times 1 = 12 \text{ cfs}.$$

c. Runoff hydrograph before mining: Based on information given in the problem statement, the lag time can be estimated from Eq. (3.53) as

$$t_L = \frac{L^{0.8}(S+1)^{0.7}}{1900Y^{0.5}}$$

$$S = \frac{1000}{CN} - 10 = \frac{1000}{62.5} - 10 = 6$$

$$L = 5000 \text{ ft}$$

$$Y = 55\%$$

$$t_L = \frac{(5000)^{0.8}(7)^{0.7}}{1900Y^{0.5}} = 0.25 \text{ hr}.$$

Use $t_L = 15$ min.

Since D should be one-fifth to one-third of t_p and 5 min is a convenient time increment, a D of 5 min is selected. The t_p can be computed from Eq. (3.26) as

$$t_p = t_L + D/2$$

$$t_p = 15 + 5/2 = 17.5.$$

For computational convenience, a t_p of 20 min is selected. The peak flow rate is estimated from Eq. (3.57) as

$$q_p = \frac{484A}{t_p} = \frac{484(100/640)}{20/60} = 227 \text{ cfs}.$$

The unit hydrograph shape is taken from the dimensionless unit hydrograph in Fig. 3.34.

The unit hydrograph ordinates are shown in Table 3.27. The volume of the unit hydrograph can be checked by using Eq. (3.46):

$$V = \frac{\Delta t \Sigma q_i}{60.5A} = \frac{5 \times 1215}{60.5 \times 100} = 1.00 \text{ in}.$$

To get an effective rainfall pattern, a 25-year, 24-hr rainfall pattern is used. With a CN of 62.5, Fig. 3.21 or the relationship $P = 0.2S$ shows that 1.20 in. of rain must occur before any runoff starts. The 25-year, 24-hr rainfall is 5.00 in. The time runoff begins is determined when 1.20 in. of rain occurs. This corresponds to a curve ordinate of $1.20/5.00 = 0.24$ or

Table 3.29 Calculation of Runoff Hydrograph for Example Problem 3.14

Time:	0	5	10	15	20	25	30	35	40	45	50	55	60	65	70	75	80	85	90	95	100	105	110	115	120			
UH:	0	34	114	197	227	204	152	98	64	43	30	18	11	9	7	5	2	0										
Rain																												
0.02	0	1	2	4	5	4	3	2	1	1	1	0																
0.02		0	1	2	4	5	4	3	2	1	1	1	0															
0.04			0	1	5	8	9	8	6	4	3	1	1	1	0													
0.17				0	6	19	33	39	35	26	17	11	7	5	3	2	2	1	1	0								
0.21					0	7	24	41	48	43	32	21	13	9	6	4	2	2	1	1	0							
0.08						0	3	9	16	18	16	12	8	5	3	2	1	1	1	1	0							
0.05							0	2	6	10	11	10	8	5	3	2	2	1	1	1	0							
0.02								0	1	2	4	5	4	3	2	1	1	1	1	0								
0.05									0	2	6	10	11	10	8	5	3	2	2	1	1	0						
0.03										0	1	3	6	7	6	5	3	2	1	1	1	0						
0.02											0	1	2	4	5	4	3	2	1	1	1	0						
0.00												0	0	0	0	0	0	0	0	0	0							
0.03													0	1	3	6	7	6	5	3	2	1	1	1				
0.00														0	0	0	0	0	0	0	0	0	0	0				
0.02															0	1	2	4	5	4	3	2	1	1	1	0		
0.00																0	0	0	0	0	0	0	0	0	0	0		
0.03																	0	1	3	6	7	6	5	3	2	1	1	1
0.02																		0	1	2	4	5	4	2	2	1	1	1
Total:	0	1	3	7	20	43	76	104	115	107	92	76	60	50	39	32	26	23	22	20	19	14	11	8	5	2	2	2

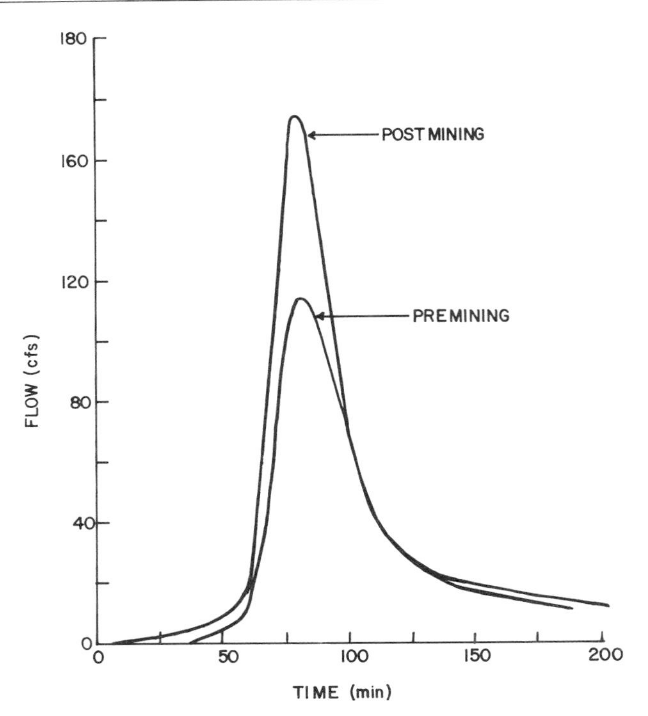

Figure 3.42 Hydrograph for Example Problem 3.14.

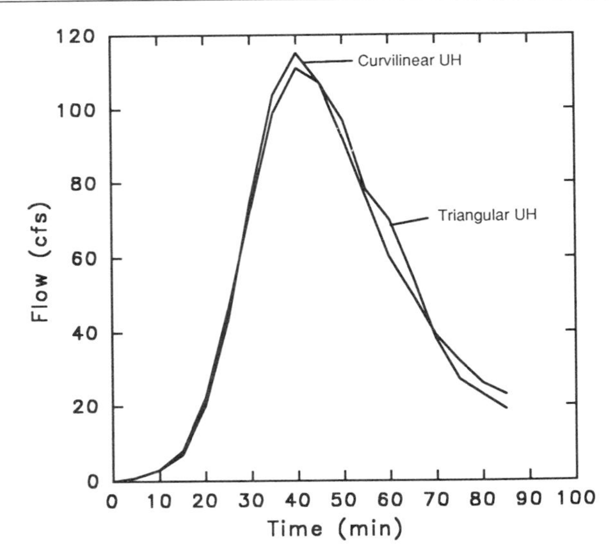

Figure 3.43 Comparison of runoff hydrographs using a curvilinear and a triangular unit hydrograph.

a time of about 11 hr on the type II curve. Table 3.28 shows the calculation of the effective rainfall for the watershed prior to mining.

Similarly, the effective rainfall pattern for the watershed after mining is shown in Table 3.28. About 1.00 in. of rain must occur before runoff starts when the CN is 66.6; 1.00/5.0 corresponds to a type II ordinate of 0.20 or a time of 10.33 hr.

In Table 3.28 the effective rainfall is shown as being calculated beginning at a time of 11.33 hr. The reason that the column for CN = 62.5 does not start at a time of 11 hr and for CN = 66.6 a time of 10.33 hr is because the values, although finite, are essentially zero for times prior to 11.33 hr.

The runoff hydrograph is calculated by applying Eq. (3.43) to the effective rainfall and unit hydrograph. Table 3.29 shows a tabular procedure for calculating the hydrograph. The unit hydrograph coordinates run across the top of the table and the effective rainfall along the left margin. The table entries are the product of the effective rainfall and the unit hydrograph ordinates. Each line in the table is moved one time increment to the right. The runoff hydrograph is obtained by summing the table entries by columns. Figure 3.42 is a plot of the resulting hydrograph.

Based on an accumulated rainfall of 3.85 in. and a CN = 62.5 or $S = 6$, Eq. (3.21) estimates the runoff volume as

$$Q = \frac{(P - 0.2S)^2}{P + 0.8S} = \frac{(3.85 - 1.2)^2}{3.85 + 4.8} = 0.81 \text{ in.}$$

Equation (3.46) applied to the runoff hydrograph results in

$$V = \frac{\Delta t \Sigma q_i}{60.5A} = \frac{5 \times 979}{60.5 \times 100} = 0.81 \text{ in.}$$

Thus the calculations in Table 3.29 are accurate.

The postmining hydrograph is similarly calculated. The unit hydrograph parameters change since the CN changes. For the postmining condition, the following parameters were used and/or determined.

CN = 66.6	$S = 5.01$ in.
$Y = 55\%$	$Q_{24} = 1.77$ in.
$L = 5000$ ft	$t_L = 13.6$ min
$D = 5$ min	$t_p = 15$ min
$A = 100$ acres	$q_p = 302.5$ cfs
$P_{24} = 5.00$ in.	$K = 3.77$

Based on these parameters, the effective rainfall of Table 3.28 and a unit hydrograph defined by Eq. (3.59), the runoff hydrograph shown in Fig. 3.42 was calculated.

To show the insensitivity of the runoff hydrograph to the shape of the unit hydrograph, t_p and q_p were held constant, and the runoff hydrograph representing the premining condition was computed using the SCS triangular unit hydrograph. The results of this calculation are compared with the results of using the curvilinear unit hydrograph in Fig. 3.43. From this figure, it is apparent that the shape of the unit hydrograph had little effect on the runoff hydrograph since the unit hydrograph had the same t_p and q_p, and several blocks of effective rainfall were included.

Example Problem 3.15. Estimation of peak discharge

Estimate the peak discharge from the watershed described in Example Problem 3.14. Use the premining watershed condition and a 25-year frequency.

Solution: Use Eq. (3.72). The ponding factor is 1.0. From previous calculations

$$t_L = 0.25 \text{ hr}.$$

From Eq. (3.52),

$$t_c = t_L/0.6 = 0.42 \text{ hr}.$$

The 25-year, 24-hr rainfall is 5.0 in. The CN is 62.5. The runoff volume from Fig. 3.21 or Eq. (3.21) is 1.47 in.

The values for C_0, C_1, and C_2 are the same as given in part b of Example Problem 3.14. q_u may be taken from Fig. 3.40 or computed from Eq. (3.73) as

$$\log q_u = 2.47 - 0.62\log(0.42) - 0.12(\log 0.42)^2 = 2.6866$$

$$q_u = 10^{2.6866} = 486.$$

From Eq. (3.72),

$$\begin{aligned} q_p &= q_u A Q F_p \\ &= 486(100/640)1.47(1) = 112 \text{ cfs}. \end{aligned}$$

Note that this peak agrees with the peak of the runoff hydrograph of 115 cfs shown in Fig. 3.42.

LONG-TERM WATER BALANCES

Often in the design of ponds and waste water disposal sites it is necessary to compute long-term water balances. Examples would include evaporation ponds, waste water disposal on vegetated surfaces, and computation of potential seepage from soil, fill, or waste disposal areas. The basic equation governing water balances is Eq. (3.1)

$$I - O = \Delta S,$$

where I and O represent all inflows to and outflows from a control volume and ΔS represents the change in the volume of water stored in the control volume. If an evaporation pond is the control volume, inflows would consist of precipitation falling on the pond, runoff entering the pond, and waste water pumped into the pond. Outflow would consist of discharges, seepage losses, and evaporation from the pond.

A control volume might also be a vegetated plot on which waste water is being sprinkled. If all subsurface flows take place in the vertical direction (no horizontal seepage into or out of the soil), then inflow to the system would be precipitation and applied waste water. Generally surface runoff would be prevented from entering or leaving the area. Outflow would be evapotranspiration, deep seepage, and runoff. The change in storage would be the change in the volume of water stored in the soil profile.

These two situations, ponds and vegetated plots, are considered in this section. Vegetated plots are considered under two conditions: (1) The soil profile is kept very wet to promote maximum evapotranspiration (ET) and (2) the soil profile is allowed to dry naturally so that ET may be limited by the soil water content.

Water balances can be computed on the basis of time intervals of minutes up to 1 year. Using long time intervals can provide quick estimates of the likely water balance for an area but will not correctly reflect storage and flow processes that have short time scales. For example, if a time interval of 1 month is used, then a rainfall of 5 in. occurring in 1 day will enter the computations in the same way as if the 5 in. were spread out over several days. Obviously the water balance of a vegetated plot will be sensitive to this difference since 5 in. of rain in 1 day could easily produce considerable runoff, while the same rain spread over several days might produce no runoff.

Two terms in the water balance equation that are very difficult to estimate are evaporation or evapotranspiration and seepage. There is no general way that seepage can be computed independent of the other water balance terms without field data. Often waste disposal sites and ponds are designed to minimize seepage. Where seepage is present, it is often computed as a residual term after all other processes in the water balance are accounted for. Thus the magnitude of seepage is taken as the flow required to balance the water balance equation.

Evaporation data in the form of pan evaporation exists for many sites in the U.S. and other parts of the world. Since evaporation pans are subject to solar heating and have more exposure than a body of water, pan evaporation generally exceeds lake or pond evaporation. Coefficients are typically applied to pan evaporation data to estimate lake evaporation. Pan coefficients vary seasonally and geographically, but a reasonable average value is 0.7 for standard U.S. Weather Bureau Class A pans.

Figures 3.44 and 3.45 show mean annual rainfall and lake evaporation for the continental U.S. For those areas where evaporation exceeds precipitation, it is theoretically possible to dispose of water through the use of evaporation ponds. The net quantity of water that can be evaporated on a long-term average is the difference in mean annual precipitation and evaporation. For example at Oklahoma City, Oklahoma, the mean annual precipitation is about 32 in. (810 mm),

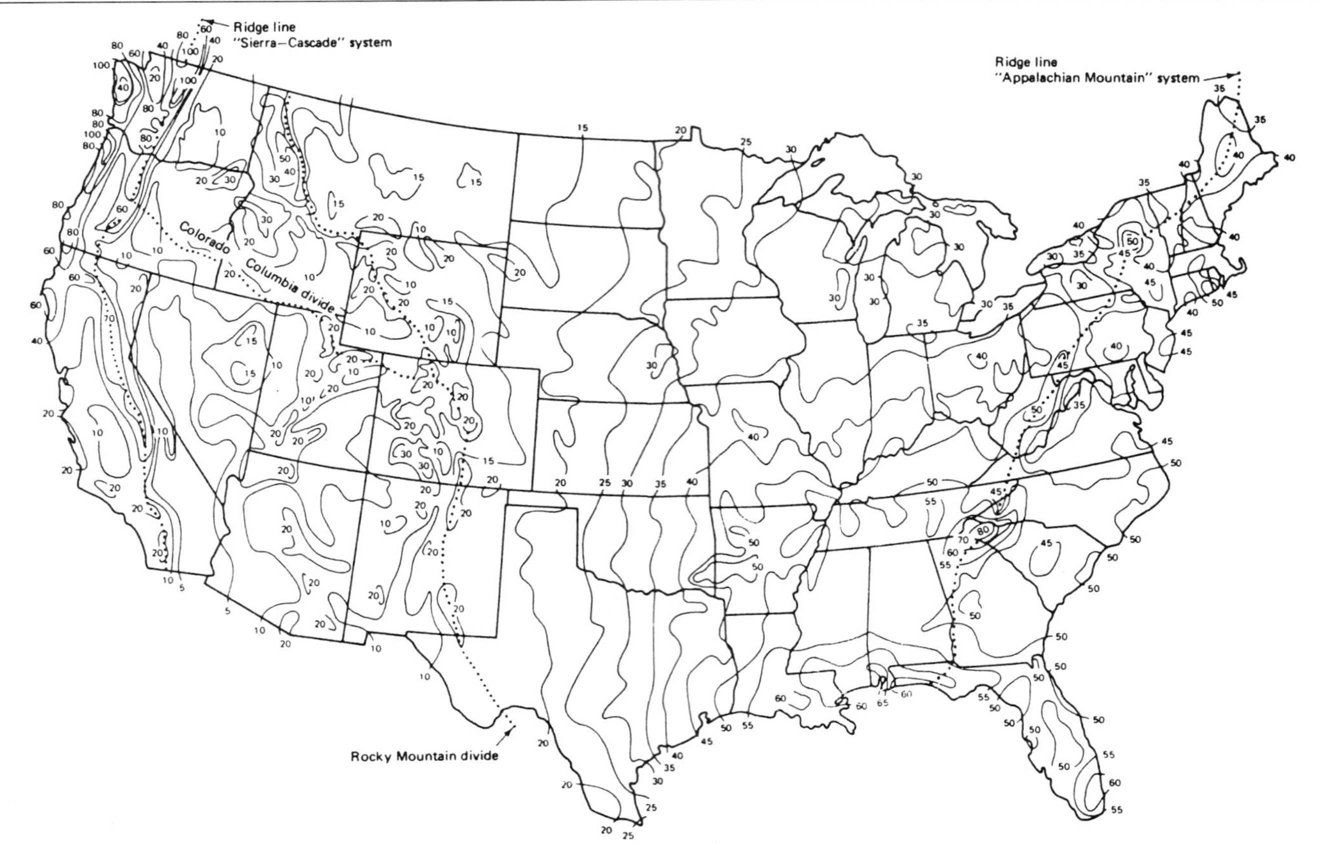

Figure 3.44 Mean annual precipitation in the U.S. (in.).

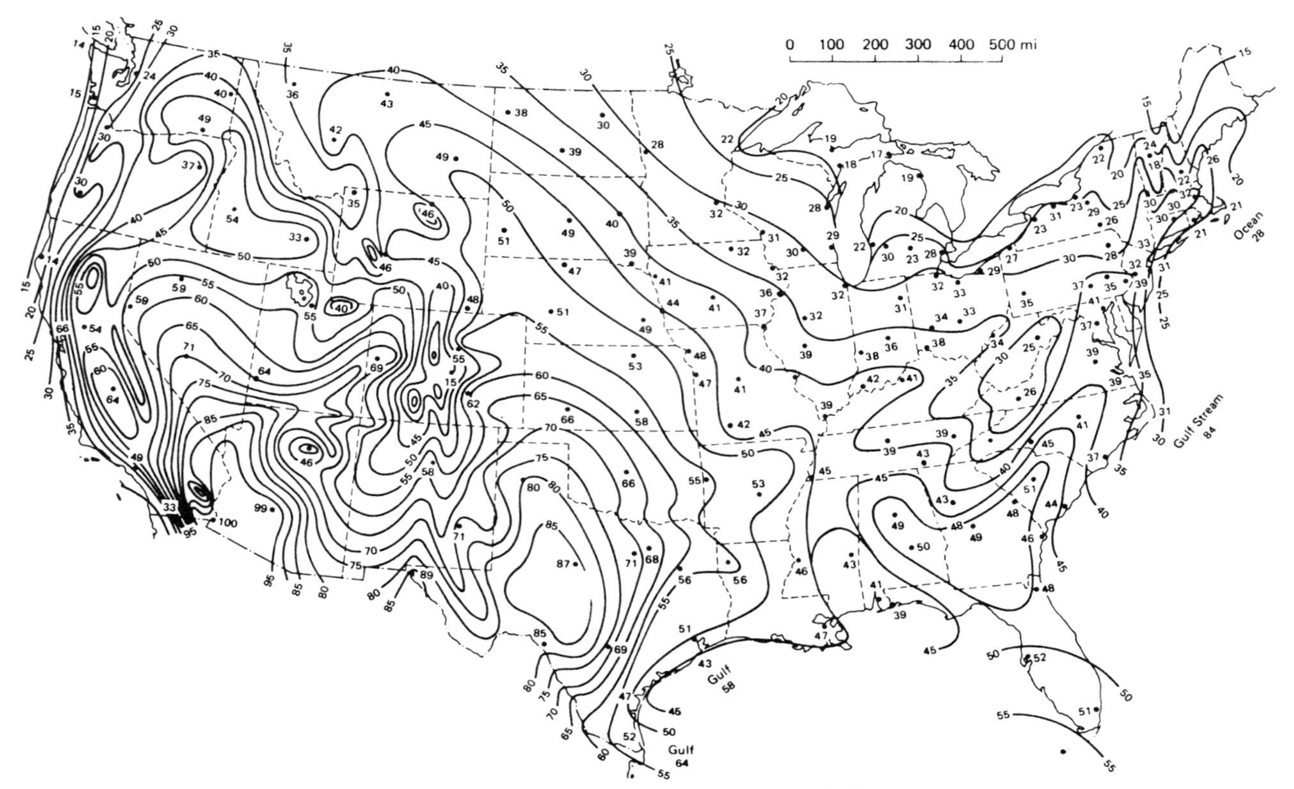

Figure 3.45 Mean annual lake evaporation in the U.S. (in.).

Table 3.30 Mean Monthly Precipitation and Lake Evaporation, Nashville, Tennessee

Month	Precipitation (in.)	Evaporation (in.)
January	5.49	0.9
February	4.51	1.3
March	5.19	1.9
April	3.74	3.3
May	3.72	4.1
June	3.25	5.1
July	3.72	5.8
August	2.86	5.4
September	2.87	4.9
October	2.32	3.7
November	3.28	2.1
December	4.19	1.1

and the mean annual lake evaporation is about 66 in. (1675 mm). Thus a net of about 34 in. (865 mm) of water is lost from the surface of small ponds and shallow reservoirs. On the other hand, near Nashville, Tennessee, the mean annual precipitation is about 45 in. (1140 mm), and the mean annual evaporation is about 39 in. (990 mm), which implies that a pond or shallow lake in this vicinity would have a net gain of water from precipitation over evaporation.

Table 3.30 shows mean monthly precipitation and evaporation for Nashville, Tennessee. This table shows that even though the average annual precipitation exceeds the average annual evaporation, the months of May through October have the opposite result with these months showing a net of 10.2 in. (259 mm) of evaporation over precipitation. This in turn implies that a 1-acre (0.40 hectare) evaporation pond could dispose of about 1500 gal/day (5675 liters/day) of waste water on the average over this 6-month period. This is calculated as

$$\frac{10.2\text{ in.}}{12\text{ in./ft}} \times \frac{43560\text{ ft}^2}{\text{acre}} \times \frac{7.48\text{ gal}}{\text{ft}^3} \times \frac{1\text{ acre}}{182\text{ days}} = 1522\text{ gpd.}$$

The stochastic nature of rainfall and evaporation makes the design of evaporation facilities based on short-term averages extremely risky. In some years, precipitation may exceed evaporation over the design period (6 months in this case). Generally, a detailed continuous simulation of the water balance in a pond should be done to assess the effectiveness of the pond as a waste disposal facility. Such a simulation could be done based on monthly precipitation and evaporation using several years of data. The results should be treated in a probabilistic manner such as flood flows were treated in Chapter 2. Thus, one might conclude that a pond of a given size would be adequate in 90% of all years. Where the probability of the disposal ponds being inadequate is appreciable, alternate means of handling the wastes may need to be available.

Various publications of the U.S. Weather Bureau contain data on precipitation and evaporation. The data in Table 3.30 were taken from Todd (1970). This type of data is very site specific. For example, average annual shallow lake evaporation varies from about 100 in. (2540 mm) in southwestern New Mexico to as low as 15 in. (380 mm) in northern Maine and northwestern Washington in the U.S.

Example Problem 3.16. Estimation of evaporation

For the months of May through October, monthly rainfall totals of 3.0, 3.5, 2.5, 2.5, 6.4, 3.4, and 4.0 were recorded at Nashville. How many gallons of waste water could be evaporated from a 1-acre pond assuming average evaporation and no other inflow or outflows?

Solution: Monthly evaporation is taken from Table 3.30. The change in storage, ΔS, is calculated from Eq. (3.1), where precipitation is I and evaporation is O.

Month	Precipitation	Evaporation	ΔS
May	3.0	4.1	−1.1
June	3.5	5.1	−1.6
July	2.5	5.8	−3.3
Aug	2.5	5.4	−2.9
Sep	6.4	4.9	1.5
Oct	4.0	3.7	0.3
Total	21.9	29.0	−7.1

A net loss of 7.1 in. of water occurs. A depth of 7.1 in. of water over 1 acre converts to 193,000 gal.

Comment: From this problem the need for alternate disposal systems or large factors of safety are apparent. If monthly rainfall had been only 1 in. greater each month, a very likely event, the disposal capacity would be reduced by 6 in. to only about 30,000 gal.

Evapotranspiration

Evaporation from vegetated surfaces is termed evapotranspiration (ET). ET is more complex than evaporation since plant and soil factors affect the process. The rate at which ET occurs from a well-watered, actively growing, completely vegetated surface is termed potential evapotranspiration (PET). PET equations are abundant and are generally based on a combination of theoretical and empirical considerations. For example, one approach to estimating PET is to set it equal to a coefficient times pan evaporation. The coefficient used may range from 0.6 to over 1.0 with the higher coefficient applicable to some forested areas. Theoretical approaches are based on energy budgets, mass transfer relationships, or a combination of these approaches. Generally, experimentally determined constants are required even for the theoretical approaches.

When a soil starts to dry, actual evapotranspiration (AET) may be reduced below PET because the rate of movement of water through the soil into plant roots may limit the transpiration rate. Thus

$$\mathrm{AET} = \mathrm{PET} \times w,$$

where w is a function of the soil water content and plant factors, including such things as stage of growth and type of plant. The factor w has been extensively investigated, yet a single expression for it has not been developed. For some deep-rooted and well-watered vegetation such as alfalfa and some trees, w may exceed 1. For many types of vegetation, the upper limit of w is unity. For any vegetation, as the soil dries, a point where soil factors become limiting and AET falls below PET or w falls below unity will be reached. The w continues to decline until it approaches a value of zero, indicating that transpiration has ceased. Some research has indicated that the point where AET becomes essentially zero is at or just below the permanent wilting point of the soil. The permanent wilting point is the soil water content at which plants wilt and cannot recover.

From this description of AET, it is apparent that w may range from above 1 at soil saturation to 0 for a dry soil. Generally the value remains near 1 until the soil dries to some value of water content that is generally below the field capacity of the soil. Field capacity is the soil water content achieved after significant vertical drainage of water from the soil due to gravity has stopped. At this point, w starts decreasing toward 0. Since the function is not well defined, some use a sigmoidal decrease. Most make the decrease in w a function of the actual soil water content. A simple model is to assign a value of 1 to w until the total soil water content in the root zone drops 1 in. below field capacity. The value of w is then linearly decreased as a function of soil water content in the root zone to 0 at the wilting point. Burman *et al.* (1983) present the relationship

$$w = \ln(\mathrm{AW} + 1)/\ln(101), \tag{3.75}$$

where AW is the available soil water as a percentage (AW = 100 at field capacity and 0 at the wilting point).

Several expressions for PET are available. Many are dependent on plant factors or solar radiation data that are not readily available. The Soil Conservation Service (1970) presents the Blaney-Criddle procedure for estimating water evapotranspiration when water is not limiting. Such plant water use is termed consumptive use and depends on temperature and length of day. Multiplying the mean monthly temperature (t) by the possible monthly percentage of daytime hours of the year (p) gives a monthly consumptive use factor (f). It is assumed that crop consumptive use varies directly with this factor when an ample water supply is available. Mathematically, $u = kf$ and U = sum of $kf = KF$, where U is the consumptive use in inches for the growing season, K is the empirical consumptive use crop coefficient for the growing season (this coefficient is crop dependent), F is the sum of monthly consumptive use factors for the growing season (sum of products of mean monthly temperature and monthly percentage of daylight hours of the year), u is the monthly consumptive use of the crop in inches, k is the empirical consumptive use coefficient for a month (varies by

Table 3.31 Monthly Percentage of Daytime Hours (p)

Latitude North	Jan	Feb	March	April	May	June	July	Aug	Sep	Oct	Nov	Dec
65	3.52	5.13	7.96	9.97	12.72	14.15	13.59	11.18	8.55	6.53	4.08	2.62
60	4.70	5.67	8.11	9.69	11.78	12.41	12.31	10.68	8.54	6.95	5.02	4.14
50	5.99	6.32	8.24	9.24	10.68	10.92	10.99	9.99	8.46	7.44	6.08	5.65
40	6.75	6.72	8.32	8.93	10.01	10.09	10.22	9.55	8.39	7.75	6.73	6.54
30	7.31	7.02	8.37	8.71	9.54	9.49	9.67	9.21	8.33	7.99	7.20	7.16
20	7.75	7.26	8.41	8.53	9.15	9.02	9.24	8.95	8.29	8.17	7.58	7.65
10	8.14	7.47	8.45	8.37	8.81	8.61	8.85	8.71	8.25	8.34	7.91	8.09
0	8.5	7.67	8.49	8.22	8.49	8.22	8.50	8.49	8.21	8.49	8.22	8.5

Table 3.32 Monthly Crop Growth Stage Coefficient (k_c)

Crop	Jan	Feb	March	April	May	June	July	Aug	Sept	Oct	Nov	Dec
Alfalfa	0.63	0.73	0.86	0.99	1.08	1.13	1.11	1.06	0.99	0.91	0.78	0.64
Grass	0.49	0.57	0.73	0.85	0.90	0.92	0.92	0.91	0.87	0.79	0.67	0.55

crop), f is the monthly consumptive use factor (product of mean monthly temperature and monthly percentage of daylight hours of the year), $f = tp/100$, t is the mean monthly air temperature in degrees Fahrenheit, and p is the monthly percentage of daylight hours in the year (Table 3.31).

The following modifications have been made to the method by the Soil Conservation Service. They define k as the product of k_t and k_c, where $k_t = 0.0173t - 0.314$ for t between 36 and 100° F. For $t < 36°$ F, k is taken as 0.3. The value of k_c may be taken from Table 3.32. The Soil Conservation Service presents a more detailed tabulation for k_c.

Example Problem 3.17. Monthly water use

Estimate the monthly water use for a well watered grass at Stillwater, Oklahoma, for the months of April through October. Stillwater is approximately $N36°$. Average monthly temperatures in degrees Farenheit are shown in Table 3.33.

Solution: The calculation are shown in Table 3.33. The factor p is from Table 3.31, f is $tp/100$, $k_t = 0.0173t - 0.314$, k_c is from Table 3.32, $k = k_t k_c$, and $u = kf$. The total water use for the period is the sum of the monthly values.

Problems

(3.1) A 927-acre catchment has been monitored for many years. It has been found that the average annual rainfall over the catchment is 37 in. and that the average annual streamflow is 1 cfs. Geologic investigations reveal a bedrock underlying the catchment that effectively prevents any deep seepage or groundwater recharge. Estimate the average evapotranspiration in inches for this catchment.

(3.2) The volume of water stored in a reservoir on July 1 was 124,000 acre-feet and on August 1 it was 122,000 acre-feet. The surface area of the reservoir on these two dates was 5650 and 5600 acres, respectively. Pan evaporation for the period was 7 in. A lake-to-pan evaporation coefficient of 0.7 is applicable. During the month of July the average stream flow into the reservoir from its 150 mile^2 drainage area was 25 cfs. Outflow was a constant 30 cfs. The catchment and reservoir received 3.5 in. of rain. Estimate the seepage loss from the reservoir assuming no inflow or outflow other than as specified in the problem.

(3.3) Develop a 3-hr, 25-year rainstorm for Chicago, Illinois, or another selected location in 15-min time increments based on (a) the appropriate SCS type curve, (b) depth–duration–frequency data, and (c) the Chicago hyetograph. Compare results by plotting them on a single chart.

(3.4) Develop a 2-hr, 10-year rainstorm for St. Louis, Missouri, or some other selected location in 10-min time increments based on (a) the appropriate SCS type curve, (b) depth–duration–frequency data, and (c) the Chicago hyetograph. Compare the results by plotting them on a single chart.

(3.5) Do problem (3.3) for New Orleans, Louisiana, and/or Seattle, Washington.

(3.6) Secure a nonlinear regression computer program and use it to estimate the coefficients of Eq. (3.4) for Chicago, Illinois, or some other location.

(3.7) For a constant duration, T, Eq. (3.4) can be written $i = K'F^x$. The coefficients K' and x can then be estimated by taking logarithms of both sides of the resulting equation and using linear regression techniques. K' will be a function of the constant value of T selected. By repeating this process for several values of T, a plot of K' versus T can be constructed. The coefficient x may also exhibit some variability as T changes. An average value of x can be chosen. Rainfall intensity for any duration and frequency can then be estimated from the equation by first estimating K'

Table 3.33 Calculations for Example Problem 3.17

Month	t	p	f	k_t	k_c	k	u
April	60	8.8	5.28	0.72	0.85	0.62	3.25
May	68	9.8	6.66	0.86	0.90	0.78	5.17
June	77	9.8	7.55	1.02	0.92	0.94	7.07
July	82	9.9	8.12	1.10	0.92	1.02	8.25
Aug	81	9.4	7.61	1.09	0.91	0.99	7.53
Sep	73	8.4	6.13	0.95	0.87	0.83	5.06
Oct	61	7.8	4.76	0.74	0.79	0.59	2.79
						Total	39.12

from the K' versus T plot and then calculating i from $i = K'F^x$. Use this approach to construct a DDF plot for the location used in problem (3.4).

(3.8) Compute the defining parameters for the rainfall patterns shown in Fig. 3.14 for a 2-hr, 10-year rainfall for St. Louis, Missouri. Use $T' = 30$ min. Plot the results on the graph of problem (3.4). What is the fundamental difference between these rainfall patterns and the patterns of problem (3.4)?

(3.9) Compute the defining parameters for the rainfall patterns shown in Fig. 3.14 for a 3-hr, 25-year rainfall at Stillwater, Oklahoma. Use $T' = 30$ min. Compare the results with Fig. 3.13 by replotting Fig. 3.13 with the results of this problem superimposed.

(3.10) Assuming you are currently inside, go to the nearest outside window. Describe what you see and estimate the amount of surface storage present as far as runoff hydrographs are concerned.

(3.11) An infiltrometer applies simulated rainfall to a plot that is 0.001 acres in size. The rainfall rate is a constant 3 in./hr. The cumulative runoff volume as a function of time is shown below. (a) Estimate the parameters of the Horton infiltration equation. (b) Estimate the parameters of the Holton infiltration equation assuming the available water capacity is 0.35 and n is 1.4. This plot is estimated to have a surface storage capacity of 0.2 in. and an unfilled soil water storage capacity of 3.0 in.

Time (min)	Cumulative runoff (in.)
0	0.00
10	0.00
20	0.00
30	0.22
40	0.56
50	0.94
60	1.35
70	1.78
80	2.21
90	2.65
100	3.09
110	3.54
120	3.99

(3.12) Based on the Horton infiltration relationship with $f_0 = 6$ iph, $f_c = 0.2$ iph, and $k = 3\ \text{hr}^{-1}$, estimate the effective rainfall pattern for the rain of problem (3.3). Surface storage is 0.25 in.

(3.13) Based on the Holton infiltration relationship with $f_c = 0.2$ iph, $n = 1.4$, $a = 1$, and $F_p = 3.5$ in., estimate the effective rainfall pattern for the rain of problem (3.3). Surface storage is 0.25 in. Assume that the soil is initially dry.

(3.14) Based on the SCS curve number equation, using a curve number of 80 estimate the effective rainfall pattern for the rain of problem (3.3).

(3.15) Based on the Green–Ampt infiltration relationship for a silty clay loam soil and 30% effective saturation, estimate the effective rainfall pattern for the rainstorm of problem (3.3). Surface storage is 0.2 in.

(3.16) Use the rainfall pattern developed for problem (3.4a) and the infiltration relation of (a) problem (3.12), (b) problem (3.13), (c) problem (3.14), (d) problem (3.15), to estimate the effective rainfall pattern.

(3.17) Use the rainfall pattern developed for problem (3.5a) and the infiltration relationship of (a) problem (3.12), (b) problem (3.13), (c) problem (3.14), (d) problem (3.15), to estimate the effective rainfall pattern.

Note: For the following problems (3.18–3.26) to be of maximum benefit and interest, two or three actual watersheds in the 5 to 1000 acre size range should be selected. A watershed that has a significant undeveloped area should be used for problem (3.26). Data on topography, soils, and land use should be gathered. If this is a class project, different students might use different approaches to the problems and then compare the results.

(3.18) Develop the 25-year, 3-hr rainstorm for the catchment using the appropriate time intervals. The DDF, SCS, or Chicago hyetograph might be used. (See note above.)

(3.19) Using an appropriate loss function, develop the effective rainfall pattern for the catchment based on the rainfall of problem (3.18). Loss functions that might be used include the SCS curve number approach or a surface storage plus infiltration with infiltration based on the Horton, Green–Ampt, or Φ index approach. (See note above.)

(3.20) Use the Santa Barbara Urban Hydrograph method to estimate the runoff hydrograph. (See note above.)

(3.21) Develop an appropriate synthetic unit hydrograph for the catchment. One of the SCS unit hydrographs, the Haan unit hydrograph, or the double triangle unit hydrograph can be used. Required parameters can be estimated based on SCS relationships, Epsey *et al.* relationships, or by other means. (See note above.)

(3.22) Combine the effective rainfall of problem (3.18) with the unit hydrograph of problem (3.21) to develop the runoff hydrograph. Plot the rainfall, effective rainfall, unit hydrograph, and runoff hydrograph on a single chart. (See note above.)

(3.23) Estimate the 25-year peak flow from the catchment without developing a runoff hydrograph. Compare this estimate to the peak of the runoff hydrograph of problem (3.22). Are the estimates different? Why? (See note above.)

(3.24) Estimate the runoff volume that will result from a 25-year, 24-hr rainfall on this catchment. Compare this volume to the volume under the runoff hydrograph of problem (3.22). Are the estimates different? Why? (See note above.)

(3.25) Estimate the runoff volume that will occur from a 100-year, 24-hr rainfall on this catchment. (See note above.)

(3.26) Assume that a substantial part (specify exactly how much) of the catchment that is currently undeveloped is going to be converted to a large shopping mall and high-density residential use. What will be the magnitude of the 25-year runoff peak and volume from the total catchment after this development? Plot the pre- and postdevelopment runoff hydrographs on the same chart. (See note above.)

(3.27) Plot the data of Table 3.4 for the type II curve along with Eq. (3.6). Is the equation a reasonable approximation to the plotted data? In what region is the deviation the greatest? What impact is this likely to have on an estimated runoff hydrograph?

(3.28) What is the 6-hr, 6-mile^2 PMP for your location? Compare this PMP estimate with the value of the greatest observed rainfall for 6-hr as estimated from Eq. (3.5).

(3.29) Examine the documentation and/or coding for an event-based computer program for estimating runoff hydrographs. Compare the techniques used in the program with those discussed in this chapter. Make reference to rainfall, abstractions, and hydrograph development.

(3.30) Assuming constant rainfall rates during each time interval, calculate the uniform loss rate or Φ index for the following storm, which produced 2.0 in. of direct runoff.

Time (hr)	Accumulated depth of rain (in.)
0	0.0
1	0.6
2	1.0
3	1.2
4	1.8
5	2.9
6	3.6
7	4.5

(3.31) The average monthly temperature in degrees Farenheit and rainfall in inches for two locations are shown below. Assuming no runoff and no deep seepage, how much waste water can be disposed of annually through evapotransporation from a 10-acre plot of grass?

	Sioux Falls, SD (Latitude 42.5° N)		Phoenix, AZ (Latitude 32.5° N)	
Month	Temperature	Rain	Temperature	Rain
Jan	15.2	0.62	49.7	0.73
Feb	19.1	0.93	53.5	0.85
March	30.1	1.54	59.0	0.66
April	45.9	2.31	67.2	0.32
May	58.3	3.38	75.0	0.13
June	68.1	4.35	83.6	0.09
July	74.3	2.84	89.8	0.77
Aug	71.8	3.59	87.5	1.12
Sep	61.8	2.61	82.8	0.73
Oct	50.3	1.25	70.7	0.46
Nov	32.6	1.00	58.1	0.49
Dec	21.1	0.74	51.6	0.85

(3.32) Do problem (3.31) for your current location.

(3.33) Develop a 90-min unit hydrograph from the following 30-min unit hydrograph.

Time (min)	Flow (cfs)
0	0
30	75
60	200
90	250
120	200
150	125
180	68
210	40
240	25
270	15
300	8
330	5
360	0

(3.34) Develop a 15 min unit hydrograph from the 30 min unit hydrograph of problem (3.33).

(3.35) Define time of concentration.

(3.36) List several methods for estimating infiltration losses from rainfall.

(3.37) What factors influence the time of concentration?

(3.38) Discuss the impact of the following catchment changes on the expected peak runoff rate, time to peak, runoff volume, and base flow. (a) Urbanization of an agricultural catchment, (b) urbanization of a natural or wildland catchment, (c) channel straightening and other hydraulic improvements within the catchment, (d) conversion of a wildland catchment to an intensive agricultural catchment, (e) installation of an improved storm drainage system in an existing urban area, (f) installation of small livestock and recreational ponds on an agricultural catchment, and (g) installation of flood detention basins on an urban catchment.

(3.39) Describe how nonrecording rain gage information can be used to supplement recording rain gage data in determining the time distribution of basin-wide rainfall.

(3.40) Based on a 1200-acre watershed, determine:

(a) the volume in acre-feet of 1.5 in. of runoff,

(b) the runoff rate in cfs equivalent to 1 in./hr,

(c) the volume in cubic feet equivalent to 0.2 in. of evapotranspiration,

(d) the rate in gallons per minute equivalent to 0.2 in. of evapotranspiration per day.

(3.41) If a steady and prolonged rain of 2 in./hr falls on a 1 mile^2 reservoir, what is the equivalent steady outflow rate in cfs?

(3.42) Develop a two-way table with location along the vertical left scale and duration along the upper horizontal scale. Tabulate the 30-min, 1-, 2-, 3-, 6-, 12-, and 24-hr rainfall for a 10-year frequency for New York City; Miami, Florida; Rapid City, South Dakota; Oklahoma City, Oklahoma; and Seattle, Washington.

(3.43) Define the probable maximum precipitation.

(3.44) What method of storm rainfall averaging would be best for determining catchment rainfall in a mountainous area? Why?

(3.45) A 1000-hectare catchment has an average annual rainfall of 990 mm and an average annual stream flow of 0.05 m^3/sec. The estimated average annual evapotransportation is 600 mm. There are no diversions of flow into or out of the catchment. Estimate the average annual recharge to groundwater.

(3.46) Develop a rainfall hyetograph in 30-min increments for a rainfall of 83 mm occurring in 5 hr. For the time distribution use (a) the SCS type II curve, (b) the curves of Fig. 3.14, and (c) the Chicago hyetograph.

(3.47) What is the maximum 1-hr and 2-hr intensities for the following rainfall?

Accumulated time (min)	Accumulated rain (in.)
0	0.00
10	0.10
15	0.25
30	0.40
60	0.45
70	0.68
90	0.92
105	1.43
110	1.79
125	2.09
140	2.47
180	2.55
210	2.65
240	2.70

(3.48) Infiltration data reveal the following:

Time (hr)	f (mm/hr)
$\frac{1}{2}$	3.52
1	2.16
5	0.51

Estimate f_0, f_c, and k in Horton's infiltration equation.

(3.49) Estimate the mean annual potential evapotransporation at a location whose latitude is 40° N and whose mean monthly temperature in degrees centigrade is:

Month:	J	F	M	A	M	J	J	A	S	O	N	D
Temp:	0	5	15	17	19	23	30	28	20	15	11	8

(3.50) A 500-acre catchment experiences a uniform rainfall of 1.8 iph for 2 hr. What is the maximum possible runoff rate in cfs? What assumptions did you make in arriving at your estimate?

(3.51) A catchment near Atlanta, Georgia, has an average overland flow distance of 350 ft. The length of the main channel is approximately 5000 ft for this 475-acre area. The predominant land use is wooded with average land slopes of 10% and an average stream slope of 0.5%. The soil is a silt loam and is rather shallow. Estimate the 25-year peak flow from the catchment under this new condition. State all assumptions that you make.

(3.52) Long-range plans call for converting the catchment of problem (3.51) to medium density urban housing. This will be accompanied by storm drains and considerable channel work on the main drainage way. Estimate the 25-year peak flow from the catchment under this new condition. State all assumptions that you make.

(3.53) The runoff hydrograph from a 1.7-mile^2 catchment can be approximated by a triangle with a peak of 600 cfs. If the volume of runoff is 1.3 in., what is the time base of the hydrograph?

(3.54) The following data on precipitation and runoff were collected over a wide range of antecedent conditions from a catchment. Estimate the AMC II SCS curve number.

P (in.)	3.4	1.7	2.8	1.9	2.2	4.7	1.5
Q (in.)	1.4	0.6	0.6	0.5	0.6	2.3	0.4

(3.55) Compare the K of Eq. (3.56) based on the SCS relationship [Eq. (3.57)] and the relationship of Eq. (3.58) approximating the exponents 0.96 and 1.07 by 1.00.

(3.56) Show that the K of Eq. (3.59) is about 3.77 for the SCS unit hydrograph.

(3.57) Compare the K of Eq. (3.56) based on the SCS relationship [Eq. (3.57)] and the relationship of Eq. (3.66).

(3.58) Develop a 3-hr, 25-year rainstorm for your location. Use 15-min time increments.

(3.59) Estimate the 6-hr, 50-year rain for your current location.

(3.60) For the rainfall shown below, develop the effective rainfall pattern assuming a curve number of 80.

Time (min)	Accumulated rain (in.)	Time (min)	Accumulated rain (in.)
0	0	105	2.48
15	0.06	120	2.63
30	0.13	135	2.74
45	0.23	150	2.82
60	0.41	165	2.89
75	0.96	180	2.96
90	2.26		

(3.61) A watershed has a time to peak of 60 min and an area of 600 acres. Write down the unit hydrograph ordinates using a time increment of 15 min and the SCS dimensionless, curvilinear hydrograph.

(3.62) Compute the runoff hydrograph resulting from the rain of problem (3.60) and the unit hydrograph of problem (3.61).

(3.63) A 400-acre watershed has the following characteristics:

A. 100 acres of forest Muskingum soil
B. 150 acres of forest Shelocta soil
C. 50 acres of surface mined Muskingum soil
D. 50 acres of surface mined Shelocta soil
E. 50 acres of bottomland Pope soil.

Calculate the appropriate SCS curve number. Convert to antecedent condition III.

References

Ardis, C. V., Jr. (1973). "Storm Hydrographs Using a Double Triangle Model." Tennessee Valley Authority, Division of Water Control Planning, Knoxville, TN.

Bernard, M. (1942). In "Hydrology" (O. E. Meinzer, ed.), Chap. II. Dover, New York.

Betson, R. P. (1964). What is watershed runoff? *J. Geogr. Res.* **69**(8):1541–1551.

Bleek, J. (1975). Synthetic unit hydrograph procedures in urban hydrology. *In* "Proceedings, National Symposium on Urban Hydrology and Sediment Control, University of Kentucky, Lexington, Kentucky, July."

Bloomsburg, C. L. (1960). A hydrograph equation, Paper presented at Pacific Northwest Section Meeting of American Geophysical Union, Moscow, Idaho, October 19–20.

Borelli, J., Hasfurther, V. R., and Burman, R. D. ed. (1983). Advances in irrigation and drainage—Surviving external pressures. *In* "Proceedings, American Society of Civil Engineers Specialty Conference, New York."

Brakensiek, D. L. (1967). A simulated watershed flow system for hydrograph prediction. *In* "Proceedings, International Hydrology Symposium, Fort Collins, CO."

Brakensiek, D. L., and Rawls, W. J. (1983). Green-Ampt infiltration model parameters for hydrologic classification of soils. *In* "Proceedings, American Society of Civil Engineers Specialty Conference, New York" (J. Borelli, V. R. Hasfurther, and R. D. Burman, eds.), pp. 226–233.

Burman, R. D., *et al.* (1983). Water requirements. *In* "Design and Operation of Farm Irrigation Systems" (Jensen, ed.), Chap. 6, Monograph 3. American Society of Agricultural Engineers, St. Joseph, MI.

Chow, V. T., ed. (1964). "Handbook of Hydrology." McGraw–Hill, New York.

Chow, V. T., Maidment, D. R., and Mays, L. W. (1988). "Applied Hydrology." McGraw–Hill, New York.

Cronshey, R. (1981). Personal communication. U.S. Department of Agriculture, Soil Conservation Service, Washington, DC.

DeCoursey, D. G. (1966). "A Runoff Hydrograph Equation." U.S. Department of Agriculture ARS 41-116, Washington, DC.

Eagleson, P. S. (1970). "Dynamic Hydrology." McGraw–Hill, New York.

Edson, C. G. (1951). Parameters for relating unit hydrographs to watershed characteristics. *Am. Geophys. Union Trans.* **32**:591–596.

Engman, E. T. (1986). Roughness coefficients for routing surface runoff. *J. Irrigat. Drain. Eng.* **112**(1):39–53.

Epsey, W. H., Jr., Altman, D. G., and Graves, C. B. (1977). Nomograph for ten-minute unit hydrographs for small urban watersheds, Technical Memo No. 32. Urban Water Resources Research Program, American Society of Civil Engineers, New York.

Frederick, R. H., Myers, V. A., and Auciello, E. P. (1977). Five to 60 minute precipitation frequency for the eastern and central United States, National Oceanic and Atmospheric Administration Technical Memorandum NWS HYDRO-35. U.S. Department of Commerce, Washington, DC.

Gray, D. M. (1961). Synthetic unit hydrographs for small watersheds. *Am. Soc. Civil Eng. Proc.* **87**(HY4):33–54.

Green, W. H., and Ampt, G. A. (1911). Studies on soil physics. I. Flow of air and water through soils. *J. Agron. Soc.* **4**:1–24.

Haan, C. T. (1970). A dimensionless hydrograph equation, File Report. Agricultural Engineering Department, University of Kentucky, Lexington, KY.

Haan, C. T. (1972). Runoff coefficients for selected small streams in Kentucky, Bulletin 713. University of Kentucky, Agricultural Experiment Station, Department of Agricultural Engineering, Lexington, KY.

Henderson, F. M. (1966). "Open Channel Flow." Macmillan, New York.

Hershfield, D. M. (1961). Rainfall frequency atlas of the United States, Technical paper 40. U.S. Department of Commerce, Weather Bureau, Washington, D.C.

Hewlett, J. D., and Hibbert, A. R. (1967). Factors affecting the response of small watersheds to precipitation in humid regions. *In* "Forest Hydrology" (W. E. Sopper and H. W. Lull, eds.). Pergamon Press, Oxford.

Hillel, D. (1971). "Soil and Water Physical Principles and Processes." Academic Press, New York.

Holtan, H. N. (1961). A concept for infiltration estimates in watershed engineering. U.S. Department of Agriculture ARS 41-51.

Horner, W. W. (1910). Modern procedure in district sewer design. *Eng. News* **64**.

Horton, R. E. (1940). Approach toward a physical interpretation of infiltration capacity. *Soil Sci. Soc. Am. Proc.* **5**:339–417.

Huff, F. A. (1967). Time distribution of rainfall in heavy storms. *Water Resources Res.* **3**(4):1007–1019.

Izzard, C. F. (1946). Hydraulics of runoff from developed surfaces. *Proc. Highway Res. Board* **26**:129–150

Keifer, C. J., and Chu, H. H. (1967). Synthetic storm pattern for drainage design. *J. Hydraulics Div., Am. Soc. Civil Eng.* **83**(HY4):1–25.

Kirpich, P. Z. (1940). Time of concentration of small agricultural watersheds. *Civil Eng.* **10**(6).

Linsley, R. K., Kohler, M. A., and Paulhus, J. L. H. (1949). "Applied Hydrology." McGraw–Hill, New York.

Linsley, R. K., Kohler, M. A., and Paulhus, J. L. H. (1982). "Hydrology for Engineers," 3rd ed. McGraw–Hill, New York.

McPherson, M. B. (1969). Some notes on the rational method of storm drain design, Technical memorandum No. 6, ASCE Urban Water Resources Research Program. American Society of Civil Engineers, New York.

Nash, J. E. (1959). Systematic determination of unit hydrograph parameters. *J. Geophys. Res.* **64**:111–115.

National Research Council (1985). "Safety of Dams—Flood and Earthquake Criteria." National Academy Press, Washington, DC.

National Oceanic and Atmospheric Administration. Precipitation-frequency atlas of the Western U.S. NOAA atlas II. Superintendent of Documents, U.S. Government Printing Office, Washington, DC.

Natural Environmental Research Council (1975). Flood studies report. Natural Environmental Research Council, London.

Norman, D. (1981). Personal communication. Bauer, Borowitz, Merchant, Inc., 2607 1/2 N. High St., Columbus, OH.

Overton, D. E., and Meadows, M. E. (1976). "Storm Water Modeling." Academic Press, New York.

Putnam, A. L. (1972). Rainfall and runoff in urban areas—A case study of flooding in the Piedmont of North Carolina. *In* "Proceedings, Urban Rainfall Management Problems, University of Kentucky, Lexington, Kentucky, April 17–18."

Rawls, W. J., Brakensiek, D. L., and Miller, N. (1983). Green-Ampt infiltration parameters from soils data. *J. Hydraulic Eng., Proc. Am. Soc. Civil Eng.* **109**(1):62–70.

Regan, R. M., and Duru, J. O. (1972). Kinematic wave nomograph for times of concentration. *Proc. Am. Soc. Civil Eng.* **98**(HY10):1765–1771.

Riedel, J. T., and Schreiner, L. C. (1980). Comparison of generalized estimates of probable maximum precipitation with the greatest observed rainfall, NOAA Technical Report NWS 25. National Oceanic and Atmospheric Administration, U.S. Department of Commerce, Silver Springs, MD.

Schwab, G. O., Frevert, R., Edminster, T. W., and Barnes, K. K. (1971). "Elementary Soil and Water Conservation Engineering." Wiley, New York.

Sherman, L. K. (1932). Streamflow from rainfall by the unit-graph method. *Eng. News Record* **108**:501–505.

Skaggs, R. W., and Khaleel, R. (1982). Infiltration. *In* "Hydrologic Modeling of Small Watersheds" (Haan *et al.*, eds.). American Society of Agricultural Engineers, St. Joseph, MI.

Snyder, F. F. (1938). Synthetic unit-graphs. *Trans. Am. Geophys. Union* **19**:447–454.

Soil Conservation Service (1970). "Irrigation Water Requirements," TR-21. U.S. Department of Agriculture, Soil Conservation Service, Washington, DC.

Soil Conservation Service (1972, 1985). "Hydrology," Sect. 4, Soil Conservation Service National Engineering Handbook. U.S. Department of Agriculture, Washington, DC.

Soil Conservation Service (1973). "A Method for Estimating Volume and Rate of Runoff in Small Watersheds," SCS-TP-149. U.S. Department of Agriculture, Soil Conservation Service, Washington, DC.

Soil Conservation Service (1975, 1986). Urban hydrology for small watersheds, Technical release No. 55. Soil Conservation Service, U.S. Department of Agriculture, Washington, DC.

Soil Conservation Service (1983). Computer program for project formulation hydrology (Draft), Technical release No. 20. Soil Conservation Service, U.S. Department of Agriculture, Washington, DC.

Strelkoff, T. (1969). One-dimensional equations of open channel flow. *Trans. Hydraulics Div. Am. Soc. Civil Eng.* **95**(HY3):861–876.

Stubchaer, J. M. (1975). The Santa Barbara urban hydrology method. *In* "Proceedings, National Symposium on Urban Hydrology and Sediment Control, University of Kentucky, Lexington, Kentucky, July 28–31."

Terstriep, M. L., and Stall, J. B. (1974). "The Illinois Urban Drainage Area Simulator, ILLUDAS," ISWS-74-Bul 58. State of Illinois, Department of Registration and Education, Urbana, IL.

Tholin, A. L., and Kiefer, C. J. (1960). Hydrology of urban runoff. *Trans. Am. Soc. Civil Eng.* **125**:1308–1355.

Thornthwaite, C. W. (1948). An approach toward a rational classification of climate. *Geogr. Rev.* **38**:55–64.

Todd, D. K., ed. (1970). "The Water Encyclopedia." Water Information Center, Water Research Building, Manhasset Isle, Port Washington, NY.

U.S. Army Corps of Engineers (1959). Flood hydrograph analysis and computation, Engineering and Design Manual EM 110-2-1405. Washington, DC.

U.S. Department of Agriculture (1968). Moisture-tension data for selected soils on experimental watersheds, Report 41-144. Agricultural Research Service, U.S. Government Printing Office, Washington, DC.

U.S. Department of Commerce (1964). Two- to ten-day precipitation for return periods of 2 to 100 years in the contiguous United States, Technical paper 49. U.S. Government Printing Office, Washington, DC.

U.S. National Weather Service (1943–1984). "National Oceanic and Atmospheric Administration Hydrometeorologic Reports," 1–55. A series of PMP reports for various areas not all of which are published, Washington, DC.

Viessman, W., Jr. (1967). A linear model for synthesizing hydrographs for small drainage areas, Paper presented at the 48th Annual Meeting American Geophysical Union, Washington, DC.

Viessman, W., Jr., Knapp, J. W., Lewis, G. L., and Harbaugh, T. E. (1977). "Introduction to Hydrology." Intext Educational Publishers, New York.

Water Pollution Control Federation (1969). Design and construction of sanitary and storm sewers, Manual of practice 9 (Am. Soc. of Civil Engineers Manual of Engineering Practice No. 37). Water Pollution Control Federation, 3900 Wisconsin Avenue, Washington, DC.

Wilkinson, L. (1987). "SYSTAT: The System for Statistics." SYSTAT, Inc., Evanston, IL.

Wilson, B. N., *et al.* (1983). A hydrology and sedimentology watershed model, Special publication SEDIMOT II design manual. Agricultural Engineering Department, University of Kentucky, Lexington, KY.

Wittenberg, H. (1975). A model to predict the effects of urbanization on watershed response. *In* "Proceedings, National Symposium on Urban Hydrology and Sediment Control, University of Kentucky, Lexington, Kentucky, July."

Wooding, R. A. (1965a). A hydraulic model for the catchment-stream problem, I. *J. Hydrol.* **3**:254–267.

Wooding, R. A. (1965b). A hydraulic model for the catchment-stream problem, II. *J. Hydrol.* **3**:268–282.

Wooding, R. A. (1966). A hydraulic model for the catchment-stream problem. III. Comparison with runoff observations. *J. Hydrol.* **4**:21–37.

Woolhiser, D. A. (1969). Overland flow on a converging surface. *Trans. Am. Soc. Agricultural Eng.* **12**(4):460–462.

Woolhiser, D. A., and Ligget, J. A. (1967). Unsteady one dimensional flow over a plane—The rising hydrograph. *Water Resources Res.* **3**(3):753–771.

Wright-McLaughlin Engineers (1969). "Urban Storm Drainage Criteria Manual." Denver Regional Council of Governments, Denver, Colorado, 1969 and 1975 revision; also a similar manual for Stillwater, OK, 1979.

Wu, I-P. (1963). Design hydrographs for small watersheds in Indiana. *Am. Soc. Civil Eng. Proc.* **89**(HY6):35–66.

Yevjevich, V. (1968). Misconceptions in hydrology and their consequences. *Water Resources Res.* **4**(2):225–232.

Open Channel Hydraulics

The subject of open channel hydraulics is extensive enough to require a complete text. Obviously, an exhaustive coverage cannot be given in one short chapter. The treatment given here is intended only to cover certain basic principles and to give the details necessary to design stable, open channels; to do simple channel routings; and to compute simple backwater profiles. The interested reader can consult several excellent texts for additional details (Chow, 1959; Henderson, 1966).

BASIC RELATIONSHIPS

Continuity Equation

When dealing with the hydraulics of open channel flow, there are three basic relationships that must be kept in mind. These relationships are the continuity equation, the energy equation, and the momentum equation. If we consider a stream with a cross section as shown in Fig. 4.1, the continuity equation may be written as

$$\text{inflow} - \text{outflow} = \text{change in storage}, \qquad (4.1)$$

where inflow represents the volume of flow across section 1 during a time interval, outflow represents the volume of flow across section 2 during this time interval, and change in storage represents the change in the volume of water stored within the section from 1 to 2.

The continuity equation may also be written in terms of flow rates as

$$\text{inflow rate} - \text{outflow rate} = \text{rate of change in storage}, \qquad (4.2)$$

where inflow rate and outflow rate represent the rate of flow across sections 1 and 2, respectively, and the rate of change in storage is the rate at which the volume of water is accumulating or diminishing within the section.

The flow rate, Q, is generally expressed in cubic feet per second (cfs) or cubic meters per second (cms) and may be written

$$Q = vA, \qquad (4.3)$$

where v is the average velocity of flow at a cross-section and A is the area of the cross section. v is generally given in feet per second (fps) or meters per second (m/sec) and A in square feet (ft^2) or square meters (m^2). Throughout this chapter, units on symbols appearing in equations will not be given unless needed for clarity. Standard units are feet and seconds or meters and seconds.

It should be kept in mind that v is the average velocity of the flow perpendicular to the cross section. The actual pattern of flow velocity can be quite com-

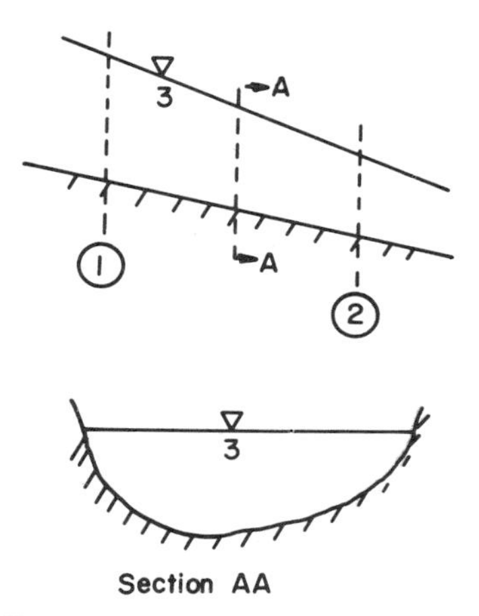

Figure 4.1 Typical channel sections.

plex. Figure 4.2 shows typical distributions of flow velocity with various channel cross sections. Figure 4.3 shows a velocity profile for an idealized situation (Chow, 1959).

Both Figs. 4.2 and 4.3 show that the actual velocity in contact with the channel boundary is quite low.

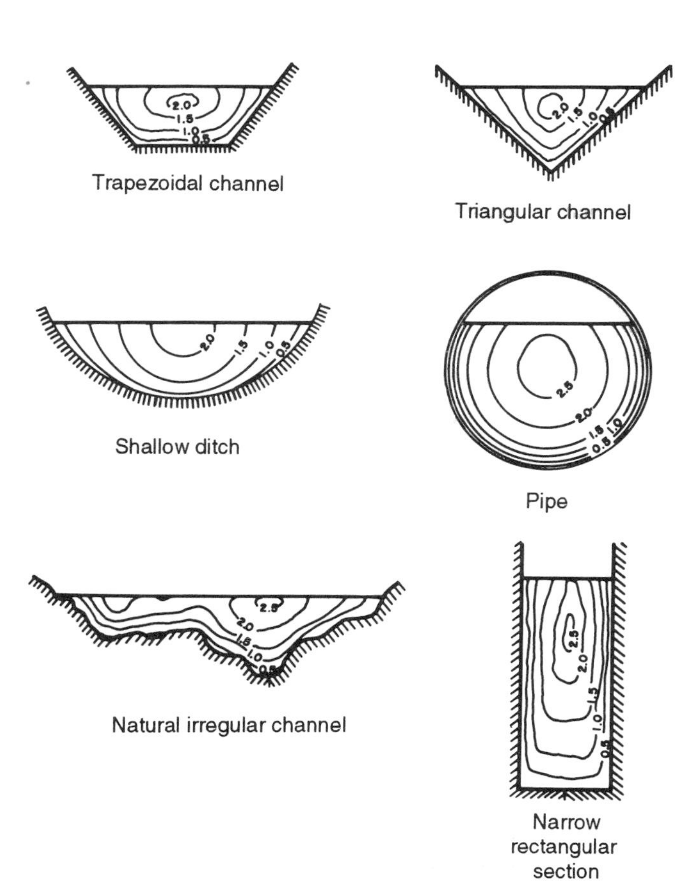

Figure 4.2 Typical velocity distributions.

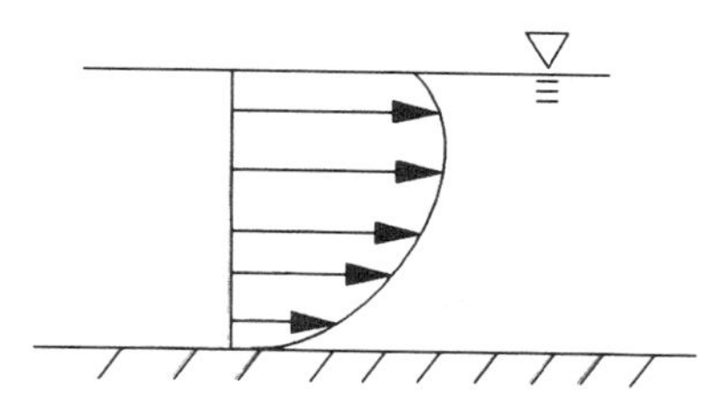

Figure 4.3 Typical velocity profile.

Theoretically, with a solid boundary, the flow velocity at the boundary is zero. Actually for natural channels it is difficult to determine precisely where the channel boundary is. The important point is that particles along the channel boundary are subjected to an actual velocity that is considerably lower than the average flow velocity of the cross-section.

Energy

In basic fluid mechanics, the energy equation is generally written in the form of Eq. (4.4). This relationship is known as Bernoulli's equation or Bernoulli's theorem:

$$\frac{v_1^2}{2g} + y_1 + z_1 + \frac{p_1}{\gamma} = \frac{v_2^2}{2g} + y_2 + z_2 + \frac{p_2}{\gamma} + h_{L,1-2}. \tag{4.4}$$

The terms in this equation are shown in Fig. 4.4. The Bernoulli equation represents an energy balance between two points along the channel. Again, v is the average flow velocity, g is the gravitational constant, y is the depth of flow, z is the elevation of the channel bottom, p is a pressure, γ is the unit weight of water, and $h_{L,1-2}$ represents the energy loss between sections 1 and 2.

Each complete term of Eq. (4.4) has the units of a length. Since the equation is an energy equation, one should consider that the terms represent energy per unit of flowing fluid. Since the units are a length, the terms are commonly associated with a "head" because

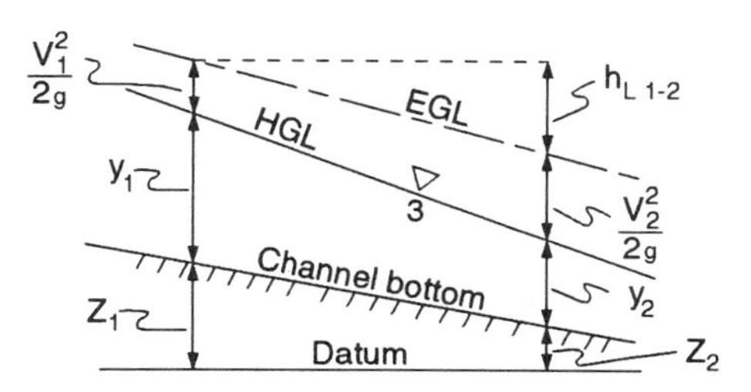

Figure 4.4 Terms in Bernoulli equation.

of the engineer's familiarity with pressure and pressure heads.

Thus, $v^2/2g$ is termed the velocity head, $y + z$ is termed the elevation head, and p/γ is known as the pressure head. Since the terms represent energy per unit of fluid, we can in a loose sense think of $v^2/2g$ as representing kinetic energy, $y + z$ as representing potential energy, and p/γ as representing stored energy.

The sum of the velocity head, elevation head, and pressure head represents the total energy. The line labeled EGL in Fig. 4.4 represents this sum and is known as the energy grade line. The sum of the elevation head and pressure head is known as the hydraulic grade line (HGL). The factor that distinguishes open channel flow from pipe flow is that in open channel flow, the free water surface is exposed to the atmosphere so that p/γ is zero. Thus, the pressure head term can generally be ignored for open channel problems, and hence, the HGL coincides with the water surface. A rather obvious fact is that the EGL must be sloping downward in the direction of flow. The EGL can only go up if external energy (through a pump for example) is supplied to the flow.

If we consider a channel section in which there is no energy loss, we can write

$$v^2/2g + y + z = \text{constant}. \tag{4.5}$$

If we take the datum elevation to be the channel bottom, we have

$$v^2/2g + y = \text{constant} = E, \tag{4.6}$$

where the constant E is known as the specific energy.

Consider now a wide rectangular channel so that the depth all across the channel cross section is y. We can then relate the flow rate on a per unit of width basis and the average flow velocity by

$$q = vy, \tag{4.7}$$

where q is the flow rate per unit of width.

Equation (4.6) can now be written as

$$q^2/2gy^2 + y = E \tag{4.8}$$

and a plot of y versus E constructed for a constant q. Figure 4.5 is such a plot and is known as a specific energy diagram. Some characteristics of the specific energy are that for a given E there are two possible depths of flow, y_1 and y_2, known as alternate depths, and there is a definite minimum E for a given q. The depth of flow corresponding to the minimum E is known as the critical depth and is denoted by y_c. The relationship between the flow rate and y_c can be determined by differentiating Eq. (4.8) and setting the

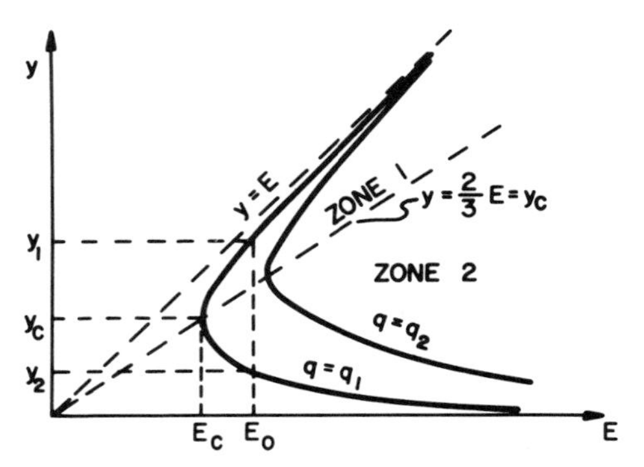

Figure 4.5 Specific energy diagrams.

differential to zero:

$$dE/dy = -2q^2/2gy^3 + 1 = 0$$

or

$$y_c = \sqrt[3]{q^2/g}. \tag{4.9}$$

Since $q = vy_c$, we can write Eq. (4.9) as

$$v/\sqrt{gy_c} = 1. \tag{4.10}$$

The term $v/\sqrt{gy_c}$ is known as the Froude number $\mathbf{F}$. Equation (4.10) shows that when $y = y_c$ or when the flow is at the critical depth, the Froude number is one. The Froude number can be used to classify the flow into subcritical, critical, and supercritical flow. When $\mathbf{F} < 1$, the flow is subcritical and $y > y_c$. This corresponds to zone 1 in Fig. 4.5. When $\mathbf{F} > 1$, the flow is supercritical and $y < y_c$. Supercritical flow is zone 2 in Fig. 4.5. $\mathbf{F} = 1$ is known as critical flow and corresponds to the line $y = y_c = 2E/3$ in Fig. 4.5. Equation (4.10) shows that for critical flow, $v_c = \sqrt{gy_c}$. This velocity corresponds to the celerity of small gravity waves in shallow water.

For nonrectangular channels, the Froude number is defined as

$$\mathbf{F} = v/\sqrt{gd_h}, \tag{4.11}$$

where d_h is the hydraulic depth. The hydraulic depth is defined as the area divided by the top width

$$d_h = A/t. \tag{4.12}$$

Since $\mathbf{F}$ is independent of slope, y_c depends only on the discharge for a given channel. For a rectangular channel, this is apparent from Eq. (4.9). In general, the relationship between Q and y_c can be determined from Eq. (4.11) for any channel by setting $\mathbf{F} = 1$ and noting from Eq. (4.3) that $v = Q/A$.

Example Problem 4.1 Critical depth

A triangular channel with side slopes of 3 : 1 is carrying 20 cfs. What is the critical depth for this channel and flow rate?

Solution: Critical depth occurs when $\mathbf{F} = 1$. Equations (4.11) and (4.12) must be used. Note that a triangular channel is a special case of a trapezoidal channel with $b = 0$. The area and top width are given by (see Fig. 4.9)

$$A = zd^2 = 3d^2$$
$$t = 2dz = 6d.$$

Therefore

$$d_h = A/t = 3d^2/6d = 0.5d.$$

From Eq. (4.11),

$$1 = \frac{v}{\sqrt{gd_h}} = \frac{Q/A}{\sqrt{0.5gd}}$$
$$1 = \frac{20/3d^2}{\sqrt{16.1d}} = \frac{1.66}{d^{5/2}}$$
$$d_c = 1.23 \text{ ft}.$$

As shown in subsequent sections of this chapter, channel roughness, velocity, discharge, and slope are interrelated. For a given discharge and roughness, the velocity can be increased and consequently, the depth of flow decreased by increasing the channel slope. When the channel slope is such that the flow depth resulting in uniform flow equals critical depth, the slope is called the critical slope, S_c. Thus for subcritical flow, the slope is less than S_c and for supercritical flow, the slope is greater than S_c. It should be pointed out that critical depth, slope, and velocity for a given section change with the discharge.

In designing channels for controlling and conveying runoff, it is generally desirable to design so that the flow is subcritical. Supercritical flow presents special problems that are not treated here.

Momentum

The momentum principle in open channel flow can be visualized by considering Fig. 4.6 and the basic relationship from mechanics

$$\Sigma F_s = \Delta(mv_s), \tag{4.13}$$

which states that the sum of the forces in the s-direction equals the change in momentum in that direction. In Eq. (4.13), F_s represents forces in the s-direction and m represents the mass. For a constant mass and a

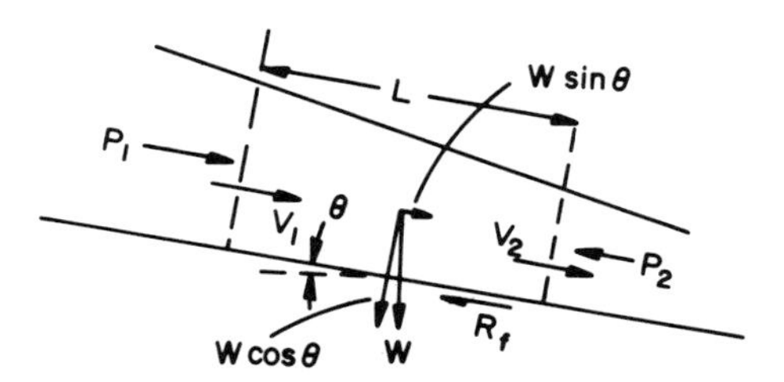

Figure 4.6 Sketch for momentum relationship.

per unit width consideration

$$\Delta(mv_s) = \rho q(v_2 - v_1).$$

The forces in the s-direction are

$$\Sigma F_s = P_1 + W \sin\theta - P_2 - R_f,$$

where P_1 and P_2 are pressure forces per unit width given by

$$P = \gamma y^2/2,$$

R_f is a frictional resistance, and $W \sin\theta$ is the s-direction component of the weight. Combining terms, we have

$$\frac{\gamma y_1^2}{2} - \frac{\gamma y_2^2}{2} + W\sin\theta - R_f = \rho q(v_2 - v_1). \tag{4.14}$$

If a short section is considered so that R_f is negligible and the channel slope is small so that $\sin\theta$ is near zero, Eq. (4.14) can be written as

$$\frac{\gamma y_1^2}{2} + \rho q v_1 = \frac{\gamma y_2^2}{2} + \rho q v_2$$

or

$$\frac{y_1^2}{2} + \frac{qv_1}{g} = \frac{y_2^2}{2} + \frac{qv_2}{g} = M, \tag{4.15}$$

where M is the specific force plus momentum and is a constant. Again it is possible to plot y versus M for a constant q in the form of a specific force plus momentum curve. Figure 4.7 is such a plot again showing two possible depths for a given M and a definite minimum

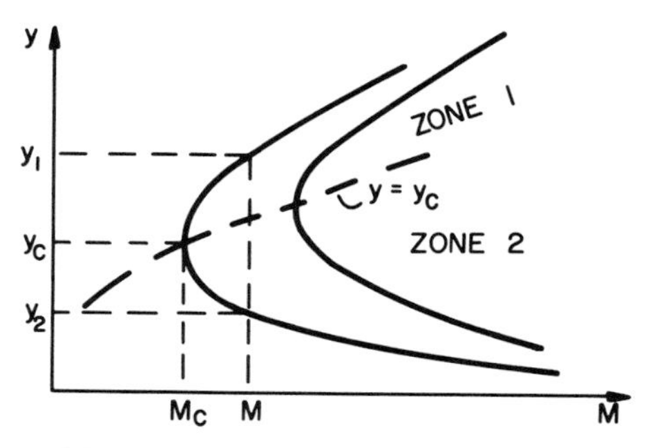

Figure 4.7 Typical specific force plus momentum curve.

M. The two possible depths for a given *M* are known as sequent depths. It can be shown that *y* corresponding to the minimum *M* is y_c. Again zone 1 represents subcritical flow and zone 2 supercritical flow.

UNIFORM FLOW

Open channel flow is generally classified with respect to changes in flow properties with time and with location along the channel. If the flow characteristics at a point are unchanging with time, the flow is said to be steady flow; otherwise the flow is unsteady. Similarly, if the flow properties are the same at every location along the channel, the flow is uniform. Flow with properties that change with channel location is nonuniform flow. In natural flow situations, the flow is generally nonsteady and nonuniform. However, in the design of most channels, steady, uniform flow is assumed with the channel design being based on some peak or maximum discharge.

When we speak of uniform flow, steady, uniform flow is generally what is considered. For uniform flow, y_1 and y_2 and v_1 and v_2 in Fig. 4.6 are equal. Thus, Eq. (4.14) reduces to

$$R_f = W \sin\theta \tag{4.16}$$

or the frictional forces are just equal to the downstream component of the weight. That is, the frictional resistance and gravitational forces are in equilibrium.

The frictional resistance to flow may be expressed as a shear, τ, per unit area times the resisting area. Neglecting the resistance generated at the surface of the flow between the water and air, the resisting area over which τ operates is the length, *L*, of a section times the wetted perimeter, *P*, of the channel. The wetted perimeter is simply the length of the boundary between the water and the channel sides and bottom at any cross section or the distance around the flow cross section starting at one edge of the channel and traveling along the sides and bottom of the channel to the other channel edge.

Thus R_f in Eq. (4.16) can be written as

$$R_f = \tau PL. \tag{4.17}$$

The weight of water in a section of the channel is simply

$$W = AL\gamma. \tag{4.18}$$

For small angles θ, $\sin\theta$ is about equal to the slope of the channel in feet per foot. Thus, Eq. (4.16) may be written as

$$\tau PL = AL\gamma S, \tag{4.19}$$

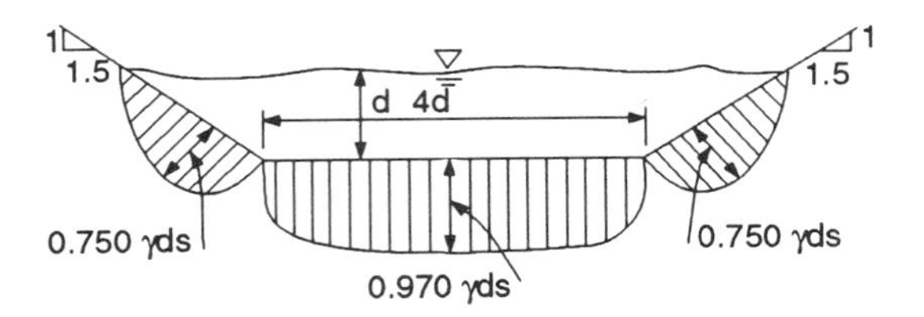

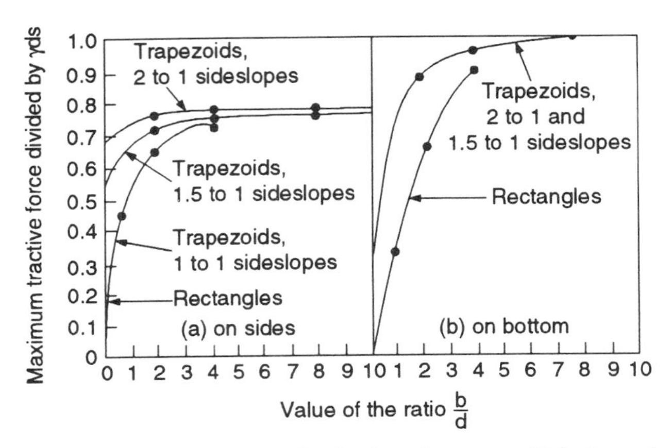

Figure 4.8 Tractive force distribution for trapezoidal channels (Lane and Carlson, 1953).

which upon rearrangement is

$$\tau = \gamma(A/P)S.$$

The term A/P represents the hydraulic radius, *R*, defined as the flow area divided by the wetted perimeter. Thus, we have

$$\tau = \gamma RS. \tag{4.20}$$

In this equation, τ represents the average shear around the periphery of the flow. At some points the actual shear will exceed τ and at other points it will be less than τ. Lane and Carlson (1953) found the shear on the periphery of a trapezoidal channel varied as shown in Fig. 4.8. The maximum shear is near γdS rather than γRS. In designing channels for stability using a critical tractive force approach as shown later, the maximum shear can be calculated as γdS.

Experimental studies on water flow in pipes has shown that τ is proportional to the Darcy–Weisbach friction factor, *f*, and the square of the flow velocity. That is

$$\tau = f\rho v^2/8 \tag{4.21}$$

or combining Eqs. (4.20) and (4.21),

$$v = \sqrt{8\gamma/f\rho}\,\sqrt{RS}\,.$$

By letting $\sqrt{8\gamma/f\rho} = C$, Chezy's equation for open channel flow is obtained as

$$v = C\sqrt{RS} = CR^{1/2}S^{1/2}, \tag{4.22}$$

where *C* is a factor related to the roughness of the channel.

An Irish engineer named Manning found that the equation

$$v = KR^{2/3}S^{1/2}$$

fit experimental data quite nicely. This equation is known as Manning's equation and differs from Chezy's equation only in the exponent on R. So that the factor related to the channel roughness would increase as roughness increased, Manning's equation is generally written as

$$v = (1/n)R^{2/3}S^{1/2}$$

in the metric system with v in meters per second and R in meters. The coefficient n is known as Manning's n. In the English system of units, Manning's equation is

$$v = \frac{1.49}{n}R^{2/3}S^{1/2}, \qquad (4.23)$$

where v is in fps, R is in feet, and S is in feet per foot. Tables of Manning's n are widely available. Table 4.1 is such a table taken from several sources, drawing heavily on Schwab *et al.* (1966, 1971). Manning's n is influenced by many factors, including the physical roughness of the channel surface, the irregularity of the channel cross section, channel alignment and bends, vegetation, silting and scouring, and obstruction within the channel. Chow (1959) displays some photographs of typical channels and the associated values for Manning's n.

Figure 4.9 contains some useful relationships for calculating the hydraulic properties of A, P, R, and top width, T, for three common channels. For natural channels, these properties are best determined from measurements based on the actual cross sections of the channel.

Table 4.1 Typical Values for Manning's n

Type and description of conduits	n Values[a]		
	Min.	Design	Max.
Channels, lined			
Asphaltic concrete, machine placed		0.014	
Asphalt, exposed prefabricated		0.015	
Concrete	0.012	0.015	0.018
Concrete, rubble	0.016		0.029
Metal, smooth (flumes)	0.011		0.015
Metal, corrugated	0.021	0.024	0.026
Plastic	0.012		0.014
Shotcrete	0.016		0.017
Wood, planed (flumes)	0.009	0.012	0.016
Wood, unplaned (flumes)	0.011	0.013	0.015
Channels, earth			
Earth bottom, rubble sides	0.028	0.032	0.035
Drainage ditches, large, no vegetation			
(a) < 2.5 hydraulic radius	0.040		0.045
(b) 2.5–4.0 hydraulic radius	0.035		0.040
(c) 4.0–5.0 hydraulic radius	0.030		0.035
(d) > 5.0 hydraulic radius	0.025		0.030
Small drainage ditches	0.035	0.040	0.040
Stony bed, weeds on bank	0.025	0.035	0.040
Straight and uniform	0.017	0.0225	0.025
Winding, sluggish	0.0225	0.025	0.030
Channels, vegetated			
(See subsequent discussion)			

Type and description of conduits	n Values[a]		
	Min.	Design	Max.
Natural Streams			
(a) Clean, straight bank, full stage, no rifts or deep pools	0.025		0.033
(b) Same as (a) but some weeds and stones	0.030		0.040
(c) Winding, some pools and shoals, clean	0.035		0.050
(d) Same as (c), lower stages, more ineffective slopes and sections	0.040		0.055
(e) Same as (c), some weeds and stones	0.033		0.045
(f) Same as (d), stony sections	0.045		0.060
(g) Sluggish river reaches, rather weedy or with very deep pools	0.050		0.080
(h) Very weedy reaches	0.075		0.150
Pipe			
Asbestos cement		0.009	
Cast iron, coated	0.011	0.013	0.014
Cast iron, uncoated	0.012		0.015
Clay or concrete drain tile (4–12 in.)	0.010	0.0108	0.020
Concrete	0.010	0.014	0.017
Metal, corrugated	0.021	0.025	0.0255
Steel, riveted and spiral	0.013	0.016	0.017
Vitrified sewer pipe	0.010	0.014	0.017
Wood stave	0.010	0.013	
Wrought iron, black	0.012		0.015
Wrought iron, galvanized	0.013	0.016	0.017

[a]Selected from numerous sources.

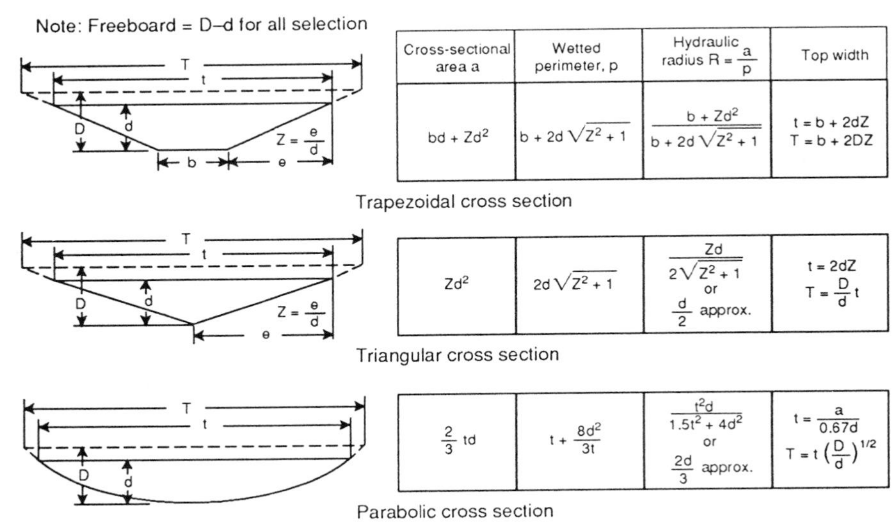

	Cross-sectional area a	Wetted perimeter, p	Hydraulic radius R = a/p	Top width
Trapezoidal cross section	$bd + Zd^2$	$b + 2d\sqrt{Z^2+1}$	$\frac{b + Zd^2}{b + 2d\sqrt{Z^2+1}}$	$t = b + 2dZ$ $T = b + 2DZ$
Triangular cross section	Zd^2	$2d\sqrt{Z^2+1}$	$\frac{Zd}{2\sqrt{Z^2+1}}$ or $\frac{d}{2}$ approx.	$t = 2dZ$ $T = \frac{D}{d}t$
Parabolic cross section	$\frac{2}{3}td$	$t + \frac{8d^2}{3t}$	$\frac{t^2d}{1.5t^2 + 4d^2}$ or $\frac{2d}{3}$ approx.	$t = \frac{a}{0.67d}$ $T = t\left(\frac{D}{d}\right)^{1/2}$

Figure 4.9 Properties of typical channels.

The expression for the hydraulic radius for wide shallow channels can be simplified from that shown in Fig. 4.9. Consider the trapezoidal channel shown in Fig. 4.10. If the trapezoid is approximated by a rectangle, one can write

$$R = \frac{A}{P} = \frac{bd}{b + 2d}.$$

If $b \gg d$, then the $2d$ in the denominator can be ignored leaving

$$R \approx bd/b = d.$$

For a parabolic channel, if $t \gg d$, then $4d^2$ in the denominator of the expression for R can be ignored leaving

$$R \approx \frac{t^2d}{1.5t^2} = \frac{2}{3}d.$$

These approximations can serve as initial estimates for d in trial and error solutions that often arise in open channel hydraulics.

The hydraulic elements of a circular conduit of diameter D can be calculated from

$$A = \frac{D^2}{8}(\theta - \sin\theta) \qquad (4.24)$$

$$R = \frac{D}{4}\left(1 - \frac{\sin\theta}{\theta}\right). \qquad (4.25)$$

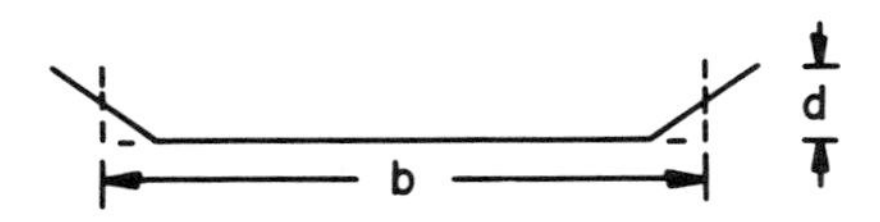

Figure 4.10 Approximation of trapezoidal channel with rectangular channel.

The angle θ is defined in Fig. 4.11 and measured in radians. Example Problems 4.2, 4.3, and 4.4 illustrate the use of Eqs. (4.24) and (4.25) to solve open channel flow problems dealing with circular conduits.

The maximum flow capacity of a circular conduit actually occurs at a depth equal to $0.938D$. Figure 4.12 shows how the hydraulic elements of a circular conduit change with depth. The subscript 0 refers to a depth

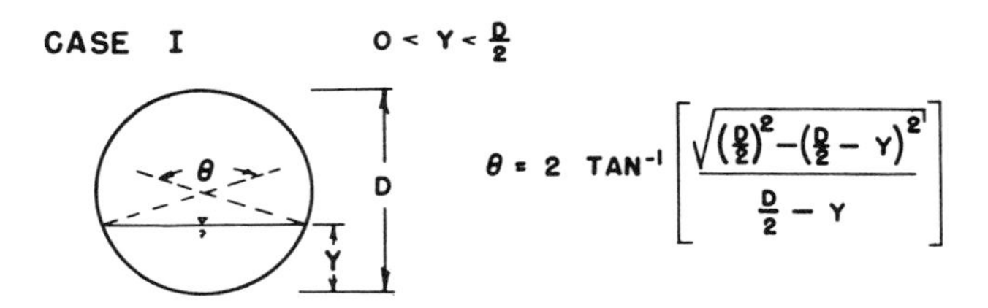

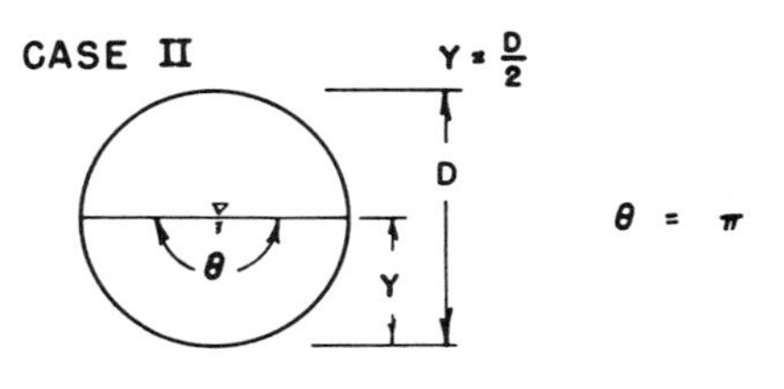

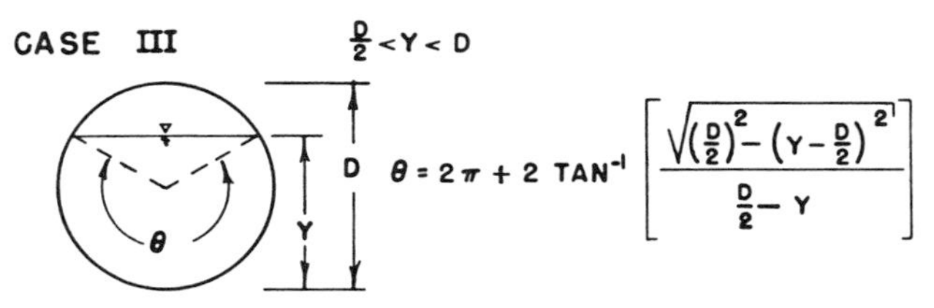

Figure 4.11 Definition of θ.

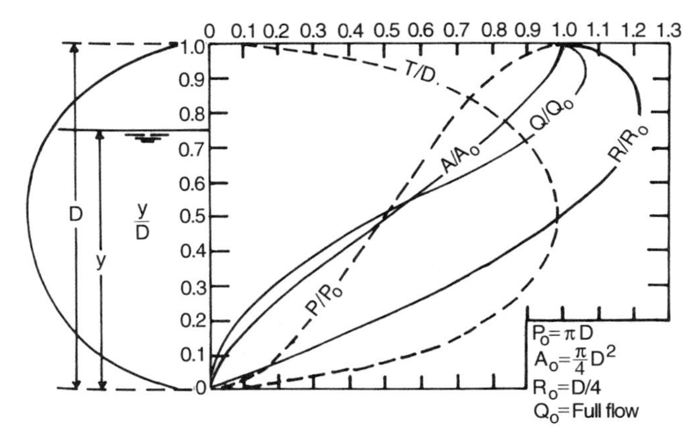

Figure 4.12 Hydraulic properties of a circular conduit.

equal to D. The line labeled Q/Q_0 assumes that n is constant with depth. Even though the maximum flow occurs at $0.938D$, it is common to design circular conduits to carry maximum flows when flowing full. Wave action and irregularities make it difficult to maintain flow at $0.938D$.

Example Problem 4.2 Flow in circular pipe 1

A circular corrugated metal pipe (CMP) that is 3 ft in diameter is flowing 1 ft deep. What is the discharge if the slope of the pipe is 4%?

Solution: Refer to Fig. 4.11 with the pipe radius, r, equal to $D/2$. Since $y < D/2$,

$$\theta = 2\tan^{-1}\left[\frac{\left[r^2 - (r-y)^2\right]^{1/2}}{r-y}\right]$$

$$= 2\tan^{-1}\left[\frac{\left[1.5^2 - (1.5-1.0)^2\right]^{1/2}}{1.5-1.0}\right]$$

$$= 2\tan^{-1}(2.828) = 2.46.$$

From Eqs. (4.24) and (4.25),

$$A = \frac{D^2}{8}(\theta - \sin\theta) = \frac{9}{8}(2.46 - \sin 2.46) = 2.06$$

$$R = \frac{D}{4}\left(1 - \frac{\sin\theta}{\theta}\right) = \frac{3}{4}\left(1 - \frac{\sin 2.46}{2.46}\right) = 0.56$$

$$Q = \frac{1.49}{n}R^{2/3}S^{1/2}A.$$

From Table 4.1, $n = 0.024$,

$$Q = \frac{1.49}{0.024}(0.56)^{2/3}(0.04)^{1/2}(2.06) = 17.4 \text{ cfs}.$$

Example Problem 4.3 Flow in circular pipe 2

A circular corrugated metal pipe that is 3 ft in diameter is flowing 2 ft deep. What is the discharge if the slope of the pipe is 4%?

Solution: Refer to Fig. 4.11. Since $y > D/2$,

$$\theta = 2\pi + 2\tan^{-1}\left[\frac{\left[r^2 - (y-r)^2\right]^{1/2}}{r-y}\right]$$

$$= 6.28 + 2\tan^{-1}\left[\frac{\left[1.5^2 - (2.0-1.5)^2\right]^{1/2}}{1.5-2}\right] = 3.81.$$

From Eqs. (4.24) and (4.25),

$$A = \frac{D^2}{8}(\theta - \sin\theta) = 5.00$$

$$R = \frac{D}{4}\left(1 - \frac{\sin\theta}{\theta}\right) = 0.87$$

$$Q = \frac{1.49}{n}R^{2/3}S^{1/2}A.$$

From Table 4.1, $n = 0.024$,

$$Q = \frac{1.49}{0.024}(0.87)^{2/3}(0.04)^{1/2}(5.00) = 56.6 \text{ cfs}.$$

Example Problem 4.4 Flow in circular pipe 3

A circular corrugated metal pipe that is 3 ft in diameter is carrying 30 cfs. How deep is the water flowing if the slope of the pipe is 4%?

Solution:

$$Q = \frac{1.49}{n}R^{2/3}S^{1/2}A$$

$$30 = \frac{1.49}{n}\left[\frac{D}{4}\left(1 - \frac{\sin\theta}{\theta}\right)\right]^{2/3} S^{1/2}\,\frac{D^2}{8}(\theta - \sin\theta).$$

After substituting $D = 3$, $n = 0.024$, and $S = 0.04$, this equation can be rearranged as

$$2.604 = \left(1 - \frac{\sin\theta}{\theta}\right)^{2/3}(\theta - \sin\theta).$$

This relationship can be solved by trial by assuming values for θ, comparing the right-hand side of the equation to the left-hand side and continuing until a match is achieved.

Trial θ	Right-hand side	
3.14	3.14	
2.50	1.58	
2.90	2.51	
2.94	2.61	OK

$\theta = 2.94$ is a solution. Since $\theta < \pi$, y must be less than r and can be obtained from

$$\theta = 2\tan^{-1}\left[\frac{\left[r^2 - (r-y)^2\right]^{1/2}}{r-y}\right]$$

$$\tan\left(\frac{\theta}{2}\right) = \frac{\left[r^2 - (r-y)^2\right]^{1/2}}{r-y}$$

$$\tan^2\left(\frac{\theta}{2}\right) = \frac{\left[r^2 - (r-y)^2\right]}{(r-y)^2}$$

$$\tan^2\left(\frac{2.94}{2}\right) = \frac{\left[2.25 - (1.5-y)^2\right]}{(1.5-y)^2}.$$

When this equation is solved for y, the result is $y = 1.35$ ft.

Example Problem 4.5 Flow in circular pipe 4

Use Fig. 4.12 to solve Example Problems 4.2, 4.3, and 4.4.

Solution:

$$Q_0 = \frac{1.49}{n} R_0^{2/3} S^{1/2} A_0.$$

$R_0 = D/4$ and $A_0 = \pi D^2/4$; therefore

$$Q_0 = \frac{1.49}{0.024}\left(\frac{3}{4}\right)^{2/3}(0.04)^{1/2}\frac{\pi 3^2}{4} = 72.4 \text{ cfs}.$$

When $y = 1$, $y/D = 0.33$. From Fig. 4.12, $Q/Q_0 = 0.23$. Therefore $Q = 0.23(72.4) = 16.7$ cfs. When $y = 2$, $y/D = 0.67$. From Fig. 4.12, $Q/Q_0 = 0.78$. Therefore $Q = 0.78(72.4) = 56.5$ cfs. When $Q = 30$, $Q/Q_0 = 0.41$. From Fig. 4.12, $y/D = 0.44$. Therefore $y = 0.44(3) = 1.32$ ft.

Natural channels often have a main channel section and an overbank section. Most flow occurs in the main channel; however, during flood events overbank flows may occur. The usual procedure for calculating such flows is to break the channel into cross-sectional parts and sum the flow calculated for the various parts. In determining the hydraulic radius for the various parts, only that part of the wetted perimeter in contact with an actual channel boundary is used. Thus

$$V_i = \frac{1.49}{n_i} S^{1/2}\left(\frac{A_i}{P_i}\right)^{2/3} \tag{4.26}$$

and

$$Q = \sum_{i=1}^{n} V_i A_i.$$

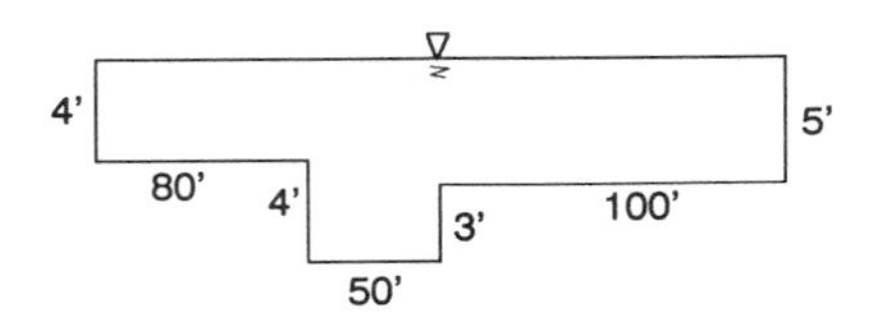

Figure 4.13 Channel section for Example Problem 4.6.

Example Problem 4.6 Compound channel

For the channel shown in Fig. 4.13, estimate the total flow for a depth of 8 ft. The channel has a slope of 0.5%. Manning's n is 0.06 for the overbank area and 0.03 for the main channel.

Solution: Use Eq. (4.26).

$A_1 = 80 \times 4 = 320$, $A_2 = 50 \times 8 = 400$,
$A_3 = 100 \times 5 = 500$
$P_1 = 80 + 4 = 84$, $P_2 = 4 + 50 + 3 = 57$,
$P_3 = 100 + 5 = 105$

$$Q = 1.49(0.005)^{1/2}\left[\frac{(320/84)^{2/3}320}{0.06} + \frac{(400/57)^{2/3}400}{0.03} + \frac{(500/105)^{2/3}500}{0.06}\right]$$

$$= 9010 \text{ cfs}.$$

DESIGN OF OPEN CHANNELS

Nonerodible Channels

The design of nonerodible open channels can be done by using Manning's equation [Eq. (4.23)]. Manning's n should be chosen carefully. Adequate consideration should be given to adding a freeboard or extra depth to the channel as a safety measure to protect against underestimates of flow or roughness and wave action. Generally a freeboard of around 20% of the depth or 0.3 to 0.5 ft, whichever is greater, should be added to the channel depth. Thus, the major consideration in the design of channels in nonerodible material is to ensure adequate capacity.

Example Problem 4.7 Flow rate concrete channel 1

Consider a concrete channel that is trapezoidal with 3 : 1 side slopes and a 6-ft bottom width. The channel is on a 0.5% slope and is flowing at a depth of 5 ft. What is the flow rate?

Solution: $S = 0.005$ and Table 4.1 gives $n = 0.015$. From Fig. 4.9,

$$R = \frac{bd + zd^2}{b + 2d\sqrt{z^2 + 1}} = \frac{6 \times 5 + 3 \times 5^2}{6 + 2 \times 5\sqrt{9 + 1}} = 2.79 \text{ ft}$$

$$v = \frac{1.49}{n} R^{2/3} S^{1/2} = \frac{1.49}{0.015} (2.79)^{2/3} (0.005)^{1/2} = 13.9 \text{ fps}$$

$$A = bd + zd^2 = 6 \times 5 + 3 \times 5^2 = 105 \text{ ft}^2$$

$$Q = vA = 13.9 \times 105 = 1459.5 \text{ cfs.}$$

Example Problem 4.8 Flow depth concrete channel 2

The channel of Example Problem 4.7 is carrying 75 cfs. How deep is the water flowing?

Solution:

$$Q = vA = \frac{1.49}{n} R^{2/3} S^{1/2} A$$

$$= \frac{1.49}{n} \left[\frac{bd + zd^2}{b + 2d\sqrt{z^2 + 1}} \right]^{2/3} S^{1/2} (bd + zd^2)$$

$$75 = \frac{1.49}{0.015} \left[\frac{6d + 3d^2}{6 + 6.32d} \right]^{2/3} (0.005)^{1/2} (6d + 3d^2)$$

$$10.68 = \left[\frac{6d + 3d^2}{6 + 6.32d} \right]^{2/3} (6d + 3d^2).$$

This last relationship must be solved by trial for a d such that the right-hand side of the equation is equal to 10.68.

Trial d	Right-hand side	
1	7.30	
1.5	15.93	
1.2	10.32	
1.22	10.65	OK

The channel is flowing 1.22 ft deep.

Example Problem 4.9 Froude number

Calculate the Froude number of the flow in example problem 4.7.

Solution:

$$\mathbf{F} = \frac{v}{(gd_h)^{1/2}}.$$

Example Problem 4.7 gives $A = 105$ ft^2; therefore

$$d_h = \frac{A}{t} = \frac{A}{b + 2dz} = \frac{105}{6 + 2 \times 5 \times 3} = 2.92$$

$$\mathbf{F} = \frac{13.9}{(32.2 \times 2.92)^{1/2}} = 1.43.$$

Thus the flow is supercritical. The high flow velocity is an early indicator of the possibility of supercritical flow.

Erodible Channels

In designing channels to be constructed in erodible materials there are two major considerations. The channel must have adequate capacity to carry the flow and it must have adequate stability to resist the erosive action of the flowing water. Erodible channels may be either vegetated or nonvegetated. Vegetation tends to protect the channel from erosion, thus permitting higher flow velocities. On the other hand, vegetation increases the roughness of the channel. The design of nonvegetated channels is considered next followed by the design of vegetated channels. Flexible linings and riprap linings are discussed in subsequent sections.

Nonvegetated Channels

Two main design procedures are used for ensuring the stability of erodible channels. One procedure is based on a limiting velocity concept and the other on a limiting tractive force (boundary shear) concept. Table 4.2 shows allowable velocities and tractive force values for several kinds of channels. This table is taken from Lane (1955) based on the work by Fortier and Scobey (1926). The values are for aged, stable channels. For newly constructed channels, the values shown in Table 4.6 should be used.

When using the limiting velocity concept, one simply sizes the channel so that it has adequate capacity and so that the average velocity does not exceed the permissible velocity.

When using the limiting tractive force concept, a channel with adequate capacity and having an average shear stress given by Eq. (4.20) that is less than the values tabulated in Table 4.2 is sought. For channels in noncohesive materials, the particles on the channel sides may move due to the combined force exerted by the flowing water and the weight component of the particles down the side of the channel. Chow (1959) should be referred to for a treatment of tractive force considerations in noncohesive materials. In cohesive materials, the cohesion generally is much greater than the gravity component so that average shear based on Eq. (4.20) can be used in design.

An alternative approach to designing stable, unlined channels is to use regime relationships. These relationships define equilibrium conditions between flow and the channel boundaries. Chapter 10 discusses this approach.

Table 4.2 Limiting Velocities and Tractive Forces for Open Channels (Straight after Aging)[a]

		For Clear Water		Water transporting colloidal silts	
Material	n	Velocity (fps)	Tractive force (psf)	Velocity (fps)	Tractive force (psf)
Fine sand colloidal	0.020	1.50	0.027	2.50	0.075
Sandy loam noncolloidal	0.020	1.75	0.037	2.50	0.075
Silt loam noncolloidal	0.020	2.00	0.048	3.00	0.110
Alluvial silts noncolloidal	0.020	2.00	0.048	3.50	0.150
Ordinary firm loam	0.020	2.50	0.075	3.50	0.150
Volcanic ash	0.020	2.50	0.075	3.50	0.150
Stiff clay very colloidal	0.025	3.75	0.260	5.00	0.460
Alluvial silts colloidal	0.025	3.75	0.260	5.00	0.460
Shales and hardpans	0.025	6.00	0.670	6.00	0.670
Fine gravel	0.020	2.50	0.075	5.00	0.320
Graded loam to cobbles when noncolloidal	0.030	3.75	0.380	5.00	0.660
Graded silts to cobbles when collodial	0.030	4.00	0.430	5.50	0.800
Coarse gravel noncolloidal	0.025	4.00	0.300	6.00	0.670
Cobbles and shingles	0.035	5.00	0.910	5.50	1.100

[a]From Lane (1955).

Example Problem 4.10 Erodible channel design

Design a channel to carry 20 cfs down a 0.5% slope. The channel material is to be an ordinary firm loan. The water will be transporting colloidal silts. The channel is to be trapezoidal with 3 : 1 side slopes. Use (a) the limiting velocity approach and (b) the limiting tractive force approach.

Solution:

(a) Limiting velocity approach. From Table 4.2, $v_p = 3.5$ fps, $n = 0.020$,

$$v_p = \frac{1.49}{n} R^{2/3} S^{1/2}$$

$$R = \left[\frac{v_p n}{1.49 S^{1/2}}\right]^{3/2} = \left[\frac{3.5(0.020)}{1.49(0.005)^{1/2}}\right]^{3/2} = 0.54$$

$$A = \frac{Q}{v_p} = \frac{20.00}{3.5} = 5.71$$

$$R = \frac{bd + zd^2}{b + 2d\sqrt{z^2 + 1}} = \frac{bd + 3d^2}{b + 6.32d} = 0.54 \qquad \text{(a)}$$

$$A = bd + zd^2 = bd + 3d^2 = 5.71. \qquad \text{(b)}$$

Substituting Eq. (b) into Eq. (a) yields

$$\frac{5.71}{b + 6.32d} = 0.54$$

or

$$b = 10.58 - 6.32d. \qquad \text{(c)}$$

Substituting this into Eq. (b) yields

$$(10.58 - 6.32d)d + 3d^2 = 5.71$$

$$-3.32d^2 + 10.58d - 5.71 = 0.00.$$

This is a quadratic equation of the form

$$ax^2 + bx + c = 0,$$

which has as a solution

$$x = \frac{-b \pm \sqrt{b^2 - 4ac}}{2a}.$$

Therefore

$$d = \frac{-10.58 \pm \sqrt{10.58^2 - 4(-3.32)(-5.71)}}{2(-3.32)}$$

$$d = \frac{-10.58 + 6.00}{-6.64} = 2.50; 0.69.$$

If $d = 2.50$, then from Eq. (c) we get

$$b = 10.58 - 6.32(2.50) = -5.22,$$

which is clearly not possible. If $d = 0.69$, we get

$$b = 10.58 - 6.32(0.69) = 6.22.$$

Therefore the channel dimensions must be

$$b = 6.22 \text{ ft}, \quad d = 0.69 \text{ ft}, \quad z = 3.0.$$

Check:

$$R = \frac{bd + zd^2}{b + 2d\sqrt{z^2 + 1}} = \frac{6.22(0.69) + 3(0.69)^2}{6.22 + 2(0.69)\sqrt{10}} = 0.54$$

$$v = \frac{1.49}{n} R^{2/3} S^{1/2} = \frac{1.49}{0.02}(0.54)^{2/3}(0.005)^{1/2} = 3.5.$$

The velocity is OK.

$$A = bd + zd^2 = 6.22(0.69) + 3(0.69)^2 = 5.72$$

$$Q = vA = 3.50(5.72) = 20.00.$$

The capacity is OK.

Add 0.3 ft of freeboard to get the final design of $b = 6.2$ ft and $d = 1.0$ ft.

(b) Critical tractive force approach. From Table 4.2, $\tau_c = 0.15$, $n = 0.020$. Figure 4.8 shows that for shallow and wide ($b/d > 8$) trapezoidal channels, the maximum bottom shear is γdS. Therefore

$$\tau_c = \gamma dS$$

$$d = \frac{\tau_c}{\gamma S} = \frac{0.15}{62.4(0.005)} = 0.48$$

$$Q = \frac{1.49}{n} R^{2/3} S^{1/2} A$$

$$R = \frac{bd + zd^2}{b + 2d\sqrt{z^2 + 1}} = \frac{0.48b + 0.69}{b + 3.03}$$

$$A = bd + zd^2 = 0.48b + 0.69$$

$$20 = \frac{1.49}{0.02}\left(\frac{0.48b + 0.69}{b + 3.03}\right)^{0.667}(0.005)^{1/2}(0.48b + 0.69)$$

$$3.797 = \left(\frac{0.48b + 0.69}{b + 3.03}\right)^{0.667}(0.48b + 0.69).$$

Solving by trial and error, b is found to be 12.4 ft. The b/d ratio is $12.4/0.48 = 26$. Thus the assumption that the maximum bottom shear is γdS is verified. If b/d had been less than 8, τ_c would have been $K\gamma dS$, where K would be approximated from Fig. 4.8.

Upon verifying that a channel with a bottom width of 12.4 ft, a depth of 0.48 ft, and 3 : 1 side slopes will have an allowable velocity and adequate capacity, a freeboard of 0.3 ft is added giving a final design of $b = 12.4$ ft, $d = 0.8$ ft, and $z = 3$.

Vegetated Channels

From the previous section it can be seen that the allowable velocities and tractive forces for nonvegetated, erodible channels are quite small, thus requiring wide shallow channels. Regime theory relationships in Chapter 10 also predict wide shallow channels for these conditions. If the channel can be protected from erosion, the allowable velocities can be increased, resulting in deeper and more narrow channels. An inexpensive and permanent form of protection is vegetation—specifically grasses. Vegetation protects the channel material from the erosive action of the flow and binds the channel material together.

Vegetated waterways generally can be used to carry intermittent flows such as storm water runoff. They are not recommended for channels having sustained base flows as most vegetation cannot survive continual submergence or continual saturation in the root zone. This means that vegetated waterways would not be used as the channel carrying the discharge from a pipe spillway in a detention basin, as this flow is likely to be a sustained one. A compound channel with a small, lined channel in the center to carry base flows and a vegetated portion to carry storm flows may be used in these situations.

Vegetated waterways are somewhat more complex to design and require more care in their establishment than nonvegetated waterways. They carry high flows at high velocities and require a minimum of maintenance if properly constructed.

The additional design consideration for vegetated waterways is the variation in roughness (Manning's n) with the height of the vegetation and with the type of vegetation. Typically a tall grass presents a great deal of flow resistance to shallow flow. As the flow depth increases, the resistance may decrease. Often the grass will lay over in the direction of flow when the flow reaches sufficient depth. With the grass in this condition, the resistance is considerably reduced as compared to the shallow flow situation.

Experimental work has shown that Manning's n can be related to the product of the flow velocity and the hydraulic radius, vR. This experimental work has also shown that different grasses have different n–vR relationships. As a matter of fact, the same grass may have a different n–vR relationship depending on the height of the grass.

Grasses have been divided into five retardance classes, designated by A, B, C, D, and E. Table 4.3 lists the retardance class for a number of grasses that are commonly used. If the grass will be mowed part of the time and long part of the time, both conditions and retardance classes must be considered. If a particular vegetation is not listed in Table 4.3, a similar vegetation might be used as a guide in selecting the retardance. In comparing vegetation, density, stem diameter, stiffness, and other physical characteristics should be considered. Information in Table 4.4 may be used to estimate the vegetal retardance if specific information on the type of vegetation is not known.

The maximum permissible velocities shown in Table 4.5 should be used for established sod in good condition. The soil erodibility factor discussed in Chapter 8 can be used to classify soils as erosion resistant or easily eroded (see pp. 126). Flow at these maximum velocities may require channel maintenance operations. If poor vegetation exists due to shade, climate, soils, or other factors, the design velocity should be about 50% of the values of Table 4.5. Data in Table 4.6 may be used to select permissible velocities when specific vegetation and erosion characteristics of soils are not known.

Figure 4.14 shows the n–vR relationship for the five retardance classes. The design procedure is to select

Table 4.3 Vegetal Retardance Classes (Soil Conservation Service, 1969)

Retardance	Cover	Condition
A		
	Reed canary grass	Excellent stand, tall (average 36 in.)
	Yellow bluestem Ischaemum	Excellent stand, tall (average 36 in.)
B		
	Smooth bromegrass	Good stand, mowed (average 12 to 15 in.)
	Bermuda grass	Good stand, tall (average 12 in.)
	Native grass mixture (little bluestem, blue grams, and other long and short midwest grasses)	Good stand, unmowed
	Tall fescue	Good stand, unmowed (average 18 in.)
	Lespedeza sericea	Good stand, not woody, tall (average 19 in.)
	Grass–legume mixture — Timothy, smooth bromegrass, or orchard grass	Good stand, uncut (average 20 in.)
	Reed canary grass	Good stand, mowed (average 12 to 15 in.)
	Tall fescue, with bird's foot trefoil or lodino	Good stand, uncut (average 18 in.)
	Blue grama	Good stand, uncut (average 13 in.)
C		
	Bahia	Good stand, uncut (6 to 8 in.)
	Bermuda grass	Good stand, mowed (average 6 in.)
	Redtop	Good stand, headed (15 to 20 in.)
	Grass–legume mixture — summer (Orchard grass, redtop, Italian ryegrass, and common lespedeza)	Good stand, uncut (6 to 8 in.)
	Centipedegrass	Very dense cover (average 6 in.)
	Kentucky bluegrass	Good stand, headed (6 to 12 in.)
D		
	Bermuda grass	Good stand, cut to 2.5 in. height
	Red fescue	Good stand, headed (12 to 18 in.)
	Buffalograss	Good stand, uncut (3 to 6 in.)
	Grass–legume mixture — fall, spring (Orchard grass, redtop, Italian ryegrass, and common lespedeza)	Good stand, uncut (4 to 5 in.)
	Lespedeza sericea	After cutting to 2 in. height, very good stand before cutting
E		
	Bermuda grass	Good stand, cut to 1.5 in. height
	Bermuda grass	Burned stubble

the vegetation, determine its retardance class and permissible velocity, and then design the channel based on the curves of Fig. 4.14. For situations where two retardance classes are applicable (for example mowed and unmowed grass), the channel should first be designed for stability based on the lower retardance and then additional depth added to the channel to accommodate the flow when the retardance increases. This procedure ensures a stable channel with adequate capacity regardless of the condition of the vegetation.

Temple *et al.* (1987) have developed the following approximation for the n–vR curves of Fig. 4.14,

$$n = \exp\left[I\left(0.01329 \ln(vR)^2 - 0.09543 \ln(vR) + 0.2971\right) - 4.16\right], \quad (4.27)$$

where the value of I is

Retardance	I
A	10.000
B	7.643
C	5.601
D	4.436
E	2.876

This relationship can be used in computer programs to make hydraulic computations for vegetated waterways. The relationships should not be used outside the range of the curves shown in Fig. 4.14.

The graphs of Fig. 4.15 are solutions to Manning's equation using the curves in Fig. 4.14. They can be used as a design aid for solving Manning's equation for all retardance classes.

Example Problem 4.11 Vegetated channel 1

Design a channel to carry 25 cfs on a 4% slope. Use a parabolic channel. The soil is easily eroded, and the grass may be mowed to 2.5 in. or it may be uncut.

Solution: Select Bermuda grass. Bermuda grass is in retardance B if unmowed and retardance D if mowed. The permissible velocity is selected from Table 4.5 as 6 fps. First design for the mowed condition

$$A = Q/v = 25/6 = 4.17 \text{ ft}^2.$$

Table 4.4 Guide to Selection of Vegetal Retardance[a]

Stand	Length of vegetation (in.)	Retardance class
Good	>30	A
	11–24	B
	6–10	C
	2–6	D
	<2	E
Fair	>30	B
	11–24	C
	6–10	D
	2–6	D
	<2	E

[a]Soil Conservation Service (1979) engineering field manual.

Table 4.5 Permissible velocities for Vegetated Channels (Ree, 1949)

	Permissible velocity (fps)					
	Erosion-resistant soils (% slope)			Easily eroded soils (% slope)		
Cover	0–5	5–10	Over 10	0–5	5–10	Over 10
Bermuda grass	8	7	6	6	5	4
Buffalo grass Kentucky bluegrass Smooth brome Blue grama Tall fescue	7	6	5	5	4	3
Lespedeza sericea Weeping lovegrass Kudzu Alfalfa Crabgrass	3.5	NR[a]	NR	2.5	NR	NR
Grass mixture	5	4	NR	4	3	NR
Annuals for temporary protection	3.5	NR	NR	2.5	NR	NR

[a]Not recommended.

Table 4.6 Permissible Velocities (fps)[a]

Soil texture	Bare channel	Retardance	Channel vegetation Poor	Fair	Good
Sand, silt	1.5	B	1.5	3	4
Sandy loam	1.5	C	1.5	2.5	3.5
Silty loam	1.5	D	1.5	2	3
Silty clay loam	2	B	2.5	4	5
Sandy clay loam	2	C	2.5	3.5	4.5
	2	D	2.5	3	4
Clay	2.5	B	3	5	6
	2.5	C	3	4.5	5.5
	2.5	D	3	4	5

[a]Soil Conservation (1979) engineering field manual.

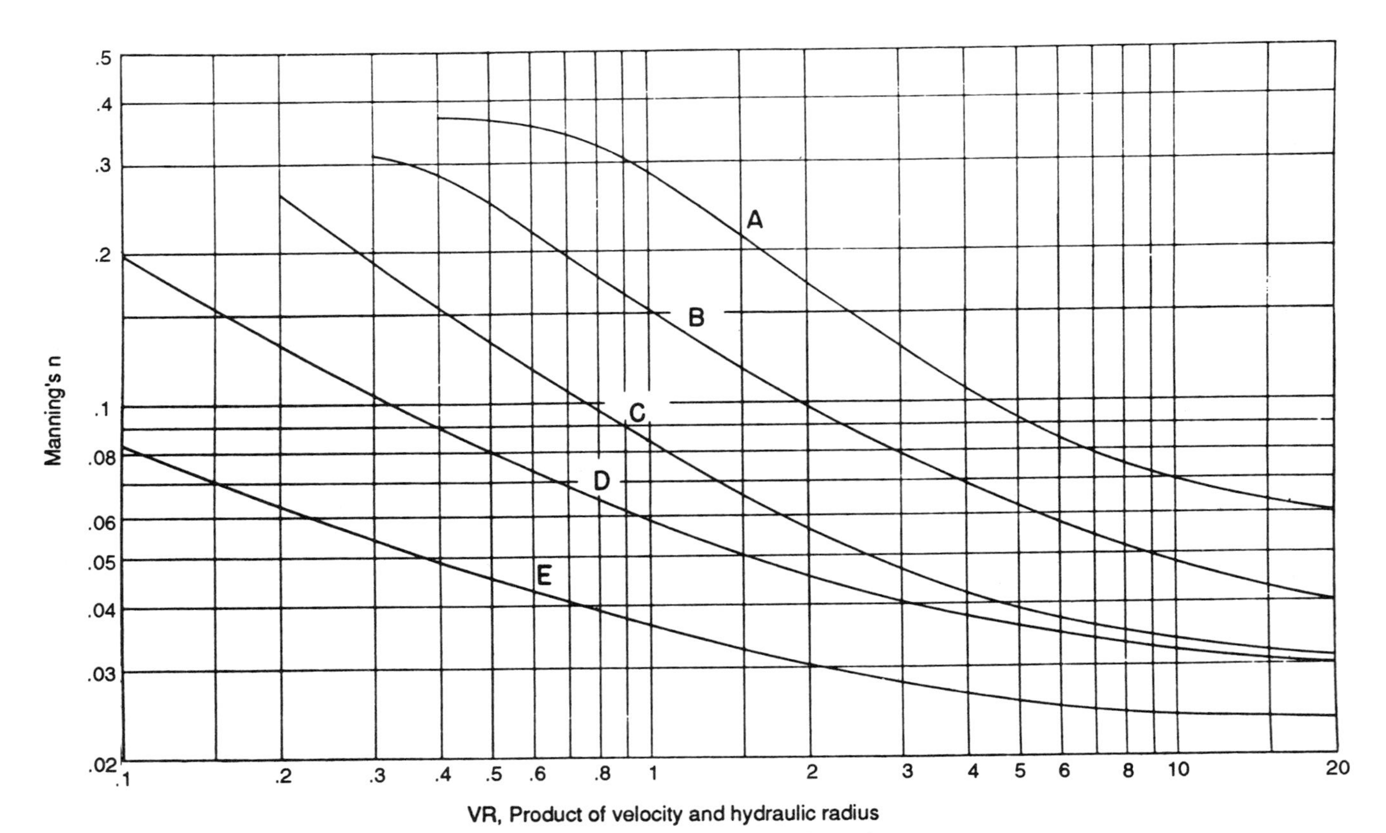

Figure 4.14 n–vR for various retardance classes.

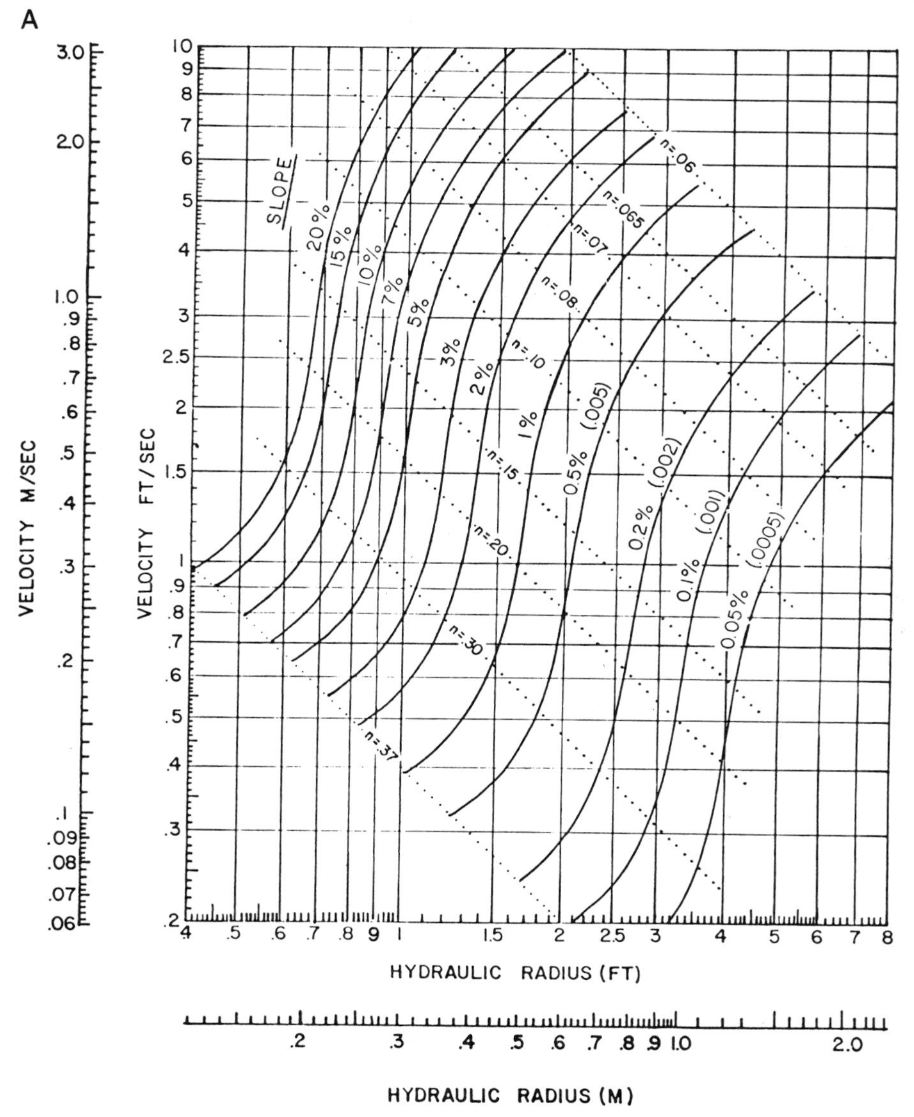

Figure 4.15a Solution to Manning's equation retardance class A.

From Fig. 4.15d for retardance class D, $R = 0.7$ ft.

$$R = 0.7 = \frac{t^2 d}{1.5t^2 + 4d^2}$$

(see Fig. 4.9)

$$A = 4.17 = \tfrac{2}{3}td.$$

For small parabolic channels, $d \approx 1.5R$. Using this approximation,

$$d = 1.05 \text{ ft}$$

$$t = \frac{3A}{2d} = \frac{3(4.17)}{2(1.05)} = 5.96 \text{ ft}.$$

Check:

$$R = \frac{(5.96)^2(1.05)}{1.5(5.96)^2 + 4(1.05)^2} = 0.646,$$

This is too small. Increase d to 1.25 feet, then

$$t = 3A/2d = 5.00 \quad \text{and} \quad R = 0.714.$$

Try $d = 1.17$ ft. Now $t = 3A/2d = 5.35$ and $R = 0.70$, which is OK.

The design for the short grass condition is

$$t = 5.35 \text{ ft}, \quad d = 1.17 \text{ ft}, \quad R = 0.7 \text{ ft}.$$

Now we must add depth using the same basic shape to get adequate capacity when the grass is long. When grass is long the retardance class is B. Try $D = 1.40$ ft. New top width

$$T = 5.35\left(\frac{1.40}{1.17}\right)^{1/2} = 5.85$$

and

$$R = \frac{t^2 d}{1.5t^2 + 4d^2} = 0.81.$$

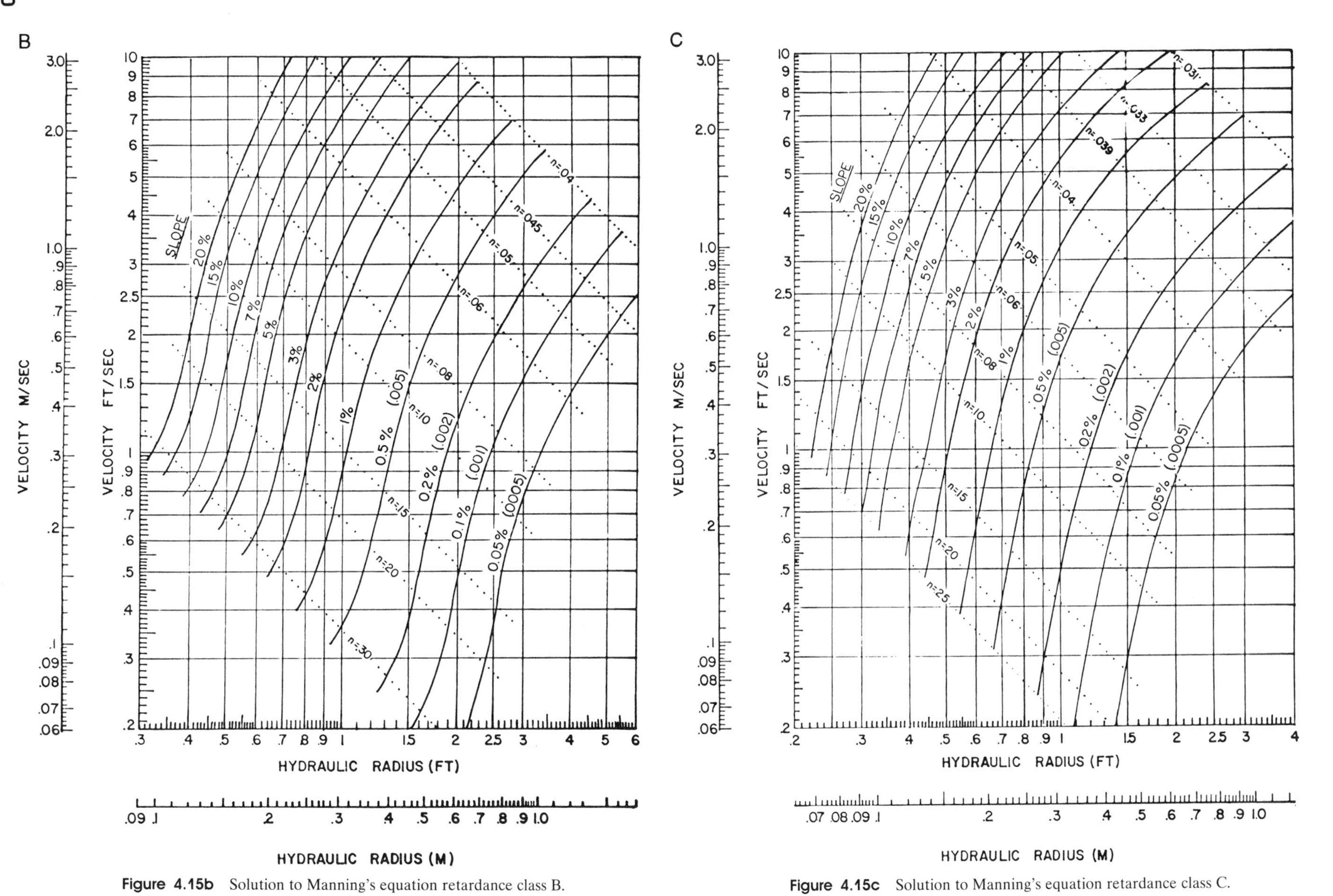

Figure 4.15b Solution to Manning's equation retardance class B.

Figure 4.15c Solution to Manning's equation retardance class C.

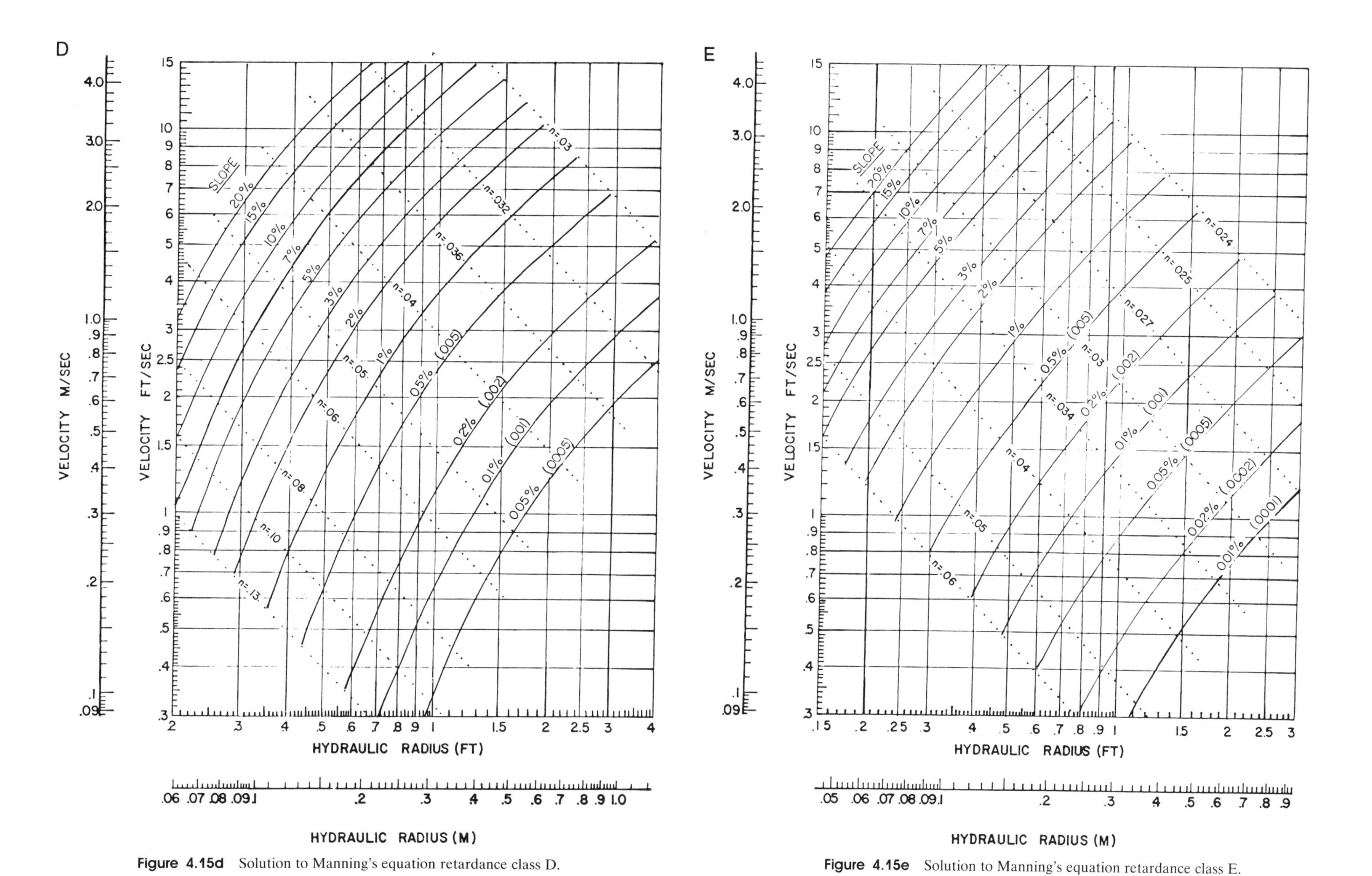

Figure 4.15d Solution to Manning's equation retardance class D.

Figure 4.15e Solution to Manning's equation retardance class E.

From Fig. 4.15b and retardance B with $R = 0.81$ and $S = 0.04$, find $v = 2.9$ fps, therefore:

$$A = 2td/3 = 5.46 \text{ ft}^2$$
$$Q = vA = 2.9 \times 5.46 = 15.8 \text{ cfs} \qquad \text{too small.}$$

Try $D = 1.75$ ft:

$$T = 5.35\left(\frac{1.75}{1.17}\right)^{1/2} = 6.54 \qquad \text{with} \quad R = 0.98$$
$$v = 4.5 \text{ ft}$$
$$A = 7.63 \text{ ft}^2$$
$$Q = vA = 35 \text{ cfs} \qquad \text{too big.}$$

Try $D = 1.6$ ft:

$$T = 5.35\left(\frac{1.6}{1.17}\right)^{1/2} = 6.26 \qquad \text{with} \quad R = 0.91$$
$$v = 3.9 \text{ fps}$$
$$A = 6.68 \text{ ft}^2$$
$$Q = 26 \text{ cfs} \qquad \text{OK.}$$

Add 0.3 freeboard to get a final design of $D = 1.9$ ft and

$$T = 5.35\left(\frac{1.9}{1.17}\right)^{1/2} = 6.8 \text{ ft.}$$

Example Problem 4.12 Vegetated channel 2

Work Example Problem 4.11 based on Eq. (4.27). Assume the grass is always mowed.

Solution:

$$v = \frac{1.49}{n}R^{2/3}S^{1/2} = \frac{1.49}{n}R^{2/3}(0.04)^{1/2} = \frac{0.298}{n}R^{2/3} = 6.$$

Assume R, compute vR, compute n, compute v, and if $v \neq 6$, repeat. For retardance D, $I = 4.436$. Assume $R = 0.8$, then $vR = 6(0.8) = 4.8$:

$$n = \exp\left[4.436\left(0.01329\ln(4.8)^2 - 0.09543\ln(4.8) + 0.2971\right) - 4.16\right] = 0.036$$

$$v = \frac{0.298}{0.036}(0.8)^{2/3} = 7.13 \qquad \text{too high.}$$

Assume $R = 0.7$, $vR = 4.2$:

$$n = \exp\left[4.436\left(0.01329\ln(4.2)^2 - 0.09543\ln(4.2) + 0.2971\right) - 4.16\right] = 0.038$$

$$v = \frac{0.298}{0.038}(0.7)^{2/3} = 6.18 \qquad \text{slightly too high.}$$

Assume $R = 0.67$, $vR = 4.02$:

$$n = \exp\left[4.436\left(0.01329\ln(4.02)^2 - 0.09543\ln(4.02) + 0.2971\right) - 4.16\right] = 0.038$$

$$v = \frac{0.298}{0.038}(0.67)^{2/3} = 6.00 \qquad \text{OK.}$$

From this point, the solution follows the procedure of Example Problem 4.11. Note the sensitivity of velocity to hydraulic radius in these calculations.

Flexible Liners

Normann (1975) presents a uniform procedure for the design of open channels using flexible liners. Liners considered are vegetation, temporary liners, and riprap. The procedure for vegetated liners is based on the procedures presented in the previous section of this book. The results for temporary liners are based on work of McWhorter *et al.* (1968), and the riprap results are largely based on Anderson *et al.* (undated) and Anderson (1973). Results are presented in the form of equations describing the maximum permissible depth of flow for a stable design

$$d_{\max} = mS^n, \tag{4.28}$$

where d is in feet and S is in feet per foot and a velocity equation of the form

$$v = aR^bS^c. \tag{4.29}$$

Vegetated Channels

Table 4.7 contains values for m and n for Eq. (4.28) for vegetated channels. Analysis of Normann's results and the results presented in the previous section of this book indicate that better agreement is obtained if $d_{\max}$ of Eq. (4.28) is replaced by the hydraulic radius, R. For example, for a grass mixture maintained at 6 to 8 in. on an erosion-resistant soil, the maximum hydraulic radius is given by

$$R = 0.12S^{-0.53}. \tag{4.30}$$

For vegetation, the velocity is determined from Figs. 4.15a–4.15e.

Example Problem 4.13 Flexible liner

Work Example Problem 4.11 using the Normann procedure.

Solution: For Bermuda grass in retardance D with an erodible soil, values of m and n are determined as $m = 0.08$

Table 4.7 *m* and *n* for Vegetation

Vegetation	Height (in.)	Retardance	Erosivity	m	n
Bermuda good stand	12	B	Resistant	0.20	−0.60
			Erodible	0.21	−0.51
	6	C	Resistant	0.13	−0.62
			Erodible	0.11	−0.59
	2.5	D	Resistant	0.10	−0.67
			Erodible	0.08	−0.63
	1.5	E	Resistant	0.072	−0.65
			Erodible	0.05	−0.67
Grass mix good stand	uncut	B	Resistant	0.20	−0.51
			Erodible	0.18	−0.48
	6–8	C	Resistant	0.12	−0.53
			Erodible	0.11	−0.50
	4–5	D	Resistant	0.084	−0.58
			Erodible	0.063	−0.60
Lespedeza	11	C	Resistant	0.12	−0.47
			Erodible	0.080	−0.53
	4.5	D	Resistant	0.13	−0.42
			Erodible	0.080	−0.47

and $n = -0.63$ from Table 4.7. Thus

$$R = 0.08(0.04)^{-0.63} = 0.61 \text{ ft.}$$

From Fig. 4.15d, v is determined as 5 fps. The solution follows the procedures of Example Problem 4.11 from this point.

Temporary Channel Linings

When vegetated linings are selected for a channel and flow can not be diverted from the channel during the establishment of vegetation, some form of temporary lining should be used to stabilize the channel during the period of vegetal establishment. This lining should be constructed of materials that will deteriorate as vegetation emerges and will not interfere with its growth.

A selected group of these linings is listed in Table 4.8 along with a brief description of the materials. Data on the effectiveness of these channel linings were collected by McWhorter *et al.* (1968) at Mississippi State University. The results of these tests indicated that the maximum allowable depth varies inversely with slope, with different relationships being given for erosion resistant and easily eroded soils. The maximum allowable depth, d_{max}, can be represented as a power function of S according to Eq. (4.28).

The data of McWhorter *et al.* (1968) as presented by Normann (1975) were analyzed, and values for m and n were determined. The results are presented in Table 4.9. McWhorter *et al.* (1968) also found that the temporary linings acted much like vegetation; hence Manning's n was not constant. An equation with a form similar to that of Manning's equation was used, but the exponents of R and S were not necessarily $\frac{2}{3}$ and $\frac{1}{2}$. The suggested form of the velocity equation is also given in Table 4.9.

In order to use the equations in Table 4.9 for design purposes, one would typically design the channel for the permanent vegetation and then select a temporary channel lining that would be stable in the channel. A lower-return period storm might be selected for the design of the temporary lining since the exposure time is short during vegetal establishment and since damages during this period can be easily repaired. Procedures for making the calculations are given in Example Problem 4.14.

Example Problem 4.14 Temporary channel liner

Select a temporary lining for use in the channel designed in Example Problem 4.11. Assume that a lower-return period storm is used for the flexible lining design since it needs to be

Table 4.8 Description of Temporary Liners (McWhorter *et al.*, 1968)

Excelsior mat

Excelsior mat is composed of 0.8 pound/yd^2 of excelsior (dried, shredded wood) covered with a fine paper net covering. The paper net, reinforced along the edges, has an opening size of approximately $\frac{1}{2} \times 2$ in. The mat is held in place by steel pins or staples at the rate of five staples per 6 linear ft of mat, with two staples along each side and one in the middle. At the start of each roll, four or five staples are spaced approximately 1 ft apart. Where more than one mat is required, the mats are butt-joined and securely stapled.

Straw and erosionet

This lining consists of straw applied at a rate of 3 tons per acre (1.25 lb/yd^2). The straw is covered with Erosionet 315 (See description following). This lining is pinned in the same manner as jute mesh, as described later in this table.

$\frac{3}{8}$ in. Fiberglass mat

This lining is fine, loosely woven glass fiber mat similar to furnace air filter material. It has a weight of 0.11 lb/yd^2. This material is not to be confused with more dense fiberglass mats used to eliminate plant growth. Steel pins or staples are placed at the rate of five staples per 6 linear ft of mat, with two staples along each side and one in the middle. At the start of each roll four or five staples are spaced approximately 1 ft apart. Where more than one mat is required, the mats are butt-joined and securely stapled.

$\frac{1}{2}$ in. Fiberglass mat

This lining is a fine, loosely woven glass fiber mat, similar to but denser than the $\frac{3}{8}$ in. fiberglass mat, as it weighs 0.35 lb/yd^2. The stapling procedure is the same as for the $\frac{3}{8}$ in. fiberglass mat.

Erosionet 315

Erosionet is a paper yarn approximately 0.05 in. in diameter, woven into a net with openings approximately $\frac{7}{8}$ in. $\times$ $\frac{1}{2}$ in. The material has little erosion prevention capability in itself and is generally used to hold other lining material in place. Erosionet weighs about 0.20 lb/yd^2 and is pinned in the same manner as jute mesh as described later in this table.

Fiberglass roving

Fiberglass roving is delivered as a lightly bound ribbon of continuous glass fibers. The material is applied to the channel bed using a special venturi nozzle driven by an air compressor, which separates the fibers and results in a web-like mat of glass fibers. The glass fibers are tacked with asphalt for adhesion to each other and to the soil. The single layer of fiberglass roving consists of one layer of blown fiberglass fibers applied at a minimum rate of 0.25 lb/yd^2 tacked with asphalt emulsion or asphalt cement at a minimum rate of 0.25 gal./yd^2. The double layer application consists of two alternating layers of fiberglass and asphalt, each layer consisting of fiberglass roving at 0.25 lb/yd^2.

Jute mesh

Jute mesh is a mat lining woven of jute yarn that varies from $\frac{1}{8}$ to $\frac{1}{4}$ in. in diameter. The mat weighs approximately 0.80 lb/yd^2, with openings about $\frac{3}{8}$ in. $\times$ $\frac{3}{4}$ in. Steel pins or staples are used to hold the jute mesh in place. The pins or staples should be spaced not more than 3 ft apart in three rows for each strip, with one row along each edge and one row alternately spaced in the center. At the overlapping edges of parallel strips, staples should be spaced at 2 ft or less. At all anchor slots, junction slots, and check slots, spacing should be 6 in. or less.

Table 4.9 Coefficients for Eqs. (4.28) and (4.29)[a]

Type lining	Erodible soil		Erosion-resistant soil		Velocity equation
	m	n	m	n	
Bare soil	0.0030	−0.687	0.0084	−0.687	$V = 22.81\,R^{0.591}S^{0.286}$
Fiberglass roving with asphalt tack (single layer)	0.0067	−0.960	0.0141	−0.960	$V = 42.45\,R^{0.667}S^{0.5}$
Fiberglass roving with asphalt tack (double layer)	0.0143	−1.01	0.027	−1.01	$V = 59.20\,R^{0.667}S^{0.5}$
Jute mesh	0.0076	−0.875	0.0202	−0.883	$V = 61.53\,R^{1.0281}S^{0.431}$
Excelsior mat	0.0572	−0.585	0.101	−0.585	$V = 32.29\,R^{1.340}S^{0.351}$
Straw and erosionet	0.052	−0.652	0.082	−0.652	$V = 70.76\,R^{1.455}S^{0.529}$
Fiberglass mat $\frac{3}{8}$ in.	0.025	−0.670	0.046	−0.670	$V = 73.53\,R^{1.330}S^{0.512}$
Fiberglass mat $\frac{1}{2}$ in.	0.048	−0.646	0.083	−0.646	$V = 14.84\,R^{1.235}S^{0.086}$
Erosionet	0.049	−0.642	0.084	−0.642	$V = 41.45\,R^{0.855}S^{0.40}$

[a]Adapted from McWhorter *et al.* (1968).

Table 4.10 Initial Calculations for Example Problem 4.14

	Maximum depth[a] (ft)	Hydraulic radius[b] (ft)	Top width[b] (ft)	Area[b] (ft^2)	Velocity[c] (fps)	Flow (cfs)
Jute mesh	0.127	0.084	1.75	0.149	1.20	0.18
Excelsior	0.376	0.241	3.02	0.758	1.55	1.17
Straw and erosionet	0.427	0.270	3.21	0.907	1.92	1.74
Fiberglass (two layers)	0.369	0.237	3.00	0.737	4.53	3.34

[a]From Eq. (4.28) and coefficients in Table 4.9.
[b]From equation in Fig. 4.9.
[c]From velocity equations in Table 4.9.

effective for only a short period of time. The design flow is found to be 10 cfs.

Solution: From Problem 4.11, the soil is easily eroded. The slope is 4%, and a parabolic channel is used with 6.8 ft top width and a depth of 1.9 ft. To facilitate selection of the lining, the values shown in Table 4.10 were calculated from Eq. (4.28) and Table 4.9.

Obviously, none of the linings are acceptable since the discharge at d_{max} is less than the design discharge. The channel will have to be redesigned for stability during the period of temporary lining. This will require an increase in the top width without increasing the total depth, thus maintaining stability. The design is made using the trial and error procedure shown in Table 4.11

What has been shown thus far is that a channel having a shape defined by a parabola with $T = 20$ and $D = 1.9$ and lined with a double layer of fiberglass will be stable enough to carry 10 cfs at a depth of 0.37 ft. A quick calculation shows that the channel if unlined will be unstable if constructed in most soils. We have seen in Example Problem 4.11 that the channel, when grass lined only, had to have a T of 6.8 ft if the D was 1.9 to safely carry 25 cfs. It thus appears that the channel need not be constructed 1.9 ft deep since the top width exceeds the required top width.

Holding the basic channel shape the same, the actual depth of flow under the long grass condition when carrying 25 cfs can be recalculated. Using retardance class B, a trial and error procedure can be used to arrive at the flow depth. Try $d = 1.00$ ft:

$$t = T\left(\frac{d}{D}\right)^{0.5} = 20\left(\frac{1.00}{1.90}\right)^{0.5} = 14.51$$

$$R = \frac{t^2 d}{1.5t^2 + 4d^2} = 0.658$$

$$v = 1.2 \text{ fps} \qquad \text{(Fig. 4.15b)}$$

$$A = 2td/3 = 9.67$$

$$Q = vA = 11.60 \qquad \text{too small.}$$

Table 4.11 Final Calculations for Example Problem 4.14

Lining type	Maximum depth (ft)	Hydraulic radius at d_{max} (ft)	Top width at d_{max} (ft)	Area at d_{max} (ft^2)	Velocity at d_{max} (fps)	Discharge at d_{max} (cfs)
Top width[a], 12 ft fiberglass, two layers	0.369	0.243	5.28	1.30	4.61	5.99 too low
Top width[a], 15 ft fiberglass, two layers	0.369	0.244	6.61	1.62	4.62	7.49 too low
Top width[a], 18 ft fiberglass, two layers	0.369	0.245	7.93	1.95	4.63	9.04 too low
Top width[a], 20 ft fiberglass, two layers	0.369	0.245	8.81	2.16	4.63	10.00 OK

[a]T at a depth of 1.9 ft.

Try $d = 1.3$, then $t = 16.54$, $R = 0.853$, $v = 3.0$, $A = 14.33$, and $Q = 43$. The channel is too large.
Try $d = 1.15$, then $t = 15.56$, $R = 0.756$, $v = 2.1$, $A = 11.93$, and $Q = 25$ cfs. This channel is OK.

Therefore, the final channel design with freeboard added would be a depth of 1.45 ft and a top width of 17.5 ft. The channel would be sprigged or seeded to Bermuda grass with a double layer of fiberglass roving with each layer tacked with asphalt to protect the channel during the establishment of the vegetation.

A similar procedure could be used to arrive at the channel design if other liners were used.

The decision to classify a soil as erodible or erosion resistant is somewhat subjective. Normann (1975) suggests that the erodibility of the soil, K in the Universal Soil Loss equation, can be used as an indicator of erosion resistance. He suggests the following classification:

$$K = 0.50 \text{ erodible}$$
$$K = 0.17 \text{ erosion resistant.}$$

For K values between 0.17 and 0.50, one would need to interpolate between the values of m and n in Table 4.9. Soil erodibility values are discussed in Chapter 8.

Riprap Linings

In situations where vegetation is not suitable, riprap is often used to stabilize channels. Riprap is generally rocks of various sizes arranged to prevent erosion of channel banks and bottom.

Rocks used for riprap should be dense and hard enough to resist deterioration due to exposure to air, water, and temperature extremes, including repeated freezing and thawing if necessary. Sometimes rock that is initially quarried may appear satisfactory but is not able to withstand weathering. If doubt exists as to the suitability of a rock source, a geologist should be consulted. Rough angular rocks are generally preferred as they interlock and resist overturning better than smooth, rounded rocks.

Surfaces on which riprap is placed should be well compacted and stable. It is especially important to ensure that the toe sections for channel bank riprap are safe from scour and sloughing, since failure of the toe may result in failure of the entire bank. Rocks should be placed in a manner that prevents segregation by size. Dumping in a manner that allows excessive rolling of the rocks in a downslope direction and spreading with a dozer potentially result in segregation. Generally front-end loaders or bucket elevators or draglines are satisfactory. Some hand work is usually required to ensure a stable and uniform riprap surface.

The design of a riprap-lined channel involves the selection of a rock size large enough that the force attempting to overturn individual rocks is less than the gravitational force holding the rocks in place. Since riprap is graded, the design procedures must also include a definition of an appropriate gradation of particle sizes such that erosion of the smaller particles on the surface will leave an armored channel that is stable. Finally, the design procedures must include a methodology for selecting appropriate underlying filters so that water flowing beneath the riprap will not erode the base material. Procedures for selecting these materials are included in this section.

Flow on a Plane Sloping Bed

At the present time, riprap design procedures are evolving. Three procedures are presented: (a) a procedure reported by the Federal Highway Administration (FHA procedure) (Norman, 1975); (b) a procedure in the Soil Conservation Service (1979) Engineering Field Manual (SCS procedure); and (c) a procedure developed at Colorado State University (CSU procedure) and reported by Stevens and Simons (1971) and Simons and Senturk (1977, 1992). The FHA and SCS procedures are similar in that a stone diameter is specified in terms of the depth of flow and channel slope. These two procedures are based on experiments and field observations. The CSU procedure includes a theoretical analysis plus laboratory and field studies. The CSU procedure is more complete and allows the specification of a safety factor. Presumably with a safety factor of 1.0, the rocks are in a state of incipient motion.

A complication in riprap design is the gradation of rock sizes. Rocks up to some particular size may be unstable in a flow, but larger rocks might tend to hold them in place. Experimental work with riprap is difficult and time consuming because of the size of the rocks involved, the many possible gradations of rocks, variation in rock shape, materials and handling costs, and the generally high flow rates required. These factors have tended to limit studies on the stability of riprap under controlled conditions.

The CSU procedure is the most theoretically complete and conservative of the three procedures. It should result in satisfactory designs. Channel sections lined with riprap should be carefully monitored, especially for the first few years after completion, to ensure that the selected riprap is stable. Any damage should be repaired immediately to prevent much more extensive damage from developing.

The FHA procedure uses a maximum stable depth of flow given by Eq. (4.28) with $n = -1.0$ and $m =$

$5D_{50}/\gamma$, where D_{50} is the riprap diameter in feet such that 50% of the stones have a diameter smaller than D_{50} and γ is the unit weight of water (62.4 lb/ft³). Thus d_{max} is given by

$$d_{max} = 5(D_{50}/\gamma S). \tag{4.31}$$

The velocity of flow is given by Manning's equation with a roughness, n, given by

$$n = 0.0395 D_{50}^{1/6} \tag{4.32}$$

so that

$$v = \frac{37.7}{D_{50}^{1/6}} R^{2/3} S^{1/2}. \tag{4.33}$$

This equation is known as the Manning–Strickler equation. Channel design is done by computing d_{max} and v for an assumed D_{50} and then determining the appropriate channel dimensions. The calculations are made easier by assuming $R = d_{max}$.

A paper by Abt *et al.* (1988) suggests that Manning's n for riprap in steep channels can be approximated by

$$n = 0.0456(D_{50}S)^{0.159}, \tag{4.34}$$

where D_{50} is in inches and S is in feet per foot. Although this relationship has not been officially adopted in any design procedures, the data presented by Abt *et al.* indicate that it better describes Manning's n than does Eq. 4.32 for the conditions they tested.

Example Problem 4.15 Riprap—FHA procedure

Determine the D_{50} riprap size required to convey 115 cfs down a 10% slope in a rectangular channel 18 ft wide. Riprap is for the bottom only. Use the FHA procedure.

Solution: Assume $R = d_{max}$, $\gamma = 62.4$, $S = 0.10$. Then

$$d_{max} = \frac{5D_{50}}{\gamma S} = 0.801 D_{50}$$

$$v = \frac{37.7}{D_{50}^{1/6}} (d_{max})^{2/3} S^{1/2} = \frac{37.7}{D_{50}^{1/6}} (0.801 D_{50})^{2/3} (0.10)^{1/2}$$

$$v = 10.28 D_{50}^{1/2}$$

$$Q = vA = 10.28 D_{50}^{1/2} d_{max} B = 10.28 D_{50}^{1/2} (0.801 D_{50})(18)$$

$$115 = 148.22 D_{50}^{3/2}$$

$$D_{50} = 0.84 \text{ ft}$$

Note:

$$d_{max} = 0.68 \text{ ft}$$

$$R = \frac{db}{2d + b} = \frac{0.68(18)}{2(0.68) + 18} = 0.63 \text{ ft}.$$

Therefore, the assumption that $R = d$ is reasonable. If the Abt relationship for n is used, the result is $v = 8.4$ fps and $D_{50} = 0.95$ ft.

The SCS procedure is based on a chart that can be approximated by

$$D_{75} = 13.5 d^{1.1} S$$

for rock diameter, D_{75}, in feet, depth of flow, d, in feet, and S in feet per foot. If D_{75} is about $1.5D_{50}$, as recommended by Simons and Senturk (1977, 1992), then

$$D_{50} = 9 d^{1.1} S$$

or

$$d_{max} = (D_{50}/9S)^{0.91}.$$

The SCS also presented a chart based on the Isbash curves, which can be approximated by

$$v = 12.84 D_{50}^{0.51}.$$

This relationship assumes $D_{100} = 2D_{50}$. An unattractive theoretical aspect of this procedure is that v is not expressed as a function of slope and thus the equation should not be considered a general result. If the expression $D_{50} = 9d^{1.1}S$ is substituted into the relationship, the result is $v = 39.4 d^{0.56} S^{0.51}$, which is analogous to Manning's equation.

Example Problem 4.16 Riprap—SCS procedure

Work Example Problem 4.15 using the SCS approximations.

Solution

$$Q = vA = 12.84 D_{50}^{0.51}(dB) = 12.84 D_{50}^{0.51} \left(\frac{D_{50}}{9S}\right)^{0.91} 18$$

$$115 = 254 D_{50}^{1.42}$$

$$D_{50} = 0.57 \text{ ft}$$

$$d_{max} = \left(\frac{0.57}{9(0.1)}\right)^{0.91} = 0.66 \text{ ft}.$$

For this problem, the FHA and SCS criteria result in similar designs with the FHA procedure resulting in larger estimates for the required D_{50}. This will generally be the case.

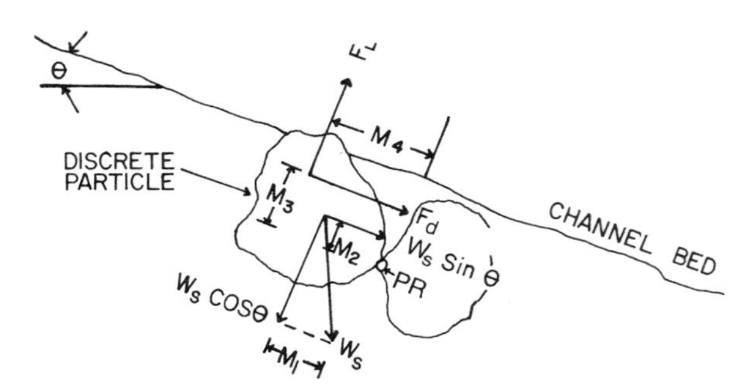

Figure 4.16 Forces on a particle in a channel bed. F_d, drag force; F_L, lift force; PR, point of rotation.

Simons and Senturk (1977, 1992) have analyzed several procedures for determining the required particle sizes for stable channel design. They present the CSU procedure, which encompasses a safety factor (SF) concept. A SF of one represents a point of incipient motion or the flow condition where forces holding particles and those tending to move particles are in exact balance. A SF of 1.5 would be preferred to add stability for particles smaller than D_{50} and to recognize statistical variability and thus prevent the initiation of localized movement, which might lead to a general failure of the riprap protection.

The FHA and SCS procedures are found to have safety factors of less than 1.0 using the CSU criteria (presented later). This indicates potential failure problems at design flows according to the CSU criteria.

The CSU procedures is developed by considering the forces on a particle on a channel bed sloping at an angle θ as shown in Fig. 4.16 along with the moment arms about the point of rotation, PR. Summing moments about PR:

$$F_L M_4 + F_d M_3 + W_s \sin\theta M_2 = W_s \cos\theta M_1. \quad (4.35)$$

These terms are defined in Fig. 4.16. The SF for a given flow situation is the ratio of the resisting moments to the overturning moments, or

$$\mathrm{SF}_b = \frac{W_s \cos\theta M_1}{W_s \sin\theta M_2 + F_L M_4 + F_d M_3}. \quad (4.36)$$

The key to a stable design is to make the safety factor greater than one. To calculate a safety factor, Eq. (4.36) must be manipulated so that it contains parameters that are readily measurable or can be determined from tables and graphs.

One readily measurable parameter is the angle of repose of a given riprap, given in Fig. 4.17. When there is no flow, the lift and drag forces are zero. Under these conditions, if the angle of the channel bottom, θ,

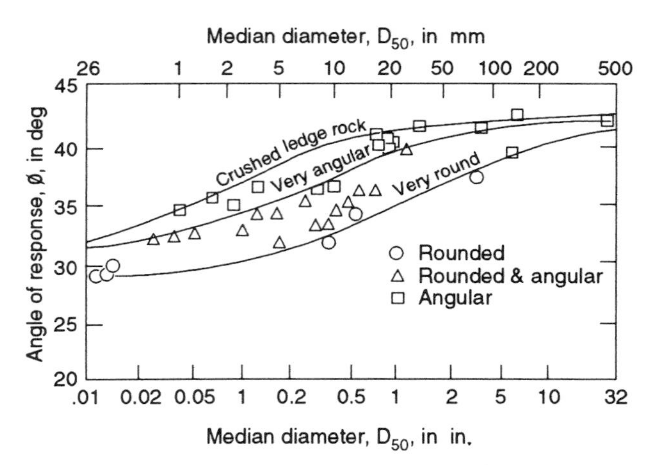

Figure 4.17 Angle of repose of dumped riprap (after Simons and Senturk, 1977, 1992).

is increased until the particles just begin to move, the particles are at their angle of repose ϕ, and the safety factor is 1.0; hence,

$$\tan\phi = M_1/M_2. \quad (4.37)$$

Using Eq. (4.37) in (4.36),

$$\mathrm{SF}_b = \frac{\cos\theta \tan\phi}{\sin\theta + (F_L M_4/W_s M_2) + (F_d M_3/W_s M_2)} \quad (4.38)$$

or

$$\mathrm{SF}_b = \frac{\cos\theta \tan\phi}{\sin\theta + \eta_b \tan\phi}, \quad (4.39)$$

where η_b is a stability parameter given by

$$\eta_b = \frac{F_L}{W_s}\frac{M_4}{M_1} + \frac{F_d}{W_s}\frac{M_3}{M_1}. \quad (4.40)$$

The nature of η_b can be determined by looking at the safety factor for a plane horizontal bed, where θ is equal to zero. Under these conditions, the safety factor becomes

$$\mathrm{SF}_{\mathrm{plane}} = \frac{W_s M_1}{F_L M_4 + F_d M_3} = \frac{1}{(F_L M_4/W_s M_1) + (F_d M_3/W_s M_1)}$$

or

$$\mathrm{SF}_{\mathrm{plane}} = 1/\eta_b. \quad (4.41)$$

For $\mathrm{SF}_{\mathrm{plane}}$ equal to 1.0, the bed is at the point of incipient motion, and the tractive force is equal to the critical tractive force. Under conditions other than incipient motion on a plane bed, it is reasonable to assume that the safety factor can be given by the ratio of critical to actual tractive force since there is no gravity component along the channel bed. Therefore

$$SF_b \Rightarrow \frac{1}{\eta_b} = \frac{\tau_c}{\tau} \tag{4.42}$$

and

$$\eta_b = \frac{\tau}{\tau_c}. \tag{4.43}$$

For fully turbulent flow, Gessler (1971) indicates that the Shield's diagram can be reanalyzed to give

$$\tau_c = 0.047\gamma(\mathrm{SG} - 1)D,$$

where SG is the specific gravity of the particles and D is the representative particle diameter, typically the average diameter. Using Gessler's analysis,

$$\eta_b = \frac{21\tau}{\gamma(\mathrm{SG} - 1)D}. \tag{4.44}$$

If τ is given by $\gamma\, dS$, these equations can be used to design a channel if the flow velocity is determined from Eq. (4.33). An illustration of the design procedure is given in Example Problem 4.17. It should be noted that these equations do not apply to channel banks, but only to the channel bottoms. Channel bank stability is considered in the following section.

Example Problem 4.17 Riprap—CSU procedure

A channel is being designed to convey a flow of 115 cfs down a 10% slope. The soil is collodial silt; hence the critical tractive force is so small that a lining is needed. Select an average diameter of riprap needed to stabilize the channel. For this example, neglect the stability problems associated with the side slopes. Assume a bottom width of 18 ft, a specific gravity of 2.65, and a rectangular cross section. Design for a safety factor of 1.5.

Solution: The solution procedure involves a trial and error approach of selecting a riprap size, calculating the depth of flow required to convey the flow, and checking the safety factor to ensure that the channel is stable. Assume a D_{50} of 2.5 ft, from Eq. (4.32).

$$n = 0.0395 D_{50}^{1/6} = 0.046.$$

From Manning's equation, assuming a wide channel,

$$Q = Av = bd\frac{1.49}{n}d^{2/3}S^{1/2}$$

$$d = \left[\frac{nQ}{1.49bS^{1/2}}\right]^{3/5} = \left[\frac{0.046(115)}{1.49(18)(0.10)^{1/2}}\right]^{3/5}$$

$d = 0.75$ ft depth required to convey the flow.

Checking for stability using Eqs. (4.44) and (4.39),

$$\tau = \gamma dS = (62.4)(0.75)(0.10) = 4.68\ \mathrm{lb/ft^2}$$

$$\eta_b = \frac{21\tau}{\gamma(\mathrm{SG} - 1)D_{50}} = \frac{21(4.68)}{62.4(2.65 - 1)2.5} = 0.382.$$

Assuming an angular riprap, Fig. 4.17 gives $\phi = 42°$. For a 10% slope, $\theta = 5.71°$. Hence, from Eq. (4.39),

$$SF_b = \frac{\cos\theta\tan\phi}{\sin\theta + \eta_b\tan\phi} = \frac{(\cos 5.71)(\tan 42)}{\sin 5.71 + 0.382\tan 42}$$

$SF_b = 2.02$ over designed.

Calculations to select a better design are contained in Table 4.12. Use a riprap with a D_{50} of 1.7 ft on the channel bed. Obviously, there is a problem with stability of the side slopes. Also the gradation of riprap must be specified and a filter blanket selected. This is covered in subsequent sections and examples.

Example Problem 4.18 Riprap—safety factor

Calculate SF for Example Problems 4.15 and 4.16.

Solution

$$SF = \frac{\cos\theta\tan\phi}{\sin\theta + \eta_b\log\phi}$$

Table 4.12 Calculations for Example Problem 4.17

D_{50} (ft)	Manning's n	Angle of repose (°)	Depth to convey flow (ft)	Tractive force τ (lb/ft^2)	Stability factor (η_b)	Safety factor (SF_b)
2.5	0.046	42	0.751	4.68	0.382	2.02
2.0	0.044	42	0.734	4.58	0.467	1.72
1.5	0.042	42	0.713	4.45	0.605	1.39
1.7	0.043	42	0.722	4.49	0.541	1.53

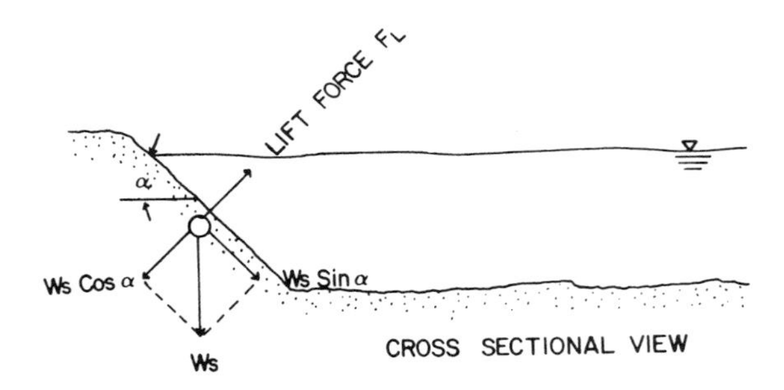

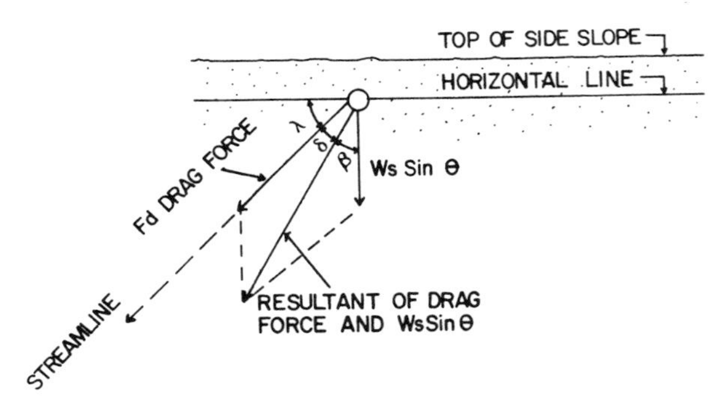

Figure 4.18 Forces on a particle on a stream channel wall.

From Problem 4.15, $d = 0.68$ ft, $D_{50} = 0.84$ ft, and $\phi = 42°$.

$$\tau = \gamma\, dS = 62.4(0.68)0.10 = 4.24 \text{ psf}$$

$$\eta_b = \frac{21\tau}{\gamma(\text{SG} - 1)D_{50}} = \frac{21(4.24)}{62.4(1.65)0.84} = 1.03$$

$$\text{SF} = \frac{\cos(5.71)\tan(42)}{\sin(5.71) + 1.03\tan(42)} = 0.87.$$

From Problem 4.16, $d = 0.66$ ft, $D_{50} = 0.57$ ft, and $\phi = 42°$:

$$\tau = 62.4(0.66)0.10 = 4.12$$

$$\eta_b = \frac{21(4.12)}{62.4(1.65)0.57} = 1.47$$

$$\text{SF} = \frac{\cos(5.71)\tan(42)}{\sin(5.71) + 1.47\tan(42)} = 0.63.$$

Based on the CSU criteria, both of these designs have SF < 1.

Channel Bank Stability

The forces on a channel bank are shown in Fig. 4.18. These forces are different from those in Fig. 4.16 for a channel bed since the drag forces are not aligned with the downslope gravitational forces. The solution of the equations describing the safety factor for this case have been given by Stevens and Simons (1971) and Simons and Senturk (1977, 1992) as

$$\text{SF} = \frac{\cos\alpha\tan\phi}{\eta'\tan\phi + \sin\alpha\cos\beta} \tag{4.45}$$

$$\beta = \tan^{-1}\left(\frac{\cos\lambda}{2\sin\alpha/\eta\tan\phi + \sin\lambda}\right) \tag{4.46}$$

$$\eta = \frac{21\tau_{max}}{\gamma(\text{SG} - 1)D_{50}} \tag{4.47}$$

and

$$\eta' = \eta\frac{1 + \sin(\lambda + \beta)}{2}, \tag{4.48}$$

where τ_{max} is the maximum shear on the channel bank.

In order to derive Eqs. (4.45) through (4.48), it was assumed that the ratio of lift to drag forces was one-half. The use of the procedures is illustrated in Example Problem 4.19.

When calculating the shear forces on a channel bank, it is desirable to take into account variations in channel shear across the channel bed. Figure 4.8 shows that for a trapezoidal channel, the maximum tractive force on the channel walls in $K\gamma\, dS$, where K is 0.74 to 0.78 depending on the channel side slope.

Example Problem 4.19 Riprap size—channel bank

Based on construction considerations and machinery limitations, side slopes of 2.5 : 1 are selected for the channel in Example Problem 4.17. Select a riprap size that will be stable on the channel sideslopes.

Solution: First the safety factor of the riprap selected in Example Problem 4.17 is calculated assuming the same material is used on the sides. From Example Problem 4.17,

$D_{50} = 1.7$ ft; $n = 0.043$; $\theta = 5.71°$; $d = 0.722$ ft.

For a trapezoidal channel, the flow depth can be calculated to be 0.72 ft, which is insignificantly smaller than 0.722 ft for the rectangular channel in Example 4.17; hence we use 0.722 ft.

From Fig. 4.8 τ_{max} is given by $0.76\gamma\, dS$:

$$\tau_{max} = (0.76)(62.4)(0.722)(0.10) = 3.41 \text{ lb/ft}^2$$

$$\eta = \frac{21\tau_{max}}{\gamma(\text{SG} - 1)D_{50}} = \frac{21(3.41)}{62.4(2.65 - 1)1.7} = 0.408.$$

Assuming uniform flow, the streamlines are parallel to the channel bottom and

$$\lambda = \theta = 5.71°.$$

Also, for a 2.5 : 1 sideslope,

$$\alpha = \tan^{-1}\frac{1}{2.5} = 21.8°.$$

Table 4.13 Calculations for Example Problem 4.19

D_{50}[a]	Manning's n	Depth to convey flow (ft)	Maximum tractive force on channel bed (lb/ft^2)	Channel bed stability factor (η_b)	Safety factor for channel bed (SF_b)	Maximum tractive force on walls (lb/ft^2)	Channel wall stability factor (η')	Channel wall safety factor (SF)
1.7	0.043	0.72	4.49	0.541	1.53	3.41	0.308	1.36
2.0	0.044	0.73	4.58	0.467	1.72	3.48	0.268	1.45
2.5	0.046	0.75	4.68	0.382	2.02	3.56	0.220	1.56
2.2	0.045	0.74	4.62	0.429	1.84	3.51	0.247	1.50

[a]Use a riprap with a D_{50} of 2.2 ft for both channel sides and bottom.

From Eq. (4.46),

$$\beta = \tan^{-1}\left[\frac{\cos\lambda}{2\sin\alpha/\eta\tan\phi + \sin\lambda}\right]$$

$$= \tan^{-1}\left[\frac{\cos(5.71)}{2\sin(21.8)/0.408\tan(42) + \sin(5.71)}\right].$$

$$\beta = 25.1°$$

From Eq. (4.48),

$$\eta' = \eta\left[\frac{1+\sin(\lambda+\beta)}{2}\right] = 0.408\left[\frac{1+\sin(5.71+25.10)}{2}\right]$$

$$\eta' = 0.308.$$

From Eq. (4.45),

$$SF = \frac{\cos\alpha\tan\phi}{\eta'\tan\phi + \sin\alpha\cos\beta}$$

$$= \frac{\cos(21.8)\tan(42)}{0.308(\tan(42)) + \sin(21.8)\cos(25.1)}$$

$$SF = 1.36.$$

Thus the riprap is stable, but does not have the required safety factor of 1.5. The selection of an acceptable riprap for the channel side slopes will be made using trial and error. The calculations are in Table 4.13. It is assumed that the riprap on the channel bed will be the same as that used on the side slopes. It would obviously be possible to vary the side slopes and channel width to obtain a smaller D_{50}. The final selection of channel dimensions and riprap size would have to be based on economics.

Selecting Proper Gradation

It is important for a riprap to have a gradation such that the voids between the larger particles are filled with smaller particles to reduce flow beneath the riprap and the formation of open pockets. A suggested gradation for riprap has been made by Simons and Senturk (1977, 1992) based on studies at Colorado State University. The proposed gradation is shown in Fig. 4.19.

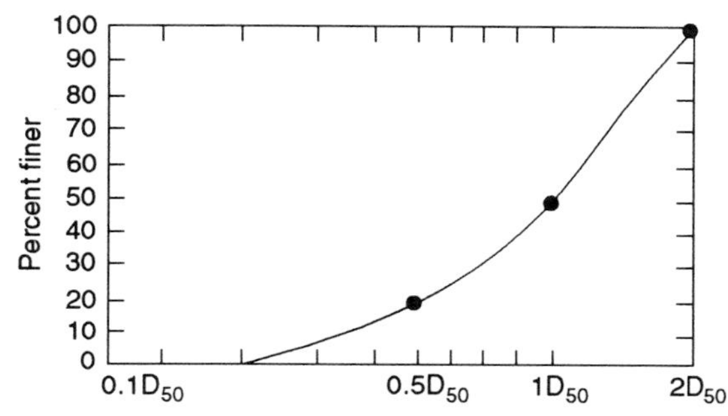

Figure 4.19 Suggested size distribution of riprap (after Simons and Senturk, 1977, 1992).

Selecting an Underlying Filter

The placement of a properly designed filter blanket underneath the riprap is necessary when the particle size of the riprap is much larger than that of the base material. The following criteria have been established for sizing the filter, based on the size distribution of the riprap and the base material:

(1) $$\frac{D_{50}(\text{filter})}{D_{50}(\text{base})} < 40 \quad \text{also} \quad \frac{D_{50}(\text{riprap})}{D_{50}(\text{filter})} < 40$$

(2) $$5 < \frac{D_{15}(\text{filter})}{D_{15}(\text{base})} < 40 \quad \text{also} \quad 5 < \frac{D_{15}(\text{riprap})}{D_{15}(\text{filter})} < 40$$

(3) $$\frac{D_{15}(\text{filter})}{D_{85}(\text{base})} < 5 \quad \text{also} \quad \frac{D_{15}(\text{riprap})}{D_{85}(\text{filter})} < 5.$$

These criteria were developed for sizing filters around drain pipe to prevent piping of the soil into the

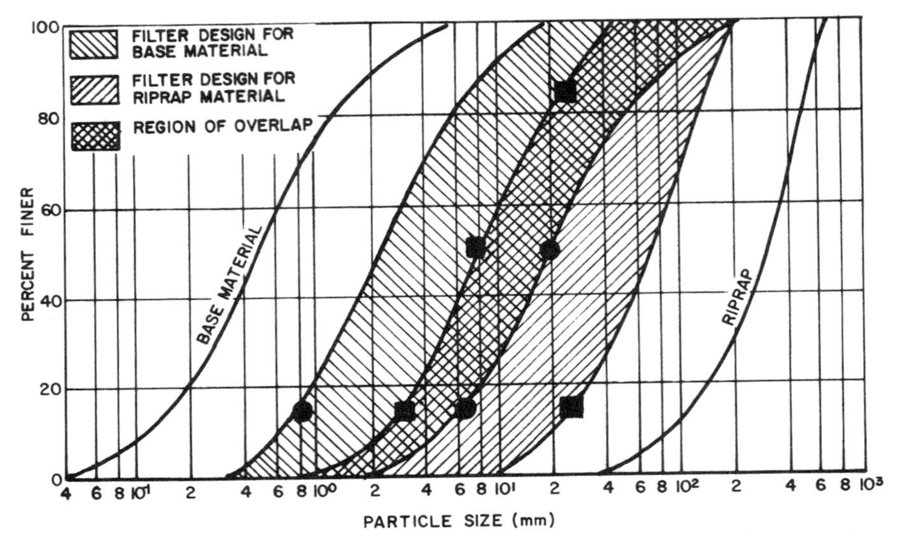

Figure 4.20 Size distribution determinations of filter material for Example Problem 4.20. The filter must have a size distribution within the region of overlap.

drain. A filter designed by the same criteria should prevent piping of the parent soil from beneath riprap. Filter thickness should be approximately one-half the thickness of the riprap, but in no case less than 6 to 9 in. An illustration of the use of these procedures is given in Example Problem 4.20. Plastic filter cloth is being used in some cases rather than granular filter materials. Normann (1975) should be consulted for details.

Example Problem 4.20 Riprap filter design

Select an appropriate riprap gradation for riprap with a D_{50} value of 1.0 ft. The base parent material on which the riprap is being placed has the properties $D_{50} = 0.5$ mm, $D_{85} = 1.5$ mm, and $D_{15} = 0.17$ mm. Select an appropriate filter blanket for the riprap.

Solution: Based on Fig. 4.19 with a D_{50} of 1.0 ft, the properties of the riprap are $D_{100} = 2.0$ ft $= 610$ mm, $D_{50} = 1.0$ ft $= 305$ mm, $D_{85} = 1.7$ ft $= 520$ mm, $D_{15} = 0.42$ ft $= 130$ mm, and $D_0 = 0.10$ ft $= 30$ mm.

These are plotted in Fig. 4.20 along with the size distribution of the parent material. Next the filter blanket must be sized. Look first at the requirements of the filter blanket with respect to the parent material:

Criterion (1)

$$\frac{D_{50}(\text{filter})}{D_{50}(\text{base})} < 40 \qquad \text{giving } D_{50}(\text{filter}) < 40 \times 0.5 = 20 \text{ mm}$$

Criterion (2)

$$\frac{D_{15}(\text{filter})}{D_{15}(\text{base})} > 5 \qquad \text{giving } D_{15}(\text{filter}) > 5 \times 0.17 = 0.85 \text{ mm}$$

and

$$\frac{D_{15}(\text{filter})}{D_{15}(\text{base})} < 40 \qquad \text{giving } D_{15}(\text{filter}) < 40 \times 0.17 = 6.18 \text{ mm}$$

Criterion (3)

$$\frac{D_{15}(\text{filter})}{D_{85}(\text{base})} < 5 \qquad \text{giving } D_{15}(\text{filter}) < 5 \times 1.5 = 7.5 \text{ mm.}$$

Therefore, with respect to the base parent material, the following criteria must be satisfied:

$$0.85 \text{ mm} < D_{15}(\text{filter}) < 6.8 \text{ mm}$$

and

$$D_{50}(\text{filter}) < 20 \text{ mm.}$$

These points are plotted as solid dots in Fig. 4.20 and curves approximating these conditions were drawn through the points.

Next, the filter must be sized relative to the riprap.

Criterion (1)

$$\frac{D_{50}(\text{riprap})}{D_{50}(\text{filter})} < 40 \qquad \text{giving } D_{50}(\text{filter}) > \frac{305}{40} = 7.6 \text{ mm}$$

Criterion (2)

$$\frac{D_{15}(\text{riprap})}{D_{15}(\text{filter})} > 5 \qquad \text{giving } D_{15}(\text{filter}) < \frac{130}{5} = 26 \text{ mm}$$

and

$$\frac{D_{15}(\text{riprap})}{D_{15}(\text{filter})} < 40 \qquad \text{giving } D_{15}(\text{filter}) > \frac{130}{40} = 3.3 \text{ mm}$$

Criterion (3)

$$\frac{D_{15}(\text{riprap})}{D_{85}(\text{filter})} < 5 \qquad \text{giving } D_{85}(\text{filter}) > \frac{130}{5} = 26 \text{ mm}.$$

Therefore the filter must also meet these criteria, or

$$D_{50}(\text{filter}) > 7.6 \text{ mm}$$

$$3.3 \text{ mm} < D_{15}(\text{filter}) < 26 \text{ mm}$$

$$D_{85}(\text{filter}) > 26 \text{ mm}.$$

These points are also plotted in Fig. 4.20 as solid boxes and curves drawn through the points. The envelope of points satisfying both criteria are crosshatched. Any material selected with a size distribution falling within the crosshatched area will satisfy the design requirements.

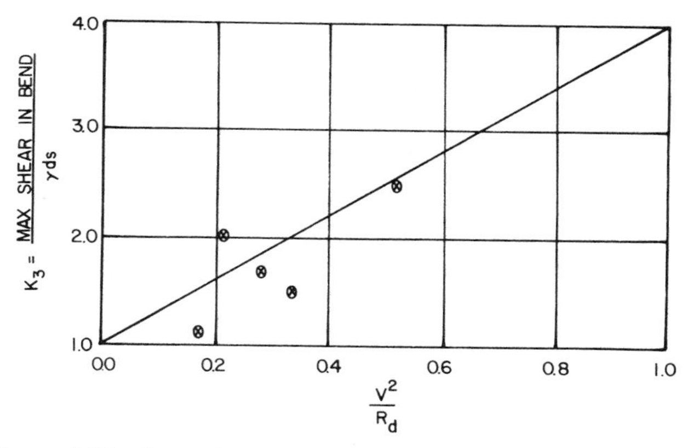

Figure 4.21 Correction factor for shear in flow in a bend (Normann, 1975).

Flow in Channel Bends

Because of the curvature in channel bends, the peak velocity typically occurs on the outside of the centerline, resulting in steeper velocity gradients and higher shear stress values on the outside banks than occur in straight channels. This extra shear must be considered when sizing riprap, vegetation, and temporary channel linings. A commonly used procedure in riprap-lined channels is to increase the riprap size in the channel bend, or in vegetated lined channels, to line sharp bends with riprap.

The location of the maximum shear varies so much within bends that it is not possible to determine the exact point at which protection is needed. Therefore, it is standard practice to protect the outside bank of the entire bend.

Data that can be used to predict shear in channel bends are not abundant. F. J. Watts [as reported by Norman (1975)] proposed that a correction factor for shear on the channels walls varying from 1.0 to 4.0 could be calculated on the basis of v^2/R_d, where v is the average flow velocity in a straight channel and R_d is the radius of the outside bank. A plot of the correction factor is given in Fig. 4.21 along with the limited verification data reported by Normann (1975). To use the relationship:

(1) Determine the velocity in a straight channel stretch.

(2) Determine the radius of curvature of the outside bank, R_d.

(3) Calculate v^2/R_d.

(4) Determine the correction factor, k_3, from Fig. 4.21.

(5) Calculate the corrected bank shear from

$$\tau = k_3 \gamma \, dS.$$

(6) Use this shear from (5) in the stability parameters η and η' and determine the required riprap size using procedures previously discussed.

It must be pointed out that these procedures have very limited verification. Their use is still somewhat speculative at this point.

General Comments

The flow range over which differing channel linings offer protection depends on channel shape and slope. An example comparison made by Normann (1975) is given in Fig. 4.22. Although the procedures used to calculate the allowable discharge for the riprap have been shown in Simons and Senturk (1977, 1992) to sometimes yield slightly unstable design, the figure gives a reasonable guide to the type of channel lining required for varying flow rates and slopes. It should be pointed out that the ranges will change based on side slopes and erodibility of underlying material.

GRADUALLY VARIED FLOW

The relationships presented in this section are for wide, open channels where the hydraulic radius may be approximated by the depth of flow. Uniform flow requires a channel of constant cross section and sufficient length for the gravitational forces to achieve a balance with the frictional resistance. At changes in slope, cross section, or roughness, the two forces will not be balanced, and the flow conditions will adjust toward equilibrium. Within the channel reach where this adjustment occurs, the flow is said to be varied flow or nonuniform flow. If the change in flow conditions occurs gradually over relatively long channel reaches, the flow is said to be gradually varied flow.

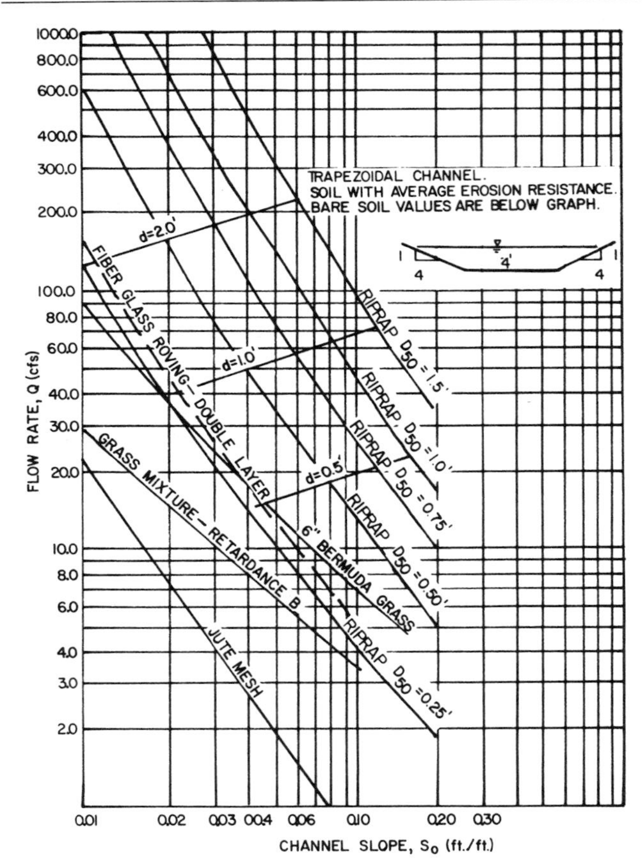

Figure 4.22 Comparison of maximum flow rate versus slope for various channel linings (Norman, 1975).

Equation (4.4) and Fig. 4.4 give the total energy, H, as

$$H = y + z + y^2/2g. \tag{4.49}$$

This may be written as

$$H = y + z + Q^2/2gA^2. \tag{4.50}$$

Differentiation with respect to x, the distance along the channel, yields

$$\frac{dH}{dx} = \frac{dy}{dx} + \frac{dz}{dx} - \frac{Q^2}{gA^3}\frac{dA}{dx}. \tag{4.51}$$

If we consider a rectangular channel or a wide channel, the last term of this equation becomes

$$\frac{Q^2}{gA^3}\frac{dA}{dx} = \frac{q^2}{gy^3}\frac{dy}{dx}.$$

The term dH/dx represents the slope of the energy grade line, S, which is by convention taken as positive downward. Similarly dz/dx is the channel slope, S_0, also positive downward. Thus

$$-S = -S_0 + \left(1 - \frac{q^2}{gy^3}\right)\frac{dy}{dx}. \tag{4.52}$$

Noting that q^2/gy^3 is $\mathbf{F}^2$ and rearranging the equation results in

$$\frac{dy}{dx} = \frac{S_0 - S}{1 - \mathbf{F}^2}. \tag{4.53}$$

This equation gives the slope of the water surface with respect to the channel bottom. If dy/dx is positive, the flow is getting deeper in the downstream direction. If dy/dx is negative, the flow is getting shallower in the downstream direction. A dy/dx of zero implies uniform flow.

A channel is said to have a mild slope if the normal depth, y_n, is greater than the critical depth, y_c. Similarly, if $y_n < y_c$, the slope is a steep slope, and if $y_n = y_c$, the slope is termed a critical slope. A slope that is negative or runs uphill in the downstream direction is known as an adverse slope. Finally a channel with no slope is said to be a horizontal channel.

In sketching gradually varied flow profiles, the profiles are conventionally labeled with the first letter of the slope type. Thus M denotes a mild slope, S a steep slope, etc.

If the flow depth exceeds both y_n and y_c, the flow is said to be in zone 1 and is denoted with the subscript 1. If the depth is between y_n and y_c (or between y_c and y_n), the zone designation is 2. A depth less than both y_n and y_c is in zone 3.

Figure 4.23 depicts possible flow profiles or backwater curves. The slope of the water surface for the various situations can be deduced from Eq. (4.53). To do this, one can approximate S as the slope calculated from Manning's equation using the actual depth of flow. S_0 is the slope in Manning's equation corresponding to y_n. The appropriate equations are

$$S = \frac{q^2n^2}{2.22y^{10/3}}$$

and

$$S_0 = \frac{q^2n^2}{2.22y_n^{10/3}}.$$

If $y_n > y$, $S_0 < S$. If $y > y_n$, $S_0 > S$. Also note that if S_0 is less than or equal to zero, y_n is not defined.

As an example of determining the slope of the water surface, consider an M_1 profile. In this situation $y_n > y_c$, $y > y_n$, and $y > y_c$. Thus $\mathbf{F} < 1$ and $S_0 > S$. This means that both the numerator and denominator of Eq. (4.53) are positive, so dy/dx is positive and the flow depth increases in the downstream direction.

As another example, consider the S_2 profile. Here $y_c > y_n$, $y > y_n$, and $y < y_c$. This means $\mathbf{F} > 1$ and $S_0 > S$. Thus, the numerator of Eq. (4.53) is positive and the denominator is negative. The S_2 curve has dy/dx negative or the depth decreases in the downstream direction.

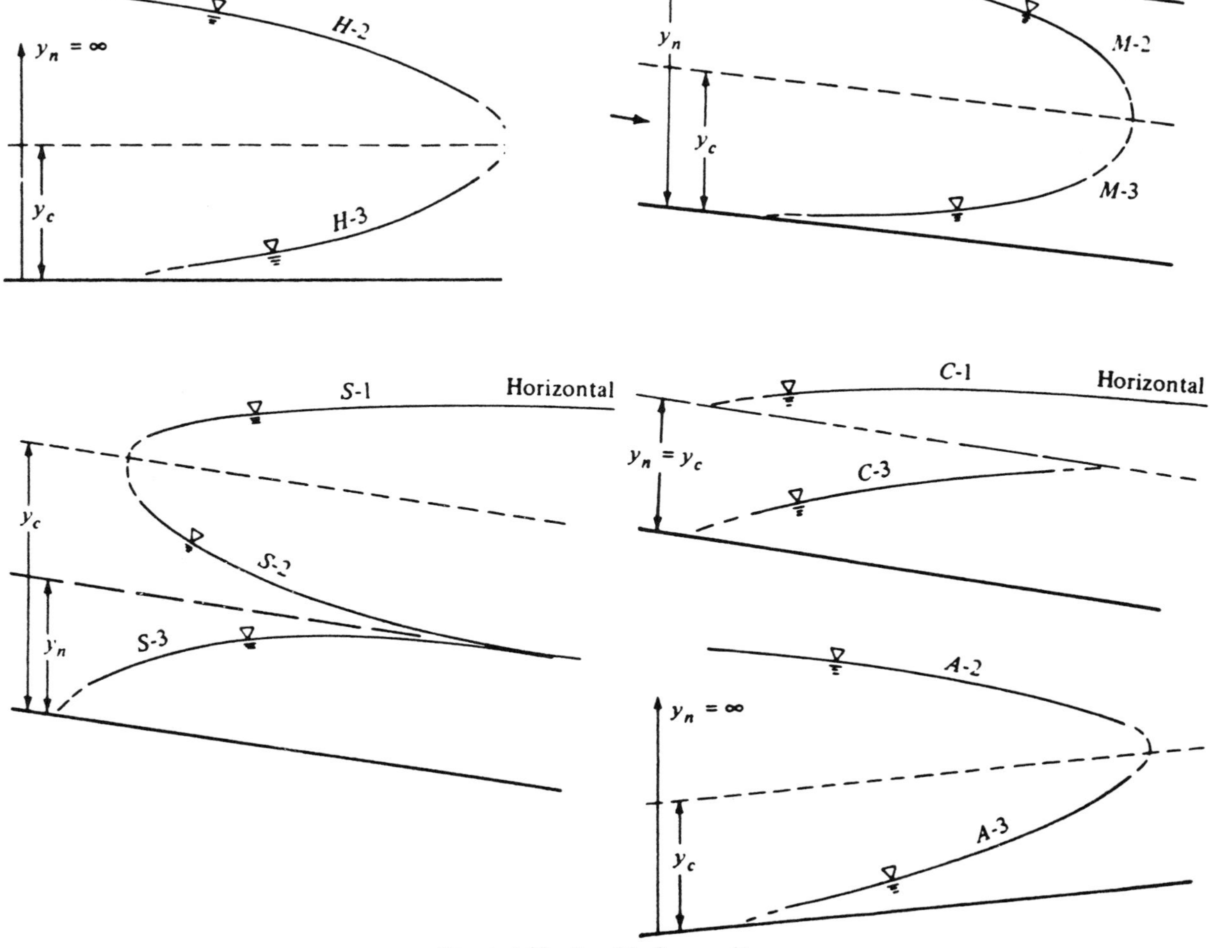

Figure 4.23 Possible flow profiles.

The above reasoning can be applied to each of the zones and profiles with the results shown in Table 4.14. Flow profiles develop at changes in channel slope, roughness, and cross section. Figure 4.24 shows some typical situations where profiles develop.

An approximate calculation of backwater profiles can be done by considering Fig. 4.4 and noting

$$E_1 + z_1 = E_2 + z_2 + h_L.$$

By definition

$$S_0 = (z_1 - z_2)/\Delta x,$$

where Δx is the length of the channel reach. Also

$$h_L = dE = S_f \,\Delta x,$$

where S_f is the friction slope or slope of the energy

Table 4.14 Slope of Water Surface Profiles with Respect to Channel Bottom

Type	Designation	Slope
Mild	M1	+
	M2	−
	M3	+
Steep	S1	+
	S2	−
	S3	+
Critical	C1	+
	C3	+
Horizontal	H2	−
	H3	+
Adverse	A2	−
	A3	+

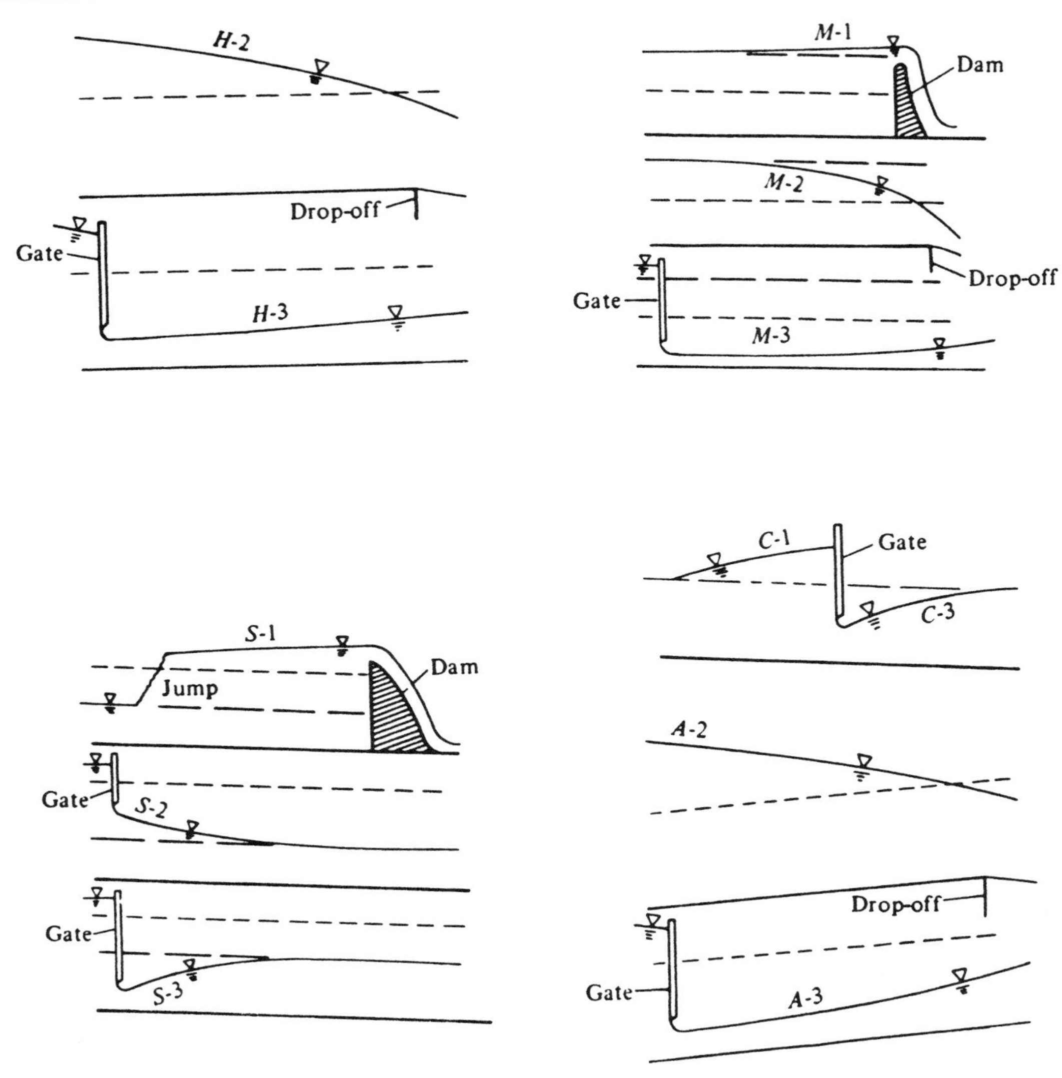

Figure 4.24 Typical flow profiles.

grade line. Combining these three equations results in

$$\Delta x = \frac{E_1 - E_2}{S_f - S_0}. \tag{4.54}$$

S_f can be approximated from Manning's equation by assuming an average flow depth for the reach. Example Problem 4.21 illustrates the computation of a backwater curve. Note that for subcritical flow, backwater curves should be computed in the upstream direction and for supercritical flow in the downstream direction. Profile calculations are started at points of known water surface elevations such as overfalls from a mild channel ($y = y_c$) or other types of control sections. Application of Eq. (4.54) is known as the direct step method.

Example Problem 4.21 Flow profile

A wide, rectangular channel is carrying 10 cfs/ft down a 0.5% slope. The channel has a Manning's n of 0.025. A 2.5-ft barrier in the channel causes flow to pass over the barrier at critical depth. Compute the flow profile upstream from the barrier to a point where the depth is within 10% of normal depth. Figure 4.25 illustrates the physical situation.

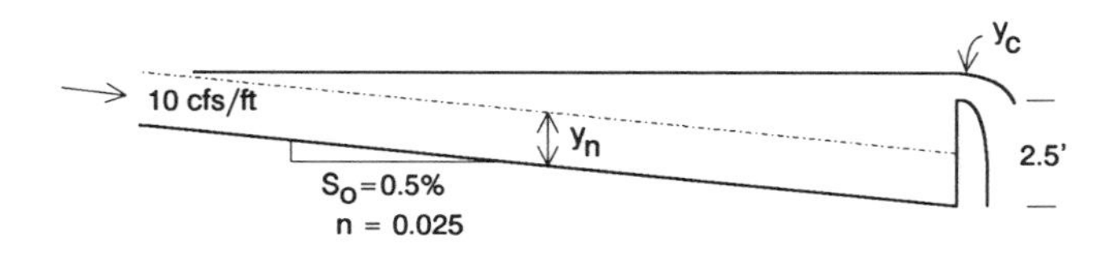

Figure 4.25 Sketch for Example Problem 4.21.

Table 4.15 Profile Calculations for Example Problem 4.21

y (ft)	v[a] (fps)	$V^2/2g$ (ft)	E[b] (ft)	y_m (ft)[c]	S_f[d]	dx (ft)[e]	x[f] (ft)
3.96	2.525	0.099	4.059				0
3.50	2.857	0.127	3.627	3.730	0.00035	–93	–93
3.25	3.077	0.147	3.397	3.375	0.00048	–51	–143
3.00	3.333	0.173	3.173	3.125	0.00063	–51	–195
2.75	3.636	0.205	2.955	2.875	0.00083	–52	–247
2.50	4.000	0.248	2.748	2.625	0.00112	–53	–300
2.25	4.444	0.307	2.557	2.375	0.00157	–56	–356
2.00	5.000	0.388	2.388	2.125	0.00227	–62	–418
1.75	5.714	0.507	2.257	1.875	0.00345	–85	–503
1.70	5.882	0.537	2.237	1.725	0.00456	–45	–548

[a] $v = q/y$.
[b] $E = v^2/2g + y$.
[c] $y_m = (y_1 + y_2)/2$.
[d] $S_f = (qn/1.49y_m^{1.67})^2$.
[e] $dx = (E_1 - E_2)/(S_f - S_o)$.
[f] $x_2 = x_1 + dx$.

Solution: From Eq. (4.9),

$$y_c = \left(\frac{q^2}{g}\right)^{1/3} = \left(\frac{100}{32.2}\right)^{1/3} = 1.46 \text{ ft.}$$

From Manning's equation (4.23) using $q = vy$,

$$y_n = \left(\frac{qn}{1.49S^{1/2}}\right)^{3/5} = \left(\frac{10(0.025)}{1.49(0.005)^{1/2}}\right)^{3/5} = 1.68.$$

The depth of flow over the brink in Fig. 4.25 is

$$y = 2.5 + y_c = 3.96 \text{ ft.}$$

The solution is carried out by assuming depths and computing Δx. Table 4.15 shows the computations.

Equation (4.54) can be used for channels where the approximation that $y = R$ is not appropriate. The calculations are somewhat more cumbersome than those illustrated in Example Problem 4.21. Fortunately, extensive computer programs are available for calculating flow profiles in natural channels. Computers are generally needed because of irregularities in natural channels and the presence of flow obstructions in the form of bridges, culverts, low dams, etc. The most widely used program in the U.S. is HEC-2, a program developed by the U.S. Army Corps of Engineers (1982) Hydrologic Engineering Center in Davis, California.

CHANNEL TRANSITIONS

Changes in channel width, shape, slope, roughness, bottom elevation, etc., cause changes in the flow regime. The location of these changes is known as the transition area. Backwater curves can be calculated to evaluate changes due to channel slope or roughness as indicated in the previous section. For smooth transitions, energy relationships can be used to evaluate the impact of the transitions. A smooth transition is one in which energy losses are minimal.

Consider the channel transition shown in Fig. 4.26. Assuming no energy loss through the transition,, Eq. (4.4) becomes

$$\frac{v_1^2}{2g} + y_1 = \frac{v_2^2}{2g} + y_2 + \Delta z, \tag{4.55}$$

showing that with a constant total energy there is a specific energy loss of Δz. A specific energy diagram can be used to visualize the flow change that occurs.

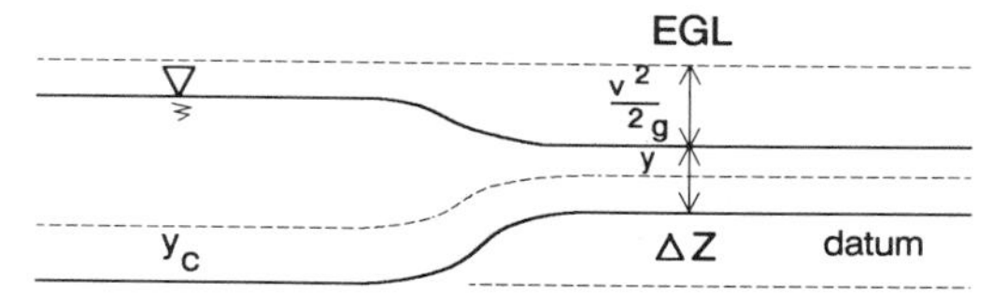

Figure 4.26 A channel transition.

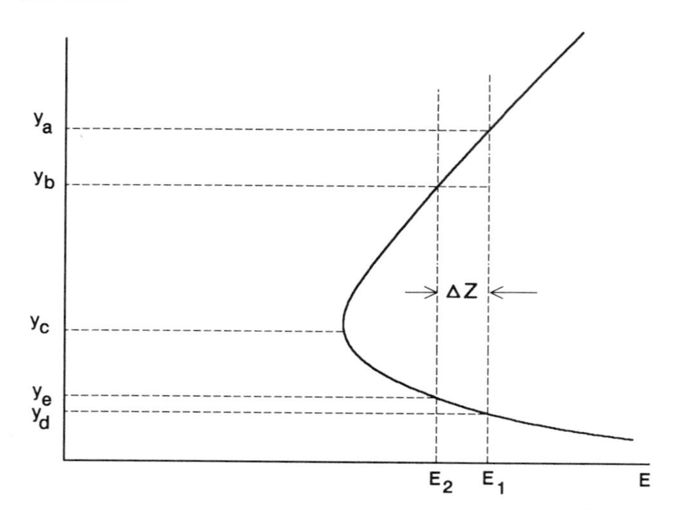

Figure 4.27 Specific energy representation in a transition.

Consider Fig. 4.27. If y_1 is subcritical and represented by y_a, y_2 must correspond to y_b so that the depth of flow due to a channel bottom rise is decreased. Conversely, if y_1 is supercritical and equal to y_d, then y_2 must correspond to y_e. It must be kept in mind that the specific energy diagram corresponds to a constant unit discharge or is based on a rectangular channel. For nonrectangular channels, Eq. (4.55) is still valid, but the specific energy representation of Fig. 4.27 can only be used conceptually, not analytically.

If the flow must pass through critical depth, the assumption of no energy loss may not be valid. This is especially true if the transition is from supercritical to subcritical flow. In such a situation, a hydraulic jump accompanied by considerable energy loss occurs. Hydraulic jumps are considered in the next section.

Example Problem 4.22 Channel transition 1

A trapezoidal channel with 2:1 side slopes and a 4-ft bottom width is flowing at a depth of 1 ft. The channel is concrete and on a slope of 0.1%. If the channel bottom is raised smoothly by 0.1 ft over a short distance, what will be the depth of flow at the exit of the transition?

Solution

$$n = 0.015 \quad \text{for concrete}$$

$$v_1 = \frac{1.5}{n} R^{2/3} S^{1/2}$$

$$A = bd + zd^2 = 4(1) + 2(1)^2 = 6 \text{ ft}^2$$

$$P = b + 2d\sqrt{z^2 + 1} = 4 + 2(1)\sqrt{2^2 + 1} = 8.47 \text{ ft}$$

$$R = A/P = 6/8.47 = 0.71 \text{ ft}$$

$$v_1 = \frac{1.5}{0.015}(0.71)^{2/3}(0.001)^{1/2} = 2.52 \text{ fps}$$

$$\mathbf{F} = \frac{v}{\sqrt{gd_h}}$$

$$d_h = \frac{A}{t} = \frac{6}{b + 2zd} = \frac{6}{4 + 2(2)(1)} = 0.75 \text{ ft}$$

$$\mathbf{F} = \frac{2.52}{\sqrt{32.3(0.75)}} = 0.51 \qquad \text{subcritical}$$

$$\frac{v_1^2}{2g} + y_1 = \frac{v_2^2}{2g} + y_2 + \Delta z$$

$$\frac{(2.52)^2}{64.4} + 1.0 = \frac{v_2^2}{64.4} + y_2 + 0.1$$

$$0.9986 = \frac{v_2^2}{64.4} + y_2$$

$$v_2 = \frac{Q}{A} = \frac{v_1 A_1}{A_2} = \frac{2.52(6)}{4y_2 + 2y_2^2}$$

$$0.9986 = \frac{3.54}{(4y_2 + 2y_2^2)^2} + y_2.$$

Solve by trial

y_2	Right-hand side	
0.90	1.03	
0.75	0.958	
0.84	0.996	OK

$$y_2 = 0.84 \text{ ft}$$

Check the Froude number:

$$v_2 = \frac{Q}{A_2} = \frac{2.52(6)}{4(0.84) + 2(0.84)^2} = \frac{15.1}{4.77} = 3.16 \text{ ft}$$

$$d_h = \frac{A}{t} = \frac{4.77}{4 + 2(2)(0.84)} = \frac{4.77}{7.36} = 0.65 \text{ ft}$$

$$\mathbf{F} = \frac{v}{\sqrt{gd_h}} = 0.69 \qquad \text{still subcritical.}$$

Solution OK.

Transitions that consist of changes in channel width can be treated similar to changes in channel bottom elevation. Again specific energy curves cannot be used directly since they are based on a constant flow per unit width, q. When the channel width changes, q must change as well.

Example Problem 4.23 Channel transition 2

A rectangular channel 10 ft wide is carrying 75 cfs. The channel smoothly narrows to 8 ft in width. The flow depth in the 10-ft section is 2.5 ft. What is the depth in the 8 ft section assuming no energy losses?

Solution

$$\frac{v_1^2}{2g} + y_1 = \frac{v_2^2}{2g} + y_2$$

$$v_1 = \frac{Q}{A} = \frac{75}{10 \times 2.5} = 3 \text{ fps}$$

$$v_2 = \frac{75}{8y_2}$$

$$\frac{3^2}{64.4} + 2.5 = \frac{(75/8y_2)^2}{64.4} + y_2 = \frac{1.36}{y_2^2} + y_2.$$

The solution may be found by trial to be $y_2 = 2.40$ ft. Thus the depth in the 8-ft section is 2.40 ft.

HYDRAULIC JUMP

An example of a flow transition that is abrupt and involves considerable energy loss is a hydraulic jump that involves a sudden transition from supercritical to subcritical flow. In looking at the flow profiles of Fig. 4.23, it can be seen that the profiles approach critical depth nearly vertically. This is also apparent from Eq. (4.53), where as y approaches y_c, **F** approaches 1 and dy/dx approaches infinity. When y approaches y_c as supercritical flow from below, a hydraulic jump may occur as shown in Figure 4.28.

A hydraulic jump cannot be analyzed using the energy equation because there is a large and unknown energy loss in the jump. By assuming that the specific force plus momentum is the same before and after a jump, Eq. (4.15) can be used:

$$\frac{y_1^2}{2} + \frac{q^2}{gy_1} = \frac{y_2^2}{2} + \frac{q^2}{gy_2}.$$

Through algebraic manipulations, it may be shown that

$$\frac{y_2}{y_1} = \frac{1}{2}\left(\sqrt{1 + 8\mathbf{F}_1^2} - 1\right) \qquad (4.56)$$

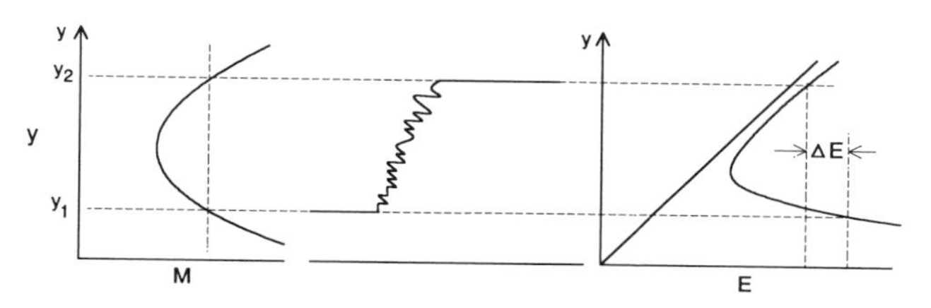

Figure 4.28 Hydraulic jump.

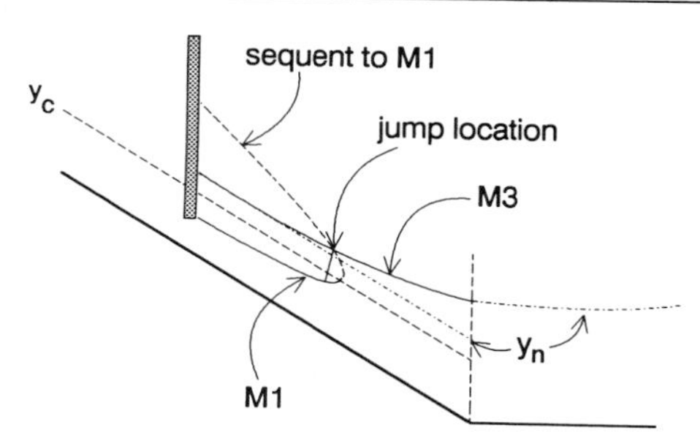

Figure 4.29 Location of a hydraulic jump.

and

$$\frac{y_1}{y_2} = \frac{1}{2}\left(\sqrt{1 + 8\mathbf{F}_2^2} - 1\right).$$

y_1 is known as the initial depth, and y_2 is the sequent depth. A hydraulic jump from $y_1 < y_c$ to $y_2 > y_c$ occurs whenever flow conditions are such that y_1 and y_2 are related by Eq. (4.15), that is momentum is conserved.

The location of a hydraulic jump can be found by plotting flow profiles and superimposing a plot of the possible sequent depth above the supercritical part of the flow. The jump occurs whenever the sequent depth line intersects the downstream flow profile (assuming the jump has zero length). Figure 4.29 illustrates the procedure.

The energy loss in a hydraulic jump can be computed directly from Bernoulli's equation as

$$E_1 = \frac{q^2}{2g}\left(\frac{1}{y_1^2} - \frac{1}{y_2^2}\right) + (y_1 - y_2).$$

Through algebraic manipulations, this relationship becomes

$$E_1 = \frac{(y_2 - y_1)^3}{4y_1 y_2}. \qquad (4.57)$$

Example Problem 4.24 Hydraulic jump

A rectangular channel is carrying 100 cfs. The channel is 10 ft wide and flowing 0.90 ft deep. Is a hydraulic jump possible? If so, what will be the sequent depth? How much energy is lost in the jump?

Solution

$$\mathbf{F}_1 = \frac{v}{\sqrt{gy}} = \frac{100/10(0.9)}{\sqrt{32.2(0.9)}} = 2.06.$$

Since the flow is supercritical, a hydraulic jump is possible.

The sequent depth is computed from Eq. (4.56) as

$$y_2 = \frac{0.9}{2}\left(\sqrt{1 + 8(2.06)^2} - 1\right) = 2.21 \text{ ft}.$$

Note that

$$\mathbf{F}_2 = \frac{100/10(2.21)}{\sqrt{32.2(2.21)}} = 0.54.$$

The depth after the jump is subcritical as it must be

$$E_1 = \frac{(y_2 - y_1)^3}{4y_1 y_2} = \frac{(2.21 - 0.9)^3}{4(2.21)(0.9)} = 0.29 \text{ ft}.$$

This loss can also be determined directly from Bernoulli's equation as

$$\begin{aligned} E_1 &= \frac{v_1^2}{2g} + y_1 - \frac{v_2^2}{2g} - y_2 \\ &= \frac{[100/10(0.9)]^2}{64.4} + 0.9 - \frac{[100/10(2.21)]^2}{64.4} - 2.21 \\ &= 0.29 \text{ ft}. \end{aligned}$$

Hydraulic jumps are accompanied by a great deal of turbulence and energy dissipation. If a hydraulic jump occurs in an erodible area of a channel, considerable degradation of the channel may occur. Hydraulic jumps are often used to provide energy dissipation below spillways and channel drop structures. To ensure that the jump occurs at a controlled location, generally on a reinforced concrete apron, stabilizing blocks are used to add drag forces to the flow. In this case,the momentum equation is modified to

$$\frac{y_1^2}{2} + \frac{q^2}{gy_1} = \frac{y_2^2}{2} + \frac{q^2}{gy_2} + \frac{F_B}{\gamma},$$

where F_B represents the drag force per unit width. The design of energy dissipation devices such as stilling basins is a special area of hydraulics and is discussed in the next chapter. Extensive model studies are often employed with the results presented in the form of dimensionless designs. These designs are then adapted to particular applications by using appropriate scaling factors. The St. Anthony Falls (SAF) stilling basin is an example.

Problems

(4.1) A trapezoidal concrete-lined ditch has a bottom width of 3 ft, a depth of 2 ft, and side slopes of 2 : 1. Estimate the discharge if the channel slope is 1.0%. What is the velocity? What is the Froude number?

(4.2) What will the depth of flow in the channel in problem (4.1) be if the flow rate is 50 cfs? What should be the freeboard?

(4.3) A channel is being designed to carry 30 cfs through a very colloidal stiff clay soil on a slope of 1%. Determine the design dimensions if the side slopes are 1 : 1 using both the tractive force and permissible velocity methods.

(4.4) The 10-year peak flow from a watershed is to be channeled through a grassed waterway of bluegrass on a slope of 4% over erosion resistant soil. The grass may be moved (2 to 5 in.) or unmowed (18 in.). The 10-year peak flow is 100 cfs. Design a grassed waterway to convey the flow.

(4.5) If a straw and erosionet liner is used in the channel of example problem (4.4), will the channel be stable before the vegetation is established under a flow of 10 cfs?

(4.6) Design a trapezoidal channel with 2 : 1 side slopes to carry 70 cfs down a 10% slope. The channel bottom width must be limited to 10 ft because of site considerations.

(4.7) A trapezoidal channel with 2 : 1 side slopes, an 8-ft bottom width, and a slope of 0.15% is flowing 1.4 ft deep. The channel is unlined and constructed in an erodible sandy loam soil. What is the flow rate? Is the channel stable at this flow rate?

(4.8) The channel of problem (4.7) is vegetated with Bermuda grass. What is the flow rate? Would there likely be any problems with this channel?

(4.9) Calculate the critical depth for the channel described in problem (4.7) if it carries 150 cfs.

(4.10) If the channel of problem 4.7 is concrete lined, what is the critical slope for the channel at a flow rate of 150 cfs?

(4.11) An elevated rectangular canal is flowing 3 ft deep. What is the horizontal force per unit length exerted by the water on the canal side walls?

(4.12) What size riprap should line the bottom of a trapezoidal channel with 4 : 1 side slopes, 10-ft bottom width, and a 7% slope? The channel is to carry 130 cfs.

(4.13) What size riprap should be used on the side slopes of the channel of problem (4.12)?

(4.14) A vegetated channel is to be used to carry 50 cfs down a 4% slope. The vegetation is to be Bermuda grass, which may be long or mowed. The soil is an easily eroded sandy loam. Design the channel.

(4.15) Will the channel of problem (4.14) be stable for a flow of 30 cfs prior to establishment of the vegetation? If not, select a temporary liner that might be used during the vegetal establishment period that will permit the safe passage of a flow of 30 cfs. Redesign the channel only if necessary.

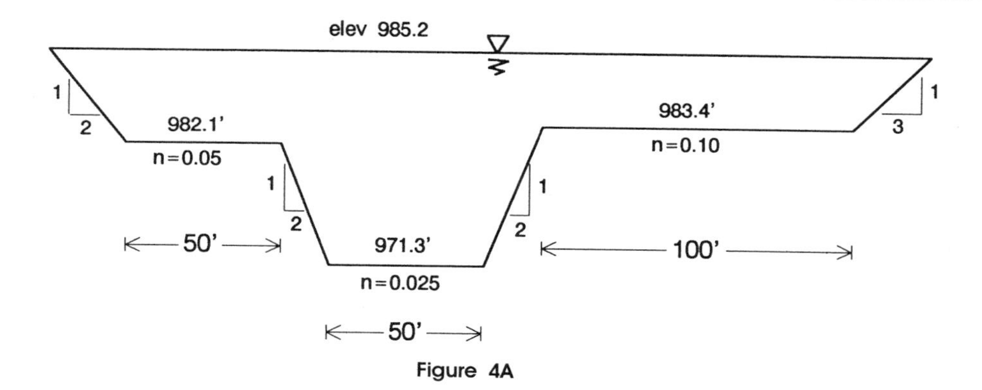

Figure 4A

(4.16) A circular, concrete storm sewer 3 ft in diameter is flowing at a depth of 2.1 ft. The sewer is on a 2% grade. What is the flow rate?

(4.17) What is the flow depth in the drain of problem (4.16) if it carries 25 cfs?

(4.18) What size circular, concrete storm drain would be required to carry 75 cfs down a 3% slope without surcharging the drain (i.e., always flowing as open channel flow)?

(4.19) Work problem (4.18) for circular corrugated metal pipe.

(4.20) Calculate the flow in the channel shown in Fig. 4A. The slope of the channel is 0.05%.

(4.21) At what depth would the channel of problem (4.20) be flowing if it were carrying 6000 cfs?

(4.22) Design a riprap-lined channel to carry 75 cfs down a 7% slope. Specify the required riprap size as well as the specifications of the filter material.

(4.23) What type of temporary lining should be used in a road ditch channel required to carry 10 cfs down a 7% slope?

(4.24) A 25 foot wide rectangular channel with a Manning's n of 0.025 is carrying 5000 cfs. The slope of the channel is 0.05%. At station 22 + 50 the slope of the channel changed abruptly to 5%. Calculate the flow profile in the upper channel from the channel break to a point where the depth is equal to 95% of normal depth.

(4.25) Calculate the flow profile in the lower channel to a point where the depth is equal to 95% of normal depth for the situation of problem 4.24.

(4.26) A wide rectangular channel has a slope of 5% and a Manning's n of 0.02. The channel slope changes abruptly to 0.04%. The flow rate is 12 cfs/foot of width. Calculate the resulting flow profile.

(4.27) A trapezoidal channel goes through a smooth transition. The flow depth is originally normal depth. The flow rate is 50 cfs. If there is no loss in energy, what will be the depth of flow immediately after the transition? The channel properties are:

	Upstream	Downstream
b	10 ft	8 ft
z	3 : 1	2 : 1
s	0.05%	0.05%

(4.28) Solve problem (4.27) as if the two channels are reversed so that the upstream channel becomes the downstream channel.

(4.29) A rectangular channel narrows from 20 ft to 15 ft. The bottom elevation simultaneously drops 2 ft. Both changes are smooth with little loss in energy. The flow rate is 400 cfs. What is the depth of flow downstream from the transition if the upstream depth is 4 ft?

(4.30) Work problem (4.29) as if the channel widens from 15 to 20 ft and the bottom elevation is raised by 2 ft.

(4.31) A hydraulic jump occurs in a wide channel where the flow is initially at a depth of 1 ft and a flow velocity of 14 fps. What is the depth after the jump? What is the energy loss within the jump?

(4.32) A rectangular channel has a Manning's n of 0.02, a slope of 0.1%, and a flow rate of 10 cfs/ft of width. Water enters the channel as supercritical flow. A hydraulic jump occurs. What must be the depth before and after the jump? How much energy is lost?

(4.33) Supercritical flow encounters some stabilizing blocks within a stilling basin. The drag force introduced by the blocks is given by $C_D \rho A v^2/2$, where C_D is a drag coefficient (use $C_D = 1$), ρ is the density of water (1.94 slugs/ft^3), and A is the cross-sectional area of the block perpendicular to the flow. The blocks are 1 ft high and occupy 75% of the flow cross section at a depth of 1 ft. Water enters the stilling well and strikes the blocks. The depth of flow is initially 2 ft with a flow rate of 25 cfs/ft. What is the downstream depth immediately after the hydraulic jump? How much energy is lost in the jump?

(4.34) Define the following terms:
(a) uniform flow
(b) supercritical flow
(c) subcritical flow
(d) steady flow
(e) gradually varied flow
(f) rapidly varied flow
(g) energy grade line
(h) velocity head
(i) pressure head
(j) flow profiles
(k) Froude number
(l) head loss.

(4.35) Flow in a wide rectangular channel encounters a barrier and an overflow spillway as shown in Fig. 4.B. Calculate the flow profile from the spillway back up to the channel to a point where the flow is within 0.2 ft of normal depth. The flow rate is 25 cfs/ft, the channel slope is 0.1%, and Manning's n is 0.015. The barrier is 3 ft high.

(4.36) Calculate the flow profiles resulting from the flow situation shown in Fig. 4.C. The underflow gate is 1000 feet upstream from the barrier. The barrier is 3 ft high. The channel slope is 0.1% slope, Manning's n is 0.015, and the flow rate is 25 cfs/ft. If a hydraulic jump will occur, locate the jump neglecting the length of the jump. The underflow gate clearance is 1.0 ft.

(4.37) What are surface water profiles used for?

(4.38) What impact might levees used to protect a particular region have on flood peaks upstream and downstream from the protected area?

(4.39) Water is flowing at 13 cfs/ft in a wide rectangular channel. What is the critical depth?

(4.40) A stream has a slope of 0.03%, a hydraulic radius of 2.2 m, and an average velocity of 1.2 m/sec. Estimate Manning's n. If the channel is 50 m wide, estimate the discharge in m^3/sec.

(4.41) A rectangular channel is carrying 10 cfs/ft of width. (a) Construct a specific energy diagram, and (b) construct a specific force and momentum diagram.

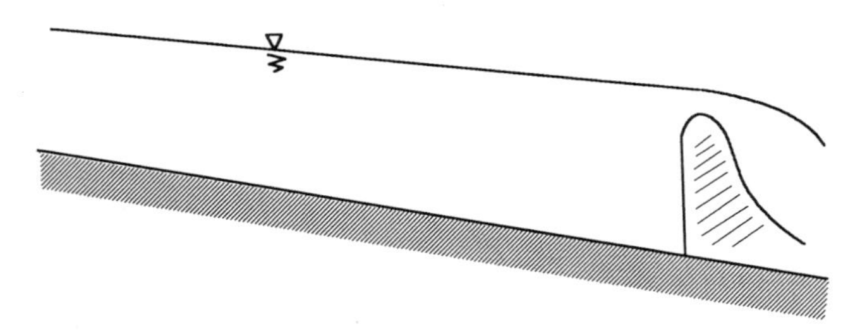

Figure 4B

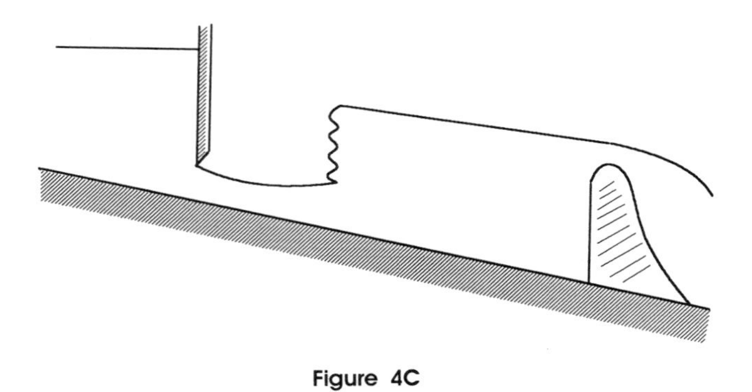

Figure 4C

(4.42) A hydraulic jump occurs in the channel of problem (4.41) with $y = 1.0$ ft. Use the diagram constructed for problem (4.41) to determine y_2 and the energy loss.

(4.43) Show that a generalized Froude number may be written as $\mathbf{F} = \sqrt{Q^2 t/gA^3}$, where t is the top width.

References

Abt, S. R., *et al.* (1988). Resistance to flow over riprap in steep channels. *Water Resources Bull.* **24**(6):1193–1200

Anderson, A. G. (1973). Tentative design procedure for riprap lined channels—Field evaluation, Project report 146. Prepared for Highway Research Board. [Original not seen, cited in Normann (1975)]

Anderson, A. G., Paintal, A. A., and Davenport, J. T. Tentative design procedure for riprap lined channels, NCHRP Report 108. Highway Research Board, National Academy of Sciences, Washington, DC. [Original not seen, cited in Normann (1975).]

Chow, V. T. (1959). "Open Channel Hydraulics." McGraw–Hill, New York.

Fortier, S., and Scobey, F. S. (1926). Permissible canal velocities. *Trans. Am. Soc. Civil Eng.* **89**:940–984.

Gessler, J. (1971). Beginning and ceasing of sediment motion. *In* "River Mechanics" (H. W. Shen, ed.), Chap. 7. Water Resources Publications, Ft. Collins, CO.

Henderson, F. M. (1966). "Open Channel Flow." MacMillan, New York.

Lane, E. W. (1955). Design of stable channels. *Trans. Am. Soc. Civil Eng.* **120**: 1234–1260.

Lane, E. W. and Carlson, E. J. (1953). Some factors affecting the stability of canals constructed in course granular materials. *In* "Proceedings, Minnesota International Hydraulics Convention, Joint meeting IAHR and Hydraulics Division, ASCE."

McWhorter, J. C., Carpenter, T. G., and Clark, R. N. (1968). Erosion control criteria for drainage channels, Study conducted for the Mississippi State Highway Department and the Federal Highwayway Administration. Department of Agricultural Engineering, Mississippi State University, State College, MS.

Normann, J. M. (1975). Design of stable channels with flexible linings, Highway engineering circular No. 15. Federal Highway Administration, U.S. Department of Transportation, Washington, DC.

Ree, W. O. (1949). Hydraulic characteristics of vegetation for vegetated waterways. *Agric. Eng.* **30**:184–189.

Schwab, G. O., Frevert, R. K., Edminister, T. W., and Barnes, J. K. (1966). "Soil and Water Conservation Engineering," 2nd ed. Wiley, New York.

Schwab, G. O., Frevert, R. K., Barnes, K. K., and Edminister, T. W. (1971). "Elementary Soil and Water Engineering," 3rd ed. Wiley, New York.

Simons, D. B., and Senturk, F. (1977). "Sediment Transport Technology." Water Resources Publications, Ft. Collins, CO.

Simons, D. B., and Senturk, F. (1992). "Sediment Transport Technology" (2nd ed.). Water Resources Publications, Ft. Collins, CO.

Soil Conservation Service (1969). "Engineering Field Manual for Conservation Practices." Soil Conservation Service, U.S. Department of Agriculture, Washington, DC.

Soil Conservation Service (1979). "Engineering Field Manual." Soil Conservation Service, U.S. Department of Agriculture, Washington, DC.

Stevens, M. A., and Simons, D. B. (1971). Stability analysis for coarse granular material on slopes. *In* "River Mechanics" (H. W. Shen, ed.), Chap. 17. Water Resources Publications, Ft. Collins, CO.

Temple, D. M., *et al.* (1987). "Stability Design of Grass-Lined Open Channels," ARS Agricultural Handbooks 667. U.S. Department of Agriculture, Washington, DC.

U.S. Army Corps of Engineers (1982). "HEC-2 Water Surface Profiles Users Manual." U.S. Army Corps of Engineers, Hydrologic Engineering Center, Davis, CA.

5 Hydraulics of Structures

INTRODUCTION

The need to measure and control flows under either open channel or pipe flow conditions has been a concern for engineers dating back hundreds of years. Whether the desire was to measure irrigation water being applied to dry croplands or control rampaging streams, structures provided potential solutions. Structures vary from very small weirs to large spillways with energy dissipators and include both inlet and outlet devices. The tremendous variability in physical size, flow capacity, and materials contribute to difficulties in designing structures.

HYDRAULICS OF FLOW CONTROL DEVICES

Introduction

Flow control devices may operate as either open channel flow in which the flow has a free water surface or pipe flow in which the flow is in a closed conduit under pressure. Although the hydraulic principles are similar for each, the application of these principles for describing the hydraulics of flow control devices varies considerably. Simple flow control devices date back at least to the ancient Egyptians who sought to control the Nile River's yearly flooding. Although the structures have changed in terms of materials and the engineer's understanding of why they work, many of the devices utilized today are remarkably similar to those that were in place hundreds of years ago.

A basic understanding of the hydraulics of flow control devices provides a basis for developing adequate design. As head increases on a structure, the flow that is discharged through the structure increases. Typical head–discharge relationships for idealized flow controls are shown in Fig. 5.1. Each of these relationships illustrates that discharge increases proportionally to the head on the flow control device. An example of a flow control structure is a principal spillway. An engineer uses a principal spillway as part of a dam design to control the rate at which water is discharged.

Weirs as Flow Control Devices

Probably the simplest and least expensive flow control device available for installation in a channel is a weir, which is simply an obstruction placed in a channel so that the flow is constricted as it goes over a crest (see Fig. 5.1). The crest is the edge of the weir over which the water flows. As the water level rises above the crest, the flow rate increases dramatically. Weirs can have a crest that is generally thin, or sharp crested. Sharp-crested weirs are constructed from sheet metal or similar thin material so that the flow over the weir, or nappe, springs free as it leaves the upstream face of

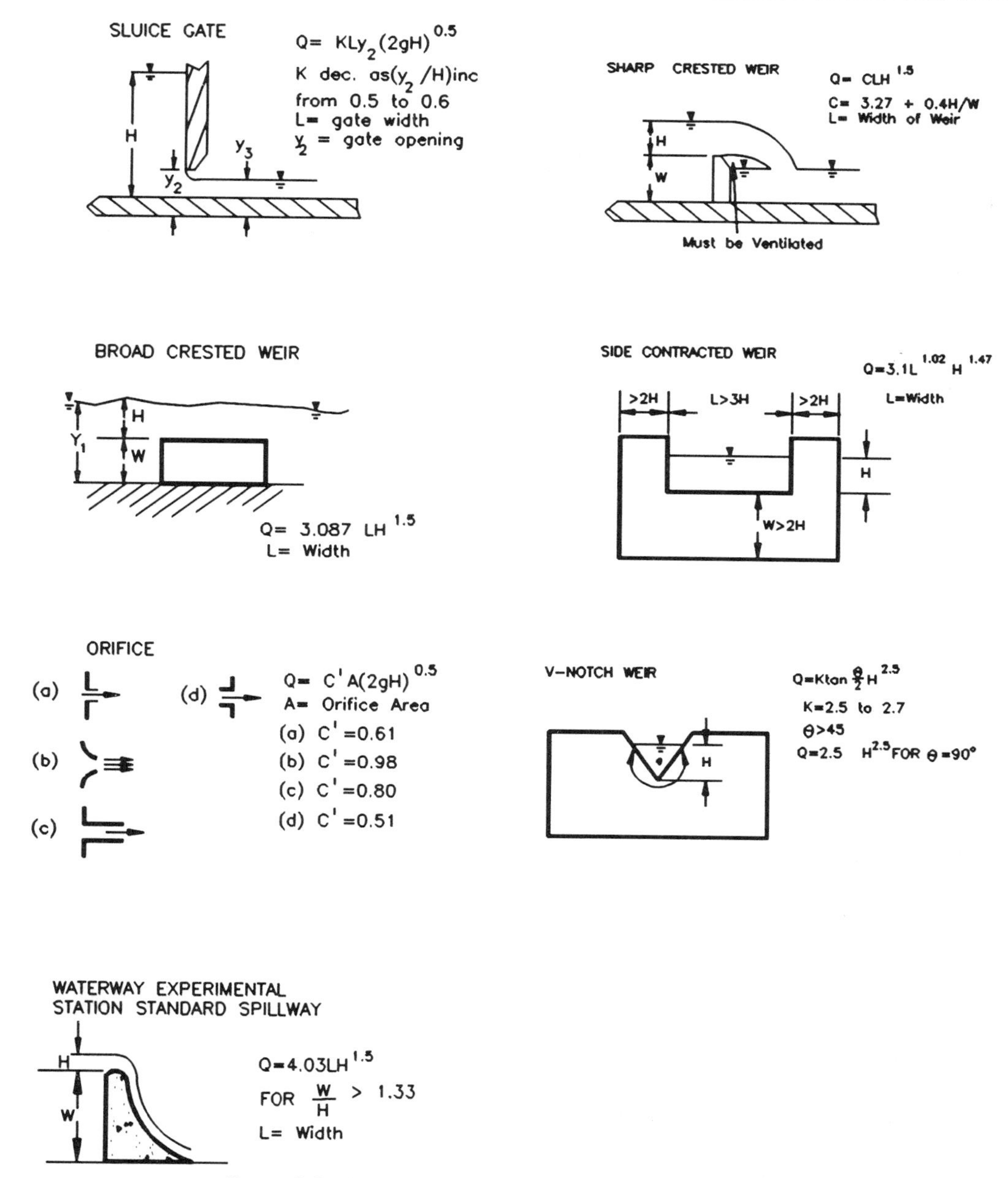

Figure 5.1 Typical head–discharge relationships (Kao, 1975).

the weir. Broad-crested weirs are sometimes used where a structure previously existed or where debris may damage a sharp-crested weir. Broad-crested weirs are discussed in a subsequent section.

Sharp-crested weirs can have several shapes, including rectangular, triangular, trapezoidal, or a combination of these, to provide the desired sensitivity at the required flow capacity. A weir is classified according to the shape of its notch. Triangular (also called V-notch) weirs have greater control under low flow conditions than do rectangular weirs and are often used where precise flow measurement is desired. Conversely, rectangular weirs have large capacity but have less sensitivity for flow measurement. A weir can vary in physical size from quite small to very large. Consequently, the controlled discharge can vary substantially. The discharge across a rectangular weir is defined by the equation

$$Q = CLH^{3/2}, \tag{5.1}$$

where Q is discharge in cubic feet per second, C is the weir coefficient (dependent upon units and weir shape),

L is weir length in feet, and H is head in feet. For a circular inlet, such as the riser on a drop inlet, L is the circumference of the pipe.

Values of C can be found in hydraulic references for many shapes. Values of C from 3.0 to 3.2 are generally used for a rectangular weir. The length L is the total length over which flow crosses the weir. If the weir coefficient and weir length are known, discharge is a function of head only. Head is the difference in elevation between the lowest point on the weir crest and the water surface elevation plus the velocity head. To avoid having to estimate velocity head, H should be measured at a location at least three times the maximum design head upstream of the weir at a point where the velocity head is negligible. The crest should be located at least two times the maximum head above the channel bottom in order to reduce the likelihood of submergence. Submergence (tailwater approaching the crest of the inlet section) can significantly decrease the capacity of the structure. The designer is cautioned to consider this possibility since it necessitates the use of different equations or field calibration of the weir under the submerged conditions. If the downstream channel is not adequate to handle the design flow, flow may be retarded and lead to submergence. Flow should spring clear of the downstream portion of the weir so that an air pocket forms beneath the nappe. Air is continuously removed from this pocket by the overflowing jet. The pressure in this pocket should be kept constant or the weir will have undesirable characteristics. French (1985) stated two such characteristics. As air pressure in the pocket decreases, the curvature of the overflowing jet increases, and the value of the coefficient of discharge will increase also. Alternatively, if the supply of air is irregular, the jet will vibrate. Flow over the weir will become unsteady, which can lead to failure of the structure.

Triangular weirs have angles ranging from 22.5° to 90° with 90° being most prevalent (see Fig. 5.1). The basic equation for discharge through a triangular weir, neglecting the velocity of approach and with no submergence, is

$$Q = C \tan(\theta/2) H^{5/2}, \qquad (5.2a)$$

where θ is the notch angle. For an angle of 90°, $\tan \theta/2$ is 1.0; hence

$$Q = CH^{5/2}. \qquad (5.2b)$$

For a 90° triangular weir, C is typically 2.5.

Example Problem 5.1 Weir flow

A 24-in. circular, vertical riser constructed from corrugated metal pipe (CMP) serves as the inlet for the principal spillway of a detention structure. Estimate the discharge if the head on the riser is 1 ft and weir flow exists. Assume the weir coefficient C equals 3.0.

Solution: Weir flow control will occur at low head. The governing equation (using $C = 3.0$) will be

$$Q = 3.0LH^{3/2},$$

where the length L is the circumference of the pipe, which equals πD or π (2 ft) or 6.3 ft. Therefore substituting L yields the equation for discharge as a function of head as

$$Q = 18.9\, H^{3/2}. \qquad (a)$$

With a head of 1 ft, the discharge Q under weir flow would be 18.9 cfs.

Grant (1978) presents an eloquent discussion of weirs, flumes, and open channel flow measurement. He discusses general requirements that lead to precise flow measurement using weirs. These criteria include construction techniques, installation, and head measurement. Grant (1978) also provides detailed information on the selection of weirs and other control devices, and he includes equations for other angles of triangular weirs, as well as other shapes.

Orifices as Flow Control Devices

An orifice is an opening through which flow occurs. Orifices can be used to control flow, as in the case of the drop inlet shown in Fig. 5.2, or they can be placed either in a pipe or at the end of a pipe to measure flow. As water flows through the opening, it can be measured because its discharged velocity through the opening is a function of head on the orifice. Orifices provide a simple means to measure pipe flow. The equation for orifice flow is

$$Q = C'A(2gH)^{1/2}, \qquad (5.3)$$

where C' is the orifice coefficient, A is the cross-sectional area of the orifice in square feet, g is the gravitational constant, and H is the head on the orifice as shown in Fig. 5.2. The leading edge of an orifice can be rounded or sharp. C' is 0.6 for sharp-edged orifices. Values for other shapes are provided in Fig. 5.1 and Hoffman (1974). Streeter (1971) presents a discussion of the theory from which Eq. (5.3) is derived. Equation (5.3) is developed in a less rigorous format in Soil Conservation Service (1984).

Example Problem 5.2 Orifice flow

Estimate the discharge through a 24-in. circular, vertical riser such as described in Example Problem 5.1 if the head is 1 ft and the riser is functioning as an orifice.

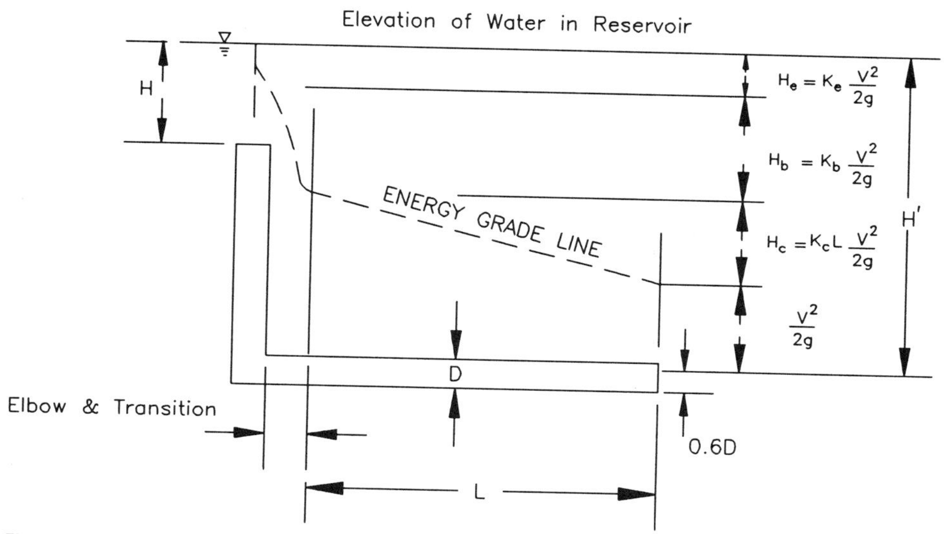

Figure 5.2 Energy losses for flow in a drop inlet spillway considering bend losses and entrance losses separately.

Solution: The discharge under orifice flow will equal

$$Q = C'a(2gH)^{1/2}.$$

The area of 24-in. pipe is 3.14 ft^2. Assuming a value of 0.6 for C' since the riser is corrugated metal pipe and substituting values including the gravitational constant, we have

$$Q = 0.6(3.14)\sqrt{2(32.2)H},$$

which reduces to

$$Q = 15.1H^{1/2}.$$

Substituting a head equal to 1 ft into the equation yields $Q = 15.1$ cfs for orifice flow.

Pipes as Flow Control Devices

A drop inlet spillway consists of a vertical pipe called a riser and a nearly horizontal pipe called a barrel. This spillway can serve as a flow control device, even when operating under pipe flow. A schematic showing energy losses with pipe flow is given in Fig. 5.2. When the water level shown in Fig. 5.2 rises to a point such that the pipe flows full, the total head causing flow is given by H' (as shown in Fig. 5.2) instead of H as it was for weir and orifice control. This head is dissipated as entrance head loss, transition head loss, bend head loss, friction head loss, and velocity head. Frequently, in pipes used to drain detention reservoirs, the only transitions and bends are at the connection between the drop inlet and the bottom pipe. If head losses are given in terms of a head loss coefficient times the velocity head, $V^2/2g$, and the transition and bend head losses are combined into a single head loss term, then the total head H' can be written as

$$H' = \frac{V^2}{2g}(1 + K_e + K_b + K_c L), \qquad (5.4)$$

where H' is the head on the pipe as shown in Fig. 5.2, K_e is the entrance head loss coefficient, K_b is the bend head loss coefficient, K_c is the head loss coefficient due to friction, L is the length of the pipe (including the riser), and V is the mean velocity in the pipe. A schematic showing the head loss terms is given in Fig. 5.2. Since discharge through the pipe is equal to velocity times area, Eq. (5.4) can be solved for discharge as

$$Q = \frac{a(2gH')^{1/2}}{(1 + K_e + K_b + K_c L)^{1/2}}, \qquad (5.5)$$

where Q is discharge and a is cross-sectional area of the pipe. Values for K_c are given in Tables 5.1 and 5.2 for circular and square pipes. Values for K_e and K_b depend on the configuration of the entrance and the bend. Typical values for K_e and K_b are 1.0 and 0.5, respectively. Brater and King (1976), as well as Hoffman (1974), can be consulted for further details.

For risers with rectangular inlets, the bend head losses are frequently combined with the entrance head losses into one term. The total head dissipated through the riser can then be written as

$$H' = \left(\frac{V^2}{2g}\right)(1 + K_e' + K_c L) \qquad (5.6)$$

Table 5.1 Head Loss Coefficients for Circular Conduits Flowing Full[a]

Head loss coefficient, K_c, for circular pipe flowing full
$K_c = 5087\ n^2/D^{4/3}$
(Note: Pipe diameter, D, is in inches)

Pipe diameter (in.)	Flow area (ft²)	Manning's coefficient of roughness, n															
		0.010	0.011	0.012	0.013	0.014	0.015	0.016	0.017	0.018	0.019	0.020	0.021	0.022	0.023	0.024	0.025
6	0.196	0.0467	0.0565	0.0672	0.0789	0.0914	0.1050	0.1194	0.1348	0.1510	0.1680	0.1870	0.2060	0.2260	0.2470	0.2690	0.2920
8	0.349	0.0318	0.0385	0.0458	0.0537	0.0623	0.0715	0.0814	0.0919	0.1030	0.1148	0.1272	0.1400	0.1540	0.1680	0.1830	0.1990
10	0.545	0.0236	0.0286	0.0340	0.0399	0.0463	0.0531	0.0604	0.0682	0.0765	0.0852	0.0944	0.1041	0.1143	0.1249	0.1360	0.1480
12	0.785	0.0185	0.0224	0.0267	0.0313	0.0363	0.0417	0.0474	0.0535	0.0600	0.0668	0.0741	0.0817	0.0896	0.0980	0.1067	0.1157
14	1.069	0.0151	0.0182	0.0217	0.0255	0.0295	0.0339	0.0386	0.0436	0.0488	0.0544	0.0603	0.0665	0.0730	0.0798	0.0868	0.0942
15	1.230	0.0138	0.0166	0.0198	0.0232	0.0270	0.0309	0.0352	0.0397	0.0446	0.0496	0.0550	0.0606	0.0666	0.0727	0.0792	0.0859
16	1.400	0.0126	0.0153	0.0182	0.0213	0.0247	0.0284	0.0323	0.0365	0.0409	0.0455	0.0505	0.0556	0.0611	0.0667	0.0727	0.0789
18	1.770	0.01078	0.0130	0.0155	0.0182	0.0211	0.0243	0.0276	0.0312	0.0349	0.0389	0.0431	0.0476	0.0522	0.0570	0.0621	0.0674
21	2.410	0.00878	0.01062	0.0126	0.0148	0.0172	0.0198	0.0225	0.0254	0.0284	0.0317	0.0351	0.0387	0.0425	0.0464	0.0506	0.0549
24	3.140	0.00735	0.00889	0.01058	0.0124	0.0144	0.0165	0.0188	0.0212	0.0238	0.0265	0.0294	0.0324	0.0356	0.0389	0.0423	0.0459
27	3.980	0.00628	0.00760	0.00904	0.01061	0.0123	0.0141	0.0161	0.0181	0.0203	0.0227	0.0251	0.0277	0.0304	0.0332	0.0362	0.0393
30	4.910	0.00546	0.00660	0.00786	0.00922	0.01070	0.01228	0.0140	0.0158	0.0177	0.0197	0.0218	0.0241	0.0264	0.0289	0.0314	0.0341
36	7.070	0.00428	0.00518	0.00616	0.00723	0.00839	0.00963	0.01096	0.0124	0.0139	0.0154	0.0171	0.0189	0.0207	0.0226	0.0246	0.0267
42	9.620	0.00348	0.00422	0.00502	0.00589	0.00683	0.00784	0.00892	0.01007	0.01129	0.0126	0.0139	0.0154	0.0169	0.0184	0.0201	0.0218
48	12.570	0.00292	0.00353	0.00420	0.00493	0.00572	0.00656	0.00747	0.00843	0.00945	0.01053	0.01166	0.0129	0.0141	0.0154	0.0168	0.0182
54	15.900	0.00249	0.00302	0.00359	0.00421	0.00488	0.00561	0.00638	0.00720	0.00808	0.00900	0.00997	0.01099	0.0121	0.0132	0.0144	0.0156
60	19.630	0.00217	0.00262	0.00312	0.00366	0.00424	0.00487	0.00554	0.00622	0.00702	0.00782	0.00866	0.00955	0.01048	0.0115	0.0125	0.0135

[a] From Soil Conservation Service (1951).

or

$$Q = \frac{a(2gH')^{1/2}}{(1 + K'_e + K_c L)^{1/2}}, \tag{5.7}$$

where K'_e is the combined entrance and bend head loss term. By providing a smooth transition, the value for K'_e can be reduced. Typical values of K'_e are given in Table 5.3.

Frequently when the drop inlet is the same size as the remainder of the pipe, orifice flow will control, and the pipe will never flow full. In this case, it may be necessary to increase the size of the drop inlet in order to utilize the full capacity of the pipe.

Example Problem 5.3 Pipe flow

An 24-in.-diameter corrugated metal pipe (CMP) is attached to the 24-in. vertical riser described in Problems 5.1 and 5.2. It is being used as the principal spillway for a detention structure. The pipe is 60 ft long and has one 90° bend. The top of the inlet riser is 15 ft above the bottom of the outlet. Assume a free outfall and estimate the discharge under pipe flow if the water elevation 30 ft from the inlet is 1 ft higher than the top of the riser.

Solution: For pipe flow, we have

$$Q = \frac{a(2gH')^{1/2}}{(1 + K_e + K_b + K_c L)^{1/2}}$$

where $K_e \approx 1.0$ for most entrances of interest and $K_b = 0.5$. Manning's n for CMP is approximately 0.024 (see Table 4.1 for a range of values for CMP). Using this value in Table 5.1, $K_c = 0.042$. Head for pipe flow is the distance from the water surface to a point $0.6D$ above the outlet as shown in Fig. 5.2 and 5.3. H' then is given in terms of the stage, H, by

$$H' = H + 15 - 0.6(2.0) = H + 13.8.$$

Table 5.2 Head Loss Coefficients for Square Conduits Flowing Full[a]

$K_c = 29.16n^2/R^{4/3}$

Conduit size (ft)	Flow area (ft^2)	Manning coefficient of roughness, n				
		0.012	0.013	0.014	0.015	0.016
2×2	4.00	0.01058	0.01212	0.01440	0.01653	0.01880
$2\frac{1}{2} \times 2\frac{1}{2}$	6.25	0.00786	0.00922	0.01070	0.01228	0.01397
3×3	9.00	0.00616	0.00723	0.00839	0.00963	0.01096
$3\frac{1}{2} \times 3\frac{1}{2}$	12.25	0.00502	0.00589	0.00683	0.00784	0.00892
4×4	16.00	0.00420	0.00493	0.00572	0.00656	0.00746
$4\frac{1}{2} \times 4\frac{1}{2}$	20.25	0.00359	0.00421	0.00488	0.00561	0.00638
5×5	25.00	0.00312	0.00366	0.00425	0.00487	0.00554
$5\frac{1}{2} \times 5\frac{1}{2}$	30.25	0.00275	0.00322	0.00374	0.00429	0.00488
6×6	36.00	0.00245	0.00287	0.00333	0.00382	0.00435
$6\frac{1}{2} \times 6\frac{1}{2}$	42.25	0.00220	0.00258	0.00299	0.00343	0.00391
7×7	49.00	0.00199	0.00234	0.00271	0.00311	0.00354
$7\frac{1}{2} \times 7\frac{1}{2}$	56.25	0.00182	0.00213	0.00247	0.00284	0.00323
8×8	64.00	0.00167	0.00196	0.00227	0.00260	0.00296
$8\frac{1}{2} \times 8\frac{1}{2}$	72.25	0.00154	0.00180	0.00209	0.00240	0.00273
9×9	81.00	0.00142	0.00167	0.00194	0.00223	0.00253
$9\frac{1}{2} \times 9\frac{1}{2}$	90.25	0.00133	0.00156	0.00180	0.00207	0.00236
10×10	100.00	0.00124	0.00145	0.00168	0.00193	0.00220

[a]From Soil Conservation Service (1951).

Table 5.3 Entrance Loss Coefficients in Drop Inlet Spillways with Rectangular Drop Inlets[a]

Description of Spillway	Minimum clear water K'_e	Maximum with debris K'_e
1. Round conduit and standard covered top riser, except with special elbow and transition		
$D \times 1.5D$ Riser	0.65	0.75[b]
$D \times 2D$ Riser	0.41	0.50[b]
$D \times 3D$ Riser	0.25	0.35[b]
$D \times 5D$ Riser	0.17	0.30[b]
2. Round conduit and standard covered top riser, with round bottom and square-edged entrance to conduit		
$D \times 3D$ Riser	0.60[b]	0.70[b]
3. Round conduit and standard rectangular open top riser, with round bottom and square-edged entrance to conduit		
$D \times 3D$ Riser	0.50[b]	0.90[b]
4. Round conduit and standard rectangular open top riser, with flat bottom and square-edged entrance to conduit		
$D \times 3D$ Riser	0.60[b]	1.10[b]
5. Round conduit and standard square open top riser, with flat bottom and square-edged entrance to conduit		
$(D + 12) \times (D + 12)$ Riser	1.20	2.00[b]

[a]Soil Conservation Service (1969).
[b]Estimated values.

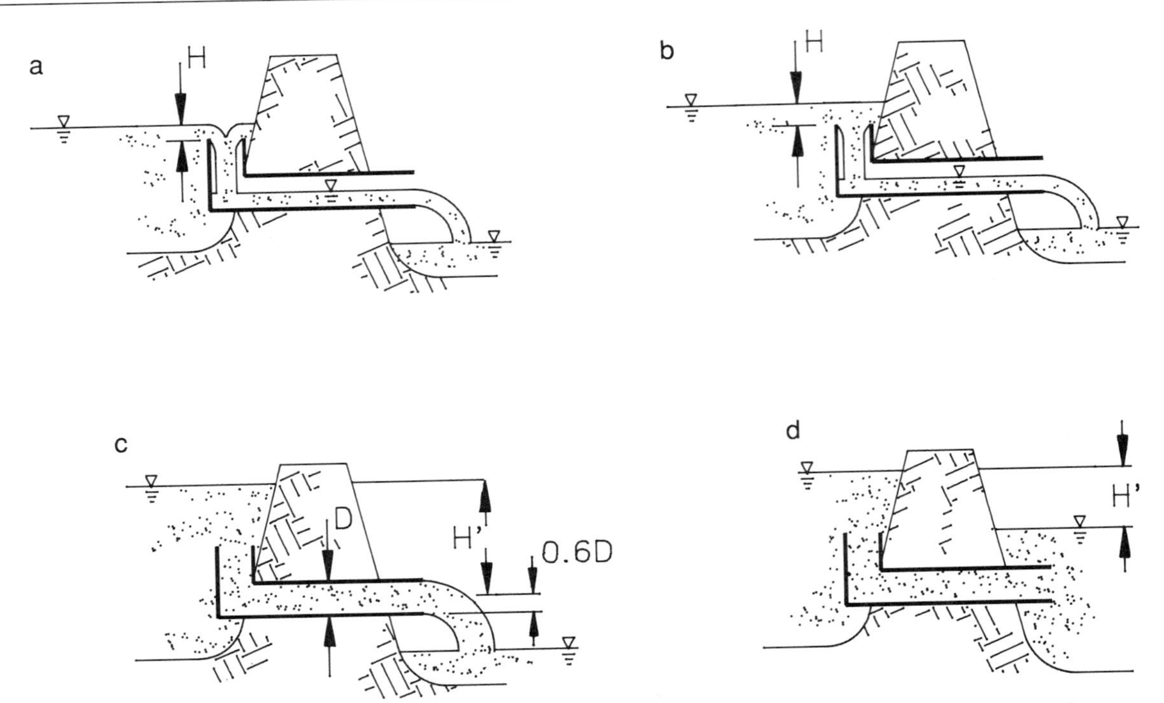

Figure 5.3 Illustration of drop inlet spillway as a flow control structure: (a) weir control, (b) orifice control, (c) pipe flow with free outfall, and (d) pipe flow with tailwater control.

The area a of the 24-in. pipe is 3.14 ft^2 (computed or found in Table 5.1). Substituting into the pipe flow equation

$$Q = \frac{3.14[2(32.2)(H + 13.8)]^{1/2}}{[1 + 1.0 + 0.5 + (0.042)(60)]^{1/2}}$$

which simplifies to

$$Q = 11.25(H + 13.8)^{1/2}.$$

Substituting a value of H equal to 1 ft yields a Q equal to 43 cfs if pipe flow exists. To determine if pipe flow exists for a drop inlet spillway, the discharge computed by pipe flow, orifice flow, and weir flow would be compared. If the pipe flow discharge is the smaller of the three, pipe flow would occur.

Using Flow Control Structures as Spillways

A given spillway can have a variety of stage discharge relationships, depending on the head. An example of the impact of flow changes as a result of changing head for a drop inlet is shown in Fig. 5.3. When the water level is just above the riser crest (very low head), the riser crest acts like a weir, and flow is weir controlled. When the flow is weir controlled, the water level inside the drop inlet is lowest near the center of the inlet. As the water level in the reservoir increases, water flowing in from all sides of the inlet interferes so that the inlet begins to act like an orifice. As the head continues to increase, the outlet eventually begins to flow full, and pipe flow prevails as shown in Fig. 5.3. A stage–discharge curve is developed by plotting Q versus H for each of the three relationships (weir, orifice and pipe flow). The minimum flow for a given head is the actual discharge used. This process is demonstrated in the next problem.

Example Problem 5.4 Stage–discharge curve

An 18-in. diameter CMP with an 18-in. vertical riser is being used as the principal spillway for a detention structure. The pipe is 50 ft long and has one 90° bend. The top of the inlet riser is 10 ft above the bottom of the outlet. Develop a stage–discharge curve. Assume a free outfall.

Solution: Weir flow control will occur first. The governing equation (using $C = 3.0$) is

$$Q = 3.0\ LH^{3/2}.$$

For an 18-in. pipe, L is the circumference of the pipe, which is 4.7 ft; therefore

$$Q = 14.1H^{3/2}. \qquad \text{(a)}$$

After a certain depth of flow occurs, the discharge may be orifice controlled or

$$Q = C'a(2gH)^{1/2}.$$

The area of an 18-in. pipe is 1.77 ft^2. Assuming a value of 0.6 for C',

$$Q = 8.51H^{1/2}. \qquad \text{(b)}$$

For pipe flow, Eq. (5.7) is used, or

$$Q = \frac{a(2gH')^{1/2}}{(1 + K_e + K_b + K_c L)^{1/2}}. \qquad \text{(c)}$$

Use $K_e = 1.0$ and $K_b = .5$. K_c is determined from Table 5.1 (using an n of 0.025) as 0.07. The head is determined for this case to be the distance from the surface to a point 0.6 diameters above the outlet as shown in Fig. 5.2. The head H' then is given in terms of the stage H by

$$H' = H + 10 - (0.6)(1.5),$$

which simplifies to

$$H' = H + 9.1.$$

Therefore

$$Q = \frac{1.77[2g(H + 9.1)]^{1/2}}{(1 + 1.0 + 0.5 + 0.07 \times 50)^{1/2}}$$

$$Q = 5.80(H + 9.1)^{1/2}. \qquad \text{(d)}$$

A stage–discharge curve is developed by plotting Q versus H for each of the relationships given in Eqs. (a), (b), and (d). These calculations can be done quite easily using a computer spreadsheet. The minimum flow for any head represents the actual flow. A plot of the stage–discharge relationship is given in Fig. 5.4.

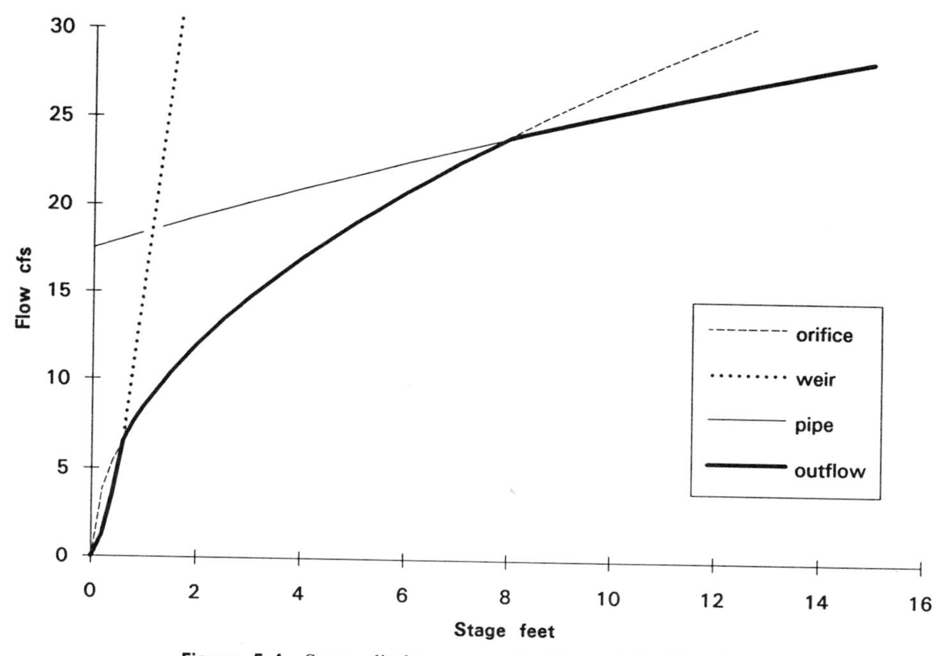

Figure 5.4 Stage–discharge curve for Example Problem 5.4.

Rockfill Outlets as Controls

Rock is by far the most abundant, and generally available, building material on earth and can often be obtained quite inexpensively. The major expenses associated with rock are grading, transporting, and placing the stone. The relative permanence of rock was recognized ages ago and rock has often been used to construct structures that are designed for long life. Rock is used as a hydraulic control for many purposes such as to construct protective channel linings and breakwaters, add stability to dams, and provide energy dissipation zones for reservoir outlets. Recently rockfill has been used as a flow control structure. As such, information on hydraulics of flow in rockfill is needed.

Analytical Procedures for Calculating Flow Hydraulics

The study of flow through rockfill has generally occurred at low Reynolds numbers in laboratory conditions. Relatively few *insitu* measurements have occurred because of the difficulties in describing the stone shapes and controlling the flow so that the desired flow conditions occur. Much of the research in this area has dealt with solution of groundwater and well problems. An example of this work is that of Stephenson (1979) who sought to correlate research of flow/head loss relationships for porous granular media over a wide range of Reynolds numbers (10^{-4} to 10^4). His work, using uniform rock sizes, showed that head loss is proportional to the flow velocity squared. Based on this effort, the head loss equation can be solved analytically for many cases in which the flow is fully developed turbulent flow.

The analysis of flow in rock media with uniform diameters utilizes a variation of the Darcy–Weisbach equation

$$\frac{dh}{dl} = f\frac{1}{d}\frac{V_p^2}{2g}, \tag{5.8}$$

where dh/dl is the gradient of head through the rock fill, f is the Darcy–Weisbach friction factor, d is the average diameter of the rock, V_p is the velocity in the pores, and g is acceleration of gravity. This is a variation of Eq. (4.21). In the form used by Stephenson (1979) for rock fill, the constant 2 is left out of the equation and a macro-velocity is substituted for pore velocity. The macro-velocity is the velocity one would have if the flow through the pores were distributed uniformly over the entire cross section (see discussion in Chapter 11) and is related to the pore velocity by the porosity, or

$$V_\mathrm{p} = V/\xi, \tag{5.9}$$

where V is the macro-velocity and ξ is the porosity. Folding the constant 2 in Eq. (5.8) into the f term, the modified Darcy–Weisbach equation becomes

$$\frac{dh}{dl} = f_k\frac{1}{d\xi^2}\frac{V^2}{g}, \tag{5.10a}$$

where f_k is the modified friction factor. Stephenson (1979) proposed that the friction factor could be given by

$$f_k = \frac{\text{const}}{R_e} + f_t, \tag{5.10b}$$

where f_t is the friction factor for fully turbulent flow and R_e is Reynolds number given by

$$R_e = \frac{Vd}{\xi\nu}. \tag{5.11}$$

Stephenson proposed that f_t equals 1, 2, or 4 for smooth polished stone, semirounded stone, or angular stone, respectively.

Recent work performed by Herrera (1989) and summarized in Herrera and Felton (1991) sought to define the hydraulics of flow through a rockfill of varying gradation using sediment-free water. They used the standard deviation of the particle diameter, σ, as the measure of gradation instead of intrinsic permeability or simple porosity as had previous workers[1]. A set of equations to predict the head loss was developed based upon a friction factor—Reynolds number relationship. Six different models were evaluated using 16 rockfill structures, with 96 tests conducted using three replications. The equations were found to predict the actual head loss data to an average error of 8%.

The standard deviation was found to be better than porosity alone as a predictor for describing the hydraulics of flow. In the original equations proposed by Herrera (1989), porosity was included as a parameter. In a later paper by Herrera and Felton (1991), porosity was deleted, since it was approximately constant at a value of 0.46 throughout all of the tests. In order to allow the relationships to be used for a wide range of conditions, porosity has been included in the equations described in this chapter. The best set of working equations for describing the hydraulics of flow and for design, considering the Herrera and Felton data, are proposed to be the following [this is Model 3 in Herrera (1989)]

1. Reynolds number equation given by

$$R_e = \frac{(d - \sigma)V}{\nu\xi} \tag{5.12}$$

2. Friction factor given by

$$f_k = \frac{gd\,\xi^2}{V^2}\frac{dh}{dl} \tag{5.13}$$

3. The friction factor–Reynolds number relationship given as

$$f_k = \frac{1600}{R_e} + 3.83 \tag{5.14}$$

[1] The average diameter, d, and the standard deviation, σ, can be obtained after a sieve analysis using the relationships

$$d = \frac{\Sigma d_i w_i}{\Sigma w_i}$$

and

$$\sigma = \left[\frac{\Sigma(d - d_i)^2 w_i}{\Sigma w_i}\right]^{1/2},$$

where all summations are from 1 to n, n is the number of sieve groupings from a sieve analysis, d_i is the average diameter of the rock obtained by averaging the sieve opening of two succeeding sieves, and w_i is the percentage by weight of the rocks assigned a diameter d_i.

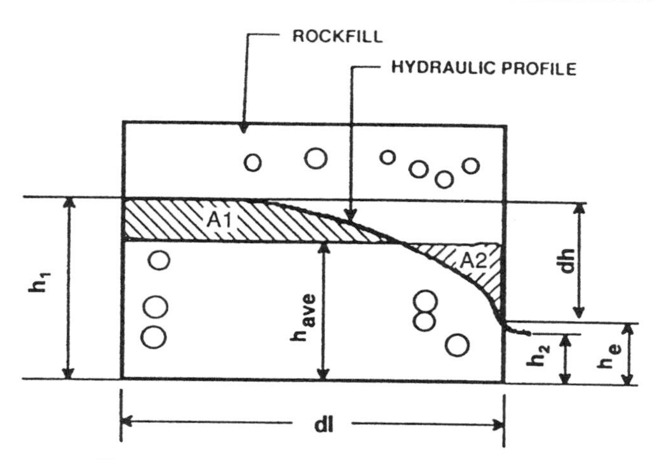

Figure 5.5 Definition sketch rockfill equations.

4. The $h_2 - h_{ave}$ relationships (see Fig. 5.5) given by

$$h_1 = h_2 + dh \tag{5.15}$$

and

$$h_{ave} = \frac{h_1 + h_2}{2}, \tag{5.16}$$

where d is average diameter of rock fill (m), ξ is porosity, σ is standard deviation (m), ν is kinematic viscosity (m/sec), h_{ave} is the average water profile inside the rockfill, h_2 is exit depth of the water in the rockfill, h_1 is the upstream depth, dh is static head drop as flow moves through the rockfill, dl is the flow length through the rockfill, f_k is the friction factor, R_e is Reynolds number, and V is average bulk velocity equal to q/h_{ave}, where q is the discharge per unit width of rockfill. Herrera suggested that a value of 0.46 be assigned to ξ in design calculations for a graded rockfill constructed by dumping.

Procedures for using the rockfill relationships are given in Example Problem 5.5.

Example Problem 5.5 Flow hydraulics through a rockfill

A rockfill dam is composed of rock having an average diameter d of 0.02 m, porosity ξ equal to 0.4, standard deviation σ of 0.001 m, and length dl equal to 1.0 m. Water with a kinematic viscosity ν of 1×10^{-6} m/sec is flowing through the rock at a rate q of 1.0 cms/m width. Downstream conditions control the exit depth of the water h_2 at 0.78 m. Find the upstream height h_1.

Solution:

1. Assume a trial h_{ave} equal to 4.6 m. The average bulk or macro-velocity equals q/h_{ave}, where q is flow per unit width. Hence

$$V = \frac{q}{h_{ave}} = \frac{1.0}{4.6} = 0.217 \text{ m/sec}.$$

2. Calculate Reynolds number. From the problem, $\sigma = 0.001$ m, $d = 0.02$ m, $\nu = 1.0 \times 10^{-6}$ m²/sec, and $\xi = 0.4$; hence Eq. (5.12) becomes

$$R_e = \frac{(d - \sigma)V}{\nu\xi} = \frac{(0.02 - 0.001)(0.217)}{1.0 \times 10^{-6}(0.4)} = 10{,}308.$$

3. Calculate friction factor using Eq. (5.14):

$$f_k = \frac{1600}{R_e} + 3.83 = \frac{1600}{10{,}308} + 3.83 = 3.98.$$

4. Calculate head loss from Eq. (5.13):

$$f_k = \frac{gd\xi^2}{V^2}\frac{dh}{dl}$$

or

$$dh = \frac{f_k V^2 dl}{gd\xi^2} = \frac{(3.98)(0.217^2)(1)}{(9.8)(0.02)(0.4^2)} = 5.98$$

5. Calculate upstream and average depth from Eqs. (5.15) and (5.16):

$$h_1 = h_2 + dh = 0.78 + 5.98 = 6.76$$

$$h_{ave} = \frac{h_2 + h_1}{2} = \frac{0.78 + 6.76}{2} = 3.77.$$

6. Additional trial calculations: Since 3.77 is less than the assumed value of 4.6, additional trials are necessary. One approach would be to use 3.77 as the new trial value. This, however, leads to oscillating values. An alternate approach is to use a second trial that is an average of the original value and the calculated value. Some oscillation occurs with this approach, but the calculated values converge. By combining Eqs. (5.12)–(5.16) using the values given in the problem, dh and h_{ave} can be related by

$$dh = \frac{1.08}{h_{ave}} + \frac{122.1}{h_{ave}^2} \tag{a}$$

and

$$h_1 = h_2 + dh = 0.78 + \frac{1.08}{h_{ave}} + \frac{122.1}{h_{ave}^2}. \tag{b}$$

Using these relationships, calculations can be quickly made. Assuming that the new trial value, $h_{ave,new}$, is the average of the previous trial value and the previous calculated value,

$$h_{ave,new} = \frac{h_{ave,trial} + h_{ave,cal}}{2} = \frac{4.6 + 3.77}{2} = 4.18.$$

From Eq. (b), using $h_{ave,new} = 4.18$ for h_{ave}, $h_1 = 8.02$ and

$$h_{ave,new} = \frac{8.02 + 0.78}{2} = 4.40.$$

The new value of 4.40 is still different from the previous value of 4.18; hence additional trials are necessary as shown below.

Trial	$h_{ave.trial}$	$h_{ave.new}$
1	4.6	3.77
2	4.18	4.40
3	4.29	4.22
4	4.26	4.27
	OK	

Thus, $h_{ave} = 4.27$. For this value, $dh = 6.95$ and $h_1 = 7.72$. Restated, this means that a head of 7.72 m on the rockfill with a thickness of 1 m will discharge 1 $m^3/sec/m$ of width if the downstream depth is 0.78 m.

Note that this problem illustrates a limitation of rockfill structures. A head of cover 7 m is required to discharge 1 cms/m. Such a structure would not be feasible. Example Problem 5.6 illustrates a more typical application of these structures.

Example Problem 5.6 Stage discharge equation

If the rock fill in Example Problem 5.5 is 2 m wide and is used as the spillway from a sediment pond, determine the stage discharge relationship up to an upstream depth of 2 m, using depths of 0.5, 1.0, 1.5, and 2.0 m. Assume that the downstream slope is such that the downstream depth is negligible.

Solution: This is also a trial and error computation process. For a given value of h_1, a value of dh can be estimated from Eq. (5.16), since h_2 is assumed to be approximately zero. Knowing dh, the unknowns in Eq. (5.14) are f_k and V. In making the calculations, a value of f_k is assumed, starting with a value of 3.83 (from Example Problem 5.5) and checked by iteration. Under these conditions, Eqs. (5.12) to (5.14) can be simplified to

$$R_e = \frac{(0.02 - 0.001)V}{(1.0 \times 10^{-6})(0.4)} = 47{,}500V \qquad \text{(a)}$$

and

$$f_k = \frac{(9.8)(0.02)(0.4^2)}{V^2}\frac{dh}{1} = 0.03136\frac{dh}{V^2}. \qquad \text{(b)}$$

Since the downstream depth, h_2 is assumed to be negligible,

$$dh = h_1. \qquad \text{(c)}$$

For an h_1 or dh of 0.5 m and an assumed f_k of 3.83, Eq. (b) can be solved for V or

$$V = \left[\frac{(0.03136)(0.5)}{3.83}\right]^{1/2} = 0.064.$$

From Eq. (a),

$$f_k = (47{,}500)(0.064) = 3040.$$

From Eq. (5.15),

$$f_k = \frac{1600}{3040} + 3.83 = 4.36.$$

Additional trials are shown below using Eq. (a), (b), and (c). Calculated f_k is used as trial f_k in the next step.

Trial No.	Trial f_k	Calculated f_k
1	3.83	4.36
2	4.10	4.37
3	4.24	4.38
4	4.31	4.38
5	4.35	4.39
		OK

Hence, for $h_1 = 0.5$ m, $f_k = 4.38$, $V = 0.06$ m/sec, $h_{ave} = 0.5/2 = 0.25$ m, q can be solved from $q = h_{ave}V = 0.0150$ m^2/sec. For a width of 2 m,

$$Q = 2q = (2)(0.0150) = 0.030 \text{ m}^3/\text{sec}.$$

Other values are tabulated below.

h_1 (m)	f_k	V (m/sec)	h_{ave} (m)	q (m^2/sec)	Q (m^3)
0.5	4.39	0.0600	0.25	0.0150	0.0300
1.0	4.22	0.0862	0.50	0.0431	0.0862
1.5	4.15	0.1065	0.75	0.0800	0.1600
2.0	4.10	0.1234	1.00	0.1234	0.2468

A plot of Q vs h_1 would constitute a stage–discharge relationship for a negligible downstream depth.

The procedure shown in Example Problem 5.6 applies only to those conditions when the downstream depth is negligible. For conditions where downstream flow occurs, the depth, h_2 would be a function of flow. In this case, it would be necessary to calculate h_2 for each discharge, using one of the flow equations in Chapter 4.

Graphical Procedure for Rockfill Flow Hydraulics

The procedures presented above require detailed computations. Computers and spreadsheets facilitate these computations considerably. When quick estimates are needed, graphical procedures are helpful. By utilizing Eq. (5.12) through (5.16), the graphical relationships shown in Fig. 5.6 were developed for predict-

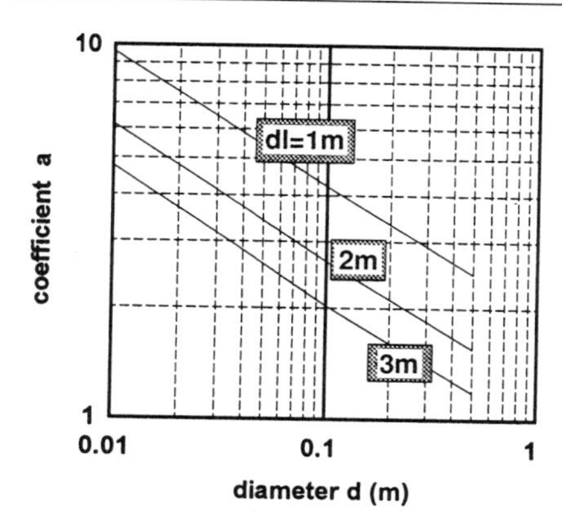

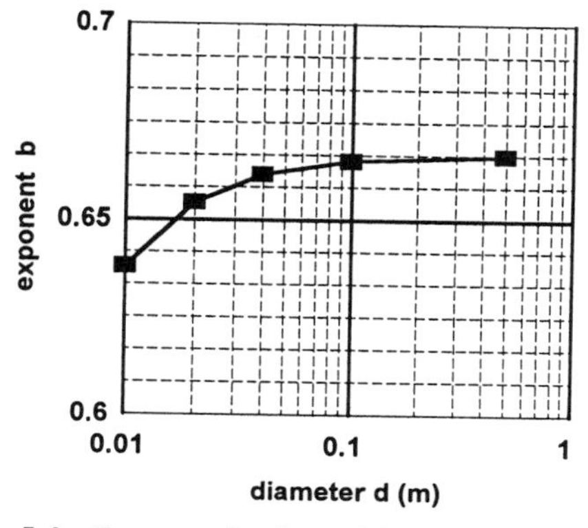

Figure 5.6 Constants for the rockfill head loss equations.

ing the average gradient through rockfill, dh/dl. These relationships can be used to develop head loss as a power function of flow, or

$$dh/dl = aq^b, \tag{5.17}$$

where a and b are constants dependent on rock size and flow path length, dl. The discharge q is flow per unit width of rockfill and has units of cms/m. The gradient, dh/dl, is dimensionless. An example showing the use of this equation is given in Example Problem 5.7.

Example Problem 5.7 Use of graphical method for rockfill outlets

Utilize the graphical method in Fig. 5.6 to calculate the stage discharge relationship in Example Problem 5.6. Assume a porosity of 0.46. Compare the results to those obtained in Example Problem 5.6.

Solution: The solution with the graphical method eliminates the trial and error involved in the process. From Fig. 5.6, $a = 7.5$ and $b = 0.653$; hence

$$dh/dl = 7.5q^{0.653}.$$

Also, since dl is 1 m,

$$dh = 7.5q^{0.653}$$

or

$$q = \left[\frac{dh}{7.5}\right]^{1/0.653}$$

Since the downstream depth is assumed to be negligible, dh is equal to h_1. The results are tabulated below along with a summary of the predicted values from Example Problem 5.6. Also included is a tabulation of values calculated using the procedures in Example Problem 5.6, but with the assumption of a porosity of 0.46 and a standard deviation of half the average diameter, or 0.01 m. These were the standard conditions used for developing Fig. 5.6.

Upstream depth h_1 (m)	Discharge q (m²/sec)		
	Graphical procedure	Example Problem 5.6 ζ=0.40, σ=0.001	New computation ζ=0.46, σ=$d/2$
0.5	0.0158	0.0150	0.0162
1.0	0.0457	0.0431	0.0474
1.5	0.0850	0.0800	0.0887
2.0	0.1321	0.1234	0.1379

The results indicate that the graphical procedure gives good agreement with the analytical procedure. It is important to note again that conditions deviating greatly from the standard conditions utilized for Fig. 5.6 could cause the procedure to be in error.

Single- and Multistage Risers

Stage–discharge relations for outlet structures are based on the physical characteristics of the outlet structure. Outlet structures can have many different entrance and exit conditions. Outlet structures may consist of a single weir or orifice, multiple weirs or orifices, or weirs and orifices used in conjunction with each other. Traditionally, basins have been designed to control the runoff from a long return event (i.e., 100-year event). A basin sized for a long return event tends to overcontrol more frequent events (i.e., 2-year events). If the design is based on frequent events, the structure tends to undercontrol long return events. These observations have led to regulations that require two or more stage risers. The development of stage–discharge curves for single-stage risers follows the procedures that were discussed previously. Two-stage risers can consist of any combination of orifices, weirs, and pipes. Most common is a riser consisting of an orifice for low flow (more frequent) events and a weir for high flow (less frequent) events. Orifices are often staggered around the circumference of the riser.

A multistage riser may be created by having more than one row of orifices on a riser. The size or number of orifices may change from one level on the riser to another. Equations used for estimating the discharge for weirs and orifices have been presented previously in this chapter.

Broad-Crested Weirs

A broad-crested weir supports the flow in the longitudinal direction (direction of flow) so that the nappe flowing across the weir does not spring free from its upstream face as shown in Fig. 5.1. Broad-crested weirs are usually calibrated in the field or by using a model. They tend to be structurally stronger than sharp-crested weirs, and they are particularly useful in locations where sharp-crested weirs suffer maintenance problems. An example is a forest area where large limbs or logs may crash into a sharp-crested weir and cause damage. Streeter (1971) described the discharge relationships for broad-crested weirs and showed that

$$Q_t = 3.09LH^{3/2}, \tag{5.18}$$

where Q_t is the theoretical discharge from a broad-crested weir of width L operating with a head of H. Streeter also noted that calibration using a broad-crested weir produced an equation of the form

$$Q_x = 3.03LH^{3/2}, \tag{5.19}$$

where Q_x is the discharge based on experimental data for a broad-crested weir having a well-rounded upstream edge. The two equations agreed within 2%. Streeter also indicated that viscosity and surface tension have a minor impact on the discharge coefficients of weirs. Equations (5.18) and (5.19) are inadequate for design of emergency spillways or road overtopping since they do not consider the effects of flow length across the crest of the spillway.

Roadway overtopping occurs when the headwater rises to the elevation of the roadway. Flows under these conditions are similar to that of a broad-crested weir with the roadway serving as the weir. The flow across the roadway is defined by

$$Q_o = C_d L(HW_r)^{3/2}, \tag{5.20}$$

where Q_o is the overtopping flow rate (ft^3/sec), C_d is the overtopping discharge coefficient, L is the length of the roadway crest (ft), and HW_r is the upstream depth (ft), measured from the roadway crest to the water surface at a location upstream of weir drawdown (ft). The overtopping discharge coefficient C_d is defined as $C_d = k_t C_r$, where C_r is the discharge coefficient and K_t is the submergence factor. The coefficients are given in Fig. 5.7. Road overtopping locations are characterized by a sagging vertical curve such that the length and elevation of the roadway crest are difficult to quantify. Federal Highway Administration (1985) suggested two methods of characterizing the sagging vertical curve.

For culvert designs that accompany road overtopping, it may be adequate to represent the sagging vertical curve by a single horizontal line. The length of this line is then taken to be the length of the weir.

The second design method involves breaking the sagging vertical curve into a series of horizontal segments. Flow across each segment is then calculated for a specified headwater using Eq. (5.20). Flows from each segment are then accumulated to obtain the total flow across the roadway. The elevation and consequent head H for each horizontal segment is that of the segment, not an average.

While calculation of the flow over the roadway is relatively simple, the difficulty is that this only represents a portion of the design flow if there is a culvert through the road fill and this culvert continues to carry flow. To determine the head for a given discharge, a trial and error procedure is required to determine the flow passing through the culvert and the amount flowing across the roadway. If the head is given and a discharge is to be calculated, the total discharge is simply the sum of that across the roadway and that through the culvert.

HYDRAULICS OF CULVERTS

Culverts are conduits that are commonly used to pass drainage water through embankments. They are employed beneath access or haul roads and perpendicular to roadside ditches. The selection of a culvert size to convey flow adequately is presented in this section. The design of ditch relief culverts is similar to that of trickle tubes and is discussed in a subsequent section. Recommended spacing of ditch relief culverts is a function of road gradient.

Culvert Classes

Chow (1959) divided culvert flow into six categories as shown in Fig. 5.8. Each of these categories is described along with methods for predicting flow for each type. Chow indicated that the entrance of an ordinary culvert will not be submerged if the outlet is not submerged unless the headwater is greater than some

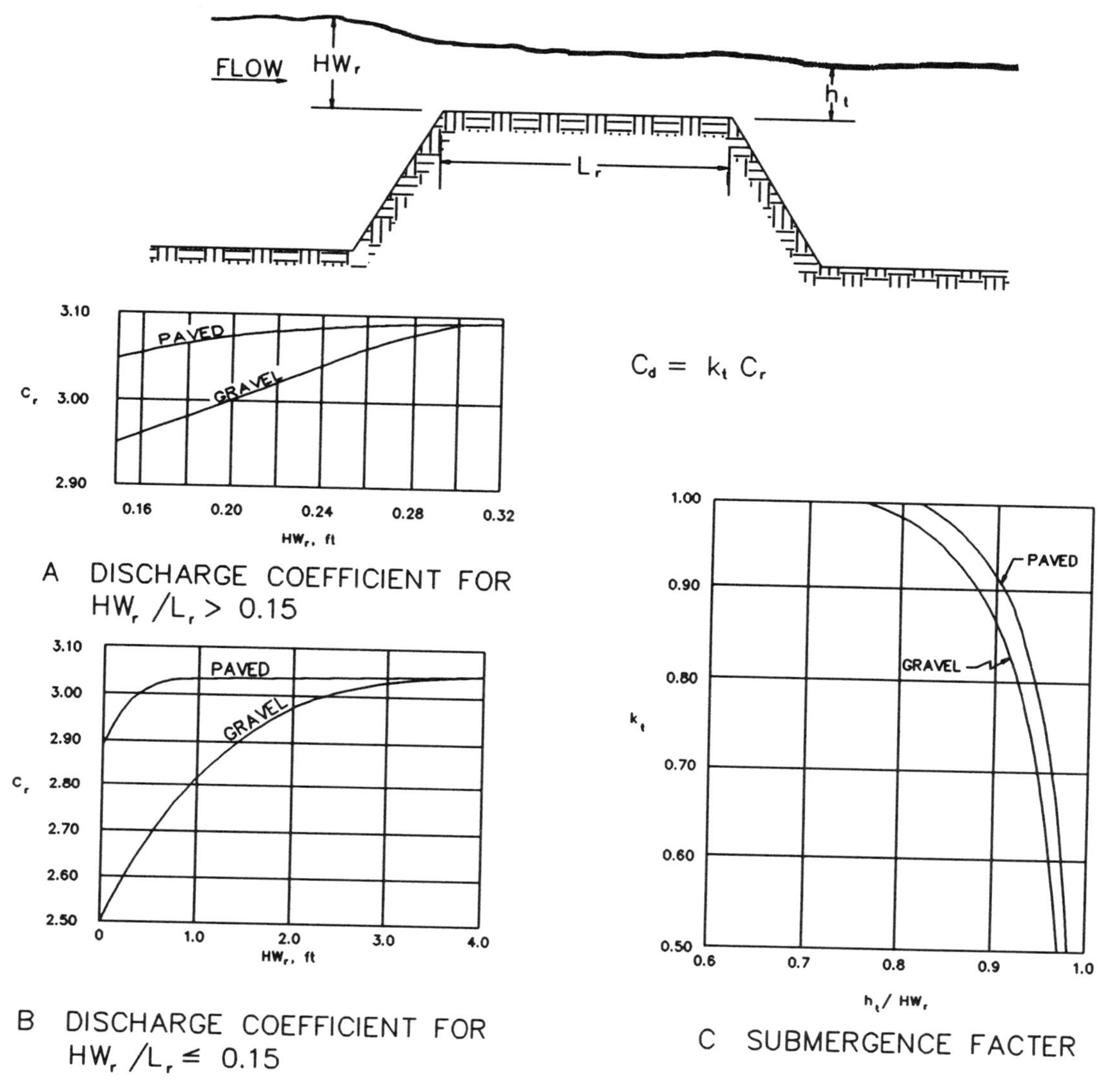

Figure 5.7 Discharge coefficients for roadway overtopping (Federal Highway Administration, 1985).

critical value, H^*. The value of H^* may vary from 1.2 to 1.5 times the culvert height as a result of entrance geometry, barrel characteristics, and approach conditions. A critical headwater height equal to 1.5 times the culvert height (diameter for circular culverts) is reasonable for preliminary analysis.

Type 1—Outlet Submerged The pipe will flow full. Discharge is calculated using Eq. (5.5).

Type 2—Outlet Not Submerged, $H > H^*$, Pipe Flowing Full This corresponds to a hydraulically long condition. Figures 5.9 or 5.10 can be checked to provide an estimate of whether proper conditions are met for corrugated pipes and concrete pipes, respectively. From Fig. 5.8, the headwater height H must be greater than the critical height H^*, and tailwater depth should be less than the height of the culvert. Calculate discharge from Eq. (5.5).

Type 3—Outlet Not Submerged, $H > H^*$, Pipe Not Flowing Full This corresponds to the hydraulically short condition. Check Figs. 5.9 or 5.10 to see if the proper conditions are met for the pipe to be classified as type 3 by locating the intersection of the bed slope and ratios in the appropriate figure. From Fig. 5.8, the headwater height H must be greater than the critical height H^*, and tailwater depth should be less than the height of the culvert. Discharge is inlet controlled and

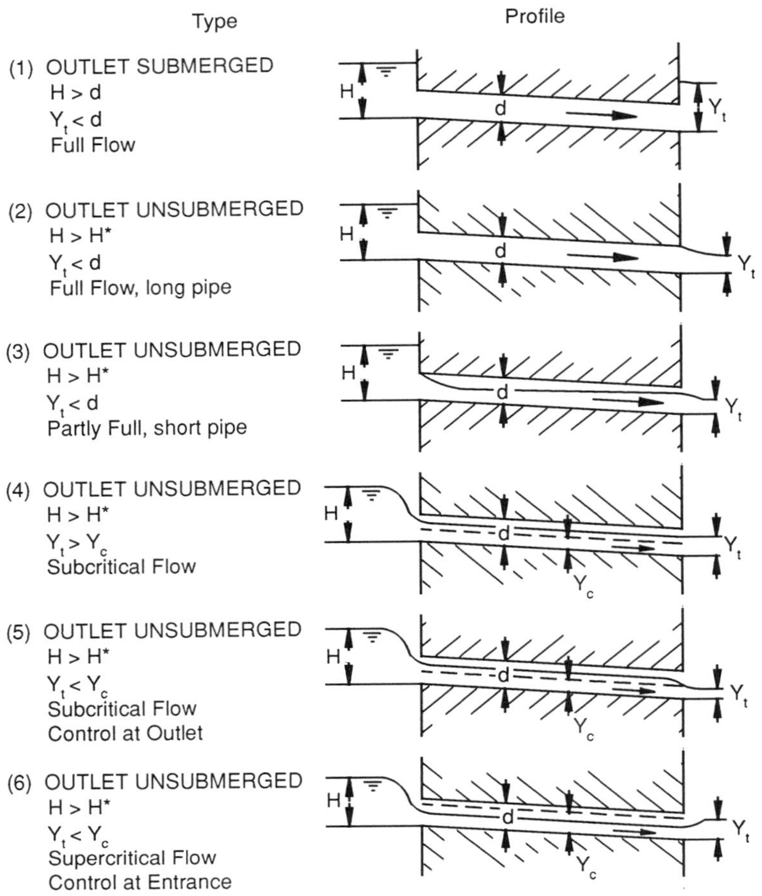

Figure 5.8 Types of culvert flow (Chow, 1959).

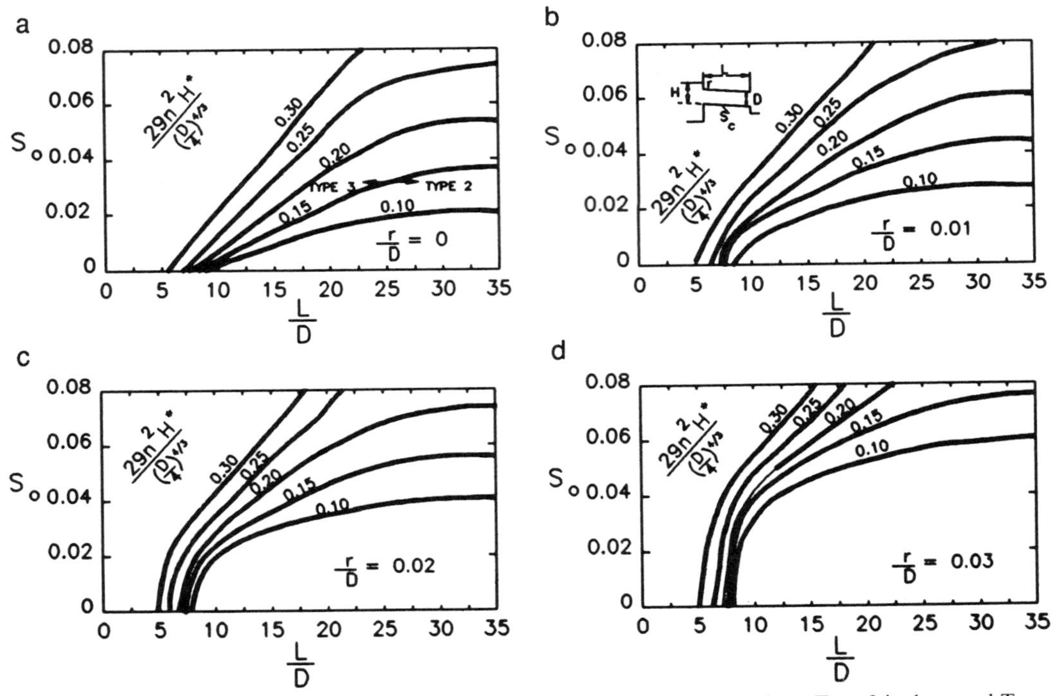

Figure 5.9 Criteria for hydraulically short and long culverts rough corrugated pipes. Type 3 is short and Type 2 is long (Carter, 1957).

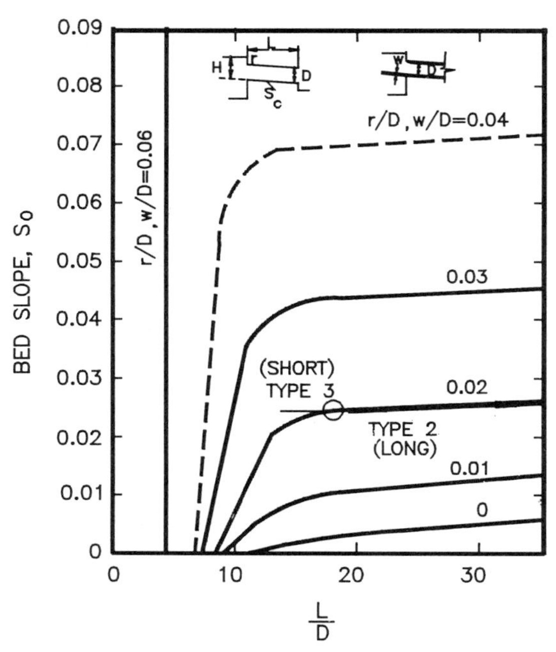

Figure 5.10 Criteria for hydraulically short and long concrete pipes (Carter, 1957).

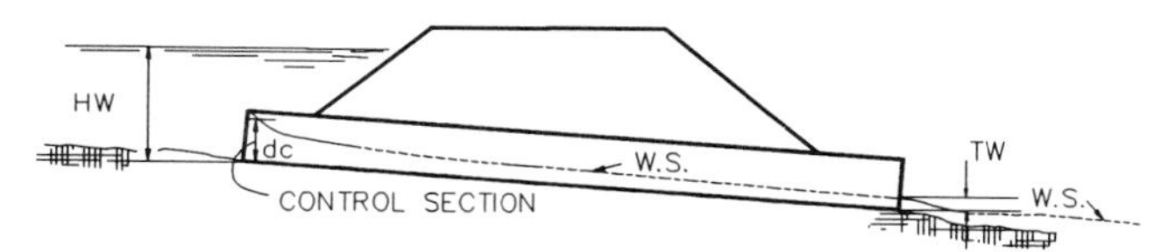

Figure 5.12 Typical inlet control flow condition (after Federal Highway Administration, 1985). HW, headwater; TW, tailwater, W.S., water surface; d_c, critical depth.

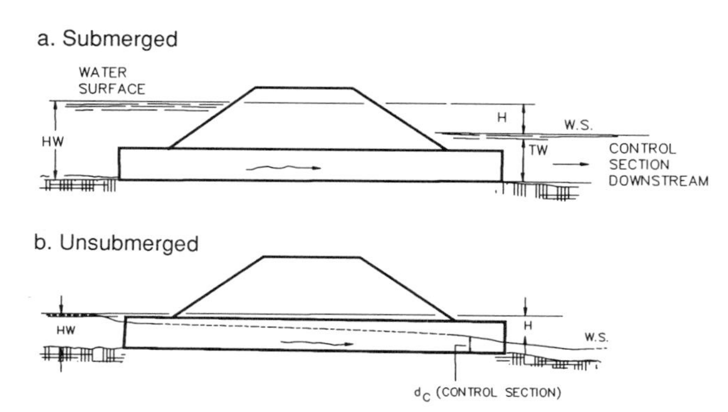

Figure 5.13 Typical outlet control flow conditions (after Federal Highway Administration, 1985). HW, headwater, TW, tailwater; W.S., water surface; H, losses through culvert.

can be determined by using dimensionless plots developed by Mavis (1942) as shown in Fig. 5.11.

Types 4–6—Outlet Unsubmerged, $H < H^*$ Under these conditions, the pipe flows as an open channel. The discharge for a given head depends on the pipe slope, entrance geometry, pipe roughness, and pipe size. To accurately predict discharge, it is necessary to develop a flow profile through the pipes. The exact shape of the profile will depend on the depth of flow at the outlet.

The factors that influence energy and hydraulic grade lines in culvert discharge are illustrated in the culvert schematic shown as Figs. 5.12 and 5.13 (inlet and outlet control, respectively). These factors are used to determine the type of control for the discharge. The flows shown are either full flow, partly full flow, or free surface flow.

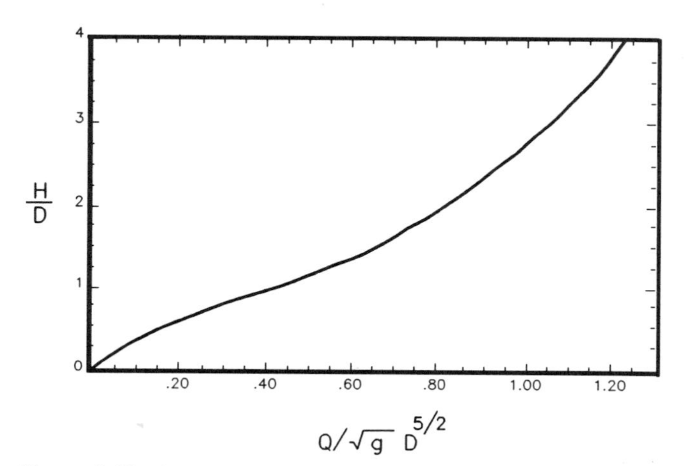

Figure 5.11 Stage–discharge relationship for a circular pipe with the control of the inlet (adapted from Mavis, 1942).

Inlet Control

Inlet control occurs when the section that controls flow is located at or near the entrance to the culvert. Discharge is dependent only on the geometry of the inlet and the headwater depth for any particular culvert size and shape. The inlet will continue to control flow as long as water flowing through the barrel of the culvert does not impede flow. If control is at the inlet, downstream hydraulic factors such as slope, length, or surface roughness will not influence capacity. However, roughness does influence the critical slope at which inlet control occurs (American Concrete Pipe Association, 1985). Smooth culverts placed on a very flat slope can have inlet control, whereas rough culverts have to be installed on a much higher slope to have inlet control. Several types of inlet control are shown in

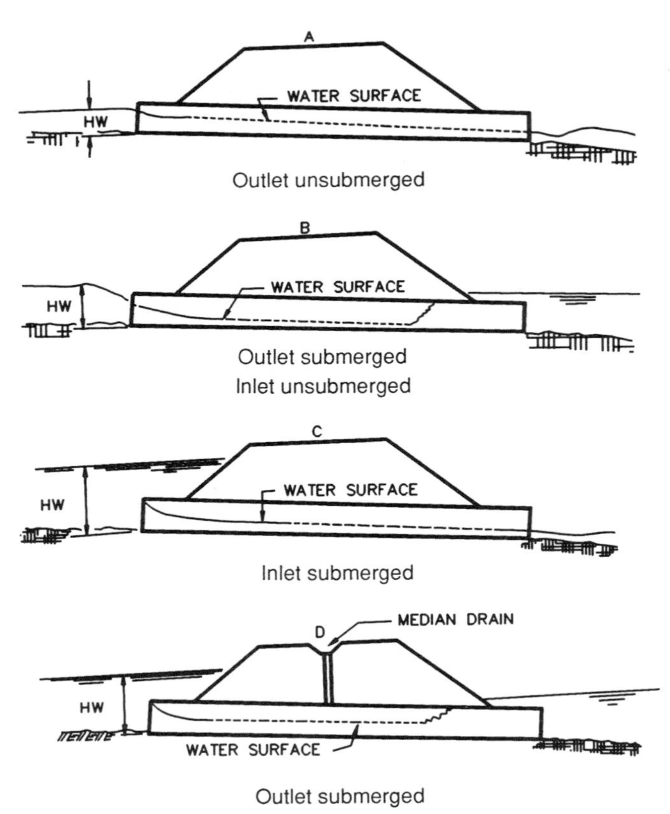

Figure 5.14 Types of inlet control (after Federal Highway Administration, 1985).

Fig. 5.14. Figures showing the critical depth for various shapes, sizes, and materials are available in many hydraulic references. If a culvert is operating under inlet control, it will not flow full throughout the entire length of the pipe.

Outlet Control

Outlet control occurs when control originates at or near the culvert outlet. In this case, discharge is dependent not only on all of the hydraulic factors upstream from the outlet including size, shape, slope, length, surface roughness, headwater depth, and inlet geometry, but on tailwater depth as well.

Culvert Selection and Design

Culvert selection techniques can range from solution of empirical formulas to comprehensive mathematical analysis of specific situations (American Concrete Pipe Association, 1985). The many hydraulic considerations involved make a precise mathematical evaluation difficult and extremely time consuming. The relatively simple empirical methods cannot account for all of the factors that impact flow, but they can be used to estimate flow through culverts for the conditions they represent. Charts in Bureau of Public Roads (1965a, b) provide a systematic design method that is based on determining the headwater depth from the charts assuming both inlet and outlet control. The condition yielding the higher headwater depth controls. Professional groups such as the American Concrete Pipe Association (ACPA) (1985) provide excellent references for many of the figures and techniques that are used in culvert selection. They can often supply detailed design specifications and design procedures for specialized products. Additionally, they are often able to supply design data on loading.

Critical Depth in Culverts

When the sum of kinetic energy plus the potential energy for a specified discharge is at its minimum, critical flow occurs. Another way to view this is that during critical flow, maximum discharge through a pipe occurs with any specified total energy head. For a given flow rate, the depth of flow and slope associated with critical flow define the critical depth and critical slope. If the culvert has an unsubmerged outlet, the maximum capacity of the culvert is established when critical flow occurs (American Concrete Pipe Association, 1985).

Culvert Nomograph Procedure

The American Concrete Pipe Association (1985) has suggested the following procedure for sizing culverts. The basic procedure for sizing a culvert for a given flow rate involves selecting a trial size culvert, determining whether inlet or outlet control prevails, and then finding the headwater required for the controlling condition. If the headwater is unacceptable (too high or low), another trial culvert is selected and the process repeated. The procedure is outlined below. The notation at this point utilizes HW as the head on the culvert instead of H used earlier, in keeping with standard terminology utilized by public agencies.

Determine Design Data.

1. Design discharge Q (cfs) for the design storm.
2. Length L (ft) of culvert.
3. Slope of culvert.
4. Allowable headwater depth HW_{al} (ft). This is the vertical distance from the culvert invert (elevation of the inside bottom of the culvert) at the inlet side of the culvert to the maximum water surface elevation permissible.

5. Flow velocities or tailwater depth in the downstream channel.

6. Size, shape and entrance type for trial culvert. A suggested trial size is a diameter (or height for non-circular culverts) of HW_{al} divided by 2.

Find Maximum Headwater Depth for Trial Culvert Under Inlet and Outlet Control Conditions

A. Inlet control

1. Given Q, D, and entrance type, select the appropriate control nomograph to find headwater depth required (Fig. 5B.1 or 5B.2).

a. Connect the given culvert diameter D and discharge Q with a straight line. Continue the line to the first HW/D scale, indicated as (1).

b. Find the HW/D scale that represents the entrance type used. If necessary, extend the point of intersection from the first line horizontally to scale (2) or (3).

c. Multiply HW/D by D to calculate HW.

2. If HW is greater or less than allowable, select another trial size until the HW is within the desired range.

B. Outlet control

1. Given Q, D, entrance type, and estimated tailwater depth TW (feet) above the outlet invert for the design flow in the downstream channel.

a. Select the outlet control nomograph for the desired culvert configuration (Fig. 5B.3 or 5B.4). Find the entrance coefficient K_e from Table 5.4.

b. Find the K_e on the length scale on the nomograph.

c. Connect the K_e point on the length scale to the size of the culvert using a straight line and mark the point where the straight line crosses the "turning line."

d. Form a straight line with the point marked on the turning line and the design Q and project to the head scale. Read H on the head scale.

2. If the tailwater, TW, elevation is lower than the top of the culvert outlet, use

$$h_0 = \frac{d_c + D}{2} \tag{5.21}$$

or TW, whichever is greater, where d_c is the critical depth (feet) determined from the corresponding critical depth chart.

3. If TW elevation is higher than or equal to the top of the culvert outlet, set h_0 equal to TW. Find HW using

$$HW = H + h_0 - S_0 L. \tag{5.22}$$

Table 5.4 Entrance Loss Coefficients (after FHA, 1985)

Outlet control, full or partly full entrance head loss

$$H_e = K_e \left[\frac{V^2}{2g}\right]$$

Type of structure and design of entrance	Coefficient K_e
Pipe, concrete	
Projecting from fill, socket end (groove end)	0.2
Projecting from fill, square cut end	0.5
Headwall or headwall and wingwalls	
Socket end of pipe (groove end)	0.2
Square edge	0.5
Rounded (radius = $\frac{1}{12}D$)	0.2
Mitered to conform to fill slope	0.7
End section conforming to fill slope[a]	0.5
Beveled edges, 33.7° or 45° bevels	0.2
Side- or slope-tapered inlet	0.2
Pipe, or pipe-arch, corrugated metal	
Projecting from fill (no headwall)	0.9
Headwall or headwall and wingwalls square edge	0.5
Mitered to conform to fill slope, paved or unpaved slope	0.7
End section conforming to fill slope[a]	0.5
Beveled edges 33.7° or 45° bevels	0.2
Side- or slope-tapered inlet	0.2
Box, reinforced concrete	
Headwall parallel to embankment (no wingwalls)	
Square edged on three edges	0.5
Rounded on three edges to radius of $\frac{1}{12}$ barrel dimension, or beveled edges on three sides	0.2
Wingwalls are 30° to 75° to barrel	
Square edged at crown	0.4
Crown edge rounded to radius of $\frac{1}{12}$ barrel dimension, or beveled top edge	0.2
Wingwall at 10° to 25° to barrel	
Square edged at crown	0.5
Wingwalls parallel (extension of sides)	
Square edged at crown	0.7
Side or slope-tapered inlet	0.2

[a]Either metal or concrete sections commonly available from manufacturers. From limited hydraulic test they are equivalent in operation to a headwall in both *inlet* and *outlet* control. Some end sections, incorporating a *closed* taper in their design, have a superior hydraulic performance.

C. Compare the headwaters required from sections A (inlet control) and B (outlet control) to determine which is higher. The higher headwater controls and is the flow control existing under the design conditions for the trial size under consideration.

D. If HW is higher than acceptable and outlet conditions control, select a larger culvert size and revise HW using section B for outlet control. It is not necessary to recheck inlet control if the smaller size culvert proved to be satisfactory in section A. It is also possible to select another shape or inlet condition if desired and repeat the procedure.

Determine Outflow Velocity

A. If outlet control exists with tailwater, outflow velocity equals the Q/A. If the outlet is not submerged, the flow area A is usually based on a flow depth equal to the average of the vertical dimension (or diameter for circular pipes) and the critical depth calculated previously.

B. If inlet flow governs, outflow velocity is approximated assuming open channel flow and using Manning's equation as described in Chapter 4.

Example Problem 5.7 Culvert size with inlet control (culvert nomograph method)

Determine the diameter of a circular concrete culvert ($n = 0.012$) that must carry a design flow of 200 cfs if the culvert length is 180 ft and the culvert slope is 0.02 ft/ft. The allowable HW equals 10 ft. Assume that the culvert has a projecting entrance, and TW equals 4 ft.

Solution:

Determine design data.

1. $Q = 200$ cfs (given).
2. $L = 180$ ft (given).
3. $S = 0.02$ ft/ft (given).
4. $HW_{al} = 10$ ft (given).
5. TW = 4 ft (given).
6. Trial size $D = HW_{al}/2 = 5$ ft.

Concrete culvert with $n = 0.012$ and projecting entrance. Find maximum headwater depth for trial culvert under inlet and outlet control conditions.

A. Inlet control

1. Given Q, D, and entrance type, the appropriate inlet control nomograph to find required headwater depth is Fig. 5B.1.

a. Connect the diameter $D = 5$ ft (60 in.), and discharge Q with a straight line. Continue the line to the first HW/D scale, indicated as (1).

b. The HW/D scale, which represents the grooved-end projecting entrance, is indicated as (3) so extend the point of intersection from the first line horizontally to scale (3) and obtain a value of $HW/D = 1.37$.

c. Multiply HW/D by $D = 5$ ft to calculate $HW = 5(1.37) = 6.85$ ft. This is the HW required for inlet control.

2. Since $HW < HW_{al}$, OK.

B. Outlet control

1. Given $Q = 200$ cfs, $D = 5$ ft, grooved end projecting, and TW = 4 ft above the outlet invert for the design flow in the downstream channel.

a. The outlet control nomograph for the circular concrete culvert is Fig. 5B.3. The entrance coefficient $K_e = 0.2$ from Table 5.4.

b. Find the K_e on the length scale on the nomograph.

c. Connect the K_e point on the length scale to the size of the culvert using a straight line and mark the point where the straight line crosses the "turning line."

d. Use the point marked on the turning line as the pivot point and connect with the design Q. Read $H = 2.80$ ft on the head scale.

2. Since the TW elevation (4 ft) is lower than the top of the culvert outlet (5 ft), obtain the critical depth d_c from the critical depth chart for 5-ft-diameter circular pipe with $Q = 200$ cfs (Fig. 5C.1) as $d_c = 4.10$ ft. Substitute d_c and D in Eq. (5.21) for h_0:

$$h_0 = \frac{d_c + D}{2} = \frac{4.10 + 5}{2} = 4.55 \text{ ft}.$$

Find HW under outlet control using Eq. (5.22)

$$HW = H + h_0 - S_0 L = 2.80 + 4.55 - 0.02(180) = 3.75.$$

C. Compare the headwaters required from sections A (inlet control) and B (outlet control) to determine which is higher. Since a higher HW is required for inlet control (HW = 6.85 ft > HW = 3.75 ft), inlet conditions control under the design conditions for the trial size under consideration.

D. Since HW = 6.85 ft is lower than the allowable HW = 10.0, the culvert is adequate. If desired, a smaller culvert size might be selected and the process repeated to check if the smaller culvert might be satisfactory.

Culvert Capacity Chart Technique

The culvert capacity chart technique offers an alternative method to determine the required culvert size or headwater depth (Portland Cement Association, 1962; American Concrete Pipe Association, 1985). Charts for numerous shapes, including circular, square, rectangular, and oval pipes, have been developed.

For specific conditions, a simple procedure is to use culvert capacity charts such as shown in Fig. 5A.1–5A.8, which are based on the data in Bureau of Public Roads (1965a). The culvert capacity charts enable a simple technique to be used for selection of culverts. The technique is somewhat limited because the pipe must be installed with the entrance and material as specified for the specific charts used.

The Portland Cement Association (1975) outlined use of these figures to obtain a direct solution of culvert size without the inlet/outlet comparison required in the nomograph procedure discussed previously.

Determine Design Data

1. Design discharge Q (cfs) for the design storm.
2. Length L (ft) of culvert.
3. Slope of culvert.
4. Allowable headwater depth HW_{al} (ft). This is the vertical distance from the culvert invert at the inlet side of the culvert to the maximum water surface elevation permissible.
5. Flow velocities in the downstream channel.
6. Size, shape, and entrance type for trial culvert. A suggested trial size is a diameter (or height for noncircular culverts) of HW_{al} divided by 2.

Find Culvert Size

1. Locate the appropriate culvert capacity chart for the culvert size that is approximately half the allowable headwater depth. A typical culvert capacity chart is shown in Fig. 5A.2. Denoted on this chart are headwater values on the ordinate, discharge quantities on the abscissa, solid curves designating inlet control, and a dash curve indicating outlet control. Some figures contained in Appendix 5A do not contain both solid and dash curves for all culvert diameters. This is an indication that the culvert would be unsuitable for the type of control that is missing from the nomograph. The horizontal dotted line denotes that the accuracy of chart values below this line are quite good, whereas prediction accuracy decreases above this line.

2. Estimate the index number by dividing L by 100 S_0. This index accounts for roughness and length effects relating to pipe flow characteristics.

a. If the calculated index value for a specific problem is less than or equal to the number shown on the solid curve for a specified culvert, the culvert will be in inlet control and the solid curve will describe culvert performance.

b. If the calculated index value is greater than the solid curve value but less than or equal to the dash curve index value, outlet control dominates and the point denoting culvert performance can be located by interpolation.

c. A calculated index value greater than a particular dash curve index value is indicative of full pipe flow, and full pipe flow conditions prevail and the culvert capacity chart should not be used. In this case, the appropriate outlet control nomograph describing full pipe flow in the nomograph procedure should be used to estimate performance of the culvert.

Determine Outflow Velocity Outflow velocity is determined in the same manner as when the nomographs are used. Culvert capacity charts permit quick estimation of headwater height given a peak flow rate and culvert size if the specified design conditions are met. The charts also can be used to select culvert size needed to meet a given headwater limitation such as to prevent roadway overtopping.

Example Problem 5.8 Culvert size with inlet control (culvert capacity chart method)

Determine the diameter of a circular concrete culvert for the conditions stated in Example Problem 5.7 using the culvert capacity chart.

Solution:

Determine design data.
1. $Q = 200$ cfs (given).
2. $L = 180$ ft (given).
3. $S = 0.02$ ft/ft (given).
4. $HW_{al} = 10$ ft (given).
5. $TW = 4$ ft (given).
6. Trial size $D = HW_{al}/2 = 5$ ft.

Concrete culvert with n = 0.012 and groove end projecting entrance.

Find culvert size.

1. Select the appropriate culvert capacity chart for the culvert size D, which is half the allowable headwater depth, or $D = HW/W = 10/2 = 5$ ft.

2. Estimate the index number, $L/100\ S_0$, which accounts for roughness and length effects relating to pipe flow characteristics:

$$\frac{L}{100\ S_0} = \frac{180}{100\ (0.02)} = 90.$$

Since the calculated index value (90) is less than the number shown on the solid curve (700) as shown in Fig. 5A.2 for a 5-ft circular concrete culvert, the culvert will be in inlet control, and the solid curve will describe culvert performance. Reading up from a $Q = 200$ cfs yields $HW = 6.6$ ft. Since 6.6 ft is less than the allowable HW of 10 ft, the culvert will carry the design flow under inlet control.

Example Problem 5.9 Headwater height required for culvert with inlet control

During construction in a subdivision, a groove-edged entrance, 42-in., 160-ft concrete culvert is placed on a 3% slope beneath a temporary road. The design storm event yields a peak runoff of 100 cfs. Find the headwater height required to pass the flow.

Solution: Select the appropriate capacity chart as Fig. 5A.2. The index is

$$\text{Index} = \frac{L}{100S_0} = \frac{160}{100(0.03)} = 53.$$

Enter the capacity chart on the abscissa with the peak flow of 100 cfs. Drawing a vertical line at 100 cfs intersects the solid line (inlet control) for the 42-in. culvert at a headwater height of 5.9 ft. The index value for inlet control given on the solid line is 400 for the 42-in. culvert size. Since the calculated index value (53) is less than the solid curve value (400), inlet control is indicated. Thus, the required headwater height HW is 5.9 ft.

Example Problem 5.10 Headwater height required for culvert with outlet control.

Use the situation described in Example Problem 5.9, except assume a slope of 0.15%.

Solution: The solution follows the same steps as in Example Problem 5.9. The difference will be the impact of the index value. The index value is

$$\text{Index} = \frac{L}{100S_0} = \frac{160}{100(0.0015)} = 1067.$$

The calculated index value, 1067, lies between the solid and dash curves, 400 and 1200, for the 42-in culvert; thus outlet control is indicated. Read the required headwater height by interpolating for the index value of 1067 between 400 and 1200, and extend a horizontal line to the left axis. The required headwater is 6.2 ft.

Example Problem 5.11 Headwater required with full pipe flow

This problem is similar to the preceding example problems except that the slope equals 0.10%.

Solution: The index value for this case is

$$\text{Index} = \frac{L}{100S_0} = \frac{160}{100(0.0010)} = 1600.$$

This value is above the dash line in Fig. 5A.2, indicating pipe full flow. Thus, the pipe full outlet control nomograph (Fig. 5B.3) must be used. Using Fig. 5B.3, assume tailwater is at the crown (top) of the 42-in. culvert so that $h_0 = 3.5$ ft.

Draw a line connecting the culvert diameter of 42 in. with the culvert length of 160 ft along the groove-edged curve ($K_c = 0.2$ in Fig. 5B.3). This located the pivot point on the turn line. Now draw a second line connecting the pivot point to the 100 cfs peak flow and extend it to read the head scale value of 3.00 ft. To determine the headwater, solve Eq. (5.22):

$$\text{HW} = H + h_0 - S_0L.$$

Substituting

$$\text{HW} = 3.0 + 3.5 - (0.0010)(160) = 6.34 \text{ ft}.$$

Example Problem 5.12 Culvert selection with headwater restriction

A corrugated metal pipe with a projecting entrance is to be located under a road during construction of an airport. The culvert length is 120 ft, and it has a slope of 0.003 ft/ft. Peak flow from the design storm is calculated to be 200 cfs. The maximum allowable headwater is only 8 ft. Estimate the culvert size that will be needed to satisfactorily convey the design flow while not exceeding the allowable headwater height.

Solution: The index value is

$$\text{Index} = \frac{L}{100S} = \frac{120}{(100)(0.003)} = 400.$$

The appropriate culvert capacity chart based on the pipe material, size, and index value is Fig. 5A.6a. The culvert size that can convey the specified drainage without exceeding the headwater constraint is selected by: (1) locating the 200 cfs discharge on the abscissa, (2) extending a line upward to the culvert size that yields a headwater value of 8 ft or greater, and (3) interpolating for the calculated index value of 400. The required culvert is a 72-in.-diameter circular corrugated metal pipe having a projecting entrance.

Trickle Tube Spillway

Trickle tube spillways are often considered to be just a variation of the common culvert that is placed in a dam. In actuality, many culverts placed in roadways act like a primary spillway for a reservoir. The major design difference is that trickle tube spillways are generally considered to have steeper slopes than typical culverts. Flow at any given head will be dependent on whether the flow is controlled by the inlet acting like a weir or an orifice, by open channel flow through the pipe, or by full pipe discharge. Chow (1959) indicated that long pipes usually flow full when the head exceeds some critical value H^*. Chow recommended a value of 1.5 times the pipe diameter for a reasonable value of H^* in order to usually have full flow. Submergence of the outlet often leads to full flow. Other hydraulic characteristics also are similar to culverts. If the outlet of a trickle tube is not submerged, the tube may not flow full even if the head exceeds the critical value H^* and is considered to be hydraulically short. In other cases, it is considered hydraulically long. Carter (1957) provided graphs (Figs. 5.9 and 5.10) that define whether a pipe is hydraulically short or long.

A detailed analysis of pipe flow control in a trickle tube spillway would require the computation of backwater curves in the pipe for a number of discharges. If the trickle tube is on a mild slope ($y_n > y_c$), flow would enter the pipe at approximately the normal depth of flow. In this case, the relationship between discharge and head could be obtained by computing the normal depth of flow for a range of discharges and then computing the required head from

$$H' = \frac{V^2}{2g} + K_e\frac{V^2}{2g} + y_n,$$

where K_e is the entrance loss coefficient and y_n is the normal depth. This method serves only as a very rough approximation and should be used with caution. Sensitive projects should always have calculations made with detailed water surface profiles.

If the pipe is on a steep slope ($y_n < y_c$), then a S_2 flow profile would develop within the pipe and the above relationship would only be approximate. It should be noted that as the pipe slope increases, the range of entrance control also increases. Example Problem 5.13 illustrates the development of a stage–discharge relationship through a trickle tube.

Example Problem 5.13 Stage–discharge curve for a trickle tube

As part of a pond design, a stage–discharge curve for an 18-in. corrugated metal trickle tube must be found. The pipe is 100 ft long, placed on a slope of 3% with a free outfall and a square entrance.

Solution: The discharges when pipe flow prevails can be found using Eq. (5.5), or

$$Q = \frac{a\sqrt{2gH'}}{\sqrt{1 + K_e + K_b + K_cL}}.$$

For a corrugated metal pipe, $n = 0.024$ (Table 4.1). From Table 5.1, K_c is 0.0621 for an 18-in. pipe with $n = 0.024$. Table 5.1 also indicates that the area is 1.770 ft^2. Assuming $K_e = 1$ and $K_b = 0$ results in

$$Q = \frac{1.770\sqrt{2 \times 32.2 \times H'}}{\sqrt{1 + 1.0 + (0.0621)(100)}} = 4.957\sqrt{H'}.$$

For a 100-ft pipe on a 3% slope, the change in elevation between the upper and lower invert is 3.0 ft. Using a point 0.6D above the outlet invert as a reference,

$$H' = H + 3.00 - 0.6 \times \tfrac{18}{12} = H + 2.10.$$

Hence,

$$Q = 4.957(H + 2.10)^{1/2}.$$

For example, with H equal to 2 ft, Q equals 10.04 cfs. For inlet control, data such as those shown in Fig. 5.11 can be used. The following table illustrates its use.

Calculation of discharge with inlet control for Example Problem 5.13

H (ft)	H/D	$Q/\sqrt{g}D^{5/2}$ (from Fig. 5.11)	Q (cfs)
0.50	0.33	.10	1.56
1.50	1.00	.40	6.25
2.00	1.33	.57	8.91
3.00	2.00	.82	12.82
4.00	2.67	.96	15.01
5.00	3.33	1.09	17.04

To determine if open channel flow controls the discharge, values of y_n are assumed, Q, A, and V determined using open channel flow relationships given in Chapter 4 for a circular conduit and calculating the required total head from

$$H' = \frac{V^2}{2g}(1 + K_e) + y_n.$$

The results are shown in the following table.

Open channel flow calculations for Example Problem 5.13

Y_n	Q	V	H'
0.15	0.21	2.24	0.31
0.30	0.86	3.43	0.66
0.45	1.93	4.33	1.03
0.60	3.32	5.03	1.38
0.75	4.93	5.58	1.72
0.90	6.62	5.98	2.01
1.05	8.25	6.24	2.26
1.20	9.63	6.36	2.45
1.35	10.50	6.27	2.57
1.50	9.86	5.58	2.47

The discharge for a given head is the minimum value obtained for any one of the different controls. Values are tabulated in the following table for pipe flow control, inlet control, or open channel control.

Comparison of pipeflow, inlet, or channel flow for Example Problem 5.13

Head (ft)	$29n^2H/R^{4/3}$	Type pipe (Fig. 5.9)	Discharges (cfs)			
			Pipe flow[a]	Inlet control[b]	Open channel[c]	Design discharge (cfs)
0.50	—	—	—	1.56	0.56	0.56
1.50	—	—	—	6.25	3.89	3.89
2.00	0.123	Long	10.04	8.91	6.62	6.62
3.00	0.185	Long	11.22	12.82	—	11.22
4.00	0.247	Long	12.27	15.01	—	12.27
5.00	0.309	Long	13.23	17.04	—	13.23

[a] $Q = 4.957(H + 2.10)^{1/2}$.
[b] Discharge with inlet control calculations.
[c] Interpolated from open channel flow calculations.

Plotting the design discharge as a function of head will yield the stage–discharge curve. In this example, flow is controlled by either open channel flow or pipe flow, depending on the head. If the slope were substantially increased, flow would have inlet control. Recall that if the trickle tube is on a hydraulically steep slope, the open channel flow calculations are only approximate. Consider the data in open channel flow calculations table and a depth of 0.75 ft. The Froude number can be determined from Eq. (4.11) as

$$\mathbf{F} = \frac{V}{\sqrt{gd_h}} = \frac{V}{\sqrt{gA/T}}.$$

For $y_n = 0.75$ ft and $D = 1.5$ ft,

$$\frac{A}{T} = \frac{\pi D^2/8}{D} = \frac{\pi D}{8} = 0.589$$

$$\mathbf{F} = \frac{5.58}{\sqrt{32.2 \times 0.589}} = 1.28.$$

The flow is supercritical, and thus the open channel calculations are only approximate. Similar calculations can be made for other flow depths.

Ditch Relief Culvert

In mining, construction, and similar land-disturbing activities, it is often desirable to divert flow away from a road ditch so that it will not attain sufficient volume, velocity, or depth to erode the ditch. Ditch relief culverts perform this function by collecting runoff in a culvert constructed across the road at a downgrade angle (typically about 30°). To further protect downstream areas, flow exiting a ditch relief culvert should be conveyed down the fill slope to a sediment trap or stabilized area by a downdrain instead of releasing it on unprotected areas.

The specific interval required between ditch relief culverts is normally regulated as a function of road grade and/or road material. Typical spacings range from 100 ft for steep slopes ($> 10\%$) to 300 ft for mild slopes ($< 5\%$).

Downdrain: Function, Use, Type, and Design

Downdrains provide erosion protection while conveying concentrated runoff from one point on a slope to another. Riprap-lined channels are limited to gentle slopes. Steep slopes require use of downdrains to route water collected by diversions and ditch relief culverts safely downstream.

Downdrains, such as those used on surface-mined lands and construction roads, can consist of chutes, flumes, and half-round rigid and flexible pipes. They may also be constructed of geotextile materials such as fiberglass, excelsior matting, jute mesh, and plastic sheeting. These geotextile materials should not be used on slopes greater 15% according to Federal Highway Administration (1975).

Two general types of downdrains can be installed, depending on the flow requirements and availability of the materials. Sectional downdrains are prefabricated sectional conduits constructed of half- or third-round pipe, corrugated metal, bituminized fiber, or other materials. Flexible downdrains are less permanent structures that are constructed of heavy-duty fabric, or other materials. Both sectional and flexible downdrains are supported by the surface profile and are constructed to carry the design storm as open channel flow using the open channel design techniques.

Corrugated metal pipe and bituminized fiber pipe are used as more permanent structures and are placed in cut and fill slopes. They are then covered much as a spillway in a dam might be. The lead section of a downdrain is normally prefabricated and placed in compacted fill protected by rock riprap with a sand bedding. The pipe slope drain is covered with approximately 1 ft of fill or spoil material. An important

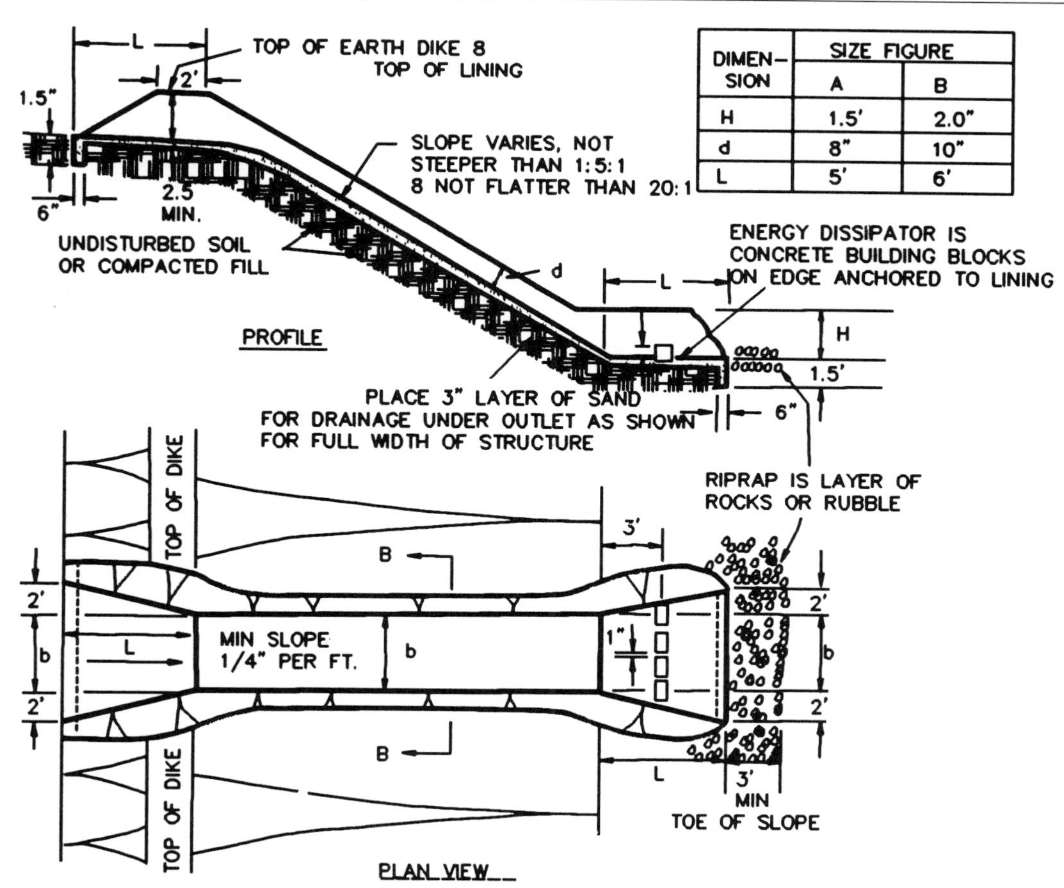

Figure 5.15 Example of a chute design (adapted from Soil Conservation Service, 1969).

component is the outlet protection consisting of sized rock riprap or a stilling basin and an energy dissipator. Such a pipe downdrain system has design requirements similar to those considered previously under trickle tubes and culverts.

Other downdrains function like a chute or flume to provide conveyance in a high-velocity, open channel suitable for carrying water to a lower elevation without erosion. Chutes or flumes are typically constructed of concrete or comparable material using either formed or freeformed methods. A schematic of a permanent chute structure is given in Fig. 5.15. Basic components are similar to those of a pipe drop downdrain. A chute can be designed using either the standard open channel procedures described previously, or it can be sized as a function of drainage area.

HYDRAULICS OF EMERGENCY SPILLWAYS

Emergency spillways are typically flat across the top with sloping inlet and exit sections as shown in Fig. 5.16. If the exit channel slope is greater than the critical slope, then flow passes through critical depth at the downstream edge of the flat crest. This point locates the control section since the velocity and depth of flow are defined. If the exit channel slope is less than the critical slope, no control section exists. Emergency spillways can be constructed both with and without control.

The hydraulics of broad-crested spillways with a control section can be analyzed with the aid of several nomographs and simple equations, which are presented in this section. The hydraulics of broad-crested spillways without control sections are not amenable to simple solution and will not be covered.

Evaluation of Head-Discharge Relations for an Emergency Spillway

The Soil Conservation Service (1968) developed a method of analysis for broad-crested spillways, such as emergency spillways. The analysis assumes that control exists at the downstream edge of the crest as summarized by Barfield *et al.* (1981). Presented are analysis procedures that can be used to:

(1) define the discharge in the spillway for a given value of head, H_p, in the reservoir, or

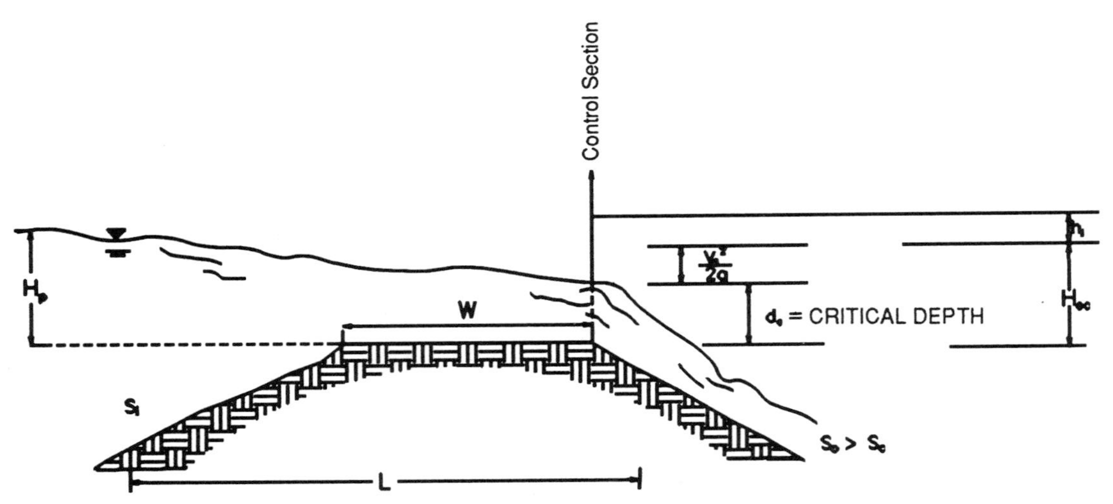

Figure 5.16 Illustration of a broad-crested spillway with a control section.

(2) define the required spillway dimension for a given discharge.

Relationship Between Discharge and H_p

When flow occurs across the crest of a spillway with a control, as shown in Fig. 5.16, critical depth and critical velocity occur at the control section. The sum of the velocity head, $V^2/2g$, and depth of flow, y_c, at this point is total critical head, H_{ec}, or

$$H_{ec} = y_c + V_c^2/2g \tag{5.23}$$

based on Eq. (4.9), where the subscript c refers to critical flow and g is the gravitational constant. For a rectangular cross section, the critical depth is given in terms of discharge as

$$y_{c,r} = (q^2/g)^{1/3}, \tag{5.24}$$

where q is discharge per unit width and the subscript r refers to a rectangular channel. For a channel of bottom width b and discharge Q, $y_{c,r}$ becomes

$$y_{c,r} = \left(\frac{Q^2}{b^2 g}\right)^{1/3}. \tag{5.25}$$

Combining Eqs. (5.23) and (4.10), it can be shown that

$$H_{ec,r} = \tfrac{3}{2} y_{c,r}. \tag{5.26}$$

Equation (5.25) can be solved for Q_r in terms of $H_{ec,r}$ and b using Eq. (5.26) to yield

$$Q_r = \left(\tfrac{2}{3}\right)^{3/2} g^{1/2} H_{ec,r}^{3/2} b \tag{5.27}$$

$$= 0.544 g^{1/2} H_{ec,r}^{3/2} b \tag{5.28}$$

for a rectangular cross section. Thus Q_r is uniquely related to $H_{ec,r}$. To define Q_r for a given value of H_p, it is necessary to relate H_{ec} and H_p.

The relationship between Q and H_{ec} for a trapezoidal section is more complex than for a rectangular channel. To develop the relationship, it is necessary to return to the energy principles discussed in Chapter 4. The total energy head for any flow is given by

$$H_e = E = V^2/2g + y \tag{5.29}$$

or in terms of discharge Q and cross-sectional area A,

$$E = Q^2/2gA^2 + y. \tag{5.30}$$

E can be differentiated with respect to y to obtain

$$\frac{dE}{dy} = 1 - \frac{Q^2}{gA^3}\frac{dA}{dy}. \tag{5.31}$$

At critical depth, dE/dy is equal to zero; hence,

$$\frac{Q^2}{gA_c^2} = \frac{A_c}{dA/dy}, \tag{5.32}$$

where A_c is the cross-sectional area at critical depth. From Fig. 4.25, dA/dy is equal to the top width; T; hence, at critical depth,

$$\frac{Q^2}{2gA_c^2} = \frac{A_c}{2T_c}, \tag{5.33}$$

where T_c is the top width at critical depth. Relationships given in Chapter 4 can be used for A_c and T_c to yield

$$\frac{Q^2}{2g(A_{c,z})^2} = \frac{V_c^2}{2g} = \frac{(b + zy_{c,z})y_{c,z}}{2(b + 2zy_{c,z})}, \quad (5.34)$$

where the subscript z refers to the side slope. Defining $H_{ec,z}$ and E at critical depth, Eq. (5.34) can be used in Eq. (5.30) to obtain

$$H_{ec,z} = \frac{(3b + 5zy_c)y_c}{2b + 4zy_c}. \quad (5.35)$$

Equations (5.34) and (5.35) can be solved simultaneously to obtain a Q–H_{ec} relationship for a trapezoidal channel. Unfortunately, the solution is implicit.

A simple approximation can be made to the $Q - H_{ec}$ problem based on the premise that the ratio of discharges of two trapezoidal channels with equal critical specific energy heads is equal to the ratio of their average widths (Soil Conservation Service, 1968), or

$$\frac{Q}{Q'} = \frac{b + zy_c}{b' + z'y_c'}. \quad (5.36)$$

When the approximation is based on a rectangular channel ($z' = 0$) of width, $b' = 100$, and using $y_c = 2/3H_{ec}$, then Eq. (5.36) can be written

$$\frac{Q}{Q'} = \frac{1.5b + zH_{ec,z}}{150}. \quad (5.37)$$

The approximate error involved in Eq. (5.37) can be determined from Fig. 5.17 or from the approximation

$$\% \text{ Error} = 2.27\,[G]^L, \quad (5.38)$$

where

$$G = zH_{ec,z}/b \quad (5.39a)$$

and

$$L = 1.1 + [\log_{10}(1/G)][0.59 - 0.136\log_{10}(1/G)]. \quad (5.39b)$$

Example Problem 5.14 illustrates the use of the procedure for calculating Q for a given value of H_{ec}.

Example Problem 5.14 Discharge for a broad-crested spillway

It has been determined that the total energy head at the control section of a broad-crested trapezoidal spillway is 5.0 ft. The side slopes are 3 : 1, and the bottom width is 100 ft. Deteremine the discharge in the spillway.

Solution: To use Eq. (5.37) to calculate Q, it is first necessary to calculate Q', the discharge in a 100-ft rectangular spillway with $H_{ec,r} = 5.0$ ft. From Eq. (5.28),

$$Q' = 0.544g^{1/2}(H_{ec,r})^{3/2}b,$$

which after substituting values becomes

$$Q' = (0.544)(32.2\text{ ft/sec}^2)^{1/2}(5\text{ ft})^{3/2}(100\text{ ft})$$
$$= 3451\text{ ft}^3/\text{sec}.$$

From Eq. (5.38),

$$Q = Q'\left[\frac{1.5b + zH_{ec,z}}{150}\right] = (3451\text{ cfs})\left[\frac{(1.5)(100) + (3)(5)}{150}\right]$$

$Q = 3796$ cfs.

The percentage error in the estimate of Q can be determined from Fig. 5.17, or from Eq. (5.38). From Fig. 5.17,

$$\% \text{ Error} = 0.13\%.$$

Using Eqs. (5.38), (5.39a), and (5.39b),

$$G = \frac{zH_{ec,z}}{b} = \frac{(3)(5\text{ ft})}{100\text{ ft}} = 0.15$$

$$L = 1.1 + [\log_{10}(1/G)][0.59 - 0.136\log_{10}(1/G)]$$
$$= 1.1 + [\log_{10}(1/0.15)][0.59 - 0.136\log_{10}(1/0.15)] = 1.49$$

$$\% \text{ Error} = 2.27[G]^L = 2.27[0.15]^{1.49} = 0.13\%.$$

The corrected value of Q is

$$Q = \frac{Q_{est}}{1 - \%e/100} = \frac{3796}{1 - 0.13/100} = 3801\text{ cfs}.$$

Although the first estimate is adequate, this will not always be the case.

The problem now remains to relate H_{ec} to H_p in order to develop a Q–H_p relationship. To do so requires the development of backwater profiles. By computing numerous profiles, the Soil Conservation Service (1968) developed nomographs of H_p versus H_{ec} for a standard reference section of $b = 100$ ft, $z = 2$, and $n = 0.04$ for the nine types of spillways shown in Fig. 5.18. The H_p versus H_{ec} curves are shown in Figs. 5D.1–5D.9 located in the appendix to this chapter.

The relationship between H_p and H_{ec} can be written[2]

$$H_p = H_{ec} + h_f, \quad (5.40)$$

where h_f is the friction head loss. The inaccuracies in

[2] H_{ec} is used here without superscripts r or z to denote a more general case.

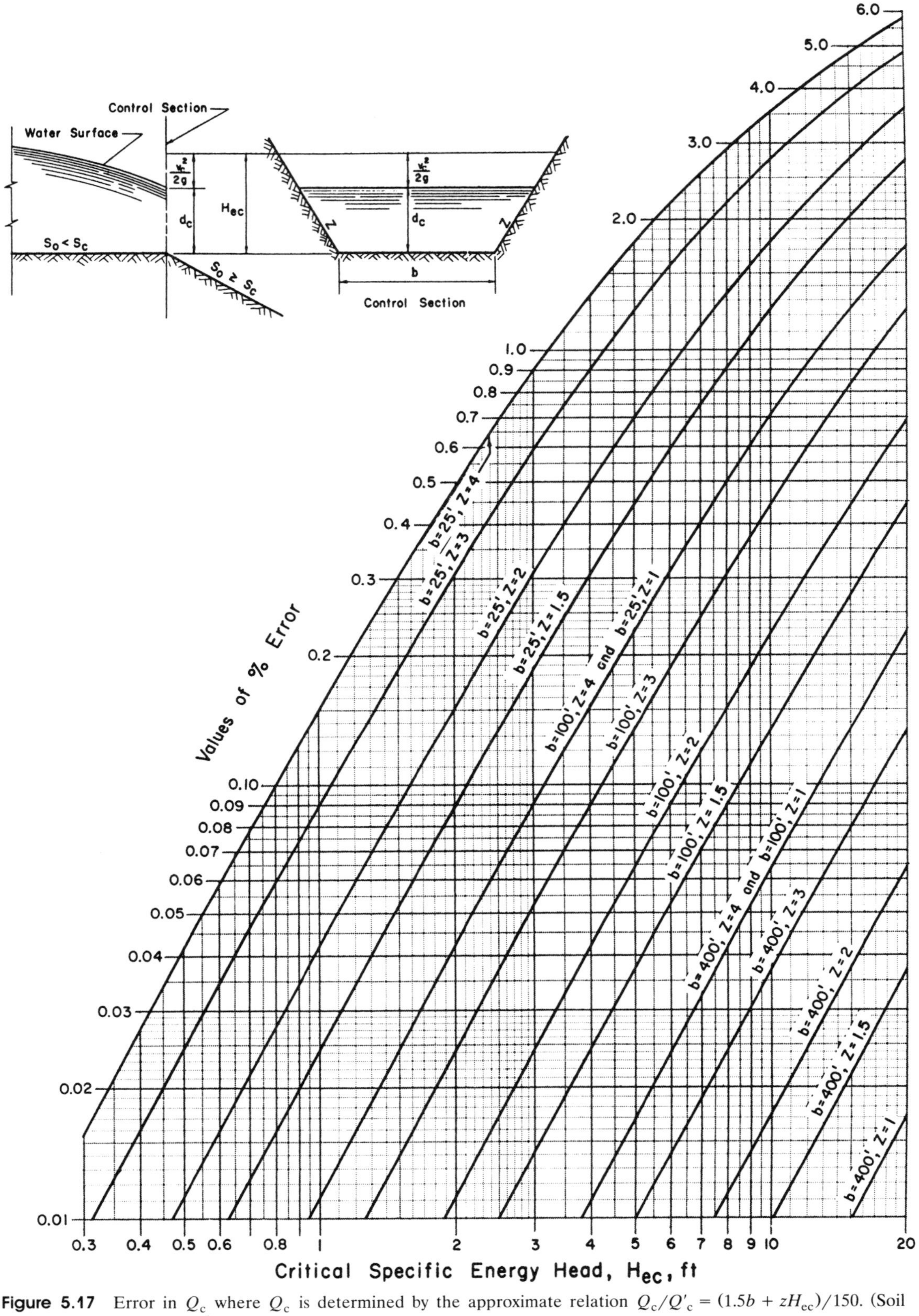

Figure 5.17 Error in Q_c where Q_c is determined by the approximate relation $Q_c/Q'_c = (1.5b + zH_{ec})/150$. (Soil Conservation Service, 1968)

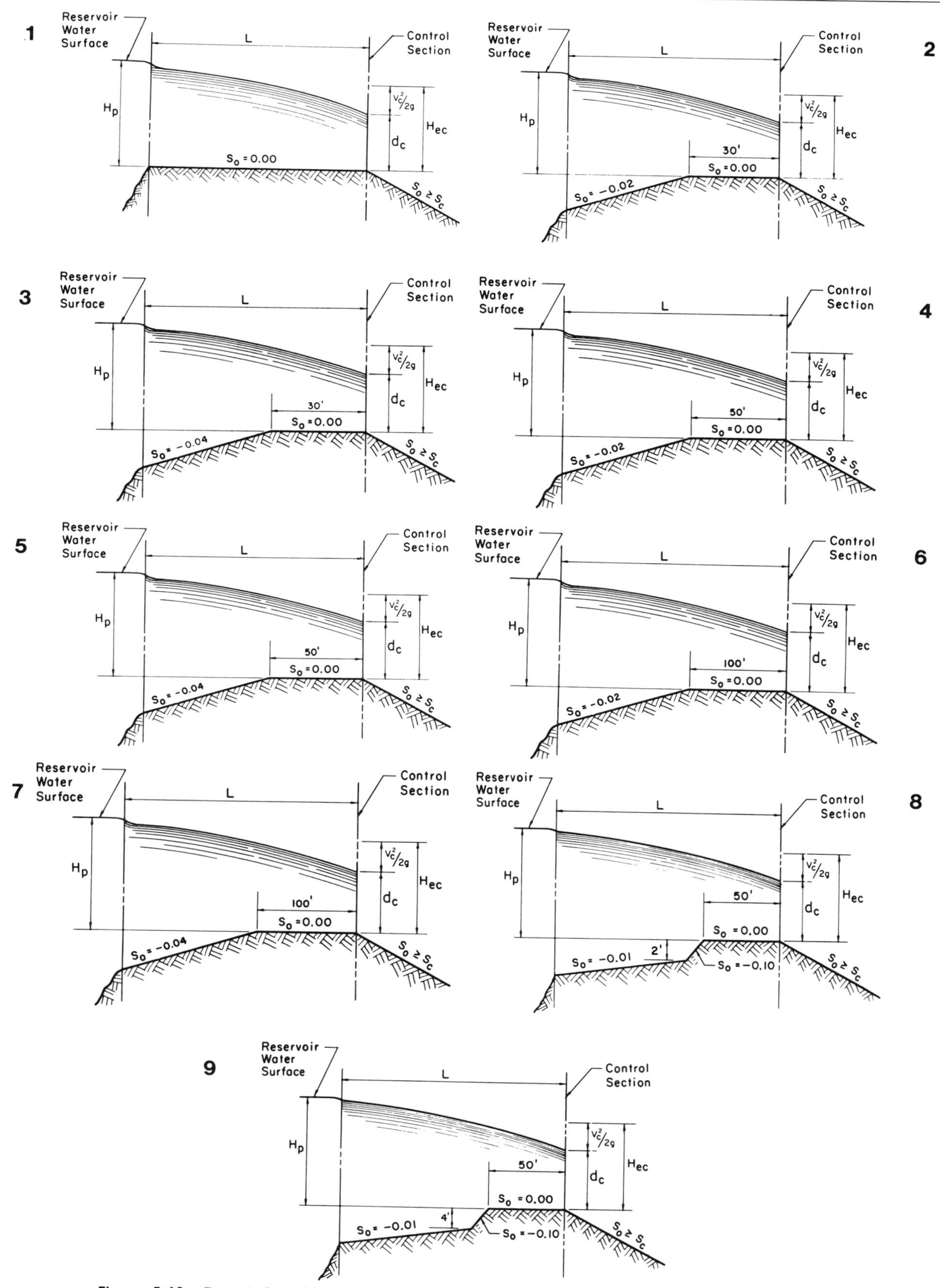

Figure 5.18 Cases 1–9 used by the Soil Conservation Service when developing curves of H_p versus H_{ec}.

Figs. 5D.1–5D.9 result from inaccuracies in h_f. These inaccuracies are given for selected cases in Figs. 5E.1–5E.5. Corrections in these figures are based on a standard condition where b is 100 ft, z is 2, and n is 0.04. Obviously, the only factor of major importance is the channel roughness n. By using Figs. 5D.1–5D.9 and the corrections in Figs. 5E.1–5E.5, the relationship between H_{ec} and H_p is established as demonstrated in Example Problem 5.15.

Example Problem 5.15 Head in reservoir with emergency spillway

Determine the value for H_p in Example Problem 5.14 if the total length of the spillway is 300 ft, the spillway is grassed so $n = 0.04$, and the section is constructed like Case 1.

Solution: Since $b = 100$, Fig. 5D.1 can be used without making corrections for channel width. For H_{ec} of 5.0 and $L = 300$ ft, the value of H_p from Fig. 5D.1 is 6.3 ft.

The effects of side slope, z, can be determined from Fig. 5E.4. From Fig. 5E.4 for a side slope of 4:1, the ratio of head loss at $z = 4$ to that of $z = 2$ is

$$H_{f,z}/h_{f,2} = 0.994;$$

hence, corrections for z are not required for a 4:1 side slope. Since a 3:1 side slope is closer to a rectangular section than a 4:1 side slope, the ratio of $h_{f,z}$ to $h_{f,2}$ would be even closer to 1.0.

Spillway Dimensions for a Given Discharge

When an emergency spillway is being selected, it is necessary to determine the design dimensions needed to transmit some expected rare event. This requires maximum values for H_p and H_{ec} be established based on channel stability in the exit section of the spillway (shown in Fig. 5.16). The maximum value of H_{ec} is selected to ensure that the velocity in the exit section does not exceed the permissible velocity. The parameter to be used will be the limiting velocity. This requires the development of backwater profiles just as in the case of Q–H_p relationships. Again on the basis of numerous computed water surface profiles, the Soil Conservation Service has developed nomographs that relate the maximum permissible H_{ec} to permissible velocity and channel slope. These nomographs are shown in the appendix to this chapter as Figs. 5E.6 and 5E.7 for Manning's n of 0.04 and 0.02, respectively. Also shown in Figs. 5E.6 and 5E.7 are curves that show the minimum slope that can be used to guarantee a control section ($S_0 > S_c$) for a discharge equal to $Q/4$. When designing an exit channel, it is important to select a slope steep enough to ensure a control section over the flow range from $Q/4$ to Q.

The procedure for selecting channel dimensions is best illustrated by example. Details are given in Example Problem 5.16. The bottom width in Example Problem 5.16 is actually quite large. It could be reduced by increasing the permissible velocity in the exit channel or by decreasing the exit slope.

Example Problem 5.16 Emergency spillway design for sediment pond

An emergency spillway is being designed for a construction site sediment pond. The design discharge is 600 cfs. The exit slope is to be grass lined with a Manning's n of 0.04 and a permissible velocity of 6.0 ft/sec. Machinery limitations require that 3:1 side slopes are the steepest that can be used. The rest of the spillway will be grassed with an n of 0.04. The Case 1 cross section in Fig. 5.18 will be used. Calculate the value of H_p needed to convey the flow, and size the emergency spillway. The spillway length, L, is 100 ft. The exit slope, S_0, is 0.03 ft/ft.

Solution: From Fig. 5E.6, the maximum permissible value for H_{ec} with $S_0 = 0.04$ ft/ft and $V_p = 6.0$ ft/sec is

$$H_{ec} \text{ max} = 1.28 \text{ ft}.$$

From Fig. 5D.1, H_p for H_{ec} of 1.28 ft, L of 100 ft, and a Manning's n of 0.04 is

$$H_{p,0.04} = 1.75 \text{ ft}.$$

Since n is 0.04 on the spillway crest, correction is not necessary. The friction head loss is therefore

$$h_{f,0.04} = 1.75 - 1.28 = 0.47 \text{ ft}.$$

If n had not been equal to 0.04, the friction head loss would be computed using Fig. 5E.1.

Corrections must also be considered for the side slopes since z is equal to 3 instead of 2. By interpolating between lines for $z = 2$ and $z = 4$ on Fig. 5E.4,

$$\frac{h_{f,n}}{h_{f,2}} = 0.999.$$

Therefore, the effects of z variations or h_f are not significant and

$$\begin{aligned} H_p &= H_{ec} + h_f \\ &= 1.28 + 0.47 \\ &= 1.75. \end{aligned}$$

The bottom width must now be determined by using Eq. (5.38). Using Eq. (5.28), the discharge for a standard 100-ft rectangular channel with $H_{ec,n}$ of 1.28 ft would be

$$Q' = 0.544 g^{1/2} H_{ec,r}^{3/2} b$$
$$= (0.544)(32.2 \text{ ft/sec})^{1/2}(1.28 \text{ ft})^{3/2}(100 \text{ ft})$$
$$= 447 \text{ cfs}.$$

From Eq. (5.38)

$$Q = Q' \frac{1.5b + zH_{ec,z}}{150}$$

substituting values and solving for b

$$600 = 447\left[\frac{1.5b + (3)(1.28)}{150}\right]$$
$$b = 131 \text{ ft}.$$

Use 130 ft. From Fig. 5.17, the error involved in using Eq. (5.38) is approximately 0.01%; hence, the bottom width is 130 ft.

Stage – Discharge Curve for Emergency Spillways

The relationships presented in this section can be used to develop a stage–discharge curve for a broad-crested emergency spillway. This can then be added to the discharge through the principal spillway to determine the total discharge.

The solution procedure involves the following steps:

(1) Calculate H_{ec} for each value of H_p.
(2) Calculate Q from the Q–H_{ec} relationship [Eq. (5.28)] for rectangular sections.
(3) Correct for a trapezoidal section [Eq. (5.37)].
(4) Add to the Q for the principal spillway.

These calculations are illustrated in Example Problem 5.17.

Example Problem 5.17 Stage–discharge curve for emergency spillways

Develop a stage–discharge curve for the spillway in Example Problem 5.16 if the side slopes are 3 : 1 and the n value is 0.04 on the crest. Assuming that the spillway crest is 12 ft above the principal spillway inlet and that the pipe used in Example Problem 5.4 is the principal spillway, calculate the total discharge from the reservoir.

Solution: From Example Problem 5.16, the spillway is trapezoidal with the following characteristics:

$$b = 130 \text{ ft}$$
$$L = 100 \text{ ft}$$
$$z = 3$$
$$n = 0.04.$$

Calculations are summarized in the following table.

Head above crest of emergency spillway[a] H_p (ft)	Total energy head[b] H_{ec} (ft)	Discharge in 100 ft rectangular spillway[c] Q' (cfs)	Discharge in trapezoidal spillway[d] Q (cfs)	Head above principal spillway[e] (ft)	Discharge in principal spillway[f] (cfs)	Total discharge[g] (cfs)
0	0	0	0	12.00	26.6	27
0.75	0.35	83	108	12.75	26.9	135
1.0	0.61	191	250	13.00	27.3	277
1.5	1.04	426	562	13.50	27.6	590
2.0	1.45	700	930	14.00	27.9	958

[a]Assumed.
[b]Figure 5D.1.
[c]Equation (5.28).
[d]Equation (5.37).
[e]$H_p + 12$ ft.
[f]From equations in Example Problem 5.4 ($Q = 5.8(H+9.1)^{1/2}$) or read from stage–discharge curve.
[g]Columns 4 + 6. (Discharge in trapezoidal spillway plus discharge in primary spillway.)

CULVERT OUTLET PROTECTION

Scour Hole Geometry

Erosion resulting from the discharge from a culvert or chute onto an unprotected, erodible material will form a hole or depression known as a scour hole. An estimation of scour hole size can best be determined through inspection of a similar existing structure located in a comparable soil environment. If this is not possible, a procedure that allows a prediction of the depth, width, length, and volume of a scour hole assuming a sandbed culvert discharge area is presented in this section. The prediction of scour hole geometry is indicative of the erosion hazard and need for soil protection. The Corps of Engineers conducted studies that indicated that scour hole geometry was related to tailwater conditions (Fletcher and Grace, 1972). They defined two tailwater categories:

(1) $$TW < \tfrac{1}{2}D, \qquad (5.41)$$

(2) $$TW \geq \tfrac{1}{2}D, \qquad (5.42)$$

where TW is tailwater depth and D is the culvert diameter.

The prediction equation needed to determine scour hole geometry is (FHA, 1975)

$$\text{Scour geometry} = \phi(D)^{\gamma}(Q/D^{2.5})^{\beta}t^{\theta}. \qquad (5.43)$$

Replacing the diameter, D, by an equivalent depth

$$y_e = (A/2)^{0.5}, \qquad (5.44)$$

where A is an area with a width that is twice the depth of flow, generalizes Eq. (5.43) to any culvert shape so that

$$\text{Scour geometry} = \alpha(y_e)^{\gamma}(Q/y_e^{2.5})^{\beta}t^{\theta}, \qquad (5.45)$$

where y_e is equivalent depth in feet, t is time of peak flow duration in minutes, and α, γ, β, and θ are coefficients that are dependent upon the desired parameter: length, width, depth, or volume of scour, as listed in Table 5.5. If the time of peak flow duration of the design storm is unknown, a maximum time of 30 min should be used in the above equation (FHA, 1975). Dimensionless rating curves for outlets of rectangular culverts (Fig. 5.19) or circular culverts (Fig. 5.20) simplify the calculation of brink depth as illustrated in the following example.

Table 5.5 Coefficients for Scour Prediction (FHA, 1975)

Maximum scour hole dimension	Coefficients			
	α	β	θ	γ
Depth (h_s)				
TW < 0.5D	0.82	0.375	0.10	1.0
TW ≥ 0.5D	0.76	0.375	0.10	1.0
Width (W_s)				
TW < 0.5D	0.55	0.915	0.15	1.0
TW ≥ 0.5D	0.39	0.915	0.15	1.0
Length (L_s)				
TW < 0.5D	1.67	0.71	0.125	1.0
TW ≥ 0.5D	2.85	0.71	0.125	1.0
Volume (v_s)				
TW < 0.5D	0.29	2.0	0.375	3.0
TW > 0.5D	0.24	2.0	0.375	3.0

Example Problem 5.18 Scour hole geometry

Determine the scour hole geometry caused by the discharge of a long 36-in. concrete circular culvert on a 0.5% slope discharging at 60 cfs.

Solution: Assume 20 min as the time duration t for the purposes of this example only. This t represents the time duration of the maximum peak discharge from a design storm.

The next step normally will be to calculate normal depth (TW) using the procedure presented in Example Problem 4.4. For this example assume TW = 1.1 ft.

Determine brink depth, y_0, using Fig. 5.20 for circular culverts to calculate

$$\frac{Q}{D^{2.5}} = \frac{60}{3^{2.5}} = 3.85 \quad \text{and} \quad \frac{TW}{D} = \frac{1.1}{3} = 0.37.$$

Then the brink depth can be found using Fig. 5.20,

$$\frac{y_0}{D} = 0.65 \therefore y_0 = 0.65(3) = 1.95 \text{ ft}.$$

Determine area of flow using Eq. (4.24) as $A = 4.86$ ft^2. Calculate the equivalent depth for nonrectangular culverts as

$$y_e = (A/2)^{1/2} = (4.86/2)^{1/2} = 1.56 \text{ ft}.$$

Noting that TW is 1.1 ft, which is less than 0.5D or 1.5 ft, the appropriate coefficients are selected from Table 5.5.

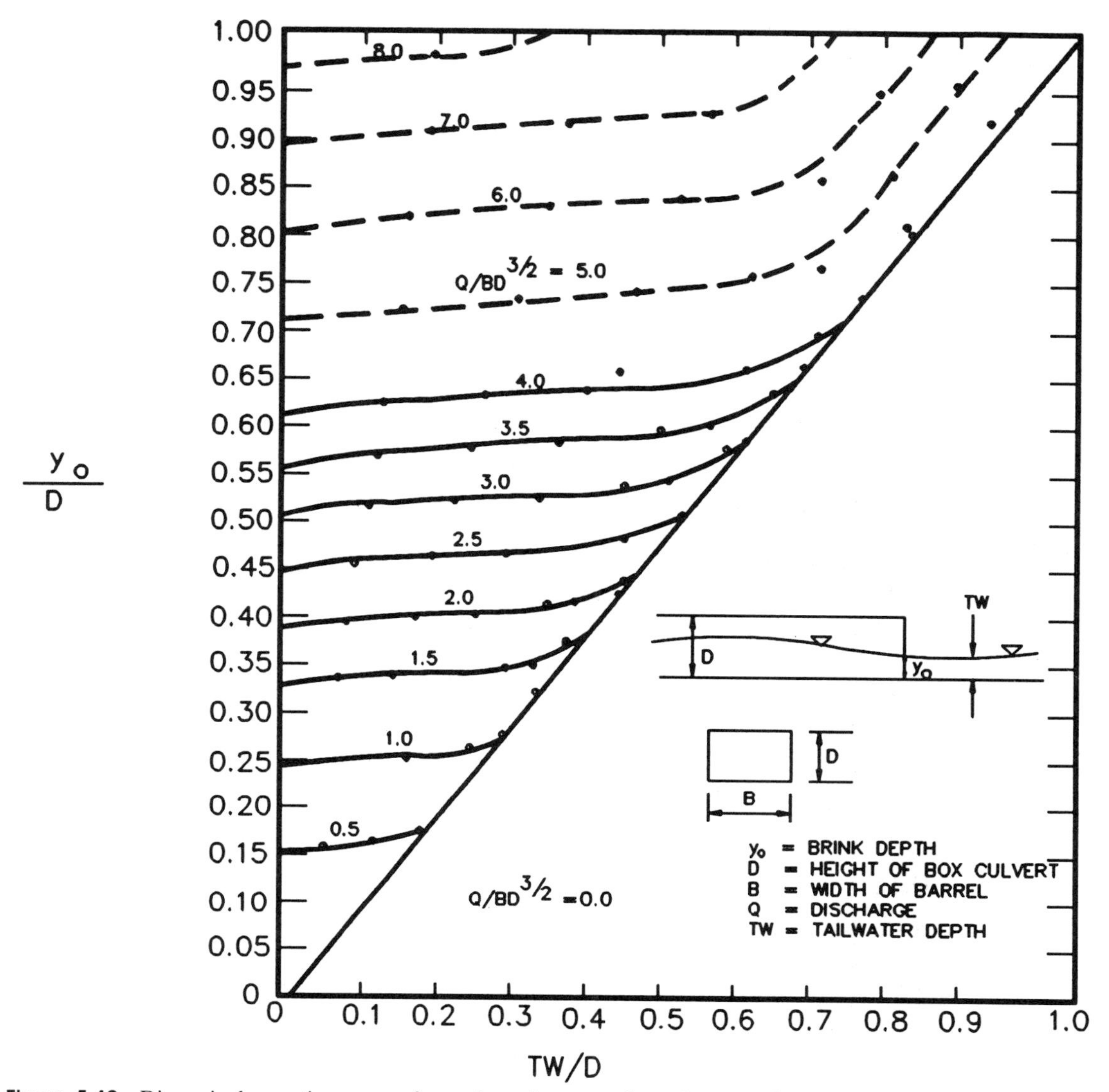

Figure 5.19 Dimensionless rating curves for outlets of rectangular culverts on horizontal and mild slopes (after Federal Highway Administration, 1975).

Calculate depth of scour using Eq. (5.45) and Table 5.5.

$$\text{Scour depth} = \alpha y_e^{\gamma}\left(Q/y_e^{2.5}\right)^{\beta} t^{\theta}$$

$$= 0.82(1.56)^{1.0}\left(60/(1.56)^{2.5}\right)^{0.375}(20)^{0.10}$$

$$= 5.3 \text{ ft}.$$

Calculate width of scour using the equation

$$\text{Scour width} = 0.55(1.56)^{1.0}\left(60/(1.56)^{2.5}\right)^{0.915}(20)^{0.15}$$

$$= 21 \text{ ft}.$$

Calculate length of scour from

$$\text{Scour length} = 1.67(1.56)^{1.0}\left(60/(1.56)^{2.5}\right)^{0.71}(20)^{0.125}$$

$$= 32 \text{ ft}.$$

Calculate volume of scour from

$$\text{Scour volume} = 0.29(1.56)^{3.0}\left(60/(1.56)^{2.5}\right)^{2.0}(20)^{0.375}$$

$$= 1319 \text{ ft}^3.$$

Energy dissipators — Rock Riprap Aprons

The design for rock riprap energy dissipators is based on research conducted at Colorado State University (FHA, 1975). The conclusion drawn from this study is that scour hole geometry is related to riprap size (d_{50}), discharge (Q), the flow depth at the discharge point

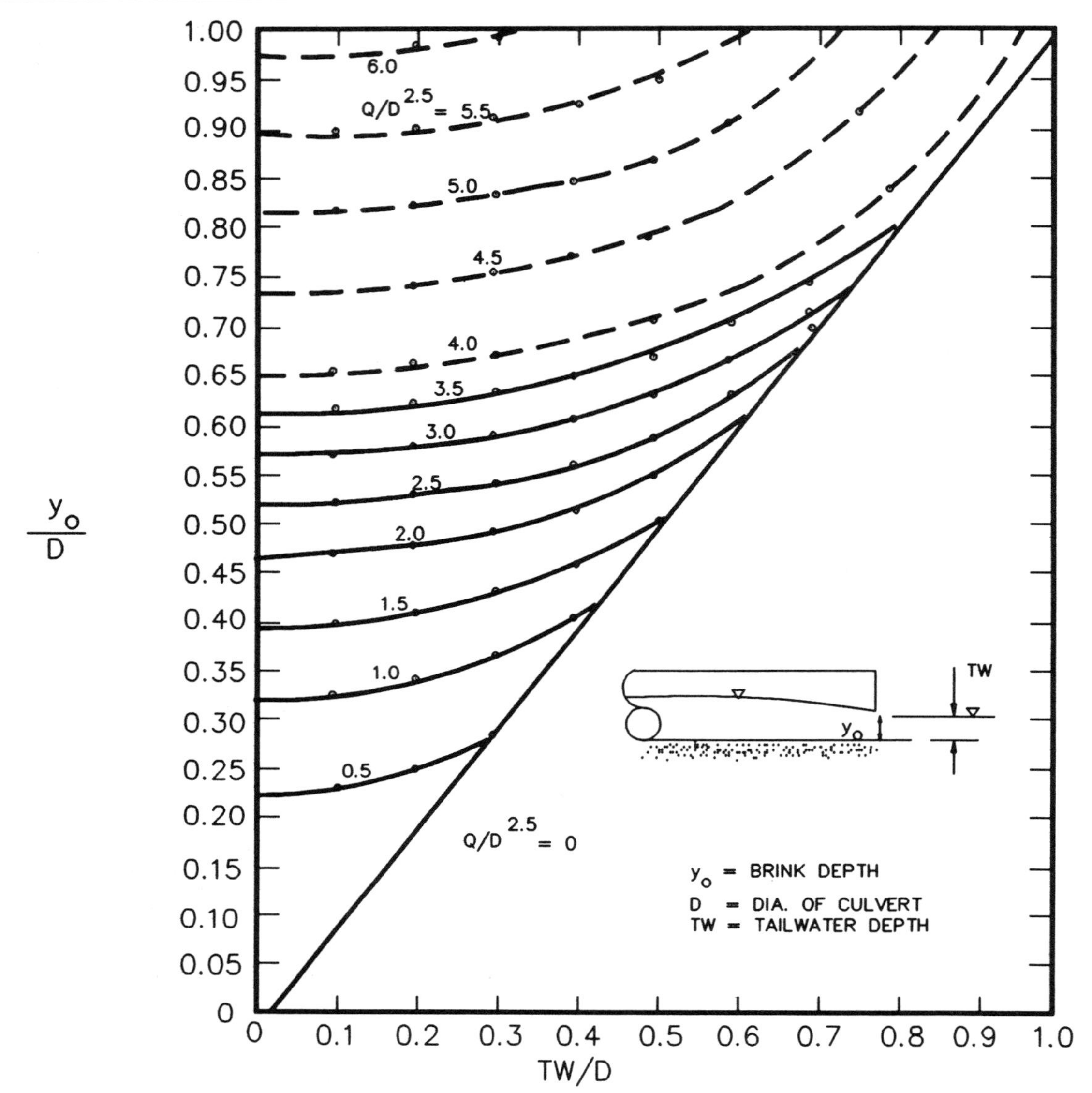

Figure 5.20 Dimensionless rating curves for outlets of circular culverts on horizontal and mild slopes (after Federal Highway Administration, 1975).

known as the brink depth (y_0), and tailwater depth (TW). Also, it was found that a riprapped scour hole functioned very efficiently as an energy dissipator if the ratio of tailwater to brink depth was less than 0.75. For greater ratios the high velocity discharge core passed through the basin creating a shallower but longer scour hole. Thus, the downstream channel required rock riprap lining.

Rock riprap basin geometry is shown in Fig. 5.21. The basin is excavated and lined with riprap. The surface of the riprapped floor is constructed at a depth h_s below the culvert exit. The ratio of scour hole depth to d_{50} should be greater than 2 and less than 4. The energy dissipator pool length is 10 h_s or 3 W_0, whichever is greater, and the overall basin length is 15 h_s or 4 W_0, whichever is greater. W_0 is equal to either the diameter for a pipe culvert or the barrel width for a box culvert.

An alternative design procedure for a rock riprap pipe outlet is to use an apron as shown in Figs. 5.22 and 5.23. The design procedure relates riprap size and apron dimensions to culvert discharge, pipe diameter, and tailwater conditions (Environmental Protection Agency, 1976). For tailwater depth less or greater than the culvert centerline, Figs. 5.24 or 5.25, respectively, should be used. The design curves are based on circu-

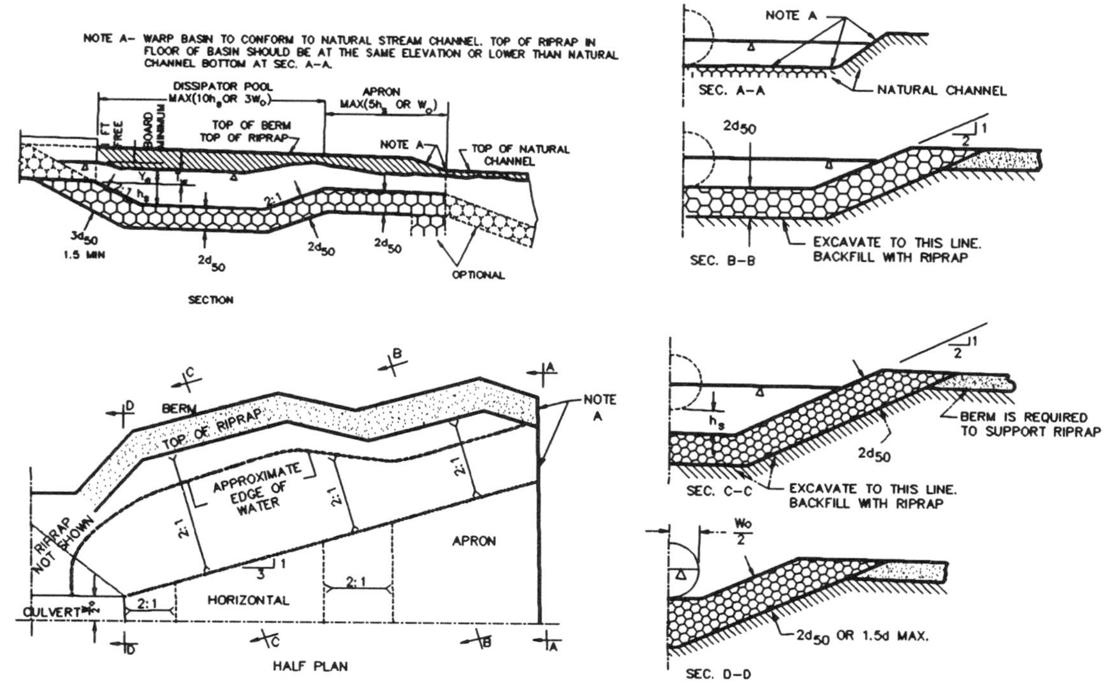

Figure 5.21 Schematic of riprapped culvert energy basin (adapted from Federal Highway Administration, 1975).

lar conduits flowing full. The design procedure is illustrated in the following examples.

Example Problem 5.19 Rock riprap selection for control of scour

Determine the rock riprap requirements for the conditions given in Example Problem 5.18.

Solution: The tailwater and brink depth from Example Problem 5.18 are 1.1 and 1.95 ft, respectively. Thus TW/$y_0 < 0.75$. The equivalent depth, y_e is 1.56 ft and flow area, A, is 4.86 ft^2. Assume rock with a d_{50} equal to 0.55 ft is available.

Compute $V_0 = Q/A = 60/4.86 = 12.3$ fps. Compute Froude Number, **F**:

$$\mathbf{F} = \frac{V_0}{\sqrt{gy_e}} = \frac{12.3}{\sqrt{32.2 \times 1.56}} = 1.74.$$

Since $d_{50} = 0.55$ ft, $d_{50}/y_e = 0.35$. From Fig. 5.23, $h_s/y_e = 1.3$ so $h_s = (1.3)(1.56) = 2.03$. Check that $2 < h_s/d_{50} < 4$; just barely OK! Depth of scour, $h_s = 2.03$ ft. Length of energy dissipation pool is the maximum of $10h_s$ or $3W_0$ or

$$\max[10h_s, 3W_0] = \max[(10)(2.03),(3)(3)].$$

Therefore, the energy dissipator pool length is 20.3 ft. The overall basin length $= \max[15h_s, 4W_0] = \max[15(2.03), 4(3)] = 30.5$ ft.

Example Problem 5.20 Rock riprap apron below culvert

A circular 36-in. concrete conduit flowing full discharges 60 cfs. The tailwater, calculated by Manning's equation, is 1.1 ft. Determine the median rock diameter by weight, d_{50}, and the appropriate length and widths of the rock riprap apron.

Solution: Calculate the apron width using either

$$W_a = D + L_a \qquad (\text{TW} < 0.5D)$$
$$W_a = D + 0.4L_a \qquad (\text{TW} > 0.5D).$$

Use the low tailwater equation since

$$\text{TW} = 1.1 < 0.5D = 1.5.$$

Hence, Fig. 5.24 is applicable. From Fig. 5.24, read both median rock diameter and minimum apron length by drawing a vertical line from the abscissa at $Q = 60$ cfs to the pipe diameter, $D = 36$ in. Read both the median rock diameter and minimum apron length from the appropriate ordinate axes. Thus $d_{50} = 0.6$ ft and $L_a = 21$ ft. Therefore the apron width using the low TW equation is

$$W_a = D + L_a = 3 + 21 = 24 \text{ ft}$$

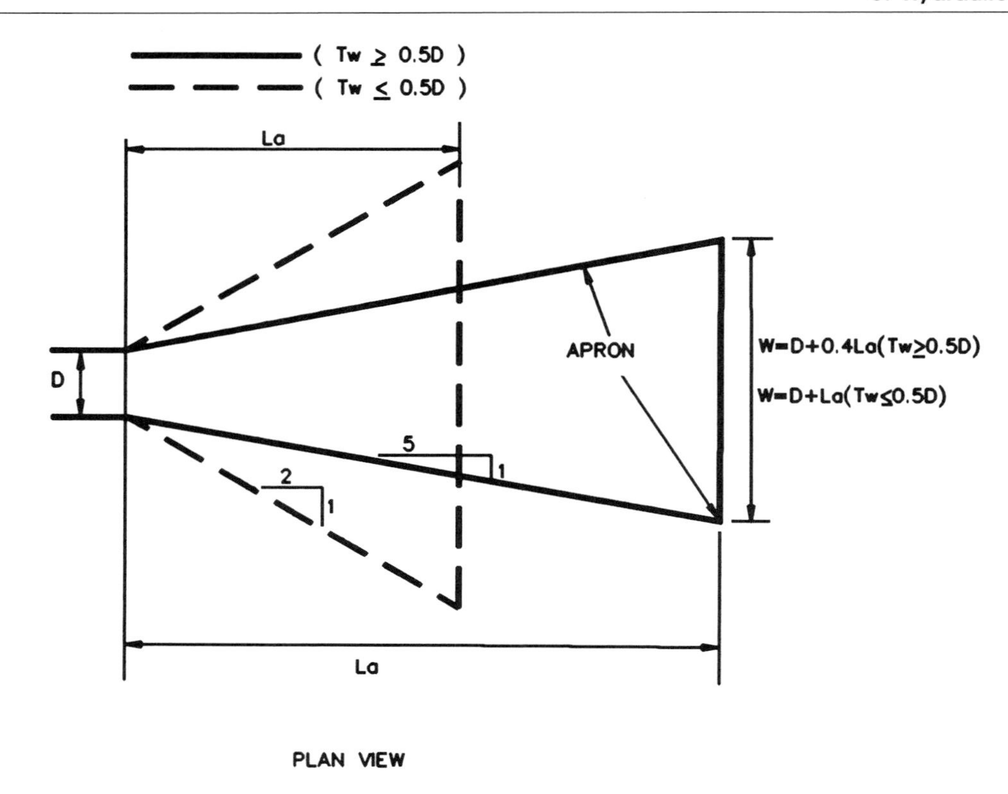

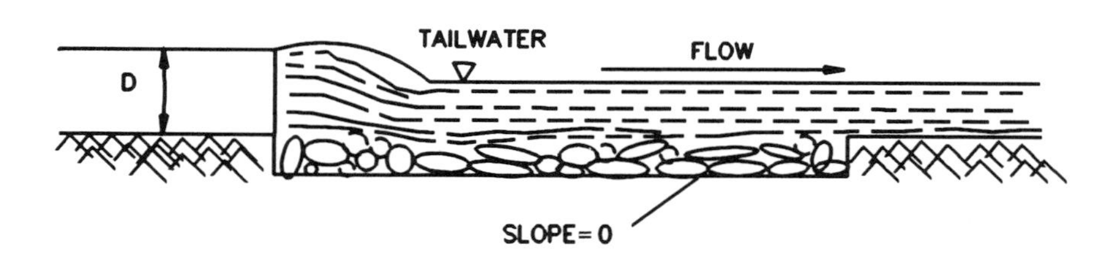

Figure 5.22 Rock riprap apron (adapted from Environmental Protection Agency, 1976).

and

$$d_{max} = 1.5 \times d_{50} = 1.5 \times 0.6 = 0.90 \text{ ft}.$$

In sizing rock riprap energy dissipators used at conduit outlets, the relative tailwater depth is a significant factor in design. Maximum scour depth is associated with low tailwater conditions, whereas maximum scour hole length is associated with high tailwater depths. For ephemeral flashy streams, tailwater depth will lag significantly during stream rise. Hence, low tailwater will always occur prior to equilibrium conditions. Computation of tailwater is difficult because of changes in channel cross section due to flood events, seasonal vegetal changes, and the inherent difficulty of measuring channel properties in poorly defined ephemeral streams. For these reasons, calculated designs should consider the worst possible tailwater condition, i.e., a low tailwater depth.

Median rock diameter, d_{50}, is predicted in the example problems. A standard practice is to prepare a well-graded mixture down to the 1-in. size particle such that 50% of the mixture by weight shall be larger than the d_{50} size (EPA, 1976). Riprap size distribution criteria are presented in Chapter 4.

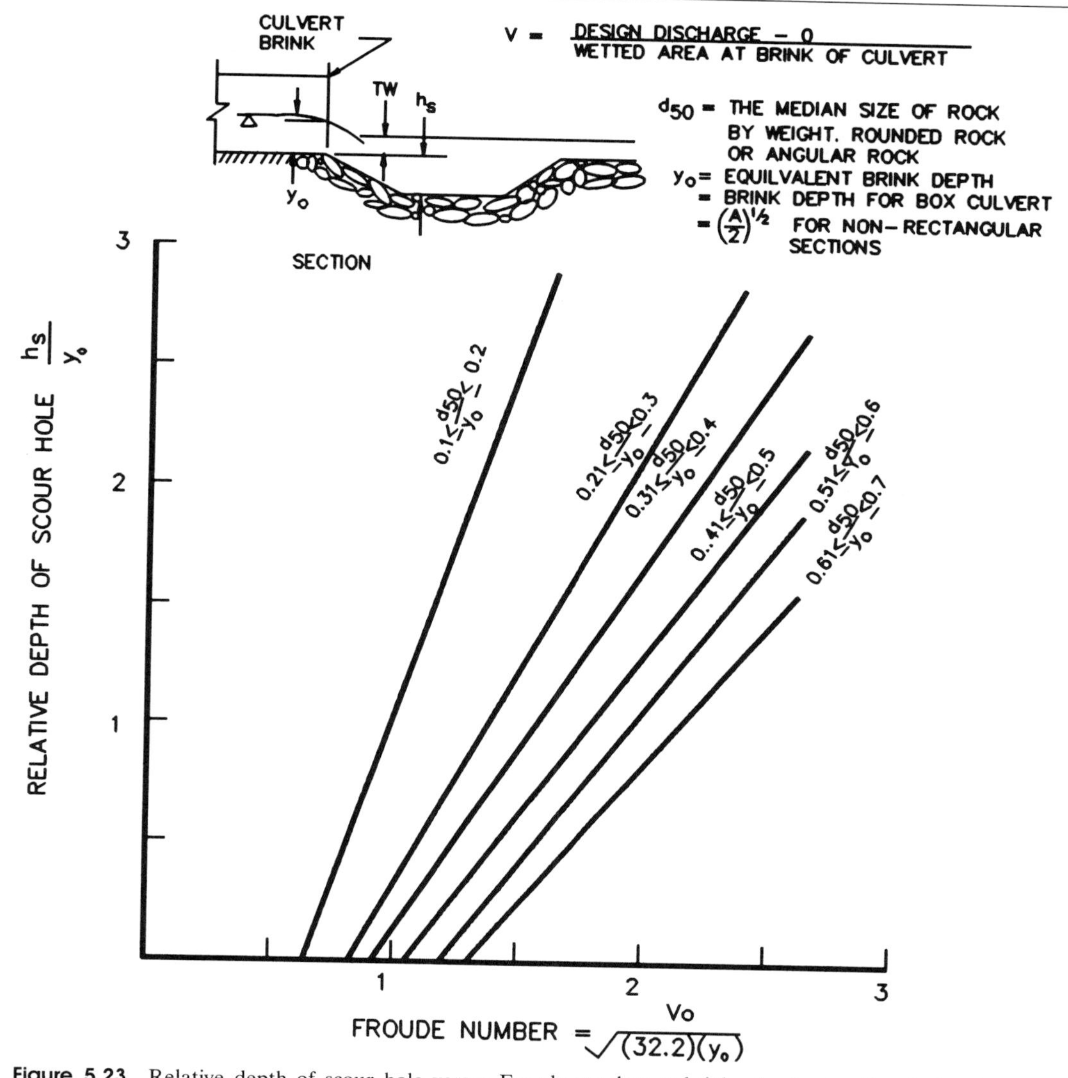

Figure 5.23 Relative depth of scour hole versus Froude number at brink of culvert with relative size of riprap as a third variable (Federal Highway Administration, 1975).

Problems

(5.1) A 36-in. circular CMP riser serves as a vertical inlet for the principal spillway of a stormwater detention structure. Estimate the discharge if the head on the riser is 1.75 ft.

(5.2) A primary spillway for a detention pond consists of a 30-in. corrugated metal pipe connected to a 48-in. vertical riser. Calculate the head required to pass a discharge of 80 cfs if weir flow controls. What if orifice flow controls?

(5.3) The spillway in the previous problem is 80 ft long and has one 90° bend where the top of the 48-in. riser joins the 30-in. pipe; the riser is 12 ft above the bottom of the outlet. Assume that there is no tailwater. Estimate the pipe discharge if pipe flow controls and the water level is 4 ft above the top of the riser.

(5.4) For the spillway indicated in problems (5.2) and (5.3), plot a stage–discharge curve. Indicate on the curve the stages at which each type of control (weir, orifice, or pipe) occurs. Do all three control types occur?

(5.5) A large stormwater detention structure is to be constructed using a 12-ft-diameter CMP for the riser and an 8-ft CMP for the barrel. The difference in elevation between the riser opening and the bottom of the downstream end of the pipe is 28 ft. The pipe is approximately 160 ft long. (a) What flow occurs through the spillway for water levels that are 1, 2, 4, 6, and 8 ft above the riser? (b) If four 1-ft-diameter holes are cut

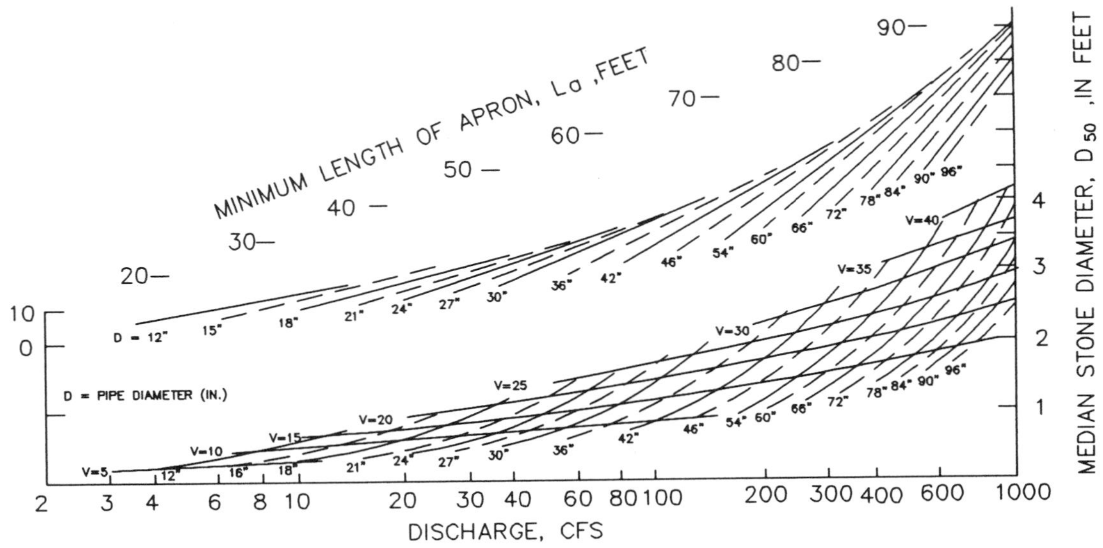

Figure 5.24 Design of outlet protection—minimum tailwater condition, $T_w < 0.5D$ (Environmental Protection Agency, 1976).

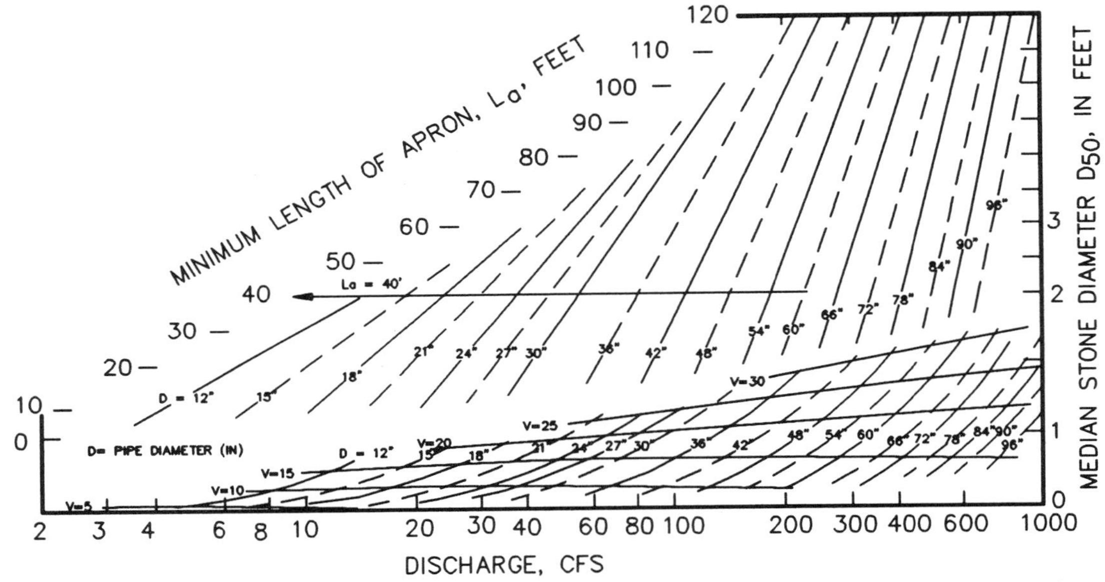

Figure 5.25 Design of outlet protection—maximum tailwater condition, $T_w \geq 0.5D$ (Environmental Protection Agency, 1976).

into the riser 3 ft below its top, what discharge will pass through the four holes with the water level at 1, 2, 4, and 8 ft above the riser? (c) What is the total discharge through the pipe? (d) How might the orifices be sized to provide better stormwater control? (e) Explain whether you would expect two rows (each consisting of four holes) of 8-in.-diameter holes to provide better results? Assume that one row is 2 ft below the riser invert and the other row is 4 ft below the riser invert.

(5.6) A gravel roadway is constructed in a low-lying area such that the roadway is frequently overtopped as a result of severe storms. The roadway is 40 ft wide, and its elevation is 36 ft. (a) If the water level upstream of the roadway is 2 ft above the crest of the roadway, what is the discharge across the roadway? (b) If the roadway is paved, what upstream depth would be required to carry the same flow? (c) Would paving reduce flooding problems?

(5.7) Compute the head loss in a 200-ft section of 2×2-ft concrete pipe flowing full and discharging 40 cfs. Assume a roughness coefficient of 0.015.

(5.8) Calculate the discharge in a 300 ft, 24-in.-diameter pipe with n equal to 0.014. The pipe is flowing full, and head on the pipe is 7 ft.

(5.9) Plot a stage–discharge curve for a trickle tube spillway (CMP) that has 24 in. diameter. The pipe is 50 ft long and is located on a 2% slope. It has a free outfall.

(5.10) Determine the critical depth and critical head for a 5000 cfs discharge across an emergency spillway. The spillway is rectangular with a bottom width of 100 ft.

(5.11) The emergency spillway in the previous problem is to be compared with a broad-crested, trapezoidal spillway having 4 : 1 sides slopes and bottom width of 150 ft. If total energy head at the control section is 4 ft, what is the discharge in the spillway?

(5.12) Determine the total energy head H_p in problem (5.10) if total length of the spillway is 240 ft and constructed as in Case 2. The spillway is grassed with tall fescue and has 3 : 1 side slopes.

(5.13) Develop a stage–discharge curve for a detention pond having a principal spillway as in problem (5.3) and an emergency spillway as in problem (5.12). Note that the discharges must be added to get total discharge. Assume that the spillway crest is 9 ft above the principal spillway inlet.

(5.14) A 48-in.-diameter circular concrete culvert draining a subdivision will have a peak discharge of 75 cfs for 20 min as a result of a 10-year storm. Determine the scour hole geometry and rock riprap requirements if the downstream channel has a 0.6% slope.

References

American Concrete Pipe Association (1985), "Concrete Pipe Design Manual." American Concrete Pipe Association, Vienna, VA.

Barfield, B. J., Warner, R. C., and Haan, C. T. (1981). "Applied Hydrology and Sedimentology of Disturbed Areas." Oklahoma Technical Press, Stillwater, OK.

Brater, E. F., and King, H. W. (1976). "Handbook of Hydraulics." McGraw–Hill, New York.

Bureau of Public Roads (1965a). Capacity charts for the hydraulic design of highway culverts, Hydraulic engineering circular No. 10. U.S. Government Printing Office, Washington, DC.

Bureau of Public Roads (1965b). Hydraulic charts for the selection of highway culverts, Hydraulic engineering circular No. 5. U.S. Government Printing Office, Washington, DC.

Carter, R. W. (1957). Computation of peak discharge at culverts, U.S. Geological Survey, Circular 376. U.S. Government Printing Office, Washington, DC.

Chow, V. T. (1959). "Open Channel Hydraulics." McGraw–Hill, New York.

Environmental Protection Agency (1976). "Erosion and Sediment Control Surface Mining in the Eastern U.S.," Vol. 2, "Design." EPA-625/3-76-006.

Federal Highway Administration (1975). Hydraulic design of energy dissipators for culverts and channels, Hydraulic engineering circular No. 14. Federal Highway Administration, Washington, DC.

Federal Highway Administration (1985). Hydraulic design of highway culverts, Hydraulic design series No. 5, Report No. FHWA-IP-85-15. Federal Highway Administration, Washington, DC.

Fletcher, B. P., and Grace, J. L., Jr. (1972). Practical guidance for estimating and controlling erosion at culvert outlets, Corps of Engineers Research Report H-72-5. Waterways Experiment Station, Vicksburg, MS.

French, R. H. (1985). "Open-Channel Hydraulics." McGraw-Hill, New York.

Grant, D. A. (1978). "Open Channel Flow Measurement Handbook," 1st ed. Instrument Specialties Co., Lincoln, NE.

Herrera, N. M. (1989). Defining the hydraulics of flow through a rockfill of varying gradation using sediment-free water, Unpublished Ph.D. dissertation. University of Kentucky, Lexington, KY.

Herrera, N. M., and Felton, G. K. (1991). Hydraulics of flow through a rockfill dam using sediment-free water. *Trans. Am. Soc. Agric. Eng.* **34**(3): 871–875.

Hoffman, C. J. (1974). Outlet works *in* "Design of Small Dams," Chap. X. U.S. Bureau of Reclamation, U.S. Government Printing Office, Washington, DC.

Kao, T. Y. (1975). Hydraulic design of storm water detention structures. *In* "Proceedings, National Symposium on Urban Hydrology and Sediment Control, UKY BU 109, College of Engineering, University of Kentucky, Lexington, KY."

Mavis, F. T. (1942). The hydraulics of culverts, Bulletin 56, The Pennsylvania State College, Engineering Experiment Station, PA.

McCuen, R. H. (1989). "Hydrologic Analysis and Design." Prentice–Hall, Englewood Cliffs, NJ.

Portland Cement Association (1962). "Culvert Design Aids: An Application of U.S. Bureau of Public Roads Culvert Capacity Charts." Portland Cement Association, Chicago, IL.

Portland Cement Association (1975). "Concrete Culvers and Conduits." Portland Cement Association, Skokie, IL.

Soil Conservation Service (1951). "Engineering Handbook," Hydraulics Section 5. U.S. Department of Agriculture, Washington, DC.

Soil Conservation Service (1968). "Hydraulics of Broad-Crested Spillways." Technical Release No. 39, Engineering Division, Soil Conservation Service, U.S. Department of Agriculture, Washington, DC.

Soil Conservation Service (1969). "Entrance Head Losses in Drop Inlet Spillways," Design Note No. 8. Engineering Division, Design Branch, Soil Conservation Service, U.S. Department of Agriculture, Washington, DC.

Soil Conservation Service (1984). "Engineering Field Manual." U.S. Department of Agriculture, Washington, DC.

Stephenson, D. (1979). "Rockfill in Hydraulic Engineering." Elsevier, New York.

Streeter, V. L. (1971). "Fluid Mechanics." 5th ed. McGraw–Hill, NY.

Channel Flow Routing and Reservoir Hydraulics

In Chapter 4, steady flow in channels was considered. Steady flow refers to situations where there is no change in flow characteristics at a point in a channel with respect to time. Similarly in Chapter 5, steady flow through hydraulic structures such as spillways, culverts, drop structures, etc., was covered. Chapter 3 was devoted to estimation of runoff hydrographs from rainfall events. In that chapter it was shown that runoff hydrographs are time-varying descriptions of flow and thus natural runoff events produce unsteady flows.

This chapter is devoted to the analysis of unsteady flow through channels and reservoirs. Storage in channels and reservoirs has a major impact on flow hydrographs. Generally, peak flows are reduced and base times of hydrographs are prolonged. These characteristics can be used to great advantage in the design and operation of systems for controlling runoff events. Consider two points, A and B, separated by several hundred feet along a stream. The flow hydrograph at point A may be known and the hydrograph at point B, downstream from A, desired. Obviously factors such as channel steepness, channel roughness, channel shape and available storage between points A and B, as well as any additional flows into the channel between points A and B, will impact the shape of the hydrograph at B.

For a steep, prismatic, smooth channel with little storage and no intermediate inflows, the hydrograph at B would be very much like the hydrograph at A. The flow would simply pass through the channel reach with little alteration in the hydrograph except a time lag equal to the travel time between A and B. If the channel between A and B was flat, irregular in shape, hydraulically rough, and had a lot of storage capacity, the hydrograph at B would be considerably different than the original hydrograph at A. In the absence of inflows between points A and B, the hydrograph at B for this latter case would likely have a longer time to peak, a longer time base, and a lower peak flow.

FLOW ROUTING

Flow routing consists of analytical techniques for determining the outflow hydrograph from a stream reach or reservoir from a known inflow hydrograph. Through flow routing, the impacts of channel and/or reservoir characteristics on hydrographs can be determined. Further the impact of channel modifications or changes in reservoir spillway characteristics on outflow hydrographs can be determined. Flow routing is essential in any storm water runoff study. Flow routing is central to designing structures to control storm water runoff events and to mitigate the impact of flood flows. Flow routing is also an integral component of sediment pond design as it provides a means of estimating the detention time and thus the settling opportunity time for sediment particles.

Flow routing is logically a topic in hydraulics, yet it is central to an understanding of hydrology. Thus flow routing is treated both in books on hydraulics, especially open channel hydraulics, and "engineering" hydrology. Flow routing procedures range in complexity from simple storage routing procedures to relatively complex procedures based on simultaneous solutions to the hydrodynamic equations dealing with the conservation of momentum and mass.

Storage routing is sometimes known as hydrologic routing since the basic equation is the continuity equation. Several methods of hydrologic routing are presented here, including basic storage routing, Muskingum routing, convex routing, and kinematic routing.

Hydraulic flow routing is different from hydrologic routing in that it is based on both the momentum and continuity equations. Hydraulic routing is usually accomplished by a numerical solution of the governing equations or by the method of characteristics. In Chapter 3, the basic hydraulic equations for flow routing were developed as Eqs. (3.36) and (3.37). The continuity equation was given as

$$\frac{\partial A}{\partial t} + A\frac{\partial v}{\partial x} + v\frac{\partial A}{\partial x} = q(x,t) \tag{6.1}$$

and the momentum equation was given as

$$\frac{\partial v}{\partial t} + v\frac{\partial v}{\partial x} + g\frac{\partial y}{\partial x} + \frac{qv}{y} - g(S_0 - S_f) = 0. \tag{6.2}$$

The relative importance of the various terms in this last equation determines to a large extent the degree of simplification warranted. This in turn determines the flow routing method(s) that is appropriate. Of course, the full momentum equation can always be used but may require more detailed information and calculation than necessary.

For example, consider a stream with a slope of 0.005 ft/ft, a change in depth of 1 ft/mile, a velocity of 5 fps, a velocity gradient of 0.5 fps/mile, a rate of change in velocity of 0.5 fps/hr, and no lateral inflow. Equation (6.2) can be written as

$$S_f = S_0 - \frac{1}{g}\frac{\partial v}{\partial t} - \frac{v}{g}\frac{\partial v}{\partial x} - \frac{\partial y}{\partial x}. \tag{6.3}$$

In looking at the individual terms and approximating partial derivatives by numerical derivatives, we have

$$S_0 = 5 \times 10^{-3}$$

$$\frac{1}{g}\frac{\partial v}{\partial t} \simeq \frac{1}{32.2\ \text{ft/sec}^2} \times \frac{0.5\ \text{ft/sec/hr}}{3600\ \text{sec/hr}} = 4.3 \times 10^{-6}$$

$$\frac{v}{g}\frac{\partial v}{\partial x} \simeq \frac{5\ \text{ft/sec}}{32.2\ \text{ft/sec}^2}\ \frac{0.5\ \text{ft/sec/mi}}{5280\ \text{ft/mi}} = 1.5 \times 10^{-5}$$

$$\frac{\partial y}{\partial x} \simeq \frac{1\ \text{ft/mi}}{5280\ \text{ft/mi}} = 1.9 \times 10^{-4}.$$

Thus from Eq. (6.3),

$$S_f = 5 \times 10^{-3} - 0.0043 \times 10^{-3}$$
$$+ 0.015 \times 10^{-3} - 0.19 \times 10^{-3} = 4.69 \times 10^{-3}$$

or S_f is 96% of S_0.

As a second example, consider a stream where changes occur more rapidly. In this case the slope might be 0.02 ft/ft, the change in depth 1 ft/200 ft, a velocity of 10 fps, a velocity gradient of 1 fps/500 ft, a rate of change of velocity of 1 fps/3 min, and no lateral inflow. The individual terms would then be

$$S_0 = 2 \times 10^{-2}$$

$$\frac{1}{g}\frac{\partial v}{\partial t} \simeq \frac{1}{32.2\ \text{ft/sec}^2}\ \frac{1\ \text{ft/sec}}{3\ \text{min} \times 60\ \text{sec/min}}$$
$$= 1.7 \times 10^{-4}$$

$$\frac{v}{g}\frac{\partial v}{\partial x} \simeq \frac{10\ \text{ft/sec}}{32.2\ \text{ft/sec}^2}\ \frac{1\ \text{ft/sec}}{500\ \text{ft}} = 6.2 \times 10^{-4}$$

$$\frac{\partial y}{\partial x} \simeq \frac{1\ \text{ft}}{200\ \text{ft}} = 5 \times 10^{-3}.$$

Thus from Eq. (6.3),

$$S_f = 0.02 - 0.017 - 0.062 - 0.500 = 0.01421$$

or $S_f = 71\%$ of S_0.

In case two the terms are more nearly of the same order of magnitude than they are in case one. In case one, a simpler routing procedure might be adopted since S_0 is nearly equal to S_f. In case two, such a simplification may add significant error. What follows is a discussion of routing techniques ranging from storage routing relying on the continuity equation to the use of the full hydrodynamic equations.

CHANNEL ROUTING

Storage Routing

The simplest form of routing is based on the continuity equation and is known as storage routing. In Chapter 3, Eq. (6.1) was shown to come from the continuity equation [Eq. (3.34)]. If this equation is written over a finite interval of time, Δt, the result is

$$\frac{I_1 + I_2}{2} \Delta t - \frac{O_1 + O_2}{2} \Delta t = S_2 - S_1. \qquad (6.4)$$

This equation may be rearranged to

$$S_2 + \frac{O_2}{2} \Delta t = S_1 - \frac{O_1}{2} \Delta t + \frac{I_1 + I_2}{2} \Delta t. \quad (6.5)$$

In these equations, the subscripts 1 and 2 refer to conditions at the beginning and end of a time interval, respectively. I represents the inflow, O the outflow, and S the storage in a channel reach. The time interval is represented by Δt.

The storage in a channel reach depends on the channel geometry and depth of flow. The flow rate may be related to the depth of flow, assuming steady, uniform flow using Manning's equation. A simple method for computing the storage is to base it on the average cross-sectional area of the reach for a given flow rate. Thus S would be determined by developing a flow rate–area relationship based on Manning's equation for each cross section. The length of the channel section multiplied by the average cross-sectional area of the channel at a given flow rate would give the storage in the reach at that flow rate.

Tributary inflows, overland flow, and ground water contributions to flow can be added to the inflow or outflow of the reach as appropriate. Normally, channel routing requires that the channel be divided into several reaches. The outflow from one reach becomes the inflow to the next reach downstream. The channel should be divided into reaches having relatively uniform hydraulic properties. The routing interval Δt should not exceed one-fifth to one-third of the time to peak of the hydrograph being routed. The routing interval should not exceed the travel time through the reach. If these guidelines on Δt are not followed, significant changes in the inflow hydrograph may occur within the selected Δt and not be reflected in the outflow hydrograph, since they may pass completely through the reach during the time interval.

For small channels, the nonuniformity of the channel may necessitate shorter reach lengths. As a general guide, the average flow velocity may be computed at bankful discharge or at a flow equal to 75% of the hydrograph peak flow.

A common method for solving Eq. (6.5) is to plot $S + O\,\Delta t/2$ and $S - O\,\Delta t/2$ versus the discharge or depth. These curves are called the characteristic curves. Example Problem 6.1 illustrates the construction of these curves and their use in routing.

Muskingum Method

A modification of storage routing that considers a linear change in depth along the reach is known as the Muskingum method. In the previous discussion of storage routing, the storage in a channel reach was related to depth of flow in the reach based on the outflow rate in the reach. In reality the flow depth would not be constant along the reach because the inflow rate would not be the same as the outflow rate. If the flow was in a rising stage, the inflow into the reach would exceed the outflow. Thus the depth of flow at the upstream end of the reach would exceed that of the downstream end of the reach. A "wedge" of storage above a uniform flow rate corresponding to the downstream flow depth would exist. To partially overcome this nonuniformity, the Muskingum method of streamflow routing makes the storage in the reach a linear function of both the inflow and outflow rate

$$S = k[xI + (1 - x)O], \qquad (6.6)$$

where k and x must be determined from channel characteristics. The coefficient k is known as the storage constant and is approximately equal to the travel time through the reach. For best results, both k and x should be based on observed hydrographs. An x of zero corresponds to reservoir storage routing. A value of $x = \frac{1}{2}$ makes the storage a function of the average flow rate in the reach based on the inflow and outflow.

In the absence of streamflow records on I and O, k may be estimated as the flow travel time in the reach, and x may be taken as about 0.25. Through manipulation of Eq. (6.5) and (6.6), one can obtain a linear expression for outflow in the form

$$O_2 = C_0 I_2 + C_1 I_1 + C_2 O_1, \qquad (6.7)$$

where

$$C_0 = \frac{-kx + 0.5\Delta t}{k(1-x) + 0.5\,\Delta t} \quad (6.8)$$

$$C_1 = \frac{kx + 0.5\,\Delta t}{k(1-x) + 0.5\,\Delta t} \quad (6.9)$$

$$C_2 = \frac{k(1-x) - 0.5\,\Delta t}{k(1-x) + 0.5\,\Delta t}. \quad (6.10)$$

Curves relating C_0, C_1, and C_2 to the outflow may be constructed. If k and x are assumed constant, the routing is greatly simplified since C_0, C_1, and C_2 then become constants.

If streamflow data on inflow and outflow to the reach of interest are available, values of k and x may be determined from these data. Equation (6.6) shows that if S is plotted against $xI + (1-x)O$, a straight line with a slope k should result. Since the inflow and outflow hydrographs are known, Eq. (6.4) can be solved for S_2. The initial value for S_1 is not important since the slope of the S vs $xI + (1-x)O$ relationship and not the intercept is being sought. With S known, values of x are assumed and S plotted against $xI + (1-x)O$. Several values of x are tried. The value that gives the smallest loop in the plotted relationship is taken as the desired x value and the slope of the plotted relationship is taken as the k value.

Generally for small catchments, measured inflow and outflow hydrographs are not available, and k and x must be approximated. Several attempts to derive procedures for estimating k and x that are better than simply setting k equal to reach travel time and x equal to some assumed constant have been made. A procedure developed by Cunge (1967) has received widespread acceptance. In the Muskingum–Cunge method, k is determined from

$$k = \Delta x / c, \quad (6.11)$$

where Δx is the reach length and c represents a flood wave celerity determined from

$$c = mv. \quad (6.12)$$

The coefficient m comes from the uniform flow equation and can be taken as 5/3 (Viessman *et al.*, 1989). The velocity, v, may be taken as the average velocity at bankful discharge. The value of x is then determined from

$$x = \frac{1}{2}\left(1 - \frac{q_0}{S_0 c\,\Delta x}\right), \quad (6.13)$$

where q_0 is the flow per unit width generally calculated at the peak flow rate and S_0 is the slope of the channel. As previously discussed, the routing interval should not exceed $t_p/5$.

Convex Routing

The Soil Conservation Service (U.S. Department of Agriculture, 1971) presents a channel routing procedure similar to the Muskingum method that is known as the Convex method. The Convex method of routing involves only inflow–outflow hydrograph relationships. The continuity equation is not directly involved, necessitating close adherence to procedures recommended by the Soil Conservation Service. In the Convex method, the routing equation is

$$O_2 = (1 - C)O_1 + CI_1, \quad (6.14)$$

where C is a parameter such that $0 \le C \le 1.0$. The parameter C can be estimated from

$$C = v/(1.7 + v), \quad (6.15)$$

where v is the average flow velocity of the reach. v may be computed at bankful discharge, at a flow equal to 75% of the peak flow or at some other appropriate value. The C value may also be approximated from the x in the Muskingum method as $C \approx 2x$ if an appropriate x is available.

The proper routing interval to use with the Convex method is

$$\Delta t = CK, \quad (6.16)$$

where K is a parameter similar to k of the Muskingum methods and may be estimated as the travel time through the reach. Generally, computing Δt from Eq. (6.16) results in an inconvenient time interval. The C value of C^* for a more convenient time interval can be calculated from

$$C^* = 1 - (1 - C)^{(\Delta t^*/\Delta t)}, \quad (6.17)$$

where Δt is from Eq. (6.16) and Δt^* is the desired time interval. The ratio $\Delta t^*/\Delta t$ should be kept as near unity as possible. Equation (6.14) is now the routing relationship with C replaced by C^*. Note that the ratio of $\Delta t^*/\Delta t$ can be varied by changing the reach length or the routing interval. Again Δt should be limited to $t_p/5$.

In 1983, the Soil Conservation Service (SCS) replaced the Convex method in their TR-20 program with a method known as the att-kin method because it combines elements from the *kin*ematic method with an *att*enuation procedure based on storage routing. The storage routing assumes

$$Q = kS^m \tag{6.18}$$

and the kinematic routing uses

$$Q = bA^m, \tag{6.19}$$

where Q is flow, S is the storage, A is the flow area, and b, k, and m are coefficients. The factor m relates velocity and wave celerity

$$c = mv. \tag{6.20}$$

Based on Manning's equation, the SCS uses an m of 5/3. The equations and repeated routings are used to assure that the peak flow resulting from the kinematic routing equals that from the storage routing and that the peak outflow from the reach occurs at the time of maximum storage.

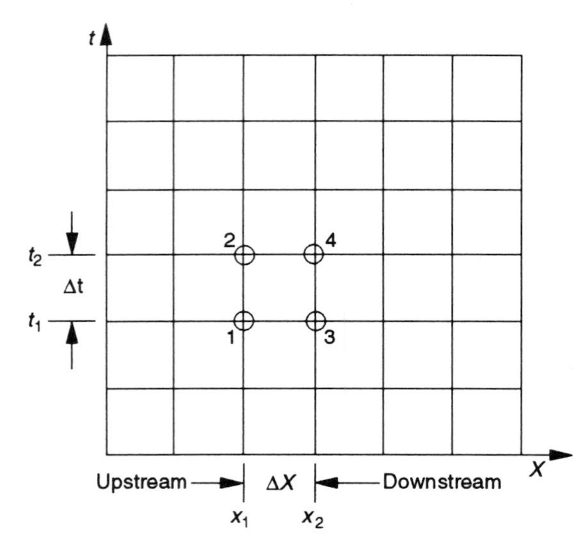

Figure 6.1 Finite difference grid.

Kinematic Routing

If all of the terms of Eq. (6.2) are neglected except the last one, the result is

$$S_f = S_0$$

or a condition of uniform flow. Kinematic routing is based on the continuity equation and a uniform flow equation. In theory, kinematic waves will not accelerate significantly and flow downstream without appreciable change in shape or loss in peak flow. Kinematic routing solves the continuity equation in the form

$$\frac{\partial Q}{\partial x} + \frac{\partial A}{\partial t} = q \tag{6.21}$$

and a flow equation or rating function

$$Q = f(A), \tag{6.22}$$

where Q is the flow rate in cfs, A is the flow area in square feet, q is any lateral inflow (+) or outflow (−) along the channel in cfs per foot, x is the distance along the channel in feet, and t is the time in seconds.

Several numerical solution techniques are available. The one that follows is based on Brakensiek (1966). Using the grid system shown in Fig. 6.1, the following approximations are made:

$$\frac{\partial Q}{\partial x} = \frac{Q_4 - Q_2}{\Delta x} \tag{6.23}$$

$$\frac{\partial A}{\partial t} = \frac{A_4 - A_3 + A_2 - A_1}{2\,\Delta t} \tag{6.24}$$

$$q = \frac{q_4 + q_2}{2} = \bar{q}. \tag{6.25}$$

Substituting these equations into Eq. (6.21) results in

$$\frac{Q_4 - Q_2}{\Delta x} + \frac{A_4 - A_3 + A_2 - A_1}{2\,\Delta t} = \bar{q}$$

or

$$\frac{\Delta t}{\Delta x} Q_4 + \frac{A_4}{2} = \frac{A_1 + A_3}{2} + \frac{\Delta t}{\Delta x} Q_2 - \frac{A_2}{2} + \Delta t \bar{q},$$

which may be written

$$\lambda Q_4 + \frac{A_4}{2} = \alpha + \beta, \tag{6.26}$$

where

$$\lambda = \frac{\Delta t}{\Delta x} \tag{6.27a}$$

$$\alpha = \frac{A_1 + A_3}{2} \tag{6.27b}$$

$$\beta = \lambda Q_2 + \Delta t \bar{q} - A_2/2. \tag{6.27c}$$

Equation (6.26) may be solved by noting that $Q = f(A)$ and rewriting the equation as

$$0 = \alpha + \beta - \lambda f(A_4) - A_4/2. \tag{6.28}$$

A value of A_4 that satisfies Eq. (6.28) is sought, and then Q_4 is calculated from Eq. (6.22). $\bar{q}$ represents lateral inflow and inflow from very small tributaries. Inflow from large tributaries would be added to Q_4 prior to routing down the next channel reach. Manning's equation can be used to define Eq. (6.22).

HYDRAULIC FLOW ROUTING

Chow (1959), Henderson (1966), Chow *et al.* (1988), and others discuss hydraulic flow routing based on equations of spatially varied (nonuniform), unsteady flow (Saint-Venant equations). These equations involve continuity and momentum as given in Eq. (6.1) and (6.2). These equations can be written as

$$A\frac{\partial v}{\partial x} + v\frac{\partial A}{\partial x} + \frac{\partial A}{\partial t} = \bar{q} \qquad (6.29)$$

and

$$\frac{\partial v}{\partial t} + v\frac{\partial v}{\partial x} + \frac{g}{A}\frac{\partial \overline{yA}}{\partial x} + \frac{v\bar{q}}{A} = g(S - S_f), \qquad (6.30)$$

where A is area, v is velocity, x is distance, $\bar{q}$ is inflow per unit length of channel, g is gravity, S is the channel slope, and S_f is the friction slope. A consistent set of units must be used in these equations.

Equations (6.29) and (6.30) may be solved using finite differences or the method of characteristics. Either treatment is beyond the scope of coverage presented here. The reader can consult one of the above references for more detailed information.

Very few computer models use the full momentum equation for routing. Situations involving backwater effects, tidal flows, surges, and flow junctions where large tributaries enter the main channel are examples of when the full momentum equation should be used. The National Weather Service DAMBRK model (Fread, 1977) uses the full equation to route the flow surge resulting from a dam failure as it passes down the channel below the dam. The National Weather Service model DEWOPER (Dynamic Wave Operational Model) (Fread, 1978) solves the full momentum equation using Newton–Raphson iterative techniques to perform channel routings.

Fread (1985) has combined DEWOPER and DAMBRK in a program known as FLDWAV adding features not found in either of the two-component programs. FLDWAV is a large, generalized, one-dimensional routing program with many features. This versatility and power comes at the expense of computer size and speed and input data requirements.

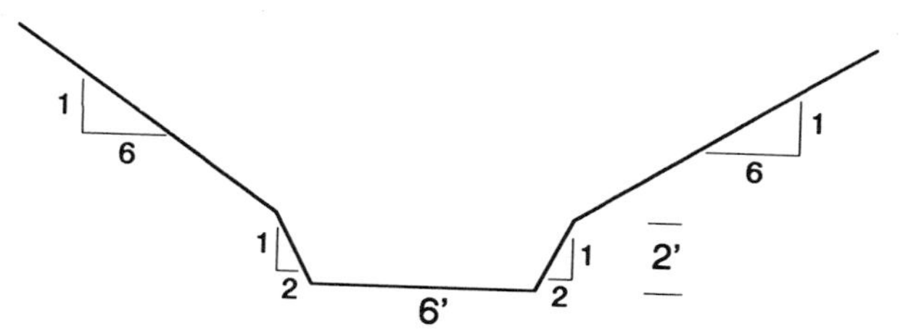

Figure 6.2 Channel cross section for Example Problem 6.1.

Example Problem 6.1 Storage routing

A channel is 2500 ft long, has a slope of 0.09%, and is clean with straight banks and no rifts or deep pools. The appropriate Manning's n is 0.030. A typical cross section is shown in Fig. 6.2. Along the length of the channel there is no lateral inflow. The inflow hydrograph to the reach is triangular in shape with a base time of 3 hr, a time to peak of 1 hr, and a peak flow rate of 360 cfs. Route the hydrograph through the channel reach using the storage routing procedure.

Solution: Table 6.1 contains the channel hydraulic properties required to compute the quantities needed for the routing using Eq. (6.5). The calculations required for the entries in this table are illustrated for a depth of 3.0 ft. The area at $y = 3$ is equal to the area at $y = 2$ plus the area between

Table 6.1 Channel Properties for Example Problem 6.1

y (ft)	A (ft^2)	wp (ft)	R (ft)	v (fps)	Q (cfs)	S (ft^3)	$S-O\Delta t/2$ (ft^3)	$S+O\Delta t/2$ (ft^3)
0	0	0	0	0	0	0	0	0
1	8	10.46	0.76	1.25	9.97	20,000	17,010	22,990
2	20	14.94	1.34	1.81	36.2	50,000	39,140	60,860
3	40	27.1	1.48	1.93	77.27	100,000	76,818	123,182
4	72	39.29	1.83	2.23	160.68	180,000	131,795	228,205
5	116	51.44	2.26	2.56	297.3	290,000	200,809	379,191
6	172	63.61	2.7	2.89	497.58	430,000	280,727	579,273

$y = 2$ and $y = 3$:

$$A_3 = A_2 + A_{2-3}$$
$$= 20 + bd + zd^2 = 20 + 14 \times 1 + 6 \times 1^2 = 40 \text{ ft}^2$$
$$wp_3 = wp_2 + wp_{2-3}$$
$$= 14.94 + 2(6^2 + 1^2)^{1/2} = 14.94 + 12.17 = 27.1 \text{ ft}$$
$$R_3 = \frac{A_3}{wp_3} = \frac{40}{27.1} = 1.48 \text{ ft}$$
$$v = \frac{1.49}{n} R^{2/3} S^{1/2} = \frac{1.49}{0.03} (1.48)^{2/3} (0.0009)^{1/2} = 1.93 \text{ fps}$$
$$Q = vA = 1.93 \times 40 = 77.27.$$

Storage in the reach, S, is computed as the product of the reach length and the cross-sectional area:

$$S = LA = 2500 \times 40 = 100{,}000 \text{ ft}^3.$$

The travel time through the reach is estimated at bankfull discharge as

$$t_t = L/v = 2500/2.9 = 862 \text{ sec} = 14.4 \text{ min}$$
$$t_p/5 = 60/5 = 12 \text{ min}.$$

Based on these two estimates, a routing interval of $\Delta t = 10$ min or 600 sec is chosen:

$$S - O\,\Delta t/2 = 100{,}000 - 77.27(600)/2 = 76{,}818 \text{ ft}^3$$
$$S + O\,\Delta t/2 = 100{,}000 + 77.27(600)/2 = 123{,}182 \text{ ft}^3.$$

A plot of $S \pm O\,\Delta t/2$ vs y yields the storage characteristic curves shown in Fig. 6.3. These curves along with equation 6.5 are used to do the routing. Table 6.2 contains the actual computations. One step in the routing is illustrated for the time interval from 40 to 50 min.

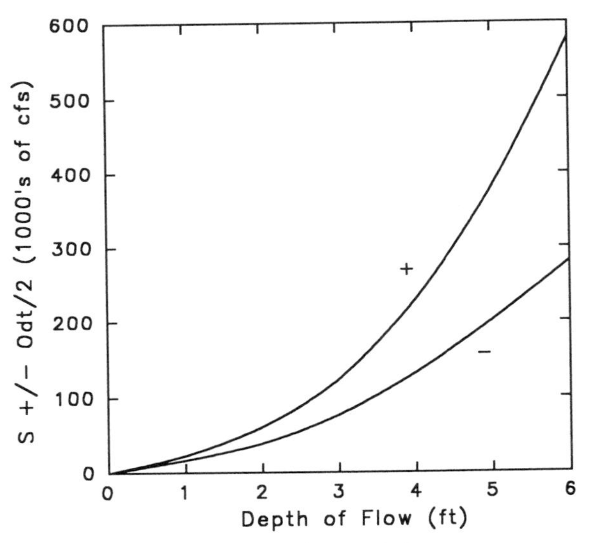

Figure 6.3 Storage characteristic curves.

$$\frac{I_1 + I_2}{2} \Delta t = \frac{240 + 300}{2} \times 600 = 162{,}000 \text{ ft}^3$$
$$y_1 = y_2 \text{ from the previous step} = 3.84 \text{ ft}.$$

$S - O\,\Delta t/2$ is read from the routing curve corresponding to $y = 3.84$ ft as 121,500:

$$S + O\,\Delta t/2 = \frac{I_1 + I_2}{2} \Delta t + S - O\,\Delta t/2$$
$$= 162{,}000 + 121{,}500 = 283{,}500.$$

y_2 is read from the routing curve corresponding to $S + O\,\Delta t/2 = 283{,}500$ as 4.40 ft.

The discharge is determined from Manning's equation as $Q = vA$, where v and A are computed for the channel corresponding to a depth of 4.40 ft. The result is $Q = 208$ cfs. Similar calculations are carried out for all of the time intervals. Figure 6.4 shows a plot of the resulting hydrograph.

Example Problem 6.2 Muskingum–Cunge routing

Work Example Problem 6.1 using the Muskingum–Cunge routing procedure.

Solution: The Muskingum–Cunge procedure relies on Eqs. (6.7) and (6.11) to (6.13). The bankful velocity is 2.89 fps; hence, from Eq. (6.12) using $m = \frac{5}{3}$,

$$c = mv = \tfrac{5}{3} 2.89 = 4.82$$
$$k = \frac{\Delta x}{c} = \frac{2500}{4.82} = 518.7$$
$$x = \frac{1}{2}\left(1 - \frac{q_0}{S_0 c\,\Delta x}\right).$$

q_0 is the flow per unit width and is approximated as a flow rate divided by the top width:

$$q_0 \simeq \frac{500}{62} = 8.06$$
$$x = \frac{1}{2}\left(1 - \frac{8.06}{0.0009(4.82)2500}\right) = 0.131$$
$$k(1 - x) + 0.5\,\Delta t = 518.7(1 - 0.131) + 0.5(600) = 750.75$$
$$C_0 = \frac{-kx + 0.5\,\Delta t}{750.75} = \frac{-518.7(0.131) + 300}{750.75} = 0.309$$
$$C_1 = \frac{kx + 0.5\,\Delta t}{750.75} = \frac{518.7(0.131) + 300}{750.75} = 0.490$$
$$C_2 = \frac{k(1 - x) - 0.5\,\Delta t}{750.75} = \frac{518.7(1 - 0.131) - 300}{750.75} = 0.201$$
$$O_2 = C_0 I_2 + C_1 I_1 + C_2 O_1$$
$$= 0.309 I_2 + 0.490 I_1 + 0.201 O_1.$$

Table 6.2 Storage Routing Computations

t (min)	I (cfs)	$(I_1+I_2)\Delta t/2$ (ft^3)	y_1 (ft)	$S-O\Delta t/2$ (ft^3)	$S+O\Delta t/2$ (ft^3)	y_2 (ft)	O (cfs)
0	0	0	0				
10	60	18,000	0	0	18,000	0.82	7
20	120	54,000	0.82	13,800	67,800	2.15	40
30	180	90,000	2.15	43,000	133,000	3.12	85
40	240	126,000	3.12	82,000	208,000	3.84	144
50	300	162,000	3.84	121,500	283,500	4.4	208
60	360	198,000	4.4	158,000	356,000	4.87	275
70	330	207,000	4.87	191,000	398,000	5.11	316
80	300	189,000	5.11	208,500	397,500	5.11	316
90	270	171,000	5.11	208,500	379,500	5	297
100	240	153,000	5	200,800	353,800	4.85	272
110	210	135,000	4.85	190,000	325,000	4.68	246
120	180	117,000	4.68	178,000	295,000	4.49	220
130	150	99,000	4.49	163,000	262,000	4.26	188
140	120	81,000	4.26	148,000	229,000	4	161
150	90	63,000	4	132,000	195,000	3.73	133
160	60	45,000	3.73	114,000	159,000	3.4	105
170	30	27,000	3.4	97,000	124,000	3.01	78
180	0	9,000	3.01	77,000	86,000	2.49	52
190	0	0	2.49	55,000	55,000	1.87	31
200	0	0	1.87	36,000	36,000	1.39	18
210	0	0	1.39	25,000	25,000	1.07	11
220	0	0	1.07	18,500	18,500	0.85	7.5
230	0	0	0.85	14,000	14,000	0.68	5
240	0	0	0.68	11,000	11,000	0.56	3.5
250	0	0	0.56	9,000	9,000	0.47	2.5
260	0	0	0.47	7,000	7,000	0.37	2
270	0	0	0.37	6,000	6,000	0.33	1.5
Sum	3240					Sum	3225

The routings based on this relationship are shown in Table 6.3. As an example calculation, the outflow at 60 min is calculated as

$$O_2 = 0.309(360) + 0.490(300) + 0.201(248) = 308.$$

The results are plotted in Fig. 6.4.

Example Problem 6.3 Convex routing

Do the routing of Example Problem 6.1 using the Convex routing method.

Solution: The solution is based on Eqs. (6.14)–(6.17) using an average bankful flow velocity of 2.8 fps:

$$O_2 = (1 - C)O_1 + CI_1$$

$$C = \frac{v}{1.7 + v} = \frac{2.8}{1.7 + 2.8} = 0.622$$

$$\Delta t = CK$$

$$K = \text{travel time} \approx L/v = 2500/2.8 = 892.9$$

$$\Delta t = 0.622(892.9) = 555 \text{ sec}$$

$$C^* = 1 - (1 - C)^{\Delta t^*/\Delta t} = 1 - (1 - 0.622)^{600/555} = 0.65$$

$$O_2 = (1 - 0.65)O_1 + 0.65I_1 = 0.35O_1 + 0.65I_1.$$

Table 6.4 contains the calculations, and Fig. 6.4 shows the results of this routing.

Example Problem 6.4 Kinematic routing

Do the routing of Example Problem 6.1 using the kinematic routing approach.

Solution: From Fig. 6.1, it can be seen that the points in the grid and the subscripts on the symbols in the kinematic routing relations have the following meaning:

Subscript	Meaning
1	Upstream end of reach at start of time interval
2	Upstream end of reach at end of time interval
3	Downstream end of reach at start of time interval
4	Downstream end of reach at end of time interval

Δt is taken as 10 min. Δx is 2500 ft. The parameter λ is given by Eq. (6.27):

$$\lambda = \frac{\Delta t}{\Delta x} = \frac{10 \text{ min} \times 60 \text{ sec/min}}{2500 \text{ ft}} = 0.24 \text{ sec/ft}.$$

Equation (6.28) can be written

$$\alpha + \beta = \lambda f(A_4) + A_4/2$$

or

$$0.24 f(A_4) + 0.5 A_4 = \alpha + \beta.$$

Figure 6.5 is a plot of this function. Points on the line of $\alpha + \beta$ or $0.24f(A_4) + 0.5A_4$ are calculated by assuming a value of y and calculating A_4 and $f(A_4)$ or Q_4. The routing is done by calculating $\alpha + \beta$ at each time step from Eq. (6.27) and then determining A_4 and Q_4 from the curves.

Table 6.5 contains the routing computations for the kinematic method. The computations are illustrated by considering the time interval from 50 to 60 min. Recall that conditions at points 1, 2, and 3 are known and conditions at point 4 are being sought. Also recall that for a prismatic channel under the assumption of uniform flow, the relationship between flow area and flow depth is the same all along the channel. From Table 6.5, A_1, A_2, and A_3 are 133.5, 126, and 111.2 ft^2, respectively:

$$\alpha = (A_1 + A_3)/2 = (133.5 + 111.2)/2 = 122.35$$

$$\beta = \lambda Q_2 - A_2/2 = 0.24(330) - 126/2 = 16.2$$

$$\alpha + \beta = 122.35 + 16.2 = 138.55.$$

From Fig. 6.5, A_4 and Q_4 are found to be 123.5 ft^2 and 321 cfs, respectively. For the next time increment, Q_1 and A_1 are Q_2 and A_2 from the previous time, Q_3 and A_3 are Q_4 and A_4 from the previous time, Q_2 is the inflow hydrograph value at the end of the time increment, and A_2 is the flow area corresponding to Q_2. Figure 6.4 contains the results of routing. It can be seen that the results are very similar to those of the other routing methods. It could be expected that the differences would increase if the routing involved several channel reaches.

Table 6.3 Routing Calculations for Example Problem 6.2

Time (min)	Inflow (cfs)	Outflow (cfs)
0	0	0
10	60	19
20	120	70
30	180	129
40	240	188
50	300	248
60	360	308
70	330	340
80	300	323
90	270	295
100	240	266
110	210	236
120	180	206
130	150	176
140	120	146
150	90	116
160	60	86
170	30	56
180	0	26
190	0	5
200	0	1
210	0	0
220	0	0
230	0	0
240	0	0
250	0	0
260	0	0
270	0	0
280	0	0
Sum	3240	3240

RESERVOIR ROUTING

Reservoir routing is generally done by using storage routing similar to the storage routing procedure explained earlier for channel routing. The continuity equation in the form of Eq. (6.4) and (6.5) is used.

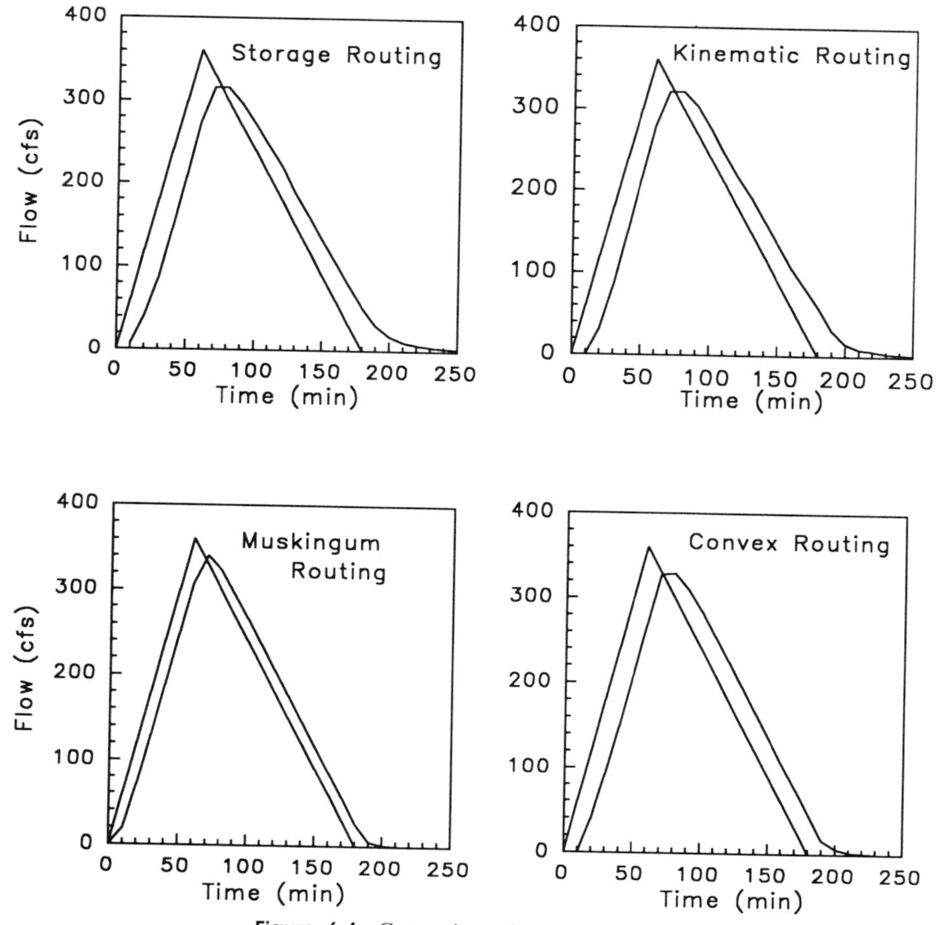

Figure 6.4 Comparison of routing results.

Outflow from the reservoir is controlled by the principal spillway at a rate depending on the height of water above the inlet. At the beginning of inflow, the depth of water above the inlet is small, and the inflow to the reservoir will exceed the outflow. As a result, the depth of water in the reservoir will increase, thereby increasing the outflow rate. This process will continue until the outflow rate equals the inflow rate. This point will occur sometime after the peak inflow rate has occurred. From this point, the outflow will exceed the inflow and the depth of water or storage in the reservoir will decrease. Figure 6.6 illustrates this process.

The relationship of storage relative to inflow and outflow hydrographs is shown in Fig. 6.7. The peak discharge of the outflow hydrograph corresponds to the point where it intersects the inflow hydrograph. When the outflow is a maximum, the storage will also be a maximum, since discharge and storage are both increasing functions of stage. The maximum storage and outflow occurs when inflow equals outflow, which is, of course, the point of intersection of the two hydrographs.

The storage volume required in the reservoir is the area between the inflow and outflow hydrographs prior to the peak outflow as shown in Fig. 6.7. An additional feature of the hydrographs is that the areas under the inflow and outflow hydrographs must be the same if the initial water level is at the spillway crest.

Graphical Routing: Puls Method

Puls method is a procedure for graphically solving the continuity equation using storage characteristic curves as was done in the case of stream routing. The stage–storage curve is developed from topographic information relative to the reservoir site, and the stage–discharge curve is based on the hydraulics of the reservoir outlet as shown in Chapter 5. The routing time interval should be 10 to 25% of the time to peak of the inflow hydrograph to ensure that the numerical

Table 6.4 Convex Routing Results

Time (min)	Inflow (cfs)	Outflow (cfs)
0	0	0
10	60	0
20	120	39
30	180	92
40	240	149
50	300	208
60	360	268
70	330	328
80	300	329
90	270	310
100	240	284
110	210	255
120	180	226
130	150	196
140	120	166
150	90	136
160	60	106
170	30	76
180	0	46
190	0	16
200	0	6
210	0	2
220	0	1
230	0	0
240	0	0
250	0	0
260	0	0
270	0	0
280	0	0
Sum	3240	3240

averaging does not diminish the impact of the peak flow. For reservoirs that store a large part of the flow, the routing interval may be increased after the peak outflow rate has occurred. If the routing time interval is changed, a new set of storage characteristic curves will be required because they depend on Δt. Extreme care must be used in any graphical routing as the errors tend to be cumulative rather than random.

Example Problem 6.5 Storage characteristic curve

Develop the storage characteristic curves for a reservoir whose stage-area and stage–discharge relationships are shown

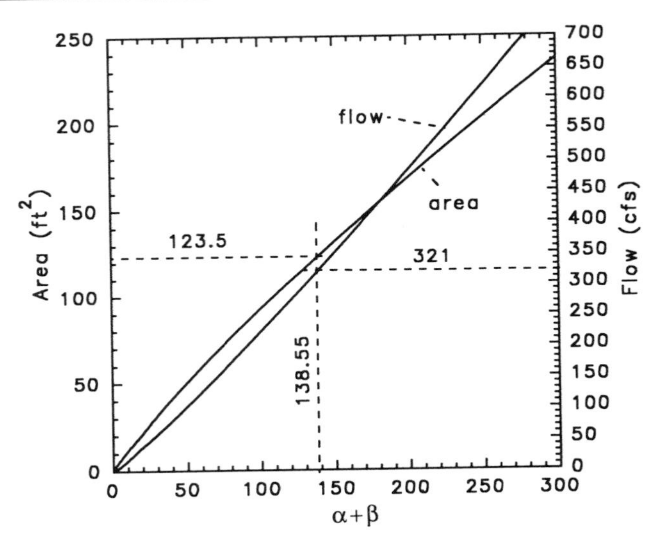

Figure 6.5 Channel properties for kinematic routing.

in Table 6.6 and Fig. 6.8 using a routing time interval of 10 min.

Solution: The results of the calculations are shown in Table 6.6. columns 1 and 3 contain topographic information relative to the site. Column 2 contains the reservoir stage assuming a datum of 0 at elevation 400.0 ft. Column 6 contains the discharge for the reservoir spillway corresponding to the stage of column 2. Column 4 is the storage volume contained between the indicated stage and the immediately preceding stage. The incremental storage is calculated from

$$\Delta S = (A_1 + A_2)(z_2 - z_1)/2,$$

where z refers to elevation. Thus for the increment from 401.5 to 402.0 ft,

$$\Delta S = (6.0 - 4.5)(402.0 - 401.5)/2 = 2.6 \text{ acre-ft}.$$

The total storage in column 5 is the sum of the incremental storages. The storage at 402.0 ft is

$$S = 4.4 + 2.6 = 7.0 \text{ acre-ft}.$$

The last two columns are based on a time increment of 10 min. The entry in column 7 for 402.0 ft is determined as

$$7.0 - 120.3(3600)(10)/60/43560/2 = 6.2.$$

The results of the calculations are plotted in Fig. 6.9.

Example Problem 6.6 Puls routing

Route the hydrograph of Example Problem 6.1 through the reservoir of Example Problem 6.5 using the Puls method.

Table 6.5 Routing Computations for Example Problem 6.4

t (min)	Q_1 (cfs)	Q_3 (cfs)	A_1 (ft^2)	A_3 (ft^2)	α (ft^2)	Q_2 (cfs)	A_2 (ft^2)	β (ft^2)	$\alpha+\beta$ (ft^2)	A_4 (ft^2)	Q_4 (cfs)
0	0	0	0	0	0	60	32.7	−1.95	−1.95	0	0
10	60	0	32.7	0	16.35	120	57.5	0.05	16.4	17.9	31
20	120	31	57.5	17.9	37.7	180	79	3.7	41.4	42.66	84
30	180	84	79	42.66	60.83	240	98	8.6	69.43	67.5	148
40	240	148	98	67.5	82.75	300	117	13.5	96.25	90.2	215
50	300	215	117	90.2	103.6	360	133.5	19.65	123.25	111.2	281
60	360	281	133.5	111.2	122.35	330	126	16.2	138.55	123.5	321
70	330	321	126	123.5	124.75	300	117	13.5	138.25	123.5	321
80	300	321	117	123.5	120.25	270	108	10.8	131.05	118	304
90	270	304	108	118	113	240	98	8.6	121.6	110	277
100	240	277	98	110	104	210	89	5.9	109.9	100	245
110	210	245	89	100	94.5	180	79	3.7	98.2	91.5	217
120	180	217	79	91.5	85.25	150	68	2	87.25	83	192
130	150	192	68	83	75.5	120	57.5	0.05	75.55	72.5	163
140	120	163	57.5	72.5	65	90	45	−0.9	64.1	63.1	135
150	90	135	45	63.1	54.05	60	32.7	−1.95	52.1	52.4	106
160	60	106	32.7	52.4	42.55	30	17.5	−1.55	41	42.5	83
170	30	83	17.5	42.5	30	0	0	0	30	32	59
180	0	59	0	32	16	0	0	0	16	17.5	31
190	0	31	0	17.5	8.75	0	0	0	8.75	10.3	15
200	0	15	0	10.3	5.15	0	0	0	5.15	7	8
210	0	8	0	7	3.5	0	0	0	3.5	5	6
220	0	6	0	5	2.5	0	0	0	2.5	3.5	3
230	0	3	0	3.5	1.75	0	0	0	1.75	2.5	2
240	0	2	0	2.5	1.25	0	0	0	1.25	2	1
250	0	1	0	2	1	0	0	0	1	1.5	1
260	0	1	0	1.5	0.75	0	0	0	0.75	1.25	1
270	0	1	0	1.25	0.625	0	0	0	0.625	1	1
280	0										0
Sum	3240										3249

Solution: The routing calculations based on Eq. (6.5) are contained in Table 6.7. The time increment from 50 to 60 min. is used to illustrate the steps involved:

$$I_1 = 360 \text{ cfs}, \qquad I_2 = 360 \text{ cfs}$$

$$\frac{I_1 + I_2}{2}\,\Delta t = \frac{300 + 360}{2} \times \frac{10(60)}{43560} = 4.55 \text{ acre-ft}.$$

y_1 is the stage at the end of the previous time interval or 1.89 ft. $S = (O/2)\,\Delta t$ is read from Fig. 6.9 corresponding to a stage of 1.89 feet as 5.12 acre-ft.

From Eq. (6.5),

$$S + \frac{O}{2}\,\Delta t = \frac{I_1 + I_2}{2}\,\Delta t + S - \frac{O}{2}\,\Delta t = 4.55 + 5.12 = 9.67.$$

From Fig. 6.9, y_2 is found to be 2.34 ft, and from Fig. 6.8, the outflow is found to be 130.2 cfs. The resulting hydrograph is shown in Fig. 6.10.

Numerical Routing

One numerical reservoir routing procedure is to solve Eq. (6.4). The unknowns in this equation are S_2

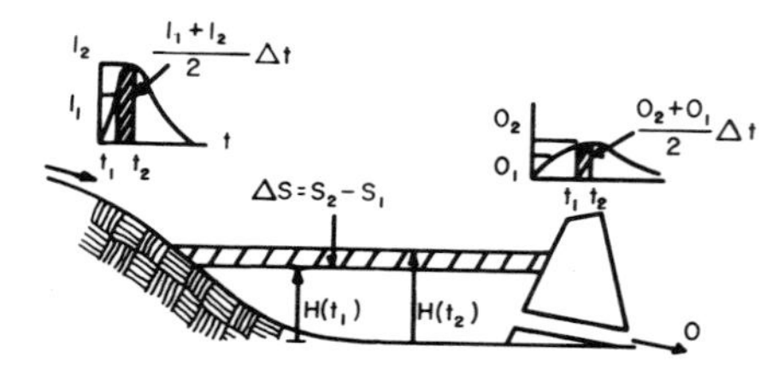

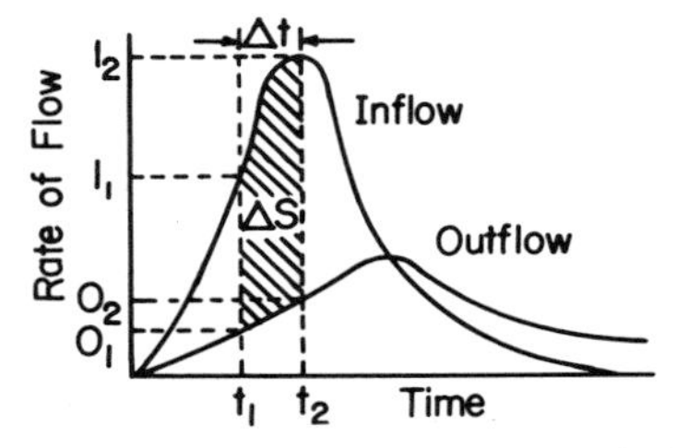

Figure 6.6 Illustration of the continuity relationship in reservoir routing.

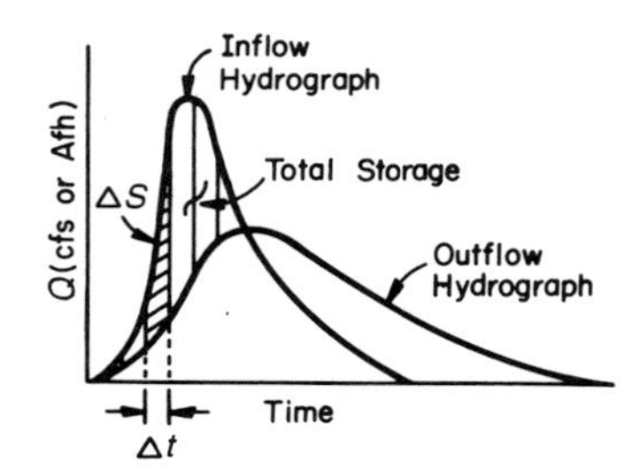

Figure 6.7 Features of inflow and outflow hydrographs.

and O_2. Both S and O are functions of the stage y. Therefore, S and O are functionally related allowing a unique solution for O_2. The solution is not explicit and thus requires an iterative solution. Fortunately, the solution converges very rapidly, and only 2 or 3 iterations are required in most instances. One solution procedure is as follows:

1. Assume $O_2 = O_1$.
2. Calculate ΔS from Eq. (6.4).
3. Calculate $S_2 = S_1 + \Delta S$.
4. Determine y_2 for S_2 and the stage–storage curve.
5. Determine O_2 for y_2 from the stage–discharge curve.
6. Repeat steps 2–5 until O_2 remains unchanged.

This procedure is illustrated in Example Problem 6.7.

Example Problem 6.7 Numerical routing

Route the inflow hydrograph of Example Problem 6.1 through a reservoir having stage–storage and stage–discharge relationships defined by

$$S = 2.1y^{1.727}$$

and

$$O = 7y^{0.498},$$

where storage, S, is in acre-feet; stage, y, is in feet; and

Table 6.6 Storage Characteristic Curves for Example Problem 6.5

Elevation (ft)	Stage (ft)	Area (acres)	Incremental volume (*a–f*)	Storage (*a–f*)	Flow (cfs)	$S–O\Delta t/2$ (*a–f*)	$S+O\Delta t/2$ (*a–f*)
399.0	0.0	0.0	0.0	0.0	0.0	0.0	0.0
399.5	0.0	0.5	0.1	0.1	0.0	0.1	0.1
400.0	0.0	1.0	0.4	0.5	0.0	0.5	0.5
400.5	0.5	2.0	0.8	1.3	49.9	0.9	1.6
401.0	1.0	3.0	1.3	2.5	85.1	1.9	3.1
401.5	1.5	4.5	1.9	4.4	104.2	3.7	5.1
402.0	2.0	6.0	2.6	7.0	120.3	6.2	7.8
402.5	2.5	7.5	3.4	10.4	134.6	9.4	11.3
403.0	3.0	9.0	4.1	14.5	147.4	13.5	15.5
403.5	3.5	12.0	5.3	19.8	153.0	18.7	20.8
404.0	4.0	15.0	6.8	26.5	158.4	25.4	27.6
404.5	4.5	17.5	8.1	34.6	163.6	33.5	35.8
405.0	5.0	20.0	9.4	44.0	168.7	42.8	45.2
405.5	5.5	24.0	11.0	55.0	173.6	53.8	56.2
406.0	6.0	28.0	13.0	68.0	178.4	66.8	69.2

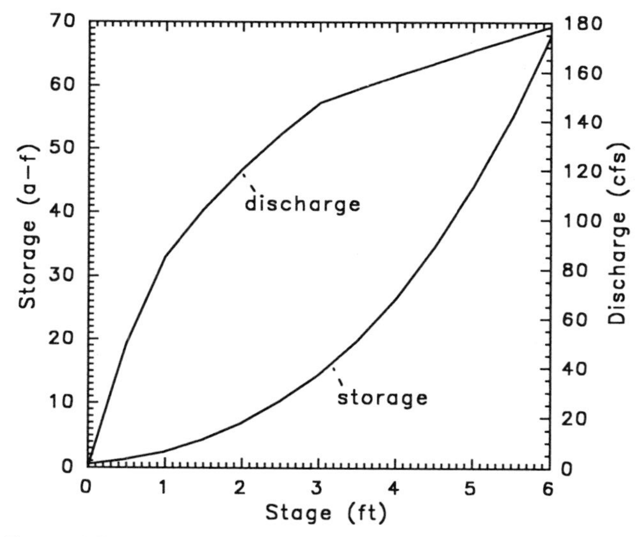

Figure 6.8 Stage–discharge and stage–storage relationship for Example Problem 6.6.

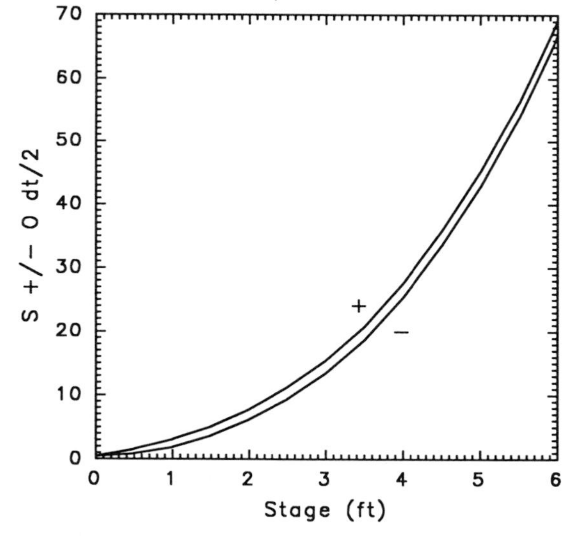

Figure 6.9 Storage characteristic curves for Example Problem 6.6.

Table 6.7 Routing Calculations for Example Problem 6.6

Time (min)	Inflow (cfs)	$I_{ave}\Delta t$ $(a\text{-}f)$	y_1 (ft)	$S\text{-}O\Delta t/2$ $(a\text{-}f)$	$S+O\Delta t/2$ $(a\text{-}f)$	y_2 (ft)	Outflow (cfs)
0	0		0.00				0.0
10	60	0.41	0.00	0.0	0.41	0.25	17.9
20	120	1.24	0.25	0.3	1.54	0.61	66.4
30	180	2.07	0.61	0.6	2.69	1.00	85.1
40	240	2.89	1.00	1.5	4.39	1.44	102.1
50	300	3.72	1.44	3.0	6.72	1.89	117.0
60	360	4.55	1.89	5.1	9.65	2.34	130.2
70	330	4.75	2.34	7.9	12.65	2.72	140.3
80	300	4.34	2.72	10.8	15.10	3.00	147.4
90	270	3.93	3.00	13.1	16.10	3.10	148.5
100	240	3.51	3.10	14.0	17.10	3.20	149.7
110	210	3.10	3.20	15.0	18.20	3.31	150.9
120	180	2.69	3.31	16.1	18.79	3.36	151.4
130	150	2.27	3.36	16.6	18.87	3.37	151.6
140	120	1.86	3.37	16.8	18.66	3.35	151.3
150	90	1.45	3.35	16.6	18.10	3.30	150.8
160	60	1.03	3.30	16.0	17.03	3.19	149.6
170	30	0.62	3.19	14.9	15.50	3.04	147.8
180	0	0.21	3.04	13.9	14.10	2.89	144.6

discharge, O, is in acre-feet per hour. Use the numerical routing method, a time increment of 10 min, and an initial stage of zero.

Solution: The solution is shown in Table 6.8. Note that after 3 time steps, the solution converges in two iterations and after about 10 time steps convergence has been achieved after the first iteration. The last column contains the outflow hydrograph.

The calculations are illustrated by considering the line corresponding to a time of 90 min. The heading numbers in parentheses refer to column numbers in Table 6.8.

(1 and 2) From inflow hydrograph

(3)

$$\frac{I_1 + I_2}{2}\Delta t = \frac{300 + 270}{2}\frac{\text{ft}^3}{\text{sec}} \times 10 \text{ min} \times \frac{60 \text{ sec}}{\text{min}} \times \frac{1 \text{ acre-ft}}{43560 \text{ ft}^3} = 3.93.$$

(4 and 5) y_1, S_1, and $O_1 = y_2$, S_2, and O_2 from previous time step (columns 15, 16, and 17).

$$(6) \quad \Delta S = \frac{I_1 + I_2}{2}\Delta t - \frac{O_1 + O_2}{2}\Delta t$$

$$= 3.93 - \left(\frac{12.08 + 12.08}{2}\right)\frac{10}{60} = 1.91.$$

$$(7) \quad S_2 = S_1 + \Delta S = 13.93 + 1.91 = 15.84.$$

$$(8) \quad y_2 = \left(\frac{S_2}{2.1}\right)^{1/1.7268} = \left(\frac{15.84}{2.1}\right)^{0.5791} = 3.22.$$

$$(9) \quad O_2 = 7(y_2)^{0.498} = 7(3.22)^{0.498} = 12.54.$$

$$(10) \quad \Delta S = \frac{I_1 + I_2}{2}\Delta t - \frac{O_1 + O_2}{2}\Delta t$$

$$= 3.93 - \left(\frac{12.08 + 12.54}{2}\right)\frac{10}{60} = 1.878.$$

$$(11) \quad S_1 = S_1 + \Delta S = 13.93 + 1.878 = 15.808.$$

$$(12) \quad y_2 = \left(\frac{S_2}{2.1}\right)^{1/1.7268} = \left(\frac{15.81}{2.1}\right)^{0.5791} = 3.218.$$

Note at this point that the y_2 calculated in column (12) agrees with the y_2 in column (8) and the solution could stop. We carry it one additional step.

$$(13) \quad O_2 = 7(y_2)^{0.498} = 7(3.218)^{0.498} = 12.528.$$

$$(14) \quad \Delta S = \frac{I_1 + I_2}{2}\Delta t - \frac{O_1 + O_2}{2}\Delta t$$

$$= 3.93 - \left(\frac{12.08 + 12.53}{2}\right)\frac{10}{60} = 1.879.$$

$$(15) \quad S_2 = S_1 + \Delta S = 13.93 + 1.879 = 15.81.$$

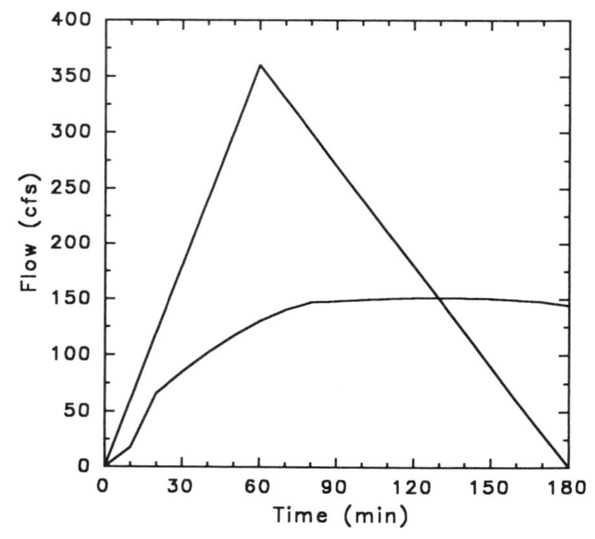

Figure 6.10 Hydrographs for Example Problem 6.6.

$$(16) \quad y_2 = \left(\frac{S_2}{2.1}\right)^{1/1.7268} = \left(\frac{15.81}{2.1}\right)^{0.5791} = 3.218.$$

$$(17) \quad O_2 = 7(3.218)^{0.498} = 12.528.$$

$$(18) \quad O_2 = 12.528 \frac{\text{acre-ft}}{\text{hr}} \times \frac{1 \text{ hr}}{3600 \text{ sec}} \times \frac{43560 \text{ ft}^3}{\text{acre-ft}} = 151.6 \text{ cfs}.$$

Other numerical techniques can be used to solve the continuity equation as it applies to reservoir routing. As in all hydrologic problems, care must be taken to ensure that a consistent set of units is used. Computer programs to carry out the routing by numerical means are widely available. These programs generally require inputs having specific units and are structured to interpolate between points on user-supplied stage–storage and stage–discharge relationships.

USES OF RESERVOIR ROUTING

Reservoirs are typically used in flood control so that some predetermined peak outflow rate is not exceeded, to delay the flow so that high runoff rates from the controlled catchment do not coincide with high flows from a second catchment, to detain water so that suspended particles may settle prior to the flow entering a stream, or for a water-use function such as recreation or water supply. Regardless of the use of the reservoir, routing is used to determine the impact

Table 6.8 Numerical Routing Example

Time	Inflow	$I_{ave}\Delta t$	y_1	S_1	Trial 1				Trial 2				Trial 3					
					O_2=O_1 (acre-ft/hr)	ΔS	S_2	y_2	O_2 (acre-ft/hr)	ΔS	S_2	y_2	O_2 (acre-ft/hr)	ΔS	S_2	y_2	O_2 (acre-ft/hr)	O (cfs)
0	0		0.00		0.00		0.00										0.00	0.0
10	60	0.41	0.00	0.00	0.00	0.41	0.41	0.39	4.37	0.05	0.05	0.11	2.32	0.22	0.22	0.27	3.63	44.0
20	120	1.24	0.27	0.22	3.63	0.63	0.85	0.59	5.39	0.49	0.70	0.53	5.11	0.51	0.73	0.54	5.16	62.4
30	180	2.07	0.54	0.73	5.16	1.21	1.94	0.95	6.84	1.07	1.80	0.91	6.69	1.08	1.81	0.92	6.71	81.2
40	240	2.89	0.92	1.81	6.71	1.77	3.58	1.36	8.17	1.65	3.46	1.34	8.08	1.66	3.47	1.34	8.09	97.9
50	300	3.72	1.34	3.47	8.09	2.37	5.84	1.81	9.40	2.26	5.73	1.79	9.35	2.27	5.73	1.79	9.35	113.2
60	360	4.55	1.79	5.73	9.35	2.99	8.73	2.28	10.56	2.89	8.63	2.27	10.52	2.89	8.63	2.27	10.52	127.3
70	330	4.75	2.27	8.63	10.52	3.00	11.62	2.69	11.47	2.92	11.55	2.68	11.44	2.92	11.55	2.68	11.44	138.5
80	300	4.34	2.68	11.55	11.44	2.43	13.98	3.00	12.09	2.38	13.93	2.99	12.08	2.38	13.93	2.99	12.08	146.2
90	270	3.93	2.99	13.93	12.08	1.92	15.84	3.22	12.54	1.88	15.81	3.22	12.53	1.88	15.81	3.22	12.53	151.6
100	240	3.51	3.22	15.81	12.53	1.42	17.23	3.38	12.84	1.40	17.20	3.38	12.84	1.40	17.20	3.38	12.84	155.3
110	210	3.10	3.38	17.20	12.84	0.96	18.16	3.49	13.04	0.94	18.15	3.49	13.04	0.94	18.15	3.49	13.04	157.8
120	180	2.69	3.49	18.15	13.04	0.52	18.66	3.54	13.14	0.51	18.65	3.54	13.14	0.51	18.65	3.54	13.14	159.0
130	150	2.27	3.54	18.65	13.14	0.08	18.73	3.55	13.16	0.08	18.73	3.55	13.16	0.08	18.73	3.55	13.16	159.2
140	120	1.86	3.55	18.73	13.16	−0.33	18.40	3.51	13.09	−0.33	18.41	3.52	13.09	−0.33	18.41	3.52	13.09	158.4
150	90	1.45	3.52	18.41	13.09	−0.73	17.61	3.43	12.94	−0.72	17.69	3.43	12.94	−0.72	17.69	3.43	12.94	156.6
160	60	1.03	3.43	17.69	12.94	−1.13	16.56	3.31	12.70	−1.11	16.58	3.31	12.70	−1.11	16.58	3.31	12.70	153.7
170	30	0.62	3.31	16.58	12.70	−1.50	15.08	3.13	12.36	−1.47	15.11	3.14	12.37	−1.47	15.11	3.14	12.37	149.6
180	0	0.21	3.14	15.11	12.37	−1.85	13.26	2.91	11.91	−1.81	13.30	2.91	11.92	−1.81	13.30	2.91	11.92	144.2

of the reservoir on hydrographs; determine reservoir design requirements, maximum water levels, and outlet hydraulic requirements; and in some cases to provide guidance in the operation of spillway gates and valves to meet downstream flow requirements for flood control or low flow augmentation.

Flood Peak Reduction

One of the most common and important uses of reservoirs is to reduce flood peaks. Based on the inflow hydrograph and the allowable downstream peak flow, the storage requirements and outlet hydraulic characteristics can be determined. For small reservoirs with noncontrolled outlets, an exact design may not be achieved because of site characteristics and available pipe sizes for spillways. SCS (1986) TR55 contains some general guidelines. The general steps in the design procedure are:

1. Determine the inflow hydrograph for the design event.
2. Determine the allowable peak outflow from the reservoir based on downstream conditions or governing regulations.
3. Estimate the storage requirement by
   a. plotting the allowable peak outflow on the falling limb of the inflow hydrograph,
   b. approximating the outflow hydrograph up to the peak outflow as either a straight line or an estimated curved line, and
   c. estimating the storage required as the volume of flow between the inflow and outflow hydrographs from the initiation of flow to the peak outflow rate.
4. Estimate the maximum stage from the stage–storage relationship.
5. Determine the spillway hydraulic configuration required to produce the peak flow with the available storage.
6. Route the inflow through the reservoir based on this spillway.
7. Adjust the spillway characteristics and repeat step 6 until a satisfactory design is achieved.

Detention Storage

Detention storage time is a measure of the average residence time of water in a reservoir. It is a useful measure of the average opportunity time for sediment to settle out of flow. The actual detention time and the theoretical detention time may differ because of dead storage in the reservoir. Dead storage represents the volume of water in the reservoir that is not displaced during the passage of flow through the reservoir. Dead storage is found in areas of the reservoir that are bypassed by the flow. In designing a sediment basin, it is desireable to minimize dead storage so that the entire volume of the basin is effective in retarding the flow and allowing sediment to settle from suspension.

The theoretical detention time in a rectangular basin with a constant flow rate is simply the average flow through time given by

$$T_d = \frac{L}{Q/A}, \tag{6.31}$$

where L is the flow length, Q is the constant discharge, and A is the cross-sectional flow area.

For flows that vary in time, the definition of detention time is more complex. One definition is based on the plug flow or first in–first out concept. To use the plug flow concept, the inflow hydrograph is divided into plugs of equal volume as shown in Fig. 6.11. These plugs are routed through the reservoir, and the detention time for each plug is determined. The theoretical detention time is the average detention time of the plugs. Obviously, plugs during the initial part of the flow have a detention time shorter than the average and the later plugs have a detention time greater than the average. Calculations based on the plug flow concept are illustrated in chapter 9.

An alternative definition of detention time is based on the centroid concept. Theoretical detention time is defined as the time difference between the centroids of the inflow and outflow hydrographs:

$$T_\mathrm{d} = T_\mathrm{mo} - T_\mathrm{mi}, \tag{6.32}$$

where T_d is the detention time and T_mi and T_mo are the time to the centroids of the inflow and outflow hydro-

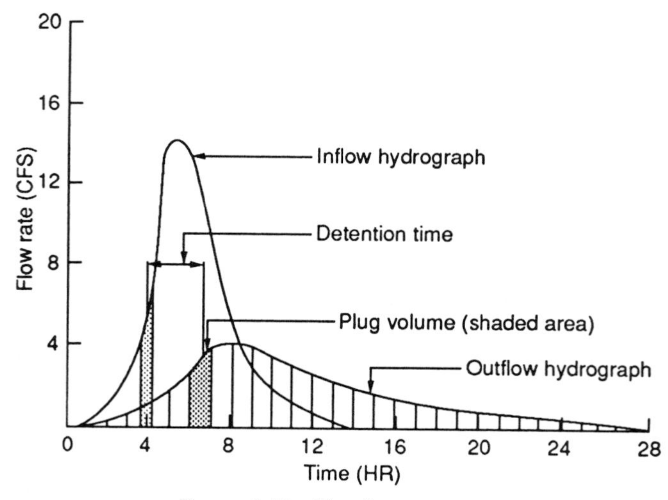

Figure 6.11 Plug flow concept.

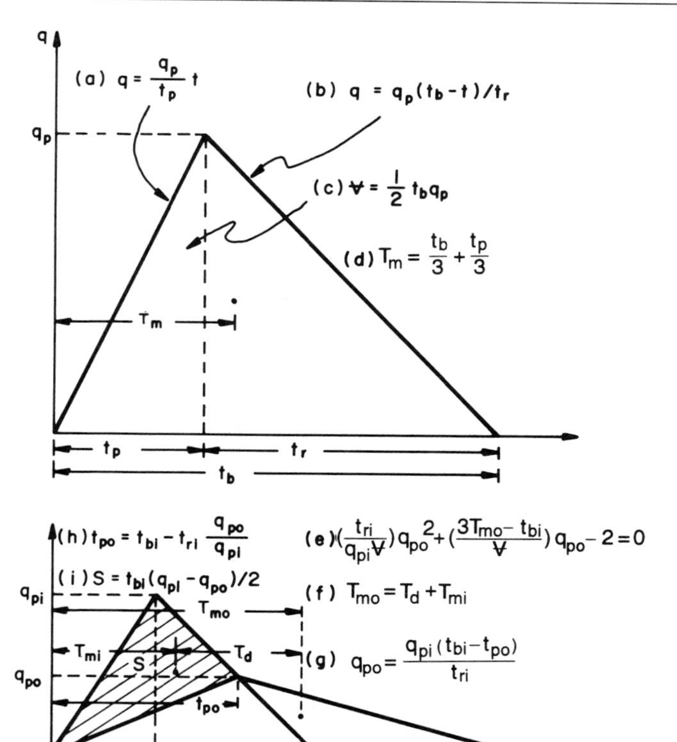

Figure 6.12 Triangular hydrograph approximations.

graphs, respectively. T_{m} can be calculated from

$$T_m = \frac{\int tq(t)\,dt}{\int q(t)\,dt}, \tag{6.33}$$

where t is the time from the beginning of the hydrograph and $q(t)$ is the flow rate at time t. A numerical approximation for T_{m} is

$$T_{\mathrm{m}} = \frac{\sum q_i t_i}{\sum q_i} \tag{6.34}$$

where t_i and q_i represent hydrograph time and flow values. The summations must be carried until q_i is negligible.

Designing a reservoir to have a specified detention time can be quite tedious. Trial outlet structures and a routing computating for each trial would have to be examined until an outlet that would provide the required detention time could be found. Using a triangular hydrograph approximation, it is possible to obtain very good estimates of the required design parameters, which can then be refined if necessary based on detailed routings. The required relationships are shown in Fig. 6.12.

Table 6.9 Computations for Example Problem 6.8

t (min)	I (cfs)	O (cfs)	tI (cfs-min)	tO (cfs-min)
0	0	0	0	0
10	48	6	480	60
20	88	21	1760	420
30	100	37	3000	1110
40	92	50	3680	2000
50	76	58	3800	2900
60	59	60	3540	3600
70	43	58	3010	4060
80	31	54	2480	4320
90	21	48	1890	4320
100	14	41	1400	4100
110	10	34	1100	3740
120	6	28	720	3360
130	4	23	520	2990
140	3	18	420	2520
150	2	14	300	2100
160	1	11	160	1760
170	1	9	170	1530
180	0	7	0	1260
190	0	5	0	950
200	0	4	0	800
210	0	3	0	630
220	0	2	0	440
230	0	2	0	460
240	0	1	0	240
250	0	1	0	250
Totals	599	595	28,430	49,920
T_{m}=	48 min	84 min	T_{d} =	36 min

Example Problem 6.8 Detention time

The inflow and outflow hydrographs shown in Table 6.9 are the result of flow from a 100-acre catchment. Compute the detention time for the reservoir.

Solution: Equation (6.34) is used to calculate the time to the centroid of the inflow and outflow hydrographs. The computations are shown in Table 6.9. Using Eq. (6.34),

$$T_{\mathrm{mi}} = 28430/599 = 48 \text{ min}$$

$$T_{\mathrm{mo}} = 49920/595 = 84 \text{ min}.$$

From Eq. (6.32),

$$T_d = T_{mo} - T_{mi} = 84 - 48 = 36 \text{ min.}$$

Example Problem 6.9 Approximate detention time

Estimate the detention time for the situation of Example Problem 6.8 using the triangular approximations.

Solution (refer to Fig. 6.12):

$$q_{pi} = 100 \text{ cfs} \approx 1 \text{ iph}$$

$$q_{po} = 60 \text{ cfs} \approx 0.6 \text{ iph.}$$

The volume of flow is found from Eq. (3.46).

$$V = \frac{\Delta t \sum q_i}{60.5A} = \frac{10(599)}{60.5(100)} = 0.99 \text{ in.}$$

The base times, t_b, for the triangular hydrographs are found from Equation (c)

$$t_{bi} = \frac{2V}{q_p} = \frac{2(.99)}{1} = 1.98 \text{ hr}$$

$$t_{bo} = \frac{2V}{q_p} = \frac{2(.99)}{0.6} = 3.30 \text{ hr.}$$

Equation (d) gives

$$t_{mi} = \frac{t_b + t_p}{3} = \frac{1.98 + 0.5}{3} = 0.83 \text{ hr}$$

for the inflow hydrograph. Equation (h) gives

$$t_{po} = t_{bi} - t_{ri}\frac{q_{po}}{q_{pi}} = 1.98 - 1.48\frac{0.6}{1.0} = 1.09 \text{ hr.}$$

Equation (d) gives

$$t_{mo} = \frac{t_b + t_p}{3} = \frac{3.30 + 1.09}{3} = 1.46 \text{ hr.}$$

Thus the detention time from Eq. (6.32) is

$$t_d = t_{mo} - t_{mi} = 1.46 - 0.83 = 0.63 \text{ hr} = 38 \text{ min,}$$

which agrees very well with the result obtained using the actual hydrographs.

Figure 6.13 shows the actual hydrographs and the triangular approximations. The figure illustrates why the triangular estimate is a reasonably good one.

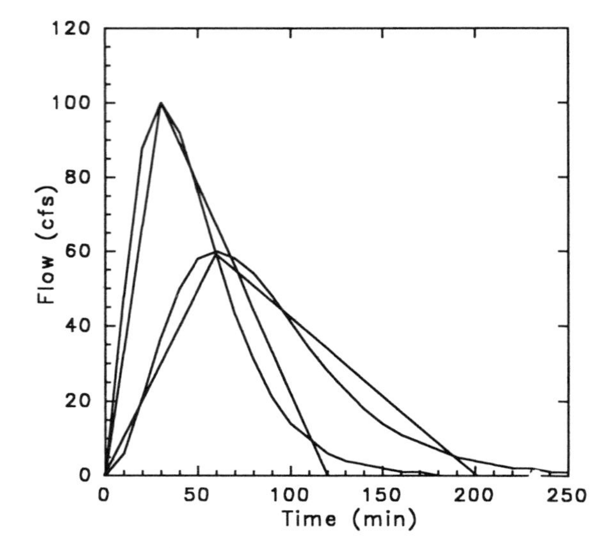

Figure 6.13 Hydrographs for Example Problem 6.9.

Example Problem 6.10 Detention storage design

Assuming that a detention time of 38 min is inadequate, estimate the storage required and the peak outflow rate for the situation of Example Problem 6.8 if a detention time of 3 hr is required. Obviously, a new outflow hydrograph is required.

Solution: Use the triangular approximation. From Eq. (f) of Fig. 6.12,

$$t_{mo} = t_d + t_{mi} = 3.00 + 0.83 = 3.83 \text{ hr.}$$

From Eq. (e),

$$\left(\frac{t_{ri}}{q_{pi}V}\right)q_{po}^2 + \left(\frac{3t_{mo} - t_{bi}}{V}\right)q_{po} - 2 = 0$$

$$\left(\frac{2 - 0.5}{1 \times 1}\right)q_{po}^2 + \left(\frac{3 \times 3.83 - 2}{1}\right)q_{po} - 2 = 0$$

$$1.5q_{po}^2 + 9.5q_{po} - 2 = 0.$$

This quadratic equation has as a solution $q_{po} = 0.204$ iph or 20 cfs. The required storage is calculated from Eq. (i) as

$$S = t_{bi}(q_{pi} - q_{po})/2 = 2(1 - 0.204)/2 = 0.8 \text{ in.}$$

The required storage is about 0.8 in. or 6.67 acre-ft.

Based on the stage–storage curve for the reservoir, the stage corresponding to 0.8 in. of storage can be determined. A principle spillway that will discharge 20 cfs at this stage can be selected. A detailed routing of the inflow hydrograph can be done, and the actual detention time calculated. If the detention time is not close enough to 3 hr, the design can be modified as needed. Generally the triangular approximations will produce very good first estimates of the required final design.

Storm Water Management

Detention basins are often used in storm water management systems to reduce peak flows. Often, regula-

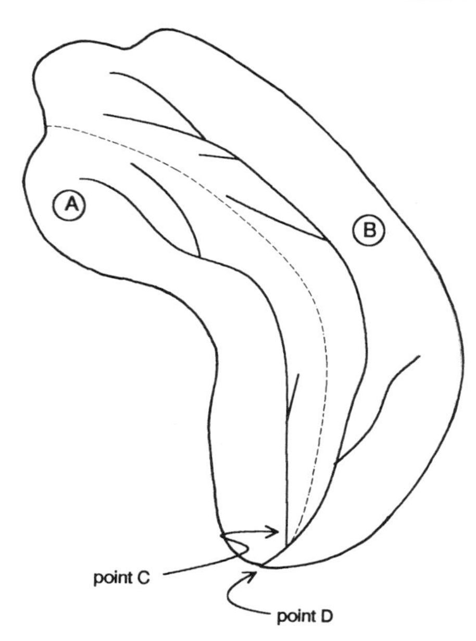

Figure 6.14 Catchments A and B with catchment A set to undergo a major land-use change.

tory authorities require that any land-use changes must be accompanied by measures that ensure that the peak discharge from a design storm after land-use changes will not exceed the peak discharge for the same event prior to the changes. Often this results in the requirement of detention storage being included in the development plans. For discussion purposes, consider that watershed A in Fig. 6.14 is going to change from a wooded area to an intensive urban development. Based on considerations in Chapter 3, it would be expected that the runoff hydrograph after the development would have a greater peak flow, a shorter time to peak, and a larger volume than the hydrograph for the same event prior to development. Figure 6.15a shows the two hydrographs of concern. Obviously the development has increased the peak flow from catchment A. To control this increase in runoff, a storm water detention basin can be installed near the outlet of the watershed. If properly designed, it is possible to have an outflow hydrograph from the basin with a peak discharge equal to the peak flow from the catchment prior to its development. Figure 6.15b shows the predevelopment and routed post development hydrographs with the same peak discharge. The runoff volume is greater after the development and is unaffected by the detention basin.

If flow as it leaves catchment A is the primary concern, the situation depicted in Fig. 6.15b could represent a satisfactory final design. If, however, flow at point D, the combined outlet of catchments A and B is of concern, further analysis is required. Figure 6.15c shows the runoff hydrograph produced by catchment A in the undeveloped state, in the developed state without detention storage, and in the developed state with detention storage. Also shown is the hydrograph from catchment B. Note that the peak of the hydrograph from catchment B occurs after the peak from catchment A. Figure 6.15d shows the combined hydrograph from catchments A and B. Both the situation where the hydrograph from catchment A is routed through a detention basin and where a detention basis is not present are shown.

A significant finding is that by not controlling the flow from catchment A, the flow from that catchment is allowed to pass from the system prior to the arrival of the high flows from catchment B. With detention storage, the flow from catchment A is delayed and caused to coincide with the flows from catchment B. The net result is that without detention storage in catchment A, the total flow at point D is no greater than it is with detention storage. Thus the expense and legal liability associated with the detention basin is largely wasted as far as flow at point D is concerned.

The purpose of this discussion is to point out that piecemeal installation of detention basins without consideration of how they fit into an overall storm water management system may not contribute to peak flow reduction at a downstream location even though they reduce peak flows immediately below and on the same stream as the detention basin. Storm water management systems must be considered on a regional basis and the placement of detention storage optimized from a regional standpoint to obtain maximum storm water control per dollar of expenditure.

A further complication in the design of detention storage arises if the detention basin does not empty to the spillway level between major runoff events. If this is the case, a continuous simulation hydrologic model may be required to compute a continuous water balance on the basin so that the proper initial stage can be used in routing a storm water runoff event through the basin. Hydrologic models are discussed in more detail in Chapter 13. A proper water balance on the detention basin would have to consider evaporation and seepage from the basin.

Problems

(6.1) A storm water drainage channel has been constructed through an urban area. The channel is trapezoidal in shape with a 15-ft bottom width, 1 : 1 side slopes, a slope of 0.3% and a Manning's n of 0.02. At the upstream end of the channel, the inflow hydrograph has a peak flow rate of 750 cfs and a time to peak of 1 hr. The hydrograph may be described using Eq. (3.59). The total volume of runoff is 1.8 in. from

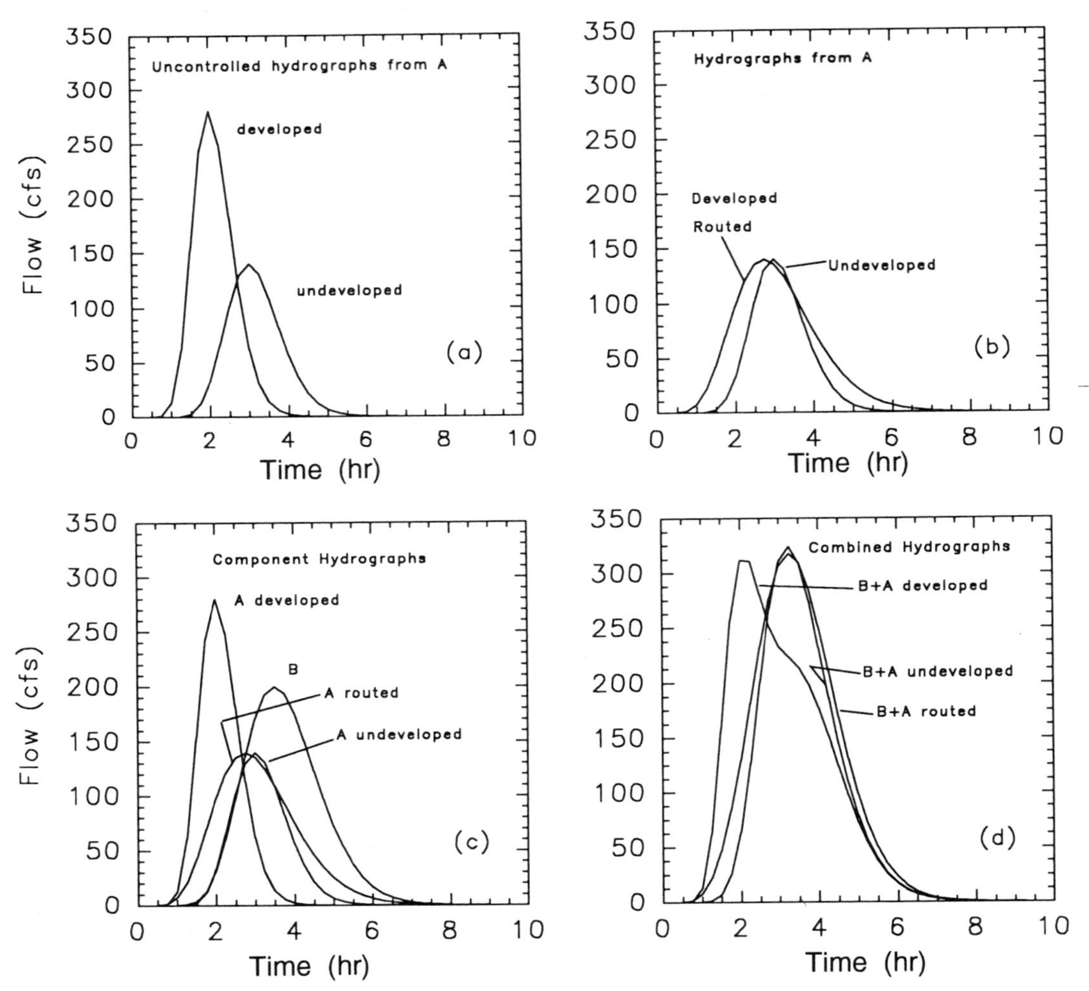

Figure 6.15 Runoff hydrographs from catchments A and B.

the 500-acre drainage area. Route the hydrograph through a 2000-ft segment of the channel.

(a) Use storage routing
(b) Use Muskingum–Cunge routing
(c) Use convex rooting
(d) Use kinematic routing.

(6.2) Route the hydrograph of problem (6.1) through 2000 ft of a trapezoidal channel having a 10-ft bottom width, 2 : 1 side slopes, a slope of 0.2%, and a Manning's n of 0.040.

(6.3) A reservoir has a stage–discharge relationship given by the equation $q = 200h^2$ and a stage–storage curve given by $S = 20h^2$, where q is cfs, h is in feet, and S is in acre-feet. The inflow hydrograph may be described by Eq. (3.59) with a peak flow rate of 1500 cfs from the 2 mile^2 drainage area and a time to peak of 75 min. The volume of runoff is 2.4 in. Estimate the outflow hydrograph.

(6.4) Using the triangular approximations of Fig. 6.12, estimate the detention time of flow of problem (6.3).

(6.5) Compute the detention time of the flow of problem (6.3) from Eqs. (6.32) and (6.34). Compare the results with the results of problem (6.4).

(6.6) Estimate a in the outflow relationship $q = ah^{1/2}$ to give a detention time of 2.5 hr for the situation described in problem (6.3). What is the required storage? Use the triangular approximation.

(6.7) Use a detailed routing to check your solution to problem (6.6). What is the actual detention time as defined by Eqs. (6.32) and (6.34)?

(6.8) A 0.5-mile^2 catchment has the following runoff characteristics from a design storm prior to and following development. Estimate the storage required for a flood control reservoir to limit the postdevelopment peak outflow to the predevelopment.

	Predevelopment	Postdevelopment
Runoff volume	3.0 in.	4.0 in.
Peak discharge	400 cfs	800 cfs
Time to peak	45 min	30 min

(6.9) A potential reservoir site has been surveyed with the results shown below. Calculate and plot the stage–storage curve using feet and acre-feet. Assume that the storage below 410 ft is zero.

Elevation	Area (acres)
410	0
411	1
412	4
413	8
414	10
415	13
416	15
417	20

References

Brakensiek, D. L. (1966). Storage flood routing without coefficients, U.S. Department of Agriculture Publication ARS 41-122.

Chow, V. T. (1959). "Open Channel Hydraulics." McGraw–Hill, New York.

Chow, V. T., Maidments, D. R., and Mays, L. W. (1988). "Applied Hydrology." McGraw–Hill, New York.

Cunge, J. A. (1967). On the subject of flood propagation method. *J. Hydraul. Res. IAHR* **7**(2):205–230.

Fread, D. L. (1977). The development and testing of a dam-break flood forecasting model. *In* "Dam-Break Flood Modeling Workshop, U.S. Water Resources Council, Washington, DC."

Fread, D. L. (1978). National Weather Service Operational Dynamic Wave Model: Verification of mathematical and physical models in hydraulic engineering. *In* "Proceedings 26th Annual Hydraulics Division Special Conference, ASCE, College Park, MD."

Fread, D. L. (1985). Channel routing. *In* "Hydrological Forecasting" (M. G. Anderson and T. P. Brent, eds.). Wiley, New York.

Henderson, F. M. (1966). "Open Channel Flow." MacMillan, New York.

Soil Conservation Service (1986). Urban hydrology for small watersheds. Technical Release No. 55. Soil Conservation Service, U.S. Department of Agriculture, Washington, DC.

U.S. Department of Agriculture (1971). Hydrology. *In* "SCS National Engineering Handbook," Sect. 4. Soil Conservation Service, Washington, DC.

Viessmann, W., Lewis, G. L., and Knapp, J. W. (1989). "Introduction to Hydrology," 3rd ed. Harper & Row, New York.

7 Sediment Properties and Transport

INTRODUCTION

Sediment is composed of many materials, including individual primary particles, aggregates, organic materials, and associated chemicals. Sediment properties impact how each individual or aggregate particle behaves in flowing water. Size, shape, and density affect the settling velocity, which in turn affects sediment transport rates and at what points particles deposit. Sediment is considered to be fully characterized when its shape, size, density, constituent texture, mineralogy, and stability are known (Martin *et al.*, 1955).

Chemical properties of a particle determine if primary particles will aggregate and form larger particles with a different shape and lighter densities. Densities of the aggregates are typically significantly less than that of primary particles (Rhoton *et al.*, 1983). Chemical properties of a sediment–water mixture may either encourage flocculation or cause dispersion of the aggregates. Organic matter, iron oxides, and carbonates can contribute to bind aggregates (Martin *et al.*, 1955).

Some of the properties discussed above are beyond the scope of this text and are appropriate for texts on soil chemistry and mineralogy. In this chapter, emphasis is placed on properties related to particle settling and sediment transport.

BASIC PRINCIPLES OF SEDIMENTATION

Types of Settling

Particle settling has been divided into four types based on concentration and the tendency of particles to hinder each other:

Discrete particle settling (Type 1) refers to settling in low-concentration solutions in which particles tend to fall independently of each other, i.e., as discrete particles.

Flocculent settling (Type 2) refers to settling in which dilute solutions of particles coalesce and form particles of larger mass and higher settling velocities.

Hindered or zone settling (Type 3) refers to settling in which particles are so concentrated that forces between particles interfere (or hinder) the settling of surrounding particles. This results in uniform settling of all particles and is characterized by a distinct liquid–solid interface at the top of the solid mass. The relatively sharp definition at the interface has led to the title zone settling.

Compression settling (Type 4) refers to settling in which the particles are so concentrated that interparticle bridging has formed a stable structure requiring compression for further settling. The force causing

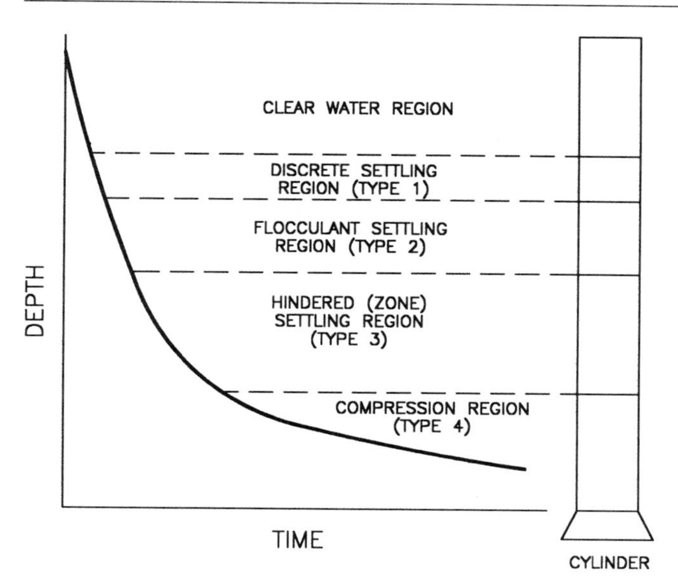

Figure 7.1 Four types of settling (after Tapp *et al.*, 1981).

compression results from the weight of additional particles settling on top of the settling mass.

Tapp *et al.* (1981) indicate that all four of these settling types can occur simultaneously in a sediment pond. The zones (Fig. 7.1) do not occur instantly, but begin to form at varying times. Since each of these types of settling can occur in a sediment pond, a detailed discussion of each type is given.

Discrete Particle Settling (Type 1)

A particle falling in turbulent free water moves in response to the difference in the submerged weight of the particle and the drag of the fluid on the particle. At steady state, the forces are in equilibrium as described by

$$C_D\left[\frac{\pi d^2}{4}\right]\left[\frac{\rho V_s^2}{2}\right] = \frac{\pi d^3}{6}(\rho_s - \rho)g, \qquad (7.1)$$

where C_D is the drag coefficient, which is normally a function of Reynold's number, d is particle diameter, ρ_s is particle density, ρ is fluid density, V_s is settling velocity of the particle, and g is acceleration of gravity.

Stokes showed that the coefficient of drag for spheres depends on Reynold's number, or

$$C_D = 24/R_e, \qquad (7.2)$$

where R_e is the Reynold's number given by

$$R_e = V_s d/\nu, \qquad (7.3)$$

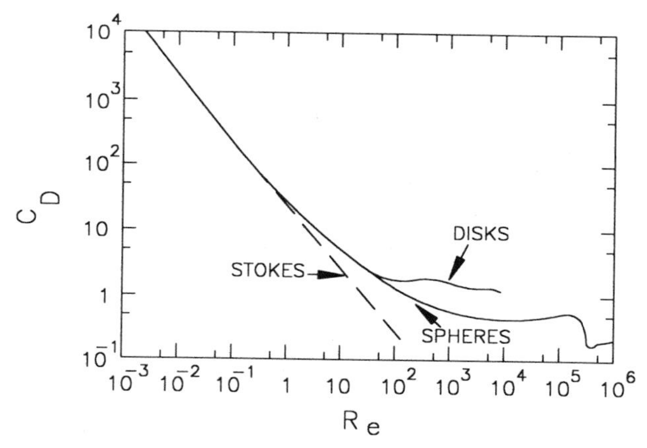

Figure 7.2 Drag coefficients versus Reynold's number for spheres and disks from experimental data (adapted from Rouse, 1950).

where ν is kinematic viscosity. Equation (7.3) is generally considered to be valid up to a value of R_e equal to 0.5.

For the Stokes' range (i.e., $R_e < 0.5$), Eqs. (7.1) and (7.2) can be simplified to yield

$$V_s = \frac{1}{18}\left[\frac{d^2 g}{\nu}(\mathrm{SG} - 1)\right], \qquad (7.4)$$

where SG is the specific gravity of the particles.

For nonspherical particles and for fall velocities outside the Stokes' range, experimental data must be relied on. A common method is to use a curve such as that shown in Fig. 7.2 for C_D to calculate fall velocities. The results define fall velocities of various diameters settling in water at specified temperatures as shown in Fig. 7.3. Equation (7.4) simplifies for SG = 2.65 and

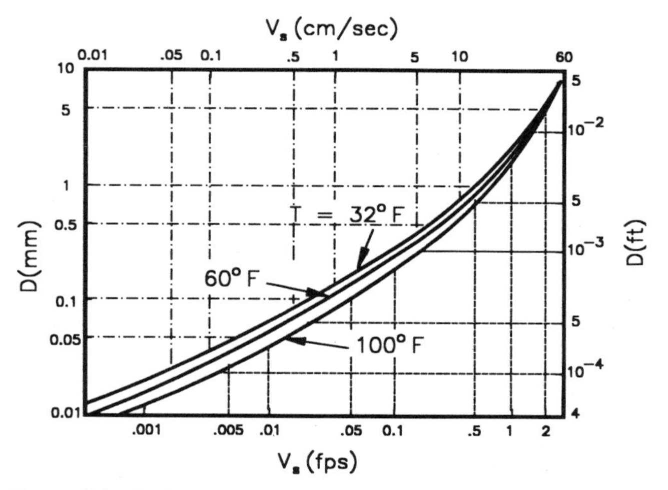

Figure 7.3 Sedimentation diameter versus fall velocity in water assuming a specific gravity of 2.65 (adapted from Rouse, 1950).

quiescent water at 68° F to become

$$V_s = 2.81d^2, \tag{7.5}$$

where V_s is given in feet per second and d is in millimeters.

The settling velocity of large particles can be estimated by a Lagrangian polynomial that is fitted through three points of a sedimentation curve such as that in Fig. 7.3. Wilson *et al.* (1982) presented on analysis that is similar to the approach used to determine Stokes' settling velocity. This analysis evaluated the drag coefficient, C_D, using experimental data for Reynold's numbers larger that 0.5. The results were

$$\log_{10} V_s = -0.34246(\log_{10} d)^2 + 0.98912 \log_{10} d + 1.14613 \quad [\text{cm/sec}] \tag{7.6a}$$

or

$$\log_{10} V_s = -0.34246(\log_{10} d)^2 + 0.98912 \log_{10} d - 0.33801 \quad [\text{ft/sec}], \tag{7.6b}$$

where d is the diameter of the particle in millimeters and V_s is the particle settling velocity in centimeters per second or feet per second, respectively. Factors that affect fall velocity of discrete particles include the following.

Particle shape. Particle shape affects the drag coefficient, which in turn affects the fall velocity. This is normally accounted for by classifying particles according to an equivalent fall diameter. Further information is contained in Graf (1971) and Simons and Senturk (1977, 1992).

Aggregation. Aggregation effectively alters the specific gravity of soil particles. Since an aggregate contains pore space, its specific gravity is less than that of the primary particles from which it was formed. Although aggregation decreases the specific gravity, the diameter of an aggregated particle is so much greater than its primary particle components that an aggregated particle typically has a higher settling velocity than its component primary particles. Site-specific fall velocity data should be collected if possible.

Turbulence. The effects of turbulence on settling velocity are somewhat mixed. Turbulence reduces the drag coefficient C_D by changing the boundary layer around the particle from laminar to turbulent. For smaller particles, this is offset by the tendency of turbulence to diffuse the particles from a zone of higher concentration near the bottom to a zone of lower concentration near the surface. No completely acceptable method for accounting for turbulence is presently available. Graf (1971) presents further discussion of these concepts.

Flocculation. Flocculation results in formation of aggregates through the satisfaction of surface charges on the sediment. The effect is very similar to that of aggregation in that particles formed have a specific gravity less than that of individual primary particles. Flocculation is discussed further in a subsequent section.

Particle settling velocities for type 1 settling are typically determined both experimentally and theoretically. The settling velocities of sand-sized and larger particles are usually determined by using the size of sieve openings and calculating a corresponding settling velocity using Stokes' law (size distribution classifications are presented in a later section). For silt and clay size particles, the size distribution must be taken by a settling analysis using a hydrometer or pipette or by more recently available electronic instrumentation. If only the primary particle size distribution is to be obtained, a hydrometer is adequate since particles are dispersed and the specific gravity is known with a relatively high degree of confidence. If an analysis of the particle size distribution of a mixture of aggregates and primary particles is being made, it is necessary to use the pipette method and calculate equivalent fall velocities since the specific gravity is not known. Since eroded sediment is typically composed of a combination of primary and aggregated particles, a pipette analysis is preferred. Obviously, a dispersing agent is not used in an aggregate analysis since this would result in dispersion of aggregates in the sample. It should be recognized that the use of a dispersing agent for primary particles is the major difference in technique that separates determination of aggregate particle size distributions from determination of primary particle size distributions. Details on hydrometer and pipette analyses can be found in standard texts dealing with soil physics and soil mechanics. The U.S. Department of Agriculture (USDA, 1979) presents substantial discussion of sieve, pipette, and hydrometer analyses.

Example Problem 7.1. Settling velocity of discrete particles

Compare the settling velocities using Eqs. (7.4) and (7.6) for 0.0002-, 0.002-, 0.02-, and 2.0-mm particles if SG = 2.65 for all particles and settling occurs in water at 68° F.

Solution: Solving for the middle size (0.02 mm) first, values for particle diameter (0.002 cm), gravitational constant (980 cm/sec^2), kinematic viscosity at 68° F (0.01003 cm^2/sec) (from the General Appendices), and particle specific gravity are substituted into Eq. (7.4):

$$V_s = \frac{1}{18}\left[\frac{d^2 g}{\nu}(\mathrm{SG} - 1)\right]$$

$$= \frac{1}{18}\left[\frac{0.002^2 \text{ cm}^2 \times 980 \text{ cm/sec}^2}{0.01003 \text{ cm}^2/\text{sec}}(2.65 - 1)\right]$$

$$= 0.0358 \text{ cm/sec}.$$

Similarly, for a particle diameter of 0.02 cm, V_s is 3.58 cm/sec, and for a diameter of 0.2 cm, V_s is 358 cm/sec. The settling velocities for 0.02- and 0.2-cm particles would be outside the Stokes' range, since Reynold's numbers would be

$$R_e = \frac{(3.58)(0.02)}{0.01003} = 7.138$$

for the 0.02-cm particle and 7138 for the 0.2-cm particle, using the calculated velocities.

Using the Wilson *et al.* (1982) relationship [Eq. (7.6)] and a particle diameter of 0.02 mm,

$$\log V_s = -0.3425(\log_{10} d_p)^2 + 0.98912 \log_{10} d_p + 1.14613$$

$$\log V_s = -0.3425(\log_{10} 0.02)^2 + 0.98912(\log_{10} 0.02) + 1.14613$$

$$\log V_s = -1.522$$

$$V_s = 0.03 \text{ cm/sec}.$$

The final estimated settling velocities for 0.00002-cm (0.0002-mm), 0.0002-cm (0.002-mm), 0.002-cm (0.02-mm), 0.02-cm (0.2-mm), and 0.2-cm (2.0-mm)-diameter particles, respectively, are:

$V_s = 3.58 \times 10^{-6}$ cm/sec, using Stokes' law [Eq. (7.5)]
$V_s = 3.58 \times 10^{-4}$ cm/sec, using Stokes' law
$V_s = 3.58 \times 10^{-2}$ cm/sec, using Stokes' law
$V_s = 1.94$ cm/sec, using Eq. (7.6a)
$V_s = 27.79$ cm/sec, using Eq. (7.6a).

Values predicted from Eq. (7.6) agree favorably with values from Fig. 7.3 and differ substantially from Stokes' law. This difference demonstrates the influence of the drag coefficient on the settling velocity at Reynold's number greater than 0.5. Equation (7.6) would be appropriate for the larger particles. When using Stokes' law, a check to assure that the Reynold's number is less than 0.5 should always be made.

Flocculent Settling (Type 2)

Colloidal particles in a dilute suspension will sometimes settle as discrete particles and sometimes coalesce and form "flocs" of numerous particles, which have sufficient mass to settle rapidly. Metcalf and Eddy (1979) reported that the degree of flocculation depends on the opportunity for particle contact, depth of the sedimentation basin, velocity gradients in the basin, concentration of particles, particle size, and interparticle forces that cause particles to repel each other. Studies associated with surface mine sedimentation indicate that flocculation can be a naturally occurring phenomena, depending on the chemistry of the water and the mineralogy of the clay particles (Tapp *et al.*, 1981; Evangelou *et al.*, 1981). Evangelou *et al.* also indicate that dispersion is a possibility under certain conditions. Dispersion may break the aggregated particles down into primary particles, thus decreasing their settling velocity. The flocculation phenomena are discussed further in this section, and the dispersion phenomena are delayed to a subsequent section of this chapter.

To begin a discussion of the flocculation process, consider a stabilized colloidal suspension. This suspension consists of dispersed or individual colloidal particles with a surface charge separated at considerable distances in an aqueous solution. For clay particles, the surface is negatively charged. Dispersed particles are said to be stabilized since they repel each other. For flocculation to occur, particles must be destabilized and brought together to allow contact and flocculation.

Three theories are often used to describe the process of destabilization—the double electric layer theory, chemical bridging, and $\mathrm{pH_{zpc}}$. The double electric layer theory considers only electrostatic attraction and diffusion. In this theory, the largest concentration of cations[1] is near the surface of the colloidal particle because of the electrical attraction of the negative charge on the colloid. The concentration of cations decreases with distance away from the colloidal surface as shown schematically in Fig. 7.4. The resulting concentration gradient of cations causes a gradient of forces that repel other clay particles that are surrounded by positively charged ions as shown in Fig. 7.5. The layer in which the concentration of counterions is increased is called the diffuse layer (See Fig. 7.4). As the concentration of ions in the aqueous phase increases, the thickness of the diffuse layer decreases, causing the magnitude of the repulsive forces to decrease more rapidly at intermediate distances from the surface of the colloidal particle. Thus the forces of repulsion depend on both the solid and aqueous phases. The forces of attraction between colloidal particles result from van der Waals energy, shown schematically in Fig. 7.5, which depends on the mass of the particle. The sum of the energy levels of repulsion and attraction is the resultant energy curve. If the concentration of counterions in the aqueous solution is low, the

[1] Positively charged ions, sometimes called counterions.

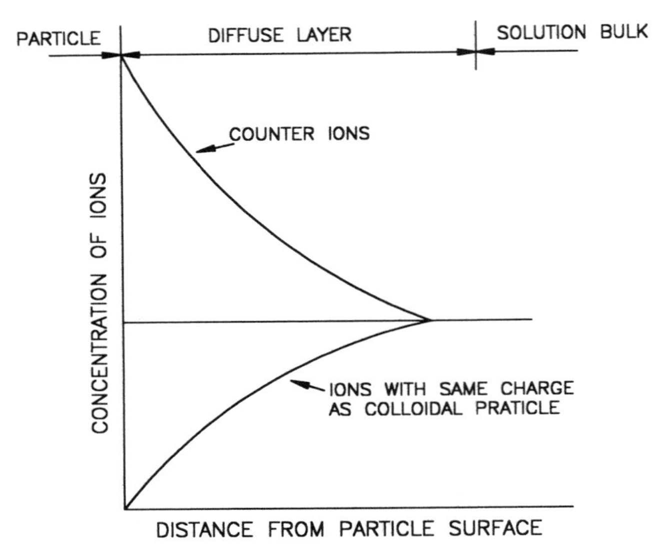

Figure 7.4 Schematic of electric double layer theory (after Tapp *et al.*, 1981).

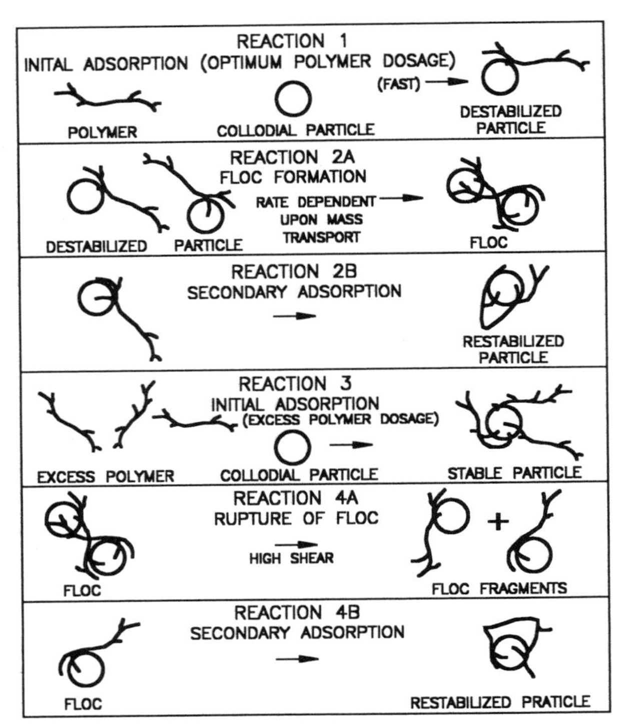

Figure 7.6 Processes involved in chemical bridging (after Tapp *et al.*, 1981).

double layer thickness is high, and a potential energy barrier exists, as shown in Fig. 7.5. As the concentration of counterions increases, the diffuse layer becomes thinner, reducing the forces of repulsion at intermediate distances. This ultimately eliminates the potential energy barrier at sufficiently high concentrations. At this point, the particle is destabilized and flocculation can occur. Charge reversal can sometimes occur as a result of overdosing when using a trivalent ion (three positive charges) such as aluminum.

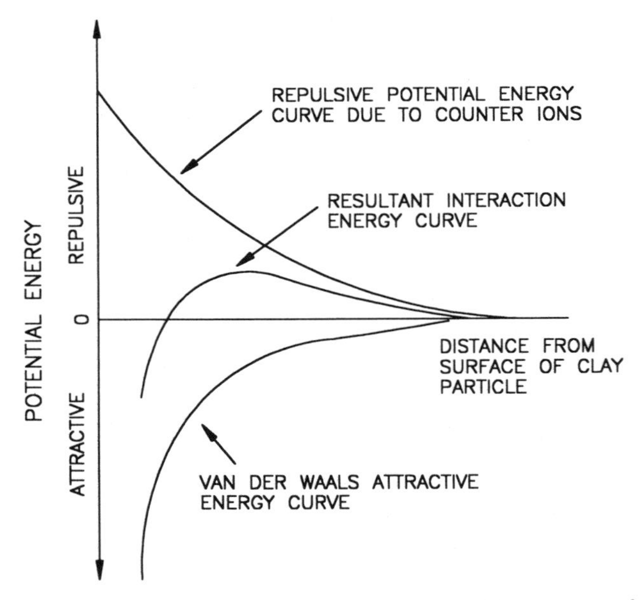

Figure 7.5 Energy levels of attraction and repulsion near a colloidal particle (after Tapp *et al.*, 1981).

In chemical bridging theory, the second flocculation mechanism, it is proposed that a polyelectrolyte (a long-chain polymer with ionizable groups) can attach itself to the surface of a colloidal particle at one or more absorption sites. Its length causes the polyelectrolyte to extend beyond the diffuse layer into the solution. This extension can thus be attached to vacant absorption sites on other colloidal particles causing flocculation. A schematic of the processes involved is given in Fig. 7.6. Further discussion can be found in Metcalf and Eddy (1979).

Procedures for predicting changes in particle size due to flocculation are limited primarily to experimental type relationships. Tapp *et al.* (1981) summarizes the equations developed by various researchers. The most promising procedure is that proposed by Argaman and Kaufman (1970). They postulate that flocs grow due to collisions between unreacted colloids and flocs that are simultaneously coming apart due to shear forces resulting from the turbulence of the flow. If n_1 represents the number of colloidal particles available for flocculation, then

$$\frac{dn_1}{dt} = -\left[\begin{array}{c}\text{rate of flocculation}\\ \text{of primary particles}\end{array}\right] + \left[\begin{array}{c}\text{rate of breakup of}\\ \text{flocs to primary particles}\end{array}\right] \quad (7.7a)$$

or

$$\frac{dn_1}{dt} = H_{1F} + B_{1T}. \quad (7.7b)$$

Argaman and Kaufman (1970) proposed that the rate of flocculation could be given by

$$H_{1F} = -4\pi K_s \alpha R_F^3 n_1 n_F [u'^2]_a, \quad (7.7c)$$

where K_s is a parameter that relates the effectiveness of the mean square turbulent velocity,[2] $[u'^2]_a$, in mixing, α is a parameter that defines the probability of a colloidal particle permanently sticking to the floc after a collision, R_F is the radius of a floc, n_1 is the number concentration of colloidal particles, and n_F is the number concentration of flocs. Argaman and Kaufman also propose that the rate of breakup is given by

$$B_{1T} = B\frac{R_F^2}{R_1^2} n_F [u'^2]_a, \quad (7.8)$$

where B is a floc breakup constant and R_1 is the radius of the colloidal particle. Combining the two expressions yields

$$\frac{dn_1}{dt} = \left(-4\pi K_s \alpha R_F^3 n_1 n_F + B\frac{R_F^2}{R_1^2} n_F\right)[u'^2]_a. \quad (7.9)$$

Argaman and Kaufman (1970) found good agreement between predicted and observed values of n_1 in a settling tank using alum as the coagulant with a suspension of clay colloids. Tapp *et al.* (1981) found that the prediction accuracy for effluent suspended solids concentration from a pilot size sediment pond was improved when Eq. (7.9) was used in conjunction with various models of pond performance. Unfortunately, values for K_s and B are not generally available for different systems, and procedures for predicting $[u'^2]_a$ are rather limited. Additional details of the Tapp *et al.* study are discussed further in a subsequent section.

[2]From turbulent flow theory, u' is the deviation of the instantaneous velocity from the time average velocity at a point. The mean square velocity is thus the time average of u'^2 and is a measure of the kinetic energy contained in the turbulence of flow.

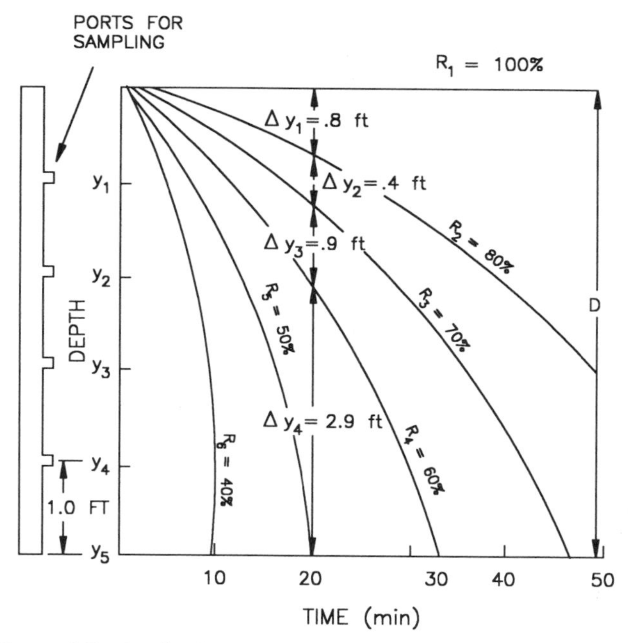

Figure 7.7 Application of settling tube for determining suspended solids removal by flocculation settling (Haan and Barfield, 1978).

In the absence of applicable procedures for predicting floc growth, most designs of settling chambers and treatment facilities in which chemical flocculants are to be added are based on settling tube analyses. A settling tube can be constructed of material having convenient diameter, typically 3–16 in. An example is shown in Fig. 7.7. Height of the tube should be equal to the height of the settling chamber. A sample containing sediment and flocculant is added to the tube; the solution is thoroughly mixed for a uniform distribution; and quiescent settling begins. After settling is initiated, samples are taken at various depths and time intervals and analyzed for total suspended solids (TSS). These are compared to the initial concentration of TSS and the percentage of particles removed calculated for each depth at a given time. Percentage removal values are then plotted as shown in Fig. 7.7. The percentage removal for an ideal clarifier with no turbulence is then the average percentage removal at a time in the settling tube equal to detention time of the clarifier.

Example Problem 7.2. Suspended solids determination using a settling tube

A sediment pond is being designed to treat drainage that is pumped from a mine. The estimated flow rate through the system and surface area are such that the pond has a detention storage time of 20 min. A settling tube test is conducted on the drainage from the mine while adding a commercial flocculant. The resulting removal curve is shown in Fig. 7.7. The depth of the pond is 5 ft. Calculate the removal efficiency of the pond.

Solution: The percentage removal is calculated by weighting the removal efficiency over given depth intervals, or $E=(1/D)\int_0^D R(y)dy \cong (1/D)\bar{R}_i\Delta y_i$ where $R(y)$ is the removal efficiency as a function of y and $\bar{R}$ is the average removal efficiency over the interval Δy_i. Using values from Fig. 7.7,

$$E = \frac{1}{5}\left[\frac{100+80}{2}(0.8) + \frac{80+70}{2}(0.4) + \frac{70+60}{2}(0.9) + \frac{60+50}{2}(2.9)\right]$$

$$E = 64\%.$$

Obviously, the trapping efficiency is low. To improve the pond's performance, the detention time would have to be increased by increasing the pond size or a better flocculating agent would have to be found.

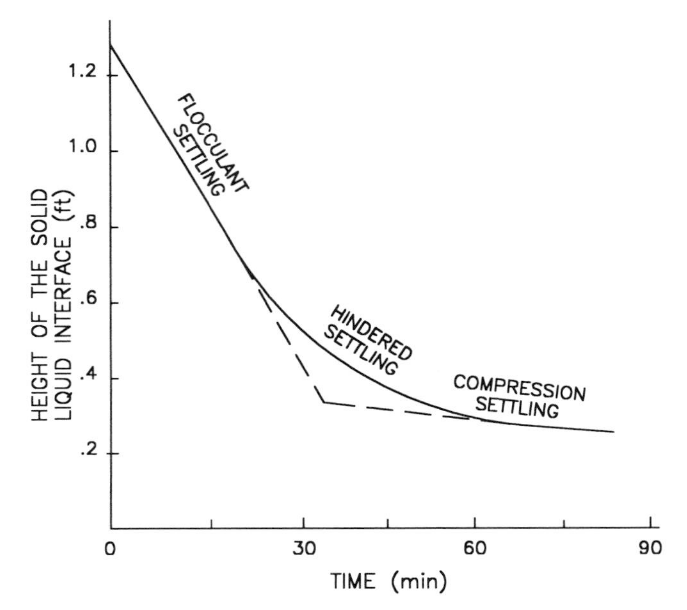

Figure 7.8 Sample settling test for hindered and compression settling.

Hindered or Zone Settling (Type 3)

As the concentration of sediment increases in density, a point will be reached such that the settling of particles is hindered. The settling of particles displaces water that flows upward through the openings between the particles. As a result, particles tend to settle as a blanket, with all of the particles maintaining the same relative velocity. This is known as hindered settling. Hindered settling would be expected near the bottom of a sediment pond.

Variability encountered with hindered settling is so great that settling tests are typically required for a design. These tests would be necessary if a system were being designed for pumping the deposited sediment (which corresponds to sludge in water treatment terminology) from a pond. Metcalf and Eddy (1979) give an excellent discussion of the topic.

Compression Settling (Type 4)

After hindered settling continues for a long period of time, a layer of particles with a definite structure begins to form at the bottom of the basin or pond. Further settling of these particles results only from compression of the layer, thus the term compression settling.

Compression settling, like hindered settling, is highly variable, usually requiring settling tests for satisfactory design information. These settling tests typically are observations of the depth of the liquid–solid interface with time. An example is shown in Fig. 7.8.

As can be seen in Fig. 7.8, the height of the sludge or compressed sediment decays exponentially with time until it reaches what might be termed a short-term constant value. This "constant" value continues to decrease, but at a slow rate. The short-term constant density will depend on the makeup of the sediment, as well as the conditions of deposition. Lara and Pemberton (1963) suggested that the short-term (or initial) weight density W of deposited sediment can be given by

$$W = W_c P_c + W_m P_m + W_s P_s, \tag{7.10}$$

where W_c, W_m, and W_s are unit weights of clay, slit, and sand, respectively, and P_c, P_m, and P_s are the fractions of clay, silt, and sand, respectively. Values for W_c, W_m, and W_s are given in Table 7.1. Miller (1953) developed a procedure for calculating how W, as calculated from Eq. (7.10), changes with time or

$$W_T = W + 0.4343\mathrm{K}\left[\frac{T}{T-1}(\ln T - 1)\right], \tag{7.11}$$

where W is the value given by Eq. (7.10), T is time in years, and K is given by

$$K = B_c P_c + B_m P_m + B_s P_s, \tag{7.12}$$

where B_c, B_m, and B_s are coefficients given in Table 7.1.

Settleable Solids

Conceptually, total suspended solids can be divided into settleable solids and those that are so small that

Table 7.1 Values of Unit Weight of Reservoir Sediments[a]

Type of reservoir operation	Values of coefficients in Eq. (7.10) and (7.12)					
	W_c	B_c	W_m	B_m	W_s	B_s
I. Sediment always submerged	26	16.0	70	5.7	97	0
II. Moderate–considerable drawdown	35	8.4	71	1.8	97	0
III. Reservoir normally empty	40	0.0	72	0.0	97	0
IV. Riverbed sediments	60	0.0	73	0.0	97	0

[a] W values from Lara and Pemberton (1963). B values from Miller (1953).

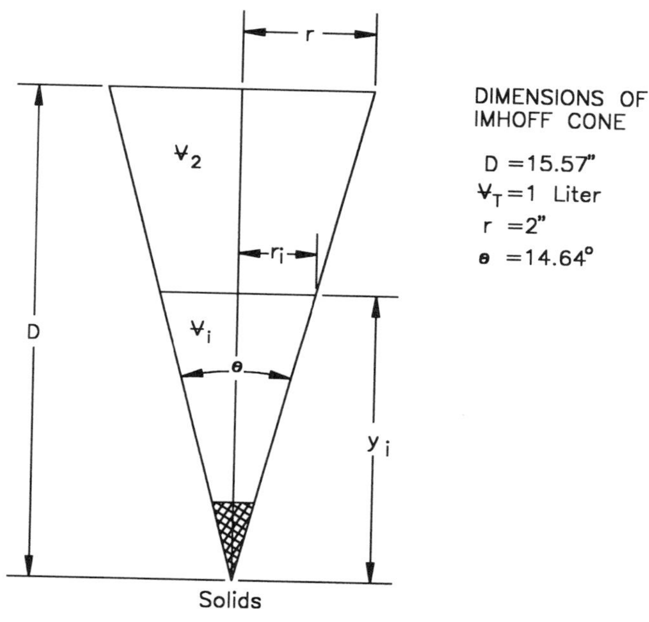

Figure 7.9 Definition sketch for settleable solids equations (after Barfield *et al.*, 1981).

they remain in suspension due to Brownian motion. Actually, settleable solids are defined as the volume of particles that settle in the bottom of an Imhoff cone in 1 hr of quiescent settling. The dimensions of an Imhoff cone are shown in Fig. 7.9. Since the displacement of an Imhoff cone is 1 liter, the units of settleable solids are typically milliliters per liter.

To convert TSS to settleable solids, the size distribution of the particles and their weight density must be used. The following derivation, developed by Barfield *et al.* (1981) and Wilson *et al.* (1982), assumes that the volume of settleable solids is small relative to the total volume. Another assumption is that the particles are uniformly distributed at the start of the test. At the end of a test of duration T hours, a particle of diameter d_i with a settling velocity V_{si} will settle through a distance

$$y_i = V_{si}T. \tag{7.13}$$

Thus all particles of size d_i will have settled out of zone Ꝟ_i in Fig. 7.9. The fraction of particles settling into the solids zone is

$$F = \text{Ꝟ}_i / \text{Ꝟ}_T, \tag{7.14}$$

where Ꝟ_T is the total volume of the Imhoff cone. From geometry

$$\text{Ꝟ}_i = \frac{\pi}{3} y_i^3 \tan^2 \frac{\theta}{2}. \tag{7.15}$$

Hence,

$$F = \frac{\tan^2(\theta/2)\, y_i^3}{r^2 D}. \tag{7.16}$$

If the velocity of a particle that will settle through D in T is given by V_0 and y_i is $V_i T$ from Eq. (7.13), then D becomes $V_0 T$. D is also $r^2/\tan^2(\theta/2)$. Therefore,

$$F_i = \frac{y_i^3}{D^3} = \frac{V_{si}^3 T^3}{V_0^3 T^3} = \frac{V_{si}^3}{V_0^3}. \tag{7.17}$$

Let ΔX_i be the fraction of particles represented by the particles of size d_i. Then the mass of sediment trapped in the size range d_i is

$$\text{mass}_i = F_i C \text{Ꝟ}_T \Delta X_i \tag{7.18}$$

or

$$\text{mass}_i = \left[\frac{\tan^2 \theta}{r^2} V_{si}^2 T^2 \left(\frac{V_{si}}{V_0} \right) \right] C \text{Ꝟ}_T \Delta X_i = \frac{V_{si}^3}{V_0^3} C \text{Ꝟ}_T \Delta X_i, \tag{7.19}$$

where C is the TSS concentration and Ꝟ_T is the total volume of the Imhoff cone.

The total mass trapped over all size ranges is

$$M_T = \forall_T C\left[(1 - X_0) + \sum_{i=1}^{n} \frac{V_{si}^3}{V_0^3}\Delta X_i\right], \quad (7.20)$$

where n is the number of size fractions of particles with a settling velocity less than V_0. The total settleable solids (SS) per unit original volume is

$$SS = M_T/WV_T, \quad (7.21)$$

where W is the dry bulk density of the solids. Based on $T = 1$ hr and the values given in Fig. 7.9 and using $V_0 = D/T$, k is defined

$$k = \frac{1}{V_0^3} = \left(\frac{T}{D}\right)^3 = \frac{3600^3 \text{ sec}^3}{(15.57/12)^3 \text{ ft}^3}$$

$$= 2.135 \times 10^{10} \text{ sec}^3/\text{ft}^3. \quad (7.22)$$

Then

$$SS = \frac{C}{W}\left[(1 - X_0) + \sum_{i=1}^{n} kV_{si}^3\Delta X_i\right]. \quad (7.23)$$

Example Problem 7.3. Use of an Imhoff cone

If the peak effluent TSS from a pond is 1000 mg/liter and the effluent sediment size distribution is given by Fig. 7.10, calculate the settleable solids concentration in the effluent.

Solution: Solving first for V_0,

$$V_0 = \frac{D}{T} = \frac{(15.57/12) \text{ ft}}{(3600 \text{ sec})} = 0.00036 \text{ ft/sec}.$$

Assuming a temperature of 68° F, then the equivalent diameter can be solved from Eq. (7.5), or

$$d = \left(\frac{V_0}{2.81}\right)^{1/2} = \left(\frac{0.00036}{2.81}\right)^{1/2} = 0.011 \text{ mm (fine silt)}.$$

Thus, one would expect to trap all of the fine silt and larger particles. The size distribution for the sediment is given in Fig. 7.10. For diameter of 0.011 mm, X_0 is 0.33. The following data are tabulated for computing settleable solids.

Diameter				
Range[a] (mm)	Average (mm)	ΔX_i[b]	V_{si}[c] (ft/sec)	$kV_{si}^3\Delta X_i$
0.0010	0.00075	0.04	1.58×10^{-6}	3.40×10^{-9}
0.0010–0.0016	0.0014	0.04	5.51×10^{-6}	1.42×10^{-7}
0.0016–0.0024	0.0021	0.04	1.24×10^{-5}	1.62×10^{-6}
0.0024–0.0034	0.0030	0.04	2.53×10^{-5}	1.39×10^{-5}
0.0034–0.0043	0.0042	0.04	4.96×10^{-5}	1.05×10^{-4}
0.0043–0.0064	0.0056	0.04	8.81×10^{-5}	5.88×10^{-4}
0.0064–0.0082	0.0070	0.04	1.38×10^{-4}	2.26×10^{-3}
0.0082–0.011	0.0095	0.04	2.54×10^{-4}	1.76×10^{-2}
	$\Sigma\Delta X_i$ =	0.33	$\Sigma kV_{si}^3\Delta X_i$ =	0.02

[a]From Fig. 7.10.
[b]From Fig. 7.10.
[c]From Eq. (7.5) or Fig. 7.3.

Using Eq. (7.23) and assuming W of 70 lb/ft³ (1120 mg/ml),

$$SS = \frac{1000 \text{ mg/liter}}{1120 \text{ mg/ml}}[(1 - 0.33) + 0.02]$$

$$SS = 0.616 \text{ ml/liter}.$$

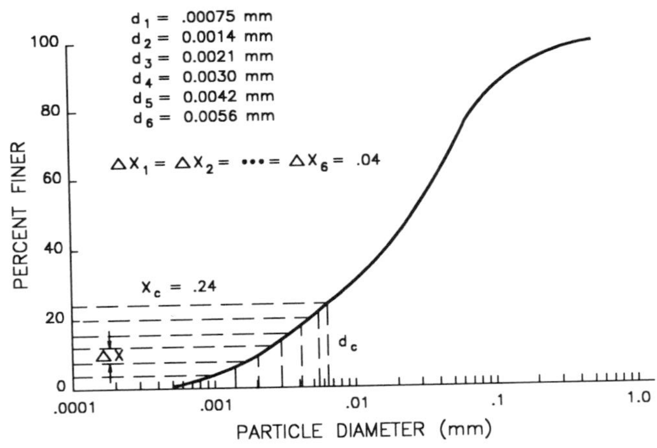

Figure 7.10 Particle size distribution for Example Problem 7.2.

The major problem involved in the estimation of SS from TSS is determining a value for W. Compression settling in the bottom of an Imhoff cone causes bulk density to change with time. Therefore when measuring SS, it is important to measure the volume of deposited material at 1 hr.

Bulk Density

Particle density and bulk density are two basic ways to express sediment weight. Particle density is defined as the mass of a unit volume of sediment solids. A simple example is that if 1 cm³ of solid material weighs 2.65 g, the particle density is 2.65 g/cm³. Particle densities generally fall between 2.60 and 2.75 g/cm³ for mineral particles. Organic matter weighs much less than an equal volume of mineral solids and often has a particle density of 1.2 to 1.4 g/cm³.

Bulk density provides a second way of defining sediment weight. Bulk density is defined as the mass of a unit volume of dry sediment, including both solids and

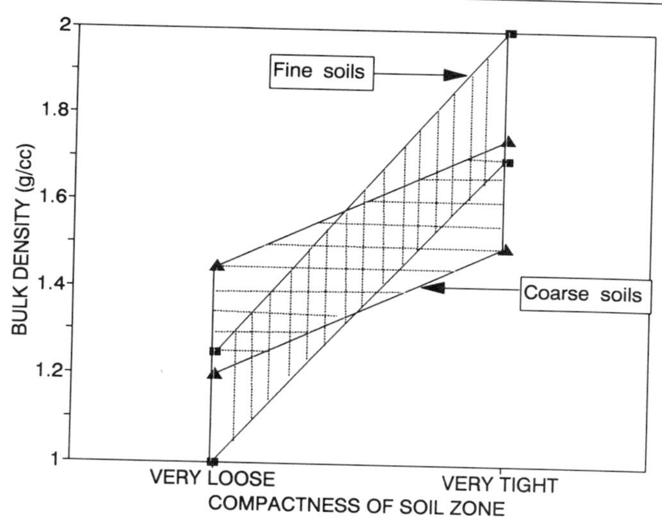

Figure 7.11 Relationship between compaction and approximate ranges of bulk densities for coarse- and fine-textured soils (after Brady, 1974).

pores. Brady (1974) discussed the factors affecting bulk density, indicating that loose and porous materials have lower bulk densities than more compact materials. Sandy soils, particularly those with low organic matter, generally fit tightly together. Soil particles in a fine-textured soil typically do not lie closely together, but instead tend to bridge between individual particles so that large pores may exist. The bulk densities of clay, clay loam, and silt loam surface soils normally range from 1.00 to about 1.60 g/cm^3, whereas sandy loams may have bulk densities ranging from 1.20 to 1.80 g/cm^3 (Brady, 1974). Figure 7.11 shows a general relationship between bulk density and compaction for both coarse- and fine-textured soils. The bulk density for fine soils is more sensitive to compaction because of large void spaces in the loose condition. Compaction allows these void spaces to be filled with solids.

A similar relationship occurs in settleable material in the bottom of a pond. Depending on the size of the material trapped in the pond, considerable range in the bulk density can occur. Also since the material is often quite loose, it is susceptible to mixing and eventual transport from the pond because of turbulence caused by runoff entering the pond. This concept must be recognized if sediment storage is to be properly accounted for when sizing a pond. Equations (7.10) and (7.11) can be used to make this evaluation.

PARTICLE SIZE CLASSIFICATIONS

Textural Classification

The size of individual grains (primary particles) is the basis for traditional approaches to classifying soils (and sediment). McKyes (1989) and the USDA (1979) discuss a number of classification systems based on individual grain sizes, indicating that the number of divisions, as well as their breakpoints, varies from one system to another. Several of the more common classification systems are shown graphically in Fig. 7.12. A major difficulty in using the classifications shown in Fig. 7.12 is that most soils do not fall into only one of the categories, but instead are combinations of several categories. One technique that is commonly used to classify these mixed soils is to use a textural triangle as shown in Fig. 7.13. To use a textural triangle, the sample is analyzed using only material that is less than 2 mm in diameter (i.e., sand or smaller). Aggregates are pulverized or soaked to separate aggregates into primary particles. The sand, or coarse material, can be sized mechanically using sieves. The material caught in each sieve is then weighed, and the fraction of the original sample is calculated. Since silt and clay particle fractions are too fine to be separated using sieves, they must be determined using either hydrometer or

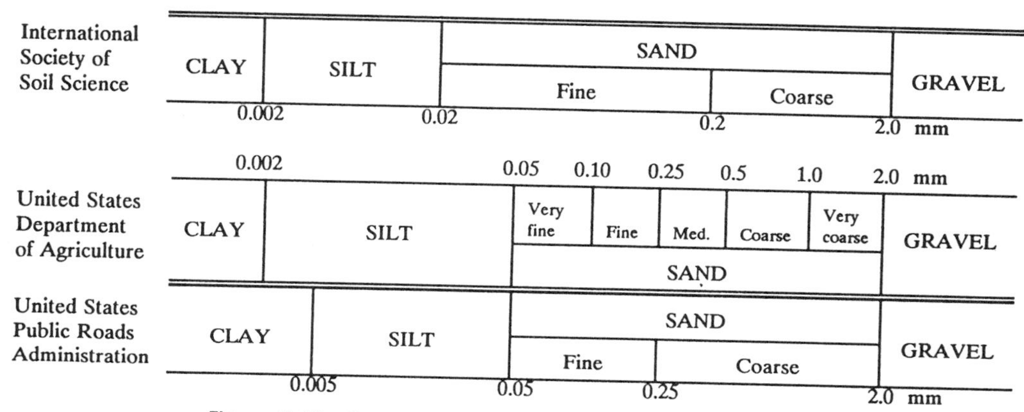

Figure 7.12 Common classification systems (after Brady, 1974).

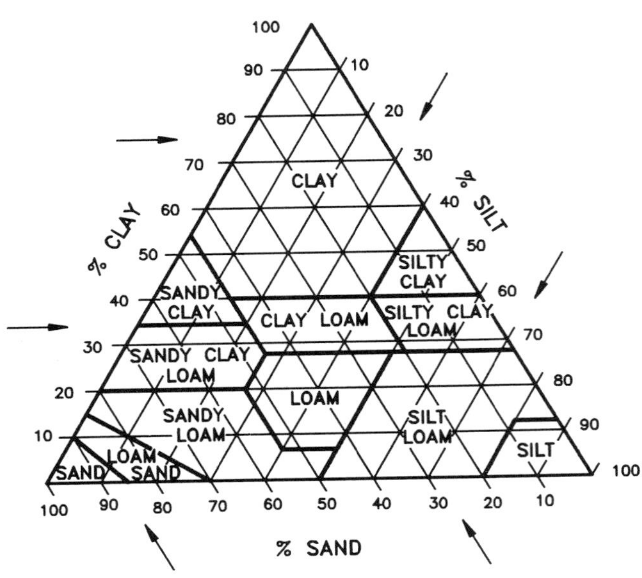

Figure 7.13 Textural triangle for soil classification (after Soil Survey Staff, 1951).

pipette analysis as described previously in this chapter. Details concerning these procedures and the equipment required for analysis can be found in ASTM (1985) and USDA (1979). The general terms used to describe soil texture in relation to the basic soil textural class names for several soils are presented in Tables 7.2 and 7.3.

Example Problem 7.4. Textural classification

Two surface soils are analyzed using a standard hydrometer test to obtain the following percentages of sand, silt, and clay particles.

Particles (%)	A	B
Sand	65	30
Silt	22	40
Clay	13	30

What is the textural classification of each soil?

Solution: Locate the point in the textural triangle (Fig. 7.13) where each of the three percentages intersect. In this case, the USDA soil classification for soil A is found to be sandy loam and for soil B is clay loam.

The Soil Conservation Service (1984) and many other agencies use the Unified Soil Classification System

Table 7.2 Grade Scale of Sediment Particle Sizes [Adapted from USDA (1979)]

	Large sizes	
Soil textural class	Metric unit (mm)	English Unit (in.)
Very large boulders	4096–2048	160–80
Large boulders	2048–1024	80–40
Medium boulders	1024–512	40–20
Small boulders	512–256	20–10
Large cobbles	256–128	10–5
Small cobbles	128–64	5–2.5
Very coarse gravel	64–32	2.5–1.3
Coarse gravel	32–16	1.3–0.6
Medium gravel	16–8	0.6–0.3
Fine gravel	8–4	0.3–0.16
Very fine gravel	4–2	0.16–0.08

Table 7.3 Size Limits for Sand and Finer Particles (USDA, 1979)

Soil textural class	Range in diameter (mm)
Very coarse sand	2.0–1.0
Coarse sand	1.0–0.5
Medium sand	0.5–0.25
Fine sand	0.25–0.10
Very fine sand	0.10–0.05
Silt	0.05–0.002
Clay	Below 0.002

(USCS) in classifying solids for engineering purposes. The USCS bases classification upon the soil's particle size, gradation, plasticity index, and liquid limit. It utilizes the material having less than 3-in. diameter. The Soil Conservation Service (1984) lists several outstanding features of USCS:

1. Test procedures are simple.
2. Important physical characteristics are described.
3. Results are realistic and consistent in properties.

Evaluation of Size Distribution Data

Considering all of the various properties of sediment, size generally has the greatest importance to the engineer working in the area of sedimentation. Size can be measured readily and also can be related to other characteristics such as particle shape, specific gravity, and settling velocity. Simons and Senturk (1977, 1992) provide a detailed discussion of the physical properties associated with sediment. One of these physical properties is particle size. It can be defined on several bases: volume, diameter, weight, fall velocity, and sieve size. All of these except volume are dependent upon shape and density of the particle. Generally particle diameter characteristics for a specific soil are expressed as a particle size distribution, either as aggregated particles or primary particles, that is presented graphically as a curve similar to the examples shown in Fig. 7.14. Each curve shows the fraction (or percentages) finer plotted versus particle size and represents a size frequency distribution. Some users plot fraction greater (instead of fraction finer) as the dependent variable. The fractions (or percentages) are based on the weight of sample that is smaller or larger than the specified size of the particles. Depending upon the soil texture and gradation, the shape of the particle size distribution can vary greatly as shown by the three curves in Fig 7.15.

Specific points in the particle size distribution are identified by Simons and Senturk (1977, 1992) as having been used by researchers to define various sediment properties. Some of the commonly used points from the distribution include the following:

D_{35}—the maximum size for the smallest 35% of the sample. Einstein (1950) specified this size to define grain size for sediment mixtures when developing his method for partitioning fluid drag. Additional information concerning this application is located in a subsequent section.

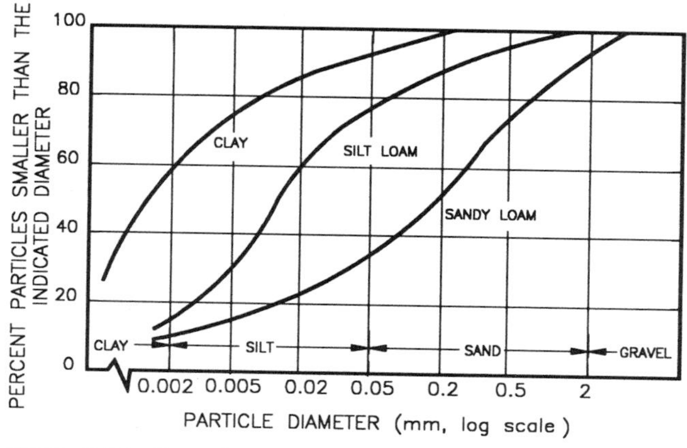

Figure 7.14 Example particle size distributions for three soils (after Brady, 1974). These examples are not intended to be typical for a given textural classification. Particle size distribution for a given soil would vary widely from these examples.

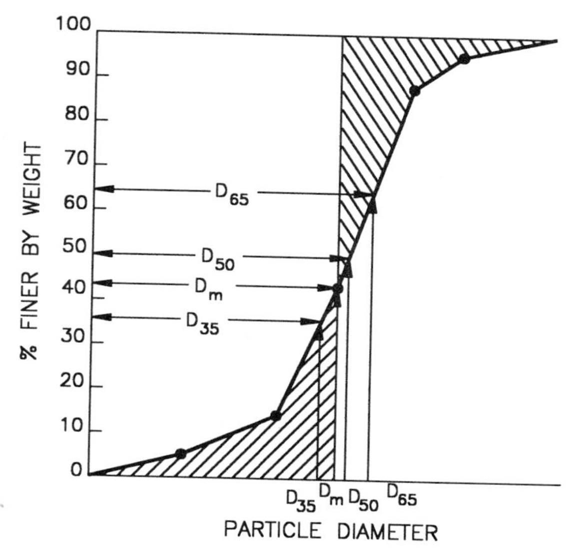

Figure 7.15 Size distribution curve showing D_{35}, D_{50}, D_{65}, and other representative diameters.

D_{50}—the maximum size for the smallest 50% of the sample and corresponds to the median diameter. This value is most naturally assumed as the best single size to represent a soil or sediment mixture, but because of various influences including the wide variation in gradation that may be in the mixture, this size is not necessarily the best size for representing a given sediment's transport and settling properties.

D_{65}—the maximum size for the smallest 65% of the sample. Researchers have often used this size to indicate roughness of a sediment mixture. It is sometimes used with D_{35} in this manner.

Example Problem 7.5. D_{35}, D_{50}, and D_{65} for three soil textures

Estimate D_{35}, D_{50}, and D_{65} for the three soil textures shown in Fig. 7.14.

Solution:

	D_{35} (mm)	D_{50} (mm)	D_{65} (mm)
Clay	0.0012	0.0017	0.0036
Silt loam	0.006	0.014	0.031
Sandy loam	0.062	0.133	0.295

Two other terms associated with particle size distributions are mean diameter and geometric mean size

(Simons and Senturk, 1977, 1992). Mean diameter is given by

$$D_{\mathrm{m}} = \frac{\sum_{i=1}^{n} \Delta_i D_i}{100}, \tag{7.24}$$

where Δ_i represents incremental percentages on the y-axis of Fig. 7.15 or 7.16, and D_i represents the mean for the sizes defined by the upper and lower values of the interval Δ_i. Mean diameter D_{m} is then computed using Eq. (7.24).

The geometric mean diameter may be approximated from

$$D_{\mathrm{gm}} = \operatorname{antilog}\left[\sum_{i=1}^{m} \frac{\Delta_i \log D_i}{100}\right]. \tag{7.25}$$

The geometric mean diameter is also not typically the same as the median diameter, D_{50}. The geometric mean and median diameter will match only if the size distribution graphs as an S-curve on a semilog scale and is symmetric about 50%.

Several of the laboratory procedures for determining the particle size distribution have been discussed previously in the sections dealing with settling. Detailed information on the laboratory tests for aggregated and primary particles can be found in soil mechanics or soil physics references or are described in ASTM Standards.[3] It must be stressed that one must recognize the difference in laboratory procedures used to obtain primary versus aggregated particle size distributions in order to properly apply the results. A field manual developed by USDA (1979) presents a thorough explanation of many traditional methods for testing soils or sediment. Recently, additional instruments have been developed. The wide diversity in detection techniques for determining particle size was discussed by Burcham (1989). Techniques range from sieving (a method that is visible in early Egyptian drawings) to sophisticated electronic methods. Methods used to delineate particle size include sieving, incremental sedimentation, microscopy, centrifugal methods, photoelectric methods, turbidimetric methods, electrical sensing zone methods, radiation scattering, adsorption, laser light extinction, sound attenuation, and particle separators. No one method is completely adequate, typically because of either a limited sampling range or difficulties in obtaining a representative sample.

[3]ASTM Standards are available for a variety of tests and are available from the American Society for Testing Materials.

Primary particle size distributions are of concern when one is interested in identifying soils, probable particle shapes, chemical contituents attached to particles, or water chemistry. Aggregated particle size distributions are of interest primarily when dealing with sedimentation because the density and size will affect both the soil erodibility in the field and the settling velocity in a sediment basin. Also, almost all soils erode as a mixture of primary and aggregated particle sizes.

Several researchers have recently studied techniques for estimating the eroded size distribution based on laboratory or field tests of soil samples. Soil aggregates have been particularly difficult to characterize in terms of parameters that are useful for engineering design of erosion control measures and sediment control structures because of the complex nature of soil aggregates (Tollner and Hayes, 1986). One of the major problems associated with soil aggregates is that they are subject to destabilizing forces both in nature and in procedures that are used to characterize aggregates. Raindrop impacts, runoff, and channel flow each contribute erosive energy, which may result in changes to the aggregate characteristics as compared to the aggregate conditions at the site prior to rainfall (Rhoton *et al.*, 1982). The stability of an aggregate describes the response of the aggregate to four inherently active soil forces (Hartge, 1978):

1. weight of the particle,
2. forces transmitted by the solid phase,
3. forces transmitted by the liquid and/or gas phase within the aggregate, and
4. forces originated by contacting interfaces.

Stability is especially crucial in soils with substantial clay content because of the dynamic nature of clays. Among researchers and field personnel, there is considerable variation in the analytical procedures utilized to characterize eroded sediment aggregates. In addition, the results of a number of studies show that considerable differences in predicted sediment aggregate size distribution can be expected with classical and analytical procedures (Tollner and Hayes, 1986). The characterization of soil or sediment properties needs to provide absolute results if they are to be useful for engineers. Many present methods provide relative results that are useful for comparison only. Basic approaches for determining aggregate size have not changed much in years. Changes that have occurred are in technology for sensing concentration and size of aggregates in solution and improved technology such as microcomputers for analyzing the information. However, standard methods are available for characterizing the size and mineralogy of primary particles. Attempts have also been made to develop methods to predict eroded size information based on primary particle size information and/or field samples.

Size Distribution after Sedimentation

In the design of sediment control structures, it is necessary to know how size distribution changes as sediment moves to a watershed exit or after significant deposition has occurred. Hayes (1979) and Hayes *et al.* (1982) describe an approach that considers the impact of trapping. The method subtracts a portion of the particle size distribution curve based upon what has been trapped. Hayes assumed that as the sediment transport capacity decreases, the largest particles would settle out first. If sediment transport capacity continues to decrease, the size of the particles remaining in flow also decreases. A simple example of this concept, as illustrated in Fig. 7.1, is that if 20% of the original sediment concentration is trapped, the largest 20% of the sediment is assumed to be trapped. To obtain an estimate of the new particle size distribution for the remaining sediment, the diameter originally corresponding to D_{80} in Fig. 7.17 becomes D_{100} for the new distribution through a normalization process. Other points along the original size distribution curve are also adjusted based on trapping until a completely new particle size distribution is obtained as shown as the intermediate curve in Fig. 7.17. The process is repeated, as necessary, to obtain distributions after additional trapping. In Fig. 7.17, the 70% curve represents an estimate of the size distribution remaining if 70% of the original sediment is trapped. Limited data have been presented, but available data show acceptable agreement between measured values and values obtained using the procedures described (Hayes *et al.*, 1982). The assumptions that are made should be recognized. Sediment deposits often contain some fine-grained particles mixed with large particles, and large particles are occasionally found far downstream from where they would be expected if the assumption were completely valid. This is likely due to the stochastic nature of turbulence and its effects on bedload, as well as the exponential nature of deposition of suspended load. A completely adequate explanation of this phenomenon is currently not available.

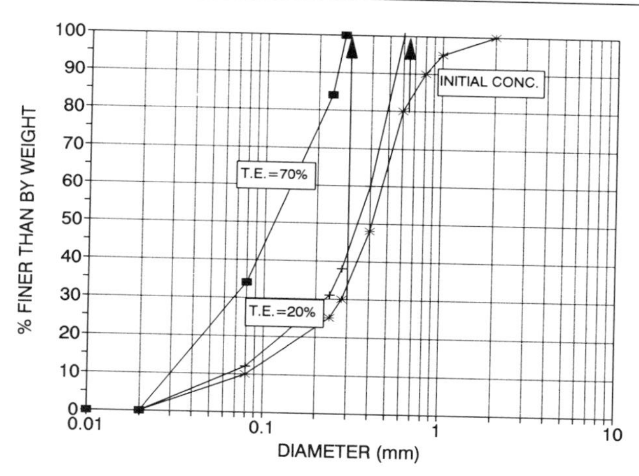

Figure 7.17 Adjusting particle size distribution for trapping.

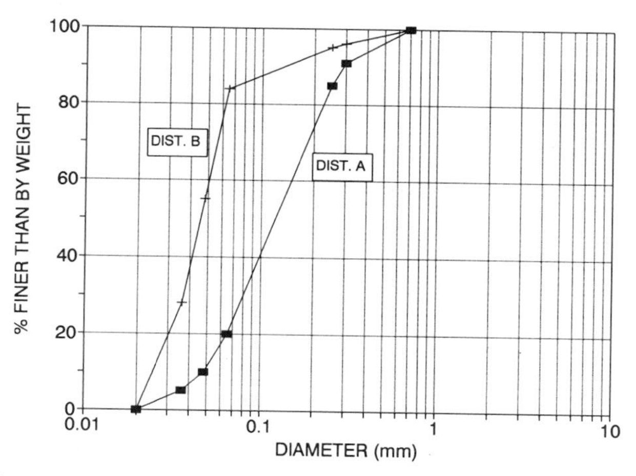

Figure 7.16 Example Problem 7.6 figure showing size distribution curves.

Barfield *et al.* (1979) also proposed a procedure based on the assumption that the largest particles in the flow settle out first when deposition occurs. Using this assumption, the size distribution of the material exiting a subwatershed contained within a larger watershed can be estimated if the eroded size distribution and the delivery ratio for the subwatershed is known. The size distribution of the percentage finer (PF) after partial deposition has occurred is given by

$$\mathrm{PF}_{2i} = \mathrm{PF}_{1i}/\mathrm{D}_{1i}; \qquad \mathrm{PF}_{2i} \leq 100, \tag{7.26}$$

where PF_{2i} and PF_{1i} are shown in Fig. 7.18, and D_{1i} is the delivery ratio for subwatershed i. An example method for calculating D_{1i} is the Modified Universal Soil Loss equation, or MUSLE (Williams, 1977), of the form

$$D_{1i} = \frac{95(Q_i \cdot q_{pi})^{0.56}}{R \cdot A_i}, \tag{7.27}$$

where Q_i is runoff volume in acre-feet, q_{pi} is peak discharge in cfs, A_i is the watershed area in acres, and R is the rainfall factor for the storm (see further discussion of the MUSLE in Chapter 8). The size distribution of material from subwatershed i that

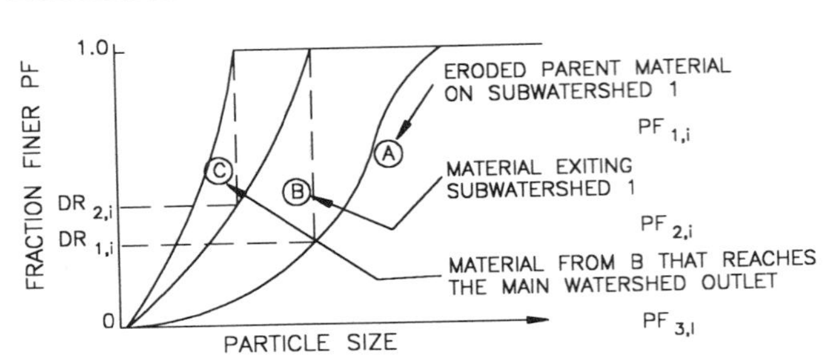

Figure 7.18 Changing particle size distribution along a flow path (after Barfield *et al.*, 1979).

reaches the main watershed exist can be estimated from

$$\mathrm{PF}_{3i} = \mathrm{PF}_{2i}/D_{2i}; \qquad \mathrm{PF}_{3i} \leq 100, \tag{7.28}$$

where D_{2i} is the fraction of material exiting subwatershed i that reaches the main watershed outlet. One method for calculating D_{2i} is the Williams routing equation [see Chapter 8, Eq. (8.106)]:

$$D_{2i} = \frac{Y_i}{Y_{0i}} = \frac{Y_{0i}e^{-BT_{ti}\sqrt{D_{50i}}}}{Y_{0i}} \tag{7.29}$$

or

$$D_{2i} = e^{-BT_{ti}\sqrt{D_{50i}}}, \tag{7.30}$$

where Y_i is the sediment from subbasin i that reaches the main watershed exit, and Y_{0i} is the yield at the subwatershed exit. A composite size distribution curve for an entire watershed can be determined by combining the curves from each subwatershed. For example, if the fraction of total yield from subwatershed i is

$$P_i = Y_i/\Sigma Y_i, \tag{7.31}$$

then the composite size distribution curve weighted according to the fraction of total yield P_i is

$$\mathrm{PF}_t = \sum_{i=1}^{n} [P_i][\mathrm{PF}_{2i}]. \tag{7.32}$$

Additional discussion of equations for estimating sediment yield from subwatersheds, main watersheds, and delivery ratios is contained in Chapter 8.

Example Problem 7.6. Percentage finer after partial deposition

Assume that distribution A shown in Fig. 7.16 represents the eroded size distribution for subwatershed i. Estimate the size distribution for material as it leaves the subwatershed if the delivery ratio for subwatershed if the delivery ratio for subwatershed i is 0.4.

Solution:

Diameter[a] (mm)	PF_{1i}[b] %	PF_{2i}[c] %
0.020	0	0
0.035	5	12.5
0.048	10	25
0.064	20	50
0.250	85	100
0.300	90	100
0.700	100	100

[a]Diameter selected from distribution A in Fig. 7.16.
[b]PF_{1i} read from points corresponding to diameter in column 1.
[c]PF_{21} calculated from Eq. (7.26) with D_{1i} equal 0.4.

Note that values of PF_{2i} must be less than 100. Partial deposition at additional downstream points could be calculated in a similar manner.

DEVELOPING SIZE DISTRIBUTION DATA

Introduction

Several radically different approaches have been considered for estimating eroded particle size distributions. One approach applies simulated rainfall onto a small sample of a representative soil. The size distribution of the particles in collected runoff from the sample is then analyzed in the laboratory to obtain an estimate of the eroded sizes. A second method seeks to simulate the impacts of erosion and transport using a series of laboratory steps involving wetting a soil sample and subjecting it to shaking. A third approach uses primary particle size data in empirical equations. Obviously, these methods cannot directly consider any deposition that occurs in the field. However, they can provide a first estimate of eroded size distribution until better methods are developed. The methods are described in the following sections.

Single-Nozzle Rainfall Simulator Method

A method that was first proposed in Barfield *et al.* (1979) utilizes a simple rainfall simulator technique to obtain a sample of the particle size distribution for eroded material. A sample of the soil of interest is

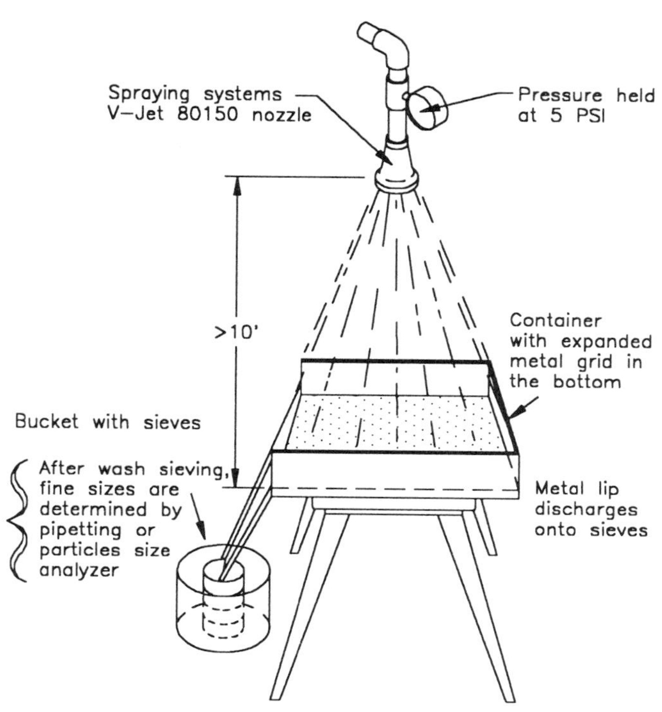

Figure 7.19 Schematic of soil erodibility test using single nozzle rainfall simulator.

placed in a small pan with an expanded metal base as shown in Fig. 7.19. Simulated rain is applied by a Spraying Systems Veejet 80150[4] nozzle located 3 m (10 ft) above the sample. As drops fall, larger drops gain speed and small drops slow so that each drop approaches its terminal velocity prior to reaching the sample. This nozzle appears appropriate based on rainfall simulator studies that have shown that it produces drop sizes and kinetic energy approximately equal to that of a 26- to 51-mm/hr rainfall (Meyer and Harmon, 1979). Similarly, a Spraying Systems Veejet 80100[4] nozzle produces drops and kinetic energy approximating those of a 2- to 13-mm/hr rainfall. The spray nozzle used is selected based on the intensity of the design storm. The sample pan is sloped so that the drainage flows onto sieves nested in an 18.9-liter (5-gal) bucket. Simulated rainfall is applied until the desired design storm's depth is reached in a rain gauge mounted adjacent to the pan. The sieves are gently washed three times to grade the sand through coarse silt fractions, fractions removed from the bucket, and dried at 105° C for 24 hr (or until constant weight is obtained). After the last sieve is removed, the volume of water in the bucket is measured; a concentration sample is obtained; and the size distribution of the fine silts and clays is determined by a pipette analysis (without dispersion). This modified wet sieve approach is simple to perform, requires very little equipment, and appears promising; but extensive data have not been presented comparing the wet sieve approach to other techniques. An apparatus for mechanizing the wet sieve process is described in Barfield *et al.* (1983).

[4]Manufactured by Spraying Systems, Inc. (Wheaton, IL).

Laboratory Method

Rhoton *et al.* (1982) proposed two methods that differ in the technique involved in wetting the soil sample. One method lets the sample soak in deaerated, distilled water. The second method allows the sample to wet by placing the soil sample on filter paper, which is placed on a saturated sponge in an enclosed tray containing distilled water. Samples equilibrate overnight and are then transferred to a flask. Distilled water is added. Each soil suspension is agitated on an orbital shaker for varying lengths of time. Immediately after agitation, size distributions are found using procedures identical to procedures used for field samples. This process includes wet sieving through a stack of five sieves with openings of 1000, 500, 250, 125, and 63 μ. Material less than 63 μ is transferred to cylinders and analyzed by pipetting after dispersing with hexametaphosphates. Discrepancies between field measured and laboratory measured percentage finers increased as sediment size decreased in Rhoton *et al.*'s results. Seventeen soils were tested, with 14 from delta and upland areas of Mississippi and 3 from Iowa. The wetting method had no significant effect on prediction of size distribution. Agitation times significantly impacted the prediction of size distribution. Best fits of the curves were obtained by varying the agitation times from 5 min for Memphis and Sharkey soils to 45 min for Loring. Most soils had a best fit in the 10 to 20 min range, and Rhoton *et al.* concluded that an agitation time of 14 min is probably appropriate for most soils. This time would generally predict the eroded size distribution within one standard deviation of the measured values from field tests.

CREAMS Equation Method

A third alternative in estimating the composition of eroded materials was proposed by Foster *et al.* (1985) and forms the basis of particle size estimation in the CREAMS and WEPP models. This technique defines five particle classes (primary clay, primary silt, small aggregate, large aggregate, and primary sand) based upon the primary particle sizes of the original soil matrix to describe the composition of the eroded mate-

Table 7.4a Representative Diameters by Classes Based on Soil Matrix Fractions [after Foster *et al.* (1985)]

Class	Representative diameter (mm)	Range limits of clay in soil matrix	Specific gravity
Clay	$D_{cl} = 0.002$		2.65
Silt	$D_{si} = 0.010$		2.65
Sand	$D_{sa} = 0.200$		2.65
Small Aggregate	$D_{sg} = 0.030$	$O_{cl} < 0.25$	1.80
	$D_{sg} = 0.2\,(O_{cl} - 0.25) + 0.030$	$0.25 \leq O_{cl} \leq 0.6$	
	$D_{sg} = 0.100$	$O_{cl} > 0.60$	
Large Aggregate	$D_{lg} = 0.30$	$O_{cl} \leq 0.15$	1.60
	$D_{lg} = 2\ \mathrm{Ocl}$	$O_{cl} > 0.15$	

Table 7.4b Fraction of Sediment by Class Based on Soil Matrix Particle Size Distributions [after Foster *et al.* (1985)]

Class	Fraction of sediment in class (mm)	Range limits of clay in soil matrix
Clay	$F_{cl} = 0.26\,O_{cl}$	
Small Aggregate	$F_{sg} = 1.8\,O_{cl}$	$O_{cl} < 0.25$
	$F_{sg} = 0.45 - 0.6\,(O_{cl} - 0.25)$	$0.25 \leq O_{cl} \leq 0.50$
	$F_{sg} = 0.6\,O_{cl}$	$O_{cl} > 0.50$
Silt	$F_{si} = O_{si} - F_{sg}$	
Sand	$F_{sa} = O_{sa}\,(1 - O_{cl})^5$	
Large Aggregate	$F_{lg} = 1 - F_{cl} - F_{si} - F_{sg} - F_{sa}$	

rial. Tables 7.4a and 7.4b summarize the equations for each classification by size range where D_{cl}, D_{si}, D_{sa}, D_{sg}, and D_{lg} in Table 7.4 are the representative diameters for clay, silt, sand, small aggregates, and large aggregates, respectively. The fractions of clay, silt, and sand in the soil matrix are denoted as O_{cl}, O_{si}, O_{sa}, respectively. Table 7.4b contains the equations for fraction of sediment by classes where F_{cl}, F_{si}, F_{sa}, F_{sg}, and F_{lg} are the fractions of sediment in clay, silt, sand, small aggregates, and large aggregates, respectively.

Example Problem 7.7. CREAMS equations used to calculate eroded size distribution

A soil has a dispersed soil matrix of 4% sand, 60% silt, and 36% clay. Estimate the fractions sand, silt, and clay using the revised CREAMS equations.

Solution: Using the equations for each class shown in Table 7.4b, the fraction of sediment by class is calculated based on soil matrix particle size distributions:

$$F_{cl} = 0.26\,O_{cl} = 0.26 \times 0.36 = 0.09$$

$$F_{sg} = 0.45 - 0.6(0.36 - 0.25) = 0.38$$

$$F_{si} = O_{si} - F_{sg} = 0.6 - 0.38 = 0.22$$

$$F_{sa} = O_{sa}(1 - O_{cl})^5 = 0.04(1 - 0.36)^5 = 0$$

$$F_{lg} = 1 - F_{cl} - F_{si} - F_{sg} - F_{sa}$$
$$= 1 - 0.09 - 0.22 - 0.38 - 0.0 = 0.31.$$

The fractions of sand-, silt-, and clay-sized particles are then calculated by recognizing that large aggregates correspond to sand particles and small aggregates with silt so that

$$\text{Fraction sand} = F_{lg} + F_{sa} = 0.31$$

$$\text{Fraction silt} = F_{sg} + F_{si} = 0.38 + 0.22 = 0.60$$

$$\text{Fraction clay} = 0.09.$$

The sum of the fractions totals 1.00 as it should. It should be noted that combining these primary particles and aggregates into sand-, silt-, and clay-sized fractions does not mean that they settle as primary sands, silts, and clays. The aggregate component has specific gravities in the range of 1.6 to 1.8 and would settle more slowly then primary particles in the same size range.

Foster *et al.*'s equations were evaluated using 28 different soils. Values of specific gravity for each of the five classes are also recommended. A specific gravity of 2.65 was suggested for the primary particle size classes (sand, silt, and clay), and values of 1.8 and 1.6 were suggested for small and large aggregates, respectively.

Little quantitative information that evaluates the accuracy of these methods is available. A limited comparison of the eroded size distributions from the Foster *et al.* (1985) equation method and the Barfield *et al.* (1979) rainfall simulator method measured values is contained in Holbrook *et al.* (1986) using data from three tillage treatments with three replications on a Piedmont soil. Rainfall for the year of interest was averaged and a 0.7-in. average depth was used for rainfall simulation with representative samples from each of the nine plots. Runoff from the rainfall simulation was wet sieved and pipetted as described previously. No dispersing agent was used. A portion of the surface soil sample was also subjected to a standard primary particle textural analysis for percentage sand, silt, and clay, which included dispersion. These data were used with Foster *et al.*'s revised CREAMS equations to account for aggregation in the eroded particle size distribution. In five of the nine tests, the equations produced values of D_{50} closer to those measured than the simulation method. The equation method showed little variation between tests, whereas the measured and rainfall simulation values covered a wide range. However, the rainfall simulator technique produced eroded size distributions that were closer to the measured values than the equation for sizes less than 0.1 mm in seven of the nine tests. Neither of the two methods considers the influence of cover or other surface stabilization measures. The procedures are, at

Table 7.5 Time of Pipette Withdrawal for Given Temperature, Depth of Withdrawal, and Diameter of Particles (after USDA, 1979)[a]

Diameter of particle (mm):	0.062	0.031	0.016	0.008	0.004	0.002
Depth of withdrawal (cm):	15	15	10	10	5	5
	Time of withdrawal					
		Min:sec				
Temperature (°C)	Sec					Hr:min
20	44	2:52	7:40	30:40	61:19	4:5
21	42	2:48	7:29	29:58	59:50	4:0
22	41	2:45	7:18	29:13	58:22	3:54
23	40	2:41	7:8	28:34	57:5	3:48
24	39	2:38	6:58	27:52	55:41	3:43
25	38	2:34	6:48	27:14	54:25	3:38
26	37	2:30	6:39	26:38	52:2	3:33
27	36	2:27	6:31	26:2	52:2	3:28
28	36	2:23	6:22	25:28	50:52	3:24
29	35	2:19	6:13	24:53	49:42	3:19
30	34	2:16	6:6	24:22	48:42	3:15

[a]Values in this table are based on particles of assumed spherical shape with an average specific gravity of 2.65, the constant of acceleration due to gravity = 980 cm/sec^2, and the viscosity varying from 0.010087 cm^2/sec at 20°C to 0.008004 cm^2/sec at 30°C.

best, a first estimate. Additional procedures are desirable and need to be developed so that sediment size distributions in runoff can be measured routinely.

Adjustment for Particle Density

In most situations, the specific gravity of sediment particles varies over a wide range. This is particularly evident in the case of mixtures that contain both aggregates and primary particles. Since the aggregates and primary particles have different specific gravities, procedures are needed to convert the data to an equivalent size distribution curve based on equivalent settling diameter. Foster *et al.* (1985) recognized this problem and suggested separating the particles into the five classes (sand, silt, clay, small aggregates, and large aggregates), which represent density, size, and composition differences. While this reduces the quantity of computations necessary to describe a particle size distribution, it is often desirable to compute equivalent diameters based on settling velocities and an assumed specific gravity. This can be accomplished quite easily by rearranging Stokes' law so that the equation is solved for the equivalent particle diameter, d, based upon an assumed specific gravity. A spherical shape and specific gravity equal to 2.65 is typically assumed. If a sample is pipetted without dispersing the sediment and samples are withdrawn according to times and depths calculated using Stokes' law, equivalent diameters are obtained that are based upon the specific gravity used to calculate the withdrawal times. Table 7.5 shows the withdrawal times and depths as a function of temperature for various equivalent diameters (USDA, 1979).

SEDIMENT TRANSPORT

Once a soil particle erodes, it becomes part of the flow and may be transported a few millimeters or hundreds of kilometers. The distance is dependent upon the sediment transport capacity of the flow. Factors controlling sediment transport capacity can be grouped into three categories: fluid properties, sediment characteristics, and hydraulic parameters associated with the flow path. Chang (1988) described how the motion of a particle is controlled by two opposing forces: the applied force and resisting force. The applied force results from hydrodynamics of flow, and the resisting force comes from the submerged weight of the particle. The particle will move if the applied force exceeds the resisting force, as discussed in Chapter 4. The threshold point for movement, or critical condition, occurs when the forces of flow are exactly balanced by the submerged weight of the particle. Variables that impact the point of incipient motion for uniform sediment on a horizontal bed are critical shear stress (τ_c), particle diameter (d), particle specific weight (γ_s), fluid specific weight (γ), fluid density (ρ), particle density (ρ_s), and kinematic viscosity (ν).

Shields (1936) recognized that these factors could be grouped into two dimensionless parameters so that

$$\frac{\tau_c}{(\gamma_s - \gamma)\, d} = fct\left(\frac{u_{*c}\, d}{\nu}\right), \qquad (7.33)$$

where u_{*c} is the critical shear velocity defined as

$$u_{*c} \equiv \left(\frac{\tau_c}{\rho}\right)^{1/2}, \qquad (7.34)$$

which is the critical friction velocity. The left side of Eq. (7.33) represents the critical Shields' stress, τ_{*c}, and the right side is the critical boundary Reynolds number, R_{e*}. Experimental data collected in flumes with a flat bed permitted graphical representation of the relationship. The original version of Shields' diagram does not apply adequately to sediment having low specific gravity and small diameter. Modifications to Shields' diagram have been suggested by numerous researchers according to Simons and Senturk (1977, 1992). Gessler (1971) rearranged dimensionless parameters and developed a graphical relationship that has the dependent variable in only one of the dimensionless terms. To overcome the difficulty with small particles, Mantz (1977) extended the range of Shields' diagram to smaller particles. This extension is shown along with the original curve in Fig. 7.20.

Example Problem 7.8 Use of Shields' diagram to determine critical shear stress

Determine the critical shear stress on the bottom of an unlined, trapezoidal channel carrying clear water. The channel has a mean flow velocity of 0.5 m/sec, and the bottom is composed of slightly rounded quartz having $d_{50} = 40$ mm. Assume that the water temperature is 68°F and the channel is wide and has a slope of 0.8%.

Solution: Base the critical shear stress criterion on Shields' diagram (Fig. 7.20) and assume that flow is in the rough zone where $R_{e*} \geq 400$. For this situation,

$$\frac{(\tau_o)_{cr}}{d(\gamma_s - \gamma)} \approx 0.06$$

Rearranging to solve for critical shear stress, the bottom can

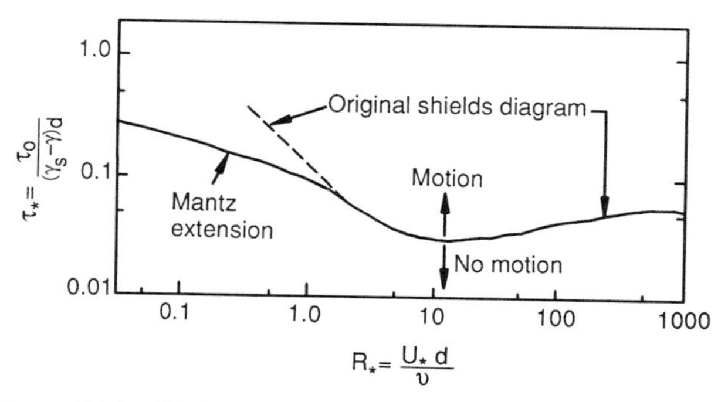

Figure 7.20 Shield's diagram as modified to include the Mantz data (after Storm *et al.*, 1990).

safely handle

$$(\tau_o)_{cr} = 0.06\, d(\gamma_s - \gamma)$$

$$(\tau_o)_{cr} = 0.06 \times 0.040\text{m} \times (2.65 - 1.0)(9.81)\frac{\text{kN}}{\text{m}^3}$$

$$= 0.039\frac{\text{kN}}{\text{m}^2}.$$

To determine if the assumption that $R_{e*} \geq 400$ is valid, the shear Reynolds number may be calculated as follows:

$$u_* = \left(\frac{\tau_c}{\rho}\right)^{1/2} = \left(\frac{0.039\text{kN/m}^2}{1000\text{kg/m}^3}\right)^{1/2}$$

$$= \left(\frac{0.039\text{kN} - \text{m}}{1000\text{ kg}} \times \frac{1000\text{ kg} - \text{m/sec}^2}{\text{kN}}\right)^{1/2}$$

$$= 0.198\text{ m/sec}.$$

(Note that 1 kN = 1000 kg − m/sec^2.)

$$R_{e*} = \frac{u_* d}{\nu} = \frac{0.198\text{ m/sec} \times 0.040\text{ m}}{1 \times 10^{-6}\text{m}^2\text{/sec}}$$

$$= 7.92 \times 10^3.$$

Since 7.92×10^3 is greater than 400, the assumption is all right.

Classification Based Transport

As a particle travels, it may be considered to be either bedload, suspended load, or washload. It should be noted that a particle's classification can change as it travels depending upon the fluid, sediment, and especially hydraulic conditions. Simons and Senturk (1977, 1992) presented extensive descriptions of each classification.

Bedload. Sediment that moves by saltation (jumping), rolling, or sliding in the flow layer just above the bed. This flow layer is generally considered to be only a few particle diameters thick.

Suspended load. Sediment that stays in suspension for some extended period of time as a result of suspension by turbulence.

Washload. Sediment that is composed of particles smaller than the channel bed material and originates from the channel bank and upslope areas.

Bedload

If the hydraulics in a channel are such that a small difference in one of the hydraulic parameters will cause the critical condition of the bed to be exceeded so that loose particles move, bedload transport occurs. Graf (1971) grouped numerous bedload equations into three classes that have slightly different approaches to bedload transport:

1. The duBoys-type equations that utilize a shear stress relationship.
2. The Schoklitsch-type equations that utilize a discharge relationship.
3. The Einstein-type equations that are grounded in statistical considerations of lift forces.

Simons and Senturk (1977, 1992) indicate that bedload amounts to about 5 to 25% of the suspended load. They emphasize that although this amount is a relatively small proportion, it controls the shape, stability, and hydraulic characteristics of the channel. Concepts presented by Einstein (1942, 1950) represent the most well-known approach to theoretically explain sediment transport. His work was summarized by Graf (1971), who recognized that Einstein's work had at least two significant differences from previous efforts.

1. The idea of defining a critical value at which sediment motion commences is demanding, and probably futile. Einstein bypassed this concept.
2. Bedload transport is related to fluctuations in velocity instead of the average velocity. The commencement and termination of particle movement is expressed statistically using probability theory to relate instantaneous hydrodynamic lift to the particle's weight forces. Experimental evidence that indicates a close association between the moving bedload and the bed was described (Graf, 1971).

Simons and Senturk (1977, 1992) also discussed some of the concepts that Einstein used to describe his theories dealing with sediment transport in open channels. Each particle moving past a cross section of a stream at a point of interest must have eroded somewhere in the watershed upstream of the cross section, and it must be transported by the flow from the point of detachment to the cross section at the point of interest. Each of these two requirements may control the sediment rate at the cross section, the availability of sediment in the watershed, and the transport capacity of the stream. Typically the finer material, which is easily carried in large amounts by flow, has limited availability in the water shed. This material is also generally considered as washload (see definition at the beginning of this section). The coarse material is much more difficult to move by flows, so its rate of movement is limited by the transport capacity of the flow. This material is designated as bed-material load and includes both bedload and suspended load. Although there is no sharp division between bed-material load and washload, washload is often considered to be silt and clay materials. This is not always appropriate because in some channels having extremely slow flows, the bed material consists of fine silts. Thus Simons and Senturk (1977, 1992) delineated a second criterion that used the D_{10} of the bed material as the dividing size between bed-material load and washload. Washload is often assumed to travel through a system by streamflow with very little deposition. Hence, very little of this small material is deposited.

Total sediment discharge (also called total sediment load) consists of both bed-material load and washload. Generally, total sediment load can only be estimated if the washload is estimated by measurement, experiment, or upland sediment yield equations because most sediment transport methods can only determine bed-material load.

Einstein concluded that there is a close relationship between bed-material load and bedload because there is a continuous exchange of particles between the bed material and bedload. In addition, particles are transported along the channel bed in a sequence of steps with the average distance of each step proportional to the particle size. It was also noted that a particular particle is not always moving, but rather, deposits on the bed after a few steps. Einstein found that the deposition rate per unit area depends on both the transport rate past a particular point and the probability that the forces on the particle are such that the particle will deposit.

Einstein's Method

Bedload. Einstein (1942) developed a relationship for sediment transport based on probability theory. Ignoring drag force effects on a particle, Einstein assumed that a particle would be dislodged from the channel bed if the lift force exceeded the submerged weight during an exchange time. The particle would move a distance of λd before striking the surface again. If the forces at the point of impact are such that lift exceeds submerged weight, the particle will take another hop of distance λd before striking the surface again. This continues until the particle comes to a point where the lift is less than the submerged weight. The actual distance a particle moves is thus random, but the average distance that a particle moves was determined by Einstein to be $A_L d$ or

$$A_L d = \frac{\lambda d}{1 - p}, \tag{7.35}$$

where p is the probability that the lift forces, L, are greater than the submerged weight, W, or

$$p = \text{Prob}(L > W). \tag{7.36}$$

Under the conditions where the number of particles of size d being eroded is equal to the number being deposited, Einstein showed that

$$\frac{p}{1 - p} = A\left(\frac{i_s}{i_b}\right)\phi, \tag{7.37}$$

where A is a constant, i_s, is the fraction of particles of size d moving in the bedload, i_b is the fraction of particles of size d in the bed material, and ϕ is given by

$$\phi = \frac{1}{F}\frac{g_s}{\gamma_s}\sqrt{\left(\frac{\gamma_s}{\gamma_s - \gamma}\right)\frac{1}{gd^3}} \tag{7.38}$$

where g_s is sediment discharge per unit width of size d, γ_s is specific weight of sediment, γ is specific weight of water, and F is a factor for settling velocity given by

$$F = \sqrt{\frac{2}{3} + \frac{36\nu^2}{gd^3(\text{SG} - 1)}} - \sqrt{\frac{36\nu^2}{gd^3(\text{SG} - 1)}}, \tag{7.39}$$

where ν is fluid viscosity and SG is sediment specific gravity. The probability that lift is greater than submerged weight is based on fluid dynamics. The lift force was assumed to be defined by the velocity at the top of a laminar sublayer formed as a result of drag due to grain roughness (see chapter 10), or

$$L = C_L \rho \frac{u_b^2}{2} A_{\text{lift}}$$
$$= C_L \frac{1}{2} \rho K_1 d^2 (11.6)^2 gR'S, \tag{7.40}$$

where S is slope and u_b is the velocity at the top of the bed layer, given by (see chapter 10)

$$u_b = 11.6U'_* = 11.6\sqrt{gR'S}, \tag{7.41}$$

where R' is the hydraulic radius due to skin friction, K_1 is a coefficient relating the cross-sectional area of a particle to the square of its diameter d and C_L is the coefficient of lift. Methods for determining R' are given in Chapter 10. Based on the assumption that the submerged weight is proportional to $(\rho_s - \rho)gd^3$, Einstein proposed that

$$p = f(\psi), \tag{7.42}$$

where ψ is referred to as the shear intensity parameter given by

$$\psi = \frac{\rho_s - \rho}{\rho} \frac{d}{R'S} \tag{7.43}$$

and B is a constant.

The combination of Eqs. (7.42) and (7.37) indicates that

$$\phi = f(\psi), \tag{7.44}$$

where the functional relationship must be defined experimentally, similarly to Fig. 7.24.

Based on data taken in a research flume, Einstein proposed that

$$0.465\phi = e^{-0.391\psi}. \tag{7.45}$$

In order to use Eq. (7.45), it is necessary to define R' values for a given water depth. Procedures for making these computations are given in Chapter 10.

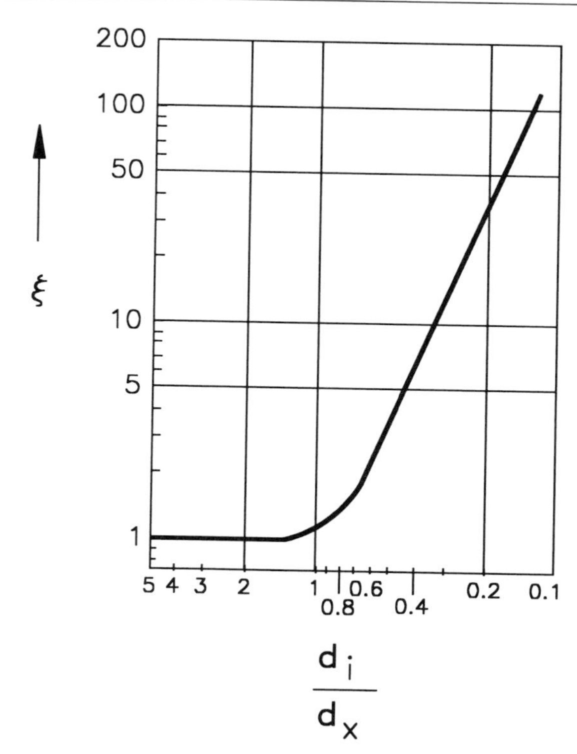

Figure 7.21 Hiding factor for use in Einstein's equation (after Einstein, 1950).

Example Problem 7.9 Bedload transport using Einstein's equation

A stream in South Carolina is estimated to have the following hydraulic data: slope of 1.5%, average depth of 1.0 m, and average velocity equal to 1.4 m/sec. The channel is wide and approximately rectangular with width of 7 m. A bed material sediment sample has been collected and analyzed, from which it was determined that d_{35} equals 13 mm. What is the rate of bedload transport?

Solution: Use Einstein's equation. To simplify computations, it will be assumed R' equals $R = A/P = 1.0$. This implies a wide channel and a smooth bed. From Eq. (7.44), the dimensionless ratios are related such that

$$\phi = f(\psi),$$

where ψ can be found by substituting into Eq. (7.43) as

$$\psi = \frac{\rho_s - \rho}{\rho} \frac{d}{R'S} = \frac{(\mathrm{SG} - 1)d}{R'S}$$

$$= \frac{(2.65 - 1)}{1} \frac{13 \times 10^{-3}\ \mathrm{m}}{(0.015\ \mathrm{m/m})(1.0\ \mathrm{m})} = 1.43.$$

From Eq. (7.45), $\phi = 1.23$, which can be substituted with other known information into Eq. (7.38) to solve for g_s after calculating F using Eq. (7.39).

To simplify calculations, first determine the common part of Eq. (7.39) as

$$\frac{36\,\nu^2}{gd^3(\mathrm{SG} - 1)} = \frac{36 \times (1 \times 10^{-6}\ \mathrm{m^2/sec})^2}{9.81\ \mathrm{m/msec^2} \times (0.013\ \mathrm{m})^3(2.65 - 1)}$$

$$= 1.01 \times 10^{-6},$$

which can be substituted under the radicals in Eq. (7.39) as

$$F = \sqrt{2/3 + (1.01 \times 10^{-6})} - \sqrt{1.01 \times 10^{-6}} = 0.817.$$

Substituting ϕ, F, and given values into Eq. (7.38),

$$1.23 = \frac{1}{0.817} \frac{g_s}{2650} \sqrt{\frac{1}{1.65} \frac{1}{9.8(13 \times 10^{-3})^3}} = 0.087 g_s.$$

Solving for g_s by rearranging yields

$$g_s = \frac{1.23}{0.087} = 14.1\ \mathrm{kg/m\text{-}sec}.$$

Finally, multiplying by the channel width provides an estimate of channel bedload transport as

$$G_s = Bg_s = 7\ \mathrm{m} \times 14.1\ \mathrm{kg/m\text{-}sec} = 99\ \mathrm{kg/sec}.$$

Note: The assumption that $R = R'$ was made to simplify computation for the example. This would not, in general, be true.

In a later modification of the procedures above, Einstein (1950) utilized a distribution of velocities at

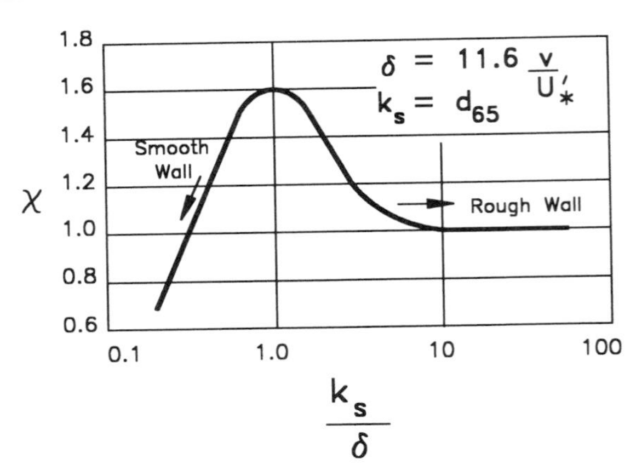

Figure 7.22 Correction factor in logarithmic velocity distribution (after Einstein, 1950).

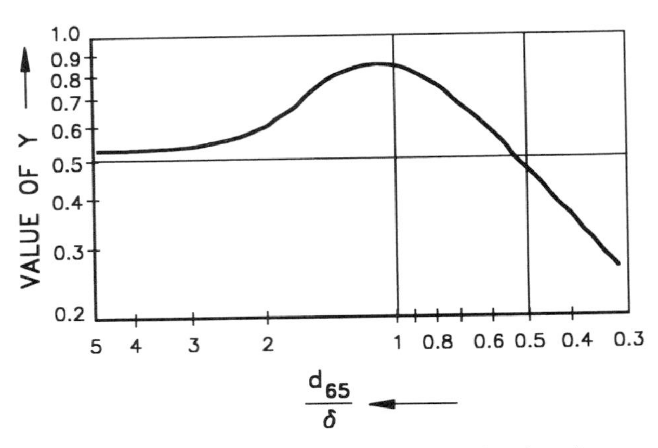

Figure 7.23 Y in Einstein's bedload function related to characteristic grain diameter (after Einstein, 1950).

the top of the laminar sublayer and proposed that

$$p = \text{Prob}\left[1 > \frac{K_2(\rho_s - \rho)gd^3}{C_L \frac{1}{2}\rho K_1 d^2 u_b^2 (1-\eta)}\right], \quad (7.46)$$

where K_2 is a coefficient relating d^3 to particle volume, and η is a parameter that defines the probability distribution of u_b, the velocity at a point $0.35X$ from the theoretical bed, where

$$\begin{aligned} X &= 0.77\Delta \qquad \text{if } \Delta/\delta > 1.80 \\ &= 1.39\delta \qquad \text{if } \Delta/\delta \leq 1.80, \end{aligned} \quad (7.47)$$

where δ is the thickness of the laminar sublayer given by

$$\delta = 11.6\nu/U'_* \quad (7.48a)$$

and Δ is a characteristic length given by

$$\Delta = \frac{k_s}{\chi} = \frac{d_{65}}{\chi}, \quad (7.48b)$$

where k_s is the characteristic particle diameter given by d_{65} and χ is Einstein's correction factor for the velocity profile, given in Fig. 7.22. Using the log velocity profile, Einstein showed that

$$u_b = 5.75^2 U^2 \log_{10}^2[10.6X/\Delta]. \quad (7.49)$$

Using Eq. (7.49) for u_b, Einstein modified Eq. (7.46) to obtain

$$p = \text{Prob}\left[1 > \left(\frac{1}{1-\eta}\right)\psi_*\right], \quad (7.50)$$

where ψ_* is given by

$$\psi_* = \frac{\xi Y(\beta/\beta_X)^2(\text{SG}-1)d}{R'S}, \quad (7.51)$$

where β and β_X are given by

$$\beta = \log_{10} 10.6 \quad (7.52a)$$

and

$$\beta_X = \log_{10}\left(\frac{10.6X}{\Delta}\right), \quad (7.52b)$$

ξ is a so-called hiding factor given by Fig. 7.21 as a function of d_i/d_x, and Y is the so-called pressure correction factor given by Fig. 7.23 as a function of d_{65}/δ.

Einstein showed that a new transport parameter could be defined as

$$\phi_* = \frac{i_s g_s}{i_b \rho_s g^{3/2} d^{3/2}\sqrt{\text{SG}-1}}. \quad (7.53)$$

Using this transport parameter and the assumption that η is normally distributed, Einstein showed that

$$p = 1 - \frac{1}{\pi^{1/2}}\int_{-B_*\psi_{*i}-1/\eta_o}^{B_*\psi_{*i}-1/\eta_o} e^{-r^2}\,dt = \frac{A_*\Phi_{*i}}{1 + A_*\Phi_{*i}}, \quad (7.54)$$

where A_*, B_*, and η_0 are constants. Based on experimental data, Einstein showed that the constants are given by 43.5, 0.143, and 0.5, respectively. Experimental data and a theoretical curve are shown in Fig. 7.24. Wilson and Barfield (1986) showed that the lower limit in Eq. (7.54) should be zero; however, the impact of the inappropriate lower limit on the constants was negligible. Procedures for calculating bedload movement with Einstein's equations are given in Example Problem 7.10.

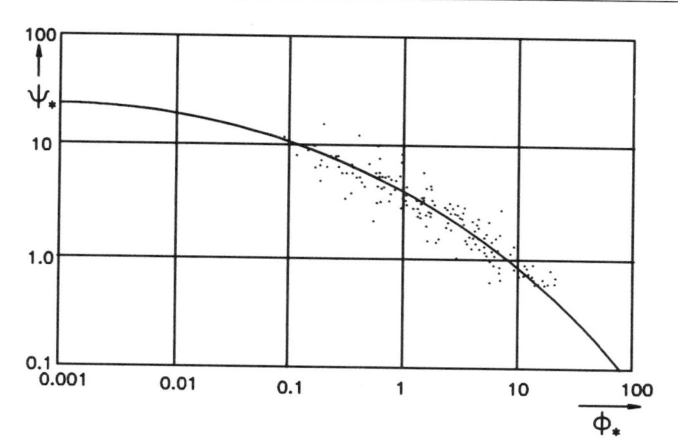

Figure 7.24 Einstein's bedload equation (after Einstein, 1950).

Suspended Load When critical conditions at the bed are exceeded so that particle motion occurs, a portion of the sediment is diffused into the main flow by the turbulence and does not move along the bed. The settling of particles tends to offset the upward diffusion, ultimately reaching a steady-state condition. Much more suspended load is transported in areas near the bed than occurs in regions near the surface. Figure 7.25 illustrates the general relationships between flow velocity and sediment concentration as a function of depth. USDA (1979) presents a figure showing discharge-weighted concentration of suspended sediment for the Missouri River at Kansas City. Figure 7.26 illustrates how concentration varies appreciably with water depth for coarse materials but is relatively uniform for fine sediments. Although the data in Fig. 7.26 are for a much larger channel than might be of general concern, the figure depicts one of the major problems in obtaining representative samples of sediment load from flowing water. A considerable lateral and vertical concentration gradient can occur in a given stream cross section. Graf (1971) also reports that the concentration of suspended load is higher near the center of a stream as compared to near the banks. The general shape of the resulting sediment concentration curve is given in Fig. 7.25. At equilibrium, the relationship between upward diffusion and downward settling is given by

$$V_s C = K \frac{\partial C}{\partial z}, \tag{7.55}$$

where V_s is the settling velocity for a particular particle diameter, K is the turbulent diffusivity for sediment, and C is sediment concentration. Based on an analogy between sediment diffusion and momentum diffusion, the momentum diffusivity can be used in Eq. (7.55),

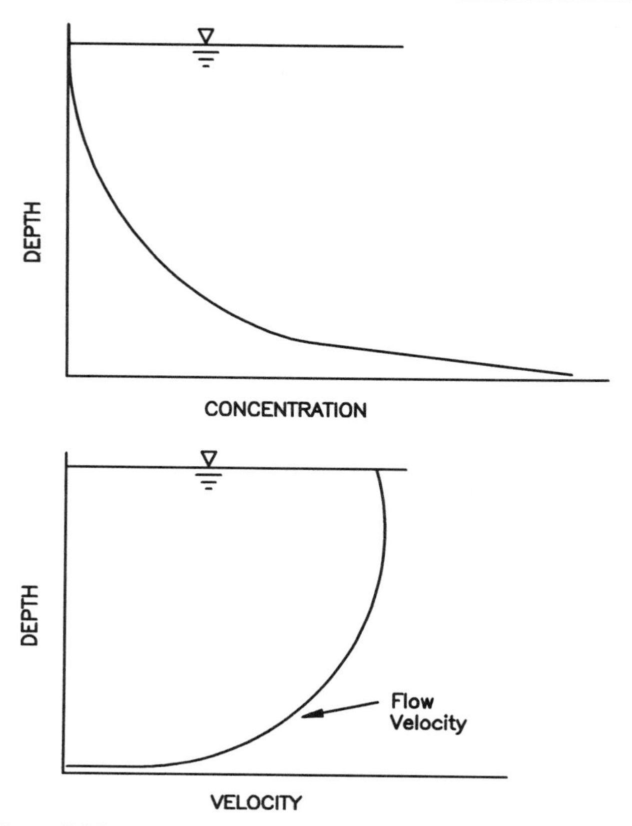

Figure 7.25 General shape of concentration and velocity profiles as a function of depth.

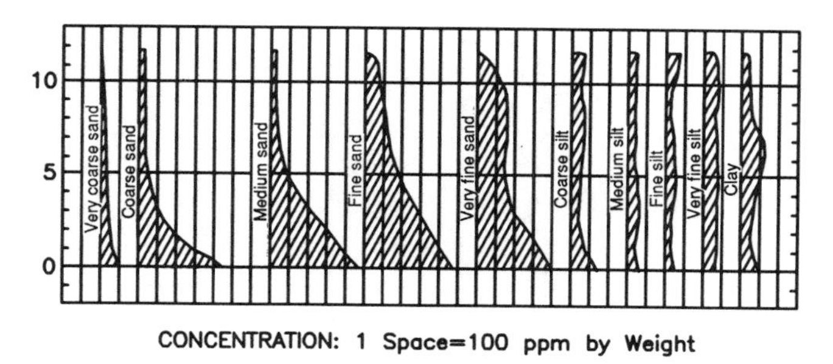

Figure 7.26 Example showing discharge weighted concentration of suspended sediment by particle size for Missouri River at Kansas City (USDA, 1979).

and the resulting concentration profile is given by

$$\frac{C}{C_a} = \left[\left(\frac{D - z}{z}\right)\left(\frac{a_E}{D - a_E}\right)\right]^{z'}, \tag{7.56}$$

where C is concentration at a depth z above the channel bed, C_a is a reference concentration measured at a distance a_E above the channel bed, D is the depth of flow, and

$$Z' = V_s / U'_* k, \tag{7.57}$$

where k is von Karman's constant (approximately 0.4 for clear water).

The total suspended load for a given particle size can be determined by the integral

$$G_{ss} = \int_{w=0}^{w=W} \int_{y=a_E}^{y=D} CU\,dy\,dx = C_m Q, \qquad (7.58)$$

where w is the stream width; D is the stream depth; C and U are concentration and velocity, respectively, at any point (x, y) above the stream bed; a_E is the reference point above the stream bed, usually a few times the mean size of the bed sediment; C_m is the mean suspended sediment concentration (discharge weighted); and Q is the stream discharge.

Using the logarithmic velocity profile of Einstein and Barbarossa (see chapter 10) for U and Eq. (7.58) for C, Eq. (7.58) can be written as

$$G_{ss} = 11.6 w C_a U'_* a_E [P_E I_1 + I_2], \qquad (7.59)$$

where

$$P_E = 2.303 \log_{10}\left(\frac{30.2 D}{\Delta}\right) \qquad (7.60)$$

$$I_1 = 0.216 \frac{A_E^{Z'-1}}{(1 - A_E)^{Z'-1}} \int_{A_E}^{1} \left[\frac{1 - z}{z}\right]^{Z'-1} dz \qquad (7.61)$$

$$I_2 = 0.216 \frac{A_E^{Z'-1}}{(1 - A_E)^{Z'-1}} \int_{A_E}^{1} \left[\frac{1 - z}{z}\right]^{Z'-1} \ln z\,dz, \qquad (7.62)$$

where $A_E = a_E/D$. The integrals I_1 and I_2 have been evaluated numerically and are presented in several texts such as Simons and Senturk (1977, 1992) and Graf (1971). Values are shown in Fig. 7.27.

The reference concentration C_a was assumed by Einstein to be the average concentration in the bed layer, a layer that he defined to be equal in thickness to two particle diameters. Assuming that the bedload is not uniformly distributed over this layer, the average concentration over this layer is divided by 11.6 to get the reference concentration, or

$$C_a = \frac{1}{11.6} \frac{i_s g_s}{U'_* a_E} \qquad (7.63)$$

Thus the suspended load can be predicted in terms of

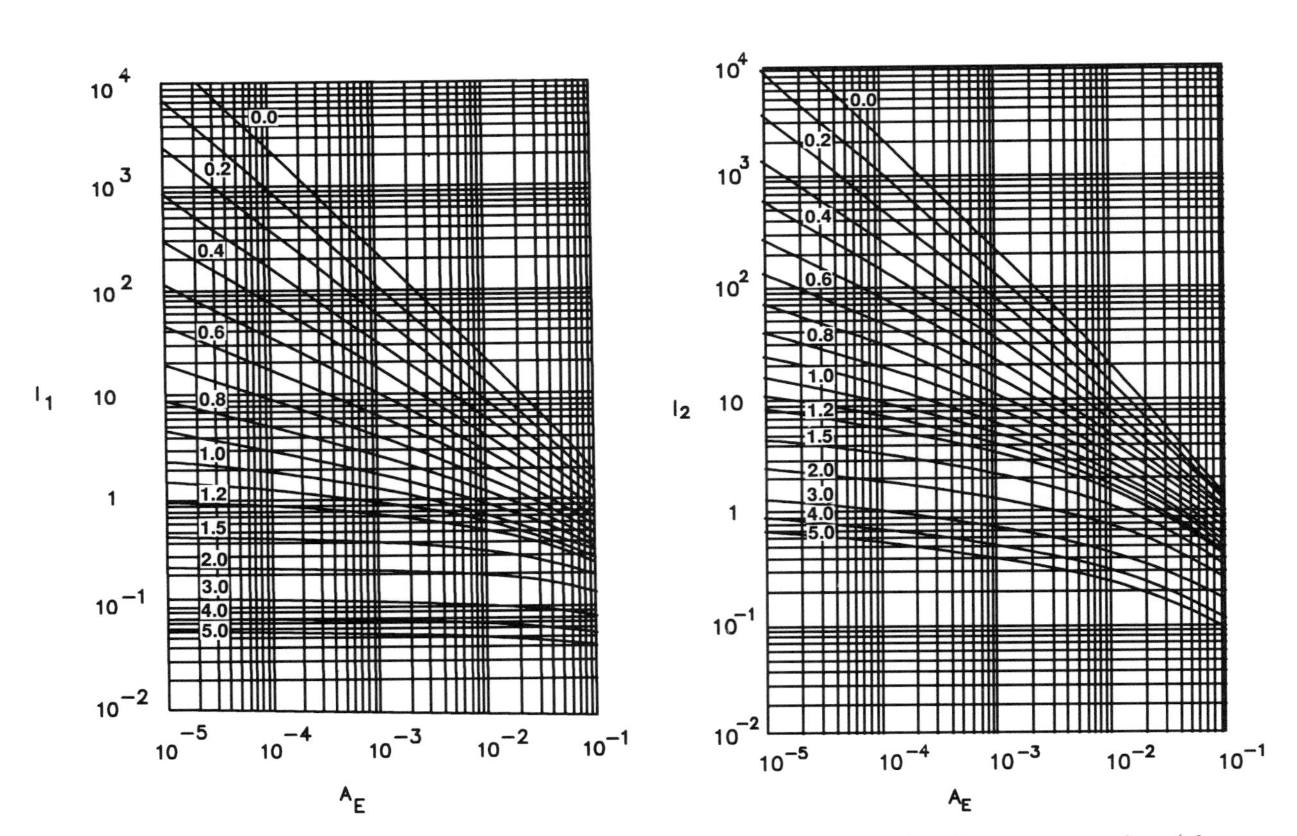

Figure 7.27 Values for the integrals I_1 and I_2 in Einstein's suspended load equation. Curve parameter is z (after Einstein, 1950).

the bedload as

$$i_{ss}G_{ss} = wi_s g_s[P_E I_1 + I_2], \quad (7.64)$$

where i_{ss} is the fraction of the particles of size d moving as suspended load.

The Total Bed-Material Load The total bed-material load, G_{st}, is the sum of suspended load and bed load corresponding to sizes existing in the channel bed. For a given particle size fraction i_{st}, this is given by

$$i_{st}G_{st} = w(i_s g_s + i_{ss} g_{ss}) \quad (7.65)$$

or

$$i_{st}G_{st} = wi_s g_s[P_E I_1 + I_2 + 1]. \quad (7.66)$$

Procedures for making the calculations required are given in Example Problem 7.10.

Example Problem 7.10 Application of Einstein's method to Big Sand Creek [adapted from Einstein (1950) and Graf (1971)]

A test reach in Big Sand Creek that can be represented by a trapezoidal cross section with 1 : 1 side slopes and a bottom width of 300 ft was selected. The channel slope was measured as 0.0007 ft/ft. A grain size distribution was obtained by averaging five samples taken to a depth of 2 ft. The sizes can be represented by four size fractions as shown in column 1 of Table 7.7, for which sediment transport is determined. These four size fractions represent 95.8% of the bed material. A flow–discharge versus sediment–discharge relation was found for the test reach. The stage–discharge relationship is shown in columns 14 and 17 of Table 7.6. (Procedures for making these calculations are illustrated in Chapter 10.) Einstein estimated that the highest discharge of interest was 20,000 cfs. Water viscosity was given as 1.0×10^{-10} ft^2/sec, and the specific gravity of the sediment as 2.65. Hydraulic radius, wetted perimeter, and cross-sectional area as a function of depth were determined on the basis of channel geometry.

Solution: Transport of each grain fraction is calculated at each given flow depth. Einstein divided the computations into hydraulic computations (Table 7.6) and bed material load calculations (Table 7.7). Explanation and references to figures and equations are included.

Einstein's bedload function was modified by Colby and Hubbell (1961) to compute total load. Difficulties with Einstein's bedload method were the complexity and detail required to solve the equations for a practical problem (Graf, 1971). Graf indicated even more difficulty with Colby and Hubbell's modified version for estimating total load because it requires bed samples, depth-integrated samples of total suspended matter, and stream flow measurements. Because of these difficulties, work has continued to find simpler, more accurate methods for defining total load.

Yalin Total Load Equation

One of the most commonly applied transport equations for small channels is the Yalin bedload equation. Yalin (1963) developed his bedload transport model for uniform, cohesionless grains over a movable bed. The model was derived using dimensional analysis and the average grain motion for uniform turbulent flow with a laminar sublayer that does not exceed the bed roughness. Yalin (1963) first presented the model as several equations that were later reduced by Alonso *et al.* (1981) to obtain an equation for sediment concentration as

$$C = 6.35 \times 10^5 \frac{\text{SG}\, dU_*}{vh} s\left[1 - \frac{1}{as}\ln(1 + as)\right], \quad (7.67)$$

where SG is sediment specific gravity, U_* is bed shear velocity in meters per second, v is average velocity in meters per second, and h is flow depth in meters. The other two terms were defined by Yalin as

$$a = 2.45 Y_{cr}^{1/2}\, \text{SG}^{-0.4}, \quad (7.68)$$

where Y_{cr} is the critical mobility factor found from Shields' diagram and

$$s = \frac{Y}{Y_{cr}} - 1, \quad (7.69)$$

where Y is a mobility number defined by

$$Y = \frac{\rho_w U_*^2}{\gamma_s\, d}. \quad (7.70)$$

To find the critical mobility factor, Shield's diagram (Fig. 7.20) is utilized to determine the value for τ_c corresponding to a given roughness Reynold's number, $U_* d/\nu$. This value for τ_c is then used to determine a critical shear velocity of $U_{*c} = (\tau_c/\rho)^{0.5}$, which is then used in Eq. (7.70) to determine Y_{cr}. Foster (1982) adapted the method to predict transport of individual particle size classes to obtain a particle size distribution.

Yang's Unit Stream Power Equation

Another commonly used sediment transport relationship for small channels is that of Yang (1972,

Table 7.6 Hydraulic Calculations for Example Problem 7.10 [Adapted from Einstein (1950) and Graf (1971)]

R'^{a}	U'^{b}	δ^{c}	k_s/δ^{d}	χ^{e}	Δ^{f}	$\bar{u}^{g}$	$\Psi*^{h}$	$\bar{u}/\bar{u}''_*{}^{i}$	$u''_*{}^{j}$	R''^{k}	R^{l}	$u_*{}^{m}$	D^{n}	A^{o}	P^{p}	Q^{q}	X^{r}	Y^{s}	$\beta_x{}^{t}$	$(\beta/\beta_x{}^{t})$	$P_E{}^{v}$
0.5	0.106	0.00110	1.05	1.61	0.00071	2.40	4.47	12.3	0.195	1.70	2.20	0.221	2.25	680	306.3	1,633	0.00153	0.84	1.36	0.57	11.49
1.0	0.150	0.00077	1.50	1.53	0.00075	3.64	2.24	18.6	0.196	1.71	2.71	0.248	2.77	845	307.8	3,075	0.00107	0.76	1.18	0.75	11.62
2.0	0.212	0.00055	2.10	1.35	0.00085	5.45	1.12	35.0	0.156	1.08	3.08	0.262	3.15	960	308.9	5,235	0.00076	0.60	0.98	1.10	11.65
3.0	0.260	0.00045	2.56	1.25	0.00092	6.90	0.75	56.5	0.122	0.66	3.66	0.286	3.75	1,140	310.5	7,860	0.00071	0.56	0.95	1.17	11.73
4.0	0.300	0.00039	2.95	1.19	0.00097	8.13	0.56	82.0	0.099	0.44	4.44	0.315	4.57	1,400	312.9	11,380	0.00075	0.54	0.91	1.27	11.90
5.0	0.336	0.00035	3.29	1.15	0.00100	9.24	0.45	115.0	0.080	0.28	5.28	0.345	5.45	1,670	315.4	15,420	0.00077	0.54	0.91	1.27	121.03
6.0	0.368	0.00032	3.60	1.12	0.00103	10.28	0.38	145.0	0.071	0.22	6.22	0.375	6.45	1,980	318.2	20,350	0.00079	0.54	0.91	1.27	12.18

[a] R', feet (hydraulic radius with respect to the grains); various values are assumed to cover the entire discharge range desired ($Q_{max} = 20{,}000$ cfs). Procedures for calculating found in Chapter 10.
[b] $U' = \sqrt{gR'S}$, fps (shear velocity with respect to grain) as shown Eq. (7.41).
[c] $\delta =$, feet (thickness of laminar sublayer), Eq. (7.48a).
[d] $k_s = d_{65}$ (roughness diameter).
[e] χ (correction in the logarithmic velocity formula, distinguishing the smooth, transition, and rough regimes; $\chi = fct(k_s/\delta)$, given with Fig. 7.22.
[f] Δ, feet (apparent roughness diameter); given in Eq. (7.48b) (7.62); $\Delta = k_s/\chi$.
[g] $\bar{u}$, fps (average velocity); $\bar{u} = (u'_*)\ 5.75 \log (12.27\ R'/\Delta)$.
[h] $\Psi*$ (intensity of shear on representative particle); $\Psi = ((\rho_s - \rho)/\rho)\ (d_{35}/R'S)$; similar to Eq. (7.43)
[i] $\bar{u}/\bar{u}''_* = fct(\Psi')$.
[j] u''_*, fps (shear velocity due to channel irregularities).
[k] R'', feet (hydraulic radius with respect to channel irregularities); from $u''_* = \sqrt{gR''S}$.
[l] R, feet (hydraulic radius); $R = R' + R'$, with no additional friction from the banks vegetation etc.; R represents the total hydraulic radius. For a detailed analysis with additional friction, see Einstein (1950).
[m] u_*, fps (shear velocity); from $u_* = \sqrt{gR_nS}$.
[n] D, feet (depth or stage); for wide channel $R \approx D$.
[o] A, square feet (cross-sectional area); obtained from description of the cross section.
[p] P, feet (wetted perimeter); obtained from description of the cross section.
[q] Q, cfs (flow discharge); $Q = \bar{u}A$; a stage–discharge relationship
[r] X, feet (characteristic distance); $X = 0.77\Delta$ for $\Delta/\delta > 1.80$ and $X = 1.39\delta$ for $\Delta/d \approx 1.80$.
[s] Y (pressure correction term); $Y = fct(k_s/\delta)$; given with Fig. 7.23.
[t] β_x (logarithmic function); $\beta_x = \log (10.6X/\Delta)$ from Eq. (7.52b).
[u] $(\beta/\beta_x)^2$; with $\beta = \log 10.6$ from Eq. (7.52a).
[v] P_E (Einstein's transport parameter); $P_E = 2.3.03\ log\ [30.2\ D/\Delta]$; further explanation in Chapter 10.

Table 7.7 Bed Material Load Calculations for Example Problem 7.10 [Adapted from Einstein (1950) and Graf (1971)]

$10^3 d$[a]	$10^2 i_b$[b]	R'[c]	Ψ[d]	d/X[e]	ξ[f]	Ψ_*[g]	Φ_*[h]	$i_s g_s$[i]	$i_s G_s$[j]	$\Sigma i_s G_s$[k]	$10^3 A_E$[l]	z[m]	I_1[n]	$-I_2$[o]	$P_E I_1 + I_2 + 1$[p]	$i_{st} g_{st}$[q]	$i_{st} G_{st}$[r]	$\Sigma i_{st} G_{st}$[s]
1.62	1.78	0.5	7.64	1.06	1.12	4.10	1.00	0.014	4.29	4.29	1.47	4.84	0.055	0.36	1.27	0.018	5.44	5.44
		1.0	3.86	1.51	1.00	2.20	2.90	0.041	12.45	12.45	1.19	3.42	0.094	0.56	1.53	0.062	19.48	19.48
		2.0	1.93	2.13	1.00	1.27	5.80	0.081	25.10	25.10	1.05	2.42	0.150	0.94	1.80	0.146	45.18	45.18
		3.0	1.29	2.28	1.00	0.85	9.10	0.127	39.50	39.50	0.89	1.97	0.215	1.35	2.17	0.276	99.53	99.53
		4.0	0.97	2.16	1.00	0.67	11.70	0.164	51.20	51.20	0.73	1.71	0.300	1.80	2.77	0.454	141.82	141.82
		5.0	0.78	2.10	1.00	0.54	14.70	0.206	65.00	65.00	0.61	1.53	0.380	2.20	3.14	0.647	204.10	204.10
		6.0	0.66	2.05	1.00	0.46	17.30	0.242	77.00	77.00	0.52	1.39	0.510	2.95	4.03	0.975	310.31	310.31
1.15	40.2	0.5	5.49	0.75	1.50	3.93	1.09	0.021	6.34	10.63	1.04	3.49	0.09	0.56	1.47	0.030	9.31	14.75
		1.0	2.74	1.08	1.11	1.73	3.90	0.074	22.84	35.28	0.85	2.47	0.15	0.96	1.78	0.132	40.66	60.14
		2.0	1.37	1.51	1.00	0.90	8.50	0.162	49.93	75.03	0.75	1.75	0.29	1.70	2.66	0.430	132.81	177.99
		3.0	0.92	1.62	1.00	0.60	13.20	0.255	79.18	11.18	0.63	1.43	0.46	2.60	3.80	0.969	300.88	400.41
		4.0	0.69	1.53	1.00	0.47	17.00	0.324	101.38	152.58	0.52	1.23	0.68	3.70	5.40	1.750	547.45	689.27
		5.0	0.55	1.50	1.00	0.38	21.00	0.400	126.16	191.16	0.44	1.10	1.05	5.00	8.62	3.448	1,087.50	1,291.60
		6.0	0.48	1.46	1.01	0.33	24.10	0.458	145.74	222.74	0.37	1.01	1.45	6.50	12.00	5.496	1,748.89	2,059.20
0.81	32.0	0.5	3.86	0.53	3.20	5.91	0.44	0.004	1.19	11.82	0.74	2.50	0.15	0.96	1.71	0.007	2.03	16.78
		1.0	1.93	0.76	1.50	1.65	4.15	0.037	11.38	46.67	0.59	1.77	0.28	1.70	1.55	0.057	17.64	77.78
		2.0	0.97	1.07	1.12	0.72	10.80	0.096	29.71	104.74	0.53	1.25	0.68	3.50	5.40	0.518	160.43	338.42
		3.0	0.65	1.14	1.08	0.46	17.30	0.154	47.82	166.50	0.44	1.02	1.35	6.20	10.60	1.632	506.89	907.30
		4.0	0.48	1.08	1.01	0.34	23.50	0.209	65.39	217.97	0.36	0.88	2.40	9.20	20.40	4.264	1,333.96	2,023.23
		5.0	0.39	1.05	1.13	0.30	26.80	0.239	75.38	266.54	0.31	0.79	3.80	13.20	32.30	7.719	2,434.77	3,726.37
		6.0	0.33	1.03	1.14	0.26	31.00	0.276	87.82	310.56	0.25	0.72	5.80	18.00	53.00	14.628	4,654.46	6,713.66
0.57	5.8	0.5	2.72	0.37	8.20	10.70	0.07	0.000	0.00	11.82	0.52	1.58	0.36	2.18	1.96	0.000	0.00	16.78
		1.0	1.36	0.53	3.20	2.48	2.45	0.002	0.74	47.41	0.42	1.12	1.00	4.80	7.82	0.016	5.79	83.57
		2.0	0.68	0.75	1.52	0.68	11.50	0.011	3.38	108.12	0.37	0.79	3.60	12.30	40.70	0.448	137.50	475.92
		3.0	0.46	0.80	1.40	0.42	19.00	0.018	5.62	172.12	0.31	0.64	8.40	23.00	76.50	1.377	429.93	1,337.23
		4.0	0.34	0.76	1.50	0.35	23.00	0.022	6.15	224.12	0.26	0.56	14.50	35.00	115.00	2.530	707.25	2,730.48
		5.0	0.27	0.74	1.55	0.29	28.00	0.027	8.45	274.99	0.21	0.50	23.40	51.00	178.40	4.817	1,507.48	5,233.85
		6.0	0.23	0.72	1.60	0.25	32.00	0.031	9.80	320.36	0.18	0.46	33.00	70.00	329.00	10.199	3,224.20	9,937.86

[a] d, feet (grain size); the representative ones are taken from a size distribution analysis (given in this problem).
[b] i_b (fraction of bed material); taken from a size distribution analysis (given in this problem).
[c] R', feet (hydraulic radius with respect to grains); taken from Table 7.6.
[d] Ψ (intensity of shear on particle); $\Psi = ((\rho_s - \rho)/\rho)\,(d/R'S)$; given in Eq. (7.43).
[e] d/X; for values of X see Table 7.6.
[f] ξ (hiding factor); $\xi = fct\,(d_i/d_x)$, given in Fig. 7.21.
[g] Ψ_* (intensity of shear on individual grain size); $\Psi_* = \xi Y(\beta^2/\beta_x^2)\Psi$. Values are given in Table 7.6.
[h] Φ_* (intensity of transport for individual grain size); $\Phi_* = fct\,(\Psi_*)$ given with Fig. 7.24.
[i] $i_s g_s$, pounds per second (ft) (bedload rate in weight per unit width and time for a size fraction); $i_s g_s = i_b \Phi_* \rho_s g^{3/2} d^{3/2}\sqrt{(SG-1)}$; from Eq. (7.53).
[j] $i_s G_s$, pounds per second (bedload rate in weight per unit time for a size fraction for entire cross section); $i_s G_s = (i_s g_s)W$.
[k] $\Sigma i_s G_s$, pound per second (bedload rate in weight per unit for all size fraction for entire cross section).
[l] $A_E = a/R$ (ratio of bed layer to hydraulic radius).
[m] z (exponent for suspension distribution); $V_g/0.4U'_*$; where V_g is particular settling velocity and U'_* is given in Table 7.6.
[n] I_1 (integral) read from Fig. 7.27; $I_1 = f(A_E, z)$; given by Eq. (7.61).
[o] I_2 (integral) read from Fig. 7.27; $I_2 = f(A_E, z)$; given by Eq. (7.62).
[p] $(P_E I_1 + I_2 + 1)$.
[q] $i_{st} g_{st}$, pounds per second (ft) (bed material load in weight per unit width and time, for a size fraction); $i_{st} g_{st} = i_s g_s (P_E I_1 + I_2 + 1)$.
[r] $i_{st} G_{st}$, pounds per second (bed material rate in weight per unit time for a size fraction for entire cross section); given by Eq. (7.66).
[s] $\Sigma i_{st} G_{st}$, pounds per second (bed material rate in weight per unit time for all size fractions for entire cross section); given by summing values.

1973), a sediment transport equation based on unit stream power. Using regression analysis, he concluded that the product of average flow velocity, V, and the slope of the energy grade line, S, is the primary indicator of flow conditions. The equation described in Yang (1972) relates total sediment concentration to unit stream power so that

$$\log_{10} C_t = \alpha + \beta \log_{10}(VS - V_{cr}S), \qquad (7.71)$$

where C_t is total sediment concentration in ppm, VS is unit stream power in meters per second, $V_{cr}S$ is the critical unit stream power required for incipient motion, and α and β are parameters.

The equation was later nondimensionalized but was only applicable to one sediment size. This limitation presents difficulties in establishing a representative particle size (Yang, 1973). In nondimensional form, Eq. (7.70) becomes

$$\log_{10} C_t = 5.435 - 0.286 \log_{10}\left(\frac{V_s d}{\nu}\right) - 0.457 \log_{10}\left(\frac{U_*}{V_s}\right) + \left[1.799 - 0.409 \log_{10}\left(\frac{V_s d}{\nu}\right) - 0.314 \log_{10}\left(\frac{U_*}{V_s}\right)\right] \times \log_{10}\left(\frac{VS}{V_s} - \frac{V_{cr}S}{V_s}\right), \qquad (7.72)$$

where V_s is particle settling velocity in meters per second, d is particle diameter in meters, ν is kinematic viscosity in square meters per second, and U_* is shear velocity in meters per second. Yang developed the equation using flume and stream data with noncohesive sand.

Yang (1984) later developed a similar equation for gravel transport in which Eq. (7.71) became

$$\log_{10} C_t = 6.681 - 0.633 \log_{10}\left(\frac{V_s d}{\nu}\right) - 4.816 \log_{10}\left(\frac{U_*}{V_s}\right) + \left[2.784 - 0.305 \log_{10}\left(\frac{V_s d}{\nu}\right) - 0.282 \log_{10}\left(\frac{U_*}{V_s}\right)\right] \times \log_{10}\left(\frac{VS}{V_s} - \frac{V_{cr}S}{V_s}\right). \qquad (7.73)$$

Hirschi (1985) provided an approach that considered nonuniform sizes.

Example Problem 7.11 Illustration of Einstein, Yalin, and Yang sediment transport equations

A channel has a channel width of 20 ft, slope of 0.1%, and overtops at a depth of 3 ft. The material in the bottom of the channel has a size distribution as shown below.

Size (mm)	PF
2.0	100
1.0	95
0.5	90
0.2	75
0.1	60
0.05	40
0.03	20
0.01	0

For the channel and size distribution given, determine the sediment discharge utilizing the Einstein, Yalin, and Yang equations.

Solution: The size distribution must be divided into size fractions. By first plotting the given distribution on graph paper, it is possible to divide the distribution into five parts, each containing 20% of the distribution. If this is done, a mean diameter and settling velocity for each fraction can be determined with a corresponding settling velocity. This information follows.

Size range (mm)	Mean diameter (mm)	Fraction (%)	V_s (ft/sec)	Mean diameter (ft)
2.0–0.26	0.7211	20	1.461	0.00236
0.26–0.1	0.1612	20	0.073	0.000528
0.1–0.05	0.0707	20	0.014	0.000231
0.05–0.03	0.0387	20	0.0042	0.000127
0.03–0.01	0.0173	20	0.0008	5.7×10^5

These size fractions and representative diameters are used in each of the methods. The Einstein method calculations follow the same procedure previously discussed in Example Problem 7.10 and shown in Table 7.7. Table 7.8 summarizes the computations used for Einstein's method.

Solutions of the problem using Yalin's and Yang's equations are illustrated in Tables 7.9 and 7.10, respectively. Detailed descriptions of the steps in the procedure are not shown. Further discussion of partitioning of the hydraulic radius and other aspects are explained and demonstrated in Chapter 10. As can be seen from comparison of Tables 7.8, 7.9, and 7.10, computation of sediment discharge for a practical problem requires solution of a series of intricate equations. Solution of the equations using computer modeling is practically a necessity because of the likelihood of calculation errors. Einstein (1950), Yalin (1963), Yang (1972), Graf (1971), and Simons and Senturk (1977, 1992) provide additional information.

Channel Armoring

All particles on a channel bed do not have the same susceptibility for lift. As a result, equations developed for critical shear stresses and velocity equations predict that smaller particles are more readily eroded than are larger ones. This results in the surface layer of the channel bed consisting of a larger fraction of coarse particles than did the original bed. This results in what is referred to as channel armoring (or bed armoring). Channel armoring refers to the protection provided by this coarse top layer to the finer particles that lie underneath. Graf (1971) presents several photographs that vividly illustrate the protection provided by armoring. Chang (1988) discusses approaches for modeling channel armoring.

Problems

(7.1) Compare the settling velocities using the Stokes equation and the Wilson *et al.* equation for 0.0004-, 0.004-, 0.01-, 0.02-, 0.04-, 0.10-, 0.2-, 0.4-, and 4.0-mm particles if SG = 2.65 for all particles and settling occurs in water at 68°F.

(7.2) Repeat problem (7.1) if the SG = 1.8. Under what conditions would you expect the SG to approach this value?

(7.3) Repeat problem (7.1) if the water is at 32 and 100°F. Are the results significantly different from those obtained in problem (7.1)?

(7.4) If the peak effluent TSS from a pond is 600 mg/liter and the sediment size distribution is given by Fig. 7.10, calculate the settleable solids concentration.

(7.5) Plot the following sediment size distribution.

Size (mm)	Percentage finer
0.0002	0
0.001	10
0.003	20
0.01	40
0.05	50
0.10	70
0.30	100

(7.6) Estimate the settleable solids concentration from a pond if the sediment size distribution is given by problem (7.4) and peak effluent TSS is 780 mg/liter.

(7.7) Two surface soils are analyzed using a standard hydrometer test to obtain the following percentages of sand, silt, and clay particles.

Particles (%)	A	B
Sand	12	40
Silt	64	25
Clay	24	35

What is the textural classification of each soil?

(7.8) Estimate D_{35}, D_{50}, and D_{65} for the sediment size distribution shown in problem (7.5).

(7.9) Determine the D_{50} and D_m representative particle sizes for the initial size distribution shown in Fig. 7.17. Are the median and mean diameters the same in each case?

(7.10) Assume that distribution A shown in Fig. 7.16 represents the eroded size distribution for subwatershed i. Estimate the size distribution for material as it leaves the subwatershed if the delivery ratio for subwatershed i is 0.3. Repeat with distribution B.

(7.11) A soil has a dispersed soil matrix of 37% sand, 50% silt, and 13% clay. Estimate the fractions sand, silt, and clay using the revised CREAMS equations. Also estimate the fractions of small and large aggregates. Plot the data as fraction finer versus size.

(7.12) What characteristic diameter is associated with each of the classes for the soil matrix shown in problem (7.11)?

(7.13) Determine the equivalent diameter for a particle having a specific gravity of 2.65 if the actual specific gravity is 2.1 and the actual particle diameter is 0.14 mm.

(7.14) Find the critical shear stress on the bottom of an unlined, trapezoidal channel carrying clear water. The channel has a mean flow velocity of 0.35 m/sec, and the bottom is composed of angular quartz having d_{50} = 25 mm. Assume that the water temperature is 68°F and the channel is wide and has a slope of 0.07%.

Table 7.8 Calculations for Example Problem 7.11 [Einstein Method]

10^3d[a]	10^2i_b[b]	R'[c]	Ψ[d]	d/X[e]	ξ[f]	Ψ_*[g]	Φ_*[h]	i_sg_s[i]	i_sG_s[j]	Σi_sG_s[k]	10^3A_E[l]	z[m]	I_1[n]	$-I_2$[o]	$P_EI_1+I_2+1$[p]	$i_{st}g_{st}$[q]	$i_{st}G_{st}$[r]	$\Sigma i_{st}G_{st}$[s]
2.36	20	0.1	389.4	0.262	19	117.	2.56E–09	0	1.54E–09	1.54E–09	6.07	203.5	0	0	1	7.14E–11	1.54E–09	1.54E–09
		0.2	194.	0.371	7.6	34.	5.08E–05	1.42E–06	3.32E–05	3.32E–05	2.77	143.9	0	0	1	1.42E–06	3.32E–05	3.32E–05
		0.5	77.8	0.587	2.3	7.	0.284	0.00794	0.204082	0.204	1.65	91.04	0	0	1	0.008	0.204	0.20482
		0.8	48.67	0.742	1.2	3.	3.322	0.0928	2.440097	2.440	1.50	71.97	0	0	1	0.092	2.440	2.440097
0.528	20	0.1	87.1	0.058	5.	1340.	0	0	7.28E–23	1.54E–09	1.35	10.17	0	0	1	0	0	1.54E–09
		0.2	43.5	0.083	388.	396.1	0	0	5.16E–16	3.32E–05	0.61	7.197	0	0	1	2.2E–17	5.16E–16	3.32E–05
		0.5	17.42	0.131	117.	79.0	8.84E–08	2.6E–10	6.71E–09	0.204	0.37	4.552	0.005	0.5	2.0	5.43E–10	1.39–08	0.204082
		0.8	10.8	0.166	63	34.3	5.25E–05	1.55E–07	4.08E–06	2.440	0.33	3.598	0.08	0.6	2.5	3.96E–07	1.04E–05	2.440107
0.231	20	0.1	38.11	0.025	8454.	5140.	0	0	0	1.54E–09	0.59	1.95	0.2	1.5	4.3	0	0	1.54E–09
		0.2	19.057	0.036	3403.	1518.	0	0	0	3.32E–05	0.27	1.384	0.5	3	9.2	0	0	3.32E–05
		0.5	7.62	0.057	1022.	303.	0	0	0	0.204	0.16	0.875	2	10	34.0	0	0	0.204082
		0.8	4.76437	0.072	551.	132.	8.62E–10	0	1.86E–11	2.440	0.14	0.692	8	23	119.0	0	2.21E–09	2.440107
0.127	20	0.1	20.95.88	0.014	40660.	13593.	0	0	0	1.54E–09	0.32	0.587	20	32	221.8	0	0	0
		0.2	10.477	0.019	16368.	4016.	0	0	0	3.32E–05	0.14	0.415	50	100	629.8	0	0	0
		0.5	4.19	0.031	4915.	801.	0	0	0	0.204	0.089	0.262	220	400	2941.	0	0	0
		0.8	2.61937	0.039	2652.	350.	0	0	0	2.440	0.080	0.207	410	600	5473.	0	0	0
5.6E–05	20	0.1	0.009	6.2E–06	2.6E + 13	3.78E + 09	0	0	0	1.54E–09	0.000144	0.117	0	0				0
		0.2	0.004	8.8E–06	1.0E + 13	1.14E + 09	0	0	0	3.32E–05	0	0.083	0	0				0
		0.5	0.0018	1.4E–05	3.1E + 12	2.28E + 08	0	0	0	0.204	0	0.052	0	0				0
		0.8	0.0011	1.8–05	1.7E + 12	9.98E + 07	0	0	0	2.440	0	0.041	0	0				0

[a] d, feet (grain size); the representative ones are taken from a size distribution analysis (given in this problem).
[b] i_b (fraction of bed material); taken from a size distribution analysis (given in this problem).
[c] R', feet (hydraulic radius with respect to grains); (assumed in this problem, see chapter 10 for more information).
[d] Ψ (intensity of shear on particle); $\Psi = ((\rho_s - \rho)/\rho)\ (d/R'S)$; given in Eq. (7.43).
[e] d/X; for values of X see Table 7.6.
[f] ξ (hiding factor); $\xi = fct\ (d_i/d_x)$, given in Fig. 7.21.
[g] Ψ_* (intensity of shear on individual grain size); $\Psi_* = \xi Y(\beta^2/\beta_x^2)\Psi$. Values are given in Table 7.6.
[h] Φ_* (intensity of transport for individual grain size); $\Phi_* = fct\ (\Psi_*)$ given with Fig. 7.24.
[i] i_sg_s, pounds per second (ft) (bedload rate in weight per unit width and time for a size fraction); $i_sg_s = i_b\Phi_*\rho_s g^{3/2} d^{3/2}\ \sqrt{(SG-1)}$; from Eq. (7.53).
[j] i_sG_s, pounds per second (bedload rate in weight per unit time for a size fraction for entire cross section); $i_sG_s = (i_sg_s)W$.
[k] Σi_sG_s, pound per second (bedload rate in weight per unit for all size fraction for entire cross section).
[l] $A_E = a/R$ (ratio of bed layer to hydraulic radius).
[m] z (exponent for suspension distribution); $z = V_s/0.4U'_*$; where V_s is particular settling velocity and U'_* is given in Table 7.6.
[n] I_1 (integral) read from Fig. 7.27; $I_1 = f(A_E, z)$; given by Eq. (7.61).
[o] I_2 (integral) read from Fig. 7.27; $I_2 = f(A_E, z)$; given by Eq. (7.62).
[p] $(P_EI_1 + I_2 + 1)$.
[q] $i_{st}g_{st}$, pounds per second (ft) (bed material load in weight per unit width and time, for a size fraction); $i_{st}g_{st} = i_sg_s(P_EI_1 + I_2 + 1)$.
[r] $i_{st}G_{st}$, pounds per second (bed material rate in weight per unit time for a size fraction for entire cross section); given by Eq. (7.66).
[s] $\Sigma i_{st}G_{st}$, pounds per second (bed material rate in weight per unit time for all size fractions for entire cross section); given by summing values.

Table 7.9 Calculations for Example Problem 7.11 (Yalin's Method)

d^a	R^b	U_*^c	R_{e*}^d	y^e	τ_c^f	U_{c*}^g	Y_{cr}^h	Y^i	a^j	q_s^k	C^l	Σq_s^m	U^n	D^o
0.000719	0.219	0.014	10.50	0.032	0.037	0.00061	0.002	0.01	0.074	0.004	185	0	0.088	0.236
	0.443	0.020	15.00	0.032	0.037	0.0061	0.002	0.02	0.074	0.022	298	0	0.145	0.519
	0.675	0.025	18.50	0.036	0.041	0.0065	0.002	0.03	0.078	0.048	208	0	0.271	0.867
	0.729	0.027	19.20	0.037	0.043	0.0066	0.002	0.04	0.079	0.055	158	0	0.371	0.958
0.000161	0.219	0.014	2.36	0.057	0.014	0.0038	0.003	0.05	0.098	0.007	296	0.006	0.088	0.236
	0.443	0.020	3.36	0.042	0.011	0.0033	0.002	0.10	0.084	0.038	513	0.037	0.145	0.519
	0.675	0.025	4.14	0.038	0.010	0.0031	0.002	0.16	0.080	0.093	403	0.093	0.271	0.867
	0.729	0.027	4.30	0.036	0.009	0.0030	0.002	0.17	0.078	0.114	326	0.114	0.371	0.958
0.00007	0.219	0.014	1.03	0.120	0.014	0.0036	0.007	0.12	0.143	0.004	180	0.009	0.088	0.236
	0.443	0.020	1.46	0.066	0.008	0.0027	0.004	0.4	0.106	0.030	399	0.068	0.145	0.519
	0.675	0.025	1.81	0.060	0.007	0.0026	0.004	0.36	0.101	0.069	297	0.162	0.271	0.867
	0.729	0.027	1.88	0.060	0.007	0.0026	0.004	0.39	0.101	0.078	224	0.192	0.371	0.958
0.000039	0.219	0.014	0.56	0.200	0.013	0.0035	0.012	0.21	0.185	0.003	126	0.012	0.088	0.236
	0.443	0.020	0.80	0.140	0.009	0.0029	0.009	0.43	0.155	0.015	198	0.082	0.145	0.519
	0.675	0.025	0.99	0.120	0.008	0.0027	0.007	0.66	0.143	0.036	155	0.198	0.271	0.867
	0.729	0.027	1.03	0.110	0.007	0.0026	0.007	0.71	0.137	0.045	129	0.237	0.371	0.958
0.000017	0.219	0.014	0.25	0.400	0.011	0.0033	0.025	0.48	0.261	0.001	76	0.014	0.088	0.236
	0.443	0.020	0.36	0.320	0.009	0.0029	0.020	0.97	0.234	0.007	94	0.089	0.145	0.519
	0.675	0.025	0.44	0.280	0.008	0.0028	0.017	1.48	0.219	0.016	70	0.214	0.271	0.867
	0.729	0.027	0.46	0.250	0.007	0.0026	0.016	1.59	0.207	0.020	60	0.258	0.371	0.958

[a]d, mm (grain size): Representative examples are taken from a size distribution analysis (given in this problem). [b]R, m (hydraulic radius with respect to grains): Values assumed in this problem. See Chapter 10 for additional information. [c]U_* (shear velocity): $U_* = \sqrt{\gamma RS}$. [d]R_{e*} (shear Reynolds number): $R_{e*} = U_* d/\nu$. [e]y: τ_* from Shield's diagram (Fig. 7.20) corresponding to R_{e*}. [f]τ_c (critical shear): Eq. 7.33 solved for τ_c. [g]U_{c*} (critical shear velocity): Eq. 7.34. [h]Y_{cr} (critical shear velocity): Eq. 7.34. [h]Y_{cr} (critical mobility factor): From Eq. 7.70 using critical shear velocity, U_{c*}. [i]Y (mobility factor): From Eq. 7.70 using shear velocity U_*. [j]a: Eq. 7.68. [k]q_s (sediment discharge for grain size d): $q_s = \frac{981\ VDC}{10^6}$ with D from column o in Table 7.9. [l]C (sediment concentration in ppm): Eq. 7.67 for grain size d. [m]Σq_s (cumulative sediment discharge): Add q_{st} + previous Σq_s for grain size d. [n]U, m/sec (average flow velocity); Typically estimated from channel geometry and cover; assumed for this example. [o]D, m (flow depth): Values up to maximum flow depth in channel.

Table 7.10 Calculations for Example Problem 7.11 (Yang's Method)

d[a]	R[b]	U_*[c]	R_{e*}[d]	V_{cr}/V_s[e]	V_s[f]	V[g]	α[h]	β[i]	C_t[j]	q_s[k]	Σq_s[l]
0.000719	0.219	0.014	10.56	3.107	0.443	0.088	5.395	1.239	—	0	0
	0.443	0.020	15.00	2.788	0.443	0.145	5.325	1.191	—	0	0
	0.675	0.025	18.5	2.634	0.443	0.271	5.283	1.163	—	0	0
	0.729	0.026	19.2	2.608	0.443	0.371	5.276	1.157	—	0	0
0.000161	0.219	0.014	2.362	7.560	0.022	0.088	5.358	1.629	—	0	0
	0.443	0.020	3.357	5.484	0.022	0.145	5.289	1.581	0.000	2.39E–08	2.39E–08
	0.675	0.025	4.141	4.752	0.022	0.271	5.247	1.552	0.024	5.67E–06	5.67E–06
	0.729	0.026	4.304	4.641	0.022	0.371	5.239	1.547	0.070	2.47E–05	2.47E–05
7.04E–05	0.219	0.014	1.033	–215	0.00424	0.088	5.338	1.844	52.58	0.001082	0.001082
	0.443	0.020	1.469	17.44	0.00424	0.145	5.268	1.796	0.146	1.09E–05	1.09E–05
	0.675	0.025	1.812	10.92	0.00424	0.271	5.227	1.76	1.896	0.00439	0.000444
	0.729	0.026	1.883	10.24	0.00424	0.371	5.219	1.762	4.3508	0.001518	0.001543
3.87E–05	0.219	0.014	0.568	–7.84	0.00128	0.088	5.324	1.999	4.3166	8.88E–05	0.001171
	0.443	0.020	0.807	–19.1	0.00128	0.145	5.254	1.952	14.444	0.00107	0.001081
	0.67	0.025	0.996	–86.7	0.00128	0.271	5.212	1.923	87.95	0.020344	0.20788
	0.729	0.026	1.03	–230.0	0.00128	0.371	5.205	1.918	301.1	0.105114	0.106657
1.73E–05	0.219	0.014	0.253	–2.84	0.00025	0.088	5.304	2.209	124.5	0.002563	0.003734
	0.443	0.020	3.360	–4.1	0.00025	0.145	5.234	2.164	373.4	0.027661	0.028742
	0.67	0.025	0.444	–5.36	0.00025	0.271	5.193	2.133	1492.0	0.3453	0.366088
	0.729	0.026	0.462	–5.66	0.00025	0.371	5.185	2.127	2973.0	1.037594	1.144251

[a] d, mm (grain size): Representative examples are taken from a size distribution analysis (given in this problem). [b] R, m (hydraulic radius with respect to grains): Values assumed in this problem. See Chapter 10 for additional information. [c] U_* (shear velocity): $U_* = \sqrt{\gamma RS}$. [d] R_{e*} (shear Reynolds number): $R_{e*} = U_* dv$. [e] V_{cr}/V_s (critical velocity over settling velocity): For $0 < R_{e*} < 70$, $V_{cr}/V_s = (2.5/(\log R_{e*} - 0.06)) + 0.66$. [f] V_s (particle settling velocity): $V_s = 2.81(d_* 1000)^2/3.28$. [g] V, m/sec (average flow velocity): Typically estimated from channel geometry and cover; assumed for this example. [h] α (parameter in Eq. 7.71): $\alpha = 5.435 - 0.286 \log_{10}(V_s d/\nu) - 0.457(\log_{10}(U_*/V_s))$. [i] β (parameter in Eq. 7.71): $\beta = 1.799 - 0.409 \log_{10}(V_s d/\nu) - 0.314 \log_{10}(U_*/V_s)$. [j] C_t (total sediment concentration in ppm): Eq. 7.72. [k] q_s (sediment discharge): $q_s = 981\ VDC_t/10^6$ with flow depths D corresponding to those from Table 7.9. [l] Σq_s (cumulative sediment discharge): Add q_{st} + previous Σq_s for grain size d.

(7.15) A stream is estimated to have the following hydraulic data: slope of 0.8%, average depth of 2.1 m, and average velocity equal to 0.6 m/sec. The channel is wide and approximately rectangular with a width of 10 m. A sediment sample has been collected and analyzed, from which it was determined that d_{35} equals 9 mm. What is the rate of bedload transport if the channel bottom is smooth?

(7.16) Collect a bedload sample from a stream at your location. Determine the particle size distribution and plot as percentage finer versus diameter. What is d_{35}? d_{50}? d_{65}? d_m?

(7.17) Estimate or measure the hydraulic data for the stream in problem (7.16) (use a flow meter if available), and estimate the rate of bedload transport using Einstein's method and the distribution data previously obtained.

(7.18) Compare the transport obtained by Yalin's, Yang's, and Einstein's methods for a stream at your location.

References

Alonso, C. V., Neibling, W. H., and Foster, G. R. (1981). Estimating sediment transport capacity in watershed modeling. *Trans. ASAE* **24**(5):1211–1226.

Argaman, Y. A., and Kaufman, W. J. (1970). Turbulence and flocculation. *J. Sanitary Eng. Div. ASCE* **96**:223.

ASTM (1985). Annual book of ASTM standards. American Society for Testing and Materials, Philadelphia, PA.

Barfield, B. J., Moore, I. D., and Williams, R. G. (1979). Sediment yield in surface mined watersheds. *In* "Proceedings, Symposium on Surface Mine Hydrology, Sedimentology and Reclamation, University of Kentucky, Lexington, KY," pp. 61–65.

Barfield, B. J. Warner, R. C., and Haan, C. T. (1981). "Applied Hydrology and Sedimentology for Disturbed Areas." Oklahoma Technical Press, Stillwater, OK.

Barfield, B. J., Barnhisel, R. I., Powell, J. C., Hirschi, M. C., and Moore, I. D. (1983). Erodibility and eroded size distribution of surface mining spoil and reconstructed topsoil, Report IMMR 84/092. Institute for Mining and Minerals Research, University of Kentucky, Lexington, KY.

Brady, N. C. (1974). "The Nature and Properties of Soils." Macmillan, New York.

Burcham, T. N. (1989). Vortical particle size distribution system, Unpublished Ph.D. dissertation. Clemson University, Clemson, SC.

Chang H. H. (1988). "Fluvial Processes in River Engineering." Wiley, New York.

Colby, B. R., and Hubbell, D. W. (1961). Simplified method for computing total sediment discharge with the modified Einstein procedure, Water Supply Paper 1593. U.S. Geol. Surv.

Einstein, H. A. (1942). Formulas for the transportation of bed-load. *Trans. Am. Soc. Civil Eng.* **107**.

Einstein, H. A. (1950). The bed-load function for sediment transportation in open channel flow, Tech. Bulletin 1026. U.S. Department of Agriculture.

Evangelou, V. P., Rawlings, F., Crutchfield, J. D., and Shannon, E. A. (1981). A simple chemo-mathematical model as a tool in managing surface mine sediment ponds. *In* "Proceedings, 1981 Symposium on Surface Mine Hydrology, Sedimentology, and Reclamation," College of Engineering, University of Kentucky, Lexington, KY.

Foster, G. R. (1982). Modeling the erosion process. *In* "Hydrologic Modeling of Small Watersheds" (C. T. Haan, ed.) Monograph No. 5. American Society of Agricultural Engineers, St. Joseph, MI.

Foster, G. R., Young, R. A., and Neibling, W. H. (1985). Sediment composition for nonpoint source pollution analyses. *Trans. ASAE* **28**(1):133–146.

Gessler, J. (1971). Beginning and ceasing of sediment motion. *In* "River Mechanics" (H. W. Shen, ed.), Chap. 7 Water Resources Publications, Fort Collins, CO.

Graf, W. H. (1971). "Hydraulics of Sediment Transport." McGraw–Hill, New York.

Haan, C. T., and Barfield, B. J. (1978). "Hydrology and Sedimentology of Surface Mined Lands." University of Kentucky, Lexington, KY.

Hartge, K. H. (1978). *In* "Structural Stability as a Function of Some Soil Properties in Modification of Soil Structure." (W. E. Emerson, R. D. Bond, and A. R. Dexter, eds.). Wiley, New York.

Hayes, J. C. (1979). Evaluation of design procedures for vegetal filtration of sediment from flowing water, Unpublished Ph.D. dissertation. University of Kentucky, Lexington, KY.

Hayes, J. C., Barfield, B. J., and Barnhisel, R. I. (1982). The use of grass filters for sediment control in strip mine drainage. III. Empirical verification of procedures using real vegetation, Technical report IMMR82/070. Institute for Mining and Minerals Research, University of Kentucky, Lexington, KY.

Hirschi, M. C. (1985). Modeling soil erosion with emphasis on steep slopes and the rilling process, Ph.D. dissertation. University of Kentucky, Lexington, KY.

Holbrook, K. F., Ligon, J. T., and Hayes, J. C. (1986). Comparison of methods for determination of eroded size distribution in sedimentology modeling, Paper No. SER-86-206 presented at the Southeast Region Meeting of ASAE, Orlando, Fl.

Lara, J. M., and Pemberton, E. L. (1963). Initial unit weight of deposited sediments. *In* "Proceedings, Federal Inter-Agency Sedimentation Conference," U.S. Department of Agriculture Miscellaneous Publication 1970. U.S. Government Printing Office, Washington, DC.

Mantz, P. A. (1977). Incipient transport in fine grains and flakes by fluids—Extended Shields' diagram. *J. Hydraulics Div. ASCE* **103**(HY6):601–615.

Martin, J. P., Martin, W. P., Paye, J. B., Raney, W. A., and DeMent, J. D. (1955). Soil aggregation. *In* "Advances in Agronomy (A. G. Norma, ed.), No. 7. Academic Press, New York.

McKyes, E. (1989). "Agricultural Engineering Soil Mechanics," Developments in Agricultural Engineering 10. Elsevier, New York.

Metcalf, and Eddy, Inc. (1979). "Wastewater Engineering, Treatment, Disposal, and Reuse." McGraw–Hill, New York.

Meyer, L. D., and Harmon, W. C. (1979). Multiple-intensity rainfall simulator for erosion research on row sideslopes. *Trans. ASAE* **22**(1):100–103.

Miller, C. R. (1953). "Determination of the Unit Weight of Sediment for Use in Sediment Volume Computation." U.S. Bureau of Reclamation, Denver, CO.

Rhoton, F. E., Meyer, L. D., and Whisler, F. D. (1982). A laboratory method for predicting the size distribution of sediment eroded from surface soils. *Soil Sci. Soc. Am. J.* **46**:1259–1263.

Rhoton, F. E., Meyer, L. D., and Whisler, F. D. (1983). Response of aggregated sediment to runoff stresses. *Trans. ASAE* **26**(5):1476–1478.

Rouse, H., ed. (1950). "Engineering Hydraulics." Wiley, New York.

Shields, A. (1936). "Application of the Theory of Similarity and Turbulence Research to the Bed Load Movement, Vol. 26, pp. 5–24. Mitt. Preuss. Vers. Wasser Schiff. [In German]

Simons, D. B., and Senturk, F. (1977). "Sediment Transport Technology." Water Resources Publications, Fort Collins, CO.

Simons, D. B., and Senturk, F. (1992). "Sediment Transport Technology." Water Resources Publications, Fort Collins, CO.

Soil Conservation Service (1984). "Engineering Field Manual." Soil Conservation Service, Washington, DC.

Soil Survey Staff (1951). "Soil Survey Manual," U.S. Department of Agriculture Handbook No. 18. USDA, Washington, DC.

Storm, D. E., Barfield, B. J., and Ormsbee, L. E. (1990). Hydrology and sedimentology of dynamic rill networks. I. Erosion model for dynamic rill networks, Research report No. 178. Kentucky Water Resources Research Institute, University of Kentucky, Lexington, KY.

Tapp, J. S., Barfield, B. J., and Griffin, M. L. (1981). Predicting suspended solids removal in pilot scale sediment ponds utilizing chemical flocculation, Technical report of Institute for Mining and Minerals Research. University of Kentucky, Lexington, KY.

Tollner, E. W., and Hayes, J. C. (1986). Measuring soil aggregate characteristics for water erosion research and engineering: A review. *Trans. ASAE* **29**(6):1582–1589.

USDA (1979). "Field Manual for Research in Agricultural Hydrology," USDA Agriculture Handbook No. 224. U.S. Government Printing Office, Washington, DC.

Williams, J. R. (1977). Sediment delivery ratios determined with sediment and runoff models. *In* "Proceedings, Erosion and Solid Matter Transport in Inland Water Symposium, IAHS, No. 122, pp. 168–179.

Wilson, B. N., and Barfield, B. J. (1986). A detachment model for non-cohesive sediment. *Trans. Am. Soc. Agric. Eng.* **29**(5):1300–1306.

Wilson, B. N., Barfield, B. J., and Moore, I. D. (1982). "A Hydrology and Sedimentology Watershed Model," Part I; "Modeling Techniques." Department of Agricultural Engineering, University of Kentucky, Lexington, KY.

Yalin, M. S. (1963). An expression for bed-load transportation. *J. Hydr. Div., Proc. ASCE* **89**(HY3):221–250.

Yang, C. T. (1972). Unit stream power and sediment transport. *J. Hydr. Div., Proc. ASCE* **98**(HY10):1805–1826.

Yang, C. T. (1973). Incipient motion and sediment transport. *J. Hydr. Div., Proc. ASCE* **98**(HY10):1679–1704.

Yang, C. T. (1984). Unit stream power equation for gravel. *J. Hydr. Div., Proc. ASCE* **110**(HY12).

8

Erosion and Sediment Yield

INTRODUCTION

Philosophy of Erosion and Sediment Control

The landscape of the earth has been greatly influenced by the processes of soil erosion and sedimentation. When uninfluenced by the activities of humans, these processes often produce picturesque landscapes such as the Grand Canyon, Niagara Falls, and other places of natural beauty. However, when the landscape is subjected to human activities with no consideration for conservation, the result is all too often a rather grotesque picture of fields riddled with gullies and muddy streams whose channels are filled with sediment.

Soil erosion and sedimentation from construction and mining activities have many similarities with those from agricultural land. This is extremely helpful since agricultural erosion has been studied for many years resulting in the development of prediction algorithms and control procedures. Economic incentives for erosion control with construction and mining, however, are drastically different from those for agriculture. Soil loss from agricultural land represents a loss in nutrients and production capability, resulting in at least some economic incentive to the landowner to reduce erosion. In the absence of regulatory penalties, soil loss from a construction site or a surface mine represents no economic loss to the operator, with the possible exception of additional cleanup costs after construction is complete. Expenditures for sediment control are therefore an economic liability. Since there are no intrinsic economic incentives to control sedimentation, it has been necessary to appeal to the altruistic interests of operators to implement controls. When this has failed, regulations have been imposed.

Soil erosion results when soil is exposed to the erosive powers of rainfall energy and flowing water. It is not possible to conduct the massive earth-moving operations necessary for major land disturbances without exposing soil to these erosive forces. It is possible, however, to plan the operation and reclamation activities so that sediment production is minimized. Through the use of properly designed and constructed conservation practices and sediment control structures, the adverse effects of land disturbances can be made minimal.

Large-scale abandoned land disturbances will eventually heal themselves by natural processes, including natural revegetation and natural armoring. Curtis (1971) analyzed the sediment yield from Appalachian watersheds during periods prior to strip mining, during active operations, and subsequent to mining. He found that sediment concentrations went from near zero prior to mining to over 46,000 mg/liter in runoff from large storms during mining operations. In the postmining period, sediment load decreased, even under poor reclamation, as the land tended to heal itself. This healing did not occur as a result of revegetation, but was a result of natural armoring by surface rocks that tended to cover the surface when the fines were washed away. These processes, along with natural revegetation, proceed at differing rates, depending on the climate, geology, and chemistry of the exposed surface. Typically, the rates are slow without humanmade enhancements, and environmental damage can be extensive prior to complete reclamation. Of course, appropriate reclamation and conservation practices dramatically enhance natural reclamation processes and can eliminate undesirable environmental impacts.

The Erosion – Sediment Yield – Deposition Process

Description of the Process

Soil erosion involves detachment, transport, and subsequent deposition (Meyer and Wischmeier, 1969). Soil is detached both by raindrop impact and the shearing force of flowing water. Sediment is transported downslope primarily by flowing water, although there is a small amount of downslope transport by raindrop splash. Runoff and resulting downslope transport do not occur until rainfall intensity exceeds infiltration rate. For this reason, soil erodibility decreases as the infiltration rate increases. Once runoff starts, the quantity and size of material transported increases with the velocity of runoff water. At some point downslope, slopes may decrease, resulting in a decreased velocity and transport capacity. At this point, sediment will be deposited, starting with the larger particles and aggregates. Smaller particles and aggregates will be carried further downslope, resulting in what is known as enrichment of fines. For this reason, the size distribution of eroded aggregates and primary particles has a major impact on soil erosion–deposition processes.

Terminology

An important part of developing an understanding of erosion and sedimentation literature is terminology. Soil eroded from exposed upland areas near the watershed divide is from rill and interrill areas. Interrill areas are those zones between small channelized flows known as rills. Interrill erosion occurs in those areas where flow is shallow overland or sheet flow and detachment forces are primarily from raindrop energy impacting exposed soil. Rill erosion occurs when flow is concentrated in microrelief channels with sufficient depth and slope to cause channel incision. The microrelief causing these small concentrated flow channels is generated by tillage or landforming operations. Since flow depth in rills is typically sufficient to absorb falling raindrop impact energy, soil detachment occurs primarily from shearing forces of the channelized flow. Tillage after a rainfall event will obliterate rills, forming a new and different microrelief. Subsequent concentrated flow channels will occur in different locations; thus rill location is generally assumed to be random.

As flow progresses further downslope and away from the watershed divide, the location of channelized flow areas ceases to be controlled by the microrelief and becomes controlled by the prevailing macrorelief. Tillage or landforming operations may obliterate any incised channels in these macrorelief drainage ways; however, channels will tend to form again in the same location unless the macrorelief is drastically changed. Soil detachment in macrorelief channels occurs primarily as a result of the shearing forces of the channelized flow, from headwall failure and from channel wall failure. If the channels form in fields that are being tilled such that the resulting incised channels are obliterated by subsequent tillage, the channels are called ephemeral gullies (Foster, 1986). If the channels are permanent features with a dynamic headwall, vertical channel walls, and dynamic features, they are called classic gullies (Harvey *et al.*, 1985). The relationship between rills and ephemeral gullies is shown in Fig. 8.1.

As flow moves still further away from the watershed divide, channels tend to have relatively stable banks and no headwalls with erosion occurring as a result of bed degradation from shear forces and channel wall failure. A discussion of erosion and sediment transport processes in these channels is given in Chapter 10.

Detachment

Erosion in interrill areas occurs primarily as a result of raindrop impact. Interrill erosion is thus relatively independent of slope length, after sufficient distance downslope to generate enough runoff to transport eroded sediment. Because interrill transport capacity from downslope splash and from overland flow does increase slightly with slope, interrill erosion is a linear function of slope steepness. Rill erosion, on the other hand, is a strong function of both slope steepness and slope length. Since rill detachment results from shearing forces of concentrated flow, increasing runoff depth and velocity will cause an increase in rill erosion. Both of these increase with distance downslope. Shear and resulting rill erosion also increase with slope steepness. The impact of slope length and steepness on rill and interrill erosion is illustrated in Fig. 8.2.

Deposition

Deposition of sediment can happen anywhere downslope of the point of erosion, occurring when

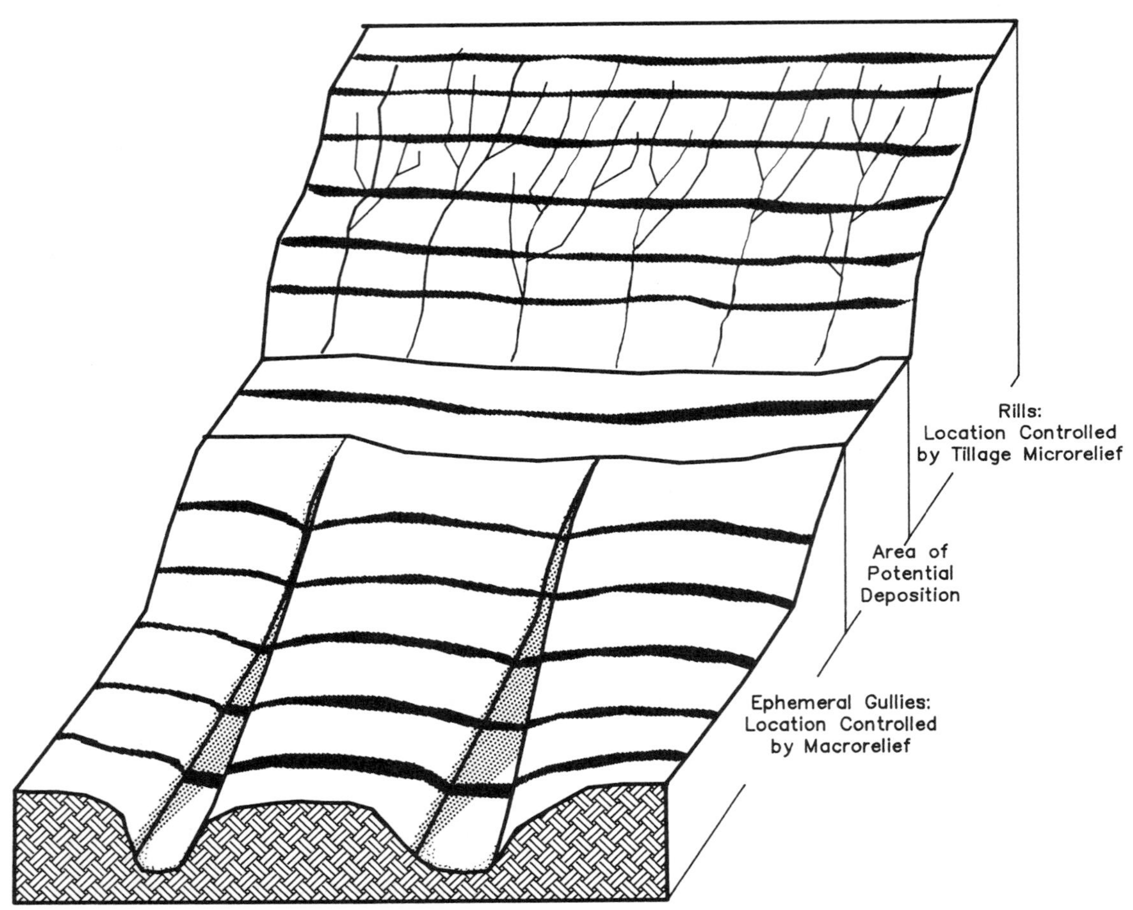

Figure 8.1 Illustration of rills and ephemeral gullies.

transport capacity of the flow is less than soil available for transport. Since transport capacity is a monotonic function of flow velocity, anything that reduces velocity in a flow segment increases deposition. Vegetal filters, terrace channels and check dams are examples of sediment control practices that reduce velocity and increase deposition. Further discussion of these practices is given in Chapter 9.

Components of the soil erosion–deposition process are interrelated as shown in Fig. 8.3. Soil available for transport at any slope segment is the sum of that carried from upslope plus that detached in the slope increment. Soil carried downslope is the lesser of transport capacity or material available for transport. Detailed information on all of the processes in Fig. 8.3 are given in varying parts of Chapters 7, 8, 9, and 10.

Erosion and Sediment Yield Modeling: A Historical Perspective

The information presented in this section is a summary from several sources but primarily that of Renard *et al.* (1993a). Development of early empirical soil erosion models started more than 50 years ago with efforts by Cook (1936), Zingg (1940), and Smith (1941) to predict the impact of slope length and steepness, cover, and supporting practices on soil erosion. Subsequently, Smith and Whitt (1948) developed a relationship for soil erosion in Missouri in which annual erosion from a claypan soil was used as a standard with corrections for slope, slope length, soil classification, and conservation practice. Similar relationships were developed for the cornbelt region. Based on a need for relationships for other regions, a national workshop was held in Ohio in 1946, which resulted in addition of a rainfall factor and reappraisal of all previously developed factors. The resulting so-called Musgrave equation (Musgrave, 1947) included factors for rainfall, surface runoff characteristics, slope steepness and length, soil characteristics, and vegetative cover.

Success of the Musgrave equation led to a desire for a truly national relationship that had universal parameters and also led to establishment of the National Runoff and Soil Loss Data Center at Purdue Univer-

A

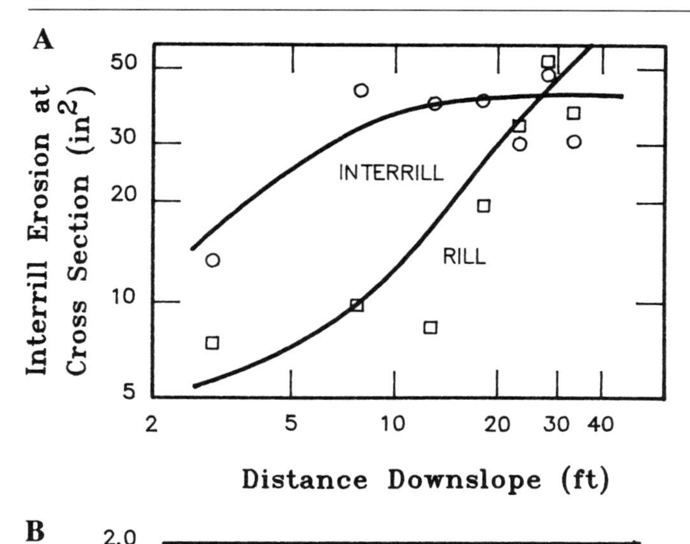

B

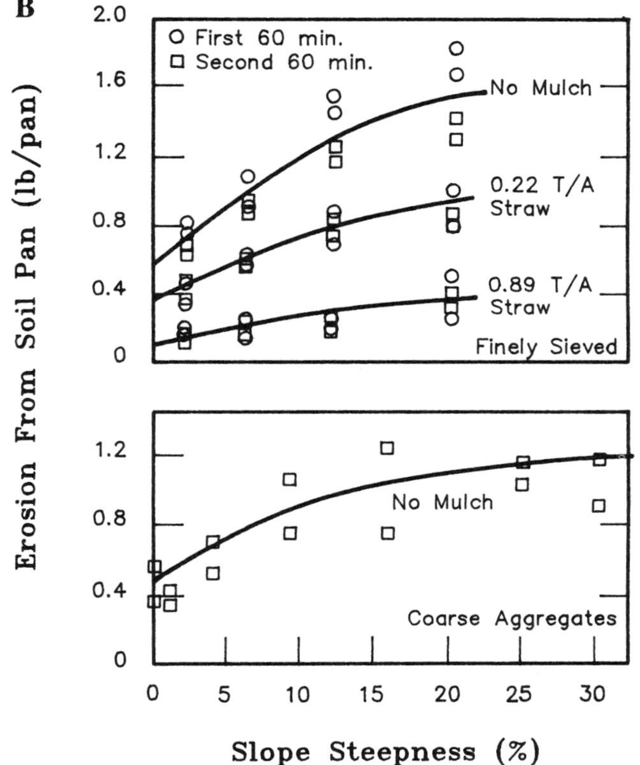

Figure 8.2 Influence of slope length and steepness on interrill and rill erosion. Pan size in B was 2.5 × 2.5 ft and rainfall intensity was 2.5 in./hr. (A) Plot length effect on interrill and rill erosion. (B) Slope steepness effect on interrill erosion as measured in 2.5 × 2.5-ft pans (after Meyer *et al.*, 1975).

sity in 1954. Federal–State cooperative research yielded 10,000 plot years of data, which were analyzed by the data center. The result was the Universal Soil Loss Equation (USLE) described by Wischmeier and Smith (1965, 1978) in Handbooks 282 and 537. The USLE describes erosion as a function of rainfall energy and intensity, soil erodibility, slope length, slope steepness, soil cover, and conservation practices, making it a valuable tool for conservation planners. Because of frequent misuse, Wischmeier (1976) strongly recommended that its use be limited to conditions for which it was originally intended.

Most of the basic USLE information was published in 1965 in Handbook 282 with additional improvements following in Handbook 537 in 1978. Subsequent to Handbook 537, a number of additional improvements have been made, specifically

- revisions in the slope and slope length factors resulting from an evaluation of the original data base,
- use of a subfactor approach for cover and management factors with major new data available to define the relationships,
- use of a time-varying relationship to account for freeze–thaw effects on erodibility,
- use of factors to account for susceptibility to rilling.

These data have been summarized in a new handbook and the revised model described as the Revised USLE or RUSLE (Renard *et al.*, 1993a). The RUSLE, in addition to containing new parameters, is available in computerized format.

Efforts to develop a more physically based approach lead to a conceptual relationship known as the Meyer model (Meyer and Wischmeier, 1969) shown in Fig. 8.3 and subsequently to the Foster–Meyer–Onstad (FMO) model (Foster *et al.*, 1977a, b). The FMO equation does not predict processes in individual rills, but rather has algorithms to estimate the sum of all rill erosion on a slope segment. Instead of a combining rill and interrill erosion estimates into one lumped prediction, interrill erosion is predicted as a function of rainfall energy, interrill erodibility, slope steepness, and interrill cover factor. Rill erosion is predicted as a function of runoff volume, peak discharge, rill erodibility, slope steepness, slope length, and rill cover and practice factors. Although separate factors were proposed for rill and interrill erosion, no data base was developed for the factors. The FMO model was used as the basic relationship for the CREAMS model (Knisel, 1980). A modification to the FMO known as SLOSS was utilized by Wilson *et al.* (1982, 1986) in the SEDIMOT II model. The SLOSS modification includes procedures for utilizing the FMO to calculate detachment and the Yang equation (Yang, 1972) to calculate sediment transport capacity.

More recent research has been oriented toward process-oriented models that consider detachment processes in individual rills. Hirschi and Barfield (1988a, b) developed a research-oriented single-storm model

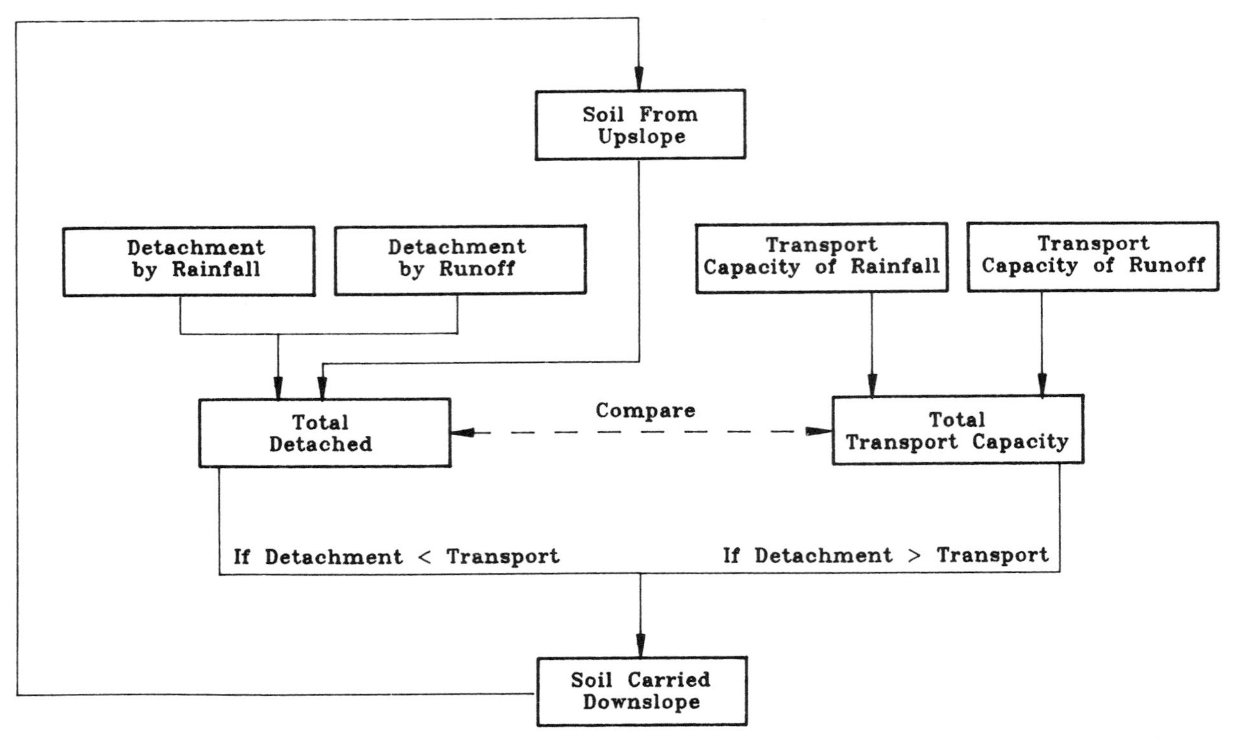

Figure 8.3 Interrelationship of detachment, transport and deposition of sediment (after Meyer and Wischmeier, 1969).

known as KYERMO, which predicted sediment detachment and transport in individual rills as a function of flow rates and shear distribution, under the assumption that rills were uniformly spaced and oriented up- and downhill. A national research team collected a large data base and developed a computerized continuous simulation model, known as WEPP, that also predicts erosion in uniformly spaced rills assuming a rill spacing of 1 m (Lane and Nearing, 1989). The WEPP data base was used to develop prediction equations for the basic erosion parameters. Further testing on the application of the model to watershed conditions is needed, particularly for construction sites. Recent research (Storm *et al.*, 1990; Lewis *et al.*, 1990) extended the fundamental erosion process approach to include evaluation of the impacts of rill density distributions on erosion.

FUNDAMENTAL EROSION MODELING

The following is intended for the reader interested in principles and concepts. Those interested primarily in applications should read subsequent sections on rill and interrill erosion modeling. An overview of the state of the art in fundamental erosion modeling is given by Storm *et al.* (1990). The following discussion draws heavily on that document.

Erosion Principles

Sediment Continuity Equation

The relationship that is basic for fundamental erosion processes is continuity of mass. For overland flow, the continuity equation is (Foster, 1982)

$$\frac{\partial q_s}{\partial x} + \rho_s \frac{\partial (cy)}{\partial t} = D_r + D_i, \tag{8.1}$$

where q_S is sediment load, x is distance downslope, ρ_S is mass density of sediment particles, c is sediment concentration, y is flow depth, t is time, D_r is rill erosion or deposition rate, and D_i is sediment delivered to the rill from interrill areas. The erosion parameters q_S, D_r, and D_i are measured per unit width of the field. The $\partial q_S/\partial x$ term represents the change in sediment flow rate along the slope, and $\rho_S \partial(cy)/\partial t$ represents the change in sediment storage over time.

For flows that are shallow and gradually varied, the $\rho_S \partial(cy)/\partial t$ storage term may be neglected, resulting in a widely used steady-state continuity equation

$$\frac{dq_S}{dx} = D_r + D_i. \quad (8.2)$$

Interaction between Sediment Load and Transport Capacity

The interaction between sediment load and transport capacity is an important consideration in modeling D_r in Eqs. (8.1) and (8.2). Foster and Meyer (1972a, 1975) proposed that rill detachment and deposition are proportional to the difference between transport capacity and sediment load, or

$$D_r = C_1(T_c - q_S), \quad (8.3)$$

where C_1 is a first-order reaction coefficient (L^{-1}), and T_c is the sediment transport capacity. If it is assumed that the maximum detachment capacity, D_{rc}, is proportional to transport capacity, or

$$D_{rc} = C_1 T_c, \quad (8.4)$$

then a relationship defining the interaction between sediment load and transport capacity can be developed, as (Foster and Meyer, 1972a)

$$\frac{D_r}{D_{rc}} + \frac{q_S}{T_c} = 1. \quad (8.5)$$

From Eq. (8.5) it can be seen that rill detachment equals detachment rate capacity when the sediment load equals zero. Conversely, when sediment load equals transport capacity, D_r approaches zero.

Foster and Meyer (1972a) utilized the stream power concept to lend credence to these concepts, claiming that the finite amount of energy in a specified flow may be used for either detaching or transporting sediment particles, but the same energy cannot be used for both. The D_r/D_{rc} term in Eq. (8.5) represents the relative amount of energy expended on sediment detachment, and the term q_S/T_c indicates the relative amount of energy expended on sediment transport. The sum of these two terms equal unity, the total relative available energy.

These concepts may also be applied to deposition. When applying Eq. (8.3) to deposition, it is clear that C_1 should be varied with sediment size to account for the differing settling velocities between particle sizes. To date, however, there are no validated relationships describing this phenomenon. On the basis of results by Einstein (1968), Foster (1982) proposed that C_1 for overland flow may be approximated by

$$C_1 = V_S/2q, \quad (8.6)$$

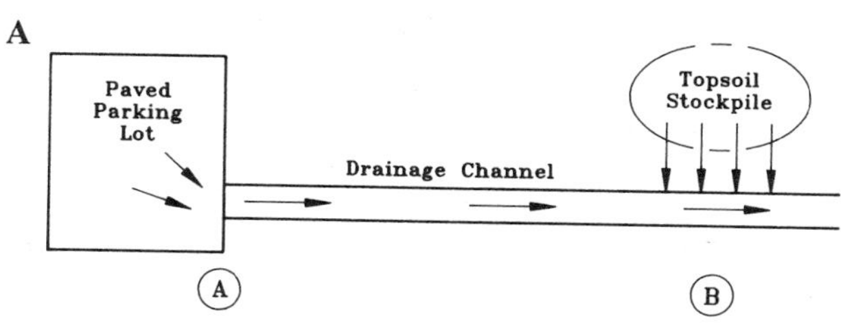

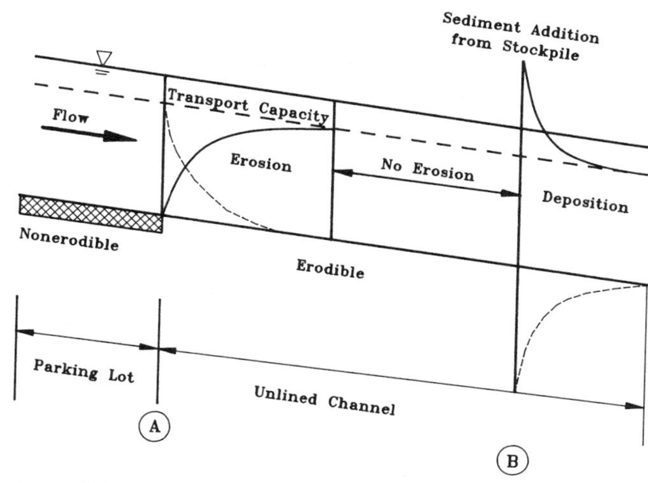

Figure 8.4 Illustration of interactions between sediment load, rill detachment and deposition, and flow transport capacity. (A) Plan view. (B) Transport rate and detachment rate changes along the channel.

where V_S is settling velocity, and q is flow discharge per unit width of plot. When Eq. (8.3) is applied to channelized flow, Foster (1982) proposed that deposition at a given flow rate would be greater than that for overland flow and that C_1 should be given by

$$C_1 = V_S/q. \quad (8.7)$$

The relationships proposed in Eqs. (8.3)–(8.6) are illustrated conceptually in Fig. 8.4. In this figure, sediment free flow is assumed to exit the parking lot and enter a channel with an erodible surface. Soil is detached at a rate that decreases exponentially with distance until sediment load equals transport capacity. When additional sediment is added to the flow at the topsoil stockpile, deposition must occur until the sediment load again equals transport capacity. This simple illustration shows how detachment changes with sediment load and transport capacity. In summary, detachment capacity or potential is based on localized flow conditions at a point, but the actual detachment rate is limited by available excess transport capacity.

Foster and Meyer Closed Form Erosion Equations

Solutions to Eqs. (8.1)–(8.7) can be developed numerically; however, Foster and Meyer (1972a, 1975) developed a closed-form erosion model by simultaneously solving these equations utilizing auxiliary relationships. They assumed steady-state rainfall, uniform shallow flow, and a steady-state rainfall excess given by

$$q = \frac{x}{L_0} q_0, \tag{8.8}$$

where q_0 is rainfall excess at $x = L_0$. In addition, potential detachment and transport rates, D_c and T_c, respectively, were defined as

$$D_c = C_D \tau^{3/2} \tag{8.9}$$

and

$$T_c = C_T \tau^{3/2}, \tag{8.10}$$

where τ is the average shear stress on the soil surface, and C_D and C_T are constants. Using auxiliary relationships from the Chezy equation, D_{c0} and T_{c0} can be defined as the detachment potential and transport capacity at $x = L_0$, leading to the relationship

$$\frac{D_r}{D_{c0}} + \frac{q_S}{T_{c0}} = g_*, \tag{8.11}$$

where

$$g_* = \frac{T_c}{T_{c0}} = \frac{D_c}{D_{c0}} \tag{8.12}$$

and

$$x_* = \frac{x}{L_0}. \tag{8.13}$$

Using Eq. (8.11) with (8.2), Foster and Meyer obtained a general solution to their equation as

$$\frac{D_r}{D_{c0}} = e^{-\alpha x_*}\left[\int\left(\frac{dg_*}{dx_*} - \theta\right)\exp^{\alpha x_*} dx_* + C_2\right], \tag{8.14}$$

where

$$\alpha = \frac{L_0 D_{c0}}{T_{c0}} \tag{8.15}$$

and

$$\theta = \frac{L_0 D_i}{T_{c0}}, \tag{8.16}$$

where x is the distance downslope, L_0 is the slope length, D_{c0} and T_{c0} are the detachment and transport capacities at $x = L_0$, respectively, and C_2 is an integration constant. If uniform slope and constant rainfall excess are assumed, the resulting relationships are

$$\frac{q_S}{T_{c0}} = x_* - (1-\theta)(1 - e^{-\alpha x_*})\alpha^{-1} \tag{8.17a}$$

$$\frac{D_r}{T_{c0}} = \frac{1}{L_0}(1-\theta)(1 - e^{-\alpha x_*}) \tag{8.17b}$$

$$\frac{D_r}{D_{c0}} = \frac{(1-\theta)(1 - e^{-\alpha x_*})}{\alpha}. \tag{8.17c}$$

The general closed-form erosion equation is applicable to both detachment and deposition, as well as several nonuniform slope conditions (Foster and Meyer, 1975). The erosion equation has been validated using field data (Foster and Meyer, 1972a, 1975) producing acceptable predictions. In addition, the closed-form equation was used in CREAMS (Foster *et al.*, 1980a, b) for identifying detachment and deposition of overland and channelized flow.

Example Problem 8.1 Illustration of the closed-form equation

At the end of a 100-ft slope, the sediment detachment potential is 0.01 lb/sec · ft^2 and the transport capacity is 2.0 lb/sec · ft. Estimate the sediment load at 2, 10, 50, 75, and 100 ft downslope, assuming steady uniform flow and an interrill detachment rate of 0.005 lb/sec · ft^2.

Solution:

1. Calculating parameters.

$$x_* = \frac{x}{L_0} = \frac{x}{100} = 0.01x$$

$$\alpha = \frac{L_0 D_{c0}}{T_{c0}} = \frac{(100\text{ ft})(0.01\text{ lb/sec}\cdot\text{ft}^2)}{2.0\text{ lb/sec}\cdot\text{ft}} = 0.5$$

$$\theta = \frac{L_0 D_i}{T_{c0}} = \frac{(100\text{ ft})(0.005\text{ lb/sec}\cdot\text{ft}^2)}{2.0\text{ lb/sec}\cdot\text{ft}} = 0.25.$$

2. Setting up the equations. From Eq. (8.17a)

$$q_S = [x_* - (1-\theta)(1 - e^{-\alpha x_*})/\alpha]T_{c0}$$

$$q_S = [x_* - (1-0.25)(1 - e^{-0.5x_*})/0.5][2.0]$$

or

$$q_S = 2.0x_* - 3.0(1 - e^{-0.5x_*}).$$

3. Tabulating the results.

x (ft)	$x_*=x/L_0$	q_s (lb/sec•ft)
2	0.02	0.010
10	0.10	0.054
50	0.50	0.336
75	0.75	0.562
100	1.00	0.819

Comments on Closed-Form Equation

The Foster and Meyer (1975) closed-form erosion model was developed using equations that describe the underlying relationships between detachment, deposition, and transport capacity. To predict erosion rates, the user must combine the closed-form equations with models of detachment capacity and transport capacity. In this combined form, the closed-form model has been used as the building block for other models such as CREAMS and the more recent WEPP model.

Interrill Erosion

Splash Erosion

Most soil detachment in interrill areas is a result of the force of falling raindrops (Young and Wiersma, 1973). For erosion to occur, detached particles must be transported from the site of detachment. Transport of detached soil particles occurs by raindrop splash as well as by overland flow. Each of these concepts is discussed separately in the following sections.

Raindrop Splash Detachment The quantity of detached soil in interrill areas is related to the kinetic energy and momentum of raindrops impacting the soil surface (Free, 1960; Quansah, 1981; Gilley and Finkner, 1985; Rose, 1960). Bubenzer and Jones (1971) proposed that splash erosion could be estimated from

$$s_S = ai^b k_e^c p_c^d \tag{8.18}$$

where s_S is the soil splash, i is rainfall intensity, k_e is the total applied rainfall kinetic energy, p_c is the soil percent clay, and a, b, c, and d are constants. Hirschi and Barfield (1988a) and Quansah (1981) suggest including the slope s as a parameter, or

$$s_S = ai^b k_e^c p_c^d s^e, \tag{8.19}$$

where s is slope, and e is a constant. Given sufficient information, Eq. (8.19) may be used to estimate soil splash from areas with various soils, topographic, and surface conditions.

Surface ponding dissipates raindrop energy; thus it has also been shown to affect significantly splash erosion (Palmer, 1965; Mutchler and Larson, 1971). In general, these studies show that splash increases up to a ponded depth of $\frac{1}{6}$ to 1 drop diameter with a decrease at deeper depths.

Raindrop Splash Transport The concentration of detached soil particles in raindrop splash is proportional to the kinetic energy of raindrops and should be equal in all directions. Thus on a level surface, the net soil transport should be zero even though the concentration of particles in the splash may be high. However, on a sloping surface, the splash trajectory goes farther downslope than upslope, just by geometric considerations. Transport by splash should, therefore, be proportional to both kinetic energy and slope, prompting Quansah (1981) to propose that

$$Q_t = ak_e^b s^c, \tag{8.20}$$

where Q_t is the net splash transport (kilogram per square meter), k_e is the total applied kinetic energy (joule per square meter), s is the soil slope, and a, b, and c are constants. In general, the transport slope exponent has been found to be much higher than the detachment slope exponent, indicating that splash transport is much more sensitive to slope than splash detachment.

Overland Flow Transport and Detachment: Sheet Erosion Overland flow transport and detachment results from shearing forces of the thin film of runoff water in interrill areas. Average shear stress in this film is the product of the surface slope and overland flow depth, a value that will be very small and typically less than a critical tractive or shear force necessary to initiate detachment (Foster and Meyer, 1975). Although the shear forces are low, which would indicate minimal transport capacity, raindrop impact on the surface increases the turbulence level and enhances transport capacity. Thus, overland flow is responsible for most of the transport of soil to rills, while detachment is primarily from raindrop impact (Young and Wiersma, 1973).

Sediment transport equations developed for streamflow conditions have been applied to overland flow conditions, with varying degrees of success. The Yalin (1963) equation is the most widely used transport equation for overland flow (Hirschi and Barfield, 1988a, b; Dillaha and Beasley, 1983). The Yang (1972) unit stream power equation has also been applied to overland flow (Moore and Burch, 1986); however, it yields better results under deep flow conditions.

Net Interrill Erosion

The quantity of sediment actually delivered to a concentrated flow network is net interrill erosion. Computationally, it is the net of detached soil particles minus deposited material. A conceptually sound method for estimating net interrill erosion would be to predict interrill detachment and transport capacity separately and use a computational framework like that in Fig. 8.3 to quantify the amount of sediment actually reaching rills (Hirschi and Barfield, 1988a, b). A frequently used alternative approach is to lump the processes together in a regression equation. Since both detachment and transport are functions of kinetic energy, which is in turn a function of rainfall intensity, Meyer (1981) proposed that net interrill erosion on bare soil could be predicted as a simple function of rainfall intensity and soil properties, or

$$D_i = aI^b, \tag{8.21}$$

where D_i is net interrill erosion, I is rainfall intensity, and a and b are constants. The b exponent increases as the percentage clay decreases, or (Meyer, 1981)

$$b = 2.1 - F_{cl}: \quad \text{clay content 20–50\%} \tag{8.22a}$$

$$b = 2.0: \quad \text{clay content} > 50\%, \tag{8.22b}$$

where F_{cl} is the fraction of clay.

Erodibility and crop factors can be added into Eq. (8.21) as was done in the USDA Water Erosion Prediction Project (WEPP) (Nearing *et al.*, 1989) resulting in

$$D_i = K_i I_e^2 C_e G_e (R_S/W), \tag{8.23}$$

where K_i is a baseline interrill erodibility for bare soil, I_e is effective rainfall intensity, C_e and G_e are dimensionless canopy and ground cover effects on interrill erosion, respectively, R_S is rill spacing, and W is a computed rill width. The effective rainfall intensity in Eq. (8.23) is calculated only for periods of rainfall excess since detached soil is not transported substantially during periods without surface runoff. Accounting only for the periods of rainfall excess, the effective rainfall intensity then becomes (Nearing *et al.*, 1989)

$$I_e^2 = \frac{1}{t_e}\int_{t_1}^{t_2} I^2\,dt, \tag{8.24}$$

where I is rainfall intensity, t is time (seconds), t_e is the cumulative time during which rainfall exceeds infiltration rate, and t_1 and t_2 are the start and end times when rainfall rate exceeds infiltration rate. The baseline erodibility factor, K_i, may be estimated from soil properties using relationships developed by Alberts *et al.* (1989). Canopy and ground effects are discussed in a subsequent section on the WEPP model.

As can be seen from the equations presented, a high degree of empiricism exists in the description of interrill erosion in current models. Operational models in the near future are likely to continue to contain such empiricism due to the complexity of the processes.

Rill Erosion

Rill Networks

Development and evolution of rills and rill networks are important erosion processes. Although significant progress has been made in describing the growth and development of individual rills, minimal progress has been made in describing rill networks. Future progress in erosion modeling will hinge on better description of the network processes in what is known as dynamic erosion models.

Network Development Rill networks are initiated on bare soil as a result of the microrelief when flow concentrates in the microrelief channels. Initially, the channels tend to be parallel (Leopold *et al.*, 1964), but overtopping of the ridges between rills as a result of changing flows results in some channels flowing at angles to the prevailing slopes and ultimate coalescence of rills into a dendritic network (Mosley, 1974). In laboratory studies, Mosley (1974) found that equilibrium sediment discharge rate for constant rainfall was related to total rill network length.

Rill Density and Its Effects on Erosion Rill density is the number of rills per unit width. Studies have shown that density varies with a number of factors such as slope steepness and length, runoff rate, soil texture, soil erodibility, and the presence or absence of rainfall (Meyer and Monke, 1965). On highly erodible soils, rill density has been shown to be high and rills have the same size from point of origin to end, indicating transport-limiting conditions (Ellison and Ellison, 1947). On less erodible soils, rill densities are less and the rills vary in width and depth from beginning to end, indicative of detachment-limiting conditions. Meyer and Monke (1965) observed that short slope lengths have higher rill densities relative to longer lengths.

Mathematical models of rill density are extremely limited. Li *et al.* (1980) developed a rill density model for laminar and turbulent flows, expressed on a unit width basis. Their model assumed that all rills were the same size for a specified distance downslope. Numerous empirical constants are required, limiting its use. Foster and Lane (1981) criticized the Li *et al.* model, indicating that their choice of a representative particle size in the Shield's diagram caused critical tractive force to be underestimated.

The effects of rill density on erosion are somewhat unclear. Using the KYERMO model, Hirschi and Barfield (1988b) performed a sensitivity analysis on the number of rills across a plot and found that the maximum sediment yield occurred at about six rills in 15 ft for their test conditions. They proposed that the decline in sediment yield at higher numbers of rills was due to lower flow rates in each rill as the surface runoff was distributed over more rills. They also showed that the effect of rill number on sediment yield is governed by the form of the rill detachment and boundary shear stress equations.

The WEPP Erosion model represents a rill network as a series of parallel rills; hence rill density is analogous to rill spacing. Based on a sensitivity analysis of the WEPP model, Nearing *et al.* (1989) indicated that WEPP predictions were somewhat insensitive to rill density; hence a default rill spacing of 1 m is used in the WEPP model. Such an assumption, however, ignores the complex interactions of rill networks. Lewis *et al.* (1990) developed a model similar to the WEPP model, but with the capability of utilizing a random distribution of rill numbers and flow in rills. Results from a sensitivity analysis from the Lewis model shows that ignoring the stochasticity of rill networks can make a significant difference in predicted erosion when a nonerodible layer is encountered, as frequently occurs in tilled soils. The presence of a nonerodible layer is not considered in WEPP.

Growth and Development of Individual Rills

Development and growth of an individual rill is governed by rill detachment potential, transport capacity, sediment load, and their interactions. The following sections discuss rill detachment potential, which includes detachment from rill incision, headwall cutting, and sidewall sloughing.

Rill Incision Shear stresses along a concentrated flow boundary will lead to incision of a channel if the shear exceeds the critical tractive force. The rate of soil detachment in rills due to rill incision is typically assumed to vary with shear excess and may be expressed as (Foster, 1982)

$$D_{rc} = a(\tau - \tau_c)^b, \qquad (8.25)$$

where D_{rc} is the maximum or potential rill detachment rate, τ is flow shear stress along the rill boundary, τ_c is the critical shear stress necessary to detach soil particles, and a and b are constants. The constant b is close to 1.0 and is typically assumed to be 1.0.

If detachment is to be estimated in a channel, it is necessary to know the shear distribution along the boundary. The average bed shear stress, τ_a, for uniform flow is given as

$$\tau_a = \gamma RS, \qquad (8.26)$$

where γ is specific weight of water, R is hydraulic radius, and S is rill bed slope. Since the distribution of the shear around the rill boundary is nonuniform, the use of average shear stress to estimate detachment potential could result in significant errors in estimating channel shape.

Rill Geometry The importance of rill shape to rill growth and development is primarily a result of its effect on shear distribution. Prior to reaching a nonerodible layer, rill shape may be approximated as a rectangle (Lane and Foster, 1980; Foster and Lane, 1983) with a width given as

$$W = aQ^b, \qquad (8.27)$$

where W is channel width, Q is discharge in the rill, and a and b are constants. Although other shapes have been suggested (Rohlf, 1981), it is likely that a rectangular cross section is developed as a result of side sloughing once the rill encounters a nonerodible layer.

Critical Shear Stress A soil's resistance to the shearing forces of concentrated flow is determined by the critical shear stress, sometimes referred to as critical tractive force. For noncohesive soils, Shields diagram (Shields, 1936) is the method most widely used to describe critical tractive force of individual particles. The original Shields diagram does not apply to sediment particles of low specific gravity and small diameter; however, Mantz (1977) extended the Shields diagram for smaller particles as shown in Chapter 7.

For cohesive materials, critical shear stress has been related to a number of soil properties including soil shear strength, soil salinity, and moisture content (Alberts *et al.*, 1989); percentage clay, mean particle size, dispersion ratio, vane shear strength, organic matter content, cation exchange capacity, and calcium–sodium ratio (Lyle and Smerdon, 1965); and plasticity index (Smerdon and Beasley, 1961). Foster (1982) recommended the equation of Smerdon and Beasley (1961) based on the dispersion ratio; however, Hirschi and Barfield (1988a) used the relationship from Smerdon and Beasley (1961) based on percentage clay.

Typical critical shear stresses range from 1 to 30 Pa. For agricultural soils, Foster and Meyer (1975) recommended an average value of 2.4 Pa. Alberts *et al.* (1989) developed regression equations using the extensive WEPP field data (including corrections from Flanagan, 1990) and found that critical shear stress of cropland soils may be predicted from very fine sand fraction, calcium carbonate fraction, sodium absorption ratio, soil specific surface area (milligrams ethylene glycol mono-ethyl ether adsorbed per gram soil), sand fraction, water-dispersible clay fraction, and clay fraction. For cropland soils with clay fraction greater than 0.30, Alberts *et al.* (1989) found that critical tractive force may be predicted from volumetric water content. Other relationships are being developed from the WEPP data set (i.e., Wilson 1993). No doubt, the final result will be considerably different from the original relationships.

Sidewall Sloughing and Headwall Cutting Sidewall sloughing and headwall cutting are significant mechanisms in the propagation of rills. Sloughing results from gravitational forces, flow hydraulics, and their combined effects. A sidewall's resistance to failure varies with slope geometry and soil properties such as cohesion, bulk density, void ratio, moisture content, and others. For rills, sidewall failure typically results from gravity forces acting on an overhang caused by undercutting, and translational or rotational slips caused by shear failure along an internal surface. Translational slips occur along a plane typically parallel to the surface slope, and rotational slips occur along a circular arc.

Hirschi and Barfield (1988a) incorporated rill sidewall stability into their erosion model based on a critical slope concept. Once the sidewall reaches a critical slope, it was assumed to slough off forming a stable slope and depositing the detached soil mass into the rill. Bradford *et al.* (1973) developed a two-dimensional rotational slip-type bank failure model. They found that the factors controlling sidewall stability are water table height, soil cohesion, and seepage rate.

The upslope propagation of rills through headcutting may contribute significant amounts of sediment. The headwall is an abrupt break in the longitudinal channel profile (Schumm *et al.*, 1984) and is the transition between wide shallow channelized flow and narrow deeper flow. Individual headcuts migrate upstream, and several headcuts may exist along the same channel. Most of the available literature on headcut development and propagation is based on drainage basin morphology.

Recent developments have included models of channel erosion that include physically based models of headwall cutting and sidewall sloughing. Fogle *et al.* (1992) summarize the recent studies on headwall cutting and channel scour and propose a model, denoted as CHANNEL, that predicts channel incision, headwall development, propagation, and washout. The model considers detachment as a function of shear excess as given in Eq. (8.25) and an interaction of transport capacity and sediment load as given by Eqs. (8.3) and (8.5). Shear downslope of the headwall is calculated by submerged jet theory or by impinging jet theory, as appropriate. Rohlf (1993) summarizes recent studies on sidewall stability. They proposed a dynamic model of channel wall failure that includes the effect of water movement into and out of the channel wall on slope stability. The analysis requires simultaneous solution of saturated–unsaturated flow equation for groundwater movement (see Chapter 11) and stress–strain dynamics for channel wall stability.

Rill Erosion Models

To date, physically based rill erosion models have been based almost exclusively on shear excess concepts. Assuming a rectangular channel cross section and erosion based on shear excess, Lane and Foster (1980) and Foster and Lane (1983) developed a deterministic channel erosion model that was incorporated into CREAMS for describing ephemeral gully growth in a tilled agricultural field. It has also been applied to rill erosion by Storm *et al.* (1990) and Lewis *et al.* (1990).

In the Foster and Lane (1983) model, channel development is partitioned into two distinct stages for steady flow rate. During the initial stage, the channel bottom erodes uniformly downward at a width equal to their so-called equilibrium width. A second stage of development occurs when the channel bottom reaches a nonerodible layer, after which lateral expansion is assumed to occur with vertical sidewall sloughing. A detailed discussion on the development and implementation of the model is presented in a subsequent section on concentrated flow modeling.

Models of rill erosion can also be developed by calculating the shear distribution around a rill (Rohlf, 1981; Hirschi and Barfield, 1988a; Fogle *et al.*, 1992). An erosion model incorporating rill network development was presented by Mossaad and Wu (1984). Their model combined a stochastic surface roughness model with a deterministic interrill and rill erosion model describing rill network development over time.

A process-based soil erosion model was developed for WEPP (Nearing *et al.*, 1989). The rill erosion model component is based on shear excess, and rill detachment depends on rill erodibility, hydraulic shear stress, surface cover, below-ground residue, consolidation, and the ratio of sediment load to transport capacity. Net deposition is assumed proportional to excess sediment load, and sediment routing is performed by applying the steady-state continuity equation.

Average shear stress for a rectangular rill is found using a rill width calculated from rill flow discharge, average slope gradient, hydraulic radius, and the ratio of friction factors for the soil and the total cross section in the rill. Transport capacity in rills is found using a simplified transport equation, calibrated with the Yalin transport equation. Further details are given in a subsequent section on theoretical models.

RILL AND INTERRILL EROSION MODELING: USLE / RUSLE EMPIRICAL MODELS

Basic Relationships

Soil erosion by water is the soil lost from a given slope, usually predicted on a per unit area basis. Sediment yield, on the other hand, is the amount of sediment that passes a given point on a watershed. Some of the sediment that leaves a given slope is deposited; hence, sediment yield and soil erosion are not the same and should not be confused as such.

The Universal Soil Loss equation, (USLE), as discussed earlier in the historical perspective, was developed to predict soil erosion, not sediment yield (Wischmeier and Smith, 1965, 1978). In fact, the term soil loss is somewhat of a misnomer. Much of the soil displaced in the erosion process is deposited subsequently in flatter and/or vegetated areas where the transport capacity is lower. Thus not all eroded soil is "lost" from the field.

The USLE is a relationship that has been widely used for planning purposes to predict the impact of land use on soil erosion. Originally developed to predict annual soil erosion averaged over long time periods, it has been modified to estimate monthly and single-storm erosion. As would be expected, the standard error of prediction increases for short-term and single-storm predictions.

Improvements to the USLE based on more recent data as well as a new evaluation of the original USLE data base have resulted in a modification known as the Revised USLE or RUSLE (Renard *et al.*, 1993a). The RUSLE/USLE are multiplicative relationships, or

$$A = RKLSCP. \tag{8.28}$$

A is the average soil loss per unit of area, expressed in units selected for K and the time period specified by R. Normal English units are tons/acre/year, but other units are used.

R is the rainfall/runoff factor, which is the number of rainfall units for rainfall energy and runoff, plus a factor for runoff from snowmelt.

K is the soil erodibility factor, which is the rate of soil loss per unit of R (erosion index units) for a given soil under continuous fallow with up and downhill cultivation on a slope of 9% with a slope length of 72.6 ft (22.1 m).

L is the slope length factor, which is the ratio of soil loss from a defined slope length relative to that from a slope length of 72.6 ft (22.1 m).

S is the slope steepness factor, which is the ratio of soil loss from a slope with a given steepness relative to that from a 9% slope.

C is the cover and management factor, which is the ratio of soil loss from an area with a given cover and management relative to that from an identical area in continuous fallow.

P is the supporting conservation practice factor, which is the ratio of soil loss from a field with a conservation support practice such as contouring relative to that with straight row farming up- and downhill.

The RUSLE was designed to be of identical form to the USLE so that the USLE parameters could be used where desirable. This interchange is particularly useful in the case of the C factor. The RUSLE approach to the C factor is sufficiently complicated to require a computer for solution of most practical problems. For many estimates, the accuracy of the USLE data base for C is adequate. In this text, the USLE data base for C factors is included along with some of the data base for the RUSLE. For simplicity, further reference to the RUSLE/USLE are simply to RUSLE; however, where distinction between the two are important, the differences will be pointed out.

Several cautions about the use of RUSLE should be considered. Predictions from RUSLE represent soil loss averaged over many storms and years. It also represents averages over a total field or disturbed area. At points on a slope or field, the soil loss will almost always be less than or greater than the average values. For example, on long slopes, the upper part of the slope will have lower erosion rates than the lower part of the slope, but the average over the entire slope over a long period of time should be approximated by RUSLE predictions. Also, the energy content of rainfall with a given intensity, as predicted by the R factor, represents a value averaged over a large number of storms over a long period of time. The value for a given storm could be much greater or less than that predicted by the R factor.

The original USLE has been extended to forest conditions and to construction applications. The RUSLE manual does not include information for these conditions, but refers the reader to Dissmeyer and Foster (1981, 1984). The Dissmeyer and Foster data are incorporated into the tables in this chapter. Also

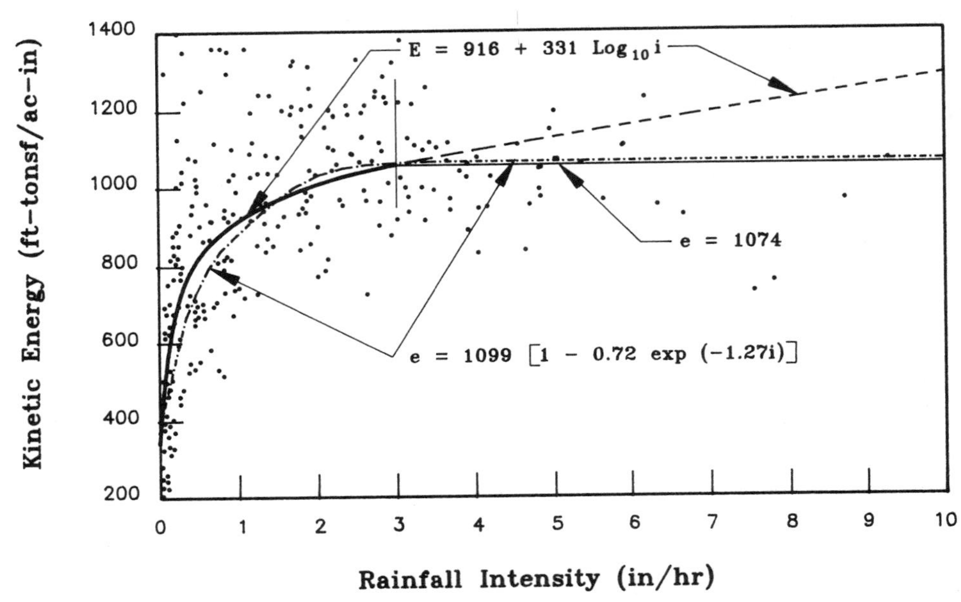

Figure 8.5 Relationship between rainfall intensity and rainfall energy. The prediction equations are those given by Eqs. (8.29a) and (8.29b). Data points are from Hirschi *et al.* (1983).

incorporated are data from other sources for construction and mining. The RUSLE is available in computerized format from the Soil Conservation Society of America.

Rainfall Energy Factor *R*

Selection of the *R* Factor

After evaluation of correlations between soil erosion and a number of rainfall parameters, the R factor selected by Wischmeier and Smith (1958) was the product of rainfall energy and maximum 30-min intensity divided by 100 for numerical convenience, known as the EI_{30} index. On an annual basis, the EI_{30} value is the sum of values over the storms in an individual year. R factors for the USLE and RUSLE are identical except for additional relationships in the RUSLE that account for ponding.

Calculations of rainfall energy require an algorithm relating energy to some measurable parameter. Up to an intensity of 3 in./hr, rainfall energy increases with storm intensity as a result of the fact that the drop size and fall velocity increase with intensity. Above 3 in./hr, the drop size reaches its maximum size and energy remains constant. Based on an analysis of Gunn and Kinzer (1949) data, Wischmeier and Smith (1958) proposed that rainfall energy is related to intensity by

$$e = 916 + \log_{10} i \qquad i \leq 3 \text{ in./hr}$$
$$e = 1074 \qquad i > 3 \text{ in./hr}, \qquad (8.29a)$$

where e is the kinetic energy in ft · tonsf/acre · in. and i is the average intensity of the storm in in./hr. Tonsf refers to force of rainfall impact as opposed to tons mass of sediment being eroded. Other relationships have been proposed for kinetic energy, but the relationship given in Eq. (8.29a) is the most widely used.

In the RUSLE, the relationship of Brown and Foster (1987) was used for the Western U.S., or

$$e = 1099[1 - 0.72\exp(-1.27i)]. \qquad (8.29b)$$

Renard *et al.* (1993a, Appendix C) recommend the use of Eq. (8.29b) in all future calculations of rainfall energy for all sections of the U.S., since it seems to better fit the data at lower intensities. To convert to total energy in a storm, e is multiplied by the depth of rainfall, or

$$E = eP, \qquad (8.30)$$

where E is total energy in a storm in ft · tonsf/acre and P is total storm depth of rainfall in inches.

Data on rainfall energy are plotted in Fig. 8.5 along with a plot of Eq. (8.29). This illustrates that the energy content predicted by Eq. (8.29) is a value averaged over a large number of storms. For any individual storm, the actual energy could be much larger or smaller than that predicted.

Using the EI_{30} index, the R factor in Eq. (8.28) is given by

$$R = \Sigma EI_{30}/100, \qquad (8.31)$$

where I_{30} is the maximum 30-min intensity (iph) for the storm and the 100 in the denominator is used for numerical convenience. Details on computation of rainfall energy computation are given in Renard *et al.* (1993a, Appendix C).

Figure 8.6 Isolines of annual R factor for the Eastern United States (after Renard *et al.*, 1993b). Maps for areas in and west of the Rockies are in Appendix 8A. R factor is in ft · tonsf · in./acre · hr · year. To convert to SI units, MJ · mm/ha · h · y, multiply by 17.02.

Units on the annual R factor are hundreds of ft · tonsf · in./acre · hr · year. The hundreds term comes from the 100 divisor in Eq. (8.31). In further reference, the units on R will be given as ft · tonsf · in./acre · hr · year for simplicity and the hundreds will be understood. The units on single storm erosivity would be ft · tonsf · in./acre · hr · storm.

To convert the annual R to SI units, MJ · mm/ha · hr · year, multiply the English Units by 17.02. A summary of other conversion factors is given in the Appendix. As a note ft · tonsf · in./acre · hr · year means

$$\frac{\text{ft tonsf in.}}{\text{acre hr year}}.$$

Average Annual R Values

Using Eq. (8.31), Wischmeier and Smith (1965) computed values for the R factor for stations east of the Rockies and extended the values to the Western States utilizing a correlation between 2-year, 6-hr rainfall and the R factor (Wischmeier and Smith, 1978). In the development of RUSLE, additional analyses were made on longer data bases and on hourly rainfall for stations in and west of the Rockies. The data are summarized in Fig. 8.6 for the eastern United States.

Corrections for Ponding

Studies have shown that ponded water on flat slopes tends to absorb rainfall energy and retard erosion

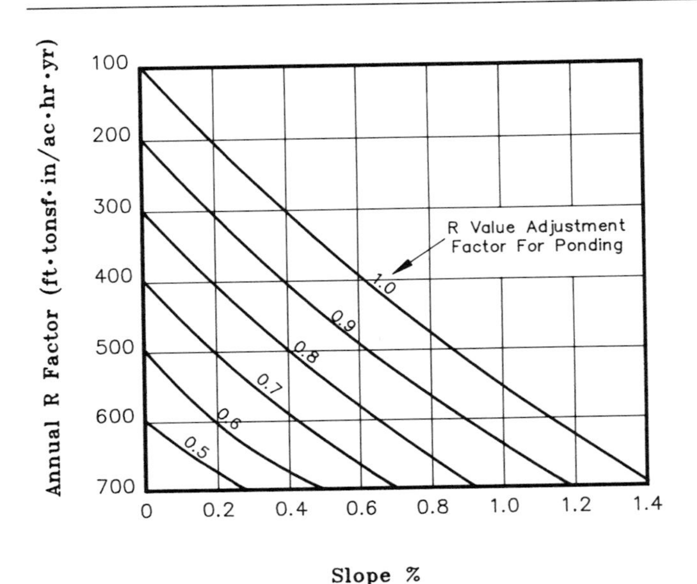

Figure 8.7 Multiplier for the *R* factor to account for ponding on flat slopes (after Renard *et al.*, 1993b).

(Mutchler and Larson, 1971). The RUSLE contains a modification for *R* that is a function of the *R* factor and slope, as shown in Fig. 8.7. The corrected *R* factor is developed by multiplying the uncorrected *R* factor from Fig. 8.6 by the correction factor in Fig. 8.7. The correction factors in Fig. 8.7 are for annual erosion. Corrections for single storms have not been developed; however, as a first estimate, one might use Fig. 8.7.

Distributions by Months

The *R* factor is not uniformly distributed over the year, but has a monthly distribution that varies widely with location. *R* factor distributions have been developed for the various zones in the U.S. These zones are shown in Fig. 8.8. Data on the cumulative percentage of annual *R* by months is given in Table 8.1 for selected zones with a full tabulation given in Appendix 8A.

Single Storm *R* Factors

Although the USLE and RUSLE were developed for predicting long-term annual soil erosion, the relationship has been used to predict single-storm erosion. In

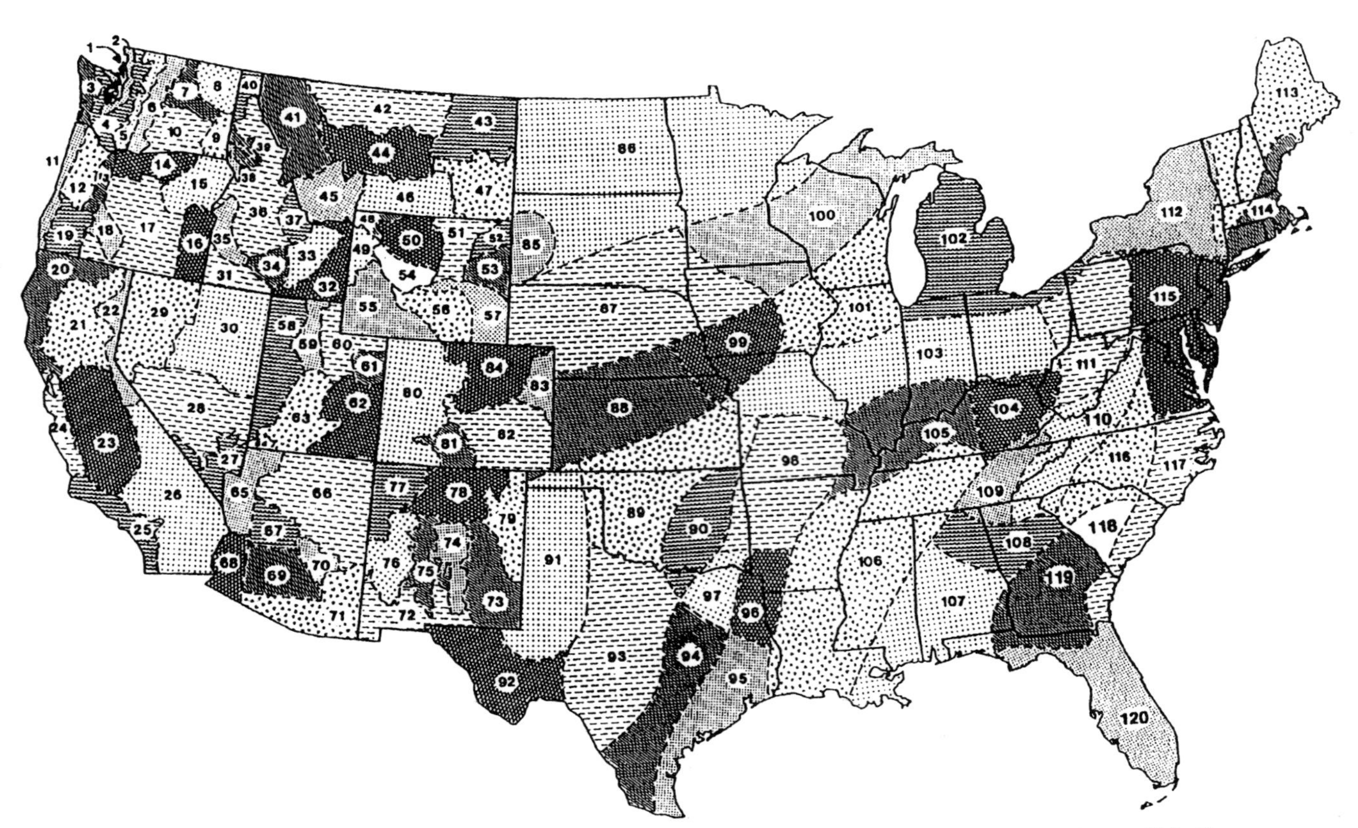

Figure 8.8 *R* factor distribution zones (after Renard *et al.*, 1993b). Zones 1–120 are for the contiguous U.S. Zones 121–139 are for Hawaii. Consult the local conservation office in Hawaii for distribution zones. Zone 140 is for Pullman, WA and Northwest dryland winter wheat.

Table 8.1 R Factor Distributions for Selected Zones Shown in Fig. 8.8 (after Renard *et al.*, 1993b)[a]

	Geographic area from Fig. 8.8.						
Date	10	25	86	104	105	112	119
1/1	0.0	0.0	0.0	0.0	0.0	0.0	0.0
1/15	0.3	9.8	0.0	2.0	1.0	0.0	2.0
2/1	0.5	20.8	0.0	3.0	3.0	0.0	4.0
2/15	0.9	30.2	0.0	5.0	6.0	1.0	6.0
3/1	2.0	37.6	0.0	7.0	9.0	2.0	8.0
3/15	4.3	45.8	0.0	10.0	12.0	3.0	12.0
4/1	9.2	50.6	1.0	13.0	16.0	4.0	16.0
4/15	13.2	54.4	2.0	16.0	21.0	5.0	20.0
5/1	18.0	56.0	3.0	19.0	26.0	7.0	25.0
5/15	22.7	56.8	6.0	23.0	31.0	12.0	30.0
6/1	29.2	57.1	11.0	27.0	37.0	17.0	35.0
6/15	39.5	57.1	23.0	34.0	43.0	24.0	41.0
7/1	46.3	57.2	36.0	44.0	50.0	33.0	47.0
7/15	48.8	57.6	49.0	54.0	57.0	42.0	56.0
8/1	51.1	58.5	63.0	63.0	64.0	55.0	67.0
8/15	57.2	59.8	77.0	72.0	71.0	67.0	75.0
9/1	64.4	62.2	90.0	80.0	77.0	76.0	81.0
9/15	67.7	65.3	95.0	85.0	81.0	83.0	85.0
10/1	71.1	67.5	98.0	89.0	85.0	89.0	87.0
10/15	77.2	68.2	99.0	91.0	88.0	92.0	89.0
11/1	85.1	69.4	100.0	93.0	91.0	94.0	91.0
11/15	92.5	74.8	100.0	95.0	93.0	96.0	93.0
12/1	96.5	86.6	100.0	96.0	95.0	98.0	95.0
12/15	99.0	93.0	100.0	98.0	97.0	99.0	97.0

[a]Values are cumulative percentage of total annual R for the day indicated. Tabulations generously provided by K. Renard of the USDA–ARS (Tucson, AZ). Additional values are given in Appendix 8A.

this computation, R is estimated for an individual storm. Wischmeier and Smith (1965, 1978) evaluated R factors for individual historic storms of all durations and summarized the information by return period. These values were not upgraded in the RUSLE, so values from the original analysis are given in Table 8.2 for selected locations. A more extensive listing is given in Appendix 8A.

Values in Table 8.2 are for historic storms. It is also useful to have values for DDF or SCS synthetic storms used for design of structures as discussed in Chapter 3. Synthetic storm R values have been determined for SCS type I, type IA, type II, and type IIA storms by Cooley (1980) or

$$R_{st} = \frac{a_1 P^{f(D)}}{D^{b_1}} \tag{8.32}$$

where R_{st} is the synthetic storm R factor, P is precipitation in inches corresponding to a duration D in hours, a_1 and b_1 are constants, and

$$f(D) = 2.119D^{0.0086}. \tag{8.33}$$

Values for a_1 and b_1 are summarized below.

Type storm	a_1	b_1
I	15.03	0.5780
IA	12.98	0.7488
II	17.90	0.4134
IIA	21.51	0.2811

Ateshian (1974) developed similar power relationships relating R_{24} to P and D; however, $f(D)$ was represented by a constant. Cooley (1980) indicates that the Ateshian equations can over- or underestimate by as much as 40% for high-intensity short-duration storms.

Cooley (1980) summarized data sets comparing energy content of historic storms to predictions from Eq. (8.32). As expected, the synthetic storm value can be much greater or less than historic storms. It is important to remember that SCS storms are design

Table 8.2 Historic Single Storm R Factors for Selected Cities (after Wischmeier *et al.*, 1978)[a]

	Index values normally exceeded once in N years				
Location	N= 1	2	5	10	20
Alabama: Montgomery	62	86	118	145	172
California: Red Bluff	13	21	36	49	65
Georgia: Columbus	61	81	108	131	152
Indiana: Indianapolis	29	41	60	75	90
Kentucky: Lexington	28	46	80	114	151
Kentucky: Middlesboro	28	38	52	63	73
Maine: Portland	16	27	48	66	88
New York: Buffalo	15	23	36	49	61
North Carolina: Raleigh	53	77	110	137	168
Oklahoma: Ardmore	46	71	107	141	179
Oregon: Portland	6	9	13	15	18
South Dakota: Rapid City	12	20	34	48	64
Tennessee: Memphis	43	55	70	82	91
Wyoming: Cheyenne	9	14	21	27	34

[a]Adapted from Wischmeier and Smith (1978). R factor values are in English units: hundreds of ft•tonsf•in/acre•hr•year. To convert to SI units, MJ•mm/ha•h•year, multiply by 17.02. Additional values are given in Appendix 8A. Isolines of 10-yr single-storm factors are given in Figs. 8A.5–8A.8.

storms and are not intended to replicate particular historic events.

Example Problem 8.2 Determining R factors

Determine the following R factors for extreme Southwest Georgia and extreme Northwestern New York: Annual R, January, April, and July R, 10-year historic storm R, and 10-year, 24-hr synthetic storm R. What would the R factor be for a specific location in Southwest Georgia if the slope at that location were 0.1%?

Solution:

1. *Annual R factor.* From Fig. 8.6, the average annual R in English units is

SW Georgia, $R = 450$
NW New York, $R = 80$.

2. *January, April, and July R.* From Fig. 8.8, SW Georgia is in zone 119 and NW New York is in zone 112. Values in Table 8.1 are cumulative percentage of total annual R. The incremental percentage change of annual R can be determined from Table 8.1 for each month and multiplied by the annual R to get the monthly value. For April in SW Georgia (zone 118), the incremental change in percentage R is from 16.0 to 25.0 or 9%. With an annual R of 450, the incremental R during April is 0.09×450 or 40.5. Tabulations for all time periods are shown below.

	January 1/1–2/1 fraction/R value	April 4/1–5/1 fraction/R value	July 7/1–8/1 fraction/R value
SW Georgia	0.04/18.0	0.09/40.5	0.20/90.0
NW New York	0.00/0.00	0.03/2.4	0.22/17.6

3. *Historic return period values.* From Table 8.2, the 10-year return period single-storm R for the nearest cities are (English units)

Columbus, GA, $R_{10} = 131$
Buffalo, NY, $R_{10} = 49$.
These values compare favorably to values in Fig. 8A.5.

4. *Synthetic storms.* Both locations utilize type II storms; hence a_1 and b_1 for Eq. (8.32) are 17.90 and 0.4134. From Appendix 3A in Chapter 3, the 10-year 24-hr storm precipitation for SW Georgia is 7.5 in. and NW New York is 3.5 in. Thus, for a duration of 24 hr, the R values can be calculated as

$$f(D) = 2.119(24)^{0.0086} = 2.18$$

$$\text{SW Georgia:} \quad R_{st} = \frac{17.90(7.5)^{2.18}}{(24)^{0.4134}} = 389$$

$$\text{NW New York:} \quad R_{st} = \frac{17.90(3.5)^{2.18}}{24^{0.4134}} = 73.8.$$

5. *Correcting R factors for ponding.* The R factor for SW Georgia needs to be corrected for ponding on flat slopes. From Fig. 8.7, the correction factor for a slope of 0.1% and an R factor of 450 is 0.70. Therefore, the corrected R value is

$$R_{cor} = (450)(0.70) = 315.$$

Special R Factors for the Pacific Northwest

In the Pacific Northwest (PNW), rain or snow falling on cropland produces erosion in excess of that expected from the R factor based on rainfall energy and maximum 30-min intensity, EI_{30}. Based on experimental data taken in the PNW, Renard *et al.*, (1993b) propose that an equivalent R can be calculated from the annual precipitation, or

$$R_{eq} = 5.9P, \tag{8.34}$$

where R_{eq} is the equivalent R factor for the cropland in ft · tonsf · in./acre · hr · year and P is annual rainfall in inches.

Special Considerations for Higher Elevation and Winter Conditions

Wischmeier and Smith (1978) proposed in Handbook 537 that precipitation falling in the form of snow could be multiplied by 1.5 and added to the R factor to account for the impacts of snowfall on erosion. The developers of the RUSLE found this unsatisfactory and do not recommend it. Other than the above-mentioned analysis for the Pacific Northwest, they recommended that no correction be made for snowfall.

At higher elevations in the West where heavy snowfall is observed, it is possible that the EI_{30} index might be too high as a result of high snowfalls that accumulate to large depths and do not generate heavy runoff rates. Predictions of erosion at these high altitudes, based on the EI_{30} index, would be correspondingly too high. When evaluating erosion under these conditions, the R factor based on total precipitation should be reduced to eliminate the snow component.

Erodibility Factor K

Definition of Erodibility

Ideally, soil erodibility is a measure of a soil's resistance to the erosive powers of rainfall energy and runoff. Practically, in the RUSLE, soil erodibility is an integration of the impacts of rainfall and runoff on soil loss for a given soil. Experimentally, soil erodibility is the soil loss per unit rainfall index on a standard erosion plot, i.e., a plot under fallow conditions on a slope of 9% with a slope length of 72.6 ft with up- and downslope tillage. Under these conditions, L, S, C, and P are all equal to 1.0, hence

$$K = \frac{\text{measured erosion}}{\Sigma EI_{30}}. \quad (8.35)$$

Practically, one seldom encounters standard conditions for a test; hence, data are taken under nonstandard conditions, and corrections are made based on accepted relationships for L, S, C, and P. Inaccuracies in any of these other parameters would be reflected in the estimated K values.

Estimating *K* Factors for Average Annual Erosion

A number of studies of soil erodibility have been made with the USLE/RUSLE format as summarized by Romkens *et al.* (1993). In the USLE, K is assumed to be constant throughout the year. Tables of K values are available from local Soil Conservation Service offices for most soils in the U.S. K values are also tabulated in the more recent soil survey manuals. In the absence of published data, a widely used relationship for predicting erodibility is a nomograph by Wischmeier *et al.* (1971), which was developed from data collected on 55 midwestern agricultural soils. Soil erodibility in the nomograph is predicted as a function of five soil and soil profile parameters:

- Percentage silt (MS; 0.002–0.05 mm).
- Percentage very fine sand (VFS; 0.05–0.1 mm).
- Percentage sand (SA; 0.1–2 mm).
- Percentage organic matter (OM).
- Structure (S_1).
- Permeability (P_1).

It is important to note that the size ranges given here are not standard for some particle classifications. Codes for structure and permeability are given in USDA soil survey manuals (Soil Conservation Service, 1983) available for most counties in the U.S. and in some foreign countries. The nomograph is shown in Fig. 8.9.

An analytical relationship for the nomograph in Fig. 8.9 is (Wischmeier *et al.*, 1971)

$$K = \frac{2.1 \times 10^{-4}(12 - \text{OM})M^{1.14} + 3.25(S_1 - 2) + 2.5(P_1 - 3)}{100}, \quad (8.36)$$

where K is soil erodibility in tons per acre per unit

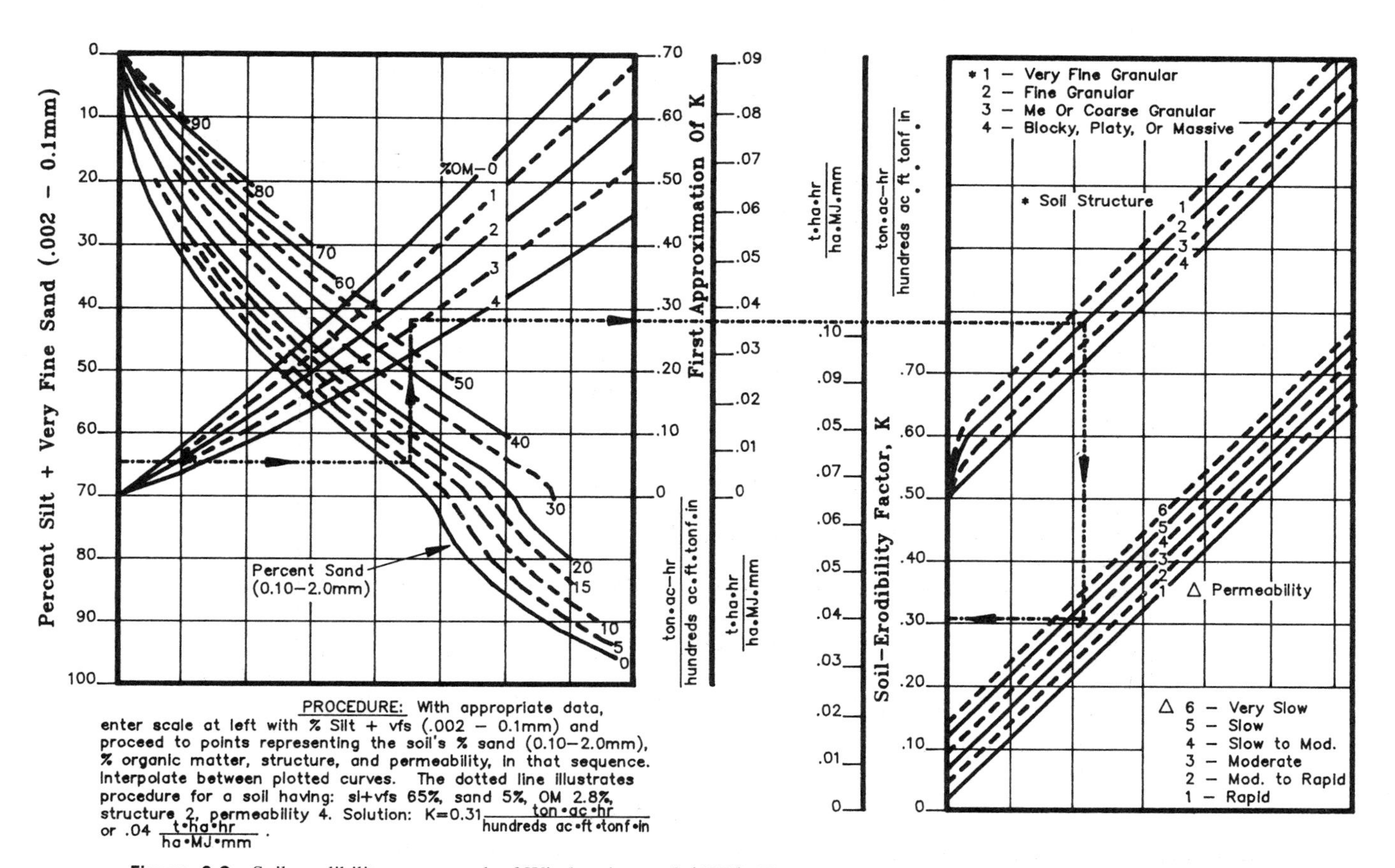

Figure 8.9 Soil erodibility nomograph of Wischmeier *et al.* (1971). The axes for K are scaled in both English and SI units.

rainfall index (tons · acre · hr/hundreds · acre · ft · tonsf · in.), OM is the percentage organic matter, P_1 is the permeability index, S_1 is the structure index, and M is a function of the primary particle size fractions given by

$$M = (\%MS + \%VFS)(100 - \%CL), \quad (8.37)$$

where %CL is percentage clay (< 0.002 mm) and other terms are as defined above. Equation (8.36) is valid for %MS + %VFS less than 70.

Since the units on K (English) are tons per acre per unit rainfall index and since the R factor is the EI_{30} index as given by Eq. (8.31), the units on K are A/R or tons · acre · hr/hundreds · acre · ft · tonsf · in., where tonsf means tons of force from the energy equation and tons are mass of sediment being eroded. The accuracy of Eq. (8.36) was evaluated by Romkens *et al.* (1993) using data from a number of sources. In general, the nomograph worked well for midwest soils, but did not work well on soils from Hawaii or for subsoils.

Seasonal Variation in K Values

A number of researchers have shown that soil erodibility varies with antecedent moisture and with freezing and thawing (Mutchler and Carter, 1983). When averaged over a number of years, freezing and thawing and

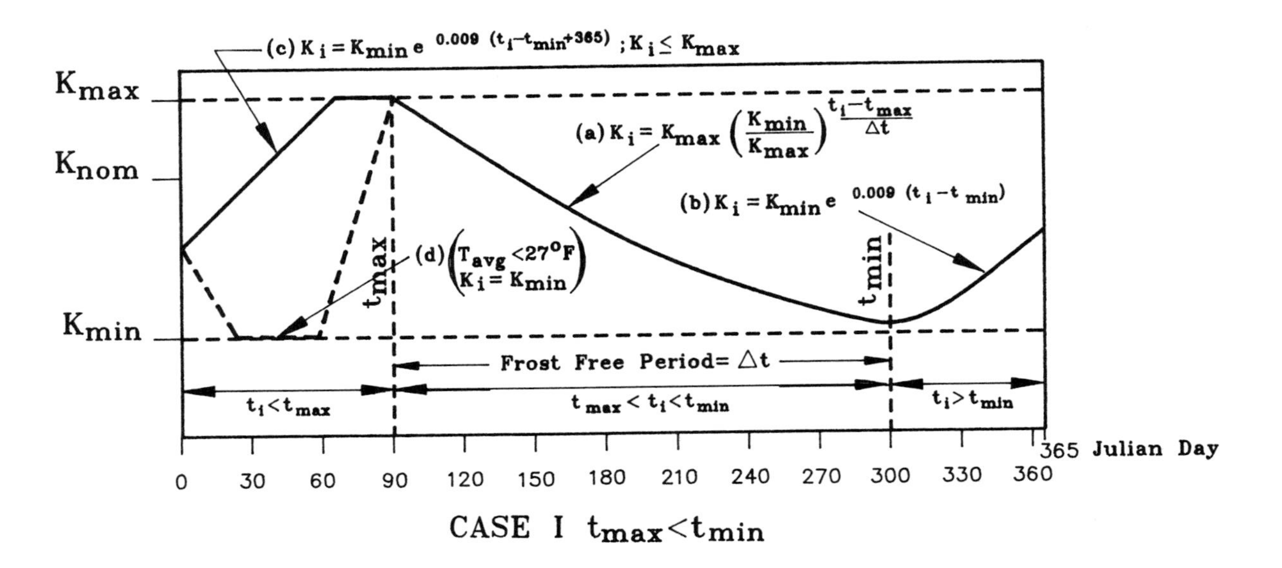

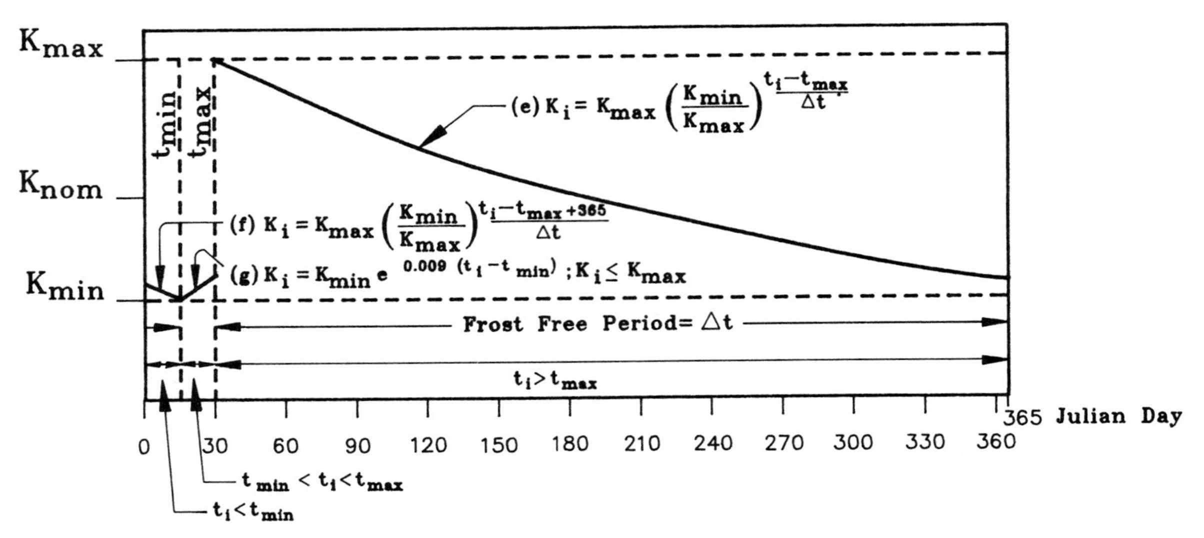

Figure 8.10 RUSLE procedure for predicting seasonal changes in K factor (after Romkens *et al.*, 1993). It is not recommended that these procedures be used for locations west of the Rockies.

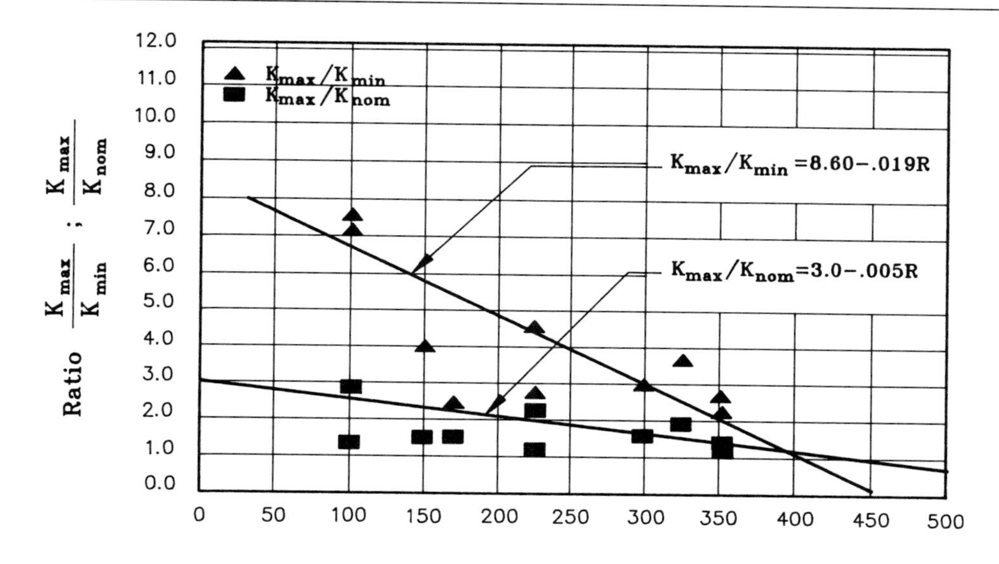

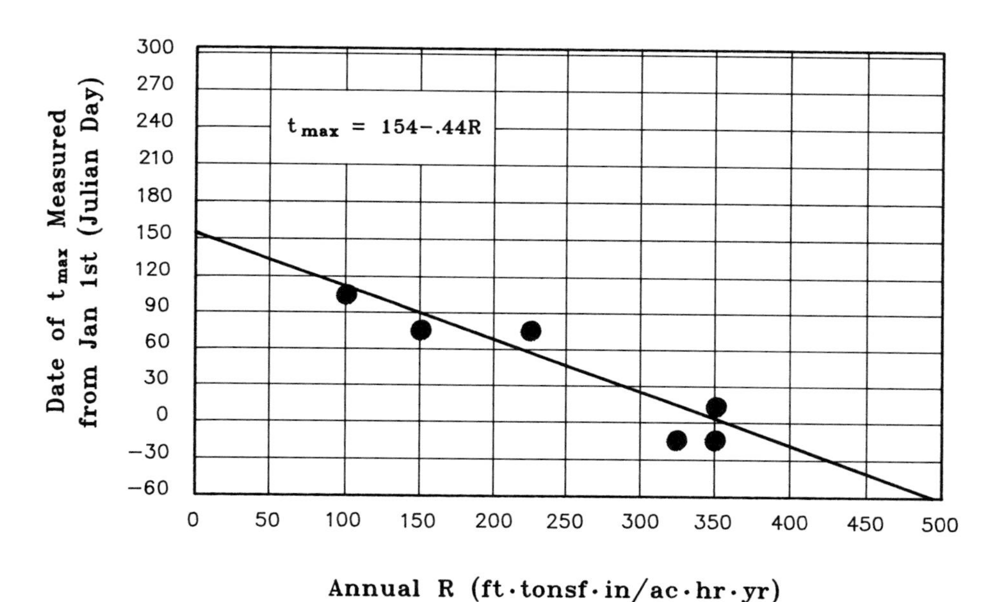

Figure 8.11 RUSLE relationship between K_{max}/K_{min}, K_{max}/K_{nom}, and the time of maximum erodibility t_{max} with the annual R factor (after Romkens *et al.*, 1993). Julian day is time measured from January 1. It is not recommended that these procedures be used for locations west of the Rockies.

high antecedent moisture conditions tend to occur on a predictable basis. Procedures were developed in the RUSLE to account for this variability, as shown in Fig. 8.10. In the RUSLE, correlations were made between the annual R factor and K_{max}/K_{min} as well as the time of maximum erodibility t_{max}, as shown in Fig. 8.11. In these relationships, t_{max} is the time (Julian days) of maximum erodibility and t_{min} is the time (Julian days) of minimum erodibility. Julian days are days numbered sequentially from 1 to 365 starting with January 1.

The procedures illustrated in Figs. 8.10 and 8.11 are divided into two categories. Case I is for t_{max} less than t_{min}, and Case II is for t_{max} greater than t_{min}. Procedures for using the relationships are illustrated in Example Problem 8.3.

Example Problem 8.3 Estimating K values

Estimate the annual K value using the Wischmeier nomograph for a soil in central Minnesota that has the following characteristics

%SA = 20% (0.1 – 2.0 mm)
%VFS = 5% (0.05 – 0.1 mm)
%MS = 45% (0.002 – 0.05 mm)
%CL = 30% (< 0.002 mm)
%OM = 2%
S_1 = soil structure = medium = 3
P_1 = permeability = rapid = 1
R = annual R factor = 100 (English units)

Correct the annual K factor, accounting for the seasonal variability in K and the variability in the EI_{30} index. The average number of frost-free days is 140. The average temperature drops below 27° F on November 16 (Julian day 320) and rises above 27° F on March 16 (Julian day 75).

Solution:

1. Nomograph value for K. For the nomograph, %MS + %VFS = 45 + 5 = 50. Using this value plus given values for %SA, %OM, S_1, and P_1 in the nomograph, the value for K read directly from the nomograph is $K = 0.21$ (English units). Comparing this to the solution from Eqs. (8.36) and (8.37)

$$M = (45 + 5)(100 - 30) = 3500$$

$$K = \frac{2.1 \times 10^{-4}(12 - 2)(3500)^{1.14} + 3.25(3 - 2) + 2.5(1 - 3)}{100}$$

$$K = 0.21.$$

2. Corrections for seasonal variability. For the location in Minnesota, the maximum erodibility would occur in the spring and the minimum in the summer or fall; thus t_{max} is less than t_{min}, which is a Case I situation for Fig. 8.10 ($t_{max} < t_{min}$ means that the date of maximum erodibility occurs before the date of minimum erodibility). Using an R factor of 100 in Fig. 8.11, the ratio K_{max}/K_{min} is 6.7, the ratio K_{max}/K_{nom} is 2.5, and t_{max} is Julian day 110. For this situation

$$\frac{K_{max}}{K_{nom}} = 2.5; \qquad K_{max} = (2.5)(0.21) = 0.525 \text{ (English Units)}$$

$$\frac{K_{max}}{K_{min}} = 6.7; \qquad K_{min} = \frac{0.525}{6.7} = 0.078.$$

The Julian day for t_{min} is t_{max} plus the frost-free period. For this location

$$t_{min} = t_{max} + \Delta t = 110 + 140 = 250 \text{ (Julian day)}.$$

For $t_{max} < t_i < t_{min}$, Julian day 110 to 250, K_i can be predicted from Eq. (a) in Fig. 8.10, or

$$K_i = K_{max}\left(\frac{K_{min}}{K_{max}}\right)^{(t_i - t_{max})/\Delta t} = 0.525\left(\frac{0.078}{0.525}\right)^{(t_i - 110)/140}$$

$$= 0.525(0.149)^{(t_i - 110)/140}; \quad 110 < t_i < 250. \qquad \text{(a)}$$

For the time periods when $t_i > t_{min}$ and in the following year when $t_i < t_{max}$, two sets of values are needed. For the periods when the temperature is below 27° F, November 16 to March 16, Eq. (d) applies, or

$$K_i = K_{min} = 0.078; \qquad t_i < 75;\ t_i > 320. \qquad \text{(d)}$$

For the periods when $T_{avg} > 27°$ F, March 16 to April 19 (Julian Day 75 to 110) and September 6 to November 16 (Julian Day 250 to 320), Eqs. (b) and (c) from Fig. 8.10 are used. For the period September 6 to November 16, Eq. (b) yields

$$K_i = 0.078 \exp[0.009(t_i - t_{min})]$$
$$= 0.078 \exp[0.009(t_i - 250)]; \qquad 250 \le t_i \le 320. \qquad \text{(b)}$$

For the period March 16 to April 19 (Julian day 75 to 110), Eq. (c) yields

$$K_i = 0.078 \exp[0.009(t_i - t_{min} + 365)]; \quad \text{subject to } K_i \le K_{max}$$
$$= 0.078 \exp[0.009(t_i - 250 + 365)]$$
$$= 0.078 \exp[0.009(t_i + 115)] \le 0.525; \qquad 75 \le t_i \le 110. \qquad \text{(c)}$$

3. Correcting for the monthly rainfall energy distribution. From Fig. 8.8, the zone for central Minnesota is 86. Utilizing the distribution in Table 8.1 for zone 86 and the relationships above for K_i, the K factor can be weighted as shown in the following table. Using the sum $\%R_iK_i$ in the last column, the average K factor becomes

$$K_{avg} = \frac{\Sigma R_i K_i}{100} = \frac{18.63}{100} = 0.186.$$

In this case, K_{avg} is only slightly lower than the nominal K. The distribution of K, however, can allow for improved calculation of erosion by correlating appropriate values of K with cover and rainfall during periods when cover may leave the soil exposed. A plot of the predicted K_i is given in Fig. 8.12.

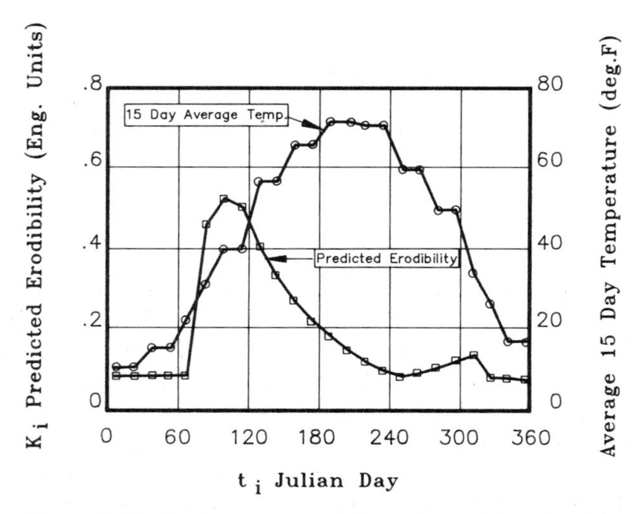

Figure 8.12 Predicted values of K_i for Example Problem 8.3.

K Value Computations for Example Problem 8.3

Time period	% annual R = $\%R$[a]	t_i Julian day[b]	T_{avg}[c] (°F)	K_i equation (in Fig. 8.10)	K_i[d] (calc)	$\%(R_i)K_i$
01/01–01/15	0	8	10	*d*	0.078	0
01/16–01/31	0	23	10	*d*	0.078	0
02/1–02/15	0	39	15	*d*	0.078	0
02/16–02/28	0	53	15	*d*	0.078	0
03/1–03/15	0	67	22	*d*	0.078	0
03/16–03/31	1	82	31	*c*	0.459	0.459
04/1–04/15	1	98	40	*c*	0.525	0.525
04/16–04/30	1	113	40	*a*	0.504	0.504
05/1–05/15	3	128	57	*a*	0.411	1.233
05/16–05/31	5	143	57	*a*	0.335	1.675
06/1–06/15	12	159	66	*a*	0.270	3.240
06/16–06/30	13	174	66	*a*	0.220	2.860
07/01–07/15	13	189	72	*a*	0.179	2.327
07/16–07/31	14	204	72	*a*	0.146	2.044
08/01–08/15	14	220	71	*a*	0.118	1.652
08/16–08/31	13	235	71	*a*	0.096	1.248
09/01–09/15	5	251	60	*b*	0.078	0.390
09/16–09/30	3	266	60	*b*	0.090	0.270
10/01–10/15	1	281	50	*b*	0.103	0.103
10/16–10/31	1	296	50	*b*	0.118	0.118
11/01–11/15	0	312	34	*b*	0.136	0
11/16–11/30	0	327	26	*d*	0.078	0
12/01–12/15	0	342	17	*d*	0.078	0
12/16–12/31	0	357	17	*d*	0.078	0
Total	100					18.63

[a]Table 8.1, zone 8b.
[b]Midpoint of half month intervals in Julian days. For example, the midpoint Julian day for the interval 2/1–2/15 is 39.
[c]Assumed average temperatures for central Minnesota.
[d]Values are given in English units (tons•acre•hr/hundreds•acre•ft•tonsf•in.).

The results in Fig. 8.12 indicate discontinuities when the air temperature drops below 27° F and rises above 27° F. This discontinuity is reasonable and occurs as a result of freezing and thawing. When frozen, soil is not very erodible. When the temperature rises above freezing, the soil will immediately return to a more erodible state.

The procedures illustrated above are for central and eastern United States and should not be used for sites in and west of the Rockies. The procedures are developed for annual erosion, not for single storms.

Soils with Rock Fragments

In the U.S., 15.6% of the soils have significant rock fragments in the soil and on the surface. These rock fragments impact erosion by absorbing impact energy and decreasing infiltration. Rocks on the surface absorb energy and decrease erosion, which must be accounted for in the C factor using the fraction of surface cover (see subsequent discussion of C factor). Rocks beneath the surface decrease infiltration rates and increase erosion, which must be accounted for in the K factor. The change in saturated hydraulic conductivity due to rocks can be estimated by

$$k_b = k_f(1 - R_w) = 2k_f\frac{(1 - R_v)}{(2 + R_v)}, \tag{8.38}$$

where k_b is the saturated hydraulic conductivity in rocky soils, k_f is the saturated hydraulic conductivity of the soil fraction, R_w is the fraction by weight of rock fragments greater than 2 mm, and R_v is the fraction by volume of rock fragments greater than 2 mm.

The impact of changes in saturated hydraulic conductivity on the K factor must be accounted for by the nomograph in Fig. 8.9. To accomplish this correction using Eq. (8.38), relationships between hydraulic conductivity and permeability classes used in Fig. 8.9 must be known. Rawls *et al.* (1982) proposed the relationship shown in Table 8.3.

Example Problem 8.4. Effects of rock fragments on K

A silty clay loam soil is classified as permeability class 5. Based on textural information, soil structure, and a permeability class of 5, K is estimated as 0.21 in English units. What would be the value for K as corrected for rock fragments if the percentage of rock fragments greater than 2 mm occupies 40% of the soil mass by weight?

Solution:

1. Impact of rock fragment on hydraulic conductivity. From Table 8.3, k_f for a silty clay loam soil is between 0.04 and 0.08 in./hr. Assume a value of 0.06 in./hr. From Eq. (8.38)

$$k_b = k_f(1 - R_w) = 0.06(1 - 0.40) = 0.036 \text{ in./hr.}$$

2. Estimating the revised permeability class. From Table 8.3, the permeability class for $k_b = 0.036$ in./hr is 6.

3. Estimating the new erodibility. Entering Fig. 8.9 with an estimated K of 0.21 for a permeability class of 5, the K value for a class 6 permeability is estimated as 0.22 (English units).

It is again important to note that this procedure corrects only for the effects of rock fragments on infiltration. Impacts on the C factor must be based on percentage ground cover, as discussed in a subsequent section.

Rough Estimates of K from Textural Information and Experimental Values for Construction and Mined Sites

The USDA–SCS has developed estimates of K based on textural classification for topsoil, subsoil, and residual materials as shown in Table 8.4. These values are first estimates only and do not include the influence of soil structure or infiltration characteristics.

A limited number of data sets have been developed for drastically disturbed lands and for reconstructed soils. A summary of the data is given in Table 8.5 along with a comparison to values from the Wischmeier *et al.* (1971) nomograph shown in Fig. 8.9. The comparison is sufficiently favorable to warrant the use of the nomograph for a first estimate of K on disturbed topsoil or A-horizon material. The comparison is not favorable for subsoil materials.

Length and Slope Factors *L* and *S*

The effects of topography on soil erosion are determined by dimensionless L and S factors, which account for both rill and interrill erosion impacts.

Slope Steepness Factor *S*

The slope steepness factor S is used to predict the effect of slope gradient on soil loss. For slope lengths

Table 8.3 Soil Water Data for the Major USDA Soil Textural Classes (after Rawls *et al.*, 1982)

Texture	Permeability class[a]	Saturated hydraulic conductivity in./hr	Saturated hydraulic conductivity mm/hr	Hydrologic soil group[b]
Silty clay, clay	6	< 0.04	<1	D
Silty clay loam, sandy clay	5	0.04–0.08	1–2	C–D
Sandy clay loam, clay loam	4	0.08–0.20	2–5	C
Loam, silt loam	3	0.20–0.80	5–20	B
Loamy sand, sandy loam	2	0.80–2.40	20–60	A
Sand	1	> 2.40	>60	A+

[a]See Soil Conservation Service National Soils Handbook (SCS, 1983).

[b]See Soil Conservation Service National Engineering Handbook (SCS, 1972, 1984).

[c]Note: Although the silt texture is missing from the NEH because of inadequate data, it undoubtedly should be in permeability class 3.

greater than 15 ft, the S factor from the USLE was modified significantly by McCool *et al.* (1987, 1993) after extensive evaluation of the original USLE data base. The modified version is

$$S = 10.8 \sin\theta + 0.03; \qquad \sin\theta < 0.09 \qquad (8.39)$$

$$S = 16.8 \sin\theta - 0.50; \qquad \sin\theta \geq 0.09, \qquad (8.40)$$

where θ is the slope angle. Based on an evaluation of data from disturbed lands with slopes up to 84%, McIssac *et al.* (1987) developed an equation similar to (8.39) and (8.40) with exponents in the same range; thus McCool *et al.* (1993) recommend that Eqs. (8.39) and (8.40) also be used for disturbed lands.

For slope lengths less than 15 ft, the S factor is not as strongly related to slope (slope exponent less than 1.0) since rilling would not have been initiated. The recommended factor is

$$S = 3.0(\sin\theta)^{0.8} + 0.56. \qquad (8.41)$$

Under conditions where thawing of recently tilled soils is occurring and surface runoff is the primary factor causing erosion (typical of the Pacific Northwest in the spring), the S factor should be (McCool *et al.*, 1987, 1993)

$$S = 4.25(\sin\theta)^{0.6}, \qquad \sin\theta \geq 0.09. \qquad (8.42)$$

For thawing soils with slopes less than 9%, Eq. (8.39) should be used.

The S factor in the RUSLE is significantly modified from the original USLE as a result of an extensive reevaluation of the original data base, addition of the factors for short slope lengths, and new values for thawing soils (McCool *et al.*, 1987). The original data base did not include values beyond 20%. When using the quadratic form of the equation for S developed for the original USLE, projections beyond 20% yielded unreasonably high values for erosion. The RUSLE equation with the linear function corrects this problem.

Table 8.4 *K* Value Estimates based on Textural Information (English Units) (Soil Conservation Service, 1978)

Texture	Estimated K value[a]
Topsoil	
Clay, clay loam, loam, silty clay	0.32[b]
Fine sandy loam, loamy very fine sand, sandy loam	0.24
Loamy fine sand, loamy sand	0.17
Sand	0.15
Silt loam, silty clay loam, very fine sandy loam	0.37
Subsoil and Residual Material	
Outwash Soils	
Sand	0.17
Loamy sand	0.24
Sandy loam	0.43
Gravel, fine to moderate fine	0.24
Gravel, medium to moderate coarse	0.49
Lacrustrine Soils	
Silt loam and very fine sandy loam	0.37
Silty clay loam	0.28
Clay and silty clay	0.28
Glacial Till	
Loam, fine to moderate fine subsoil	0.32
Loam, medium subsoil	0.37
Clay loam	0.32
Clay and silty clay	0.28
Loess	0.37
Residual	
Sandstone	0.49
Siltstone, nonchannery	0.43
Siltstone, channery	0.32
Acid clay shale	0.28
Calcareous clay shale or limestone residuum	0.24

[a]These values are typical based only on textural information. Values for an actual soil can be considerably different due to different structure and infiltration.

[b]Units on K in this table are English units (tons•acre•hr/hundreds•acre•ft•tonsf•in.). To convert to metric units (t•ha•h/ha•MJ•mm), multiply K values by 0.1317.

Slope Length Factor

The slope length factor was developed by McCool *et al.* (1989, 1993) from the original USLE data base augmented with theoretical considerations. The L factor retains its original form

$$L = \left[\frac{\lambda}{72.6}\right]^m, \qquad (8.43)$$

where λ is the slope length in feet, 72.6 ft is the length of a standard erosion plot, and m is a variable slope length exponent. Slope length, λ, is the horizontal projection of plot length, not the length measured along the slope. The difference in horizontal projections and slope lengths becomes important on steeper slopes.

The slope length exponent is related to the ratio of rill to interrill erosion, β (Foster *et al.*, 1977b; McCool *et al.*, 1989, 1993), by

$$m = \frac{\beta}{1+\beta}. \qquad (8.44)$$

Table 8.5 Experimental K Value Estimates for Disturbed Lands (English Units)

Reclaimed soil or residual material	Location of experimental site	K Exp[a]/Nomo[b]	Reference
Hosmer silt loam	Indiana	0.387/0.485[c]	Stein *et al.* (1983)
Alfred silt loam	Southern Indiana	0.812/0.485	
Ava silt loam	Southern Indiana	0.842/0.478	
Graded overburden	Southern Indiana	0.197–0.835/ 0.250–0.478	
Clinton silt loam[d]	Western Illinois	0.370/0.360	Mitchell *et al.* (1983)
Tama silty clay loam[d]	Westem Illinois	0.210/0.310	
Hosmer silt loam[d]	Southern Indiana	0.450–0.650/ 0.470	
Sadler silt loam (A horizon)	Western Kentucky	0.415/0.385	Barfield *et al.* (1988)
Sadler silt loam (B horizon)	Western Kentucky	0.380/0.640	
Shale spoil material	Western Kentucky	0.140/0.180	

[a]Values measured experimentally with rainfall simulators.
[b]Values calculated from Wischmeier *et al.* (1971) nomograph shown in Fig. 8.9.
[c]Values in English units of tons•acre•hr/hundreds•acre•ft•tonsf•in. To convert to metric units of t•a•h/ha•MJ•mm, multiply by 0.1317.
[d]The dominant soil series. Some mixing occurred with other series.

For soils that are classed as being moderately susceptible to erosion, McCool *et al.* (1989) proposed that

$$\beta_{\text{mod}} = \frac{11.16 \sin \theta}{3.0(\sin \theta)^{0.8} + 0.56}, \tag{8.45}$$

where θ is the slope angle. Thus, the slope exponent is a function of the slope angle θ.

Soils in the RUSLE are classed as having low, moderate, or high susceptibility to rill erosion. Equation (8.45) is for soils that are moderately susceptible to erosion. Conversions for soils that have low or high susceptibility to erosion are given in Table 8.6. Values in Table 8.6 are based on the assumption that moderately erodible soils have a β defined by Eq. (8.45), soils highly susceptible to rilling have a β that is twice that given by Eq. (8.45), and soils with low susceptibility to rilling have a β that is defined by half that given by Eq. (8.45).

For soils in the Pacific Northwest, or other soils that are exposed to runoff during thawing without sufficient rainfall energy to cause interrill erosion, the values in Table 8.6 should not be used. Instead, McCool *et al.* (1989) recommend that a slope length exponent of 0.5 be used for all slopes. When runoff on thawing soils is exposed to rainfall sufficient to cause significant interrill erosion, the slope length exponent for the low rill to interrill erosion ratio should be used (column 1 in Table 8.6). For rangeland soils, the use of a low rill to interrill erosion ratio is proposed. Selection of the appropriate column to use in Table 8.6 requires professional judgement. The assistance of a soil scientist may be helpful.

Combined Length and Slope Factors

Combined slope length and slope steepness factors were calculated using the factors from Eqs. (8.39) to (8.43). These combination factors are given in Fig. 8.13 for all susceptibilities and for thawing soils.

Irregular and Segmented Slopes

Soil loss is strongly impacted by slope shape (Foster and Huggins, 1979). A convex shape will have greater erosion than a uniform slope by as much as 30%. A concave slope will have less erosion than a uniform slope. Foster and Wischmeier (1974) developed a procedure for evaluating the impact of irregular slopes by dividing the slope into segments. The soil loss per unit area from the ith segment is

$$A_i = RK_iC_iP_iS_i\left[\frac{\lambda_i^{m+1} - \lambda_{i-1}^{m+1}}{(\lambda_i - \lambda_{i-1})72.6^m}\right], \tag{8.46}$$

where λ_i and λ_{i-1} are the slope lengths at the start and end of segment i, and K_i, C_i, P_i, and S_i are USLE factors for segment i. Equation (8.46) can be used for each segment i. The total erosion from each segment

Table 8.6 Slope Length Exponent m in Eq. (8.43) (after McCool *et al.*, 1993)[a]

Percentage slope	Rill/interrill ratio		
	Low[b]	Moderate[c]	High[d]
0.2	0.02	0.04	0.07
0.5	0.04	0.08	0.16
1.0	0.08	0.15	0.26
2.0	0.14	0.24	0.39
3.0	0.18	0.31	0.47
4.0	0.22	0.36	0.53
5.0	0.25	0.40	0.57
6.0	0.28	0.43	0.60
8.0	0.32	0.48	0.65
10.0	0.35	0.52	0.68
12.0	0.37	0.55	0.71
14.0	0.40	0.57	0.72
16.0	0.41	0.59	0.74
20.0	0.44	0.61	0.76
25.0	0.47	0.64	0.78
30.0	0.49	0.66	0.79
40.0	0.52	0.68	0.81
50.0	0.54	0.70	0.82
60.0	0.55	0.71	0.83

[a]Values in table are not applicable to thawing soils. See text for explanation.
[b]$\beta = 1/2$ value from Eq. (8.45) in Eq. (8.44).
[c]$\beta = 1 \times$ value from Eq. (8.45) in Eq. (8.44).
[d]$\beta = 2 \times$ value from Eq. (8.45) in Eq. (8.44).

would be $A_i(\lambda_i - \lambda_{i-1})$, and the average erosion per unit area over the entire slope length would be

$$A = R \sum_{i=1}^{n} K_i C_i P_i S_i \frac{\left[\lambda_i^{m+1} - \lambda_{i-1}^{m+1}\right]}{\lambda_e 72.6^m}, \qquad (8.47)$$

where λ_e is the total slope length. Equation (8.47) can also be used to evaluate the effects of variation in K, C, and P over the slope length.

An alternate method for evaluating irregular slopes is the use of a slope length adjustment factor (SAF). If the slope is divided into n increments of equal length ΔX, then

$$A = R \sum_{i=1}^{n} K_i C_i P_i S_i \frac{\left[(i\,\Delta X)^{m+1} - ([i-1]\,\Delta X)^{m+1}\right]}{n\,\Delta X 72.6^m}. \qquad (8.48)$$

Dividing by n times the soil loss from a uniform slope of equal length and assuming constant values of K_i C_i P_i along the slope, a slope adjustment factor can be developed for each segment, or

$$\mathrm{SAF}_i = \frac{A_i}{A} = \frac{i^{m+1} - (i-1)^{m+1}}{n^m}, \qquad (8.49)$$

where n is the number of segments and SAF is the slope adjustment factor. The sum of the SAF_i for a given slope is equal to the number of segments n; thus the average erosion over the slope is

$$A = \frac{R}{n} \sum_{i=1}^{n} K_i C_i P_i S_i L_i (\mathrm{SAF})_i. \qquad (8.50a)$$

where L_i is the slope length factor calculated from Eq. (8.43) using the m value corresponding to the segment steepness. In the development of a SAF relationship, R, K, C, and P remain constant over all segments; thus Eq. (8.50a) can be solved for an equivalent LS factor

$$LS = \frac{1}{n} \sum_{i=1}^{n} S_i L_i (\mathrm{SAF})_i. \qquad (8.50b)$$

Factors calculated from Eq. (8.50b) are given in Table 8.7. An example of its use is given in Example Problem 8.5.

Example Problem 8.5. Estimating *LS* factors

A soil that is very susceptible to rilling has a slope length of 210 ft and an average slope of 15%. Estimate the LS factor if:

(1) the slope is uniform
(2) the slope is convex with slopes of 10, 15, and 20% on segments 1, 2, and 3
(3) the slope is concave with slopes of 20, 15, and 10% on segments 1, 2, and 3.

Assume that the soil is not freezing and thawing.

Solution:

1. Uniform slope. The slope angle is

$$\theta = \tan^{-1} 0.15 = 8.53^\circ.$$

From Eq. (8.45) for soils moderately susceptible to rilling,

$$\beta = \frac{11.16 \sin 8.53}{3.0(\sin 8.53)^{0.8} + 0.56} = 1.37.$$

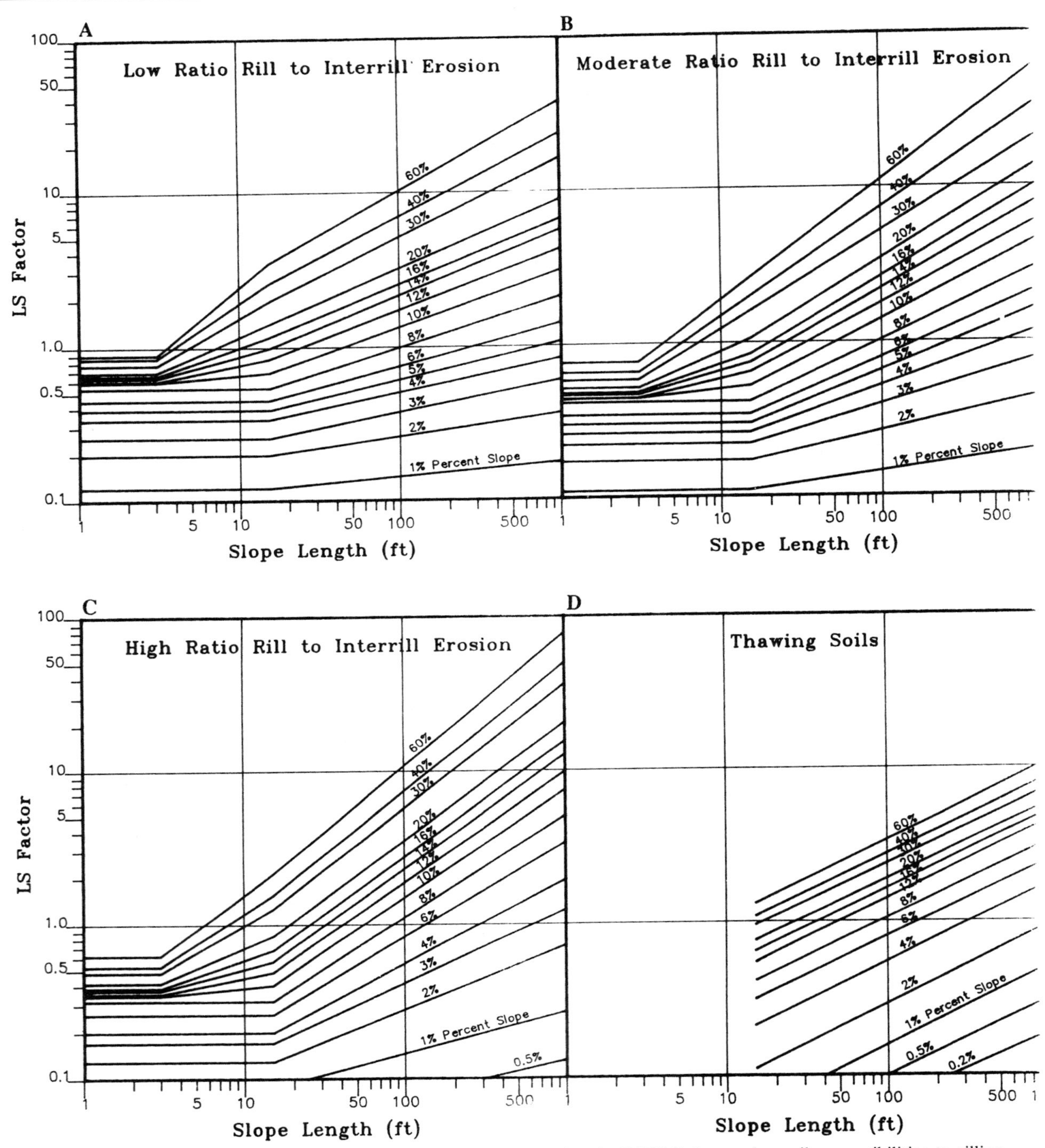

Figure 8.13 Combined slope length and steepness factors, LS, for the RUSLE for varying soil susceptibilities to rilling and for thawing soils.

For a soil highly susceptible to rilling, β is doubled before estimating m; thus $\beta = 2.74$. From Eq. (8.44)

$$m = \frac{2.74}{1 + 2.74} = 0.73,$$

which is equal to the value in Table 8.6. From Eqs. (8.40) and (8.43)

$$LS = \left(\frac{\lambda}{72.6}\right)^m (16.8 \sin\theta - 0.50)$$

$$= \left(\frac{210}{72.6}\right)^{0.73} (16.8 \sin 8.53 - 0.50)$$

$$= 4.33.$$

Table 8.7 Adjustment Factors to LS to Estimate Soil Loss on Irregular Slopes (after McCool *et al.*, 1993)[a]

Number of segments	Sequential number of segments	Slope length exponent								
		0.05	0.1	0.2	0.3	0.4	0.5	0.6	0.7	0.8
$n=$ 2	$i=$ 1	0.97	0.93	0.87	0.81	0.76	0.71	0.66	0.62	0.57
	2	1.03	1.07	1.13	1.19	1.24	1.29	1.34	1.38	1.43
3	1	0.95	0.90	0.80	0.72	0.64	0.58	0.52	0.46	0.42
	2	1.01	1.02	1.04	1.05	1.06	1.05	1.05	1.04	1.03
	3	1.04	1.08	1.16	1.23	1.30	1.37	1.43	1.50	1.55
4	1	0.93	0.87	0.76	0.66	0.57	0.50	0.44	0.38	0.33
	2	1.00	1.00	0.98	0.96	0.94	0.92	0.88	0.85	0.82
	3	1.03	1.05	1.09	1.13	1.16	1.18	1.20	1.22	1.23
	4	1.04	1.08	1.17	1.25	1.33	1.40	1.48	1.55	1.62
5	1	0.92	0.85	0.73	0.62	0.53	0.45	0.38	0.32	0.28
	2	0.99	0.97	0.94	0.90	0.86	0.82	0.77	0.73	0.69
	3	1.01	1.03	1.04	1.05	1.06	1.06	1.06	1.05	1.03
	4	1.03	1.06	1.12	1.17	1.21	1.25	1.29	1.32	1.35
	5	1.05	1.09	1.17	1.26	1.34	1.42	1.50	1.58	1.65

[a]Soil loss factors = $[i^{1+m} - (i-1)^{1+m}]/n^m$, where i is the sequential number of segment, m is the slope length exponent, and n is the number of segments. Values are forced to give a factor total equal to n.

2. *Convex slope.* Values are tabulated below.

Slope segment	Slope percentage	Slope expon. Table 8.6	θ (°)	Uniform slope *LS* factor[a]	*SAF,* Table 8.7	Segment *LS* factor (Column 5 × Column 6)
1	10	0.68	5.71	2.41	0.47	1.13
2	15	0.73	8.53	4.33	1.04	4.50
3	20	0.76	11.31	6.26	1.53	9.58
						Σ=15.21

[a]Equation (8.40) x Eq. (8.43) with λ=210 ft.

From Eq. (8.50b)

$$LS = \Sigma/n = 15.21/3 = 5.07.$$

3. *Concave slope.* Values are tabulated below.

Slope segment	Slope percentage	Slope expon. Table 8.6	θ (°)	Uniform slope *LS* factor[a]	*SAF,* Table 8.7	Segment *LS* factor (Column 5 × Column 6)
1	20	0.76	11. 31	6.26	0.44	2.75
2	15	0.73	8.53	4.33	1.04	4.50
3	10	0.68	5.71	2.41	1.49	3.59
						Σ=10.84

[a]Equation (8.40) x Eq. (8.43) with λ=210 ft.

From Eq. (8.50b)

$$LS = \Sigma/n = 10.84/3 = 3.61.$$

4. Summary.

Concave $LS = 3.61$
Uniform $LS = 4.33$
Convex $LS = 5.07$.

Thus the convex shape has the highest LS factor.

Estimating Slope Lengths for Watersheds

Slope length estimates for fields or watersheds with nonplanar surfaces require considerable professional judgment. A large number of slope lengths occur in any given real watershed. Erosion can be estimated for each of these and area weighted to determine average erosion on a watershed.

Slope length is defined as the slope distance from the point of origin of overland flow to the point of

Table 8.8 Selected USLE C Values for Construction, Mining, and Forest Lands.

Condition	C factor	References
1. Bare soil conditions		
Undisturbed except scraped	0.66–1.30	a
Compacted		
Smooth	1.00–1.40	a,b
Root raked	0.90–1.20	a
Disk tillage		
Fresh	1.00	a
After one rain	0.89	a
2. Mulch		
Straw		
0.5 tons/ac	0.30	a,d
1.0 tons/ac	0.18	a,d
2.0 tons/ac	0.09	a,d
4.0 tons/ac	0.02	a,d
Wood chips		
0.5 tons/ac	0.90	a,d
2.0 tons/ac	0.70	a,d
4.0 tons/ac	0.42	a,d
6.0 tons/ac	0.22	a,d
3. Chemical binders		
Asphalt emulsion, 605 gal/ac	0.14-0.52	a
Aquatan, Terra-tack	0.67	a
4. Seedings		
No prepared seedbed		
New planting	0.64	a
After 60 days	0.54	a
Prepared seedbed		
New planting	0.40	a
After 60 days	0.05	a
5 . Undisturbed forest		
100–75% canopy, 100–90% litter	0.0001–0.001	c
35–20% canopy, 70–40% litter	0.003–0.009	c
6. Permanent pasture and brush cover		
0% canopy, 80% ground cover		
Grass	0.013	c
Weeds	0.043	c
50% Brush, 80% ground cover		
Grass	0.012	c
7. Mechanically prepared woodland sites		
Burned, 10% cover at ground		
Good soil	0.240	c
Poor soil	0.360	c
Burned, 0% cover at ground		
Good soil	0.260	c
Poor soil	0.450	c
Disked, 0% cover at ground		
Good soil	0.720	c
Poor soil	0.940	

Note. Additional values are given in Appendix 8B.
[a]Transportation Research Board (1980).
[b]Barfield *et al.* (1988).
[c]Wischmeier and Smith (1978).
[d]Meyer *et al.* (1972). C factors for mulch vary depending on slope length and steepness. Slope length limits apply (see Appendix 8B, Table 8B.5).

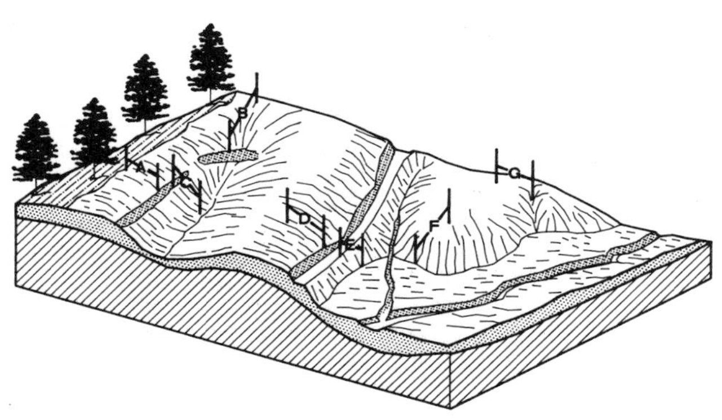

Figure 8.14 Example slope lengths (after Dissmeyer and Foster, 1984). Slopes: (A) If undisturbed forest soil above does not yield surface runoff, the top of slope starts with edge of undisturbed forest soil and extends down slope to windrow of brush if runoff is concentrated by windrow; (B) point of origin of runoff to windrow if runoff is concentrated by windrow; (C) from windrow to flow concentration point; (D) point of origin of runoff to road that concentrates runoff; (E) from road to flood plain where deposition would occur; (F) on nose of hill, from point of origin of runoff to flood plain where deposition would occur; (G) point of origin of runoff to slight depression where runoff would concentrate.

concentrated flow or until deposition occurs. A field evaluation is helpful to determine points at which flows become concentrated and points at which deposition occurs. Example slope lengths are given in Fig. 8.14.

Table 8.9 Selected USLE C Values for Cropland (after Wischmeier and Smith, 1978)[a]

Crop stage[b]	Corn HP[c]	Corn LP[d]	Corn NT[e]	Soybeans HP[f]	Small Grain[g]
Fal	0.44	0.65	—	0.45	—
SB	0.65	0.78	0.08	0.69	0.29
1	0.53	0.65	0.08	0.57	0.24
2	0.38	0.45	0.08	0.38	0.19
3	0.20	0.26	0.07	0.18	0.06
4	—	—	0.19	—	—

[a]For a complete listing, see Handbook 537, Wischmeier and Smith (1978).

[b]Period Fal (Rough fallow), inversion plowing to secondary tillage. Period SB (seedbed), secondary tillage for seedbed preparation until 10% of canopy cover, Period 1 (establishment), 10 to 50% canopy cover. Period 2 (development), 50 to 75% cover. Period 3 (maturing crop), 75% cover to harvest. Period 4 (residue or stubble), harvest to plowing or seeding.

[c]Corn, fall turnplowing, high productivity.

[d]Corn, fall turnplowing, low productivity.

[e]Corn, minimum tillage, plant in stubble.

[f]Soybeans after corn, fall turnplow, high productivity.

[g]Small grain, after corn, average productivity, 40% cover after emergence.

Cover Factor C

Processes Being Considered in Cover Factor

The cover factor accounts for the effects of cover above the ground, ground cover, root mass, incorporated residue, surface roughness, and soil moisture on soil erosion. In general, these factors can be divided into three categories:

- Above-ground effects.
- Surface effects, including surface cover and surface roughness.
- Below-surface effects, including active root growth, incorporated residue from previous crops, and effects of tillage and land disturbances on structure and consolidation.

Each of these effects are discussed below.

Above-Ground Cover. Above-ground cover intercepts rainfall and absorbs the raindrop energy. Intercepted water moves to the surface by either stem or drip flow. Stemflow is that intercepted rainfall that moves down the main stem of a plant to the ground while drip flow is intercepted rainfall that collects on leaves and drips to the ground. Stemflow causes no interrill erosion; thus conversion of rainfall to stemflow typically reduces erosion. The quantity of stem flow depends on the type of vegetation, with grasses having a high ratio of stem to drip flow and broad leaf plants such as soybean or oak having a low ratio. Drip flow, if coming from leaves high on trees, can contain as much energy as undisturbed rainfall. Thus, conversion of rainfall to drip flow does not always lead to reduced erosion. In forest watersheds, of course, the litter will absorb much of the energy of drip flow and protect the soil surface.

Surface Effects. The impact of surface cover on raindrop energy impact and shearing forces of runoff is included in surface effects. In addition, surface effects include the impact of surface conditions on transport capacity of runoff. These processes are affected by residue on the surface, surface roughness, and to an extent the residue incorporated in the soil.

Subsurface Effects. Subsurface effects include the effects of live root mass, residual root mass, incorporated residue, consolidation, compaction, prior land disturbance, and soil moisture on erosion. These effects are the most difficult to quantify due to significant changes during a growing season and interactions between the components.

Tabulated Values for C

The simplest approach to defining the C factor is to use tabulated values for C that lump together all of the factors discussed above. An extensive data base of lumped parameters collected for the USLE is available for use. Selected values are included in Table 8.8. for construction, mining, and forest lands and in Table 8.9 for agricultural lands. Additional data are given in Appendix 8B.

C factors change with land-use cover, resulting in varying erosion rates over the course of a season, even with invariant month-to-month rainfall energy. This variation can be accounted for by weighting according to the fraction of R factor received during a given month, following the weighting procedure used for the K factor in Example Problem 8.3. This weighting procedure is illustrated in Example Problem 8.6.

Example Problem 8.6. Estimating erosion with tabulated C factors on disturbed lands

A 40-acre field near Middlesboro, Kentucky, has a K value of 0.35 (English units tons · acre · hr/hundreds · acre · ft · tonsf · in.). The field is subjected to the following sequence of operations:

1/1–4/1	Dense forest (75% effective canopy)
4/1–6/1	Stripped of vegetation (no cover, poor soil, burned)
6/1–8/1	Active mining
8/1–9/1	Regraded (no cover, disked, poor soil)
9/1–11/1	Permanent seeding (first 60 days)
11/1–12/31	Permanent seeding (remainder of the year).

The soils have a high tendency to rill, and slopes are 50% with slope lengths of 150 ft. Estimate the weighted annual C factor using the tabulated values in Table 8.8, and calculate the average annual erosion. Assuming that a 10-year storm occurs during the most susceptible period, estimate the 10-year single-storm erosion. Compare these predictions to the erosion rate from a corn field in West Kentucky with the same soil on a slope of 8% and slope length of 150 ft if the following sequence of operations is followed:

4/15	Seedbed preparation and planting
5/15	10% canopy
6/15	50% canopy
7/1	75% canopy
10/1	Harvest
10/15	Turnplow.

Assume that the corn crop has a low yield. In both cases, assume that $P = 1.0$

Solution:

East Kentucky Strip Mine

1. Estimating a weighted C factor. Computations for the C factors and weighting factors are given in the table below. Middlesboro, Kentucky, is in zone 104 in Fig. 8.8. Using the sum of the values in Column 6, the weighted C factor is

$$C = \Sigma \text{ column } 6/\Sigma \text{ column } 4 = 63.84/100 = 0.64.$$

Example Problem 8.5 Weighted C Factors—Eastern Kentucky Strip Mine

Time period[a]	Activity[a]	Cumulative percentage annual R^b	Percentage annual R during period[c]	Cover factor C_i	Weighted cover factor[i]
1/1–4/1	Dense forest	13	13	0.001[d]	0.013
4/1–6/1	Bare soil, burned	27	14	0.45[e]	6.300
6/1–8/1	Active mining	63	36	1.0[f]	36.000
8/1–9/1	Regraded (bare)	80	17	0.94[g]	15.980
9/1–11/1	Perm. seeding (first 60 days)	93	13	0.40[h]	5.200
11/1–12/31	Perm. seeding (remainder of year)	100	7	0.05[h]	0.350
		Σ=	100		63.840

[a]Specified in the problem.
[b]Value from Table 8.1 (Zone 104 in Fig. 8.8).
[c]Incremental value from column 3.
[d]Table 8.8, item 5.
[e]Table 8.8, item 7.
[f]Construction value = 1.0.
[g]Table 8.8, item 7.
[h]Table 8.8, item 4 (seedbed was disked in regrading operation).
[i]Column 4 × Column 5.

2. Estimating LS factors. From Table 8.6, $m = 0.82$ for 50% slope and high tendency to rill. For a 50% slope, $\theta = \tan^{-1} 0.5 = 26.6°$. Using Eqs. (8.43) and (8.40)

$$LS = \left[\frac{\lambda}{72.6}\right]^m [16.8 \sin\theta - 0.5]$$

$$= \left[\frac{150}{72.6}\right]^{0.82} [16.8 \sin(26.6) - 0.5] = 12.73.$$

3. Calculating average annual erosion rates. From Fig. 8.6, the average annual R factor is 175. Using Eq. (8.28)

$$A = RKLSCP = (175)(0.35)(12.73)(0.64)$$
$$= 499 \text{ ton/acre} \cdot \text{yr}.$$

4. Calculating single storm erosion rates. The most sensitive period is with a C factor of 1.0. The 10-year single-storm R factor for Middlesboro is given in Table 8.2 as 63 (a similar value is obtained from Appendix 8A). Using the LS factor calculated above and the C factor of 1.0

$$A = RKLSCP = (63)(0.35)(12.73)(1.00)$$
$$= 280.7 \text{ ton/acre} \cdot \text{storm}.$$

Western Kentucky Cropland

1. Estimating a weighted C factor. Western Kentucky is in zone 105 in Fig. 8.8. C factors and weighting factors are tabulated below. The weighted C factor is

$$C = \Sigma \text{ column } 6/\Sigma \text{ column } 4 = 50.08/100 = 0.50.$$

2. Estimating LS factors. From Table 8.6, $m = 0.65$ for an 8% slope and a high tendency to rill. For an 8% slope, $\theta = \tan^{-1} 0.08 = 4.6°$. Using Eqs. (8.43) and (8.39)

$$LS = \left[\frac{\lambda}{72.6}\right]^m [10.8 \sin\theta + 0.03]$$

$$= \left[\frac{150}{72.6}\right]^{0.65} [10.8 \sin(4.6) + 0.03] = 1.44.$$

3. Calculating average annual erosion rates. From Fig. 8.6, the average annual R factor is 250. Using Eq. (8.28)

$$A = RKLSCP = (250)(0.35)(1.44)(0.50)$$
$$= 63.0 \text{ ton / acre} \cdot \text{year}$$

4. Calculating single-storm erosion rates. The most sensitive period is during the seedbed period with a C factor of 0.78. The 10-year single-storm R factor for the city closest to west Kentucky given in Table 8.2 would be for Memphis, Tennessee or 82, which compares favorably with Fig. 8A.5 in Appendix 8A. Using the LS factor calculated above and the C factor of 0.78,

$$A = RKLSCP = (82)(0.35)(1.44)(0.78)$$
$$= 32.2 \text{ ton / acre} \cdot \text{storm}$$

Comparison of Predicted Erosion Values

	Eastern KY strip mine	Western KY cropland
Annual (ton/acre•year)	499	63
10-year storm (ton/acre•storm)	281	32

Example Problem 8.5 Weighted C Factors—Western Kentucky Corn

Time period[a]	Activity[a]	Cumulative percentage annual R[b]	Percentage annual R during period[c]	Cover factor C_i	Weighted cover factor[f]
1/1–4/15	Rough fallow	21	21	0.65[d]	13.65
4/15–5/15	Seedbed (0–10%)	31	10	0.78	7.8
5/15–6/15	Period 1 (10–50%)	43	12	0.65	7.8
6/15–7/1	Period 2 (50–75%)	50	7	0.45	3.15
7/1–10/1	Period 3 (75%–hv)	85	35	0.26	9.1
10/1–10/15	Period 4 residue	88	3	0.26[e]	0.78
10/15–12/31	Rough fallow	100	12	0.65	7.8
		Σ=	100		50.08

[a]Specified in the problem.
[b]Value from Table 8.1 (zone 105 in Fig. 8.8).
[c]Incremental value from Column 3.
[d]Table 8.9, Column 2.
[e]Assumed the same as period 3.
[f]Column 4 × Column 5.

The strip mine obviously has a much larger erosion rate by almost a factor of 10. This is due partially to the greater C factors, but primarily due to the higher LS factor resulting from much steeper slopes prevalent in that area.

Cover Factors: Subfactor Approach for Agricultural Lands

The reader not interested in the refinement offered by subfactors may wish to skip to the supporting practice factor section and read the section on tabulated P factors.

Introduction to the Subfactor Approach

An alternative to the tabulated values is the subfactor approach proposed by Wischmeier (1975) and Mutchler *et al.* (1982), which allows for more detailed evaluation of the interacting processes affecting cover than possible with the tabulated values in Tables 8.8 and 8.9. Originally proposed as a tool to develop composite C factors for the USLE, the subfactor approach is used totally in the RUSLE. The actual algorithm for the RUSLE follows that of Laflen *et al.* (1985). Subfactors are used to account for prior land use, canopy, surface cover, surface roughness, and soil moisture. In this text, procedures for estimating cover factors are presented for three land uses: agricultural and rangeland, disturbed forest, and construction and mining. In each case, the cover factor is developed as a function of several subfactors to account for the parameters listed above as well as special parameters for construction and forest lands. For cropland, the subfactors for above- and below-ground effects would change rapidly, but with rangeland, the subfactors would change slowly.

The subfactors presented in this section are those of the RUSLE, as presented in the draft documentation available as this text went to press. The original procedures, proposed in a 1991 draft are likely to change. Reference should be made to the final document (Renard *et al.* 1993) for further information.

Using the subfactor analogy, the C factor for agricultural and rangelands is defined by

$$C = C_{plu} C_{cc} C_{sc} C_{sr} C_{sm}, \tag{8.51}$$

where C_{plu} is the prior land-use factor, C_{cc} is the canopy cover subfactor, C_{sc} is the surface cover subfactor, C_{sr} is the surface roughness subfactor, and C_{sm} is the soil moisture subfactor. These subfactors depend on cropping and management and can be expressed as a function of residue cover, canopy cover, canopy height, surface roughness, below-ground root mass and residue, prior cropping, and time. For cropped areas, Yoder *et al.* (1993) recommend that the values be estimated for 15-day periods of time.

Canopy Cover Subfactor

The effects of canopy cover and height on energy reduction of falling rain are given by the canopy subfactor. Raindrops either fracture into smaller drops with less energy or drip from leaf edges. The canopy cover subfactor is

$$C_{cc} = 1 - F_c e^{-0.1H}, \tag{8.52}$$

where F_c is the fraction of surface covered by canopy and H is the average canopy height in feet. This is the original relationship proposed graphically by Wischmeier and Smith (1978) in which it was assumed that the fraction of rainfall intercepted is equal to the fraction of canopy cover. It was also assumed that intercepted rainfall leaves the canopy at height H with a drop size of 0.1 in. Quinn and Laflen (1983) reported that the relationship gave satisfactory results for cover although the assumptions were not exactly correct. The recommended values for H and F_c are listed in Table 8.10A for selected crops.

Surface Cover Subfactor

The impacts of surface cover include a reduction in soil exposed to rainfall energy, reduction in transport capacity, and deposition in ponded areas. Included in surface cover is residue, rocks, and other material in contact with the ground surface. The surface cover factor is

$$C_{sc} = \exp\left[-bR_c\left(\frac{6}{6+R_G}\right)^{0.08}\right], \tag{8.53}$$

where R_C is the fraction of ground cover, R_G is a variable to account for the effects of surface roughness on the effectiveness of mulch, and b is a constant. Recommended values for b are given in Table 8.10B. Residue cover, R_C is estimated from

$$R_C = 1 - e^{-a_W R_W}, \tag{8.54}$$

where R_C is the fraction of residue cover, R_W is residue weight (pounds per acre) and a_W is the ratio of area covered to mass of residue (acres per pound). Example values for a_W are given in Table 8.10.

The random roughness parameter in Eq. (8.53), R_G, accounts for the impact of rainfall and buried residue on random roughness and its further impact on the

Table 8.10 RUSLE Parameters for Selected Crops, Rangeland, and Tillage Practices[a]

A. Recommended values for root mass, canopy cover, and canopy height for selected crops

	Root Mass (R_{sr}) Mass of roots (lb • acre) in upper 4 in. of soil					Canopy cover (F_c) Cover fraction of land surface covered by canopy (decimal fraction)					Canopy height, H (ft)				
Days after planting	Corn	Soybean/cotton	Sorghum	Winter small grain	Spring small grain	Corn	Soybean/cotton	Sorghum	Winter small grain	Spring small grain	Corn	Soybean/cotton	Sorghum	Winter small grain	Spring small grain
15	0	0	0	92	92	0	0	0	0	0	0	0	0	0	0
30	92	0	92	180	180	0.1	0.1	0.1	0.1	0.1	0.5	0.2	0.5	0.2	0.2
45	180	92	180	364	452	0.5	0.3	0.5	0.3	0.4	1.2	.5	1	0.5	1
60	272	180/272	272	544	724	0.8	0.7	0.8	0.4	0.9	2.5	1	1.5	0.7	2
75	544	364	544	544	908	1	1	1	0.4	1	5	1.7	2.9	0.7	2.5
90	544	364	544	544	908	1	1	1	0.4	1	6	2.5	3.7	0.7	2.5
120	544	364	544	544	908	1	.5	1	0.4	1	6	2.5	3.7	0.7	2.5
195	—	—	—	908	—	—	—	—	1.0	—	—	—	—	2.5	—

B. Recommended values for coefficients b in Eq. (8.53) and C_1 in Eq. (8.63)

Condition	C_1	b
Irrigation or snowmelt (rill erosion dominates)	0.00102	5
Interrill erosion dominates	0.00053	2.5
Typical cropland erosion	0.00088	3.5
Rangeland values	[a]	4.5

[a] C_{plu} = 1.0 for rangeland.

C. Residue decomposition parameter values [Eqs. (8.54), (8.58)]

Crop	Residue/grain ratio	U/R	C_N	a_w	Yield
Alfalfa	0.15	0.0040	30	0.00056	6 tons/acre
Bromegrass	0.15	0.0040	80	0.00056	5 tons/acre
Corn	1	0.0017	62	0.00038	130 bu/acre
Cotton	1	0.0022	40	0.00022	900 lb/acre
Sorghum	1.0	0.0022	60	0.00034	65 bu/acre
Soybean	1.5	0.0022	31	0.00058	35 bu/acre
Wheat-S	1.7	0.0018	107	0.00060	30 bu/acre

D. Rangeland random roughness values [Eq. (8.55)]

Condition	Random roughness (in.)
California annual grassland	0.2
Tallgrass prairie	0.3
Natural shrub	0.8
Cleared	0.7
Pitted, cleared and pitted	1.0
Root plowed	1.3
Clipped and bare	0.6
Seeded rangeland drill	0.8

E. Recommended root mass for meadows [for Eq (8.57)]

Type of grass	R_{sr} 0–4 in. root zone
	lb/acre
Clover, orchard grass	1200
Lespedeza, Bahia, and Bermuda Grass	2100
Alfalfa, Dallis grass, Fescue	2600
Reed Canary grass	6100

F. Parameter values for estimating below-ground root mass for grasses and prairie vegetation [Using Eq. (8.56)]

	n_i		a_i	
Vegetation type	Best estimates	Range	Best estimates	Range
Northern mixed prairie	0.34	0.22–0.77	30.0	0.64–119.6
Tallgrass prairie	0.74	0.73–0.75	7.4	0.23–20.3
Shortgrass prairie	0.41	0.24–0.64	3.2	1.12–10.7
Sagebrush, bunch grass	0.38	0.35–0.41	28.8	27.30–29.6
Sagebrush, herbaceous interspaces	0.45	0.41–0.45	10.2	0.93–27.6
Cold desert shrubs	0.46	NA	5.0	4.09–11.0
California annual grassland	0.33	NA	3.0	NA

[a] Additional data are available in Appendix 8C. (Adapted from Yoder *et al.* 1993).

Table 8.11 Selected Field Operations and Associated Parameter Values for RUSLE (after Yoder *et al.*, 1993)

Field operations	R_R, random roughness (in.)	Percentage surface residue buried (%)	Tillage depth (in.)	Soil surface disturbed (%)
Chisel (2-in. shovels)	0.9	25	8	100
Cultivator, row	0.6	30	4	85
Disk, 1-way (18- to 24-in. disks)	1.1	40	4	100
Drill, conventional	0.4	10	2	80
Harrow (tine)	0.4	5	4	100
Moldboard (8 in. deep)	1.9	90	8	100
Planter, row	0.4	15	4	20

mulch effects. The factor is given by

$$R_G = (25.4R_R - 6)(1 - e^{-0.0015R_S})e^{-0.14P_T};$$
$$R_G \geq 0.0 \tag{8.55}$$

where R_R is the total random roughness (inches) after a field operation, P_T is the total rainfall (inches) after the last field operation, and R_S is the total root and buried residue after tillage in the top 4 in. of soil (pounds per acre). Selected values for the live root mass component of R_S, R_{sr}, are given in Table 8.10A. The buried residue component of R_S, R_{br} is discussed under the prior land-use subfactor. Total random roughness is the standard deviation of land surface elevation after furrows and slopes are removed from calculations. Example values for random roughness are given in Table 8.10D for rangeland and in Table 8.11 for tillage operations.

Below-Ground Root Mass Example data are given in Table 8.10A and E for below-ground live root mass, R_{sr}. For those crops where data are not available, below-ground live root mass can be estimated from the above-ground root mass by

$$R_{sr} = B_{AG} n_i a_i, \tag{8.56}$$

where R_{sr} is the root mass in the upper 4 in. (pounds per acre), B_{AG} is the above-ground biomass (pounds per acre), a_i is the ratio of root mass (pounds per acre) to above-ground biomass (pounds per acre), and n_i is the ratio of root mass in the upper 4 in. to that in the total root zone. Values for n_i and a_i are given in Table 8.10F for selected plant communities in the Western U.S.

The total below-surface mass in the upper 4 in., R_s, is the sum of root mass and buried residue, or

$$R_s = R_{sr} + R_{br}, \tag{8.57}$$

where R_{sr} and R_{br} are live root mass and buried residue, respectively (pounds per acre). The mass of buried residue at any time depends on the initial mass of residue at harvest, the rate of decomposition of residue, and tillage sequences after harvest. In cases where the initial mass of residue is not known, it can be estimated from the grain yield and residue-to-grain ratio in Table 8.10C. Residue decomposition in draft documentation available for RUSLE at the writing of this text was estimated by relationships from Gregory *et al.* (1985), or

$$P_{Ri} = \left[1 - 14.1D_i\left(\frac{U}{R}\right)(T_{ai} - 30)\frac{A_{mi}}{C_N}\right]^2, \tag{8.58}$$

where P_{Ri} is the fraction of initial mass remaining after D_i days, U is a constant for the given crop producing the residue, R is the average radius of a residue stem, T_{ai} is air temperature in degrees Farenheit during the D_i days, A_{mi} is an antecedent moisture term, and C_N is the carbon to nitrogen ratio for the residue. For periods where $T_{ai} < 30°$ F little decomposition occurs; hence $P_{Ri} = 1.0$. Selected values for U/R

and C_N are given in Table 8.10C. The effects of selected tillage practices on buried residue are given in Table 8.11. Values for A_m are estimated on an annual basis for the Western U.S. by

$$A_m = 0.018 R_{AN}, \tag{8.59}$$

where R_{AN} is the annual rainfall in inches. For areas east of the Rockies, A_m is estimated for 15-day periods by

$$A_{mi} = 0.22 R_{15,i}, \tag{8.60}$$

where $R_{15,i}$ is the rainfall in 15-day increments. The cumulative value of residue decomposition effects, P_{RC}, at any time is the product of P_{Ri} for the 15-day increments after residue was produced, or

$$P_{RC} = \prod_{i=1}^{N_P} P_{Ri}, \tag{8.61}$$

where N_P is the number of 15-day increments evaluated.

At the writing of this text, the authors of RUSLE (Renard *et al.*, 1993) were planning to change the form of the residue decay relationship in Eq. (8.58). The changes are not expected to drastically change the results for most situations.

Surface Roughness Subfactor

The direct impact of surface roughness on erosion is given by the surface roughness subfactor. The indirect impacts of surface roughness on the effectiveness of mulch and residue as a surface cover are included under the surface cover factor in Eq. (8.53). The surface roughness subfactor is given by

$$C_{sr} = e^{-0.026 R_G}, \tag{8.62}$$

where R_G is defined by Eq. (8.55). Rainfall decreases surface roughness and thus decreases its impact on erosion. Rainfall impacts are included in computation of R_G in Eq. (8.55).

Soil Moisture Subfactor

The soil moisture subfactor accounts for the effects of antecedent moisture on infiltration. In general, the effects of antecedent moisture on annual soil erosion are accounted for by the seasonal variation in the K factor. For single storms, a correction may be needed. When the soil is near field capacity, the soil moisture subfactor, C_{sm}, is 1.0. When soil moisture is near the wilting point to a depth of 6 ft, the value for C_{sm} is 0.0. A conservative estimate is to assume a value of 1.0.

In the Western U.S., particularly the Pacific Northwest, K values are not varied with season; hence, soil moisture corrections are in order. Information is given in Appendix 8C, Table 8C.3, on replenishment and depletion rates for these lands. Moisture balance computations are made on 15-day increments and compared to field capacity and wilting point values to determine C_{sm}. Yoder *et al.* (1993) recommend that a linear relationship be used between 1.0 at field capacity and 0.0 at the wilting point. Soil moisture factors are not used for rangelands.

Below-Ground Effects: Prior Land Use (PLU) Factor

The subfactor for prior land use (PLU) is used to estimate the impact of prior cropping, tillage practices, soil consolidation, time, and biological activity on erosion. For example, when a change is made from a meadow to a cropland, there is a residual beneficial effect of the meadow on erosion for 2 years. When the change is made from cropland to meadow, there is a residual detrimental effect for 3 years. The PLU factor is generalized to consider both type effects through the relationship

$$C_{plu} = D_{en} e^{-C_1 R_s}, \tag{8.63}$$

where D_{en} is a density variable related to tillage practices, C_1 is a constant (Table 8.10B), and R_s is the amount of live roots and buried residue in the upper 4 in. of soil (pounds per acre). Values for the buried residue must be estimated on the basis of: (1) tillage impacts on buried residue as given in Table 8.11 and (2) residue decay. Procedures for making the estimates are given in Example Problem 8.7.

The parameter D_{en} is used to account for the effects of changes in surface density that occur as a result of tillage practices. Tillage breaks soil particle bonds and reduces density, thus increasing the potential for erosion. With the passage of time and natural reconsolidation, these bonds reform and erosion potential is reduced. Yoder *et al.* (1993) refer to the work of Dissmeyer and Foster (1981) to show that D_{en} should vary exponentially from 1.0 for freshly tilled soil to 0.45 after 7 years with no tillage. The Dissmeyer and Foster data are presented in Fig. 8.15 as an estimator of D_{en}.

Use of RUSLE to Estimate Soil Erosion

The RUSLE procedure allows the systematic evaluation of the interactions between soil and plant properties that effect soil erosion; however, the procedures are computationally intensive and require a computer

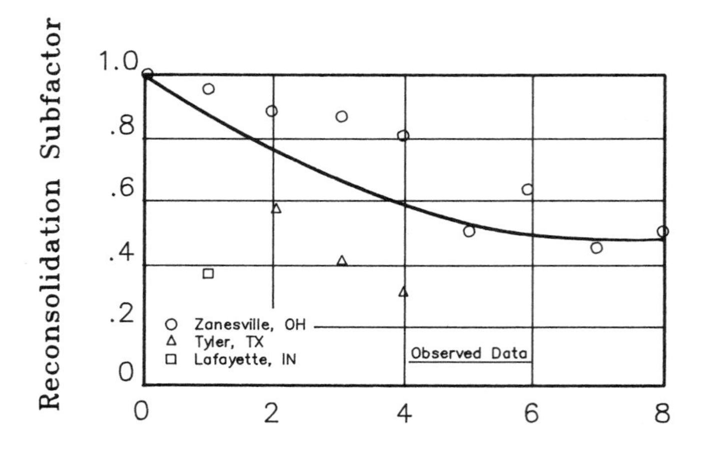

Figure 8.15 Effects of reconsolidation on erodibility as reported by Dissmeyer and Foster (1981). These values are recommended as estimates of D_{en} in the prior land-use factor.

or a major spreadsheet for any extensive use. A computerized version is available from the Soil and Water Conservation Society of America.

Example Problem 8.7 Use of RUSLE subfactors to estimate the cover factor for agricultural land

Estimate the C factor on April 15 for the following conditions:

> Soybeans following corn. Corn harvested on October 1 with a yield of 6200 lb/acre; soybeans planted on March 1. Monthly average temperatures: October, 60° F; November, 50° F; December, 42° F; January, 28° F; February, 32° F; March, 45° F; and April, 50° F. Soil moisture is at field capacity. Moldboard plowing on November 1; row planter on March 1. Precipitation during the period from October 1 to April 15 is: October, 3 in.; November, 5 in.; December, 4 in.; January, 4 in.; February, 5 in.; March, 7 in.; April, 10 in.

Assume that the soil is such that erosion is a typical rill and interrill mixture.

Solution:

1. Prior land-use factor [Eq. (8.63)]. From Fig. 8.15, D_{en} for freshly tilled soil (45 days after tillage) is still approximately 1.0. From Table 8.10B, the prior land-use parameter C_1 is 0.00088. The prior land-use factor C_{plu} is for the effects of crops grown in previous years. Hence, the C_{plu} residue parameters are for corn residue; whereas active growing roots will be soybeans. April 15 is 45 days after planting; hence from Table 8.10A, the growing root mass, R_{sr} = 92 lb/acre for soybeans. The below-ground and above-ground residue from previous corn must be estimated from residue decomposition and tillage incorporation information. Parameters for Eq. (8.58) for predicting decay of corn residue left at harvest are taken from Table 8.10C, $U/R = 0.0017$ and $C_N = 62$. To utilize Eq. (8.58), the value for P_{Ri} must be estimated for each time period and used multiplicatively. Calculations are summarized below by 15-day increments. Monthly precipitation is divided equally among 15-day periods.

Time period	T_{ai} (°F)	$R_{15,i}$ (in.)	A_{mi}[a]	P_{Ri}[b]
10/1–10/15	60	1.5	0.33	0.888
10/15–11/1	60	1.5	0.33	0.888
11/1–11/15	50	2.5	0.55	0.876
11/15–12/1	50	2.5	0.55	0.876
12/1–12/15	42	2.0	0.44	0.940
12/15–1/1	42	2.0	0.44	0.940
1/1–1/15	28	2.0	0.44	1.00
1/15–2/1	28	2.0	0.44	1.00
2/1–2/15	32	2.5	0.55	0.987
2/15–3/1	32	2.5	0.55	0.987
3/1–3/15	45	3.5	0.77	0.870
3/15–4/1	45	3.5	0.77	0.870
4/1–4/15	50	5.0	1.10	0.761

[a]Equation (8.60) $A_{mi} = 0.22\, R_{15,i}$.
[b]Equation (8.58), using 15-day increments for D, $P_{Ri} = \left[1-(14,1)(15)(0.0017)(T_{ai}-30)\left(\frac{A_{mi}}{62}\right)\right]^2 = [1-0.0058\,(T_{ai}-30)\,(A_{mi})]^2$.

From Eq. (8.61)

$$P_{RC} = \prod_{i=1}^{13} P_{Ri} = 0.30.$$

The total residue mass present at April 15 is thus 0.300 times that left at harvest on October 15. From Table 8.10C, the residue-to-grain ratio for corn is 1.0; thus the residue produced equals the weight of grain or 6200 lb/acre. The total residue present at April 15, accounting for decay by P_{RC}, is thus

$$\text{Residue} = (0.300)(6200) = 1860 \text{ lb/acre}.$$

Of this residue, moldboard plowing on October 1 buried 90% to a depth of 8 in. (Table 8.11). This would leave 45% in the upper 4 in. and 45% below 4 in., assuming a uniform burial distribution. On March 1, row planting placed 15% (Table 8.11) of the mulch remaining on the surface (10% of original) into the 0 to 4-in. zone. Thus the fraction of original surface residue that is in the 0- to 4-in. zone on April 15 is

$$\left|\begin{matrix}\text{fraction of}\\ \text{original surface}\\ \text{residue in upper 0– 4 in.}\end{matrix}\right| = [0.45 + 0.15(1 - 0.90)]$$

$$= 0.465.$$

On a mass basis, the buried residue in this zone is thus

$$R_{br} = (0.465)(P_{RC})(6200)$$
$$= (0.465)(0.300)(6200) = 865 \text{ lb/acre}.$$

Since planting incorporates only 15% of the surface residue (Table 8.11), 85% of the residue at planting remains on the surface. Thus, the fraction remaining on the surface on April 15 is 85% of the 10% left initially after turnplowing or (0.85)(0.1) = 0.085. The above ground residue, R_W, is therefore 0.085 times the residue left, after accounting for decay, or

$$R_W = (0.085)(P_{RC})(6200)$$
$$= (0.085)(0.300)(6200)$$
$$= 158 \text{ lb/acre}.$$

The total mass in the upper 0–4 in. of soil on a per inch basis is the sum of the incorporated residue after accounting for decomposition, R_{br}, and the below-ground root mass, R_{sr}, or

$$R_s = R_{sr} + R_{br} = 92 + 865 = 957 \text{ lb/acre}.$$

Using $R_s = 957$, $D_{en} = 1.0$ and $C_1 = 0.00088$ in Eq. (8.63) for C_{plu}

$$C_{plu} = D_{en}e^{-C_1 R_s} = (1.0)e^{-0.00088(957)}$$
$$= 0.431.$$

2. *Canopy cover factor* C_{cc}. From Table 8.10A, the canopy height and canopy cover for soybeans at 45 days is $H = 0.5$ ft and $F_c = 0.3$. The cover factor from Eq. (8.52) is thus

$$C_{cc} = 1 - (0.3)e^{-0.1(0.5)} = 0.715.$$

3. *Surface cover factor* C_{sc}. Parameters needed for surface cover factor predictions in Eq. (8.53) are the fraction of ground cover, R_c, the roughness parameter, R_G, and the coefficient b. From Table 8.10B, $b = 3.5$ for typical cropland erosion.

Fraction of residue ground cover R_c. From earlier calculation in part 1 of this problem, the above-ground residue weight R_W is 158 lb/acre of corn residue. From Table 8.10C, a_W for corn residue is 0.00038. From Eq. (8.54)

$$R_c = 1 - e^{-0.00038(158)} = 0.058.$$

Surface roughness factor. Equation (8.55) is used to calculate R_G, which is then further used in Eq. (8.53) to calculate the effects of random roughness on both the surface cover subfactor and a surface roughness subfactor. From Table 8.11, the random roughness is 0.4 in. after row planting on March 1. (Note that row planting would obliterate roughness elements remaining from earlier turnplowing.) The precipitation between March 1 and April 15 from the tabulation in part 1 is 3.5 + 3.5 + 5.0 or 12.0 in. Also from part 1, $R_s = 1057$ lb/acre. Thus from Eq. (8.55) with a random roughness of 0.4 in.,

$$R_G = [25.4R_R - 6][1 - e^{-0.0015R_s}]e^{-0.14P_T}$$
$$= [(25.4)(0.4) - 6][1 - e^{-0.0015(957)}]e^{-0.14(12)}$$
$$R_G = 0.591.$$

Calculating surface cover subfactor. Equation (8.53) is used with R_c of 0.058, R_G of 0.616, and b of 3.5 to obtain

$$C_{sc} = \exp\left[(-3.5)(0.058)\left(\frac{6}{6 + 0.591}\right)^{0.08}\right]$$
$$= 0.818.$$

4. *Surface roughness subfactor [Eq. (8.62)].*

$$C_{sr} = e^{-0.026R_G} = e^{-0.026(0.591)} = 0.985.$$

5. *Soil moisture subfactor.* Soil moisture is at field capacity; thus

$$C_{sm} = 1.0.$$

6. *Calculating C factor [Eq. (8.51)].*

$$C = C_{plu}C_{cc}C_{sc}C_{sr}C_{sm}$$
$$= (0.431)(0.715)(0.818)(0.985)(1.0)$$
$$= 0.248.$$

Cover Factors: Subfactor Approach for Construction and Mined Lands

Introduction

Soils that are drastically disturbed and/or reconstructed have erosion characteristics that are different from agricultural lands. With passage of time, the forces of weathering and reconsolidation tend to move these characteristics more toward agricultural lands.

No published subfactor procedure has been developed for construction and mined lands, but a modification of Eq. (8.51) seems to be in order. In addition to the factors that are included in Eq. (8.51), other important factors include effects of compaction and timing after reconstruction. No published comprehensive procedures have been developed for all the subfactors; hence procedures proposed below are somewhat speculative, especially for below-ground biomass. The C factor for construction and mined lands could be represented by

$$C = C_T C_D C_{plu} C_{cc} C_{sc} C_{sr} S_{sm}, \quad (8.64)$$

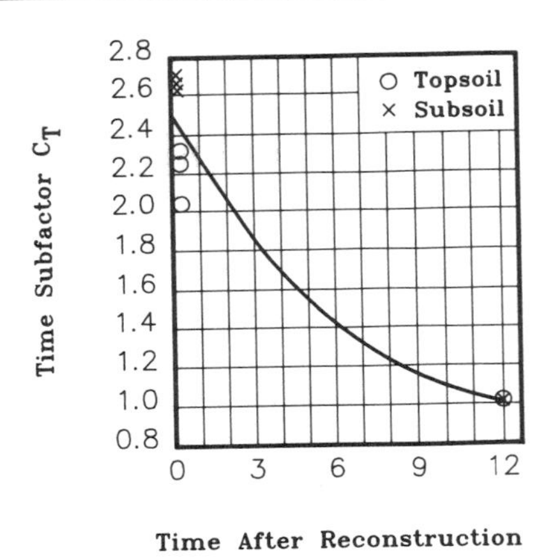

Figure 8.16 Subfactor to account for the effects of time after reconstruction on the C factor (after Barfield *et al.*, 1988). These results are based on studies of reconstructed surface mine soils.

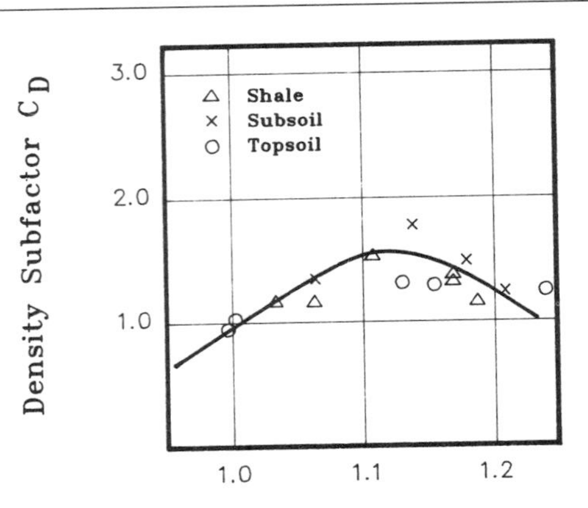

Figure 8.17 Subfactor for the effects of compacted density on the C factor (after Barfield *et al.*, 1988). These results are based on studies of reconstructed surface mine soils.

where C_T is the subfactor for time and C_D is the subfactor for density and other factors as previously described for Eq. (8.51) for agricultural lands.

Time Subfactor

After soil reconstruction, there is an apparent loss of soil structure resulting in greatly enhanced erosion, even under compacted conditions (Barfield *et al.*, 1988). With the passage of time and weathering, soil structure recovers and erosion rates decrease. The impacts of such changes have been evaluated by Barfield *et al.* (1988) who proposed the relationship shown in Fig. 8.16.

Density Subfactor

As soil is compacted artificially, its inherent resistance to erosion decreases. This is in contrast to the effects of reconsolidation after tillage, where the increase in density causes a decrease in erosion (Barfield *et al.*, 1988; Israelson *et al.*, 1980). In the latter case, the increase in density occurs as soil aggregate bonds are enhanced, resulting in increased erosion resistance. Mechanical compaction apparently does not create such resistance. A subfactor for the effects of density changes due to mechanical compaction is given in Fig. 8.17.

Prior Land-Use Factor

When mulch is applied and the surface is planted to vegetation, Eq. (8.63) can be used to estimate C_{plu}. Since vegetation does not grow as rapidly on reconstructed land, the root mass estimates from Tables 8.10 should be decreased significantly. In the absence of long-term data on erosion on reconstructed lands, the value for D_{en} in Eq. (8.63) should be conservatively set at 1.0.

Other Subfactors, C_{cc}, C_{sc}, C_{sr}, C_{sm}

The prior land-use factor discussed above for construction and mined lands should be different from that of agricultural lands. Other factors are a function of what occurs on the surface and above ground and should be estimated the same as for agricultural lands, given the appropriate crop parameters. However, since growth in the early years after reclamation can be considerably different from agricultural lands, the crop parameters given in Table 8.10 should be adjusted accordingly based on experience in the area. An experienced agronomist or reclamation specialist could be of assistance.

Example Problem 8.8 Cover Factor for construction sites

Estimate the C factor for a reconstructed surface-mined land 6.5 months after regrading and reseeding with rye and fescue on October 1. One ton per acre of straw mulch is applied at seeding. During regrading, the soil is compacted to 1.1 times its loose tilled density due to compaction effects of reclamation machinery. Assume no seedbed preparation and adequate moisture. Assume the same temperature and moisture conditions as in Example Problem 8.7.

Solution:

1. *Subfactors for density and timing.* Density subfactor for relative density of 1.1 (Fig. 8.17)

$$C_D = 1.5.$$

Timing subfactor for time after reconstruction of 6.5 months (Fig. 8.16)

$$C_T = 1.36.$$

2. Subfactor for surface cover. As a first approximation, assume no mulch decomposition and utilize values from Table 8.8, item 2,

$$C_{sc} = 0.18.$$

This would be an acceptable value for short periods after application of mulch and for longer periods if a rough estimate were acceptable. As an alternative, Eqs. (8.53) through (8.60) can be utilized. First, estimate the fraction of ground cover by mulch (residue) using Eq. (8.54). For this equation, $a_W = 0.00060$ from Table 8.10C (assuming wheat straw), and $b = 3.5$ from Table 8.10B. Using a mulch rate of 1 ton/acre or 2000 lb/acre and a residue decay P_{RC} of 0.300 from Example Problem 8.7,

$$R_W = (0.300)(2000) = 600 \text{ lb/acre}.$$

Use of P_{RC} from Example Problem 8.7 is justified since the same temperature and moisture conditions were assumed. From Eq. (8.54),

$$R_c = 1 - e^{-a_W R_W} = 1 - e^{-0.00060(600)}$$
$$= 0.302.$$

From Eq. (8.53), assuming zero random roughness ($R_G = 0$),

$$C_{sc} = \exp\left[-bR_c\left(\frac{6}{6+R_G}\right)^{0.08}\right]$$
$$= \exp\left[-(3.5)(0.302)\left(\frac{6}{6+0}\right)^{0.08}\right]$$
$$= 0.347.$$

A C_{sc} of 0.347 is the more appropriate value since it takes into account residue decay.

3. Canopy subfactor. This computation requires an estimate of canopy cover. At 6.5 months, the primary canopy will be rye, which is a small grain. Since planting was in the fall, the crop is classed as winter grain. From Table 8.10A, the canopy cover and height for small grains on agricultural lands after 6.5 months would be 1.0 and 2.5 ft. For an unprepared seedbed, the values would be smaller. For a first approximation half these values will be assumed, or $F_c = 0.50$ and $H = 1.25$ ft. In actual practice, reclamation specialists should be consulted for information on crop stands under reclaimed conditions. Using these values and Eq. (8.52),

$$C_{cc} = 1 - F_c e^{-0.1H} = 1 - 0.50e^{-0.1(1.25)} = 0.559.$$

4. Other factors. Since the surface roughness is zero, and moisture is at field capacity C_{sr} and $C_{sm} = 1.0$. Immediately after reconstruction, there will be no crop residuals; hence

$$C_{plu} = 1.0.$$

5. Calculating C factor. From Eq. (8.62),

$$C = C_D C_T C_{plu} C_{cc} C_{sc} C_{sr} C_{sm}$$
$$= (1.5)(1.36)(1.0)(0.559)(0.347)(1.0)(1.0)$$
$$= 0.395.$$

Cover Factors: Subfactor Approach for Disturbed Forest and Woodlands

Factors for Undisturbed Forests and Grassed Forests

As discussed previously, first estimates of C factors for undisturbed forests and grassed forest lands are best estimated with tabulations developed by Wischmeier and Smith (1978) as summarized in Table 8.8 and Appendix 8B. In this section, cover factors are addressed for forests and woodlands that are disturbed to some degree.

Introduction to Forest and Woodland Subfactors

Cover factors for forest and woodlands with some disturbance can be estimated using procedures developed by Dissmeyer and Foster (1984). Using a subfactor analogy, the C factor is estimated from a product of factors to account for bare soil, fine root mat, soil reconsolidation, canopy, steps, depression storage, and contour tillage. Dissmeyer and Foster (1984) evaluated the accuracy of the method on four plots and 35 watersheds in the southeast. The correlation between predicted and observed values was 0.90 and the average prediction error was ±71%.

Bare Soil and Fine Root Mat Subfactor

Undisturbed soils in forested areas tend to have infiltration rates greater than rainfall rates. Thus, bare soil areas become the primary source of sediment and runoff. A 0% cover in forest lands frequently results in zero sediment yield as a result of the root mass, whereas zero cover for agricultural land will normally have some sediment yield. Thus the cover factors for forest land use are different from agricultural land uses.

The bare soil and root mass subfactors are given in Table 8.12 for forest lands. The reconsolidation factor discussed in the following section is also incorporated into these values.

Soil Reconsolidation Subfactor

After a tillage operation, soil reconsolidates naturally as a result of wetting and drying cycles. Unlike mechanical compaction, this reconsolidation improves soil particle bonding with a resultant increase in erosion resistance and decrease in erosion rates. Plots maintained in a fallow condition at Zanesville, Ohio,

Table 8.12 Selected Subfactors for Disturbed Forest Lands (after Dissmeyer and Foster, 1984). Additional Data Are Available in Appendix 8D

A. Effect of bare soil, fine root mat of tree roots, and soil reconsolidation on *C* factor

1. Untilled soils

Percentage bare soil	Percentage of bare soil with dense mat of fine roots in top 3 cm of soil		
	100	60	0
0	0.0000	0.0000	0.0000
1	0.0004	0.0007	0.0018
10	0.005	0.009	0.0123
40	0.023	0.042	0.104
100	0.099	0.180	0.450

2. Tilled soils with good initial fine root mat in topsoil. Subsoil has good structure and permeability

Percentage bare soil	Time (months) since tillage		
	0	12 and 72+	24+ thru 60
0	0.0000	0.0000	0.0000
1	0.0014	0.0018	0.0020
10	0.019	0.023	0.026
40	0.083	0.104	0.117
100	0.360	0.450	0.510

3. Tilled soil with poor initial fine root mat in topsoil. Subsoil has good structure and permeability

Percentage bare soil	Time (months) since tillage		
	0	12 thru 36	72+
0	0.0000	0.0000	0.0000
1	0.0021	0.0025	0.0018
10	0.027	0.033	0.023
40	0.122	0.144	0.104
100	0.530	0.630	0.450

B. Step effect on soil erosion

Percentage slope	Percentage of total slope in steps			
	0	30	50	100
5	1.00	0.98	0.96	0.92
10	1.00	0.81	0.68	0.36
15	1.00	0.75	0.59	0.18
30+	1.00	0.72	0.53	0.06

C. Contour tillage subfactors for forestlands

Percentage slope	On contour	Degrees off contour		
	0	15	45	90
1.0	0.80	0.88	0.94	1.00
10.0	0.80	0.88	0.94	1.00
19+	1.00	1.00	1.00	1.00

had a decreased erosion rate with consolidation after tillage, decreasing to 0.45 of the original tilled value after 7 years. Most of this decrease occurred prior to 3 years. Subfactors for reconsolidation are given in Table 8.12A, in combination with bare soil and fine root subfactors. When considered alone, the soil reconsolidation subfactor proposed by Dissmeyer and Foster is given in Fig. 8.15.

Canopy Subfactor

A canopy of forest vegetation will have an impact similar to that of agricultural crops if the canopy heights are similar. Thus, the above-ground canopy subfactor should be estimated from Eq. (8.52).

Depression Storage Subfactor

Water stored in depressions cannot transport sediment off site; thus erosion is reduced by depression storage. The depression storage subfactor is estimated by selecting the appropriate factor from Fig. 8.18. For example, if the surface is composed primarily of 6-in. clods, the factor would be 0.5. Likewise, if the surface is bare soil chopped with 1-in.-deep slits along the contour, the factor would be 0.8.

Step Subfactor

In forest lands, debris washed by surface runoff will form small dams with subsequent ponding and deposition. The final result is a series of steps that have the appearance of mini-terraces, as shown in Fig. 8.19. Based on observation from 100 steps throughout the southeast, Dissmeyer and Foster (1984) found that the deposition behind the slopes occurred at a slope of 3%, as shown in Fig. 8.19. Using this measurement, the Foster and Wischmeier (1974) relationship for irregular slopes, and the assumption that steps were small

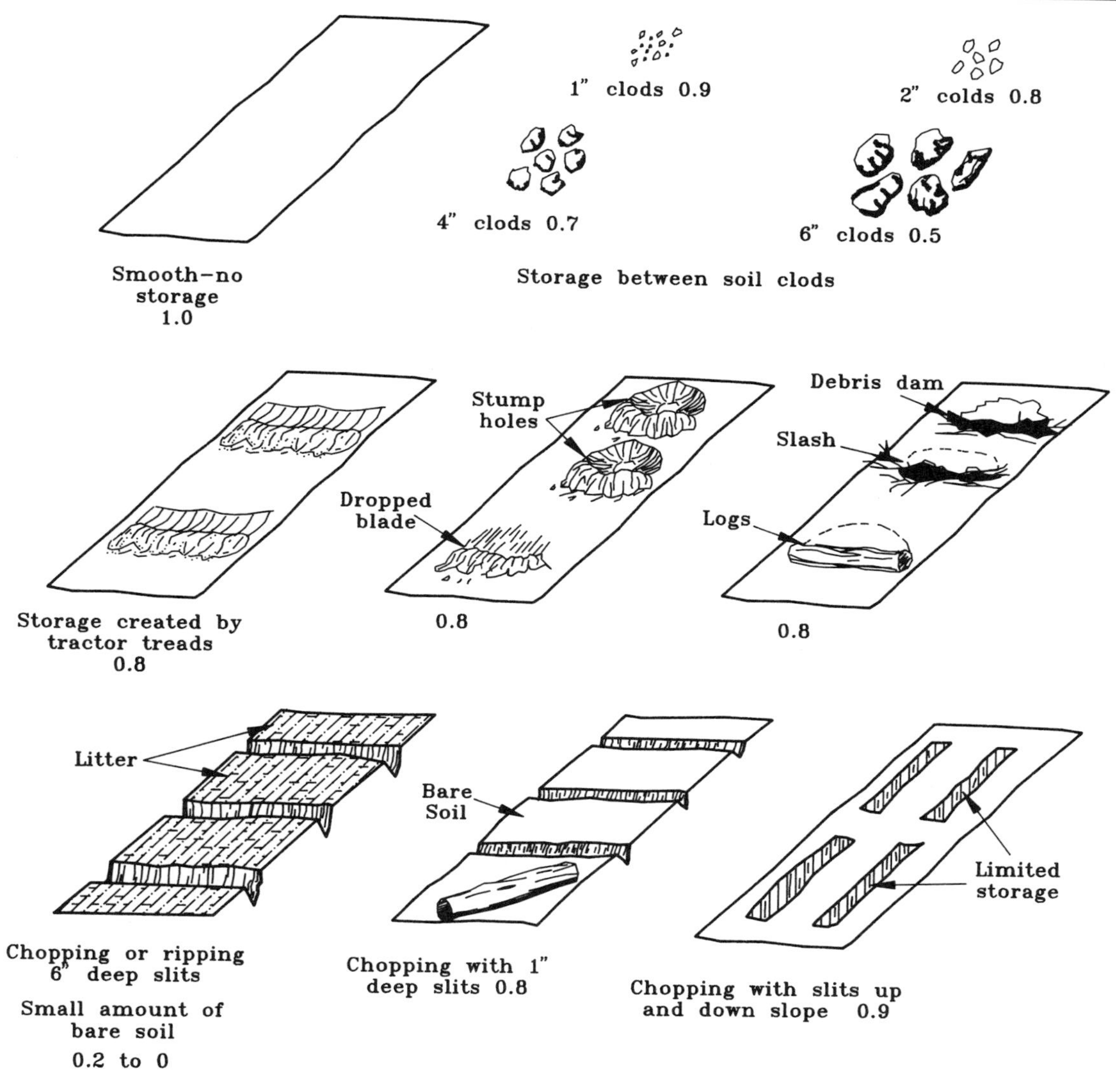

Figure 8.18 Subfactors for on-site depression storage in forests (after Dissmeyer and Foster, 1984).

and randomly distributed, Dissmeyer and Foster (1984) developed the relationship for steps given in Table 8.12B. In general, this relationship should be used when forest lands are being disturbed for logging or other operations.

Contour Tillage Subfactor

Contour disking and plowing in forests that have been disturbed will generally reduce erosion. Such operations are most effective when conducted on the contour. When this has not been accomplished, corrections are necessary.

Dissmeyer and Foster (1984) modified the USLE factors for contour plowing to account for disking on the contour, deviations from the contours, and the fact that ridges from disking are not as high as typical agricultural rows. These factors are given in Table 8.12C.

Example Problem 8.9 Estimating forest lands cover factors with subfactors

Estimate the cover factor for a forest land that has been subjected to logging operation. In the logging operation, 40% of the surface area is disturbed. Tracked vehicles that generate random slits 1 in. deep were used. The remaining vegetation provides only 30% above-ground canopy cover at a height of 20 ft. Ground slope is 10%. Estimate the C factor assuming: (A) a fine root mat covers the entire area and (B) logging removes the root mat.

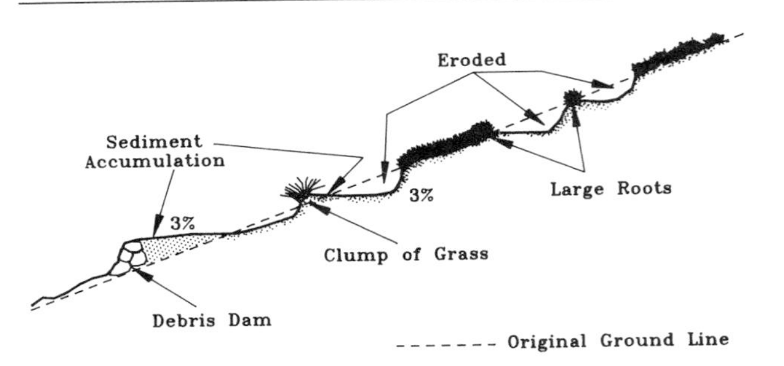

Figure 8.19 Illustration of step formation in forest lands (after Dissmeyer and Foster, 1984).

Solution:

With Fine Root Mat

1. Subfactor for reconsolidation, bare area, and fine root mass. Using Table 8.12A, with 100% fine root mass, 40% bare soil, untilled

$$\text{Subfactor } 1 = 0.023.$$

2. Canopy subfactor. From Eq. (8.52) with a canopy cover of 30% at a height of 20 ft,

$$\text{Subfactor } 2 = C_{cc} = 1 - F_c e^{-0.1H}$$
$$= 1 - 0.3e^{-(0.1)(20)} = 0.96.$$

3. Depression storage subfactor. From Fig. 8.18, slits up- and downslope would have a subfactor of 0.9 and across slope would be 0.8. Using an average

$$\text{Subfactor } 3 = 0.85.$$

4. Step subfactor. Assume that all disturbed areas (40%) are in steps. From Table 8.12B, with a ground slope of 10%, and 40% of the area in steps,

$$\text{Subfactor } 4 = 0.75.$$

5. Contour tillage subfactor. Disturbed areas are not tilled; therefore, contour tillage factor is 1.0.

6. Calculating C factor.

$$C = (0.023)(0.96)(0.85)(0.75)(1.0)$$
$$= 0.014.$$

Fine Root Mat Removed

The only change in the solution is to subfactor 1 above. With zero fine root mat in the disturbed area, subfactor 1 for the effect of fine root mat and soil reconsolidation from Table 8.12A with 40% bare area would be 0.104. Thus the C factor becomes

$$C = (0.104)(0.96)(0.85)(0.75)(1.0) = 0.064,$$

Conservation Support Practice P Factor

Introduction

The conservation practice factor, P, by definition is the ratio of soil loss from any conservation support practice to that with up- and downslope tillage. It is used to evaluate the effects of contour tillage, stripcropping, terracing, subsurface drainage, and dryland farm surface roughening. The effects of sod-based crop rotations, minimum tillage, residue management, and humid area surface roughening are included in the C factor, as discussed in the previous section. The P factor is typically used only for agricultural lands and rangelands, but could be used with some caution on construction and disturbed lands.

A first approximation to the P factor can be developed by using tabulated values from the USLE data base. This can be used by the practitioner interested in first or rough estimates, particularly for planning purposes. For more refined estimates that allow for detailed consideration of a variety of combined practices, a subfactor approach from the RUSLE would be preferred. Both procedures are presented here.

Tabulated P Factors from the USLE

Tabulated total values for the P factor are available for contouring, strip cropping, and contour terracing, as given in Table 8.13. A detailed discussion of the values is beyond the scope of this text. A few comments, however, are in order. First, it should be noted that some of the practices have limits within which they are assumed to be totally effective. These limits decrease with increasing slope. For example, P values increase with slope, and the maximum effective slope length for contouring decreases with slope. Limits on contouring result from the decreasing surface storage with steeper slopes and the greater tendency to form rills at steeper slopes.

The RUSLE Subfactor Approach to the P Factor

An alternative to the tabulations in Table 8.13 for the P factor is to use a subfactor analogy from the RUSLE, or

$$P = P_c P_{st} P_{ter}, \tag{8.65}$$

where P_c is the contour subfactor, P_{st} is the strip cropping subfactor, and P_{ter} is the terracing subfactor. Use of this subfactor analogy allows a more detailed evaluation of factors affecting P, particularly when considering a combination of practices. Also, it allows a correction for the impact of large storms on contouring.

At the writing of this text, the RUSLE was in draft form. Revisions are likely to be made in some of the subfactors before final printing.

Contour Support Factor P_c The contour support factor accounts for the impact of tillage on the contour on soil erosion. If the surface is tilled up- and downslope or is relatively smooth, a drainage pattern that allows eroded sediment to be readily transported downslope

Table 8.13 Selected USLE *P* Factors (after Wischmeier and Smith, 1978)

1. *P* values and slope length limits for contouring

Land slope percentage	*P* value	Maximum length[a] (ft)
1 to 2	0.60	400
3 to 5	0.50	300
6 to 8	0.50	200
9 to 12	0.60	120
13 to 16	0.70	80
17 to 20	0.80	60
21 to 25	0.90	50

2. *P* values, maximum strip widths, and slope length limits for contour stripcropping

Land slope Percentage	P values[b]			Strip width[c] (ft)	Maximum length (ft)
	A	B	C		
1 to 2	0.30	0.45	0.60	130	800
3 to 5	0.25	0.38	0.50	100	600
6 to 8	0.25	0.38	0.50	100	400
9 to 12	0.30	0.45	0.60	80	240
13 to 16	0.35	0.52	0.70	80	160
17 to 20	0.40	0.60	0.80	60	120
21 to 25	0.45	0.68	0.90	50	100

3. *P* values for contour-farmed terraced fields

Land slope Percentage	Farm planning		Computing sediment yield	
	Contour factor	Stripcrop factor	Graded channels sod outlets	Steep backslope underground outlets
1 to 2	0.60	0.30	0.12	0.05
3 to 8	0.50	0.25	0.10	0.05
9 to 12	0.60	0.30	0.12	0.05
13 to 16	0.70	0.35	0.14	0.05
17 to 20	0.80	0.40	0.16	0.06
21 to 25	0.90	0.45	0.18	0.06

[a]Limit may be increased by 25% if residue cover after crop seedings will regularly exceed 50%.

[b]*A*, for 4-year rotation of row crop, small grain with meadow seeding, and 2 years of meadow. *B*, for 4-year rotation of 2 years row crop, winter grain with meadow seeding, and 1-year meadow. *C*, for alternate strips of row crop and small grain.

[c]Adjust strip-width limit, generally downward, to accomodate widths of farm equipment.

develops. If tillage is on the contour, flow collects in the furrows between tillage ridges, allowing significant amounts of deposition. The effectiveness of contouring depends on the ability of the tillage marks to store runoff and is obviously impacted by the size or roughness of the tillage system, the slope of the system, the amount of runoff, as well as the peak intensity. When contour tillage marks have a cross-contour component, as they frequently do, the effectiveness of contouring is also reduced. As flows move downslope, the quantity of runoff increases, reducing the effectiveness of a given contour tillage system. Thus a critical slope length is

Table 8.14 Selected P Subfactors for the RUSLE (after Foster *et al.*, 1993)

A. P_b, Contouring subfactors for tillage along the contour when slope lengths are less than critical[a]

Downhill slope	Ridge of oriented roughness height (in.): Low 1–3	Moderate 3–5	Ridge system >5	Critical slope length (ft)
0.5	1.0	0.8	0.8	1000+
3.0	0.9	0.5	0.3	630
5.0	0.8	0.5	0.2	323
10.0	0.8	0.6	0.2	125
15.0	1.0	0.6	0.3	78
20.0		0.8	0.4	57
25.0		1.0	0.7	45[b]
32.0			1.0	30[b]

[a]To determine the contour factor value, use the base P_b value from above in Fig. 8.20. For locations with a 10-yr storm EI less than 50 or greater than 160, use 50 or 160. The P_s value obtained in Fig. 8.20 is used to determine the contour P_c factor value in Fig. 8.21.

[b]Projections beyond the values presented by Foster *et al.* (1991).

B. P_b effective, contouring subfactors for tillage along the contour when slope lengths are greater than critical[a]

(Upslope P value)/(Slope length exponent m)[b]

Critical slope:	0.2			0.4			0.6			0.8			1.0
Length ratio[c]:	0.1	0.5	1.0	0.1	0.5	1.0	0.1	0.5	1.0	0.1	0.5	1.0	All
0.5	0.63	0.72	0.80	0.72	0.79	0.85	0.81	0.86	0.90	0.91	0.93	0.95	1.0
00.6	0.54	0.63	0.71	0.66	0.72	0.78	0.77	0.81	0.86	0.89	0.91	0.93	1.0
0.7	0.46	0.53	0.61	0.59	0.65	0.71	0.73	0.77	0.80	0.86	0.88	0.90	1.0
0.8	0.37	0.43	0.49	0.53	0.57	0.62	0.69	0.71	0.74	0.84	0.86	0.87	1.0
0.9	0.29	0.32	0.35	0.47	0.49	0.51	0.64	0.66	0.68	0.82	0.83	0.84	1.0
1.0	0.20	0.20	0.20	0.40	0.40	0.40	0.60	0.60	0.60	0.80	0.80	0.80	1.0

[a]Use these values instead of P_b values in Item A when the slope length exceeds the critical slope length from Item A.

[b]Slope length exponent in the RUSLE slope length factor [see Table 8.6 or Eq. (8.44)].

[c]Ratio of critical slope length to total slope length.

C. P_t subfactor for terracing with conservation planning[a]

Horiz. Ter. int. (ft)	Closed outlets[b]	Open outlets with percentage grade[c] (%): 0.1–0.3	0.4–0.7	>0.8
<110	0.5	0.6	0.7	1.0
150	0.7	0.8	0.9	1.0
300	1.0	1.0	1.0	1.0

[a]Multiply values by other P subfactor values for practices on terrace interval.

[b]Values also apply to terraces with underground outlets and to level terraces with open outlets.

[c]Channel grade measured on the 300 ft closest to the outlet or one-third of the channel length, whichever is less.

D. Terracing impact on sediment yield.[a]

Terrace grade	Delivery subfactor
Closed outlet[b]	0.05
0 (level)	0.10
0.2	0.13
0.4	0.17
0.6	0.29
0.7	0.49
0.8	0.83
0.9	1.0

[a]To be used only to evaluate the impact of terracing on sediment yield from the field. Not to be included in gross erosion estimate or conservation planning. Sediment deposited in the terrace channel outlet is not effective in maintaining soil productivity.

[b]Includes terraces with underground outlets.

E. P_{sc} subfactor for stripcropping[a]

System[b]	P_{sc} for conservation planning[c]	Impact on sediment delivery from field[d]
Rotation stripcropping		
RC-WSG-M1-M2	0.78	0.53
RS-SSG-RC-SSG	0.91	0.75
RC-RCrt-RCrt-M1	0.84	0.65
RC-WSG	0.86	0.71
0.1 Filter	0.91	0.24
Buffer strip[f]	0.67	0.15
Buffer strip[g]	0.75	

[a]Used to evaluate the impact for selected stripcropping, buffer and filter strips.

[b]Rotation stripcropping is cropping in strips parallel to the contour. RC, row crop; WSG, winter small grain; SSG, spring small grain; M1, 1st year meadow; M2, second year meadow; C, corn; SB, soybeans; rt, reduced tillage; nt, no till.

[c]Use as an estimate of the impact of practice on gross erosion value.

[d]Use as an estimate of the impact on sediment yield from the field erosion.

[e]Permanent meadow filter strip that covers 10% of slope below row crop.

[f]Permanent meadow buffer strips located at 40–50% and 90–100% slope length.

[g]Permanent meadow buffer strip located at 40–50% slope length.

typically defined, beyond which contouring effectiveness is decreased.

The size of the ridge–furrow storage system, slope, the degree of the cross-contour component of the tillage marks, the runoff amount, and peak runoff rate impact the effectiveness of contour tillage. Using the limited data base available plus simulations with the CREAMS model, Foster *et al.* (1993) developed *P* subfactors for the impacts of contouring, which include these parameters. Base values for contouring, P_b, are given in Table 8.14A for slopes that do not exceed the critical slope length. For slopes that exceed the critical slope length, values for P_b are given in Table 8.14B. Corrections for cross contouring and storm intensity were developed using the CREAMS model and are given in Figs. 8.20 and 8.21. In revisions being planned to the initial draft of RUSLE, equations will be used in lieu of Figs. 8.20 and 8.21.

Foster *et al.* determined critical slope lengths from a simplification of a mulch stability analysis, or

$$\lambda_c = \frac{10471 n_t^{1.5}}{(EI_{10})^{0.31} S^{1.667} r}, \tag{8.66}$$

where λ_c is the critical slope length, n_t is Manning's n, EI_{10} is the 10-year storm EI index in English units, S is slope in feet per foot, and r is a dimensionless runoff reduction factor to account for infiltration.

Strip Cropping Conservation Support Factor P_{sc} Strip cropping is the use of alternating strips of close growing vegetation such as grasses and legumes between strips of clean-tilled or nearly clean-tilled row crops, all planted on the contour. Typically, the strips are rotated.

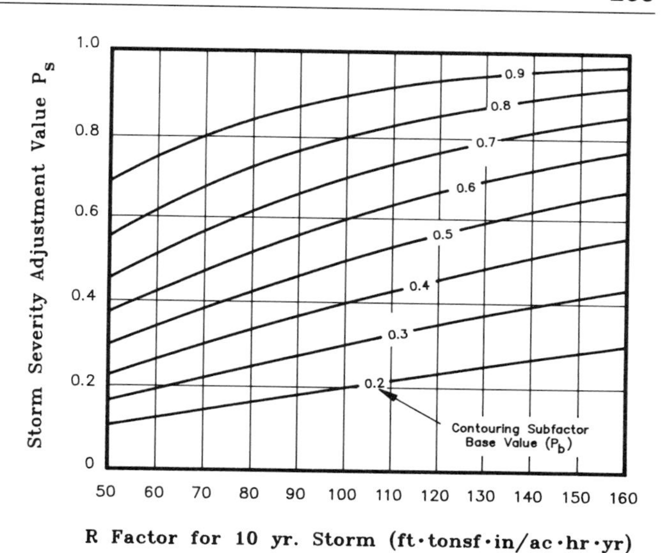

Figure 8.20 Effect of storm erosivity on the contouring subfactor (after Foster *et al.*, 1993).

An alternative to rotating strips is to include permanent strips of grass on the slopes. If the grass strips are located within the growing crops, the strips are referred to as buffer strips. If they are located along the bottom of the slope, they are referred to as filter strips. Buffer strips, being permanently located, do not contribute as heavily as strip cropping to soil conservation as the soil is not filtered in a location where it is used for further production of the row crops. Thus, if one is considering conservation planning, less credit would be

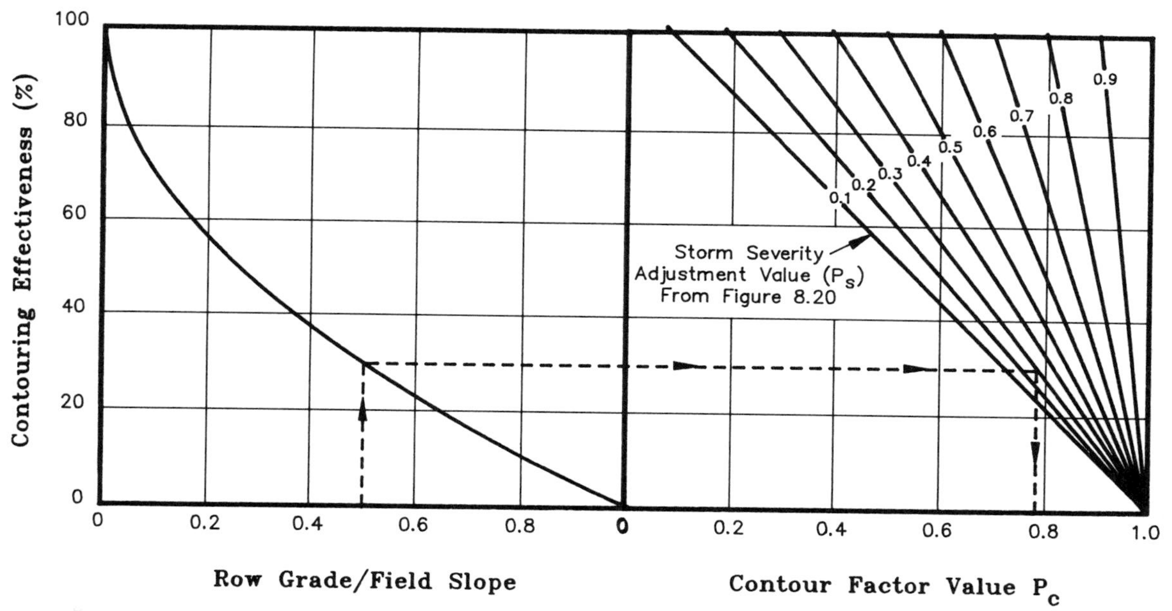

Figure 8.21 Adjustment to contour subfactor to account for grades off the contour and storm severity (after Foster *et al.*, 1993).

given for the filter strip than if one is considering sediment yield and off-site effects. The filter strips, being located permanently below the cropland, would not contribute at all to maintaining soil productivity; hence they would be considered ineffective as a conservation practice, but would contribute to controlling or reducing sediment yield.

As flow moves from the clean-tilled strip through the grass strip, transport capacity is decreased and sediment may be deposited. The effectiveness of the strip depends on strip width, slope, and type of tillage. Using the CREAMS equation, Foster *et al.* (1993) developed a computational procedure that is included in the RUSLE computer program. Foster *et al.* also published typical subfactor values for selected strip crops and buffer and filter strips. Selected values are given in Table 8.14E.

Terracing Support Factor, P_{ter} Terraces collect flows from slopes and divert them to a stabilized waterway or to closed outlets, thus preventing long slope lengths. In addition, flat sloped terrace channels cause deposition. Foster and Highfill (1983) developed factors to account for the effects of terraces on soil loss, also accounting for deposition in channels. The net soil loss is that lost from the slopes minus channel deposition. Values for the conservation practice factor for terracing, P_{ter}, are given in Table 8.14C. In addition to using P_{ter} subfactor, the slope length used in the RUSLE, should be the terrace interval.

Terraces reduce sediment yield in two ways, by decreasing the slope length and by allowing deposition in the terrace channels. These two factors are considered separately in the RUSLE. The impact of slope length change is included in the *LS* factor by reducing the slope length in the calculation. The effects of deposition are included in the terracing factor in Table 8.14C. The factor P_{ter} accounts for deposition in the terrace channel. The impact of terraces on the loss of soil from the slope is reflected in the slope length factor calculation based on the terrace spacing.

Unlike the USLE data base, the impact of contouring is not included in the *P* subfactor for terracing. Hence, to get a *P* factor for both contouring and terracing (the normal combination), the two RUSLE subfactors must be multiplied together as given in Eq. (8.65).

Rangeland Support Practices Rangeland treatment practices include surface roughening by tillage or pitting, contouring, and terracing. The impact of terracing and contouring must be evaluated with the support factors discussed above. The effectiveness of other practices is given in Foster *et al.* (1993).

Example Problem 8.10 Estimating conservation practice factor

A planner is interested in the impact of a proposed conservation plan on agricultural production. The plan includes contour strip cropping with conventional tillage sloped at 2% to the contour, and tile outlet terraces, spaced at 100 ft. Terrace channel slope is 1.0% and general downhill slope is 8%. The EI_{30} index (*R* factor) for a 10-year storm is 100 (English units) and ridge heights from tillage are moderate (4 in.). The slope has equal width strips of alternate row crop and winter small grain that are rotated.

Solution:

1. *Contouring subfactor.* From Table 8.14A, the uncorrected contouring subfactor for 4-in. ridge heights (moderate) and 8% slope is $P_b = 0.55$. Also, from the same table, the critical slope length is greater than 100, so the value of 0.55 is appropriate. From Fig. 8.20, the storm severity adjustment factor P_s for a 10-year *R* of 100 and P_b of 0.55 is $P_s = 0.55$. The ratio of tillage slope to downhill slope is $0.02/0.08 = 0.25$. From Fig. 8.21 with P_s of 0.55 and a grade-to-slope ratio of 0.25,

$$P_c = 0.76.$$

2. *Strip-cropping subfactor.* From Table 8.14E, the strip cropping subfactor for RC-WSG rotation is

$$P_{sc} = 0.86.$$

Since the interest is in conservation planning (effects on crop production), this is the appropriate number.

3. *Terrace subfactor.* From Table 8.14C for terrace spacing of 100 ft and closed outlets

$$P_{ter} = 0.5.$$

4. *Calculating support practice factor.* From Eq. (8.65),

$$P = P_c P_{st} P_{ter} = (0.76)(0.86)(0.5) = 0.33.$$

Prediction of Annual Erosion Using RUSLE

Estimating annual erosion with the RUSLE can be quite tedious, due to a need to consider so many variables that change with time. The procedure, however, is readily adapted to a spreadsheet or to a computer algorithm. As discussed earlier, computerized version of the RUSLE is available from the Soil and Water Conservation Society of America.

RILL AND INTERILL EROSION MODELING: COMMENTS ON PROCESS-BASED MODELS

Recent developments have led to process-based models such as CREAMS (Foster *et al.*, 1980a, b) and

WEPP (Lane and Nearing, 1989). These models are sufficiently complex to require a computer for solution; thus illustration of model applications is beyond the scope of this chapter. However, since these models are being widely used, a discussion of the basic concepts seems appropriate and is given in a section at the end of this chapter.

CALCULATING CONCENTRATED CHANNEL FLOW EROSION

Background

Concentrated flow erosion is classed as channel erosion when the location of channels is controlled by the macrorelief. As discussed earlier, concentrated flow erosion is classed as rill erosion when the location of channels is controlled by the tillage microrelief. Concentrated flow erosion, when controlled by the macrorelief may be classed as ephemeral gully erosion, classical gully erosion, or channel erosion. These channels erode by three mechanisms:

- channel bed degradation due to shear
- channel wall failure
- knickpoint (headwall) advance.

A characteristic of concentrated flow channels is that there is no smooth transition from a zone of wide shallow flow to an incised channel with deeper flow. Rather, the transition occurs abruptly at a knickpoint or headwall. In fact, several such transitions may occur in a concentrated flow channel. Eventually, the knickpoints merge, making one incised channel.

Physically based methods for locating knickpoints are not well developed. Present analytical methods are based primarily on geomorphological studies with limited transferability to other climatic regions. In general, these studies indicate that some threshold flow normally exists above which the channel is incised.

Harvey *et al.* (1985) summarized much of the work on gully and channel erosion. Based on the existing literature, they concluded the following:

(1) A gully may develop in a short time due to exceeding an intrinsic or extrinsic threshold.
(2) The response of the system to gullying is complex; secondary responses complicate the adjustment to change.
(3) Empirical data bases for a homogeneous region can be used to estimate thresholds.
(4) All incised channels follow the same evolutionary trend: initiation, headwall migration, channel widening, channel slope reduction, reduction of bank angle, sediment deposition, and vegetation establishment.
(5) The nature of the sediment eroded and transported affects the morphology of the channel and the nature of the channel adjustment.

Models of the channel headwall migration and channel bank failure are currently being developed, but are not available at this writing.

In general, channel erosion estimates are highly empirical, relying on field surveys. One exception would be the use of the model of Foster and Lane (1983) and its derivatives for concentrated flow erosion in terraces and ephemeral gullies. These models are discussed below.

Foster and Lane Model

The Foster and Lane (1983) model is composed of four major elements:

- An equilibrium channel width model
- A model for conveyance function
- Channel erosion model prior to reaching a nonerodible layer
- Channel erosion model after reaching a nonerodible layer.

Foster and Lane developed the model for steady-state flow, but use has extended to varying flow rates. An overview of the model is given below. Storm *et al.* (1990) and Foster (1982) discuss the model further.

Equilibrium Channel Geometry

The basic detachment relationship is the shear excess concept, or

$$D_{rc} = K_r(\tau - \tau_c), \tag{8.67}$$

where D_{rc} is detachment rate potential (kg/m · sec), K_r is rill erodibility (m/sec), and τ and τ_c are actual and critical tractive force respectively (Pa). The data base for the WEPP model (Lane and Nearing, 1989) can be used to estimate K_r and τ_c. In development of their equilibrium channel width model, Foster and Lane (1983) assumed that a symmetrical distribution of shear exists, given by

$$\tau_* = \frac{\tau}{\tau_a} = 1.35\left[1 - (1 - 2X_*)^{2.9}\right]; \qquad X_* < 0.5, \tag{8.68}$$

where τ_* is a dimensionless shear, τ is actual shear, τ_a is the average shear stress given by

$$\tau_a = \gamma RS, \tag{8.69}$$

and X_* is a normalized distance along the wetted

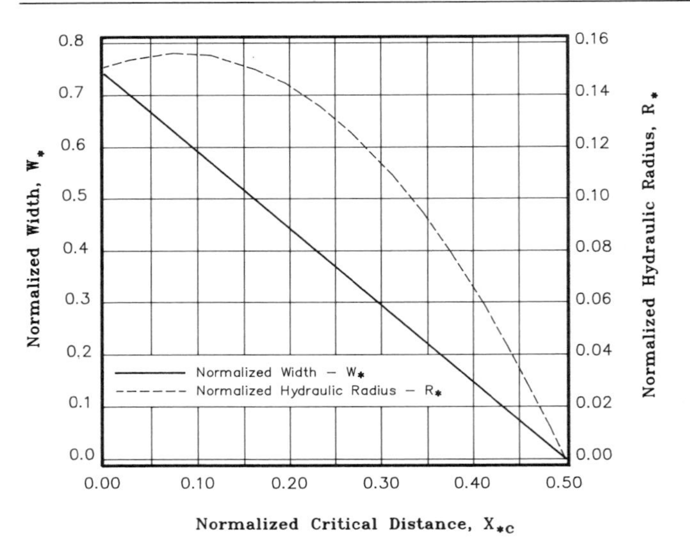

Figure 8.22 Normalized equilibrium characteristics for an eroding channel (adapted from Lane and Foster, 1980). Tabulations are given in Appendix F, Table 8F.1.

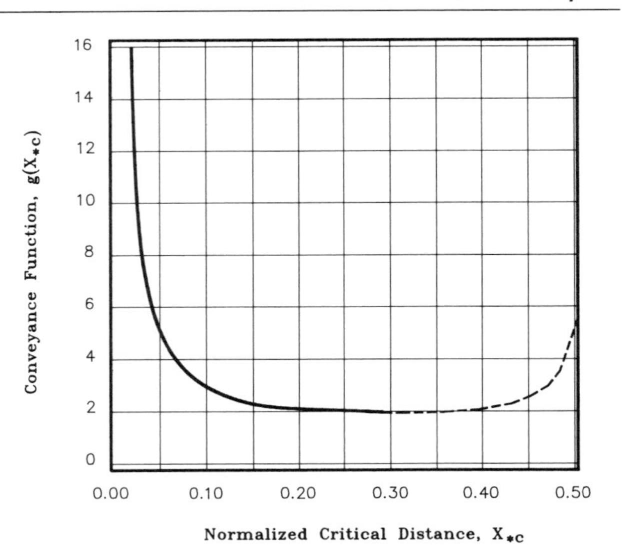

Figure 8.23 Conveyance function of Foster and Lane (1983). Tabulations are given in Appendix F, Table 8F.1.

perimeter starting at the water surface. X_* is given by

$$X_* = X/\mathrm{WP}, \tag{8.70}$$

where WP is wetted perimeter. Equations (8.68) and (8.69) can be used to show that the maximum shear on the channel bottom is 1.35 γRS at $X_* = 0.5$. The distribution given by Eq. (8.68) is symmetrical about $X_* = 0.5$.

Foster and Lane (1983) also defined normalized channel hydraulic radius and channel width as

$$R_* = R/\mathrm{WP} \tag{8.71}$$

and

$$W_* = W_{\mathrm{eq}}/\mathrm{WP}, \tag{8.72}$$

where W_{eq} is the equilibrium channel width. A normalized parameter known as X_{*c} was also defined as X_*, where $\tau = \tau_c$. Using procedures defined below, the relationship between X_{*c} and R_* and W_* were developed as shown in Fig. 8.22.

To define the channel geometries in Fig. 8.22, the assumption was made that the channel erodes vertically, rapidly adjusting to an equilibrium width. The rate of vertical movement was assumed to be constant across the entire cross section at a rate defined by the maximum shear stress. The rate of movement normal to the channel boundary at any point was defined by the actual shear stress. The angle between the normal and vertical movement would be defined by

$$\cos\alpha = \frac{\tau - \tau_c}{1.35\tau - \tau_c} = \frac{\tau_* - \tau_{*c}}{1.35\tau_* - \tau_{*c}}. \tag{8.73}$$

Utilizing Eqs. (8.67)–(8.72), Foster and Lane (1983) developed relationships for normalized channel geometry by: (1) selecting a value for X_{*c} (X_* when $\tau_* = \tau_{*c}$), (2) dividing the boundary between $X_* = 0.5$ and $X_* = X_{*c}$ into 50 segments of length ΔX_*, (3) determining an average X_*, τ_*, and thus α for each segment, (4) determining the final coordinates of the segment, knowing α, ΔX_*, and the starting coordinates, and (5) starting with $X_* = 0.5$, repeating (3)–(5) until $X_* = X_{*c}$. At X_{*c}, the boundary is vertical to $X_* = 0.0$. Knowing the coordinates of X_{*c}, the values for W_* and R_* were determined. By varying X_{*c}, the relationship in Fig. 8.22 was developed.

Conveyance Function

W_* can be determined once X_{*c} is known. Relationships are needed to estimate X_{*c} and equilibrium width, W_{eq}. To accomplish this, Foster and Lane (1983) developed a conveyance function to predict X_{*c}. Starting with Manning's equation and Eqs. (8.69)–(8.72), they showed that

$$g(X_{*c}) = \frac{1}{\tau_{*c} R_*^{3/8}} = \frac{\gamma S}{\tau_c}\left[\frac{nQ}{\sqrt{S}}\right]^{3/8}, \tag{8.74}$$

where $g(X_{*c})$ is a function only of X_{*c} as defined by Fig. 8.23. If $g(X_{*c})$ is greater than 35, a value of 35 is typically used. The development of Eq. (8.74) is illustrated in Example Problem 8.11.

Example Problem 8.11 Development of conveyance function for Foster and Lane model

Starting with Manning's equation, show that the conveyance function of Foster and Lane (1983) given by Eq. (8.74) is correct.

Solution: Manning's equation [Eq. (4.23)] can be written in SI units as

$$Q = \frac{A}{n} R^{2/3} S^{1/2}. \tag{a}$$

Using the definition $R = A/\text{WP}$ and $R_* = R/\text{WP}$,

$$A = \frac{R^2}{R_*}. \tag{b}$$

Hence

$$Q = \frac{R^{8/3} S^{1/2}}{n R_*} \tag{c}$$

and

$$R = \left[\frac{nQ}{S^{1/2}}\right]^{3/8} R_*^{3/8}. \tag{d}$$

From the definition

$$\tau_{*c} = \frac{\tau_c}{\tau_a} = \frac{\tau_c}{\gamma RS}, \tag{e}$$

a definition for R can be written as

$$R = \frac{\tau_c}{\gamma S \tau_{*c}}. \tag{f}$$

Equating Eqs. (d) and (f),

$$\frac{1}{\tau_{*c} R_*^{3/8}} = \frac{\gamma S}{\tau_c}\left[\frac{nQ}{\sqrt{S}}\right]^{3/8}. \tag{g}$$

Since both τ_{*c} and R_* can be predicted as functions of X_{*c}, then Eq. (g) can be written

$$g(X_{*c}) = \frac{1}{\tau_{*c} R_*^{3/8}} = \frac{\gamma S}{\tau_c}\left[\frac{nQ}{\sqrt{S}}\right]^{3/8}, \tag{h}$$

which is Eq. (8.74). Using Fig. 8.22 to define the relationship between X_{*c} and R_* and Eq. (8.68) to define the X_{*c}–τ_* relationship, the functional relationship between X_{*c} and $g(X_{*c})$ given in Fig. 8.23 was developed

---

To determine the normalized parameters R_* and W_*, X_{*c} can be determined for given values of Q, S, γ, and τ from Fig. 8.23 and Eq. (8.74). It should be noted that X_{*c} is undefined for $g(X_{*c})$ values less than 1.8. This corresponds to a flow condition with insufficient shear force to cause erosion and channel incision. Also, over certain ranges of $g(X_{*c})$, X_{*c} is double valued. The conservative procedure for estimating erosion is to select the lower values.

Given X_{*c}, values for the normalized parameters R_* and W_* are determined from Fig. 8.22. Relationships are then needed to convert these normalized quantities to channel geometry. By assuming a rectangular geometry with an equilibrium width, Foster and Lane (1983) showed that

$$\text{WP} = \frac{W_{eq}^2}{W_{eq} - 2R}, \tag{8.75}$$

where W_{eq} is the equilibrium width. Utilizing Manning's equation, it can be shown that

$$\text{WP} = \left[\frac{nQ}{\sqrt{S}}\right]^{3/8} R_*^{-5/8} \tag{8.76}$$

and from Eq. (8.72)

$$W_{eq} = W_* \text{WP} = \left[\frac{nQ}{\sqrt{S}}\right]^{3/8} W_* R_*^{-5/8}. \tag{8.77}$$

Thus, knowing the conveyance function, values for X_{*c}, R_* and W_* can be determined. Knowing R_* and W_*, a value can be determined for WP and W_{eq}. Knowing R_* and WP, R can be determined from

$$R = R_* \text{WP}. \tag{8.78}$$

It is significant to point out that Eq. (8.77) would predict that W_{eq} is a power function of discharge for a given slope, critical tractive force, and bulk density. This is in keeping with empirical relationships given in Chapter 10 relating channel width and flow rate.

The relationships developed to this point are for channel geometry. Procedures discussed in the next sections are used to convert these geometrical relationships to erosion rates. The discussion is divided into erosion prior to reaching a nonerodible layer and erosion after reaching a nonerodible layer.

Stage 1: Channel Erosion Prior to Nonerodible Layer Prior to reaching the nonerodible layer, the channel is assumed to erode vertically at a width equal to W_{eq} and at a potential rate defined by the maximum tractive force (see Fig. 8.24A) or

$$E_{rc} = D_{rc} W_{eq} = K_r(1.35\tau_a - \tau_c) W_{eq}, \tag{8.79}$$

where E_{rc} is the potential rate of vertical erosion, K_r is the soil erodibility, and τ_a is given by γRS [Eq. (8.69)]. Using Manning's equation for R (Eq. (d), Example Problem 8.11)

$$\tau_a = \gamma S \left[\frac{nQR_*}{\sqrt{S}}\right]^{3/8}. \tag{8.80}$$

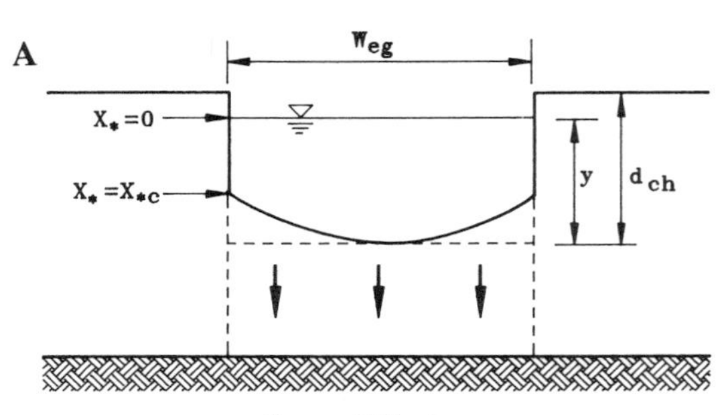

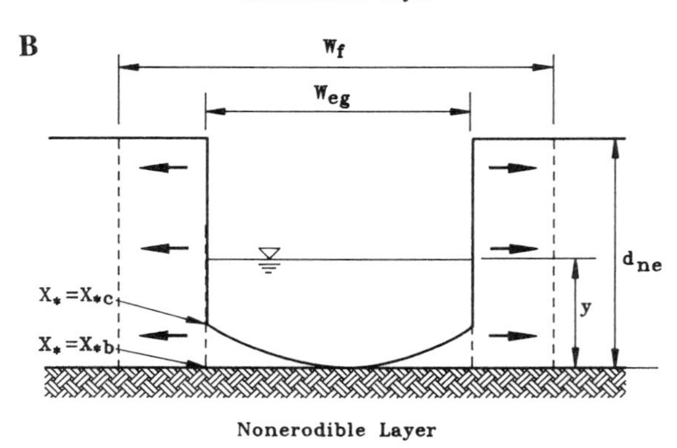

Figure 8.24 Equilibrium geometry for (A) stage 1 and (B) stage 2 channel erosion.

The maximum rate of downward movement will be defined by

$$M_{\mathrm{rc}} = \frac{E_{\mathrm{rc}}}{W_{\mathrm{eq}} \rho_{\mathrm{b}}}, \tag{8.81}$$

where ρ_{b} is soil bulk density.

Actual detachment will be limited by the ratio of sediment load to transport capacity, as given in Eq. (8.5), or

$$M_{\mathrm{r}} = \frac{E_{\mathrm{rc}}}{W_{\mathrm{eq}} \rho_{\mathrm{b}}} \left(1 - \frac{q_{\mathrm{s}}}{T_{\mathrm{c}}}\right). \tag{8.82}$$

Stage 2: Channel Erosion After Reaching a Nonerodible Layer After reaching a nonerodible layer, the channel is assumed to expand laterally at a rate defined by shear at the intersection of the erodible channel wall and the nonerodible bed, or $X_{*\mathrm{b}} = y/\mathrm{WP}$ (See Fig. 8.24B). Thus

$$\frac{dW}{dt} = K_{\mathrm{r}} \frac{(\tau_{\mathrm{b}} - \tau_{\mathrm{c}})}{\rho_{\mathrm{b}}}, \tag{8.83}$$

where τ_{b} is shear at the intersection of the channel bottom and channel wall. The erosion rate is given by

$$E_{\mathrm{rc}} = \rho_{\mathrm{b}} \frac{dW}{dt} d_{\mathrm{ne}} = d_{\mathrm{ne}} K_{\mathrm{r}} (\tau_{\mathrm{b}} - \tau_{\mathrm{c}}), \tag{8.84}$$

where d_{ne} is the depth to nonerodible layer.

The final channel width, W_{f}, will be reached when $X_{*\mathrm{c}}$ is such that the shear on the channel wall is everywhere less than τ_{c}. The value of X_* corresponding to this condition will be denoted as

$$X_{*\mathrm{cf}} = \frac{y_{\mathrm{f}}}{\mathrm{WP}}, \tag{8.85}$$

where y_{f} is the depth of flow when $W = W_{\mathrm{f}}$. By assuming a rectangular channel shape, it can be shown that the final width can be given by

$$\frac{W_{\mathrm{f}}}{\mathrm{WP}} = 1 - 2X_{*\mathrm{cf}} \tag{8.86}$$

and

$$\frac{W_{\mathrm{f}}}{\mathrm{WP}} = \frac{R_*}{X_{*\mathrm{cf}}}. \tag{8.87}$$

Using Eq. (8.87) in (8.86) yields

$$R_* = X_{*\mathrm{cf}}(1 - 2X_{*\mathrm{cf}}), \tag{8.88}$$

and substituting Eq. (8.88) into Eq. (8.74) for R_*, one can obtain

$$g(X_{*\mathrm{cf}}) = \frac{1}{\tau_{*\mathrm{cf}}[X_{*\mathrm{cf}}(1 - 2X_{*\mathrm{cf}})]^{3/8}} = \frac{\gamma S}{\tau_{\mathrm{c}}} \left[\frac{nQ}{\sqrt{S}}\right]^{3/8}. \tag{8.89}$$

Equation (8.89) must be solved implicitly for $X_{*\mathrm{cf}}$. Tabulated values for $g(X_{*\mathrm{cf}})$ versus $X_{*\mathrm{cf}}$ are given in Appendix 8F, Table 8F.1. Solution procedures are illustrated in Example Problem 8.12. At first glance, it may seem that Fig. 8.23 could be used to solve Eq. (8.89). Such is not the case. Figure 8.23 and the $g(X_{*\mathrm{c}})$ relationship in Appendix 8F are based on Eq. (8.74) and cannot be used to solve Eq. (8.89) as the two equations are different.

From Example Problem 8.11, Eq. (d), $R = [nQ/S^{1/2}]^{3/8} R_*^{3/8}$. Using the known value for $X_{*\mathrm{cf}}$ from Eq. (8.89), the final hydraulic radius is

$$R = \left[\frac{nQ}{\sqrt{S}} X_{*\mathrm{cf}}(1 - 2X_{*\mathrm{cf}})\right]^{3/8}. \tag{8.90}$$

Using Eq. (8.87), Eq. (8.90), and the fact that $R_* = R/\mathrm{WP}$, the final width is given by

$$W_{\mathrm{f}} = \left[\frac{nQ}{\sqrt{S}} \frac{(1 - 2X_{*\mathrm{cf}})}{X_{*\mathrm{cf}}^{5/3}}\right]^{3/8}. \tag{8.91}$$

In the transition between initial width and final width, Foster (1982) defined a dimensionless time and width from

$$t_* = \frac{t(dW/dt)_{in}}{W_f - W_{in}} \qquad (8.92)$$

and

$$W'_* = \frac{W - W_{in}}{W_f - W_{in}}, \qquad (8.93)$$

where $(dW/dt)_{in}$ is the initial value for dW/dt and W_{in} is the initial channel width upon reaching a nonerodible layer. For steady-state flow, W_{in} would be equal to W_{eq}, the equilibrium width prior to reaching the nonerodible layer.

The rate of change of channel width in the Foster and Lane model was expressed by

$$dW'_*/dt_* = e^{-t_*} \qquad (8.94)$$

or

$$W'_* = 1 - e^{-t_*} \qquad (8.95)$$

Knowing W'_*, the actual width at any time is determined from

$$W = W'_*(W_f - W_{in}) + W_{in}. \qquad (8.96)$$

The use of the Foster and Lane model is illustrated in Example Problem 8.12.

The Ephemeral Gully Erosion Model

The complexity of the Foster and Lane model discussed above led to a desire for an explicit prediction equation. Using a data base generated from the Foster and Lane model, regression equations known as the Ephemeral Gully model were developed for W_{eq}, W_f, and τ (Watson *et al.*, 1986), or

$$W_{eq} = 2.66(q_p^{0.396})(n^{0.387})(S^{-0.16})(\tau_c^{-0.24}) \qquad (8.97)$$

and

$$W_f = 179(q_p^{0.552})(n^{0.556})(S^{0.119})(\tau_c^{-0.476}), \qquad (8.98)$$

where q_p is peak discharge (m^3/sec), n is Manning's roughness, S is slope (m/m), and τ_c is critical tractive force (Pa). When estimating the erosion rate, the maximum shear force on the channel bottom ($1.35\gamma RS$) is estimated by

$$\tau = 4867(q_p^{0.375})(n^{0.375})(S^{0.811}). \qquad (8.99)$$

In application of the Ephemeral Gully model, Watson *et al.* (1986) recommend that peak discharge be used to predict channel equilibrium width and rate of growth of the gully. In using the Ephemeral Gully model, it should be remembered that the regression equations were developed from a data base generated by the Foster and Lane model and that Fig. 8.23 or Table 8F.1 in Appendix 8F would indicate an undefined value for X_{*c} when $g(X_{*c})$ is less than 1.8. As stated earlier, this would correspond to insufficient shear strength to cause channel erosion and an incised channel.

A comparison between the Foster and Lane predictions and the Ephemeral Gully model predictions is given in Table 8.15 for selected input parameters.

Table 8.15 Comparison between Predictions from the Foster and Lane Concentrated Flow Model and the Ephemeral Gully Model.

Flow rate (m^3/sec)	Slope (m/m)	Manning's n	Critical tractive force τ_c (Pa)	$g(X_{*c})$	Equilibrium width W_{eq} (ft) FLM Eq. (8.77)	Equilibrium width W_{eq} (ft) EGM Eq. (8.97)	Final width W_f (ft) FLM Eq. (8.91)	Final width W_f (ft) EGM Eq. (8.98)	Maximum shear τ_{max} (Pa) FLM[b]	Maximum shear τ_{max} (Pa) EGM Eq. (8.99)
0.1950	0.02	0.03	1.35	43.98	0.734	0.624	4.186	5.623	39.45	29.65
0.1500	0.02	0.03	2.00	26.91	0.648	0.512	3.068	4.035	35.88	26.88
0.0050	0.02	0.03	2.00	7.52	0.170	0.133	0.434	0.617	10.09	7.51
0.0010	0.02	0.03	2.00	4.11	0.085	0.070	0.164	0.254	5.53	4.10
0.0005	0.02	0.03	2.00	3.17	0.062	0.053	0.106	0.173	4.27	3.17
0.0005	0.01	0.03	2.00	1.80	0.045	0.060	0.062	0.159	2.21	1.80

[a] FLM, Foster–Lane model; EGM, ephemeral gully model.
[b] $1.35\gamma RS$ [see Eqs. (8.68) and (8.69)].

Example Problem 8.12 Computing concentrated flow erosion by Foster and Lane Model

A construction area has been regraded to include a 20-m long drainage channel on a slope of 2%. In channel construction, the soil is graded to a depth of 0.5 m over a rock layer. Prior to stabilizing the channel, a storm occurs with a peak discharge of 0.195 m^3/sec. If the soil in the unstabilized state has an erodibility of 0.005 sec/m, a critical tractive force of 1.35 Pa, and a bulk specific gravity of 1.3, estimate the erosion potential per unit length of channel and the total erosion for the entire channel. The storm has a duration of 6 hr and a runoff volume of 1728 m^3.

Solution:

1. *Conveyance factors [Eq. (8.74)].* Assume a smooth soil $n = 0.03$ using the values in Appendix 8E, Table 8E.2. A frequently used factor in the Foster and Lane equations is

$$\left(\frac{nQ}{\sqrt{S}}\right)^{3/8} = \left(\frac{(0.03)(0.195)}{\sqrt{0.02}}\right)^{3/8} = 0.303.$$

Using $\gamma = 9803\ N/m^3$, the conveyance factor given by Eq. (8.74) is

$$g(X_{*c}) = \left(\frac{nQ}{\sqrt{S}}\right)^{3/8}\frac{\gamma S}{\tau_c} = (0.303)\frac{(9803)(0.02)}{1.35} = 44.0.$$

For $g(X_{*c}) > 35$, use 35.

2. *Determining X_{*c}, W_*, and R_*.* From Fig. 8.23 or Appendix 8F, Table F.1,

$$g(X_{*c}) = 35.0;\ X_{*c} = 0.0;\ W_* = 0.744;\ R_* = 0.151.$$

3. *Wetted perimeter [Eq. (8.76)].*

$$WP = \left[\frac{nQ}{\sqrt{S}}\right]^{3/8} R_*^{-5/8} = (0.303)(0.151)^{-5/8} = 0.988\ m.$$

4. *Equilibrium width [Eq. (8.77)].*

$$W_{eq} = WP\ W_* = (0.988)(0.744) = 0.735\ m.$$

5. *Equilibrium hydraulic radius [Eq. (8.78)].*

$$R = R_* WP = (0.151)(0.988) = 0.149\ m.$$

6. *Stage 1 erosion rate and maximum downward movement.* Using Eq. (8.79)–(8.81) with ρ_b of 1300 kg/m^3

$$\tau_a = \gamma S\left(\frac{nQ}{\sqrt{S}}\right)^{3/8} R_*^{3/8} = (9803)(0.02)(0.303)(0.151)^{3/8}$$
$$= 29.2\ Pa$$

$$E_{rc} = K_r(1.35\tau_a - \tau_c)W_{eq}$$
$$= (0.005)[(1.35)(29.2) - 1.35](0.735)$$
$$= 0.140\ kg/sec \cdot m$$

$$M_{rc} = \frac{E_{rc}}{W_{eq}\rho_b} = \frac{0.140}{(0.735)(1300)}$$
$$= 1.46 \times 10^{-4}\ m/sec \text{ or } 0.527\ m/hr.$$

7. *Time to reach nonerodible layer (d_{ne} = 0.5 m).*

$$t_{ne} = \frac{d_{ne}}{M_{rc}} = \frac{0.5\ m}{0.527\ m/hr} = 0.95 = 1\ hr.$$

From geometry for a rectangular channel, the depth of flow y is $\frac{1}{2}(WP - W_{eq})$. Thus, when erosion has reached the nonerodible layer, the following calculations can be made for y, the dimensionless distance X_* corresponding to y, τ_* corresponding to y, and the actual shear at y, τ_b, by using Eqs. (8.68), (8.69), and (8.70),

$$y = \frac{WP - W_{eq}}{2} = \frac{0.988 - 0.735}{2} = 0.1265\ m$$

$$X_* = \frac{y}{WP} = \frac{0.1265}{0.988} = 0.128$$

$$\tau_* = 1.35\left[1 - (1 - 2X_*)^{2.9}\right]$$
$$= 1.35\left(1 - [1 - (2)(0.128)]^{2.9}\right) = 0.777.$$

From earlier calculations, $\tau_a = 29.2$ Pa, which does not change until the nonerodible layer is reached, at which point the flow depth will change. When the nonerodible layer is reached, the nondimensional shear corresponding to y would be $\tau_* = \tau_b/\tau_a$, hence

$$\tau_b = \tau_*\tau_a = (0.777)(29.2) = 22.69\ Pa.$$

From Eq. (8.83),

$$\left(\frac{dW}{dt}\right)_{in} = K_r\frac{\tau_b - \tau_c}{\rho_b} = \frac{(0.005)(22.69 - 1.35)}{(1300)}$$
$$= 8.21 \times 10^{-5}\ m/sec \text{ or } 29.55\ cm/hr.$$

8. *Initial erosion rate after reaching nonerodible layer [Eq. (8.84)].*

$$E_{rc} = \rho_b\left(\frac{dW}{dt}\right)_{in} d_{ne}$$
$$= (1300\ kg/m^3)(8.21 \times 10^{-5}\ m/sec)(0.5\ m)$$
$$= 0.053\ kg/m \cdot sec.$$

9. *Final width [Eqs. (8.89), (8.90), and (8.91)].* Calculating X_{*cf},

$$g(X_{*cf}) = \frac{1}{\tau_{*cf}\left[X_{*cf}(1 - 2X_{*cf})\right]^{3/8}} = \frac{\gamma S}{\tau_c}\left[\frac{nQ}{\sqrt{S}}\right]^{3/8}.$$

This must be solved by iteration for X_{*cf}. From item 1 above, the RHS of the equation is 44.0. Thus, a value for X_{*cf} that makes $g(X_{*cf})$ equal to 44.0 must be found. Since $g(X_{*cf})$ is solved analytically for X_{*cf}, no cutoff of 35.0 is used, as in the case with $g(X_*)$. For a first trial, let $X_{*\,cf} = 0.05$; then from Eqs. (8.68) and (8.89),

$$\tau_{*\,cf} = 0.355 \qquad \text{and} \qquad g(X_{*\,cf}) = 9.0 < 44.0.$$

Other trials are given below.

X_{*cf}	$g(X_{*cf})$
0.05	9.0
0.02	29.21
0.015	42.80
0.014	46.94
0.0147	43.98 OK

Using $X_{*\,cf} = 0.0147$ in Eq. (8.91),

$$W_f = \left[\frac{nQ}{\sqrt{S}}\right]^{3/8}\left[\frac{1 - 2X_{*\,cf}}{(X_{*\,cf})^{5/3}}\right]^{3/8}$$

$$= 0.303\left[\frac{1-(2)(0.0147)}{(0.0147)^{5/3}}\right]^{3/8} = 4.19 \text{ m}.$$

10. Dimensionless time and width [Eqs. (8.92) and (8.93)].

$$t_* = \frac{t(dW/dt)_{in}}{W_f - W_{in}} = \frac{(t)(8.21 \times 10^{-5})}{4.19 - 0.735} = 2.376 \times 10^{-5} t$$

$$W'_* = \frac{W - W_{in}}{W_f - W_i} = \frac{W - 0.735}{4.19 - 0.735} = 0.289(W - 0.735).$$

11. Potential detachment rates after reaching nonerodible layer [Eq. (8.84)]. From the chain rule of differentiation, Eq. (8.94) and the definitions given in item 10,

$$\frac{dW}{dt} = \frac{dW}{dW'_*}\frac{dW'_*}{dt_*}\frac{dt_*}{dt} = \left(\frac{dW}{dt}\right)_{in} e^{-t_*}.$$

From item 10 above, $t_* = 2.376 \times 10^{-5}\, t$, and from item 7 $(dW/dt)_{in} = 8.21 \times 10^{-5}$ m/sec; hence

$$\frac{dW}{dt} = 8.21 \times 10^{-5} e^{-2.376\times10^{-5}t} \text{ m/sec}.$$

Finally, using Eq. (8.84),

$$E_{rc} = \rho_b \frac{dW}{dt} d_{ne} = (1300 \text{ kg/m}^3)\frac{dW}{dt}(0.5 \text{ m})$$

$$= 650\frac{dW}{dt} \text{ kg/m} \cdot \text{sec}$$

or

$$E_{rc} = (650)(8.21 \times 10^{-5})e^{-2.37\times10^{-5}t}$$

$$= 0.0533e^{-2.376\times10^{-5}t} \text{ kg/m} \cdot \text{sec}.$$

Values for E_{rc} are tabulated below.

Time after start of storm (hr)	Time after reaching d_{ne} (hr)	t (secs)	E_{rc} (kg/m•sec)	Hourly average E_{rc} (kg/m•sec)
0.0	—	—	0.140	
0.5	—	—	0.140	0.140
1.0	0.0	0	0.053	
1.5	0.5	1800	0.051	0.051
2.0	1.0	3600	0.049	
2.5	1.5	5400	0.047	0.047
3.0	2.0	7200	0.045	
				0.043
4.0	3.0	10,800	0.041	
				0.040
5.0	4.0	14,400	0.038	
				0.037
6.0	5.0	18,000	0.35	
			ΣE_{rc} =	0.358

12. Total detachment potential (length = 20 m). Total erosion is given by

$$E_{tot} = \sum E_{rc,i}\Delta t_i \times 20 \text{ m}.$$

Using Δt of 1 hr,

$$E_{tot} = (0.358 \text{ kg/sec} \cdot \text{m})(1 \text{ hr})(3600 \text{ sec/hr})(20 \text{ m})$$
$$= 25{,}776 \text{ kg}.$$

13. Converting to sediment concentration. Conversion of total erosion to an average concentration can be made by noting that the average sediment concentration is simply the mass of sediment divided by the runoff mass, which is runoff volume times density of water. In this case, runoff mass is 1728 m^3 × 1000 kg/m^3; hence the average concentration is

$$C = \frac{25{,}776 \text{ kg}}{1728 \text{ m}^3 \times 1000 \text{ kg/m}^3}$$
$$= 0.0149 \text{ kg/kg or } 14{,}901 \text{ mg/liter}.$$

These values represent what could happen if transport capacity is not limiting. Corrections for the effects of transport capacity are done in a following section.

The DYRT Model for Concentrated Flow Erosion

A recent modification of the Foster and Lane (1983) model was developed by Storm *et al.* (1990). Known as DYRT, this model allows for variable flow rates and can be modified for layers of varying density and tractive force. To account for variable flow rates, procedures were developed to account for flow changes. Computations of channel width are made numerically rather than using the exponential form of Eq. (8.95), or

$$W_{t+\Delta t} = W_t + \frac{dW}{dt}\Delta t, \qquad (8.100)$$

where W_t, $W_{t+\Delta t}$ are channel widths at time t and $t + \Delta t$, and Δt is the solution time interval.

The use of Eq. (8.100) with the Foster and Lane (1983) model for equilibrium channel width allows for computation of rill erosion under varying flow rates. Procedures for the computation are given in Storm *et al.* (1990).

Potential versus Actual Channel Erosion

The concentrated flow models presented in this section predict detachment potential in a channel. Potential detachment can be translated to actual detachment by using a modification of Eq. (8.5) or

$$D_r = D_{rc}(1 - q_s/T_c), \qquad (8.5)$$

where D_{rc} is detachment potential and D_r is actual detachment. This can be combined with the continuity equation [Eq. (8.1)] to predict sediment load at any point on a channel. The procedure is illustrated in Example Problem 8.13.

Example Problem 8.13 Effects of transport capacity on detachment

Calculate the total channel erosion in Example Problem 8.12, assuming that the channel drains a regraded area and that the ratio of sediment load to transport capacity, q_s/T_c, is 0.7 in the runoff from the regraded area.

Solution:

1. Converting potential to actual detachment. From Eq. (8.5),

$$D_r = D_{rc}(1 - q_s/T_c).$$

D_r represents actual and D_{rc} potential erosion as calculated in Example Problem 8.12.

2. Correcting initial erosion rates and time to reach nonerodible layer. Corrections are needed for the rate of actual erosion to determine the time to reach the nonerodible layer. From item 6 in Example Problem 8.12, $E_{rc,pot} = 0.140\ \text{kg/m}\cdot\text{sec}$ and $M_{rc,pot} = 1.46 \times 10^{-4}$ m/sec. Correcting for the effects of transport capacity, an analogy can be made to equation 8.5, or

$$E_{rc,act} = E_{rc,pot}\left(1 - \frac{q_s}{T_c}\right) = (0.140\ \text{kg/m}\cdot\text{sec})(1 - 0.7)$$
$$= 0.042\ \text{kg/m}\cdot\text{sec}$$

and

$$M_{rc,act} = M_{rc,pot}\left(1 - \frac{q_s}{T_c}\right) = (1.46 \times 10^{-4}\ \text{m/sec})(1 - 0.7)$$
$$= 4.38 \times 10^{-5}\ \text{m/sec}.$$

The time to reach the nonerodible layer ($d_{ne} = 0.5$ m) is thus

$$t_{ne} = \frac{d_{ne}}{M_{rc,act}} = \frac{0.5\ \text{m}}{4.38 \times 10^{-5}\ \text{m/sec}}$$
$$= 11{,}415\ \text{sec or } 3.17\ \text{hr}.$$

3. Correcting the rate of widening for actual erosion. The equilibrium and final widths would not be affected by sediment load, but the rate of change would be. Hence from items 4 and 9 in Example Problem 8.12, $W_f = 4.19$ m and $W_{eq} = 0.735$ m. Also, from items 7 and 8, the potential rate of widening and erosion rate after reaching the nonerodible layers are $(dW/dt)_{in,pot} = 8.21 \times 10^{-5}$ m/sec and $E_{rc,pot} = 0.053\ \text{kg/m}\cdot\text{sec}$. Actual rates would be [using an analogy to Eq. (8.5)]

$$E_{rc,act} = E_{rc,pot}\left(1 - \frac{q_s}{T_c}\right)$$
$$= 0.053(1 - 0.7)$$
$$= 0.016\ \text{kg/sec}\cdot\text{m}$$

and

$$\left(\frac{dW}{dt}\right)_{in,act} = \left(\frac{dW}{dt}\right)_{in,pot}\left(1 - \frac{q_s}{T_c}\right)$$
$$= (8.21 \times 10^{-5}\ \text{m/sec})(1 - 0.7)$$
$$= 2.46 \times 10^{-5}\ \text{m/sec}.$$

The actual erosion rate would be given by

$$E_{rc,act} = \rho_b\left(\frac{dW}{dt}\right)_{act} d_{ne}$$
$$= (1300\ \text{g/m}^3)\left(\frac{dW}{dt}\right)_{act}(0.5\ \text{m})\ \text{kg/m}\cdot\text{sec}.$$

The dimensionless time, t_* is corrected for actual erosion rate by

$$t_* = \frac{t(dW/dt)_{in,act}}{W_f - W_i} = \frac{2.46 \times 10^{-5}}{4.19 - 0.735}t$$
$$= 7.12 \times 10^{-6}t.$$

From item 11 of Example Problem 8.12,

$$\left(\frac{dW}{dt}\right)_{act} = \left(\frac{dW}{dt}\right)_{in,act} e^{-t_*}.$$

Using the value calculated above for $(dW/dt)_{in,act}$

$$\left(\frac{dW}{dt}\right)_{act} = 2.46 \times 10^{-5} e^{-7.12\times 10^{-6}t}.$$

Hence

$$E_{rc,act} = (1300\ \text{kg/m}^3)(0.5\ \text{m})(2.46 \times 10^{-5}\ \text{m/sec})e^{-7.12\times 10^{-6}t}$$
$$= 0.016e^{-7.12\times 10^{-5}t}\ \text{kg/m}\cdot\text{sec}.$$

Values are tabulated below.

Example Problem 8.12

4. Corrected erosion rates.

Time after start of storm (hr)	Time after reaching d_{ne} (hr)	t (sec)	E_{rc} (kg/m•sec)	Hourly average E_{rc} (kg/m•sec)
0	0.00	0		
1.00	0.00	0	0.0420[a]	0.0420
2.00	0.00	0	0.0420[a]	0.0420
3.00	0.00	0	0.0420[a]	0.0420
3.16	0.00	0	0.0420[a]	
3.17	0.00	0	0.0160	
4.00	0.83	2988	0.0157	0.0225
5.00	1.83	6588	0.0152	0.0155
6.00	2.83	10,188	0.0149	0.0151
			$\Sigma E_{rc,act}$ =	0.1791

[a]Values determined by previous computation for erosion prior to reaching the nonerodible layer. The last value at 3.16 hr is just prior to reaching the nonerodible layer.

Total detachment is

$$E_{tot} = 20 \text{ m} \sum E_{rc,act} \Delta t_i$$
$$= (20 \text{ m})(0.1791 \text{ kg/m} \cdot \text{sec})(1 \text{ hr})(3600 \text{ sec/hr})$$
$$= 12{,}895 \text{ kg}.$$

The ratio of actual to potential erosion, using the estimated potential erosion of 25,776 kg in Example Problem 8.12, is 12,895/25,766 = 0.50. Thus the effect of having q_s/T_c equal to 0.7 was to reduce the erosion by 50%. Obviously, the effect is not linear.

5. Calculating concentration. To calculate concentration, the sediment load entering the ditch must be combined with the erosion rate to get total sediment. Since only the ratio of q_s/T_c was given in the problem, sediment concentration at the end of the channel cannot be calculated.

ESTIMATING SEDIMENT YIELD

Erosion – Sediment Delivery Ratio Method

The Concept of a Sediment Delivery Ratio

The soil loss equations described earlier are useful tools for predicting the amount of soil loss from a field, referred to as gross erosion. Some of the models, such as RUSLE, assume no deposition. Between the field and point of final deposition, sediment will normally have numerous opportunities to be deposited, reducing the sediment yield accordingly. To quantify the amount of deposition occurring, a sediment delivery ratio has been defined as

$$D = \frac{Y}{\text{gross erosion} \times \text{watershed area}}, \tag{8.101}$$

where Y is the sediment yield from a watershed and gross erosion would be the erosion per unit area occurring on the watershed. Gross erosion is composed of rill and interrill erosion, gully erosion, and stream erosion. On disturbed areas, sheet and rill erosion are the principal components of gross erosion. In the previous section, procedures were presented for predicting rill and interrill erosion as well as ephemeral channel erosion. Procedures are now needed for determining the delivery ratio in order to obtain sediment yield.

In this section, empirical delivery ratio methods are discussed. It should be pointed out that the degree of understanding of sediment delivery ratios is probably less than any other area of sedimentation.

Application of a delivery ratio to estimated erosion should be done with careful consideration of the manner in which the model considers deposition. For example, the USLE/RUSLE does not include procedures for evaluating deposition, whereas the CREAMS and WEPP models (discussed in a subsequent section) consider deposition on hillslopes and in ephemeral gullies. Thus, the delivery ratio for the USLE/RUSLE would need to evaluate deposition that occurs in overland flow prior to reaching concentrated flow channels.

Graphical Methods for Predicting Delivery Ratios

A number of methodologies have been proposed to predict the sediment delivery ratio. These include simple estimates by an areal relationship and a relief–length ratio. Also, the accounting of many on-site factors such as water available for overland flow; texture of eroded material; ground cover; slope shape, gradient, and length; surface roughness; and additional site-specific factors have been recommended by the Forest Service (1980). Additionally, delivery ratio concepts based on storm modeling techniques have been advanced by Williams (1977).

Area–Delivery Ratio Relationship A first approximation of the delivery ratio may be obtained by the area effects illustrated in Fig. 8.25. Erosion estimates used in developing delivery ratio curves were based on the simple USLE. Therefore the curves should only be used with erosion estimates from the USLE or RUSLE. Considerable scatter is present in the data; however, the general trend indicates a strong effect of area on delivery ratio. Differences exist in the form of curves for different areas. Therefore, it is desirable, where possible, to obtain a curve specific for a given area. As

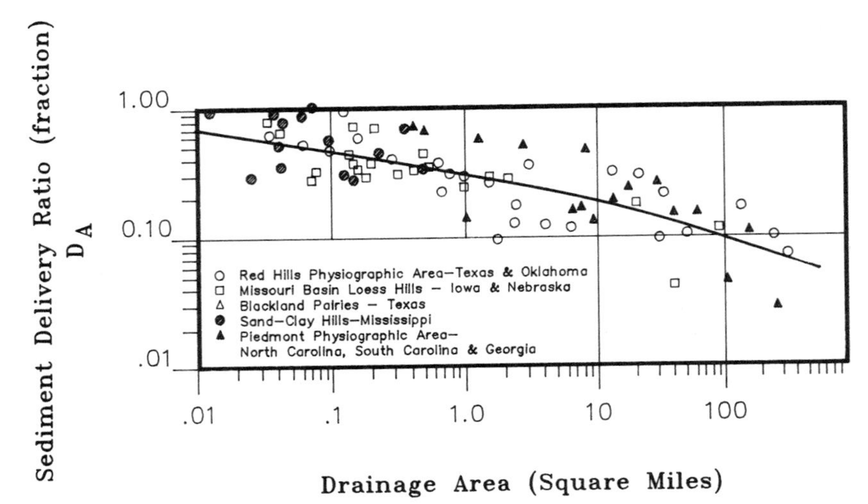

Figure 8.25 Sediment delivery ratio versus drainage area size for use with USLE/RUSLE (after Boyce, 1975).

expected, a method as simple as this yields only a rough approximation and has limited applicability in determining the expected sediment impact of alternative surface treatment techniques. In particular, the method is not recommended for single storms.

The Effects of Channelization on Delivery Ratio The degree of channelization affects how efficiently eroded sediment can be transported through a watershed channel system. A well-channelized watershed will transport most eroded material out of the watershed, whereas a poorly channelized watershed will transport the sediment slowly, leaving many opportunities for deposition. One measure of channelization is known as the relief–length ratio, calculated as

$$\left[\frac{\text{elevation difference between watershed divide at the main stem and the watershed outlet}}{\text{length of flow path along the main stem}}\right].$$

An example of the effects of relief–length ratio on the sediment delivery ratio is shown in Fig. 8.26 for the Red Hills area of Oklahoma and Texas (Renfro, 1975). It should be applied with caution to other areas, although the shape of the curve should be similar for most areas.

Forest Service Sediment Delivery Index Model

The Forest Service (1980) developed a methodology for predicting sediment delivery ratio that can be used for a single storm. The method uses a stiff diagram to predict delivery ratio as a function of (1) delivery distance from the slope to stream, (2) slope shape, (3) percentage of ground cover, (4) texture of eroded material, (5) surface runoff, (6) slope gradient, and (7) surface roughness. The parameters are shown on Fig. 8.27 and discussed below.

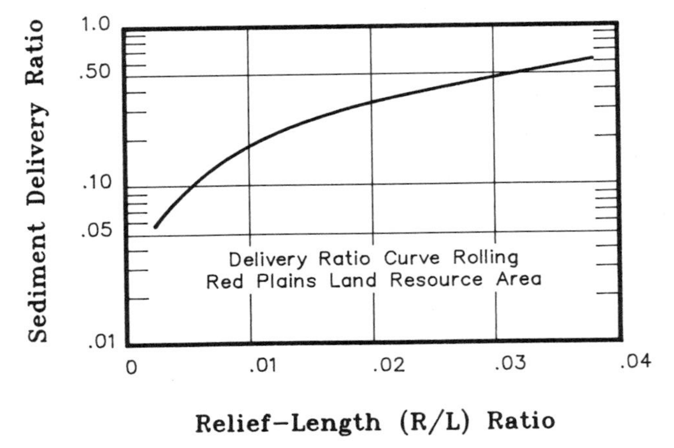

Figure 8.26 Effect of relief-length ratio on delivery ratio for use with the USLE/RUSLE (from Renfro, 1975).

Runoff Factor The surface runoff factor, given by peak discharge in cfs/ft, defines the quantity of water available to transport sediment for a storm. In the absence of peak discharge information, the magnitude of the runoff factor can be estimated by

$$F = 2.31 \times 10^{-5}\,\sigma L, \tag{8.102}$$

where F is runoff rate per foot of slope width (cfs/ft), σ is rainfall excess (in./hr), and L is the length of the disturbed area (ft). Values greater than 0.1 are assumed to be 0.1.

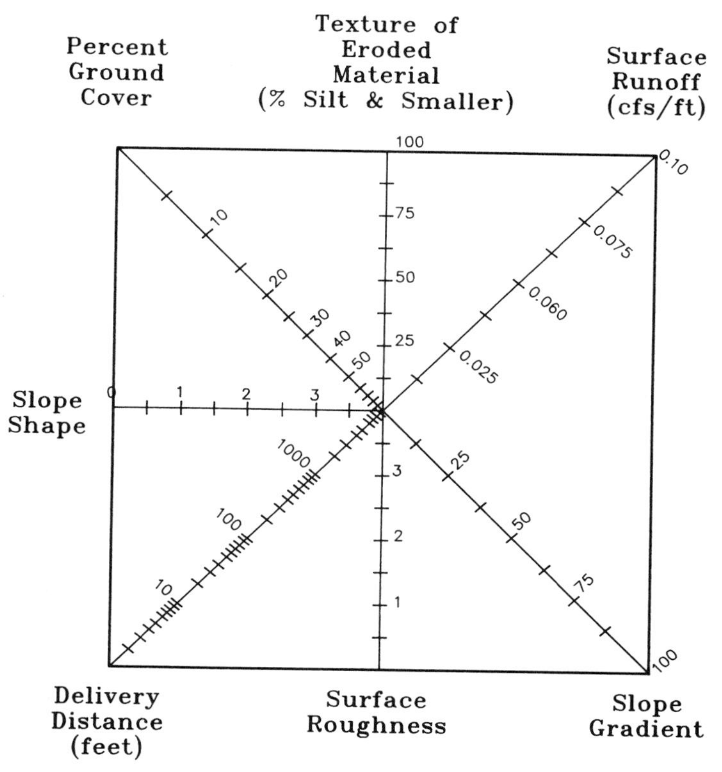

Figure 8.27 Stiff diagram for estimating sediment delivery with the USLE/RUSLE (after Forest Service, 1980).

Texture of Eroded Sediment Texture of eroded sediment is a parameter used to define the impact of particle size on delivery. The parameter is the percentage finer than 0.05 mm (silt size and finer). A value of 100%, for example, means that all particles are silt size and smaller.

Ground Cover Factor The ground cover factor refers to the percentage ground cover in the flow path between the source area and the stream. Ground cover, in this case, is defined as cover such as litter that is in contact with the surface. Zero indicates no cover and 100 means complete cover.

Slope Shape Factor Slope shape is the factor that accounts for the impact of concave or convex surfaces on a sediment delivery. The shape refers to slope shape between the source area and channel. A factor of zero represents a convex shape and four a concave shape.

Slope Gradient Slope gradient is the average slope between the source area and the receiving channel.

Delivery Distance Delivery distance is the flow distance in feet between the source area and channel.

Surface Roughness Surface roughness is a subjective index of the impact of roughness on sediment delivery. A value of 0 indicates a smooth surface and 4 is a very rough surface.

Use of the Forest Service Model The forest service method should only be used with the USLE/RUSLE estimates of gross erosion. To use the method, all of the parameters must be estimated and plotted on the appropriate axis in Fig. 8.27. All of the points are then connected, forming a polygon. The ratio of the area within the polygon to the total area of the rectangle is used in Fig. 8.28 to predict the sediment delivery index (delivery ratio). Procedures are illustrated in Example Problem 8.14.

Example Problem 8.14 Predicting delivery ratio with Forest Service method

A construction operation is disturbing a large area in a forest watershed upslope of a sensitive stream. The regulatory authority requires that a riparian forest zone be left undisturbed for 200 ft on either side of the stream. During a storm, the sediment yield from the disturbed area to the edge of the forest is estimated from the RUSLE to be 30 tons/acre with a size distribution that includes approximately 50% silt plus clay. The disturbed area has a slope length of approximately 720 ft. The peak rainfall excess from the site is 4.50 in./hr. The riparian zone between the disturbed area and the stream has the following characteristics:

Slope shape, moderately convex
Change in elevation, 50.0 ft
Average slope, 25%
Ground cover, 100%
Surface roughness, rough
Flow distance, 200 ft.

Estimate the sediment yield to the stream, both with and without the riparian zone, if the disturbed area extends 1000 ft along both sides of the stream. If the runoff volume is 3.0 in., estimate the average sediment concentration of the flow entering the stream.

Solution:

1. *Area of polygon on stiff diagram.* From the data above, the texture index is percentage silt plus clay or 50; the slope gradient is 25.0; surface roughness value is 3.0; delivery distance is 200 ft, slope shape is 1, and percentage ground cover is 100. Peak flow rate is

$$q_p = 2.31 \times 10^{-5}\sigma L = 2.31 \times 10^{-5}(4.50)(720)$$

$$= 0.075 \text{ cfs/ft}.$$

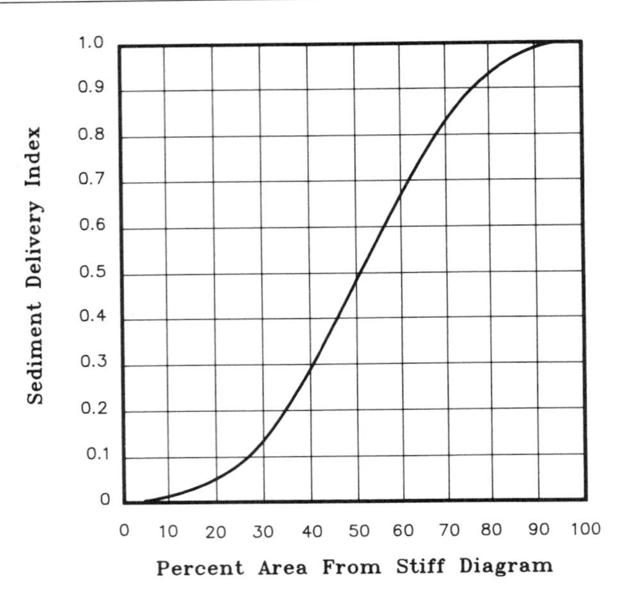

Figure 8.28 Diagram to convert percentage of area in Fig. 8.27 to sediment delivery ratio (after Forest Service, 1980).

Using these values in Fig. 8.27, a polygon is plotted as shown in Fig. 8.29 and the relative area determined to be 0.10 or 10%.

2. *Delivery ratio.* Using 10% in Fig. 8.28,

$$D = 0.02.$$

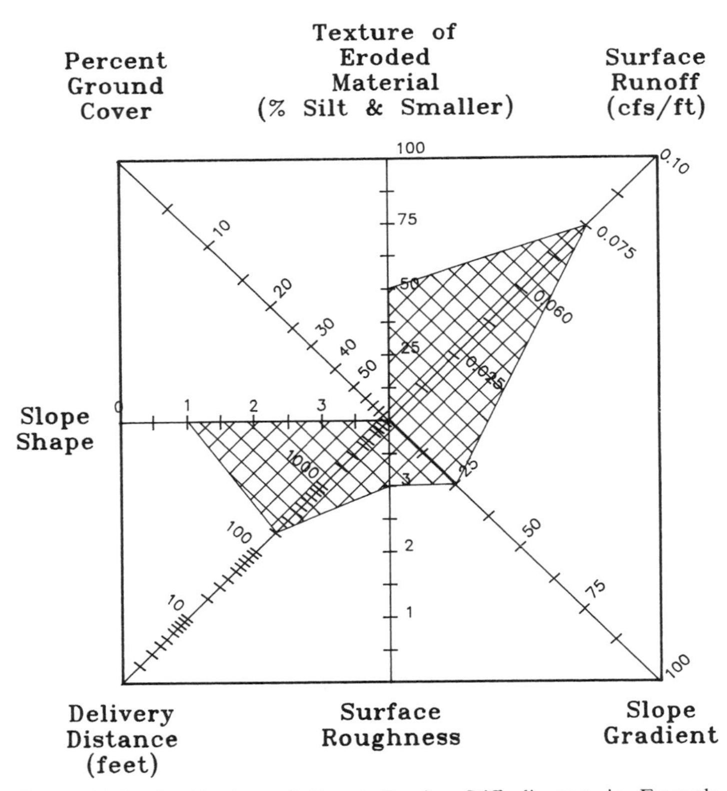

Figure 8.29 Application of Forest Service Stiff diagram in Example Problem 8.14.

3. Sediment yield.

$$Y = D \times (\text{gross erosion})$$

$$= (0.02)(30\ \text{tons/acre})(720\ \text{ft})(1000\ \text{ft})\left(\frac{1\ \text{acre}}{43{,}560\ \text{ft}^2}\right)$$

$$Y = 9.91\ \text{tons},$$

4. Estimating concentration. The runoff volume is

$$Q = \text{runoff depth} \times \text{area}$$

$$= \left(\tfrac{3}{12}\ \text{ft}\right)(720 \times 1000\ \text{ft}^2)$$

$$= 180{,}000\ \text{ft}^3 = 4.13\ \text{acre} \cdot \text{ft}.$$

Sediment concentration is the mass of sediment divided by the mass of water or

$$C = \frac{(9.91\ \text{tons})(2000\ \text{lb/ton})}{(62.4\ \text{lb/ft}^3)(180{,}000\ \text{ft}^3)}$$

$$= 0.00176\ \text{lb/lb or } 1760\ \text{mg/liter}.$$

This is the concentration entering the stream.

5. Sediment yield without the riparian zone. Without the riparian zone, the sediment yield would be

$$Y = \frac{(30\ \text{tons/acre})(720\ \text{ft})(1000\ \text{ft})}{43{,}560\ \text{ft}^2/\text{acre}} = 496\ \text{tons}.$$

The average concentration would be

$$C = \frac{(496\ \text{tons})(2000\ \text{lb/ton})}{(62.4\ \text{lb/ft}^3)(180{,}000\ \text{ft}^3)}$$

$$= 0.088\ \text{lb/lb or } 88{,}000\ \text{mg/liter}.$$

Obviously, riparian zones have a major impact on sediment yield.

Reservoir-Survey Method of Estimating Sediment Yield

One of the most common methods of estimating the volumes of sediment produced is by the Reservoir-Survey method. The Soil Conservation Service maintains a program in which thousands of reservoirs are surveyed annually to determine the quantity of sediment deposited. Information is also obtained, if available, on watershed characteristics. When a new reservoir site is proposed in an area, the SCS geologist simply finds the record of the closest surveyed reservoir that has similar watershed characteristics. The measured loss of storage capacity in acre-feet per year per acre watershed for the existing reservoir now becomes the design sediment storage volume.

Caution should be exercised in selecting record years for which the survey was made to assure that the rainfall over the watershed was near normal during the period of observation. Otherwise, the design capacity may be far too small or too large. To select a comparison reservoir, it is important to select a site that has similar characteristics insofar as they affect sediment yield. The more important characteristics to compare are relief ratio, area, vegetative cover, and fraction of area disturbed.

A problem often encountered with the reservoir survey method for small watersheds is that many sedimentation surveys do not adequately describe the characteristics of the drainage area. The sediment yield expected from a highly disturbed watershed cannot be adequately extrapolated from the yield measured during an existing nondisturbed land use. Simply adjusting factors in the USLE/RUSLE, such as K, C, and P, will not be adequate since the entire watershed drainage pattern and flow characteristics of individual streams have been drastically modified through increased peak flows and varied sediment loads as a result of disturbance. Thus, caution should be exercised in the application of this method.

Estimating Sediment Yield with Modified Universal Soil Loss Equation (MUSLE)

Research in recent years has generated new models of soil erosion and sediment yield with varying degrees of applicability. Some of these models are application oriented, while others have not reached the point of application. Some of the models have been combined and computerized along with hydrology models and reservoir models to yield a complete computational package of watershed hydrology and sedimentology.

A detailed discussion of all the models is beyond the scope of this chapter. One of the less complex relationships is the Modified USLE.

Williams (1976) proposed that the rainfall energy term, EI_{30} index, in the USLE could be replaced with a runoff energy term in the USLE to predict sediment yield directly. Procedures were developed for homogeneous watersheds using a lumped parameter approach and for nonhomogeneous watersheds using sediment routing procedures. A lumped parameter approach is one in which the entire watershed is represented by one characteristic parameter.

MUSLE Lumped Parameter

Williams (1977) and Williams and Brendt (1972) developed the MUSLE using data from 778 storms on watersheds near Reisel, Texas, and Hastings, Nebraska. The drainage areas ranged from 2.7 to 4380 acres and the average slope and slope lengths ranged from 0.9 to 5.9% and 258 to 570 ft. He replaced the R factor in the USLE with various parameters and used the resulting equation to predict the sediment yield from the watersheds. The parameter that gave the best estimate was Qq_p, the product of runoff volume and peak discharge. Williams denoted this term runoff energy. The resulting equation is

$$Y = 95(Q \times q_p)^{0.56}\{K\}_a\{LS\}_a\{CP\}_a, \qquad (8.103)$$

where Y is single-storm sediment yield in tons, Q is runoff volume in acre-feet, q_p is peak flow in cfs, and $\{K\}_a$, $\{LS\}_a$, and $\{CP\}_a$ are area weighted average USLE/RUSLE parameters for the watershed. Obviously, from the definition of delivery ratio, the single-storm delivery ratio is

$$D = \frac{95(Q \times q_p)^{0.56}}{R \times \text{area}}, \qquad (8.104)$$

where area is the watershed area in acres, R is the single-storm value of the rainfall factor in English units, and Q and q_p are acre-feet and cfs.

MUSLE Routing Procedures

The use of the MUSLE as given by Eq. (8.103) should be limited to those watersheds that are relatively homogeneous. Watersheds that are in the process of being disturbed are typically very heterogeneous. Williams (1975, 1978) proposed the following procedure to account for watershed heterogeneity:

(1) Divide the watershed into n homogeneous watersheds as shown in Fig. 8.30 and determine the travel time T_{ti} from each subwatershed to the watershed outlet.

(2) Determine the sediment yield Y_{0i} for each homogeneous subwatershed using Eq. (8.103).

(3) Determine the D_{50}, average diameter of sediment, exiting each subwatershed.

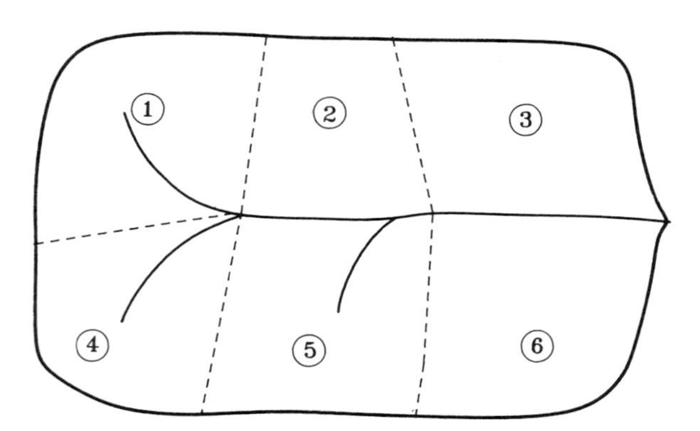

Figure 8.30 Example watershed division for MUSLE routing procedures.

(4) Calculate the amount of sediment from each subwatershed that reaches the exit by assuming that the rate of deposition is proportional to the particle size, amount of sediment, and travel time, or

$$\frac{dY_i}{dt} = -BY_i\sqrt{D_{50_i}}, \qquad (8.105)$$

where Y_i is the yield at any point downstream of the subwatershed exit, B is a routing coefficient, and D_{50} is the average diameter of sediment. Equation (8.105) can be integrated to obtain

$$Y_i = Y_{0i}e^{-BT_{ti}\sqrt{D_{50_i}}}, \qquad (8.106)$$

where Y_{0i} is the yield at the subwatershed exit, Y_i is the sediment from watershed i that reaches the main watershed exit, and T_{ti} is the travel time from the subwatershed to the main watershed exit.

(5) Calculate the total sediment yield for the entire watershed by

$$Y = \sum_{i=1}^{n} Y_{0i}e^{-BT_{ti}\sqrt{D_{50_i}}}. \qquad (8.107)$$

(6) Determine the routing coefficient from

$$(Q \times q_p)_{ws}^{0.56} = \sum_{i=1}^{n} (Q_i \times q_{pi})^{0.56} e^{-BT_{ti}\sqrt{D_{50_i}}}, \qquad (8.108)$$

where $(Q \times q_p)_{ws}$ is the runoff energy term for the entire watershed.

Williams presented verification data from five storms on a 4380-acre watershed with an average slope of 2% in Reisel, Texas, and proposed that the relationship had acceptable accuracy.

Williams' routing procedure is simple to use but does not predict a time distribution of sediment (sedigraph) or a size distribution of sediment. Neither is flood plain erosion or channel erosion considered. Williams (1978) presented a more complex procedure for routing that considered both channel erosion and deposition. The procedures have not been verified nor widely adopted.

Size Distribution of Sediment

To use Eq. (8.108), it will be necessary to have a D_{50} value for the sediment exiting the subwatershed. Barfield *et al.* (1979) proposed a simple procedure based on the assumption that all of the larger particles in the flow settle out first in a deposition process. Based on this assumption, the size distribution of material exiting the subwatershed can be estimated from the size distribution of the percentage finer (PF) of eroded parent material. These procedures are discussed in Chapter 7.

Foster *et al.* (1985) divided eroded soil into clay, silt, and sand fractions plus large and small aggregates. The fraction of these components in runoff is predicted by regression equations based on a limited amount of data. The independent variables in the regression equation is the fraction of the components in the parent material. These procedures are also discussed in Chapter 7.

PREDICTING THE TIME DISTRIBUTION OF SEDIMENT: A SEDIGRAPH

In the sediment yield computational procedures presented thus far, only the total storm sediment yield was calculated. A procedure is needed to translate this yield to a sedigraph, a time distribution of sediment yield. The simplest procedure is to make the assumption that concentration is a power function of water discharge or

$$c = kq^{a}, \tag{8.109}$$

where k and a are constants. Based on a value for a, k can be calculated from runoff and sediment yield information as discussed below.

Since the mass flow rate of sediment is given by

$$q_{\mathrm{s}} = cq, \tag{8.110}$$

then

$$q_{\mathrm{s}} = kq^{a+1}. \tag{8.111}$$

To obtain the total sediment yield, Eq. (8.111) must be integrated over the entire storm, or

$$Y = \int_0^{D_{\mathrm{st}}} kq^{a+1}\, dt, \tag{8.112}$$

where D_{st} is the storm duration. Since k is a constant, it can be determined from

$$k = \frac{Y}{\int_0^{D_{\mathrm{st}}} q^{a+1}\, dt} = \frac{Y}{\sum_{i=1}^{n} q_i^{a+1}\, \Delta t_i}, \tag{8.113}$$

where the storm is divided into n increments for the purpose of evaluating the integral. Thus given a runoff hydrograph and total sediment yield, a sedigraph can be calculated. Values for a are typically near 0.5 to 1.0.

Other models for developing sedigraphs have been presented (Williams, 1979; Rendon-Herrero, 1974; Renard and Laursen, 1975; Bruce *et al.*, 1975), but are much more complex than the procedures above.

PROCESS-BASED EROSION MODELS: CREAMS SEMITHEORETICAL RILL AND INTERRILL MODEL

Basic Equations

Erosion Model

The CREAMS (chemicals runoff erosion in agricultural management systems) interrill and rill erosion model (Foster *et al.*, 1980a) is based on a semitheoretical relationship known as the Foster–Meyer–Onstad (FMO) equation (Foster *et al.*, 1977a). Foster *et al.* start with a partition of erosion into rill and interrill components

$$G = G_{\mathrm{r}} + G_{\mathrm{i}}, \tag{8.114}$$

where G_{r} is the cumulative rill erosion (mass/width · time) and G_{i} is cumulative interrill erosion (mass/width · time) at any distance X down the hillslope. Relationships for predicting rill and interrill erosion in the FMO are based on detachment as a function of surface shear for rill erosion and rainfall energy for interrill erosion. Starting with the assumption that interrill erosion is proportional to slope and rainfall intensity and that detachment in rills is proportional to tractive force to the $\frac{3}{2}$ power, the FMO equation for cumulative erosion at any point X downslope is given by

$$G = X^2 K_{\mathrm{r}} a' (\sin\theta)^{e'} F_{\mathrm{t}} C_{\mathrm{r}} P_{\mathrm{r}} + X K_{\mathrm{i}} (b' \sin\theta) I_{\mathrm{t}} C_{\mathrm{i}} P_{\mathrm{i}}, \tag{8.115}$$

where K_{r} and K_{i} are rill and interrill erodibilities, θ is slope angle, C_{r} and C_{i} are rill and interrill cover factor, P_{r} and P_{i} are rill and interrill conservation practice factors, F_{t} and I_{t} are rill and interrill erosivities, and a', b', and e' are constants. Foster *et al.* (1977a) proposed that

$$F_{\mathrm{t}} = 15 Q q_{\mathrm{p}}^{1/3} \tag{8.116}$$

$$I_{\mathrm{t}} = EI_{30}, \tag{8.117}$$

where E is rainfall energy, I_{30} is the maximum 30-min intensity, Q is runoff volume in watershed feet, and q_{p} is peak discharge in volume per area per time (feet per second). By equating the FMO equation and the USLE, Foster *et al.* (1977b) developed the following form for the rill and interrill erosion components

$$D_{\mathrm{i}} = 0.210\left(\frac{EI_{30}}{100}\right)(S + 0.014) K_{\mathrm{i}} C_{\mathrm{i}} P_{\mathrm{i}} \left[\frac{\sigma_p}{V_u}\right] \tag{8.118}$$

$$D_{\mathrm{rc}} = 37983 m V_{\mathrm{u}} \sigma_{\mathrm{p}}^{1/3} \left[\frac{X}{72.6}\right]^{m-1} S^2 K_{\mathrm{r}} C_{\mathrm{r}} P_{\mathrm{r}} \left[\frac{\sigma_{\mathrm{p}}}{V_{\mathrm{u}}}\right], \tag{8.119}$$

where D_i and D_{rc} are denoted as interrill and rill erosion potential in lb/ft^2 · sec, $S = \sin\theta$, σ_p is the peak flow rate in ft/sec, V_u is runoff volume in feet, K_r and K_i are rill and interrill erodibility in English units, and m is a slope length exponent. The use of the potential instead of actual erosion definition allows the relationship to be used where deposition occurs, as is discussed later. Equations (8.118) and (8.119) are the basic erosion relationships for CREAMS. In the usage developed for CREAMS, the only time P is used is for contouring. Since separate K, C, and P values for rill and interrill erosion are not available using the USLE modeling format, the interrill and rill values are assumed to be the same and equal to the USLE values.

Erosivity of Storm

Storm erosivity is calculated from rainfall distribution information. For breakpoint rainfall (rainfall distributed over the duration of the storm) energy is calculated from Eq. (8.31). If the rainfall distribution is not available, the storm EI_{30} index is estimated from daily rainfall, or

$$\frac{EI_{30}}{100} = 8.0V_R^{1.51}, \qquad (8.120)$$

where V_R is the 24-hr rainfall in inches.

Slope Exponent *m*

To limit excessive erosion estimates on long steep slopes, the slope exponent is defined as

$$\begin{aligned} m &= 1.0 + \frac{5.011}{\ln X} \qquad X > 150 \text{ ft} \\ m &= 2.0 \qquad X < 150. \end{aligned} \qquad (8.121)$$

By combining rill and interrill components, Foster *et al.* (1980a) point out that the effective exponent is a function of slope, slope length, and ratio $EI_{30}/V_u\sigma_p^{1/3}$.

Transport Capacity, Particle Size Distribution, and Erosion Interactions

Transport capacity in the CREAMS model is calculated by the Yalin equation using the CREAMS eroded size distributions, as discussed in Chapter 7. The interaction between detachment and transport is defined by the closed-form erosion solution, given in Eq. (8.17). The closed-form equation is applied to four different cases

Case I —Deposition over the entire segment
Case II —Detachment by flow on the upper end and deposition in the lower end
Case III —Deposition on the upper end and detachment by flow on the lower end
Case IV —Detachment over the entire slope.

Specific equations defining these cases are given in the CREAMS manual (Knisel, 1980).

Transport capacity calculations with Yalin's equation requires a value of shear stress acting on the soil surface. In CREAMS, shear stress on the soil is estimated from

$$\tau_{so} = \gamma y_{bs} S \left(\frac{n_{bs}}{n_{cov}} \right)^{0.9}, \qquad (8.122)$$

where τ_{so} shear stress on the soil in pounds per square foot, γ is the weight density of water in pounds per cubic foot, y_{bs} is the depth of flow relative to bare soil (i.e., the flow depth assuming a bare soil), S is the sine of the slope angle, n_{bs} is Manning's n for bare soil (assumed to be 0.01), and n_{cov} is the total Manning's n. Values for the parameters are given in Appendix 8E, Tables 8E.1 and 8E.2. Depth of flow relative to bare soil is estimated from Manning's equation

$$y_{bs} = \left[\frac{q n_{bs}}{1.5 S^{1/2}} \right]^{0.6}, \qquad (8.123)$$

where q is water flow rate in cfs/ft.

Application of the Model

The equations used in CREAMS are steady state. To use them on a storm basis, Foster *et al.* (1980a) recommend that calculations be made with peak discharge as a flow rate. This should be used to estimate a sediment concentration. The concentration is then used with the runoff volume to estimate total sediment yield.

CREAMS is a model of erosion for a field-sized area and is not intended to estimate watershed erosion, although it has been used for that purpose. Computation of soil erosion with the CREAMS model is too complex to be made without a computer; hence the model is available in a computer format.

PROCESS-BASED EROSION MODELS: WEPP THEORETICAL RILL AND INTERRILL MODEL

Background

A group of hydrologic and erosion scientists, primarily in the USDA–ARS, conducted a massive data collection and modeling effort with the objective of developing a process-oriented model that incorporates modern erosion and hydrologic science (Lane and Nearing, 1989; Foster and Lane, 1987). The hydrologic and sediment algorithms were developed at several erosion laboratories and calibrated on data collected on crop lands, rangelands, and forest lands research sites dispersed throughout the 48 states. Results from the studies included a data base on soil erodibility parameters, rill and interrill hydraulic parameters, and effects of buried residue on erosion parameters.

The WEPP erosion model is included as a component of an overall continuous simulation model including a rainfall generator, runoff predictor, plant growth simulator, and sediment yield model. The final result is a process-oriented continuous simulation computer model for predicting runoff and soil erosion on hillslopes as affected by soils, climate, and management. Based on the experimental results, prediction equations for hydrologic and erosion parameters have been developed and evaluated. Further evaluation of the parameter prediction relationships continues. All units in the model are SI. The model will be available as a personal computer program in a hillslope version and an upcoming so-called watershed version. The watershed version, like the CREAMS model, is intended for small upland watersheds and will include impoundment channel erosion elements.

The following information covers erosion fundamentals, presented here to show how theoretical concepts are combined in WEPP to calculate sediment yield. The WEPP model is still under development, hence the alogrithms presented here are tentative. Most algorithms have reached a final condition as of the writing of this text with a final "freezing" of the algorithms expected in early 1994.

Basic Governing Erosion Relationships

Continuity

The basic continuity relationship is the steady-state spatially varied sediment continuity equation, given earlier by Eq. (8.2) or

$$dq_s/dx = D_i + D_r,$$

where D_i and D_r are interrill and rill erosion rates (kg $\sec^{-1}$ m^{-2}), q_s is sediment load (kg/sec · m), and x is distance downslope (m). In this formulation, D_i and D_r are detachment rates per unit slope area, not per unit length of rill. Net soil detachment is related to detachment potential and transport capacity by Eq. (8.5), or

$$D_r = D_{rc}(1 - q_s/T_c), \tag{8.124}$$

where D_{rc} is detachment potential (kg/sec · m) and T_c is transport capacity (kg/sec · m). Detachment potential is defined by shear excess given in Eq. (8.25) or

$$D_{rc} = K_r(\tau - \tau_c), \tag{8.125}$$

where rill erodibility K_r replaces a in Eq. (8.25), τ is stress acting on the soil particles (Pascal), and τ_c is critical tractive force (Pascal). When $\tau < \tau_c$, detachment is zero.

Net deposition is defined by a modification of Eqs. (8.6) and (8.3), or

$$D_r = \beta_T \frac{V_s}{q}(q_s - T_c), \tag{8.126}$$

where β_T is a parameter defining the impact of turbulence on settling. For WEPP, $\beta_T = 0.5$. Storm *et al.* (1990) showed that Eq. (8.126) is a modification of the overflow rate discussed in Chapter 9. The formulation in Eq. (8.126) is the laminar form of the overflow rate, but the relationship approaches that of a turbulent form because (1) the slope is discretized into segments and (2) the assumption is made that flow is completely mixed at the beginning of each segment. Storm *et al.* (1990) showed that discretizing the slope into 20 segments will ensure that the deposition model approached that of a fully turbulent model.

Equilibrium Channel Width

Lane and Foster (1980) showed that concentrated flow erosion tends to develop channels with vertical walls and with an equilibrium width that is proportional to flow rate. In WEPP, a rectangular shape is assumed with an equilibrium width given by

$$W_e = c_w q^{d_w} \tag{8.127}$$

where W_e is the equilibrium width (meters), q is flow rate at the end of the slope (cubic meters per second), and c_w and d_w are equilibrium parameters that are functions of soil and vegetation. Unlike the Foster and Lane (1983) model presented in an earlier section, the WEPP model does not provide for the presence of a nonerodible layer. Presently, Gilley *et al.* (1990) recommend that universal values of $c_w = 1.13$ and $d_w = 0.303$ be used. The Foster and Lane (1983) concentrated flow model, discussed earlier, could be used to correct c_w and d_w for soil properties. The flow rate for Eq. (8.127) is a steady-state flow at the end of a slope, assumed in WEPP to be given by the peak runoff rate, or

$$q = I_p X S_{rs}, \tag{8.128}$$

where I_p is the peak rate of rainfall excess (meters per second), X is the slope length (meters), and S_{rs} is the rill spacing (meters). In the present formulation of WEPP, S_{rs} is assumed to be 1 m.

Flow depth is calculated using the Darcy Weisbach friction factor [Eq. (4.21)], rill width, and average slope. With the calculated depth and width, hydraulic radius is determined and shear stress on the soil calculated from

$$\tau_e = \gamma S_a R(f_s/f_t), \tag{8.129}$$

where τ_e is shear stress on the soil at the end of an average uniform slope (Pa), γ is density of water

(N/m^3), S_a is the average channel slope, R is hydraulic radius, f_s is friction factor for soil, and f_t is total friction factor. Methods for estimating f_t and f_s are given in Lane and Nearing (1989).

Sediment Transport Capacity Sediment transport is calculated by a relationship developed by Finkner *et al.* (1989), or

$$T_c = k_t \tau^{3/2}, \quad (8.130)$$

where T_c is transport capacity, τ is shear stress on the soil surface, and k_t is a calibration coefficient calculated from T_c and τ estimated at the end of the slope. Details on calculation of k_t are given in Finkner *et al.* (1989).

Normalized Parameters Slopes are not necessarily uniform, but can be divided into segments. For each segment, a normalized slope is defined as

$$S_* = S/S_a = aX_* + b, \quad (8.131)$$

where

$$X_* = X/L. \quad (8.132)$$

L is the slope length, S_a is the average slope measured over L, and a and b are coefficients set to make the model fit the land slope.[1] An equivalent transport capacity T_{ce} at the end of the uniform slope is defined as a function of the shear on the uniform slope τ_e, or

$$T_{ce} = k_{t1} \tau_e^{3/2}. \quad (8.133)$$

Shear stress versus distance downslope can be derived using the Darcy–Weisbach equation

$$\tau = \gamma \left[\frac{I_p}{C} XS \right]^{2/3}, \quad (8.134)$$

where C is the Chezy coefficient given by $C = (8g/f_t)^{0.5}$ is the friction factor for rills and X is the downslope distance. Likewise, τ_e can be defined from slope length L and S_a. Defining τ_* as τ/τ_e,

$$\tau_* = \left[\frac{X}{L} \frac{S}{S_a} \right]^{2/3} \quad (8.135)$$

or, since $S/S_a = aX_* + b$,

$$\tau_* = \left[aX_*^2 + bX_* \right]^{2/3}. \quad (8.136)$$

Sediment load and transport capacity are also normalized to transport capacity at the end of the uniform slope, or

$$q_s^* = \frac{q_s}{T_{ce}} \quad (8.137)$$

[1] Given two points on a slope segment, a and b can be determined for Eq. (8.131).

and

$$T_c^* = \frac{T_c}{T_{ce}}. \quad (8.138)$$

Thus, since $T_{c1} = k_{t1} \tau_e^{3/2}$, then

$$T_c^* = k_{tr}(ax_*^2 + bx_*), \quad (8.139)$$

where k_{tr} is k_t/k_{t1}.

Normalized Rill and Interrill Detachment and Deposition Parameter

WEPP utilizes dimensionless detachment and deposition equations and parameters in making calculations. Using dimensionless parameters, rill detachment is corrected for consolidation, freeze–thaw, and below-ground root mass. The dimensionless rill parameters are

$$\eta = LK_r K_{rc} K_{rbr} \tau_e / T_{ce} \quad (8.140)$$

and

$$\tau_{cn} = \tau_c \tau_{cc} / \tau_e, \quad (8.141)$$

where η is a dimensionless rill detachment parameter, K_{rc} is a dimensionless subfactor for soil consolidation and freeze–thaw effects, K_{rbr} is a dimensionless subfactor for root effects, τ_{cn} is dimensionless tractive force, and τ_{cc} is a dimensionless subfactor to account for consolidation and freeze–thaw. To apply the subfactor correction to Eq. 8.125, multiply k_r by k_{rc} and k_{rbr} while τ_c is multiplied by τ_{cc}. The standard condition for WEPP, used to define K_r, is for bare soil immediately after tillage. A dimensional deposition parameter is given by

$$\phi = \beta_T (V_s / I_p). \quad (8.142)$$

where V_s is settling velocity (m/sec). A nondimensional interrill detachment parameter is given by

$$\varepsilon = LD_i / T_{ce}, \quad (8.143)$$

where D_i is interrill erosion rate given by

$$D_i = K_i I_e^2 S_f C_c C_{gc} (S_{rs}/W_e), \quad (8.144)$$

where S_f is the interrill slope adjustment factor given by

$$S_f = 1.05 - 0.85 e^{-4 \sin \Omega}, \quad (8.145)$$

where Ω is the interrill slope angle, K_i is the baseline interrill erodibility, I_e is the equivalent rainfall intensity (m/sec), and C_c is the canopy factor, C_{gc} is the ground cover factor, S_{rs} is rill spacing (meters), and W_e is equilibrium rill width (meters) defined by Eq. (8.127). The canopy effect is given by a relationship similar to that in RUSLE, or

$$C_c = 1 - F_c e^{-0.34H}, \quad (8.146)$$

where F_c is the fraction of ground cover and H is crop height (meters). The ground cover factor is given by

$$C_{gc} = e^{-2.5 F_{gc}}, \quad (8.147)$$

where F_{gc} is fraction of ground cover in the interrill area.

Normalized Erosion Equation

Detachment and Deposition Equation

Utilizing Eqs. (8.2), (8.4), and (8.124) along with WEPP parameters, a fundamental erosion equation (detachment) can be written as

$$\frac{dq_s^*}{dx_*} = \eta(\tau_* - \tau_{cn})\left(1 - \frac{q_s^*}{T_c^*}\right) + \varepsilon, \qquad (8.148)$$

where η, τ_{cn}, and ε are the normalized parameters defined earlier. The dimensionless parameters τ_c^*, q_s^*, and τ_* are normalized functions of X_* defined earlier. The equation must be solved numerically. Utilizing Eqs. (8.2) and (8.126), the normalized deposition equation is

$$dq_s^*/dx_* = (\phi/x_*)(T_c^* - q_s^*) + \varepsilon, \qquad (8.149)$$

where ϕ and ε are the normalized erosion parameters defined earlier. In application, ϕ can be calculated for each particle class, if desired.

Sediment Yield

Sediment load is calculated from normalized sediment load by

$$q_s = q_s^* T_{ce}[W_e/S_{rs}], \qquad (8.150)$$

where q_s is sediment load in kilograms per second per unit width of hillslope. Equations (8.148)–(8.150) would be used for the first segment of a slope where the inflow is zero at the top of the slope. Equations for nonzero inflow are given in the following section.

Evaluating Downslope Variability with Inflow at the Top of the Slope

Downslope variability in flow rate and erosion parameters can be evaluated with WEPP. In WEPP, the flow paths are divided into sections with homogenous erosion properties. Those sections are elements that may have complex topography, but the erosion and cover parameters would be constant over a section. Each section is treated separately with a lateral inflow defined at the top of the section. With lateral inflow, the nondimensional shear stress becomes

$$\tau_* = \left(A_0 x_*^2 + B_0 x_* + C_0\right)^{2/3}, \qquad (8.151)$$

where A_0, B_0, and C_0 are coefficients defined by

$$A_0 = \frac{a}{q_0^* + 1} \qquad (8.152)$$

$$B_0 = \frac{a q_0^* + b}{q_0^* + 1} \qquad (8.153)$$

$$C_0 = \frac{b q_0^*}{q_0^* + 1} \qquad (8.154)$$

$$q_0^* = \frac{q_0}{I_p L} \qquad (8.155)$$

and q_0 is the inflow of water (meters per second) at the top of the homogeneous section. Nondimensional transport capacity is given by

$$T_c^* = k_{tr}\left(A_0 x_*^2 + B_0 x_* + C_0\right). \qquad (8.156)$$

The form of the detachment equation, Eq. (8.148), must remain the same. The form of the deposition equation, Eq. (8.149), is changed by replacing (ϕ/x_*) with $(\phi/x_* + q_0^*)$. Equation (8.156) would be used on any slope segment with inflow at the top of the slope.

Use of the WEPP Model

As indicated earlier, the WEPP erosion model presented in this chapter is but one component of a large hydrology and sedimentology continuous simulation model that includes crop growth models, hydrology models, and parameter generation relationships. Its use obviously requires a computer.

Problems

(8.1) Discuss the difference between rill erosion, interrill erosion, ephemeral gully erosion, and channel erosion. Enumerate the processes and forces involved in each type of erosion.

(8.2) On a uniform slope of 150 ft, three different mulch rates are being considered. The resulting interrill detachment rates would be 0.008, 0.005, and 0.002 lb/sec · ft. If the detachment capacity and transport capacity at the end of the slope is 0.02 lb/sec · ft and 1.8 lb/sec · ft, respectively, estimate the sediment load at 25, 50, 75, 100, 125, and 150 ft downslope for each of the mulch rates. Assume steady uniform flow. Plot the results, showing the difference between sediment loads.

(8.3) A researcher measures rainfall energy of 1300 ft · tonsf/acre · in. in a storm with an intensity of 2.0 in./hr. Is this a reasonable value? Justify your answer. Compare this value to predictions from equations of Wischmeier and Smith and Brown and Foster.

(8.4) Estimate the energy content (ft · tonsf/acre · in.) of a storm with the following rainfall intensities

Time period (min)	Intensity (in./hr)	Time period (min)	Intensity (in./hr)
0–15	0.6	30–45	4.0
15–20	1.4	45–60	0.8
20–30	3.0		

What is the EI_{30} index? If this storm occurred in Memphis, Tennessee, what would be the return period, based on the EI_{30} index? What would be the return period if the storm occurred in your area?

(8.5) Determine the following R factors for Los Angeles, California and Rapid City, South Dakota: Average annual R factor, 10-year, 24-hr historic storm R factor, and 10-year, 24-hr synthetic storm R factor. If the slope being considered has a slope steepness of 0.1%, make a correction for ponding.

(8.6) If you are scheduling construction operations, how would the difference in distribution of R factor affect your scheduling decision for Los Angeles and South Georgia?

(8.7) Estimate the average annual K using the Wischmeier nomograph if the soil has the following characteristics.

Property		Value
% SA	0.1–2.0 mm	25%
% VFS	0.05–0.1 mm	4%
% MS	0.002–0.05 mm	40%
% CL	<0.002 mm	31%
% OM		1%
Structure		Medium
Permeability		Moderate
Annual R factor	English units	80

Correct the annual K factor accounting for seasonable variations in K and the variability in the EI_{30} index. The average number of frost-free days is 120, and the site is located in central New York. The average temperature drops below 27°F on December 1 and rises above 27°F on March 1. Make a plot of the resulting K factor versus time of year. Indicate the nominal K value on the plot.

(8.8) If the soil in Problem (8.7) has rock fragments greater than 2.0 mm that occupy 30% of the soil by weight, what would be the resulting K value?

(8.9) Estimate the LS factor for the RUSLE for a slope of 10% under the conditions of low, moderate, and high ratio of rill to interrill erosion and a slope length of 100 ft.

(8.10) Estimate the LS factor for a 250-ft slope length with an average slope of 15% under the following assumptions: (1) Uniform slope; (2) convex slope divided into 5 equal segments with slopes of 5, 10, 15, 20, and 25% moving from top to bottom of the slope; (3) concave slope divided into 5 equal segments of 25, 20, 15, 10, and 5% moving from top to bottom; and (4) S-shaped slope with slopes of 10, 15, 25, 15, and 10% moving from top to bottom. Assume a moderate ratio of rill to interrill erosion.

(8.11) A 180-ft-long slope has the following characteristics.

Section	Parameter	Value
1	Length	60 ft
	Steepness	25%
	Erodibility	0.2
	Cover factor	0.6
	Practice factor	1.0
2	Length	60 ft
	Steepness	15%
	Erodibility	0.3
	Cover factor	0.2
	Practice factor	1.0
3	Length	60 ft
	Steepness	5%
	Erodibility	0.2
	Cover factor	0.1
	Practice factor	1.0

Estimate the erosion from a single storm with an R factor of 100 for this site.

(8.12) Estimate the average annual C factor for corn grown in upstate New York under the following assumptions: (1) high production conventional tillage; fall turnplowing on Oct 15, seedbed preparation on May 1, planting on May 15, 10% cover on Jun 5, 50% cover on July 1, 75% cover on July 15, harvest on October 1 (assume that the C factor for period 4 is the same as period 3); (2) minimum tillage with the same planting and harvest dates and same dates for percentage of cover.

(8.13) If the K in Problem (8.12) is 0.3, slope lengths are 150 ft, and slope steepness are 8%, estimate the average annual erosion assuming a moderate ratio of rill to interrill erosion and an average R factor of 175.

(8.14) If the field in Problem (8.12) is the same as that in Problem (8.7), estimate the average annual erosion, accounting for seasonal variation in both K and C. Assume a slope length of 150 ft and a slope steepness of 8%.

(8.15) If the bulk density of the soil in Problem (8.14) is 1.2, what depth of soil is being eroded annually by the crop.

(8.16) For a typical soil and field in your location, estimate the average annual erosion for a crop that would routinely be grown. Correct for annual variation in C and K. Based on information from the soil survey on allowable annual erosion, develop a cropping management plan to keep the erosion rate below the allowable value.

(8.17) A strip mine in Southwestern Wyoming has the following sequence of operations.

Time period	Description
1/1–3/1	Rangeland, 25% cover
3/1–6/1	Bare soil, stripped of vegetation (no cover, poor soil)
6/1–9/1	Active mining
9/1–10/1	Regraded
10/1–12/31	Seeded to range grass

(1) Estimate the C value for the site. (2) If the soil has an erodibility of 0.4, a slope steepness of 12%, a slope length of 250 ft, and a low ratio of rill to interrill erosion, estimate the average annual erosion rate.

(8.18) Estimate the C factor on March 15 for the following conditions: soybeans following corn, corn harvested October 15 with a yield of 5500 lb/acre, soil moisture at field capacity, moldboard plowing on November 1, row planting on March 1, and a moderate mixture of rill to interrill erosion. The following monthly average temperatures and precipitations are applicable

Month	Monthly average temperature (°F)	Monthly average precepitation (in.)
Oct	55	2.5
Nov	50	4.0
Dec	40	3.0
Jan	25	3.0
Feb	30	4.0
Mar	40	6.0

(8.19) If the RUSLE is to be used to estimate erosion, an estimate of the quantity of residue on the surface and below the surface must be made for every 15-day period. This means that the effects of each tillage operation on the distribution of above- and below-ground residue must be calculated. Develop an algorithm to evaluate that distribution of residue, using a spreadsheet.

(8.20) Estimate the C factor for a reconstructed surface mine soil on March 15, assuming that the soil is regarded on October 1, reseeded to winter small grain, and mulched with 2 tons/acre of wheat straw mulch. During reconstruction, the soil is compacted to 1.2 times its loose density. Assume the same temperature and moisture as in Problem (8.18).

(8.21) If you live in an area where timber harvesting occurs, take a trip to an active operation. Estimate the percentage of bare soil, the percentage of soil with root mat, the canopy height and cover, and the percentage of total slope in steps. Using this information, estimate the C factor. Collect information on the soil, slope length, steepness, and R factor, and estimate the average annual erosion.

(8.22) Take a trip to a farming operation with and without conservation practices. Estimate the annual C and P factors using the USLE and RUSLE approach. Using these factors and the RKLS factors, estimate the average annual erosion.

(8.23) Take a trip to a farm with a crop in the development stage. From visual observations and your best estimates, develop the subfactors for the RUSLE C factor.

(8.24) Starting with Manning's equation, show that Eq. (8.76) is correct.

(8.25) An ephemeral gully contains a flow rate of 0.005 m^3/sec a slope of 0.015, n of 0.025, bulk density of 1.4, and a τ_e of 1.8 Pa. Estimate the equilibrium width, final width, initial detachment rate, and detachment rate at reaching a nonerodible layer located at a depth of 0.3 m. Use both the Foster and Lane model and the Ephemeral Gully model and compare the results.

(8.26) If the flow for Problem (8.25) is from a disturbed area with a sediment load equal to half the transport capacity, what would be the detachment rates prior to reaching the nonerodible layer.

(8.27) A 20-acre watershed is being disturbed to develop a major shopping center. The planning commission requires that the average sediment concentration in the drainage shall be no greater than 1000 mg/liter in a 10-year, 24-hr type II storm. The developer proposed that a 100-ft-wide riparian zone of natural vegetation would solve the problem. Is the developer correct? The riparian zone is wooded with 80% grass at the surface. The slope shape is concave, the slope gradient is 12.5%, and the surface is moderately rough. The average watershed slope length for the contributing area is 200 ft with a 6% slope and a K of

0.32. The sediment in the runoff from the contributing area is 50% silt and clay. For the site being considered, the 10-year, 24-hr rainfall is 5.0 in., and the curve number is 80 during construction. The surface will be bare during construction. The peak runoff rate for the watershed is 800 cfs/mile^2 · in. runoff. (Hint, utilize Fig. 8.27.)

(8.28) For the postmining hydrograph in Fig. 3.42, determine a corresponding sediment graph if the constant a in Eq. (8.109) is 1.0 and the sediment yield is 2000 tons. Express the sediment graph in terms of concentration (milligrams per liter).

(8.29) A watershed is divided into three homogeneous subareas as shown below. Estimate the total yield from each subwatershed and the combined watersheds using the MUSLE procedures.

Sub watershed	Hydraulic parameters	USLE parameters	Average particle diam (mm)
1	Q (acre•ft) = 5.1	$K = 0.3$	0.11
	q_p (cfs) = 81.2	Length 175 ft	
	T_t (hr) = 0.5	Stpns 4.3%	
		$C = 0.45$	
		$P = 1.0$	
2	Q (acre•ft) = 6.3	$K = 0.46$	0.09
	q_p (cfs) = 102.0	Length 220 ft	
	T_t (hr) = 0.32	Stpns 8.2%	
		$C = 0.84$	
		$P = 0.7$	
3	Q (acre•ft) = 4.1	$K = 0.26$	0.32
	q_p (cfs) = 66.2	Length 160 ft	
	T_t (hr) = 0.25	Stpns 10.0%	
		$C = 0.02$	
		$P = 1.0$	

(8.30) For watershed 1 in Problem (8.29), estimate the interrill and rill detachment potential using the CREAMS model, Eqs. (8.118)–(8.124).

References

Alberts, E. E., Laflen, J. M., Rawls, W. J., Simanton, J. R., and Nearing, M. A. (1989). Soil component. *In* "USDA Water Erosion Prediction Project: Hillslope Profile Model Documentation," Chap. 6, NSERL Report No. 2. National Soil Erosion Laboratory, USDA-ARS, W. Lafayette, IN.

Ateshian, J. K. H. (1974). Estimation of rainfall erosion index. *Proc. Am. Soc. Civil Eng.*, **100**(IR3): 293–307.

Barfield, B. J., Moore, I. D., and Williams, R. G. (1979). Sediment yield in surface mined watersheds. *In* "Proceedings, Symposium on Surface Mine Hydrology, Sedimentology and Reclamation," University of Kentucky, Lexington, KY," pp. 83–92.

Barfield, B. J., Barnhisel, R. I., Hirschi, M. C., and Moore, I. D. (1988). Compaction effects on erosion of mining spoil and reconstructed topsoil. *Trans. Am. Soc. Agric. Eng.* **3**(2): 447–452.

Boyce, R. C. (1975). Sediment routing with sediment delivery ratios. *In* "Present and Prospective Technology for Predicting Sediment Yields and Sources," Publication ARS-S40, pp. 61–65. USDA-Agricultural Research Service.

Bradford, J. M., Farrell, D. A., and Larson, W. E. (1973). Mathematical evaluation of factors affecting gully stability. *Soil Sci. Soc. Am. Proc.* **37**(1):103–107.

Brown, L. C., and Foster, G. R. (1987). Storm erosivity using idealized intensity distributions. *Trans. Am. Soc. Agric. Eng.* **30**(2):293–307.

Bruce, R. R., Harper, L. A., Leonard, R. A., Snyder, W. M., and Thomas, A. W. (1975). A model for runoff of pesticides from small uplane watersheds. *J. Environ. Quality* **4**(4):541–548.

Bubenzer, G. D., and Jones, B. A. (1971). Drop size and impact velocity effects on the detachment of soils under simulated rainfall. *Trans. Am. Soc. Agric. Eng.* **14**(4):625–628.

Cook, H. L. (1936). The nature and controlling variables of the water erosion process. *Soil Sci. Soc. Am. Proc.* **1**:60–64.

Cooley, K. R. (1980). Erosivity "R" for individual design storms. *In* "CREAMS—A Field Scale Model for Chemicals, Runoff and Erosion from Agricultural Management Systems," Vol. III, Chap. 2, USDA-SEA Conservation Report No. 26, pp. 386–397.

Curtis, W. R. (1971). Strip mining, erosion and sedimentation. *Trans. Am. Soc. Agric. Eng.* **14**(3):434–436.

Dillaha, T. A., and Beasley, D. B. (1983). Distributed parameter modeling of sediment movement and particle size distributions. *Trans. Am. Soc. Agric. Eng.* **26**(6):1766–1777.

Dissmeyer, G. E., and Foster, G. R. (1981). Estimating the cover-management factor (C) in the universal soil loss equation for forest conditions. *J. Soil Water Conserv.* **36**:235–240.

Dissmeyer, G. E., and Foster, G. R. (1984). "A Guide for Predicting Sheet and Rill Erosion on Forest Land," Forest Service Technical Publication RA-TP6. United States Department of Agriculture.

Einstein, H. A. (1968). Deposition of suspended particles in a gravel bed. *Proc. Am. Soc. Civil Eng.* **94**(HY5):1197–1205.

Ellison, W. D., and Ellison, O. T. (1947). Soil erosion studies. IV. Soil detachment by surface flow. *Agric. Eng.* **28**(9):402–408.

Finkner, S. C., Nearing, M. A., Foster, G. R., and Gilley, J. E. (1989). A simplified equation for modeling sediment transport capacity. *Trans. Am. Soc. Agric. Eng.* **32**(5):1545–1550.

Flanagan, D. C. (1990). WEPP Second Edition, NSERL Report No. 4. National Soil Erosion Lab, United States Department of Agriculture, Agricultural Research Service, Lafayette, IN.

Fogle, A. W., Barfield, B. J. (1992). CHANNEL, a model of channel erosion by shear, scour, and channel headwall advance. Part I. Model Development. Research Report 186 Water Resources Research Institute, Univ. of Kentucky, Lexington.

Forest Service (1980). "An Approach to Water Resources Evaluation of Non-point Silvicultural Sources," a procedural handbook. EPA-600/8-80-012, Environmental Protection Agency, Washington, DC.

Foster, G. R. (1982). Modeling the erosion process. *In* "Hydrology Modeling of Small Watersheds" (Haan, Johnson, and Brakensiek, eds.), Monograph No. 5. American Society of Agricultural Engineers, St. Joseph, MI.

Foster, G. R. (1986). Understanding ephemeral gully erosion. In "Soil Conservation, Assessing the National Resources Inventory," Vol. 2. National Academy Press, Washington, DC.

Foster, G. R., and Highfill, R. E. (1983). Effect of terraces on soil loss: USLE P factors for terraces. *J. Soil Water Conser.* **38**(1):48–51.

Foster, G. R., and Lane, L. J. (1981). Modeling rill density–Discussion. *Proc. Am. Soc. Civil Eng.* **107**(IR1):109–112.

Foster, G. R., and Lane, L. J. (1983). Erosion by concentrated flow in farm fields. *In* "Proceedings, D. B. Simons Symposium on Erosion and Sedimentation," Colorado State University, Ft. Collins, CO, pp. 9.65–9.82.

Foster, G. R., and Lane, L. J. (1987). User requirements, USDA-water erosion prediction project (WEPP), NSERL Report No. 1. National Soil Erosion Research Laboratory, United States Department of Agriculture, Agricultural Research Service, West Lafayette, IN.

Foster, G. R., and Meyer, L. D. (1972a). A closed-form soil erosion equation for upland areas. *In* "Sedimentation: Symposium to Honor Prof. H. A. Einstein" (Shen, ed.), Chap. 12, pp. 12.1–12.19. Colorado State University, Ft. Collins, CO.

Foster, G. R., and Meyer, L. D. (1975). Mathematical simulation of upland erosion by fundamental erosion mechanics. *In* "Present and Prospective Technology for Predicting Sediment Yields and Sources," ARS-S-40, pp. 190–207. USDA-Agricultural Research Services.

Foster, G. R., and Wischmeier, W. H. (1974). Evaluating irregular slopes for soil loss prediction. *Trans. Am. Soc. Agric. Eng.* **17**(2):305–309.

Foster, G. R., Meyer, L. D., and Onstad, C. A. (1977a). An erosion equation derived from basic erosion principles. *Trans. Am. Soc. Agric. Eng.* **20**(4):678–682.

Foster, G. R., Meyer, L. D., and Onstad, C. A. (1977b). A runoff erosivity factor and variable slope length exponents for soil loss estimates. *Trans. Am. Soc. Agric. Eng.* **20**(4):683–687.

Foster, G. R., Lane, L. J., Nowlin, J. D., Laflen, L. M., and Young, R. A. (1980a). A model to estimate the sediment yield from field-sized areas: Development of model. *In* "CREAMS—A Field Scale Model for Chemicals, Runoff, and Erosion from Agricultural Management Systems," Vol. I, "Model Documentation," Chap. 3, USDA-SEA Conservation Report No. 26, p. 36–64.

Foster, G. R., Lane, L. J., and Nowlin, J. D. (1980b). A model to estimate the sediment yield from field-sized areas: selection of parameter values. *In* "CREAMS—A Field Scale Model for Chemicals, Runoff, and Erosion from Agricultural Management Systems," Vol. II, USDA-SEA Conservation Report No. 26, pp. 193–281.

Foster, G. R., Young, R. A., and Neibling, W. H. (1985). Sediment composition for nonpoint source pollution analysis. *Trans. Am. Soc. Agric. Eng.* **28**(1):133–146.

Foster, G. R., Weesies, G. A., Renard, K. G., Yoder, D. C., and Porter, J. P. (1991). Conservation practice factor. *In* "Estimating Soil Erosion by Water—A Guide to Conservation Planning with the Revised Universal Soil Loss Equation (RUSLE) (Renard *et al.*, eds.), Chap. 6, ARS publication, U.S. Department of Agriculture, in press.

Free, G. R. (1960). Erosion characteristics of rainfall. *Agric. Eng.* **41**(7):447–455.

Gilley, J. E., and Finkner, S. C. (1985). Estimating soil detachment caused by raindrop impact. *Trans. Am. Soc. Agric. Eng.* **28**(1):140–146.

Gilley, J. E., Woolhiser, D. A., and McWhorter, D. B. (1985a). Interrill soil erosion. I. Development of model equations. *Trans. Am. Soc. Agric. Eng.* **28**(1):147–159.

Gilley, J. E., Woolhiser, D. A., and McWhorter, D. B. (1985b). Interrill soil erosion. II. Testing and use of model equations. *Trans. Am. Soc. Agric. Eng.* **28**(1):154–159.

Gilley, J. E., Kottwitz, E. R., and Simanton, J. R. (1990). Hydraulic characteristics of rills. *Trans. Am. Soc. Agric. Eng.*, **33**(6):1900–1906.

Gregory, J. M., McCarty, T. R., Ghidey, F., and Alberts, E. E. (1985). Derivation and evaluation of a residue decay equation. *Trans. Am. Soc. Agric. Eng.* **28**(1):98–105.

Gunn, R., and Kinzer, G. D. (1949). Terminal velocity of fall for water droplets in stagnant air. *J. Meteorol.* **6**(4):243–248.

Harvey, M. D., Watson, C. C., and Schumm, S. A. (1985). Gully erosion, Technical note 366. Bureau of Land Management, U. S. Department of Interior, Denver Service Center, Denver, CO.

Hirschi, M. C., and Barfield, B. J. (1988a). KYERMO—A physically based research erosion model. I. Model development. *Trans. Am. Soc. Agric. Eng.* **31**(3):804–813.

Hirschi, M. C., and Barfield, B. J. (1988b). KYERMO—A physically based research erosion model. II. Model sensitivity analysis and testing. *Trans. Am. Soc. Agric. Eng.* **31**(3):814–820.

Hirschi, M. C., Barfield, B. J., and Moore, I. D. (1983). Modeling erosion on long steep slopes with emphasis on the rilling process, Research report No. 148. Water Resources Research Institute, University of Kentucky.

Horton, R. E. (1945). Erosional development of streams and their drainage basins, hydrological approach to quantitative morphology. *Geol. Soc. Am. Bull.* **56**:275–370.

Huang, C., Bradford, J. M., and Cushman, J. H. (1982). A numerical study of raindrop impact phenomena: The rigid case. *Soil Sci. Soc. Am. J.* **46**(1):14–19.

Huang, C., Bradford, J. M., and Cushman, J. H. (1983). A numerical study of raindrop impact phenomena: the elastic deformation case. *Soil Sci. Soc. Am. J.* **47**:855–861.

Israelson, C. E., Clyde, C. G., Fletcher, J. E., Israelson, E. K., Haws, F. W., Parker, P. E., and Farmer, E. E. (1980). Erosion control during highway construction: Manual on principles and practices, Report 221. Transportation Research Board, National Research Council, Washington, DC.

Kelly, W. E., and Gularte, R. C. (1981). Erosion resistance of cohesive soils. *Proc. Am. Soc. Civil Eng.* **107**(HY10):1211–1224.

Kilinc, M., and Richardson, E. V. (1973). Mechanics of soil erosion from overland flow generated by simulated rainfall, Hydrology paper No. 63. Colorado State University, Ft. Collins, CO.

Knisel, W. G., ed. (1980). "CREAMS—A Field-Scale Model for Chemicals, Runoff, and Erosion from Agricultural Management Systems," Conservation Research Report No. 26. U.S. Department of Agriculture.

Laflen, J. M., Foster, G. R., and Onstad, C. A. (1985). Simulation of individual storm soil loss for modeling the impact of soil erosion on productivity. *In* "Soil Erosion and Conservation" (El-Swaify *et al.*, eds.), pp. 285–295. SCS of Am., Ankeny, IA.

Lane, E. W. (1953). Progress report on studies on the design of stable channels of the Bureau of Reclamation. *In* "Proceedings, American Society of Civil Engineers, Irrigation and Drainage Division," Separate No. 280.

Lane, E. W. (1955). Design of stable channels. *Am. Soc. Civil Eng. Trans.* **120**:1234–1279.

Lane, L. J., and Foster, G. R. (1980). Concentrated flow relationships. *In* "CREAMS—A Field Scale Model for Chemicals, Runoff, and Erosion from Agricultural Management Systems," Vol. III, Supporting Documentation, Chap. 11, USDA-SEA Conservation Report No. 26, pp. 474–485.

Lane, L. J., and Nearing, M. A., ed. (1989). Water erosion prediction project: Hillslope profile model documentation, NSERL Report No. 2. USDA-ARS National Soil Erosion Research Laboratory, West Lafayette, IN.

Larson, W. E., Holt, R. F., and Carlson, C. W. (1978). Residue for soil composition. *In* "Crop Residue Management Systems" (Oshwald *et al.*, eds.), pp. 1–15. American Society of Agronomy, Madison, WI.

Leopold, L. B., Wolman, M. G., and Miller, J. P. (1964). "Fluvial Processes in Geomorphology." Freeman, San Francisco.

Lewis, S. M. (1990). PRORIL—A probabilistic physically based erosion model, Unpublished master thesis. Department of Agricultural Engineering, University of Kentucky, Lexington, KY.

Lewis, S. M., Barfield, B. J., and Storm, D. (1990). An erosion model using probability distributions for rill flow and density, Paper No. 90-2623. American Society of Agricultural Engineers, St. Joseph, MI. [In press for 1994 Transaction ASAE]

Li, R., Ponce, V. M., and Simons, D. B. (1980). Modeling rill density. *Proc. Am. Soc. Civil Eng.* **106**(IR1):63–67.

Lundgren, H., and Jonsson, I. G. (1964). Shear and velocity distribution in shallow channels. *Proc. Am. Soc. Civil Eng.* **8**(3):419–422.

Lyle, W. M., and Smerdon, E. T. (1965). Relation of compaction and other soil properties to erosion resistance of soils. *Trans. Am. Soc. Agric. Eng.* **8**(3):419–421.

Mantz, P. A. (1977). Incipient transport of fine grains and flakes of fluid—Extended shields diagram. *Proc. Am. Soc. Civil Eng.* **103**(HY6):601–615.

McCool, D. K., Brown, L. C., Foster, G. R., Mutchler, C. K., and Meyer, L. D. (1987). Revised slope steepness factor for the Universal Soil Loss Equation. *Trans. Am. Soc. Agric. Eng.* **30**(5):1387–1396.

McCool, D. K., Brown, L. C., Foster, G. R., Mutchler, C. K., and Meyer, L. D. (1989). Revised slope length factor for the Universal Soil Loss Equation. *Trans. Am. Soc. Agric. Eng.* **32**(5):1571–1576.

McCool, D. K., Foster, G. R., and Weesies, G. A. (1993). Slope length and steepness factor. *In* "Predicting Soil Erosion by Water —A Guide to Conservation Planning with the Revised Universal Soil Loss Equation (RUSLE)" (Renard *et al.*, eds.), Chap. 4, USDA-ARS Special Publication, in press.

McIssac, G. F., Mitchell, J. F., and Hirschi, M. C. (1987). Slope steepness effects on soil loss from disturbed lands. *Trans. Am. Soc. Agric. Eng.* **30**(4):1005–1013.

Meyer, L. D. (1981). How rain intensity affects interrill erosion. *Trans. Am. Soc. Agric. Eng.* **24**(6):1472–1475.

Meyer, L. D., and Monke, E. J. (1965). Mechanics of soil erosion by rainfall and overland flow. *Trans. Am. Soc. Agric. Eng.* **8**(4):572–580.

Meyer, L. D., and Wischmeier, W. H. (1969). Mathematical simulation of the process of soil erosion by water. *Trans. Am. Soc. Agric. Eng.* **12**(6):754–758, 762.e

Meyer, L. D., Johnson, C. B., and Foster, G. R. (1972). Stone and wood chip mulches for erosion control on construction sites. *J. Soil. Water Conserv.* **27**(6):264–269.

Meyer, L. D., Foster, G. R., and Romkens, M. J. M. (1975). Source of soil eroded by water from upland slopes. *In* "Present and Prospective Technology for Predicting Sediment Yields and Sources," ARS-S-40, pp. 177–179. USDA-Agricultural Research Service.

Mitchell, J. K., Moldenhauer, W. C., and Gustavson, D. G. (1983). Erodibility of selected reclaimed surface mine spoil. *Trans. Am. Soc. Agric. Eng.* **26**(5):1413–1417, 1421.

Moore, I. D., and Burch, G. J. (1986). Sediment transport capacity of sheet and rill flow: Application of unit stream power theory. *Water Resources Res.* **22**(8):1350–1360.

Mosely, M. P. (1974). Experimental study of rill erosion. *Trans. Am. Soc. Agric. Eng.* **17**(5):909–913, 916.

Mossaad, M. E., and Wu, T. H. (1984). A stochastic model of soil erosion. Int. J. Numer. Analyt. Methods Geomech. **8**:201–224.

Musgrave, G. W. (1947). The quantitative evaluation of the factors in water erosion, a first approximation. J. Soil Water Conserv. **2**(3):133–138.

Mutchler, C. K., and Carter, C. E. (1983). Soil erodibility variation during the year. Trans. Am. Soc. Agric. Eng. **26**(4):1102–1104, 1108.

Mutchler, C. K., and Larson, C. L. (1971). Splash amounts from waterdrop impact on a smooth surface. Water Resources Res. **7**(1):195–200.

Mutchler, C. K., Murphree, C. E., and McGregor, K. C. (1982). Subfactor method for computing C factors for continuous cotton. Trans. Am. Soc. Agric. Eng. **25**(2):327–332.

Nearing, M. A., Foster, G. R., Lane, L. J., and Finkner, S. C. (1989). A process-based soil erosion model for USDA-water erosion prediction project technology. Trans. Am. Soc. Agric. Eng. **32**(5):1587–1593.

Palmer, R. S. (1965). Waterdrop impact forces. Trans. Am. Soc. Agric. Eng. **8**(1):69–70.

Quansah, C. (1981). The effect of soil type, slope, rain intensity, and their interactions on splash detachment and transport. J. Soil Sci. **32**:215–224.

Quinn, M. W., and Laflen, J. M. (1983). Characteristics of raindrop throughfall under corn canopy. Trans. Am. Soc. Agric. Eng. **26**(5):1445–1450.

Rawls, W. J., Brakensiek, D. L., and Saxton, K. E. (1982). Estimation of soil water properties. Trans. Am. Soc. Agric. Eng.

25(5):1316–1320.

Renard, K. G., and Laursen, E. M. (1975). Dynamic behavior model of ephemeral stream. Proc. Am. Soc. Civil Eng. **101**(HY5):511–528.

Renard, K. G., Foster, G. R., Weesies, G. A., McCool, D. K., and Yoder, D. C. (1993a). "Predicting Soil Erosion by Water—A Guide to Conservation Planning with the Revised Universal Soil Loss Equation RUSLE," Publication ARS. U.S. Department of Agriculture, Washington, DC, in press.

Renard, K. G., McCool, D. K., Cooley, K. R., Mutchler, C. K., and Foster, G. R. (1993b). Rainfall-runoff erosivity factor (R). Chapter 2 *In* Renard, Foster and Weesies (eds) "Predicting Soil Erosion by Water—A Guide to Conservation Planning with the Revised Universal Soil Loss Equation RUSLE," Publication ARS. U.S. Department of Agriculture, Washington, DC, in press.

Rendon-Herrero, O. (1974). Estimation of washload produced on certain small watersheds. Proc. Am. Soc. Civil Eng. **100**(HY7):835–848.

Renfro, G. W. (1975). Use of erosion equations and sediment delivery ratios for predicting sediment yield. *In* "Present and Prospective Technology for Predicting Sediment Yields and Sources," Agricultural Research Service Publication ARS-S40, pp. 23–45. U.S. Department of Agriculture, Washington, DC.

Rohlf, R. A. (1981). Development of a deterministic mathematical model for interrill and rill runoff and erosion, M.S. thesis, University of Kentucky, Lexington, KY.

Rohlf, R. A. (1993). Groundwater flow and elastoplastic stress–strain modeling of channel bank failure. PhD dissertation, Department of Civil Engineering, University of Kentucky, Lexington.

Romkens, M. J., Young, R. A., Poesen, J. W., McCool, D. K., ElSwaify, S. A., and Bradford, J. M. (1993). Soil erodibility factor K. "Predicting Soil Erosion by Water—A Guide to Conservation Planning with the Revised Universal Soil Loss Equation RUSLE" (Renard, Foster, and Weesies, eds.), Chap. 3, Publication ARS. U.S. Department of Agriculture, Washington, DC, in press.

Rose, C. W. (1960). Soil detachment caused by rainfall. Soil Sci. **89**(1):28–35.

Schumm, S. A., Harvey, M. D., and Watson, C. C. (1984). "Incised Channels: Morphology, Dynamics and Control." Water Resources Publication, Littleton, CO.

Shields, A. (1936). Application of the theory of similarity and turbulence research research to the bed load movement. *In* "Anwendung der Aehnlichkeitsmechanik und der turbulenzforschung auf die geschiebebewengung," Vol. 26, pp. 5–24. Wasserbau der Preussischen Versuchsanstalt fur Wasserbau und Schiffbau, Heft, Berlin. [A translation of this paper by Ott and van Uchelen is on file with the Engineering Societies Library]

Smerdon, E. T., and Beasley, R. P. (1961). Critical tractive forces in cohesive soils. Agric. Eng. **42**(1):26–29.

Smith, D. D. (1941). Interpretation of soil conservation data for field use. Agric. Eng. **22**:173–175.

Smith, D. D., and Whitt, D. M. (1948). Estimating soil losses from field areas. Agric. Eng. **29**:394–396.

Soil Conservation Service (1978). "Water Management and Sediment Control for Urbanizing Areas." U.S. Department of Agriculture, Soil Conservation Service, Columbus, OH.

Soil Conservation Service (1983). "National Soils Handbook." Soil Conservation Service, U.S. Department of Agriculture, Washington, DC.

Stein, O. R., Roth, C. B., Moldenhauer, W. C., and Hahn, D. T. (1983). Erodibility of selected Indiana reclaimed strip mine soils. *In* "Proceedings, 1983 Symposium on Surface Mining, Hydrology, Sedimentology and Reclamation, College of Engineering, University of Kentucky, Lexington, KY."

Storm, D. E., Barfield, B. J., and Ormsbee, L. E. (1990). Hydrology and sedimentology of dynamic rill networks. I. Erosion model for dynamic rill networks, Research report No. 178. Kentucky Water Resources Research Institute, University of Kentucky, Lexington, KY.

Transportation Research Board (1980). Erosion control during highway construction. Manual on principles and practices. Report 221. National Cooperative Highway Research Program. Transportation Research Board, National Research Council, Washington, DC.

Watson, D. A., Laflen, J. M., and Franti, T. G. (1986). Estimating ephemeral gully erosion, Paper No. 86-2020. American Society of Agricultural Engineers, St. Joseph, MI.

Williams, J. R. (1975). Sediment routing for agricultural watersheds. Water Resources Bull. **11**(5):965–974.

Williams, J. R. (1976). Sediment yield prediction with Universal Equation using runoff energy factor. *In* "Present and Prospective Technology for Predicting Sediment Yields and Sources," Publication ARS-S-40. Agricultural Research Service, U.S. Department of Agriculture, Washington, DC.

Williams, J. R. (1977). Sediment delivery ratios determined with sediment and runoff models. *In* "Proceedings, Erosion and Solid Matter Transport in Inland Water Symposium," IAHS No. 122, pp. 168–179.

Williams, J. R. (1978). A sediment yield routing model. *In* "Proceedings, American Society of Civil Engineers Conference," Verification of mathematical and physical models in hydraulic engineering, pp. 662–670. American Society of Civil Engineers, New York.

Williams, J. R. (1978). A sediment graph model based on an instantaneous sediment graph. Water Resources Res. **14**(4):659–664.

Williams, J. R., and Brendt, A. D. (1972). Sediment yield computed with Universal Equation. Proc. Am. Soc. Civil Eng. **98**(HY12):2087–2098.

Wilson, B. N., and Haan, C. T. (1993). Development of a fundamentally based detachment model. Trans. Am. Soc. Agr. Eng. **36**(4):1105–1114.

Wilson, B. N., Barfield, B. J., and Moore, I. D. (1982). A hydrology and sedimentology watershed model. I. Modeling technique. Department of Agricultural Engineering, University of Kentucky, Lexington, KY.

Wilson, B. N., Barfield, B. J., and Warner, R. C. (1986). Simple models to evaluate non-point sources and controls. *In* "Agricultural Nonpoint Source Pollution: Model Selection and Application" (Giorgini and Zingales, eds.), pp 231–263. Elsevier, New York.

Wischmeier, W. H. (1975). Estimating the soil loss equations cover and management factor for undisturbed lands. *In* "Present and Prospective Technology for Predicting Sediment Yields and Sources," USDA-ARS Publication ARS-S-40, pp. 118–124. U.S. Department of Agriculture, Washington, DC.

Wischmeier, W. H. (1976). Use and misuse of the universal soil loss equation. J. Soil Water Conserv. **31**:5–9.

Wischmeier, W. H., and Smith, D. D. (1958). Rainfall energy and its relationship to soil loss. Trans. Am. Soc. Am. Geophys. Union **39**(2):285–291.

Wischmeier, W. H., and Smith, D. D. (1965). Predicting rainfall-erosion losses from cropland east of the Rocky Mountains—A guide for selection of practices for soil and water conservation, Agricultural Handbook No. 282. U.S. Department of Agriculture, Washington, DC.

Wischmeier, W. H., and Smith, D. D. (1978). Predicting rainfall erosion losses—A guide to conservation planning, Agricultural Handbook No. 537. U.S. Department of Agriculture, Washington, DC.

Wischmeier, W. H., Johnson, C. B., and Cross, B. V. (1971). A soil erodibility nomograph for farmland and construction sites. J. Soil Water Conserv. **26**:189–193.

Yalin, M. S. (1963). An expression for bed-load transportation. Proc. Am. Soc. Civil Eng. **89**(HY3):221–250.

Yang, C. T. (1972). Unit stream power and sediment transport. Proc. Am. Soc. Civil Eng. **98**(HY10):1805–1826.

Yoder, D. C., Porter, J. P., Loften, J. M., Simenton, J. R., Renard, K. G., McCool, D. K., and Foster, G. R. (1993). Cover management factor (c). *In* "Predicting Soil Erosion by Water—A Guide to Conservation Planning with the REvised Universal Soil Loss Equation (RUSLE) Publication ARS" (Renard, Foster, and Weesises, eds.), Chapter 5. U.S. Department of Agriculture, Washington, DC.

Young, R. A., and Wiersma, J. L. (1973). The role of rainfall impact in soil detachment and transport. Water Resources Res. **9**(6):1629–1636.

Zingg, A. W. (1940). Degree and length of land slope as it affects soil loss in runoff. Agric. Eng. **21**:59–64.

9

Sediment Control Structures

INTRODUCTION

From an environmental standpoint, erosion prevention and on-site control are the most desirable strategies for controlling sediment pollution. Such strategies keep soil at the source, thus maintaining its value as a natural resource and eliminating expensive cleanup requirements for sediment materials deposited off-site. Thus, on-site control is typically considered to be the first line of defense in any sediment control plan.

After exhausting on-site control as an option, the next line of defense is off-site controls. During the massive land disturbances resulting from construction, surface mining, and some types of farming operation, unprotected soil will invariably be exposed to the erosive powers of rainfall and surface runoff in spite of best control technologies. Some type of off-site control technique is typically needed as an adjunct to on-site controls; thus a well-designed sediment control plan should include both on- and off-site technologies. A summary of common control techniques is given in Table 9.1.

Any evaluation of the effectiveness of sediment control techniques should be made on a systems basis

Table 9.1 Erosion and Sediment Controls

On-site overland flow controls	**Channel erosion controls**
Chiseling	Vegetative controls
Surface roughing	Riprap
Vegetation	Gabions
Mulching	Energy dissipator
Straw	Drawdown tubes
Hay	Culverts
Cellulose	
Fiberglass	**Small structures**
Sawdust	Filter fence
Hydromulch	Straw bales
Wood cellulose	Level spreader
Wood chips	Swirl concentrator
	Vegetative filter
Contouring	Sediment traps
Strip crops	Riprap outlets
Crop rotation	Vegetated outlets
Conservation tillage	Filter fabric
Runoff diversions (swales)	**Sediment basins**
Vegetated	Dewatered
Bare soil	Permanent pond
Riprap	First flush filter
Combination	
Temporary lining	**Wetlands**

rather than by an evaluation of individual components, because of the nonlinearity of the system. For example, a sediment pond controlling runoff from a bare soil may be designed to trap 90% of the sediment. After revegetation, however, the vegetation will tend to prevent erosion of the larger particles resulting in a finer particle size distribution. The trapping efficiency may drop drastically as a result of the change in inflow particle size caused by revegetation. As shown in a later section, the reduction in trapping efficiency due to a finer particle size distribution can be demonstrated theoretically. The incoming sediment load, however, would also be reduced as a result of vegetation. Therefore, the pond sediment effluent would be reduced, even though the trapping efficiency is decreased by revegetation. This is illustrated in Example Problem 9.1. A conclusion from the above discussion is that pond effectiveness cannot be evaluated in isolation from a hydrologic and sedimentation analysis of the entire watershed.

Example Problem 9.1 Illustration of problems with trapping efficiency

Sediment yield in a given storm has been estimated to be 200 tons for bare soil and 5 tons after vegetation is established. The trapping efficiencies for the same pond are estimated to change from 90% for the bare soil to 70% for the vegetated case as a result of size distribution changes. Estimate the sediment discharged from the pond in both cases.

Solution: Sediment discharge rate would be the product of incoming load times $(1 - E)$, where E is trapping efficiency (fraction). Results are tabulated below.

Watershed status	Incoming load (tons)	Trapping efficiency (%)	Sediment outflow (tons)
Bare soil	200	90	20.0
Revegetated	5	70	1.5

Although the pond had a lower trapping efficiency when the watershed was revegetated, the sediment discharge was less. This points out the difficulty of using trapping efficiency alone as a design criteria.

One can conclude from Example Problem 9.1 that trapping efficiency alone is not an adequate measure of a sediment control system's performance. A summary of other descriptors of performance is given in Table 9.2.

Table 9.2 Summary of Measures of Performance of Sediment Control Techniques

1. *Trapping efficiency.* The easiest parameter to predict. Effective only in determining the amount of sediment trapped. Not a good measure of the impact of a control structure on the environment.
2. *Effluent concentration.* Can be either peak or average storm effluent concentration. Most difficult to predict. A small error in predicted trapping efficiency can have a large impact on predicted effluent concentration. Considers all particle sizes. Good measure of a structure's effect on total turbidity.
3. *Settleable solids.* Considers only those particles that would settle out of an Imhoff cone in 1 hr; thus it generally is only those particles 0.01 mm and larger. Good measure of a structure's impact on those particles likely to settle in downstream conveyance systems and small reservoirs.

SEDIMENT DETENTION BASINS

Introductory Comments

The most commonly used off-site control is a sediment detention basin (pond). Any size pond or reservoir can serve as a sediment detention basin; however, sediment detention basins are typically small structures. In general, basins that are designed primarily for sediment trapping have storage volumes that are equal to or less than the runoff volume in a design storm. In that situation, thermal stratification and vertical mixing are not likely to be major problems. For larger reservoirs, such is not the case. The procedures presented in this chapter apply primarily to the smaller reservoirs.

Some authorities have issued regulations requiring the sizing of sediment detention structures by a fraction of watershed area disturbed, detention storage time, size of storm to be stored, or an effluent standard. The material presented in this chapter is not oriented toward any one design requirement. Instead, procedures are presented for evaluating the performance of a sediment control system regardless of the design methodology. In addition, a discussion of the limitations of some of the design methodologies is given.

The major factors that control sediment transport through a detention basin include:

- Physical characteristics of the sediment.
- Hydraulic characteristics of the basin.
- Inflow sedimentgraph.
- Inflow hydrograph.
- Basin geometry.
- Chemistry of the water and sediment.

A discussion of these parameters is given in this section, along with a discussion of the models more commonly used for evaluating sediment trapping in ponds. Finally, special design considerations are discussed. No attempt is made to present information on structural and geotechnical design.

Sediment detention basins can be of the permanent pool or self-dewatering type as shown in Fig. 9.1. In any detention basin, a volume, $Ꝟ_s$ is included for sediment storage. A storage volume, $Ꝟ_D$, is also provided for detention storage to detain the design storm long enough to allow adequate sedimentation. In some cases, additional permanent pool volume, $Ꝟ_p$, is added above the sediment storage volume as illustrated in Fig. 9.1C; this additional volume protects stored sediment from resuspension and provides for additional detention time for first flush flow. A final flood storage volume, $Ꝟ_f$, is added for storage of sufficient water to prevent overtopping and resulting dam failure during rare events. Potential downstream damage will dictate the return period for which the flood storage volume is designed. Procedures for determining $Ꝟ_s$, $Ꝟ_D$, and $Ꝟ_f$ are introduced in subsequent sections.

Factors Affecting Pond Performance

The background information in this section is included to assist in understanding the process involved in reservoir sedimentation and to assist in developing inputs to models of pond performance. The reader interested primarily in mathematical models of pond performance might wish to read the section on modeling pond performance prior to reading this section.

Particle Size Distribution and Pond Performance

Particle size distribution effects on trapping efficiency of a reservoir are reflected primarily by particle fall velocities or settling velocities. The importance of settling velocity to reservoir trap efficiency can be illustrated by a simple model. If steady-state flow is assumed in a rectangular channel, the distance required for settling of 1 ft can be calculated for varying particle sizes as shown in Fig. 9.2. Obviously, the smaller the particle, the longer the flow path required through the reservoir to trap the sediment. The importance of particle size and particle size distribution is illustrated in Example Problem 9.2 for a simple rectangular reservoir with constant flow rate.

Example Problem 9.2. Simple illustration of impact of particle size on pond trapping

Water is being pumped into a sediment detention reservoir at a constant rate resulting in an average flow through

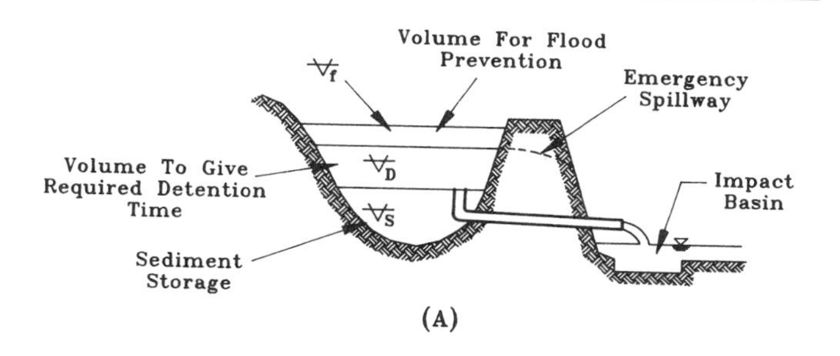

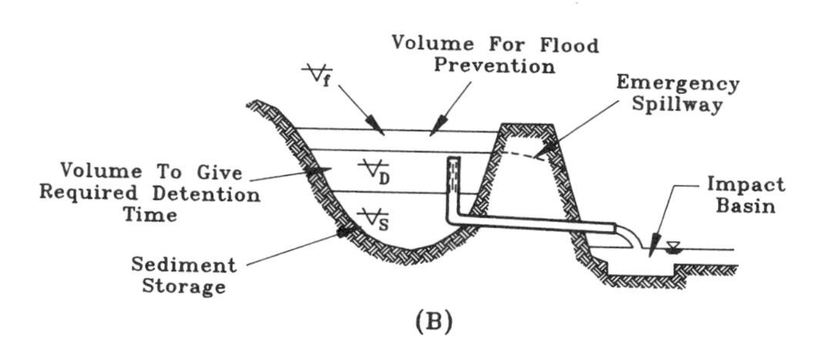

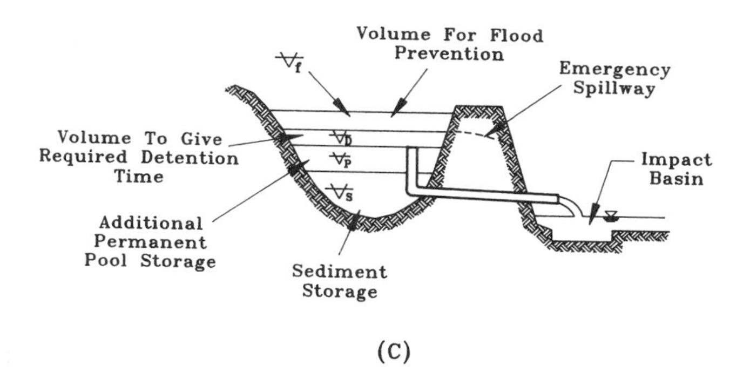

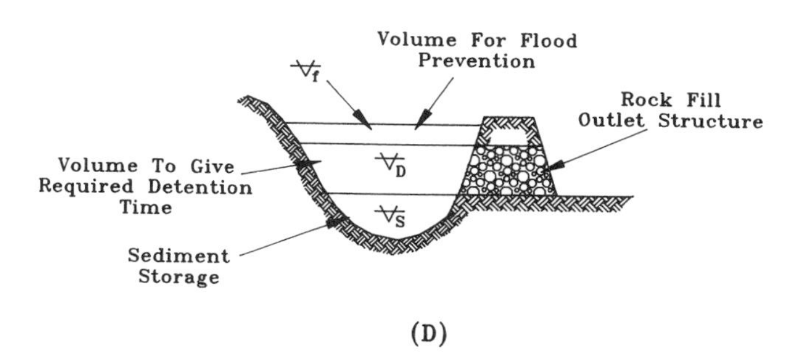

Figure 9.1 Illustration of different types of sediment detention structures. (A) Self-dewatering-type sediment detention basin with inlet to principal spillway located at top of sediment storage. A trickle tube will display similar hydraulic characteristics. (B) Self-dewatering-type sediment detention structure with slotted principal spillway. (C) Permanent pool-type sediment detention structure illustrating the additional storage volume. (D) Rock fill outlet type sediment detention structure.

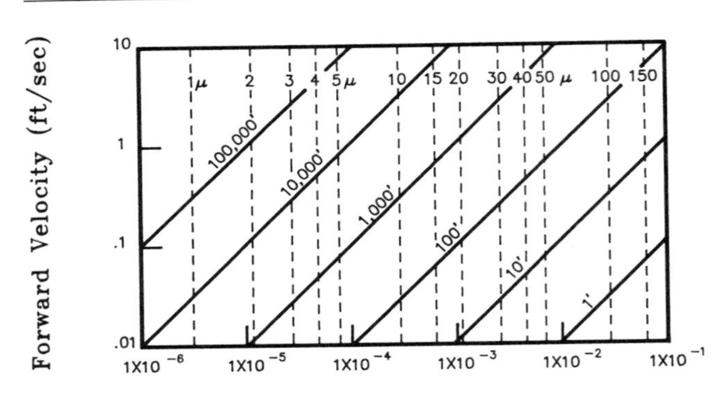

Settling Velocity (ft/sec)

Figure 9.2 Pond length for quartz particles to settle 1 ft at various forward velocities, using Stokes' law and assuming SG = 2.65 (after Bondurant *et al.*, 1975).

velocity of 1.0 ft/sec. The inlet to the outflow riser is situated so that the average settling depth is 2 ft. Estimate the flow distance required to trap 90% of the two sediments shown below.

Sediment A		Sediment B	
Particle size (mm)	% Finer	Particle size (mm)	% Finer
0.002	10	0.02	10
0.02	50	0.2	50
0.2	100	2.0	100

Solution: Figure 9.2 can be entered either with fall velocity or equivalent settling diameter. We assume a SG of 2.65 and spherical particles and enter the figure with particle size. As a rough approximation, it is assumed that 90% or more of the sediment will be trapped when the particle size corresponding to 10% finer settles from the surface to the bottom. To settle 2 ft, the flow path will be twice that for 1 ft.

For sediment A, the D_{10} is 0.002 mm (2 μm). This is the upper limit of the clay size particles. The flow path to settle 1 ft is 100,000 ft, so the flow path to settle 2 ft is 200,000 ft.

For sediment B, the D_{10} is 0.02 mm (20 μm). This the upper limit of the silt size particles. The flow path to settle 1 ft is 1000 ft, so the flow path to settle 2 ft is 2000 ft.

It should be noted that the rough estimate of trapping efficiency used here is for steady flow rates and only accounts for trapping of those particles that could settle all the way from the surface to the bottom of the reservoir in the flow through time. A portion of particles that settle only a fraction of that distance will also be trapped. This effect on trapping efficiency is considered in a later section.

It is obvious that size distribution of inflowing sediment is a major factor determining trap efficiency. Thus, any sediment pond model or design procedure should consider particle sizes of sediment reaching the reservoir, and model users should emphasize accurate estimates of inflow particle size distributions.

Another parameter of importance to trapping efficiency is flow velocity. By changing the flow velocity to 0.1 ft/sec, the distance changes in Example Problem 9.2 from 200,000 ft to 20,000 ft and 2000 ft to 200 ft. The effect of flow velocity is reflected in detention storage time and overflow rate, two concepts discussed in a subsequent section.

In addition to particle size and flow velocity, particle shape, particle density, turbulence levels, and sediment concentration have influences on settling velocities. A discussion of the effect of these parameters on settling velocity is given in Chapter 7.

Pond Hydraulic Response and Reactor Models

Hydraulic characteristics of a pond are represented by a variety of parameters including detention storage time, dead storage, and short-circuiting. Typically, hydraulic effects are defined by either a hydrodynamic model or a reactor model that desribes mixing processes. Hydrodynamic models utilize conservation of mass or momentum principles to derive partial differential equations that are usually difficult to solve. Reactor theory models divide the pond into conceptual chambers or reactors in which complete mixing, plug flow, or a combination of mixing processes is assumed. Alternately, some of the material may be assumed to bypass the pond entirely. Equations defining the hydraulic response of the pond are derived by conducting a mass balance on each reactor to account for the time variation of concentration of any tracer. Because of their relative simplicity, only reactor theory models are discussed in this chapter.

In the following discussions, models defining the hydraulic mixing of a pond are developed on the basis of the assumption of a tracer with no settling velocity. The principles developed are then applied to sediment-laden flows in a subsequent section. Reactor model discussion is presented for two reasons:

- to develop background information for pond models, and
- to illustrate numerical procedures for analyzing tracer studies of pond mixing processes.

Reactor models illustrated in this section include:

- single continuous stirred tank reactor
- continuous stirred tank reactors in series (CSTRS)
- plug flow reactors
- combinations of reactors (hybrid).

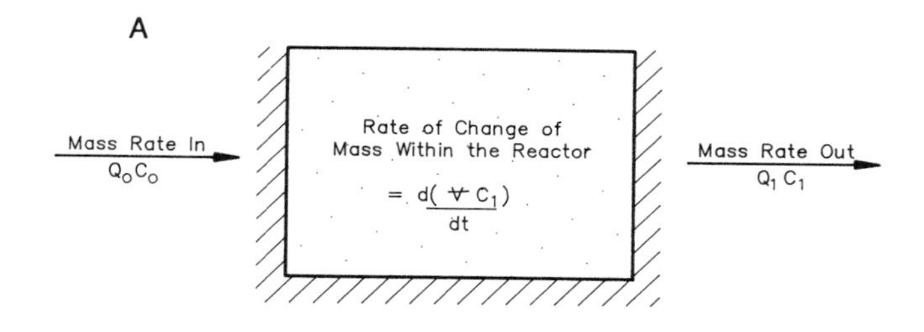

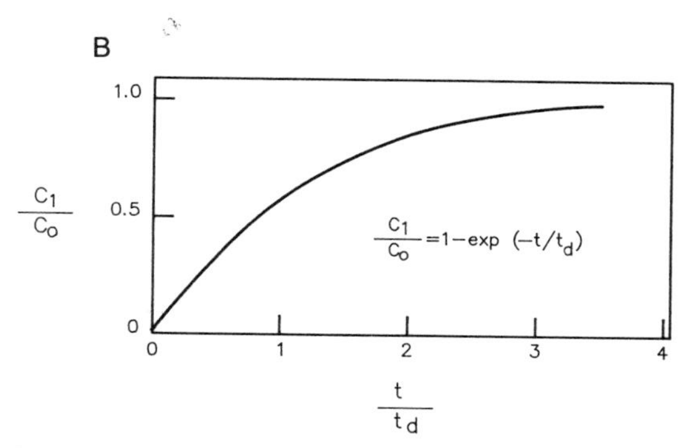

Figure 9.3 A single continuous stirred reactor. (A) Single reactor mass balance, (B) Outflow concentration for a single reactor with continuous influent concentration.

A detailed derivation of the single continuous stirred reactor is given. Final equations are presented for other reactors. Steady-state flows are assumed in development of the reactor models.

Single Continuous Stirred Tank Reactor Using steady-state assumptions, a mass balance can be conducted on the completely mixed continuous stirred reactor shown in Fig. 9.3. Assuming a tracer of concentration C_0 introduced continuously into the reactor with a constant inflow and outflow volumetric flow rate of Q, the mass balance can be written as

$$\begin{bmatrix}\text{Mass}\\\text{rate}\\\text{in}\end{bmatrix} - \begin{bmatrix}\text{Mass}\\\text{rate}\\\text{out}\end{bmatrix} = \begin{bmatrix}\text{Rate of change}\\\text{of mass in}\\\text{the reactor}\end{bmatrix} \tag{9.1}$$

or

$$QC_0 - QC_1 = d\forall C_1/dt, \tag{9.2}$$

where C_1 is concentration of tracer in the reactor and $\forall$ is reactor volume. In a continuous stirred reactor, the assumption that the reactor volume is completely mixed is typically made. This means that the outflow concentration is equal to the concentration in the reactor, as shown in Eq. (9.2).

Using the assumption of steady state, i.e., Q and $\forall$ are constant, Eq. (9.2) becomes

$$\frac{Q}{\forall}(C_0 - C_1) = \frac{dC_1}{dt} \tag{9.3}$$

or

$$\int_0^{C_1} \frac{dC_1}{C_0 - C_1} = \int_0^t \frac{Q}{\forall}\, dt. \tag{9.4}$$

For an initial concentration of zero and a constant inflow concentration C_0 starting at time zero, Eq. (9.4) can be integrated to

$$\ln \frac{C_0 - C_1}{C_0} = -\frac{Q}{\forall}t \tag{9.5}$$

or

$$\ln\left(1 - \frac{C_1}{C_0}\right) = -\frac{Q}{\forall}t. \tag{9.6}$$

The equation can be simplified by the concept of a theoretical detention time, defined as

$$t_d = \forall/Q, \tag{9.7}$$

where t_d is the time required for a flow of Q to completely displace a reactor volume $\forall$ assuming no mixing. Using this definition of theoretical detention time, Eq. (9.6) becomes

$$F(t) = \frac{C_1}{C_0} = 1 - e^{-t/t_d}, \tag{9.8}$$

where C_1/C_0 is defined as $F(t)$ to denote that C_1/C_0 is a function of t. It should be understood that F is dimensionless and is a function of the dimensionless time ratio t/t_d, but is written as $F(t)$ for simplicity.

The effluent concentration curve shown in Fig. 9.3 corresponds to Eq. (9.8). It should be noted that an assumption of idealized flow, i.e., no dead storage or short-circulating, was made in the development. These concepts are discussed subsequently.

An alternate method to continuous injection for defining mixing is the use of an instantaneous slug injection of dye at time zero. The resulting effluent concentration curve defines what is known as a residence time distribution (RTD), which is essentially a probability distribution for residence time in a reactor. Since inflow concentration is zero for all times greater than zero for an instantaneous slug injection, the mass balance given by Eq. (9.1) will become

$$-QC_1 = \frac{d\forall C_1}{dt} \tag{9.9}$$

or

$$\int_{C_i}^{C_1} \frac{dC_1}{C_1} = -\int_0^t \frac{\forall}{Q}\, dt, \tag{9.10}$$

where the initial concentration, C_i is given by

$$C_i = M/\forall \tag{9.11}$$

and M is the mass of tracer injected in the slug. Thus, Eq. (9.10) becomes

$$C_1/C_i = e^{-t/t_d}. \tag{9.12}$$

To be a probability distribution, the relationship must integrate to 1.0; thus, Eq. (9.12) must be modified to

$$\frac{C_1}{C_i t_d} = \frac{1}{t_d} e^{-t/t_d} \tag{9.13}$$

or

$$E(t) = \frac{C_1}{M/Q} = \frac{1}{t_d} e^{-t/t_d}, \tag{9.14}$$

where M is the mass in the injector slug and $C_1/M/Q$ is defined as $E(t)$. Thus, a plot of $E(t)$ versus t/t_d is a probability distribution known as the RTD, which defines the probability distribution of residence times for any fluid particle entering a reactor.

A special relationship exists between $E(t)$ and $F(t)$. From Eqs. (9.8) and (9.14), it can be seen that

$$F(t) = \int_0^t E(t)\, dt. \tag{9.15}$$

Thus, $E(t)$ is the probability density function and $F(t)$ is the cumulative distribution function for residence time. Specifically, the probability distribution for a completely mixed reactor is an exponential distribution with parameter $1/t_d$. A further discussion of the RTD concept is given in Levenspiel (1972).

Series of Continuous Stirred Tank Reactors (CSTRS) In the development of Eqs. (9.8) and (9.14) for the single continuous stirred reactor, it was assumed that the tracer was instantly mixed throughout the reactor. Obviously, a time lag is needed to allow the introduced tracer to reach the outlet. This can be accounted for by introducing the concept of a series of continuous stirred reactors as shown in Fig. 9.4. The inflow concentration for a given reactor is the effluent from the previous reactor. By conducting a mass balance in the reactors, it can be shown (Levenspiel, 1972) that the $F(t)$ and

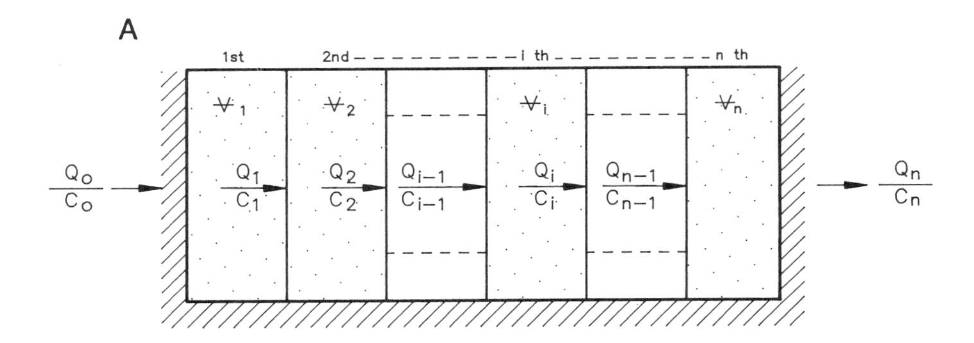

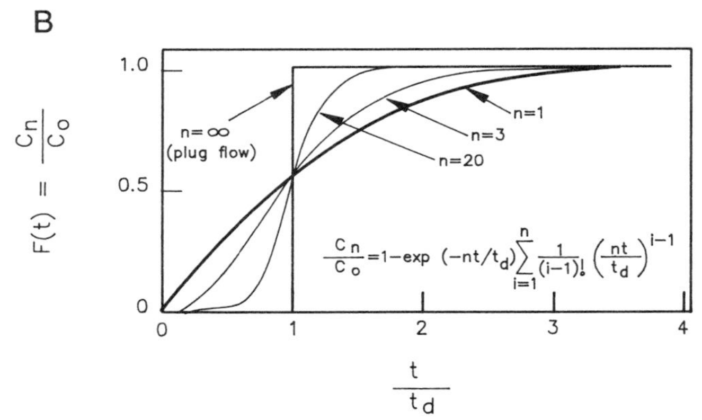

Figure 9.4 A series of continuous stirred reactors (after Wilson *et al.*, 1982). (A) n continuous shirred reactors. (B) Outflow concentration for n CSTR in series.

$E(t)$ distribution can be given by

$$F(t) = \frac{C_n}{C_0} = 1 - \exp\left(-\frac{nt}{t_d}\right) \sum_{i=1}^{n} \frac{1}{(i-1)!}\left(\frac{nt}{t_d}\right)^{i-1} \tag{9.16}$$

and

$$E(t) = \frac{C_n}{M/Q} = \left(\frac{n}{t_d}\right)\exp\left(-\frac{nt}{t_d}\right)\frac{1}{(n-1)!}\left(\frac{nt}{t_d}\right)^{i-1}, \tag{9.17}$$

where n is the number of reactors, and zero factorial is defined as one. A comparison of the RTDs for varying numbers of reactors is given in Fig. 9.4. As can be seen, the CSTRS model can be used to predict a reservoir's mixing responses, ranging from a completely mixed reactor to a nonmixed plug flow system discussed below. The number of reactors that should be used is discussed in a subsequent section on estimating reactor parameters.

Plug Flow Reactor As the number of reactors in a CSTRS system approaches infinity, the volume of each reactor ($\forall_i = \forall/n$) approaches zero. In this case, the inflow tracer is not mixed with any of the reactor volume, thus yielding a tracer concentration curve as shown in Fig. 9.4 for n of infinity. Therefore, if an inflow tracer is added to a plug flow reactor, the outflow concentration will equal the inflow concentration after a time delay required for the steady-state flow to displace the reactor volume. This time, defined earlier as the theoretical detention time, is given by $t_d = \forall/Q$. The process is illustrated in Fig. 9.5. Mathematically, the effluent concentration for a plug flow reactor would be defined by

$$\begin{aligned} F(t) &= C_1/C_0 = 0, \qquad t < t_d \\ F(t) &= C_1/C_0 = 1.0, \qquad t > t_d. \end{aligned} \tag{9.18}$$

From this discussion, it can be seen that an obvious advantage of the CSTRS model is that it can be used to evaluate mixing as well as time delay effects of plug flow.

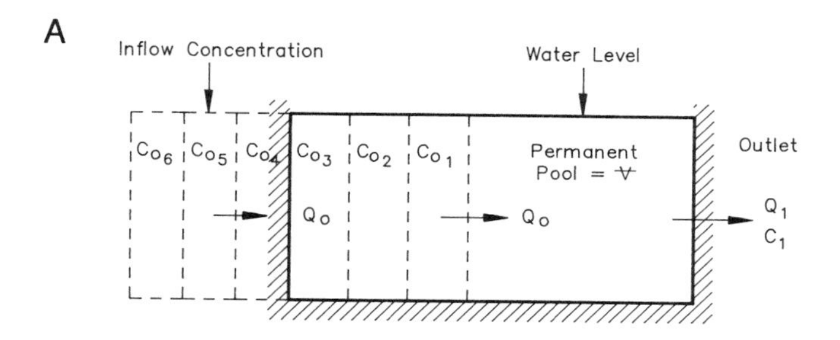

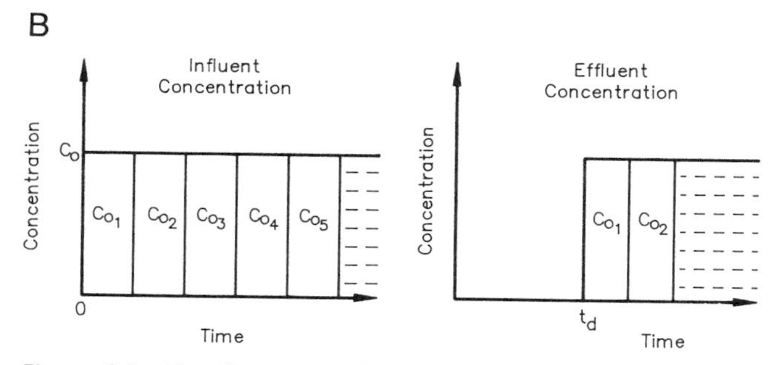

Figure 9.5 Plug flow reactor (after Wilson *et al.*, 1982). (A) Conceptualized view of a plug flow reactor. (B) Influent concentration and resulting effluent concentration for a plug flow reactor.

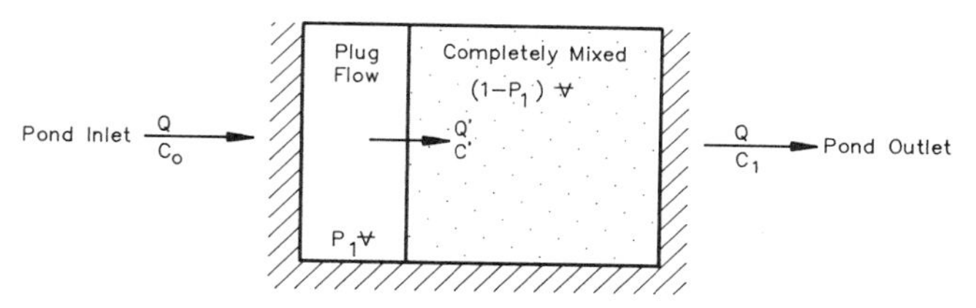

Figure 9.6 A plug flow and CSTR in series (after Wilson *et al.*, 1982). P_1, Portion of the total pond volume that is plug flow.

Hybrid Reactors An alternate to the CSTRS model is the use of hybrid reactor models, in which pure mixing reactors and plug flow reactors are combined in parallel or put in series as shown in Fig. 9.6. Other possible combinations exist. For the system shown in Fig. 9.6, the resulting steady-state effluent concentration curves would be given by

$$\begin{aligned} F(t) &= C_1 = 0, \qquad 0 < t < (1 - P_1)t_d \\ F(t) &= \frac{C_1}{C_0} \\ &= 1 - \exp\left[\frac{-[t - (1 - P_1)t_d]}{P_1 t_d}\right], \\ &\qquad t \geq (1 - P_1)t_d, \end{aligned} \tag{9.19}$$

where P_1 is the reactor fraction attributed to the completely mixed reactor and $(1 - P_1)$ is the fraction assigned to a plug flow reactor. The value for P_1 would need to be developed from tracer studies or experience, as discussed subsequently. In Eq. (9.19), the dead storage space is assumed to be zero.

Dead Storage and Short-Circuiting in Reactor Models Dead storage and short-circuiting are two concepts used to explain nonideal behavior of a reactor. Application of these concepts is illustrated with a single continuous stirred reactor as illustrated in Fig. 9.7. Equations for other reactors are summarized.

Conceptually, short-circuiting is viewed as that portion of the flow that bypasses the reactor volume entirely, thus short-circuiting directly to the outlet. Dead storage is conceptualized as that portion of the reactor that does not mix with the inflow. Using the definitions given in Fig. 9.7, f_1 is the fraction of the flow that moves through the reactor while $(1 - f_1)$ short-circuits and f_2 is the fraction of the reactor that is active while $(1 - f_2)$ is dead storage. Using these definitions and a mass balance; the effluent from the active portion of the reactor becomes

$$\frac{C_1}{C_0} = 1 - \exp\left(-\frac{f_1}{f_2}\frac{t}{t_d}\right), \tag{9.20}$$

and the total concentration C_T from the combined short-circuiting and active flow is given by

$$\frac{C_T}{C_0} = 1 - f_1 \exp\left(-\frac{f_1}{f_2}\frac{t}{t_d}\right). \tag{9.21}$$

The fractions f_1 and f_2 would need to be estimated from dye tracer studies or experience. Equations for other reactor models are given in Table 9.3.

Estimating Parameters for Reactor Models Parameter estimation must be based on experience, previous analysis, or dye tracer studies. Dye tracer studies are conducted by injecting a tracer, such as a fluorescent dye of known concentration, into the inlet and observing its concentration at the outlet. From the observed concentration curve, parameters are estimated by obtaining a best fit of the data to the particular model selected. Obviously, the model selected will have an effect on the value of the best fit parameter. Thus, if parameters are being determined for use in a pond sedimentation analysis, it is important that the reactor model used in parameter estimation be the same as that which forms the basis for the sediment or water quality model.

Table 9.3 Equations for Reactor Models with Dead Storage and Short Circuiting[a]

Model	Equation	No.
Single continuous stirred reactor	$\frac{C_T}{C_0} = 1 - f_1 \exp\left(-\frac{f_1}{f_2}\frac{t}{t_d}\right)$	(9.22)
Continuous stirred reactors in series	$\frac{C_T}{C_0} = 1 - f_1 \exp\left(-n\frac{f_1}{f_2}\frac{t}{t_d}\right)$	(9.23)
	$\sum_{i=1}^{n} \frac{1}{(i-1)!}\left(n\frac{f_1}{f_2}\frac{t}{t_d}\right)^{i-1}$	
Plug flow reactor	$\frac{C_T}{C_0} = (1 - f_1) \quad t < f_2 t_d$	(9.24)
	$\frac{C_T}{C_0} = 1, \quad t < f_2 t_d$	(9.25)
Plug flow–continuous stirred reactor	$\frac{C_T}{C_0} = (1 - f_1) \text{ for } t < t^1 = \frac{f_2}{f_1} P_1 t_d$	(9.26)
	$\frac{C_T}{C_0} = 1 - f_1 \exp\left[-\left(\frac{f_1}{f_2}\frac{1}{(1-P_1)}\frac{(t-t^1)}{t_d}\right)\right], \quad t > t^1$	(9.27)
Diffusion plug flow reactor	$E(t) = \frac{1}{2(\pi N_F N_D)^{0.5}} \exp\left[\frac{-(1-N_F)^2}{4N_F N_D}\right]$	(9.28)
	$N_F = \frac{Qt}{V} = \frac{t}{t_d}$	(9.29)
	$N_D = \frac{D}{UL}$	(9.30)

[a]Definition of terms: C_T ; total concentration of short circuited flow plus reactor flow; C_o, inflow concentration; t, time; t_d, theoretical detention time = V/Q; V, reactor volume; f_1, fraction of flow going to reactor; f_2, fraction of reactor that is active volume; $1\text{-}f_2$, fraction of reactor that is dead storage; P_1, fraction of hybrid reactor that is plug flow; D, turbulent diffusivity; U, mean flow velocity; L, length of the reactor.

Time[a] (min)	C_1 (ppm)	t/t_d	$F(t) = C_1/C_0$	$E(t) = dF/dt$	$1\text{-}C_1/C_0$
0.0	0.000	0.00	0.00	0.0029	1.00
3.0	0.006	0.16	0.01	0.0392	0.99
6.0	0.160	0.31	0.24	0.0686	0.76
9.0	0.286	0.47	0.42	0.0485	0.58
12.0	0.358	0.62	0.53	0.0304	0.47
15.0	0.410	0.78	0.60	0.0189	0.40
18.0	0.435	0.93	0.64	0.0208	0.36
21.0	0.495	1.09	0.73	0.0297	0.27
24.0	0.556	1.24	0.82	0.0238	0.18
27.0	0.592	1.40	0.87	0.0088	0.13
30.0	0.592	1.55	0.87	0.0039	0.13
33.0	0.608	1.71	0.89	0.0108	0.11
36.0	0.636	1.87	0.94	0.0108	0.06
39.0	0.652	2.02	0.96	0.0059	0.04
42.0	0.660	2.18	0.97	0.0039	0.03

[a]For all except first and last values, $E_i = (F_{i+1} - F_{i-1})/2\Delta t = (F_{i+1} - F_{i-1})/6$ min.

The plug flow model requires somewhat different estimating procedures. Assuming no short-circuiting, Griffin *et al.* (1985) proposed that dead storage for the plug flow model could be estimated from

$$\text{DS} = 1 - t_g/t_d \tag{9.31}$$

where t_g is the time to center of mass (centroid) of the $E(t)$ curve. Methods for estimating t_g are given in Example Problem 9.3.

An additional relationship for evaluating nonideal behavior is known as the plug-flow-diffusion model, which combines axial diffusion with plug flow concepts. The $E(t)$ RTD for the plug flow diffusion model is given by Eqs. (9.28)–(9.30) in Table 9.3. Assuming no dead storage or short-circuiting, Levenspiel and Smith (1957) showed that the term N_D in Eq. (9.30) could be estimated from

$$N_D = \tfrac{1}{8}\left(\sqrt{8\sigma^2 + 1} - 1\right), \tag{9.32}$$

where σ is the standard deviation of the $E(t)$ curve. Procedures for calculation of parameters from dye tracer tests are illustrated in Example Problem 9.3. Suggestions for parameters in the absence of such data are given in a subsequent section.

Example Problem 9.3. Calibration of reactor models and estimating dead storage

Results from a continuous injection dye test on a laboratory pond are given below. The constant inflow concentration was 0.68 ppm, the theoretical detention time was 19.3 min, and the steady-state flow rate was 38.1 liters/min. Estimate the best-fit parameters for the CSTRS model, plug flow model, and plug flow diffusion model.

Solution: Note: The solutions to this problem were developed with a spreadsheet, and results are rounded for inclusion in tables. Values developed with a calculator will not agree precisely.

1. *CSTRS model*. The model to be used is Eq. (9.23) in Table 9.3. A direct solution is available only for an assumption of one reactor. The problem here is to find both the dead storage and the number of reactors. To do this, Eq. (9.23) is solved for all values of time in the data set and the difference between observed and predicted values calculated. The sums of squared differences defined by

$$\text{DEV} = \sum_{j=1}^{m} (\log O_i - \log P_i)^2$$

are calculated where O_i and P_i are observed and predicted values of $(1 - C_T/C_0)$ for a given value of t/t_d and m is the number of observations available. Following Wilson *et al.* (1982), observed and predicted values are log transformed to account for the exponential nature of Eq. (9.23). After making the computations, the values of DEV are tabulated in the table below for varying values and dead storage fraction, $(1 - f_2)$, and number of reactors, n. The minimum least-squares value occurs at

$$n = 2$$
$$1 - f_2 = 15\%.$$

These would be the best-fit values for the parameters. Obviously, if the log transformation had not been made, the results would have been different. Values in the table below were calculated with base 10 logarithms.

Dead space $(1-f_2)$ (%)	Number of reactors (n)				
	1	2	3	4	5
0.0	0.985	0.529	0.271	0.223	0.376
5.0	0.788	0.317	0.139	0.250	0.631
10.0	0.599	0.149	0.117	0.473	1.193
15.0	0.425	0.052	0.254	0.979	2.195
20.0	0.276	0.060	0.624	1.896	3.829
25.0	0.168	0.224	1.337	3.406	6.375
30.0	0.124	0.621	2.553	5.785	10.250
35.0	0.176	1.366	4.513	9.446	16.082

2. *Plug flow model*. Dead storage for the plug flow model will be calculated from Eq. (9.31) or

$$\mathrm{DS} = 1 - t_g/t_d,$$

where t_g is the centroid of the area beneath the $E(t)$ curve. By definition, the centroid of the $E(t)$ is given by

$$t_g = \frac{\int_0^\infty tE(t)\,dt}{\int_0^\infty E(t)\,dt} = \frac{\int_0^\infty t(dF/dt)\,dt}{1.0} = \int_0^\infty t\,dF.$$

As illustrated in Fig. 9.8, let t' be given by

$$t' = t/t_d.$$

Hence

$$t = t' t_d$$

and

$$t_g = t_d \int_0^\infty t'\,dF = t_d t'_g.$$

From Fig. 9.8, it can be seen that t'_g can be estimated from

$$t'_g = \int t'\,dF = \int [1 - F(t')]\,dt'$$

$$= \sum [1 - F(t')]\Delta t' = \Delta t' \sum [1 - F(t')].$$

For this case

$$\Delta t' = \Delta\left(\frac{t}{t_d}\right) = \frac{3 \text{ min}}{19.3 \text{ min}} = 0.155$$

and

$$t_g = t'_g t_d.$$

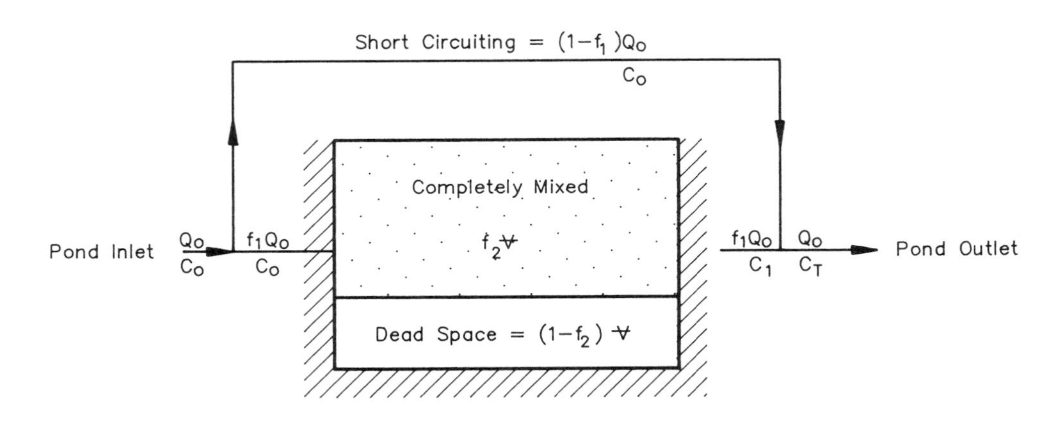

Figure 9.7 A single CSTR with short-circuiting and dead space (after Wilson *et al.*, 1982). f_1, fraction of discharge entering CSTR. f_2, fraction of completely mixed reactor that is active volume.

Values for $F(t')$ and $[1 - F(t')]$ are tabulated below.

Computation of $F(t')$ and $1 - F(t')$ for Example Problem 9.3

$t' = t/t_d$	$F(t')$	$1 = F(t')$
0.00	0.00	1.00
0.16	0.01	0.99
0.31	0.24	0.76
0.47	0.42	0.58
0.62	0.53	0.47
0.78	0.60	0.40
0.93	0.64	0.36
1.09	0.73	0.27
1.24	0.82	0.18
1.40	0.87	0.13
1.55	0.87	0.13
1.71	0.89	0.11
1.87	0.94	0.06
2.02	0.96	0.04
2.18	0.97	0.03
		$\Sigma = 5.51$

$$t'_g = \Delta t' \sum [1 - F(t')] = (0.155)(5.51) = 0.854$$

$$t_g = t'_g \times t_d = 0.854 \times 19.3 \text{ min} = 16.48 \text{ min}.$$

Using Eq. (9.31),

$$DS = 1 - \frac{t_g}{t_d} = 1 - \frac{16.48 \text{ min}}{19.3 \text{ min}}$$

$$DS = 0.146.$$

In this case, dead storage is 0.15 for the plug flow model as well as for the CSTRS model. This will not always be true.

3. *Plug flow–diffusion model.* The task here is to estimate the axial dispersion numbers, N_D and N_F. Estimating N_D from Eq. (9.32),

$$N_D = \tfrac{1}{8}\left(\sqrt{8\sigma^2 + 1} - 1\right),$$

where σ^2 is the variance of the $E(t)$ curve, given by

$$\sigma^2 = \int_0^{\infty} E(t')\left[t' - t'_{avg}\right]^2 dt'$$

$$= \sum_{i=1}^{m} E_i\left[t'_i - t'_{avg}\right]^2 \Delta t',$$

where $t' = t/t_d$ and t'_{avg} is the value of t/t_d at which $F(t/t_d) = 0.5$. Using the data set, t'_{avg} can be found by interpolation, or

$$t'_{avg} = 0.579.$$

Calculations are summarized below.

			Average over time interval			
$t'_i = (t/t_d)_i$	C_T	$F(t_i')$	$E(t_i') = \Delta F/\Delta t'$[a]	E_j[b]	t'_j[c]	$E_j(t'_j - t_{avg})^2$
0.0000	0.000	0.0	0.057	0.407	0.078	0.1002
0.1554	0.006	0.01	0.757	1.041	0.233	0.1246
0.3109	0.160	0.24	1.324	1.130	0.389	0.0408
0.4666	0.286	0.42	0.937	0.762	0.544	0.0009
0.6218	0.358	0.53	0.587	0.475	0.699	0.0068
0.7772	0.410	0.60	0.364	0.383	0.855	0.0292
0.9326	0.435	0.64	0.402	0.487	1.010	0.0905
1.0880	0.495	0.73	0.572	0.516	1.166	0.1778
1.2435	0.556	0.82	0.459	0.315	1.321	0.1734
1.3990	0.592	0.87	0.170	0.123	1.477	0.0992
1.5544	0.592	0.87	0.076	0.142	1.632	0.1575
1.7098	0.608	0.89	0.208	0.208	1.788	0.3040
1.8652	0.636	0.94	0.208	0.161	1.943	0.2995
2.0207	0.652	0.96	0.114	0.095	2.098	0.2192
2.1762	0.660	0.97	0.076	$\Sigma = 6.244$		$\Sigma = 1.826$

[a] Based on Eq. (9.15), $E(t_i') = [F(t'_{i+1}) - F(t'_{i-1})]/2\Delta t'$. Calculations were made with a spreadsheet with $F(t'_i) = C_T/C_0$ and not from rounded values in the third column.

[b] $E_j = (E_{i+1} + E_i)/2$ = average over time interval.

[c] $t'_j = (t'_{i+1} + t'_i)/2$.

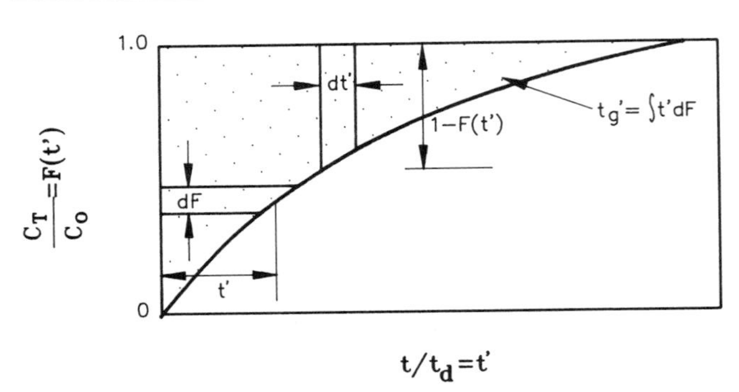

Figure 9.8 Illustration of procedures for calculating t'_g.

Check on E_i distribution

$$\sum E_i \, \Delta t' = (6.244)\left(\frac{3}{19.3}\right) = 0.97 \cong 1.0.$$

Calculating σ^2,

$$\sigma^2 = \sum (E_j)(t'_j - t'_{avg})^2 \, \Delta t'$$

$$= (1.826)\left(\frac{3 \text{ min}}{19.3 \text{ min}}\right) = 0.284$$

$$\sigma = 0.532$$

$$N_D = \tfrac{1}{8}\left(\sqrt{8\sigma^2 + 1} - 1\right) = \tfrac{1}{8}\left(\sqrt{8(0.284) + 1} - 1\right)$$

$$N_D = 0.101.$$

Estimating N_F from Eq. (9.29),

$$N_f = \frac{t}{\forall/Q} = \frac{t}{t_d}$$

$$= \frac{t}{19.3} = 0.052t.$$

4. *Summary*. Using the data supplied, three reactor models have been calibrated to the RTD. For all models the assumption of no short-circuiting, $f_1 = 1.0$, was made. Putting the calibrated parameters into the reactor models, the following equations were obtained:

CSTRS: $n = 2$; $\quad (1 - f_2) = 0.15$ or $f_2 = 0.85$

$$F(t) = 1 - f_1 \exp\left(-n\frac{f_1}{f_2}\frac{t}{t_d}\right) \sum_{i=1}^{n} \frac{1}{(i-1)!}\left(n\frac{f_1}{f_2}\frac{t}{t_d}\right)^{i-1}$$

$$= 1 - \left[1.0 \exp\left((-2)\frac{1}{(0.85)}\frac{t}{(19.3)}\right)\right]$$

$$\cdot \left[\frac{1}{0!}\left(2\frac{1}{0.85}\frac{t}{19.3}\right)^0 + \frac{1}{1!}\left(2\frac{1}{0.85}\frac{t}{19.3}\right)^1\right]$$

$$= 1 - [\exp(-0.122t)][1 + 0.122t].$$

Plug flow DS $= 0.14$; $f_2 = 0.86$,

$$F(t) = 0, \qquad t/t_d < 0.86$$
$$F(t) = 1.0, \qquad t/t_d > 0.86.$$

Plug flow diffusion $N_D = 0.101$; $N_F = 0.052t$,

$$E(t) = \frac{1}{2\sqrt{\pi N_F N_D}} \exp\left[\frac{-(1 - N_F)^2}{4N_F N_D}\right]$$

$$= \frac{1}{2\sqrt{\pi(0.052t)(0.101)}} \exp\left[\frac{-(1 - 0.052t)^2}{4(0.101)(0.052t)}\right]$$

$$= 3.89t^{-0.5} \exp\left[\frac{-(1 - 0.052t)^2}{0.0210t}\right].$$

Values are plotted in Fig. 9.9. Based on visual observation of this data set, the CSTRS model provided the best fit to the data.

Circulation Patterns and Reactor Models. In the reactor models presented in Table 9.3, dead storage is visualized as pond volume that is bypassed entirely by incoming flow. Such is not the case. An example of a circulation pattern in a pond with large dead storage is given in Fig. 9.10. Studies by Griffin *et al.* (1985) and by Noe and Barfield (1990) show that a significant portion of the pond flow bypasses the outlet and forms a recirculation pattern, moving into the area typically considered to be dead storage. This volume in the recirculation pattern is not as effective in the sedimentation process because it it initially bypassed by the inflow. At any point in time, the flow being discharged from a pond is a combination of flow that moves directly across the pond and that which has recirculated.

Detention Storage Time and Pond Performance

Detention storage time, which is a measure of flow through time in a structure, was shown in Example Problem 9.2 to have a significant impact on trapping efficiency. For steady-state systems, detention storage time was defined by Eq. (9.7) as

$$t_d = \forall/Q. \tag{9.7}$$

Physically, it is the time required for a given flow, Q, to displace the stored water $\forall$.

For nonsteady flows typical of stormwater, the definition of detention storage time is not quite so simple. Detention storage time for nonsteady flows, as discussed in Chapter 6, is the average time that a given flow resides in a pond. The concept of detention time for nonsteady flows using the plug flow concept is illustrated in Fig. 9.11 for a reservoir with and without dead storage. Example computations of plug flow are

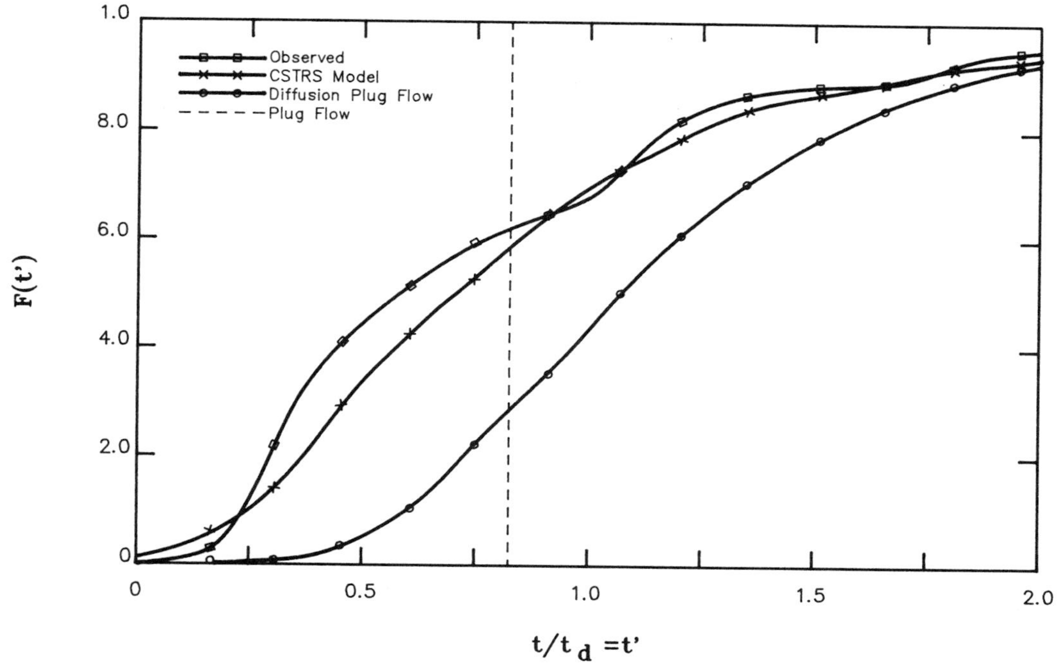

Figure 9.9 Predicted and observed values for $F(t')$ for three reactor models in Example Problem 9.3. Values of F for the diffusion plug flow model were determined by numerical integration of the $E(t)$ equation.

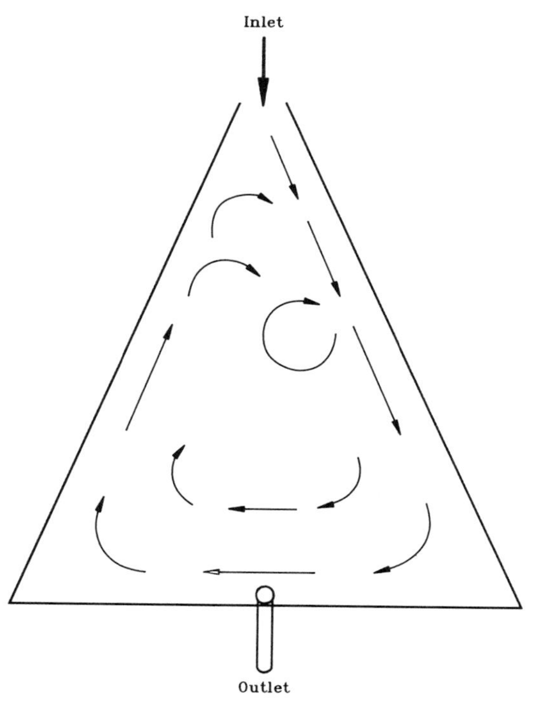

Figure 9.10 Example circulation pattern for a model pond with a large dead storage estimate from dye tracer tests (after Griffin *et al.*, 1985).

given later in Example Problem 9.7. Alternatively, detention storage time for reservoirs that are completely dewatered can be estimated by the time difference between centers of mass of the inflow and outflow hydrographs as illustrated in Chapter 6. For reservoirs with a permanent pool, such estimates from centers of mass can be misleading. Water stored in the permanent pool will have a detention time equal to the time since the last storm. Much of this water will be displaced by the incoming flow, giving a detention time greater than that estimated from hydrograph centroids or from the average detention of a plug, as illustrated in Fig. 9.11B.

Once a detention time is assigned to the permanent pool volume, the total storm detention time, using the centroid concept, becomes

$$T_d\left[(\forall_p - \forall_{ds})T_{dp} + (\forall - \forall_p + \forall_{ds})(T_{mo} - T_{mi})\right]/\forall \tag{9.33}$$

where $\forall_p$ is the volume in the permanent pool, $\forall_{ds}$ is dead storage, T_{dp} is the detention time assigned to the permanent pool, $\forall$ is the storm runoff volume, and $T_{mo} - T_{mi}$ is the time between the center of mass of the inflow and outflow hydrographs.

The problem with using the centroid concept for computing detention time for ponds with permanent pools is that of assigning an acceptable detention time

A

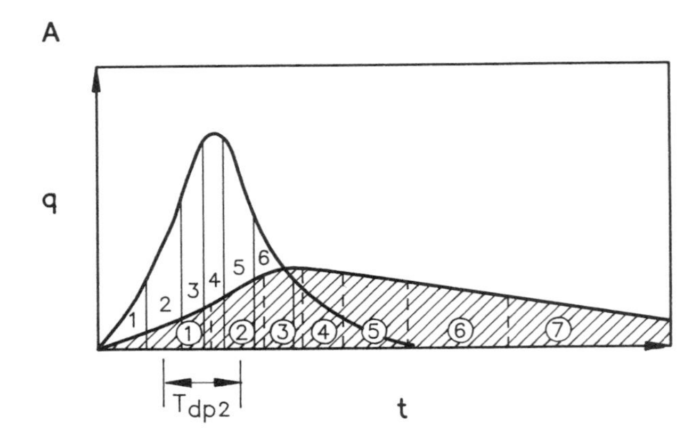

B

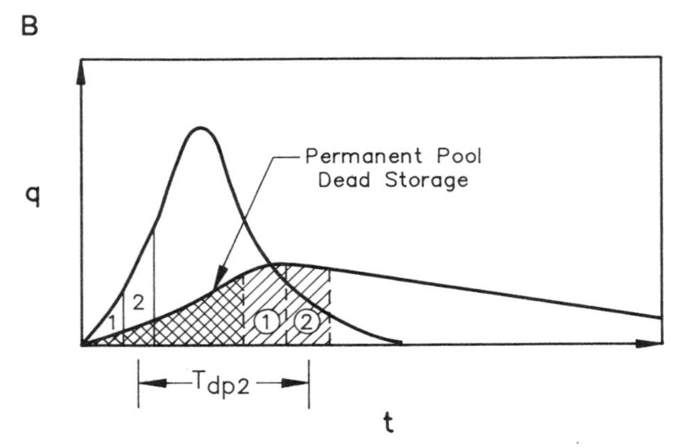

Figure 9.11 Illustration of plug flow detention time concept in a reservoir both with (A) and without (B) permanent-pool storage.

to the water displaced from the permanent pool. Studies using a computer model of reservoir sedimentology known as DEPOSITS (Ward *et al.*, 1977, 1979) indicate that a conservative value is 2.5 to 2.7 times $T_{mo} - T_{mi}$. This value is actually the detention time of the last plug of outflow in a storm.

To circumvent the problem of stored permanent-pool water versus stormwater the Environmental Protection Agency (Driscoll *et al.*, 1986) has proposed a model that evaluates trapping separately for stormwater and permanent-pool water. The EPA model is discussed in a later section.

Example Problem 9.4. Plug flow computation of theoretical detention time accounting for permanent pool

A reservoir is being designed for a detention time of 24 hr. The detention time assigned to the permanent pool is 2.5 ($T_{mo} - T_{mi}$). The runoff volume for a 10-year, 24-hr storm is 2.4 watershed in. A permanent pool is to be constructed to contain 25% of the runoff volume. What is the required time between the center of mass of the inflow and outflow hydrographs. Assume $\forall_{ds} = 0.25\forall_p$.

Solution: Calculating permanent pool and dead storage volumes.

$$\forall_p = (0.25)(2.4) = 0.6 \text{ in.}$$

$$\forall_{ds} = (0.25)(0.6) = 0.15 \text{ in.}$$

Solving for $T_{mo} - T_{mi}$. From Eq. (9.33), using $T_{dp} = 2.5$ $(T_{mo} - T_{mi})$,

$$\forall T_d = (2.4)(24) = (0.6 - 0.15)2.5(T_{mo} - T_{mi}) + (2.4 - 0.6 + 0.15)(T_{mo} - T_{mi}).$$

Hence

$$T_{mo} - T_{mi} = 18.7 \text{ hr}.$$

Using the procedures discussed in Chapter 6, a pond could be designed to give a required detention time of 18.7 hr.

Although detention storage time is frequently proposed as a design criteria, it should be pointed out that it has severe limitations as a design parameter. Particle size distribution, as shown by Example Problem 9.2, has a major effect on trapping efficiency. Even when evaluating relative trapping efficiency within a given size distribution, McBurnie *et al.* (1990) showed that trapping efficiency varied widely with surface area for a given detention time, as shown in Fig. 9.12.

Reservoir Shape and Dead Storage

Reservoir shape has a major influence on how effectively the pond volume is utilized in sedimentation. It is frequently assumed that some areas of a pond, referred to as dead storage, are bypassed and are therefore totally ineffective in the settling process. Classic examples of reservoir shapes that have large dead storage volumes are shown in Fig. 9.13.

To minimize dead storage, the Environmental Protection Agency (1976) recommends that the ratio of average length of flow path to the effective width of the reservoir ($L:W$ ratio $= L/W_e$) be greater than 2.0. W_e is an effective width calculated from

$$W_e = A/L, \tag{9.34}$$

where A is the surface area of the reservoir. The effective length of the reservoir can be increased by installing baffles in the structure as shown in Fig. 9.14.

Estimation of dead storage volume is difficult at best. As discussed earlier, dye tracer studies are typically used in the analysis. Since dead storage zones can occur in the vertical as well as horizontal plane, three-dimensional evaluations are needed to make visual observations of dead storage. Visual observations of the tracers in two dimensions can be made, but the

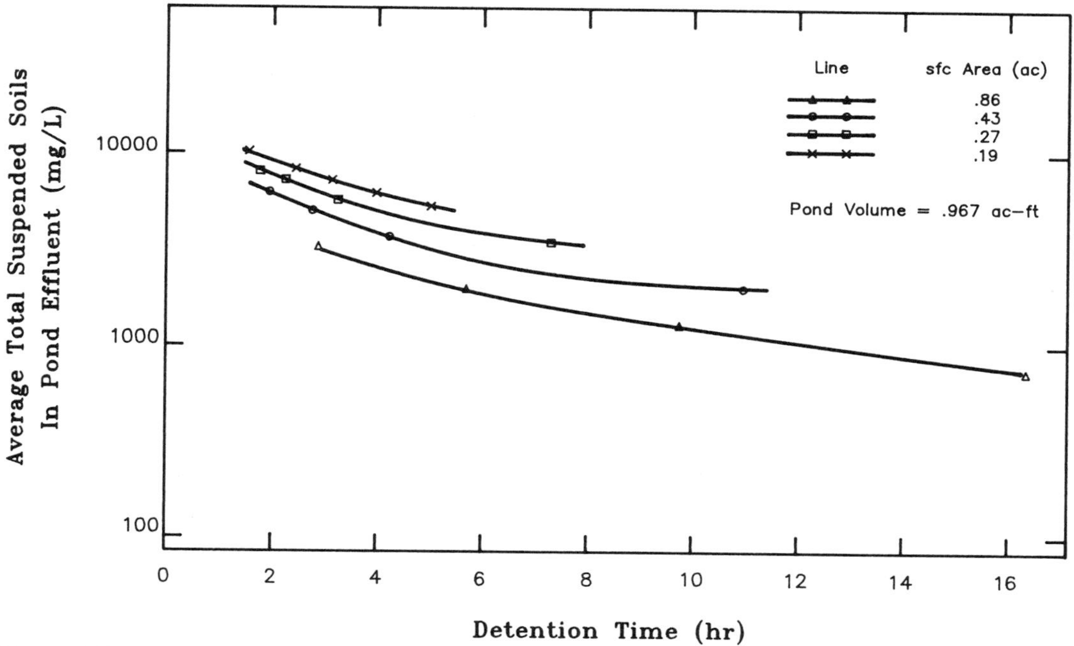

Figure 9.12 Effects of detention time on effluent TSS for a hypothetical reservoir in Maryland's Coastal Plains. Predicted values are from simulations using SEDIMOT II with ponds of varying surface area but the same volume (after McBurnie *et al.*, 1990).

three-dimensional picture of flow is not available with any reasonable expenditure of effort. Because of this problem, dead storage estimates are typically made from measured dye concentrations at the pond outlet, resulting either from a slug of dye or continuous injection of dye placed at the pond inlet. If an appropriate reactor model is used, as discussed earlier, these dye concentrations can be translated into a dead storage volume.

Dye injection studies in a model sediment pond were used by Griffin *et al.* (1985) to estimate dead storage using plug flow and CSTRS models. The results are shown in Fig. 9.15.

The results shown in Fig. 9.15 clearly show that dead storage depends on the length-to-width ratios and not inflow momentum. In general, short ponds with length-to-width ratios of less than 2.0 have a dead storage of 25%. Longer ponds with length-to-width ratios greater than 2.0 have a dead storage of 15%.

Dead storage values for the CSTRS model vary with the number of reactors assumed. Using the CSTRS reactor model, Griffin *et al.* (1985) showed that the optimum number of reactors for a best fit to the dye tracer data was 2.0.

It is important to note that dead storage calculations from tracer studies will vary widely depending on the model used. This is due partly to the fact that the term dead storage is somewhat misleading. In Fig. 9.13, it is implied that no flow enters the dead storage volume. While there may be no movement of sediment or tracer into dead storage due to the mean flows, there is a diffusion of sediment or tracer into and out of dead storage zones due to turbulent eddies and recirculation flows that may occur. These nonideal flow phenomena tend to make the dead storage volume part of the active volume of the reservoir. Diffusion and recirculating flow may or may not be accounted for, depending on the model adopted to interpret the tracer study.

Reservoir Type and Pond Performance

Reservoirs are typically classified as permanent-pool reservoirs and reservoirs without permanent pool. A permanent-pool reservoir, as shown schematically in Fig. 9.1, is one in which a permanent pool of water is kept below the crest of the principal spillway. Theoretically, the permanent pool has two major functions, to shield the deposited sediment to prevent resuspension from large storms and to provide a body of water that will be clarified by settling over the period between runoff events. In the plug flow concepts used in some models of pond performance, it is assumed that the inflow storm water displaces the clearer permanent-pool water in a first-in first-out concept. Thus, the permanent-pool water is the first flow discharged, resulting in a higher quality effluent for the first flush of stormwater.

The relative effectiveness of permanent-pool volume depends on reservoir shape. As shown earlier, ponds

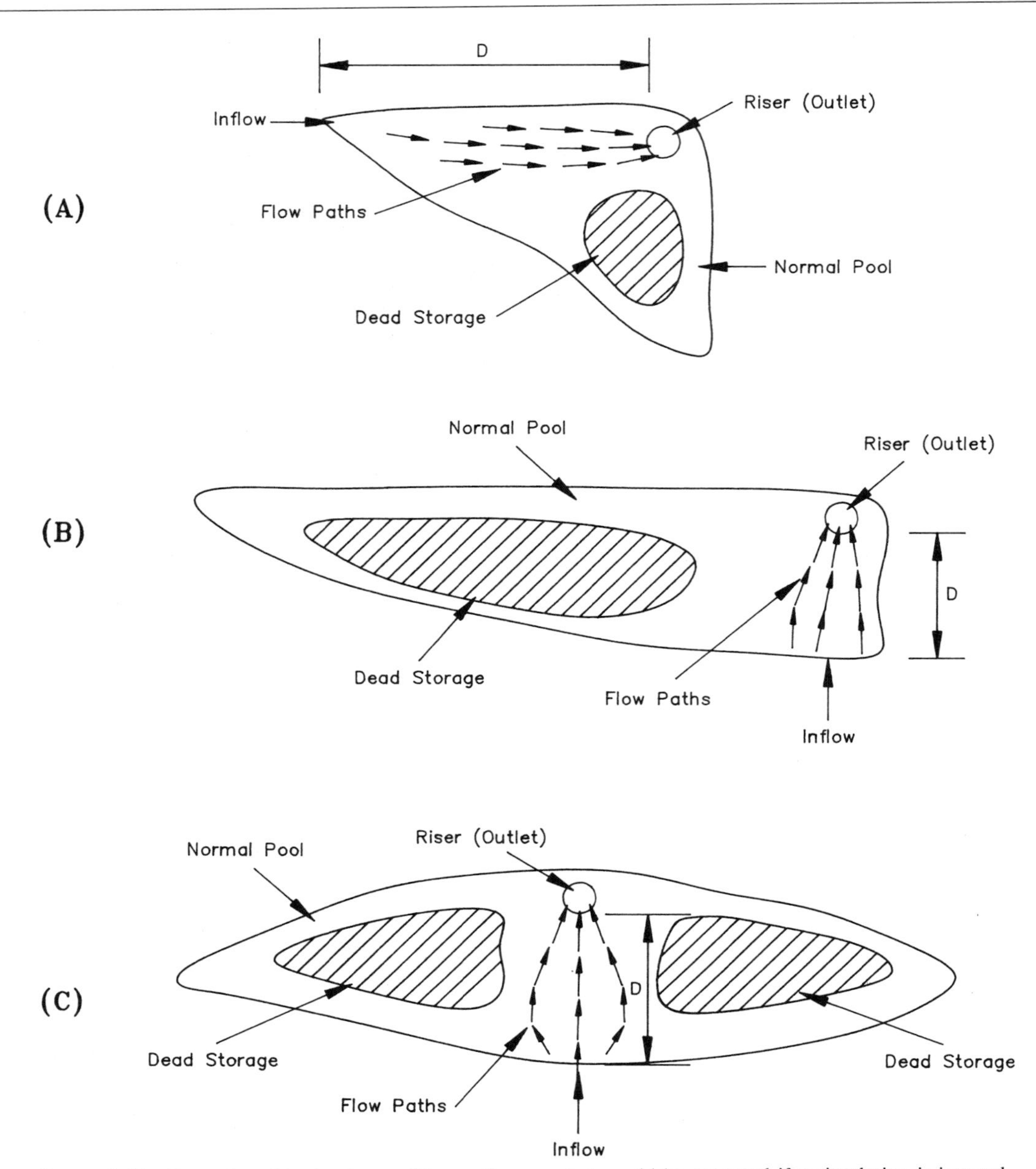

Figure 9.13 Examples of basin shapes. Flow paths are what would be expected if recirculation is ignored. Shaded areas are conceptualized dead storage areas (after Environmental Protection Agency, 1976).

with a low length-to-width ratio have higher dead storage volume (Griffin *et al.*, 1985).

Water Chemistry and Pond Performance

Some ponds are relatively clear within a few days after a runoff event and others will not be clear when observed months later. Studies indicate that runoff water chemistry is one of the major factors causing this difference (Tapp *et al.*, 1981; Tapp and Barfield, 1986; Evangelou *et al.*, 1981), primarily due to its influence on flocculation or dispersion.

Flocculation or dispersion processes can be explained by the double layer theory or by particle bridging. As shown in Chapter 7, the presence of a potential energy barrier (PEB) will prevent two particles from flocculating. If the ionic strength of cations (concentrations of cations) is increased, the double layer thickness will decrease and the magnitude of the PEB will decrease. At some point, the ionic strength will reach a value such that the barrier is eliminated and particles will flocculate. Thus, ionic strength has an important effect on flocculation.

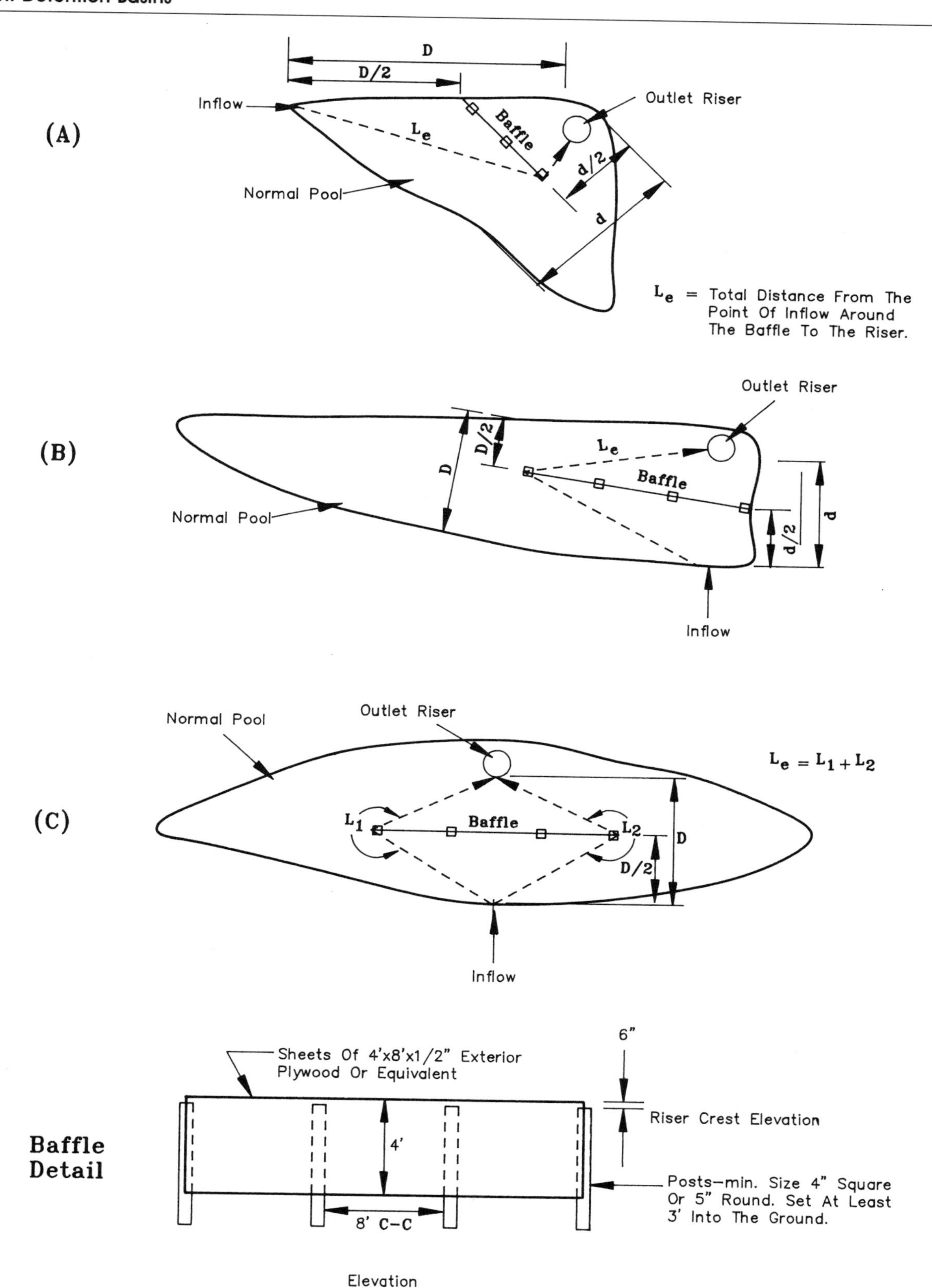

Figure 9.14 Sediment basin baffles for reducing dead storage (after Environmental Protection Agency, 1976).

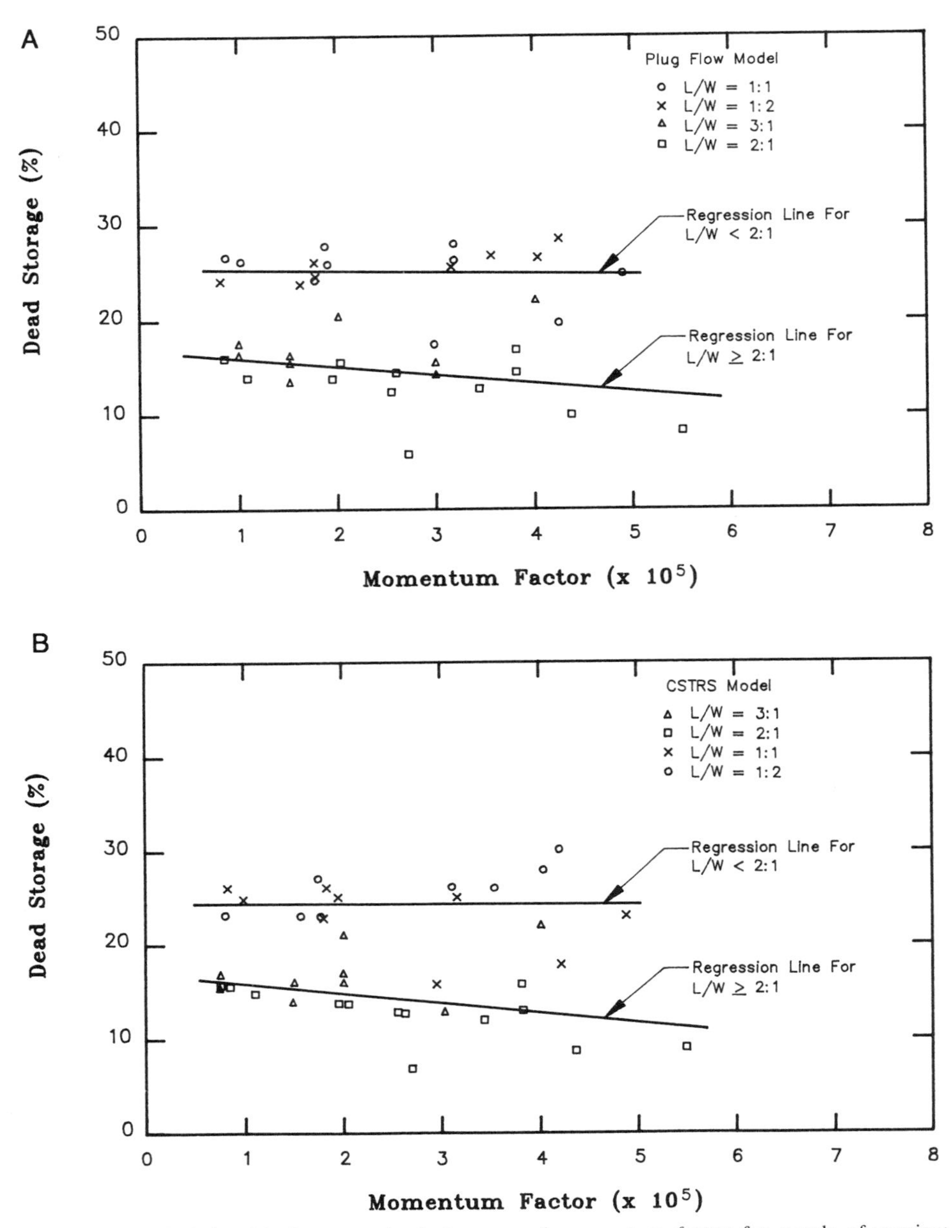

Figure 9.15 Relationship between dead storage and momentum factor for ponds of varying length-to-width ratio. Computations were made with the plug flow and CSTRS models (after Griffin *et al.*, 1985). The momentum factor is the inflow momentum divided by the weight of fluid in the pond. (A) Optimum dead storage value for plug flow model. (B) Optimum dead storage value for CSTR model with two reactors.

In addition to ionic strength, the type of absorbed cation is also important. Calcium (Ca^{2+}) and Magnesium (Mg^{2+}) are both divalent ions with a strong electric charge. Thus, when absorbed on the colloidal surface, the charge density goes from negative to positive, resulting in a thick double layer and probable flocculation. Sodium (Na^{+}) on the other hand is a monovalent ion with a weak charge and is highly hydrated (surrounded by water molecules) such that it is weakly absorbed. The charge density for Na^{+} does not change as rapidly from negative to positive as in the case of Ca^{2+}, resulting in a PEB until very high concentrations of sodium are reached. Therefore, the relative magnitude of Na^{+} to $Ca^{+} + Mg^{2+}$ is also important to the determination of flocculation potential. Two measures are used to describe the relative magnitude of these ions. One measure known as the exchangeable sodium percentage (ESP) is the fraction of absorption sites on the colloidal surface (exchange phase) filled by Na^+. The other measure, known as the sodium absorption ratio (SAR), is the fraction of ions in solution that are sodium. For soils that are heavy clays, with high numbers of exchange sites (high cation exchange capacity), the ESP and SAR are highly related. In soils that are composed of kaolonitic colloids, the number of exchange sites with a double negative charge are low as compared with heavy clays. As a result; the monovalent Na^+ ion is more easily absorbed on the kaolonite; thus the SAR and ESP values are typically not equal.

In an effort to predict the occurrence of flocculation or dispersion, Evangelou *et al.* (1981) proposed that the thickness of the double layer could be used as a prediction parameter. Using electrical conductivity as a measure of ionic strength, Evangelou *et al.* analyzed a kaolonitic soil with a SAR of less than 5.0 and a solution phase composed primarily of $Ca^{2+} + Mg^{2+}$. They proposed that a measure of the double layer thickness could be estimated from

$$\mathrm{DLT} = \frac{1}{[(\mathrm{EC})(0.014)]^{1/2}}, \tag{9.35}$$

where DLT is thickness in arbitrary units and EC is the electroconductivity of the water in millimhos per centimeter. For DLT values less than 15, Evangelou *et al.* indicated that flocculation would occur. Data for other clay types and other dissolved chemicals were not evaluated, limiting the use of Eq. (9.35) for other situations.

Sediment Scour in Ponds

When water flows over deposited sediment, the lift and drag forces will attempt to move particles out of the bed. The critical velocity that will cause motion in a rectangular settling tank with constant flow rate was analyzed by Camp (1946) using Shield's parameters, or

$$V_{\mathrm{H}} = \left[\frac{8k(\mathrm{SG}-1)gd}{f}\right]^{1/2}, \tag{9.36}$$

where V_{H} is the critical velocity causing scour of a particle of diameter d, g is acceleration of gravity, SG is particle specific gravity, f is the Darcy–Weisbach friction factor, and k is Shield's parameter discussed in Chapter 7. Typical values for f are 0.02 to 0.03 for settling chambers and k is 0.4 for granular materials and 0.06 for particles that have cohesive properties. Units on g and d will dictate the units on V_{H}. In practice, the actual velocity would be designed to be less than V_{H} to prevent scour.

Relationships more complex than Eq. (9.36) are available, but require complex computer models for their execution. Wilson and Barfield (1985, 1986a) describe such a model, known as BASIN, which evaluates sediment resuspension using the Einstein equation. BASIN is discussed further in a subsequent section.

Modeling Pond Performance: Theoretically Based Predictors

Reservoir and pond sedimentation is not a new field. The first theories on reservoir trapping efficiency were probably developed by Seddon (1989). Since that time, many models have been developed. In the following section a brief overview of the most commonly used models is given, starting first with steady-state and moving then to non-steady-state models.

Steady-State Overflow Rate Models—Quiescent Flow

An analysis of the trapping efficiency of rectangular basins with steady-state inflows and outflows can be made using settling velocities, flow rates, and surface areas. Based on the trajectory of particles through a settling basin, a critical settling velocity, V_{c}, that will just allow a particle to settle to the bottom in its trajectory through the basin can be selected as shown in Fig. 9.16. The critical settling velocity can be given by

$$V_{\mathrm{c}} = D/T, \tag{9.37}$$

where D is the depth of the basin and T is the flow through time. If it is assumed that flow is quiescent (turbulent free) and that no resuspension occurs, then all particles with a settling velocity greater than V_{c} will be trapped. Based on geometry, it can be shown that the fraction of particles trapped with a settling velocity

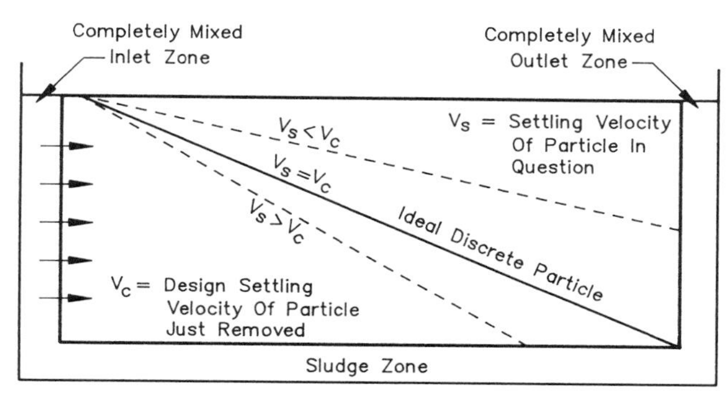

Figure 9.16 Illustration of sediment flow trajectories in an ideal rectangular sediment pond.

V_s less than V_c is given by

$$F = V_s/V_c. \tag{9.38}$$

Equations (9.37) and (9.38) can be converted to the familiar overflow rate equation by noting that the flow through time, T, is given by effective length of the basin L divided by the flow through velocity V, or

$$F = \frac{V_s}{V_c} = \frac{V_s}{D/L/V} = \frac{V_s}{DV/L}. \tag{9.39a}$$

Further modification can be made to yield

$$F = \frac{V_s}{WDV/(LW)}, \tag{9.39b}$$

where W is the width of the chamber. The term WDV is simply the discharge Q and LW is the surface area A. Hence

$$F = \frac{V_s}{Q/A} = \frac{V_s}{V_c} \leq 1.0, \tag{9.40}$$

which is the familiar overflow rate equation. The quantity Q/A is equal to V_c and is known as the overflow rate. Equation (9.40) illustrates that the trapping fraction for a given particle is independent of the depth of the settling basin under steady-state conditions. It should again be noted that the following assumptions were made:

- quiescent flow
- no resuspension of sediment
- rectangular shaped reservoir
- steady flow.

It was also implicit that inlet and outlet zones were completely mixed and that settling was discrete particle settling as described in Chapter 7.

If X represents the fraction of particles finer than a given size and dX represents a differential element of X, then the total trapping efficiency for a basin integrated over all sizes is

$$E = \int_0^\infty F\,dX.$$

If the fraction of particles with settling velocity less than V_c is given by X_c, the total trapping efficiency of a basin can be calculated from

$$\begin{aligned} E &= (1 - X_c) + \int_0^{X_c} \frac{V_s}{V_c}\,dX \\ &\cong (1 - X_c) + \sum_{i=1}^{n} \frac{V_{si}}{V_c}\,\Delta X_i. \end{aligned} \tag{9.41}$$

The use of Eq. (9.41) is illustrated in Example Problem 9.5.

The effluent size distribution can also be estimated with the overflow rate concept. Since the fraction

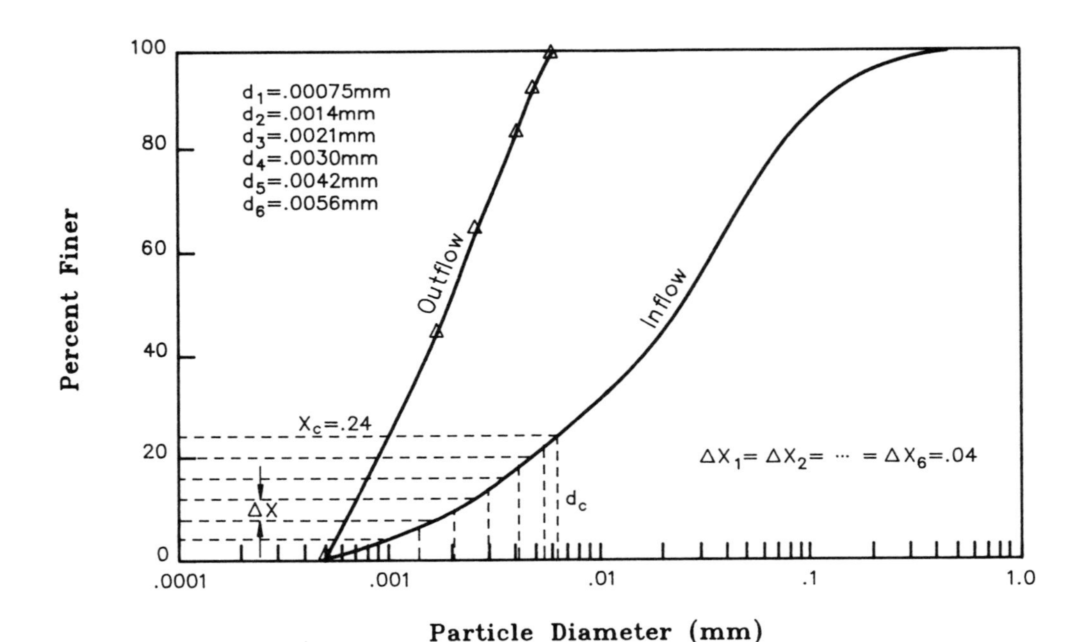

Figure 9.17 Size distributions used for Example Problem 9.5.

trapped of a given size is V_{si}/V_c, the fraction discharged is $1 - V_{si}/V_c$. If the total mass in the storm is M_s, the mass discharged for a given size, i, is $(1 - V_{si}/V_c)\,\Delta X_i\,M_s$, and the fraction of total discharged mass of size i is

$$\Delta FF_{o,i} = \frac{(1 - V_{si}/V_c)\,\Delta X_i\,M_s}{\Sigma(1 - V_{si}/V_c)\,\Delta X_i\,M_s} = \frac{(1 - V_{si}/V_c)\,\Delta X_i}{\Sigma(1 - V_{si}/V_c)\Delta X_i}. \tag{9.42}$$

The fraction finer than size j, $FF_{o,j}$, is simply the sum of $\Delta FF_{o,i}$ for all smaller particles.

This approach has the inherent limitations of the overflow rate concept. Computations are illustrated in Example Problem 9.5.

Example Problem 9.5. Calculating trapping efficiency and effluent size distribution with the overflow rate concept

A rectangular reservoir has a steady inflow and outflow rate of 5 cfs and a surface area of 1.0 acre. Calculate the overflow rate. If sediment inflow has a size distribution as given in Fig. 9.17, calculate the trapping efficiency of the structure and estimate the effluent size distribution.

Solution:

1. Calculating the overflow rate. The overflow rate is given by

$$V_c = \frac{Q}{A} = \frac{5.0 \text{ ft}^3/\text{sec}}{(1.0 \text{ acre})(43{,}560 \text{ ft}^2/\text{acre})}$$
$$= 1.15 \times 10^{-4} \text{ ft/sec}.$$

Assuming a temperature of 68° F, the corresponding equivalent diameter sphere corresponding to V_c can be determined from Eqs. (7.5) (or (7.6) for larger particles) or

$$d_c = \left(\frac{V_c}{2.81}\right)^{1/2} = \left(\frac{1.15 \times 10^{-4}}{2.81}\right)^{1/2} = 0.0064 \text{ mm}.$$

2. Calculating trapping efficiency. From Fig. 9.17, the fraction of particles smaller than d_c is 0.24. The evaluation of the integral in Eq. (9.41) is given in columns 1–5 in the table below. From Eq. (9.41),

$$E = (1 - X_c) + \sum_{i=1}^{n} \frac{V_{si}}{V_c}\Delta X_i$$
$$= (1 - 0.24) + 0.0634 = 0.8234.$$

Hence the trapping efficiency is approximately 82%.

3. Calculating effluent size distribution. The fraction of effluent smaller than particle size j, $\Delta FF_{o,j}$, is calculated from Eq. (9.42) and the fraction finer than size j is

$$FF_{o,i} = \sum_{i=1}^{j} \Delta FF_{o,j},$$

where j is the number of particle sizes smaller than i. The computations are summarized in columns 1–3 and 7–9 in the table below and the computed effluent size distribution plotted on Fig. 9.17. The size corresponding to $FF_{o,j}$ is the larger value of the size range in column 1.

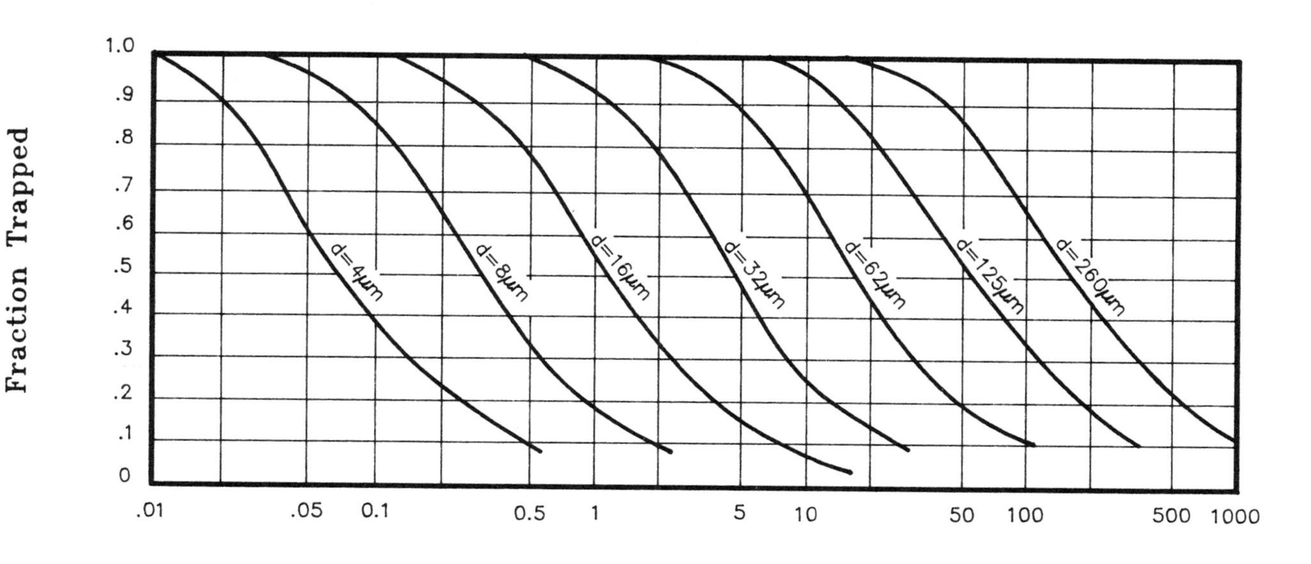

Figure 9.18 Sedimentation as a function of overflow rate for various particle sizes (1 μm = 10^{-3} mm).

Tabulations for Example Problem 9.5

Particle size range[a] (mm)	Diameter d^a (mm)	ΔX^a	V_{si}[b] (ft/sec)	$(V_{si}/V_c)\Delta X_i$[c]	$(1 - V_{si}/V_c)\Delta X_i$[d]	Column 6/Σ (Column 6)[e]	$FF_{o,j}$ Σ (Column 7)[f]	Particle size for (Column 8)[g]
0.0001–0.00095	0.00075	0.04	1.58×10^{-6}	0.0005	0.039	0.222	0.222	0.00095
0.00095–0.0016	0.0014	0.04	5.51×10^{-6}	0.0019	0.038	0.216	0.438	0.0016
0.0016–0.0025	0.0020	0.04	1.24×10^{-5}	0.0043	0.036	0.205	0.643	0.0025
0.0025–0.0036	0.0030	0.04	2.53×10^{-5}	0.0088	0.031	0.176	0.819	0.0036
0.0036–0.0045	0.0042	0.04	4.96×10^{-5}	0.0173	0.023	0.131	0.950	0.0045
0.0045–0.0064	0.0056	0.04	8.81×10^{-5}	0.0306	0.009	0.050	1.000	0.0064
		Σ = 0.24		Σ = 0.0634	Σ = 0.176	Σ = 1.001		

[a]Taken from Fig. 9.17 for each diameter.
[b]Min of $V_{si} = 2.81\ d^2$ and Eq. (7.6).
[c]$V_{si}\Delta X_i/1.15 \times 10^{-4}$.
[d]$(1 - V_{si}/1.15 \times 10^{-4})\Delta X_i$.
[e]$\Delta FF_{o,i}$ = (Column 6)/$\Sigma(1 - V_{si}/1.15 \times 10^{-4})\Delta X_i$ = (Column 6/Σ Column 6) = Column 6/0.176.
[f]$FF_{o,j} = \Sigma \Delta FF_{o,i}$.
[g]Upper limit of size range in column 1.

From Example Problem 9.5, it is clear that particle size and overflow rate are important parameters affecting the performance of sediment ponds. This is further illustrated in Fig. 9.18. Since the horizontal axis (overflow rate) in Fig. 9.18 decreases as area increases, an increase in surface area, for a given particle size and discharge, will increase trapping. Thus, for steady-state quiescent flow, the optimum rectangular reservoir is one that maximizes surface area subject to the constraint that particles not be suspended. This will typically minimize depth. Also, from Example Problem 9.5 and Fig. 9.18, it is obvious that discrete particles 4 μm and smaller are very difficult to trap with any reasonably sized basin. As is shown in following sections, the same conclusion can be drawn about turbulent flow and non-steady-state flow.

Steady-State Overflow Rate Models: Turbulent Flow

A theoretical analysis of turbulent flow is much more complex than that required for quiescent settling as a result of upward diffusion of sediment by turbulence. An analysis of turbulent diffusion of sediment should start with the equilibrium case in which the upward turbulent diffusion of sediment of a given size d just equals the rate at which particles are settling (see Fig. 9.19A), or

$$\rho V_s C = -\rho \varepsilon \frac{\partial C}{\partial z},$$

where ρ is density of water, V_s is settling velocity of a particle diameter of size d, C is the concentration (mass sediment/mass water) of particles of size d, and ε is the turbulent diffusivity for sediment of size d. Simplifying

$$\varepsilon \frac{\partial C}{\partial z} + V_s C = 0. \tag{9.43}$$

To solve Eq. (9.43), it is necessary to have an estimate of the eddy diffusivity ε. By assuming that the eddy diffusivity for momentum and sediment are equal and by using the logarithmic velocity profile to describe velocity, the eddy diffusivity becomes

$$\varepsilon = kU_*\left(1 - \frac{z}{D}\right)z, \tag{9.44}$$

where U_* is the so-called shear velocity, D is depth of flow, and k is von Karmon's constant given by 0.4 in clear water. For uniform flow in open channels, shear velocity is defined as

$$U_* = \sqrt{\tau_0/\rho} = \sqrt{gDS}, \tag{9.45}$$

where τ_0 is shear on the channel bottom and S is the energy gradient.

Using Eq. (9.44), Eq. (9.43) can be integrated to yield

$$\frac{C}{C_a} = \left[\frac{D/z - 1}{D/a - 1}\right]^{z_*}, \tag{9.46}$$

where C_a is the concentration of d-sized particles

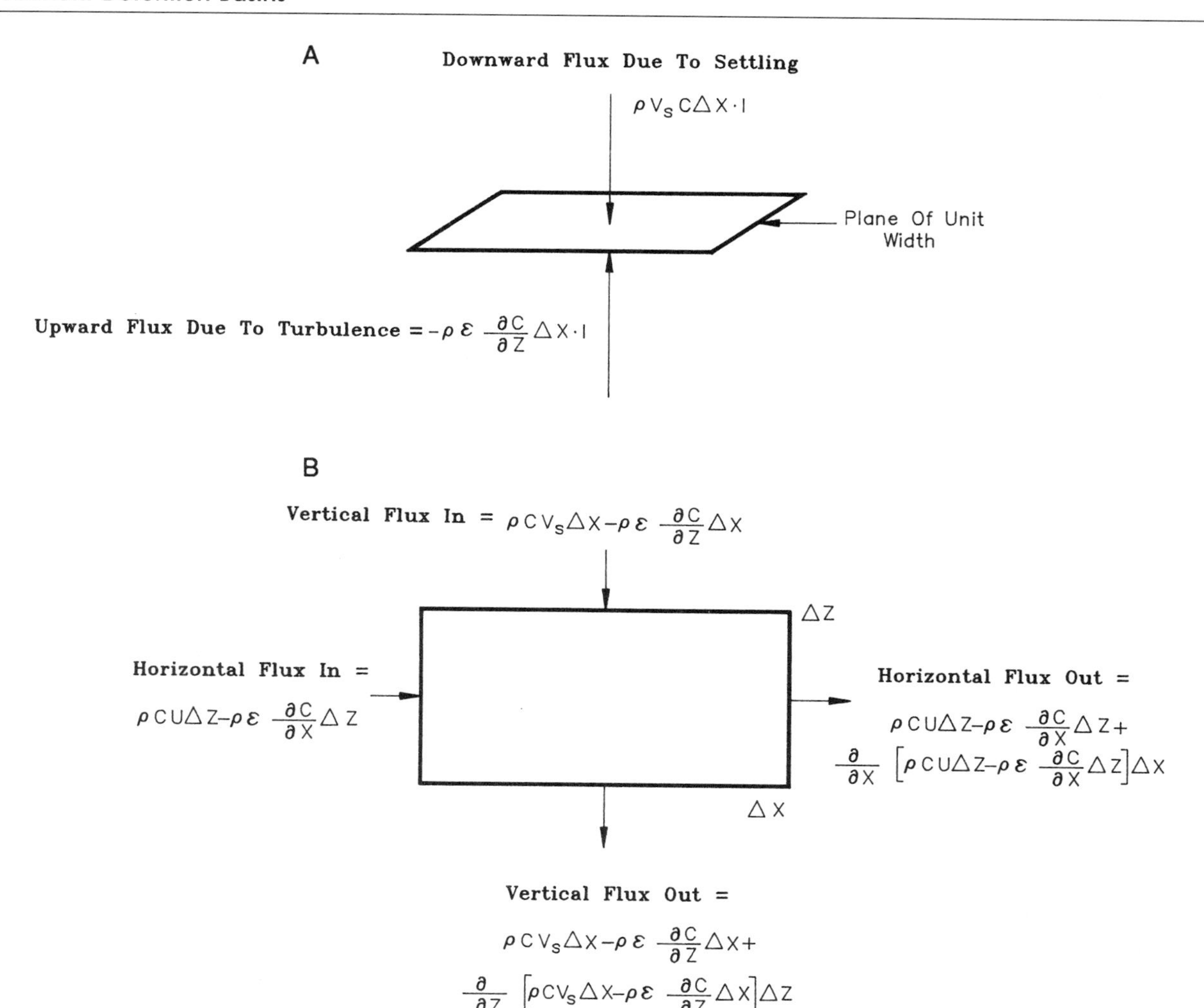

Figure 9.19 Schematics for deriving sediment trapping efficiency equation for turbulent flow. (A) Schematic showing equilibrium sediment movement. The minus sign on the upward flux due to turbulence accounts for the fact that sediment diffuses from high to low concentration. (B) Control elements for deriving non-steady-state sediment transport equation.

measured at a distance a above the channel bed and

$$z_* = V_s/kU_* . \tag{9.47}$$

Equation (9.46) describes the case where the rate of resuspension of sediment at the channel bed equals the rate of settling; hence there are no changes of concentration in the horizontal flow direction. In any small sediment pond this will not be the case; therefore an equation that considers changes in the mean flow direction, must be derived. Such a derivation will require a mass balance on a fluid element.

A schematic of a mass balance of a fluid element is shown in Fig. 9.19B. In such a mass balance, the rate of inflow of mass into the element minus the rate of outflow of mass from the element must equal the rate of change of mass in the element. Performing the mass balance using the terms in Fig. 9.19B results in

$$\begin{aligned}\frac{\partial \rho\, C\, \Delta x\, \Delta z}{\partial t} &= -\frac{\partial}{\partial x}\rho C U \Delta x\, \Delta z + \frac{\partial}{\partial x}\rho\varepsilon\frac{\partial C}{\partial x}\Delta x\, \Delta z \\ &\quad - \frac{\partial}{\partial z}\rho C V_s\, \Delta x\, \Delta z + \frac{\partial}{\partial z}\rho\varepsilon\frac{\partial C}{\partial z}\Delta x\, \Delta z .\end{aligned} \tag{9.48}$$

After dividing by $\rho\, \Delta x\, \Delta z$ and assuming that U and V_s are constant,

$$\begin{aligned}&\frac{\partial C}{\partial t} + U\frac{\partial C}{\partial x} + V_s\frac{\partial C}{\partial z} \\ &\quad = \frac{\partial}{\partial x}\varepsilon\frac{\partial C}{\partial x} + \frac{\partial}{\partial z}\varepsilon\frac{\partial C}{\partial z} .\end{aligned} \tag{9.49}$$

The general solution to Eq. (9.49) will require the use of numerical procedures. Dobbins (1944) developed an analytical solution for the special case in which ε is a constant by using what is known as a product series solution. Camp (1946) manipulated Dobbins solution for the special case in which there is no scour to obtain

$$F = 1 - 8B_1^2 e^{B_1} \sum_{n=1}^{\infty} \left[\frac{\alpha_n^2 H_n e^{B_2}}{(B_1^2 + \alpha_n^2 + 2B_1)(B_1^2 + \alpha_n^2)^2} \right], \tag{9.50}$$

where

$$B_1 = \frac{V_s D}{2\varepsilon} \tag{9.51}$$

$$B_2 = -(B_1^2 + \alpha_n^2)\left(\frac{V_s}{2B_1 V_c}\right), \tag{9.52}$$

and $\alpha_1, \alpha_2, \alpha_3, \ldots, \alpha_n$ are the real positive roots of the transcendental equation

$$2\,\mathrm{Cot}\,\alpha = \frac{\alpha}{B_1} - \frac{B_1}{\alpha}. \tag{9.53}$$

In Eq. (9.50), H_n is $+1$ when α is in the first and second quadrants and -1 when α is in the third and fourth quadrants. Dobbins verified his equations experimentally using carefully controlled laboratory studies.

Numerical computations using Eq. (9.50) require the use of up to 40 terms of the infinite series. To simplify the analysis procedures, Camp (1946) modified Eq. (9.50) to predict trapping efficiency of a rectangular reservoir and developed the graphical solution shown in Fig. 9.20. To use Fig. 9.20, it is necessary to have an estimate of ε. Camp (1946) proposed that ε could be estimated from

$$\varepsilon = 0.75 U_* D, \tag{9.54}$$

where U_* is shear velocity and D is depth of the rectangular reservoir. In this case, ε is the average diffusivity over the entire flow cross section as opposed to the point value given earlier in Eq. (9.44). Brown (1950) proposed that U_* be estimated from a modification of Manning's equation, or

$$U_* = \frac{nV\sqrt{g}}{1.5D^{1/6}}, \tag{9.55}$$

where V is the average horizontal velocity in the reservoir and n is Manning's roughness. Using Eq. (9.54)

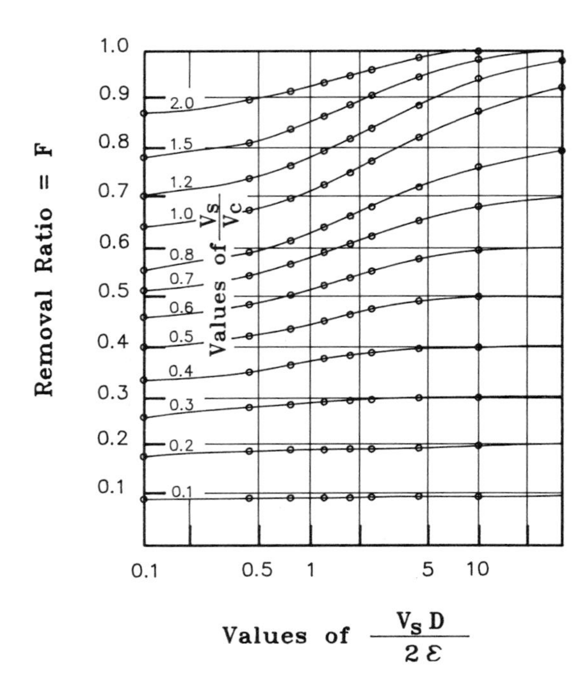

Figure 9.20 Sediment trapping efficiency of a rectangular reservoir with steady-state turbulent flow (after Camp, 1946).

and (9.55), the horizontal axis of Fig. 9.20 becomes

$$\frac{V_s D}{2\varepsilon} = \frac{10 V_s D^{1/6}}{nV\sqrt{g}}. \tag{9.56}$$

Wilson and Barfield (1986b) summarize other methods for estimating turbulent diffusivities in sediment ponds. Details are given in Appendix 9A.

The effect of turbulence on basin trap efficiency is apparent from Fig. 9.20. When turbulence levels are low, the value of $V_s D/2\varepsilon$ is large and trapping efficiency approaches that of quiescent settling given in Eq. (9.40) as V_s/V_c. When turbulence levels are high, ε is large, $V_s D/2\varepsilon$ is small, and trapping efficiency is reduced. From Fig. 9.20 it can also be seen that trapping efficiency for small values of V_s/V_c is relatively unaffected by turbulence.

Since trapping efficiency in the turbulent flow condition is dependent on $V_s D/2\varepsilon$, it is not independent of flow depth as in the case of quiescent settling. However, it is interesting to note how a doubling of depth would affect trapping. Since Eq. (9.56) indicates $V_s D/2\varepsilon$ is proportional to $D^{1/6}$, then changing the depth by a factor of 10 would cause a change in $V_s D/2\varepsilon$ of $10^{1/6}$ or a factor of 1.5 (a 50% change). From Fig. 9.20, a 50% change in $V_s D/2\varepsilon$ would change F by less than 0.05 in all cases. This weak dependence of F on D prompted Camp (1946) to infer that the optimum settling tank design is one that maximizes surface area and minimizes depth, without causing

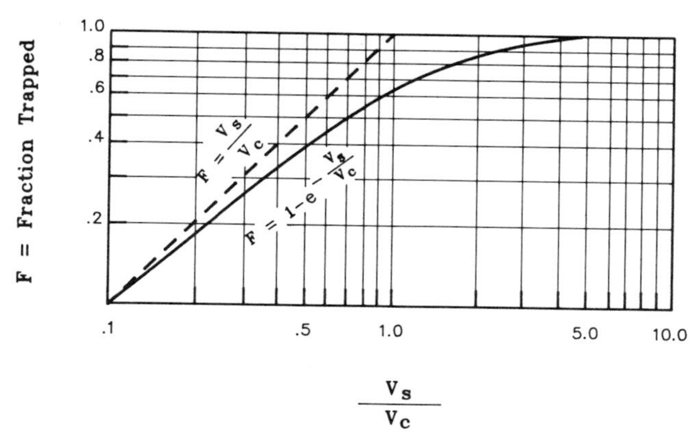

Figure 9.21 Trap efficiency versus ratio of settling velocity to overflow rate for a high-turbulence model (after Chen, 1975).

scour. This is the same conclusion reached earlier for quiescent flow.

Chen (1975) defined a high turbulence flow condition as one in which $V_s D/2\varepsilon$ is equal to 0.01 and determined values of trapping efficiency versus V_s/V_c for the condition. These values are plotted in Fig. 9.21 along with those predicted for quiescent settling. Chen utilized Vetter's (1940) equation and proposed that the turbulence values can be approximated by

$$F = 1 - \exp(-V_s/V_c). \tag{9.57}$$

Total trapping efficiency for quiescent settling was found earlier by integrating Eq. (9.40) over all size ranges to yield

$$E = (1 - X_c) + \int_0^{X_c} \frac{V_s}{V_c}\, dx.$$

The total trapping efficiency for fully turbulent flow can be found by integrating Eq. (9.40) over all size ranges to yield

$$E = 1 - \int_0^1 \exp\left(-\frac{V_s}{V_c}\right) dx. \tag{9.58}$$

These models are illustrated in Example Problem 9.6.

Example Problem 9.6. Comparison of the turbulent and quiescent flow models

Estimate the trapping efficiency of the reservoir in Example Problem 9.5 using the turbulent flow procedures of Chen and those of Dobbin and Camp. Compare the results to the quiescent flow model (overflow rate) of Camp. Assume that the reservoir is 5 ft deep and 10 ft wide.

Solution: From Example Problem 9.5,

$V_c = Q/A = 1.15 \times 10^{-4}$ ft/sec
$d_c = 0.0064$ mm
$Q = 5.0$ cfs
$A = 1.0$ acre.

First, the turbulence parameter of Dobbins must be calculated. In the discussion of reservoir scour in this chapter following Eq. (9.36), it was stated that the friction factor for a reservoir is 0.02 to 0.03. Using a value of 0.03, it can be shown that Manning's n can be given by (see Eqs. (4.21)–(4.23))

$$n = 1.49 \frac{R^{1/6}}{\sqrt{8g/f}},$$

where R is the hydraulic radius. Since R for a rectangular section is

$$\frac{WD}{2D + W} = \frac{(10)(5)}{2 \times 5 + 10} = \frac{50}{20} = 2.5,$$

then

$$n = \frac{1.49(2.5)^{1/6}}{\sqrt{(8 \times 32.2)/0.03}} = 0.019.$$

The forward flow velocity is

$$V = \frac{Q}{A_{cs}} = \frac{5}{(5 \times 10)} = 0.1 \text{ fps},$$

where A_{cs} is the cross-sectional area perpendicular to flow. Dobbin's parameter is, therefore,

$$\frac{V_s D}{2\varepsilon} = \frac{10 V_s D^{1/6}}{nV\sqrt{g}} = \frac{10(5)^{1/6} V_s}{0.019(0.10)\sqrt{32.2}} = 1.213 \times 10^3 V_s.$$

Calculations for predicting trapping efficiency by Dobbin's and Camp's procedures as well as those of Chen are summarized in the Table 9.4.

Using the results from Example Problem 9.5 for quiescent settling along with calculations from Table 9.4, the following comparison can be made:

Procedure	Trapping efficiency
Quiescent settling Eq. (9.41)	0.823
Dobbin–Camp (Fig. 9.20)	0.794
High-turbulence model (Eq. (9.58)	0.789

Comments: The differences in predicted trapping efficiencies are relatively small. As expected, the quiescent value is the highest, the fully turbulent value (Chen) the lowest, and the Dobbin–Camp model intermediate. An explanation for this lack of a major difference can be developed from Fig. 9.21.The maximum deviation between quiescent settling and fully turbulent flow occurs at $V_s/V_c = 1.0$. For this problem,

Table 9.4 Tabulations for Example Problem 9.6

Particle size range[a] (mm)	Average diameter d_i^a (mm)	ΔX_i^a	V_{si}^b (ft/sec)	Fully turbulent model [Eq. (9.57)] V_{si}/V_c^c	$F_i =$ 1-exp $(-V_{si}/V_c)$	$F_i\Delta X_i$	Dobbin and Camp model (Fig. 9.20) $\frac{10V_{si}D^{1/6}}{nVg^{1/2}d}$	F_i^e	$F_i\Delta X_i$
0.110–0.500	0.170	0.10	8.1×10^{-2}	706	1.0	0.100	98.5	1.0	0.10
0.072–0.110	0.088	0.10	2.17×10^{-2}	189	1.0	0.100	26.0	1.0	0.10
0.048–0.072	0.058	0.10	9.45×10^{-3}	82	1.0	0.100	11.5	1.0	0.10
0.035–0.048	0.040	0.10	4.50×10^{-3}	39.1	1.0	0.100	5.5	1.0	0.10
0.026–0.035	0.030	0.10	2.53×10^{-3}	22.0	1.0	0.100	3.1	1.0	0.10
0.17–0.026	0.022	0.10	1.36×10^{-3}	11.8	1.0	0.100	1.6	1.0	0.10
0.009–0.017	0.012	0.10	4.05×10^{-4}	3.52	0.97	0.097	0.49	1.0	0.10
0.004–0.009	0.0065	0.10	1.19×10^{-4}	1.03	0.64	0.064	0.14	0.65	0.065
0.002–0.004	0.0034	0.10	3.25×10^{-5}	0.28	0.24	0.024	0.04	0.25	0.025
<0.002	0.0012	0.10	4.05×10^{-6}	0.04	0.04	0.004	0.00	0.04	0.004
						$\Sigma = 0.789$			$\Sigma = 0.794$

[a]From Fig. 9.17.
[b]$V_{si} = 2.81\ d_i^2$. This overestimates V_s for larger particles, but refined estimates would also yield $F = 1.0$ for these particles.
[c]$V_{si}/1.15 \times 10^{-4}$.
[d]$1.23 \times 10^3\ V_{si}$.
[e]From Fig. 9.20.

only 10% of the particles fell in that range. If 30 to 40% had been in that range, the difference between the quiescent and fully turbulent model would have been 10 to 15% which is a significant difference.

Variable Flow Rate—Plug Flow Model

Overflow rate models have found widespread acceptance in the design of settling tanks for water and sewage treatment systems. They can also be applied to the analysis of sediment ponds that have constant inflows as might well be the case with pumping from quarries or deep mines. The majority of the sediment ponds, however, must handle surface runoff and inflow rates that vary over several orders of magnitude during a runoff event.

One variable flow rate procedure that has found widespread acceptance in the analysis and design of sediment ponds is a computer model known as DEPOSITS. The DEPOSITS Model (*de*tention *p*erformance *o*f *s*ediments *i*n *t*rap *s*tructures) was developed to study the sedimentation process in small reservoirs in the hopes of providing an insight into the development of a better design method for sediment structures (Ward *et al.*, 1977, 1979). DEPOSITS is one of the pond options in SEDIMOT II (Wilson *et al.*, 1982).

To make the model sufficiently general to be applicable to most sediment basins, the flow within the basin is idealized by the plug flow concept (see Fig. 9.11). Plug flow assumes delivery of the flow on a first-in, first-out basis and allows no mixing between plugs. Although this concept does not account for short-circuiting or turbulent flow, provision for a correction factor to account for these phenomena has been incorporated into the model.

Computational procedures for DEPOSITS are summarized in Ward *et al.* (1977) and are illustrated in Example Problem 9.7. A description of the model computational procedures follows, but can probably best be understood after working through the example.

In DEPOSITS, an outflow hydrograph is generated first by continuity routing through the reservoir and a total cumulated flow hydrograph determined for both inflow and outflow. Cumulative flow is calculated by summing the area under the hydrograph. The inflow and outflow hydrographs are then divided into plugs of equal volume by equally dividing the cumulative inflow and outflow hydrograph into equal volumes. Once the hydrograph is divided into plugs, corresponding plugs are identified on the inflow and outflow hydrograph, the detention time for each plug is calculated, and an average depth of flow and surface area for the plug residence time is determined. The volume of each plug is divided into four layers for sedimentation computations. Using settling velocities calculated from Stokes Law, the amount of sediment in each layer is calculated at the end of the plug detention time. Particles are considered to be trapped as soon as they reach the basin bed. Withdrawal can be specified as uniform, or distributed among the four layers at the user's option. The outflow concentration for each plug of flow is calculated from the mass of sediment and the mass of water in the plug as it exits the pond.

Calculations of effluent size distributions follow similar procedures to those of the overflow rate method illustrated in Example Problem 9.5, but must be computed for each layer. Thus, procedures are more complex. The procedure is developed in Example Problem 9.9. Further details are given in Ward *et al.* (1977) and Wilson *et al.* (1982).

Example Problem 9.7 Illustration of DEPOSITS model computations

Using the plug flow concept, divide the hydrograph in Fig. 9.22A into 15 plugs. For plug 10, determine the average flow depth and plug detention time. Divide the average depth into four layers, and calculate the particle size that will settle out of plug 10 during the plug detention time. Finally, assuming uniform withdrawal, calculate the effluent concentration for plug 10 assuming that the inflow concentration for plug 10 is 19,270 mg/liter and the inflow particle size distribution is given in Fig. 9.23B.

Solution:

1. *Dividing hydrographs into plugs.* Subdivision into plugs is done from a cumulative inflow and outflow hydrograph where

$$\mathrm{IC}(t) = \frac{\int_0^t I(t)\,dt}{\int_0^\infty I(t)\,dt} = \frac{\sum_{i=1}^{n} I(t_i)\,\Delta t_i}{Q_{\mathrm{TR}}},$$

where IC(t) is relative cumulative inflow at time t, $I(t)$ is inflow rate at any time t, and Q_{TR} is the total inflow volume. Using this approach, the inflow and outflow hydrographs in Fig. 9.22A are transformed to the cumulative hydrographs in Fig. 9.22B. To divide the hydrograph into 15 plugs, the vertical axis is simply divided into 15 equal increments. The tenth plug is emphasized in Fig. 9.22B.

2. *Plug detention time.* Detention time for the tenth plug is shown in Fig. 9.22B as the time interval between plug inflow and plug outflow, or

$$T_{\mathrm{D},10} = 7.62 - 4.0 \text{ hr} = 3.62 \text{ hr or } 217.2 \text{ min}.$$

A

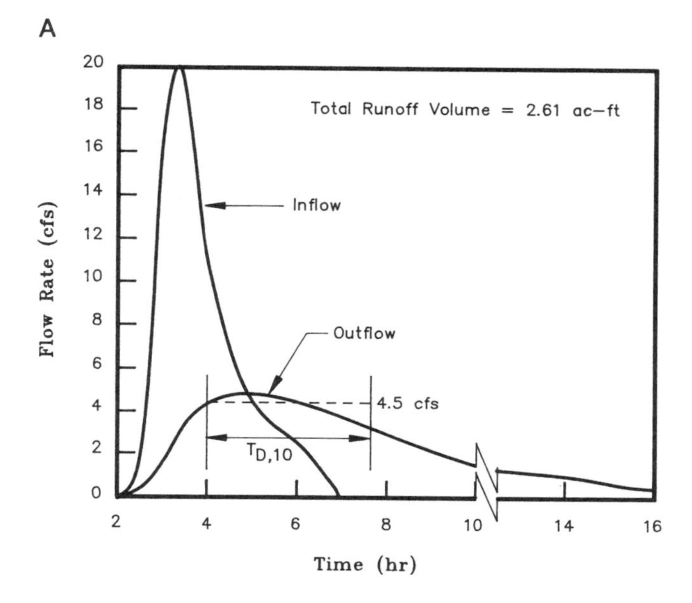

B

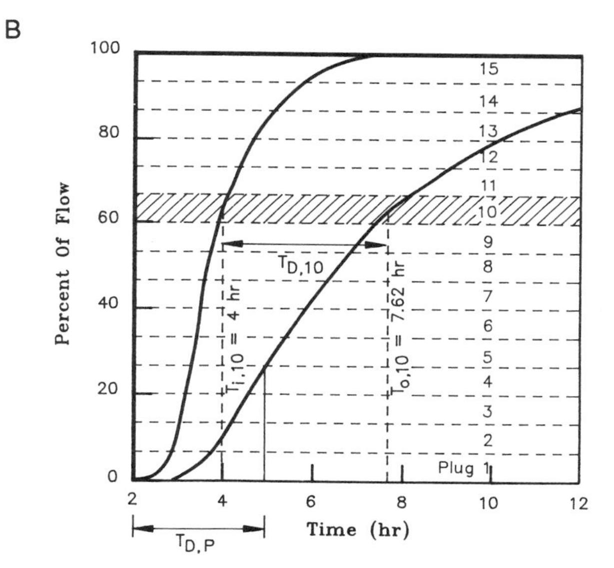

Figure 9.22 Hydrographs for Example Problems 9.7 and 9.8. (A) Inflow and outflow hydrographs, (B) Cumulative flow graphs.

3. *Average depth for plug* 10. From Fig. 9.22B, the inflow and outflow times for the tenth plug are

$$t_{i,10} = 4.00 \text{ hr}$$

$$t_{\mathrm{o},10} = 7.62 \text{ hr}.$$

Using the average reservoir depth versus time curve in Fig. 9.23 and $t_{i,10}$ and $t_{\mathrm{o},10}$ given above, the average depth can be determined by numerically integrating the depth–time curve and dividing by 3.62 hr to be

$$D_{10} = 1.51 \text{ ft}.$$

A

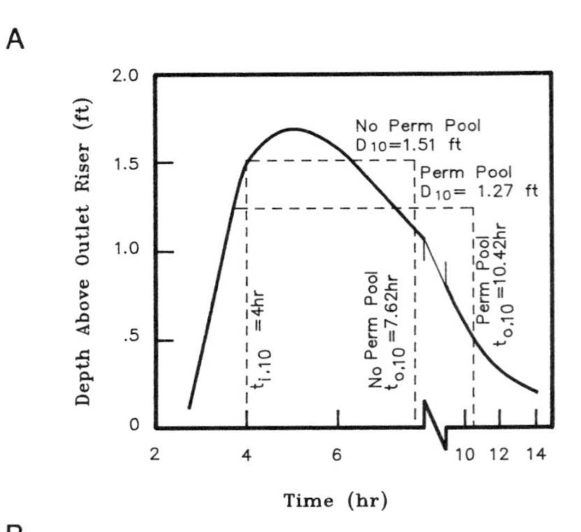

B

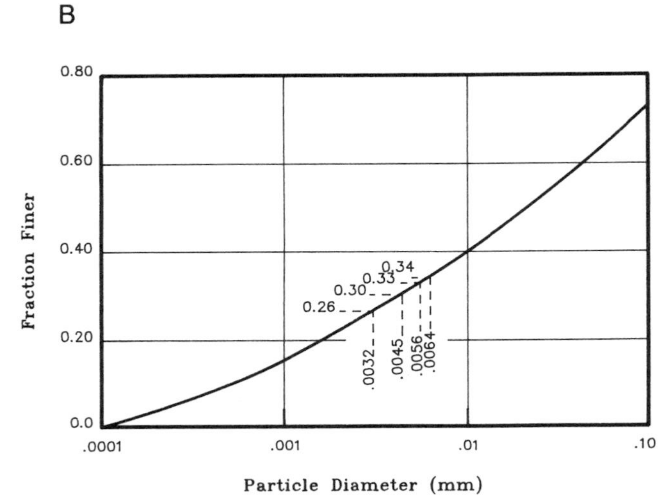

Figure 9.23 Average depth and size distributions for Example Problems 9.7 and 9.8. (A) Depth–time relationship. (B) Particle size distribution.

4. *Calculating particles that will settle from each layer*. The pond depth, D, is divided into four layers. To settle out of layer one, a particle would settle through a depth of $D/4$, to settle out of layer two, a particle must settle through a depth of $2D/4$, layer three, $3D/4$, and layer four, $4D/4$.

The required settling velocities and particle diameters are summarized below.

Layer	T_D, 10 (min)	Settling depth, SD (ft)	Required Settling velocity $SD/T_{D,10}$ (ft/sec)	Diameter $d_{c,j}$ from Eq. (7.5) and Fig. 7.3 (mm)
1	217.2	$0.378 = D/4$	0.0000290	0.0032
2	217.2	$0.755 = D/2$	0.0000579	0.0045
3	217.2	$1.133 = 3D/4$	0.0000869	0.0056
4	217.2	$1.510 = D$	0.0001158	0.0064

5. *Calculating effluent concentration*. Once d_c, the particle diameter that will just settle out of a layer, is determined, this can be transformed into a percentage finer by the following assumptions:

- All particles larger than d_c are out of the layer and all particles smaller than d_c are in the layer.
- Inflow sediment is uniformly mixed.

Using d_c calculated above and the inflow size distribution given in Fig. 9.23B, the following fractions finer than d_c were determined.

Layer j	$d_{c,j}$ (mm)	Fraction finer, FF $(d_{c,j})$
1	0.0032	0.26
2	0.0045	0.30
3	0.0056	0.33
4	0.0064	0.34

The outflow concentration $C_{o,j}$ for layer j is given by

$$C_{o,j} = C_i \mathrm{FF}(d_{c,j}),$$

where C_i is the inflow concentration. Given the inflow concentration of 19,270 mg/liter, the effluent concentration for each layer is

Layer j	$C_{o,j}$ (mg/liter)
1	5010
2	5781
3	6359
4	6552

The average outflow concentration is

$$C_o = \frac{C_{o,1} + C_{o,2} + C_{o,3} + C_{o,4}}{4}$$

$$C_o = \frac{5010 + 5781 + 6359 + 6552}{4} = 5926 \text{ mg/liter}$$

The computations in Example Problem 9.7 are for a totally dewatered reservoir, i.e., no permanent pool. Procedures for considering the effects of a permanent pool are given in Example Problem 9.8.

Example Problem 9.8 Effects of permanent pool on DEPOSITS prediction

Assume that the reservoir in Example Problem 9.7 has a permanent pool volume of 0.91 acre · ft and a dead storage of 25%. To develop this permanent pool, the crest of the riser is left at the same elevation, and a 1-ft-deep pool is excavated. How will this impact the effluent concentration of plug 10 in Example Problem 9.7?

Solution: Adding permanent pool is typically accomplished by raising the elevation of the spillway. This modifies the outflow hydrograph and the average depth time relationship. To accurately evaluate such a situation, the analysis given in Example Problem 9.7 must be redone. In this problem, it is assumed that the permanent pool is developed by excavation in order to simplify computation. For a first approximation with this small permanent pool, it will be further assumed that the outflow hydrograph is not appreciably affected.

The permanent-pool water displaced is equal to the permanent pool volume minus the dead storage, or

$$\text{Volume displaced} = \mathbf{V}_{\mathrm{p}} - \mathbf{V}_{\mathrm{ds}}$$
$$= 0.91 - (0.25)(0.91) = 0.68 \text{ acre} \cdot \text{ft}.$$

From Fig. 9.22A, the total volume of runoff is 2.61 acre · ft. Thus the runoff displaced represents 0.68/2.61 or 26% of the total volume. From Fig. 9.22B, the time required to displace 26% of the total flow is shown as $T_{\mathrm{D,p}} = 4.8 - 2.0$ hr $=$ 2.8 hr. (Note that the starting time for Fig. 9.22B is 2 hr). Thus the permanent pool effectively has the impact of adding 2.8 hr to the detention time of each plug. From example Problem 9.7, $T_{\mathrm{D,10}} = 3.62$ hr; thus the effective detention time of plug 10 with a permanent pool becomes

$$T_{\mathrm{D,10}} = 3.62 + 2.80 = 6.42 \text{ hr}.$$

The value for average outflow time, $t_{i,10}$, remains the same as in Example Problem 9.7, or 4.0 hr. The value for average outflow time, $t_{0,10}$, thus becomes 4.0 + 6.42 or 10.42 hr. Other computations change also. From Fig. 9.23A, the average depth above the outlet riser for the period from 4.0 to 10.42 hr is 1.27 ft. Including the 1.0-ft permanent pool, the average depth is 2.27 ft. Using this depth and the detention time, the following concentrations are calculated.

Layer	FF ($d_{\mathrm{c},j}$)	$C_{\mathrm{o},j}$ (mg/liter)	Average C_{o} (mg/liter)
1	0.23	4432	
2	0.29	5588	
3	0.31	5974	5588
4	0.33	6359	

Computational details for the above table follow those of Example Problem 9.7 and are left to the reader.

Effluent concentrations for plug 10 in Example Problems 9.7 and 9.8 are similar for the totally dewatered pond as compared to the pond with a permanent pool, although the detention time is increased from 127 to 385 min. The increase in settling depth offsets the increase in the detention time. The permanent pool does allow effluent early in the storm to be near zero as the stored permanent-pool water is discharged prior to any stormwater being discharged. Calculation with a computerized DEPOSITS model would yield a trapping efficiency for Example Problem 9.7 (no permanent pool) of 63.2% and for Example Problem 9.8 (with a permanent pool) the trapping efficiency would be 68.82%. The plug flow model, however, does not always predict that permanent pool will increase trapping efficiency.

Computational details are quite tedious for using this model. Software for making these computations is available from the authors.

Example Problem 9.9 Size distribution equations for plug flow model

Develop a routine to predict the particle size distribution for sediment in the discharge from a plug of flow, using the DEPOSITS model approach. Assume uniform withdrawal and a completely mixed inflow.

Solution: Let the inflow particle size distribution be defined by $\mathrm{FF}_i(d_{\mathrm{k}})$, where d_{k} is particle diameter. For a layer j, the particle size that just settles out of the layer is $d_{\mathrm{c},j}$. All particles in the effluent from layer j are equal or smaller than $d_{\mathrm{c},j}$. If $FF_{\mathrm{o},j}(d_{\mathrm{k}})$ represents the size distribution of the effluent from layer j, then $\mathrm{FF}_{\mathrm{o},j}(d_{\mathrm{c},j})$ is 1.00. For diameters smaller than $d_{\mathrm{c},j}$,

$$\mathrm{FF}_{\mathrm{o},j}(d_{\mathrm{k}}) = \frac{\mathrm{FF}_{\mathrm{i}}(d_{\mathrm{k}})}{\mathrm{FF}_{\mathrm{i}}(d_{\mathrm{c,j}})} \leq 1.0. \qquad \text{(A)}$$

This equation effectively normalizes the effluent size distribution from layer j to the range of 0 to 1.0. The mass corresponding to particles finer than the size d_{k} would be

$$M_{\mathrm{o},j}(d_{\mathrm{k}}) = \mathrm{FF}_{\mathrm{o},j}(d_{\mathrm{k}})C_{\mathrm{o},j}q_j, \qquad \text{(B)}$$

where $C_{\mathrm{o},j}$ is the effluent concentration from layer j and q_j is the water discharge rate from layer j. For uniform withdrawal,

$$q_j = q/4, \qquad \text{(C)}$$

where q is the total discharge rate for the plug. Hence

$$M_{o,j}(d_k) = \frac{q}{4} C_{o,j} FF_{o,j}(d_k). \qquad \text{(D)}$$

Summing over all layers to get the total mass finer than d_k, $M_o(d_k)$,

$$M_o(d_k) = \frac{q}{4} \sum_{j=1}^{4} C_{o,j} FF_{o,j}(d_k). \qquad \text{(E)}$$

The total mass discharged is

$$M_T = \frac{q}{4} \sum_{j=1}^{4} C_{o,j}. \qquad \text{(F)}$$

The fraction finer than d_k summed over all four layers is therefore

$$FF_o(d_k) = \frac{\Sigma C_{o,j} FF_{o,j}(d_k)}{\Sigma C_{o,j}}. \qquad \text{(G)}$$

Equations (A) and (G) make up the size distribution algorithm for the plug flow model.

Variable Flow Rate—Modified Overflow Rate Models

Because of the no-mixing assumption made in the overflow rate derivation given earlier, these type relationships could be classed as plug flow models. To apply the overflow rate concept to variable flow models, a flow rate and surface area must be selected to define the overflow rate for each time increment. Several different approaches have been utilized as described below.

Early EPA Model The early EPA Model is essentially an application of the overflow rate equation to the non-steady-state system (Hill 1976). In the EPA modification of the overflow rate method, surface area, A is set equal to the basin surface area at the top of the outlet riser, and Q is set equal to the peak discharge from the reservoir; hence

$$F = \frac{V_s A}{1.2\, q_{po}}, \qquad (9.59)$$

where q_{po} is the peak outflow and the factor 1.2 is added to account for nonideal settling. The EPA method provides a poor indication of the effects of basin geometry, inflow hydrograph, and sedigraph shape on sediment trapping; however, because of its simplicity, it has been used with some regularity in the analysis of surface-mined sediment ponds. To use the EPA procedures, the argument V_s/V_c in Eq. (9.41) is replaced with $V_s A/1.2 q_{po}$, and X_c is calculated from the settling velocity corresponding to $1.2 q_{po}/A$. This effectively makes V_c equal to $1.2 q_{po}/A$. The calculation procedures for using the EPA model are

1. The particle size distribution is divided into intervals, and the medium diameter in each interval is used to compute the settling velocity, V_{si}, for that interval.
2. The basin overflow rate, V_c, is calculated from the routed peak outflow and the effective pond surface area at the permanent pool using $V_c = 1.2 q_{po}/\text{A}$. The effective area is the area at the permanent pool minus the dead storage fraction.
3. The removal fraction, F, is calculated for each size interval and the trapping efficiency, E, determined using procedures shown in Example Problem 9.5.

Tapp Method 1 The Tapp Method 1 (Tapp *et al.*, 1981) is a modification of the steady-state overflow rate equation. The method is based on the discharge and effective surface area at a given time. The basin overflow rate for each time interval is calculated as

$$(V_c)_i = \frac{C_{OR} Q_o}{A_i - C_A A_i}, \qquad (9.60)$$

where $(V_c)_i$ is the basin overflow rate at a given time, C_{OR} is a constant, A_i is the basin surface area at a given time, C_A is the fraction of surface area that does not contribute to settling, and Q_o is the basin outflow rate. The calculation procedure is summarized after the following section.

Tapp Method 2 The previous two methods were based on Eq. (9.41), which uses surface area in calculating the basin overflow rate. The overflow rate method was originally developed assuming a basin with vertical sidewalls. A more theoretically palatable equation for an irregular geometry like that found in a sediment pond, would be to consider settling depth and volume in calculating the basin overflow. The equation used in this method is

$$(V_c)_i = \frac{C'_{OR} Q_o D}{\forall - C_\forall \forall}, \qquad (9.61)$$

where $(V_c)_i$ is the basin overflow rate at a given time, C'_{OR} is a constant, Q_o is the outflow rate, D is the settling depth, $\forall$ is the volume in the basin at a given time, and $C_\forall$ is the fraction of dead space volume accounting for dead storage. The quantity $(\forall - C_\forall \forall)$ would be the effective volume. The calculation procedure would be the same as the Tapp Method 1, except the removal ratio would be calculated using the above equation for basin overflow rate.

All three of the modified overflow rate methods described may be used with quiescent settling or turbulent settling computations.

Example Problem 9.10 Illustration of modified overflow rate models for variable flow

Estimate the trapping efficiency of the reservoir in Example Problem 9.7 for the time interval given by plug 10 using Tapp method 1. A stage–area–average depth curve is given in Fig. 9.24 for the reservoir.

Solution:

1. *Determining average surface area for plug 10*. From Fig. 9.23, the average depth for plug 10 is 1.51 ft. Using Fig. 9.24, the average surface area is given as 0.90 acres or 39,204 ft^2.

2. *Calculating overflow rate*. From Fig. 9.22, the average discharge rate for plug 10 is 4.5 cfs. Assuming a dead storage volume of zero, the overflow rate becomes

$$(V_c)_{10} = \frac{Q_{o,10}}{A_{10}} = \frac{4.5}{39{,}204} = 0.000115 \text{ ft/sec}.$$

3. *Calculating trapping efficiency*. Using Stokes' law [Eq. (7.4)], the diameter corresponding to the overflow rate is

$$d_c = \sqrt{\frac{V_s}{2.81}} = \sqrt{\frac{0.000115}{2.81}} = 0.0064 \text{ mm}.$$

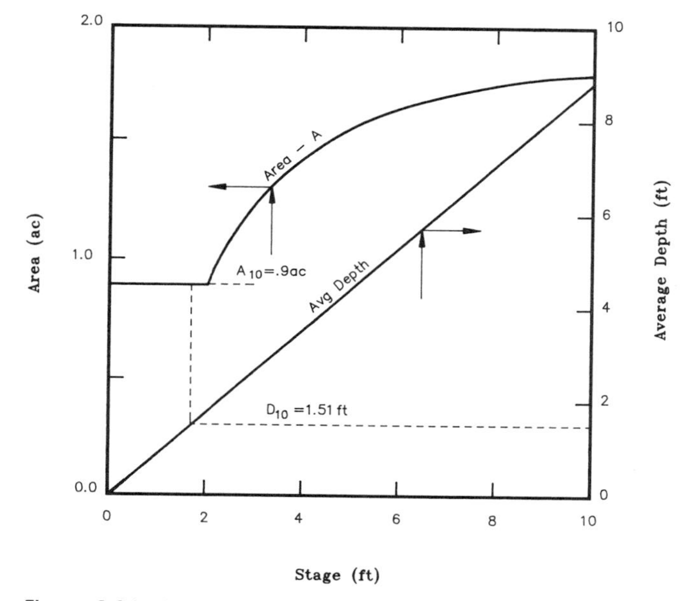

Figure 9.24 Stage–area–average depth curves for Example Problem 9.10. Average depth is an area-weighted depth over the entire reservoir.

Using this diameter in Fig. 9.23B, the value for X_c is 0.34. Following procedures given in Example Problem 9.5, the trapping efficiency for plug 10 is

$$E = (1 - X_c) + \sum_{i=1}^{n} \frac{V_{si}}{V_c} \Delta X_i.$$

Dividing the particle sizes below 0.0064 mm into intervals of $\Delta X = 0.04$, and following procedures from Example Problem 9.5, the summation in the RHS is 0.047. Therefore trapping efficiency for plug 10 is

$$E_{10} = (1 - 0.34) + 0.047 = 0.707.$$

The effluent concentration for plug 10 would thus be (assuming the same $C_{i,10}$ as Example Problem 9.7)

$$C_{o,10} = (1 - E_{10})C_{i,10} = (1 - 0.707)(19{,}270)$$
$$C_{o,10} = 5646 \text{ mg/liter},$$

which is similar to the plug flow model. This will not always be true.

Variable Flow Rate—CSTRS Model

Plug flow concepts are useful for defining simple models, but do not describe the mixing known to exist in sediment basins. As will be seen in a subsequent discussion of model accuracy, plug flow models predict trapping efficiencies with reasonable accuracy, but do not accurately predict timing or magnitudes of sediment concentrations. To overcome this deficiency, Wilson and Barfield (1984) modified the CSTRS concept for sediment pond modeling. A schematic of a reservoir conceptualized as CSTRS is given in Fig. 9.25.

Effluent concentrations are predicted by conducting a mass balance on each reactor, or

$$\begin{bmatrix}\text{Mass rate into}\\ \text{reactor } i\end{bmatrix} - \begin{bmatrix}\text{Mass rate out}\\ \text{of reactor } i\end{bmatrix} - \begin{bmatrix}\text{Deposition rate}\\ \text{in reactor } i\end{bmatrix} = \begin{bmatrix}\text{Rate of change of}\\ \text{mass in reactor } i\end{bmatrix}.$$

For a continuous stirred reactor, effluent concentration is equal to concentration in the reactor. Hence, the mass balance for reactor i can be written as

$$Q_{i-1}C_{i-1} - Q_iC_i - DR_i = \frac{d(\maltese_i C_i)}{dt}, \quad (9.62)$$

where Q_{i-1} and C_{i-1} are the water inflow rate and the influent sediment concentration, respectively, Q_i and C_i are the water discharge and the effluent sediment concentration, respectively, DR_i is the deposition rate

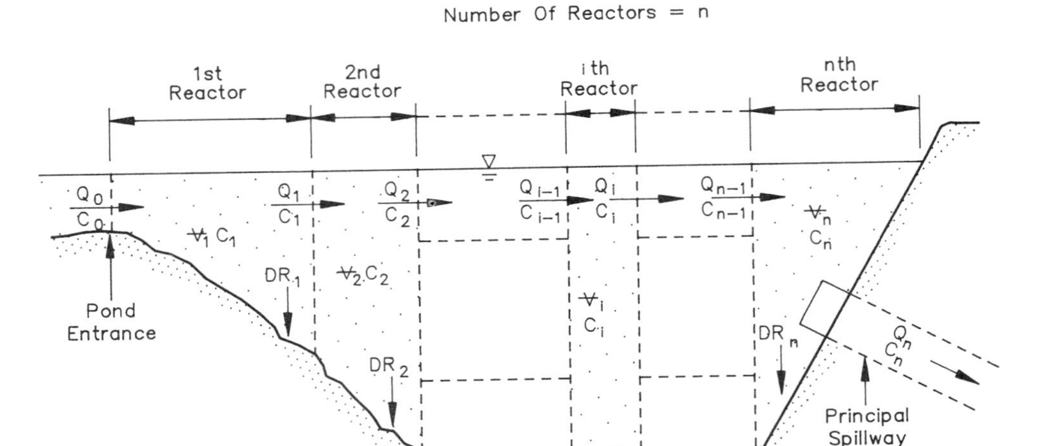

Figure 9.25 Pond divided into a series of CSTRS (after Wilson and Barfield, 1985). n, number of reactors.

of sediment in the reactor, $\forall_i$ is the volume of the reactor, and t is time.

Knowing the concentration in all the reactors at the start of a time step (C_i^0 for $i = 1$ to n) and the volume of each reactor at the start of the time step ($\forall_i^0$ for $i = 1$ to n), the concentration in the ith reactor, C_i at the end of the time step is determined by using a finite difference approximation

$$C_i^n = \frac{\Delta t\left(\{Q_{i-1}C_{i-1}\}_{\text{avg}} - \{Q_i\}_{\text{avg}}C_i^0/2\right) - \text{DEP} + C_i^0\forall_i^0}{\forall_i^n + \Delta t\{Q_i\}_{\text{avg}}/2} \tag{9.63}$$

where Δt is the time increment used to route the sediment-laden flow, $\{Q_{i-1}C_{i-1}\}_{\text{avg}}$ is the average inflow mass rate during Δt, $\{Q_i\}_{\text{avg}}$ is the average outflow discharge rate during Δt, $\forall_i^0$ and $\forall_i^n$ are the reactor volumes at the beginning and end of the time increment, respectively, C_i^0 and C_i^n are the effluent/reactor concentrations at the beginning and the end of the time increment respectively, DEP_i is the mass deposited within Δt, and the other terms are as previously defined. In Eq. (9.63), $\{Q_{i-1}C_{i-1}\}_{\text{avg}}$ is defined from the previous reactor or from the inflow sediment graph (for the first reactor) and C_i^0 is a known value calculated by the previous time step.

The average discharge, $\{Q_i\}_{\text{avg}}$, is estimated by interpolating between the average discharge at the inlet and the outlet of the pond, or

$$\{Q_i\}_{\text{avg}} = \{Q_0\}_{\text{avg}} + (i-1)\frac{\{Q_n\}_{\text{avg}} - \{Q_0\}_{\text{avg}}}{n}, \tag{9.64}$$

where $\{Q_i\}_{\text{avg}}$ is the average inflow rate for the ith reactor, $\{Q_0\}_{\text{avg}}$ is the average flow rate at the pond's inlet, $\{Q_n\}_{\text{avg}}$ is the average discharge at the pond's outlet, and n is the number of reactors.

The reactor volumes in Eq. (9.64) ($\forall_i^0$ and $\forall_i^n$) are defined as

$$\forall_i^{(t)} = [\text{PV}(t) - \text{DEAD}]/n, \tag{9.65}$$

where $\text{PV}(t)$ is the pond volume at time t, DEAD is the volume of the permanent pool that is dead space, and n is the number of reactors. $\text{PV}(t)$ is determined and stored by a hydrograph routing procedure. Equation (9.65) divides the pond volume equally among each reactor.

To understand the methodology that is used to estimate the deposition term DEP_i of Eq. (9.63), it is useful to discuss the differences between the sedimentology of sediment particles of a plug flow model, such as DEPOSITS and that of a mixing model such as CSTRS. In a plug flow model, each particle of a given inflow slug[1] of sediment has the same detention time.

[1]A slug of sediment is defined in this context as the amount of sediment inflow into the 1st reactor during the time step.

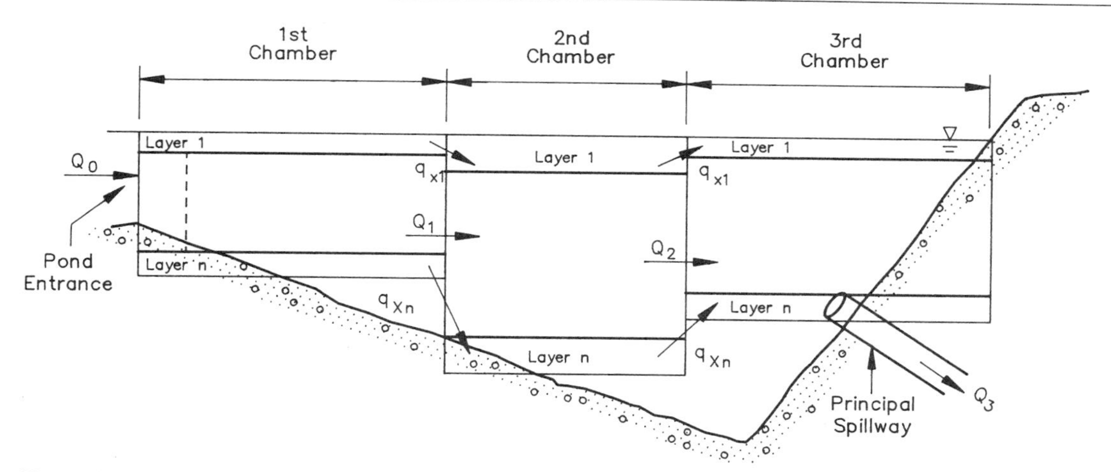

Figure 9.26 Conceptualization of a reservoir represented by the BASIN model (after Wilson and Barfield, 1985).

As such, a single time parameter can be used to characterize the fall time of this sediment. For a CSTRS model, inflow sediment mixes with sediment from previous inflow slugs; therefore, when sediment is discharged from the pond, it contains some mass from the current inflow slug as well as previous inflow slugs. Consequently, a single time parameter cannot be used to represent the residence time of sediment particles contained in each inflow slug since some sediment particles remain in the pond for a long time, whereas other particles are discharged almost immediately. To handle a variation in the residence times of individual particles, the CSTRS model maintains a mass balance for each inflow slug of sediment.

Computational algorithms for the CSTRS model are described in detail in Wilson and Barfield (1984). These procedures are too complex for hand calculations and must be solved with computer software. The CSTRS model is a component of the SEDIMOT II hydrology and sedimentology computer model (Wilson *et al.*, 1982, 1986) and SEDCAD (Warner and Schwab, 1992).

Variable Flow Rate—Basin Model

Use of the CSTRS concept allows the effects of mixing to be evaluated, but cannot be used to evaluate bed scour and resuspension. To overcome that deficit, a model of reservoir sedimentation, known as BASIN was developed (Wilson and Barfield, 1985), in which scour is predicted using a modification of the entrainment equation from Einstein (1950) and resuspension of scoured material predicted by a coupling of reactor and diffusion theory. Reactor theory is used to predict residence times of particles, and diffusion theory is used to model settling characteristics due to gravity and vertical diffusion due to turbulence. A conceptualization of a reservoir represented by BASIN is given in Fig. 9.26.

In the BASIN procedures, the inflow size distribution is divided into a maximum of nine particle size classes. Using the assumption that each layer is completely mixed horizontally with vertical diffusion only, a differential equation for concentration, C, of any particle size class at any level z in reactor i can be written as

$$\frac{\partial C}{\partial t} = \frac{Q_i U_f(z)}{V_{ei}}[C_p - C] + V_s\frac{\partial C}{\partial z} + \frac{\partial}{\partial z}\left(\varepsilon\frac{\partial C}{\partial z}\right), \tag{9.66}$$

where C_p is the concentration flowing into a layer from the previous reactor, V_{ei} is the effective volume of reactor i (actual volume minus dead storage), V_s is settling velocity for the particle class (taken as positive in the negative z direction), ε is turbulent diffusivity, Q_i is inflow to reactor i, and $U_f(z)$ is the fraction of flow through the reactor that moves through a given level of the reactor, or

$$U_f(z) = U(z)/U_{avg}, \tag{9.67}$$

where $U(z)$ is velocity at level z and U_{avg} is average velocity for the section. $U_f(z)$ is a user input to the model that can vary between reactors. It can be estimated from a known velocity distribution at the inlet to the reservoir. For the special case where the reservoir water surface is rising slowly, Wilson and Barfield (1985) recommend that $U_f(z)$ in each reactor i be related to that in the previous reactor by

$$U_{f,i}(z) = 1 - \frac{Q_{i-1}}{Q_i}[1 - U_{f,i-1}(z)], \tag{9.68}$$

where subscripts i and $i-1$ refer to values in reactors i and $i-1$. Q_i and Q_{i-1} can be estimated from Eq. (9.64). To start the computation, $U_{f,0}$ can be estimated from the log velocity profile of Einstein (1950) given by Eq. (10.15), or

$$U_{avg} = \frac{U_*}{k}\log\left(12.27\frac{D}{k_s}\chi\right) \tag{9.69}$$

and

$$U(z) = \frac{U_*}{k}\log\left(30.27\frac{Z}{k_s}\chi\right), \tag{9.70}$$

where U_* is shear velocity given by $(gDS)^{0.5}$, g is gravity, D is reservoir depth, S is channel slope, k_s is channel roughness given by the sediment diameter d_{65}, χ is Einstein's log velocity constant from Fig. 10.10, and k is von Karmon's constant ($\simeq 0.4$). Hence $U(z)/U_{avg}$ can be approximated by

$$U_{f,0}(z) = \frac{U(z)}{U_{avg}} = \frac{\ln(30.2\, z/k_s)}{\ln(12.27\, D/k_s)}, \tag{9.71}$$

where D is the depth in the inflow channel.

Boundary conditions are needed for the solution of Eq. (9.66) at the inlet to the reservoir, at the water surface, and at the bed of the reservoir. At the inlet to the reservoir, sediment concentration can be given by

$$C(z) = k_1 \exp(-V_s z/\varepsilon), \tag{9.72}$$

where k_1 is a constant of integration based on mass continuity or

$$k_1 = \frac{DC_{avg}}{\int_0^D(-V_s(z/\varepsilon))U_f(z)\,dz}. \tag{9.73}$$

ε is the turbulent diffusivity at the inlet, C_{avg} is the average inflow concentration at a given time at the inlet to the reservoir, and D is flow depth at the inlet.

The boundary condition at the water surface is defined by the statement of zero flux, i.e., upward diffusion equals downward settling, or

$$V_s C\Big|_{z=D} = -\varepsilon\frac{\partial C}{\partial z}\Big|_{z=D}. \tag{9.74}$$

At the reservoir bed, the rate of settling would not necessarily be equal to upward diffusion and the boundary condition is defined by

$$S_c = -\varepsilon\frac{\partial c}{\partial z}\Big|_{z=0}, \tag{9.75}$$

where S_c is the rate of scour due to turbulence. S_c is evaluated by Einstein's (1950) scour function. Coincidentally, the net of scour and settling, F_n, is

$$F_n = \left[-V_s C - \varepsilon\frac{\partial c}{\partial z}\right]_{z=0}, \tag{9.76}$$

where, again, V_s is positive in the negative z direction.

Variable Flow Rate Models—Evaluation of Accuracy

Evaluations of the accuracy of variable flow rate models have been made with laboratory and field data. The DEPOSITS model has been evaluated the most extensively on both laboratory and field data. In general, estimates of trapping efficiency were reasonably accurate for both field and laboratory data, but timing of effluent concentration and magnitudes of effluent concentration were poorly predicted. Based on laboratory model studies, the modified overflow rate equations were no more accurate than the DEPOSITS model in predicting trapping efficiencies. When compared to the DEPOSITS model utilizing laboratory data, both the CSTRS and BASIN models were more accurate in predicting the shape of the effluent sedimentgraph and peak concentration. Nevertheless, neither CSTRS nor BASIN models are more accurate than DEPOSITS in predicting trapping efficiencies. In general, BASIN was no more accurate than the CSTRS model in predicting effluent concentration. The BASIN model, however, can be utilized to predict resuspension, whereas the CSTRS model cannot. Detailed comparisons of model predictions are given in Wilson and Barfield (1985).

Example Problem 9.11. Comparison of CSTRS and DEPOSITS model predictions

For the storm and reservoir in Example Problem 9.8, compute the trapping efficiency and peak effluent concentration for the storm using the plug flow model (DEPOSITS) and CSTRS models. Compare the models. Assume a dead storage volume of 20%.

Solution: The solution to the problem requires the use of computer models due to the complexity of computations. Input data required by the models are summarized below

1. *Storage area discharge data (required of all models).*

Stage	Area (acre)	Discharge (cfs)
0.00	0.91	0
1.00	0.91	0
3.00	0.91	5.84
6.00	1.50	6.22
11.00	1.80	6.82
16.00	2.20	7.36

2. Special input.

Model	No. reactors	Inflow distribution	Withdrawal distribution
DEPOSITS	0[a]	Mixed	Uniform
CSTRS	2	Mixed	Uniform

[a]Zero reactors actually correspond to an infinite number of reactors using the CSTRS model (see Fig. 9.4).

3. Computed values. Trapping efficiencies and peak effluent concentrations were predicted with computerized algorithms. Values are summarized below.

Model	Trapping efficiency (%)	Peak effluent concentration (mg/liter)
Plug flow (DEPOSITS)	68.8	52,527
CSTRS	72.4	17,492

Modeling Pond Performance: 1986 EPA Urban Methodology

Model Description

The EPA methodology (Driscoll *et al.*, 1986) attempts to evaluate long-term sediment trapping in a reservoir. The method accounts for trapping under storm flow (dynamic) conditions and subsequent settling under the quiescent conditions that occur in ponds after most stormwater is discharged. To make such an evaluation, the model must account for variations in stormflow and variations in the duration of quiescent conditions. To predict settling during quiescent periods, the model must account for the portion of sediment already deposited and that remaining in suspension at the time all stormwater is discharged.

The EPA methodology extends the single-storm concept to long-term trapping in a reservoir. Stochastic concepts are used for developing stormwater and sediment discharges into a pond along with a reservoir sedimentation model to predict trapping under quiescent conditions. The overall concept is to utilize the single-storm model along with rainfall and runoff statistics to predict trapping during stormflow and to use a quiescent model plus statistics on the interarrival times between storms for trapping after storm flow ceases. Perhaps the weakest link in the approach is a lack of physical basis for the models used for trapping.

The single-storm reservoir model used for predicting trapping of a given particle under dynamic (stormwater) conditions is an empirical relationship

$$F = 1 - \left[1 + \frac{1}{\beta}\frac{V_s}{V_c}\right]^{-\beta}, \qquad (9.77)$$

where V_s is settling velocity of the particle, V_c is overflow rate defined as Q/A in Eq. (9.40), and β is a turbulence or short-circuiting parameter reflecting the nonideal performance of the pond. Recommended values for β are

$\beta = 1$, very poor performance
$\beta = 2$, average performance
$\beta = 3$, good performance
$\beta > 5$, very good performance.

For $\beta = 1$, Eq. (9.77) reduces to

$$F = \frac{V_s}{V_c} - \left(\frac{V_s}{V_c}\right)^2 + \left(\frac{V_s}{V_c}\right)^3 - \cdots.$$

For small values of V_s/V_c, this becomes equal to the quiescent overflow rate equation, but deviates for larger values. For $\beta = \infty$, the equation becomes

$$F = 1 - e^{-V_s/V_c},$$

which is the fully turbulent model given earlier by Eq. (9.57). A comparison of the models is given in Fig. 9.27, showing that the EPA model does not predict trapping efficiencies greater than the fully turbulent model of Eq. (9.57), even under their so-called "ideal" conditions.

To estimate total sediment removed for a single storm, Eq. (9.77) would be applied to all particle sizes in a procedure like that applied to the overflow rate procedure (Example Problems 9.5 and 9.6).

To predict long-term trapping, the EPA model combines Eq. (9.77) with stochastically generated flows. The analysis is divided into dynamic (stormwater) flows and quiescent flows between stormwater events. Dynamic flows are assumed to be gamma distributed and characterized by a mean flow and a coefficient of

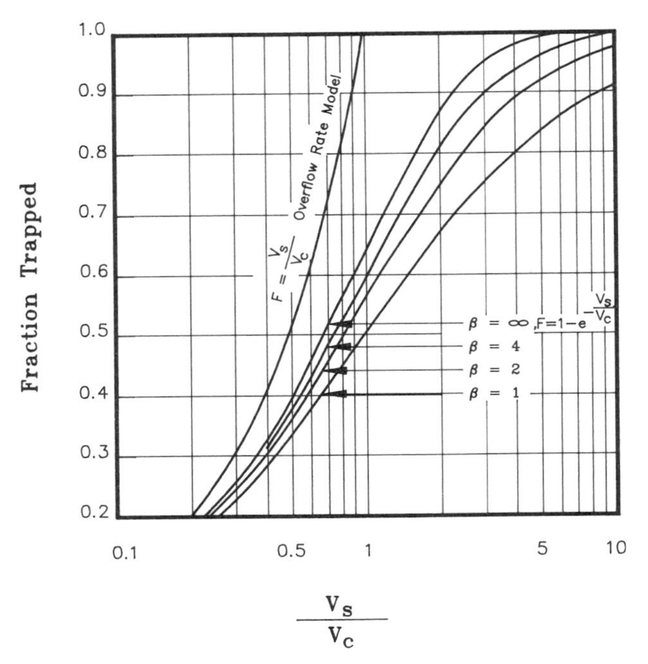

Figure 9.27 Comparison of the EPA and overflow rate models.

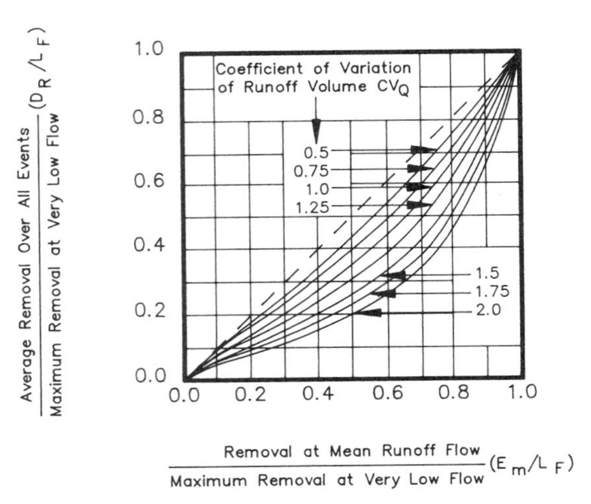

Figure 9.28 Long-term sediment trapping in sediment ponds under storm water flow (after Driscoll *et al.*, 1986).

variation for flow, CV_Q. Using the gamma distribution, Driscoll *et al.* (1986) proposed that the total removal efficiency could be calculated by

$$D_R = L_F\left[\frac{1/CV_Q^2}{1/CV_Q^2 - \ln(E_m/L_F)}\right]^{\frac{1}{CV_Q^2}+1}, \tag{9.78}$$

where D_R is long-term dynamic removal fraction for stormwater, L_F is the removal ratio (fraction) for very low flow rates, E_m is mean storm removal fraction [typically estimated from Eq. (9.77)], and CV_Q is the coefficient of variation of flows. Equation (9.78) is illustrated graphically in Fig. 9.28. Values for L_f and E_m must be determined for single storms, as is discussed subsequently.

Driscoll *et al.* (1986) recommended that removal efficiencies for quiescent conditions between stormwater flows be estimated from

$$Q_R = V_s A_Q, \tag{9.79}$$

where Q_R is quiescent removal rate, V_s is settling velocity, and A_Q is surface area under quiescent conditions. For an average condition, the model defines a removal ratio as

$$RR = \frac{T_{IA} Q_R}{\forall_R}, \tag{9.80}$$

where T_{IA} is the average time interval between storms, $\forall_R$ is mean runoff volume, and Q_R is the removal rate defined by Eq. (9.79). Values for T_{IA} are given in Table 9B.1 in Appendix 9B under the column labeled "Interval."

The storage volume under quiescent conditions in the EPA model is not assumed to be a fixed quantity, but varies between storms. This is accounted for in the EPA model in Fig. 9.29 by relating the ratio of effective pond storage volume to mean runoff volume, $\forall_E/\forall_R$, to the ratio of storage volume (empty) to mean runoff volume, $\forall_B/\forall_R$, and the removal ratio from Eq. (9.80). The effective storage volume ratio from Fig. 9.29, $\forall_E/\forall_R$ is then used in Fig. 9.30 along with the coefficient of variation of runoff volume to estimate the percentage removal under quiescent conditions. Computational procedures are illustrated in Example Problem 9.12.

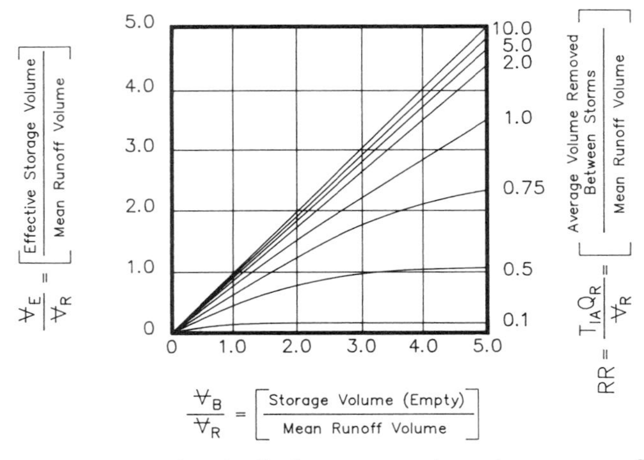

Figure 9.29 Ratio of effective storage volume to mean runoff volume (after Driscoll *et al.*, 1986).

A relationship for combined conditions, i.e., stormwater trapping plus trapping during quiescent intervals between storms, is needed in order to estimate the total long-term fraction removed. As stormwater moves through the pond, part of the sediment is retained at the end of the storm as a consequence of settling, and part is discharged. Of that remaining, some is in suspension in the permanent pool and some has already settled to the bottom. Of that remaining in suspension, some will continue to settle out during the quiescent period between storms. Any sediment remaining in suspension at the start of a subsequent storm will be discharged. To accurately account for this combination of dynamic and quiescent settling, a complex computer model would be required. The EPA model recommends a simple alternative

$$E_T = 1 - (1 - E_D)(1 - E_Q), \qquad (9.81)$$

where E_T is total removal efficiency, E_D is total dynamic removal efficiency, and E_Q is total quiescent removal efficiency.

To use the EPA model, special rainfall statistics are needed. Values recommended by Driscoll *et al.* (1986) are given in Appendix 9B. Computational procedures for the model are illustrated in Example Problem 9.12.

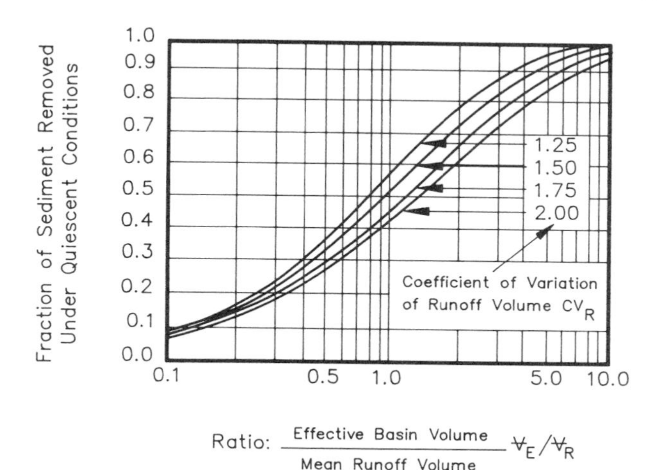

Figure 9.30 Long-term removal ratio for sediment ponds under quiescent conditions between storms (after Driscoll *et al.*, 1986).

Example Problem 9.12 Illustration of the EPA model

A reservoir is being constructed with the following characteristics:

Parameter	Value
Average surface area (principal spillway to emergency spillway) ft^2	10,000
Storage volume ft^3	40,000
Average depth ft	4.0

The reservoir is located in an area that has the following rainfall characteristics (see Appendix 9B)

Parameter	Mean	Coef. var.
Average rainfall volume (in.)	1.14	1.3
Intensity (in./hr)	0.3	1.3
Duration (hr)	4.0	1.2
Interval between storms T_{IA} (hr)	100	1.0

The watershed has an area of 80 acres, a curve number of 90, and a Rational Equation C factor of 0.8. The particle size distribution for sediment in an average storm is such that the settling velocities and percentages in the table below are applicable. Calculate the total fraction trapped for both stormflow and quiescent flow. Assume that the reservoirs have good hydraulic performance.

Particle class i	Percentage finer	Settling velocity, V_{si} (ft/hr)
1	10	0.02
2	30	0.20
3	60	1.00
4	80	8.00
5	100	50.00

Solution:

1. *Calculating mean runoff parameters*. From the curve number method in Fig. 3.21, the runoff volume for a rainfall of 1.14 in. and curve number of 90 is 0.412 in. The peak discharge for the average strom using the rational method [Eq. (3.70)] is

$$q_p = CIA_w = (0.8)(0.3\text{ in./hr})(80\text{ acre})$$
$$= 19.2\text{ cfs or } 69{,}120\text{ ft}^3/\text{hr}.$$

It will be assumed that the coefficient of variation of runoff is the same as rainfall; thus

$$CV_Q = 1.3$$
$$CV_R = 1.3.$$

2. *Calculating removal under dynamic conditions*. The overflow rate for the mean storm will be conservatively estimated using the peak flow, or

$$V_c = \frac{Q}{A} = \frac{q_p}{A} = \frac{69{,}120 \text{ ft}^3/\text{hr}}{10{,}000 \text{ ft}^2} = 6.91 \text{ ft/hr}.$$

Since the pond will have good performance, $\beta = 3$. For particle class 1, $V_s = 0.02$ ft/hr. From Eq. (9.77),

$$F = 1 - \left[1 + \frac{1}{3}\frac{0.02}{6.91}\right]^{-3} = 0.003.$$

Values for the other particle classes are given in the table below. Average trapping over all storms is given by Eq. (9.78). It will be assumed conservatively that the average trapping is given by Eq. (9.77), or $E_m = F$. From Eq. (9.78) using $CV_Q = 1.3$ and assuming that low flow trapping efficiency, L_F, is 1.0,

$$D_R = 1.0\left[\frac{1/1.3^2}{1/1.3^2 - \ln(0.003/1.0)}\right]^{1+1/1.3^2} = 0.023.$$

Trapping by each particle class averaged over all dynamic conditions is estimated in the table below.

Particle class [a]	Percentage in class[a]	Settling velocity V_{si} [a] (ft/hr)	Single-storm storm trapping F[b]	Fraction removed over all storms D_R [c]	Column 5 × Column 2
1	10	0.02	0.003	0.023	0.23
2	20	0.20	0.028	0.045	0.90
3	30	1.00	0.132	0.094	2.82
4	20	8.00	0.624	0.393	7.86
5	20	50.0	0.975	0.936	18.72
Total	100				$E_D = 30.53$

[a]Given in problem statement.
[b]Equation (9.77), using $\beta = 3$ for good performance, $F = 1 - \left[1 + \frac{1}{3}\frac{V_{si}}{V_c}\right]^{-3}$.
[c]Equation (9.78), assuming that $L_F = 1.0$ and $CV_Q = 1.3$, $D_R = 1.0\left[\frac{1/CV_Q^2}{(1/CV_Q^2 - \ln F)}\right]^{1+1/CV_Q^2}$.

The fraction trapped under dynamic conditions is thus

$$E_D = \frac{30.53}{100} = 0.305.$$

3. *Fraction trapped under quiescent conditions*. The mean runoff volume is runoff depth (0.412 in.) times watershed area or

$$V_R = \left(0.412 \text{ in.} \times \frac{1 \text{ ft}}{12 \text{ in.}}\right)\left(80 \text{ acre} \times 43{,}560 \frac{\text{ft}^2}{\text{acre}}\right)$$

$$= 119{,}645 \text{ ft}^3.$$

The reservoir volume is given as 40,000 ft^3; hence the ratio of reservoir volume to runoff volume is

$$\frac{V_B}{V_R} = \frac{40{,}000 \text{ ft}^3}{119{,}645 \text{ ft}^3} = 0.334.$$

The removal ratio from Eq. (9.80) is

$$RR = \frac{T_{IA} Q_R}{V_R} = \frac{T_{IA} V_s A_Q}{V_R}.$$

Assuming that A_Q is the average surface area, and knowing that T_{IA} is given by the problem statement as 100 hr,

$$RR = \frac{(100 \text{ hr})(10{,}000 \text{ ft}^2)\, V_s}{(119{,}645 \text{ ft}^3)} = 8.36\, V_s,$$

where V_s is feet per hour. For a given V_s, RR can be calculated and used along with V_B/V_R in Fig. 9.29 to determine the ratio of effective storage volume to runoff volume. This is then used in Fig. 9.30 along with CV_R to determine the fraction removed under quiescent conditions. Computations are summarized in the following, assuming that the size distribution does not change between storm flow and quiescent flow.

Particle class	Percentage in class	Settling velocity V_s (ft/hr)	Removal ratio, RR[a] (ft^3/hr)	Effective volume ratio, V_E/V_R[b]	Fraction removed under quiescent conditions[c]	Column 6 × column 2
1	10	0.02	0.17	0.100	0.09	0.09
2	20	0.20	0.167	0.334	0.25	5.0
3	30	1.00	0.836	0.334	0.23	7.5
4	20	8.00	66.86	0.334	0.23	5.0
5	20	50.0	418.00	0.334	0.23	5.0
Total	100					$\Sigma = 23.4$

[a] $RR = 8.36\ V_s$.
[b] Figure 9.29 with $V_B/V_R = 0.334$.
[c] Figure 9.30 with $CV_R = 1.3$.

Thus

$$E_Q = \frac{23.4}{100} = 0.234.$$

4. *Total fraction trapped under combined conditions.*

$$\begin{aligned} E_T &= 1 - (1 - E_D)(1 - E_Q) \\ &= 1 - (1 - 0.305)(1 - 0.234) \\ &= 0.468. \end{aligned}$$

Comments on EPA Model

The rainfall data provided by the EPA contain information for small storms where runoff is unlikely; therefore, designers are urged to make conservative decisions and consider doubling flow rates and volumes to be conservative. Another option is to evaluate the rainfall data base for a given locale, eliminating all small rainfall events not likely to produce runoff, and develop a modified set of values for rainfall duration, intensity, etc.

The graphical relationships and equations presented by the EPA represent, at best, a rough estimate of long-term trapping efficiency in a pond. It does, at least, represent a useful framework with which to approach the problem. To accurately predict long-term trapping, a detailed model of reservoir performance such as the CSTRS model (Wilson and Barfield, 1984) should be combined with a continuous stormwater and sediment runoff model to develop estimates.

Other Models of Pond Performance

Several researchers have proposed empirical procedures for evaluating reservoir trapping efficiency. Some of the procedures were developed from reservoir survey data and typically do not reflect many of the factors that are known to affect pond performance. One of the procedures is based on data generated by a simulation model.

Brune's Method

Figure 9.31 shows the empirical curves developed by Brune (1953) with data collected from 44 large reservoirs. Brune relates trap efficiency to the capacity–inflow (C/I) ratio described by Hazen (1904). Considerable scatter exists in the data, and there appears to be no correlation with estimates for semidry reservoirs that are typical of sediment structures.

The C/I ratio used as an independent variable in Brune's method does not account for sediment characteristics, inflow sedimentgraph, outlet discharge curves, and outlet discharge distribution variation with depth. The capacity of a basin is a very poor indicator of the geometry and anticipated flow conditions within the basin.

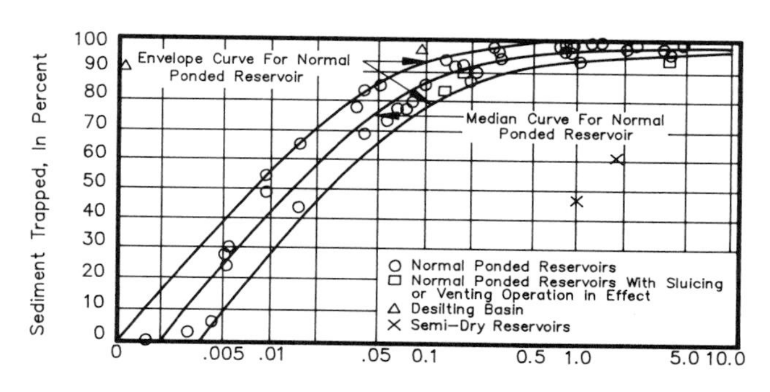

Figure 9.31 Brune's trap efficiency curves (after Brune, 1953).

Churchill's Method

Churchill (1948) developed a method based on results obtained from several TVA reservoirs in which a sedimentation index is related to trap efficiency as shown in Fig. 9.32. The sedimentation index is the detention time divided by the mean velocity of the flow through the basin.

While the method does not account for varying sediment sizes, many of the other factors affecting sediment transport are directly incorporated in the sedimentation index. The small amount of scatter indicates a high correlation with trap efficiency for the large reservoirs analyzed.

Based on a theoretical analysis using the overflow rate concept, Chen (1975) showed that both Churchill's and Brune's method overpredicted trapping efficiency for small particles and underpredicted trapping efficiency for large particles.

The Maryland Model

A simulation analysis of sediment trapping in reservoirs used in sediment control for construction sites was conducted by McBurnie *et al.* (1990) on several watersheds typical of conditions across Maryland. Using the SEDIMOT II model, they predicted sediment trapping in reservoirs with varying pond volumes and surface areas. Inflows were simulated for disturbed watersheds with varying soil types. The results indicated that the primary variables controlling trapping efficiency were soil classification and the surface area to peak discharge ratio. The results of the study, shown in Fig. 9.33, can be used for a very rough estimate of trapping in sediment detention ponds. The surface area in the horizontal axis is the area at the crest of the principal spillway.

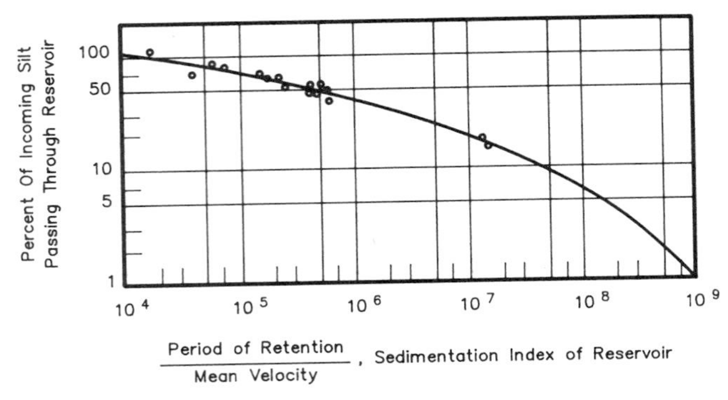

Figure 9.32 Churchill's trap efficiency curves (adapted from Brune, 1953).

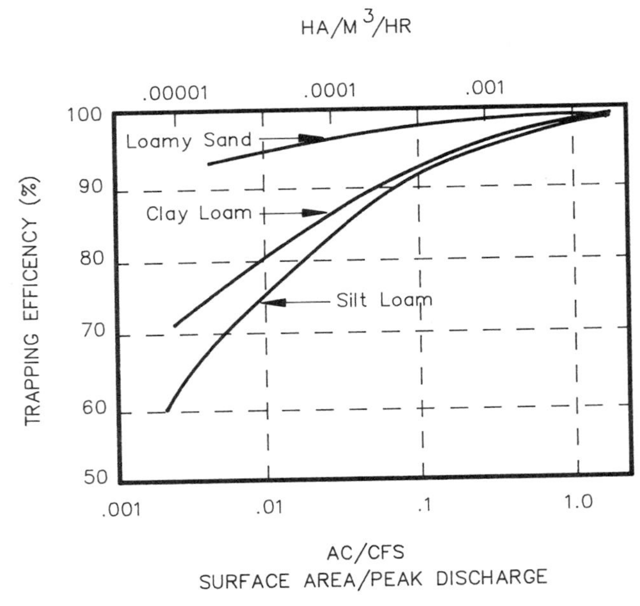

Figure 9.33 Maryland model of trapping efficiency in a pond (after McBurnie et al., 1990)

Predicting Pond Performance with Chemical Flocculation

It has been demonstrated clearly that chemical flocculants can be used to obtain sediment trapping efficiencies near 100% if the pond is properly sized and adequate mixing is obtained (Tapp *et al.*, 1981; Tapp and Barfield, 1986; Janiak, 1979). Prediction of pond performance when using flocculants is a difficult problem since floc sizes are changing due to flocculation in the reservoir. Procedures developed for predicting sediment removal with flocculation in steady-state flow will not work with variable-flow rates.

The performance of a given chemical used to enhance flocculation is typically determined by a procedure known as "jar tests." A jar test apparatus consists of a stirring platform on which solution in beakers, typically 1000 ml, is stirred with variable-speed paddle mixers. Sediment–water mixtures are placed in the jars and varying dosages of a given flocculant added. The mixture is typically stirred at a high speed of 100 rpm for 2 min to properly distribute the flocculant, and then at 60 rpm for 10 min followed by 40 rpm for 5 min. The purpose of the lower speeds is to give a high probability of particle contact and resulting flocculation without creating such large shear forces that flocs break apart. The jars are then tested for clay size particles remaining in suspension. The dosage that gives the minimum concentration of clay-sized particles is the optimum dosage. For the optimum dosage, a particle size distribution can be developed by withdrawing samples at varying depths and times as shown in Table 9.5. This size distribution information can then be used as input to any of the pond models described earlier for a first estimate of trapping efficiency. As indicated earlier,

Table 9.5 Sampling Time and Depths and Corresponding Overflow Rates and Particle Diameters

Time (min-sec)	Depth(cm)	Particle settling velocity (ft/hr)	Overflow rate[a] (gal/day•ft^2)	Equivalent particle diameter[b] (mm)
2–21	10	8.37	1500	0.029
3–31	10	5.59	1000	0.024
5–18	7.5	2.79	500	0.017
8–49	5	1.11	200	0.011
17–40	5	0.56	100	0.007
35–20	5	0.28	50	0.005

[a]1 gal/day•ft^2 = 40.7 liters/day•m^2.
[b]Equivalent particle diameter is the diameter of spherical particle that, according to Stoke's law, settles the indicated depth in the given time at 20°C and at a particle specific gravity of 2.65.

this would not account for the change in particle size distribution that occurs with flocculation in the pond.

Tapp *et al.* (1981) showed that the use of the size distribution determined from these jar test procedures in any of the variable flow models overpredicted the trapping efficiency of a pilot-scale sediment pond. The probable reason for the overprediction was the lack of proper mixing in the pilot-scale apparatus. The jar tests optimize the mixing process; hence maximum size flocs would be obtained. In the pilot-scale apparatus, mixing was conducted using techniques that would be available at a sediment pond located at a remote site without power. This consisted of running the water–sediment–flocculant mixture through a baffle system for the low-energy mixing with a total contact time of 1 min. Such mixing would not be optimum; hence the flocs likely did not reach the maximum size. It should be pointed out that although the flocs in the pilot-scale apparatus did not reach optimize size, the effluent-suspended solids concentration for the flocculant tests were at least one order of magnitude lower than those from identical tests without flocculant.

Because of the growth of flocs in the pond, the settling characteristics of the sediment change with time. Tapp and Barfield (1986) modified Eq. (7.7b) to account for floc growth. Since mean square turbulent velocity is required as an input to Eq. (7.7b), Tapp used the equations of Li *et al.* (1980) to predict turbulence. The use of flocculation routines improved the accuracy of the predictions in ponds; however, the procedures of Li *et al.* are not applicable to nonrectangular reservoirs. Wilson and Barfield (1986b) compared methods for predicting turbulence in sediment ponds, as summarized in Appendix A.

It can be concluded that the use of chemical flocculants will improve the performance of sediment ponds, but procedures for predicting effluent concentrations when using flocculants are not highly accurate.

Sediment Pond Design Procedures

The following discussion is an overview of procedures for designing a sediment pond. Specific attention is paid to designing each component of the ponds shown in Fig. 9.1. A typical reservoir site plan is shown in Fig. 9.34.

The procedures presented are not specific to any one set of regulations, but are general enough to be adapted to specific design requirements.

Designing for Sediment Storage Volume, $\forall_S$

The first decision to be made is the selection of an appropriate sediment storage volume. This volume is simply the storage occupied by the sediment deposited over the given design period. The design periods may be the life of the reservoir or the time between clean outs. Using a computed sediment yield and a bulk density of deposited sediment, the sediment storage volume is

$$\forall_S = \frac{Y_D}{W \times 43{,}560}, \qquad (9.82)$$

where $\forall_S$ is sediment storage volume in acre · feet, Y_D is sediment deposited over the design period in pounds, and W is weight density of deposited sediment in pounds per cubic foot. W can be determined from

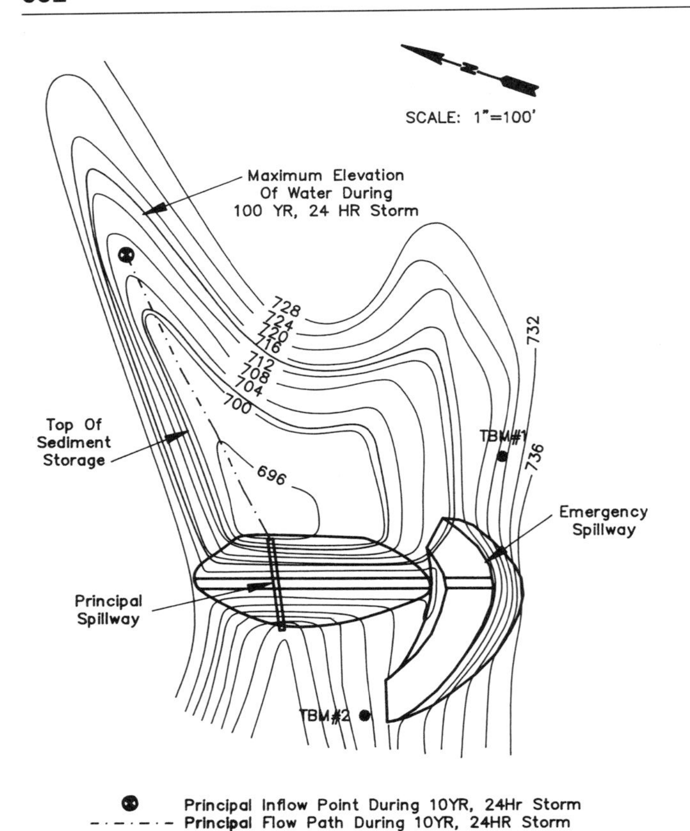

Figure 9.34 Example of a sediment pond site plan.

Eq. (7.10). Y_D is the product of the sediment yield over the design period and the trapping efficiency. This volume is assumed to be stored beneath the crest of the principal spillway as shown in Fig. 9.1.

Determining Peak Outflow Rate and Storage Volume

The key problem in sizing a sediment detention structure is determining the peak outflow rate and storage volume to meet a given standard. A number of design criteria are utilized in various regions of the world. Three criteria are discussed in this section.

1. Sizing the structure and outlet to meet a required detention time.
2. Sizing the structure and outlet to meet a required effluent concentration, either suspended solids or settleable solids.
3. Sizing the structure and outlet so that the peak outflow during or after disturbance does not exceed the predisturbance peak.

Each of the criteria will be discussed subsequently. The design should be based on the maximum storage and minimum outflow required by the standard as set by the appropriate regulatory authority.

Designing for Detention Storage Time To design for detention storage time, T_d, a reservoir storage volume and peak outflow discharge that gives the specified detention storage time must be selected. Procedures for making these calculations were discussed earlier. First estimates can be obtained from the triangular approximation given in Fig. 6.12, but should be checked by a detailed routing. If the detention storage time is too short, the outflow structure must be reduced in size and a new routing calculated to determine a new outflow hydrograph.

Credit for permanent pool volume can be calculated from Eq. (9.33). Caution should be exercised in selecting permanent pool volumes that approach the design storm runoff volume where there is no detention volume; i.e., the design storm does not appreciably increase the elevation of water in the pond. In this case, the principal spillway must be sized large enough to pass the peak of the design storm without an appreciable increase in water surface elevation. Short-circuiting could be a major problem with the large pipe sizes that result from such a design.

If a series of reservoirs in tandem are being used, no acceptable definition of detention storage time has been established. In this case, it will be necessary to select a series of reservoirs that give an equivalent trapping to one large reservoir with a given detention storage time. One can show from the overflow rate method that it is more effective to have one large reservoir than two smaller ones with a combined volume equal to that of the larger one.

Designing for Outflow Concentration To design for a maximum outflow concentration, a design storm must be specified. In addition, a hydrograph, sedimentgraph, and particle size distribution must be determined. If these are used as input to one of the non-steady-state sediment pond models, predictions can be made directly of suspended solids concentration in the effluent. If the DEPOSITS or CSTRS model that is in SEDIMOT II (Wilson *et al.*, 1982) is used, the model can also be used to predict settleable solids.

The above models require computer capability. In the absence of a computer, an average effluent concentration can be computed from

$$C = \frac{Y(1 - E)}{\gamma \forall}, \tag{9.83}$$

where C is average effluent concentration, Y is sediment yield during the storm, E is computed trap efficiency of the reservoir (must be calculated from one of the previous procedures), γ is weight density of water and $\forall$ is the volume of runoff. The concentration predicted will be the average concentration in a storm. One major problem with Eq. (9.83) is the effect of small errors in predicting trap efficiency. For example, with an inflow concentration of 100,000 mg/liter (not unusual for construction) a 1% error in E can make a 1000 mg/liter error in the predicted concentration. This inherent limitation in the method must be recognized.

After the effluent concentration is computed, this can be converted to a settleable solids concentration, if the effluent size distribution is known. Procedures for calculating the effluent size distribution are given in Example Problem 9.5 for the overflow rate and in Example Problem 9.9 for the plug flow model. Procedures for converting an effluent size distribution into settleable solids are given in Example Problem 7.2.

If the predicted effluent concentration is too high, the storage volume will need to be increased and the outlet pipe size decreased. Another option is to modify the size of permanent pool. As shown in Example Problem 9.8, an increase in permanent pool does not always increase the overall trapping efficiency; however, permanent pool does improve the water quality of the effluent from the first flush of flow.

Selecting the correct combination of permanent pool and temporary detention is a trial and error process that should become easier with experience. No optimization algorithms are available.

Design for Peak Discharge Reduction Design for peak discharge reduction was discussed in Chapter 6 under the title Flood Peak Reduction. Approximate procedures have been developed by the Soil Conservation Service (1975, 1986), in what is known as TR55, for determining storage volumes and peak outflow discharges. Details for making exact computations with routing are given in Chapter 6. Computer models such as TR20 (Soil Conservation Service, 1983) and SEDIMOT II (Wilson *et al.*, 1982) can also be used.

Sizing Emergency Spillways

The return period of a design storm for sediment control will typically be between 2 and 10 years. A reservoir that has only sufficient volume to safely pass this storm will overtop if an event with a much longer return period occurs during the life of the system. Such overtopping could result in a dam breach with subsequent downstream flooding. To prevent overtopping, an emergency spillway is designed to pass the rare event through a channel constructed in stabilized soil. This stabilized channel should not be constructed through the reservoir embankment.

The return period required for the emergency spillway design storm depends on the hazard classification of the reservoir. Example hazard classifications used by the Soil Conservation Service are given in Table 9.6 along with typical associated design storm return periods. Classifications and return periods are set by regulatory authorities in each state.

Once a return period precipitation is selected for the emergency spillway, it is translated into an inflow hydrograph. If the reservoir is small, the emergency spillway is usually sized to safely pass the peak inflow rate. If the reservoir is large, the flood should be routed

Table 9.6 Dam Hazard Classification and Return Period Storm for Emergency Spillway Design[a]

Hazard classification[b]	Description	Height–storage volume relationship[c]	Typical design storm for emergency spillways[d]
A	Dams in rural and agricultural areas where failure may damage farm buildings, agricultural lands, townships, or country roads	$S \times H < 3,000$	P_{25}
		$3,000 < S \times H < 30,000$	$P_{100} + 0.12(\text{PMP} - P_{100})$
		$30,000 < S \times H$	$P_{100} + 0.26\ (\text{PMP} - P_{100})$
B	Dams in predominantly rural and agricultural lands where failure may damage isolated houses, main highways, or minor railroads, or cause interruption of relatively important public utilities.		$P_{100} + 0.40\ (\text{PMP} - P_{100})$
C	Dams located where failure may cause loss of life, serious damage to homes, industrial and commercial buildings, important main highways or railroads.		PMP

[a]Adapted from Earth Dams and Reservoirs, TR 70, Soil Conservation Service.
[b]Hazard classifications and return periods are specified by regulatory agencies. Description and values given here are typical.
[c]S, storage in acre feet. H, dam height in feet.
[d]P_{25} = 25 year, 24-hr precipitation. P_{100} = 100 year, 24-hr precipitation. PMP, probable maximum precipitation.

through the reservoir to take advantage of the reservoir storage. The size of the emergency spillway for the routed storm will be less than that required to pass the peak flow. Procedures illustrated in Chapter 5 can be used to size the spillway.

Geotechnical Considerations

In the design of a reservoir, attention must be given to *seepage* through and under the dam and slope stability. To prevent this seepage, *cutoff trenches* are constructed below the dam and *antiseep collars* are located along the principal spillway. In addition, the dam must be properly compacted from appropriate materials. Details are given in appropriate soil mechanics texts.

Design of Ponds for Combination Sediment and Stormwater Control

In some situations, ponds are used for both sediment and storm water management. In this case, the final design must accomplish both objectives. Typically, sediment control standards are set for a more frequent return period, i.e., 2- to 10-year storm and storm water control for a more rare event, i.e, a 10- to 100-year storm. As shown in Chapter 10, the channel forming event is a 2-year storm; thus some regulatory authorities require basin design for storm water control for 2-, 10-, 25-, and 100-year events. This means that peak storm water discharges after land disturbances must match those of predisturbed conditions for each of the return periods. Design procedures for multiple return period storms, therefore, must assure that the peak discharges from the basin for each of the return period events do not exceed those prior to disturbance. In addition, if the basin is to be used for sediment control, the design must be checked for sediment effluent.

Designing for First Flush Filtration

After a disturbed area has been developed and stabilized, the highest concentration of sediment and other pollutants frequently occurs during the first flush of runoff when dry deposition materials are being washed from the unsaturated surface. A technique for controlling sediment in postdevelopment runoff is to use first-flush filtration. In this process, the first $\frac{1}{2}$ to 1 in. of runoff is diverted to a sedimentation/filtration basin and the remainder of the runoff goes directly into a storm water detention basin as shown conceptually in Figs. 9.35 and 9.36. An example of a first-flush diverter is shown in Fig. 9.37. As storm water flows into the inlet channel in Fig. 9.36, the first-flush flows through the slots on the isolation baffle into the sedimentation basin shown in Fig. 9.36. This flow continues until the

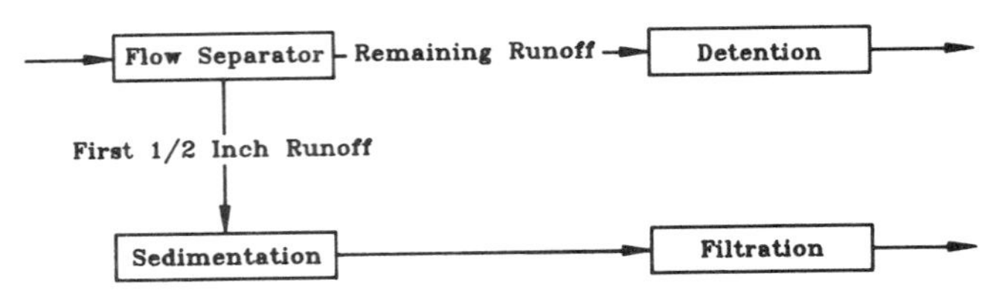

Figure 9.35 Flow chart for a first-flush filtration system.

sedimentation basin is filled with the required volume for the first flush. In design calculations, the elevation corresponding to this first-flush volume would be set as the elevation of the crest of the diversion weir in Fig. 9.37. At this point, water overtops the diversion weir and most of the subsequent storm flow diverts directly to the storm water detension basin.

Water in the sedimentation basin is detained long enough to allow coarse particles to settle out. The remaining finer particles are discharged along with the first-flush volume onto a filtration basin with a sand bed filtration system and an underdrain piping system as shown in Fig. 9.36. The volume of the sedimentation basin is set by the first-flush runoff to be treated, typically $\frac{1}{2}$ to 1 in. The surface area of the sedimentation basin is set by overflow rate and particle size trapping requirements. If it is assumed that 100% of particles with settling velocity V_{sr} are required to be trapped in the sedimentation basin, then the overflow rate given by Eq. (9.40) can be used to determine the required surface area, or

$$\frac{V_{sr}}{Q/A} = 1.0$$

and

$$A = \frac{Q}{V_{sr}}. \tag{9.84}$$

The flow rate can be approximated by a steady-state rate averaged over the flow time t_f, or

$$A = \frac{D_{FF} A_w}{t_f V_{sr}}, \tag{9.85}$$

where D_{FF} is the depth of runoff to be diverted, t_f is the drawdown time, and A_w is the watershed area. Appropriate units should be used. A value for V_{sr} should be selected on the basis of a minimum particle size to be trapped. To prevent clogging of the filter beds, one would anticipate trapping most of the coarse sediment in the sedimentation basin and the remainder of the fines in the filtration basis. Thus a particle diameter that represents a large fraction of the sedi-

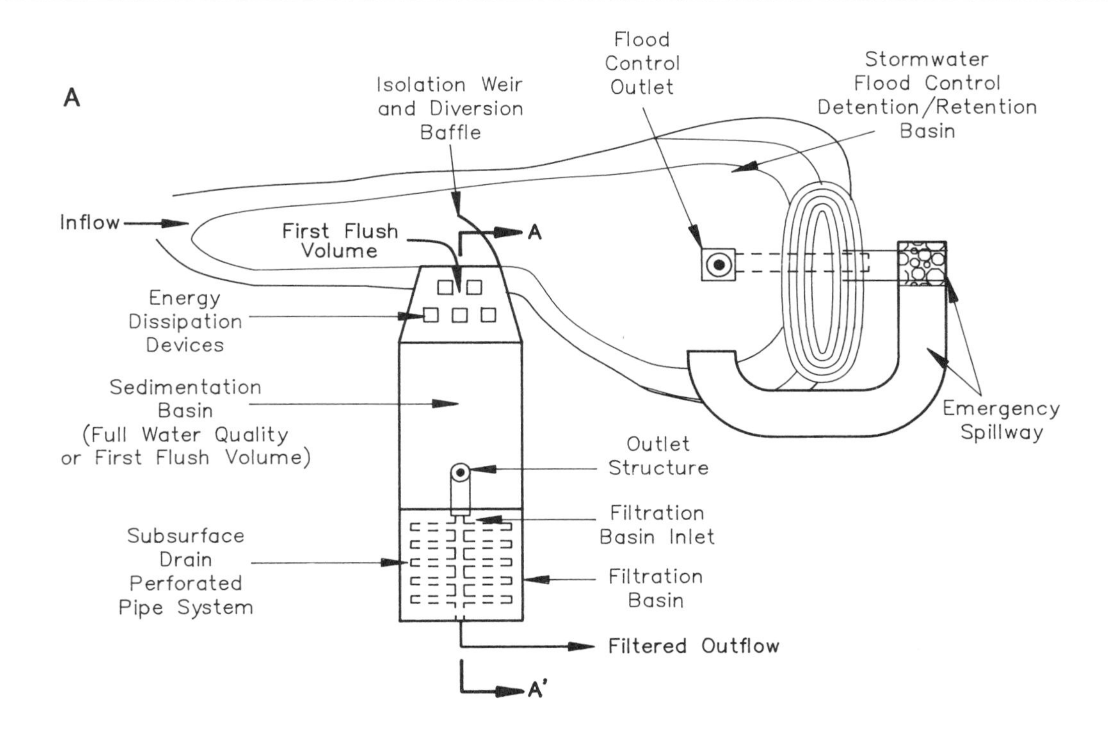

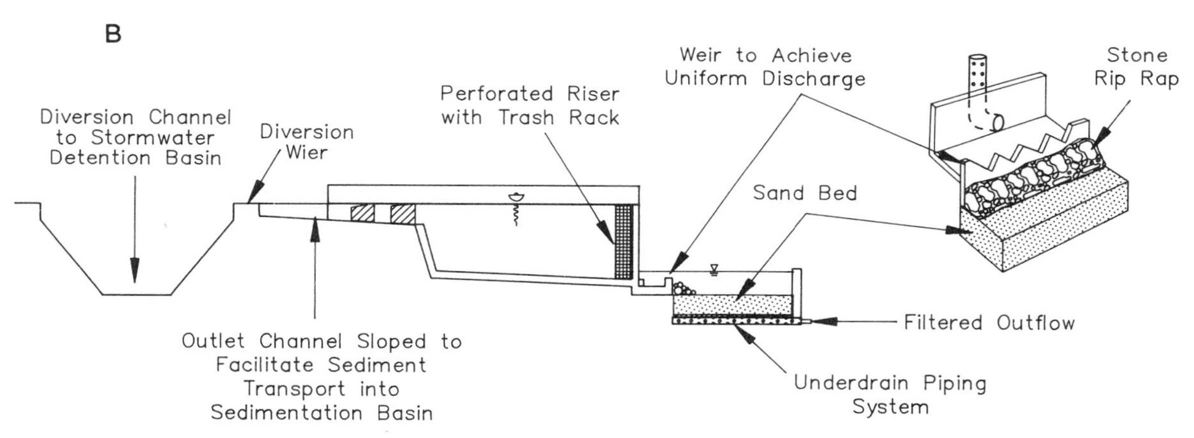

Figure 9.36 Example of a complete first-flush filtration system. (A) Conceptual plan view. (B) Elevation $A - A'$.

ment in a postdevelopment situation should be selected. The city of Austin, Texas, recommends a value of 20 μm (Austin, 1988) based on data from the Environmental Protection Agency (Driscol *et al.*, 1986). Using a diameter of 20 μm or other site-specific measurements, a corresponding settling velocity V_{sr} can be calculated from the minimum value of Eqs. (7.5) and (7.6). The drawdown time, t_f, should be selected on the basis of interarrival time between storms as well as safety and site characteristics. Values between 24 and 60 hr are currently being used (Austin, 1988). Flow controls for the drawdown should not be based on pipe size, but should be manually adjustable. Pipe sizes small enough to regulate the flows would likely plug from debris.

The first-flush volume should be the runoff that contains the first flush of material washed from the surface. The first flush is often called water quality volume and should include runoff from all impervious surfaces such as roadways, rooftops, and parking lots, plus all runoff from previous areas that drain onto impervious areas. Typically this is assumed to be $\frac{1}{2}$ in. of runoff (Austin, 1988).

The surface area of the filter chamber is controlled by the permeability and depth of the sand bed. A minimum depth of 18 in. is typically recommended

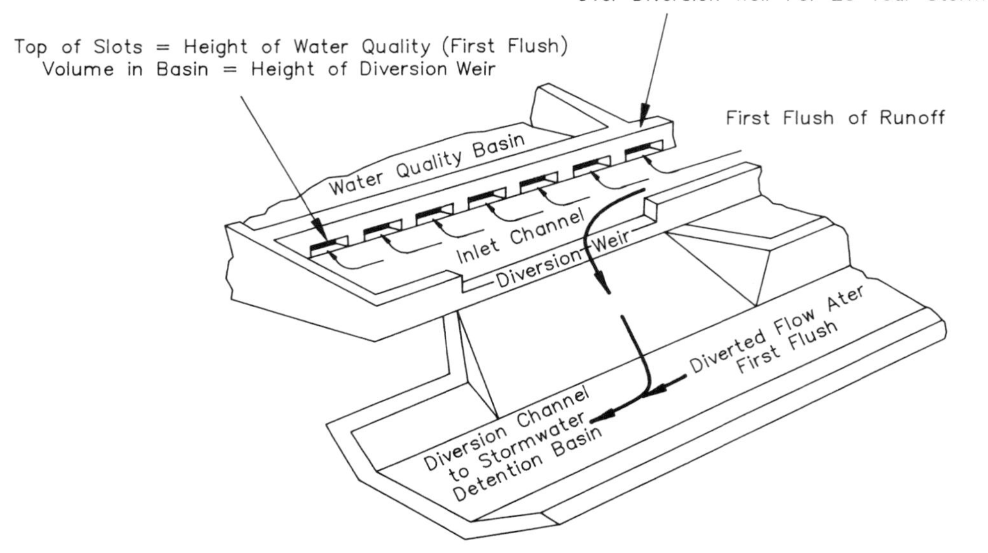

Figure 9.37 Details of a system to divert the first flush of storm water. The elevation of the crest of the diversion weir is set to allow overflow as soon as the first-flush volume is satisfied.

(Austin, 1988). Using Darcy's law for flow, Eq. (11.5), the velocity of drainage flow through the filter is given by

$$V_{\text{per}} = K\frac{\Delta H}{\Delta L}, \tag{9.86}$$

where V_{per} is flow through or percolation velocity (feet per second), K is hydraulic conductivity (feet per second), ΔH is head loss in flow through the filter, and ΔL is the distance over which the head is dissipated (see Fig. 9.38). If the depth of ponding on the filter at any time is given by H and the thickness of the filter is given by L_{f}, then ΔH is given by $H + L_{\text{f}}$ and ΔL is given by L_{f}. The flow through velocity then becomes

$$V_{\text{per}} = K\frac{H + L_{\text{f}}}{L_{\text{f}}}. \tag{9.87}$$

The surface area is set to allow the ponded water to drain in a specified time t_{dr}. To determine the depth that can be removed, one must utilize the continuity equation. For the purposes of developing equations to compute drawdown times, assume a column of water initially at height H_{o} as shown in Fig. 9.38, with no additional inflow. For this no-inflow situation, the rate of outflow per unit area of the water column, V_{per}, must equal the change in height of the column of

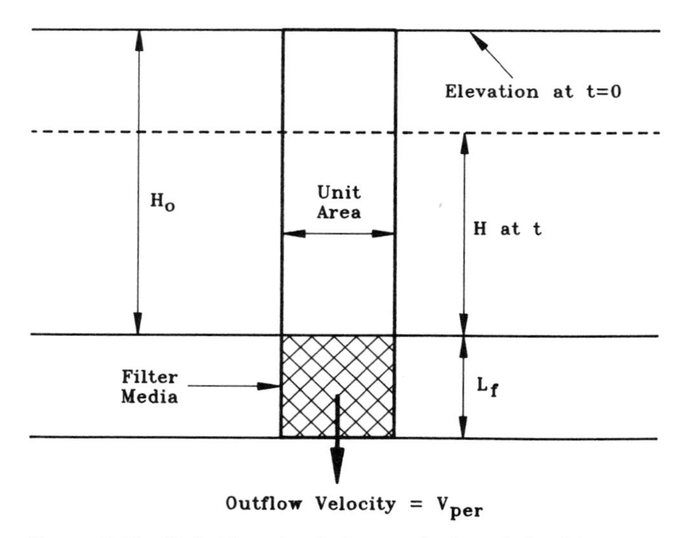

Figure 9.38 Definition sketch for continuity relationship as applied to filtration chamber.

water, or

$$dH/dt = -V_{\text{per}}. \tag{9.88}$$

Using Darcy's law, Eq. (9.87), for V_{per},

$$\frac{dH}{dt} = -K\frac{H + L_{\text{f}}}{L_{\text{f}}}. \tag{9.89}$$

Separating variables and integrating yields

$$H + L_f = (H_o + L_f)e^{-Kt/L_f}. \qquad (9.90)$$

The total volume of water, $C_{\forall,\text{per}}$, per unit area that has passed through the filter at any time t is

$$C_{\forall,\text{per}} = \int_0^t V_{\text{per}}\,dt \qquad (9.91)$$

or using Darcy's law for V_{per},

$$C_{\forall,\text{per}} = \int_0^t K\frac{(H + L_f)}{L_f}\,dt. \qquad (9.92)$$

Substituting Eq. (9.90) for $H + L_f$ and simplifying

$$C_{\forall,\text{per}} = (H_o + L_f)(1 - e^{-Kt/L_f}). \qquad (9.93)$$

If $C_{\forall,\text{per}}$ is set equal to the total volume above the filter, H_o, then Eq. (9.93) becomes

$$H_o = L_f(e^{Kt/L_f} - 1). \qquad (9.94)$$

Equation (9.94) defines the depth of water that can be drained through a filter of thickness L_f with hydraulic conductivity, K in time t. This can be used to define the filter area required. If D_{FF} is the depth of runoff from watershed area A_w, which is diverted, then the filter surface area A_f is given by

$$A_f = D_{FF}A_w/H_o, \qquad (9.95)$$

where H_o would be defined by the required drainage time. A maximum constraint for H_o would also typically be defined based on safety and site requirements. Computational procedures for first-flush filtration design are given in Example Problem 9.13.

Example Problem 9.13 Illustration of first-flush filtration

Runoff from a 20-acre watershed is to be controlled by a storm water detention basin with first-flush filtration capabilities, including a sedimentation basin and a filtration basin. Regulatory requirements dictate that the first 0.50 in. of runoff be diverted. The filter is to be 18 in. thick with sufficient underdrain capacity to remove peak flows. A sand that has a hydraulic conductivity of 3.5 ft/day is to be used. The sedimentation basin is to be drawn down in 24 hr or less. The filtration basin is to be drawn down in 36 hr. Water depth in the filtration and settling basins must not exceed 6.0 ft. Determine the area of the sedimentation basin and the filtration basin. Following the recommendations of the City of Austin, Texas, assume that 20-μm and larger particles should be trapped in the sedimentation basin.

Solution:

1. *Sedimentation basin design.* The runoff volume to be diverted is $\forall_{\text{div}} = D_{FF}A_w$, where A_w is the watershed area, or

$$\forall_{\text{div}} = \left(\frac{0.5}{12}\text{ft}\right)\left(20 \text{ acre} \times 43{,}560\ \frac{\text{ft}^2}{\text{acre}}\right) = 36{,}300 \text{ ft}^3.$$

Assume that a 20 μm (0.02 mm) particle is to be trapped in the basin, the required settling velocity is determined from Eq. (7.6) in centimeters per second as

$$\begin{aligned}\log_{10}V_{sr} &= -0.342(\log_{10} d)^2 + 0.989\log_{10} d + 1.15\\ &= -0.342(\log_{10}0.02)^2 + 0.989\log_{10}(0.02) + 1.15\\ &= -1.518\\ V_{sr} &= 10^{-1.518} = 0.030 \text{ cm/sec or } 3.54 \text{ ft/hr}.\end{aligned}$$

Using Eq. (9.85) to define surface area for the required drawdown time and settling velocity

$$A = \frac{D_{FF}A_w}{t_f V_{sr}} = \frac{36{,}300 \text{ ft}^3}{(24 \text{ hr})(3.54 \text{ ft/hr})}$$

$$A = 427 \text{ ft}^2 \text{ or } 0.01 \text{ acre}.$$

Thus, the settling basin should have a surface area of at least 427 ft^2. The required depth would be

$$D = \frac{36{,}300 \text{ ft}^3}{427 \text{ ft}^2} = 85 \text{ ft} > 6.0 \text{ ft}.$$

Use $D = 6.0$ ft based on the safety requirements set by the problem statement. The surface area becomes

$$= \frac{36{,}300 \text{ ft}^3}{6.0 \text{ ft}} = 6050 \text{ ft}^2 \text{ or } 0.14 \text{ acre}.$$

Since the surface area is much grater than that required to trap a 20-μm particle, the discharge time can be modified. Using Eq. (9.85), the minimum discharge time now becomes

$$t_f = \frac{D_{FF}A_w}{AV_{sr}} = \frac{36{,}300 \text{ ft}^3}{(6050 \text{ ft}^2)\cdot(3.54 \text{ ft/hr})} = 1.69 \text{ hr}.$$

An outlet must be sized to allow the sedimentation basin to completely discharge in 1.69 hr or more. This would typically be a slotted riser with an orifice on the outlet barrel. The riser would be wrapped with filter cloth to prevent coarse material from entering.

2. *Filtration basin.* The maximum head on the basin to allow discharge in 36 hr is given by Eq. (9.94) as

$$H_o = L_f(e^{Kt/L_f} - 1)$$

$$\frac{Kt}{L_f} = \left(\frac{3.5 \text{ ft/day}}{24 \text{ hr/day}}\right)(36 \text{ hr})/1.5 \text{ ft} = 3.5$$

$$H_o = 1.5 \text{ ft}\,(e^{3.5} - 1) = 48.2 > 6 \text{ ft}.$$

The maximum allowable head was stated to be 6.0 ft; thus

use H_o of 6.0 ft and reduce the drawdown time. The new drawdown time (which does not need to be controlled) is solved from Eq. (9.94), or

$$6 \text{ ft} = 1.5 \text{ ft}(e^{Kt/L_f} - 1).$$

Solving for Kt/L_f,

$$Kt/L_f = \ln\left(\frac{6}{1.5} + 1\right) = 1.609$$

$$t = \frac{(1.609)(1.5 \text{ ft})}{3.5 \text{ ft/day}}$$

$$= 0.690 \text{ days or } 16.6 \text{ hr}.$$

The required surface area is

$$A = \frac{D_{FF} A_w}{H_o} = \frac{36{,}300 \text{ ft}^3}{6.0 \text{ ft}} = 6050 \text{ ft}^2 \text{ or } 0.14 \text{ acre}.$$

The total surface area for the sedimentation basin and filtration basin is 0.14 + 0.14 or 0.28 acre.

No routing time delays are accounted for in the procedures above. The time delay for flow from the sedimentation basin along with the discharge time from the filtration basin during inflow periods would make the above analysis conservative. A more detailed analysis using routing techniques could be used to refine the design and reduce the surface areas.

Provisions would need to be made to clean the filters or provide for replacing the filter material.

Effect of Sediment Basins on Water Quality

A portion of the sediments deposited in a reservoir will be in the clay fraction and have surfaces with a charge density. This charge density, which varies with clay mineralogy, is known as the exchange phase of a soil and attracts ions from the solution. A substantial fraction of the total dissolved solids load (TDS) in a flow will be tied up on the exchange phase and hence will be trapped in a basin along with sediment. Present technology is not sufficiently developed to theoretically predict the amount of dissolved chemicals trapped in a reservoir, although Evangelou *et al.* (1987) propose procedures based on equilibrium chemistry for a limited number of metals. Hence, one must resort to empirical studies to define the effects of sedimentation basins on water quality.

Grizzard *et al.* (1986) recommend the following criteria for dry basins designed for water quality:

(1) Detain the runoff volume from an average annual storm for 24 hr.
(2) The runoff volume detained should be larger than the volume of runoff from an average rainstorm.
(3) The full volume of the basin should be drained in 40 hr or less.

Table 9.7 Range of Measured Long–Term Pollutant Removal for Sediment Detention Basins (after Stahre and Urbonos, 1990)

Item	Removal percentage
Total suspended solids (TSS)	50–70
Total phosphorous (TP)	10–20
Nitrogen	10–20
Organic matter	20–40
Lead	75–90
Zinc	30–60
Hydrocarbons	50–70
Bacteria	50–90

Stahre and Urbonos (1990) summarize the data from Grizzard *et al.* (1986), Occoquan Watershed Monitoring Laboratory (1986), and Whipple and Hunter (1981) and propose the range of trapping percentages for extended detention basins as given in Table 9.7. The values given are long-term averages. Considerable variation exists between storms. In some cases, a negative trapping efficiency will exist for certain stroms, resulting from resuspension of sediment and/or diffusion of chemicals from previously deposited sediment. Other data are given in Randall (1982) and Reed (1978).

CONSTRUCTED WETLANDS

Native Wetlands

Native Wetlands have been defined by Hammer and Bastian (1989) as

> an *ecotone*, an edge habitat, a transition space between dry land and deep water, an environment that is neither clearly terrestrial nor clearly aquatic. (p. 5)

The U.S. Fish and Wildlife Service (FWS) also recognizes wetlands as a transition zone with a water table that is periodically at or near the surface or covered by shallow water, meeting one or more of the following criteria (Cowardin *et al.*, 1979):

(1) Areas that primarily support hydrophyles.
(2) Areas with predominantly undrained hydric soils (wet enough for long enough to produce anaerobic conditions that limit the type plants).
(3) Areas with nonsoil substrate (such as sand or gravel) that are saturated or covered with shallow water at some time during the growing season.

This description by the FWS encompasses areas that are traditionally called swamps, marshes, and bogs. Swamps are wetlands that contain primarily woody species, marshes are areas that contain soft-stemmed plants, and bogs contain predominantly mosses (Hammer and Bastian, 1989).

The natural dynamics of wetlands lead to alternative wetting and drying as the area receive, hold, and recycle nutrients and runoff from upland areas. The wetting and drying tends to enhance native productivity. Continually wet areas have a different species than alternatively wet and dry areas.

According to Hammer and Bastian (1989), the most important and least understood function of wetlands is in water quality improvement. As a result of complex chemical and biological interactions, these areas can remove or convert large quantities of chemicals from a variety of point and nonpoint sources, including sediment, metals, nutrients, and organic matter. Some chemicals are completely immobilized and some are broken down into simpler forms.

Constructed Wetlands

Constructed wetlands are defined by Hammer and Bastian (1989) as

> a designed and man-made complex of saturated substrates, emergent and submergent vegetation, animal life and water that simulates natural wetlands for human use and benefits. (p. 12)

Other terms used for constructed wetlands include artificial, manmade, and engineered.

The most promising type of wetland for construction purposes is a marshland, due to its adaptability to fluctuating water levels and rapidly growing vegetation (Gearthart *et al.*, 1983). Vegetation that is likely to be successfully grown in manmade wetlands includes cattail, bulrush, rush, and giant reed, all of which tend to be adapted to high pollutant concentrations as well as fluctuating water levels (Small, 1976).

Elements of constructed wetlands include (Hammer and Bastian, 1989)

(1) subsurface flow zone with varying rates of hydraulic conductivity
(2) plants adapted to anaerobic conditions
(3) a water column above the surface
(4) invertebrates and vertebrates
(5) an aerobic and anaerobic population of microbes.

The subsurface zone includes the porous zone plus the zone of deposited litter and debris. To survive insect/pest outbreaks, plant species should be mixtures instead of monocultures.

The plant types present have the ability to transport enough oxygen to their roots to survive anaerobic conditions. The film aerobic region, called the rhizosphere, that is produced around the roots is essential to the microbial population that modifies nutrients, metallic ions, and other compounds. The microbial population is essential to the operation of wetlands and is, fortunately, ubiquitous, naturally occurring in most waters. They are fast growing and genetically plastic and thus can be rapidly adapted to a variety of incoming pollutants. Because of this plasticity, wetlands have been used to treat urban wastewater, acid mine drainage, highway runoff with high lead concentration, manganese, and runoff laden with organics and nutrients. They provide a simple alternative to physical treatment (see Fig. 9.39).

Constructed wetlands should not be used as a primary settling system for sediment since sediment particles will tend to cause deltas and clog flow systems. If sediment loads are high, pretreatment is necessary. Constructed wetlands are best used for removing contaminants other than sediment from the flow, and thus should be used downstream of a sediment control structure. Hammer (1989) edited a book that contains an excellent summary of the state of the art in wetland design.

VEGETATIVE FILTER STRIPS AND RIPARIAN VEGETATION

Description of Vegetative Filter Strips

Vegetative filter strips (VFS) are zones of vegetation through which sediment and pollutant-laden flow are directed before being discharged to a concentrated flow channel. This control technique has been described by a number of different terms including vegetative filter strips, grass filters, grass filter strips, buffer strips, riparian vegetation buffer strips, and constructed filter strips. For the purpose of this text, two main classifications will be used with subclassifications for each:

Constructed filter strips: Filter strips that are constructed and maintained to allow for primarily overland flow through the vegetation. Vegetation is grass-like plants with density approaching that of tall lawn grass.

Natural vegetative strips: Any natural vegetative area through which sediment-laden flow is directed, including riparian vegetation around drainage channels. Flow is typically *not* broad overland sheet flow,

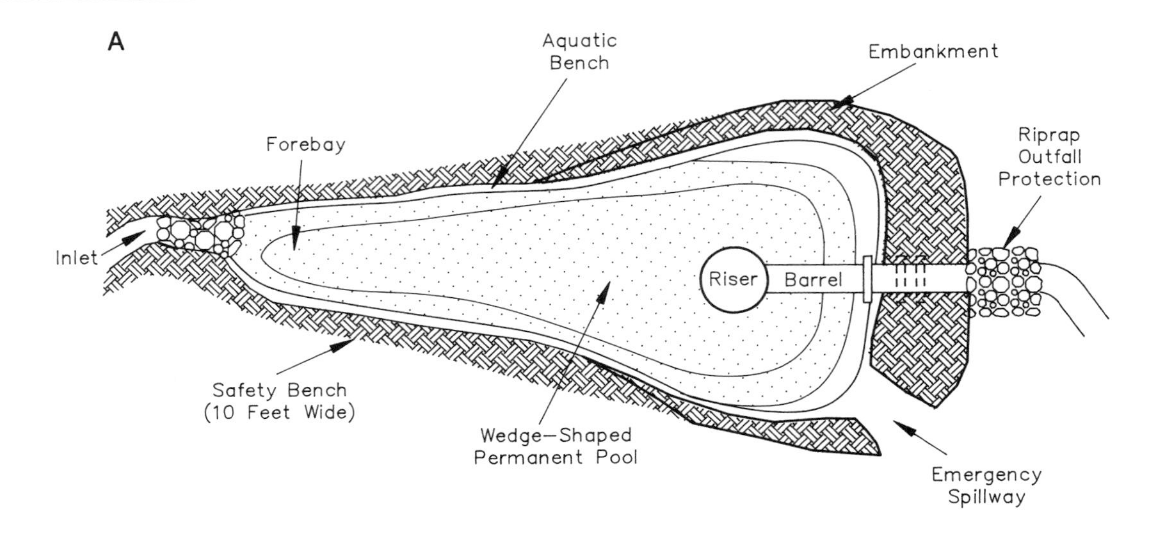

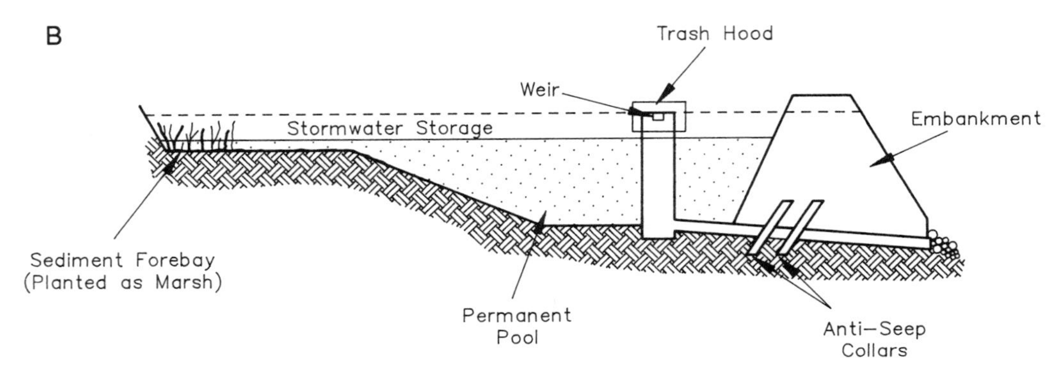

Figure 9.39 An example of a constructed wetland integrated as a part of a sedimentation basin. (A) Plan view. (B) Section view.

but occurs in small concentrated flow channels or flow zones. These channels occur as a result of channelization resulting from the natural topography as well as a result of the deposition delta that frequently forms at the leading edge of the vegetation. Vegetation can range from grass-like plants to brush or trees with ground litter.

Riparian vegetative buffer strips are strips of vegetation that grow along stream and concentrated flow channels. The vegetation may be constructed or natural.

To be effective, the VFS will normally be located on the contour perpendicular to the general direction of flow. A schematic of a typical VFS is given in Fig. 9.40.

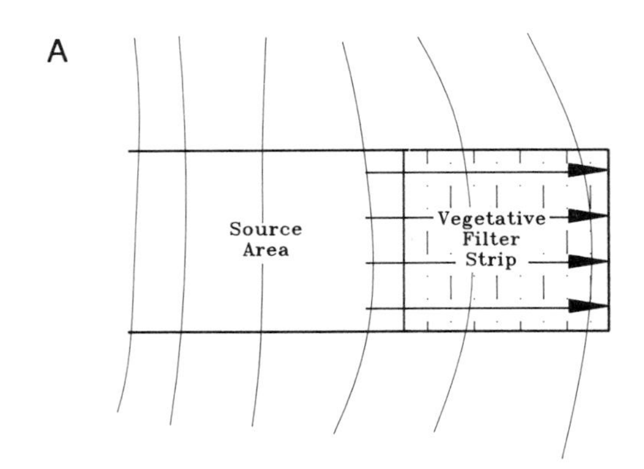

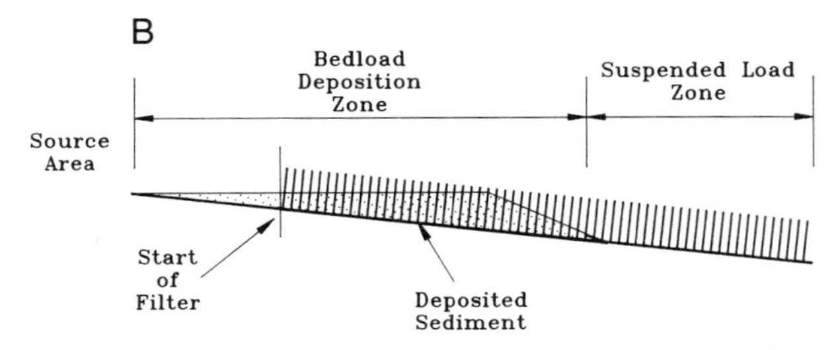

Figure 9.40 Schematic of a typical vegetative filter strip. (A) Plan view. (B) Profile view.

Effectiveness of VFS in Trapping Sediment

It has been established that VFS are effective as a nonpoint pollution control device when properly utilized. Measurements on naturally occurring vegetative strips have shown that they are very effective in removing sediments and some dissolved solids, but the degree of the effectiveness is not clearly established except for a few specific cases (Dillaha *et al.*, 1986, 1988; Cooper and Gilliam, 1987). Operational models to evaluate the effectiveness of naturally occurring strips have not been developed, although a research oriented model is available for natural grassed areas (Inamdar, 1993; Barfield *et al.*, 1993). Conversely, measurements and prediction methods are fairly well defined for constructed VFS (Hayes *et al.*, 1984) and evaluations can be made of their effectivness. These techniques can also be used for design, as discussed subsequently. Differences between constructed and naturally occurring VFS are discussed in this section, showing why the measurement and prediction techniques can be be projected from one to the other without modification. In the absence of any specific data, a discussion is given of how one might use experience and predictions from constructed VFS to give a first estimate of the effectiveness of natural VFS.

Whether designed and constructed or occurring naturally, VFS remove solids primarily by three mechanisms:

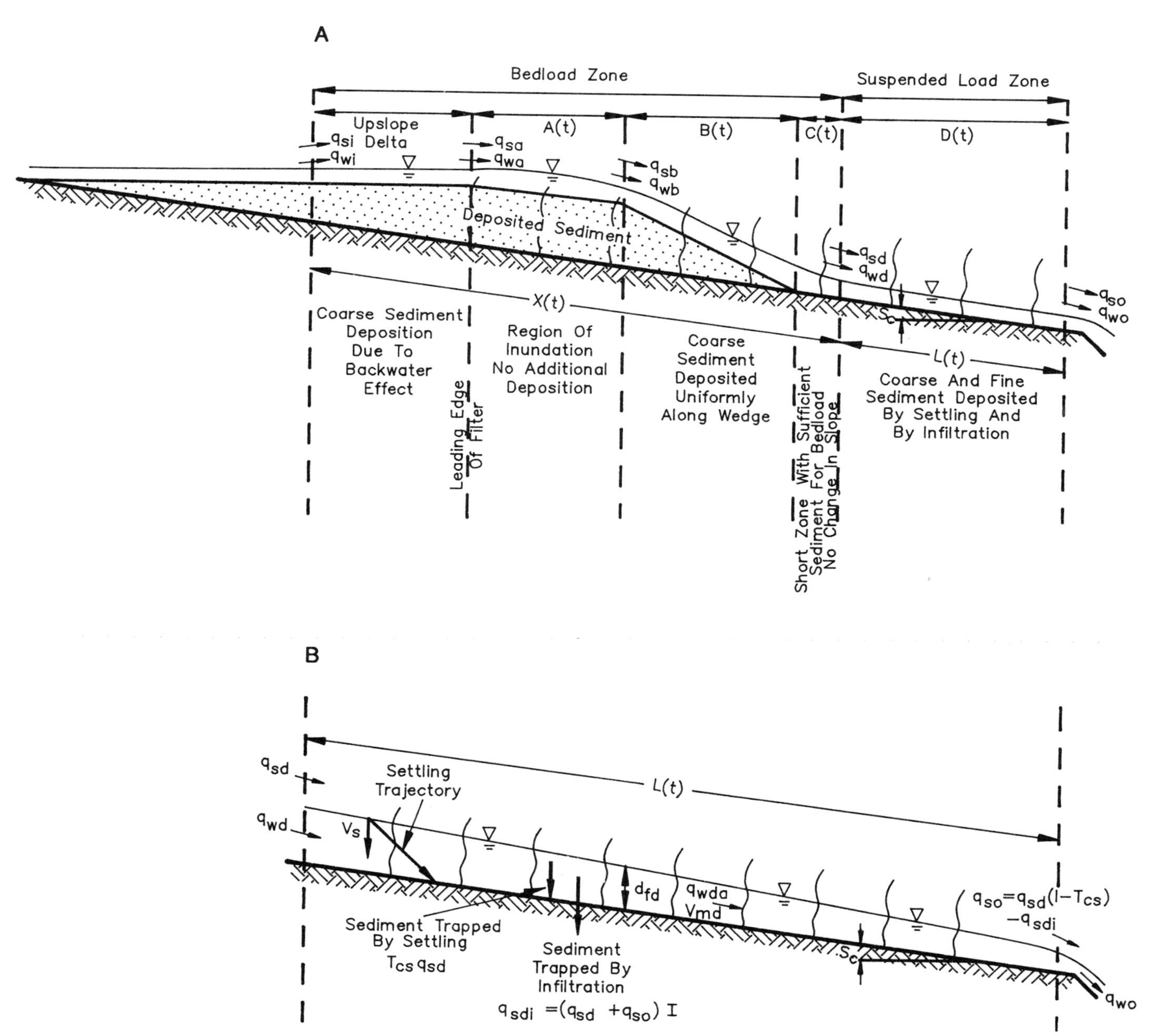

Figure 9.41 Illustration of the trapping mechanisms in VFS. Note that $C(t)$ is a short zone and its length is typically ignored. (A) Overview of all mechanisms. (B) Trapping mechanisms in zone $D(t)$.

(1) Deposition of bedload material and its attached chemicals as a result of decreased flow velocities and transport capacity. Such deposition occurs in a deposition wedge, either at the leading edge of the VFS or in a ponded area upslope of the VFS.

(2) Trapping of suspended solids in the litter collected at the surface. When suspended solids settle to the bed, they are trapped in the litter at the soil surface instead of being resuspended as would occur in a concentrated flow channel. When the litter becomes inundated with sediment, trapping no longer occurs as a result of this mechanism.

(3) Trapping of suspended material that moves into the soil matrix along with infiltrating water. This is the primary mechanism by which dispersed colloidal particles are trapped.

These trapping mechanisms are shown schematically in Fig. 9.41. The degree to which a VFS traps sediment thus depends on a number of factors. A factor of considerable importance is the degree of channelization that occurs. If runoff occurs as shallow overland flow spread fairly uniformly across the width of the filter, more sediment will be trapped than if the flow occurs in one or more small concentrated flow channels. In channelized flow, depth of flow will be greater, the transport capacity greater, and hence the amount of sediment transported through the filter will be much greater (trapping efficiency will be less) than for the case of shallow overland flow. Also, if flow is concentrated, the area where infiltration is occurring is reduced, which reduces the trapping of suspended sediment.

Table 9.8 Factors Affecting Trapping Efficiency for Vegetative Filter Strips

Parameter	Effect on trapping
Flow rate	Trapping inversely related to flow rate.
Size distribution	Trapping directly related to particle size. Larger particles are more easily trapped. Sand size and larger particles can be trapped with short filters on relatively steep slopes. Clay size particle are trapped only by infiltration that occurs on long relatively flat slopes and highly permeable soils.
Slope	Trapping inversely related to slope.
Type of vegetation	Vegetation that grows in clumps tends to be less effective than uniform growth.
Density of vegetation	Trapping directly related to the density of vegetation.
Stiffness of vegetation	Trapping related to stiffnes and height of vegetation. If the vegetation is not sufficiently stiff to remain erect under storm flow conditions, trapping will be greatly reduced as vegetation lays flat.
Height of vegetation	Theoretically, trapping increases with height, assuming that vegetation remains erect. Actually, tall vegetation does not naturally remain dense near the ground, resulting in a decreased trapping based on the number of vegetal elements per unit area at the ground surface. Also taller vegetation tends to lie flat during storm flow, which reduces trapping.
Infiltration rate	Trapping of suspended sediment, particularly colloidal and small silt particles, is directly proportional to infiltration rate.
Mass of litter	The depth of surface litter is directly related to the amount of sediment that can be trapped before the filter looses its effectiveness. For tree or brush like vegetation, trapping of sediment typically occurs as a result of the litter; thus trapping efficiency is directly related to mass of litter.
Degree of channelization	Trapping is inversely related to the degree to which flow is channelized. Channelized flow is deeper with higher velocities than overland flow, resulting in higher transport capacities and reduced trapping efficiencies.

Based on studies conducted by Hayes *et al.* (1978) and Tollner *et al.* (1976), it was found that the transport of sediment, and hence trapping of sediment, in VFS that are designed and constructed to minimize or eliminate concentrated flow is a function of a number of variables as shown in Table 9.8. Equations that predict sediment trapping as a function of these variables are presented in a later section.

Examples of the measured trapping efficiencies of constructed and natural VFS are given in Table 9.9, showing that the ability of the VFS to trap sediment is quite high. From this information, it is obvious why VFS are recommended practices, particularly riparian VFS.

Analysis of Flow in VFS

Flow velocities can be predicted in VFS by using a specially calibrated form of Manning's equation (Hayes *et al.*, 1978; Tollner *et al.*, 1976) in which an analogy is made between flow in vegetation with a spacing of S_S and flow in a rectangular channel with a flow depth of d_f, or

$$V_m = \left(\frac{1.5}{xn}\right) R_s^{2/3} S_c^{1/2}\ (\text{ft/sec}) \tag{9.96a}$$

or

$$V_m = \left(\frac{1.0}{xn}\right) R_s^{2/3} S_c^{1/2}\ (\text{m/sec}), \tag{9.96b}$$

where xn is a calibrated value for Manning's roughness, S_c is the slope of the channel, and R_s is a hydraulic radius based on average spacing of the media elements and the flow depth, or

$$R_s = \frac{S_s\, d_f}{2d_f + S_s}. \tag{9.97}$$

Table 9.9 Examples of Measured Trapping Efficiency for Vegetative Filter Strips

Study location and type vegetation	Description of site	Sediment trapping efficiency	Impact on water quality	Reference
University of Kentucky Fescue (KY 31)	Constructed filter field site, length approximately 30 m slopes 3–20%	87–99%	Not measured	Hayes *et al.* (1984)
University of Kentucky, SE KY Fescue (KY 31)	Constructed filter at strip mine site, 70 m slope 20%, monitored natural rainfall for 1 year	70–90%	Not measured	Barfield and Albrecht (1982)
Mississippi State University Fescue (KY 31)	Contructed filter for construction site, high clay soil, monitored natural rainfall for 1 year, slope 3%, 30 m long.	> 90%	Not measured	Hayes and Harriston (1983)
North Carolina State University Wetland Vegetation	Natural riparian forested area in Middle Coastal Plains of North Carolina	> 50%	50% Total P	Cooper *et al.* (1987)
Virginia Polytechnic University Orchard Grass	Nine constructed VFS controlling feedlot runoff. Lengths 4.6 to 9.1 m with 5 to 15% slopes	81–91%	50–69% Total P 60–74% Total N	Dillaha *et al.* (1986, 1988)
	Same type plots, controlling runoff from cropland	70–84%	61–79% Total P 54–73% Total N	Dillaha *et al.* (1989)
North Carolina State University Crab Grass and Bermuda	4.3- and 5.3-m plots controlling runoff from croplands	70%	50% Ortho P 26% Total P 50% Total N	Parsons *et al.* (1991)
Maryland KY 31 Fescue	4.6- and 9.2-m filters controlling drainage from plots spread with poultry manure and liquid nitrogen	66%	27% Total P 0% Total N	Magette *et al.* (1989)

Table 9.10 Values for Calibrated Manning's Roughness *xn* Vegetative Density and Vegetative Stiffness for Various Vegetative Types

Vegetation	Retardance class[a] unmowed/mowed	Density[b] (stems/ft^2/ spacing in.) $**/S_c$	Max. height before mowing[c] (in.)	Calibrated Manning roughness, *xn*	Stiffness[d] MEI unmowed/mowed N/m^2	Type stand
Vegetation typically recommended for VFS						
Yellow bluestem	A/D	250/0.76	NA	NA	300/0.1	Good
Tall fescue	B/D	360/0.63[f]	15	0.056	20/0.1	Good
Blue grama	B/D	350/0.64	10	0.056	20/0.1	Good
Ryegrass (perennial)	B/D	360/0.67[f]	7	0.056	20/0.1	Good
Weeping lovegrass	B/D	350/0.64	12	NA	20/0.1	Good
Bermudagrass	B&C/D	500/0.54	10	0.074	9/0.1	Good
Bahiagrass	C/D	NA	8	0.056	5/0.1	Good
Centipedegrass	C/D	500/0.54	6	0.074	5/0.1	Good
Kentucky bluegrass	C/D	350/0.64	8	0.056	5/0.1	Good
Grass mixture[e]	C/D	200/0.85	7	0.05	5/0.1	Good
Buffalograss	D/D	400/0.60	5	0.056	0.1/0.1	Good
Vegetation not typically recommended for VFS						
Alfalfa	B/E	100/1.20[g]	14	0.037[h]	20/0.05	Good
Sericea lespedeza	C/E	60/1.55[g]	16	0.037[h]	5/0.05	Good
Common lespedeza	D/E	30/2.19[g]	5	0.037[h]	0.1/0.05	Good
Sudangrass	D/E	10/3.80[g]	NA	0.037[h]	0.1/0.05	Good

[a]Best estimate by authors based on values from Table 4.3.

[b]Taken from Temple *et al.* (1987) except as noted. To convert densities for good stand to other stands, Temple *et al.* recommend multiplying the given densities by $\frac{1}{3}$, $\frac{2}{3}$, 1, $\frac{4}{3}$, and $\frac{5}{3}$ for poor, fair, good, very good, and excellent covers, repectively. The first number in the column is density in stems/ft^2. The second number is stem spacing in inches.

[c]Maximum heights recommended before mowing. Based on a value that would typically keep the vegetation erect. Checks should be made with shear velocity as shown in the text. Mowing should be routine to promote a dense growth, even if the maximum height given here is not reached. Heights of VFS after mowing would be 3 to 5 in., based on use of a standard tractor drawn field rotary mower. Adjustments should be made for other mowing implements.

[d]Adapted from Kouwen *et al.* (1981). Values are based on retardance class rather that calibration values given by Kouwen *et al.* For retardance class D, an average of the calibration value was used, as the retardance value appeared unrepresentative. For stands other than good stands, the retardance class changes for a given height; hence, MEI changes. Judgement will need to be used to estimate this change.

[e]Values could vary widely, depending on mixture. If a given grass type predominates, values for that species should be used.

[f]Adapted from Table V.1, Hayes *et al.* (1982).

[g]Not recommended for VFS. Studies by the authors have shown that spacings greater than 1 in. tend to cause scour and are hence not recommended. These numbers are one-fifth those in Temple *et al.* (1987). For VFS, the spacing of interest is that close to the surface. Values given by Temple *et al.* are for grass-lined channels where one is interested in an effective spacing that relates the retardance of the vegetation to shear force; hence Temple *et al.* multiplied the actual spacings for legumes by five to get an effective spacing.

[h]Based on grain sorghum, Table V.1, Hayes *et al.* (1982).

From continuity, the flow per unit width, q_w, is given as

$$q_w = V_m d_f. \qquad (9.98)$$

Equations (9.96)–(9.98) characterize the flow in VFS. Values for the calibrated xn are given in Table 9.10 for some of the various vegetal classes shown for grass waterways in Table 4.3. The user is cautioned that the values for xn are *not* standard values for Manning's n for overland flow, but are specially calibrated values, based on an analogy to flow in a rectangular channel with a width equal to the spacing of grass blades. Standard Manning's n values for overland flow should not be used in this equation.

Flow velocity equations presented above are valid only if the vegetation remains erect. Kouwen *et al.* (1981) evaluated the erectness of vegetation by comparing the shear velocity of the flow to a critical shear velocity based on the stiffness of the grass. He developed two relationships to predict the critical shear velocity, or

$$U_{c1}^* = 0.091 + 20.76\,(\text{MEI})^2 \qquad (9.99)$$

and

$$U_{c2}^* = 0.754\,(\text{MEI})^{0.106}, \qquad (9.100)$$

where U_{c1}^* is the critical shear velocity (ft/sec) of grass that is elastic, U_{c2}^* is the critical shear velocity (ft/sec) of grass that is stiff, and MEI is the stiffness of the grass (Nm^2) as given in Table 9.10. Kouwen *et al.* (1981) selected the lesser of the two values to compare to the actual shear velocity. Actual shear velocity, of course, is given by

$$U^* = \sqrt{g d_f S_c}, \qquad (9.101)$$

where g is the gravitational constant with other terms as described previously. If U^* is greater than the lessor of U_{c1}^* and U_{c2}^*, then the vegetation will not remain erect.

Example Problem 9.14 Calculation of flow depth and velocity in vegetative filter strips

Estimate the flow depth and velocity for a constructed VFS with overland flow if the vegetation is perennial KY 31 fescue on a slope of 8% and the flow rate is 0.074 cfs/ft width. Determine whether or not the grass will remain erect. Estimate the Reynold's number.

Solution:

1. *Parameters for calculation*. From Table 9.10, assuming a good stand of tall fescue

MEI = 20 unmowed, 0.1 mowed
$xn = 0.056$
$H = 15$ in. or 1.25 ft unmowed
$H = 4$ in. or 0.333 ft mowed (footnote C, Table 9.10)
$S_s = 0.63$ in. or 0.0525 ft.

2. *Calculating velocity and depth of flow*. Using Eqs. (9.96)–(9.98)

$$V_m = \frac{1.5}{xn} R_s^{2/3} S_c^{1/2}$$

$$R_s = \frac{S_s d_f}{2d_f + S_s}$$

$$q_w = V_m d_f.$$

Solving the equations simultaneously,

$$q_w = d_f \frac{1.5}{xn} R_s^{2/3} S_c^{1/2}$$

$$0.074 = \frac{1.5}{0.056} d_f^{5/3} \left(\frac{0.0525}{2d_f + 0.0525} \right)^{2/3} (0.08)^{1/2}$$

Collecting d_f terms into a term denoted as $f(d_f)$,

$$f(d_f) = \frac{d_f^{5/3}}{(2d_f + 0.0525)^{2/3}}$$

$$= \frac{(0.074)(0.056)}{(1.5)(0.08)^{1/2}(0.0525)^{2/3}} = 0.069.$$

The solution is by trial and error, or

Trial 1: $d_f = 0.15$ ft, $f(d_f) = 0.085$, high
Trial 2: $d_f = 0.10$ ft, $f(d_f) = 0.054$, low
Trial 3: $d_f = 0.12$ ft, $f(d_f) = 0.066$, low
Trial 4: $d_f = 0.125$ ft, $f(d_f) = 0.069$, OK.

Calculating velocity, spacing hydraulic radius from d_f, and subsequently checking the flow rate,

$$R_s = \frac{(0.125)(0.0525)}{(2)(0.125) + 0.0525} = 0.022 \text{ ft}$$

$$V_m = \frac{1.5}{0.056}(0.022)^{2/3}(0.08)^{1/2} = 0.594 \text{ ft/sec}$$

$$q_w = (0.594)(0.125) = 0.074 \text{ cfs/ft} \quad \text{OK.}$$

3. *Estimating the Reynold's Number*. Assuming that $\nu = 1.0 \times 10^{-5}$ ft²/sec,

$$R_e = \frac{V_m R_s}{\nu} = \frac{(0.594)(0.022)}{10^{-5}} = 1306.$$

4. *Checking to see if vegetation will remain erect.* Critical shear velocities [Eqs. (9.99) and (9.100)] Unmowed − MEI = 20

$$U_{c1}^* = 0.091 + 20.76\,(\mathrm{MEI})^2$$
$$= 0.091 + 20.76(20)^2 = 8304 \text{ ft/sec}$$
$$U_{c2}^* = 0.754\,(\mathrm{MEI})^{0.106}$$
$$= 0.754(20)^{0.106} = 1.04 \text{ ft/sec.}$$

Use $U_c^* = 1.04$ ft/sec.
Mowed − MEI = 0.1

$$U_{c1}^* = 0.091 + 20.76\,(\mathrm{MEI})^2$$
$$= 0.091 + 20.76(0.1)^2 = 0.299 \text{ ft/sec}$$
$$U_{c1}^* = 0.754\,(\mathrm{MEI})^{0.106}$$
$$= 0.754(0.1)^{0.106} = 0.591 \text{ ft/sec.}$$

Use $U_c^* = 0.591$ ft/sec.
Actual shear velocity [Eq. (9.101)]

$$U^* = (g d_f S_c)^{1/2} = [(32.2)(0.125)(0.08)]^{1/2}$$
$$= 0.567 \text{ ft/sec.}$$

Therefore, the grass will remain erect in either the mowed or unmowed condition as

$$U^* < U_c^*.$$

Sediment Trapping in VFS

Constructed VFS

As described earlier, constructed VFS are designed for broad overland flow without significant channelization. Mechanisms for trapping were illustrated earlier in Fig. 9.41. The representation of the zones as functions of time (t) is used to illustrate the fact that the length and location of the zones change with time. Based on studies by Tollner *et al.* (1976, 1978) and Hayes *et al.* (1984), equations for predicting sediment deposited and transported through each zone have been developed. Actual application to a design situation would normally require the use of a computer program.

An alternative approach to the use of the complete set of equations is to assume that the length of the deposition wedge is negligible and calculate the trapping that occurs in the suspended load zone, given as zone $D(t)$ in Fig. 9.41. Using this approach, the following steps would be taken:

(1) Calculate q_{sd}, the equilibrium sediment load entering zone $D(t)$
(2) Calculate the size distribution entering zone $D(t)$
(3) Calculate the fraction trapped in zone $D(t)$.

The steps are explained in the following sections. Subsequently, a summary is given in Appendix 9C of the relationships utilized to calculate the length of the deposition wedge.

Definitions of deposition and transport can best be given by moving from the downstream zones back to the upstream edge. This is the approach taken in the following discussion. The parameters discussed in the following sections are shown on Fig. 9.41. Although the length of zone $D(t)$ is treated as a constant in this discussion, it is denoted $L(t)$ for the reader who might want to use the relationships in Appendix 9C and calculate the variation in $L(t)$ with the formation of a deposition wedge.

In the following section, subscripts c and d refer to inflow to zone $C(t)$ and $D(t)$. When average parameters are needed on a segment, they are given the additional subscript a. The subscript o refers to the exit from zone $D(t)$. For example, q_{so} is sediment load leaving zone $D(t)$, q_{sd} is sediment load entering zone $D(t)$, and q_{sda} is average sediment load on zone $D(t)$.

In the discussions that follow, relationships for transport and deposition are given for a single particle size, although computations are usually made for a range of particle sizes. A discussion is given, after presentation of transport and deposition equations, of procedures for partitioning sediment load among particle classes and for estimating representative particle diameters. The discussion given below starts first with deposition in zone $D(t)$ and then moves to zone $C(t)$.

Deposition in Zone $D(t)$ Zone $D(t)$ represents that area within the filter in which the layer of litter on the bed has not been totally filled with sediment; thus bedload transport would be zero. Tollner *et al.* (1976) proposed that the trapping efficiency in this layer was a function of the number of times a particle could settle to the bed and of the flow Reynold's number. They experimentally determined that

$$T_s = \frac{q_{sd} - q_{so}}{q_{sd}} = \exp\left(-1.05 \times 10^{-3} R_e^{0.82} N_f^{-0.91}\right), \tag{9.102}$$

where T_s is the trapping efficiency in zone $D(t)$, R_e is the flow Reynold's number, and N_f is the fall number. Values for R_e were defined using the spacing hydraulic radius in Eq. (9.97), or

$$R_e = V_{mda} R_{sda}/\nu, \tag{9.103}$$

where ν is kinematic viscosity. Addition of the subscript da to V_m, R_s, and d_f in Eqs. (9.96)–(9.98) indicates that these values are estimated in zone $D(t)$ at the midpoint. Values for the fall number N_f are given by

$$N_f = V_s L(t)/V_{mda} d_{fda}, \tag{9.104}$$

where V_s is the settling velocity of the sediment and $L(t)$ is the total length of zone $D(t)$ in Fig. 9.41. Values for settling velocities are dependent on particle size, as defined in Chapter 7; thus the trapping efficiency changes with particle size. In the approximation being utilized here, $L(t)$ is a constant. If one wished to calculate the advance distance of the deposition wedge, $X(t)$, utilizing equations in Appendix 9C, $L(t)$ could be represented as a function of time by

$$L(t) = L_T - X(t), \tag{9.105}$$

where L_T is the total initial length of the filter prior to deposition and $X(t)$ is the distance the deposition wedge in zone $B(t)$ has advanced downstream. It is assumed in Eq. (9.105) that the length of zone $C(t)$ is small, as shown by Hayes *et al.* (1982, 1984).

In the development of Eq. (9.102), it was assumed that all particles reaching the bed are trapped in the litter. While this is a good assumption initially, it becomes questionable with time and deposition; thus, Wilson *et al.* (1982) proposed a correction function to the trapping efficiency, or

$$C' = 0.5\exp[-3D_{ep}] + 0.5\exp[15(0.2D_{ep} - D_{ep}^2)], \tag{9.106}$$

where D_{ep} is the average depth of sediment deposited in zone $D(t)$. The corrected trapping efficiency now becomes

$$T_{cs} = C'T_s \tag{9.107}$$

and the outflow sediment load (without considering infiltration losses) becomes

$$q_{so} = q_{sd}(1 - T_{cs}). \tag{9.108}$$

Sediment load in zone $D(t)$ is further reduced by infiltration. Hayes *et al.* (1984) evaluated the impact of infiltration on sediment load by assuming that:

(1) the difference in flow rate between the inlet of zone $D(t)$, q_{wd}, and outlet, q_{wo}, is only a result of infiltration, and
(2) the mass of sediment contained in a given infiltration volume is either transported into the soil matrix by infiltration or is trapped on the surface.

Based on these assumptions, the sediment deposited as a result of infiltration, q_{sdi}, is

$$q_{sdi} = (q_{sd} + q_{so})I, \tag{9.109}$$

where I is a dimensionless term related to the average infiltration rate. To simplify further equations, I is taken as half the infiltration rate to offset the fact that $(q_{sd} + q_{so})$ is not divided by two, or

$$I = \frac{q_{wd} - q_{wo}}{2q_{wda}}, \tag{9.110}$$

where q_{wd} and q_{wo} are flow rates entering and exiting zone $D(t)$ and q_{wda} is the average flow rate in zone $D(t)$ given by

$$q_{wda} = \frac{q_{wd} + q_{wo}}{2}. \tag{9.111}$$

Thus, the sediment load exiting zone $D(t)$ for a given particle size, when considering infiltration, becomes

$$q_{so} = (q_{sd} - q_{sdi})(1.0 - T_{cs})$$

or

$$q_{so} = [q_{sd} - (q_{sd} + q_{so})I][1.0 - T_{cs}]. \tag{9.112}$$

Solving for q_{so},

$$q_{so} = \frac{q_{sd}(1.0 - I)(1.0 - T_{cs})}{1.0 + I(1.0 - T_{cs})}. \tag{9.113}$$

Converting to a total fraction trapped, f_d, becomes

$$f_d = \frac{T_{cs} + 2I(1 - T_{cs})}{1 + I(1 - T_{cs})}. \tag{9.114}$$

Thus, sediment discharge rate from zone $D(t)$ for a given filter slope, length, and media spacing can be calculated, if the inflow rate of water and sediment, the infiltration rate, and the sediment particle size are known. The following calculations must be made

(1) Knowing the inflow rate of water, q_{wd}, the effective length of zone $D(t)$, and the infiltration rate, the outflow rate of water is given by

$$q_{wo} = q_{wd} - iL(t),$$

where i is the infiltration rate.
(2) The average flow rate, q_{wda}, on $D(t)$ is determined from Eq. (9.111) and the infiltration parameter from Eq. (9.110).
(3) Calculate the average flow depth, d_{fda}, and velocity, V_{mda}, from q_{wda} and simultaneous solution of Eqs. (9.96), (9.97), and (9.98).
(4) Calculate the fall number, N_f, the Reynold's number, R_e, from Eqs. (9.103) and (9.104).
(5) Calculate the trapping efficiency for particles settling to the bed by Eq. (9.102).

(6) If a cumulative deposition of D (inches) has occurred, calculate a corrected trapping efficiency by Eq. (9.106).
(7) Calculate the total sediment discharge corrected for infiltration by Eq. (9.113).

Use of these procedures is illustrated in a subsequent example.

Equilibrium Transport in Zone $C(t)$. Zone $C(t)$ is the zone in the filter where there is sufficient sediment deposition to allow bedload transport, but not sufficient deposition to alter the bedslope. Based on observations by Hayes *et al.* (1982), the length of zone $C(t)$ is small, and changes in flow rate and sediment load in the zone are small. Thus the equilibrium transport in zone $C(t)$ is the transport into zone $D(t)$. The subscript d, therefore, is used to represent values in zone $C(t)$.

Tollner *et al.* (1982) developed a calibrated version of the Einstein bedload function to predict sediment transport in this zone, or

$$\psi = 1.08\phi^{-0.28} \tag{9.115}$$

where ψ is Einstein's shear intensity given by

$$\psi = (\mathrm{SG} - 1)\frac{d_{\mathrm{pd}}}{S_{\mathrm{c}} R_{\mathrm{sd}}} \tag{9.116}$$

and ϕ is Einstein's transport rate function given by

$$\phi = \frac{q_{\mathrm{sd}}}{\gamma_{\mathrm{s}}\sqrt{(\mathrm{SG} - 1)\, g d_{\mathrm{pd}}^{3}}}, \tag{9.117}$$

where S_{c} is slope, SG is particle specific gravity, γ_{s} is particle weight density (pounds per cubic foot in the fps system), and d_{pd} is the representative particle size for bedload transport. For computational purposes, Eq. (9.115) can be rearranged to

$$q_{\mathrm{sd}} = \frac{K(R_{\mathrm{sd}} S_{\mathrm{c}})^{3.57}}{d_{\mathrm{pd}}^{2.07}}, \tag{9.118}$$

where K is a constant given by

$$K = (1.08)^{3.57}\gamma_{\mathrm{w}} g^{1/2}\mathrm{SG}(\mathrm{SG} - 1)^{-3.07}. \tag{9.119a}$$

If q_{sd} is in pounds per second per foot width, R_{sd} is in feet, and d_{pd} is particle diameter in millimeters, then

$$K = 6.462 \times 10^{7}\mathrm{SG}(\mathrm{SG} - 1)^{-3.07}. \tag{9.119b}$$

Because q_{sd} is a function of particle diameter, calculations must be made based on an average particle diameter. Procedures for estimating an average diameter are discussed in a following section.

To determine the sediment discharge rate from zone $C(t)$, and hence inflow into zone $D(t)$, for a given filter slope, length, media spacing, inflow rate of water, and representative sediment size, the following calculations must be made:

(1) Calculate the flow depth, velocity, and spacing hydraulic radius as in zone $D(t)$, using the flow rate q_{wd} instead of q_{wda}. Assume no infiltration in zones $A(t)$ through $C(t)$; hence

$$q_{\mathrm{wd}} = q_{\mathrm{wi}}.$$

(2) Calculate the sediment discharge rate, q_{sd}, for the representative particle size from Eq. (9.118).

Procedures for determining a representative particle diameter are discussed in a subsequent section.

Particle Size Distributions, Effective Particle Diameters, and Sediment Load Partitioning. Sediment transport and deposition for areas upstream of the VFS and in zones $A(t)$ through $D(t)$ are illustrated in Fig. 9.41. Zone $D(t)$ is the only one discussed here. Zones $A(t)$ and $B(t)$ are discussed in Appendix 9C. Discussions to this point have focussed on a single particle size. The use of a single particle size, however, to calculate deposition could indicate complete trapping of a mixture, although it is unlikely that fines would be trapped, particularly in the deposition wedge. To apply these equations to the various zones, it is necessary to make calculations for a variation of particle sizes and adjust the particle size distribution at various points in the VFS.

The following analysis is that presented by Hayes *et al.* (1982, 1984). Based on their observations, it was apparent that particles smaller than 0.037 mm were not trapped in the deposition wedge and that particles smaller than 0.004 mm were not trapped by settling in zone $D(t)$. Thus the particle size distribution was divided into three classes:

(a) $d_{\mathrm{pi}} > 0.037$ mm
(b) 0.004 mm $< d_{\mathrm{pi}} <$ 0.037 mm
(c) 0.004 mm $> d_{\mathrm{pi}}$.

For computational purposes, the particles greater than 0.037 mm can be divided into multiple size ranges.

Using trapping efficiencies and the size ranges described above, the size distribution was modified at three points in the VFS as shown in Fig. 9.42. For transport in the deposition wedge, zones $A(t)$ and $B(t)$, observations indicated that the particles being transported as bedload were primarily those greater than 0.037 mm (Hayes *et al.*, 1984); therefore, it was

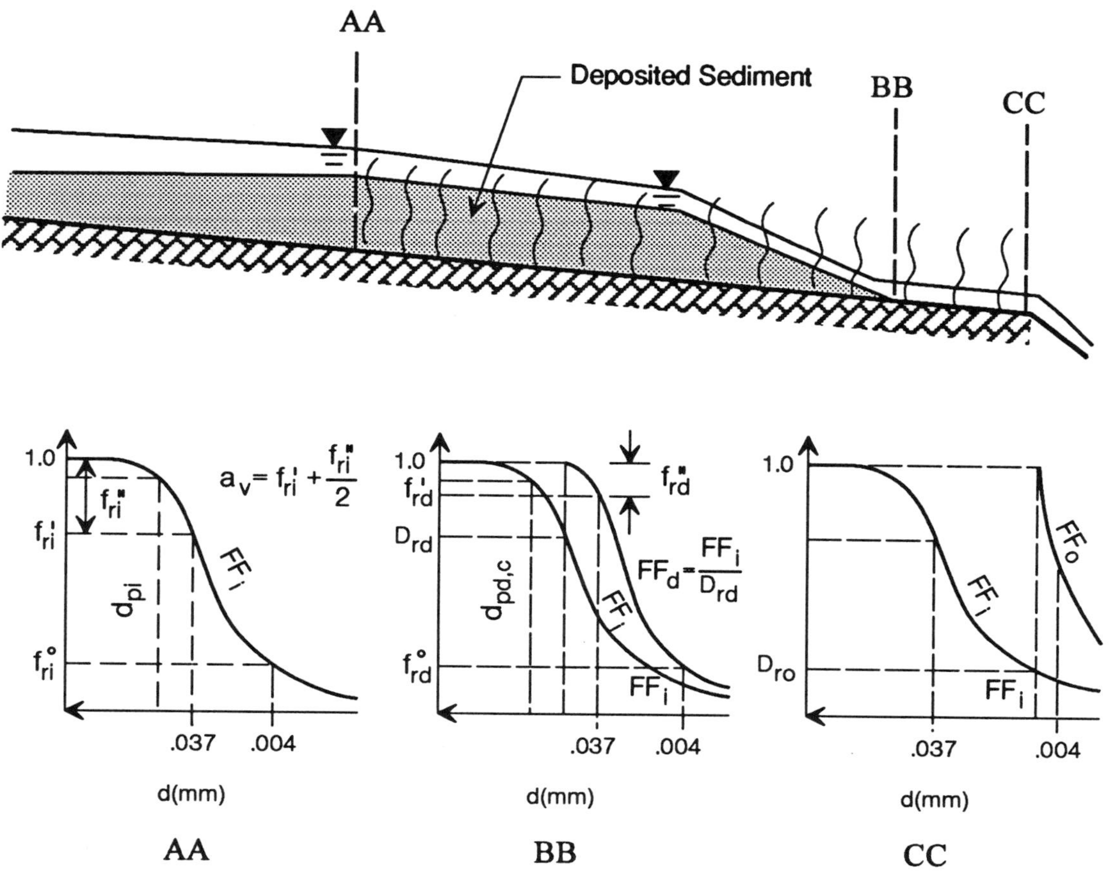

Figure 9.42 Illustration of particle size distribution computations.

assumed that all particles smaller than 0.037 mm were transported to zone $D(t)$ and that the representative diameter for bedload transport d_{pd} (also d_{pc}), was the average diameter of all particles coarser than 0.037 mm. Using these definitions, the total sediment load incoming to the upstream edge of the VFS is divided into two groups, or

$$q^{t}_{si} = q^{c}_{si} + q^{f+m}_{si}, \tag{9.120}$$

where q^{t}_{si} is the total load, q^{c}_{si} is the coarse material transport rate, and q^{f+m}_{si} is the fine material (silt and clay) transport rate. Further,

$$q^{f+m}_{si} = q^{t}_{si} f^{1}_{ri} \tag{9.121}$$

and

$$q^{c}_{si} = q^{t}_{si}(1.0 - f^{1}_{ri}), \tag{9.122}$$

where f^{1}_{ri} is the fraction of particles smaller than 0.037 mm, the lower limit of coarse material. Values for f^{1}_{ri} are determined from the input size distribution, using logarithmic interpolation as necessary. For bedload transport, the mean diameter of particles at the inlet is not the d_{50} of the entire size distribution, but the mean diameter of particles coarser than 0.037 mm. This is the size corresponding to a fraction finer given by

$$a_{v} = \frac{f^{11}_{ri}}{2} + f^{1}_{ri}, \tag{9.123}$$

where f^{11}_{r} is the fraction of coarse material. Given a_{v}, the value for d_{pi} is determined from the input size distribution, as shown in distribution AA in Fig. 9.42.

The next task is to calculate the size distribution at location BB in Fig. 9.42. Using the value for $d_{pd} = d_{pi}$, the transport capacity for bedload can be calculated for zone $C(t)$ from Eq. (9.118). The fraction of sediment trapped in the upstream deposition wedge can be calculated using

$$f = \frac{q^{c}_{si} - q^{c}_{sd}}{q^{c}_{si}}. \tag{9.124}$$

Using this value for f, the fraction of sediment delivered to zones $C(t)$ and $D(t)$ is

$$D_{\mathrm{rd}} = 1 - f \cdot f_{\mathrm{ri}}^{11}. \tag{9.125}$$

Using D_{rd}, the size distribution at the inlet, $\mathrm{FF_i}$, can be converted to that at BB, $\mathrm{FF_d}$, by

$$\mathrm{FF_d} = \mathrm{FF_i}/D_{\mathrm{rd}} \tag{9.126}$$

with the constraint

$$\mathrm{FF_d} \leq 1.0.$$

The resulting size distribution is shown in Fig. 9.42 at section BB. As is obvious from the figure, f_{r}^{1} and f_{r}^{11} at BB are different from that at the inlet.

In zone $D(t)$, sediment is trapped by settling and by infiltration as predicted by Eq. (9.114). When considering total sediment trapped in zone $D(t)$, it is necessary to divide the particles into three size ranges given above and calculate f_{d}, Eq. (9.114), for each size range. The following sizes are used for calculation

(a) Coarse particles: $d_{\mathrm{p}} > 0.037$ mm. Calculate d_{pda} using size distribution at BB.
(b) Medium (silt) particles: $0.004 \text{ mm} < d_{\mathrm{p}} \leq 0.037$ mm, $d_{\mathrm{pda}} = 0.012$ mm.
(c) Small (clay) particles: $d_{\mathrm{p}} \leq 0.004$ mm, $d_{\mathrm{pda}} = 0.002$ mm.

The total fraction trapped for each size is calculated from Eq. (9.114) and designated as $f_{\mathrm{d}}^{\mathrm{f}}$, $f_{\mathrm{d}}^{\mathrm{m}}$, and $f_{\mathrm{d}}^{\mathrm{c}}$ for clay, silt, and coarse particles, respectively. Using these values, the total trapping efficiency for the entire filter is then

$$\begin{aligned} f_{\mathrm{to}} = {} & [f + (f_{\mathrm{d}}^{\mathrm{c}})(1-f)][(1 - f_{\mathrm{ri}}^{1})] \\ & + f_{\mathrm{d}}^{\mathrm{m}}(f_{\mathrm{ri}}^{1} - f_{\mathrm{ri}}^{\mathrm{o}}) + f_{\mathrm{d}}^{\mathrm{f}} f_{\mathrm{ri}}^{\mathrm{o}}, \end{aligned} \tag{9.127}$$

where $f_{\mathrm{ri}}^{\mathrm{o}}$ is the fraction of inflow sediment smaller than 0.004 mm, and the total fraction of sediment reaching the outlet of the VFS is

$$D_{\mathrm{ro}} = 1 - f_{\mathrm{to}}. \tag{9.128}$$

The particle size distribution at section CC in Fig. 9.42, $\mathrm{FF_o}$, is given by

$$\mathrm{FF_o} = \mathrm{FF_i}/\mathrm{D_{ro}} \tag{9.129}$$

with the constraint

$$\mathrm{FF_o} \leq 1.0.$$

These procedures are illustrated in Example Problem 9.15.

Example Problem 9.15 Trapping efficiency of vegetative filter strips

A 50-ft-long vegetated filter strip is being built to control sediment from a disturbed area with an average sediment discharge of 0.5 lb/ft/sec and a size distribution as given in Fig. 9.43. The average inflow rate is 0.074 cfs/ft, the vegetation to be used is KY 31 fescue, and the slope is 8%. Estimate the trapping efficiency of the filter at the start of the storm. Assume steady-state flow conditions, and a specific gravity of sediment of 2.65, and a steady infiltration rate of 0.30 in./hr.

Solution:

1. *Input parameters for filter (same as Example Problem 9.14).* Assume a good stand, unmowed conditions are

$$xn = 0.056;\ S_{\mathrm{s}} = 0.63 \text{ in. or } 0.0525 \text{ ft};\ H = 15 \text{ in. or } 1.25 \text{ ft}.$$

Unmowed height is

$$H = 4 \text{ in. or } 0.333 \text{ ft}.$$

2. *Estimating flow depth and velocities in zone $D(t)$.* Calculations same as Example Problem 9.14,

$$R_{\mathrm{sd}} = 0.022 \text{ ft};\ V_{\mathrm{md}} = 0.594 \text{ ft/sec};\ d_{\mathrm{fd}} = 0.125 \text{ ft}.$$

3. *Estimating average size distribution for transport capacity in zones $C(t)$ and $D(t)$.* From Fig. 9.43, the fraction of particles entering the VFS are $f_{\mathrm{r}}^{1} = 0.36$ and $f_{\mathrm{r}}^{11} = 0.64$. From Eq. (9.123), the fraction corresponding to the average diameter of coarse particles (> 0.037 mm) is

$$a_{\mathrm{v}} = f_{r\mathrm{i}} + \frac{f_{r\mathrm{i}}^{11}}{2} = 0.36 + \frac{0.64}{2} = 0.68.$$

Thus from Fig. 9.43, the average diameter becomes

$$d_{\mathrm{pd}} = 0.17 \text{ mm}.$$

4. *Estimating bedload sediment transport capacity in zone $C(t)$.* From Eq. (9.118), using the symbol $q_{\mathrm{sd}}^{\mathrm{c}}$ for the coarse fraction of q_{sd},

$$q_{\mathrm{sd}}^{\mathrm{c}} = \frac{K(R_{\mathrm{sd}} S_{\mathrm{c}})^{3.57}}{d_{\mathrm{pd}}^{2.07}}$$

$$\begin{aligned} K &= 6.462 \times 10^{7}\,\mathrm{SG}(\mathrm{SG} - 1)^{-3.07} \\ &= 6.462 \times 10^{7}\,(2.65)(2.65 - 1)^{-3.07} = 3.68 \times 10^{7}. \end{aligned}$$

From item 2 above, $R_{\mathrm{sd}} = 0.022$; hence

$$\begin{aligned} q_{\mathrm{sd}}^{\mathrm{c}} &= \frac{(3.68 \times 10^{7})[(0.022)(0.08)]^{3.57}}{(0.17)^{2.07}} \\ &= 0.211 \text{ lb/sec} \cdot \text{ft}. \end{aligned}$$

5. *Estimating fraction of bedload trapped in upstream deposition wedge.* Using Eq. (9.124)

$$f = \frac{q_{\mathrm{si}}^{\mathrm{c}} - q_{\mathrm{sd}}^{\mathrm{c}}}{q_{\mathrm{si}}^{\mathrm{c}}}.$$

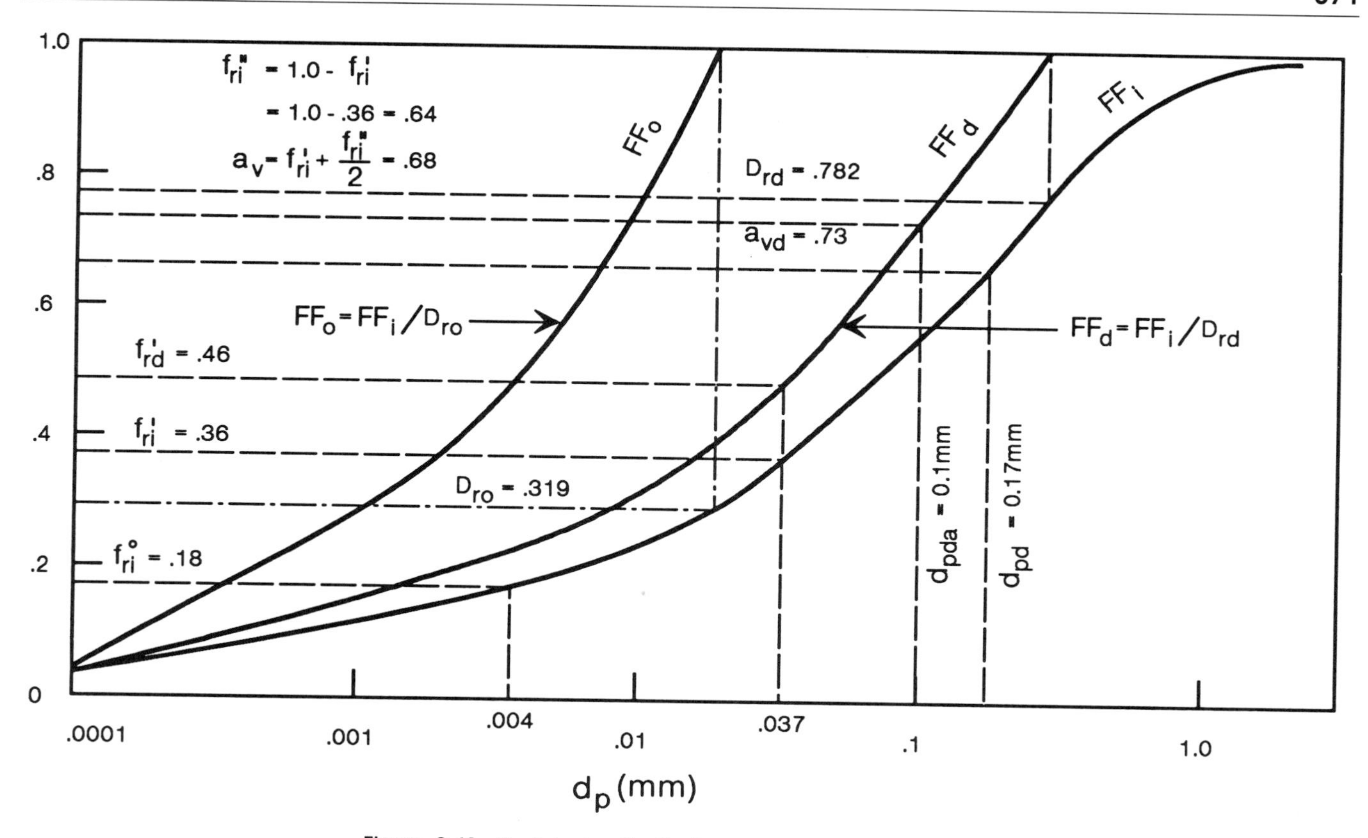

Figure 9.43 Particle size distributions for Example Problem 9.15.

The incoming sediment load that is coarse material (> 0.037 mm) is

$$q_{si}^{c} = q_{si} f_{ri}^{11} = (0.50 \text{ lb/sec} \cdot \text{ft})(0.64)$$
$$= 0.32 \text{ lb/sec} \cdot \text{ft}$$

and

$$f = \frac{0.32 - 0.211}{0.32} = 0.341.$$

6. *Estimating size distribution at BB (see Fig. 9.42).* From Eq. (9.125), the fraction delivered to zone $D(t)$ is

$$D_{rd} = 1 - f \cdot f_{r}^{11} = 1 - (0.341)(0.64)$$
$$= 0.782.$$

The size distribution at BB is given by

$$FF_{d} = \frac{FF_{i}}{D_{rd}} = \frac{FF_{i}}{0.782} \leq 1.0.$$

This distribution is shown in Fig. 9.43. From this distribution, $f_{rd}^{1} = 0.46$ and $f_{rd}^{11} = 1 - 0.46 = 0.54$. Using Eq. (9.123), the new value for a_{v} is denoted as a_{vd} and is

$$a_{vd} = 0.46 + \frac{0.54}{2} = 0.73.$$

Using a_{vd} with FF_{d} in Fig. 9.43, the average diameter for coarse particles in zone $D(t)$, d_{pda}, is 0.1 mm.

7. *Estimating total sediment trapped.* Ignoring the effect of the deposition wedge on the effective length of zone $D(t)$,

$$L(t) = L_{T} = 50 \text{ ft}.$$

The average diameters for clay and medium size particles in the model are 0.002 and 0.012 mm, respectively. The average diameter for coarse particles, d_{pda}, was determined under item 6 to be 0.1 mm. From Chapter 7, settling velocities are

$$V_{s,c} = 0.021 \text{ ft/sec} \quad \text{(Fig. 7.3)}$$

$$V_{s,m} = 2.81 d_{p}^{2} \quad \text{[Eq. (7.5)]}$$
$$= (2.81)(0.012)^{2} = 4.05 \times 10^{-4} \text{ ft/sec}$$

$$V_{s,f} = 2.81 d_{p}^{2} \quad \text{[Eq. (7.5)]}$$
$$= 2.81(0.002)^{2} = 1.124 \times 10^{-5} \text{ ft/sec}.$$

Trapping, settling, and infiltration must be estimated at the average flow condition for zone $D(t)$; thus the infiltration volume must be calculated. Knowing the infiltration rate i

$$q_{wo} = q_{wd} - iL(t)$$
$$iL(t) = \left(\frac{0.30}{12}\,\text{ft/hr}\right)(50\ \text{ft})\left(\frac{1\ \text{hr}}{3600\ \text{sec}}\right)$$
$$= 0.00035\ \text{cfs/ft}.$$

Hence

$$q_{wo} = 0.074 - 0.00035 = 0.07365\ \text{cfs/ft}.$$

Thus infiltration has no significant impact on flow. For purposes of calculating average depths of flow, it is therefore assumed that $q_{wo} = q_{wd} = 0.074$ cfs/ft and the average depths of flow and velocities are the same as those at the inlet to zone $D(t)$, or

$$V_{mda} = V_{md} = 0.594\ \text{ft/sec}$$
$$d_{fda} = d_{fd} = 0.125\ \text{ft}$$
$$R_{sda} = R_{sd} = 0.022\ \text{ft}.$$

Using Eq. (9.103) for Reynold's number and $\nu = 1.0 \times 10^{-5}$ ft^2/sec,

$$R_e = \frac{V_{mda} R_{sda}}{\nu} = \frac{(0.594)(0.022)}{10^{-5}} = 1306.$$

Using Eq. (9.104), the fall number is

$$N_f = \frac{V_s L(t)}{V_{mda} d_{fda}} = \frac{V_s(50)}{(0.594)(0.125)}$$
$$= 673\ V_s.$$

Calculating individual values

$$\text{Coarse } N_f^c = (673)(0.021) = 14.13$$
$$\text{Medium (silt) } N_f^m = (673)(4.05 \times 10^{-4}) = 0.2726$$
$$\text{Clay } N_f^f = (673)(1.124 \times 10^{-5}) = 0.00756.$$

From Eqs. (9.110) and (9.111), the infiltration parameter is

$$I = \frac{q_{wd} - q_{iso}}{2q_{wda}} = \frac{q_{wd} - q_{wo}}{q_{wd} + q_{wo}} = \frac{0.00035}{0.074}.$$
$$= 0.00473$$

From Eq. (9.102), the fraction trapped by settling, T_s is

$$T_s = \exp\left[-1.05 \times 10^{-3} R_e^{0.82} N_f^{-0.91}\right]$$
$$= \exp\left[-1.05 \times 10^{-3} (1306)^{0.82} N_f^{-0.91}\right]$$
$$= \exp\left[-0.377\ N_f^{-0.91}\right].$$

Calculating values for individual particle classes,

$$\text{Coarse } T_s^c = 0.967$$
$$\text{Medium } T_s^m = 0.291$$
$$\text{Clay } T_s^f = 0.0$$

From Eq. (9.114), the total sediment trapped in zone $D(t)$ including infiltration is

$$f_d = \frac{T_{cs} + 2I(1 - T_{cs})}{1 + I(1 - T_{cs})}.$$

Assuming no prior deposition, $T_{cs} = T_s$. From the above calculations, $I = 0.00473$; hence

$$f_d = \frac{T_{cs} + (2)(0.00473)(1 - T_{cs})}{1 + (0.00473)(1 - T_{cs})}.$$

Values are tabulated below.

Particle class	T_{cs}	Parameter	Calculated f_d
Coarse	0.967	f_d^c	0.967
Medium	0.292	f_d^m	0.298
Fine	0.000	f_d^f	0.009

The total trapping efficiency [Eq. (9.127)]

$$f_{to} = [f + f_d^c(1 - f)](1 - f_{ri}^1) + f_d^m(f_{ri}^1 - f_{ri}^o) + f_d^f f_{ri}^o.$$

Separating the incoming sediment into coarse (sand), medium (silt), and clay-sized particles is done from Fig. 9.43, or

$$\text{Coarse } f_{ri}^{11} = 0.64$$
$$\text{Medium + fine } f_{ri}^1 = 0.36$$
$$\text{Clay } f_{ri}^o = 0.18.$$

The effects of the length of the deposition wedge (zones $A(t)$ and $B(t)$) are being ignored in calculations of trapping in zone $D(t)$ as affected by the effective length of the zone. However, bedload is actually being trapped there and with a fraction trapped, f, of 0.341, as calculated under item 5. Hence, from Eq. (9.127) cited above, the total trapping in the filter will be

$$f_{to} = [0.341 + 0.967(1 - 0.341)](1 - 0.36)$$
$$+ 0.298(0.36 - 0.18) + 0.009(0.18)$$
$$f_{to} = 0.681.$$

Thus the filter traps 68.1% of the incoming sediment.

8. *Calculating outflow sediment load*. By definition of the total fraction trapped,

$$q_{so} = q_{si}(1 - f_{to})$$
$$= 0.5(1 - 0.681)$$
$$= 0.160\ \text{lb/sec} \cdot \text{ft}.$$

9. *Particle size of outgoing sediment*. From Eq. (9.125), the total delivery ratio of sediment at the filter exit is

$$D_{ro} = 1 - f_{to} = 1 - 0.681 = 0.319.$$

From Eq. (9.126), the particle size distribution at the filter exit is

$$FF_o = \frac{FF_i}{D_{ro}} = \frac{FF_i}{0.319} < 1.0.$$

This particle size distribution is shown in Fig. 9.43.

10. *Summary*.

$$q_{si} = 0.5 \text{ lb/ft} \cdot \text{sec}$$

$$q_{so} = 0.160 \text{ lb/ft} \cdot \text{sec}$$

Total trapping efficiency equals 68.1%.

Natural VFS

Trapping in natural VFS differs from that in constructed VFS due to differences in types of vegetation and to the effects of channelized flow. Corrections needed for each effect are discussed below.

Effects of Channelized Flow Channelization causes an increase in flow rate per unit width and a decrease in trapping. Such channelization is normally randomly distributed across a VFS, making it difficult to quantify. A first estimate of the effects of channelization on trapping could be made by a correction to the inflow rate and sediment load, or

$$q'_{wi} = q_{wi}/C_{cf} \quad (9.130)$$

$$q'_{si} = q_{si}/C_{cf}, \quad (9.131)$$

where C_{cf} is a correction factor for channelized flow. A possible correction factor would be the fraction of the VFS through which flow occurs. For example, if, as a result of channelization, a given VFS has flow through only 40% of its area, then a first estimate for C_{cf} might well be 0.4. After making the correction to flow rate and sediment load, sediment trapping could be evaluated by Eqs. (9.96)–(9.129).

Effects of Vegetation Type Impacts of vegetation type on trapping are related to growth patterns and type of litter. Grass-like vegetation that grows in clumps tends to be less effective as a sediment trap than uniformly growing grass. Tall grass-like vegetation tends to be less dense at the surface than low-growing-type vegetation and hence has a reduced trapping efficiency. These effects can be accounted for in the spacing parameter S_s in Eqs. (9.96)–(9.129).

For tree- and brush-like VFS, sediment trapping normally occurs in the litter at the ground surface, since the spacing of trees and brush are typically too large to have a significant impact on shallow overland flow. The flow equations described earlier were developed on the basis of the analogy between flow in a rectangular channel with a width equal to the media spacing and flow in a VFS. Conversely, flow in litter tends to be more analogous to turbulent flow in a porous media. Equations for such transport have not been developed.

Design of VFS

In the design of VFS, one must select a vegetative type, a groundslope, and a length of the filter (distance parallel to the flow path). The groundslope should be as close to the natural slope as possible. Slopes less than 1% and greater than 10% should be avoided, if possible. In both cases, maintenance becomes crucial. VFS should be located on the contour to minimize channelization. Vegetation should be selected to be dense, lawn-like grass in order to minimize channelization. Turf specialists in an area should be contacted for information on desirable species.

The length of the filter is an important design parameter. Typically, the length should be such that excessive deposition will not occur during the design life. In selecting the filter length, the following steps can be used.

(1) Select the design life and a maximum allowable deposition. Ten years and 15 cm are recommended by Hayes and Dillaha (1991).

(2) Estimate the long-term sediment yield entering the filter (see Chapter 8) and a design single-storm yield. Hayes and Dillaha (1991) recommend a 10-year storm.

(3) Calculate the required trapping efficiency to meet a water quality standard. If a water quality standard is not required, make a first estimate of trapping efficiency.

(4) Estimate the length necessary to prevent deposition above that allowable over the life of the system.

(5) Use the filter length to calculate the trapping efficiency for the design storm, ignoring the deposition wedge (zones $A(t)$ and $B(t)$ in Fig. 9.41).

(6) Repeat (4) and (5) until the lengths match.

(7) Estimate, if desired, the trapping efficiency for the design storm, considering deposition in zones $A(t)$–$C(t)$ using relationships in Appendix 9C.

Example Problem 9.16 Design of a vegetative filter strip

Estimate the length of a VFS needed in Western Kentucky to reduce the long term sediment yield to less than 1 ton/acre/yr if the actual sediment yield to the VFS is 5 tons/acre/yr with the size distribution given by FF_i in Fig.

9.43. Flows and vegetation are the same as in Example Problem 9.14 and 9.15. The average flow length from the disturbed area is 200 ft.

Solution:

1. *Selecting a vegetation type*. The vegetation assumed in Example Problem 9.15 was KY 31 fescue, which is acceptable for growing conditions in Western Kentucky.

2. *Select a design life*. Based on a desired frequency of repair, use a design life of 10 years.

3. *First estimate the required filter length, based on depth of deposition*. As recommended by Hayes and Dillaha (1991), assume that the maximum allowable deposition is 15 cm and a bulk density of 1.5. Also for the first approximation, assume that the width of the filter (distance parallel to the slope) equals the width of the disturbed area. To reduce the sediment load from 5 tons/acre to 1 ton/acre, the trapping efficiency must be

$$f_{\mathrm{to}} \geq \tfrac{4}{5} = 0.8.$$

The total sediment to be trapped, using the given length of 200 ft for the disturbed area, is

$$\begin{aligned}&(0.8)(5\ \mathrm{tons/acre\cdot yr})(10\ \mathrm{years})(1\ \mathrm{acre}/43{,}560\ \mathrm{ft}^2)\\&\quad\times(2000\ \mathrm{lb/ton})(200\ \mathrm{ft})\\&\quad= 367\ \mathrm{lb/ft\ filter\ width}.\end{aligned}$$

The units here are pounds per foot measured along the contour. The depth of sediment to be trapped per unit area is the total mass of sediment per unit length of filter divided by the bulk density. The bulk density is

$$\gamma_{\mathrm{sb}} = \mathrm{SG} \times 62.4 = (1.5)(62.4) = 93.6\ \mathrm{lb/ft^3}.$$

Hence the depth of deposition is

$$\begin{aligned}\mathrm{Depth} &= \frac{367\ \mathrm{lb/ft}}{L_{\mathrm{T}}(\mathrm{ft})\gamma_{\mathrm{sb}}}\\&= \frac{367\ \mathrm{lb/ft}}{L_{\mathrm{T}}(\mathrm{ft})(93.6\ \mathrm{lb/ft^3})} = \frac{3.92}{L_{\mathrm{T}}}\ \mathrm{ft}.\end{aligned}$$

To keep this depth below 15 cm or 0.49 ft,

$$L_{\mathrm{T}} \geq \frac{3.92}{0.49} = 8.0\ \mathrm{ft}.$$

4. *Calculate the trapping efficiency*. To be conservative, all deposition in the deposition wedge (zones $A(t)$ and $B(t)$) will be ignored). Thus the size fraction and diameters are given as $\mathrm{FF_i}$ in Fig. 9.43. Using Fig. 9.43, the values in the following table were determined in Examples Problem 9.15.

Particle class	Fraction, $\mathrm{FF_i}$	Average diameter, d_{pd} (mm)	Fall Velocity, V_{s} (ft/sec)
Coarse (sand)	0.64	0.170	0.021
Medium (silt)	0.18	0.012	4.05×10^{-4}
Fine (clay)	0.18	0.002	1.124×10^{-5}

For the flow rate of 0.074 cfs/ft, the depth of flow is determined the same as in Example Problem 9.14. The results are $d_{\mathrm{f}} = 0.125$ ft, $R_{\mathrm{s}} = 0.022$ ft/sec, $S_{\mathrm{s}} = 0.0525$ ft, $xn = 0.056$, $V_{\mathrm{m}} = 0.594$ ft/sec, and $R_{\mathrm{e}} = 1306$. Hence, since L_{T} equals L, the fall number becomes

$$N_{\mathrm{f}} = \frac{V_{\mathrm{s}}L_{\mathrm{T}}}{V_{\mathrm{m}}d_{\mathrm{f}}} = \frac{V_{\mathrm{s}}L_{\mathrm{T}}}{(0.594)(0.125)} = 13.5\, V_{\mathrm{s}}L_{\mathrm{T}}.$$

For $L_{\mathrm{T}} = 8$ ft,

$$\begin{aligned}\mathrm{Coarse}\ N_{\mathrm{f}}^{\mathrm{c}} &= 2.27\\\mathrm{Medium}\ N_{\mathrm{f}}^{\mathrm{m}} &= 0.44\\\mathrm{Fine}\ N_{\mathrm{f}}^{\mathrm{f}} &= 0.001.\end{aligned}$$

Trapping efficiency for each particle size

$$\begin{aligned}f_{\mathrm{d}} = T_{\mathrm{s}} &= \exp\left[-1.05\times10^{-3}R_{\mathrm{e}}^{0.82}N_{\mathrm{f}}^{-0.91}\right]\\&= \exp\left[-1.05\times10^{-3}(1306)^{0.82}N_{\mathrm{f}}^{-0.91}\right]\\&= \exp\left[-0.377\,N_{\mathrm{f}}^{-0.91}\right].\end{aligned}$$

Total trapping efficiency

$$\begin{aligned}\mathrm{Coarse}\ f_{\mathrm{d}}^{\mathrm{c}} &= \exp\left[(-0.377)(2.27)^{-0.91}\right] = 0.836\\\mathrm{Medium}\ f_{\mathrm{d}}^{\mathrm{m}} &= 0.002\\\mathrm{Fine}\ f_{\mathrm{d}}^{\mathrm{f}} &= 0.0.\end{aligned}$$

Using Eq. (9.127), assuming $f = 0$,

$$\begin{aligned}f_{\mathrm{to}} &= f_{\mathrm{d}}^{\mathrm{c}}\left(1 - f_{\mathrm{ri}}^{1}\right) + f_{\mathrm{d}}^{\mathrm{m}}\left(f_{\mathrm{ri}}^{1} - f_{\mathrm{ri}}^{\mathrm{o}}\right) + f_{\mathrm{d}}^{\mathrm{f}}f_{\mathrm{ri}}^{\mathrm{o}}\\&= 0.836(0.64) + 0.002(0.36 - 0.18) + (0.0)(0.18)\\&= 0.535.\end{aligned}$$

To reduce yield to 1 ton/acre, the trapping efficiency must be 0.80; thus the length must be increased.

5. *Calculating the length required based on trapping efficiency*. The length must be greater than 8 ft, which will make $f_{\mathrm{d}}^{\mathrm{c}}$ approach 1.0, as shown by the above computations. Based on the settling velocity, $f_{\mathrm{d}}^{\mathrm{f}} = 0$, for any reasonable length. Therefore using Eq. (9.127), the trapping efficiency becomes

$$\begin{aligned}f_{\mathrm{to}} &= \left(f_{\mathrm{d}}^{\mathrm{c}}\right)\left(1 - f_{\mathrm{ri}}^{1}\right) + f_{\mathrm{d}}^{\mathrm{m}}\left(f_{\mathrm{ri}}^{1} - f_{\mathrm{ri}}^{\mathrm{o}}\right)\\0.80 &= (1.0)(0.64) + f_{\mathrm{d}}^{\mathrm{m}}(0.18)\\f_{\mathrm{d}}^{\mathrm{m}} &= \frac{0.80 - 0.64}{0.18} = 0.889.\end{aligned}$$

Keeping $f_{\mathrm{d}}^{\mathrm{m}} \geq 0.889$ will ensure a total trapping efficiency of 0.80. From item 4 above,

$$f_{\mathrm{d}}^{\mathrm{m}} = \exp\left[-0.377(\mathrm{N_f^m})^{-0.91}\right] = 0.889$$

$$N_{\mathrm{f}}^{\mathrm{m}} = \left[\frac{\ln(0.889)}{-0.377}\right]^{-1/0.91} = 3.595.$$

Also, from item 4 above,

$$N_f^m = 13.5\, V_{sm} L_T = (13.5)(4.05 \times 10^{-4}) L_T$$

$$L_T = \frac{3.595}{(13.5)4.05 \times 10^{-4}} = 657 \text{ ft}.$$

Thus a total length of 657 ft of KY 31 fescue should reduce the sediment load to the desired level and remain effective for the design life of 10 years. This is obviously an excessive length. An alternative would be to utilize some cultural practices to reduce on-site erosion.

Impact of VFS on Water Quality

Data sets on the impacts of VFS on water quality are limited for anything other than sediment. The trapping will be highly related to the fraction of sediment in the clay size range since chemicals are attached to the exchange phase of the clay. As shown by the previous examples, colloidal particles not flocculated or aggregated are not likely to be trapped by settling, but could be trapped by infiltration with high infiltration rates. Thus a knowledge of the fraction of colloidal sediment tied up in aggregates is important. This is discussed in Chapter 7.

A summary of selected data sets available on the impacts of VFS on water quality is given in Table 9.9. The quantities of dissolved solids trapped varies widely. The complete mechanisms involved are not well understood or modeled at the present time.

POROUS STRUCTURES: CHECK DAMS, FILTER FENCES, AND STRAW BALES

Introduction

Porous structures are finding wide use as sediment control techniques. Examples include rock fill check dams in a stream, rock over sand check dams, filter fabric fence (filter fence), filter fabric over brush, gabions, and brush barriers. Many of these structures are simply built with the materials on hand.

Experimental data on the trapping efficiency on porous structures are extremely limited. In general, it is expected that the trapping efficiency will be low, particularly for rock-filled check dams. Reed (1978) measured a trapping efficiency of less than 5% for a check dam associated with highway construction. Depending on particle size, theoretical considerations indicate that the trapping efficiency of filter fence could be relatively high, as indicated in a subsequent example.

Evaluation of the sediment trapping by porous structures can be made by utilizing basic principles of sedimentation and hydraulics. These principles are discussed in the next section, followed by a description of the types of porous structures.

Predicting the Trapping Efficiency of Porous Structures

A limited number of prediction equations have been developed for defining the sediment trapping efficiencies of porous structures (Hirschi, 1981; Warner and Hirschi, 1983), as discussed in the section.

Porous structures trap sediment by three mechanisms.

1. Reduced total load transport capacity resulting in deposition (primarily bedload) behind the dam.
2. Mechanical filtration inside the porous media.
3. Trapping of fines as a result of infiltration into the deposited sediment.

The relative importance of the three mechanisms varies between types of porous structures, as discussed below.

Porous structures trap sediment only if the flow does not overtop the structure. Once overtopping occurs, the trapping efficiency effectively becomes zero.

Modeling Sediment Trapping by Settling

Hirschi (1981) and Warner and Hirschi (1983) developed equations defining sediment trapping as a result of reduced transport capacity behind porous structures. The trapping equations are based on the following assumptions (see Fig. 9.44)

(1) A quiescent settling zone occurs upstream of the sediment structure with a length calculated from a single-step backwater curve.
(2) The settling depth is the average of the depth of 1.1 times the normal flow depth and the depth at the dam.
(3) The flow through velocity is either
  (a) given as a function of porosity, or
  (b) defined by slurry flow rate.

The following derivations are given for both assumptions (3a) and (3b).

Trapping Efficiency—Velocity Defined by Porosity In this derivation, equations are developed only for triangular channels with side slopes of 2:1. The assumption is made that the discharge is given by

$$Q = V_n A_n = V_d A_d \epsilon, \qquad (9.132)$$

where V_n is the normal velocity, A_n is the area corresponding to normal depth, V_d is the velocity at the dam, A_d is the cross-sectional area at the dam, and ϵ is the porosity of the dam. Using Manning's equation

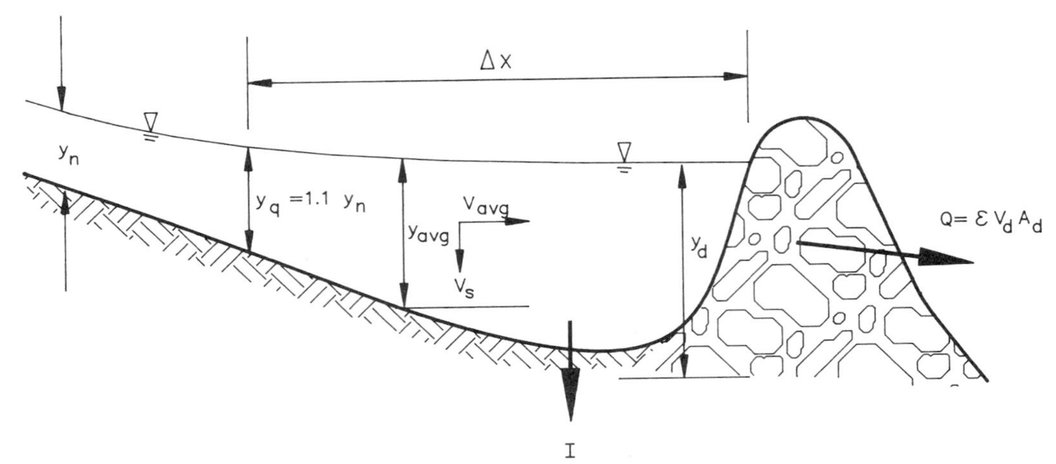

Figure 9.44 Backwater profile behind porous dam.

to define normal flow depth for a triangular channel, the normal depth becomes (assuming that the hydraulic radius is half the depth)

$$y_n = 1.024\left(\frac{nQ}{ZS_c^{0.5}}\right)^{3/8}, \qquad (9.133)$$

where Q is discharge in cfs, y_n is normal depth in feet, n is Manning's roughness, Z is the channel side slope, and S is channel slope in feet per foot. If the assumption is made that the zone of quiescent flow starts at a depth of $y_q = 1.1\ y_n$, then

$$y_q = 1.126\left(nQ/ZS_c^{0.5}\right)^{3/8}. \qquad (9.134)$$

Using Eq. (9.132) for continuity, and $A_d = Zy_d^2$,

$$y_d = [A_d/Z]^{1/2} = [Q/\epsilon V_d Z]^{1/2}. \qquad (9.135)$$

If further assumption is made that

$$V_d = \epsilon V_n, \qquad (9.136)$$

then

$$y_d = y_n/\epsilon. \qquad (9.137)$$

Next, the length of quiescent settling zone can be estimated from a single-step backwater curve, using Eq. (4.54), or

$$\Delta x = \frac{E_d - E_q}{S_c - S_f}, \qquad (9.138)$$

where E_d and E_q refer to total energy at the dam and at the start of the quiescent zone and S_f and S_c are the slope of the energy grade line and channel, respectively. If it is assumed that S_f is given by Manning's equation using the velocity at the dam, then

$$\Delta x = \frac{y_d + V_d^2/2g - y_q - V_q^2/2g}{S_c - \left[nV_d/1.49\ R_d^{2/3}\right]^2}, \qquad (9.139)$$

where R_d is the hydraulic radius at the dam. For Z greater than 2.0,

$$R_d \cong y_d/2; \qquad (9.140)$$

hence

$$\Delta x = \frac{y_n/\epsilon + Q'\epsilon^2 - 1.1y_n - 0.683\ Q'}{S - \left(2.27gn^2\epsilon^{3.33}/y_n^{1.33}\right)Q'}, \qquad (9.141)$$

where

$$Q' = Q^2/2gy_n^4Z^2. \qquad (9.142)$$

Since the constant used for Manning's equation in Eq. (9.141) is 1.49, all inputs must be in feet per second and feet. The gravitational constant must be 32.2 ft/sec^2.

Hirschi (1981) proposed that the trapping efficiency for a given particle size could be given by

$$T_s = V_s t_D/y_{avg}, \qquad (9.143)$$

where V_s is the particle settling velocity, t_D is the average flow through or detention time, and y_{avg} is the average depth in the quiescent zone. The value of y_{avg} can be given by

$$y_{avg} = \frac{y_d + y_n}{2} \qquad (9.144)$$

or using Eq. (9.137),

$$y_{avg} = \frac{y_n}{2}\left(1 + \frac{1}{\epsilon}\right). \tag{9.145}$$

The flow-through time is determined by using the average of V_n and V_d as an average velocity; hence

$$t_D = \frac{2\,\Delta x}{V_n + V_d}. \tag{9.146}$$

Since $V_d = \epsilon V_n$,

$$t_d = \frac{2Zy_n^2\,\Delta X}{Q(1+\epsilon)}. \tag{9.147}$$

The trapping efficiency for a given particle size is determined by simultaneously solving Eqs. (9.133), (9.141), (9.143), (9.145), and (9.147). The use of these equations is illustrated in Example Problem 9.17. Example predictions for a variety of conditions is also given in Fig. 9.45.

Example Problem 9.17. Porosity-defined trapping efficiency of porous structure

Calculate the trapping efficiency for a 0.10-mm particle behind a 60% porosity check dam located in a triangular channel with 3 : 1 side slopes, a bottom slope of 1%, and a Manning's roughness of 0.015. The flow rate is 5 cfs. What would be the required dam height?

Solution:

1. *Calculate the normal depth from equation 9.133.*

$$y_n = 1.024\left(\frac{nQ}{ZS_c^{0.5}}\right)^{3/8} = 1.024\left[\frac{(0.015)(5)}{3(0.01)^{0.5}}\right]^{3/8} = 0.609\text{ ft.}$$

2. *Calculate the depth of water at the dam from Eq. (9.137).*

$$y_d = \frac{y_n}{\epsilon} = \frac{0.609}{0.6} = 1.015\text{ ft.}$$

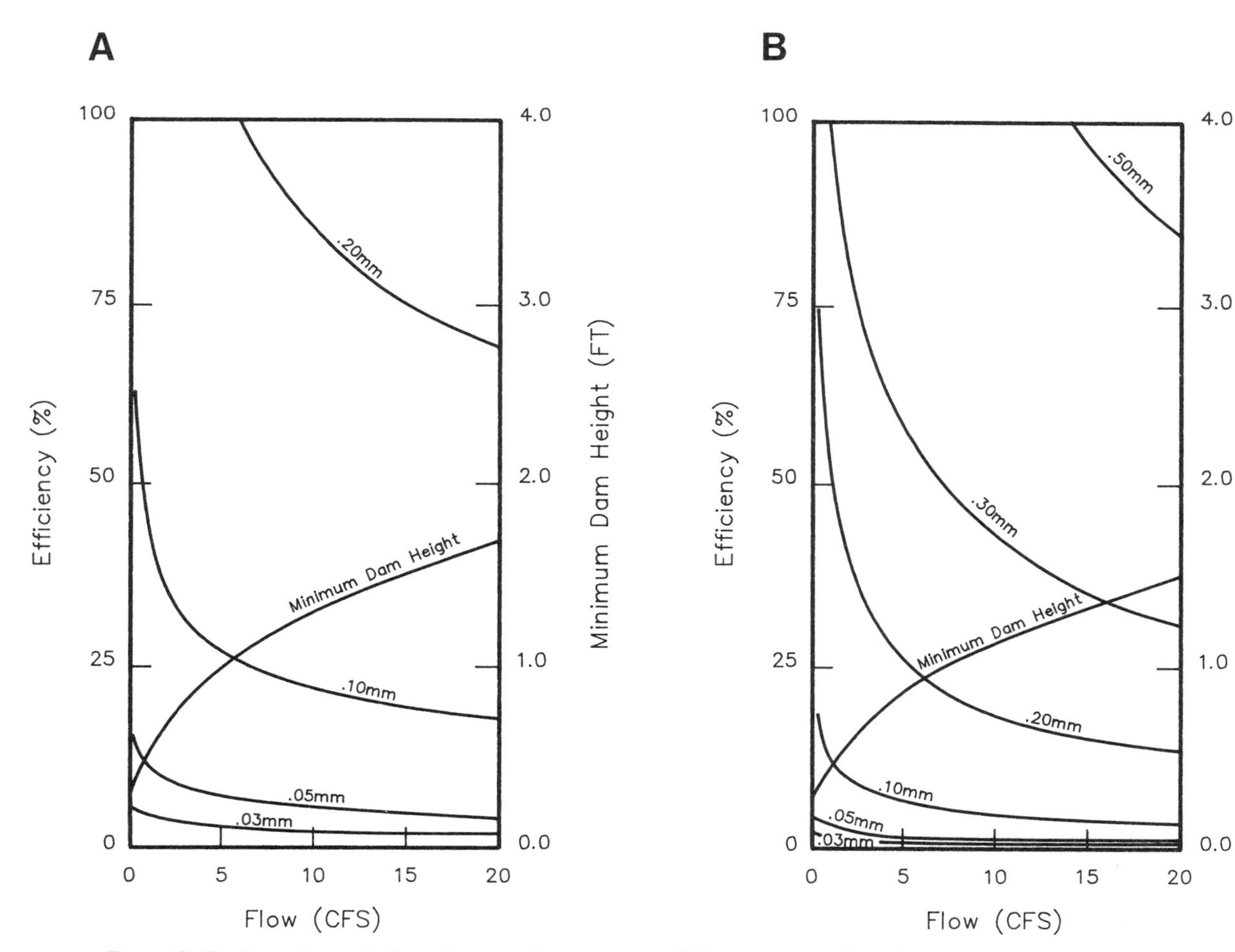

Figure 9.45 Example prediction of porous dam trapping efficiency for two channel slopes and a porosity of 60%. (A) $S_c = 0.01$. (B) $S_c = 0.02$. The maximum dam height corresponds to dam overtopping for a given flow rate.

3. *Calculate the length of the quiescent zone, Eq. (9.141).* From Eq. (9.142), the parameter Q' is

$$Q' = \frac{Q^2}{2gy_n^4Z^2} = \frac{(5)^2}{(2)(32.2)(0.609)^4(3)^2} = 0.314.$$

From Eq. (9.141), Δx is

$$\Delta x = \frac{y_n/\epsilon + Q'\epsilon^2 - 1.1y_n - 0.683Q'}{S_c - (2.27gn^2\epsilon^{3.33}/y_n^{1.33})Q'}$$

$$= \frac{0.609/0.6 + (0.314)(0.6)^2 - (1.1)(0.609) - (0.683)(0.314)}{0.01 - (2.27)(32.2)(0.015)^2(0.6)^{3.333}(0.314)/(0.609)^{1.333}}$$

$= 29.78$ ft; use 30 ft.

4. *Calculate V_n.*

$$V_n = \frac{1.49}{n}\left(\frac{y_n}{2}\right)^{2/3} S_c^{1/2} = \frac{1.49}{0.015}\left(\frac{0.609}{2}\right)^{2/3}(0.01)^{1/2}$$

$= 4.50$ ft/sec.

5. *Calculate t_D from Eq. (9.147).*

$$t_D = \frac{2Zy_n^2\,\Delta X}{Q(1+\epsilon)} = \frac{(2)(3)(0.609)^2(30)}{(5)(1+0.6)} = 8.34 \text{ sec.}$$

6. *Calculate the setting velocity.* From Fig. 7.3, V_s is 0.022 ft/sec for a 0.10-mm particle.

7. *Calculate average depth from Eq. (9.145).*

$$y_{avg} = \frac{y_n}{2}\left(1 + \frac{1}{\epsilon}\right) = \frac{0.609}{2}\left(1 + \frac{1}{0.6}\right) = 0.812 \text{ ft.}$$

8. *Calculate T_s from Eq. (9.143).*

$$T_s = \frac{V_s t_D}{y_{avg}} = \frac{(0.022)(8.34)}{0.812} = 0.226 \text{ or } 22.6\%.$$

This is approximately the same value one obtains from Fig. 9.45A.

Example Problem 9.18 Effects of size distribution on porous structure trapping efficiency

Using the same flow conditions and channel as Example Problem 9.17 estimate the fraction of sediment trapped by the porous dam if the sediment has the following size distribution.

Diameters (mm)	Class %
0.004	20
0.01	30
0.03	20
0.05	20
0.10	5
0.30	5

Solution:

These conditions fit those in Fig. 9.45A; thus T_s values can be read directly. The total trapping efficiency summed over all particle sizes is then

$$T_{ts} = \sum (T_{si})(\Delta FF_i).$$

Calculations are tabulated below.

Size (mm)	ΔFF_i	T_{si}[a]	$T_{si}\,\Delta FF_i$
0.004	0.20	0.00	0.000
0.01	0.30	0.01	0.003
0.03	0.20	0.03	0.006
0.05	0.20	0.07	0.014
0.10	0.05	0.26	0.013
0.30	0.05	1.00	0.050
			$\Sigma = 0.086$

[a]From Fig. 9.45A. Note: If the flow conditions do not match those in Fig. 9.46A, T_s would need to be calculated using Eq. (9.143).

Thus

$$T_{cs} = 0.086 \quad \text{or} \quad 8.6\%.$$

It can be seen from this example that the trapping efficiency for porous dams is typically small. This is a result of the small detention time.

An important point should be made about the value of Manning's n for the above equations. The roughness of the channel will be dependent on the equivalent roughness of the deposited material. Relationships given in Chapter 4 and Chapter 10 can be used to predict n for a given particle size.

Trapping Efficiency—Defined Flow Through Velocity The porosity definition of velocity is difficult to apply to flow through filter fence-type material. As an alternative, the flow through velocity, known as slurry flow

Table 9.11 Slurry Flow Rates Through Porous Filter Material

Material	Slurry flow rate		Reference
	gpm/ft^2	ft/sec	
Straw bale	5.6	0.0125	[a]
Burlap (10 oz.)	2.4	0.0053	[a]
Synthetic fabric	0.3	0.000674	[a]

[a]Virginia Soil and Water Commission Conservation (1980).
[a]Maryland Water Resources Administration (1983).

rate, is defined for the material. Example slurry flow rates are given in Table 9.11, as recommended by state regulatory agencies. The value for synthetic fabric is recommended by several states; however, studies by Fisher and Jarrett (1984) and Waynt (1980) indicate that the slurry flow rate of 0.5 gpm/ft^2 is very low. A further discussion is given in the following section on filter fences.

Warner, as reported in Warner and Hirschi (1983) and Barfield *et al.* (1981), derived equations defining sediment trapping behind a porous structure based on a defined slurry flow rate, V_{sl}. Channel shapes considered are triangular and wide rectangular. Procedures followed are similar to those based on porosity. In addition to defined velocities, the equations are based on normal velocities and depths from Manning's equation and a single-step backwater profile.

Triangular Channel For a triangular-shaped channel, using a defined slurry flow rate of V_{sl} (feet per second), calculations of normal depth and depth at the dam are made with Manning's equation and continuity, given earlier as Eq. (9.133), or

$$y_n = \left[\frac{nQ}{1.49\,ZS_c^{1/2}}\right]^{3/8}.$$

Given that the area at the dam is $A_d = Zy_d^2$ for a triangular-shaped channel and $A_d = Q/V_{sl}$, then

$$y_d = \left[\frac{Q}{ZV_{sl}}\right]^{0.5}. \tag{9.148}$$

Using a single-step backwater profile defined by Eq. (9.139) and ignoring terms with powers of V_{sl} greater than one,

$$\Delta x = \frac{(Q/ZV_{sl})^{0.5} - y_n - Q^2/2gZ^2y_n^4}{S_c}. \tag{9.149}$$

The average velocity, of course, is

$$V_{avg} = \frac{V_n + V_d}{2}. \tag{9.150}$$

The travel time is

$$t_D = \frac{\Delta x}{V_{avg}}, \tag{9.151}$$

and the trapping efficiency is

$$T_s = \frac{V_s t_D}{y_{avg}}, \tag{9.152}$$

where

$$y_{avg} = \frac{y_d + y_n}{2}. \tag{9.153}$$

Rectangular Channels For a wide rectangular channels of width b, where $R \cong y_n$, and $A = by_n$, the equations are

$$y_n = \left[\frac{nQ}{1.49S_c^{1/2}b}\right]^{3/5} \tag{9.154}$$

$$y_d = \frac{Q}{bV_{sl}} \tag{9.155}$$

$$\Delta X = \frac{Q/V_{sl}b - y_n - Q^2/2gb^2y_n^2}{S_c} \tag{9.156}$$

$$V_{avg} = \frac{Q/by_n + V_{sl}}{2} \tag{9.157}$$

$$y_{avg} = \frac{y_n + y_d}{2} \tag{9.158}$$

$$t_D = \frac{\Delta X}{V_{avg}} \cong \frac{2\,\Delta X by_n}{Q} \tag{9.159}$$

$$T_s = \frac{V_s t_D}{y_{avg}} \le 1.0. \tag{9.160}$$

A relationship between slurry flow rates and porosity for porous rock fill dams is given in Chapter 5 along with example computations.

Procedures for using Eqs. (9.154)–(9.160) are illustrated in Example Problem 9.19.

Example Problem 9.19 Trapping efficiency defined by slurry flow rates

Estimate the trapping efficiency due to settling behind a straw bale filter assuming that a flow of 5 cfs is spread over a width of 200 and 75 ft, the slope is 1%, and the channel has a roughness of 0.015. Assume a rectangular channel. The size distribution is the same as that in Example Problem 9.18. Compare those results to trapping by a filter fence with slurry flow rates of 5 to 10 gpm/ft^2.

Solution: From Table 9.12, the slurry flow rate for straw bales is V_{sl} = 5.6 gpm/ft^2 (0.012 ft/sec). Slurry flow rates for the filter fence are given in the problem statement. To solve for trapping efficiencies, one must make calculations for y_n from Eq. (9.154), y_d from (9.155), ΔX from (9.156), V_{avg} from (9.157), y_{avg} from (9.158), t_D from (9.159) and T_s from (9.160). Values are tabulated below.

Material	$V_{sl} \times 10^2$ (ft/sec)	y_d (ft)	y_n (ft)	ΔX (ft)	V_{avg} (ft/sec)	y_{avg} (ft)	t_D (sec)	T_s
Width = 200 ft								
Straw bales	1.25	2.00	0.0275	196	0.461	1.01	425	$421V_s$
V_{sl} = 5.6 gpm/ft^2								
Filter fabric								
V_{sl} = 0.3 gpm/ft^2	0.0674	37.09	0.0275	3705	0.455	18.56[a]	8143	$439V_s$
V_{sl} = 10 gpm/ft^2	2.23	1.12	0.0275	108	0.466	0.57	232	$407V_s$
Width = 75 ft								
Straw bales								
V_{sl} = 5.6 gpm/ft^2	1.25	5.33	0.0497	526	0.677	2.69	777	$289V_s$
Filter fabric								
V_{sl} = 0.3 gpm/ft^2	0.0674	98.91	0.0497	9893	0.671	49.48[a]	14743	$298V_s$
V_{sl} = 10 gpm/ft^2	2.23	2.99	0.0497	291	0.682	1.52	427	$280V_s$

[a]Unrealistic situation.

Trapping efficiencies now must be calculated for each particle size and summed over all particles classes as shown below.

Trapping Efficiency for a Particular Size, T_{si}[a]

Particle Size				Straw bale		Filter fabric			
		V_{si}[b]	V_{sl}:	5.6 gpm/ft^2		0.3 gpm/ft^2		10 gpm/ft^2	
(mm)	ΔFF_i	(ft/sec)	b:	200	75	200[c]	75[c]	200	75
0.004	0.20	4.5×10^{-5}		0.02	0.01	0.02	0.01	0.02	0.01
0.01	0.30	2.8×10^{-4}		0.12	0.08	0.12	0.08	0.11	0.08
0.03	0.20	0.00253		1.00	0.73	1.00	0.73	1.00	0.71
0.05	0.20	0.0062		1.00	1.00	1.00	1.00	1.00	1.00
0.10	0.05	0.025		1.00	1.00	1.00	1.00	1.00	1.00
0.30	0.05	0.13		1.00	1.00	1.00	1.00	1.00	1.00
		$T_{ts} = \Sigma\, T_{si}\Delta FF_i$ =		0.54	0.47	0.54	0.47	0.54	0.47

[a]T_{si} from Eq. (9.160.).
[b]Minimum of Eq. (7.5) or Fig. 7.3.
[c]Since the values for y_d for this case are unrealistic, i.e. (overtopping occurs), these trapping fractions are presented for illustrative purposes only.

Table 9.12 Results of Tests by Fisher and Jarrett (1984) on Selected Filter Fabric

Fabric	Equiv. opening size (Sieve No.)	Slurry flow rates (gpm/ft²) for flows with sediment consisting of				Measured mechanical filtration trapping efficiency		
		Clear water	Sand	Coarse silt	Silt-clay	Sand	Coarse silt	Silt-clay
Cerex 34	NA	131	27	4.5	99	100	82	2
Cerex 68	NA	94	22	4.5	3	100	91	22
Supac 139	80/120	111	21	10.5	75	100	72	5
Supac 407	70/100	111	29	40.5	110	100	48	2
Typar 64	4/70	37	12	33.0	44	95	6	2
Mirafi 100	30	15	5	16.5	5	90	9	1

From Example Problem 9.19, one can conclude that trapping is very much a function of flow rate per unit width and only weakly a function of slurry flow rate. In the prediction equations, the decreased travel time resulting from higher slurry flow rates are offset by decreased depth of flow, requiring less settling time to reach the bottom. Again, it should be pointed out that the trapping predicted here results from settling behind the dam. Mechanical filtration is a separate issue, discussed subsequently.

Also, it should be pointed out that the depth of flow behind the filter fence is excessive for slurry flow rates of 0.3 gpm/ft^2, the value typically recommended for filter fence fabric. A value of 5 to 10 gpm/ft^2 gives a more reasonable flow depth.

Mechanical Filtration

Studies by Fisher and Jarrett (1984) indicate that mechanical filtration is very much dependent on the type fabric, equivalent opening size (EOS) and the sediment size distribution. As sediment flows through the filter cloth, holes are plugged by sediment in the size range of the EOS, if such sizes are present. As a consequence, the fraction of sediment trapped by mechanical filtration and the slurry flow rates are highly related to the size distribution of the sediment as well as the filter material. Fisher and Jarrett (1984) indicate that slurry flow rates and fraction of sediment trapped are related to flow orientation, thickness of the fabric, and accumulation of sediment on the upstream side of the fabric. Results for the fabrics tested by Fisher and Jarrett are shown in Table 9.12.

Studies by Waynt (1980) on 15 materials show slurry flow rates much smaller and more in the range of 0.3 gpm/ft^2. His results also show greater mechanical filtration. Waynt's results are based on low head filtration. He did not test the same materials as Fisher and Jarrett (1984).

Procedures for estimating trapping by mechanical action have not been developed. In general, one has to use empirical data for a given size fraction, such as those in Table 9.12.

Infiltration Impacts

Sediment trapping by infiltration through the previously deposited sediment also accounts for some of the trapping. As infiltration occurs, suspended sediment is transported to the soil matrix. This would be the primary mechanism for trapping clay size particles. Since infiltration rates are typically much less than the slurry flow rate, this mechanism occurs primarily after the filter fabric is clogged and sediment laden flow drains through previously deposited sediment.

Types of Porous Structures

Equations for trapping efficiency were given in the previous section. The following description gives some general details on design and installation.

Rock Fill Dam and Gabion

A commonly used porous structure is the rock fill check dam, simply constructed by end dumping rock in a channel as shown in Fig. 9.46B. In some cases, the rock may cover a sand filter core. Sometimes these structures are used for grade stabilization as well as sediment trapping. When used for grade stabilization, the heights of the dams and the distance between dams is set by the slope, as shown in Fig. 9.47

Sediment trapping with rock fill check dams occurs as a result of all the mechanisms listed in the introduc-

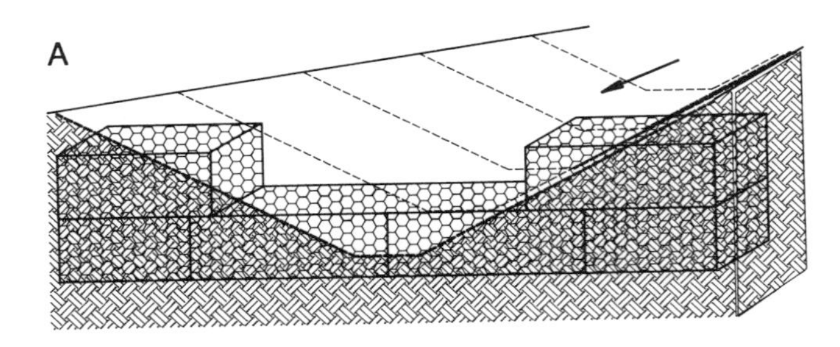

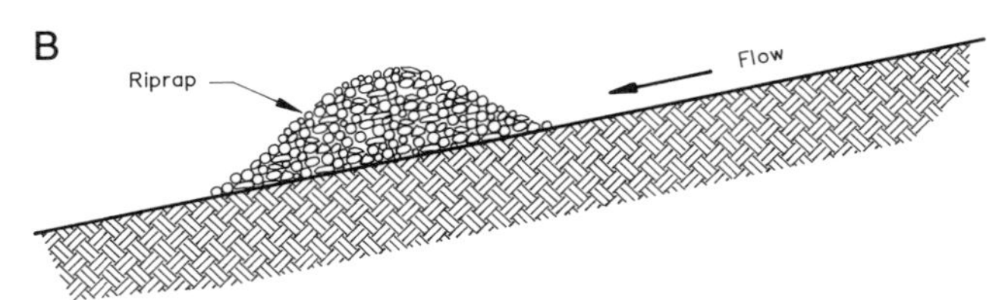

Figure 9.46 Schematic of rock fill check dam and a gabion check dam. (A) Gabion and (B) end-dumped riprap check dam.

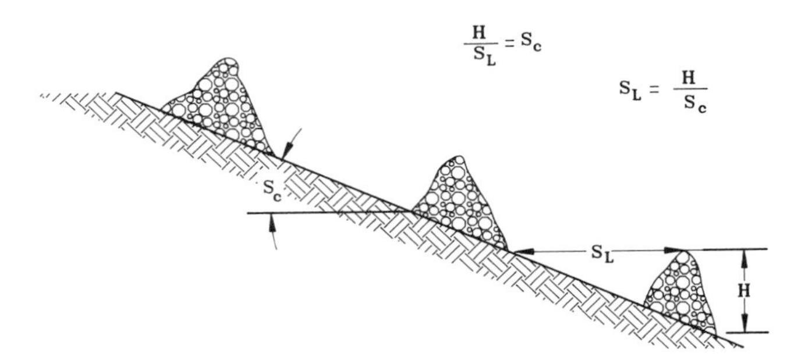

Figure 9.47 Height and spacing relationships for check dams used for grade stabilization and sediment trapping.

tion, but it appears that reduced total transport capacity behind the dam is the primary trapping mechanism. Trapping within the dam itself could also be occurring, but to a lesser extent. Such trapping would be expected with small rocks and small pores. The degree of trapping inside the pores has not been evaluated experimentally. Thus, rock fill check dams would not be expected to trap particles other than the larger bed-load materials.

The design of rock fill structures is based on the intended use. If designed as a grade control structure, the relationship between dam height and spacing of check dams is based on slope as shown in Fig. 9.47. When designed for trapping efficiency, calculation of trapping efficiency is made as shown in Example Problems 9.17 to 9.19.

Straw Bale Check Dams

Straw bales are frequently used to control overland flow from disturbed areas as shown in Fig. 9.48. Properly installed, they can be effective in removing sediment. Proper installation includes placing the bales in a shallow trench and securing them in place with stakes. In some cases, straw bales form a core of a rock fill check dam. The rock fill is placed over the straw bales, resulting in improved anchoring.

Design calculations are limited to making checks to assure that the flow does not overtop the straw bales and that the desired trapping efficiency can be met. Multiple-layer straw bale dams are not recommended. Flow and sediment trapping calculations should follow those in Example Problem 9.19.

Filter Fence

A filter fence performs in much the same manner as a porous rock fill check dam or a straw bale check dam. It is normally used to control overland flow. A filter fabric, typically a reinforced geotextile, is placed in the flow path, and trapping is provided by the three mechanisms described in the introduction to this section.

Design of the filter fence should require calculation to ensure that the fence does not overtop and that the required trapping efficiencies are met. Calculation procedures should follow those in Example Problem 9.19 with modification to account for mechanical trapping, if empirical data are available for the particular fabric.

In the installation of the filter fences, particular care should be taken to assure that the toe of the fabric is buried, as shown in Fig. 9.49. As much as possible, the fence should be installed on the contour to prevent excessive concentration of flow at any one point. A filter fence installation is shown in Fig. 9.49.

Brush Barrier

Organic litter and spoil material from site clearing can be used effectively on a site by pushing or dumping a mixture of limbs, small vegetation, and root mat into windrows along the toe of any slope where accelerated erosion and runoff are expected. Anchoring a filter fabric over the brush enhances the filtration capacity of the vegetation. Maintenance requirements are small.

When covered with filter cloth, the upstream edge of the filter cloth should be buried just upstream of the barrier brush. A brush barrier installation is illustrated in Fig. 9.50.

The trapping efficiencies of brush barriers without filter fabric has not been evaluated experimentally, nor have theoretical relationships been developed to predict their effectiveness. Thus, design cannot be based on predicted trapping efficiences. For brush with filter cloth cover, the effectiveness can be determined, as a first approximation, by the effectiveness of the filter material.

In general, the recommended height of the barrier should be at least 3 ft at construction, and the base should be at least 5 ft. Some consolidation will occur with time and vegetative decay, reducing the effective height. The effective life of the brush barrier is short, limited by the vegetative decay.

SEDIMENT TRAPS

A sediment trap is a small temporary excavated basin that intercepts sediment-laden flow and detains it long enough to allow sedimentation. Such traps are, in fact, small sediment ponds, but typically have ill-defined outlet structures. Outlets can be open channels, riprap-lined weirs, riprap porous dams, or any other of a large variety of controls. The outlet is typically selected to control the volume of water detained and not the flow rate. Because of the lack of attention to the outlet design, a maximum drainage area of 5 acres is recommended (Clar *et al.*, 1981).

The trapping efficiency of sediment traps can be defined by one of the several sediment pond models described earlier, if the stage-discharge curve is de-

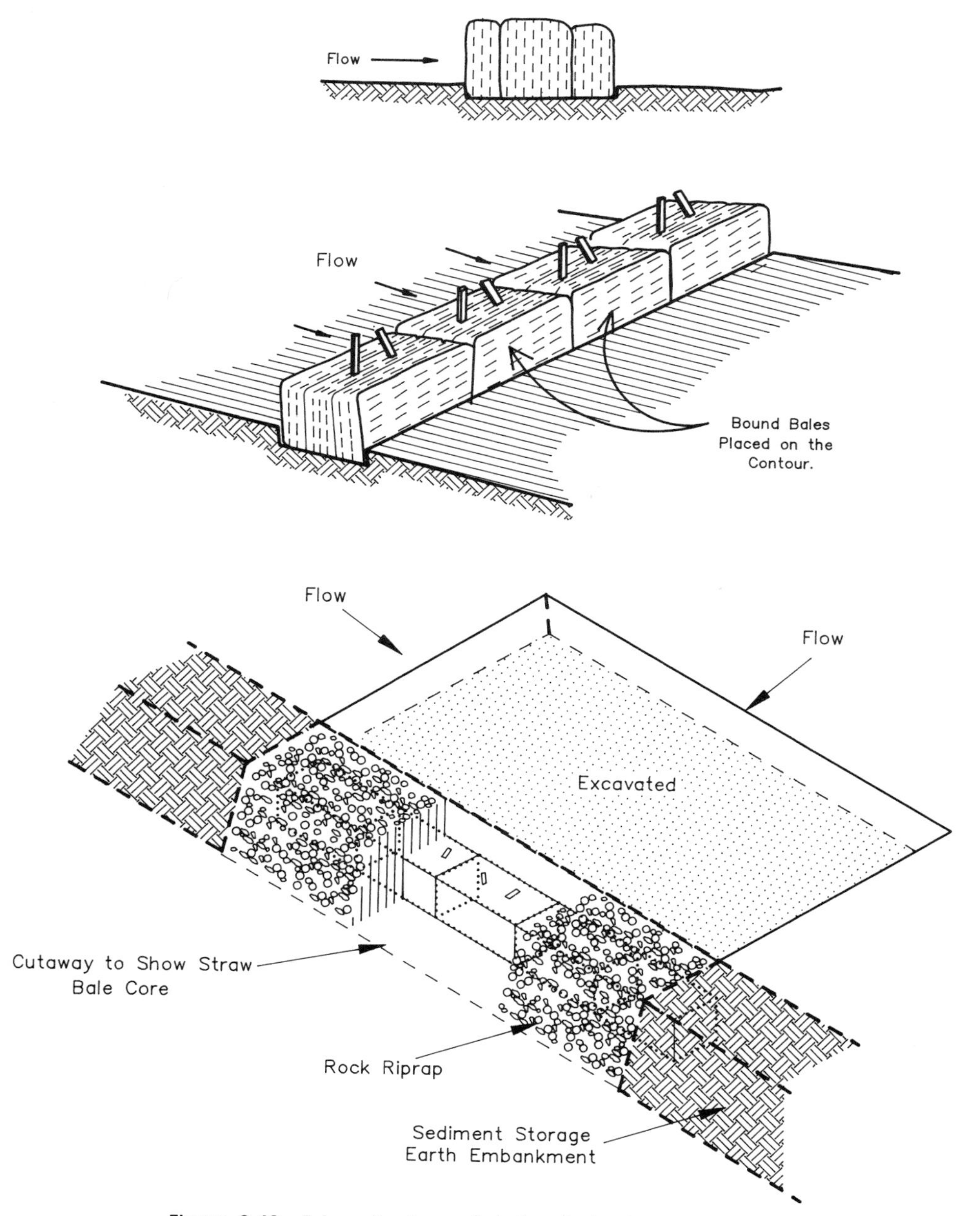

Figure 9.48 Schematic of straw bale installation for sediment control.

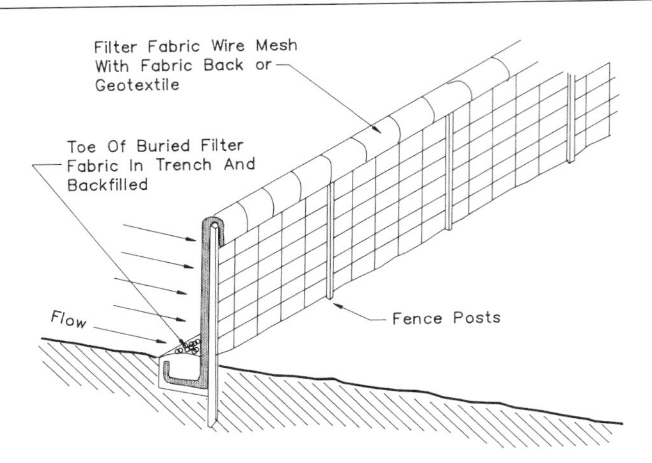

Figure 9.49 Schematic of a filter fence installation.

fined. As an alternative, Hansen (1973) utilized the fully turbulent form of the overflow rate equation, given by Vetter (1940) and Eq. (9.57) or

$$F = 1 - e^{-V_s/V_c},$$

where V_c is the overflow rate given earlier by

$$V_c = Q/A.$$

In the above equations, F is the fraction trapped, V_s is the settling velocity, Q is the flow rate, and A is the surface area.

Design of a trap must consider both volume and surface area. Surface area is selected to give a desired trapping efficiency. Calculations of trapping efficiency are made using Eq. (9.57). Volume is based on the need to store trapped sediment and should theoretically equal some fraction of the sediment expected to be trapped over the life of the trap. If the total volume of trapped sediment cannot be stored, clean out will be required.

INERTIAL SEPARATION: THE SWIRL CONCENTRATOR

The swirl concentrator is a device designed to use the centrifugal force of the flow to separate sediment from water in storm runoff. The effluent sludge is transmitted to a small basin while the treated flow is discharged directly to a stream. A schematic is shown in Fig. 9.51. Flow enters through an inlet pipe at the outer edge of the concentrator and discharges through an overflow section at the center of the device. Sludge is discharged from the lower section. The device is portable and can be reused.

Sullivan *et al*, (1976) presented curves, as shown in Fig. 9.52, for predicting the capture efficiency of a swirl concentrator for the size unit shown in Fig. 9.51. Warner and Dysart (1983) used these curves in conjunction with the discharge from a 100-year storm in

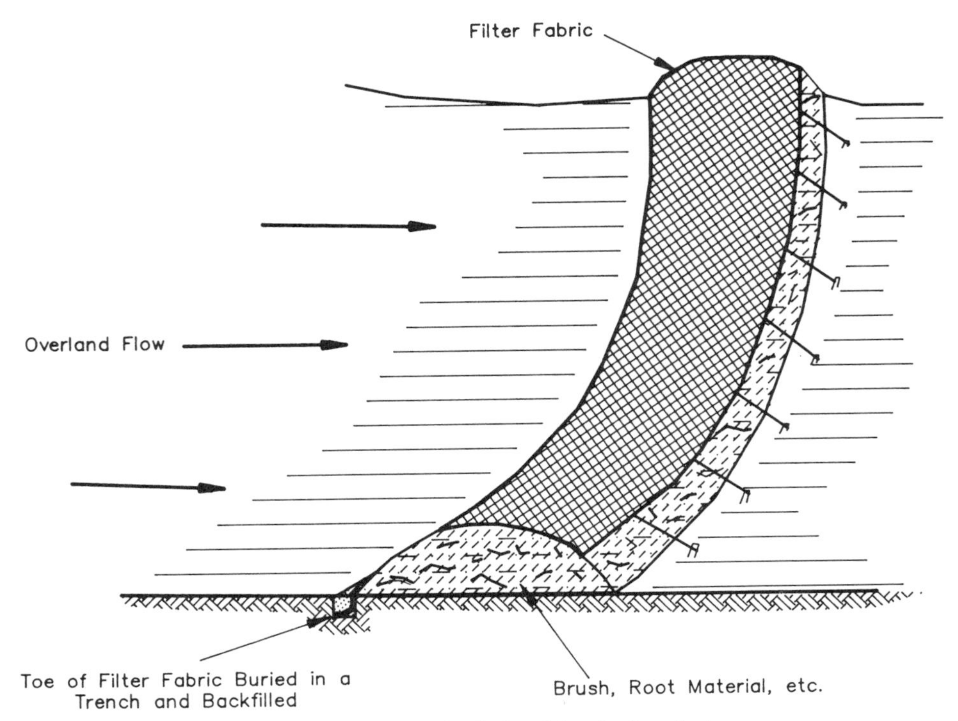

Figure 9.50 Schematic of a brush barrier installation.

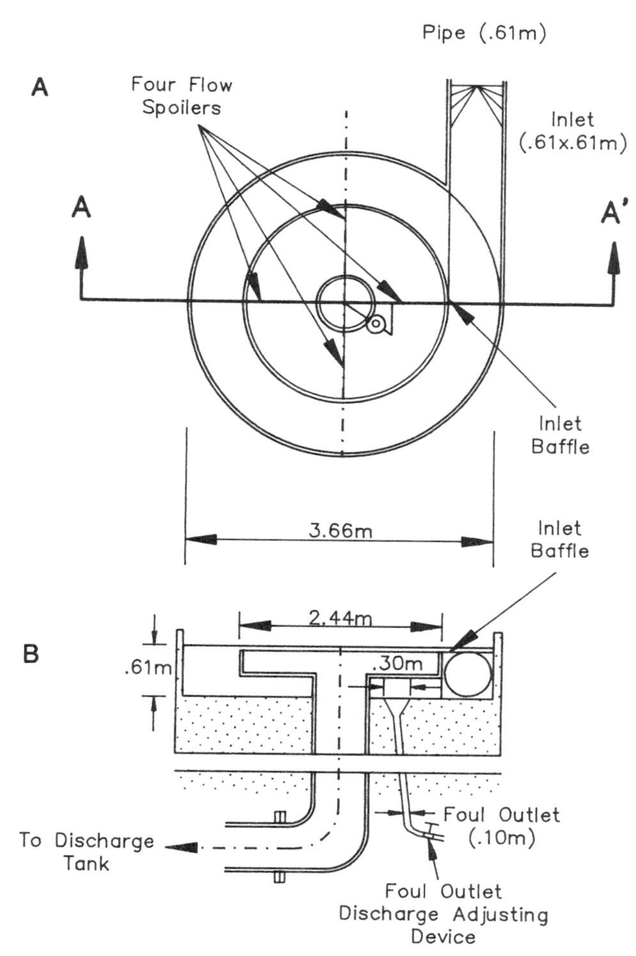

Figure 9.51 Schematic of swirl concentrator (after Sullivan *et al.*, 1976). (A) Plan view. (B) Elevation, section $A - A'$.

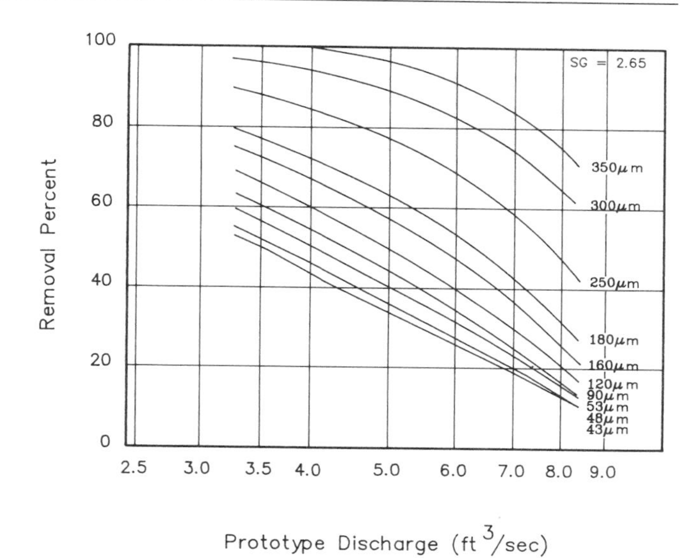

Figure 9.52 Swirl concentrator device particle size capture efficiency for a 10% draw off (after Sullivan *et al.*, 1976).

central Appalachia and predicted a trapping efficiency of 33% for a surface mine site. They proposed that the trapping in more common events would be much higher. Caruccio and Buxton (1984) present results for a limited series of field tests in which they experienced problems generating vortex action in the separator. Field studies are needed to settle the controversy.

SYSTEMS APPROACH TO SEDIMENT CONTROL

Sediment control by any one of the technologies described in this chapter or previous chapters will be less effective than a combination of these systems together along with on-site controls. Example combinations may include the use of diversions, mulching, check dams, and grass filters. The effectiveness of a combination of systems will not be as great as a linear combination of each of the systems. For example, a check dam in a flow channel will have a lower trapping efficiency if the upslope areas are covered with straw mulch, than if the upslope areas are bare. This decrease is a result of the fact that the large particles will be trapped by the mulch. The combination of the two will, however, be more effective than either of the practices used alone.

The effectiveness of a system of controls is highly dependent upon site specific parameters, including the specific combination of control techniques. An evaluation of a proposed system, due to the nonlinearity of the combined systems, will need to be made with some type of watershed model such as SEDIMOT II (Wilson *et al.*, 1982, 1986). Such a model must be capable of evaluating systems of controls used in various combinations.

An example analysis was made by Barfield and Wells (1981) using SEDIMOT II and reported in part by Warner *et al.* (1982b), which illustrated the effectiveness of a variation of erosion control techniques on the same location. The predicted parameter in the study was the single storm (10-year, 24-hr) sediment yield at varying stages in the mining operation. An example of the results is shown in Fig. 9.53. From inspection of the results, it is obvious that the effects of a combination of controls is nonlinear and depends on the location. Although total suspended solids (TSS) were not summarized in this study, it would have been possible to do so. TSS concentrations would respond in the same manner as single-storm sediment yields.

Problems

(9.1) Describe the criteria used to evaluate the effectiveness of sediment control structures. Which crite-

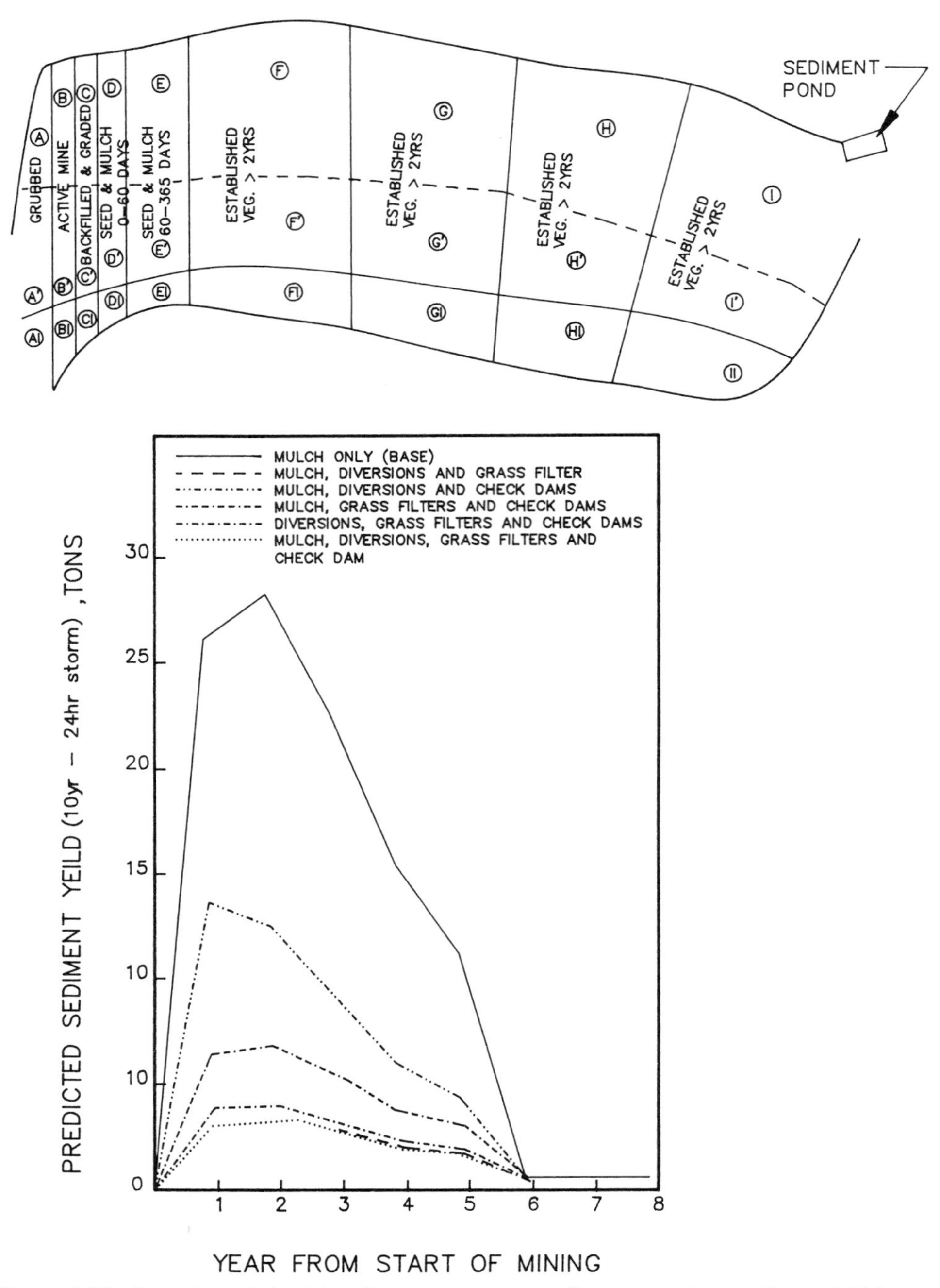

Figure 9.53 Example analysis of the effects of a system of sediment controls on sediment yield from an area surface mine in Northern Appalachia. Computations were made with the SEDIMOT II model (after Warner *et al.*, 1982b).

ria would you recommend as a design standard for sediment control structures for surface-mined lands, construction sites, and storm water detention structures for urbanized areas following construction?

(9.2) A pond is being used to control sediment from pumped discharge from a deep mine operation. The storage volume is 1.0 acre feet and the pump rate is 3.0 cfs. Dye was pumped into the pond such that the inflow concentration was 0.334 ppm. If the outflow concentration varies as shown below, estimate the best-fit parameters for the CSTRS, plug flow, and plug flow diffusion models.

Time (min)	Concentration (ppm)	Time (min)	Concentration (ppm)
0	0	320	0.278
40	0.003	360	0.278
80	0.086	400	0.304
120	0.143	440	0.318
160	0.179	480	0.326
200	0.205	520	0.330
240	0.217	560	0.332
280	0.247	600	0.334

(9.3) A rectangular reservoir has a steady inflow and outflow rate of 3 cfs and a surface area of 1.3 ac. Calculate the overflow rate.

(9.4) If the reservoir in Problem (9.3) has a size distribution given by the inflow distribution in Fig. 9.17, estimate the trapping efficiency. If the inflow concentration of sediment is 25,000 mg/liter, estimate the effluent concentration.

(9.5) Estimate the outflow size distribution for Problem (9.4).

(9.6) The mine operator for the pond in Problems (9.3) to (9.5) finds that he cannot physically locate the pond as described. Therefore, he proposes to construct two ponds in series that have a combined volume equal to that of the originally proposed pond, but with a surface area of 0.65 acres for each pond. What will be the impact on trapping efficiency? Estimate the total trapping efficiency of the two ponds and the effluent size distribution for each pond. If the inflow concentration is 5000 mg/liter, what will be the outflow concentration for Problems (9.3) to (9.5) and for this problem?

(9.7) If the reservoir in Problems (9.3) to (9.5) has a depth of 2.0 ft and a width of 10 ft, estimate the effects of turbulence on trapping efficiency. What would be the outflow concentration if the inflow concentration is 2000 mg/liter?

(9.8) Work Example Problem 9.7 for plug 12.

(9.9) Work Example Problem 9.8 for plug 12.

(9.10) Work Example Problem 9.10 for plug 12.

(9.11) A reservoir is being constructed with the following characteristics:

Parameter	Value
Average surface area (principal spillway to emergency spillway) ft^2	15,000
Storage volume ft^3	52,500
Average depth ft	3.5

The watershed is located in an area that has the following rainfall characteristics:

Parameter	Mean	Coef. var.
Average rainfall volume (in.)	1.4	1.2
Intensity (in./hr)	0.5	1.5
Duration (hr)	5.0	1.2
Interval between storms T_{IA} (hr)	80	5.0

The watershed area is 80 acre with a curve number of 85 and a Rational Equation C factor of 0.6. The particle size distribution for sediment in an average storm is shown below. Calculate the total fraction trapped for both stormflow and quiescent flow. Assume that the reservoirs have good hydraulic performance.

Particle class i	% finer	Settling velocity, V_{si} (ft/hr)
1	15	0.01
2	25	0.25
3	55	1.49
4	90	7.32
5	100	22.00

(9.12) If one wanted to extend the sampling time and depth in Table 9.5 to a particle diameter of 0.001 mm, what would be the sampling time if a depth of 5 cm is used?

(9.13) A first-flush filtration system is proposed for a 30-acre watershed with established apartments and houses. Regulatory requirements specify that the first 0.4 in. of runoff must be diverted and filtered. Filters must be at least 18 in. thick with a hydraulic conductivity of 3.5 ft/day. The sediment basin and filter must be drawn down in 24 hr or less and water depths must not exceed 4 ft. Determine the surface area and depth of the sedimentation basin and filter basin. Assume that 18 μm and larger particles are to be trapped in the sedimentation basin. Make a sketch for the system.

(9.14) A developer proposes that sediment basins are improperly constructed and that the deepest portion should be at the inlet and the shallow portion near the outlet. What is your response to that proposal?

(9.15) A vegetated filter strip (VFS) is constructed on a slope of 5% and seeded to perennial rye grass. If the flow rate is 0.02 cfs/ft, estimate the flow depth and determine whether or not the vegetation will remain erect.

(9.16) If the VFS in Problem (9.15) has a 100-ft flow length and is subjected to a sediment load of 1.0 lb/ft · sec with a size distribution as given in Fig. 9.43, estimate the trapping efficiency and the outflow concentration. Assume a steady infiltration rate of 0.20 in./hr.

(9.17) A VFS is being used to control sediment near Atlanta, Georgia. If the sediment yield must be less than 1.5 tons/acre · year and the actual sediment yield to the filter is 6 tons/acre · year, estimate the length of filter required. Assume that the flow and vegetation is the same as those in Problems (9.15) and (9.16) and that the average flow length from the disturbed area is 150 ft.

(9.18) Calculate the trapping efficiency for a 0.15-mm particle behind a check dam in a triangular channel with 3 : 1 side slopes, slope 1%, $n = 0.015$, $Q = 3$ cfs, and porosity 80%.

(9.19) What would be the overall trapping efficiency for the check dam in Problem (9.18) if the sediment has the following size distribution:

Diameter (mm)	Percentage in class	Diameter (mm)	Percentage in class
0.006	15	0.05	15
0.01	25	0.10	10
0.03	30	0.40	5

(9.20) If the dam in Problem (9.19) is a synthetic fabric with a slurry flow rate of 5 gpm/ft^2, what would be the trapping efficiency?

(9.21) What are some possible reasons for differences in the slurry flow rates in Tables 9.11 and 9.12?

(9.22) Estimate the trapping efficiency, effluent concentration, and effluent size distribution if a swirl concentrator of the size shown in Fig. 9.51 is used to control the flows in Problems (9.18) and (9.19). Assume that the inflow concentration is 10,000 mg/liter and that the trapping efficiency is zero for particle sizes less than 30 microns.

References

Austin, TX, City of (1988). "Environmental Criteria Manual." Department of Public Works, City of Austin, TX.

Barfield, B. J., Warner, R. C. and Haan, C. T. (1981). Hydrology and Sedimentology of Disturbed Lands. Oklahoma Technical Press, Stillwater, OK.

Barfield, B. J., and Wells, C. G., (1981). Unpublished data associated with studies for Hittman, Inc. (Available from senior author).

Barfield, B. J., Fogle, A. W., Carey, D. I., Inamdar, S. P., Blevins, R. L., Madison, C. E., and Evangelou, V. P. (1992). Water quality impacts of natural riparian grasses. Research Report No. 184. Kentucky Water Resources Research Institute, University of Kentucky, Lexington.

Barfield, B. J., and Albrecht, S. (1982). Use of a vegetative filter zone to control fine grained sediment from surface mine. *In* "Proceedings, 1982 Symposium on Surface Mine Hydrology, Sedimentology and Reclamation," UKY BU 129, pp. 481–490. College of Engineering, University of Kentucky, Lexington, KY.

Barfield, B. J., and Hayes, J. C. (1988). Design of Grass Waterways for channel stabilization and sediment filtration. *In* "Handbook of Engineering in Agriculture," Vol II, "Soil and Water Engineering." CRC Press, Boca Raton, FL.

Bondurant, J. A., Brockway, C. E., and Brown, M. J. (1975). Some aspects of sedimentation pond design. *In* "Proceedings, 1975 National Symposium on Urban Hydrology and Sediment Control," UK BU 109, pp. 116–122. College of Engineering, University of Kentucky, Lexington, KY.

Brown, C. B (1950). Sediment transport. *In* "Engineering Hydraulics" (Hunter Rouse, ed.). Wiley, New York.

Brune, G. M. (1953). Trap efficiency of reservoirs. *Trans. Am. Geophy. Union* **34** (3):407–418.

Camp, T. R. (1946). Sedimentation and the design of settling tanks. *Trans. Am. Soc. Civil Eng.* **111**:895–958.

Caruccio, F. T., and Buxton, H. (1984). An evaluation of the swirl concentrator: Its effectiveness in clarifying sediment laden drainage. *In* "Proceedings, 1984 Symposium on Surface Mining, Hydrology, Sedimentology and Reclamation," UK BU 136, pp. 169–178. College of Engineering, University of Kentucky, Lexington, KY.

Chen, C. (1975). Design of sediment retention basins. *In* "Proceedings, National Symposium on Urban Hydrology and Sediment Control," UK BU 109, pp. 285–298. College of Engineering, University of Kentucky, Lexington, KY.

Churchill, M. A. (1948). Discussion of "Analysis and use of reservoir sedimentation data" by L. C. Gottschalk. *In* "Proceedings, Federal Interagency Sedimentation Conference, Washington, DC, pp. 139–140.

Clar, M. L., Das, P., Ferrandino, J. J., and Barfield, B. J. (1981). Handbook of erosion and sediment control measures for coal mines, Unpublished final report on Contract No. J5104049 to U. S. Office of Surface Mining. Hittman Associates, Columbia, MD.

Cooper, J. R., and Gilliam, J. W. (1987). Phosphorous redistribution from cultivated fields to riparian areas. *Soil Sci. Soc. Am. J.* **51** (6):1600–1604.

Cooper, J. R., Gilliam, J. W., Daniels, R. B., and Robarge, W. P. (1987). Riparian areas as filters for agricultural sediment. *Soil Sci. Soc. Am. J.* **51** (2):416–420.

Cowardin, L. M., Carter, V., Golet, F. C., and LaRoe, E. T. (1979). Classification of wetlands and deep water habitats of the United States. U.S. Department of the Interior Publication FWS/OBS-79/31.

Dillaha, T. A., Sherrard, J. H., Lee, D., Shanholtz, V. O., Mostaghimi, S., and Magette, W. L. (1986). Use of vegetative filter strips to minimize sediment and phosphorous losses from feed lots: Phase I experimental plot study, Bulletin 151. Virginia Water Resources Research Institute, Virginia Tech University, Blacksburg, VA.

Dillaha, T. A., Sherrard, J. H., Lee, D., Mostaghami, S., and Shanholtz, V. O. (1988). Evaluation of vegetative filter strips as a best management practice for feed lots. *J. Water Pollut. Control Fed.* **60** (7):1231–1238.

Dillaha, T. A., Reneau, R. B., Mostaghami, S., and Lee, D. (1989). Vegetative filter strips for agricultural nonpoint source pollution control. *Trans. Am. Soc. Agric. Eng.* **32** (2):513–519.

Dobbins. W. E. (1944). Effect of turbulence on sedimentation. *Trans. Am. Soc. Civil Eng.* **109**:629–678.

Driscoll, E. D., DiToro, D., Gaboury, D., and Shelley, P. (1986). Methodology for analysis of detention basins for control of urban runoff quality, Report No. EPA 440/5-87-01 (NTIS No. PB87-116562). U.S. Environmental Protection Agency, Washington, DC.

Einstein, H. A. (1950). The bed-load function for sediment transportation in open channel flows, Technical Bulletin No. 1026. U.S. Department of Agriculture, Washington, DC.

Environmental Protection Agency (1976). Erosion and sediment control—Surface mining in the Eastern U.S., Vol. I and II, EPA-615/2-76-006. U.S. Environmental Protection Agency, Washington, DC.

Evangelou, V. P., Rawlings, F., Crutchfield, J. D., and Shannon, E. A. (1981). A simple chemo-mathematical model as a tool in managing surface mine sediment ponds. *In* "Proceedings, 1981 Symposium on Surface Mining Hydrology, Sedimentology, and Reclamation," UKY BU 126, pp. 49–58. College of Engineering, University of Kentucky, Lexington, KY.

Evangelou, V. P., Barnhisel, R. I., Barfield, B. J., and Garyotis, C. L. (1987). Modeling release of chemical constituents in surface mine runoff. *Trans. Am. Soc. Agric. Eng.* **30**(1):82–89.

Fisher, L. S., and Jarrett, A. R. (1984). Sediment retention efficiency of synthetic filter fabric. *Trans. Am. Soc. Agric. Eng.* **27**(2):429–436.

Gearheart, R. J., Wilbur, S., Williams, J., Hull, D., Finney, B., and Sunburg, S. (1983). Final report City of Arcata Marsh pilot project effluent quality results—System design and management, Project report C-06-2270. City of Arcata, Department of Public Works, Arcata, CA.

Gottschalk, L. C. (1965). Trap-efficiency of small floodwater retarding structures. *In* "Proceedings, ASCE Water Resources Engineering Conference, Mobile, AL."

Griffin, M. L., Barfield, B. J., and Warner, R. C. (1985). Laboratory studies of dead storage in sediment ponds. *Trans. Am. Soc. Agric. Eng.* **28**(3):799–804.

Grizzard, T. L., Randall, C. W., Weand, B. L., and Ellis, K. L. (1986). Effectiveness of extended detention ponds. Urban Runoff Quality, American Society of Civil Engineers.

Hammer, D. A., ed. (1989). "Constructed Wetlands for Wastewater Treatment: Municipal, Industrial and Agricultural." Lewis, Chelsea, MI.

Hammer, D. A., and Bastian, R. K. (1989). Wetlands ecosystems: Natural water purifiers? *In* "Constructed Wetlands for Wastewater Treatment: Municipal, Industrial and Agricultural" (D. A. Hammer, ed.) pp. 5–20. Lewis, Chelsea, MI.

Hansen, E. A. (1973). In stream sedimentation basins-a possible tool for trout habitat management. U. S. Department of Agriculture, Forest Service, North Central Forest Experimental Station, East Lansing, MI.

Hayes, J. C., and Dillaha, T. (1991). Procedure for the design of vegetative filter strips, Report prepared for the USDA-Soil Conservation Service, Washington, DC. [Available from J. C. Hayes, Department of Agricultural and Biological Engineering, Clemson University, Clemson, SC]

Hayes, J. C., and Harriston, J. (1983). Modeling the long term effectiveness of vegetative filters as onsite sediment controls, Paper 83-2081. American Society of Agricultural Engineers, St. Joseph, MI.

Hayes, J. C., Barfield, B. J., and Barnhisel, R. I. (1978). Evaluation of grass characteristics related to sediment filtration, Paper 78-2513. American Society of Agricultural Engineers, St. Joseph, MI.

Hayes, J. C., Barfield, B. J., and Barnhisel, R. I. (1982). The use of grass filters for sediment control in strip mine drainage. III. Empirical verification of procedures using real vegetation. Report No. IMMR82/070. Institute for Mining and Minerals Research, University of Kentucky, Lexington, KY.

Hayes, J. C., Barfield, B. J., and Barnhisel, R. I. (1984). Performance of grass filters under laboratory and field conditions. *Trans. Am. Soc. Agric. Eng.* **27**(5):1321–1331.

Hazen, A. (1904). On sedimentation. *Trans. Am. Soc. Civil Eng.* **980**:45–87.

Hill, R. D. (1976). Sedimentation ponds—A critical review. *In* "Proceedings, 6th Symposium on Coal Mine Drainage Research, Louisville, KY."

Hirschi, M. C. (1981). Efficiency of small sediment controls. Unpublished Agricultural Engineering file report. University of Kentucky, Lexington, KY.

Inamdar, S. P. (1993). Modeling sediment trapping in riparian vegetative filter strips. Masters Thesis, Agricultural Engineering Department, Univ. of Kentucky, Lexington.

Janiak, H. (1979). Purification of waters discharged from Polish lignite mines, EPA-600/7-79-099. U.S. Environmental Protection Agency Industrial Research Laboratory, Cincinnati, OH.

Kathuria, D. V., Nawrocki, M. A., and Becker, B. C. (1976). Effectiveness of surface mine sedimentation ponds, Report No. EPA-600/2-76-117. U.S. Environmental Protection Agency, Cincinnati, OH.

Kouwen, N., Li, R. M., and Simons, D. B. (1981). Flow resistance in vegetated waterways. *Trans. Am. Soc. Agric. Eng.* **24**(3):684–690.

Levenspiel, O. (1972). "Chemical Reaction Engineering." Wiley, New York.

Levenspiel, O., and Smith, W. K. (1957). Notes on the diffusion-type model for the longitudinal mixing of fluids in flow. *Chem. Eng. Sci.* **6**:227–233.

Li, R. M., Schall, J. D., and Simons, D. B. (1980). Turbulence prediction in open channel flow. *Proc. Am. Soc. Civil Eng.* **106**(HY4):575–587.

Magette, W. L., Brinsfield, R. B., Palmer, R. E., and Wood, J. D. (1989). Nutrient and sediment removal by vegetated filter strips. *Trans. Am. Soc. Agric. Eng.* **32**(2):663–667.

Maryland Water Resources Administration (1983). Maryland standards and specifications for soil erosion and sediment control. Maryland Water Resources Administration, Annapolis, MD.

McBurnie, J. C., Barfield, B. J., Clar, M. L., and Shaver, E. (1990). Maryland sediment detention pond design criteria and performance. *Appl. Eng. Agric.* **6**(2):167–173.

Noe, S., and Barfield, B. (1990). Unpublished results of dye tracer studies in a model sediment pond. Department of Agricultural Engineering, University of Kentucky, Lexington, KY.

Occoquan Watershed Monitoring Laboratory (1986). Final Contract Report: Washington Area NURP Project, Prepared for the Metropolitan Washington Council of Governments.

Parsons, J. E., Daniels, R. B., Gilliam, J. W., and Dillaha, T. A. (1991). The effect of vegetation filter strips on sediment and nutrient removal from agricultural runoff. *In* "Proceedings, Environmentally Sound Agriculture Conference, April, Orlando, FL."

Randall, C. W. (1982). Stormwater detention ponds for water quality control, stormwater detention facilities—Planning design operation and maintenance. *In* "Proceedings, Engineering Foundation Conference, ASCE, 1982."

Reed, L. A. (1978). Effectiveness of sediment control structures used in highway construction in Central Pennsylvania, Water Supply Paper 2054. U.S. Geological Survey.

Seddon (1889). Cleaning water by settlement (original not seen, cited Hazen, A., 1904). *J. Am. Assoc. Eng. Soc.*

Small, M. M. (1976). Data report-marsh/pond systems, USERDA Report BNL 50600.

Soil Conservation Service (1975, 1986). Urban hydrology for small watersheds, Technical release No. 55. Soil Conservation Service, U.S. Department of Agriculture, Washington, DC.

Soil Conservation Service (1983). Computer program for project formulation hydrology, Technical Release No. 20. Soil Conservation Service, U.S. Department of Agriculture, Washington, DC.

Stahre, P., and Urbonos, B. (1990). "Stormwater Detention for Drainage, Water Quality and CSO Management." Prentice–Hall, Englewood Cliffs, NJ.

Sullivan, R. H., Cohn, M. M., Ure, J. E., Parkinson, F. E., and Zielinski, P. E. (1976). The swirl concentrator for erosion runoff treatment, EPA 600/2-76-271. U.S. Environmental Protection Agency, Washington, DC.

Tapp, J. S., and Barfield, B. J. (1986). Modeling the flocculation process in sediment ponds. *Trans. Am. Soc. Agric. Eng.* **29**(3):741–747.

Tapp, J. S., Barfield, B. J., and Griffin, M. L. (1981). Predicting suspended solids removal in pilot size sediment ponds using chemical flocculation, Institute for Mining and Minerals Research Report IMMR 81/063. University of Kentucky, Lexington, KY.

Temple, D. M., Robinson, K. M., Ahring, R. M., and Davis, A. G. (1987). Stability design of grass-lined open channels, Agriculture Handbook Number 667. U.S. Department of Agriculture, Agricultural Research Service.

Tollner, E. W., Barfield, B. J., Haan, C. T., and Kao, T. Y. (1976). Suspended sediment filtration capacity of simulated vegetation. *Trans. Am. Soc. Agric. Eng.* **19**(4):678–682.

Tollner, E. W., Barfield, B. J., Vachirakornwatana, C., and Haan, C. T. (1977). Sediment deposition patterns in simulated grass filters. *Trans. Am. Soc. Agric. Eng.* **20**(5):940–944.

Tollner, E. W., Hayes, J. C., and Barfield, B. J. (1978). The use of grass filters for sediment control in strip mine drainage. I. Theoretical studies on artificial media, Report No. IMMR 35-RRR2-78. Institute for Mining and Minerals Research, University of Kentucky, Lexington, KY.

Tollner, E. W., Barfield, B. J., and Hayes, J. C. (1982). Sedimentology of erect vegetal filters. *Proc. Am. Soc. Civil Eng.* **108**(HY12):1518–1531.

Vetter, C. P. (1940). Technical aspects of the silt problem on the Colorado River. *Civil Eng.* **10**:698–701.

Virginia Soil and Water Conservation Commission (1980). "Virginia Erosion and Sediment Control Handbook." Richmond, VA.

Ward, A. D., Haan, C. T., and Barfield, B. J. (1977). Simulation of the sedimentology of sediment detention basins, Research report No. 103. Water Resources Research Institute, University of Kentucky, Lexington, KY.

Ward, A. D., Haan, C. T., and Tapp, J. S. (1979). "The DEPOSITS Sedimentation Pond Design Manual," OISTL. Institute for Mining and Minerals Research, University of Kentucky, Lexington, KY.

Warner, R. C., and Dysart, B. C. (1983). Potential use of the swirl concentrator for sediment control on surface mined lands. *In* "Proceedings, 1983 Symposium on Surface Mining Hydrology, Sedimentology and Reclamation." University of Kentucky, College of Engineering, Lexington, KY.

Warner, R. C., and Hirschi, M. C. (1983). Modeling check dam trap efficiency, Paper No. 83-2082. American Society of Agricultural Engineers, St. Joseph, MI.

Warner, R. C., and Schwab, P. J. (1992). "SEDCAD + Version 3 training Manual." Civil Software Design, Ames, IO.

Warner, R. C., Wilson, B. N., Barfield, B. J., Logsdon, D. S., and Nebgen, P. J. (1982a). A hydrology and sedimentology watershed model. II. Users' manual. Special Publication. Department of Agricultural Engineering, University of Kentucky, Lexington, KY.

Warner, R. C., Wilson, B. N., Barfield, B. J., and Wells, L. G. (1982b). Evaluation of alternative on site sediment controls using the SEDIMOT II Model. *In* "Proceedings, Modeling and Simulation Conference, University of Pittsburg, 1982."

Waynt, D. C. (1980). Evaluation of filter fabric for use as silt fences. VA Highway and Transportation Research Council, Richmond, VA.

Whipple, W., Jr., and Hunter, J. V. (1981). Settleability of urban runoff pollution. *J. Water Pollut. Control Fed.* **53**(12):1726–1732.

Wilson, B. N., and Barfield, B. J. (1984). A sediment detention pond model using CSTRS mixing theory. *Trans. Am. Soc. Agric. Eng.* **27**(5):1339–1344.

Wilson, B. N., and Barfield, B. J. (1985). Modeling sediment detention ponds using reactor theory and advection-diffusion concepts. *Water Resources Res.* **21**(4):523–532.

Wilson, B. N., and Barfield, B. J. (1986a). A detachment model for non-cohesive sediment. *Trans. Am. Soc. Agric. Eng.* **29**(2):445–449.

Wilson, B. N., and Barfield, B. J. (1986b). Predicted and observed turbulence in detention ponds. *Trans. Am. Soc. Agric. Eng.* **29**(5):1300–1306.

Wilson, B. N., Barfield, B. J., and Moore, I. D. (1982). A simulation model of the hydrology and sedimentology of surface mined lands. I. Modeling techniques, Special publication. University of Kentucky Agricultural Engineering Department.

Wilson, B. N., Barfield, B. J., and Warner, R. C. (1986). Simple modles to evaluate nonpoint source and controls. *In* "Agricultural Nonpoint Source Pollution: Model Selection and Application." (Giorgini and Singales, eds.), pp. 231–263. Elsevier, New York.

10

Fluvial Geomorphology: Fluvial Channel Analysis and Design

INTRODUCTION

The Fluvial System

Definition of a Fluvial System

A fluvial system can be divided, for convenience sake, into zones of production, transfer, and deposition of sediment, as shown in Fig. 10.1. Given that these three processes occur in all three zones, the division is rather artificial. However, it is possible to speak of individual zones where one of the three processes predominate.

The processes occurring in zone 1, the zone of production, are of interest to hydrologists and soil erosion scientists. In this zone, processes are very dynamic, responding to short-term climatic conditions, and channels tend to be unstable with rapid changes. Hydrologic and sedimentation processes occurring in zone 1 have been discussed in detail in Chapters 2, 7, and 8.

The processes occurring in zones 2 and 3 are of interest in this chapter. Processes in these zones tend to be the most stable with the best defined configurations. Channel systems in these two zones remain dynamic and occasionally experience abrupt changes; however, the specific processes tend to be amenable to prediction. Some of the deposition processes have been discussed in Chapter 9.

Time Frames of Fluvial Systems

An important distinction needs to be made about steady-state concepts as applied to fluvial systems. Fluvial channels tend to be dynamic, as a result not only of changing flows but also of continued erosion and deposition, which occur with steady flows. Fluvial channels are thus always in a state of change, requiring that the concept of a stable channel be applied to conditions averaged over a period of time and not to a particular instant of time. If a channel system, over a period of years, has developed hydraulic properties such that the sediment load entering the system is transmitted through the system, it is said to be graded or in regime (Mackin, 1948). Such a condition is a delicate balance and would be modified by changes in any of the upstream zones in Fig. 10.1. For example, a channel-straightening operation would increase the channel slope, resulting in a short-term increase in sediment load, upstream erosion, a wider channel, and possible changes in the channel classification. A further discussion of the impacts of such changes on channel geometry and properties is given in a subsequent section.

The choice of a time period for analysis determines whether a variable is independent or dependent. For example, over a short time period, the rate of sediment transports tends to be a function of channel flow rate and the channel hydraulic properties of slope, width, and bed material composition, as discussed in Chapters 4 and 7. Over long periods of time, however, these channel hydraulic properties are determined by flow rate and incoming sediment load to zone 2. In fact, these channel hydraulic properties must ultimately assume values that allow the long-term sediment inflow to equal the outflow. Therefore, the channel hydraulic properties become dependent variables and sediment load becomes the independent variable. In an analysis of the impact of either manmade or natural phenomena on channel geomorphology, it is important to consider the time frame of interest. The absolute duration of the time is not important. What is important is the concept that a variable can be independent or dependent, depending on the time frame. An illustration of the changes of parameters from independent to dependent with time is given in Table 10.1.

The rate of change of channel hydraulic properties is also variable for fluvial systems, tending to be episodic in nature. This led Schumm (1977) to propose the concept of thresholds for change, indicating that changes in flow do not immediately lead to a change in channel properties until some threshold level is reached. At that point, a major change in channel properties is observed.

Rigid versus Fluvial Channels

A detailed discussion is given in Chapter 4 of flow in rigid channels, where the channel banks and bed do not change their configuration with flow. With the exception of vegetative channels, the roughness was also assumed to remain constant. This is in contrast to channels in alluvial material where channel width, depth, slope, curvature, and roughness change with flow rate and sediment load. Adjustment in these channel properties may be slow or may occur in an episodic nature, as suggested by the threshold concept. For purposes of this chapter, the following definitions are

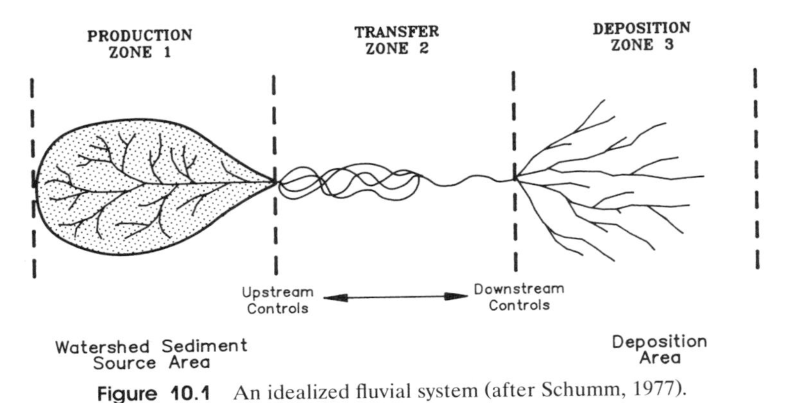

Figure 10.1 An idealized fluvial system (after Schumm, 1977).

Table 10.1 Independent and Dependent Variables for Different Time Spans (Adapted from Schumm, 1971a,b)

	Status of Variable[a]		
Variable	Steady (short term)	Graded (long term)	Geologic (very long term)
Geology	*I*	*I*	*I*
Valley morphology (slope, width, and depth)	*I*	*I*	*D*
Water discharge	*I*	*I*	*X*
Sediment inflow rate	*I*	*I*	*X*
Channel slope, width, and depth	*I*	*D*	*X*
Bed sediment size	*I*	*D*	*X*
Channel sinuosity	*I*	*D*	*X*
Channel roughness	*D*	*D*	*X*
Observed sediment load and flow rate	*D*	*X*	*X*

[a]*I* ; independent variable; *D* ; dependent variable; *X* ; indeterminate.

used:

Rigid channel: A channel whose slope, width, depth, roughness and curvature are assumed to remain constant.

Fluvial channel: A channel whose slope, width, depth, roughness, and curvature are functions of flow rate and sediment load.

In the discussion in this chapter, a special section is devoted to those alluvial channels whose bed is composed of gravel and large roughness elements. Typically, the size of the roughness elements in these channels approaches that of the flow depth, making the flow relationships more complicated. In addition, the bed material is either fixed or only mobile in larger runoff events, as contrasted to the more common moveable beds where bed material is in transport in most runoff events.

Channel response in a fluvial channel is a complex interaction of many factors, not always completely predictable, but amenable to some analytical procedures. The major physical factors involved are water discharge, sediment load, channel slope and shape, soil and geologic characteristics of channel banks and beds, vegetative effects, and human activities.

Overview of the Chapter

The purpose of the discussion in this chapter is to develop a rational system of channel classification, to discuss information on factors controlling channel properties, and to discuss changes that occur in the transfer and deposition zone as a result of natural or human actions. Finally, a discussion of both simple and complex analytical models for predicting such changes is given. Each of these topics could fill an entire text; therefore, the coverage in this chapter is limited to an overview.

CHANNEL CLASSIFICATION

Geomorphic

Davis (1899) proposed that streams could be classified based on stream age, using three age categories: youth, maturity, and old age. This system of classification would allow a stream to go from youth in the headwaters to old age in the delta.

A youthful stream is one in the initial state of development. Such streams are typically V shaped and very irregular, such as headwater streams in mountain tributaries. Mature streams occur in broader valleys where channel slopes are flatter and downward cutting has essentially ceased, being replaced by bank widening. Meanders tend to occur across the entire valley. Mature stream channels, in their natural state, tend to be graded. Old streams occur in flatter flood plains and have flatter gradients. Stream meanders do not typically occupy the entire valley. Natural levees occur near the stream with swamp-like areas adjacent to the levees. Tributaries flow for long distances parallel to the main stream channel before entering the channel at a natural break in the levee. The channels of most interest in small catchments fall in the youthful to mature classification.

Schumm (1971a, b) classified channels by discharge and sediment load into stable, depositing, and eroding channels. The form of the sediment load (bedload versus suspended load), along with flow rate, determines how the stream will shape its channel.

Planform

Based on a plan view of a river network, rivers can be classed as braided, straight, or meandering, or some combination of the three, as shown in Fig. 10.2 (Leopold and Wolman, 1957). If the channel walls and bed are alluvial, straight channels are typically artificial situations or, at most, unstable and short lived. Even under laboratory conditions, artificially formed straight channels become naturally sinuous with time. An exception results from anthropological activity where meandering and resulting sinuosity can be prevented to some extent by channel bank controls.

Braided channels are typically wide with ill-defined banks and multiple islands. Inherently unstable, these channels experience rapid changes in shape, resulting in a braided pattern, thus allowing the channel to establish a sort of equilibrium between flow rates and sediment load. Due to the inherent instability, it is difficult to predict the precise nature of changes in channel morphology resulting from changes in inputs.

Lane (1957) proposed that the mechanisms causing braiding are: (1) steep slopes that result in wide shallow channels where multiple islands can form and (2) excessive sediment loads resulting in deposition islands. Simons, Li and Associates (1982) propose that easily eroded banks can also be a cause of braided channels. In general, braided streams have steep slopes and excessive sediment loads composed primarily of bedload materials.

Meandering channels have an S-shaped appearance resulting from multiple meanders. Within the main channel itself is a narrow deep section known as the thalweg with a more sinuous shape than the main channel. The deepest part of the channel exists in the pool at the apex of the meander. The shallow crossing can occur in the straight section between meanders.

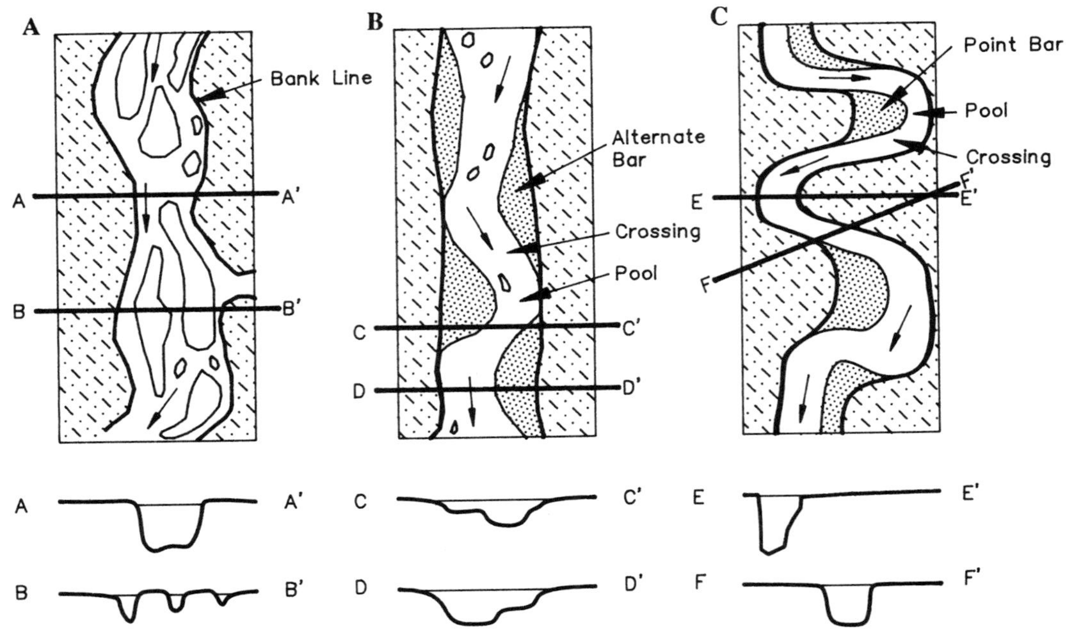

Figure 10.2 Illustration of planform patterns for stream channels (after Simons *et al.*, 1975; Simons, Li and Associates, 1982). (A) Braided, (B) straight, (C) meandering.

Brice (1983) classified channels into four planforms: sinuous canaliform, sinuous point bar, sinuous braided, and nonsinuous braided channels. In general, the slope increases with a change from sinuous canaliform to sinuous point bar to braided channels. The primary difference between the sinuous canaliform and sinuous point bar is in the frequency of point bars. The sinuous canaliform river has broad uniformly wide channels with high sinuosity and a lack of distinct crossings and pools.

Description

Simons, Li and Associates (1982) credit an additional classification to Culbertson *et al.* (1967), who use vegetation, sinuosity, and bank characteristics to characterize streams. Bank characteristics include presence of oxbow lakes, meander scroll patterns, type of bank, bank height, type of natural levee, type of floodplain, and type of vegetal patterns on the bank. These subclassifications assist in determining possible changes that might occur with changes in input parameters.

CHANNEL MORPHOLOGY

Hydraulic Geometry and Shape of Channels

Water discharge and sediment load combine to determine stream channel width, depth, velocity, and planform. The interrelationship is known as hydraulic geometry. Numerous empirical relationships have been developed to predict this interrelationship, with equations generally divided into those that predict geometry at a point and those that predict changes as one moves from one point to another on a stream.

As discussed earlier, gravel bed streams and streams with large roughness elements frequently have roughness elements of magnitudes equal to or greater than the flow depth. Equations developed in this section do not apply to those streams. A special section is included later in the chapter to deal with the morphology of such streams and their resistance to flow.

Hydraulic Geometry at a Point

For fluvial channels, surface width B, mean depth D, mean velocity U, and suspended sediment load Q_s, are all related to water discharge, Q. Leopold and Maddock (1953) proposed that each where related to water discharge by a power function, or

$$B = C_a Q^a \tag{10.1}$$

$$D = C_b Q^b \tag{10.2}$$

$$U = C_c Q^c \tag{10.3}$$

$$Q_s = C_d Q^d, \tag{10.4}$$

where C_a, C_b, C_c, C_d, a, b, c, and d are constants,

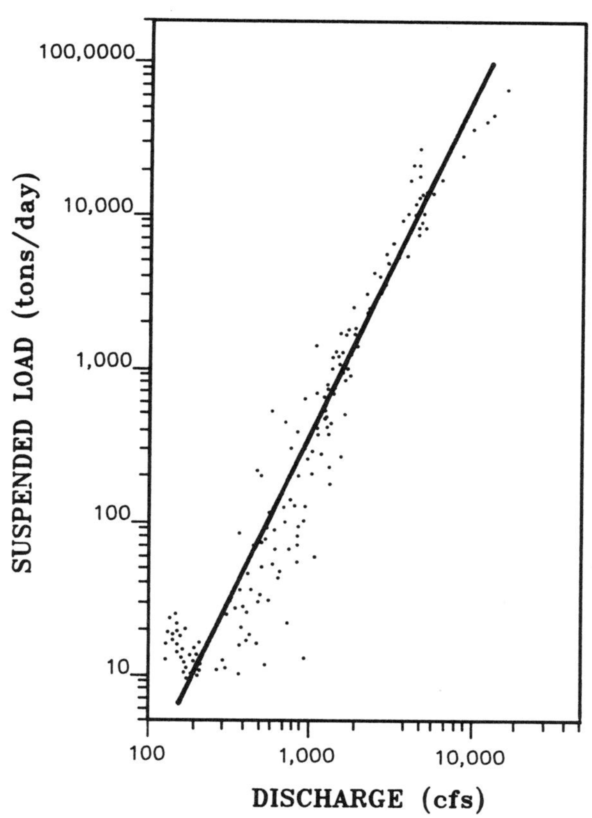

Figure 10.3 Suspended sediment load versus discharge for Brandywine Creek at Wilmington, Delaware (after Wolman, 1955).

interrelated by flow. If the average flow is defined as $Q = BDU$, then a logical solution for the constants is

$$C_a C_b C_c = 1 \quad (10.5)$$

and

$$a + b + c = 1. \quad (10.6)$$

Leopold and Maddock (1953) obtained values for a, b, and c of 0.26, 0.40, and 0.34, respectively, for rivers in the Great Plains and the Southwest. Considerable scatter in the data exists among rivers. They also found that values for d were between 2 and 3, indicating that sediment load increases with discharge more rapidly than B, D, and U. Values for d vary widely among rivers, as a result of varying bed material and channel characteristics. An example of a relationship between sediment load and discharge for a given channel and location is given in Fig. 10.3 As can be seen from the scatter, there is considerable variation among events. Procedures for estimating a relationship between sediment load and water flow-rate for a given channel and location where monitoring data are not available are given in Chapter 7. One of these procedures should be used to develop specific sediment discharge curves.

It can be concluded from the relationships presented that channel width, depth, average velocity, and sediment load all increase with water discharge at a point. An example is given in Fig. 10.4A for Brandywine Creek in Pennsylvania (Wolman, 1955).

Hydraulic Geometry along a Reach

Studies similar to those of hydraulic geometry at a point have been developed for hydraulic geometry along channel reaches. An example is shown in Fig. 10.4B for Brandywine Creek in Pennsylvania. Chang (1988) presents the results of studies from twenty investigators worldwide as summarized by Ming (1983). His results show that the slopes of the regression lines of B, U, and D versus discharge along a channel reach are remarkably similar. Values of the exponents are:

$$a = 0.39\text{–}0.60, \quad b = 0.29\text{–}0.40, \quad c = 0.09\text{–}0.28.$$

These exponents, again, indicated that width, depth, and flow velocity increase with increasing discharge.

The coefficients C_a, C_b, and C_c in Eqs. (10.1)–(10.4) depend on a number of variables, including the size of the bed material and the type of channel bank. When prediction equations that correlate hydraulic geometry to flow and sediment load are developed, the resulting equations are known as regime relationships. The range of their usefulness is limited since they do not include many of the important parameters affecting hydraulic geometry. Examples of additional parameters include bed material size and type of bank material. Schumm (1968) evaluated hydraulic geometry of sand channels in the Great Plains of the United States and the Riverine Plains of New South Wales, Australia. Considerable variation in width, depth, and velocity was present in streams with the same discharge. All channels studied were stable with no major adjustments in 10 years. Schumm determined that the shape of the channel, as determined by the width to depth ratio, F, was related to the percentage of silt and clay, M, in the bank material. Silt and clay were measured as the percentage of sediment smaller than 0.074 mm. Schumm found that

$$F = 255\, M^{-1.08}. \quad (10.7)$$

Studies by Schumm and Khan (Schumm, 1968; Khan, 1971; Schumm and Khan, 1972) also show that F is related to the composition of the sediment load as well as the total load. In general, width increases and flow becomes more shallow as sediment load increases. In studies by Schumm (1960) on ephemeral channels, flows with a high percentage of clays and silts tended to have vertical side walls and small width-to-depth

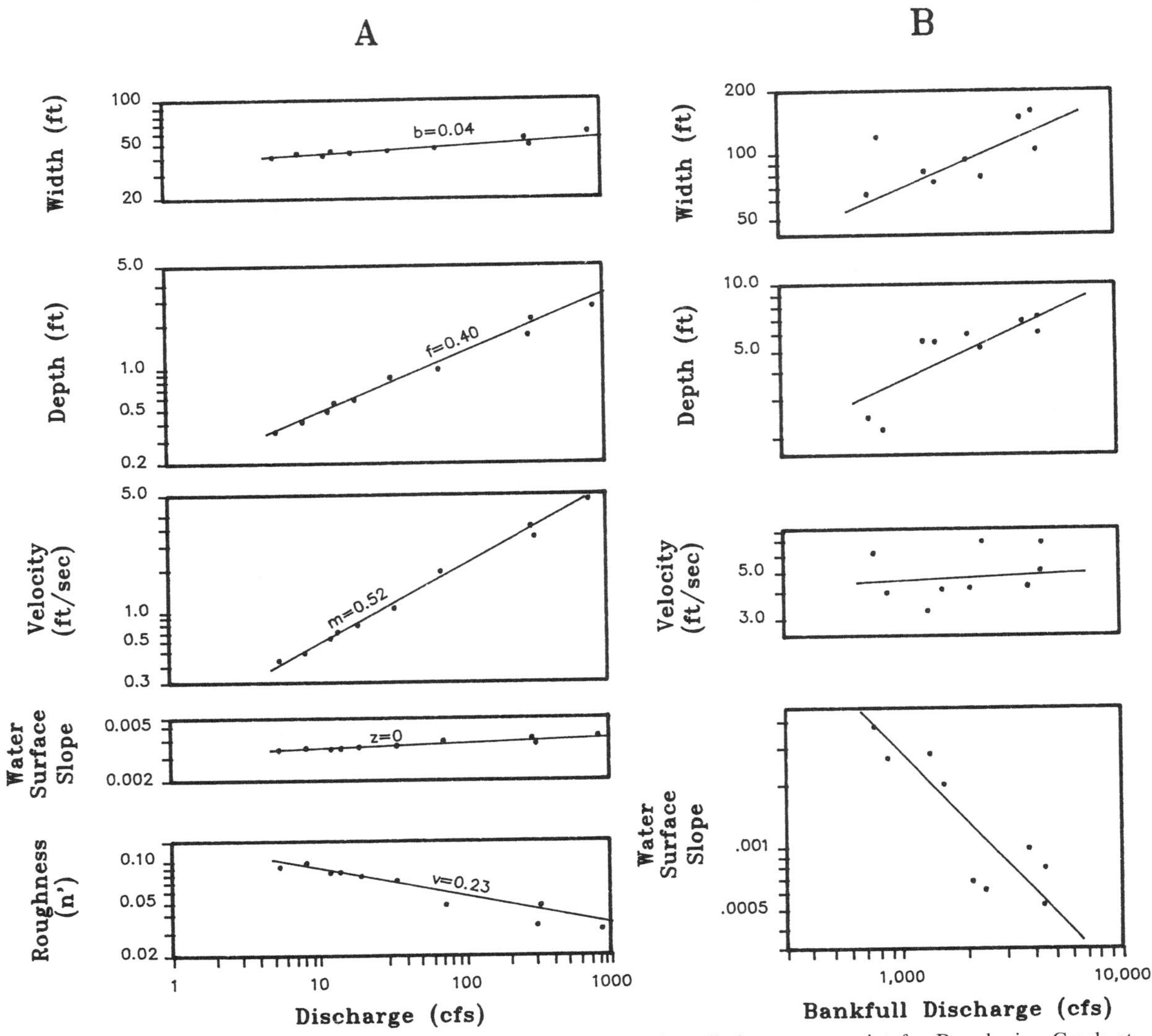

Figure 10.4 (A) Relationship between hydraulic parameters related to discharge at a point for Brandywine Creek at Cornog, Pennsylvania. (B) Relationship between hydraulic parameters and discharge along a 25-mile stretch of Brandywine Creek, Pennsylvania (after Wolman, 1955).

ratios. Channels with primarily sandy sediment load tended to have shallow flow depths and large width-to-depth ratios.

Specific regime relationships used for design of fluvial channels are given in a subsequent section of this chapter.

Channel-Forming Discharge

Bankfull discharge is typically considered to be the channel-forming event. Flows lower than bankfull discharge transport less sediment and cause less scour. Flows grater than bankfull discharge are absorbed by the overflow areas, without major increases in depth of flow and resulting scour. Thus, the bankfull discharge is assumed to be the event that forms the channel. Leopold *et al.* (1964) found that the bankfull discharge had a return period of 1.5 years for 13 stations in the Eastern United States. Williams (1978) utilized 233 sets of data and found that the recurrence interval was not consistent. Chang (1979b) evaluated several data sets and developed the relationship in Fig. 10.5. Dury (1973) analyzed data for the U.S. and showed that the return period of the bankfull discharge is 1.58 years.

Channel Gradient

Channel gradients generally decrease with distance downstream in association with increased discharge

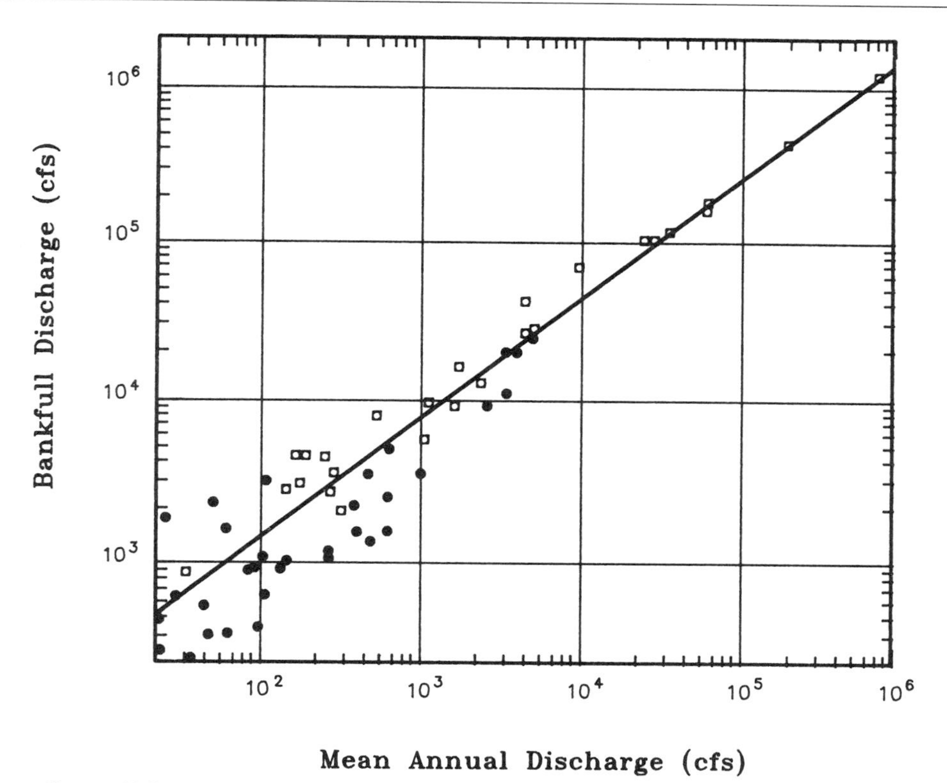

Figure 10.5 Bankfull discharge and mean annual discharge relationship (Chang, 1979b).

and decrease in stream bed particle size. In general, the decrease is exponentially related to distance downstream, or

$$S = S_0 e^{-\alpha X} \tag{10.8}$$

and

$$d_p = d_{p0} e^{-\beta X}, \tag{10.9}$$

where S and d_p are slope and particle diameter at any X; S_0 and d_{p0} are slope and particle diameter at X equal to zero; and α and β are constants. Regression equations for slope as a function of mean annual discharge, average particle diameter, and percentage of silt and clay in the total sediment load have been developed, but the relationships are not universally applicable.

Meandering and Channel Sinuosity

As previously stated, a straight channel is inherently unstable and will tend to develop meanders unless artificially confined. Laboratory studies by Schumm and Khan (1972) as shown in Fig. 10.6A, showed that the degree of sinuosity for a laboratory sand channel with steady flow of 0.15 cfs increased with increasing slope up to a point and then declined until becoming a braided channel at high slopes. Similar results were found for the Mississippi River, in the early 1900s before channelization occurred, which cut off some of the meanders and increased slope gradients. These results are also shown in Fig. 10.6B. One would expect, at any point on the channel, to observe wide variations in the sinuosity as a result of cutoffs and the changes in channel slope that occur. The scatter in the Mississippi river data is likely a result of this changing sinuosity and length of channels as cutoffs occur.

The results in Fig. 10.6 point to a threshold value for slope at which a change occurs and the channel goes from meandering to braided. The presence of such a threshold has been verified by Land (1957) and Leopold and Wolman (1957). The analysis of Lane is shown in Fig. 10.7 indicating that changing the slope of a channel can lead to a transition from a meandering pattern to a braiding pattern, or vice versa.

ALLUVIAL CHANNEL BEDFORM

Description of Channel Bedforms

Channel bedforms have long been of interest because of their impact on flow and sediment transport. As bedforms change, resistance, flow depth, and width

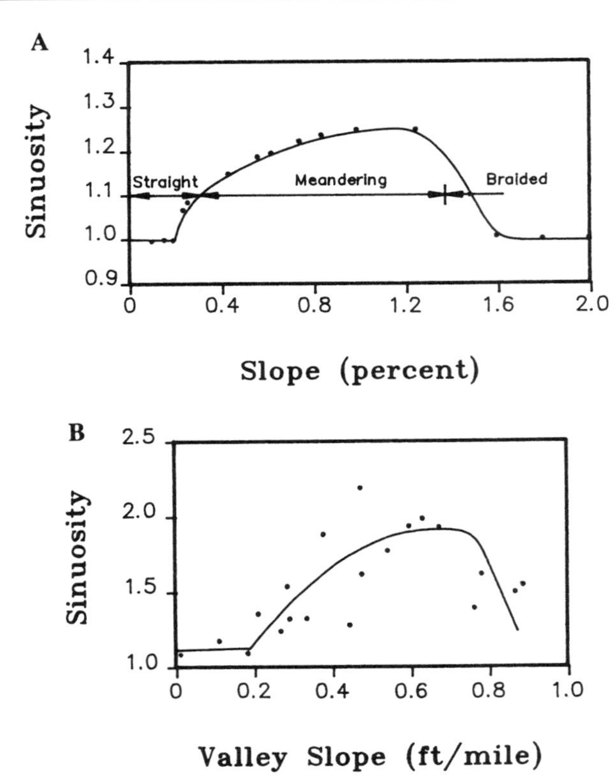

Figure 10.6 (A) Laboratory studies. Relationship between slope and sinuosity for laboratory studies at a flow rate of 0.15 cfs. The relationship would vary, dependent on flow rate and channel properties (after Schumm and Khan, 1972). (B) Relationship between slope and sinuosity for the Mississippi River between Cairo, Illinois, and Head of Passes, Louisiana. Data obtained between 1911 and 1915 surveys before cutoffs (after Simons, Li and Associates, 1982).

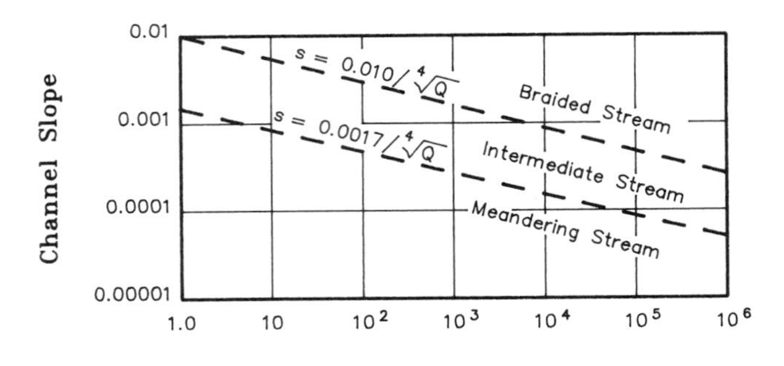

Figure 10.7 Lane's (1957) analysis of meandering as related to annual discharge and slope. The regression lines developed by Lane are taken from streams ranging from highly meandering to braided.

also change. Bedforms, however, are induced by flow rates and shear forces: hence, there is a nonlinear relationship between flow and channel geometry for fluvial channels.

As flow rates and bed shear forces change, there is resulting change in bedforms as shown in Fig. 10.8. A description of the characteristics of each bedform follows. The bedforms follow an increase of stream power, $\tau_0 U$, where τ_0 is shear on the channel bed and U is the mean velocity.

Flat bed: A channel with no bedforms occurring at low stream power. Alternately, it is called a plane bed and a smooth bed.

Ripples: As stream power increases, roughness elements known as ripples form that have no consistent wavelength. Average wavelengths are typically less than 1 ft long and heights are less than 0.1 ft.

Dunes: As stream power increases further, bedforms known as dunes occur that are out of phase with the water surface. Dunes tend to be triangular with upslopes approximately at the angle of repose. Dunes may have ripples superimposed on their surface.

Transition or plane bed: As stream power continues to increase, the roughness elements start to wash out and a plane bed again forms. This form is also called washed out dunes or sand waves. The bedform is unstable, frequently alternating between dunes, plane bed, and antidunes.

Antidunes: With a continued increase in stream power, a bedform known as antidunes appears. Antidunes have an apparent upstream movement, although the net movement of sediment is downstream. The surface water profile is in phase with the antidunes. The amplitude of the antidunes increases as stream power increases and surface waves reach a point of breaking, leading to the term antidunes with breaking waves.

Chutes and Pools: With a further increase in stream power, the bedform becomes chutes and pools. Sediment deposits in large mounds, connected by chutes with supercritical flow and pools that may be either supercritical or subcritical.

Characteristics of each of the bedforms are given in Table 10.2. Of particular interest is the roughness, which increases with increasing stream power and bedform, but decreases significantly as the flow transitions from lower regime to upper regime.

Prediction of Bedforms

A number of studies of bedforms have been conducted using flume data as well as that from canals and natural channels (Simons and Richardson, 1966; Athaullah, 1968; van Rijn, 1984). Perhaps the most widely used relationship is that of Simons and Richardson (1966), shown in Fig. 10.9. To predict the bedform,

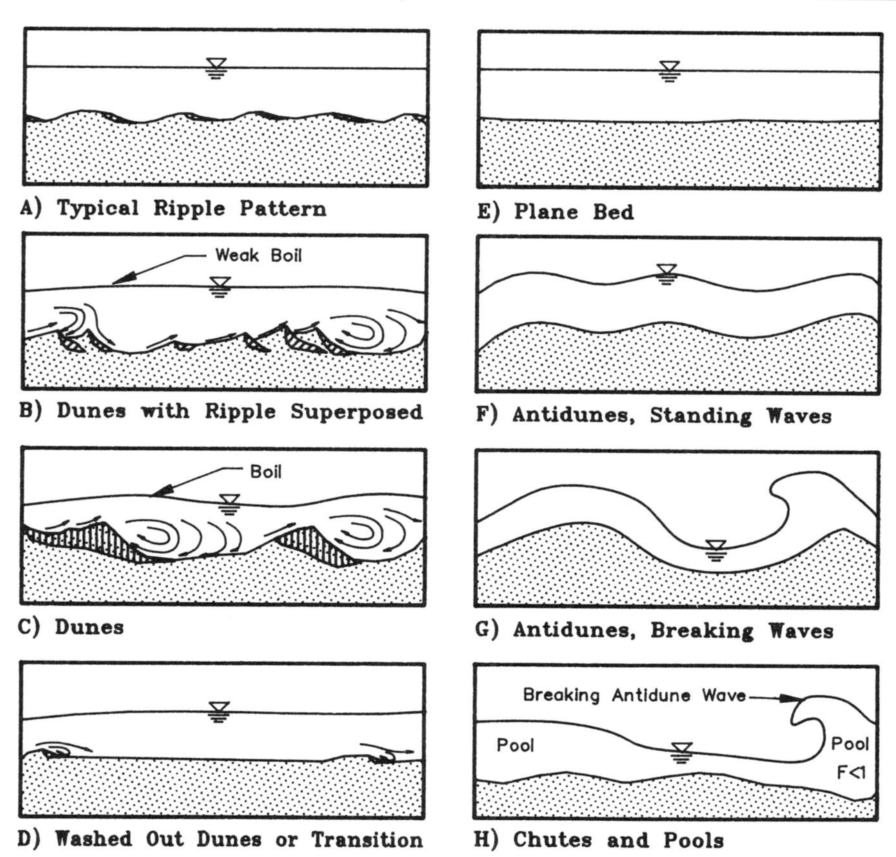

Figure 10.8 Bedforms in alluvial channels (after Simons and Richardson, 1966).

Table 10.2 Bedform Characteristics (after Simons *et al*, 1965; Simons and Richardson, 1966)

Regime	Bedform	Bed material concentration (ppm)	Mode of sediment transport	Type of roughness	Friction factor	Phase relation between bed and water surface
Lower flow regime	Ripples	10–200	Discrete steps	Form roughness dominates	0.05–0.13	Out of phase
	Dunes with ripples	—			—	
	Dunes	100–2000	—		0.05–0.15	
Transition zone	Alternates between dunes, plane bed, and antidunes	1000–4000	—	Variable	0.03–0.08	Variable
Upper flow regime	Plane beds, antidunes, chutes, and pools	1500–3000 > 5000	Continuous	Grain roughness dominates	0.02–0.03 0.03–0.08 0.07–0.10	In phase

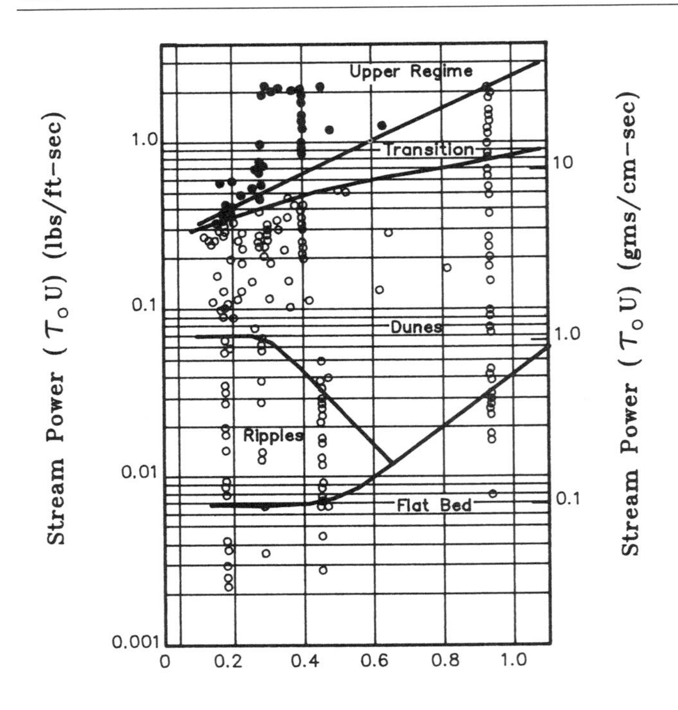

Figure 10.9 Bedforms as related to stream power and median particle diameter (after Simons and Richardson, 1961).

one needs only to know the stream power and median particle diameter.

Two observations can be made from Fig. 10.9. First, for diameters greater than 0.64 mm, the bedform goes directly from flat to dunes without the intermediate ripple form. Also, based on the slopes of the transition line between bedforms, dune bedforms will disappear as one gets to large gravel-sized particles.

FLOW RESISTANCE

Channel roughness is composed of two components, grain roughness and form roughness. Grain roughness is related to the shear forces and form roughness is related to the changes in pressure that result from interactions of bedforms with the flow. To understand roughness of moveable beds, it is first necessary to evaluate roughness of fixed beds.

Fixed Bed Roughness

If channels are lined with unmoveable uniform sand or gravel elements of size d, the resistance is a result of grain roughness. One widely used relationship for predicting roughness is the Strickler (1923) formula, which relates Manning's n to particle diameter, or

$$n = d^{1/6}/21.1, \qquad (10.10)$$

where d is diameter in meters. For d in feet, Eq. (10.10) becomes Eq. (4.32), or

$$n = d^{1/6}/25.7. \qquad (10.11)$$

Using either Eq. (10.10) or (10.11) in Manning's equation, a relationship known as the Manning–Strickler equation results [Eq. (4.33)], or

$$U/U_* = 6.74(R/d)^{1/6}, \qquad (10.12)$$

where U is mean velocity and U_* is shear velocity given by

$$U_* = \sqrt{gRS}\,. \qquad (10.13)$$

When the channel is lined with mixtures of particles, the average diameter is obviously not the diameter that controls roughness. Meyer-Peter and Muller (1948) proposed that d_{90} be used and that

$$n = d_{90}^{1/2}/26, \qquad (10.14)$$

where d_{90} is given in meters.

Equations (10.10)–(10.14) are for fully turbulent flow and a fully turbulent boundary layer. For conditions where the boundary layer is not turbulent, i.e., a laminar sublayer exists, other relationships are needed. Einstein (1950) proposed that the logarithmic velocity profile be applied for both turbulent and laminar sublayer conditions and proposed the following relationships:

$$\frac{u}{U_*} = 5.75\log_{10}\left(30.2\frac{z}{\Delta}\right) \qquad (10.15)$$

and

$$\frac{U}{U_*} = 5.75\log_{10}\left(12.27\frac{D}{\Delta}\right), \qquad (10.16)$$

where u is the velocity at some distance z above the channel bottom, U is average velocity, D is depth of flow, and Δ is an apparent roughness. Δ is related to the roughness of the channel boundary, k_s, by

$$\Delta = k_s/\chi, \qquad (10.17a)$$

where k_s is channel roughness (d_{65}) and χ is a factor given in Fig. 10.10 that is used to correct the logarithmic velocity profile as transition occurs from a smooth to a rough boundary layer. The parameter χ in Fig. 10.10 is given as a function of k_s/δ, where δ is the

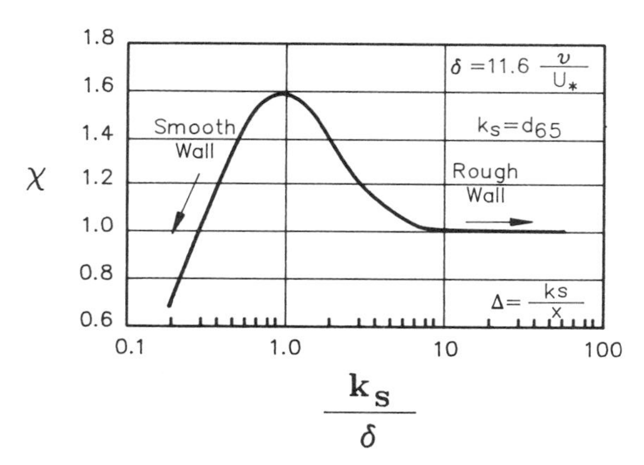

Figure 10.10 Einstein's correction factor χ for the logarithmic velocity profile (After Einstein, 1950).

thickness of the laminar sublayer given by

$$\delta = 11.6\, \nu/U_*, \tag{10.17b}$$

where ν is kinematic viscosity (L^2T^{-1}).

For flow in gravel bed channels, the relationships used here need to be specially calibrated. The Federal Highway Administration (1975) evaluated several data sets and developed an equation for rock riprap as

$$\frac{ng^{1/2}}{D^{1/6}} = 0.225\left(\frac{d_{50}}{D}\right)^{1/6}, \tag{10.18}$$

which yields Eq. (4.32). Each of the terms in Eq. (10.18) must have consistent dimensions.

Fluvial Bed Roughness

Predictors of roughness and velocities for beds with form roughness are based on partitions of drag between form and friction drag. One approach to partitioning drag includes partitioning the energy slope, or

$$S = S' + S'', \tag{10.19}$$

where S is total slope, S' is the slope required to overcome the friction drag, and S'' is the slope required to overcome the form drag. Partitioning can also be based on the hydraulic radius, or

$$R = R' + R'', \tag{10.20}$$

where R' and R'' are the hydraulic radius relative to friction and form drag respectively. If Eq. (10.20) is multiplied by γS, then a shear stress relationship results, or

$$\tau_0 = \tau_0' + \tau_0'', \tag{10.21}$$

where τ_0 is total shear, τ_0' is shear due to grain roughness, and τ_0'' is shear due to form roughness. If Eq. (10.21) is divided by ρU^2, then a friction factor relationship results, or

$$f = f' + f'', \tag{10.22}$$

where f, f', and f'' are friction factors for the channel, grain roughness, and form roughness, respectively. In the following discussion, methods for partitioning the hydraulic radius and channel slope are given. Methods for partitioning friction factors are given in Chang (1988) and other references.

Partitioning Hydraulic Radius – Einstein Barbarossa Method

Einstein and Barbarossa (1952) divided total resistance into friction and form drag by partitioning the hydraulic radius. They assumed that velocity could be predicted by using a modification of the logarithmic velocity profile given by Eq. (10.16). In this partitioning, the hydraulic radius for predicting mean velocity in Eq. (10.16) was assumed to be that for grain roughness, R', or

$$\frac{U}{U_*'} = 5.75 \log_{10}\left(12.27\frac{R'\chi}{k_s}\right), \tag{10.23}$$

where

$$U_*' = \sqrt{gR'S} \tag{10.24}$$

and χ in Fig. 10.10 is given as a function of k_s/δ', with k_s defined by d_{65}. In calculating δ', of course, U_*' is used and in the calculation of U_*', R' is used. For a given R', the mean velocity and hence discharge can be calculated. The difficulty in using Eq. (10.23) comes in determining the value for R'' and hence R for a given R'. To make this possible, Einstein and Barbarossa proposed that

$$\frac{U}{U_*''} = fct(\psi_{35}), \tag{10.25}$$

where

$$\psi_{35} = \left[\frac{(\rho_s - \rho)}{\rho}\right]\frac{d_{35}}{R'S} = (\mathrm{SG} - 1)\frac{d_{35}}{R'S}, \tag{10.26}$$

where SG is the specific gravity. The functional relationship for Eq. (10.25) is shown in Fig. 10.11. Procedures for using the relationship are given in a subsequent example.

Partitioning Channel Shear—Engelund Method

Another widely used resistance formula is that of Engelund (1966), in which the slope is divided into

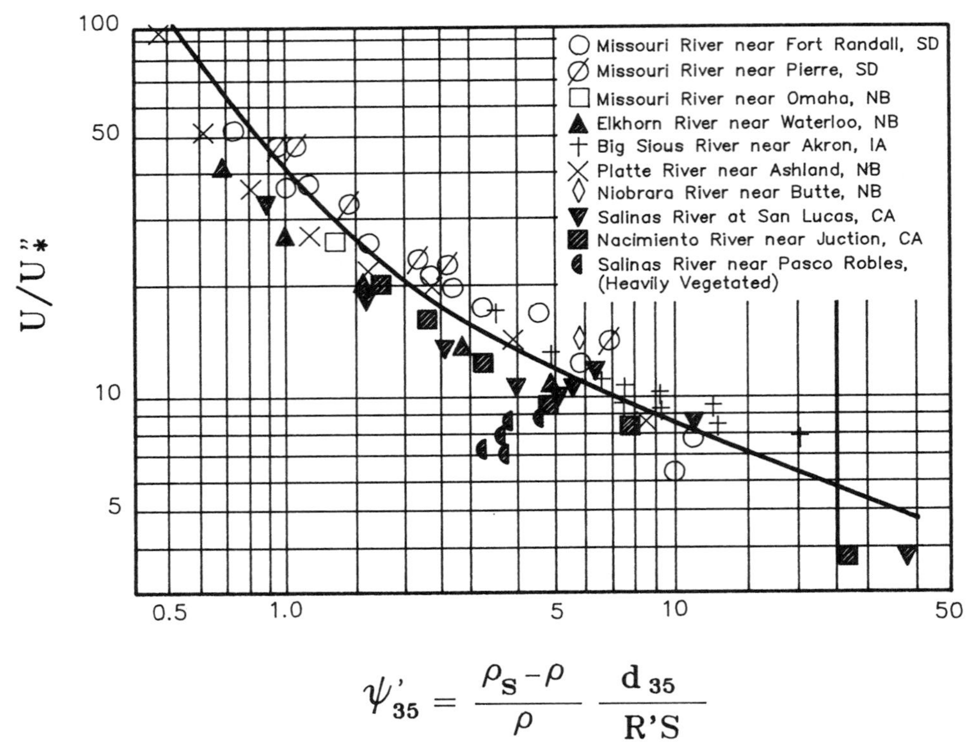

$$\psi'_{35} = \frac{\rho_s - \rho}{\rho} \frac{d_{35}}{R'S}$$

Figure 10.11 Einstein and Barbarossa relationship for form resistance (after Einstein and Barbarossa, 1952).

form and grain roughness components. By evaluating the expansion head loss that results as flow moves over a bedform element, Engelund proposed that the head loss due to flow over a form roughness element of length λ and height h is caused by expansion loss, $\Delta H'' = \alpha(\Delta U)^2/2g$, where ΔU is the flow change over the roughness element. Since U is discharge per unit width over depth (q/D) and the depth on the form roughness in the lee of the form roughness element is $D - h/2$ and $D + h/2$, then the slope could be divided into grain roughness and form roughness components by

$$S = S' + \frac{\alpha}{2g\lambda}\left[\frac{q}{D - h/2} - \frac{q}{D + h/2}\right]^2, \quad (10.27)$$

where D is depth of flow, S' is slope due to grain roughness, α is a loss coefficient, h and λ are the height and average wavelength, respectively, of the form roughness elements, and U is the mean velocity of the flow. The second term on the right-hand side is $\Delta H/\lambda$. Equation (10.27) can be modified to

$$S \cong S' + \frac{\alpha U^2}{2g\lambda}\left[\frac{h}{D}\right]^2. \quad (10.28)$$

Equation (10.27) can be further modified by multiplying by $\gamma R/[(\gamma_s - \gamma)d_{50}]$ and by assuming that $\gamma RS'$ equals $\gamma R'S$ to obtain

$$\tau_* = \tau'_* + \tau''_*, \quad (10.29)$$

where

$$\tau_* = \frac{\gamma RS}{(\gamma_s - \gamma)d_{50}} \quad (10.30)$$

$$\tau'_* = \frac{\gamma R'S}{(\gamma_s - \gamma)d_{50}} \quad (10.31)$$

and

$$\tau''_* = \frac{\gamma R''S}{(\gamma_s - \gamma)d_{50}} = \left(\frac{\alpha}{2}\right)\frac{\gamma h^2}{(\gamma_s - \gamma)\lambda d_{50}}\frac{U^2}{gR}. \quad (10.32)$$

Shear due to grain roughness is based on the logarithmic formula (Engelund and Hansen, 1967)

$$\frac{U}{U_*} = 6 + 2.5\ln\left[\frac{R'}{2.5d_{65}}\right]. \quad (10.33)$$

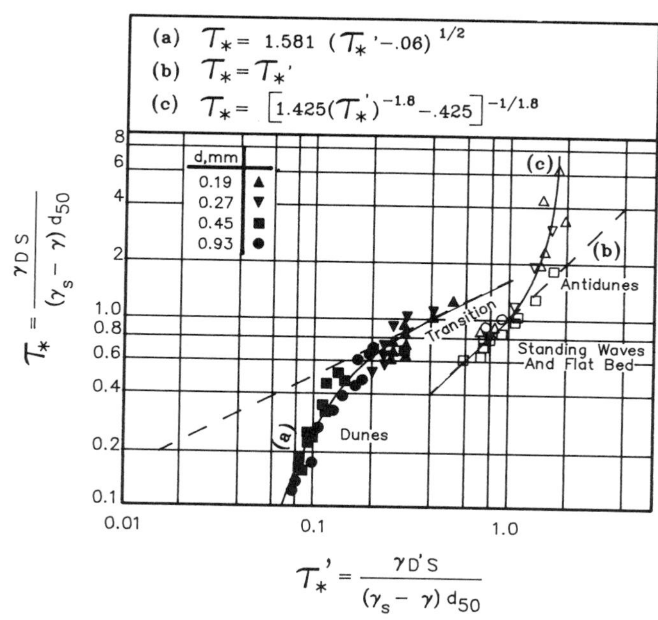

Figure 10.12 Engelund's relationship for total shear and grain shear.

Using this model, Engelund and Hansen (1967) developed the relationship in Fig. 10.12. According to Chang (1988), the relationship is generally satisfactory for predicting stage–discharge relationships, but should not be used for channels with very coarse sand beds. Procedures for making the calculations are illustrated in a subsequent problem.

From Fig. 10.12, it can be seen that a discontinuity exists in the roughness curve at τ_* of 1.0, resulting from a transition from the lower regime to the upper regime. Such a discontinuity has been supported by flume data as shown in Fig. 10.13, but it is less common in natural channels due to the variation in flow regimes that exist across a cross section of varying depth.

Example Problem 10.1 Calculation of stage–discharge curves

A channel has a flow width of 20 ft and a maximum depth of 4.0 ft before overtopping the banks. The channel slope is 0.0005 ft/ft. Bed material has the following particle size information: $d_{35} = 0.15$ mm (0.000492 ft), $d_{50} = 0.3$ mm (0.000984 ft), $d_{65} = 0.7$ mm (0.00229 ft). Develop a stage–discharge curve for depths up to bankfull discharge using the Einstein—Barbarossa and the Engelund methods. Assume that the kinematic viscosity is 10^{-5} ft^2/sec and that the channel is rectangular.

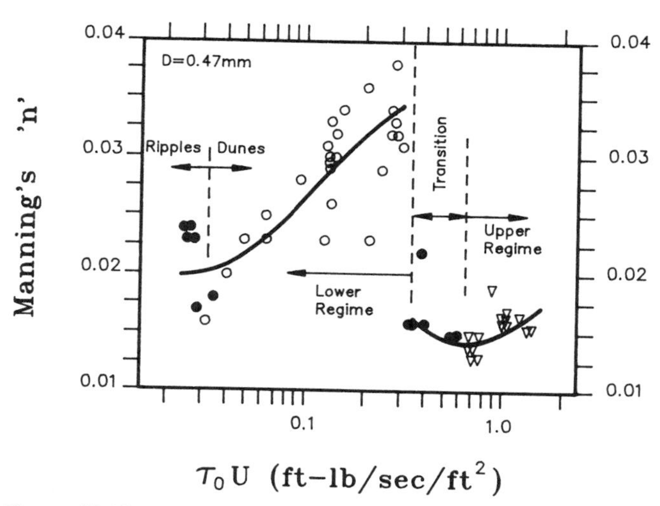

Figure 10.13 Flume data showing a relationship between bed roughness, bedform, and stream power (after Chang, 1979b).

Solution:

Einstein–Barbarossa Method

1. *Assume a value for R′ and calculate a trial velocity.* Assume that $R' = 1.0$ to start the computation. From Eqs. (10.23), (10.24), and (10.17) using R' instead of R,

$$U_*' = (gR'S)^{1/2} = (32.2 \times 1.0 \times 0.0005)^{1/2} = 0.127 \text{ ft/sec}$$

$$\delta' = \frac{11.6\nu}{U_*'} = \frac{(11.6)(10^{-5}\text{ ft}^2/\text{sec})}{0.127 \text{ ft/sec}} = 0.000913 \text{ ft}$$

$$\frac{k_s}{\delta'} = \frac{0.00229 \text{ ft}}{0.000913 \text{ ft}} = 2.51.$$

From Fig. 10.10 with $k_s/\delta' = 2.51$, $\chi = 1.25$. Using Eq. (10.23) with $k_s = d_{65} = 0.00229$ ft,

$$U = U_*' 5.75 \log_{10}\left(\frac{12.27R'\chi}{k_s}\right)$$

$$= (0.127)(5.75)\log_{10}\left(\frac{12.27 \times 1.0 \times 1.25}{0.00229}\right) = 2.79 \text{ ft/sec}.$$

2. *Calculating R″.* Calculate a value for ψ_{35}, assuming that R' is 1.0 ft. Utilizing a value of 2.65 for specific gravity,

$$\psi_{35} = \frac{(SG - 1)d_{35}}{R'S} = \frac{(2.65 - 1)(0.000492)}{(1.0)(0.0005)} = 1.62.$$

From Fig. 10.11, $U/U_*'' = 26$; hence

$$U_*'' = \frac{U}{26} = \frac{2.79 \text{ ft/sec}}{26} = 0.107 \text{ ft/sec}$$

$$R'' = \frac{U_*''^2}{gS} = \frac{(0.107)^2}{(32.2)(0.0005)} = 0.711 \text{ ft}.$$

3. *Calculate R, A, and Q.*

$$R = R' + R'' = 1.0 + 0.711 = 1.711 \text{ ft}.$$

For a rectangular channel $D = BR/(B - 2R) = 20R/(20 - 2R)$,

$$D = \frac{(20)(1.711)}{20 - (2)(1.711)} = 2.06$$

$$A = BD = (20)(2.06) = 41.2 \text{ ft}^2$$

$$Q = AU = (41.2)(2.79) = 114.9 \text{ cfs}.$$

4. *Tabulation for other values of R'.* Other values are tabulated and shown below.

(1) R'	(2) U_*'	(3) k_s/δ'	(4) χ	(5) U	(6) Ψ_{35}	(7) U/U_*''
0.1	0.040	0.79	1.50	0.67	16.2	7.0
0.5	0.090	1.78	1.50	1.86	3.24	15.0
1.0	0.127	2.51	1.25	2.79	1.62	26.0
1.5	0.155	3.06	1.15	3.53	1.08	37.0
2.0	0.179	3.53	1.12	4.20	0.81	51.0
3.0	0.219	4.32	1.08	5.34	0.54	85.0
4.0	0.254	5.01	1.02	6.34	0.41	150.0

(1) R' (ft)	(8) U_*'' (ft/sec)	(9) R'' (ft)	(10) R (ft)	(11) D (ft)	(12) Q (cfs)
0.1	0.096	0.572	0.672	0.72	9.6
0.5	0.124	0.955	1.455	1.70	63.2
1.0	0.107	0.711	1.711	2.06	114.9
1.5	0.095	0.560	2.060	2.59	182.9
2.0	0.082	0.418	2.418	3.19	268.0
3.0	0.062	0.239	3.239	4.79	511.6
4.0	0.042	0.110	4.110	6.98	885.1

Column entries in this table are determined from

(1) Assumed value for R'

(2) $U'_* = \sqrt{gR'S} = \sqrt{32.2R'(0.0005)}$ ft/sec, Eq. (10.24)

(3) $\dfrac{k_s}{\delta'} = \dfrac{d_{65}}{11.6\nu/U'_*} = 19.74U'_*$, Eq. (10.17)

(4) $\chi = Fct\left(\dfrac{R'}{\delta'}\right)$, Fig. 10.10

(5) $U = U'_* 5.75 \log_{10}\left(12.27\dfrac{R'\chi}{k_s}\right)$, Eq. (10.23)

(6) $\psi_{35} = (SG - 1)\dfrac{d_{35}}{(R'S)} = \dfrac{(1.65)(0.000492 \text{ ft})}{(R')(0.0005)} = \dfrac{1.62}{R'}$, Eq. (10.26)

(7) $\dfrac{U}{U''_*} = Fct(\psi_{35})$, Fig. 10.11

(8) $U''_* = \dfrac{U}{U/U''_*}$

(9) $R'' = \dfrac{U''^2_*}{gS} = \dfrac{U''^2_*}{(32.2)(0.0005)} = 62.11U''^2_*$

(10) $R = R' + R''$

(11) $D = \dfrac{BR}{B - 2R} = \dfrac{20R}{20 - 2R}$

(12) $Q = BDU$.

Engelund Method

1. *Assume a value for R' and calculate τ'_*.* Assume $R' = 1.0$ ft. From Eq. (10.31),

$$\tau'_* = \frac{\gamma R'S}{(\gamma_s - \gamma)d_{50}}$$

$$= \frac{R'S}{(SG - 1)d_{50}} = \frac{(1.0)(0.0005)}{(1.65)(0.000984)} = 0.307.$$

2. *Determining τ_*.* From Fig. 10.12, since $\tau'_* < 0.55$, lower flow regime is assumed. Using Eq. (a) in Fig. 10.12 for $\tau'_* = 0.307$,

$$\tau_* = 1.581(\tau'_* - 0.06)^{0.5} = 0.786.$$

3. *Calculate depth, velocity, and discharge.* From definition of τ. [Eq. (10.30)],

$$R = \frac{\tau_*(\gamma_s - \gamma)d_{50}}{\gamma S} = \frac{\tau_*(SG - 1)d_{50}}{S}$$

$$= \frac{(0.786)(1.65)(0.000984)}{0.0005} = 2.55.$$

For a 20-ft-wide channel,

$$D = \frac{BR}{B - 2R} = \frac{(20)(2.55)}{20 - (2)(2.55)} = 3.42.$$

From Eq. (10.33) for mean velocity,

$$U = U'_*\left[6 + 2.5\ln\left(\frac{R'}{2.5d_{65}}\right)\right]$$

$$U'_* = \sqrt{gR'S} = \sqrt{(32.2)(1)(0.0005)} = 0.127$$

$$d_{65} = 0.00229 \quad \text{(given)}.$$

Hence

$$U = 0.127\left[6 + 2.5\ln\left(\frac{1.0}{(2.5)(0.00229)}\right)\right]$$

$$= 2.40 \text{ ft/sec}.$$

From continuity for a rectangular channel,

$$Q = BDU = (20)(3.42)(2.40) = 164.2 \text{ cfs}.$$

4. *Tabulations for other R′ values.*

(1) R' (ft)	(2) U_*' (ft/sec)	(3) U (ft/sec)	(4) τ_*'	(5) τ_*	(6) R (ft)	(7) D (ft)	(8) Q (cfs)
0.5	0.090	1.55	0.154	0.485	1.58	1.88	58.3
1.0	0.127	2.40	0.307	0.786	2.55	3.42	165.1
1.5	0.155	3.09	0.461	1.001	3.26	4.84	297.3
2.0	0.179	3.69	0.614	0.614	2.00	2.50	184.5
3.0	0.219	4.74	0.921	0.921	3.00	4.29	406.7
4.0	0.254	5.68	1.228	1.381	4.50	8.18	929.2

Column entries in this table determined from

(1) Assume a value for R'

(2) $U'_* = \sqrt{gR'S} = \sqrt{(32.2)R'(0.0005)}$

(3) $U = U'_*\left[6 + 2.5\ln\left(\frac{R'}{2.5d_{65}}\right)\right]$, Eq. (10.33)

(4) $\tau'_* = \frac{\gamma R'S}{(\gamma_s - \gamma)d_{50}} = 0.307R'$, Eq. (10.31)

(5) τ_* from Fig. 10.12

(6) $\tau_* = \frac{\gamma RS}{(\gamma_s - \gamma)d_{50}} = \frac{RS}{(SG - 1)d_{50}}$, Eq. (10.30)

$R = \tau_*/0.307$

(7) $D = \frac{BR}{(B - 2R)} = \frac{20}{(20 - 2R)}$

(8) $Q = BDU = 20\ DU$.

Note: The stage–discharge curve results from a plot of Q versus D.

CHANNELS IN REGIME

The Regime Concept

A channel is assumed to be in regime if it has not changed its characteristics over a long period of time. This would imply that the channel hydraulic geometry was in a state of equilibrium with the incoming flow rate and sediment load such that the flow was just carrying the sediment load introduced into the stream. Such a concept would be significantly different for a channel with widely varying flows as compared to a canal with a reasonably stable discharge. Both flow situations are discussed.

Regime Equations for Canals and Channels with Steady Discharge

Regime theory originated with analyses on irrigation canals in India and Pakistan in an attempt to develop stable designs for canals transporting steady discharges of water with high sediment loads. Early strictly empirical relationships were developed by Lacey (1930) followed by Blench (1970) and Simons and Albertson (1960). More recently, Chang (1980b, 1985b) developed what is described as a rational basis for a regime concept based on minimization of stream power. Chang (1979b, p. 311) proposed as his hypothesis for stream power minimization:

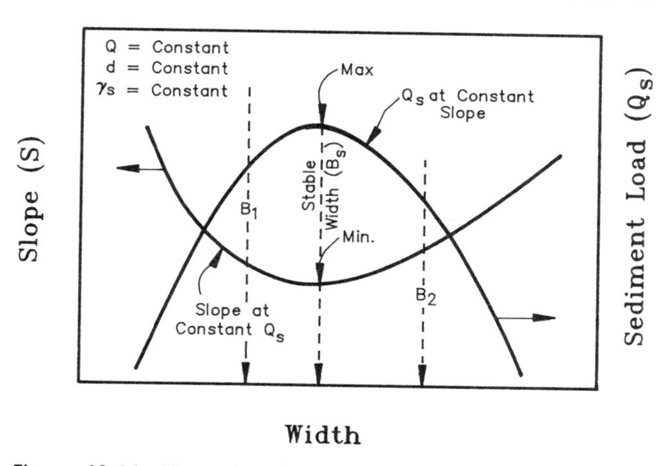

Figure 10.14 Illustration of the minimum stream power and maximum sediment efficiency as applied to stable width determination. For a constant Q, the minimum value for S also corresponds to the minimum stream power, γQS, and the maximum sediment transport (after Chang, 1988; copyright © 1988. Reprinted by permission of John Wiley & Sons, Inc.)

> For an alluvial channel, the necessary and sufficient condition of equilibrium occurs when the stream power per unit channel length γQS is a minimum subject to given constraints. Hence, an alluvial channel with water discharge Q and sediment load Q_s as independent variables tends to establish its width, depth and slope such that γQS is a minimum. Since Q is a given parameter, minimum γQS also means minimum channel slope S.

Application of this principle is illustrated in Fig. 10.14. A family of curves with varying Q_s could be plotted like that of slope versus width in Fig. 10.14. For a constant width, Q_s would increase with slope. Thus, for a constant slope, Q_s versus width would display the shape known, illustrating that the width corresponding to the minimum stream power is also the most efficient section for sediment transport.

The methods of Blench, Simons and Albertson, and Chang are illustrated in this section. In addition to the information presented on these methodologies, Chang (1988) points out that experience on canals in India and Pakistan indicates that Froude numbers greater than 0.3 make maintenance of straight channels difficult, even when designed with regime equations. At higher Froude numbers, a thalweg is formed with a tendency to meander.

Simons and Albertson's Method

Using data collected in India and Pakistan as well as the Western United States, Simons and Albertson (1960) divided channels into five types: (1) sand beds and banks, (2) sand beds and cohesive banks with no sediment load, (3) sand beds and cohesive banks with sediment loads between 2000 and 8000 mg/liter, (4) cohesive bed and banks, and (5) coarse noncohesive material. Simons and Albertson developed three groups of graphical relationships and equations to define stable width and wetted perimeter, stable depth and hydraulic radius, and stable velocity.

Chang (1988) grouped the Simons and Albertson (1960) equations into three categories and presented equations for the graphical relationships. Equations for stable wetted perimeter and stable width (foot–pound–seconds) are of the form

$$P = K_b Q^{1/2} \quad (10.34)$$

$$B_a = 0.9P \quad (10.35)$$

and

$$B_s = 1.087B_a + 2.17, \quad (10.36)$$

where P is the wetted perimeter in feet, K_b is a constant dependent on type of channel, Q is discharge in cfs, B_a is average channel width in feet, and B_s is top width in feet. Equations for stable flow depth and hydraulic radius are of the form

$$R = K_d Q^{0.36} \quad (10.37)$$

$$D = 1.21R; \qquad R < 7 \text{ ft} \quad (10.38)$$

$$D = 2 + 0.93R; \qquad R \geq 7 \text{ ft}, \quad (10.39)$$

where R is the hydraulic radius and K_d is a coefficient dependent on channel type. Equations for stable flow rates are of the form

$$U = K_u(R^2S)^{\varepsilon} \quad (10.40)$$

and

$$\frac{U^2}{gDS} = K_r\left[\frac{UB_a}{\nu}\right]^{0.37}, \quad (10.41)$$

where U is average velocity in the channel, ν is kinematic viscosity, and K_u, K_r, and ε are constants dependent on channel type. Values for the constants were developed by Chang (1988) from graphical relationships presented by Simons and Albertson (1960).

Table 10.3 Coefficient Values for Simons and Albertson Regime Equations (after Chang, 1988)

Coefficient	Channel type[a] 1	2	3	4	5
K_b	3.50	2.60	1.70	2.20	1.75
K_d	0.52	0.44	0.34	0.37	0.23
K_u	13.90	16.90	16.00	—	17.90
K_r	0.33	0.54	—	0.87	—
ξ	0.33	0.33	0.29	—	0.29

[a]Channel types are: (1) Sand beds and banks, (2) sand beds and cohesive banks with no sediment load, (3) sand beds and cohesive banks with sediment loads between 2000 and 8000 mg/liter, (4) cohesive bed and banks, and (5) coarse non-cohesive material.

These constants are given in Table 10.3. Equation (10.40) is only applicable to the channel types 1, 2, and 4 as identified in Table 10.3.

Calculation of discharge from Eqs. (10.34)–(10.41) is accomplished by continuity, or

$$Q = UB_aD \quad (10.42)$$

For classification purposes, sediment sizes, channel banks, and bed are assumed to be sand bed channels if the bed and bank materials are medium to fine sands. Cohesive bend materials are those finer than sand size and coarse bed materials are those with medium particle sizes between 20 and 82 mm.

Blench Regime Method

Blench (1970) developed a regime method that utilizes a bed factor, a side factor, and a flow resistance equation to develop relationships for stable width, depth, and slope. His relationships are

$$B_a = \left(\frac{F_bQ}{F_s}\right)^{1/2}, \quad (10.43)$$

where B_a is average width in feet, F_b and F_s are bed and side factors, and Q is discharge in cfs. Average depth in feet is determined as

$$D = \left(\frac{F_sQ}{F_b^2}\right)^{1/3} \quad (10.44)$$

and stable slope is

$$S = \frac{F_b^{5/6}F_s^{1/12}}{K_1Q^{1/6}(1 + C/2330)}, \quad (10.45a)$$

where

$$K_1 = \frac{3.63g}{\nu^{1/4}}. \tag{10.45b}$$

ν is kinematic viscosity in square feet per second, g is the gravitational constant in feet per second squared, and C is sediment concentration in ppm by weight. Values for the bed and side factors were determined by Blench to be

$$F_b = 1.9 d_{50}^{1/2} \tag{10.46}$$

and

$F_s = 0.1$ slightly cohesive banks (very friable)
$= 0.2$ moderately cohesive banks (silty clay loam)
$= 0.3$ highly cohesive banks (tough clay) (10.47)

where d_{50} is median diameter of sediment in millimeters. Blench also proposed that F_b increases with sediment concentration by 12% for every 1% increase in C.

Chang's Rational Method

Chang (1980b, 1985b) utilize the concept of minimum stream power, γQS, to develop stable channel width, depth, and slopes for channels with constant discharges. Using sediment discharge relationships and stage flow rate predictors, Chang developed a model known as FLUVIAL that determines the width corresponding to a minimum stream power. A flow diagram of FLUVIAL is given in Fig. 10.15.

Based on computations using FLUVIAL, Chang (1985b) developed the graphical relationship in Fig. 10.16 for side slopes of 2:1, but subsequently suggested applying the results to 1.5:1 channels also (Chang, 1988). Similar curves could have been obtained for other sides slopes.

The critical slope for bedload movement in Fig. 10.16 is given by

$$\frac{S_c}{\sqrt{d_{50}}} = 0.00238 Q^{-0.51}, \tag{10.48}$$

where d_{50} is average particle size in millimeters and Q is flow rate in cfs. Empirical relationships for surface width B and depth D are

$$B = 4.17\left(\frac{S}{\sqrt{d_{50}}} - \frac{S_c}{\sqrt{d_{50}}}\right)^{0.05} Q^{0.5} \tag{10.49}$$

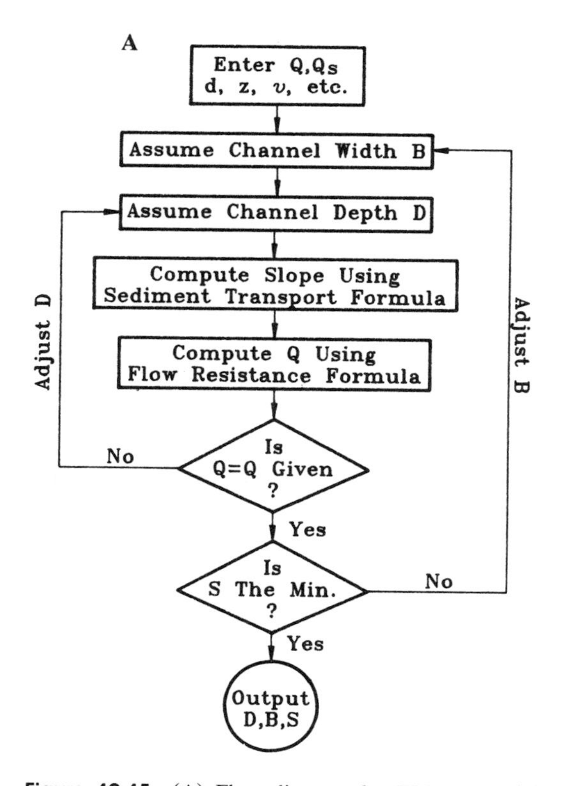

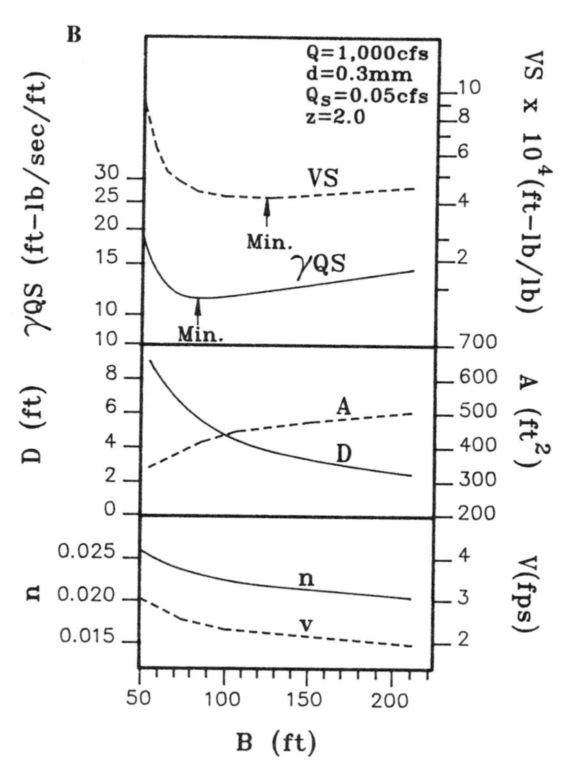

Figure 10.15 (A) Flow diagram for FLUVIAL. (B) Illustration of the change in stream power (γQS) and other parameters with changes in channel width (after Chang, 1980b).

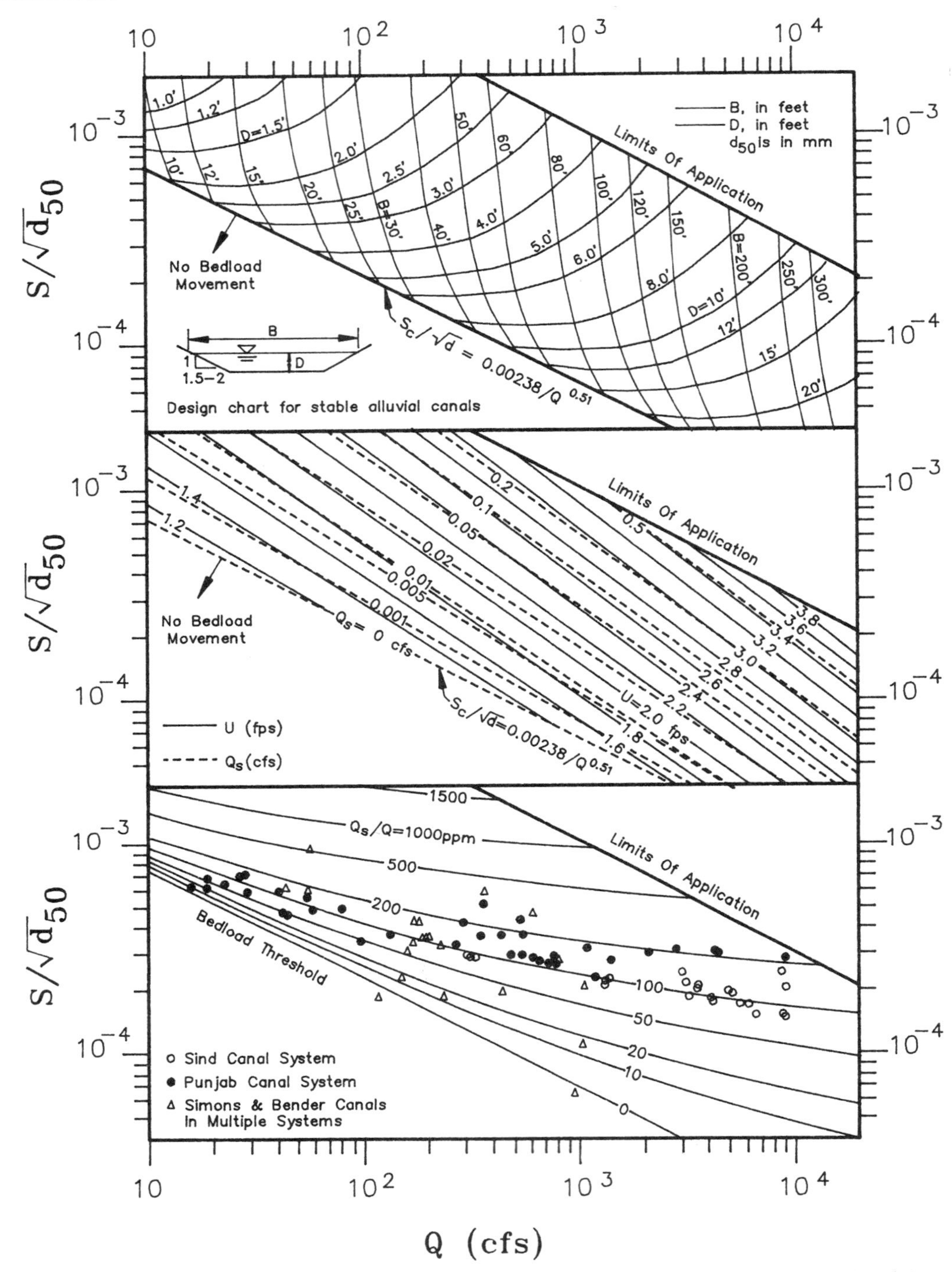

Figure 10.16 Stable alluvial canal (or steady flow channel) design chart for side slopes of 1.5–2 : 1 (after Chang, 1985b). Also shown is the bed load and velocity relationship for the stable section.

and

$$D = 0.055\left(\frac{S}{\sqrt{d_{50}}} - \frac{S_c}{\sqrt{d_{50}}}\right)^{-0.3} Q^{0.3}, \quad (10.50)$$

where B and D are in feet. An empirical equation for bedload is

$$\frac{S}{\sqrt{d_{50}}} = 0.433\frac{Q_s}{Q^{0.789}} + \frac{S_c}{\sqrt{d_{50}}}. \quad (10.51)$$

where Q_s is volumetric bedload flow rate in cfs.

Example Problem 10.2 Illustration of regime equations

A flow of 50 cfs and a bed sediment load of 3.12 lb/sec is to be conveyed in a canal in alluvial material with a median bed particle size of 0.40 mm. Determine the stable width, depth, and slope by the Blench, Simons and Albertson, and Chang methods if the channel banks are slightly noncohesive with 1.5 : 1 side slopes. Assume a viscosity of 10^{-5} ft^2/sec. If the actual groundslope is 0.0003 ft/ft, what would be recommended?

Solution:

Blench Method

1. *Bed and side factors*. From Eq. (10.46) with d_{50} = 0.4 mm,

$$F_b = (1.9)(0.4)^{0.5} = 1.20.$$

From Eq. (10.47) for slightly cohesive banks,

$$F_s = 0.1.$$

2. *Average width and depth*. From Eq. (10.43) for width

$$B_a = \left[\frac{(1.20)(50)}{0.1}\right]^{0.5} = 24.5 \text{ ft}.$$

From Eq. (10.44) for depth

$$D = \left[\frac{(0.1)(50)}{1.20^2}\right]^{1/3} = 1.51 \text{ ft}.$$

3. *Average stable slope*. Concentration of sediment

$$C = \frac{Q_s}{\gamma Q} = \frac{3.12}{(62.4)(50)} = 0.001 \text{ lb/lb} = 1000 \text{ ppm}.$$

From Eq. (10.45) for stable slopes

$$K_1 = \frac{(3.63)(32.2)}{(10^{-5})^{1/4}} = 2079$$

$$S = \frac{(1.2)^{5/6}(0.1)^{1/12}}{(2079)(50)^{1/6}(1 + 1000/2330)}$$
$$= 0.0001685.$$

Simons and Albertson Methods

1. *Coefficients for equations*. Channel type is similar to sand bed and banks (type 1). From Table 10.3,

$$K_b = 3.5, \quad K_d = 0.52, \quad K_u = 13.9,$$
$$K_r = 0.33, \quad \varepsilon = 0.33.$$

2. *Stable R values from Eqs.* (10.37) *and* (10.38).

$$R = K_d Q^{0.36} = 0.52(50)^{0.36} = 2.126.$$

Since $R < 7$ ft, use Eq. (10.38) for D, or

$$D = 1.21, \quad R = 2.57.$$

3. *Average channel width* [*Eqs*. (10.34) *and* (10.35)].

$$B_a = 0.9\,P = 0.9\,K_b Q^{1/2} = 0.9(3.5)(50)^{1/2} = 22.27.$$

4. *Average velocity and slope*. From continuity, the required velocity is

$$U = \frac{Q}{A} = \frac{Q}{B_a D} = \frac{50}{(22.27)(2.57)} = 0.874 \text{ ft/sec}.$$

From Eqs. (10.40) and (10.41) for stable slopes,

$$S = \frac{[U/K_u]^{1/\varepsilon}}{R^2} = \frac{[0.874/13.9]^{1/0.33}}{(2.13)^2} = 0.000050387$$

or

$$S = \frac{U^2}{gDK_r(UB_a/\nu)^{0.37}}$$

$$= \frac{(0.874)^2}{(32.2)(2.57)(0.33)((0.876)(22.2)/10^{-5})^{0.37}} = 0.000132$$

The two slopes are based on Reynold's number-type relationships and on tractive force relationships.

Chang's Method

1. *Calculating a stable slope*. Chang's method for analyzing canals does not lend itself to the calculations shown above. A slope must be specified. Starting with sediment load, a slope that will just transport the sediment can be defined. From the calculations above, C = 1000 ppm. From Fig. 10.16 with C = 1000 ppm and Q = 50 cfs,

$$S/\sqrt{d} = 1.52 \times 10^{-3}.$$

2. *Stable channel dimension*. From Fig. 10.16 with $S/d^{0.5} = 1.52 \times 10^{-3}$ and Q = 50 cfs,

$$B_a = 21 \text{ ft}$$
$$D = 1.3 \text{ ft}$$
$$U = Q/(B_a D) = 1.8 \text{ ft/sec}$$
$$S = (1.52 \times 10^{-3})\sqrt{d} = (1.52 \times 10^{-3})\sqrt{0.4} = 0.00096.$$

Using Eqs. (10.48)–(10.51) would have yielded similar results.

Summary of Predictions

	Blench	Simons and Albertson	Chang
Avg. width (ft)	24.5	22.3	21.0
Avg. depth (ft)	1.5	2.57	1.3
Stable slope (ft/ft)	0.00017	0.000132	0.00096

It should not be surprising that the slopes are drastically different. Slopes are highly variable in flat terrain. What is significant is the impact of sediment load. The Simons and Albertson relationship does not consider sediment loads, whereas sediment load is considered in both the Blench and Chang relationships. The Chang relationship is based on theoretical relationships and has been experimentally verified. It would thus be preferred in the absence of field data.

Recommendation for a Ground slope of 0.0003 ft/ft

1. *Discussion of options.* Comparing the ground slope to predicted stable slopes where sediment load is considered, it is obvious that Chang slopes are greater than the ground slope. This indicates that the ground slope is insufficient to transport the sediment load and provide stable channel dimensions. In the absence of constructed works, deposition would be expected with the resulting increase in channel slope. The most reasonable action would be to develop a sediment removal facility to reduce sediment load to the transport capacity.

2. *Analysis of stable channel using Chang's relationship.* For $S = 0.0003$,

$$\frac{S}{\sqrt{d}} = \frac{0.0003}{\sqrt{0.4}} = 0.000474.$$

From Fig. 10.16 with $S/d^{0.5} = 0.000474$ and $Q = 50$ cfs, the stable sediment load would be

$$C = 50 \text{ ppm}.$$

Also from Fig. 10.16, the stable channel geometry would be

$$B = 17 \text{ ft}$$
$$D = 2.5 \text{ ft}.$$

3. *Recommendations.* Construct a sediment diversion system to reduce sediment load to 50 ppm and construct a channel with average width of 17 ft and average flow depth of 2.5 ft on the prevailing ground slope.

Rational Regime Relationship for Natural Channels

Natural channels are significantly different from constant flow canals in that their flow rates and sediment loads change frequently. Under this concept, it is important to think of a natural channel in regime as one that is stable over a long period of time, discharging a sediment load that equals the incoming load. To accomplish this end, the channel geometry is self-forming. Channel characteristics such as width, depth, slope, sinuosity, bank slope, and transverse slope in bends are dependent variables that allow the channel considerable degrees of freedom to come into regime with inflow rate and sediment load.

Early studies of stable channel morphology, presented previously in this chapter, were based on statistical analyses of field data, yielding only marginal insight into the mechanics of channel change. More recent developments in modeling channel resistance to flow, sediment transport, flow in channel bends, and channel bank behavior have given rise to complete channel models that can give a reasonable approximation to natural channel morphology.

Chang (1985a) combined a number of basic relationships into an overall model of flow and sediment transport to develop a model of stable channel geometry for sand bed rivers. The analysis is based on the use of bankfull discharge as the stable flow event. In the analysis, physically based equations were used, including relationships for: (1) flow continuity, (2) sediment transport, (3) flow resistance as related to flow rate, (4) minimum stream power for a cross section and minimum stream power for a channel reach, and (5) relationships for transverse sediment movement and flow in channel bends. Utilizing these relationships, Chang (1979a, 1985a, 1986) developed the curves shown in Fig. 10.17, which relate discharge, slope, average diameter, average depth, and average bottom width for rivers. Chang (1985a) quoted experimental data indicating general agreement with theoretical relationships.

In Chang's method the channels are classified into several regions.

Region 1

These channels tend to have flat slopes, low sediment loads, and low velocities. Channel depth is sensitive to slope, but width is insensitive to slope. Flow resistance is such that the channels are in the lower flow regimes. Channels would generally be classed as sinuous canaliform with constant depths and widths throughout the meanders. Regime canals normally fall in this region, indicating that it is possible to have long straight channels. When flows are such that the channel moves intermittently to region 2, channel sinuosity could result.

Empirical equations for surface width and depth in Region 1 are

$$B = 3.49\left(\frac{S}{\sqrt{d_{50}}} - \frac{S_c}{\sqrt{d_{50}}}\right)^{0.02} Q^{0.47} \quad (10.52)$$

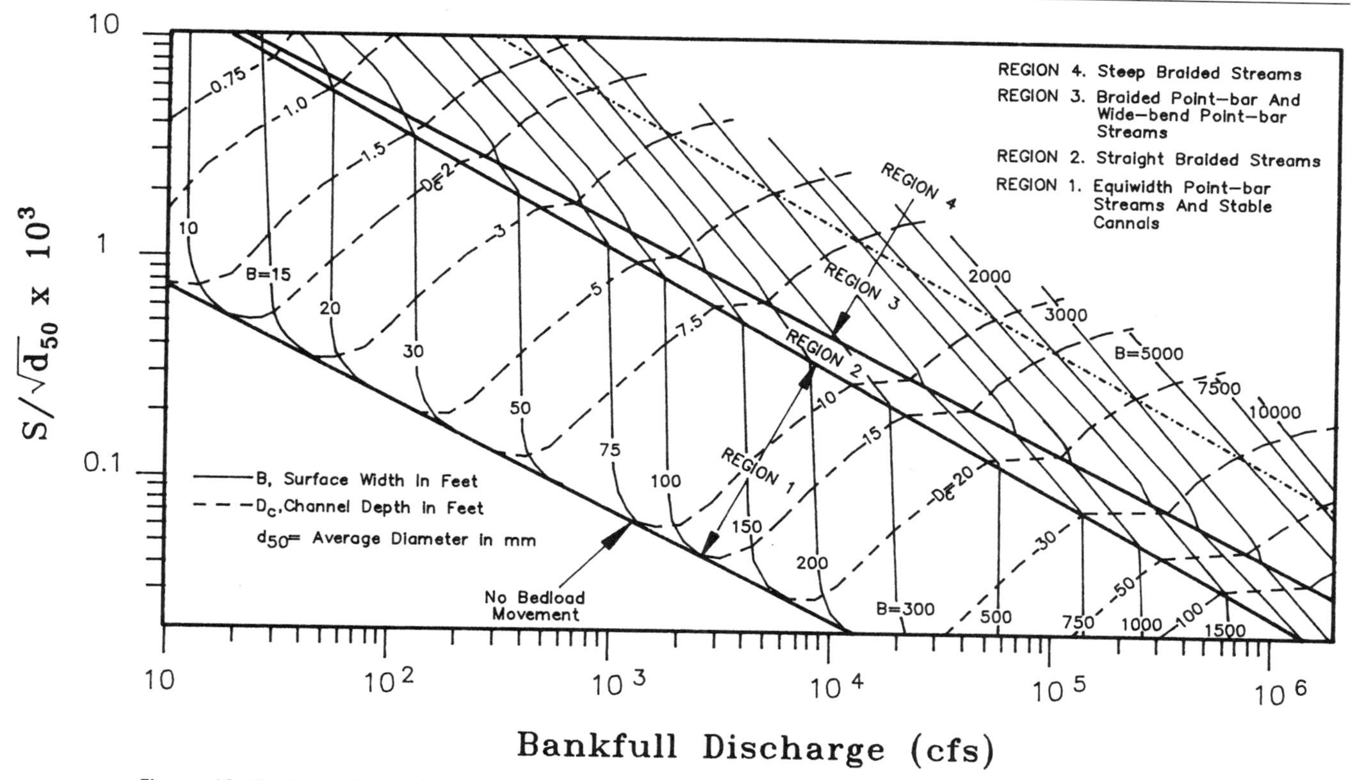

Figure 10.17 Stable (regime) channel geometry for sand bed rivers with bankfull discharge (after Chang, 1985a).

and

$$D_c = 0.51Q^{0.47}\exp\left[-0.38\left(\frac{S}{S_c} - 1\right)^{0.4}\right], \quad (10.53)$$

where D_c is the center depth in feet, B is surface width in feet, d_{50} is average particle diameter in millimeters and Q is bankfull discharge in cfs. For B and D in meters and Q in cms, replace the constant 3.49 with 5.68 and 0.51 with 0.83.

Region 2

This is a region that occurs over a rather narrow range of flow conditions. A minimum stream power exists in the lower regime and in the upper regime, but with a more defined value at the lower regime. Width-to-depth ratios are usually large, leading to frequent braided channels. Width and depth are sensitive to slope. Few natural channels occur in this region.

No empirical equations are available for Region 2.

Region 3

Channels in this region have widths and depths that are very sensitive to slope. As a result, changes in slope can quickly lead to braiding. Stream power–channel width plots have minimums in both lower and upper regimes. If the lower regime minimum occurs at slopes equal to the valley slope, the channel can be straight. If the lower regime minimum occurs at slopes lower than the valley slopes, this implies the possibility of both riffles and pools to maintain the channel at a slope less than the valley slope (Chang, 1985a). These channels tend to be alternating point bar or sinuous braided. For discharges less than bankfull, streams can become Region 1 or 2 channels with significantly different channel forms.

Empirical equations for B and D_c for Region 3 are

$$B = 33.2Q^{0.93}\left(\frac{S}{\sqrt{d_{50}}}\right)^{0.84} \quad (10.54)$$

and

$$D_c = \left[0.015 - 0.025\ln Q - 0.049\ln\left(\frac{S}{\sqrt{d_{50}}}\right)\right]Q^{0.45}, \quad (10.55a)$$

with B and D_c in feet, d_{50} in millimeters and Q in cfs. For B and D_c in meters and Q in cms, replace the constant 33.2 with 278 in Eq. (10.54) and use the following expression for D_c

$$D_c = \left[-0.112 - 0.0379 \ln Q - 0.0743 \ln\left(\frac{S}{\sqrt{d_{50}}} \right) \right] Q^{0.45}. \quad (10.55b)$$

Region 4

These channels are similar to those in Region 3, but have steeper slopes with a greater tendency to braiding. Width-to-depth ratios are typically greater than 100; thus the braided channels tend to be straight. Prediction equations for B and D_c are not available.

Chang (1988) utilized a sediment transport equation by Engelund and Hansen (1967) to develop the bed material transport rate for Regions 1, 3, and 4, or Region 1

$$\frac{S}{\sqrt{d_{50}}} = 9.66 \times 10^{-5} C^{0.58} Q^{-0.17} \quad (10.56)$$

Regions 3 and 4

$$\frac{S}{\sqrt{d_{50}}} = 7.28 \times 10^{-6} C^{0.87}, \quad (10.57)$$

where C is the bed material load in ppm by weight, d is in millimeters, and Q is in cfs. If Q is in cms, then the constant in equation (10.56) is 5.27×10^{-5}. Since Q is not a parameter in Eq. (10.57), the constant is unchanged with the use of Q in cms. Relationships were not given for Region 2.

Example Problem 10.3. Illustration of Chang's regime channel geometry relationship for natural channels

A bankfull flow of 600 cfs and a bed sediment load concentration of 1000 ppm is flowing in a channel with a median size of 0.4 mm. Determine the stable width, depth, and slope by Chang's relationships for regime sand bed streams.

Solution:

1. *Stable channel slope*. Assuming Region 1, Eq. (10.56) can be used for slope with Q in cfs, C in ppm, and d in millimeters:

$$\frac{S}{\sqrt{d_{50}}} = 9.66 \times 10^{-5} C^{0.58} Q^{-0.17}$$

$$= 9.66 \times 10^{-5} (1000)^{0.58} (600)^{-0.17} = 0.00179 \text{ mm}^{-1/2}$$

$$S = 0.00179\sqrt{0.4} = 0.00113.$$

2. *Stable channel width and center depth*. From Fig. 10.17 with $S/d_{50}^{0.5} = 0.00179$ and $Q = 600$ cfs.

$$D = 3.5 \text{ ft}$$
$$B = 65 \text{ ft}.$$

Use of Eq. (10.52) and (10.53) yields similar values.

3. *Channel form*. From Fig. 10.17, the channel is on the border of Regions 1 and 2. Region 1 is indicated; therefore, the channel is a sinuous point bar stream with the potential to be a braided point bar steam.

GRAVEL CHANNELS

Resistance to Flow

Gravel channels and channels with large roughness elements typically do not follow the same resistance relationships that are used for channels with smaller median diameter bed materials due to the highly three-dimensional flow characteristics around roughness elements. In addition, the major fluid drag with large roughness elements results from grain roughness and not the presence of mobile bedforms that exist with smaller particles. The empirical data available indicate that relationships available for more moveable beds do not apply to these coarser materials.

Because of its dimensionless form, Bathurst (1985) utilized the Darcy–Weisbach friction factor to evaluate existing data. For uniform flow, it can be shown that the friction factor is given by

$$f = \frac{8gRS}{U^2} \quad (10.58)$$

or since $U_*^2 = gRS$,

$$\frac{U}{U_*} = \sqrt{\frac{8}{f}}. \quad (10.59)$$

Using data from 16 field sites in the United Kingdom, Bathurst (1985) evaluated U/U_* for a number of flows and developed the relationships in Fig. 10.18. For slopes greater than 0.004, the scatter in the data is quite large for any relative submergence, D/d_{84}. Bathurst suggests that no accurate relationship exists for all situations, but proposed the following relationship

$$\frac{U}{U_*} = \sqrt{\frac{8}{f}} = 5.62 \log_{10}\left(\frac{D}{d_{84}} \right) + 4.0 \quad (10.60)$$

as a compromise where D is the depth of flow and d_{84} is the diameter for which 84% of the roughness ele-

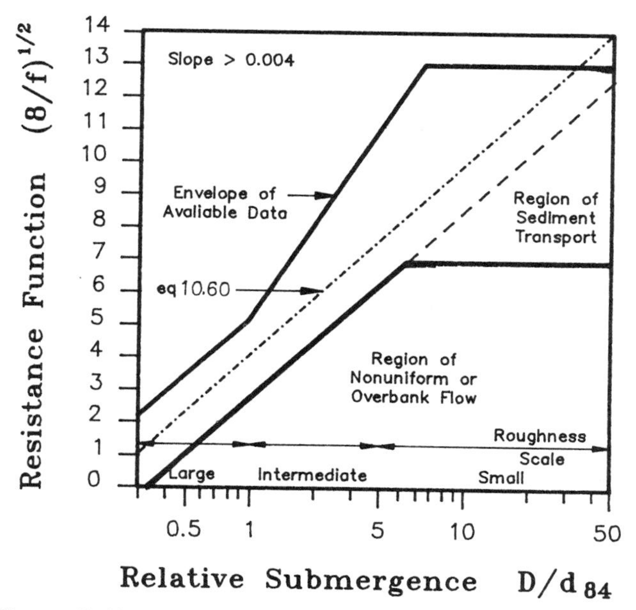

Figure 10.18 Resistance function for large roughness elements and steep slopes (after Bathurst, 1985).

ments are smaller. Bathurst suggests that the expected errors associated with using Eq. (10.60) are likely to be on the order of 25 to 35%. For D/d_{84} greater than 4, no acceptable relationship is available.

Regime Relationships for Stable Gravel Bed Streams

Regime relationships for gravel bed streams are different from those presented earlier. Parker (1979) analyzed empirical data and proposed that

$$\frac{B}{d_{50}} = 4.4(Q_*)^{1/2}, \qquad (10.61)$$

where B is the surface width normalized by d_{50}, the median bed particle diameter, and Q_* is a dimensionless flow parameter given by

$$Q_* = \frac{Q}{\left[(\mathrm{SG}-1)gd_{50}^5\right]^{1/2}}, \qquad (10.62)$$

where SG is specific gravity of a particle and Q is bankfull flow rate. Consistent units are needed in order for B/d_{50} and Q_* to be dimensionless. Parker (1979) extended the relationships using theoretical calculations to obtain relationships for D and S, or

$$\frac{D_a}{d_{50}} = 0.253Q_*^{0.415} \qquad (10.63)$$

and

$$S = 0.223Q_*^{-0.410}, \qquad (10.64)$$

where D_a is average depth. Parker used the assumption that shear stress at bankfull discharge exceeded the critical shear stress by 20% in developing Eqs. (10.63) and (10.64).

Chang (1980a) utilized the FLUVIAL model with its concept of minimum stream power, as illustrated in Fig. 10.14, to develop a rational regime diagram for gravel bed streams in a manner similar to that used to develop Fig. 10.16. In this development, a modification of the Einstein bed load equation, calibrated for gravel channels, was used for sediment transport calculations. The results are shown in Fig. 10.19.

Empirical equations for B and D_a in Fig. 10.19 are

$$B = \left[1.905 + 0.249\left(\ln\frac{0.001065d_{50}^{1.15}}{SQ^{0.42}}\right)^2\right]Q^{0.47} \quad S > S_c \qquad (10.65)$$

and

$$D_a = \left[0.2077 + 0.0418\ln\frac{0.000442d_{50}^{1.15}}{SQ^{0.42}}\right]Q^{0.42}, \quad S > S_c, \qquad (10.66)$$

where B is surface width in feet, D_a is average depth in feet, and Q is flow in cfs.

Example Problem 10.4. Illustration of regime relationships for gravel channels

A major development is being planned in the headwaters of a mountain stream with bed material of average diameter of 200 mm. The annual storm flow downstream from the development is expected to change from 100 to 200 cfs as a result of the development. The stable surface width and slope prior to development are 23 ft and 0.024 ft/ft. What would be the expected change in channel geometry as a result of the increase in Q.

Solution:

1. *Determining expected dimensions prior to development.* For a d_{50} of 200 mm, the vertical axis in Fig. 10.19 is

$$S\left(\frac{50}{d_{50}}\right)^{1.15} = 0.024\left(\frac{50}{200}\right)^{1.15} = 0.0048,$$

From Fig. 10.19 for $Q = 100$ cfs and $S\,(50/d_{50})^{1.15} = 0.0048$, the channel is in a zone where no transport is expected. The channel is close to the threshold where $B_a = 19$ ft and $D_a = 1.5$ ft. Therefore a width of 23 ft would be expected to be stable.

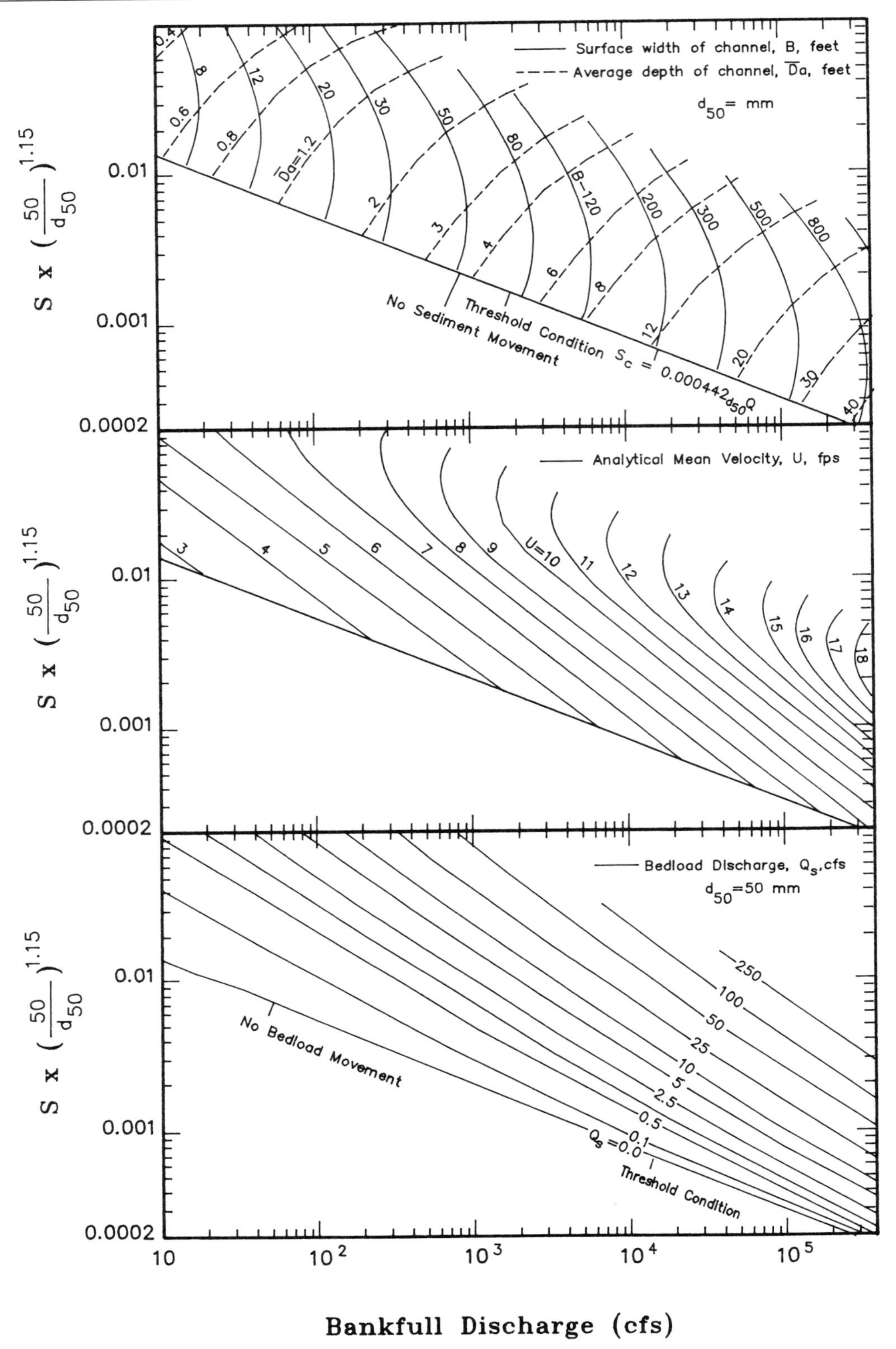

Figure 10.19 Rational regime relationship for geometry, mean velocity, and bed load for gravel bed channels. d_{50} is in millimeters (after Chang, 1980a; copyright © 1988. Reprinted by permission of John Wiley & Sons, Inc.)

2. *Expected change in dimensions with flow change.* From Fig. 10.19 with $Q = 200$ cfs,

$$B = 25 \text{ ft}$$
$$D_a = 1.9 \text{ ft}$$
$$Q_s \cong 0.005 \text{ cfs of sediment.}$$

The sediment load is a volume flow rate, but is very small. Therefore the new channel would be expected to have little sediment transport.

3. *Expected change.* It would be reasonable to assume that the original channel had come into regime with its flow. Assuming that Fig. 10.19 could be used in this situation to predict relative changes, then the ratio of width after development and prior to development is $25/19 = 1.316$. The surface width, after development, would be expected to be

$$B = (1.316)(23) = 30 \text{ ft.}$$

Since the sediment load is very small, the slope would not be expected to change; hence $S = 0.024$ is still a stable slope.

MODELING CHANNEL RESPONSE TO CHANGE

Over a period of time, channels will tend to come into regime with flows, sediment loads, and properties of the geologic parent material through which the channel flows. These characteristics can change as a result of nature through climatic shifts or tectonic activity, and by human interference through surface mining, urbanization, deforestation, channel straightening, reservoir construction, etc. Any of these changes will result in new equilibrium conditions and a change in channel properties. Examples of such changes are abundant.

As a part of any development project, it would be desirable to predict the impact of development on channel properties. Such a prediction may be a simple qualitative projection of new equilibrium conditions or may be specific detailed predictions of change with time. The needs of the project will dictate the type of analysis conducted. The following discussion outlines the methodology for each analysis and gives examples.

Qualitative Predictors

Lane's Geomorphic Relationship

Lane (1955) developed a well-known geomorphic relationship to predict a balance between the sediment load, particle diameter, discharge, and slope in an alluvial channel, or

$$QS \sim Q_s d_{50}, \qquad (10.67)$$

where the symbol $\sim$ means that QS changes monotonically with $Q_a d_{50}$. This principle is illustrated graphically in Fig. 10.20. A change in one variable requires a corresponding change in another variable to maintain equilibrium. For example, an increase in sediment load, Q_s, resulting from deforestation with no appreciable change in flow or average particle size, would require an increase in slope to maintain equilibrium. The increased slope would result from deposition in the channel of the excess of sediment load over the transport capacity. The deposition would continue until the slope is great enough to transport the increased sediment load.

Example Problem 10.5. Illustration of Lane's geomorphic relationship

A channel-straightening project is proposed to eliminate stream resistance to flow and minimize flooding in a subdivision. The channel is in relatively deep alluvial material. What would be the impact of the project on equilibrium channel properties in the proposed project area and downstream areas.

Solution:

1. *Impact in the project area.* We assume that flow, sediment load into the straightened reach, and particle diameter are unchanged. As a result of the straightening activity, channel slope will increase. Thus, we can write Lane's relationship [Eq. (10.67)] as

$$Q^0 S^+ \sim Q_s^+ d_{50}^0$$

where the superscript 0 implies no change, + implies an increase, and − a decrease. Since Q and d_{50} are unchanged, Q_s must increase to offset the increase in slope. The incoming sediment load to the reach is not increased; thus the increase in Q_s in the reach must come from detachment in the stream bed. Therefore, it is anticipated that channel erosion will increase in the section.

2. *Downstream changes.* In the downstream section that has not been straightened, there will be no increase in S, Q, or d_{50} as a result of human activity. However, the sediment load coming into the section has increased from upstream channel erosion. Therefore, slope must increase to offset the increased Q_s. The increased slope will actually result from aggradation occurring downstream. As seen in the next section, an increase in sediment load results in a decrease in depth, an increase in channel width, and an increase in meandering and sinuosity. From Fig. 10.7, there is a tendency toward braiding as the slope is increased. Therefore, the net result of channel straightening is detailed below.

Local change.

1. Increased slope
2. Channel erosion and increased sediment load
3. Increased channel width

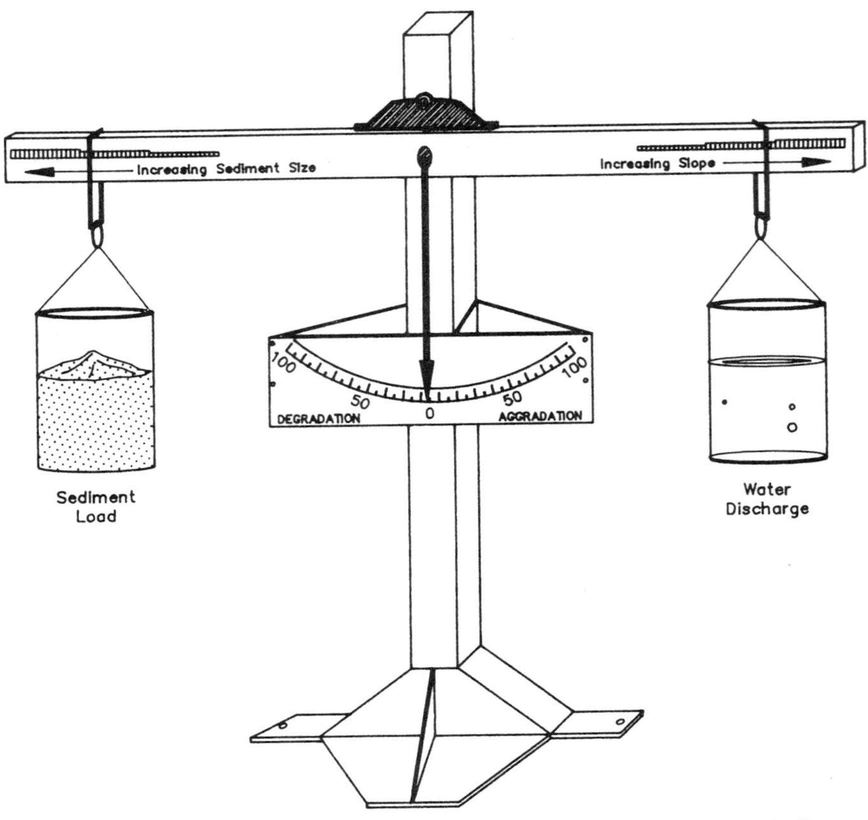

Figure 10.20 Graphical illustration of Lane's (1955) geomorphic model of channel adjustment.

4. Decreased depth of flow
5. Tendency toward braiding.

Downstream change.

1. Aggradation and decreased slope
2. Decreased meandering
3. Increased channel width
4. Decreased depth of flow
5. Tendency toward braiding.

To prevent these unwanted features, it will be necessary to use channel stabilization in the straightened section such as grade stabilization structures and channel bank stabilization.

Schumm's Qualitative Model

Schumm (1969) combined Lane's (1955) geomorphic response as given by Eq. (10.67) with other regime equations and developed a summary of qualitative relationships, as summarized in Table 10.4, that include not only the parameters of Q, Q_s, d_{50}, and S, but channel depth, ratio of channel width-to-depth, average wavelength of meanders, and sinuosity. These relationships along with Eq. (10.67) can be used to provide a fairly complete qualitative evaluation of channel metamorphosis as inputs change.

In some cases, it is not possible to predict the impact of a change in Q or Q_s on hydraulic geometry without resorting to detailed models of channel response. These are indicated in Table 10.4 by a $\pm$ superscript. Detailed models are addressed in the following section.

Example Problem 10.6. Illustration of Schumm's qualitative model

A reservoir proposed in an arid area is to be built on a stream that receives constant flow from snow melt. Assuming that the discharge after the reservoir fills is approximately the same as that prior to filling and that the sediment discharge is essentially zero, estimate the qualitative impact of the reservoir on the downstream channel.

Solution: From Schumm's relationship in Table 10.4, the following response

$$Q_s^- \sim B^- D^+ F^- \lambda^- S^- S_u^+$$

is proposed, which implies that average width decreases, depth increases, width-to-depth ratio decreases, meander wavelength decreases, slope decreases, and sinuosity in-

Table 10.4 Schumm's Qualitative Model of Channel Metamorphosis (after Schumm, 1969, 1971a, b)

Increase in discharge alone $Q^+ \sim B^+D^+F^+\lambda^+S^-$	Decrease in discharge alone $Q^- \sim B^-D^-F^-\lambda^-S^+$
Increase in bed material discharge alone $Q_s^+ \sim B^+D^-F^+\lambda^+S^+S_u^-$	Decrease in bed material discharge alone $Q_s^- \sim B^-D^+F^-\lambda^-S^-S_u^+$
Discharge and bed material discharge increase together (example: during urban construction) $Q^+Q_s^+ \sim B^+D^{\pm}F^+\lambda^+S^{\pm}S_u^-$	Discharge and bed material discharge decrease together (example: discharge from reservoir) $Q^-Q_s^- \sim B^-D^{\pm}F^-\lambda^-S^{\pm}S_u^+$
Discharge increase, bed material decrease (example: urbanization after construction) $Q^+Q_s^- \sim B^{\pm}D^+F^-\lambda^{\pm}S^-S_u^+$	Discharge decrease, bed material increase (example: diversion for irrigation with sediment laden return flow) $Q^-Q_s^+ \sim B^{\pm}D^-F^+\lambda^{\pm}S^+S_u^-$

[a]Q, water flow rate; Q_s, bed material flow rate; B, average channel width; D, average depth; F, ratio of channel width to depth; λ, average wavelength of meanders; S, channel slope; S_u, sinuosity.

creases. The most important impact is an increase in depth and decrease in slope. Since the sediment load is decreased, slope must decrease to come into regime. This must result from channel degradation immediately downstream of the dam, resulting in higher banks. At water withdrawal points, this translates into increased pumping charges. There is also an impact on tributaries, resulting from increased slope at the junction with the main channel.

Chang's Quantitative Model

As described earlier, Chang developed predictors of channel morphology in sand bed streams by combining physically based equations for: (1) flow continuity, (2) sediment transport, (3) flow resistance as related to flow rate, (4) minimum stream power for a cross section and minimum stream power for a channel reach, and (5) relationships for transverse sediment movement and flow in channel bends. Utilizing these relationships, Chang developed the curves shown earlier in Fig. 10.17, which relate discharge, slope, average diameter, average depth, and average bottom width. Using the Engelund–Hansen (1967) sediment transport equation, he also developed equations for sediment concentration for the equilibrium shapes. With these graphical relationships and associated equations, it is possible to develop first estimates of changes in channel regime characteristics corresponding to changes in input. This is illustrated in the following example.

Example Problem 10.7. Utilization of Chang's regime channel geometry relationship for predicting equilibrium changes in natural channels

In Example Problem 10.3, a bankfull flow of 600 cfs and a bed sediment load of 1000 ppm flowing in a channel with a median size of 0.40 mm yielded a channel with a bed slope of 0.00113 ft/ft, a width of 65 ft, and an average flow depth of 3.5 ft. If the channel is straightened and the slope increases to 0.002 ft/ft, what would be the resulting channel width, flow depth, and sediment load? How does this compare quantitatively to the Lane and Schumm relationships? Explain any difference.

Solution:

Chang's Analysis

1. Change in width and depth of flow.

$$\frac{S}{\sqrt{d_{50}}} = \frac{0.002}{\sqrt{0.4}} = 0.00316.$$

2. *Width and depth*. From Fig. 10.17 with $Q = 600$ cfs and $S/d^{0.5} = 0.00316$

$$B_a = 100 \text{ ft}$$
$$D = 2.5 \text{ ft}.$$

3. *Sediment load*. After the modification, flow is in Region 3 (braided channel); therefore, Eq. (10.57) is used

$$C = \left[\frac{S/\sqrt{d_{50}}}{7.28 \times 10^{-6}}\right]^{1/0.87} = \left[\frac{0.00316}{7.28 \times 10^{-6}}\right]^{1/0.87}$$

or

$$C = 1076 \text{ ppm}.$$

4. *Conclusions*. Due to changes in slope the following changes are anticipated

	Before	After
Width B_a	65 ft	100 ft
Depth	3.5 ft	2.5 ft
Sediment concentration	1000 ppm	1076 ppm
Classification	Sinuous	Braided

5. *Comparison to Lane and Schumm analysis*. Quantitatively, the local changes with straightening are the same as those indicated from Lane qualitative model in Eq. (10.67), or

$$Q^0 S^+ \sim Q_s^- d_{50}^0.$$

Since there is an increase in S with Q and d_{50} remaining constant, there must be an increase in Q_s to balance the relationship. From Table 10.4, it is seen that an increase in sediment load alone leads to an increase in width, decrease in depth, and a decrease in sinuosity. These changes are consistent with predictions from Chang's relationship.

DYNAMIC MODELS OF CHANNEL CHANGE

As has already been pointed out in this chapter, channels in alluvial material are self-adjusting as they change characteristics in response to changes in the environment. Adjustments in width, flow depth, slope, meander length, sediment load, and channel form are possible to allow a channel to adjust so that it can balance the sediment load.

Early studies of channel change were limited to physical model evaluations, which are often expensive and time consuming. More recently, improvements in our understanding of fluvial processes has led to development of dynamic models of fluvial channels. Dawdy and Vanoni (1986) review the existing models, including the U.S. Army Corps of Engineers (1977), HEC-6 model FLUVIAL-11 by Chang and Hill (1976), and Chang (1982). A summary of the requirements of adequate models is given in the following section.

There is a difference between dynamic models of channel behavior and the regime models discussed earlier. The dynamic models respond to short-term fluctuations, whereas the regime models are for long-term equilibrium conditions. Thus, the models have different purposes. The dynamic models, however, can be used with long-term simulations to develop regime relationships.

Requirements of Dynamic Models

Water Routing

Water routing must account for variations in resistance resulting from bedform changes, for transverse as well as longitudinal flow, and for upstream and downstream boundary conditions. Water routing is discussed in Chapter 6 and is typically accomplished using numerical solutions to the continuity and momentum equations, or

$$\frac{\partial A}{\partial t} + \frac{\partial Q}{\partial X} - q_1 = 0 \qquad (10.68)$$

and

$$\frac{1}{A}\frac{\partial Q}{\partial t} + g\frac{\partial Z}{\partial x} + \frac{1}{A}\frac{\partial}{\partial x}\left(\frac{Q^2}{A}\right) + gS_e - \frac{Q}{A^2}q_1 = 0, \qquad (10.69)$$

where Q is water discharge, A is the cross-sectional area, x is the streamwise coordinate, q_1 is lateral inflow of water per unit channel length, g is gravitational constant, S_e is slope of the energy gradient, and Z is water surface elevation. Upstream boundary conditions would be the inflow hydrograph, and downstream boundary conditions would be a water surface level or a stage–discharge relationship. Numerical solution possibilities are included in a number of standard references (i.e., U.S. Army Corps of Engineers, 1982).

A resistance relationship such as the Engelund curves in Fig. 10.12 is needed to relate channel roughness to flow rate. Such relationships must account for the changes that occur in form roughness with changing flows.

Total energy gradient must account for both lateral flow and transverse flow. Transverse flow must account for changes in secondary currents and transverse slopes that occur in channel bends. Relationships for making these computations are too complex to be included here, but can be reviewed in Chang (1988).

Sediment Load Calculations

A relationship must be available for predicting sediment transport capacity over a range of particle sizes. A number of relationships are available, as summarized in Chapter 7. The relationship used can then be combined with a bed armoring function and a sediment continuity equation to predict actual sediment discharge rates. A sediment armoring function should allow for interactions of flow with the bed to detach smaller particles and leave behind larger particles. These larger particles will form an armor over smaller particles in the bed, preventing detachment as flow rates decrease. An example of an armoring relationship is that of Borah *et al.* (1982).

Sediment calculations should also account for the diffusion phenomenon that moves suspended load into the flow and prevents suspended load from coming to rapid equilibrium with flow changes.

The basic sediment continuity relationship is

$$(1 - P)\frac{\partial A_{cb}}{\partial t} + \frac{\partial Q_s}{\partial x} - q_s = 0, \qquad (10.70)$$

where P is porosity of bed material, A_{cb} is bed surface area, Q_s is bed material discharge, and q_s is lateral inflow of sediment including that detached from the bed. Equation (10.70) must be solved numerically along with Eq. (10.69). The output needed for further calculations is the change in bed area, ΔA_{cb} over a routing interval Δt. This can be used to determine the change in channel width and depth.

Changes in Channel Width and Depth

Because channel width and depth are not constrained, it is necessary to select rules for computations based on additional principles. For example, it is possible to have simultaneous erosion or deposition in banks and stream beds, or erosion in one and deposition in the other. In the FLUVIAL model (Chang, 1982), for example, the direction of width adjustment is such that the channel is moving toward uniform stream power along the reach. If a channel reach is steeper than adjacent sections, its width is decreased to reduce the gradient, and vice versa. The rate of widening in FLUVIAL is controlled by the banks. If a channel is widening, the rate is controlled by the erodibility of the banks. If it is decreasing in width, it decreases at a rate controlled by deposition along the banks.

Once the rate of widening is determined, and subtracted from ΔA_{cb}, the remainder can be used to determine bed profile adjustments. For example, in the FLUVIAL model, Chang (1982) adjusts the bed elevation by using the shear excess concept, or

$$\Delta Z = \frac{(\tau_0 - \tau_c)^m}{\sum_B (\tau_0 - \tau_c)^m \Delta y}\Delta A_{cb}, \qquad (10.71)$$

where ΔZ is a bed channel elevation correction, τ_0 is the local tractive force, τ_c is critical tractive force, m is an exponent, y is the horizontal coordinate, and the summation over B indicates that the denominator is summed over the channel width.

Changes Due to Curvature Effect

A final correction must be made for transverse sediment transport, which creates a transverse channel slope. Available relationships correlate transverse sediment load to streamwise sediment load as a function of the bottom current in the transverse direction (related to the curvature) and the transverse slope of the bed. Sediment continuity is also followed in the transverse direction, or

$$\frac{\partial Z}{\partial t} + \frac{1}{1 - P}\frac{1}{r}\frac{\partial}{\partial r}(rq_s') = 0, \qquad (10.72)$$

where r is the radial coordinate at any streamwise position x, and q_s' is the transverse sediment load.

Application of the Models

The models that result from the relationships discussed above are very complex. Input parameters are not known with a high degree of certainty; particularly bed roughness parameters, bank and bed erodibility parameters, and sediment transport relationships. Successful application of the model to a specific situation requires at least some field data for calibration runs specific to the location. Examples of successful applications of such models are included in Simons, Li, and Associates (1982) and Chang (1988).

Problems

(10.1) In an attempt to improve navigation in a very sinuous channel, the public works department in Outer Slobovia proposes to eliminate a series of meanders. If the stream channel is erodible, what will be the local impacts and downstream impacts? How might these impacts be reduced?

(10.2) A large forested watershed is to be developed into a urban and industrial complex over a period of 10 years, leaving more than 50 acres disturbed in any 1 year. What would be the probable impact on the main channel draining the watershed (a) during construction and (b) after construction is complete.

(10.3) List the differences between a rigid boundary and moveable boundary (fluvial) channel. For each type channel, indicate how the following parameters would change with increasing flow rate: depth, width,. slope, width to depth ratio, and sinuosity.

(10.4) For your local geographic region, select and describe in detail natural channels that are sinuous canaliform, sinuous point bar, sinuous braided, and nonsinuous braided. Include in your description such things as boundary vegetation, channel slope, valley slope, sinuosity, etc.

(10.5) Explain to the best of your ability why flow depth decreases and width increases as sediment load increases.

(10.6) Describe the changes one experiences in channel morphology as slope increases. What causes the phenomenon of braiding?

(10.7) Describe the changes in bedform that are observed as stream power increases from low to high values. It is often observed in the transition zone that the bedform oscillates between dunes, plane bed, and antidunes. Explain why this might occur.

(10.8) A riprap-lined channel is being proposed with an average diameter of 0.25 ft and a d_{90} of 0.5 ft. If the slope is 1.0% and the flow depth is 2.0 ft, what is the Manning's n according to the Strickler equation and the Federal Highway Administration equation? What would be the average velocity and discharge per unit width (assume a wide channel)?

(10.9) The flow in a 10-ft-wide rectangular channel with a slope of 0.1% is 50 cfs. If the average roughness of the channel, k_s, is 0.01 ft, what is the depth of flow according to Einstein's equation? (Hint: Replace D in Eq. (10.16) with hydraulic radius.)

(10.10) A 40-ft wide channel has a maximum depth of 5.0 ft before overtopping the banks. The channel slope is 0.0004 ft/ft and the bed material has the following characteristics: $d_{35} = 0.12$ mm, $d_{50} = 0.25$ mm, $d_{65} = 0.65$ mm. Develop a stage–discharge curve using (a) the Einstein–Barbarossa method and (b) the Englelund method. Assume that the water temperature is 20°C and that the channel is approximately rectangular.

(10.11) For Problem (10.10), what will be the bedform at a depth of 3.0 ft?

(10.12) Explain the difference between a regime relationship for a steady flow channel and a channel conveying naturally varying runoff.

(10.13) A flow of 75 cfs is being diverted from a river with an average suspended load of 10,000 ppm of sediment. Define a stable channel width, depth, and slope by the Blench, Simons and Albertson, and Chang methods. The actual groundslope is 0.001 ft/ft. What would you recommend?

(10.14) Why would the bankfull discharge be classed as the channel-forming event? Why would the annual storm (2-year event) be approximately equal to the bankfull discharge?

(10.15) As a result of urbanization, the 1.5-year storm is expected to change from 400 to 800 cfs in the drainage from a watershed. During construction, the 1.5-year storm is expected to be 600 cfs. Prior to development, the channel draining the watershed has a bed material with an average diameter of 0.4 mm, a slope of 0.0002 ft/ft, and a surface width of 45 ft. No information on the sediment load is available. During development, it is estimated that the average bed material concentration will be 10,000 ppm and after construction the incoming bed material load will be near zero. What will be the expected channel characteristics (width, depth, slope) during and after construction? Explain how this will happen, giving a sequence of events.

(10.16) To allow for navigation on the Arkansas River, a series of locks and dams were built up to Tulsa, Oklahoma. What changes in channel characteristics would you expect in areas where the channel bed is erodible? (Hint: Utilize the slope of the energy gradient instead of the slope of the channel in your analysis.)

(10.17) As a result of a series of events in the 1930s, most of the cotton fields in the Eastern Piedmont areas of South Carolina were converted from agricultural lands to forest with a major decrease in sediment yield and runoff rates. What impact would that have on stream channel characteristics in the area? Justify your answer.

(10.18) A gravel channel has a d_{84} of 0.3 ft, a slope of 0.01 ft/ft, and a width of 30 ft. Estimate the stage–discharge curve up to a relative submergence of 5 using Bathurst's relationship. Use depths of 0.3, 0.5, 0.75, 1.0 ft, etc.

(10.19) A recreational development in Appalachia is expected to change the annual flow in a stream channel from 150 to 400 cfs. Prior to development, the surface width in the annual storm is estimated to be 30 ft with a slope of 0.03 ft/ft. What would be the expected change in channel characteristics as a result of the development? The average diameter of the channel material is 300 mm.

References

Athaullah, M. (1968). Prediction of bed forms in erodible channels, Ph.D. thesis. Department of Civil Engineering, Colorado State University, Ft. Collins, CO.

Bathurst, J. C. (1985). Flow resistance estimation in mountain rivers. *J. Hydraul. Eng.* **111**(4): 625–643.

Blench, T. (1970). Regime theory design of canals with sand beds. *Proc. Am. Soc. Civil Eng.* **96**(IR2): 205–213.

Borah, D. K., Alonso, C. V., and Prasad, S. N. (1982). Routing graded sediments in streams: Formations. *Proc. Am. Soc. Civil Eng.* **102**(HY12): 1486–1503.

Brice, J. C. (1983). Planform properties of meandering rivers. River Meandering. *In* "Proceedings, 1983 Rivers Conference, American Society of Civil Engineers, New York."

Chang, H. H. (1979a). Geometry of rivers in regime. *Proc. Am. Soc. Civil Eng.* **105**(HY6): 691–706.

Chang, H. H. (1979b). Minimum stream power and river channel patterns. *J. Hydrol.* **41**: 303–327.

Chang, H. H. (1980a). Geometry of gravel streams. *Proc Am. Soc. Civil Eng.* **106**(HY9): 1443–1456.

Chang, H. H. (1980b). Stable alluvial canal design. *Proc. Am. Soc. Civil Eng.* **106**(HY5): 873–891.

Chang, H. H. (1982). Mathematical model for erodible channels. *Proc. Am. Soc. Civil Eng.* **108**(HY5): 678–689.

Chang, H. H. (1984). Modeling of river channel changes. *Proc. Am. Soc. Civil Eng.* **110**(HY2): 157–172.

Chang, H. H. (1985a). River morphology and thresholds. *J. Hydraul. Eng.* **111**(HY3): 503–519.

Chang, H. H. (1985b). Design of stable alluvial canals in a system. *J. Irrigat. Drain. Eng.* **111**(1): 36–43.

Chang, H. H. (1986). River channel changes: Adjustments of equilibrium. *J. Hydraul. Eng.* **112**(1): 43–55.

Chang, H. H. (1988). "Fluvial Processes in River Engineering." Wiley–Interscience, New York.

Chang, H. H., and Hill, J. C. (1976). Computer modeling of erodible flood channels and deltas. *Proc. Am. Soc. Civil Eng.* **102**(HY10): 1461–1477.

Culbertson, D. M., Young, L. E., and Brice, J. C. (1967). Scour and fill in alluvial channels, U.S. Geological Survey, Open File Report. [Quoted by Simons, Li, and Associates, 1982; original not seen].

Davis, W. M. (1899). The geographical cycle. *Geogr. J.* **14**: 481–504.

Dawdy, D. R., and Vanoni, V. A. (1986). Modeling alluvial channels. *Water Resources Res.* **22**(9): 71S–81S.

Dury, G. H. (1973). Magnitude-frequency analysis and channel morphology. *In* "Fluvial Geomorphology" (Morisawa, ed.), Publications in Geomorphology, pp. 91–121. SUNY, New York.

Einstein, H. A. (1950). The bed load function for sediment transportation in open channels, Technical bulletin 1026. Soil Conservation Service, U.S. Department of Agriculture, Washington, DC.

Einstein, H. A., and Barbarossa, N. (1952). River channel roughness. *Trans. Am. Soc. Civil Eng.* **117**: 1121–1146.

Engelund, F. (1966). Hydraulic resistance of alluvial streams. *Proc. Am. Soc. Civil Eng.* **92**(HY2): 315–326.

Engelund, F., and Hansen, E. (1967). "A Monograph on Sediment Transport in Alluvial Channels." Teknisk Vorlag, Copenhagen, Denmark.

Federal Highway Administration (1975). Design of stable channels with flexible liners, Hydraulic engineering circular No. 15. U.S. Department of Transportation, Washington, DC.

Foster, G. R., and Lane, L. J. (1983). Erosion by concentrated flow in farm fields. *In* "Proceedings, D. B. Simons Symposium on Erosion and Sedimentation, Colorado State University, Ft. Collins, CO," pp. 9.65–9.82.

Khan, H. R. (1971). Laboratory studies of alluvial river morphology, Ph.D. dissertation. Civil Engineering Department, Colorado State University, Ft. Collins, CO.

Lacey, G. (1930). Stable channels in alluvium. *Proc. Inst. Civil Eng.* **229**: 259–384.

Lane, E. W. (1955). The importance of fluvial geomorphology in hydraulic engineering. *Proc. Am. Soc. Civil Eng.* **81**: 1–17.

Lane, E. W. (1957). A study of shape of channels formed by natural streams flowing in erodible material, M. R. D. sediment series No. 9. U.S. Army Engineering Division, Missouri River, US Army Corps of Engineers, Omaha, NE.

Leopold, L. B., and Maddock, T. (1953). The hydraulic geometry of stream channels and some physiographic implications, USGS professional paper No. 252.

Leopold, L. B., and Wolman, M. G. (1957). River channel patterns: Braided, meandering and straight, USGS professional paper 282-B; pp. 45–62.

Leopold, L. B., Wolman, M. G., and Miller, J. P. (1964). "Fluvial Processes in Geomorphology." Freeman, San Francisco, CA.

Mackin, J. H. (1948). Concept of the graded river. *Geol. Soc. Am. Bull.* **59**: 463–512.

Meyer-Peter, E., and Muller, R. (1948). Formulas for bed-load transport. *In* "Proceedings, 2nd Meeting International Association Hydraulic Research," Paper No. 2; pp. 39–64.

Ming, Z. F. (1983). Hydraulic geometry of alluvial channels. *J. Sediment Res.* **4**: 75–84.

Parker, G. (1979). Hydraulic geometry of active gravel rivers. *Proc. Am. Soc. Civil Eng.* **105**(HY9): 1185–1201.

Richards, K. (1982). "Rivers, From and Process in Alluvial Channels." Methuen, New York.

Schumm, S. A. (1960). The effect of sediment type on the shape and stratification of some modern fluvial deposits. *Am. J. Sci.* **258**: 177–184.

Schumm, S. A. (1968). River adjustment to cultured hydrologic regimes, Murrumbridgee River and Palco Channels, Australia, USGS professional paper No. 598.

Schumm, S. A. (1969). River metamorphosis. *Proc. Am. Soc. Civil Eng.* **95**(HY1): 255–273.

Schumm, S. A. (1971a). Fluvial geomorphology, historical perspective, *In* "River Mechanics" (H. H. Shen, ed.). P.O. Box 606, Ft. Collins, CO.

Schumm, S. A. (1971b). Channel adjustment and river metamorphosis, *In* "River Mechanics" (H. H. Shen, ed.). P.O. Box 606, Ft. Collins, CO.

Schumm, S. A. (1977). "The Fluvial System." Wiley, New York.

Schumm, S. A., and Khan, H. R. (1972). Experimental study of channel patterns. *Geol. Soc. Am. Bull.* **83**: 1755–1770.

Simons, D. B., and Albertson, M. L. (1960). Uniform water conveyance channels in alluvial material. *Proc. Am. Soc. Civil Eng.* **86**(HY5): 33–71.

Simons, D. B., and Richardson, E. V. (1961). Forms of bed roughness in alluvial channels. *J. Hydraul. Div. Am. Soc. Civil Eng.* **88**(HY3): 87–105.

Simons, D. B., and Richardson, E. V. (1966). Resistance to flow in alluvial channels, USGS professional paper 4-22-J.

Simons, D. B., Richardson, E. V., and Nordin, C. F. (1965). Sedimentary structures generated by flow in alluvial channels, Special publication No. 12. American Association of Petroleum Geologists.

Simons, D. B., Lagasse, P. F., Chen, Y. H., and Schumm, S. A. (1975). The river environment—A reference document. U.S. Department of Interior, Fish and Wildlife Service, Twin Cities, MN.

Simons, Li and Associates (1982). "Engineering Analysis of Fluvial Systems." Simons, Li and Associates, Ft. Collins, CO.

Strickler, A. (1923). Some contributions to the problem of the velocity formula and roughness factors for rivers, canals and closed conduits, No. 16. Mitteilungen des eidgenossicschen Amtes fur Wasserwirtschaft, Bern, Switzerland.

U.S. Army Corps of Engineers (1977). HEC-6, scour and deposition in rivers and reservoirs, users manual. Hydrologic Engineering Center, Davis, CA.

U.S. Army Corps of Engineers (1982). HEC-2, water surface profiles, users manual. Hydrologic Engineering Center, Davis, CA.

van Rijn, L.C. (1984). Sediment transport. III. Bed forms and alluvial roughness. *J. Hydraul. Eng. Am. Soc. Civil Eng.* **110**(12): 1733–1754.

Williams, G. P. (1978). Bankfull discharge of rivers. *Water Resources Res.* **14**(6): 1141–1154.

Wolman, M. G. (1955). The natural channel of Brandywine Creek, Pennsylvania, U.S. Geological Survey professional paper 271. U.S. Geological Survey, Reston, VA.

11

Ground Water

INTRODUCTION

Most of the fresh water in the continental United States is ground water. Ground water refers to all of the water beneath the earth's surface. Ground water may be found in zones that are either saturated or unsaturated. Traditionally, the water of interest to hydrologists is contained in saturated layers similar to those shown in Figs. 11.1 and 11.2. The unsaturated zone is also known as the vadose zone or the phreatic zone. Ground water in the saturated zone is considered to be water contained in underground formations in a saturated or near-saturated condition and under a pressure greater than atmospheric. Water stored under pressures less than atmospheric such as water in the vadose zone is sometimes thought of as underground water, not as ground water.

Ground water in saturated zones is located in aquifers. Aquifers are water-bearing formations that are saturated (pore spaces are filled with water). They are generally considered to be those soil and rock formations that have enough hydraulic conductivity to allow water to be pumped out through wells at useable rates. The Soil Conservation Service (1984) indicated that although there is water under most of the earth's surface, it may not be practical to utilize because it is too difficult to pump or is held too tightly. For example, a heavy clay may contain a high percentage of water by volume, but the water may be held so tightly by the clay that it cannot be removed effectively by pumping.

The exact depth at which the saturated zone occurs varies by location (spatially) and through time (temporally). The depth to the saturated zone can vary from zero at the soil surface to hundreds, even thousands, of feet. In some areas, it is almost impossible to locate the saturated zone because of rock formations or the extremely large depth at which ground water is located. Typically, there is a region above the ground water that consists of soil and other unconsolidated materials containing some water, but the larger pore spaces are not filled with water. The water in this zone is typically not available for pumping because it is generally held under tension. However, if sufficient water is available, it will move through this unsaturated zone by gravity drainage and become part of the saturated zone (Fig. 11.2). Thus, the unsaturated zone is important because it is a zone of transmittal to the saturated zone. As such, it becomes a source or sink for any chemicals associated with it as it moves toward the saturated zone.

Although ground water is conventionally classified as in either the saturated zone or vadose zone, this division of waters is somewhat artificial. There is no difference in physical or chemical properties of the water or in the basic physical laws governing the movement

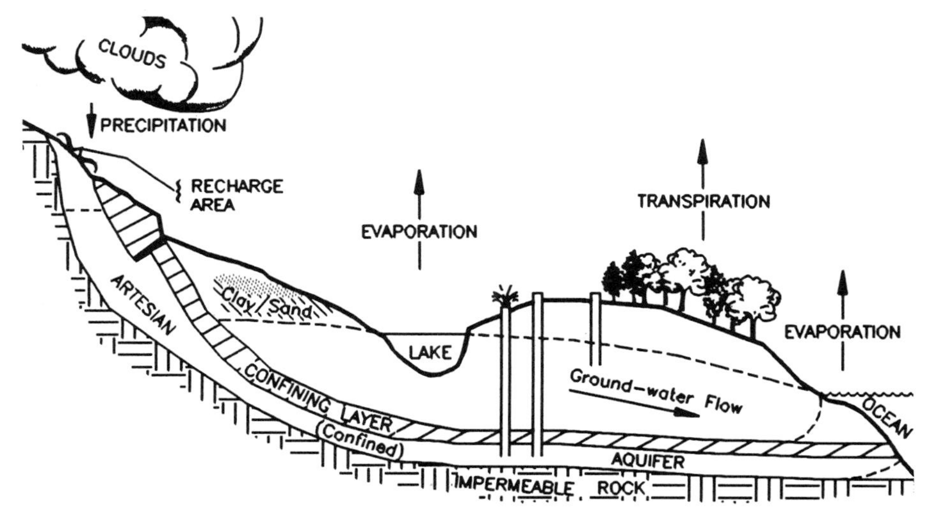

Figure 11.1 Ground water's role in the hydrologic cycle (after Soil Conservation Service, 1984).

of water in these two zones. Water is transferred freely between the zones. Under unconfined (water table) conditions, a saturated zone occupied by ground water may become a part of the vadose zone as the water table drops. At a later time, as the water table rises, the zone may again become a part of the saturated water zone. Virtually the same water may be in the saturated zone at some time, vadose zone at a later time, and in the saturated zone once again at a still later time.

Ground water becomes surface water when it emerges as a spring or seep or when it directly enters a stream, pond, or lake. Ground water discharge to streams, springs, and seeps generally forms the base flow for small streams between major runoff-producing events. The pathways taken by water as it infiltrates and percolates to become ground water and then to emerge as baseflow to become surface water has a major impact on the quality of that water. An indication of the impact of geologic factors on quality of ground water is provided in Table 11.1.

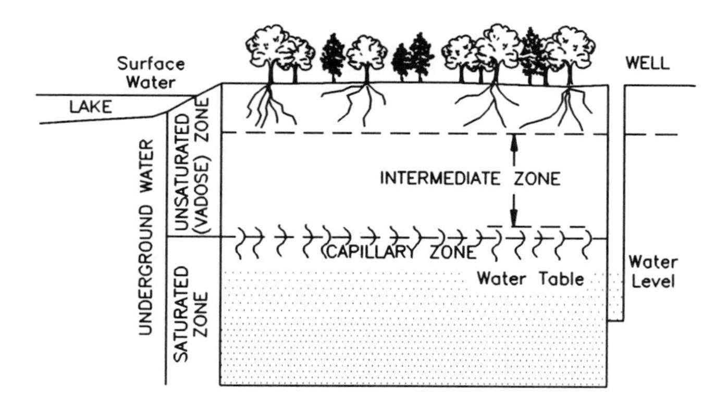

Figure 11.2 Saturated and unsaturated zones in an unconfined aquifer (after Heath, 1982).

The interchange between ground water, vadose water, and surface water points to the need to consider the entire hydrologic system in assessing the impact of alterations within a catchment on any aspect of the hydrology of that catchment.

Ground water is of interest to hydrologists and engineers for several reasons. The first arises because much of the drinking water used by humans and domesticated animals comes from ground water. Thus, there is considerable interest in any activity that has potential for impacting either the quantity or quality of ground water. Particular concern about the possibility of chemicals entering ground water because of the extreme difficulty of removing contaminants from ground water zones has been expressed. Recently, considerable attention has been directed toward determining potential for certain activities to cause ground water contamination (CAST, 1985; National Research Council, 1984). A second interest, primarily to engineers, occurs when construction, mining, or other activities involve ground water aquifers or recharge areas or where ground water restricts these activities (National Research Council, 1990a, b). Additionally, ground water provides a large portion of the flow to streams. Ground water, like almost all of the waters of the earth, is part of the hydrologic cycle as described in Chapter 3 and illustrated in Fig. 11.1.

The speed with which ground water moves from the surface to other aspects of the hydrologic cycle can

Table 11.1 Naturally Occurring Inorganic Chemicals that Pollute Ground Water (Adapted from Heath, 1982)

Substance	Major natural sources
Bicarbonate and carbonate	Solution of carbonate rocks, limestone, and dolomite by water
Calcium and magnesium	Rocks containing limestone, dolomite, or gypsum
Chloride	Either geologic seawater or seawater in contact with aquifers
Fluoride	Sedimentary or igneous rocks
Iron and manganese	Iron present in most soils or rocks, manganese less widespread
Sodium	Similar to chloride
Sulfate	Gypsum, pyrite, or other rocks containing sulfur

vary greatly. In one location, water may move rapidly through the hydrologic cycle as it goes from precipitation to percolation into a rock fracture or a highly structured soil. Then it may quickly flow to a spring or stream and into an ocean. At another location, water may infiltrate and percolate into a slowly permeable formation and require many years before it emerges and reenters the more active parts of the cycle through surface water flow or evapotranspiration.

Surface mining, construction, and similar land-disturbing activities may alter many of the hydrologic processes related to ground water (National Research Council, 1990a). Infiltration, overland flow, surface runoff, surface storage and detention, interception, evapotranspiration, percolation, ground water flow, and stream flow can all be impacted. As a matter of fact, the only part of the cycle not generally considered to be potentially impacted by human activities is precipitation, and even that assumption is currently under serious challenge.

The flow and storage processes shown in Fig. 11.1 are unsteady—that is, the flow rate and volume of water in any particular form of storage are constantly changing with time. The time rate of change of some processes such as precipitation and surface runoff may be rapid, while the rate of change of ground water storage and discharge may be very slow.

LOCATION OF GROUND WATER PROVINCES

The U.S. Geological Survey identified significant ground water provinces in the United States based on the area of coverage of water-bearing formations. Figure 11.3 delineates these ground water provinces al-

Figure 11.3 Geographic delineation of ground water provinces (after Soil Conservation Service, 1984).

phabetically. The Soil Conservation Service (1984) presents a brief description of each province as follows.

A. Atlantic and Gulf Coastal Plain Province

Water is derived in rather large quantities from sands and gravels interbedded with clay. Large supplies are obtained from alluvial gravels in the Mississippi Valley and adjacent areas. The province includes extensive areas of artesian flow. In mineral content, the ground water varies from low to high.

B. Northeastern Drift Province

Ground water comes principally from glacial drift. The till yields small supplies to many springs and shallow wells (less than 60 m); the outwash gravels yield large supplies. Many drilled rock wells receive small supplies, chiefly from joints in crystalline rocks or in Triassic sandstone. Ground water is generally soft and low in mineral content.

C. Piedmont Province

Water generally low in mineral content is supplied in small quantities by the crystalline rocks and locally by Triassic sandstone. Many shallow dug wells are supplied from surface deposits or from the upper decomposed part of the bedrock. Many moderately deep (60–300 m) drilled wells are supplied from joints in the crystalline rocks. Some wells in Triassic sandstone yield large supplies.

D. Blue Ridge–Appalachian Valley Province

This is a region of rugged topography with numerous springs that generally yield good-quality water from Paleozoic strata, pre-Cambrian crystalline rocks, or post-Cambrian intrusive rocks. The water is derived chiefly from springs, spring-fed streams, and shallow wells.

E. Southcentral Paleozoic Province

The principal water sources are the Paleozoic sandstones and limestones. In many of the valleys, large supplies are obtained from alluvial sands and gravels.

F. Northcentral Drift–Paleozoic Province

Most water is derived from glacial drift, where it is generally hard but otherwise good. Numerous drilled wells produce large supplies from glacial outwash or from gravel interbedded with till. Many drilled wells end in Paleozoic sandstone or limestone and receive ample water.

G. Wisconsin Paleozoic Province

Most of the water is from wells of moderate depth drilled into Cambrian or Ordovician sandstone or limestone. These wells as a rule yield ample supplies of hard but otherwise good water. In many of the valleys, artesian flows are obtained from the Paleozoic aquifers. The region has no water-bearing drift except in the valleys, where there are water-bearing outwash gravels.

H. Superior Drift–Crystalline Province

In most parts of this province, satisfactory water supplies are obtained from glacial drift. Where the drift is thin, water is generally scarce, because the pre-Cambrian crystalline rocks in most places yield only meager supplies, and as a rule there are no intervening Paleozoic, Mesozoic, or Tertiary formations thick enough to yield much water. The drift and rock waters range from soft waters in Wisconsin to highly mineralized waters in the western and northwestern parts of the province.

I. Dakota Drift–Cretaceous Province

The two important sources of ground water are the glacial drift and the Dakota sandstone. The drift supplies numerous wells with hard but otherwise good water. The Dakota sandstone has extensive areas of artesian flow that supply many strong flowing wells, a considerable number of which are more than 300 m deep. The Dakota sandstone waters are highly mineralized but are used for domestic supplies.

J. Black Hills Cretaceous Province

The conditions in this province are, on the whole, unfavorable for shallow water supplies, because most of the province is underlain by the Pierre shale or by shales of the White River group. The principal aquifer is the Dakota sandstone, which underlies the entire region except the Black Hills. This sandstone will probably yield water wherever it occurs, and over considerable parts of the province it will give rise to flowing wells; however, throughout much of the province it is far below the surface. In the Black Hills, water is obtained from a variety of sources, ranging from pre-Cambrian crystalline rocks to Cretaceous or Tertiary sedimentary rocks.

K. Great Plains Pliocene–Cretaceous Province

The principal aquifers of this province are the late Tertiary sands and gravels (Ogahalla and related formations) and the Dakota sandstone. The Tertiary deposits underlying the extensive smooth and uneroded plains supply large quantities of water to shallow wells. The Dakota sandstone underlies nearly the entire province and gives rise to various areas of artesian flow. Throughout much of the province, however, it lies too far below the surface to be a practical source of water. Where the Tertiary beds are absent or badly eroded and the Dakota sandstone is buried beneath thick beds of shale, as in parts of eastern Colorado, developing even small water supplies may be difficult. Many of the valleys contain Quaternary gravels, however, which supply large quantities of good water. Considerable Tertiary and Quaternary sections can yield a supply suitable for irrigation.

L. Great Plains Pliocene–Paleozoic Province

The principal aquifers of this province are the late Tertiary and Quaternary sands and gravels, which give the same favorable conditions as those in the Great Plains Pliocene–Cretaceous province. The Tertiary deposits are underlain practically throughout the province by Permian or Triassic deposits, which in most places yield little or only highly mineralized water. Where the Tertiary deposits are thin or absent, or where they have been eroded, the ground water conditions are generally unfavorable.

M. Trans-Pecos Paleozoic Province

The bedrock consists of Carboniferous, Permian, and Triassic strata, including limestone, gypsum, red beds of shale and shaly sandstone, and some less shaly sandstone. In most of the province these rocks yield only meager supplies of highly mineralized waters to deep wells. In the Pecos Valley, however, Carboniferous limestones and sandstones yield large supplies to numerous flowing wells; the water is very hard but good enough for irrigation, domestic, and livestock purposes. Locally the bedrock is overlain by Quaternary water-bearing gravels.

N. Northwestern Drift–Eocene–Cretaceous Province

Ground water is obtained from glacial drift and from underlying Eocene and Upper Cretaceous formations. Where the drift is absent or not water bearing, wells are sunk into the underlying formations, with variable success. The Eocene and latest Cretaceous, which underlie most of the eastern part of the province, generally include water-yielding strata or lenses of sand, gravel, or coal. The Cretaceous formations in the western part consist chiefly of alternating beds of shale and sandstone. The sandstones generally yield water, but the shales are unproductive, and where a thick shale formation immediately underlies the drift or is at the surface, successful wells may be difficult to obtain. In certain localities, upland gravels yield water to shallow wells.

O. Montana Eocene–Cretaceous Province

Enough fairly good water for domestic and livestock supplies and even for small municipal supplies is obtained from strata and lenses of sand, gravel, and coal in the Fort Union (Eocene) and Lance (late Cretaceous or Eocene) formations that underlie most of this province. These formations usually rest on the Pierre shale, a thick, dense Upper Cretaceous shale that yields only meager amounts of generally poor-quality water or none at all. Hence, locally, where the Fort Union and Lance are absent or do not yield enough, satisfactory water supplies are very difficult to obtain. The northern part of the province has a little water-bearing glacial drift.

P. Southern Rocky Mountain Province

In this mountain province, underlain mostly by crystalline rocks, water is obtained chiefly from springs, from streams fed by springs and melted snow, or from very shallow wells near streams.

Q. Montana–Arizona Plateau Province

This large area is mostly an arid to semiarid plateau underlain by sedimentary formations ranging in age from Paleozoic to Tertiary. The formations are not violently deformed, but they are warped and broken enough that the presence of ground water is closely related to rock structure, and conditions vary over short distances. On the whole, water is neither plentiful nor of very satisfactory quality. Where thick formations of nearly impervious material are at the surface, or where the plateau is greatly dissected, as in the Grand Canyon region, water is scarce. Locally, however, sandstone aquifers can be developed and may yield very satisfactory supplies—in some places giving rise to flowing wells. There are also local deposits of water-bearing Quaternary gravels.

R. Northern Rocky Mountain Province

This is a relatively cold region, chiefly mountainous but with extensive valleys and plains. It is underlain by a wide variety of rocks with complicated and diverse structure. As in other mountain regions, water is obtained largely from mountain springs and streams. Considerable water is available in places made up of ordinary alluvial sand and gravel and the outwash deposits of mountain glaciers. Water is also obtained from wells drilled into various pre-Cambrian and Tertiary rock formations.

S. Columbia Plateau Lava Province

The principal aquifers of this province are the widespread Tertiary and Quaternary lava beds and interbedded or associated Tertiary sand and gravel. In general, the lava yields abundant supplies of good water. It gives rise to many large springs, especially along the Snake River in Idaho. Locally, the lava or the interbedded sand and gravel give rise to flowing wells.

However, much of the lava is so permeable and the relief of the region is so great that in many places the water table can be reached only by deep wells. In certain parts of the province, glacial outwash and ordinary valley fill are also important water sources.

T. Southwestern Bolson Province

The principal source of water in this arid province is the alluvial sand and gravel of the valley fill underlying the numerous intermountain valleys. In the elevated marginal parts of the valleys, the water table may be far below the surface or ground water may be absent; in the lowest parts, underlain by clayey and alkaline beds, ground water may be scarce and of poor quality; at intermediate levels, however, large supplies of good-quality water are generally found. Most of the water in the valleys of this province is recovered by means of pumping wells, but there are many springs and areas of artesian flow. In mountain areas of the province, many springs, small streams, and shallow wells furnish valuable supplies.

U. Central Valley of California Province

Good-quality ground water is found chiefly in alluvial cones formed by streams emerging from the Sierra Nevada, although water can be obtained throughout the valley. The yield of cones flanking the Coast Range is small. Poor-quality water generally comes from the south and central sections, and somewhat better quality from the north. Underlying piedmont deposits consist of marine, lacustrine, and alluvial formations. Highly mineralized water is found in deep strata throughout the valley and near the ground surface in the center of the valley. Extensive irrigation in the valley depends on ground water pumped from wells.

V. Coastal Ranges of Central and Southern California Province

The principal ground water bodies are in the mountain valley and piedmont plains draining to the Pacific Ocean. Aquifers consist of valley fill and alluvial sand and gravel deposits. Locally, good water supplies are developed from underlying younger Tertiary sandstones. Heavy development of ground water along the coast for municipal and irrigation needs has caused sea water to enter and contaminate aquifers in several valley mouths.

W. Willamette Valley–Puget Sound Province

A large body of alluvium fills the structural trough forming this province. Abundant supplies of surface water have delayed investigation and exploitation of the extensive ground water resources of the area.

X. Northern Coast Range Province

Ground water is found in the alluvial fill of the valleys draining to the Pacific Ocean. A small area in the southern part of the province contains heated ground water, hot springs, and geysers. Because surface water is abundant and the province is relatively undeveloped, little detailed information on ground water conditions is available.

BASIC CONCEPTS OF GROUND WATER HYDRAULICS

Conservation of Mass

The principle referred to as conservation of mass indicates that over some time interval, the difference in the volume of water entering and leaving a control element must equal the change in volume of water stored in the element. Mathematically this is expressed as

$$I\Delta t - O\Delta t = \Delta S, \tag{11.1}$$

where $I\Delta t$ is the inflow volume in time Δt, $O\Delta t$ is the outflow volume in time Δt, and ΔS is the change in storage. As a rate relationship, the continuity of mass is given by

$$I - O = dS/dt, \tag{11.2}$$

where I and O are the inflow and outflow rates (volume per unit or time) and dS/dt is the time rate of change in storage within the control element.

Over short intervals of time (up to a few years), the change in inflow, outflow, or the volume of water stored in the control element may be substantial. In an undisturbed control element, in the absence of some natural, cataclysmic event, and for long time intervals (several years), the change in the average value of these quantities is relatively small, and the inflow and outflow volumes are nearly equal. Under these conditions, a state of dynamic equilibrium exists.

The relationship between the inflow, outflow, and change in storage for any control element defines the hydrologic balance of the element. Any factors that alter inflows, outflows, or storage characteristics potentially alter the hydrologic balance and thus ground water. Such factors include impacts of construction, mining, and other land-disturbing activities.

The size of the control element must also be considered in evaluating changes to the hydrologic balance.

A small control element (a few acres in areal extent) may have its hydrologic balance altered by minor disturbances within the element. More substantial alterations would be required to alter the hydrologic balance on large control elements that contain many acres.

Hydrologic impacts of changes within a control element may be transient. A disturbance upsets the state of dynamic equilibrium existing within the control element. A substantial time period, often several years, may have to pass for a new state of dynamic equilibrium to come into existence. During this transient period, the hydrologic balance is certainly altered. After the transient period, the new state of dynamic equilibrium must be compared to the original state to determine if permanent changes to the hydrologic balance have occurred.

A control element that encompasses only water in the saturated zone may be defined. The inflow to this element could be flow from outside the element, percolation, and infiltration. Outflow from the element could be saturated zone flow, flow becoming surface water via springs, seeps or direct discharge to surface water bodies, unsaturated flow to the vadose zone generally from a water table, and transpiration via plants drawing water directly from the saturated zone.

Occurrence and Movement of Ground Water

The material that constitutes the earth's outer mantle is composed of solid material and void spaces. The solid material may be in the form of individual particles or more massive rock formations. The void spaces are occupied by air or water. In the saturated zone, of course, the voids are filled with water (some entrapped air may be present).

Water-bearing formations may be either consolidated or unconsolidated. Except for rock outcroppings, the earth's surface is covered by a layer of unconsolidated material that may range in thickness from a few centimeters to several thousand meters. Consolidated material always underlies the unconsolidated material at some depth. Alternating strata of consolidated and unconsolidated material may exist above the final consolidated rock.

Unconsolidated material consists of individual mineral particles derived from the breakdown of consolidated rock. Individual particles may range from clay-sized particles measuring fractions of a millimeter in diameter to rocks and boulders measuring several meters across. Coarser materials, such as sand and gravel deposits, often make excellent aquifers because they allow water to move with relative ease.

Consolidated material consists of mineral particles that have been fused together by heat and pressure or by chemical reactions to form solid masses. They generally consist of sedimentary and igneous rocks. Rocks of importance to ground water are limestone, dolomite, shale, siltstone, sandstone, conglomerate, granite, and basalt.

Void spaces may consist of primary voids that were formed at the same time as the rock or secondary voids formed by fracturing of consolidated materials and chemical action-forming solution channels.

One characteristic of an aquifer is porosity, n, which is defined as the fraction (or percentage if preferred) of volume that is occupied by voids divided by the total volume of the media so that

$$n = \frac{V_t - V_s}{V_t} = \frac{V_v}{V_t}, \tag{11.3}$$

where V_t, V_s, and V_v are the total volume, solid volume, and void volume, respectively. Materials with high porosity contain considerable water when saturated. Porosity provides an upper limit that represents the volume of water contained in a material when saturated. All of the water contained in a formation will not drain solely due to gravity either because the pores are not connected or because some water is held too tightly to the individual particles. The amount of water that will drain from a saturated material due to gravity is referred to as the specific yield, while the amount of water retained is known as the specific retention. For a confined aquifer, water yield is defined as the volume of water an aquifer takes in or discharges per unit surface area per unit change in head normal to the surface. This yield is termed the storage coefficient.

The sum of the specific yield and the specific retention is the porosity of the material as illustrated mathematically by

$$n = S_y + S_r, \tag{11.4}$$

where S_y is specific yield and S_r is specific retention. Table 11.2 contains typical values for porosity, specific yield, and specific retention (Heath, 1982) for various materials. As shown, fine-grained material tends to have higher porosity. Although clay has high porosity because of the small size of individual particles, it has a very low specific yield. This property is undesirable for an aquifer used as a water supply. Coarse materials, sands, and gravels form most of the highly productive aquifers and can yield up to 80% of their water as shown from the porosity and specific yield percentages shown in Table 11.2.

Estimates of ground water flow velocities for several aquifer classes and selected hydraulic gradients are provided in Table 11.3. (Dunne and Leopold, 1978).

Table 11.2 Selected Values of Porosity, Specific Yield, and Specific Retention (Heath, 1982)[a]

Material	Porosity	Specific yield	Specific retention
Soil	55	40	15
Clay	50	2	48
Sand	25	22	3
Gravel	20	19	1
Limestone	20	18	2
Sandstone (semiconsolidated)	11	6	5
Granite	0.1	0.09	0.01
Basalt (young)	11	8	3

[a]Values in percentage by volume.

Table 11.3 Approximate Velocities, V, for Different Soil Classes and Hydraulic Gradients, i (Adapted from Dunne and Leopold, 1978)

Soil class	k (ft/day)	i (ft/ft)	V (ft/year)
Clean gravel	2.4×10^4	0.001	8,860
		0.05	443,000
Clean sand	2.4×10^2	0.001	89
		0.005	4,430
Very fine sand	2.4×10^{-2}	0.001	0.0089
		0.05	0.044
Unweathered clay	2.4×10^{-5}	0.001	8.9×10^{-6}
		0.05	4.4×10^{-4}

Figures 11.1 and 11.4 illustrate three types of aquifers—unconfined, confined, and perched. An unconfined aquifer, also known as a water table aquifer, has the free water surface as its upper boundary. The free water surface is also called the water table. With an unconfined aquifer, the upper surface of the saturated zone is subject to rise or fall. A confined aquifer (Fig. 11.1) has as an upper boundary a relatively impervious layer that restricts the height to which water can rise. The pressure potential of the water in contact with this confining layer is greater than atmospheric and defines the piezometric surface. The confining layer may be impermeable or only slightly permeable. A buildup of pressure results so that if a well is drilled into the aquifer, water will rise in a pipe to the height of the piezometric surface above the upper boundary of the aquifer. Confined aquifers that are under pressure are also known as artesian aquifers. A well piercing an artesian aquifer will flow without pumping if the piezometric surface is above the ground surface. A perched aquifer (Fig. 11.4) is an unconfined aquifer of limited areal extent that retains water against gravity because of an underlying restricting layer such as clay. Perched aquifers may be seasonal or permanent.

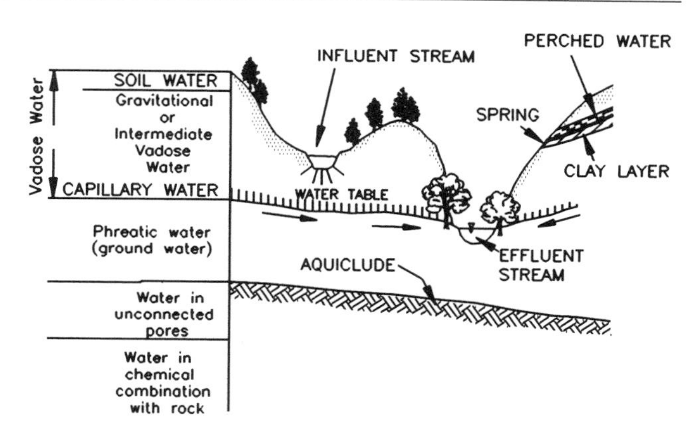

Figure 11.4 Schematic showing relationships between influent and effluent streams and ground water zones (after Vesilind *et al.*, 1988).

Darcy's Law

Water moves within an aquifer due to hydraulic gradients or water potential gradients created by gravity. Ground water flow may be in any direction, even upward. The actual flow patterns may be difficult to ascertain because of difficulties in defining underground formations. Although gravity is the dominant force in ground water flow, surface topography often does not match the water table of an unconfined aquifer or the piezometric surface of a confined aquifer. The hydraulic gradient, and not surface topography, controls ground water movement. An example situation that is indicative of the complexity of ground water flow problems is shown in Fig. 11.5. The figure shows a pumping well and stream located in an unconfined aquifer. The equipotential lines delineate points that have the specified potential. The equipotential lines show drawdown occurring near the pumping well and near the stream. If the water level in the stream was greater than the water level in the aquifer, the stream would recharge the aquifer. It should also be noted that the direction of ground water flow is perpendicular to the equipotential lines in isotropic aquifers. In some cases, it is desirable to draw equipotential lines so that the same drop in head occurs across adjacent equipotential lines. Then flow lines are drawn so that they form curvilinear squares with the equipotential lines, and the flow is equal between adjacent flow lines.

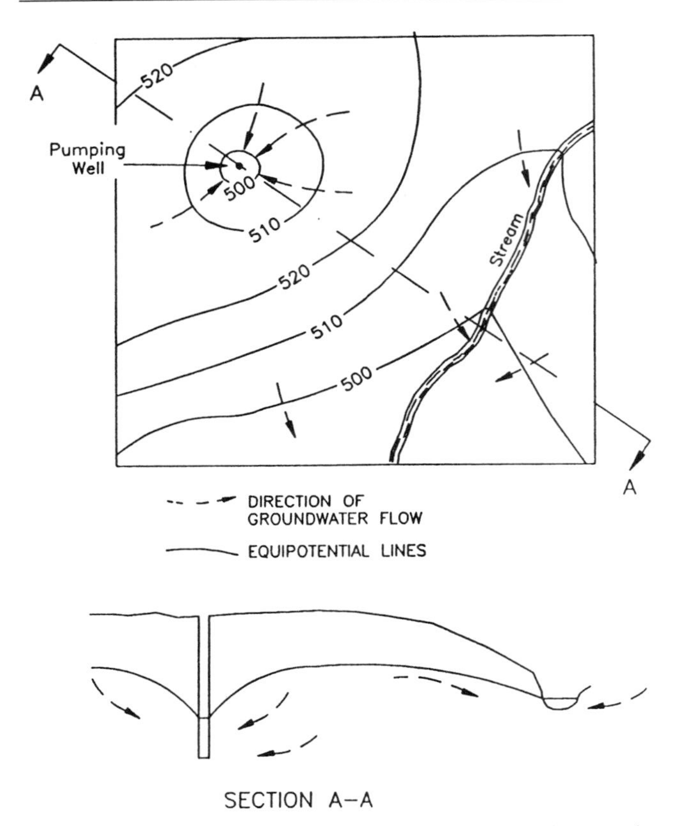

Figure 11.5 Schematic of example equipotential lines for pumping well and stream.

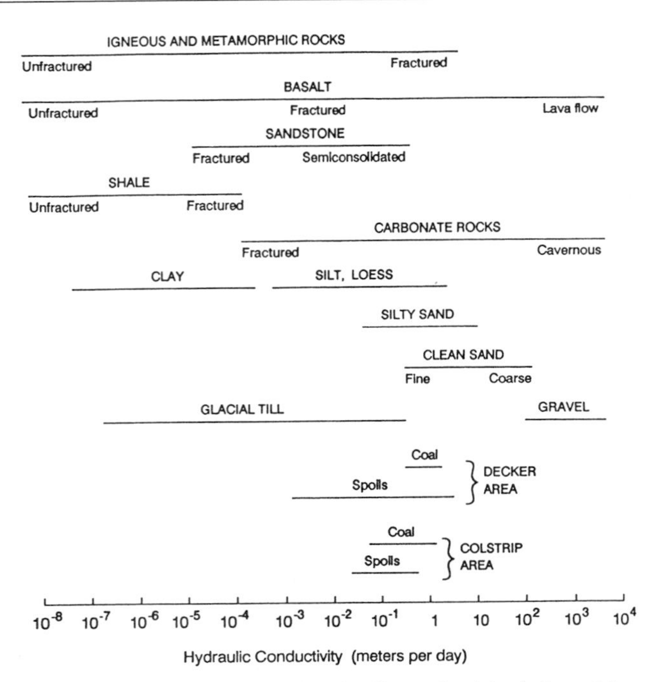

Figure 11.6 Typical values of hydraulic conductivity (adapted from Heath (1982) and Van Voast and Reiten (1988).

The combination of equipotentials and flow lines is referred to as a flow net.

Henri Darcy applied a theory for water flow in capillary tubes proposed by Hagen and Poiseville to ground water (Vesilind *et al.*, 1988). (Darcy's law is discussed in Chapter 3 with respect to infiltration using other mathematical forms of the equation.) The rates and direction of flow between two points are described by Darcy's law, or

$$Q = -KA(dh/dL), \qquad (11.5)$$

where Q is the rate of flow through the media, K is the hydraulic conductivity of the material, A is the total cross-sectional area of the porous medial including both pores and particles, h is the head, and L is the length of the porous media. According to Darcy's law, the rate of ground water movement is proportional to the hydraulic gradient, dh/dL. The proportionality factor, K, is known as the hydraulic conductivity. High values of hydraulic conductivity mean that the material readily transmits water. Among other things, hydraulic conductivity depends on the size, shape, and connectivity of pores and fractures in the aquifer. Figure 11.6 shows the ranges of typical values for hydraulic conductivity. The equation

$$Q = VA \qquad (11.6)$$

can be combined with Darcy's equation to obtain a mean flow velocity for the entire cross-sectional area. This velocity is a macrovelocity. The actual flow velocity through the pores would be much larger than this velocity since the true flow area is much smaller than the total cross-sectional area.

Example Problem 11.1. Velocity of water movement in porous media

Estimate the actual velocity of ground water movement through an aquifer composed of coarse sand and through an aquitard that confines it. Assume that dh/dL equals 1 m/1000 m for the aquifer and dh/dL equals 1 m/10 m for the aquitard. Assume $K = 50$ m/day for the aquifer and 0.0001 m/day for the aquitard.

Solution:

Equations (11.5) and (11.6) can be combined to obtain the average velocity of the entire cross-sectional area as

$$V = -K(dh/dL),$$

which is also known as the Darcian velocity. Since water only flows through the pores in the media, a porosity term must be included to obtain

$$V_a = -K/n(dh/dL),$$

where V_a is pore velocity. Table 11.2 indicates that porosity of a sandy material (aquifer) is approximately 25%. Substituting n into the previous equation along with the hydraulic conductivity K and hydraulic gradient dh/dL yields

$$V_a = -\frac{50}{0.25}\left(\frac{1}{1000}\right) = -0.2 \text{ m/day}.$$

Similarly, Table 11.2 contains a porosity for clay of 50%. Substituting with other values for the aquitard yields

$$V_a = -\frac{0.0001}{0.50}\left(\frac{1}{10}\right) = -0.00002 \text{ m/day}.$$

The negative velocity simply indicates that water flows in the opposite direction of the hydraulic gradient or from high potential to low potential. Thus the flow is expected to travel 10,000 times faster in the aquifer under these conditions. Movement in locations that contain limestone caverns or rock fractures may be much faster. While actual flow velocities are difficult to accurately quantify, the potential for spreading a pollutant over a large area once it reaches an aquifer is dependent on V_a.

Example Problem 11.2. Flowrate using Darcy's law in porous media

Estimate the flowrate through an aquifer if the aquifer is 30 m thick and 200 m wide and has hydraulic conductivity equal to 0.8 m/day. Observation wells have been placed 1200 m apart in a direction parallel to the flow. The head in one well is 28 m, and the head in the second is 20 m.

Solution: Since the observation wells are in the direction of water movement, this is a direct application of Darcy's law such that substituting yields

$$Q = -KA\frac{dh}{dL} = -\frac{0.8 \text{ m/day}(30 \text{ m})(200 \text{ m})\ (28 \text{ m} - 20 \text{ m})}{1200 \text{ m}},$$

which simplifies as

$$Q = -32 \text{ m}^3/\text{day}.$$

A variety of methods for measuring saturated hydraulic conductivity are available. Dorsey *et al.* (1990) compared four field methods that are commonly employed for near-surface measurements. These included the Guelph permeameter, the velocity permeameter, a pumping test procedure, and the auger hole method. They found that the Guelph permeameter produced significantly lower estimates than the other methods. They indicated that if the instrument is to be used on heavy clay soils, the procedure and equipment would need to be modified. The other three methods produced comparable results. Dorsey *et al.* (1990) suggested that selection of the best method for a specific application is dependent upon the soil water conditions.

Probably the simplest and most widely used method for measuring saturated hydraulic conductivity in the presence of a water table is the auger hole method. A hole is augered to a depth below the water table, and the water is allowed to rise until equilibrium is reached. The water is then removed, and the water level recorded. Subsequent changes in water level as a function of time are recorded. The water is again removed, and the process is repeated several times. The method is probably the best alternative for shallow water table conditions.

Hydraulic conductivity of aquifers is difficult to measure directly but can be determined indirectly if the depth of the aquifer is known and the transmissivity is estimated using a well test. Transmissivity is defined as

$$T = KD, \tag{11.7}$$

where K is hydraulic conductivity of the aquifer and D is the aquifer thickness. Transmissivity is a good measure of the total ability of an aquifer to transmit water under a unit hydraulic gradient.

Laplace's Equation

Darcy's law is useful for understanding the theory of fluid movement in porous materials, but it should be recognized that it is limited in direct application because it deals only with movement in one general direction. Water movement in aquifers is quite dynamic. It does not occur in only one direction, but may change directions depending upon head. For general situations where the flow is not one-dimensional, Darcy's equation must be combined with the equation of continuity to provide the basis for a general differential equation. Consider a rectangular volume element $\Delta x \, \Delta y \, \Delta z$ of soil having sides parallel to an x, y, z coordinate system as shown in Fig. 11.7. Let the velocity of water flowing in the x direction through a unit area perpendicular to the x-direction be V_x. The flow of water per unit time into the face $\Delta y \, \Delta z$ is then $V_x \, \Delta y \, \Delta z$. On the opposite face, the flow out is $V_x + (\partial V_x/\partial x)\, \Delta x$, where the term $(\partial V_x/\partial x)$ represents the change in flow across the distance Δx. Analysis of flow in the other two directions yields similar terms. The inflow minus outflow in each direction represents the net volume of water per unit time accumulating in the

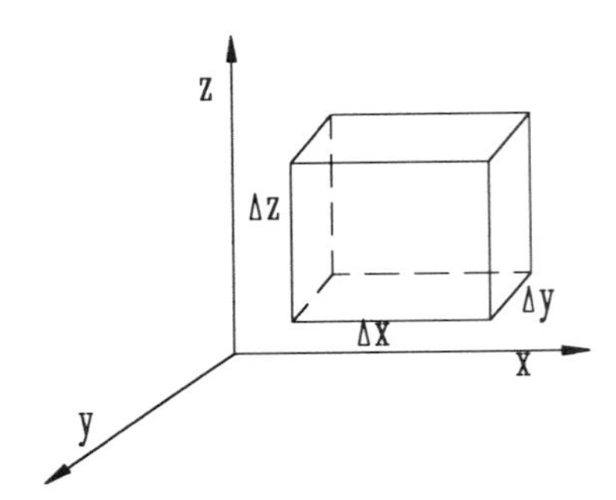

Figure 11.7 Definition of volume for Laplace's equation derivation.

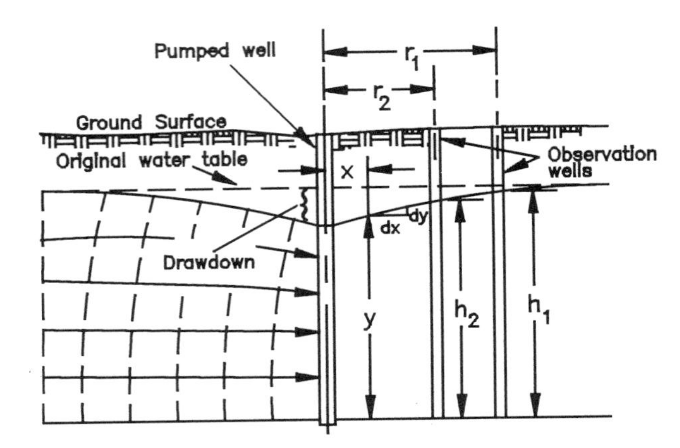

Figure 11.8 Schematic defining a simple well flow problem (after Linsley *et al.*, 1982).

rectangular volume element from flow in the x-direction as

$$\text{Inflow}_x - \text{Outflow}_x = V_x\,\Delta y\,\Delta z - (V_x + \Delta x\,\partial V_x/\partial x)\,\Delta y\,\Delta z,$$

which simplifies for the x-direction to

$$\text{Inflow}_x - \text{Outflow}_x = -(\partial V_x/\partial x)\,\Delta x\,\Delta y\,\Delta z.$$

The net flow accumulating in the volume from the y-direction and z-direction is found similarly so that the water accumulating in the volume element is

$$\frac{\partial\theta}{\partial t}\Delta x\,\Delta y\,\Delta z = -(\partial V_x/\partial x + \partial V_y/\partial y + \partial V_z/\partial z)\,\Delta x\,\Delta y\,\Delta z,$$

where V_x, V_y, and V_z are the velocities in the x-, y-, and z-directions and θ is the water content on a volumetric basis. Eliminating the volume element from each side of the equation leaves the general differential equation

$$-\left(\frac{\partial V_x}{\partial x} + \frac{\partial V_y}{\partial y} + \frac{\partial V_z}{\partial z}\right) = \frac{\partial\theta}{\partial t}. \tag{11.8}$$

Since the equation cannot be solved directly for the velocities, modification using Darcy's law allows the equation to be written in the form

$$\frac{\partial}{\partial x}\left(-K_x\frac{\partial h}{\partial x}\right) + \frac{\partial}{\partial y}\left(-K_y\frac{\partial h}{\partial y}\right) + \frac{\partial}{\partial z}\left(-K_z\frac{\partial h}{\partial z}\right) = \frac{\partial\theta}{\partial t}, \tag{11.9}$$

where K_x, K_y, and K_z are the hydraulic conductivities in the x-, y-, and z-directions, respectively; h is the hydraulic head at a point; θ is the water content on a volumetric basis; and t is time. If θ is a constant, such as is the case under saturated conditions, and if K_x, K_y, and K_z are equal and constant such that they can be replaced by K (i.e., homogeneous and isotropic conditions), then the general equation can be reduced to

$$\frac{\partial^2 h}{\partial x^2} + \frac{\partial^2 h}{\partial y^2} + \frac{\partial^2 h}{\partial z^2} = 0, \tag{11.10}$$

which is referred to as Laplace's equation. Although Laplace's equation appears relatively simple, in actual ground water flow problems, it is difficult to apply because of limitations associated with its simplifying assumptions.

Well Hydraulics under Equilibrium Conditions

A simplified well problem is shown in Fig. 11.8, which shows a well in a homogeneous and isotropic, unconfined aquifer (i.e., the hydraulic conductivity does not vary with location or direction) of infinite areal extent having water that initially is moving horizontally toward the well. For water to enter the well, there must be a drawdown at the well that forms a cone of depression. However, if drawdown is small compared to the total thickness of the aquifer, and if the well is fully penetrating the aquifer, flow streamlines may be assumed to be horizontal so that an approximation of well discharge as a function of aquifer characteristics can be obtained. Dupuit proposed a solution technique based on Darcy's law and the assumption that flow through a cylindrical surface at radius x from the well must equal the discharge of the well. With this assumption,

$$Q = -2\pi xyK\frac{dy}{dx}, \tag{11.11}$$

where $2\pi xy$ is the cylinder's area and dy/dx is the

slope of the water table, which is the driving force behind the water movement. Integration can be performed with respect to x from r_1 to r_2 and y from h_1 to h_2 to obtain

$$Q = \frac{\pi K(h_2^2 - h_1^2)}{\ln(r_2/r_1)}, \tag{11.12}$$

where h_1 and h_2 are the heights of the water table above the bottom of the aquifer at distances r_1 and r_2 from the pumped well. A similar equation for a confined aquifer is

$$Q = \frac{2\pi bK(h_2 - h_1)}{\ln(r_2/r_1)}. \tag{11.13}$$

Equations (11.12) and (11.13) represent the interrelationship between Q, K, h, and r for steady-state (or equilibrium) conditions. Linsley *et al.* (1982) emphasized that low ground water flow velocities cause true equilibrium conditions to occur only after pumping a very long time at a constant rate.

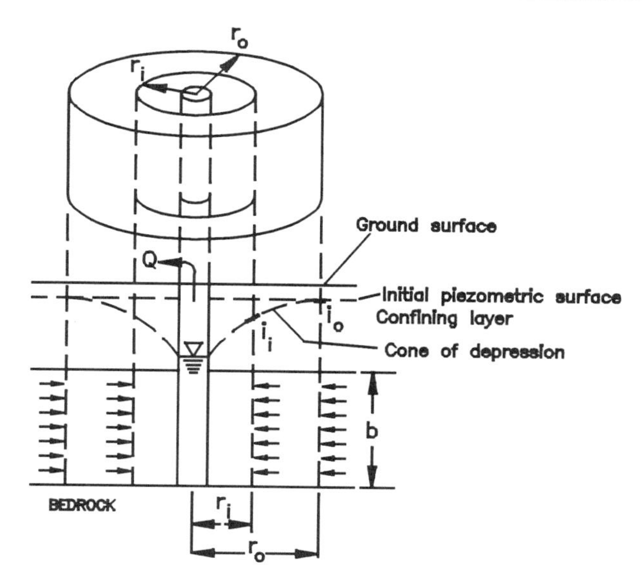

Figure 11.9 Cone depression and radius of influence for single well (after Linsley *et al.*, 1982).

Generally, ground water movement is not nearly as simple as indicated in the preceding analysis. Ground water moves in the direction of decreasing total head, which may or may not be the same as that of the decreasing pressure head. Theis sought to account for the effect of time and storage coefficients of the aquifer because he realized that only a fraction of the total aquifer depth provides flow to a well. Heath (1982) describes a technique involving three wells to determine the movement of ground water. The method requires that the wells be arranged in a triangular pattern so that relative location and distance between wells is known along with the total head at each well.

Another difficulty that occurs in defining flow is that flow in fractured systems is dependent on the extent of fracturing, the interconnectivity of the fractures, and the mechanisms available for water to enter the fracture systems. All of these factors are highly variable and site specific. Figure 11.6 shows the range of values experienced for hydraulic conductivity in consolidated materials. This variability is largely due to the nature of the secondary porosity of the material. Highly fractured rock may have quite high hydraulic conductivities and thus be able to rapidly transmit water.

Karst systems are also difficult to define mathematically. These systems contain numerous conduits that vary in size from a few centimeters to several meters. Caves are common. It is not unusual to find the major flow in a conduit moving opposite to the flow direction dictated by the general piezometric surface.

To achieve better understanding about ground water availability, ground water supplies are often evaluated by drilling test wells into the ground and pumping the water at a known rate. Test wells can supply much information about ground water availability, including the position and thickness of aquifers and hydraulic features, such as hydraulic conductivity, transmissivity, storage, and specific capacity. For flow to occur to a well, there must be a gradient to the well. This gradient forms a cone of depression similar to the idealized shape shown in Fig. 11.9 for a confined aquifer. In a confined, artesian aquifer, the actual water level does not drop. In an unconfined aquifer, the water surface corresponds to the cone of depression. The cone of depression is dependent upon the pumping rate and aquifer characteristics. An increased pumping rate or low transmissivity will increase the depth to the cone of depression. If more than one well is in an area, cones of depression may overlap and increase the depth to water in regions between wells as shown in Fig. 11.10 for an unconfined aquifer. Such well interference in-

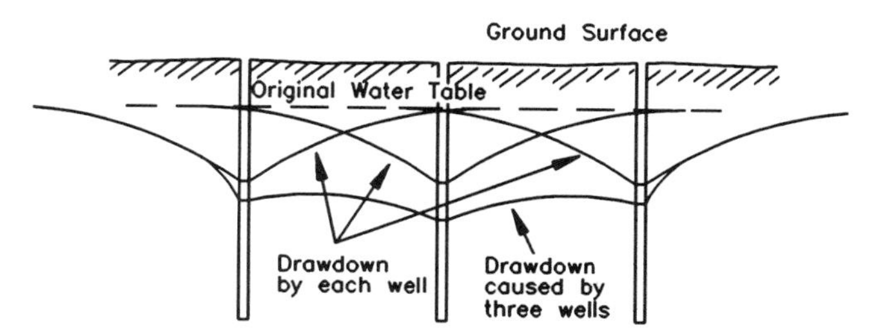

Figure 11.10 Combining cones of depression using superposition and showing well interference for multiple wells (after Linsley *et al.*, 1982).

fluences the available drawdown and also reduces the maximum yield of a well. If a large well is installed, the resulting cone of depression may cause previously operating wells to go dry because the water level drops below the screens of these wells. The piezometric surface of confined, artesian aquifers will drop in a similar manner as a result of well interference.

Specific Capacity and Transmissivity

Transmissivity, as given by Eq. (11.7), represents the rate of flow through a section of aquifer of unit thickness under a unit head. Specific capacity of a well is the flow per unit drop of water level in the well. Specific capacity of a well is dependent upon both an aquifer's hydraulic characteristics and those of the well itself. Heath (1982) listed several components that control specific capacity for ground water wells:

(A) Transmissivity of the zone where the water enters the well. (This may be much less than transmissivity of the aquifer depending on the size of the screen.)
(B) Storage coefficient of the water-bearing formation.
(C) Duration of pumping.
(D) Effective radius of the well.
(E) Pumping discharge.

All of these factors except pumping discharge can be evaluated using the Theis method as discussed below. Since pumping discharge influences well loss, it can only be estimated from an aquifer test in which drawdowns are measured in both pumping and observation wells.

Theis Equation

Theis developed an equation to relate drawdown to transmissibility of a confined aquifer (Viessman *et al.*, 1989). The equation can also be simplified if the pumping continues for a relatively long time. Heath (1982) presented Theis's equation as

$$T = \frac{W(u)}{4\pi}\frac{Q}{s}, \qquad (11.14)$$

where T is the transmissivity, Q/s is the specific capacity (Q is the pumping discharge and s is drawdown), and $W(u)$ is the well function such that

$$u = \frac{r^2 S_c}{4Tt}, \qquad (11.15)$$

where r is the effective radius of the well, S_c is the storage coefficient, and t is the length of pumping time prior to determination of specific capacity. The well function is given by

$$W(u) = -0.577216 - \ln(u) + u - \frac{u^2}{2 \times 2!} + \frac{u^3}{3 \times 3!} - \frac{u^4}{4 \times 4!} + \cdots. \qquad (11.16)$$

The form of the Theis equation is such that it cannot be solved directly. To overcome this problem, Theis developed a procedure that involves plotting the type curve and test data using logarithmic graph paper. A type curve is a log–log plot of u versus $W(u)$. Values of observed r^2/t versus s are plotted on log–log scales. The two curves are superimposed and moved about until segments coincide with the axes parallel. The coincident points determine u, $W(u)$, r^2/t, and S_c. Equations (11.14) and (11.15) can then be used to determine the transmissibility and storage coefficient.

One potential problem in applying this procedure is that Theis assumed that the discharging well is fully penetrating the aquifer. Sometimes it is not possible, or desirable, to fully penetrate the aquifer. The impact of partial penetration on drawdown must then be considered. The Theis method has been the basis for other methods that are more easily utilized. However, the assumptions for the Theis method are not nearly as restrictive as for other methods.

Computers permit trial and error estimation of formation constants using pumping test data. Estimated values of the formation constants can be substituted into the Theis method and optimized until the test data are approximated by the simulation.

Jacob's Method

Other methods for analyzing aquifer test data have been developed. One that is somewhat easier to use was developed by Jacob (Heath, 1982).

Jacob's method utilizes data collected after a long pumping time. As time passes, the shape of the cone and the drawdown rate vary. Initially after beginning a test, the cone of depression is rapidly changing. At a later time, the cone changes more slowly. Jacob's method works only for times such that the terms beyond $\ln(u)$ in Eq. (11.16) are negligible. This condition is considered met if

$$u \le 0.05, \qquad (11.17)$$

Substituting 0.05 into Theis's equation for u and making units conversions, the minimum time at which Jacob's equation applies is given by using the form

$$t_c = \frac{7200 r^2 S_c}{T}, \qquad (11.18)$$

where t_c is the time in minutes when steady-state conditions develop at a distance r in feet from the pumping well, S_c is the dimensionless storage coefficient, and T is the transmissivity in square feet per day.

By setting $W(u)$ equal to $-0.577216 - \ln(u)$, Jacob solved for T as

$$T = \frac{2.3Q}{4\pi \Delta s} \log_{10} \frac{t_2}{t_1}, \qquad (11.19)$$

where Δs is the drawdown in the interval between t_2 and t_1. By setting the ratio t_2/t_1 equal to an order of magnitude (i.e., one log cycle), T becomes

$$T = \frac{2.3Q}{4\pi \Delta s} \qquad (11.20)$$

and the storage coefficient becomes

$$S_c = \frac{2.25Tt_0}{r^2}, \qquad (11.21)$$

where Q is the discharge (cfs), Δs is the drawdown (feet) measured across one log cycle in the straight line portion of a plot of drawdown versus lag time, t_0 is the time where the straight line intersects the zero drawdown line, and r is distance (feet) between the pumping and discharge well.

To apply Jacob's method, drawdown is plotted on the vertical arithmetic axis versus time on a logarithmic horizontal axis. A straight line is approached after steady conditions are reached as shown in Fig. 11.11. The straight line has a slope that is proportional to the pumping rate and the transmissivity. The procedures are illustrated in Example Problem 11.3. One feature of Jacob's method that makes it easier is its use of semilog graph paper instead of logarithmic graph paper. Under ideal conditions, data plot as a straight line instead of as a curve. A difficulty with Jacob's method is, unlike Theis's, it is applicable only in those situations in which $u \leq 0.05$, whereas Theis's procedure is more generally applicable.

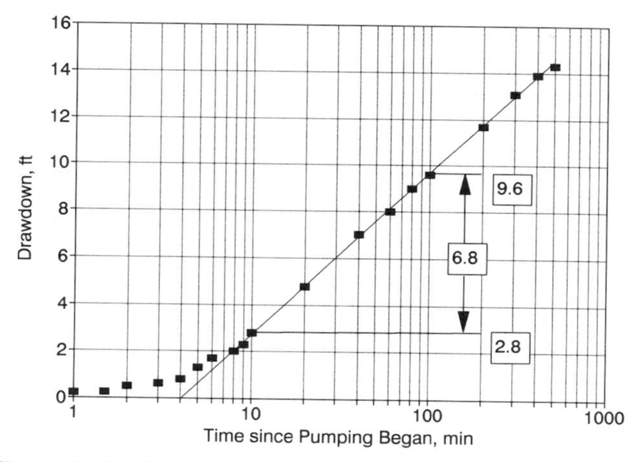

Figure 11.11 Pump test data plotted for drawdown problem.

Example 11.3. Jacob's method

Use the data shown in Fig. 11.11 to find the formation constants, transmissivity, and storage coefficient for an aquifer. The pumping rate is 600 gpm, and the observation well is located 500 ft from the pumped well.

Solution: First plot a straight line through the data points that are linear on the semilog plot. The points that fall along this line were obtained after steady state was approached. Next find Δs, the difference in drawdown over one log cycle, by reading the difference in drawdown using the straight line over one log cycle. Thus Δs is 6.8 ft. Then use Eq. (11.20) with appropriate substitution and unit conversions to obtain

$$T = \frac{2.3(600 \text{ gpm})}{4\pi(6.8 \text{ ft})(7.48 \text{ gal/ft}^3)} = 2.16 \text{ ft}^2/\text{min}.$$

Equation (11.21) can then be solved by reading the straight line at a point corresponding to zero drawdown to obtain t_0 equal to 4 min. Substitute this value and other knowns into Eq. (11.21) as

$$S_c = \frac{2.25\left(2.16\dfrac{\text{ft}^2}{\text{min}}\right)(4 \text{ min})}{(500 \text{ ft})^2} = 7.8 \times 10^{-5}$$

A similar procedure is described in Heath (1982), which can be used to estimate distance versus drawdown relationships. In either set of well tests, three observation wells are preferred. They should be located in a triangular arrangement with varying distances from the pumping wells. Limited information can also be obtained from a single observation well, or even a production well. Heath (1982) provides procedural information on this subject.

Ground Water Recharge

Ground water recharge is the quantity of water added to the ground water reservoir. Natural recharge occurs as a result of the natural movement of water through the hydrologic cycle as shown in Fig. 11.1. Natural conditions tend to have much higher recharge rates than less pervious areas such as cities. As a result of this, there has recently been considerable interest in the concept of providing artificial recharge in developing areas in order to increase ground water supplies and provide storm water control by reducing surface runoff. Natural recharge to ground water aquifers may range from negligible amounts up to 40 or 50% of precipitation. Artificial recharge occurs when water is added to the ground water reservoir that would not have naturally reached the reservoir. Artificial recharge can result from recharge basins, artificial impoundments, recharge wells, applying water to the land surface through irrigation, waste disposal, and other means. Recharge enhancement refers to activities that increase the rate of natural recharge. Such activities as land treatment to increase infiltration could constitute recharge enhancement. Care must be taken to ensure that artificial recharge does not pollute the aquifer. If recharge water contains pollutants, they may be carried into the aquifer and be difficult to eliminate.

Some methods of providing artificial recharge suggested by Linsley *et al.* (1982) include:

(A) Holding storm water in reservoirs located over permeable areas.
(B) Containing surface runoff in reservoirs and releasing it into stream channels at rates that match the percolation rate of the channel.
(C) Discharging streamflow into infiltration areas located over permeable formations.
(D) Constructing recharge basins that reach permeable formations.
(E) Forcing water down wells into an aquifer.
(F) Overapplying irrigation water over permeable formations.

Todd (1980) described many applications where recharge methods are being used. Applications discussed include irrigation, spreading basins, overland flow, and recharge wells.

The combination of hydrologic and geologic settings that contribute to natural ground water recharge are many and varied. Some of the major settings include general infiltration and percolation over large areas, percolation from bodies of surface water, such as the influent stream recharging a confined aquifer, inflow of ground water from other sources, and rapid movement of water from the surface through fractures and solution channels. Each of these mechanisms is individually discussed.

Recharge depends on the availability of water for recharge, the physical characteristics of soils and rock material that the water must pass through, and the ability of the ground water reservoir to accept the recharge water. Any one of these three major factors may be limiting and thus define the actual recharge. Obviously the recharge rate cannot exceed the rate at which water is available to supply the recharge process. In the case of surface water bodies, water availability will not be limiting as long as surface water is present. Deep percolation of infiltrated precipitation is a common and widespread means of natural recharge. Hydrologic processes at the surface and in the vadose zone largely determine the quantity of water that becomes deep percolation. Rainfall amounts, timing, and intensities are influential. Large quantities of rainfall occurring at low intensities during high infiltration periods will maximize the water available for recharge via percolation. Evapotranspiration removes a large fraction of the infiltrated water before it can become deep percolation. Climatic, plant, and soil factors govern evapotranspiration so that it is often nearly equal to precipitation. Large and infrequent precipitation events contribute most of the water to deep percolation. In humid regions, deep percolation can be a significant part of the hydrologic budget and may account for several percent of the annual precipitation.

Thus recharge via deep percolation is governed to a large extent by the hydrologic processes that take place in the near-surface zone. This zone generally constitutes the root zone of any actively growing vegetation. The character of the vegetation is a determinant of the amount of recharge. In evaluating evapotranspiration, the impact of vegetation through such things as type of vegetation, density of vegetation, leaf area index, root density, and growing season must be considered. Factors that reduce evapotranspiration tend to increase recharge if all other factors remain the same.

The permeability of the material above an aquifer is often an important determinant of the recharge rate. Highly permeable materials that allow rapid movement of water vertically are primary contributors to recharge. If the aquifer is part of a layered system, the least permeable layer will govern the recharge rate.

Fracture zones and solution channels through rock material may increase recharge if they are located so that they come in contact with water having a pressure potential equal to or greater than atmospheric pressures. In such cases they may act as localized sinks and rapidly transmit water under conditions of free surface and pressure flow to the ground water reservoir.

Localized areas overlying an aquifer may contribute much of the recharge to an aquifer. Such areas may have more favorable conditions for allowing the relatively rapid movement of water from the ground surface to the ground water. These conditions may be the result of very permeable soils, solution channels, highly fractured rock, or shallow ground water conditions. Such areas are often termed recharge areas even though recharge may be occurring at slower rates over other parts of the aquifer. Disturbances of these recharge areas have a great potential for impacting the actual recharge rate of an aquifer.

Streams, lakes, and ponds may be a source of recharge for some aquifers. In humid regions, the water level of surface water bodies is often a reflection of the ground water level with the slope of the ground water surface being downward toward the surface water. In such instances the surface water is being augmented by subsurface or ground water flow. In semiarid and arid conditions, the slope of the ground water surface is often away from the surface body of water indicating that the surface water is a recharge source for the ground water. Streams that contribute water to ground water are often known as influent streams, while streams that gain water from ground water are known as effluent streams, as shown in Fig. 11.4. Some prefer the terminology gaining and losing streams rather than effluent and influent streams. A particular stream may be a gaining stream over a part of its length and a losing stream over another part of its length. A stream may also be a gaining stream part of the time at a particular location and a losing stream at the same location at another time. The determining factor as to whether a surface water body is gaining or losing is the relative elevations of the surface water and the ground water. Losing streams are frequently ephemeral; that is, they go dry during droughty periods because percolation depletes the flow.

Aquifer characteristics may limit water recharge in instances where the potential recharge rate exceeds the rate at which the water is transmitted away from the recharge area resulting in the buildup of a ground water mound. This mound would continue to build until the hydraulic gradients in the aquifer were sufficient to cause lateral flows in the aquifer equal to the recharge rate or until the mound limited the recharge rate itself.

Changes in any of the factors influential in governing recharge rates may result in an alteration of the actual recharge of ground water. Actual recharge will be reduced if the factor currently limiting recharge is altered so as to be more restrictive to recharge or if a factor not currently limiting is changed so as to become the limiting factor.

Topography also impacts recharge because it affects the time available for precipitation to infiltrate. Steeply sloping land will provide less opportunity time than will flat land under the same cover conditions. In recharge applications that utilize overland flow, a mild slope is desirable.

FRACTURE ROCK HYDROLOGY

Flow systems in fracture zones are very difficult to quantify. The controlling factors are the extent, size, distribution, and degree of interconnectedness of the fractures. A highly fractured material may allow rapid transmission of water and thus promote recharge of ground water as shown by the hydraulic conductivities in Fig. 11.6. Hydraulic conductivity of fractured rock may be orders of magnitude higher than in unfractured rock. If fractures are not interconnected, they cannot serve as conduits for water movement. Slightly fractured systems are thus not likely to allow significant movement of water, whereas highly fractured systems may serve as major conduits.

Fracturing of rock is brought about by stresses applied to and released from rock formations. Stress relief fractures are common in the Appalachian area where overlying soil material has gradually eroded away and removed part of the compression load on the underlying rock. As this load is relieved, the rocks tend to expand and fracture. A fractured zone is very common on the upper layer of rock in the Appalachian region. This fractured zone may be up to 80 ft thick and can provide pathways for significant movement of water.

Once water enters a fracture system, it tends to continue its generally downward movement. Fracture flow results in ground water recharge, hillside seepage, or seepage into tributary streams. For water to enter any but the smallest fractures, it must be at or above atmospheric pressure. An example situation where such flow may occur is that water from saturated materials can move readily into fracture systems. Infiltration basins located over a fractured zone can provide large quantities of recharge.

Thomas and Phillips (1979) described several instances where water movement in macropores would produce noticeably different impacts than would Darcian (diffuse) flow. An estimate for the Missouri Ozarks is that water travelling through macropores contributes five times as much to ground water recharge and spring flow than does Darcian flow. When much of the water flows through macropores, pollutants can move from the surface at a speed that is much greater than is expected using Darcian theory. They can easily add contaminants to saturated zones since the contaminants do not have an opportunity to bond or chemically react with the soil profile. Obviously, potential movement of contaminants into ground water is of concern to many people for a variety of reasons.

MOVEMENT OF POLLUTANTS

Pollution of ground water is a serious problem because of the difficulties in correcting the problem once it occurs. Ground water often contains large amounts of dissolved solids. This is a result of the ability of water to dissolve some of almost any substance it contacts combined with the extremely long residence time of water after it enters the ground water. Table 11.1 lists several natural inorganic substances that are commonly dissolved in water that may affect its use. Other substances that are not naturally occurring may also cause contamination of ground water as shown in Table 11.4. This table also shows the relative importance of the different sources. Obviously, the pollutants of significance vary considerably from region to region. Since pollutants cannot easily be removed from ground water and ground water is the source of much of the drinking water in the United States, maintaining or improving water quality is receiving increasing attention from regulatory personnel, environmental groups, and individuals. Chemical pollution has recently been detected in locations that were considered to be free of pollution problems only a few years ago. Leakage from underground petroleum tanks has received widespread concern, and legislation has been passed that seeks to alleviate this particular source of ground water pollution. Deterioration of ground water as a result of other human activities is also of concern. Pollution of ground water also results from inadequate disposal of wastes on the land surface and improper application of fertilizers or other agricultural chemicals. Septic tanks may contribute bacteria and nutrients to ground water. Industrial wastes are a particularly difficult problem because many chemicals are toxic in extremely small concentrations. Coastal locations often have problems with saltwater intrusion. This results from the increased pumping of ground water necessary to supply water to rapidly increasing populations. Landfills represent another potential source of ground water pollution.

One aspect of the ground water pollution problem is that it can be difficult to detect. Some activities, such as industrial plants and wastewater treatment facilities, are point sources for pollution. Others, such as agricultural and silvicultural applications, are diffused over large areas and are referred to as nonpoint sources. Chemicals associated with pollution are often diluted so that a small trace of the chemical may impair large

Table 11.4 Sources of Ground Water Pollution in the U.S. by Region[a] (Adapted from National Research Council, 1984)

Source	NE	NW	SE	SC	SW
Natural pollution					
Mineralization from soluble aquifers				1[b]	1
Aquifer interchange					3
Ground water development					
Overpumping/land subsidence				1	4
Underground storage/artificial recharge	4		1	4	
Water wells	4				
Saltwater encroachment	3	4	3		1
Agricultural activities					
Dryland farming		1			
Animal wastes, feedlots		4			
Pesticide residues	4	3	2		4
Irrigation return flow		1		2	1
Fertilization	4	3	2		2
Mining activities	2	2	3		2
Waste disposal					
Septic tanks/cesspools	1	1	2	2	1
Land disposal, municipal and industrial wastes			3	2	2
Landfills	1	3	1		
Surface impoundments		2	1	3	
Injection wells		2		4	2
Miscellaneous					
Accidental spills	2	3	1	3	2
Urban runoff					3
Highway deicing salts	1	4	4		
Seepage from polluted surface waters	3		4		

[a]Northeast includes NY, NJ, PA, MD, DE, and New England; Southeast includes AL, FL, GA, MS, NC, SC; Northwest includes CO, ID, MT, OR, WA, WY; Southwest includes AZ, CA, NV, UT: South central includes AR, LA, NM, OK, TX. Reports not completed for Great Lakes and North Central regions, AK, and HI.

[b]Numbers indicate degree of contamination: 1, high; 2, medium high; 3, medium low; 4, low.

quantities of water. For example, a pollutant from a point source may enter the ground water at a very specific point. It then moves laterally and longitudinally as the aquifer carries it along. Detection of the pollutant in observation wells typically shows a plume that can be traced to locate the source area for the pollutant. Such is the case with oil-related substances. Since petroleum products are less dense than the water and do not easily mix with water, they tend to disperse as a thin film over widespread areas.

The potential for ground water contamination problems necessitates the need for careful consideration of operations that may contribute to ground water pollution. Heath (1982) described several factors that can be used to avoid ground water pollution in the selection of waste disposal sites:

(A) Select a site having significant depth of unsaturated clay and/or organic material.
(B) Locate near a point of natural ground water discharge.
(C) Divert surface runoff and minimize surface infiltration.

To avoid potential contamination of ground water, wells must be situated so that they are not surrounded by areas prone to surface or subsurface pollutants. Such areas include locations adjacent to septic tanks, agricultural fields and feedlots, and waste disposal sites. Wellhead protection is especially important because leakage around the outside of a pipe going into ground water will provide easy access for pollutants to the water below. Grouting around the pipe reduces the opportunity for pollutants to travel down the outside of the pipe to the aquifer or to travel from one aquifer to another. A similar problem occurs if confining layers leak water between aquifers. If one aquifer gets polluted, a leaky aquifer can provide an easy route to another aquifer. A bibliography and numerous references dealing with human-caused ground water pollution is contained in Todd and McNulty (1974).

The need to protect ground water from pollution is evident. Some measures only allow for control of bacterial contamination. In most instances, additional protection and conservation considerations are appropriate (Soil Conservation Service, 1984):

(A) Measures should be designed to prevent cross-contamination where wells penetrate two or more aquifers.
(B) Discharge from a flowing artesian aquifer should be controlled so that large quantities of water are not wasted.
(C) An overall management scheme should include analysis of the amount that can be pumped economically, the importance of its use, and expected recharge.
(D) Close proximity with currently operating systems may lower the water levels such that some systems do not operate.
(E) Wellheads should be properly constructed and protected.
(F) Recharge area should be protected from contamination.

Ground water flow has been modeled for many years using physical models such as sand tank (porous media) models, analog models using heat or electricity, and membrane models. A variety of these models are described in Todd (1980).

Recently, mathematical ground water flow and transport models that assess the movement of potential pollutants into and with ground water have been developed. Most mathematical models use the finite difference method. This computational method divides an aquifer into a grid and analyzes the flow using a variation of the equation of continuity (Anderson and Woessner, 1992). Additional information on several specific models that have recently been developed can be found in Bedient and Huber (1992). This reference contains detailed descriptions of the equations and examples of the code used for several ground water flow problems.

Increasing availability of computer hardware and software has led to rapid growth in the use of models to anticipate the potential movement of pollutants or track them to a source. A variety of models are available for use ranging from nonpoint-source pollution models, such as GLEAMS developed by the U.S. Department of Agriculture, to commercially available models for pollutant plume analysis. A major difficulty with many of these models is the lack of an adequate description of the underground media. Further growth in these applications is anticipated. Additional information summarizing the basic concepts of ground water hydrology and a list of references grouped by topic can be found in Heath (1982). The National Research Council (1990b) has presented a comprehensive discussion of ground water modeling.

Problems

(11.1) Describe the prevailing ground water province at your location. Are either unconfined or confined formations commonly used for drinking water?

(11.2) At your location, what naturally occurring inorganic chemicals would you expect to be most likely to pollute ground water?

(11.3) Ground water is the source for baseflow in a small stream located 100 m from a well. A highly water-soluble chemical is spilled at the wellhead and leaks into the ground water via an improperly grouted casing. How long would you expect it to take for the pollutant to reach the stream if the aquifer is composed of good sand and the hydraulic gradient is 1 m/25 m? Impermeable rock is located below the aquifer.

(11.4) Referring back to Problem (11.3), could the well be managed to reduce the damage from the pollutant? Explain.

(11.5) If the aquifer in Problem (11.3) is located above rock fractures, would you expect a different result?

(11.6) Estimate the Darcian velocity for water flowing through a column having an area of 80 cm^2, hydraulic conductivity of 0.1 m/day, and hydraulic gradient of 0.05. What material(s) would you expect the column to contain?

(11.7) Use the same data as shown in Fig. 11.11 to find the formation constants, transmissivity, and storage coefficient for an aquifer. The pumping rate is 480 gpm, and the observation well is located 300 ft from the pumped well.

(11.8) Calculate the relative pore velocity for a sand aquifer that has a Darcian velocity of V and a porosity equal to 37%.

(11.9) Many aquifers are composed of more than one horizontal layer, each individually isotropic, but with different thicknesses and hydraulic conductivities. Consider a case where there are two horizontal layers that have thickness z_1 and z_2 and hydraulic conductivities K_1 and K_2, respectively. Derive an equation for the hydraulic conductivity in the horizontal direction. Assume a homogeneous system and a unit width.

(11.10) Generalize the equation derived in Problem (11.9) for n horizontal layers. This equation defines the equivalent horizontal hydraulic conductivity for a stratified material.

(11.11) Consider again the two-layered, stratified aquifer. Derive an equation for the equivalent vertical hydraulic conductivity, K_z, for the system. Assume homogeneous conditions and a unit width. Generalize the equation for n horizontal layers.

(11.12) An alluvial flood plain contains an unconfined aquifer with ground water flowing parallel to the stream channel. The depth from the water table to the aquitard below the aquifer is 40 m, and the flood plain is approximately rectangular with a width perpendicular to the stream of 1200 m. Estimate the flow in the aquifer in cubic meters per day and cubic meters per second if the aquifer has a K equal to 80 m/day and a hydraulic gradient equal to 0.01. How does the flow in the aquifer compare to that of a stream?

(11.13) An aquitard seperates an unconfined aquifer from an underlying aquifer. The water table is higher than the piezometric surface, and water leaks vertically downward from the unconfined aquifer to the confined aquifer through the aquitard. Using the bottom of the aquitard as a datum, the water table is at an elevation of 40 m, the piezometric surface is at 35 m, and the top of the aquitard is at 8 m. Hydraulic conductivity of the unconfined aquifer is 12 m/day and for the aquitard is 0.3 m/day. Assume steady-state conditions, and determine the leakage velocity in m/day.

(11.14) Plot the pumping test data shown below on semilogarithmic paper. Fit a straight line through the points. The test data were collected at a pumping rate of 2000 m^3/day with drawdowns measured at an observation well located 50 m away. Estimate the transmissivity and storage coefficient.

t (min)	Drawdown (m)	t (min)	Drawdown (m)
0	0	10	0.46
1	0.20	15	0.50
2	0.27	20	0.54
3	0.32	50	0.64
4	0.36	100	0.72
5	0.38	500	0.90
8	0.43		

(11.15) A well that pumped at 350 m^3/hr for 8 hr was shut down. Measurements of the drawdown as it recovered were made in an observation well located 30 m away. Calculate the transmissivity of the system. (Hint: Plot the residual drawdown s' as a function of t/t' on semilog paper where t' is minutes since shutdown and t is time since pumping began.)

t' (min)	s' (m)
1	0.62
2	0.57
3	0.54
4	0.53
5	0.51
10	0.46
15	0.44
20	0.42
30	0.39
60	0.35
90	0.32
120	0.31
240	0.27

References

Anderson, M. P., and Woessner, W. M. (1992). "Applied Groundwater Modeling—Simulation of Flow and Advective Transport." Academic Press, San Diego, CA.

Bedient, P. B., and Huber, W. C. (1992). "Hydrology and Floodplain Analysis," 2nd ed. Addison–Wesley, Reading, MA.

Bouwer, H. (1978). "Ground Water Hydrology." McGraw–Hill, New York.

CAST (1985). Agriculture and ground water quality, Agricultural Science and Technology, Council for Report No. 103. Ames, Iowa.

Dorsey, J. D., Ward, A. D., Fausey, N. R., and Bair, E. S. (1990). A comparison of four field methods for measuring saturated hydraulic conductivity. *Trans. Am. Soc. Agric. Eng.* **33**(6): 1925–1931.

Dunne, T. and Leopold, L. B. (1978). "Water in Environmental Planning." Freeman, San Francisco, CA.

Heath, R. C. (1982). Basic ground-water hydrology, U.S. Geological Survey water supply paper 2220. U.S. Geological Survey, Denver, CO.

Linsley, R. K., Kohler, M. A., and Paulhus, J. L. H. (1982). "Hydrology for Engineers," 3rd ed. McGraw–Hill, New York.

National Research Council (1984). "Ground Water Contamination." National Academy Press, Washington, DC.

National Research Council (1990a). "Surface Coal Mining Effects on Ground Water Recharge." National Academy Press, Washington, DC.

National Research Council (1990b). "Ground Water Models." National Academy Press, Washington, DC.

Soil Conservation Service (1984). "Engineering Field Manual," 4th printing. Department of Agriculture, U.S. Soil Conservation Service, Washington, DC.

Thomas, G. W., and Phillips, R. E. (1979). Consequences of water movement in macropores. *J. Environ. Quality* **8**(2): 149–152.

Todd, D. K. (1980). "Ground Water Hydrology." Wiley, New York.

Todd, D. K., and McNulty, D. E. O. (1974). "Polluted Groundwater." Water Information Center, Huntington, New York.

Van Voast, W. A., and Reiten, J. C. (1988). Hydrogeologic responses: Twenty years of surface coal mining in Southeastern Montana, Memoir 62. Montana Bureau of Mines and Geology, Montana College of Mineral Science and Technology, Butte, MT.

Vesilind, P. A., Peirce, J. J., and Weiner, R. (1988). "Environmental Engineering," 2nd ed. Butterworths, New York.

Viessman, W., Jr., Lewis, G. L., and Knapp, J. W. (1989). "Introduction to Hydrology," 3rd ed. Intext Educational Publishers, New York.

Walton, W. C. (1970). "Ground Water Resource Evaluation." McGraw–Hill, New York.

12

Monitoring Hydrologic Systems

A large part of this book presents techniques for estimating various hydrologic quantities in the absence of any actual measurements of these quantities at the location(s) of interest. Only a small fraction of small catchments throughout the world are actually monitored. For the vast majority of these catchments some type of model must be used to estimate the quantities of interest. Generally the estimation technique used has been developed and tested on only a small subset of the small catchments that are actually monitored. The net result is that estimation techniques used on small catchments have only been tested on a very low number of the total existing catchments. Finally, no estimation technique has been found that is 100% accurate for any catchment. Thus, it is frequently desirable to monitor hydrologic and water quality variables on a local and site-specific basis.

UNCERTAINTY

When a model is selected for application to a small catchment, at least two points of uncertainty regarding the model exist. The first is the uncertainty associated with the model itself. No matter how rigorous, every model has associated with it uncertainty regarding how well the actual catchment processes are being represented. The second source of uncertainty is associated with the values of the parameters used in the model. Every model has parameters that are used to characterize the particular catchment of concern. The true values of these parameters are not known and must be estimated. There is much uncertainty associated with the parameter estimates.

Actual data on the physical process of concern from a catchment can be extremely valuable in reducing model and parameter uncertainty. Data provide a basis for estimating model parameters reflective of the actual catchment and for testing the model and the selected parameters. Without any actual data on the catchment of concern, one can only infer model performance and model parameter values based on other (and hopefully similar) small catchments. Even very limited data can prove to be quite useful in parameter evaluation.

Municipalities and agencies may well find it very cost effective to operate moderate catchment monitoring programs for the purpose of determining locally applicable models and for determining model parameters applicable to local conditions. For instance, if a municipality consistently overestimates storm water runoff by 20%, overexpenditure on storm water-related facilities will be on this same order of magnitude. On the other hand, if a consistent underestimation is made, excessive costs may be incurred due to flood damage and maintenance costs.

Good data are the best source of information on hydrologic response. Good data are preferred to the same information generated from a model. Poor data may be misleading and can be inferior to model results. To promote similarity in the type of data collected and the manner they are collected, many agencies of the U.S. Government joined efforts to produce a "National Handbook of Recommended Methods for Water-Data Acquisition" (U.S. Geological Survey, 1977; and revisions). This handbook contains recommendations for collecting data on precipitation, surface and ground water, water quality and sediment, soil water, evaporation, hydrometeorology, snow and ice, and catchment characteristics. The U.S. Department of Agriculture (Brakensiek *et al.*, 1979) has also prepared a comprehensive manual for field measurement of hydrologic data. This manual is divided into sections on precipitation, runoff, climate, sedimentation, geology and soil conditions, and watershed characteristics. Similarly, the U.S. Weather Bureau has guides for the installation of various hydrometeorologic instruments. In any data collection effort, it is important to use standard techniques so that the data obtained will be consistent with other data sources.

All measurements are subject to an unknown amount of error. Two major types of errors are random errors and systematic errors. Random errors result in errors that are both positive and negative and have a mean value of zero. Random errors may result from insensitivity of an instrument to the phenomena being measured, errors in reading scales, or pulsating conditions. Random errors are generally chance errors. Since these errors are random with a zero mean, repeated sampling or measurement may be used to reduce this source of error.

Systematic errors introduce a bias into data in that the mean systematic error is not zero. Improper instrumentation, always reading on the high or low side of a scale, and faulty calibrations are frequently the cause of systematic errors. Repeated sampling cannot overcome systematic errors.

Data with small random errors are said to have high *precision* in that the data are repeatable. Data with small systematic errors are said to have high *accuracy* in that they are representative of the true data value. Obviously, minimizing both random and systematic errors is desirable.

Faulty calibration may be the source of error in some instruments and may produce both random and systematic errors. For example, streamflow is often related to flow depth. Theoretically, there is a one-to-one relationship between flow rate and depth for uniform flow in a prismatic channel with constant properties. Natural flow may not be uniform and channel properties change. Thus, for a given depth, flow may be above or below the value indicated by the calibration curve. This could be a random error. Similarly, the data or procedure used to define the depth–flow relationship may not be representative of the true relationship over the entire range of interest. Nonuniformities may introduce error into the result in certain depth ranges. These errors are systematic and may lead to over- or underestimation of the flow for certain depth ranges.

INSTRUMENTS

For many years, instruments used to measure hydrologically important variables were generally of the mechanical type using spring wound clocks, floats attached to mechanically activated pens, or weighing devices attached to mechanically driven pens. Although many of these devices are still in use, they are being replaced by electrical and electronic devices. Modern data recorders use digital clocks and microelectronic technology to receive and/or transmit data to a central receiving station. Data are now stored on magnetic tapes, magnetic disks, and optical disks.

Many recording stations are battery powered, with solar energy recharge capability greatly reducing the time required to actually visit the sites and service the instruments. Battery-powered sites can transmit data to central receiving stations for retransmission or storage, thus eliminating the need to change charts. Instrumentation technology is changing rapidly. For this reason, this chapter addresses the principles of measurement of hydrologic variables but does not detail the actual instruments used.

Reliability of instruments and training personnel that service the instruments must be considered in selecting the type of instrumentation to be used for a particular application. Often the most valuable data, the data of real concern, occur during the most adverse weather and flooding conditions when instrumentation failure is the most likely. Simple, reliable instruments that function properly without observer attention are a must under these conditions.

SOURCES OF DATA (U.S.)

Hydrologic data are available from a variety of state and federal agencies. The primary source of weather-related data, such as precipitation and temperature, is the National Weather Service, which has a data center in Asheville, North Carolina. The U.S. Geological Survey, with offices in Reston, Virginia, is a primary source of data on surface and ground water quantity and quality as well as geology, topography, and aquifer characteristics. Soils and land-use data may generally be obtained from the Soil Conservation Service of the U.S. Department of Agriculture headquartered in Washington, D.C. These large Federal agencies have state offices responsible for data pertaining to particular states. The Soil Conservation Service also has many local offices scattered throughout the U.S.

Other Federal agencies that collect data of value to hydrology and sediment studies are the U.S. Forest Service and Agricultural Research Service of the U.S. Department of Agriculture; the U.S. Army Corps of Engineers; the U.S. Bureau of Reclamation; the Environmental Protection Agency; the Office of Surface Mining; the U.S. Park Service; and the Bureau of Land Management of the U.S. Department of Interior.

State and local agencies should not be overlooked as a source of data. Often universities, especially in conjunction with agricultural experiment stations, conduct special hydrologic studies producing valuable data. State water agencies and environmental agencies also may be a source of data on water quantity and quality. Local governments often conduct special studies of hydrology and may have data applicable to their particular areas of responsibility.

Private firms are also archiving data collected by federal and state agencies in the form of microcomputer-compatible disks. These firms also are vendors of software useful in analyzing the raw data.

PRECIPITATION

Precipitation may occur in the forms of rain, drizzle, freezing rain, snow, and ice. The only active form of precipitation generally considered in the design of storm water or erosion control facilities is rain. Snow melt may contribute appreciably to runoff and erosion during the melting phase. Snow melt may be an important erosive agent on small catchments but rarely produces flood flows on them. By contrast, snow melt may be an important contributor to floods on larger catchments.

Precipitation may be measured directly with a collecting gage of some type or estimated via radar. Collector-type rain gages are generally cylindrical with diameters ranging from 2 to 12 in. (5 to 30.5 cm) or more. Studies have shown that the diameter of the gage has little impact on gage accuracy. Collector gages may be recording or nonrecording. Nonrecording gages are simply designed to store the collected precipitation until it can be manually measured. The standard nonrecording raingage in the U.S. Weather Service has an 8- in. (20.3 cm) opening and is read once every 24 hr.

Recording raingages are of three main types—weighing, tipping bucket, and float. Data from recording gages are required for storm water computations from small catchments since the time distribution of rainfall is as important as the volume of rainfall for these catchments. The timing of the gages should be such that rainfall time increments reflective of the hydrologic response time of the catchment can be determined. These time increments are discussed in Chapter 3 and may be as short as 5 min.

As the name implies, weighing raingages respond to the weight of precipitation passing through the collector opening. This weight is converted to an equivalent depth over the area of the collector. The cumulative weight (depth) as a function of time is recorded. The slope of the depth versus time relationships is the intensity of the precipitation. Weighing gages may be suitable for estimating snowfall if a suitable method of melting the snow without allowing it to bridge over the gage openings is used.

Tipping-bucket gages have two small "buckets" that rotate on an axis below a collection orifice. The buckets rotate after collecting a designed and very small volume of rainfall. When one bucket tips, a second, empty bucket is rotated in place under the collector orifice. Thus one bucket is always in place under the collector orifice. During the actual tipping process, some error may be introduced into the recording. For light rainfalls, this error will be small and generally not exceed 2%. The water spilled from the tipping bucket is collected and measured so that the tipping recordings can be adjusted as required to reflect the correct total volume of rainfall.

Float-type gages have a cylindrical chamber into which the precipitation is directed. A float in this chamber transmits the water depth to the recording device so that a continuous record of rain depth is obtained.

All types of gages should be protected from freezing by using an appropriate antifreeze solution. Similarly, evaporation may be controlled through the use of some type of oil as an evaporation suppressant.

Raingage catch is very sensitive to wind turbulence. Gages must be located to minimize this source of error. It is generally recommended that the horizontal distance from any vertical obstructions be at least twice the height of the obstructions and preferably four times the height. Gages should also be located on horizontal or nearly horizontal areas. Up- and down-slope drafts can appreciably impact the catch of gages located on slopes. Gages should not be located on roofs. The gages should be installed so that the top of the collecting cylinder is horizontal. A carpenter's level can be used to periodically check this condition. It is also important that the opening of the collector be cylindrical and undamaged. Any dents in the opening should be repaired. Standard gage height in the U.S. is generally 30 in. (76.2 cm) above the ground.

Raingages should be distributed uniformly. The density of the gage network depends on the purpose of the data and the prevalent types of rainfall. When storms are cyclonic, widespread, and of low intensity, a sparse network may be adequate. A more dense gage network is required to characterize convective thunderstorms of high intensity and variable areal extent. In mountainous areas, orographic effects must be considered in the layout of the gage network.

Often a combination of recording and nonrecording gages can be used to improve network density and storm definition. The network might consist of four nonrecording gages for every recording gage. At least two recording gages and preferably three should be included in each network.

Radar had become a standard instrument in meteorological networks and is especially valuable in providing real-time information on the location, intensity, type, vertical and horizontal extent, and movement of rainstorms. Quantitative estimates of rainfall amounts and intensities can be enhanced by incorporating recording raingages with data transmission capabilities. Correlation of data transmitted from the gages and the radar echos can produce reasonable estimates of rainfall as it occurs. This type of information is valuable in the real-time operation of storm water management facilities.

RUNOFF

The measurement of runoff from small catchments is a complex topic because of the wide range of flow conditions that might exist. The references by the U.S. Geological Survey (1977) and Brakensiek *et al.* (1979) have excellent coverage of the topic and should be consulted for many details associated with the measurement of runoff.

Flow measurements may be done by:

(1) Measuring stream stage and relating stage to discharge.
(2) Measuring flow velocity and calculating discharge based on velocity and area.
(3) Using precalibrated weirs and flumes.
(4) Using hydraulic characteristics of channels and flow control structures.
(5) Using model studies of flow control structures.
(6) Using dye dilution techniques.
(7) Using a tipping bucket.

Flow measurement is fraught with difficulties. Problems that must be anticipated are debris, ice, sediment, vegetation, erosion, stream alterations, and vandalism. Stream-gaging stations require constant attention, maintenance, and recalibration to ensure the integrity of the measurement.

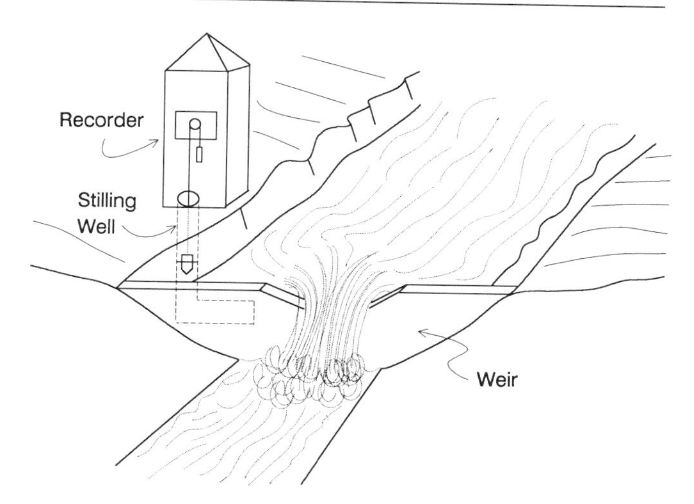

Figure 12.1 Stream-gaging arrangement.

Stream Stage Determination

Stream stage is the depth of water above some arbitrary datum. The reference datum should be selected so that negative stages are not possible. For weirs and flumes, the reference is often selected so that a stage of zero corresponds to no flow, and any positive stage results in some flow. Stage may be measured with a staff gage, float-activated recorder, pressure-activated recorder, wire weight gage, or a crest stage gage. Figure 12.1 shows a typical stream gaging setup using a float recorder.

Staff gages may be used as the primary means of measuring stage or as a check on other types of measurements. A staff gage is a scale placed in the flow so that the stage can be read directly. Staff gages may be placed vertically or inclined. Inclined gages must be calibrated to provide correct vertical stage reading for the particular angle of inclination involved. Staff gages should be placed where they can be easily read from stream banks or bridges and where they are protected from damage due to debris in the flow.

Float-activated recorders are the most common device used to measure stage. Generally the installation for a recorder of this type has an instrument shelter and stilling well located off to the side of a channel with an intake pipe extending into the flow. The shelter must be located so that it is above the highest flow level to be recorded, is accessible during flood flows, does not interfere with the flow, and is not damaged by floating debris and ice. The intake to the stilling well must be located so that it will not become clogged with sediment or debris such as leaves, rags, paper, plastic, and other trash. The intake must also be large enough so that the water level in the stilling well corresponds to the water level in the stream. The suitable opening size depends on the area of the stilling well and the rate of change of stage in the stream.

Once the recorder is installed, care must be taken to ensure the chart readings are properly related to stream stage, that the recorder will not be subject to vibrations from flow, wind, traffic, etc., and that the float and counterweight can operate freely through the entire range of the anticipated stage.

A bench mark should be established near the gage so that the reference elevation of the gage can be checked periodically. It is also desirable to check the recorded stage against the actual water stage at several different water surface elevations to ensure correct data over the range of the expected stages.

A pressure-activated gage operates on the principle that the depth of water above a given point is directly related to the water pressure at that point. By measuring the pressure, the stage can be determined. The most common method of pressure measurement is to bubble a gas into the flow and record the required gas pressure. Bubble gage installations are often less-expensive than installation of a stilling well and can easily measure a 50-ft (15 m) range in stage. Bubble gages can also be moved from one location to another with relative ease. The orifice through which the gas is bubbled must not be located where it will be covered with mud or debris or in highly turbulent flow. The most common bubble gage uses a mercury manometer for recording pressure. Continuous records can be obtained using pressure transducers and automatic recorders.

A wire-weight gage consists of a weight suspended from a wire or cable wrapped on a cylinder. The weight is lowered until it contacts the water surface at which point a counter on the cylinder indicates the length of wire required. This in turn is related to water stage. Bridges are frequently used to house wire-weight gages as they make excellent platforms from which to make measurements.

A crest-stage gage is used to determine the maximum stage reached during a runoff event. A popular crest-stage gage consists of a vertical 2 in. (5 cm) pipe that houses a wooden stick or piece of thin-walled conduit. Burnt cork is placed within the pipe, which is perforated around the bottom. The cork floats up with the water in the pipe and tends to cling to the measuring stick when the water level drops. By removing the measuring stick and recording the height of the deposited cork, the maximum stage can be determined. Measuring sticks made of buoyant materials must be prevented from floating. If the stick is of the proper length, the cap on the pipe may serve this purpose.

One of the most frustrating flow-monitoring problems is the loss of data during extreme but rare events. By definition, a 100-year event is expected to occur just once every 100 years. An event of this frequency produces large flows with a lot of erosive action and carrying a lot of debris. Installations that are not well constructed with adequate foundations and located above the flood level will invariably fail during critical events, and the most valuable data of the gaging program will be lost, to say nothing of the instrumentation itself.

Velocity Determination

Flow velocity may be determined using rotating element current meters, floats, and tracers. Occasionally, electromagnetic and acoustical flow meters, pitot tubes, or optical methods are used. These latter methods are more prevalent in laboratory research than they are in field hydrologic studies.

Rotating element current meters may be of the propeller or cup type. The rotating element is held in proper orientation to the flow by a streamlined body and flow vanes. The rotations are electronically or manually timed and related to flow velocity through a calibration curve. The velocity obtained is an estimate of the velocity at one point in the flow. Figures 4.2 and 4.3 show that the velocity varies throughout the cross section of the channel. Thus several measurements must be obtained.

The general procedure is to divide the stream width into a number of sections, determine the average velocity for each section, compute the discharge for each section as the product of the section area and average velocity, and finally sum the section discharges to determine the total discharge.

If a logarithmic velocity profile is assumed, the average velocity of a vertical profile is approximately equal to the average of the velocities at 0.2 and 0.8 times the flow depth. If the flow depth is too shallow to reliably determine the 0.2 and/or 0.8 depth velocity, the average velocity may be taken as the velocity at 0.6 times the depth. Occasionally, the velocity at all three depths is averaged. Field studies of actual velocity profiles have verified these approximations.

On small catchments, especially in urban areas, the depth of flow may be changing rapidly, making it difficult to make very many velocity measurements without a substantial change in stage. In this event, relatively few vertical profiles can be sampled and the 0.6 depth reading should be used.

Floats may be used to approximate average flow velocities. Difficulties encountered with floats include the variation in downstream velocity across the channel and the fact that surface and not average velocity is measured. A correction factor of 0.85 times the float velocity is often used to approximate the average profile velocity. Correction factors from 0.80 to 0.95 may be applicable depending on the particular flow conditions.

Tracer techniques for determining flow velocities consist of injecting a slug of tracer into the flow and measuring the time it takes for the centroid of the concentration–time curves to pass two points a known distance apart. Very little error is introduced if the time between peak concentrations rather than centroids is used. Occasionally, visual observation of the time of passage of a dye tracer is used; however, actual concentration measurements give more reliable results. A fluorescent dye such as Rhodamine WT can be detected with good accuracy at low concentrations using a fluorometer.

The amount of Rhodamine WT 20% solution required for a time of travel study can be estimated from (Kilpatrick, 1970),

$$V_{\mathrm{d}} = 2.0 \times 10^{-3}\left(Q_{\mathrm{m}}\frac{L}{V}\right)^{0.93} C_{\mathrm{p}}, \qquad (12.1)$$

where V_{d} is the volume of dye (liters), Q_{m} is the maximum discharge (cubic meters per second), L is the length of the reach (kilometers), V is the estimated mean velocity (meters per second) and C_{p} is the desired peak concentration (micrograms per liter) at the downstream end of the reach. C_{p} generally should not exceed 5 μg/liter.

Regardless of the method employed, velocity measurements should be made in straight and uniform sections of a channel. This makes the determination of discharge based on velocity and area more accurate. By combining measurements on stage and discharge, the stage–discharge or rating curve for a stream may be determined. Frequently the relationship may be expressed in the form

$$Q = aH^{b}, \qquad (12.2)$$

where Q is discharge and H is stage. For some locations, a separate rating curve is required for the rising and falling stages of the hydrograph. The rating of streams from larger catchments, especially in the presence of alluvial channels, is a specialized topic beyond the scope of the coverage presented here. For small catchments, problems with shifting controls and erosion and sedimentation within the channel are not as severe. A major problem with small catchments is the rapid rise and fall of the hydrograph, making it difficult to complete discharge measurements.

Precalibrated Weirs and Flumes

Precalibrated weirs and flumes are commonly used to measure runoff from small research catchments. Such weirs do not generally require field calibration. U.S. Geological Survey (1977) and Brakensiek *et al.* (1979) contain excellent treatments of this topic. In Chapter 5, the use of weirs for flow control is discussed. Weirs and flumes make excellent flow-measuring devices because they have a one-to-one relationship between discharge and stage.

A weir is an overflow structure placed across the flow. The edge over which the flow occurs is called the crest of the weir. Stage is generally measured relative to the weir crest. The crest of a weir may be sharp (metal plates) or broad (small dams or humps). If the overflow from the weir is unobstructed by the water level downstream from the weir, a free overfall is said to exist. If the weir overflow is partially under water, the weir is termed submerged and a unique stage discharge relationship may no longer exist. Weirs may have a free overfall under low flows and become submerged under high flows. For flow measurement, submergence of the weir should be avoided.

The stage–discharge relationships for some typical sharp-crested weirs are given in Chapter 5. Sharp-crested weirs are difficult to maintain in natural streams because the edge becomes nicked and bent from debris in the flow. To overcome this problem, the U.S. Department of Agriculture (Brakensiek *et al.*, 1979) has developed a series of broad-crested, V-notch weirs that have been widely used on research catchments. Rating tables and dimensions of weirs for measuring discharges up to 1300 cfs are presented.

Weirs present problems if the flow is transporting sediment or debris. Sedimentation tends to occur immediately upstream from the weir causing shifts in the stage–discharge relationship and requiring frequent clean out. This problem may be largely overcome by using a flume. Flumes are designed to cause flow to pass through critical depth thus producing a unique stage–discharge relationship. They also are designed to allow the flow to pass without ponding and without a bottom obstruction so that sediment and debris tend to pass without clogging the flume.

The common types of flumes are H-flumes and Parshal flumes (Fig. 12.2). Brakensiek *et al.* (1979) present design information on H-flumes having capacities up to 117 cfs and Parshal flumes up to 3000 cfs. H-flumes provide excellent accuracy for very small catchments. Parshal flumes have the advantage of low

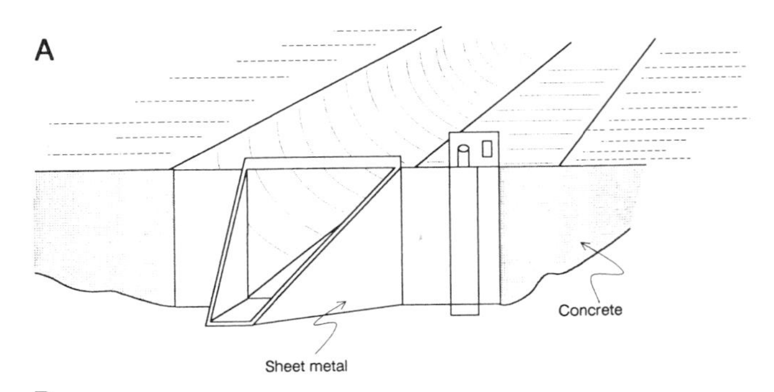

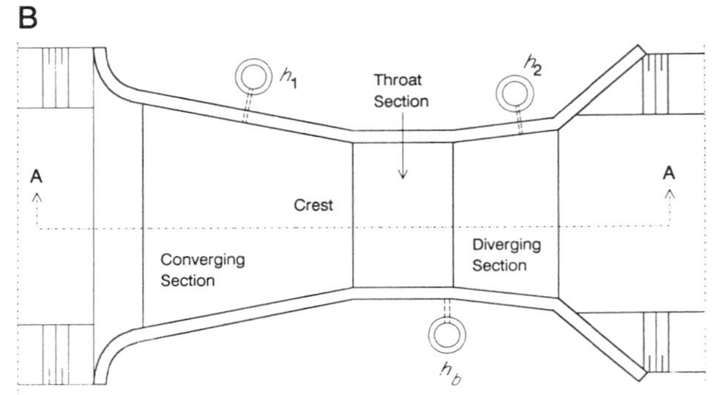

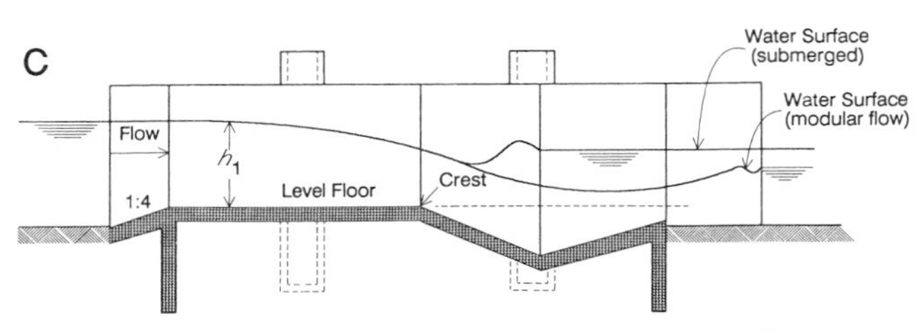

Figure 12.2 Flumes used for flow measurements. (A) H-flume, (B) Parshal flume.

head requirements—about a fourth of that required for a weir having the same crest length—and can thus be used in channels of low slopes. Parshal flumes are more expensive and may not provide satisfactory measurement of low flows. For example, a Parshal flume having a maximum capacity of 1000 cfs has a minimum free-flow capacity of 10 cfs. Depending on the purpose of the gaging program, it may be necessary to install an additional weir for low-flow measurements on some catchments. Weirs and H-flumes are much easier to construct than Parshal flumes. H-flumes and Parshal flumes are available commercially in various sizes and materials.

Flow Control Structures

The hydraulic characteristics of some flow control structures may be used to develop a stage–discharge relationship and thus serve as a flow-measuring device. For example, in Chapter 5 the hydraulics of culverts was discussed. Frequently these hydraulic relationships are used to develop a rating curve for culverts, which then is used to determine the discharge. A stage-recording device can be used to monitor flow continuously or high-water marks can be used to estimate peak discharges.

Occasionally, flow hydraulics may not be amenable to analytic considerations, and model studies may be required to develop the needed stage–discharge relationship. Properly conducted model studies are expensive and are generally limited to important installations that are going to be in operation over a long period of time. Laboratory facilities and the expertise required for hydraulic similitude model studies are available in a limited number of places—generally government or university facilities. Often these facilities will contract for outside studies.

Dye Dilution

Tracer dyes may be used to make flow measurements. If a tracer is injected at a constant rate and allowed to completely mix with the stream, discharge may be estimated from

$$Q = \left(\frac{C_1 - C_2}{C_2 - C_b}\right) q, \tag{12.3}$$

where Q is the stream discharge, q is the rate of tracer injection, C_b is the background tracer concentration in the stream, C_1 is the tracer concentration in the injection stream, and C_2 is the measured tracer concentration in the stream. This relationship assumes that an equilibrium condition exists.

Dye dilution techniques require that the dye be thoroughly mixed with the flow at the point of sampling. A manifold system may be required for injecting the dye so that complete mixing occurs in a relatively short distance. Care must be taken not to inject the dye in a concentrated fashion so that the dye stream is not completely mixed with the stream flow. Incomplete mixing will render the concentration C_2 nonrepresentative and will invalidate the results. If the flow contains significant quantities of sediment with an exchange phase with the dye, significant error in flow estimation may result.

GROUND WATER

The most fundamental quantity measured when monitoring ground water is the elevation of the water table or piezometric surface. Other quantities of interest include physical characteristics of aquifers and measures of ground water quality. Ground water monitoring sites should be selected to be as representative as possible of the factors being monitored. They should be free of influences that are extraneous to the purpose of the monitoring program. For example, if a ground water level monitoring program is desired to measure the impact of land treatment on aquifer recharge, the monitoring sites should not be located in a municipal well field where drawndown due to pumping would obscure the desired data. Monitoring sites should be located where they are accessible under all expected weather conditions. They should also be secure from external disturbance and damage.

Although existing wells may be used for monitoring wells, it is generally preferable to install observation wells specifically for data collection. When an observation well is installed, the following data should be collected (U.S. Geological Survey, 1977):

(1) Aquifer(s) tapped.
(2) Aquifer hydrologic characteristics
(3) Lithologic and geophysical logs.
(4) Well depth, size and type of casing or finish, location, and type of perforations.
(5) Elevation of land surface and measuring point.
(6) Diagram and photograph of well showing access to well and measuring point.
(7) Date the well was drilled.
(8) Well-response data for unpumped well (specific capacity tests).
(9) Local well name and owner.
(10) Location by legal description such as latitude and longitude coordinates.
(11) Significant features near well that could affect the water level.
(12) Use of the well.

Ground water monitoring sites might consist of existing wells, observation wells, piezometers, or exposed features such as pits, ponds, springs, or streams. Existing wells may prove satisfactory as monitoring locations; however, pumping from the well may interfere with the monitoring program. It is also essential that data of the type listed above be available. For example, the aquifer from which the water is originating must be known if more than one aquifer exists in the locality. Abandoned wells may also be satisfactory; however, the reason for abandonment should be known. If clogging or well collapse or some other problem is present, faulty data may result.

Specially drilled observation wells are the best ground water monitoring sites. As the well is drilled, detailed lithologic and geologic data can be collected. Data are easy to collect because there is no pump to interfere with the measurements. Observation wells must be of sufficient diameter, when finished, to accommodate measuring and sampling equipment. Observation wells must be carefully installed to ensure that the resulting measurements are reflective of the specific aquifer of interest. Precautions must be taken to prevent leakage and possible contamination of one aquifer by another.

Piezometers are much like observation wells except they are generally smaller in diameter and designed to collect water at one location rather than throughout the depth of an aquifer, as in the case of an observation well. Piezometers are most frequently used in fine packed material where changes in the piezometric surface occur very slowly.

Pits, ponds, streams, etc., may be used as an indicator of ground water levels in the immediate vicinity. Water quality determinations from these sources may not be reflective of ground water quality because of the high likelihood of surface water contamination. If these surface water sources are used as indicators of ground water levels, care must be taken to determine the aquifer that is in hydraulic contact with them. A pit, for example, may contain water that has entered from the surface and reflects a temporary, perched water table rather than the water table elevation of an expansive aquifer.

If a network of monitoring locations is established, the elevation of a measuring point or reference point should be established at each location. The elevations should all be referenced to a common datum. The reference points should be permanent and easily relocated.

Water-Level Measurements

The most common methods of water-level measurement are graduated steel tape, electrical measuring line, air lines, and float-activated recorders.

Water-level measurements with a steel tape are straightforward. The tape is simply lowered into the well and a reading taken at the reference mark. The tape is withdrawn and the wet line on the tape determined. The difference in these two readings is the depth below the reference mark to the water surface. A dark tape is easier to use than a bright one. Frequently the tape is coated with chalk to aid in determining the wet line reading. A lead weight should be attached to the down hole end of the tape to ease lowering the tape and to ensure that the tape remains straight.

The electrical method consists of an electrical probe that is lowered into the well until electrical continuity is established. The probe may be one wire serving as a positive lead with the well casing serving as the negative lead. Alternatively, the probe may contain both a positive and negative lead. A battery is connected to an ammeter or light and to the electrical probe. When the probe touches the water, the electrical circuit is completed and the ammeter or light activated. The electric probe should be weighted to ensure that it hangs straight in the well. The probe should also be shielded to protect against false readings that might occur if the probe contacts the walls of the well. The probe wire is marked so that when the circuit is complete, the depth from the reference point can be easily determined.

An air line consists of a small diameter tube to which a compressed air source is attached. The tube is lowered into the well. The depth to the water surface is obtained by subtracting the length of the submerged tubing from the total length of the tubing. The length of the submerged tubing is obtained by measuring the air pressure on the tube and converting this pressure to a hydrostatic head ($h = p/\gamma$). This calculation assumes no significant pressure loss in forcing the air through the tubing. Thus air flow rates must be kept small, and clogging of the tubing must be avoided.

The piezometric surface elevation for flowing wells can be determined by capping the well and measuring the water pressure in the capped well. The pressure must be allowed to stabilize before this determination is made. If the pressure in the well is small, a short piece of tubing may be used by attaching the tubing to the capped well and raising the tubing until the flow stops. For higher pressures, a mercury manometer or pressure gage can be used.

For continuous records of water level, a float-actuated recorder may be installed on the well. If this approach is used, care must be used to ensure that the float and the counterweight do not interfere and that the tape does not drag along the side of the well.

The frequency that water level measurements are made depends on the purpose of the measurements and the dynamics of the ground water system being monitored. Rapidly responding systems may require daily measurements, while systems that respond slowly may be adequately characterized by weekly or even monthly measurements. In a cyclic system such as a pumped well field for municipal or irrigation water, sampling frequency must be such that the cyclic response of the water table can be completely defined.

Geophysical Measurements

A number of geophysical methods to assist in determining the makeup of geologic formations have been developed. These methods are useful in locating and mapping the extent of aquifers. Geophysical methods may be divided into surface and subsurface measurements. Surface methods rely on the differential physical properties of various geologic formations to transmit electrical signals and elastic disturbances. Many geophysical methods have been developed by petroleum geologists and are more appropriate to that field than to geohydrology.

A common electrical geophysical method is to apply a direct current or very low-frequency alternating current to two electrodes. Between these two electrodes, the voltage drop across two additional electrodes is recorded. These measurements permit an estimate to be made of the apparent resistivity of the underlying strata. Repeated measurements are made as the electrode spacing is increased allowing the electrical influence to penetrate more deeply. As the electrode spacing is increased, the apparent resistivity of deeper layers is determined.

The resistivity of geologic materials varies widely ranging from 10^{-6} Ω-m for graphite to 10^{-2} Ω-m for quartzite. Dry materials have a higher resistivity than the same material when wet. Clays have smaller resistivities than gravels. Thus, based on apparent resistivities, an estimate of underlying formations can be made. If resistivity data are correlated to well log data, the interpretation of the underlying geology can be greatly improved.

Seismic methods rely on the differential travel time of seismic waves through geologic materials. Loose unconsolidated material transmits the waves more slowly than consolidated material. The method imposes a sudden disturbance either by an explosive device or by a heavy hammer blow to a steel plate lying on the ground. The time it takes for the resulting seismic wave to travel various distances is determined by a series of geophones. By studying these travel times, the depth to rock layers can be determined. The petroleum industry makes extensive use of seismic reflection methods, while hydrogeologists often find seismic refraction methods more satisfactory. Seismic methods use a travel time versus distance curve for the seismic waves. The slope of various portions of this curve can be used to determine velocities. The intercept can be used to determine depths.

The variations in gravimetric and magnetic fields may also be related to changes in the unconsolidated material such as buried stream channels, which may prove to be productive aquifers.

As with electrical methods, seismic, gravimetric, and magnetic field data are especially useful when used in conjunction with well logs. Geophysical well logging gives direct access to information on subsurface materials. Well logs may be constructed from visual observation of material produced as a well is drilled, by recording the speed of penetration of the drilling process, or by more sophisticated logging techniques such as electrical, acoustical, or nuclear logging.

Generally good borehole geophysical data are obtained from boreholes used in conjunction with the petroleum industry. Only large, municipal water wells are generally logged by other than visual means. Domestic water wells are often logged visually. Many states require well logs to be submitted to a central state agency where they are filed and available for others to inspect. The following descriptions of borehole techniques are summarized from Fetter (1980).

Caliper logs: Measures the diameter of uncased boreholes. Hole diameter may vary because of caving or solution-enlarged bedding planes and joints in carbonate aquifers.

Temperature logs: Best done after fluid in borehole regains temperature equilibrium after boring. May reveal geothermal gradients. Aquifers may also have distinct temperatures.

Single-point resistance: One electrode is lowered into borehole on insulated cable. The other electrode is grounded at the surface. The resistance of all material between the lower electrode and the surface is determined. If the borehole fluid is homogeneous, some of the variation in resistance will be due to lithology near the borehole.

Resistivity: Two electrodes are lowered and the resistance between the electrodes determined. Resistivity logs can often be calibrated and used quantitatively.

Spontaneous potential: Measures the natural electrical potential that develops between the formation and the borehole fluid. One use is to distinguish shale from sandstone lithology. Shale has a positive spontaneous potential and sandstone a negative one if the salinity of the formation fluid is greater than that of the borehole fluid.

Nuclear logging: Measures either natural radioactivity of the rocks and fluid or their attenuation of induced radiation. Nuclear logging may be done in cased or uncased wells and the results are not affected by the type of drilling mud. Natural gamma radiation, neutron logging, and gamma-gamma radiation are used.

WATER QUALITY

A good water quality monitoring program requires that representative samples be collected, that samples be handled in a manner that will maintain their integrity, and that proper analytical procedures be used during the analysis phase. The weakest link in this triad determines the confidence that can be placed in the results of the monitoring program.

Surface Water

Sample sites must be accessible and must produce samples representative of the objectives of the sampling programs. Generally they must be located where flows are well mixed. Certain types of instrumentation may require external electrical power while others may be battery operated. Samples must be taken from points in the flow that are representative of the general flow situation. For example, if the objective is to sample sediment concentrations, several samples at different depths at the same location may be required because of the variation in sediment concentration with depth. Ideally the sampling location should be such that regardless of where within the stream cross section the sample is collected, the same results would be obtained.

Sampling immediately below a concentrated source of lower quality water may result in samples that are not representative of the catchment as a whole. On the other hand, if the purpose is to monitor potential water quality degradation due to a concentrated source of pollutant, sampling sites might be located immediately below and above the impacted site.

If automatic, remote samples are used, they must be located where they will not be impacted by high flows or by vandalism. The samplers must also be located where they can be easily inspected and where the samples that are collected can be properly handled.

Sampling locations at or near stream-gaging stations are desirable if information on total loads of various constituents is to be determined. Sometimes correlations between concentrations and flow rates can be made so that stream-gaging information is useful in estimating concentrations prior to the sampling program and between the actual sampling times.

Sampling frequency is dependent on the variability in the constituents being sampled. Rapid changes in concentration require frequent sampling, while monthly samples may be sufficient where concentrations are relatively stable. If concentrations are dependent on flow rates, a sampling schedule tied to the flow rate or stage in the stream may be desirable. Some commercial automatic samplers can be set to sample on a flow basis or on a time interval basis. The purpose of the study and the variability in the constituents of interest will determine the most appropriate sampling scheme. Since the "first flush" of runoff water is often rich in pollutants, it is generally desireable to include it in the sampling program.

Types of samples range from grab samples to sophisticated automatic monitoring. With grab samples, a collecting jar is lowered into the flow, uncovered, and the sample collected. Sample containers should be rinsed in the flow prior to this procedure. Plastic containers are generally preferable to glass containers unless the constituents of interest react with or are absorbed by plastic. Glass containers are preferred for organic components. Clean sample containers are essential. Good laboratory procedures should be used to ensure that the sample containers are not contaminated.

A grab sample is a point sample. A sampling procedure that takes water from all depths at a particular location (depth-integrated sample) is generally preferred to a point sample. Some constituents exhibit definite concentration profiles. Sediments and constituents transported by sediments generally have higher concentrations near the bottom of a channel and lower concentrations near the surface of the flow. A true depth-integrated sample will collect water at a rate proportional to the flow velocity.

Automatic pumping samplers are frequently used to collect water quality samples from small catchments. The expense of these samplers generally means that only one, fixed location in that stream is sampled. Obviously for valid sampling, this single point must be representative of the flow in general. Often small flumes are installed to measure the flow rate and to improve flow mixing so that a point sampled by an automatic pumping sampler is more representative. Automatic samplers can generally be programmed to take samples on a set time interval or on a flow proportional basis.

Samples collected over time may be treated as individual samples or may be composited. The choice is dependent on the purpose of the study and the resources available for sample analysis.

Continuous sensing of certain parameters such as temperature, pH, and conductivity is possible. Sensors may be located directly in the flow, or water may be continuously pumped into a sensing chamber of limited volume.

Sediment sampling in streams is an especially difficult task (see Chapter 7). The difference in density between water and sediment particles results in the particles settling toward the bottom of the stream. Turbulence acts to suspend particles. The net result is a sediment concentration gradient with increasing concentration near the stream bottom. There is also a separation of particle sizes with the larger particles predominating along the stream bottom.

Sediment is transported in streams as suspended load and bed load. In reality, there is not a sharp boundary between the two transport mechanisms. They grade into each other. Devices for sampling the suspended load are better developed than are bedload samplers. Brakensiek *et al.* (1979) should be consulted for a discussion of the various types of sediment samplers.

In sampling for sediment load, it is important to obtain representative samples of the entire cross section in both a horizontal and vertical context. Using a pumping sampler with a single intake location is not likely to yield a representative average sediment concentration.

Ground Water

Ground water samples must be handled with the same precautions as surface water samples for accurate and meaningful determination of water quality parameters. The siting of sampling points and the frequency of sampling will be governed by the purpose of the sampling program. If the purpose of the monitoring is to evaluate the overall water quality associated with a region or an aquifer, sampling points must be selected to be representative of the region or the aquifer.

If the sampling program is designed to monitor ground water quality impacts from a waste site, manufacturing facility, or feedlot, then the monitoring wells must be located downgradient from these facilities, where they will truly sample the ground water that might be impacted. Ideally, both upgradient and downgradient locations would be sampled to better quantify impacts of the object of the monitoring program.

Well-casing material may impact water quality. Shallow wells may be cased with plastic. For deeper wells cased with metal, sufficient water should be pumped so that none of the water standing in the well and possibly interacting with the casing is included in the sample.

Multiple aquifers in the same location require special care. Ideally, a series of wells should be used so that a particular well draws water from only one aquifer. Fetter (1980) outlines a procedure whereby a single well can be used to sample water quality from multiple aquifers. The procedure described would be useable only one time as it involves sampling an aquifer and then increasing the depth of the borehole to the next aquifer.

Ground water quality generally does not change very rapidly. Unnatural disturbances to the system can, however, result in rapid changes. The arrival of a pollutant plume, for example can signal a step change in some water quality parameters. Spills and other sudden contaminations may also cause abrupt changes in ground water quality. The time between the spill and the detection of the impact in ground water depends on the distance the sampling location is from the spill and the rate of movement of ground water through the intervening formations.

For deep aquifers, relatively infrequent sampling may suffice as a check on water quality. A frequency of one sample per year may be satisfactory. Shallower aquifers and recharge areas for deep aquifers may require more frequent sampling since the quality in those areas can be more easily impacted. Wells used for ground water sampling should be properly grouted if permanent and filled and sealed if not permanent to prevent surface water contamination of the aquifer.

Sample Handling

Sufficient volume of sample must be collected to meet the requirements of all of the analytical tests that are to be performed. For certain constituents, immediate field testing of the sample must be done to obtain reliable results. For these constituents, changes occur rapidly and render a delayed analysis of little value. Temperature, pH, specific conductance, and dissolved gases are examples of constituents requiring immediate field analysis.

For some constituents it is best to chill or possibly freeze the sample for later analysis. Other constituents may require the addition of a stabilizing agent such as nitric acid. Biologic activity is greatly reduced by chilling the sample to 0.5°C. Frozen samples may yield unreliable results for certain dissolved organic and inorganic constituents. Samples for metal analysis treated with nitric acid may be stored for several months prior to analysis. Refrigerated samples are generally stable in the absence of sediments. Samples should always be stored in a cool place out of direct sunlight. Regardless of storage life, samples should be analyzed as soon as possible after collection for best results. All samples should be clearly and permanently marked with sufficient information to identify all pertinent variables such as location of sample site, date and time of sampling, name of observer, flow rate or gage height, and other needed information.

Analytical Procedures

Standard analytical techniques should be used for all analyses. Some references to accepted analytical procedures are the American Society for Testing Materials (1981 and revisions), American Public Health Association (1980 and revisions), U.S. Environmental Protection Agency (1979 and revisions), and U.S. Geological Survey (1977 and revisions). These standard techniques must be complemented by good laboratory management and record keeping.

All instruments should be routinely calibrated. Analytic techniques should be constantly checked through duplicate sample analysis, spiked samples, alternate techniques, and/or the use of alternate laboratories. Standard reference materials are available from the Office of Standard Reference Materials of the U.S. National Bureau of Standards in Washington, D.C., or the Quality Assurance Branch of the U.S. Environmental Protection Agency in Cincinnati, Ohio.

Problems

(12.1) The following stream-gaging data have been collected. Estimate the flow rate. Stations are referenced to the left bank and measured perpendicular to the stream.

Station (ft)	Depth (ft)	Velocity (fps) at	
		0.2 day	0.8 day
0 + 00	0.00		
0 + 25	3.91	2.5	2.3
0 + 50	8.62	3.1	2.8
0 + 80	10.87	3.9	3.4
1 + 20	12.14	4.1	3.6
1 + 80	10.42	3.8	3.3
2 + 20	9.58	3.7	3.2
2 + 50	5.17	3.2	2.9
2 + 75	2.34	2.5	2.3
2 + 90	0.00		

(12.2) The following are stage–discharge data for Crooked Creek near Nye, Kansas. Plot the data on log paper. Estimate the coefficients a and b in Eq. (12.2). Estimate the discharge for a stage of 8.0 ft. Briefly discuss the confidence you have in this estimated discharge.

Stage (ft)	Discharge (cfs)	Stage (ft)	Discharge (cfs)
2.03	118	6.28	2880
3.68	1360	6.72	4370
4.65	2310	7.40	7400
4.78	2530	7.59	10000
5.66	3970	5.47	3070
6.13	4950	4.49	1550
4.18	2080	5.98	3730
3.72	1400	5.68	3210
3.89	1610	5.10	2370
5.12	3330	4.47	1320
6.93	7100	8.01	13600
5.28	3490	4.25	2140
6.82	5970	4.07	1840
4.65	2150	4.21	1840
5.00	3570	4.56	2290
4.89	1820	7.94	13200
6.20	3930	7.00	5680
6.50	4660	8.98	12000
7.15	6360	9.00	12100
6.70	4910	5.53	2280
6.08	2980		

(12.3) A float requires 38 sec to travel 150 ft. The width of the stream at the measuring point is estimated to 25 ft and the average depth 5.5 ft. Estimate the dischange.

(12.4) A float requires 29 sec to pass through a 100-ft wide rectangular bridge opening. The road bed is 20 ft wide. The depth of flow is estimated at 3 ft. Estimate the flow rate.

(12.5) A stream reach 1200 m long is available for flow measurement using a Rhodamine WT fluorescence dye. The maximum desired dye concentration is 4 μg/liter. The estimated mean velocity and maximum discharge are 1.4 m/sec and 150 m^3sec, respectively. How many liters of dye are required?

(12.6) A tracer solution injected at a rate of 1 liter/min into a stream that is initially free of the tracer. The tracer solution is 50% tracer and 50% water. Downstream a distance sufficient for complete mixing, the tracer concentration is found to be 2×10^{-6} liter/liter. Estimate the stream discharge.

(12.7) Derive Eq. 12.3.

References

American Public Health Association, American Water Works Association, and Water Pollution Control Federation (1980 and revisions). "Standard Methods for the Analysis of Water and Wastewater," 15th ed. American Public Health Association, Washington, DC.

American Society for Testing Materials (1981 and revisions). "Annual Book and ASTM Standards," Part 31. Water. Philadelphia.

Brakensiek, D. L., Osborn, H. B., and Rawls, W. J., coordinators (1979). "Field Manual for Research in Agricultural Hydrology," Agricultural Handbook 224. U.S. Department of Agriculture, Washington, D.C.

Fetter, C. W., Jr. (1980). "Applied Hydrogeology." Merrill, Columbus, OH.

Kilpatrick, F. A. (1970). Dosage requirements for slug injections of Rhodamine BA and WT dyes, U.S. Geological Survey Professional Paper 700-B, pp. 250–253.

U.S. Environmental Protection Agency (1979 and revisions). "Methods for Chemical Analysis of Water and Wastes," EPA 600/4-79-020, 3rd ed. U.S. Environmental Protection Agency Environmental Monitoring and Support Laboratory, Cincinnati, OH.

U.S. Geological Survey (1977 and revisions). "National Handbook of Recommended Methods for Water-data Acquisitions," proposed under the sponsorship of the office of Water Data Coordination, Geological Survey. U.S. Department of Interior, Reston, VA.

13 Hydrologic Modeling

Hydrologic modeling has become commonplace over the past 25 years. Virtually all hydrologic design is based on the results of applying a hydrologic model. The ready availability of models and computers and the "user-friendly" nature of many hydrologic models ensures continued and virtually absolute reliance on such models. Haan *et al.* (1982) present a detailed treatment of hydrologic modeling of small watersheds.

The availability of sophisticated hydrologic models has greatly improved our ability to perform complex, detailed hydrologic analyses. Many different designs can be evaluated at minimal cost once baseline data are collected. Software that not only performs hydrologic analyses but also suggests appropriate model parameters in the form of "pop up" screens on microcomputers is available today. Neat, professional-looking reports can be prepared almost automatically.

Models are used for a variety of hydrologic studies. Possibly the most common use is to evaluate the impact of some physical change within a catchment on the hydrology of that catchment. For a model to be useful in this mode, it must contain parameters that are sensitive to the catchment changes that are taking place. If an internal channel system is modified and the model has no way of reflecting this modification, the model obviously will not be able to define the hydrologic impact of the channel modifications.

Concerns about nonpoint-source pollution have led to regulatory requirements for showing how control practices will mitigate adverse water quality impacts. Models are generally required to evaluate the potential effectiveness of various control efforts such as best management practices. Once a model is operational for a particular catchment, various combinations of storage, channel modifications, and land-use changes can be evaluated at very little incremental cost. Often a particular combination of designs can be found to meet a hydrologic and economic objective that will result in considerable savings over the life of the project. These cost savings will generally greatly exceed the cost of the modeling effort.

If a planning body has a development plan for a particular area and a timetable for the plan, a hydrologic model can be used to determine the type and timing of various storm water control works required to meet an agreed on flow objective. For example, it may be possible to determine that 10 years into the development it will be necessary to install a detention basin at a particular location to provide the required level of storm water control. Financial scheduling can be done to ensure that the available funds are in hand for the project. The land might be secured prior to it becoming inflated in value due to the development.

There is a great tendency for a developer or municipality to consider each development unit as a separate entity and design conveyance and storage facilities independently for that particular unit. In Chapter 6, the advantages of considering storm water management on a regional basis were discussed. This approach requires the use of a hydrologic model. The hydrologic model makes it feasible to evaluate the impact of a structure locally and its regional impact as well.

A well-conducted model study requires detailed knowledge of the system being modeled. The model does not replace system knowledge. A model simply carries out computations. By being relieved of computations, however, the hydrologist or engineer should be more willing to do less lumping of parameters and to give more attention to variability that exists within a catchment. Small, nonhomogeneous areas can be included as they are rather than lumping them with larger units. Models also allow and encourage the use of "what if" scenarios and encourage the use of innovation.

To limit the scope of this chapter somewhat, it is necessary to define what is meant by a hydrologic model. The term "model" brings to mind different things to different individuals. Websters dictionary provides the following as definitions of model: "a generalized, hypothetical description, often based on an analogy, used in analyzing or explaining something." If we substitute for the word "something" the word "hydrology," we have a reasonably good definition of a hydrologic model.

Hydrologic models have been classified in many ways. Some of the terms that have been used in model classification are deterministic, parametric, statistical, stochastic, physically based, empirical, blackbox, lumped, linear, nonlinear, distributed, theoretical, predictive, operational, research, design, similarity, iconic, analog, numerical, regression, event, continuous simulation, and conceptual.

This chapter is limited to a discussion of mathematical models. Mathematical models range from single prediction equations to complex computer simulation algorithms. The mathematical basis for a model may be theoretical or empirical. A completely theoretical model would contain only relationships derived entirely from basic physical laws. Such laws are the conservation of mass, conservation of energy, laws of thermodynamics, etc. Empirical relationships are based on observations and/or experimentation. Manning's equation for uniform open channel flow is an empirical equation. Often empirical equations become so well accepted and entrenched in usage that they are viewed as physical laws. The application of Darcy's equation for flow through a porous media is an example of this in hydrology.

There exists no completely theoretical operational model in hydrology. All hydrologic models contain empirical relationships. Thus we generally liberalize the definition of theoretical models to be models that include both a set of general or theoretical principles and a set of statements of empirical circumstances. A strictly empirical model is one that is based on no basic physical laws but contains only a representation of data (empirical results). Admittedly this is not a very clear distinction, but such a distinction is only of academic interest anyway. Suffice it to say that no matter how sophisticated and detailed, *all* hydrologic models rely on empirical results to some extent.

For our purposes a hydrologic model will be defined as:

> a collection of physical laws and empirical observations written in mathematical terms and combined in such a way as to produce hydrologic estimates (outputs) based on a set of known and/or assumed conditions (inputs).

There are many ways of "collecting" physical laws and empirical observations and of "combining" them to produce a model. Certain of these ways result in a formulation in which the use of a computer is desirable. Such models are called computer models. Computer models are the focus of the remainder of the discussion in this chapter.

Regardless of how models are classified, they can generally be represented as

$$\mathbf{O} = \mathbf{f}(\mathbf{I}, \mathbf{P}, t) + \mathbf{e}, \tag{13.1}$$

where $\mathbf{O}$ is an $n \times k$ matrix of hydrologic responses to be modeled, $\mathbf{f}$ is a collection of l functional relationships, $\mathbf{I}$ is an $n \times m$ matrix of inputs, $\mathbf{P}$ is a vector of p parameters, t is time, $\mathbf{e}$ is an $n \times k$ matrix of errors, n is the number of data points, k is the number of responses, and m is the number of inputs.

Responses in $\mathbf{O}$ may range from a single number, such as a peak flow or a runoff volume, to a continuous record of flow, soil water content, evapotranspiration, and other quantities.

Model classification refers to the nature of $\mathbf{f}$. The distinction between $\mathbf{I}$ and $\mathbf{P}$ is not always clear and not of extreme importance to the discussion here. Generally $\mathbf{I}$ represents inputs, some of which are time varying, such as rainfall, temperature, and land use, while $\mathbf{P}$ represents coefficients particular to a watershed that must be estimated from tables, charts, correlations, observed data, or some other means.

The error term, $\mathbf{e}$, represents the difference between what actually occurs, $\mathbf{O}$, and what the model predicts, $\mathbf{O}_p$.

$$\mathbf{O}_p = \mathbf{f}(\mathbf{I}, \mathbf{P}, t) \tag{13.2}$$

$$\mathbf{e} = \mathbf{O} - \mathbf{O}_p. \tag{13.3}$$

A parametric model is a model having parameters that must be estimated in some fashion. The parameters may be estimated based on observed data (calibration), tables and/or charts (Manning's n or curve numbers), correlation-type relations (regional analysis), analysis of site-specific information (water-holding capacities of soils, etc.), experience, an edict from an agency, or by some other means.

An empirical model is a model containing any empirical relationship. Empirical here means data based or based on observation. Even Darcy's law and Manning's equation are empirical equations.

A lumped model describes processes on a scale larger than a point. A completely distributed model would be one in which all processes are described at a point and then integrated over three-dimensional space and time taking into account variations in space and time to produce the total watershed response.

Based on these somewhat restrictive definitions, *all* hydrologic models are to some degree parametric, empirical, and lumped.

A BRIEF LOOK BACK

Many of the current developments in hydrologic modeling depend more heavily on computational improvements than on improved representations of hydrologic processes when compared to models of 25 years ago.

The 1960s might be thought of as the golden years of hydrologic modeling. Digital computers were becoming widely available and hydrologic researchers began taking advantage of their power. The Stanford Project in Hydrologic Simulation was initiated in 1959 and, under the general leadership of R. K. Linsley, developed approaches to hydrologic modeling that continue to play a major role in the field to this day. An important development from this program was the initiative it provided in fostering research in the development of hydrologic models in several locations throughout the U.S.

The most famous of the models developed at Stanford was the Stanford Watershed Model, SWM (Crawford and Linsley, 1962, 1966). In their 1966 report, Crawford and Linsley stated:

> The objective of the research is to develop a general system of quantitative analysis for hydrologic regimes. The most effective way for doing this has been to establish continuous mathematical relationships between elements of the hydrologic cycle. The operation of these mathematical relationships is observed and improved by using digital computers to carry the calculations forward in time. ... As mathematical relationships are developed, every attempt is made to realistically reproduce physical processes in the model. Experimental results and analytical studies are used wherever possible to assist in defining the necessary relationships.

This statement of their research objective has been paraphrased and repeated by others in many locations throughout the world.

The SWM defined watershed flow and storage processes in mathematical terms and combined them so that a continuous record of estimated streamflow was generated in response to hourly rainfall and daily evapotranspiration. In developing the governing relationships, many model parameters were defined and a conceptual basis assigned to them. Estimation of the many parameters generally relied on a streamflow record and manually adjusting the parameters until a satisfactory estimate of streamflow was obtained.

In 1966, Huggins and Monke (1966) presented the first developments of what has become known as ANSWERS (Beasley and Huggins, 1981). This model is representative of a class of models that breaks a watershed into a grid system, simulates the hydrology of each grid and the interaction of the grids with their neighbors, and integrates these results over the watershed to produce the total watershed response. Generally with models of this type, the goal is to use physically based parameters, since a set of parameters must be defined for each grid within the watershed.

In addition to general purpose models applicable to a wide range of watersheds, hydrologic models have been developed to address specific questions. The model developed at Iowa State University (Haan and Johnson, 1968a, b; DeBoer and Johnson, 1971; Campbell *et al.*, 1974) to evaluate the impact of subsurface drainage on flood flows in north central Iowa represents such a model.

Since the 1960s there have been thousands of papers, articles, and books written dealing with hydrologic modeling. In the 1960s, hydrologic modelers spent a great deal of time justifying why hydrologic models should be developed and their applicability to certain problems. Today the situation is reversed. Everyone wants to use models for every conceivable hydrologic problem, and applications are frequently made well outside the verified domain of the model being used. Many modern hydrologic investigations would not be possible without computer models.

MODEL SELECTION

The availability of a large number of models, strong advocates of the various models, in some cases vocal detractors from certain modeling approaches, optimistic claims by model developers, and conflicting modeling objectives all contribute to a great deal of confusion when it comes to model selection for a particular application. There are at least three possible approaches to modeling:

(1) Use of an existing model.
(2) Modification of an existing model.
(3) Development of a new model.

These options should be considered in the order they are listed. A great deal of time and money is involved in model development and testing. Developing a new model can only be justified for large and important projects. If an existing model is modified, reports based on the modified model should clearly indicate the modification and should not imply that the original, unmodified model was used.

The choice of the best model depends on:

(1) The problem to be solved.
(2) Computer facilities available.
(3) Likelihood of other applications for the model.
(4) Available documentation and other forms of assistance.
(5) Characteristics of the model.
(6) Users modeling experience.

The Problem to Be Solved

The model selected should be matched to the problem to be solved. The problem should not be redefined to fit a model that is available. The problem is fixed; models are variable. In evaluating the problem to be solved, the required flow characteristics must be defined. Some models estimate only peak flow or runoff volumes. Other models generate runoff hydrographs from single events. Continuous simulation of streamflow over long periods of time is also possible with certain models. Water quality parameters may be required. One must be certain that the selected model can provide an estimate of the flow characteristics that are of importance. Finally, the value of the decision to be made should be considered. There is little justification in spending a great deal of money to arrive at a flow to size a culvert if simply using a larger size culvert is cheaper (than the modeling cost) and is hydrologically acceptable.

Computing Facilities Available

This consideration is of less concern than in the past but still must be considered. Most hydrologic modeling on small catchments is currently done using microcomputers. The model selected must be compatible with the computer system in terms of input and output of data and results, operating systems, internal memory, disk storage, and graphics. For some it may be best to hire computer services if a limited need for a particular type of system is at hand. Required peripheral devices, such as modems, pointers, plotters, graphics, and monitors, must be considered.

Likelihood of Other Applications for Model

If the likelihood of frequent use of a model is apparent, more resources can be used to obtain the required hardware and software. One temptation that must be resisted is that of redefining problems to match an available modeling system so as to recover more of the investment in the system.

Available Documentation and Other Forms of Assistance

Well-documented models are a must. The documentation should not only explain how to run the model, but should explain or reference the algorithms used in the model. The availability of help screens and training sessions also enhances a model and helps a user initially get started with the model. Guidance in parameter estimation and error checking should be provided.

Characteristics of the Model

Obviously, a model must be able to predict the desired quantities with the accuracy required for a particular study. Model evaluation is discussed later in this chapter. The simplest model that will meet the requirements should be selected. One should not confuse model complexity with model accuracy. The parameters of the model selected should be sensitive to the purpose of the modeling effort.

Users Modeling Experience

As one gains experience in the use of hydrologic models, parameter selection, model evaluation, error detection, and other modeling attributes are accomplished more quickly. An inexperienced user can be overwhelmed by the number of parameters and complexity of some of the more detailed models. Because of parameter correlations and the interrelationships

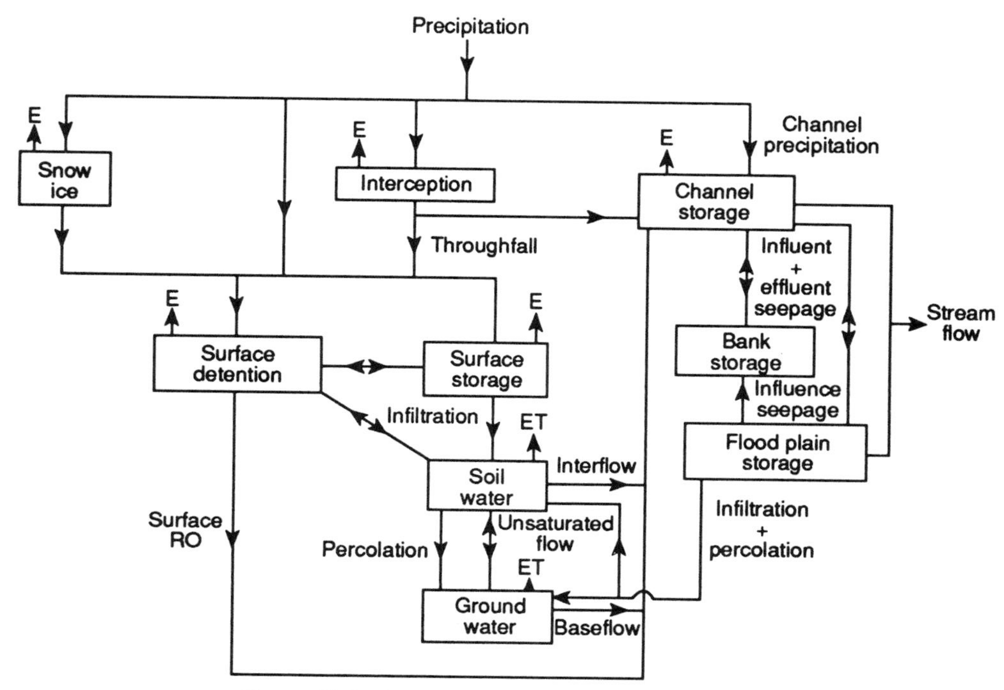

Figure 13.1 A representation of the hydrologic cycle.

programmed into most models, it generally takes considerable time and effort to become familiar with some of the more complete and complex hydrologic models.

BASIC MODELING APPROACHES

One way of differentiating hydrologic models is based on the time scale of importance. This leads to two approaches—event simulation and continuous simulation. Event simulation refers to modeling the hydrologic response to a single, isolated storm. Continuous simulation refers to modeling the ongoing hydrology of a catchment over long periods of time, such as years. Event-based models are commonly used in designing storm water control facilities for small catchments. Continuous simulation models are used where long-term flow volumes and storage considerations are important.

Hydrologic models are generally written in terms of flow and storage processes. Figure 13.1 represents the hydrologic cycle on a catchment in these terms with the boxes representing storage processes and the arrows representing the flow processes. Note that in this diagram, only the inflow and outflow of water to the system is being considered. Other factors such as solar radiation and air masses play an important role in that they import energy to drive evaporative processes and play a major role in governing vegetative growth, which also influences evapotranspiration. This discussion is limited to the water flow and storage processes.

In building a hydrologic model around a diagram, such as Fig. 13.1, the factors that must be defined include the capacity limits in the various storages, the rate of release of water from the storage, and the rate of movement of water in the various flow phases. It is desirable to define the processes in as physically based a manner as possible so that model parameters can be conceptually, if not actually, related to physical, catchment parameters. For example, soil water storage might be related to the available water-holding capacity of the soil in the root zone and soil water flow processes might be related to the water-transmitting properties of the soil.

Figure 13.1 could serve as a basis for developing a continuous simulation hydrologic model. Some models omit some of the components shown, some include additional components, and some expand on the components shown.

For models that attempt to describe the same process, event-based models generally require less detail than continuous simulation models. Often subsurface flow processes and abstractions with the exception of infiltration, evapotranspiration, and baseflow are neglected in event-based models based on time scale and order of magnitude arguments. Figure 13.2 might serve

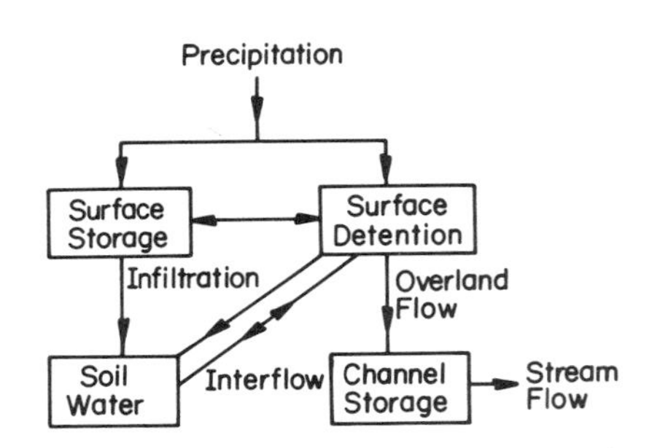

Figure 13.2 A representation of an event-based model.

as a basis for an event model to predict the hydrologic response of a catchment to a design storm. In Chapter 3, the steps in estimating a runoff hydrograph were given as (a) defining the design storm, (b) deducting losses (abstractions) from the design storm, (c) routing the rainfall excess to the channel system, and (d) routing the channel flow to the catchment outlet. These are the processes that are depicted in Fig. 13.2. The rainfall excess is the overland flow arrow and is composed of water released from surface storage. This is the amount in excess of infiltration, surface storage capacity, and interception.

In Chapter 3, it was shown that there are many ways of representing these processes. Building an event-based model consists of combining algorithms representing flow and storage processes in a logical manner and providing a means of estimating the various parameters that are required as a result of the representations.

From Figs. 13.1 and 13.2, it can be seen that a continuous simulation model may be considerably more complex than an event model. The complexity is increased as the time step is decreased. If a time step of 1 day is used in a small catchment, then detailed overland flow routing and infiltration calculations cannot be done. In this case, the storage process becomes dominant in the model. Generally for an event-based model, a short time increment is used. Typically, for small catchments the time increment will be in minutes. If a time step of minutes is used in the continuous case, the event model becomes basically a submodel to the overall model with other components of the model keeping track of antecedent storage. For an event model, antecedent conditions in terms of soil water storage are either specified or become an implicit assumption of the model.

Another model classification concerns the manner in which physical processes are represented. In some models, processes and parameters are defined as average representations over the entire catchment, while in other models the processes are defined at points or within cells and then integrated over the watershed to define the total catchment response. The former approach is termed lumped and the latter distributed. In actuality, distributed models represent averages over some finite area, often a hectare or so, and thus are lumped to some extent. Similarly, lumped models may be applied to very small catchments with the results combined and routed. In this way, lumped models may become distributed models.

It is apparent that regardless of whether a model is event based or continuous, lumped or distributed, it will contain parameters that must be estimated to define the flow and storage processes.

PARAMETER ESTIMATION

As the name implies, parameter estimation is the process by which the parameters of a hydrologic model are estimated for a particular application. Rational parameter estimation must be tied to some criterion if a unique parameter set is to be found. Some criteria that might be used include (1) personal judgment of goodness of fit of simulated hydrographs to observed hydrographs, (2) direct measurement of physical properties in the field or in the lab, (3) indirect measurement of physical properties through their relationship with other hydrologic processes and watershed characteristics, (4) optimization of some objective function either computationally or by trial and error, (5) satisfaction of agency requirements, and (6) compliance with published tables and charts.

Some of the things that make parameter estimation for hydrologic models difficult are (1) specification of appropriate criteria for parameter selection, (2) correlation among parameters, (3) amount of computation involved in many models, (4) restrictions on appropriate values for some of the parameters, (5) nonuniqueness of parameter sets for certain objective functions, (6) thresholding in some of the model relationships, and (7) errors in data. Parameter estimation is made more difficult by increasing the number of parameters to be estimated, the lack of correspondence between individual parameters and measurable physical properties of the catchment, multiple objectives, limited data, and pronounced seasonality in hydrologic regimes.

Problems in parameter estimation have been recognized for some time. Dawdy and O'Donnell (1965) investigated the possibility of obtaining an efficient automatic procedure for finding numerical values of the various parameters of an overall watershed model. Beard (1967) and DeCoursey and Snyder (1969) addressed computer procedures for finding optimal values of parameters for a hydrologic model. Jackson and Aron (1971) reviewed parameter estimation techniques in hydrology. Most of the earlier papers approached parameter estimation from a mathematical rather than statistical point of view. An objective function, generally a minimization of a sum of squares, was defined, and search techniques were employed to find the parameter set that optimized the objective function.

Sorooshian (1983) reviewed parameter estimation techniques for hydrologic models. In his review the shift from a deterministic, mathematical interpretation of parameters toward a stochastic, statistical interpretation can be noted. Increasingly, causal models are being viewed as "somewhat structured empirical constructs whose elements are regression coefficients with physical sounding names" (Klemes, 1982). Adoption of this viewpoint leads to parameter estimation in a statistical framework and focuses attention on treating a parameter as a random variable (rv) with a probability density function (pdf). Troutman (1985) made an extensive investigation of parameter estimation using this approach.

To this point, a vast majority of the research on parameter estimation in hydrology has been for the case where a single objective is to be met. For example, this objective might relate to prediction of peak flows, storm runoff volumes, *or* daily streamflow. Diskin and Simon (1977) investigated 12 such univariate objective functions. Attempts to find parameter sets that meet multiple objectives such as peak flows, runoff volumes, *and* daily streamflow in some optimal sense have been scarce.

The use of multiple objective criteria for parameter estimation permits more of the information contained in a data set to be used and distributes the importance of the parameter estimates among more components of the model. For example, if a continuous flow simulation model is optimized based on peak flows, parameters related to evapotranspiration (ET) may be poorly estimated. If the estimation criteria included both peak flow and ET, it is likely that the precision of the ET parameters would be greatly improved without an adverse impact on the peak flow parameters.

Edwards (1988) reported an example of improved parameter estimates using multiobjective optimization criteria when applied to the Soil Conservation Service (SCS) runoff model. When model parameters were estimated based solely on a minimization of prediction error sum of squares for peak flows, the resulting parameters would do a good job of predicting peak flows but gave very poor estimates of runoff volume. Changing the estimation criteria to include a measure of error sum of squares on peaks and volumes and their interactions had no appreciable impact on predictions of peaks but greatly improved runoff volume estimates. If interest lies only in runoff peaks, then the choice of the simpler univariate objective function may be appropriate; however, potential users of the model or the estimated parameters may rightfully be skeptical of the univariate optimization and/or the model itself when the poorly estimated SCS curve number parameter (which governs runoff volume) is compared to more conventional estimates of this parameter as commonly found in tables. Inclusion of a measure of prediction errors on volumes in the objective function overcame this problem and resulted in curve number estimates that were in agreement with conventional estimates.

Runoff typically accounts for only 10 to 35% of annual rainfall. Estimation of all model parameters based solely on runoff ignores 65 to 90% of the processes accounting for water loss from a catchment. The assumption that if runoff can be predicted well then all model parameters must have been adequately determined has no clear justification. Including some measure of performance in the optimization that reflects some of the flow and storage processes occurring within a watershed in addition to runoff should improve the stability (reduce the variance) and accuracy (reduce the absolute error) of the estimated parameters.

Traditionally, parameter estimation criteria have been tied to some measure of how well predicted streamflow agreed with observed streamflow. Especially for continuous flow simulation, these criteria are difficult to apply. An alternative measure of performance and thus a basis for parameter estimation is how well the model performs in a design situation. For example, one might select the parameters of a daily flow model so that the estimated capacity of a reservoir to meet some demand would be as close as possible to the capacity estimated on the basis of observed streamflow data using the same capacity estimation algorithm.

Parameter Estimation Criteria

Parameter estimation techniques can be divided into two major categories—personal and objective. Personal parameter estimation relies solely on the judgment of the modeler in arriving at parameter values. Objective parameter optimization generally deals with some function of the error term, **e**, of Eq. (13.1) and thus requires some observed data on the quantity being modeled. Generally, a probability density function for **e**, as well as other properties such as independence, constant variance, and zero mean, is assumed. Equations (13.1)–(13.3) show that **e** is a function of the parameters, **P**.

Personal Parameter Estimation

Possibly the most commonly used parameter estimation technique relies on the personal judgment of the modeler. Parameters are initially assigned on the basis of judgment, published guides, and physical properties and characteristics of the catchment. These initial parameters are then adjusted again based on judgment as to the appropriateness of the model results. If observed data are available, the parameters may be adjusted several times in an effort to obtain a "satisfactory" fit to the observations. In the absence of observed data, parameter adjustment depends entirely on judgment. This method of parameter estimation has the advantage of allowing the user to weight mentally the importance of various flow components such as peak flows, low flows, and runoff volumes, whereas most objective estimation techniques can focus only on a single objective. The method has the disadvantage of sole reliance on the judgment of the modeler and generally a poorly defined objective function in the mind of the modeler. Different hydrologists would arrive at different parameter estimates on the basis of the same model using the same data.

Least Squares

In the least-squares procedure, a sum, S, is defined as the sum of squares of the $\mathbf{e}_i$:

$$S = \sum_{i=1}^{n} e_i^2. \tag{13.4}$$

Values of the p parameters that minimize S are sought. For hydrologic models, numerical search techniques are generally employed in p-dimensional space.

Minimization of Absolute Errors

The minimization of absolute errors requires a sum, A, be computed as

$$A = \sum_{i=1}^{n} |e_i|.$$

Values of the parameters that minimize A in p-dimensional parameter space are sought, generally through numerical search techniques.

Method of Moments

The method of moments requires equating the first p sample moments of $\mathbf{e}_p$ with the first p population moments of the pdf of $\mathbf{e}$. In general the population moments will be a function of the p model parameters. Thus p equations in p unknowns, which must be solved for values of the p unknown parameters, result.

Maximum Likelihood

The likelihood function, L, of $\mathbf{e}$ is written as

$$L = \prod_{i=1}^{n} p(e_i|\mathbf{P}), \tag{13.5}$$

where $p(e_i|\mathbf{P})$ is the pdf of $\mathbf{e}$ given $\mathbf{P}$. Values of $\mathbf{P}$ that maximize L are sought. Again for hydrologic models, numerical search techniques are generally required.

Arbitrary Objective Functions

Any objective function or criterion function, C, can be used to find $\mathbf{P}$. In general,

$$C = G[f_1(\mathbf{O}), f_2(\mathbf{O}_\mathrm{p})] = G[f_1(\mathbf{O}), f_3(\mathbf{P})], \tag{13.6}$$

where G is the arbitrary function, f_1 is a function of the observed values of $\mathbf{O}$, and f_2 is a function of the estimated values of $\mathbf{O}_\mathrm{p}$. Note that f_2 may be transformed to a function of the parameters, f_3, through Eq. (13.2). Using numerical search techniques, $\mathbf{P}$ that optimizes C is sought. An example of this approach might be setting $f_1(O) = \ln(O)$, $f_2 = \ln(O_\mathrm{p})$, and $G = \Sigma[f_1(O) - f_2(O_\mathrm{p})]^2$ and finding P that minimizes G. This would be a minimization of the sum of squares of the differences in the logarithms of the observed and predicted outputs.

For all of the above four estimation procedures, $\mathbf{e}$ is a $n \times 1$ vector, point estimates for the parameters are obtained, and a single objective is used.

Bayesian Estimation

Bayesian estimation is fundamentally different from the above procedures in that it evolves from probabilistic considerations rather than some arbitrarily specified objective function. Bayesian estimation is concerned with the probability distribution of the parameters rather than point estimates. Point estimates, however, may be derived as the mode of the resulting distribution.

Some references to Bayesian estimation are Box and Tiao (1973), Kuczera (1983), Vicens *et al.* (1975), Edwards (1988), and Edwards and Haan (1988, 1989). Bayesian estimation has the advantage that multiple objectives may be incorporated into the analysis. Box and Tiao (1973) show that the point estimates for $\mathbf{P}$ may be found by minimizing the determinant of $\mathbf{S(P)}$, $|\mathbf{S(P)}|$, with respect to $\mathbf{P}$, where

$$\mathbf{S(P)} = \mathbf{e}'\mathbf{e} \tag{13.7}$$

for the case where $\mathbf{e}$ is a $n \times k$ matrix of errors as in Eq. (13.1). $\mathbf{S(P)}$ is $k \times k$ matrix of sums of squares and cross products of errors. Since $|\mathbf{S(P)}|$ is simply a number (the determinant of $\mathbf{e}'\mathbf{e}$), numerical search procedures can be used to find the $\mathbf{P}$ that minimizes $|\mathbf{S(P)}|$. Note that if $\mathbf{e}$ is a $n \times 1$ vector, Eqs. (13.4) and (13.7) are identical. In this case, the Bayesian approach provides a statistical justification for the least-squares procedure.

Other Considerations

James and Burges (1982) present an excellent discussion of parameter estimation and estimation criteria. If a model structure is such that only a subset of parameters impact one particular aspect of model output, those parameters may be estimated on the basis of an optimization criteria related to that particular aspect of model output. For example a p-parameter model may be structured so that a subset p_1 of the p parameters governs runoff volumes and a second nonoverlapping subset p_2 governs peak flows. The parameters in p_1 could then be estimated on the basis of a criterion related to volumes while the parameters in p_2 could be estimated on the basis of a criterion related to peaks. If in this case the parameters in p_1 and p_2 are all estimated on the basis of peaks, the parameters in p_1 will be poorly determined and likely will have a large variance. While strict division of influence between parameter sets and outputs generally does not occur, it is common for some parameters to have relatively little influence on optimization criteria and thus be poorly determined. James and Burges (1982) define sensitivity coefficients that can help identify this possibility. If the model is then used in a situation where the aspect of the model that is highly dependent on the poorly defined parameters is important, a questionable design may result. From the preceding situation, designing a storage reservoir with a model whose parameters were optimized on peaks would be a situation where a potentially poor design could result. Use of Eq. (13.7) partially overcomes this problem since prediction errors on both peaks and volumes can be incorporated into parameter estimation.

As an alternative to Eq. (13.7) to include multiple objectives, a weighted criterion function could be used. Such a function might be

$$C = w_1 \sum e_1^2 + w_2 \sum e_2^2, \tag{13.8}$$

where the subscripts refer to different objectives (i.e., peaks and volumes), w is a weight, and e is the error. The selection of the weights is arbitrary and may be done to reflect the relative importance of the two objectives. Equation (13.8) differs from Eq. (13.7) in that interaction between the two objectives is not included. For the case where $k = 2$, Eq. (13.7) becomes

$$C = \sum e_1^2 \sum e_2^2 - \left(\sum e_1 e_2 \right)^2. \tag{13.9}$$

Equation (13.8) can be generalized to any number of objectives. It is not necessary that all e_i used in an equation like (13.7) or (13.8) be related to flow. For example, e_2 might refer to some measure of soil water content if observed data were available and the model provided estimates of soil water.

EVENT MODELING

To illustrate event models and modeling, the construction of an event model based on concepts discussed in Chapters 3, 4, and 6 is illustrated. As the model is developed, it will become apparent that the bookkeeping of modeling is often as difficult as the modeling itself. In the simple model to be developed, the bookkeeping or sequence management will be left to the modeler and not incorporated into the model.

Recall that event models are used to describe the response of a catchment to a single hydrologic event or rainstorm. Thus components of the model must deal with the characteristics of the input rainstorm, abstractions from the rainfall, routing of overland and shallow subsurface flow to the channel system, and finally the routing of the channel flow to the catchment outlet. Detailed procedures for carrying out these steps have been presented in earlier chapters. Thus model development becomes one of choosing algorithms for representing the various model components and combining them in a logical and functional manner to produce a runoff hydrograph.

The model to be used as an illustration will rely on the SCS type II rainfall distribution to define the storm input, the curve number approach to determine abstractions from rainfall, the SCS curvilinear unit hydrograph for overland and shallow subsurface flow routing, and the Muskingum routing procedure for channel routing.

The catchment to be modeled is shown in Fig. 13.3. Based on land use, soils, and stream configuration, the overall catchment is divided into four subcatchments. Physical properties of these catchments are shown in Table 13.1. Table 13.2 shows the identification system used in differentiating the various hydrographs. Catchment D is to undergo urbanization. It is desired to evaluate the feasibility of using a detention basin just below the total catchment to mitigate the impact of this urbanization on downstream flooding. Figure 13.4 shows the basic model unit (BMU) that constitutes the main routine of the model to be developed. The BMU consists of the type II distribution, curve number procedure, and unit hydrograph procedure. The BMU must be applied to each of the catchments. Figure 13.5 shows how the BMUs must be combined to produce the total catchment hydrograph. The schematic of Fig. 13.5 represents the portion of the model that manages output from the BMU. This management aspect is highly dependent on the particular catchment being

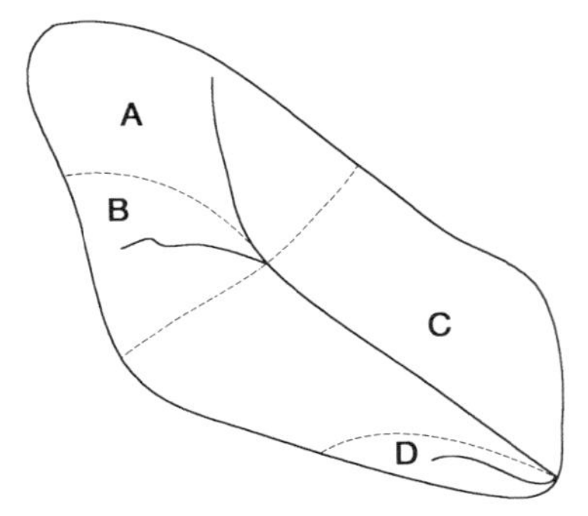

Figure 13.3 Watershed schematic.

Table 13.2 Catchment Identification System

Code	Catchment or hydrograph description
A	A
B	B
C	C
D	D before development
DP	D after development
E	C and D lumped
EP	C and DP lumped
AB	A added to B
ABR	AB routed to outlet of C
ABRC	ABR added to C
ABRE	ABR added to E
ABREP	ABR added to EP
ABRCD	ABRC added to D
ABRCDP	ABRC added to DP

modeled. A good hydrologic model must have an easily used management scheme.

The event to be modeled is a 6-in. rain in 24 hr. The time distribution for the rain will be taken as the type II distribution of the SCS. The modeling approach is to determine the runoff hydrograph from catchments A and B, add these hydrographs together, route the combined hydrographs through stream segment C, and add the hydrographs from catchment C and catchment D to produce the total runoff hydrograph. Figure 13.6 shows the component hydrographs and the total resulting hydrograph. The same procedure is followed after the development of catchment D. The only change is in the curve number representing catchment D. Figure 13.7 shows that the impact of the development on the hydrograph from catchment D is an increase in the runoff volume and peak flow rate. Figure 13.8 indicates, however, that the impact of the increase in flow from catchment D is of no significance as far as total catchment flow is concerned. A detention basin would not be required. The reasons for this are (1) the relatively small part of the total catchment that is represented by catchment D and (2) catchment D in the undeveloped state has a relatively high curve number and development of the catchment does not have a large impact on this curve number.

In a modeling effort such as this, the manner in which catchments are defined can have a pronounced influence on the conclusions reached. In this particular example, it might be logical to define catchment E as the combination of catchments C and D (C and D lumped). A predevelopment hydrograph could then be developed by applying the BMU to catchment E and adding the result to the routed sum of the flows from catchments A and B to produce the total runoff hydrograph. The total runoff hydrograph after development could be determined as before by considering catchments C and D separately. Figure 13.9 shows the result of this approach. If Fig. 13.9 was the only information available, one might conclude that development of

Table 13.1 Catchment Properties

	Catchment				
Property	A	B	C	D	DP
Soil HSG	B	B	C	C	C
CN	72	65	79	79	85
Area (acres)	60	75	80	20	20
Land slope (%)	5	6	7	10	10
Stream slope (%)	1	1	0.5	2	2
Max flow length (ft)	3200	3500	6000	1000	1000
Stream seg. length			3000		

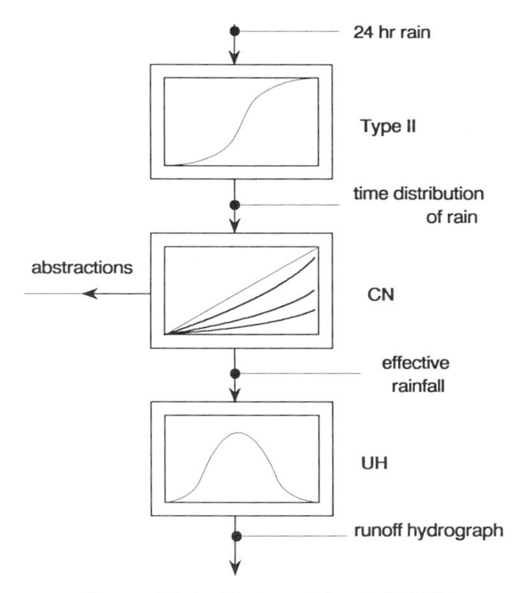

Figure 13.4 Basic model unit (BMU).

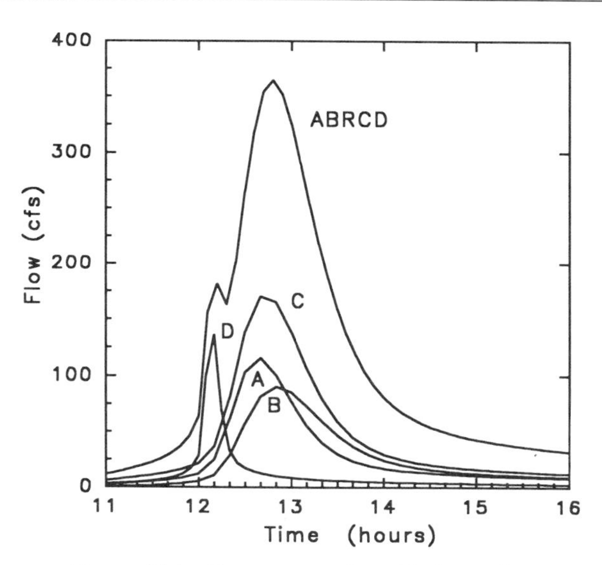

Figure 13.6 Component and total hydrographs.

catchment D produced a quick hydrologic response so that runoff from D was nearly over before the flow from the bulk of the catchment reached the catchment outlet. Further, one might conclude that development of catchment D actually reduced the total runoff peak flow. These conclusions are, of course, contrary to the conclusions reached on the basis of Fig. 13.8. The reason for this disparity is that in Fig. 13.9 in the undeveloped case, C and D were lumped, while in the developed case, C and D were treated separately. Figure 13.10 shows this lump/no lump impact when D is undeveloped. Clearly both modeling approaches cannot be correct.

A third alternative would be to lump C and D both prior to and after development and generate runoff hydrographs. Figure 13.11 shows the result of this approach from which one would conclude development on catchment D has only minor (negligible) impact on the total runoff hydrograph.

One purpose of this discussion has been to illustrate the importance of catchment definitions and the impact of lumping. If the effect of a small change on a large watershed is to be investigated, the lumping of the area undergoing change with the rest of the catchment will often mask any impacts. Generally, catchments with widely varying characteristics should not be lumped. Catchments with different land uses, soils, surface topography, stream characteristics, etc., should

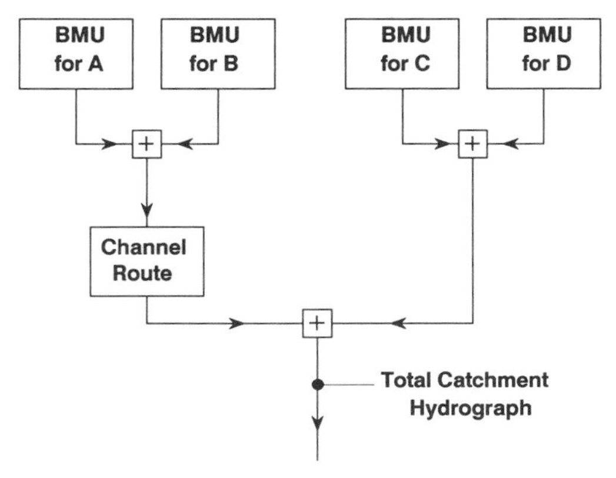

Figure 13.5 Management module.

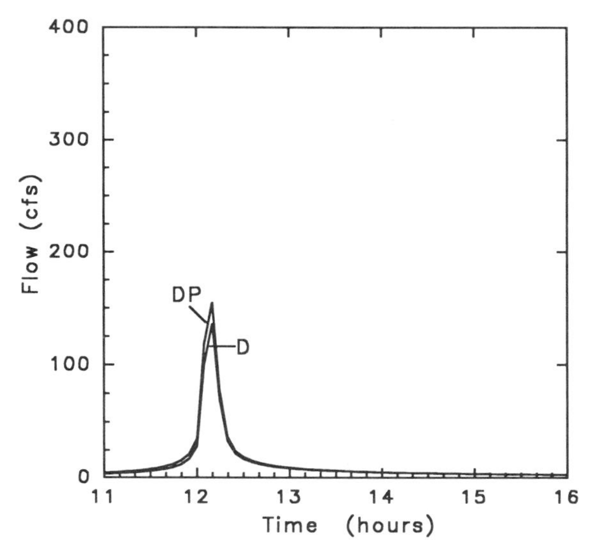

Figure 13.7 Impact of development on catchment D.

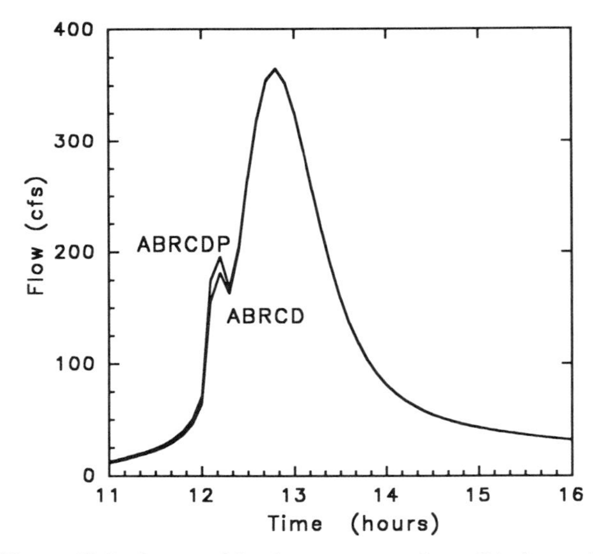

Figure 13.8 Impact of development on total runoff hydrograph.

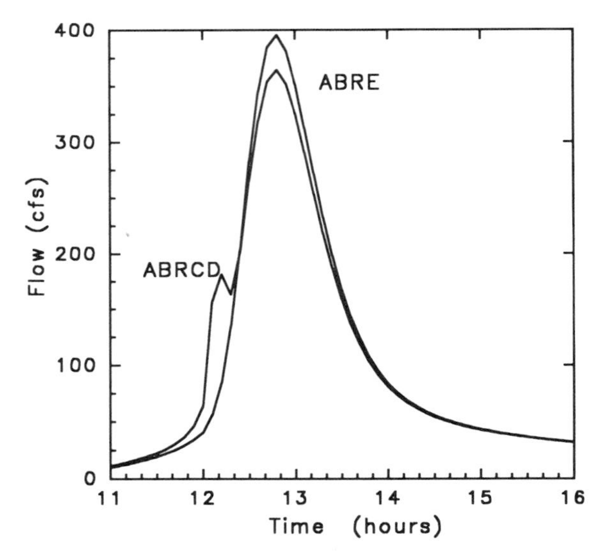

Figure 13.10 Effect of using different catchment definitions on runoff hydrographs.

be identified and treated separately with the flows combined through a routing process. Certainly different degrees of lumping should not be used to delineate some hydrologic impact. The approach of Fig. 13.9 is clearly invalid. Arguments may develop over the preference for the approach of Fig. 13.8 or Fig. 13.11. Possibly Fig. 13.8 is more valid as it tends to more sharply identify changes due to developments on catchment D. This illustration points out that consistently applied models may be used to evaluate relative impacts, but overreliance on the absolute numerical predictions should be avoided, especially for uncalibrated

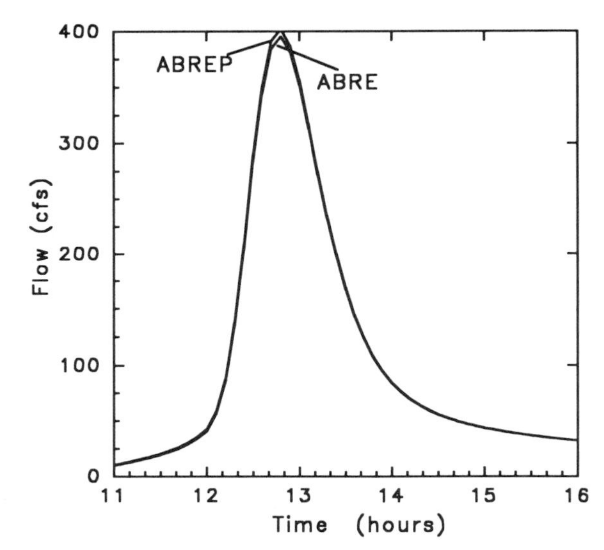

Figure 13.11 Impact of development C and D lumped.

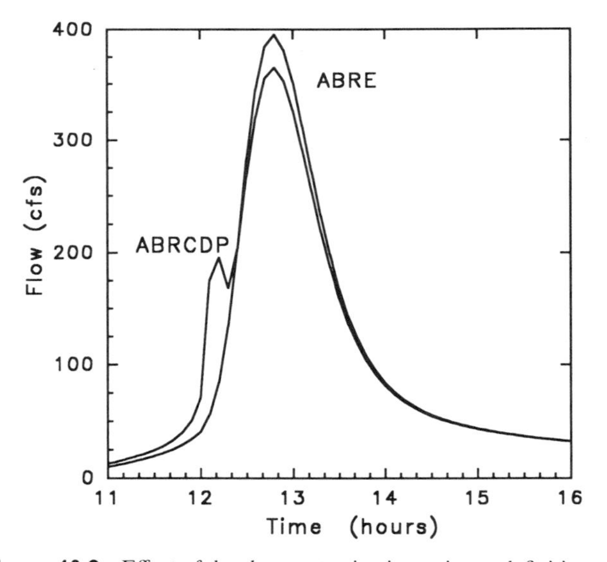

Figure 13.9 Effect of development using inconsistant definitions of catchments.

model applications (no observed data used to temper parameter estimates). Hydrologic models produce hydrologic estimates. For comparisons between estimates to be valid, a consistent modeling approach must be used.

A Note on Parameter Estimation

The BMU for this event model requires estimation of watershed physical characteristics and the runoff curve number. Assuming the physical characteristics can be estimated from maps and other data sources, the curve number, CN, becomes the only parameter that must be estimated.

In the absence of any streamflow data, CN will necessarily be estimated from tables such as those in Chapter 3 and from the judgment of the modeler. Obviously a good estimate for this parameter is important since it will be called on to reflect changes in hydrology due to development of catchment D. Since there are no streamflow records to use in validating either the model or the estimates for the various CNs, the modeler must rely on previous validation of the model for other catchments and assume the model will be equally valid for this particular catchment. Changes in land use, such as development, must be reflected in the estimates for CN. Without validation data, the modeler can assume that the relative changes in flow from the pre- to postdevelopment state are accurate if the model is a valid one and is properly used, but cannot assume that the quantitative estimates are absolutely correct.

Proper estimation of model parameters is absolutely essential to valid model studies. A model that has parameters that can be estimated from available sources of data, including streamflow if streamflow data are available, must be selected.

If streamflow data are available at the outlet of the total catchment, these data can be a valuable source of information about the correct values for the parameters. One could go through the data and pick out isolated storm events. From each of these storm events, a CN could be determined on the basis of Eqs. (3.21) and (3.22). This CN could then serve as a guide for estimating the individual CNs for the four individual catchments. For models where the parameter(s) cannot be explicitly determined, parameter value (s) that minimizes S of Eq. (13.4) can be sought. The e_i would represent the difference between observed and predicted runoff volumes. Again if the total catchment is considered as one unit, a single value for the CN can be estimated rather easily by a straightforward minimization of S.

If it is desired to estimate a CN for each of the four catchments that make up the total catchment, then the estimation process becomes more involved. It still remains to find the set of CNs that minimize S; however, the problems discussed in the parameter estimation section of this chapter begin to creep in. These are problems of nonuniqueness, errors in data, parameter correlations, etc. In the example given here, a minimization routine that would allow minimizing S over four dimensions (one CN estimate for each component catchment) would be required. For this particular model and parameter set, one may well find that several parameter sets produce nearly identical values of S, making the determination of the single "best" set difficult. This relates to the nonuniqueness problem.

There are many examples of currently available event models. Some of the most widely used in the U.S. are the SCS TR-20 model of the Soil Conservation Service (U.S. Department of Agriculture, 1973); the University of Kentucky's SEDIMOT II (Wilson *et al.*, 1983); the USGS model (Carrigan, 1973); the Storm Water Management Model, SWMM, of the Environmental Protection Agency (Huber *et al.*, 1981); and the U.S. Army Corps of Engineers HEC-1 model (U.S. Army Corps of Engineers, 1985). These models are all lumped parameter models. Two examples of distributed parameter, event models are ANSWERS (Beasley and Huggins, 1981) and FESHM (Ross *et al.*, 1979). All of these event models combine components discussed in Chapters 3, 4, and 6. Effective use of the models requires a basic understanding of the individual model components.

CONTINUOUS SIMULATION MODELS

Continuous simulation models differ from event models in that they attempt to simulate the hydrologic response of a catchment over long periods of time. Continuous input streams of precipitation and frequently temperature or solar radiation are required. Continuous models should represent more of the hydrologic processes that are occurring on a catchment than event models. Generally, a continuous accounting of soil water content and ground water storage is required. A representation of the interaction between soil water content, evaporative demand, and stage of plant growth is often required. In general, many more parameters are required by continuous simulation models than by event models. Some continuous simulation models require that 30 or more parameters be estimated before the model can be run. Obviously a large number of parameters implies a fairly complex model structure to incorporate their individual impacts. It is not only difficult to visualize the effect of so many parameters, but it is difficult to determine a unique set of parameters to fulfill some objective function.

To illustrate the working of a continuous simulation model, a simple four-parameter model for simulating monthly water yield from small catchments will be used. Again this is not an attempt to promote any particular model. This model has been chosen for illustrative purposes so that the basic process of continuous simulation modeling can be illustrated without getting involved in trying to understand the multitude of relationships required in the more complex models having several parameters.

This four-parameter water yield model was developed at the University of Kentucky (Haan, 1972). The model is representative of many such models. One attractive feature of the model is that it has a built in parameter estimation scheme that determines the values for the four parameters of the model that minimizes the sum of squares between the observed and predicted monthly flows.

The water yield model shown schematically in several ways in Fig. 13.12 divides the water-holding capacity of the soil into a volume M_r, which is readily available for evapotranspiration, and a volume M_l which is less readily available for evapotranspiration. The maximum capacity of M_r is 1 in. of water. The maximum capacity of M_l is MLC, a model parameter. M_r and M_l are used to denote the current volume of water stored in the readily and less readily available (for ET) soil water zones, respectively.

Precipitation input into the model is daily totals. This daily total is then divided into hourly rainfall on the basis of the appropriate SCS type curve as shown in Chapter 3. Hourly precipitation is divided into infiltration and runoff. The infiltration rate, f, is determined from

$$\begin{aligned} f &= f_{max} \quad \text{for } P > f_{max} \quad \text{and} \quad M_r < 1 \quad \text{or } M_l < \text{MLC} \\ f &= P \quad \text{for } P < f_{max} \quad \text{and} \quad M_r < 1 \quad \text{or } M_l < \text{MLC} \\ f &= 0 \quad \text{for } M_r = 1 \quad \text{and} \quad M_l = \text{MLC}, \end{aligned} \tag{13.10}$$

where f_{max} is the maximum possible infiltration rate and P is the precipitation rate. All infiltrated water is stored in M_r until the entire 1-in. capacity is filled, at which point any additional infiltrated water is added directly to M_l. When both storage are filled to their capacity, all precipitation is assumed to become runoff.

The surface runoff volume SRO is determined from

$$\begin{aligned} \text{SRO} &= (P - f)t \quad \text{for } P > f \\ \text{SRO} &= 0 \quad \text{for } P < f, \end{aligned} \tag{13.11}$$

where t is the time increment involved.

The daily evapotranspiration ET is determined from

$$\begin{aligned} \text{ET} &= E_p \quad && \text{for } P_d = 0; \quad && 0 < M_r < 1.00 \\ \text{ET} &= E_p \frac{M_l}{\text{MLC}} \quad && \text{for } P_d = 0; \quad && M_r = 0 \\ \text{ET} &= 0.5E_p \quad && \text{for } P_d > 0.01; \quad && 0 < M_r < 1.00 \\ \text{ET} &= 0.5E_p \frac{M_l}{\text{MLC}} \quad && \text{for } P_d > 0.01; \quad && M_r = 0, \end{aligned} \tag{13.12}$$

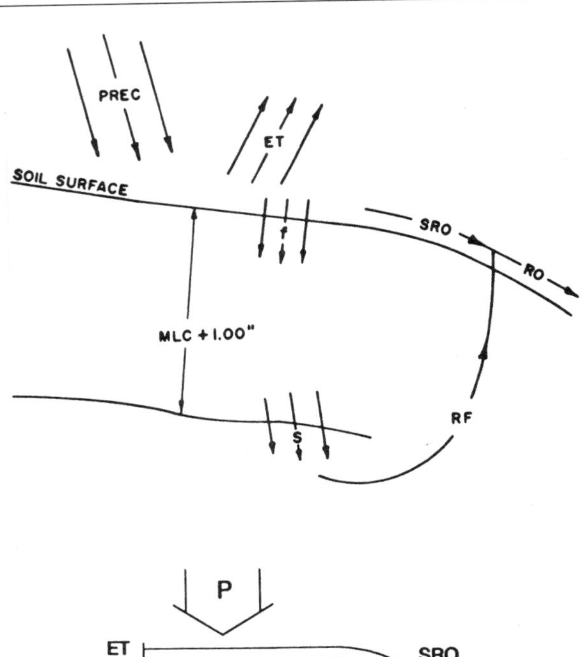

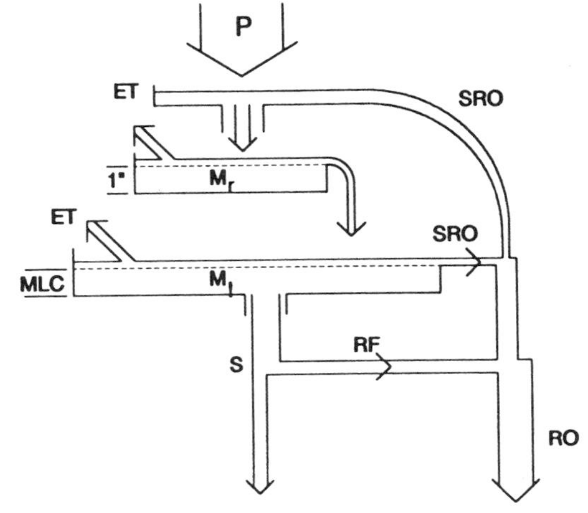

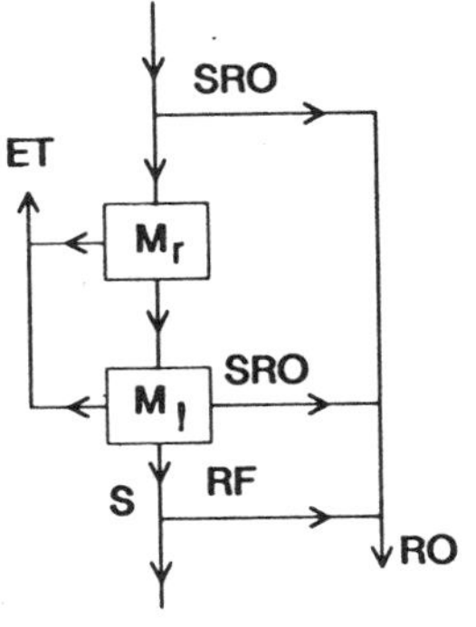

Figure 13.12 Schematic of four-parameter water yield model.

where E_p is the potential daily evapotranspiration and P_d is the depth of rainfall (inches) that occurred on the day in question.

Evapotranspiration is equal to potential evapotranspiration as long as water is contained in the readily available zone and then is reduced by the ratio of M_l

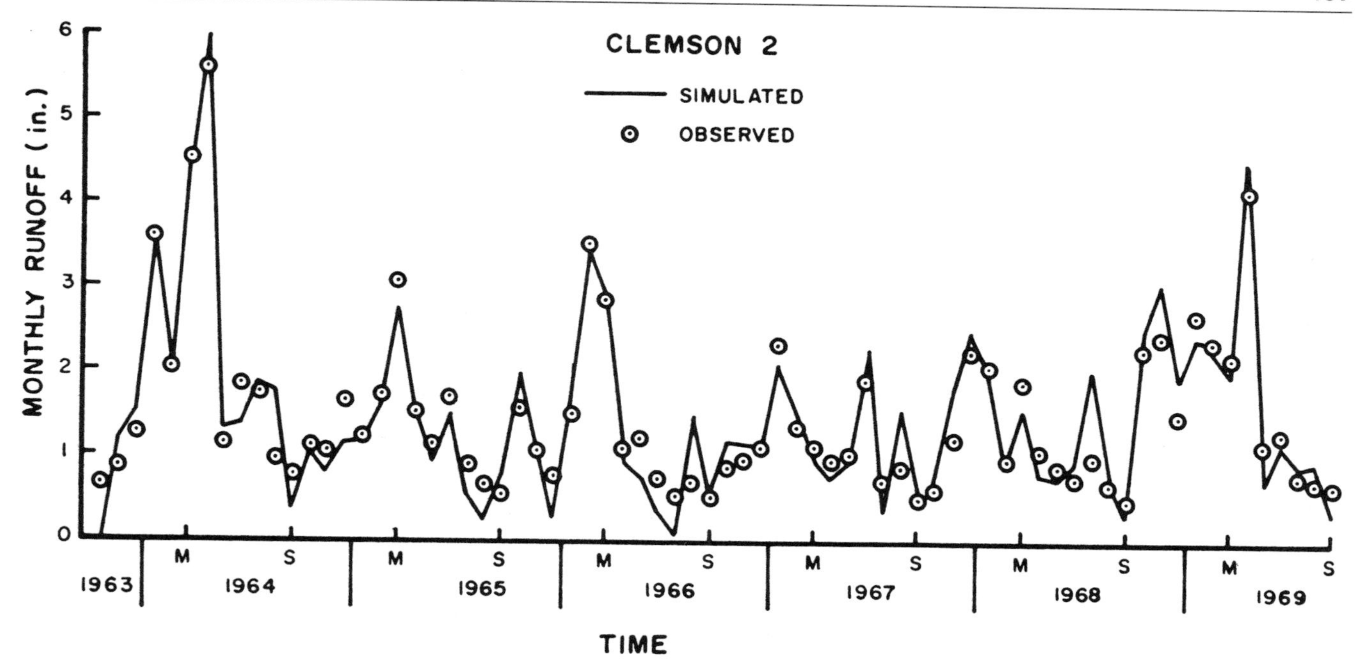

Figure 13.13 Results of simulation on Clemson 2 watershed.

to MLC. On days when precipitation occurs, the evapotranspiration rate is reduced by a factor of 2 to account for cloudy conditions and low solar radiation.

Deep seepage S or water that does not appear as streamflow within the watershed is determined from

$$S = S_{max} \frac{M_1}{MLC}, \tag{13.13}$$

where S_{max} is the maximum possible seepage rate in inches per day. A certain amount of return flow RF is allowed within the catchment and is calculated from

$$RF = F \times S, \tag{13.14}$$

where F is a constant defining the fraction of seepage that becomes runoff.

The total runoff RO is then equal to the sum of surface runoff and the return flow

$$RO = SRO + RF. \tag{13.15}$$

Thus there are four model parameters to be estimated:

S_{max}, the maximum deep seepage rate
MLC, the soil water storage capacity of the less readily available soil zone
F, the fraction of the seepage that becomes runoff
f_{max}, the maximum infiltration rate.

Figure 13.13 shows a time series of the results of applying this model to a 561-acre catchment near Clemson, South Carolina. The first 2 years of the 6-year record were used to obtain optimal parameter values. The results were

$$f_{max} = 0.30 \quad \text{(inches per hour)}$$
$$S_{max} = 0.185 \quad \text{(inches per day)}$$
$$MLC = 3.75 \quad \text{(inches)}$$
$$F = 0.40 \quad (-).$$

These parameters were used to simulate the entire 6-year period of record shown in Fig. 13.13.

Figure 13.14 shows a scatter plot of observed versus predicted monthly streamflow. If the model were doing a perfect job of prediction, all the points would fall on the line of equal values, the correlation coefficient would be 1.00, and the slope of the regression line between observed and predicted flows would be 1.00. Figure 13.14 demonstrates a common procedure for evaluating hydrologic models if observed data are available. In using this comparison, it is important to evaluate both the correlation coefficient and the slope of the regression line since either a low correlation coefficient or a slope significantly different from 1.00 can indicate a poor fit of the model. Neither measure, when used alone, provides adequate information since a slope of near 1.00 and a low correlation coefficient or a correlation coefficient of near 1.00 and a slope very different from 1.00 could indicate poor modeling results. The slope provides a measure of the bias in the estimates made by the model. The correlation coeffi-

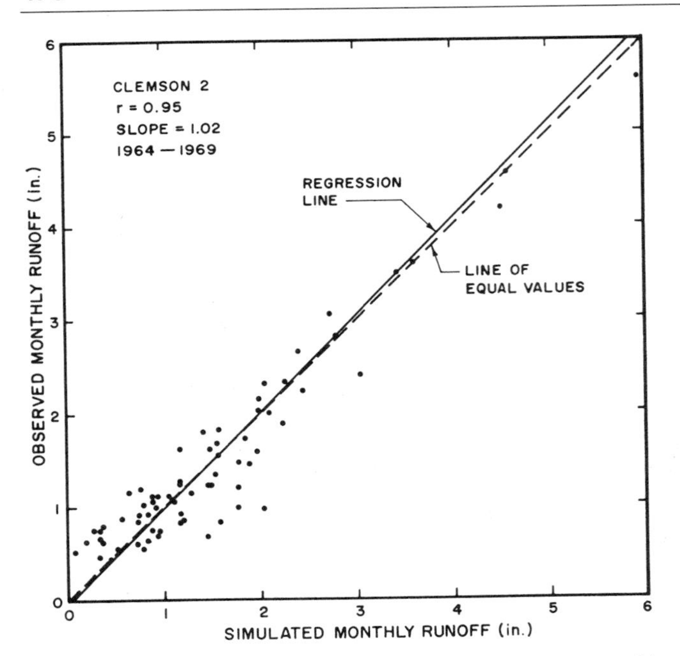

Figure 13.14 Comparison of observed and simulated monthly runoff for Clemson 2 watershed.

cient is a measure of how the model results track the observed results in a relative sense (model predictions go up or down when observed results go up or down, respectively).

There is no generally accepted, simple criteria to use in evaluating continuous simulation models. Evaluation of these models has always relied to a certain extent on the judgment of the modeler based on comparisons of observed and predicted hydrographs (such as Fig. 13.13) and based on how the model predicts in the face of extreme or changing conditions in comparison to the modelers expectation.

Some quantitative comparisons that can be made include means; variances; correlation between observed and predicted, standard error of estimate, and first-order autocorrelations of flows integrated over some finite time period such as 1 day or 1 month; flood peaks for various frequencies, such as the 5-year peak flow; and/or low-flow duration—frequency estimates such as the 10-day, 2-year low-flow volume. Another valuative approach is to compare design or operational decisions that would be made based on model results and observed data. Such things as required storage capacity in a detention basin, storage to meet a projected water supply demand, or size of bridge opening or culvert to meet a given criteria might be compared. Certainly if the model-predicted flows agree very well with observed flows, then design decisions based on modeled and observed flows should also agree. Certain design decisions are more sensitive to certain flow quantities than others. Thus what appears to be a good representation of a flow hydrograph may produce a substantially different design if the critical flow quantity is not properly modeled. An example would be the importance of modeling the persistence in low flows for water supply storage requirement determinations.

There are several currently available continuous flow simulation models. Continuous flow models are especially popular as a foundation for water quality models because of the importance of antecedent conditions at the time of a storm event.

The Stanford Watershed Model (Crawford and Linsley, 1966) and its many derivatives and the precipitation—runoff modeling system (PRMS) of the U.S. Geological Survey (Leavesley *et al.*, 1983) represent two popular continuous simulation models that can generate daily streamflow hydrographs and storm flow hydrographs.

These and other continuous simulation models require numerous computations in the form of water budgets for the various storages in the model. Parameter estimation, if based on observed data, requires running the model using a number of different potential parameter sets, and selecting the parameter set that optimizes the desired objective function [i.e., minimizes S of Eq. (13.4)]. PRMS has an option for automatic parameter estimation based on Eq. (13.4). Experience with PRMS and other models has shown that for some catchments, the model results are not very sensitive to the values used for some of the parameters. This makes estimation of these parameters difficult—a difficulty that may not be of importance unless some watershed modification or change over time whose hydrologic impact should be reflected in the insensitive parameter(s) is being evaluated. This shortcoming is common to all continuous simulation models, not just PRMS.

Use of a model such as PRMS or SWM or one of its derivatives requires considerable time to become familiar with the model, to fully understand all of the internal workings of the model, and to comprehend the interactions and interrelationships of the various flow components. Before a model is selected, the user's manual should be studied to determine the adequacy of the documentation for the model and whether the model is suited for the intended purpose. A complex model with inadequate documentation should not be used unless one is thoroughly familiar with the model. The documentation should clearly state the capability and the limitations of the model and provide guidance on parameter estimation and model evalution. Inadequate documentation is often a stumbling block to proper and efficient use of a model.

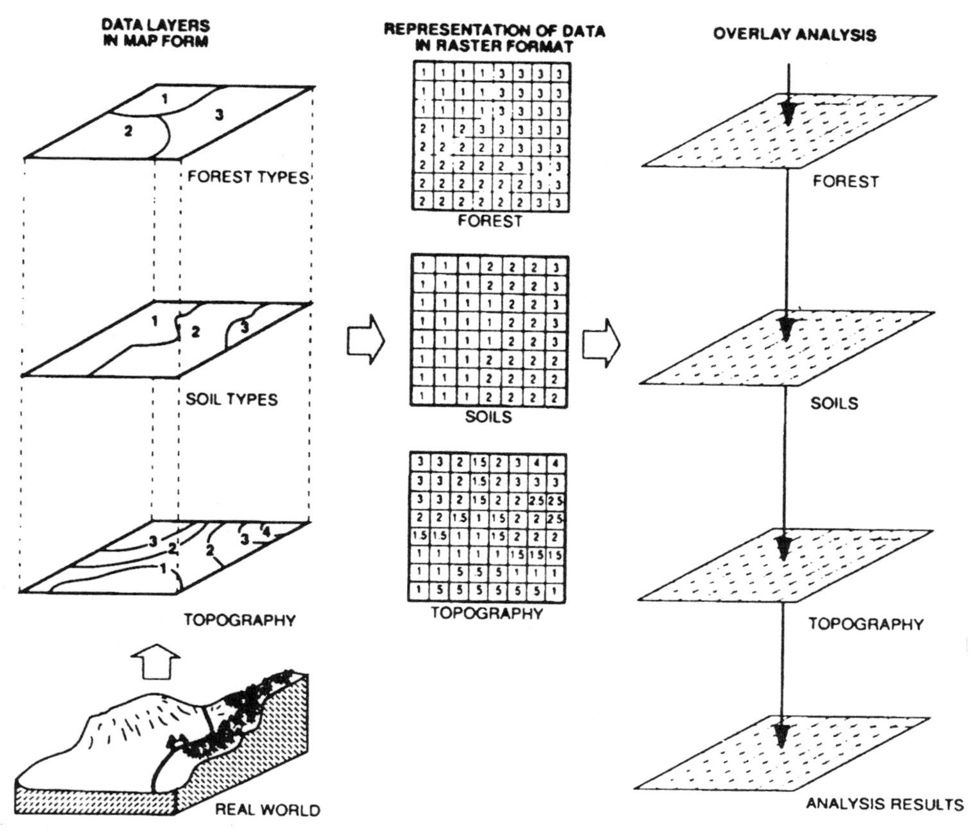

Figure 13.15 A representation of GIS.

INFORMATION SYSTEMS TECHNOLOGIES

Advances in microcomputer technology has been paralleled by developments in software that promise to revolutionize hydrologic and water quality modeling. Three of these technologies are geographical information systems (GIS), expert systems (ES), and visualization techniques (VT).

Geographical Information Systems

A characteristic of sophisticated, distributed parameter, hydrologic models is the large amount of spatially oriented data that they require. GIS is a technology for manipulating spatial data (Zhang *et al.*, 1990a, b). Data of various types are collected in the form of layers. Typical layers might be soils, topography, and land use. The ability to extract, overlay, and delineate land characteristics makes GIS eminently suited to the delineation of hydrologically homogeneous subareas.

A complete GIS includes the hardware and software used to perform geographic analyses as well as the data bases and the people who use the system to meet a specific set of objectives (Brown, 1986). A GIS will, in general, have a means of encoding and converting geographically (or spatially) oriented data and the capacity to retrieve and display the information through a variety of media including computer screens, printers, and plotters.

Two basic techniques of representing spatial data are employed. The first is a vector format where the basic unit of data is a single vector or map line. Vectors form polygons that enclose like information within a data layer. A second data model is a raster format made up of cells or pixels. The raster format is commonly called a grid-cell system. Figure 13.15 depicts the basis of a raster GIS system. The figure shows how the data layers are converted to cells having attribute values. These layers can then form the required data base for input into a hydrologic model or they can be combined by overlaying techniques to generate the input data.

A hydrologic model can perform an analysis on the basis of the imported data and export results by cell back to the GIS. The GIS can then prepare visuals or maps of the resulting hydrologic analysis. Zhang *et al.* (1990b) have interfaced a root-zone chemical transport model with a GIS to produce maps showing the probability of the applied chemical exceeding the health advisory limit of the U.S. Environmental Protection Agency at a depth of 1 m.

Most GIS are data managers and do not themselves contain hydrologic modeling capability. Hydrologic models must be interfaced with the GIS. The GIS manages data for the model. The model in turn provides the GIS with the results of an analysis that can be made into a GIS layer and mapped in various ways.

GIS application will grow rapidly over the next several years as data bases become more readily available and the concepts and software become better understood.

Expert Systems

Expert systems are a subset of "artificial intelligence." ES are computer software that offer advice to the software user based on its own store of knowledge and the users response to a number of if—then rules or questions. ES obtain their knowledge from the developers of the software and the users of the software. Some ES have a learning ability in that they accumulate knowledge in response to the experience of their users.

ES are potentially valuable in hydrologic and water quality modeling as a means of selecting appropriate modeling approaches based on available data and of estimating the appropriate values for model parameters. ES, when properly formulated and used, make it possible for a novice to make modeling decisions similar to those of a modeling expert. For this to happen, the software making up the ES must have captured within it expert knowledge. Often this knowledge is gained by interviewing experts and coding their responses to key questions in a form that captures the required knowledge. The best ES are based on the knowledge of more than one expert.

ES or any modeling tool should not be totally relied upon. The user of any model, including those containing ES, has the ultimate responsibility for ensuring that the modeling is done correctly using appropriate model parameters.

Visualization Technology

Visualization technology is the use of computer graphics to enhance the understanding of computer-generated results. Television weather forecasters make good use of visualization technology when they show storm systems sweeping across the country and use this visual impression to explain the weather forecast. Similarly, techniques that will show the detailed movement of a severe storm across a catchment along with the resulting anticipated flood flow elevations and areal extent of flooding along streams draining the catchment are under development.

Visualization technology will encompass GIS and hydrologic models along with high-speed parallel processing to computer generate the required graphic information. This technology will be useful in flood forecast centers and in explaining emergency operational plans to citizens. It may also help show citizens the expected extent of flooding as a storm develops and thus the likelihood of their particular property being impacted by the flood.

Problems

(13.1) Investigate mathematical techniques for finding the maximum or minimum of an objective function with respect to a single unknown parameter. Consider both analytical and numerical approaches. Discuss the applicability and relative merits of the approaches to hydrologic modeling.

(13.2) Same as Problem (13.1) but with multiple unknown parameters.

(13.3) (a) Write computer coding for the hydrologic model shown schematically as Fig. 13.12. (b) Select a hydrologic record of at least 1 year in length from a humid region catchment and estimate the parameters for this model. (c) Discuss quantitatively and qualitatively how well the model describes the hydrology of the selected catchment.

(13.4) Discuss the merits of using the model depicted in Fig. 13.12 for evaluating the hydrologic impact of forest clear cutting on stream hydrology for a 250-acre (100-ha) catchment. Include in your discussion how the model might be used, your opinion as to whether the model would produce reasonable results, and the hydrologic quantities (water yield, peak flow, etc.) that likely could and could not be evaluated with this model. What aspects of the model would be the most important in this application? How are these important aspects reflected in the model in terms of parameters and model structure?

(13.5) Select a hydrologic model and a catchment. Estimate the parameters for the model and the catchment. Select four of the parameters of the model. Vary the values of the parameters by 10, 20, and 50% from their estimated values, and run the model using these parameter values. Vary the parameters individually. Discuss the sensitivity of the parameters with respect to hydrologic estimates that might be made with the model.

(13.6) Do Problem (13.5) except vary the parameters simultaneously in pairs, triplicates, and all simultaneously.

(13.7) Select a hydrologic model. Discuss the basic structure of the model, the number of parameters, how the parameters can be estimated in the absence of stream flow data, situations where the model could and could not be expected to produce reliable hydrologic estimates.

(13.8) Prepare the computer coding for the basic model unit of Fig. 13.4.

(13.9) Apply the coding developed for Problem (13.8) to a selected catchment of around 50 acres (20 ha).

(13.10) How would the impact of a land-use change, such as surface mining on runoff hydrographs, be reflected in the model depicted in Fig. 13.4?

(13.11) For a selected hydrologic model, discuss the approach used to properly sequence the hydrology of subwatersheds (i.e., discuss the model management approach for combining and routing hydrographs).

(13.12) Select a particular catchment for which a streamflow record is available. Without any reference to the streamflow, use a hydrologic model to estimate the hydrologic record for the same period as the available record. Discuss the difficulties encountered. Discuss how well the estimated records resemble the actual record of streamflow.

(13.13) Use the available streamflow record of Problem (13.12) to improve the estimates of the model parameters and repeat the estimation and discussion. Are the estimated flows more in agreement with the observed flows after modifying the parameters? Why?

(13.14) Discuss the similarities and differences among deterministic, parametric, and stochastic hydrologic models. Under what conditions would each of these modeling approaches be the most appropriate?

(13.15) Discuss the procedure that one might use to verify the results of an application of an event-based hydrologic model (a) with a good record of streamflow and (b) in the absence of any streamflow data.

(13.16) Discuss the procedure that one might use to verify the results of an application of a continuous simulation hydrologic model (a) with a good record of streamflow and (b) in the absence of any streamflow data.

(13.17) Describe the basic mathematical structures of a selected hydrologic model such as the SWM or PRMS.

(13.18) Discuss the advantage of objective parameter estimation based on a mathematical fitting criteria as compared to reliance on the judgment of the model user.

(13.19) Under what condition would parameter estimation based on personal judgment be preferred over an objective mathematical fitting criteria?

(13.20) Describe desirable characteristics of a hydrologic model that is going to be used as a framework for a water quality model.

(13.21) Describe at least two potential modeling approaches for generating runoff hydrographs from impervious parking lots. What are the advantages and disadvantages of each approach. Which approach do your prefer? Why?

(13.22) Develop computer coding for one of the models described for Problem (13.21). Test the coding by simulating the runoff from a hypothetical parking lot.

References

Beard, L. R. (1967). Optimization techniques for hydrologic engineering. *Water Resources Res.* **3**(3):809–815.

Beasley, D. B., and Huggins, L. F. (1981). "ANSWERS Users Manual," EPA-905/9-82-01. U.S. Environmental Protection Agency Region V, Chicago, IL.

Box, G. E. P., and Tiao, G. C. (1973). "Bayesian Inference in Statistical Analysis." Addison–Wesley, Reading, MA.

Brown, C. (1986). Implementation of Geographic Information System. What makes a new site a success? *In* "Proceedings, Geographic Information System Workshop, Bethesda, MD." Am. Soc. Photogrammetry and Remote Sensing.

Campbell, K. K., Johnson, H. P., and Melvin, S. W. (1974). Mathematical modeling of drainage watersheds, American Society of Civil Engineers preprint 2373. Annual and National Environmental Engineering Convention, Kansas City, MO.

Carrigan, P. H. (1973). Calibration of the U.S. Geological Survey Rainfall-Runoff Model for peak flow synthesis—Natural basins, U.S. Geological Survey Computer Report. U.S. Department of Commerce NTIS.

Crawford, N. H., and Linsley, R. K. (1962). Synthesis of continuous streamflow hydrographs on a digital computer, Technical report 12. Stanford University Department of Civil Engineering, Stanford, CA.

Crawford, N. H., and Linsley, R. K. (1966). Digital simulation in hydrology: Stanford Watershed Model IV, Technical report 39. Stanford University Department of Civil Engineering, Stanford, CA.

Dawdy, D. R., and O'Donnell, T. (1965). Mathematical models of catchment behavior. *Am. Soc. Civil Eng. Proc.* **91**(HY4):123–127.

DeBoer, D. W., and Johnson, H. P. (1971). Simulation of runoff from depression characterized watersheds. *Trans. Am. Soc. Agric. Eng.* **14**(4):615–620.

DeCoursey, D. G., and Snyder, W. M. (1969). Computer oriented method of optimizing hydrologic model parameters. *J. Hydrol.* **9**:34–56.

Diskin, M. H., and Simon, E. (1977). A procedure for the selection of objective functions for hydrologic simulation models. *J. Hydrol.* **34**:129–149.

Edwards, D. R. (1988). Incorporating parametric uncertainty into flood estimation methodologies for ungaged watersheds and watersheds with short records, Ph.D. dissertation. Oklahoma State University Library, Stillwater, OK.

Edwards, D. R., and Haan, C. T. (1988). Confidence limits on peak flow estimates for ungaged watersheds. *In* "Proceedings, International Symposium on Modeling Agricultural, Forest, and Rangeland Hydrology." American Society of Agricultural Engineers, St. Joseph, MI.

Edwards, D. R., and Haan, C. T. (1989). Incorporating uncertainty into peak flow estimates. *Trans. Am. Soc. Agric. Eng.* **32**(1):113–119.

Haan, C. T. (1972). A water yield model for small watersheds. *Water Resources Res.* **8**(1):58–68.

Haan, C. T., and Johnson, H. P. (1968a). Hydraulic model of runoff from depressional areas. I. General considerations. *Trans. Am. Soc. Agri. Eng.* **11**(3):364–367.

Haan, C. T., and Johnson, H. P. (1968b). Hydraulic model of runoff from depressional areas. II. Development of model. *Trans. Am. Soc. Agric. Eng.* **11**(3):368–373.

Haan, C. T., Johnson, H. P., and Brakensiek, D. L. (1982). Hydrologic modeling of small watersheds, ASAE Monograph 5. American Society of Agricultural Engineers, St. Joseph, MI.

Huber, W. C., *et al.* (1981). "Storm Water Management Model users manual," Version III, EPA-600/2-84-109a. Environmental Protection Agency, Cincinnati, OH.

Huggins, L. F., and Monke, E. J. (1966). The mathematical simulation of the hydrology of small watersheds, Technical Report 1. Purdue University Water Resources Research Center.

Jackson, D. R., and Aron, G. (1971). Parameter estimation in hydrology: The state of the art. *Water Resources Bull.* **7**(3):457–472.

James, L. D., and Burges, S. J. (1982). Selection, calibration and testing of hydrologic models. *In* "Hydrologic Modeling of Small Watersheds" (Haan *et al.*, eds.), pp. 437–472. ASAE Monograph 5, American Society of Agricultural Engineers, St. Joseph, MI.

Klemes, V. (1982). Empirical and causal models in hydrology. *In* "Scientific Basis of Water Resources Management," pp. 95–104. National Academy Press, Washington DC.

Kuczera, G. (1983). Improved parameter inference in catchment models 1: Evaluating parameter uncertainty. *Water Resources Res.* **19**(5):1151–1162.

Leavesley, G. H., Lichty, R. W., Troutman, B. M., and Saindon, L. G. (1983). "Precipitation-Runoff Modeling System: User's Manual," Water Resources Investigations Report 83-4238. U.S. Geological Survey, Denver, CO.

Ross, B. B., Contractor, D. N., and Shanholtz, V. O. (1979). A finite element model of overland and channel flow for assessing the hydrologic impact of land use changes. *J. Hydrol.* **41**:11–30.

Sorooshian, S. (1983). Surface water hydrology: On line estimation. *Rev. Geophys. Space Phys.* **21**(3):706–721. [U.S. National Report to International Union of Geodesy and Geophysics 1979–1982.]

Troutman, B. M. (1985). Errors and parameter estimation in precipitation-runoff modeling 1: Theory. *Water Resources Res.* **21**(8):1195–1212; 2: Case Study. *Water Resources Res.* **21**(8):1214–1222.

U.S. Army Corps of Engineers (1985). "HEC-1 Flood Hydrology Package Users Manual." Hydrologic Engineering Center, Davis, CA.

U.S. Department of Agriculture, Soil Conservation Service (1973). "Computer Program for Project Formulation Hydrology." Technical Release 20, Washington, DC.

Vicens, G. J., Rodriguez-Iturbe, I., and Shaake, J. C. (1975). A Bayesian framework for the use of regional information in hydrology. *Water Resources Res.* **11**(3):405–414.

Wilson, B. N., Barfield, B. J., and Moore, I. D. (1983). "A Hydrology and Sedimentology Watershed Model. I. Modeling Techniques," Special publication. Agricultural Engineering Department, University of Kentucky, Lexington, KY.

Zhang, H., Haan, C. T., and Nofziger, D. L. (1990a). Hydrologic Modeling with GIS: An Overview. *Appl. Eng. Agric.* **6**(4):453–458.

Zhang, H., Nofziger, D. L., and Haan, C. T. (1990b). Interfacing a root-zone transport model with GIS, ASAE Paper 903034 presented at the 1990 International Summer Meeting of Am. Soc. Agr. Engrs., Columbus, OH, June 24–27, 1990.

Appendix 2
Cumulative Standard Normal Distribution

Appendix 2
Cumulative Standard Normal Distribution

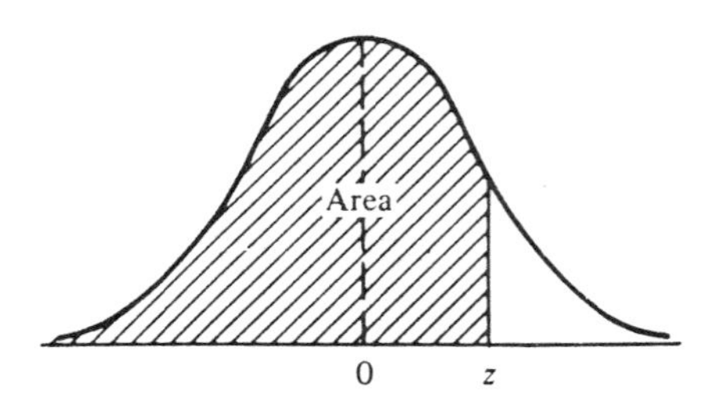

z	0.00	0.01	0.02	0.03	0.04	0.05	0.06	0.07	0.08	0.09
−3.4	0.0003	0.0003	0.0003	0.0003	0.0003	0.0003	0.0003	0.0003	0.0003	0.0002
−3.3	0.0005	0.0005	0.0005	0.0004	0.0004	0.0004	0.0004	0.0004	0.0004	0.0003
−3.2	0.0007	0.0007	0.0006	0.0006	0.0006	0.0006	0.0006	0.0005	0.0005	0.0005
−3.1	0.0010	0.0009	0.0009	0.0009	0.0008	0.0008	0.0008	0.0008	0.0007	0.0007
−3.0	0.0013	0.0013	0.0013	0.0012	0.0012	0.0011	0.0011	0.0011	0.0010	0.0010
−2.9	0.0019	0.0018	0.0017	0.0017	0.0016	0.0016	0.0015	0.0015	0.0014	0.0014
−2.8	0.0026	0.0025	0.0024	0.0023	0.0023	0.0022	0.0021	0.0021	0.0020	0.0019
−2.7	0.0035	0.0034	0.0033	0.0032	0.0031	0.0030	0.0029	0.0028	0.0027	0.0026
−2.6	0.0047	0.0045	0.0044	0.0043	0.0041	0.0040	0.0039	0.0038	0.0037	0.0036
−2.5	0.0062	0.0060	0.0059	0.0057	0.0055	0.0054	0.0052	0.0051	0.0049	0.0048
−2.4	0.0082	0.0080	0.0078	0.0075	0.0073	0.0071	0.0069	0.0068	0.0066	0.0064
−2.3	0.0107	0.0104	0.0102	0.0099	0.0096	0.0094	0.0091	0.0089	0.0087	0.0084
−2.2	0.0139	0.0136	0.0132	0.0129	0.0125	0.0122	0.0119	0.0116	0.0113	0.0110
−2.1	0.0179	0.0174	0.0170	0.0166	0.0162	0.0158	0.0154	0.0150	0.0146	0.0143
−2.0	0.0228	0.0222	0.0217	0.0212	0.0207	0.0202	0.0197	0.0192	0.0188	0.0183
−1.9	0.0287	0.0281	0.0274	0.0268	0.0262	0.0256	0.0250	0.0244	0.0239	0.0233
−1.8	0.0359	0.0352	0.0344	0.0336	0.0329	0.0322	0.0314	0.0307	0.0301	0.0294
−1.7	0.0446	0.0436	0.0427	0.0418	0.0409	0.0401	0.0392	0.0384	0.0375	0.0367
−1.6	0.0548	0.0537	0.0526	0.0516	0.0505	0.0495	0.0485	0.0475	0.0465	0.0455
−1.5	0.0668	0.0655	0.0643	0.0630	0.0618	0.0606	0.0594	0.0582	0.0571	0.0559
−1.4	0.0808	0.0793	0.0778	0.0764	0.0749	0.0735	0.0722	0.0708	0.0694	0.0681
−1.3	0.0968	0.0951	0.0934	0.0918	0.0901	0.0885	0.0869	0.0853	0.0838	0.0823
−1.2	0.1151	0.1131	0.1112	0.1093	0.1075	0.1056	0.1038	0.1020	0.1003	0.0985
−1.1	0.1357	0.1335	0.1314	0.1292	0.1271	0.1251	0.1230	0.1210	0.1190	0.1170
−1.0	0.1587	0.1562	0.1539	0.1515	0.1492	0.1469	0.1446	0.1423	0.1401	0.1379
−0.9	0.1841	0.1814	0.1788	0.1762	0.1736	0.1711	0.1685	0.1660	0.1635	0.1611
−0.8	0.2119	0.2090	0.2061	0.2033	0.2005	0.1977	0.1949	0.1922	0.1894	0.1867
−0.7	0.2420	0.2389	0.2358	0.2327	0.2296	0.2266	0.2236	0.2206	0.2177	0.2148
−0.6	0.2743	0.2709	0.2676	0.2643	0.2611	0.2578	0.2546	0.2514	0.2483	0.2451
−0.5	0.3085	0.3050	0.3015	0.2981	0.2946	0.2912	0.2877	0.2843	0.2810	0.2776
−0.4	0.3446	0.3409	0.3372	0.3336	0.3300	0.3264	0.3228	0.3192	0.3156	0.3121
−0.3	0.3821	0.3783	0.3745	0.3707	0.3669	0.3632	0.3594	0.3557	0.3520	0.3483
−0.2	0.4207	0.4168	0.4129	0.4090	0.4052	0.4013	0.3974	0.3936	0.3897	0.3859
−0.1	0.4602	0.4562	0.4522	0.4483	0.4443	0.4404	0.4364	0.4325	0.4286	0.4247
−0.0	0.5000	0.4960	0.4920	0.4880	0.4840	0.4801	0.4761	0.4721	0.4681	0.4641
0.0	0.5000	0.5040	0.5080	0.5120	0.5160	0.5199	0.5239	0.5279	0.5319	0.5359
0.1	0.5398	0.5438	0.5478	0.5517	0.5557	0.5596	0.5636	0.5675	0.5714	0.5753
0.2	0.5793	0.5832	0.5871	0.5910	0.5948	0.5987	0.6026	0.6064	0.6103	0.6141
0.3	0.6179	0.6217	0.6255	0.6293	0.6331	0.6368	0.6406	0.6443	0.6480	0.6517
0.4	0.6554	0.6591	0.6628	0.6664	0.6700	0.6736	0.6772	0.6808	0.6844	0.6879
0.5	0.6915	0.6950	0.6985	0.7019	0.7054	0.7088	0.7123	0.7157	0.7190	0.7224
0.6	0.7257	0.7291	0.7324	0.7357	0.7389	0.7422	0.7454	0.7486	0.7517	0.7549
0.7	0.7580	0.7611	0.7642	0.7673	0.7704	0.7734	0.7764	0.7794	0.7823	0.7852
0.8	0.7881	0.7910	0.7939	0.7967	0.7995	0.8023	0.8051	0.8078	0.8106	0.8133
0.9	0.8159	0.8186	0.8212	0.8238	0.8264	0.8289	0.8315	0.8340	0.8365	0.8389
1.0	0.8413	0.8438	0.8461	0.8485	0.8508	0.8531	0.8554	0.8577	0.8599	0.8621
1.1	0.8643	0.8665	0.8686	0.8708	0.8729	0.8749	0.8770	0.8790	0.8810	0.8830
1.2	0.8849	0.8869	0.8888	0.8907	0.8925	0.8944	0.8962	0.8980	0.8997	0.9015
1.3	0.9032	0.9049	0.9066	0.9082	0.9099	0.9115	0.9131	0.9147	0.9162	0.9177
1.4	0.9192	0.9207	0.9222	0.9236	0.9251	0.9265	0.9278	0.9292	0.9306	0.9319
1.5	0.9332	0.9345	0.9357	0.9370	0.9382	0.9394	0.9406	0.9418	0.9429	0.9441
1.6	0.9452	0.9463	0.9474	0.9484	0.9495	0.9505	0.9515	0.9525	0.9535	0.9545
1.7	0.9554	0.9564	0.9573	0.9582	0.9591	0.9599	0.9608	0.9616	0.9625	0.9633
1.8	0.9641	0.9649	0.9656	0.9664	0.9671	0.9678	0.9686	0.9693	0.9699	0.9706
1.9	0.9713	0.9719	0.9726	0.9732	0.9738	0.9744	0.9750	0.9756	0.9761	0.9767
2.0	0.9772	0.9778	0.9783	0.9788	0.9793	0.9798	0.9803	0.9808	0.9812	0.9817
2.1	0.9821	0.9826	0.9830	0.9834	0.9838	0.9842	0.9846	0.9850	0.9854	0.9857
2.2	0.9861	0.9864	0.9868	0.9871	0.9875	0.9878	0.9881	0.9884	0.9887	0.9890
2.3	0.9893	0.9896	0.9898	0.9901	0.9904	0.9906	0.9909	0.9911	0.9913	0.9916
2.4	0.9918	0.9920	0.9922	0.9925	0.9927	0.9929	0.9931	0.9932	0.9934	0.9936
2.5	0.9938	0.9940	0.9941	0.9943	0.9945	0.9946	0.9948	0.9949	0.9951	0.9952
2.6	0.9953	0.9955	0.9956	0.9957	0.9959	0.9960	0.9961	0.9962	0.9963	0.9964
2.7	0.9965	0.9966	0.9967	0.9968	0.9969	0.9970	0.9971	0.9972	0.9973	0.9974
2.8	0.9974	0.9975	0.9976	0.9977	0.9977	0.9978	0.9979	0.9979	0.9980	0.9981
2.9	0.9981	0.9982	0.9982	0.9983	0.9984	0.9984	0.9985	0.9985	0.9986	0.9986
3.0	0.9987	0.9987	0.9987	0.9988	0.9988	0.9989	0.9989	0.9989	0.9990	0.9990
3.1	0.9990	0.9991	0.9991	0.9991	0.9992	0.9992	0.9992	0.9992	0.9993	0.9993
3.2	0.9993	0.9993	0.9994	0.9994	0.9994	0.9994	0.9994	0.9995	0.9995	0.9995
3.3	0.9995	0.9995	0.9995	0.9996	0.9996	0.9996	0.9996	0.9996	0.9996	0.9997
3.4	0.9997	0.9997	0.9997	0.9997	0.9997	0.9997	0.9997	0.9997	0.9997	0.9998

Appendix 3A
Rainfall Maps

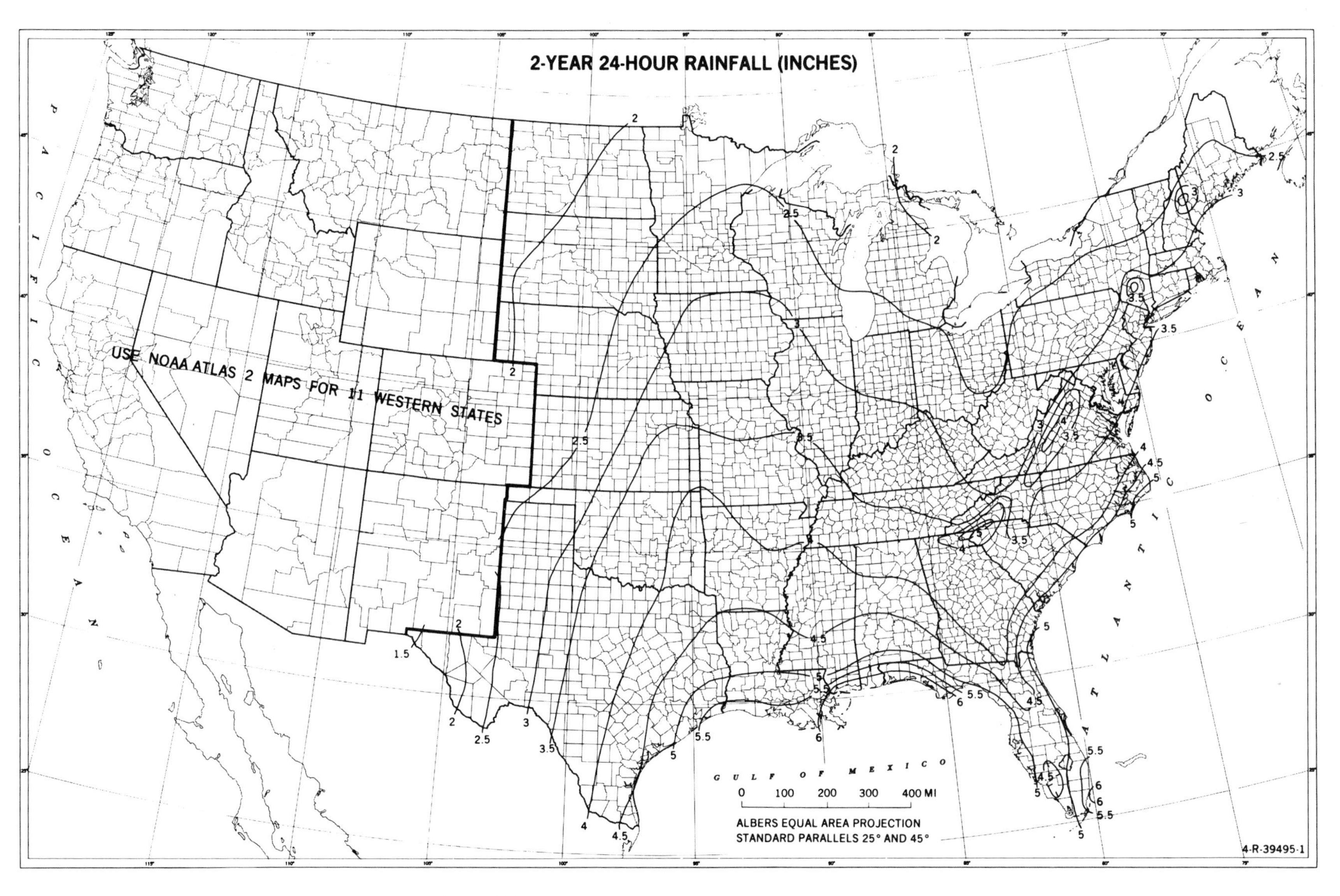

Figure 3A.1 Two-year, 24-hr rainfall. From Soil Conservation Service (1986).

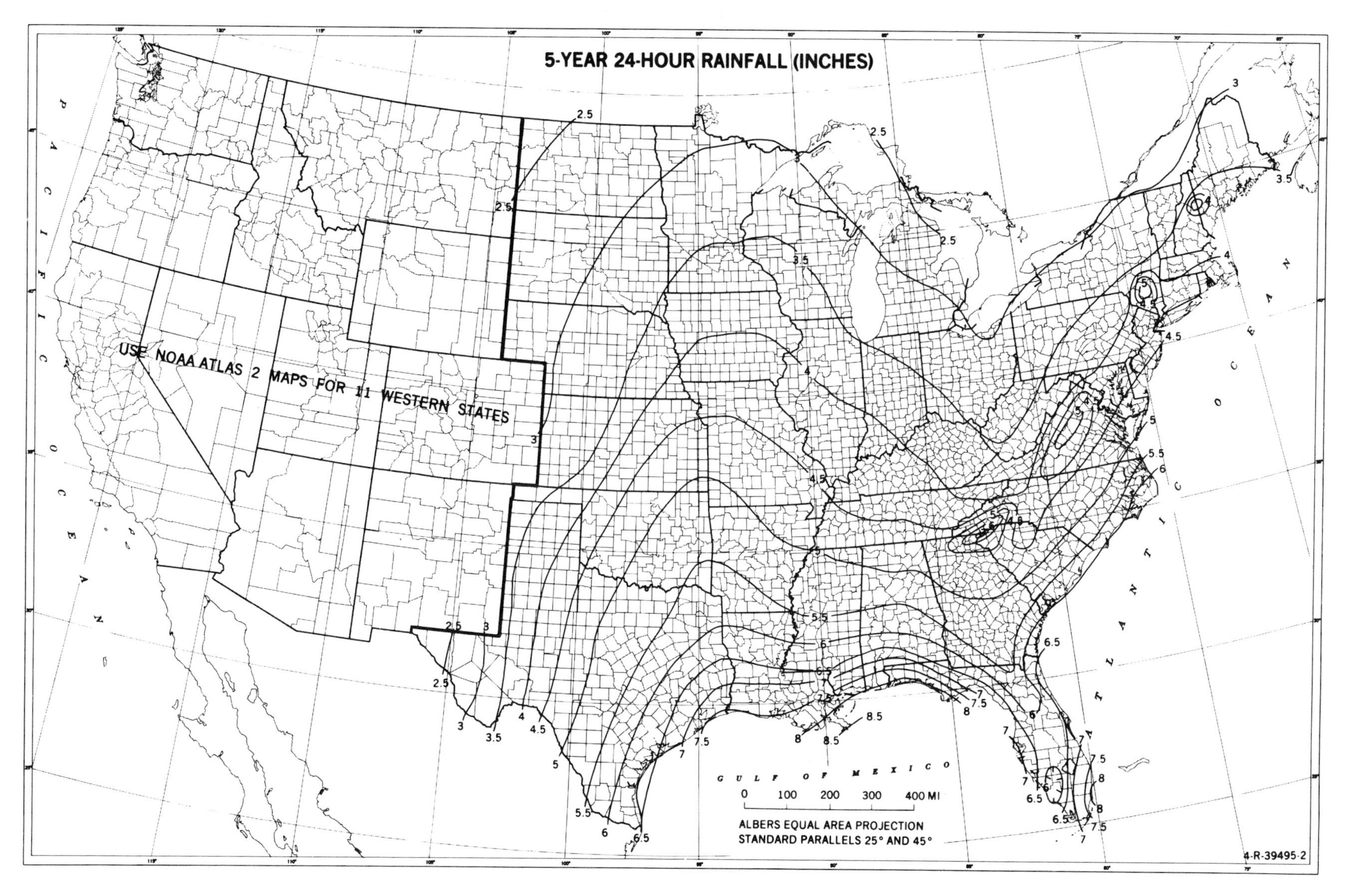

Figure 3A.2 Five-year, 24-hr rainfall. From Soil Conservation Service (1986).

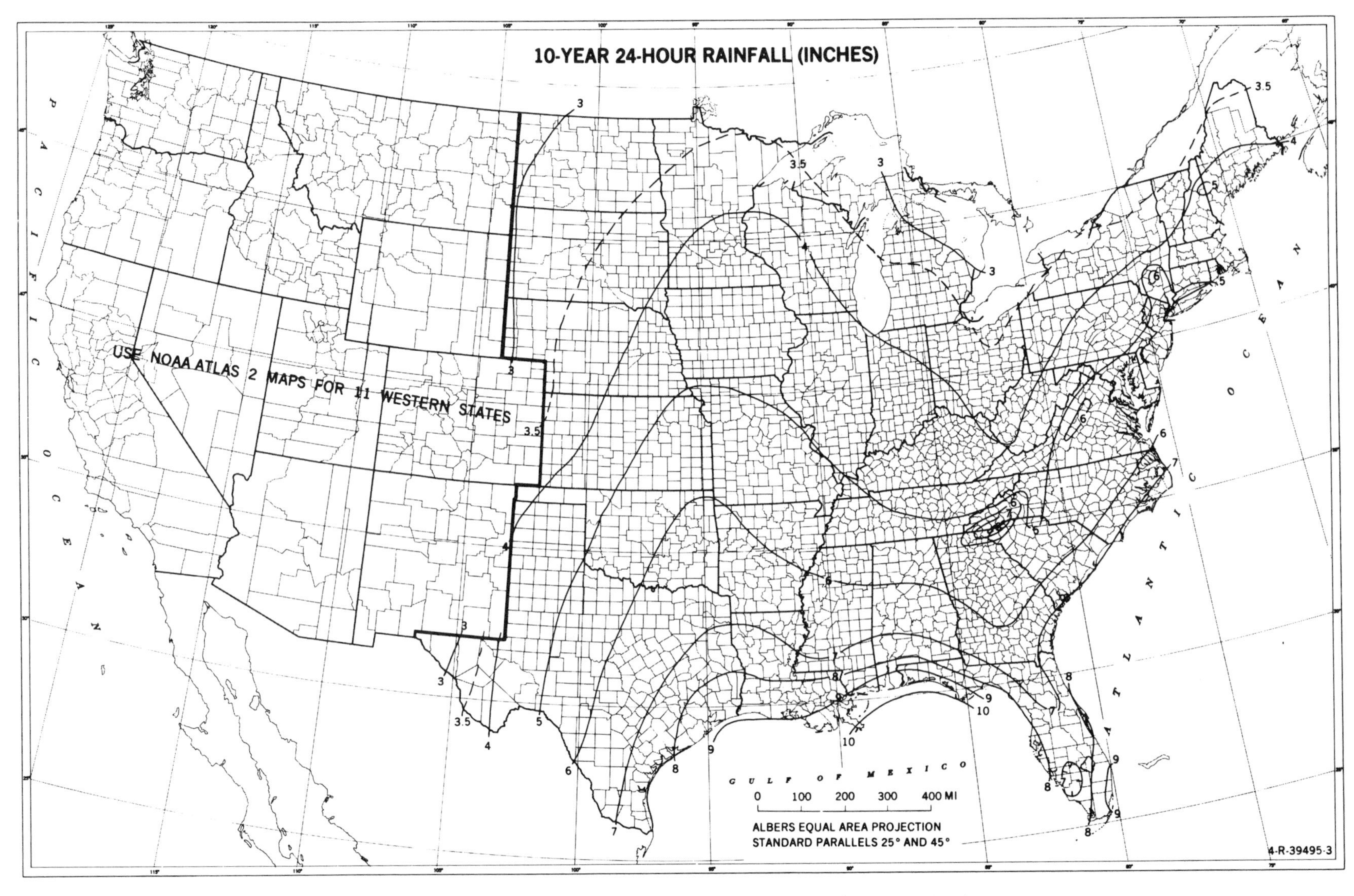

Figure 3A.3 Ten-year, 24-hr rainfall. From Soil Conservation Service (1986).

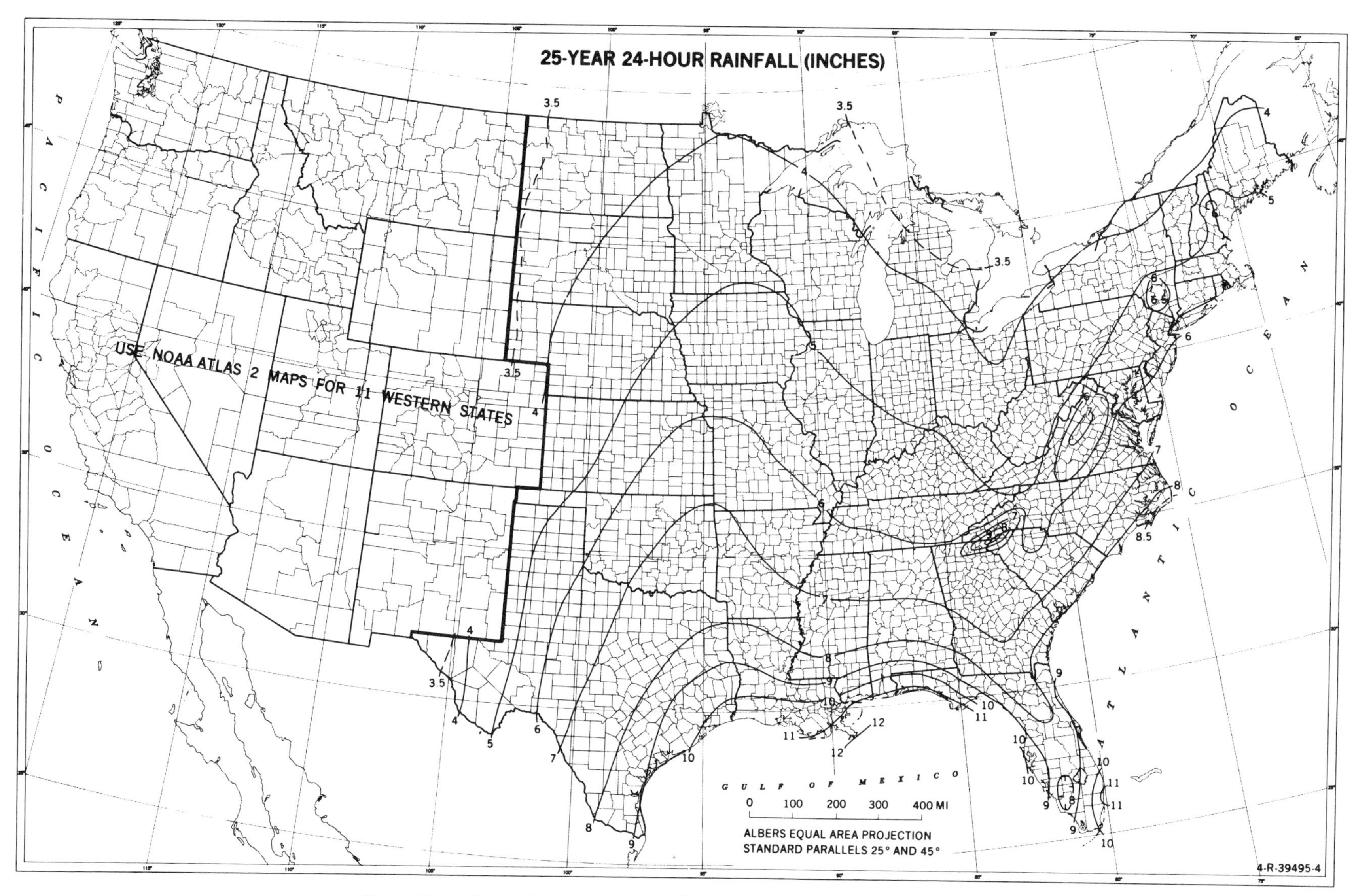

Figure 3A.4 Twenty-five-year, 24-hr rainfall. From Soil Conservation Service (1986).

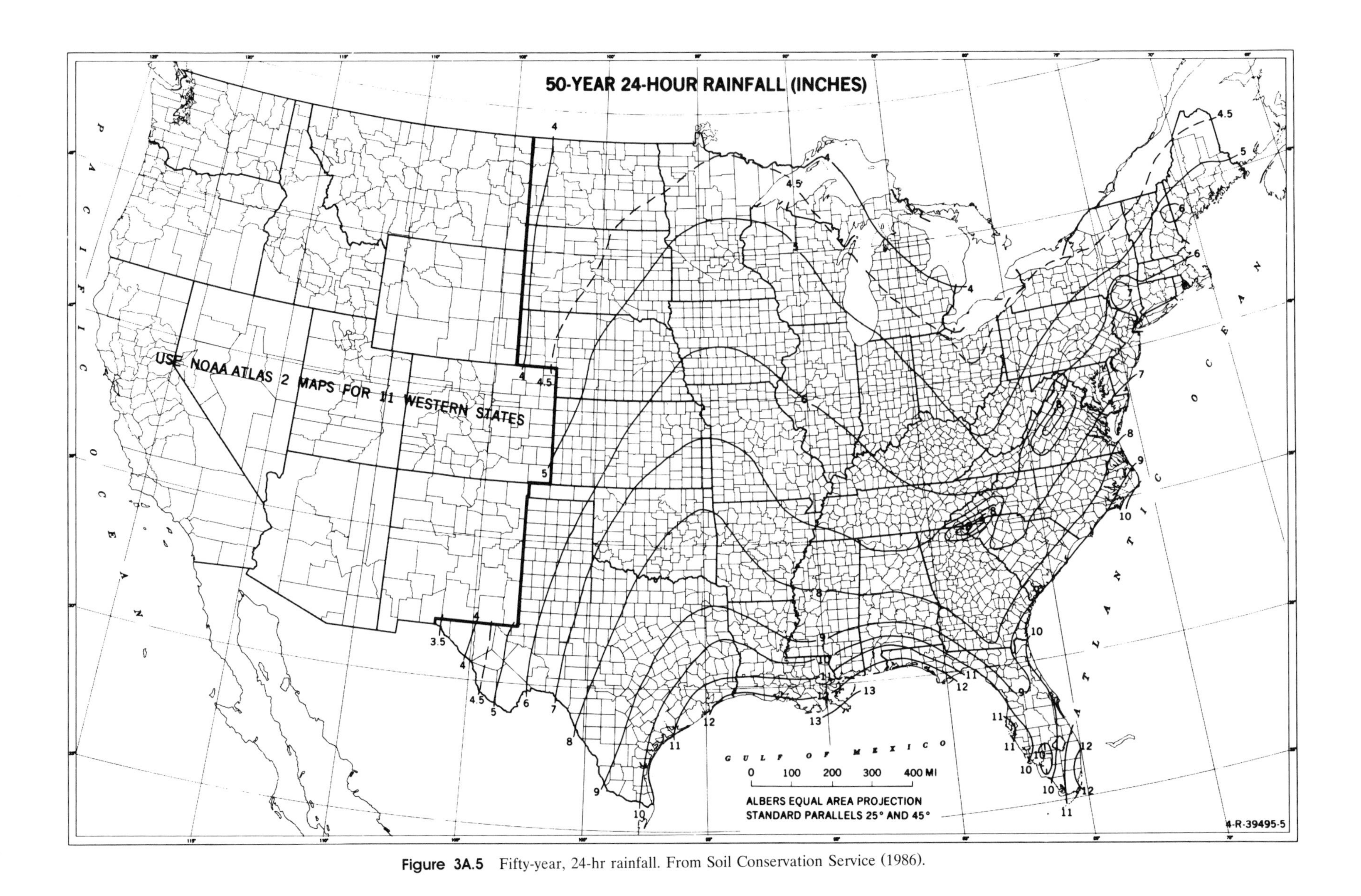

Figure 3A.5 Fifty-year, 24-hr rainfall. From Soil Conservation Service (1986).

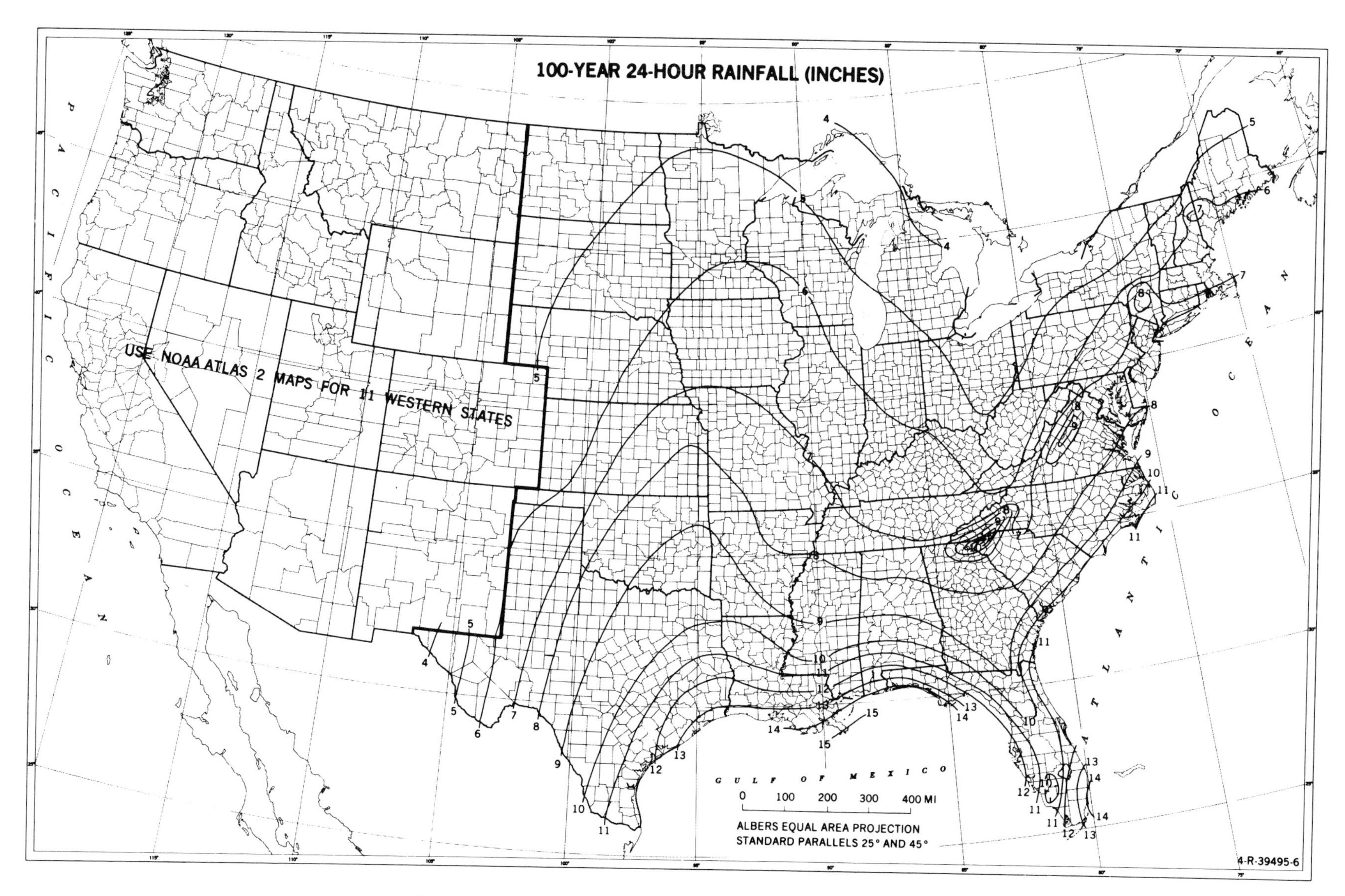

Figure 3A.6 One-hundred-year, 24-hr rainfall. From Soil Conservation Service (1986).

Appendix 3B
Hydrologic Soil Groups

Aasted B
Aberdeen D
Abilene C
Abington B
Acadia D
Acme B
Acton B
Adair D
Adams A
Adamville C
Ade A
Adel B
Adelanto B
Adelphia B
Adler C
Adolph D
Afton D
Agar B
Agate C
Agawam B
Agency C
Agnew B
Aguadilla A
Aguilita C
Ahmeek B
Ahnberg C
Aiken B
Airmont B
Akaka B
Akan C
Akan C
Akaska B
Akaska B
Alachus B
Alaeola B
Alama C
Alamance C
Alamosa C
Albaton D
Albermarle* B
Albertville C
Albia C
Albion B
Alcester B
Alcoa B
Alden D
Alderwood B
Aldino D
Aleen C
Alexandria B
Alexis B
Alford B
Algarrobo D
Algiers D
Alicel B
Allard B
Alleghany B
Allen B
Allendale C
Allenwod B
Allie C
Alligator D
Allison C
Allovers B
Alma C
Almena C
Almirante D
Almo D
Almy B
Alonso C
Alpe C
Altamaha D
Altamont C
Altavista B
Alto C
Alton B
Altoona C
Altura B
Altvan B
Alvin B
Alvira C
Amalu D
Amarillo B
Amenia B
Americua A
Amerlia C
Ames C
Amite B
Amity C
Amsterdam B
Andover D
Andrea B
Angie C
Angola C
Ankeny A
Annandale B
Anoka A
Anselmo B
Antanus C
Anthony B
Antigo B
Apache B
Apakuie A
Apishapa C
Apison C
Applegate C
Appling D
Arch B
Archer C
Arenzville B
Argyle B
Ark C
Arkport B
Arland B
Armagh D
Armer C
Armour B
Armuchee C
Arnold A
Arnot C
Arredondo A
Artesia C
Arvada D
Arveson D
Arzell D
Asa B
Asboy B
Asbury B
Ascalon B
Ash Springs C
Ashby C
Ashe B
Ashkum C
Ashley C
Ashton B
Ashuelot C
Ashwood D
Asotin B
Assumption B
Astoria B
Athelwold B
Athena B
Atherton D
*Athol B
Atkins D
Atterberry B
Atwood B
Au Cree C
Auburn C
Auburndale D
Augusta C
Aurora C
Austin B
Ava C
Avalanche B
Avery B
Avon B
Avonburg D
Axtell D
Ayr B
Babylon A
Baca B
Bagnell D
Baigh C
Bainville B
Baker C
Balch A
Baldock B
Baldwin C
Balfour B
Balm A
Balmorhea D
Bancombe A
Bangor B
Banks A
Bannerville C
Barabook B
Barbour B
Barbourville B
Barclay C
Barnard C
Barnes B
Barneston A
Barnet D
Barney A
Barnhardt A
Barnstead B
Barrancas D
Barron C
Barronett C
Barth B
Bartle D
Basher B
Bass B
Baster B
Bastrop B
Batavia B
Bates B
Bath C
Baudette B
Bayard A
Bayboro D
Bayside C
Baysmon B
Beadle C
Bear Lake D
Bear Prairie B
Bearden C
Beardstown C
Beatty B
Beaucomp C
Beauford D
Beaumont D
Beauregard C
Beaver B
Beaverhead B
Beaverton B
Beckett C
Beckton D
Beckwith C
Bedford C
Bedington B
Beecher C
Beechy D
Belfast D
Belfore C
Belgrade B
Belknap C
Bell D
Belle B

Bellinghouse D
Belmont B
Beltrami B
Beltsville C
Belvoir C
Benefield C
Benevola B
Benld D
Bennington C
Benoit D
**Benson C
Bentonville C
Beotia B
Berg C
Bergland D
Berka C
Berkley C
Berkshire B
Bermudian B
Bernard D
Bernardston C
Berrien B
Berthoud B
Bertle C
Bertolotti A
Bertrand B
Berwyn C
Bethany C
Bethel D
Beulah B
Beverly A
Bewleyville B
Bibb D
Bickleton B
Biddeford D
Bienville B
Big Horn C
Biggs A
Biggsville B
Billett A
Billings D
Binnaville
Bippus B
Birds C
Birdsall D
Birdsboro B
Birkbeck B
Birkhardt B
Biscay D
Bitterroot C
Blacklock D
Blackwater D
Bladen D
Blago D
Blain C
Blair C
Blairton C
Blakeland A
Blakely B
Blanc C
Blanchard A
Blanco B
Blandford C
Blanding B
Blanket D
Blanton A
Blencoe C
Blichton C
Blockton C
Blodgett B
Blomford B
Bloomfield A
Bloomington B
Blount C
Blue Earth D
Bluffton D
Bluford D
Bobtail C
Bodine B
Bogota C
Bohemian B
Bold B
Bolivia B
Bolton B
Bombay C
Bonaccord D
Bonaparte A
Bonham B
Bonilla B
Bonita D
Bonner B
Bonneville A
Bonnie D
Bono D
Bonpas B
Boomer C
Boone A
Bordeaux B
Bosket B
Boswell D
Bow D
Bowamnville D
Bowdoin D
Bowdre C
Bowie B
Boyd D
Boyer B
Boynton C
Bozarth C
Bozeman C
Braceville B
Bracken D
Brackett C
Braddock B
Braden C
Bradenton C
Bradley B
Brady B
Braham B
Brallier A
Brandon B
Brandywine C
Branford B
Brashear C
Bratton B
Braxton B
Brayton C
Brazito A
Brazoe B
Brecknock C
Breece B
Breese D
Bremo C
Brennan B
Brennar C
Brenton B
Bresser B
Brewater C
Brewer B
Brewster D
Briacoe B
Brianfield C
Brickton C
Bridgehampton B
Bridgeport B
Bridger B
Bridgeville B
Briggs A
Briggsdale C
Brill B
**Brimfield
Brimley B
Brinkerton D
Brittain C
Broadbook C
Broadview C
Brockport C
Brooke C
Brookfield B
Brookings B
Brooklyn D
Brookston B
Brooksville D
Broughton D
Broward C
Brownfield A
Browning B
Brownlee C
Bruno A
Bryce D
Bub C
Buchanan C
Buckingham C
Buckland C
Buckley D
Buckner A
Bucks B
Bucoda D
Bude C
Bunkerville D
Burchard B
Burdett C
Burgess B
Burgin D
Burke B
Burleson D
Burnham D
Burnside D
Burnsville B
Burnt Fork B
Burrell C
Burton B
Buse B
Butler D
Butlertown C
Butte B
Buxin D
Buxton C
Byars D
Byrds D
Cabinet C
Cabo Rojo C
Cabot C
Cacapon B
Cacapon B
Caddo C
Cagey D
Caguas D
Cahaba B
Cajon A
Calais B
Caldwell B
Calhoun D
Califon C
Calloway C
Calvert D
Calverton C
Calvin C
Camaguey D
Camas A
Cambridge C
Camden B
Cameron C
Camillus C
Camp B
Campo C
Camroden C
**Canaan C
Canadaigua C
Canadian B
Canadice C
Canaserage B
Cane B
Caneadea C
Caneyville C
Canfield B
Cannett D
Canoncito D
Canyon C
Cape Fear D
Capeshaw C
Capron B
Captina C
Capulin B
Carbo C
Cardiff B
Cardinton C
Carey B
Caribou B
Carisle D
Carlsborg A
Carlton B
Carnegie B
Carnero C
Caroi B
Caroline C
Carrington B
Carrizo A
Carroll D
Carson D
Carstairs C
Carver A
Carytown D
Casa Grande C
Cascade C
Casco A
Casey C
Cashmere A
Cashton D
Cass A
Cass B
Cassville B
Castana B
Castle D
Castner C
Catalina C
Catalpa C
Catano A
Cataula C
Cathcart C
Catlett C
Catlin B
Catoctin C
Catron D
Catskill C
Cattaraugus B
Cave C
Cavode C
Cavot B
Cavour D
Cayagus c
Caylor B
Cayucos D
Cayuga C
Cazenovia C
Cecil B
Celina B
Center C
Centerton B
Central A
Chagrin B
Chalfont D
Chama B
Chamber D
Chamberlino C
Chamokane B
Chandler C
Channahon B
Chariton C
Charleston C
Charloe B
Charlotte D
Charlton B
Chaseburg B
Chastain D
**Chatfield C
Chatsworth D
Chattahoochee B
Chauncey C
Chehalis B

Chelsea A
Chemawa B
Chenango B
Cheney B
Chenoweth B
Cherette A
Cherokee D
Cherry B
Cherryhill B
Cheshire B
Chester B
Chesterfield B
Chetek A
Chewacla C
Chewelsh B
Cheyenne B
Chichasha B
Chiefland A
Chigley C
Chilgren C
Chilhowie D
Chili B
Chillisquaque C
Chillum C
Chilmark C
Chilo D
Chipeta D
Chippewa D
Chiricahua D
Choctaw B
Choptank A
Choteau C
Christian B
Christiana C
Christianburg D
Churchill D
Cialee B
Cialitoe C
Ciane D
Ciapus A
Cicero C
Cie Elum B
Cincinnati B
Cinebar A
Cintrona D
Clackamana C
Claiborne B
Clallam C
Clarence D
Clareville C
Clarinda D
Clarion B
Clark Fork B
Clarksburg C
Clarksdale C
Clarkson C
Clarksville B
Clary B
Clatsop D
Claverack B
Clawson C
Clayton C
Cleaver D
Cleburne B
Cleman B
Clement C
Cleora B
Clermont D
Clifton B
Climax D
Clinton B
Clio C
Clipper C
Clodine B
Cloquallum C
Cloquet B
Cloud D
Clover Creek C
Clovia B
Clyde C
Clymer B
Coamo C
Cobb B
Cochise B
Cocoa B
Cody A
Coeburn C
Cogswell C
Cokedale C
Coker D
Cokesbury D
Colbert D
Colby B
Colden D
Coldwater D
Colebrook Fine
Colebrook Loamy
Fine Sand A
Coleman B
Colemantown D
Colfax C
Colinas C
Collamer B
Collington B
Collins C
Collinsville C
Colo C
Coloao D
Coloase A
Coloma A
Colonie A
Colp D
Colrain B
Colta Neck B
Colton A
Columbia B
Colville B
Colwood D
Colyar C
Comfrey C
Comly C
Commerce C
Comoro A
Compton B
Conassuga C
Concord D
Condit D
Condon B
Conent B
Conestoga B
Conesus C
Congaree B
Conley C
Conottom B
Conover B
Conowingo C
Constable A
Continental C
Conway C
Cook D
Cookeville B
Cookport C
Coolidge B
Coolville C
Cooney C
Cooper B
Copaka B
Copas C
Copeland C
Copelia C
Coplay D
Coral B
Corcege D
Corduroy B
Corkindale C
Corley C
Cornutt D
Corvalis B
Corwin B
Corydon C
Cossayuna C
Cotaco C
Coto C
Cottonwood C
Cougar C
Couparle D
Coupeville B
Course C
Courtland B
Courtney D
Cove D
Coveland C
Coverytown C
Covington D
Cowden D
Cowiche B
Cowling B
Crago B
Craig C
Crandon B
Crane B
Craven C
Crawford D
Creal D
Creedmoor C
Cresbard C
Crescent B
Crestmore C
Crete D
Crevasse A
Crider B
Crockett D
Crofton B
Croghan B
Croom C
Crosby C
Crossville B
Croton D
Crow C
Crowder D
Crowley D
Crown B
Crystal B
Culleoka B
Cullo C
Culpeper B
Culvers C
Cumberland B
Curran C
Curtis B
Cushman B
Custer D
Cut Bank B
Cuthbert C
Dade A
Daggett B
Dakota B
Dalbo B
Dale C
Dalhart B
Dalton C
Dana B
Dandridge C
Daniela B
Danley C
Dannermors C
Danvers C
Dariem C
Darling C
Darnell C
Darret B
Darwin D
Davidson B
Davie D
Dawes C
Dayton D
De Soto C
Deachutes A
Deary C
Decatur B
Deckerville D
Decorrs B
Defiance D
*Dekalb B
Del Rey C
Delanco C
Delfina C
Dell C
Dellrose B
Delphi B
Delpine D
Delray D
Demers C
Denham D
Dennis C
Denny D
Denrock D
Denson C
Denton C
Depew C
Derby A
Descalabrado B
Detour C
Detroit B
Dewart B
Dewey B
Dexter B
Dick A
Dickey A
Dickinson Fine
Sandy Loam A
Dickinson Loam B
Dickson C
Dill B
Dilldown B
Dillinger B
Dillon D
Dilman
Dimmick D
Disco B
Dishlo D
Dixmont B
Dodgeville
Shallow Phase C
Dodgeville Deep
Phase B
Doland B
Dominguito D
Dominic B
Donerail C
Donlonton C
Dorchester B
Dorsey D
Dos Cabezas B
Doty B
Dougherty B
Douglas B
Dover C
Dowelton D
Dowling D
Downs B
Doylestown D
Dragston C
Drake B
Dreeden B
Driping Springs D
Drummer B
Drummond D
Drury B
Dryad C
Du Page B
Duane B
Dubbe B
Dubois C
Dubuque B
Dubuque Deep
Phase B
Dubuque
Shallow Phase C
Duffield B

Duffy C
Dukes A
Dulac C
Dunbar C
Duncan C
Duncannon B
Duncom D
Dundas C
Dundee C
Dune Sand A
Dunellen B
Dungeness B
Dunham B
Dunkirk B
Dunlap C
Dunmore C
Dunning D
Duplin B
Dupo C
Dupont D
Durant D
Durham B
Durkee C
*Dutchess B
Dutson C
Duval B
Dwight D
Dwyer A
Dyke B
Easton C
Eastonville A
Ebbe B
Ebbert D
Ebeys B
Eckman B
Ector C
Edalgo D
Eddy C
Eden C
Edenton C
Edge D
Edgeley C
Edgemont B
Edgewick A
Edgington C
Edina D
Edinburg C
Edisto C
Edith A
Edmonds D
Edna D
Edneyville B
Edom C
Edwards D
Eel C
Efland C
Egam C
Egeland B
Eibbard C
Elbert D
Elburn B
Elco B
Eld B
Eldon C
Eldorado C
Elfort C
Elfrida C
Elioak B
Elk B
Elkina D
Elkinsville B
Elkton D
Ellery D
Elliber A
Elliott C
Ellis D
Ellison B
Ellsberry C
Ellsworth D
Elmmert A
Elmo C
Elmore C
Elsinboro B
Elsmere A
Elswood C
Elwha C
Emmet B
Emory B
Empey C
Empeyville C
Enders C
Enfield B
Englund D
Ennis B
Enon C
Ensenada C
Ensley D
Enstrom B
Enterprise A
Enuclaw B
Ephrata A
Epping D
Era B
Eram D
Erie C
Ernest C
Escondido C
Eskin B
Espinosa B
Esquatzel B
Essex C
Estacion B
Estellin B
Estevan C
Estherville B
Esto C
Etowah B
Ettrick D
Eubanks B
Eufaula A
Eulonia C
Eustis A
Eutaw D
Evans B
Evendale C
Everett A
Everson C
Ewing A
Exline D
Eylar D
Faceville B
Fahey B
Fairfax B
Fairhaven B
Fairhope C
Fairmount D
Fajardo C
Falaya C
Falcon B
Falkner D
Fall B
Fallbrook B
Fallsburg C
Fallsington D
Falun B
Fannin B
Fargo D
Farland B
**Farmington C
Farragut C
Farum C
Fauguier B
Faunce A
Fawcett C
Faxon D
Fayette B
Fe D
Felda D
Felida B
Fellowship C
Fergua C
Fidalgo C
Fillmore D
Fincastle C
Fitch A
Fitchville C
Fitzhugh B
Flamingo D
Flander C
Flandreau B
Flanagan B
Flasher A
Flathead B
Fleetwood B
Fletcher C
Flint C
Flora D
Florence C
Florsheim D
Floyd B
Fluvanna C
Foard D
Foley C
Folsom C
Fordney A
Fordville B
Fore D
Forestdale D
Forrest D
Fort Collins B
Fort Lyon B
Fort Meade A
Fort Pierce B
Fortuna D
Fox B
Foxhome B
Frankford D
Franklinton C
Frankstown B
Fraternidad D
Frederick B
Fredon C
Freehold B
Freeland C
Freeon B
Freer C
Fremont C
Frenchtown D
Freneau D
Frio B
Frost D
Fruita B
Frye C
Fullerton B
Fulton C
Gage D
Gainesville A
Gale B
Galen B
Galestown A
Gallatin D
Gallion B
Galveston A
Galvin B
Gann B
Gara C
Gardnerville C
Garfield D
Garner D
Garrison A
Garwin D
Gasconade D
Gaviota A
Gayville D
Gearhart A
Geary B
Geer C
Geiger D
Gem C
Genesee B
Genoa D
Genola B
Georgetown B
Georgetown C
Gerald D
Germania B
Geronimo B
Gila B
Gilcrest B
Gilead C
Giles A
Gilford D
Gilligan B
Gilman B
Gilpin C
Gilson B
Gilt Edge D
Ginat D
Gird B
Givin C
Glasagow C
Glenbar C
Glencoe D
Glendale C
Glendive A
Gleneig B
Glenfield C
Glenford C
Glenoma B
Glenville C
Gloucester B
Godwin B
Goessel D
Gogebic B
Goldridge A
Goldsboro B
Goldston C
Goldvein C
Goliad C
Gooch C
Gore D
Gorua B
Goshen B
Gosort C
Gothard D
Gowen C
Grady D
Graham C
Grail C
Granby D
Grande Roda C
Grant B
Grantsdale B
Granville B
Grayling A
Great Bend B
Greeham B
Greeley B
Green Bluff B
Green River B
Greenbush B
Greendale B
Greenfield B
Greenport C
Greensboro B
Greenville B
Greenwater A
Greer C
Grenada C
Grenville B
Greybull C
Greys B
Griffin C
Grimstad C
Groeclose C
Groton A
Grove A
Groveland B
Grover B
Groveton B

Grundy C
Guadalupe B
Guanica D
Guyayabo C
Guaysma C
Guckeen C
Gudrid B
Guelph B
Guernsey C
Guin A
Guthrie D
Gypremort C
Habersham B
Haccke D
Hackers B
Hackettstown B
Hadley B
Hagener A
Hagerstown B
Haig C
Haiku C
Hainea B
Halawa C
Haleakala A
Halewood B
Half Moon C
Halfway C
Halii C
Haliimaile B
Hall B
Halsey D
Hamburg B
Hamilton B
Hamlin B
Hammerly C
Hammond D
Hampshire C
Hampton B
Hanaka C
Hanalei C
Hanceville B
Hand B
Hanford B
Hanipoe B
Nannahatchee B
Hanover B
Hanson B
Harbin B
Harbourton C
Harlem B
Harley C
Harlingen D
Harmon D
Harmony C
Harpeter B
Harriet D
Harris D
Harrisburg C
Harrison C
Harstine A
Hartford B
Hartland B
Hartleton B
Hartsburg B
Hartselle B
Harwood B
Hasel C
Hasen B
Haskill A
Hassel D
Hastings C
Hatchie C
Haven B
Havre B
Haxtun A
Hayden B
Hayesville B
Haymond B
Haynie B
Haytar B
Heach B
Heath B
Hebo D
Hector B
Hedville C
Heisler B
Heitt C
Helena C
Hemmni C
Hempstead B
Henderson D
Hennepin B
Henry D
Henshaw C
Herando B
Herbert B
Herkinser B
Hermiston B
Hermitage B
Hermon B
Hermosa C
Herdon C
Hero B
Herrick C
Hershal B
Hesseltin A
Hesson C
Hialeah D
Hiawatha A
Hibbing B
Hickory C
Hicks B
Hidalgo B
Hidewood C
Highfield Higley B
Hiko Springs D
Hildsboro B
Hilgar B
Hilliard B
Hillsdale B
Hilo B
Hilton C
Hinckley A
Hinman D
Hiwassee B
Hiwood A
Hixton B
Hobble B
Hoble B
Hockley C
Hoffman C
Hogansburg B
Hoko C
Holbrook D
Holcomb D
Holdrege B
Hollad B
Hollinger C
**Hollis
Hollister C
Holloway B
Holly C
Hollywood D
Holmdel B
Holston B
Holt B
**Holyoke C
Homer C
Hondo C
Honeoye B
Honokas B
Honolua B
Honomanu C
Honoulfuli C
Hood B
Hoodsport A
Hooker B
Hoosic B
Hopewell C
Hopper B
Hoquiam B
Hord B
Hornell C
Hortman C
Horton B
Hosmer C
Houdexk B
Houghton D
Houlka D
Housetonic C
Houseville C
Houston Black B
Houston D
Houton C
Hovde B
Hoven D
Howard B
Howell B
Hoye C
Hoyleton C
Hoypus A
Hubbard A
Huckabee A
Huckleberry C
Hudson C
Huey D
Huff B
Huffine B
Huggins C
Hugo B
Huikau A
Humacao C
Humbarger B
Humeston C
Humphreys B
Hunt D
Huntaville B
Hunters B
Hunters B
Huntington B
Hurst D
Hutchinson C
Hyde D
Hymon C
Hyrley D
Hysattsville B
Iberia D
Ida B
Idena C
Ilion C
Ilion C
Illiopolis B
Ima B
Immokalee
Imperial D
Ina C
Independence A
Inglefield C
Ingomar D
Inman C
Inola D
Io B
Iola A
Iona B
Ipeva B
Irdell D
Irion D
Irish D
Iron River B
Irurena D
Irvia D
Irving D
Irvington C
Isagore C
Isanti D
Iso B
Isolte B
Isom B
Issaquah B
Istokpoga D
Iuka C
Ivanhoe D
Ives B
Jacana C
Jackson B
Jacob D
Jaffreyu A
Jaucas A
Jeanerette C
Jefferson B
Jerome D
Jersuld D
Jessup D
Joe Creek B
Johnston D
Joliet C
Jonesville A
Joplin B
Josefa D
Josephine C
Joy B
Juana Diaz B
Judith B
Judson B
Jules B
Juliaette A
Juncoe C
Juniata B
Junius C
Juno A
Kaena D
Kahana B
Kaines B
Kalamzoo B
Kalihi D
Kalispell B
Kalkaska A
Kalmia B
Kaloko D
Kamnanui C
Kanab C
Kanapaha B
Kaneohe B
Kapapela B
Karnak D
Karro B
Kars B
Kasota C
Kasson C
Kato B
Katy D
Kaufman D
Kawaihae C
Kawsihapai A
Keansburg D
Keating C
Kedron C
Keelakekua B
Keene C
Keith B
Kelly D
Lelso C
Kelton A
Kempsville B
Kempton B
Kenansville B
Kendaia C
Kendall B
Kennebec B
Kennedy B
Kenney A
Kenspur B
Kent D
Keomah C
Kerby B
Keri C
Kerrtown B
Kershaw A
Kessau D
Kettle B

Kettleman C
Keyport C
Keysport C
Keystone A
Kibbie C
Kickerville B
Kilauea A
Kilbourne A
Kimbrough D
Kinghurst B
Kings C
Kinrose D
Kipling D
Kipp B
Kipson D
Kirkland D
Kirvin C
Kistler B
Kitsap C
Kittitas B
Kittson B
Kiwanis B
Klaberg C
Klamath C
Klaus A
Klej B
Kline A
Klinesville C
Knappa B
Knight C
Know B
Koch C
Koehler B
Kohala B
Kokokahi D
Kokomo D
Konokti A
Koolsu D
Kopish D
Kosmos D
Koster C
Kranzburg B
Krause B
Kreamer C
Kresson C
Krum C
Kukaisu
Kunia B
Kutztown C
La Brier C
La Palma B
La Prairie B
La Rose B
La Verkin B
LaBelle C
Labette C
Labounty C
Lacamas D
Lackawanna B
Ladd C
LaDelle B
Ladoga C
Ladysmith D
Lafe D
Lagonda D
LaGrande C
LaHogue B
Lahontan D
Laidig C
Laidlaw A
Laie D
Lairdsville B
Lajas C
Lake Charles D
Lake Creek B
Lakehurst A
Lakeland A
Lakemont C
Lakeville Loam B
Lakeville Sandy
Loam A
Lakewood A
Lakin A
LaLande B
Lamington D
Lamont Fine
Sandy Loam A
Lamont Loam B
Lamonta C
Lamoure C
Lamson C
Lanark B
Lancaster B
Land C
Landes B
Landisburg C
Lane C
Lane C
Langford C
Langley b
Lanham D
Lanrgrell B
Lansdale B
Lansdowne C
Lansing B
Lantz D
Lapine A
Lapon D
Laporte D
Laredo B
Lares C
Largent C
Largo C
Larimer B
Larkia B
Larry D
Las Animas A
Las Casa C
Las Lucas D
Las Piedras D
Las Vegas D
Lashley B
Lassen C
Latah C
Lauderdale C
Laurel B
Lauren A
Laveen B
Lawhorn D
Lawrence C
Lawrenceville
Clubbock C
Lawson B
Lawton B
Lax C
Leadvale C
Leaf D
Leavenworth A
Leavitt B
Leavittville B
Lebanon D
LeBar B
Leck Kill B
Lee B
Lee D
Leeds B
Leel B
Leetonia B
Legore C
Lehigh C
Leicester C
Lela D
Lempster D
Lena D
Lenoir Fine
Sandy Loam B
Lenoir Fine
Sandy Loam C
Lenox C
Leon C
Leona D
Leonardtown D
Leota C
Leshara B
Lester B
LeSueur B
Letcher D
Letort B
Levan A
Lewisberry B
Lewiston C
Lewisville B
Lexington B
Lick B
Lick Creek B
Lickdale D
Lightning D
Lignum C
Lihen A
Likes A
Lima
Limerick C
Lincoln A
Lindley C
Lindsborg D
Lindside C
Linecroft A
Linganore C
Link B
Linker B
Linneus B
Lino C
Lintonia B
Lisbon B
Lismas D
Lismore B
Little Horn B
Littlefield D
Littleton B
Litz C
Liverty C
Livingston D
Llave C
Lloyd B
Lobdell B
Lobelville C
Lockhard C
Lockhart B
Lockport C
Locust C
Lodi B
Lofton C
Logan C
Logendale C
Lolo B
Lomax B
Lone Rock B
Lonepine B
Longford C
Longrie C
Lonoka B
Lookout C
Loon A
Lorain C
Lordstown C
Lorella C
Lorenso A
Loring B
Lorsdale B
Los Guineoa D
Los Osoe C
Loudon C
Loudonville C
Louisa B
Louisburg B
Loup D
Lowell C
Loy D
Loysville C
Lualualei D
Lucas C
Lucien C
Ludlow C
Lufkin D
Lukin C
Lumni C
Lun C
Lunt C
Lupton D
Lura D
Luray C
Luton D
Luverne B
Luzena D
Lycoming C
*Lyden
Lynchburg C
Lynden A
Lynndyl C
Lyons D
Lystair A
Mabi D
Machete C
Mack B
*Macomber c
Macon D
Madalin C
Maddock A
Maddox B
Madison B
Madras C
Madrid B
Maginnis C
Magnolia B
Mahaska B
Mahnomen B
Mahoning D
Maiden C
Maile B
Makalapa D
Makawso B
Makena B
Malago B
Malays C
Maleza B
Mamala C
Manalapan D
Manana C
Manassa B
Manassas B
Manatash C
Manatee D
Manchester A
Mangua C
Manhattan B
Manhelm C
Manor B
Mansfield D
Mansic C
Mansker B
Mantachie C
Manteo D
Manvel B
Maple D
**Mapleton c
Marble A
Marcus D
Marcy D
Mardin C
Marengo C
Marhsall B
Mariana C
Marias D
Marina A
Marion D
Marissa C
Markland C
Marksboro B
Marlboro B
Marletta B

Marlow C
Marlton C
Marna D
Marquette B
Martha C
Martin Pena D
Martinsdale B
Martinton C
Masada B
Mason B
Massena C
Massillon B
Matansas C
Matapeake B
Matawan C
Matlock D
Matmon C
Matney B
Mattapex C
Maumee D
Maunabo D
Maury B
Maverick D
May B
Mayhew D
Maynard Lake B
Mayo
Mayodan B
Maytown C
Mazeppe B
McAfee B
McAllister C
McBride B
McDonald C
McDowell
McEwen C
McGary C
McKamie C
McKay D
McKenna B
Mckenzie D
McLain C
McMurray A
McNeal C
McPherson D
Meadin A
Meadowville
Meadville B
Mecklenburg C
Meda B
Medary C
Medford C
Medina B
Medio C
Meeteetee C
Mehlhorn C
Meigs C
Melbourne C
Mellenthin B
Melrose C
Melvern C
Melvin D
Memphis B
Menahga A
Mench B
Menfro B
Menlo D
Mentor B
Mercedita D
Mercer C
Mereta D
Meridian B
Meros A
Merrimac B
Mertz B
Mesa B
Meskill D
Metes B
Methow B
Metolum A
Mexico D
Mhoon C
Miami B
Micres A
Middlebury B
Midland D
Mifflinburg B
Miquel D
Milaca B
Milam B
Miles B
Milford c
Mill Creek B
Millbrook B
Miller D
Millington B
Millsdale B
Milo D
Milroy D
Mimosa C
Minatare D
Minco B
Mineola B
Miner D
Minnequa B
Minora C
Minvale B
Miota D
Mission C
Mitchell B
Moca D
Modale C
Moenkopie C
Moffat B
Mohave C
Mohawk B
Moiese B
Moira C
Mokena C
Molena A
Molina D
Monarda C
Monee D
Monmouth C
Monona B
Monongahala C
Monroeville C
Montalto B
Montara D
Montell D
Monteola D
Montesano B
Montevallo C
Montgomery D
Monticello B
Montoya D
Moody B
Moreau D
Moree D
Morley C
Morman Mesa
Moro Cojo A
Moro May D
Morocco C
Morrill B
Morris C
Morrison B
Morrow C
Morton B
Moscow B
Moshannon B
Mossyrock A
Mottsville A
Mount Carroll B
Mount Lucas C
Mountview B
Mucars B
Muir B
Muirkirk B
Mukilteo A
Mullins D
Munising B
Munuacong D
Murrill B
Muscatine B
Muse B
Muskingnum C
Muskogee C
Musselshell B
Myatt D
Myersville B
Naches B
Nacimiento C
Macogdoches C
Naiwa B
Nakelele B
Nantucket C
Nanum B
Napa D
Napier B
Nappanee D
Naranjito
Narcisse B
Narragansett C
Nasel C
Naslehu B
Nason C
**Nassau C
Natalie C
Natches B
Natchitoches D
National A
Navajo D
Navesink B
Navesota D
Naylor C
Nebish B
Neble A
Needmore C
Negley B
Nehalem B
Nellie B
Neosho D
Neptune A
Nereson B
Neshaminy B
Nesika B
Nester D
Neubert B
Nevada D
Neville B
New Cambria C
Newark C
Newart B
Newberg B
Newbery C
Newfane B
Newkirk B
Newport B
Newton D
Newtonia B
Nicholson B
Nicholville B
Nickel D
Nicollet B
Niles C
Ninigret B
Ninrod C
Nipe B
Nisqually A
Nixa
Nixon B
Nixonton B
Nobescot A
Noble B
Nodaway B
Nogalee C
Nohili D
Nolan C
Nolichucky B
Nolo C
Nonopahu C
Nookachamps D
Nooksack C
Nora B
Norden B
Norfolk B
Norge C
Norma C
North Powder C
Northport C
NorthumberlndC
Northville D
Norton B
Norwich D
Norwood B
Nosbig D
Nuby C
Nuckolls C
Nueces A
Nunda C
Nunn C
Nutley D
Nymore A
Oahe B
Oakford B
Oakland B
Oasie C
O'Fallon D
O'Neill B
Ochlockonee B
Ochopee D
Ockley B
Oconee C
Odessa C
Odin C
Ogemaw B
Okav D
Okeechobee D
Okeelanta D
Okemah C
Okenee D
Okoboji B
Oktibbeha D
Olequa C
Olinda B
Olivier C
Olmitz B
Olmsted C
Olsa B
Olympic B
Omaha B
Omega A
Ona C
Onalaska B
Onamia B
Onarga B
Onaway B
Ondawa B
Oneida C
Onslow B
Ontario B
Ontonagon C
Onyx B
Ookala B
**Oquaga C
Oquawka A
Ora C
Oracle B
Orange D
Orangeburg B
Orcas A
Orchard B
Ordway D
Orelia D
Orells D
Orient B
Orienta B
Orio C
Orion C

Orlando High Phase A
Orlando Low Phase B
Orman D
Orrville C
Ortello A
Orting C
Osage D
Osceola D
Oshawa D
Oshtemo B
Osmund B
Oso A
Ostrander B
Otero B
Othello D
Otisville A
Otsego C
Ottawa A
Otter D
Otterhold B
Ottokee A
Otway D
Ovid C
Owaneco D
Owen Creek C
Owens D
Ozona C
Pace B
Paden C
Page D
Pahranagat C
Pahroc D
Paia B
Painesville B
Paiute B
Palestine B
Paletine C
Palm Beach A
Palmas Atlas D
Palmdale C
Palmyra B
Palouse B
Palso C
Pandura B
Panton D
Papago B
Papakating D
Papineau C
Parishville C
Parkdale A
Parker B
Parks B
Parkwood C
Parnell D
Parr B
Parsons D
Pasco B
Pasloe B
Paso Seco C
Pasquotank D
Pasuhau B
Patent C
Patit Creek B
Patoutville C
Patrick B
Patton C
Paulding D
Pauwela C
Pawlet B
Pawnee D
Paxton C
Paymaster B
Payne D
Peace River D
Peacham D
Pearman
Pearson C
Pecetonics B
Pedernales C
Pekin C
Pelan B
Pella B
Pence B
Penn C
Pennington B
Penoyer C
Penrose C
Penwood A
Peoh C
Peona C
Peotone C
Pequea C
Perkinsville B
Perks A
Perrine D
Perry D
Persayo D
Pershing C
Peru C
Peshastin A
Patoskey A
Petrolis D
Pettis B
Pewbroke B
Pheba C
Phelpe B
Phillips D
Philo C
Piagah B
Picacho D
Pickford D
Pickway C
Pickwick B
Pierce B
Pierre D
Pilchuck A
Pilot B
Pilot Rock C
Pilhonua B
Pima C
Pinal D
Pinckney C
Pinones D
Pinson B
Pintura A
**Pittsfield C
Pittstown C
Pittwood B
Placentia D
Plainfield A
Plano B
Plata C
Platea C
Platner C
Plattsmouth B
Plattville B
Pledger D
Plummer D
Plymouth B
Pocomoke D
Podunk B
Poinsett B
Poland B
Polson C
Pomello A
Pompano D
Pomroy B
Pope B
Port B
Port Bryon B
Portales B
Porters B
Portland D
Portsmouth D
Portugues D
Poskin C
Poso Blanco C
Post D
Potamo D
Potter C
Pottsville D
Poultney B
Poverty C
Powder B
Powell C
Poygan D
Prather C
Pratt A
Prentiss C
Prescott D
Presque Isle B
Preston A
Prewitt C
Prieta B
Princeton B
Pring B
Proctor B
Progresso B
Promise D
Prospect B
Prosser B
Providence C
Provo B
Prowers B
Ptarmingan B
Puget B
Puhi B
Pulaski B
Pulehu A
Pullman D
Puna B
Purdy D
Purgatory D
Puu Oo B
Puu Pa B
Puysllup B
Quamba D
Quandahl B
Quay C
Quicksell C
Quincy A
Quinlan B
Quonset A
Raber C
Rabun B
Racine B
Racoon D
Radford B
Radnor D
Ragnar D
Rago C
Raina D
Rainbow C
Ralston B
Ramona C
Ramsey B
Randall D
Ranger C
Rankin C
Rantoul D
Rapidan C
Rarden C
Raritan B
Raub B
Rauville D
Ravalli D
Ravenna C
Ravola B
Ray B
Rayne B
Reagan C
Reaton B
Reaville C
Rebuck C
Red Bay B
Red Hook C
Redfield B
Redington C
Redlands B
Redmond B
Reed D
Reeser C
Reeves C
Regent C
Regnier D
Reinach B
Reliance C
Renfrow D
Reno D
Renohill C
Renova B
Renshaw B
Renslow B
Rentide C
Reparada D
Retsof C
Rex C
Reynolds B
Rhinebeck C
Rhoades D
Richfield B
Richland C
Richview C
Richwood B
Ridgebury C
Ridgely B
Ridgeville B
Riesel D
Riffe A
Riga C
Riggs D
Riley B
Rillito B
Rimer C
Rinard D
Ringling C
Ringold B
Rio Arribe D
Rio Canas C
Rio Lajas A
Rio Piedras D
Ritchey B
Rittman C
Ritzville B
Riverside A
Riverton C
Roane C
Roanoke D
Robbe D
Robertsville D
Robinsonville B
Roby C
Rockaway B
Rockbridge B
Rockdale D
Rockmart C
Rockport B
Rockton B
Rockwood B
Rocky Ford B
Rodman A
Roe B
Roebuck D
Rogers D
Rohrersville D
Rokeby D
Rolfe C
Romeo C
Rosario C
Roscoe C
Roscommon D
Rosebud B
Rosedell D
Roselms D
Rosemount B
Roseville B
Rositas A
Rosyln C

Ross B
Rosschi D
Rossmoyne C
Rosulus D
Round Butte D
Routon D
Rowe D
Rowland C
Rowley C
Rox B
Roy B
Royalton C
Roza D
Rozetta B
Ruark C
Rubicon A
Rubio C
Rucker B
Rudyard C
Rumford B
Rumney C
Rupert A
Rushtown A
Rushville D
Ruskin C
Russell B
Russellville C
Ruston B
Rutlege D
Ryder C
Sabana C
Sabana Seca D
Saco D
Saffell B
Sage D
Sagemoor C
Salal B
Salem B
Salemsburg B
Salisbury D
Salix B
Salkum D
Salmon B
Salol D
Saltillo C
Saluvia C
Salvisa D
Samish D
Sammamish B
Sams B
San Antonio D
San German C
San Jose B
San Josquin D
San Juan B
San Saba D
Sango C
Santa C
Santa Clara D
Santa Isabel D
Santa Lucia C
Santiago B
Sargeant D
Sarpy A
Sassafras B
Sauble A
Saugstuck C
Sauk B
Savage C
Savannah C
Sawaill C
Sawyer C
Saybrook B
Scandia B
Scantic C
Scarboro D
Schapville C
Schoharie C
Schooley D
Schumacher C
Schuylkill B
Scio B
Sciotoville C
Scipio D
Scituate C
Scobey B
Scott D
Scott Lake B
Scowlale B
Scranton C
Searing B
Seaton B
Sebeka D
Sebewa D
Sebring D
Sedan C
.Sediu C
Segal D
Segno B
Selah C
Selkirk D
Selle A
Selma B
Semiahmoo C
Sequatchie B
Sequois C
Serrano D
Sewtooth C
Sexton D
Seymour C
Shannon B
**Shapleigh C
Sharkey D
Sharon B
Sharpeburg B
Shavano B
Shelburn C
Shelby C
Shelbyville c
Shelmadine c
Shelocata B
Shelton B
Sheppard A
Sheridan B
Sherman B
Shiloh C
Shoals C
Shook B
Shoshone B
Shouna B
Shrewsbury D
Shubuta C
Shuvah C
Sicily B
Sidell B
Sierrs C
Sifton B
Signal C
Siler A
Silerton B
Silver Creek D
Sima D
Simcoa B
Simla B
Sinai C
Sinclair B
Singsaas B
Sioux B
Sipple B
Siskiyou B
Sites C
Skaggs C
Skagit B
Skalkho B
Skamanis B
Skames C
Skerry C
Skiyou B
Skokomish A
Skyberg C
Skykomish A
Sleeth C
Sloan D
Slocum D
Smoky Butte C
Smolan C
Snow B
Soda Lake B
Sodus C
Soga C
Soller D
Solomon D
Somers B
Somerset B
Sonoita B
Sontag D
Souva D
Sparta A
Spearfish B
Spencer B
Sperry C
Spilo D
Spooner C
Spottswood B
Spring Creek D
Spring D
Springer A
Springfield D
Springtown B
Spur B
St. Albans B
St. Charles B
St. Clair D
St. Helens A
St. Joe B
St. Johns D
St. Lucie A
St. Marys C
St. Paul C
Staatsburg C
Stacum D
Stambaugh B
Stamford D
Stanfield C
Stanton D
Starks C
Starr B
Staser B
State B
Steekee C
Steinauer B
Steinsburg B
Stendal C
Stephensburg C
Stephenville B
Stetson B
Stevenson B
Steward D
Stidham B
Stimmon
Stissing C
Stockbridge C
Stockland B
Stockton B
Stoneham B
Stonington B
Stono C
Stookey B
Storden B
Story C
Stough C
Stoy D
Strasburg C
Strauss C
Strawn B
Stronghurst B
Stryker D
Stukal C
Stumpp D
Sudbury B
Suffield C
Sula B
Sulphurs C
Sultan B
Sumas C
Sumit C
Summerville C
Sumner A
Sumter D
Sun D
Sunbury B
Suncook A
**Sunderland C
Sunniland C
Sunnyside B
Sunrise B
Sunsweet C
Superstition A
Surry B
Susquehanna D
Sutherlin C
Sutphen D
Sutton B
Swaim C
Swanton C
Swantown A
Swartswood C
Sweden B
Sweeney C
Sweetwater D
Swims B
Switzer D
Swygert C
Sylvan B
Symerton B
Tabernash B
Tabler D
Tabor D
Taft C
Tohoe C
Talanta D
Talbott D
Talcot D
Talihina D
Talladega C
Tallula B
Tally B
Taloka D
Tama B
Tammas C
Tanama C
Tanberg D
Taneum C
Tanwax A
Taos B
Tarrant D
Tate B
Tatum B
Taylor C
Teague D
Teas C
Tedrow A
Teja C
Tell B
Teller B
Tellico B
Tenino A
Tepee D
Teresa D
Terril B
Terry B
Tescott B
Teton B
Tetonka C
Thackery B
Thatuna C
Thayer B
Thomasville C
Thompson A
**Thorndike C

Thornton D
Thornwood A
Thoroughfare A
Thorp C
Thurman A
Thurmont B
Thurston B
Tiburones D
Tice C
Tickfaw D
Tijeras B
Tilden C
Tillman B
Tilsit C
Timmer B
Timmerman A
Timpahute D
Timula B
Tinton B
Tiocano D
Tioga B
Tippah D
Tipparary A
Tippecanoe B
Tipton B
Tirah B
Tisbury B
Tisch C
Tishominto C
Tiston Tofton B
Titusville C
Tiwoli A
Toa C
Tobin B
Tobosa D
Todd B
Toddville B
Tokul A
Toledo D
Tolley B
Tolo B
Toltec B
Tombigbee A
Tonawanda C
Tongue River C
Tonopah B
Toppenish C
Topton B
Torres C
Tortugas D
Tours C
Toutle A
Tower D
Townsbury B
Townsend B
Toxaway D
Toyah B
Toyey B
Traer C
Traller B
Transylvania B
Trapper B
Travessilla B
Travessilla D
Travis C
Treadway D
Trego C
Trempealeau B
Trenary B
Trent B
Trexler C
Trinity D
Tripp B
Tromp B
Trout River A
Trowbridge B
Troxel B
Troy C
Truman B
Trumbull D
Tuacumbia D
Tubac D
Tucker D
Tucumcari B
Tuffit B
Tughill D
Tujunga A
Tuller D
Tully C
Tumacacori B
Tumbez D
Tumwater A
Tunica D
tunkhannock B
Tupelo D
Turbotville C
Turis C
Turkey Creek B
Turnbow D
Turner B
Turnerville C
Tuscan B
Tuscarora C
Tuscola B
Tusquites B
Tuxedo D
Twin Creek B
Twin Lakes B
Two Dot B
Tyler C
Udolpho C
Uinta B
Ulen B
Ulm B
Ulupalakum B
Ulysses B
Umapine B
Una D
Unadilla B
Uncompahgre B
Ungers B
Union C
Unison B
Unity A
Upehur C
Urbana B
Urbo D
Ursula D
Usine D
Utica B
Utuado B
Uvalde B
Vader B
Vaiden D
Vale B
Valentine A
Valers C
Vallecitos C
Valois C
Vance C
Vandalia C
Vanderville C
Vanoss B
Varna C
Vaucluse C
Vayma D
Vebad B
Vega Alta C
Vega Baja D
Vekok C
Velma B
Venango D
Venedy D
Verdel D
Verdigris B
Verdun D
Vergennes D
Verhalen D
Vernon D
Verona B
Veyo D
Via C
Vicksburg B
Victor B
Victoria D
Vienna B
Vilas A
Viola D
Vira B
Virden C
Virgil B
Virgin River D
Virtue C
Vista B
Viston A
Vivee C
Vivi B
Vlassaty C
Volga D
Volin B
Volke C
Volney B
Volperite C
Voluaia C
Vona A
Vrooman B
Wacoasta C
Wade D
Wadell D
Wadena B
Wadesboro B
Wadsworth D
Wagner D
Waha C
Wahee C
Wahiawa B
Wahtum C
Waialua B
Waikaloe B
Waikapu B
Wailes B
Waimanalo D
Waimea B
Waipahu C
Waiska B
Waits B
Wakeland B
Wakonda C
Walla Walla B
Wallace B
Waller C
Wallington C
Wallkill C
Wallpack B
Walpole C
Walsh B
Walters D
Walton C
Wampeville B
Wann A
Wapato C
Wapping B
Ward D
Warden B
Warman D
Warne D
Warners C
Warrenton A
Warrior C
Warsaw B
Wartrace C
Warwick B
Washjburn D
Washington B
Washoe C
Washougal C
Washtenaw C
Wassaic C
Wassuk D
Wagauga B
Watchaug B
Waterboro D
Wateloo C
Waterville C
Watsaka A
Watson C
Watsonvile D
Watt D
Watton C
Waubay B
Waugh C
Waukeesha B
Waukegan B
Waukon B
Waumbek B
Wausson D
Waverly D
Wayland C
Wayne B
Waynesboro B
Wea B
Weaver C
Webb C
Webster C
Weeksville B
Wehadkee D
Weikert C
Weinbach C
Weir D
Weld C
Weller D
Wellington D
Wellman C
Wellsboro C
Wemple B
Wenaa C
Wenatchee B
Wesley C
Wessington B
West Point D
Westbury C
Westfall C
Westland D
**Westminister C
Westmoreland C
Weston C
Westphalia B
Westport A
Westville C
Wethersfield C
Weymouth C
Whalan B
Wharton C
Whatcom D
Whately D
Wheeling B
Whidbey A
Whippany C
White House C
White Store D
White Swan C
Whitefish B
Whiteford C
Whitelaw
Whitesburg C
Whiteson D
Whitetail B
Whitlock A
Whitman D
Whitson D
Whitwell C
Wibaux C
Wibaux C
Wichita C
Wickersham C
Wickham B
Wickiup A
*Wilbraham C
Wilcox D
Wildwood D

Wilkee C
Wilkeson B
Will D
Williamette B
Willard C
Williams B
Williamsburg C
Williamson B
Willoughby C
Willow Creek B
Wilson D
Winchester A
Wind River A
Windom B
Windsor A
Windthorst C
Winema B
Winfield C
Wingville B
Winifred D
Winlock D
Winnett D
Winooski B
Winslow B
Winston A
Winterset C
Witt B
Wolcottsburg C
Woldale D
Wolf B
Wolfever C
Wood River D
Woodbridge C
Woodglen D
Woodinville B
Woodlyn C
Woodson D
Woodstown C
Woodward B
Wooster B
Woostern B
Worland B
Worsham D
Worth C
Worthen B
Worthington B
Wortman C
Wrightsville D
Wurtsboro C
Wykoff B
Wynoose D
Xenia B
Yabucoa D
Yadkin B
Yahola B
Yakim A
Yale C
Yauco D
Yeoman B
Yoder B
Yonaba C
Yordy B
York C
Yunes D
Zaca D
Zahl B
Zaleeki B
Zaneis C
Zapeta D
Zell B
Zimmerman A
Zion C
Zipp C
Zita B
Zook D
Zuber B
Zwingle D

Appendix 3C

Runoff Curve Numbers

Table 3.2a Runoff Curve Numbers for Urban Areas[a]

Cover description		Curve numbers for hydrologic soil group			
Cover type and hydrologic condition	Average percent impervious area[b]	A	B	C	D
Fully developed urban areas (vegetation established)					
Open space (lawns, parks, golf courses, cemeteries, etc.)[c]					
Poor condition (grass cover < 50%)		68	79	86	89
Fair condition (grass cover 50% to 75%)		49	69	79	84
Good condition (grass cover > 75%)		39	61	74	80
Impervious areas					
Paved parking lots, roofs, driveways, etc. (excluding right-of-way)		98	98	98	98
Streets and roads:					
Paved; curbs and storm sewers (excluding right-of-way)		98	98	98	98
Paved; open ditches (including right-of-way)		83	89	92	93
Gravel (including right-of-way)		76	85	89	91
Dirt (including right-of-way)		72	82	87	89
Western desert urban areas					
Natural desert landscaping (pervious areas only)[d]		63	77	85	88
Artificial desert landscaping (impervious weed barrier, desert shrub with 1- to 2-inch sand or gravel mulch and basin borders)		96	96	96	96
Urban districts					
Commercial and business	85	89	92	94	95
Industrial	72	81	88	91	93
Residential districts by average lot size					
1/8 acre or less (town houses)	65	77	85	90	92
1/4 acre	38	61	75	83	87
1/3 acre	30	57	72	81	86
1/2 acre	25	54	70	80	85
1 acre	20	51	68	79	84
2 acres	12	46	65	77	82
Developing urban areas					
Newly graded areas (pervious areas only, no vegetation)		77	86	91	94
Idle lands (CN's are determined using cover types similar to those in Table 3.2c).					

Note. Source: Soil Conservation Service (1986).

[a]Average runoff condition, and $I_a = 0.2S$.

[b]The average percent impervious area shown was used to develop the composite CN's. Other assumptions are as follows: impervious areas are directly connected to the drainage system, impervious areas have a CN of 98, and pervious areas are considered equivalent to open space in good hydrologic condition. CN's for other combinations of conditons may be computed using Eq. 3.24.

[c]CN's shown are equivalent to these of pasture. Composite CN's may be computed for other combinations of open space cover type.

[d]Composite CN 's for natural desert landscaping should be computed using Eq. 3.24 based on the impervious area percentage (CN = 98) and the pervious area CN. The pervious area CN's are assumed equivalent to desert shrub in poor hydrologic condition.

[e]Composite CN's to use for the design of temporary measures during grading and construction should be computed using Eq. 3.24, based on the degree of development (impervious area percentage) and the CN's for the newly graded pervious areas.

Table 3.2b Runoff Curve Numbers for Cultivated Agricultural Lands[a]

Cover description			Curve numbers for hydrologic soil group			
Cover type	Treatment[b]	Hydrologic condition[c]	A	B	C	D
Fallow	Bare soil	—	77	86	91	94
	Crop residue cover (CR)	Poor	76	85	90	93
		Good	74	83	88	90
Row crops	Straight row (SR)	Poor	72	81	88	91
		Good	67	78	85	89
	SR + CR	Poor	71	80	87	90
		Good	64	75	82	85
	Contoured (C)	Poor	70	79	84	88
		Good	65	75	82	86
	C + CR	Poor	69	78	83	87
		Good	64	74	81	85
	Contoured & terraced (C&T)	Poor	66	74	80	82
		Good	62	71	78	81
	C&T + CR	Poor	65	73	79	81
		Good	61	70	77	80
Small grain	SR	Poor	65	76	84	88
		Good	63	75	83	87
	SR + CR	Poor	64	75	83	86
		Good	60	72	80	84
	C	Poor	63	74	82	85
		Good	61	73	81	84
	C + CR	Poor	62	73	81	84
		Good	60	72	80	83
	C&T	Poor	61	72	79	82
		Good	59	70	78	81
	C&T +CR	Poor	60	71	78	81
		Good	58	69	77	80
Close-seeded or broadcast legumes or rotation meadow	SR	Poor	66	77	85	89
		Good	58	72	81	85
	C	Poor	64	75	83	85
		Good	55	69	78	83
	C&T	Poor	63	73	80	83
		Good	51	67	76	80

Note. Source: Soil Conservation Service (1986).
[a]Average runoff condition, and $I_a = 0.2S$.
[b]*Crop residue cover* applies only if residue is on at least 5% of the surface throughout the year.
[c]Hydrologic condition is based on combination of factors that affect infiltration and runoff, including (a) density and canopy of vegetative areas, (b) amount of year-round cover, (c) amount of grass or close-seeded legumes in rotations, (d) percent of residue cover on the land surface (good ≥ 20%), and (e) degree of surface roughness. *Poor:* Factors impair infiltration and tend to increase runoff. *Good:* Factors encourage average and better than average infiltration and tend to decrease runoff.

Table 3.2c Runoff Curve Numbers for Other Agricultural Lands[a]

Cover type	Hydrologic condition	A	B	C	D
Pasture, grassland, or range—continuous forage for grazing[b]	Poor	68	79	86	89
	Fair	49	69	79	84
	Good	39	61	74	80
Meadow—continuous grass, protected from grazing and generally mowed for hay.	—	30	58	71	78
Brush—brush-weed-grass mixture with brush the major element[c]	Poor	48	67	77	83
	Fair	35	56	70	77
	Good	30[d]	48	65	73
Woods—grass combination (orchard or tree farm)[e]	Poor	57	73	82	86
	Fair	43	65	76	82
	Good	32	58	72	79
Woods[f]	Poor	45	66	77	83
	Fair	36	60	73	79
	Good	30	55	70	77
Farmsteads—buildings, lanes, driveways, and surrounding lots.	—	59	74	82	86

Cover description; curve numbers for hydrologic soil group (A–D).

Note. Source: Soil Conservation Service (1986).

[a]Average runoff condition, and $I_a = 0.2S$.

[b]*Poor:* <50% ground cover or heavily grazed with no mulch. *Fair:* 50 to 75% ground cover and not heavily grazed. *Good:* >75% ground cover and lightly or only occassionally grazed.

[c]*Poor:* <50% ground cover. *Fair:* 50 to 75% ground cover. *Good:* >75% ground cover.

[d]Actual curve number is less than 30; use CN = 30 for runoff computations.

[e]CN's shown were computed for areas with 50% woods and 50% grass (pasture) cover. Other combinations of conditions may be computed from the CN's for woods and pasture.

[f]*Poor:* Forest litter, small trees, and brush are destroyed by heavy grazing or regular burning. *Fair:* Woods are grazed but not burned, and some forest litter covers the soil. *Good:* Woods are protected from grazing, and litter and brush adequately cover the soil.

Table 3.2d Runoff Curve Numbers for Arid and Semiarid Rangelands[a]

Cover type	Hydrologic condition[b]	A[c]	B	C	D
Herbaceous—mixture of grass, weeds, and low-growing brush, with the minor element	Poor		80	87	93
	Fair		71	81	89
	Good		62	74	85
Oak-aspen—mountain brush mixture of oak brush, aspen, mountain mahogany, bitter brush, maple, and other brush	Poor		66	74	79
	Fair		48	57	63
	Good		30	41	48
Pinyon-juniper—pinyon, juniper, or both; grass understory	Poor		75	85	89
	Fair		58	73	80
	Good		41	61	71
Sagebrush with grass understory	Poor		67	80	85
	Fair		51	63	70
	Good		35	47	55
Desert shrub—major plants include saltbrush, greasewood, creosotebrush, blackbrush, bursage palo verde, mesquite, and cactus	Poor	63	77	85	88
	Fair	55	72	81	86
	Good	49	68	79	84

Cover description; curve numbers for hydrologic soil group (A–D).

Note. Source: Soil Conservation Service (1986).

[a]Average runoff condition, and $I_a = 0.2S$. For range in humid regions, use Table 3.2c.

[b]*Poor:* <30% ground cover (litter, grass, and brush overstory). *Fair:* 30 to 70% ground cover. *Good:* >70% ground cover.

Appendix 5A

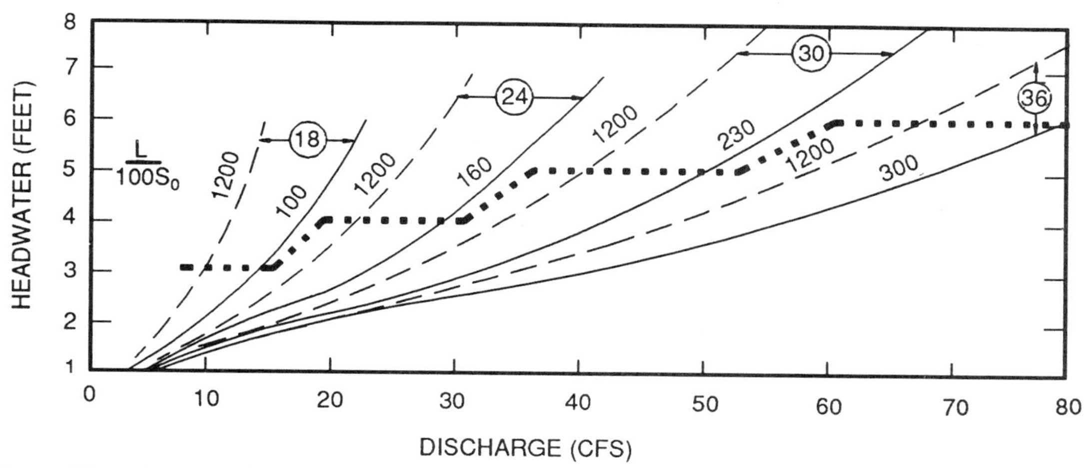

Figure 5A.1 Culvert capacity of circular concrete pipe, groove-edged entrance, 18–36 in. diameter (after Bureau of Public Roads, 1965a).

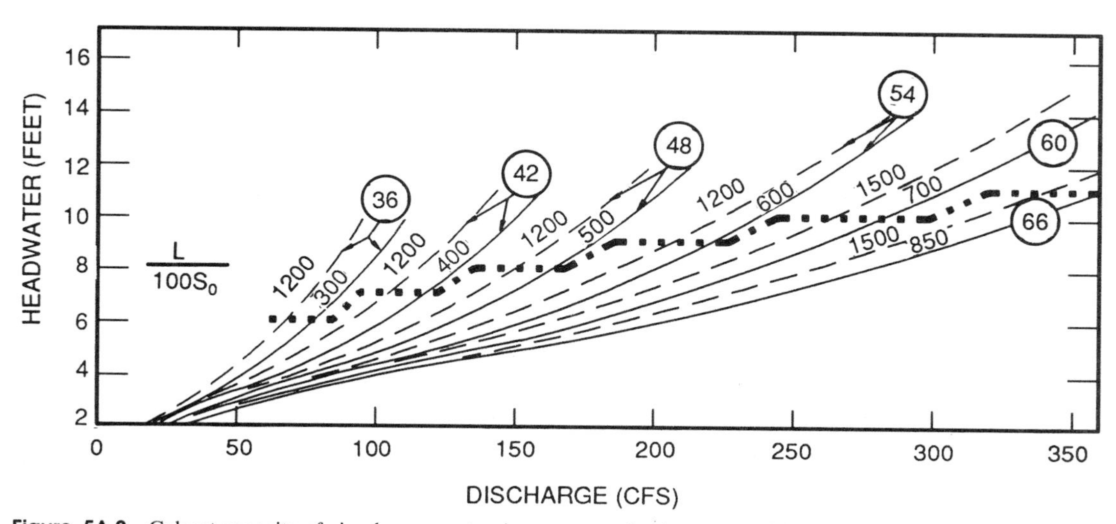

Figure 5A.2 Culvert capacity of circular concrete pipe, groove-edged entrance, 36–66 in. diameter (after Bureau of Public Roads, 1965a).

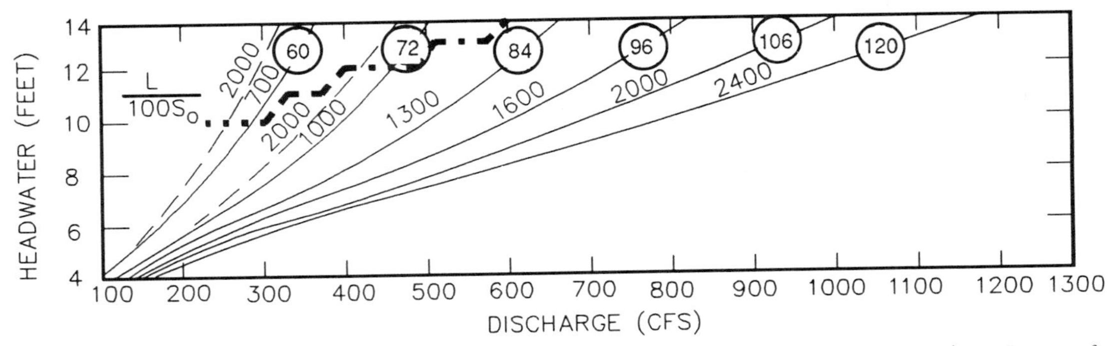

Figure 5A.3 Culvert capacity of circular concrete pipe, groove-edged entrance, 60–120 in. diameter (after Bureau of Public Roads, 1965a).

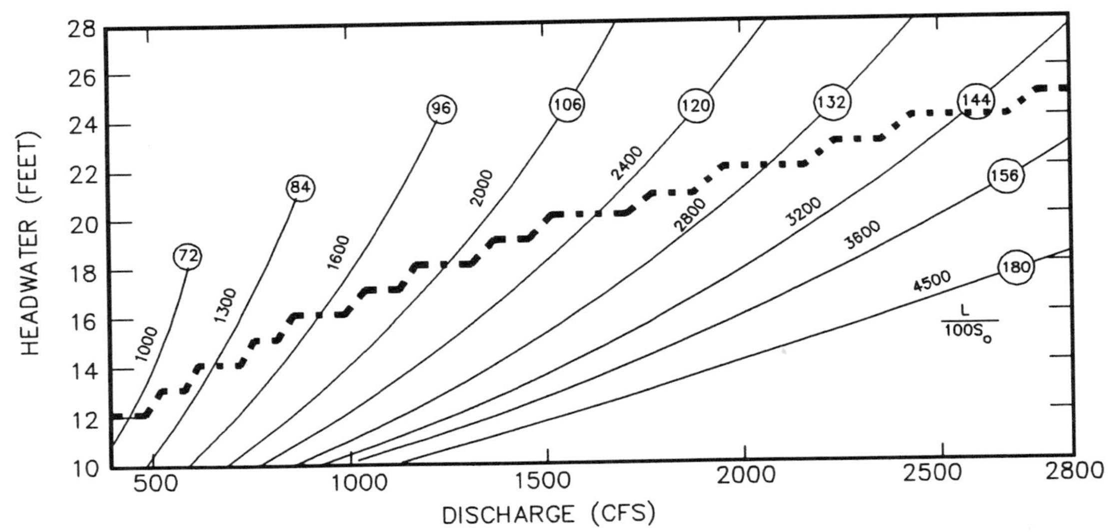

Figure 5A.4 Culvert capacity of circular concrete pipe, groove-edged entrance, 72–180 in. diameter (after Bureau of Public Roads, 1965a).

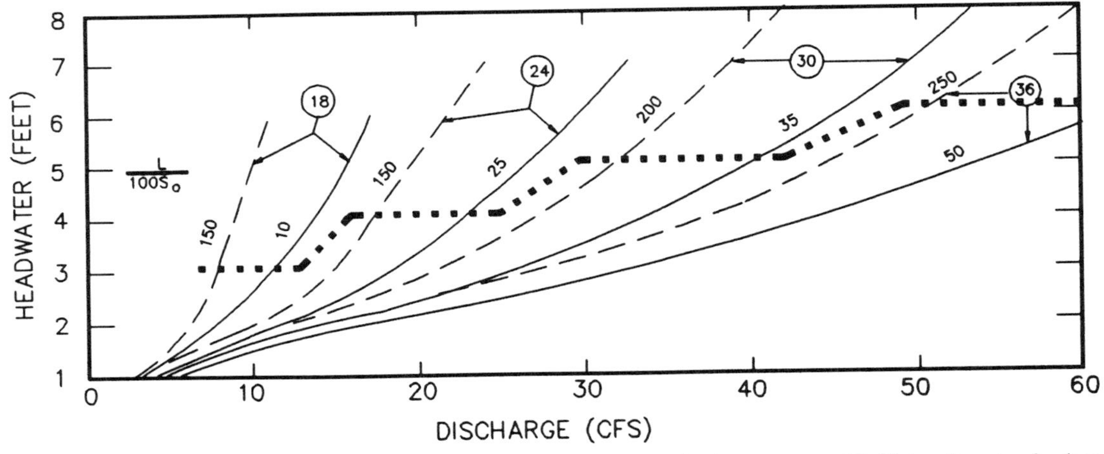

Figure 5A.5a Culvert capacity of circular corrugated metal pipe, projecting entrance, 18–36 in. diameter for low values of $L/1000S_0$ (after Bureau of Public Roads, 1965a).

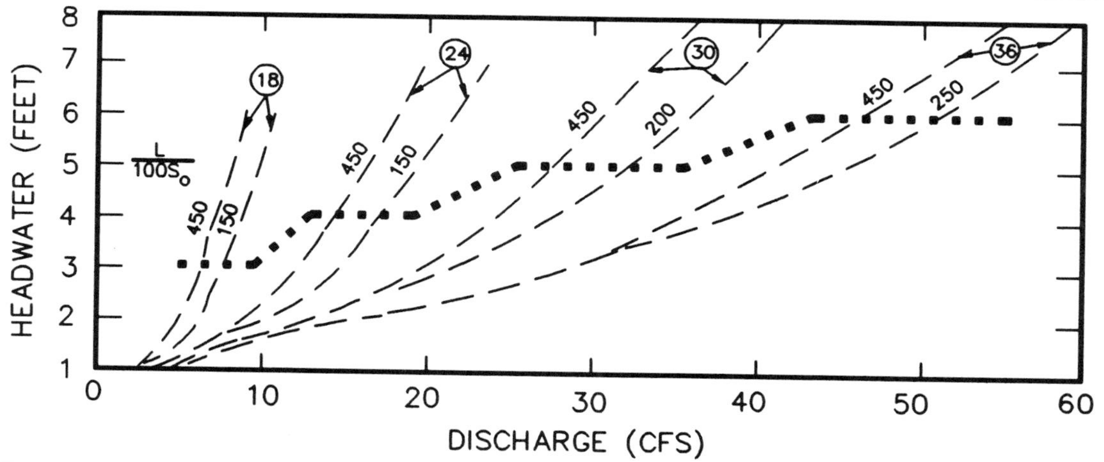

Figure 5A.5b Culvert capacity of circular corrugated metal pipe, projecting entrance, 18–36 in. diameter for high values of $L/1000S_0$ (after Bureau of Public Roads, 1965a).

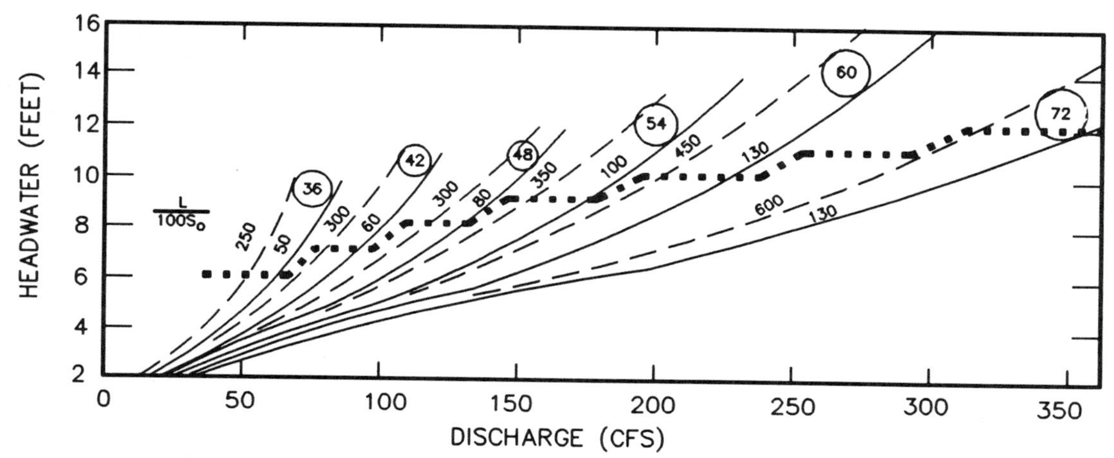

Figure 5A.6a Culvert capacity of circular corrugated metal pipe, projecting entrance, 36–72 in. diameter for low values of $L/1000S_0$ (after Bureau of Public Roads, 1965a).

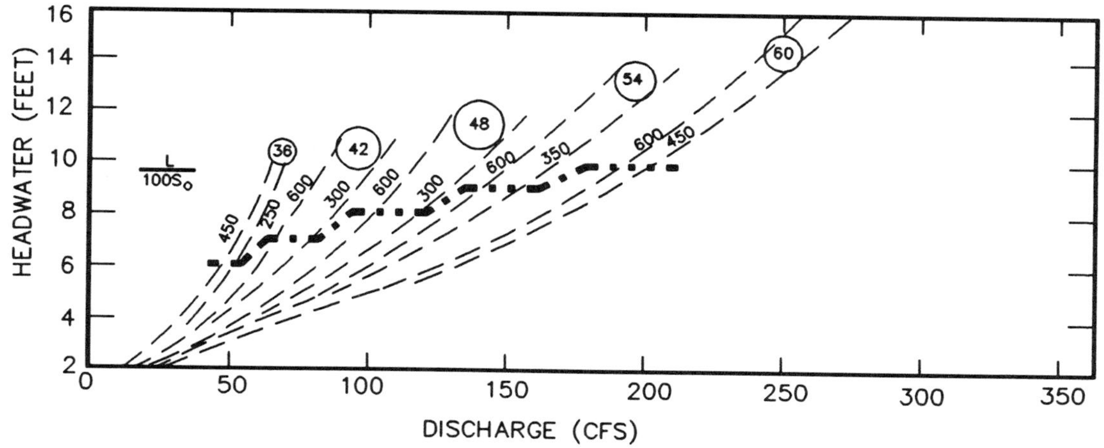

Figure 5A.6b Culvert capacity of circular corrugated metal pipe, projecting entrance, 36–72 in. diameter for high values of $L/1000S_0$ (after Bureau of Public Roads, 1965a).

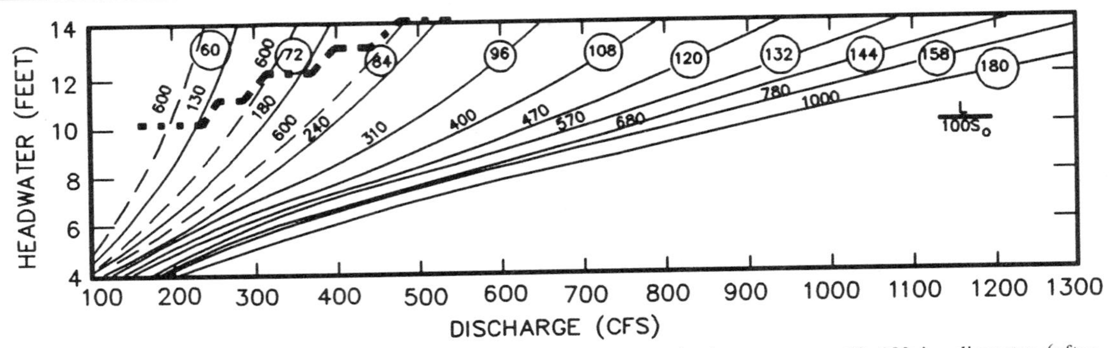

Figure 5A.7 Culvert capacity of circular corrugated metal pipe, projecting entrance, 60–180 in. diameter (after Bureau of Public Roads, 1965a).

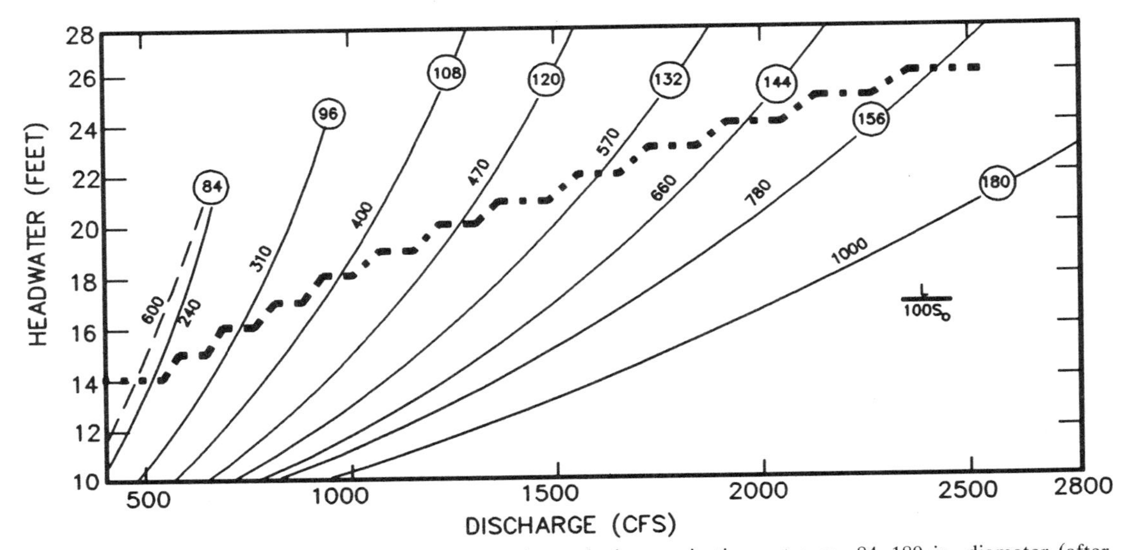

Figure 5A.8 Culvert capacity of circular corrugated metal pipe, projecting entrance, 84–180 in. diameter (after Bureau of Public Roads, 1965a).

Appendix 5B

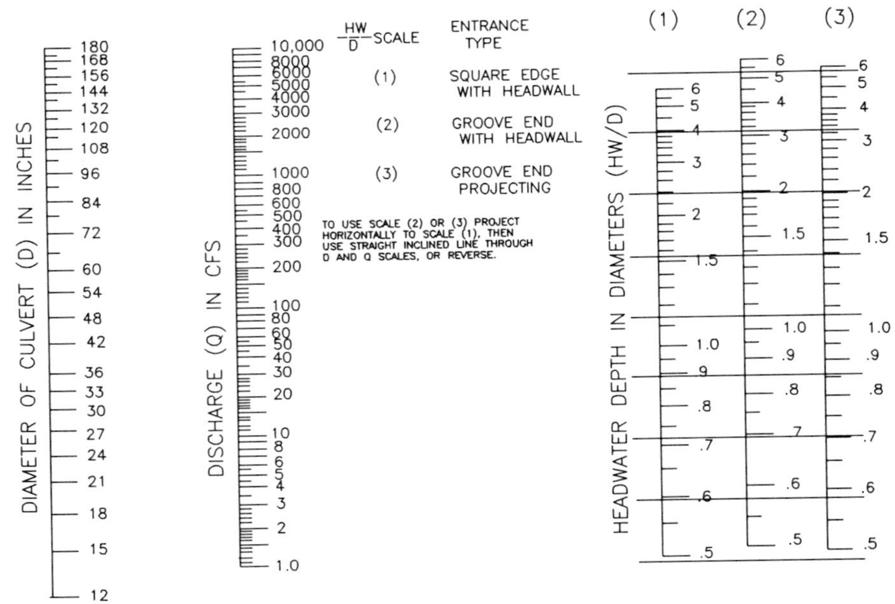

Figure 5B.1 Headwater depth for circular concrete pipe culverts with inlet control (after Federal Highway Administration, 1985).

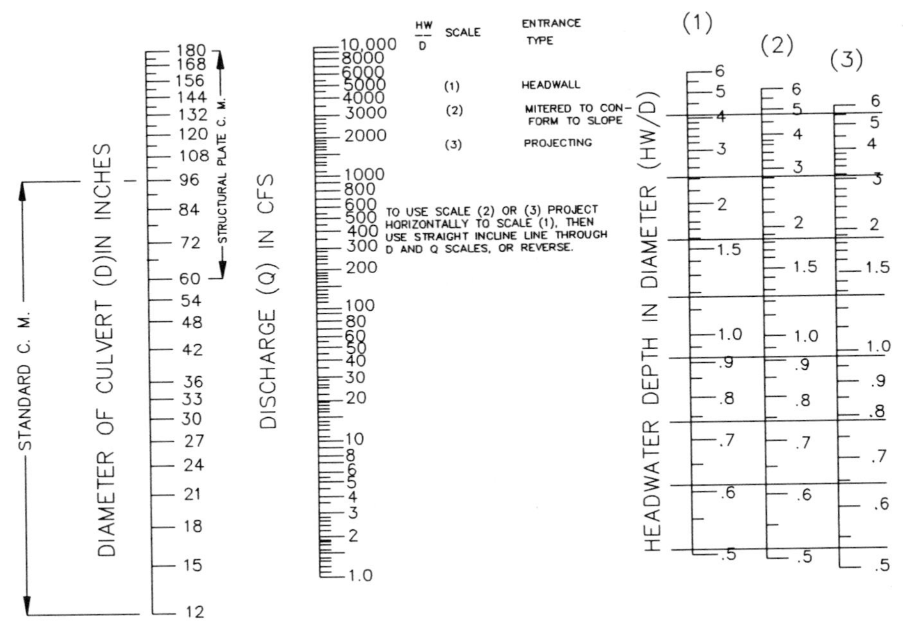

Figure 5B.2 Headwater depth for circular corrugated metal pipe culverts with inlet control (after Federal Highway Administration, 1985).

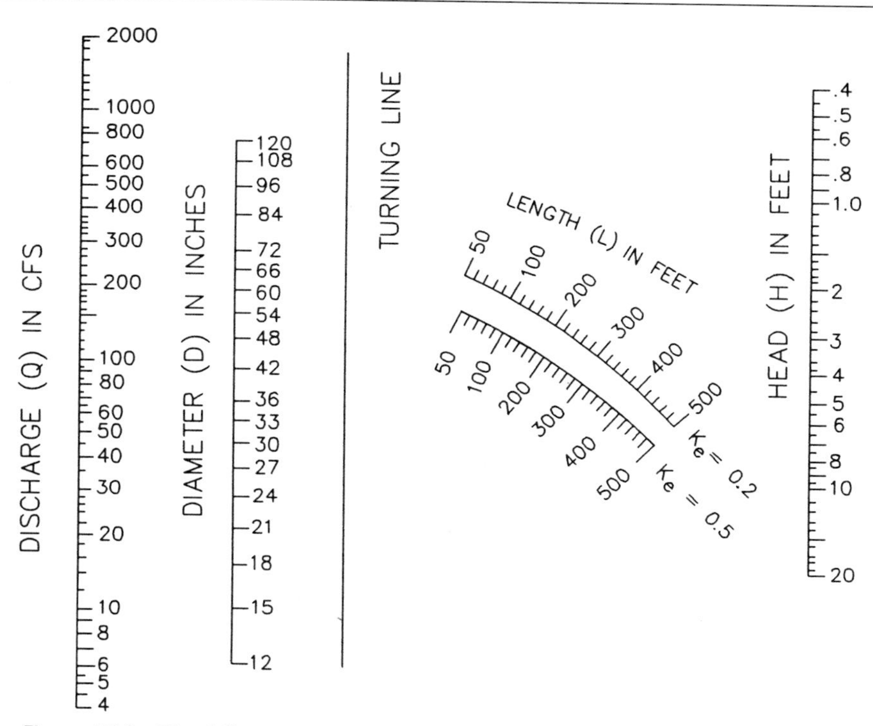

Figure 5B.3 Pipe full nomograph for concrete pipe culverts flowing full with $n = 0.012$ (after Federal Highway Administration, 1985).

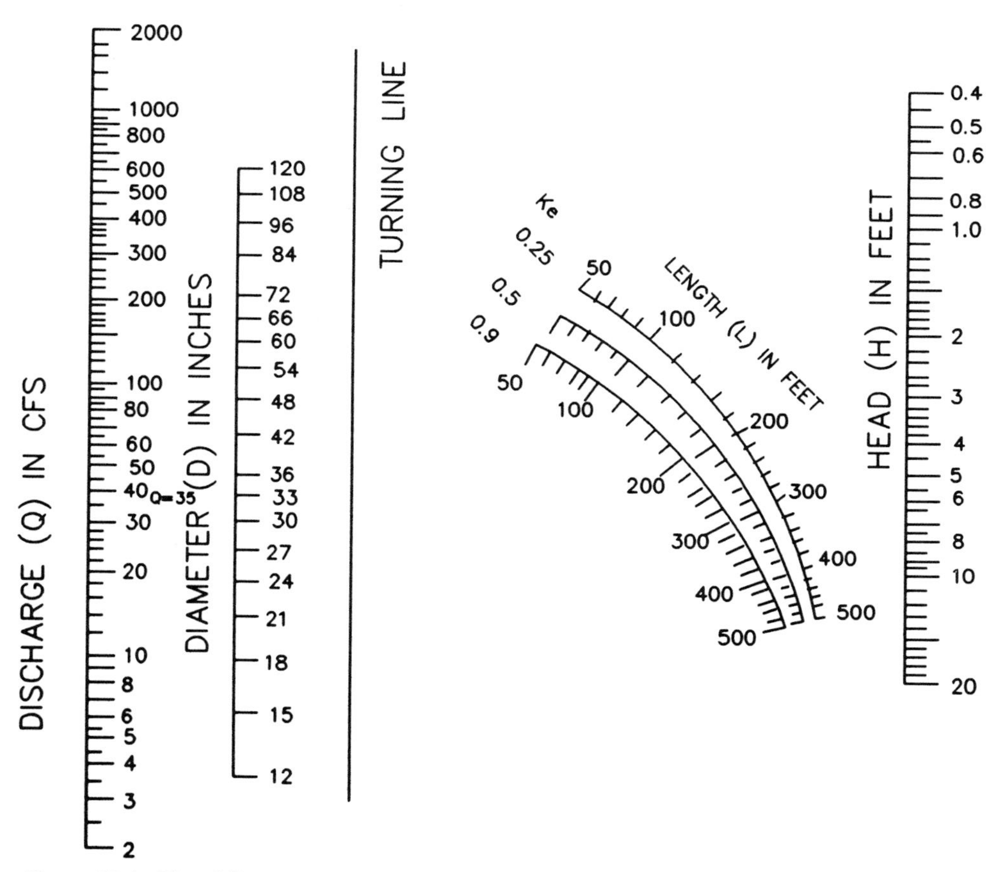

Figure 5B.4 Pipe full nomograph for corrugated metal pipe culverts flowing full with $n = 0.0024$ (after Federal Highway Administration, 1985).

Appendix 5C

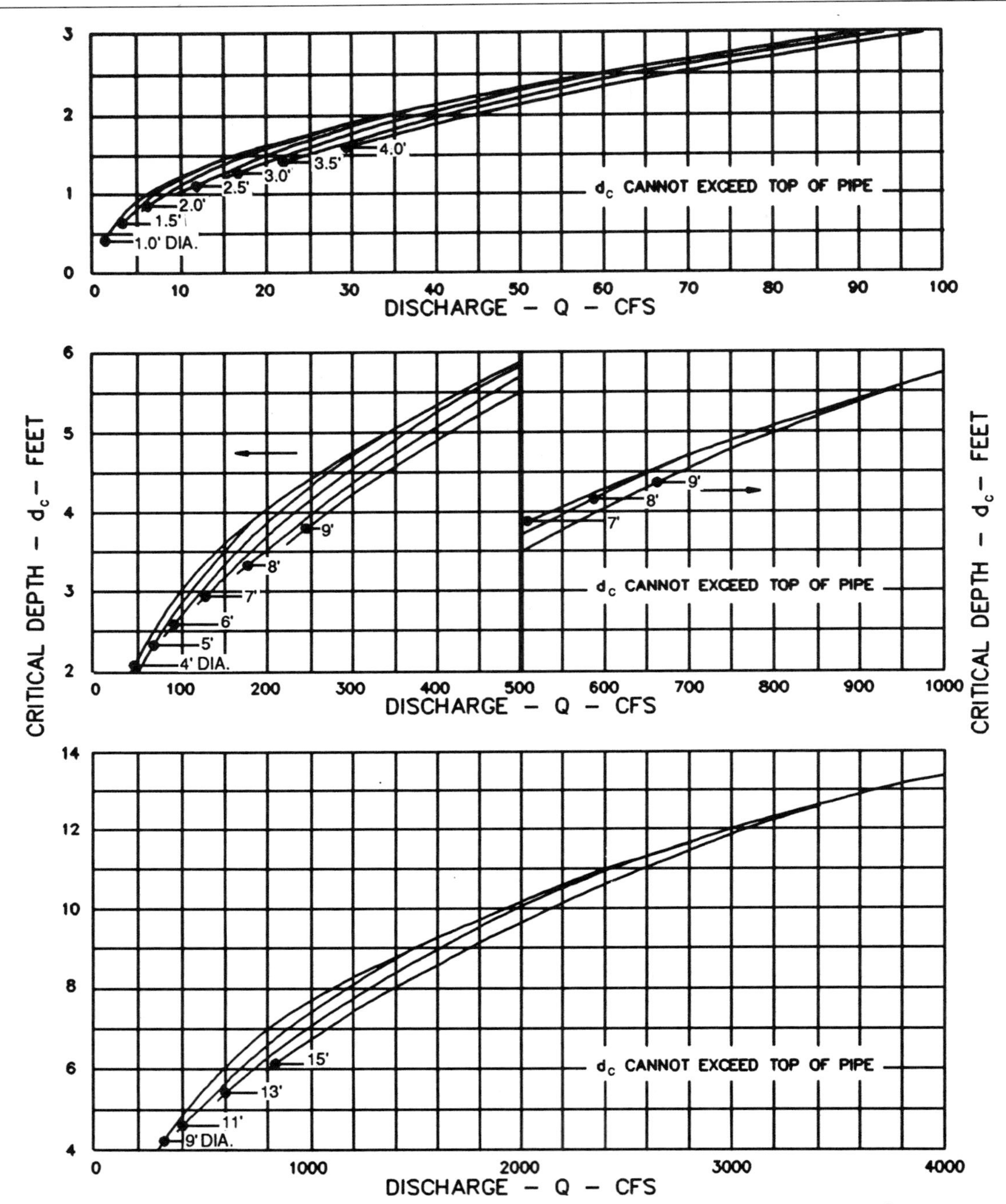

Figure 5C.1 Critical depth for circular pipe (after Federal Highway Administration, 1985).

Appendix 5D

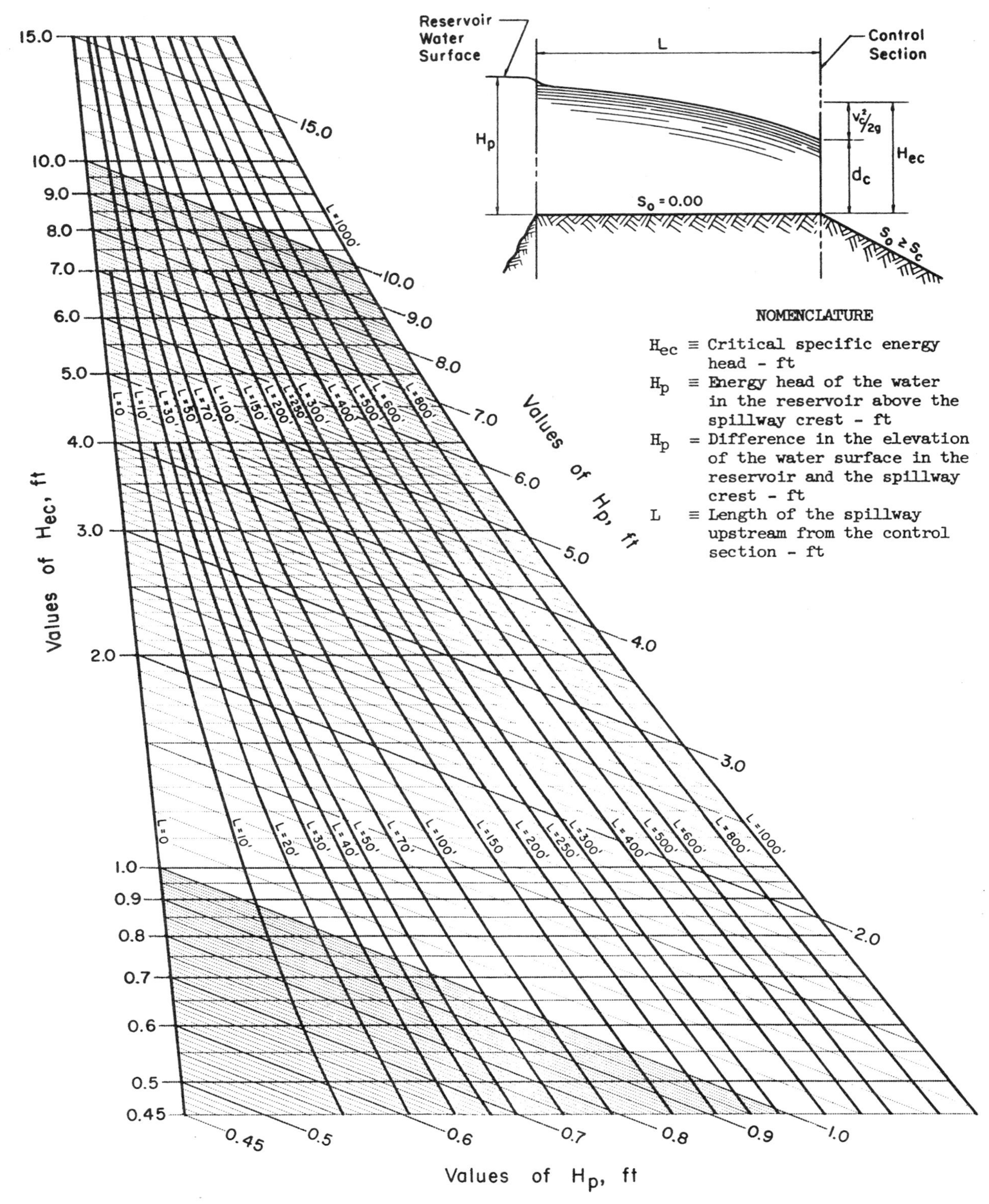

Figure 5D.1 H_{ec} versus H_p for emergency spillways of various lengths L. Case 1, b = 100 ft, z = 2, n = 0.04. (Soil Conservation Service, 1968).

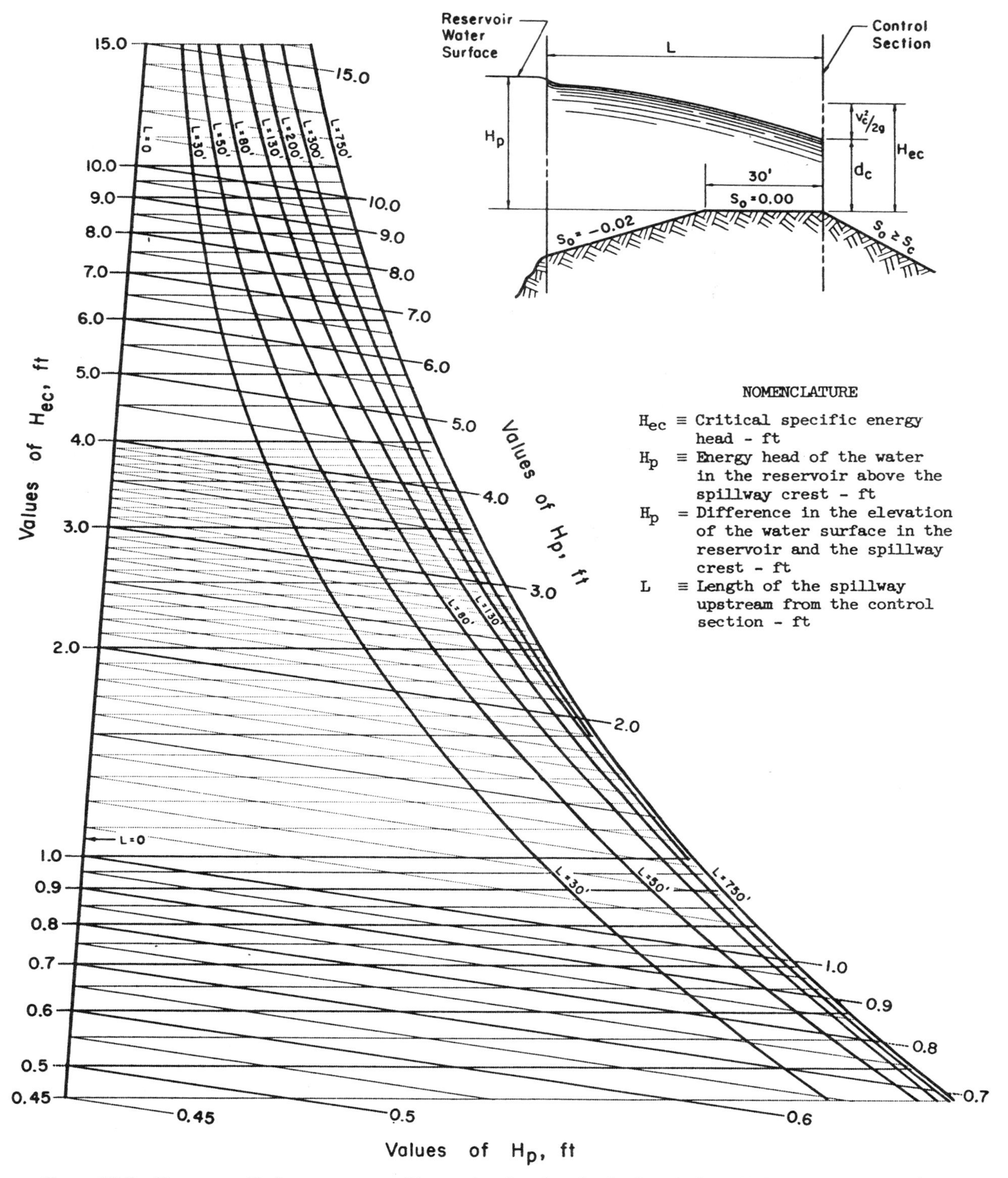

Figure 5D.2 H_{ec} versus H_p for emergency spillways of various lengths L. Case 2, b = 100 ft, z = 2, n = 0.04. (Soil Conservation Service, 1968).

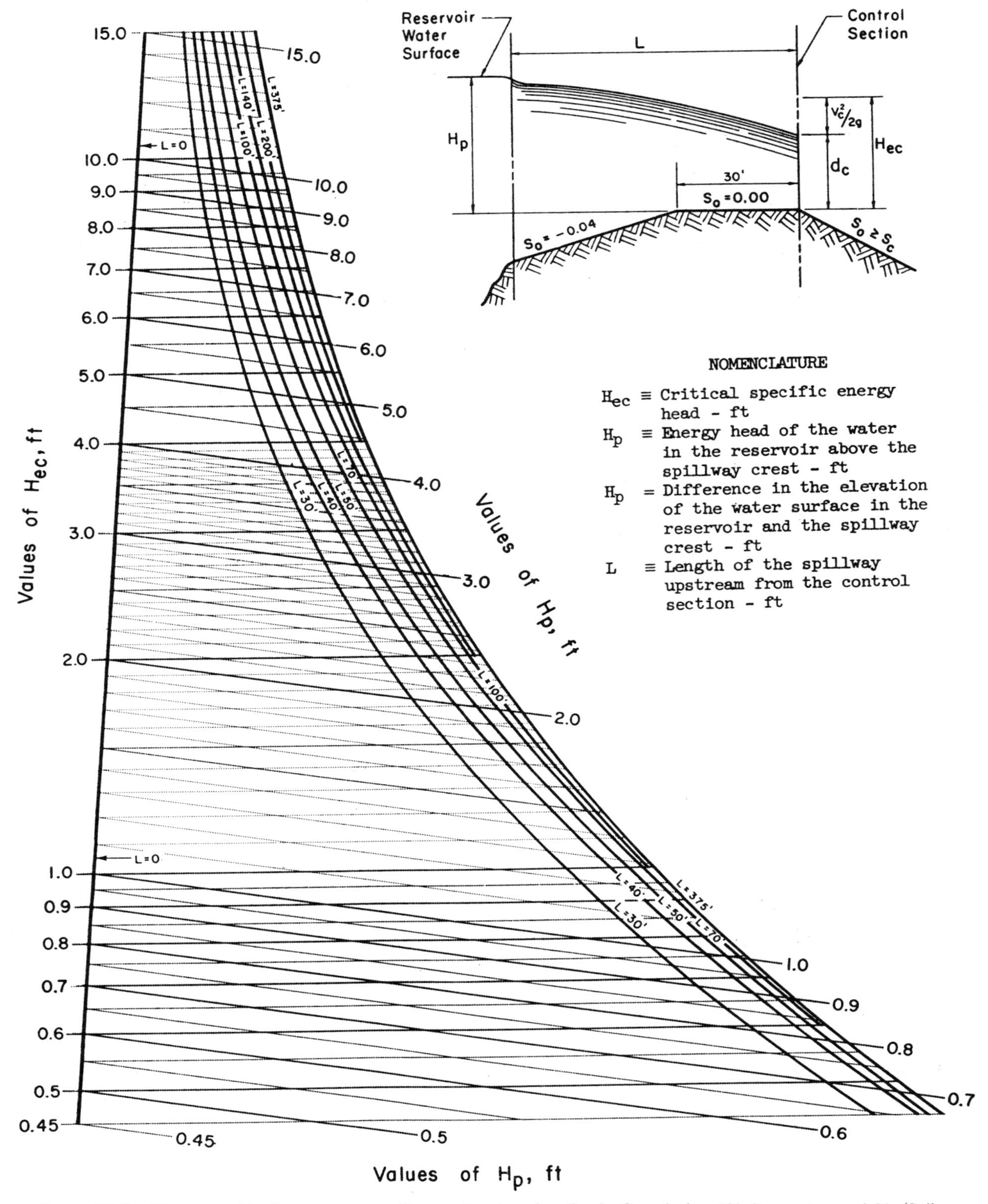

Figure 5D.3 H_{ec} versus H_p for emergency spillways of various lengths L. Case 3, b = 100 ft, z = 2, n = 0.04. (Soil Conservation Service, 1968).

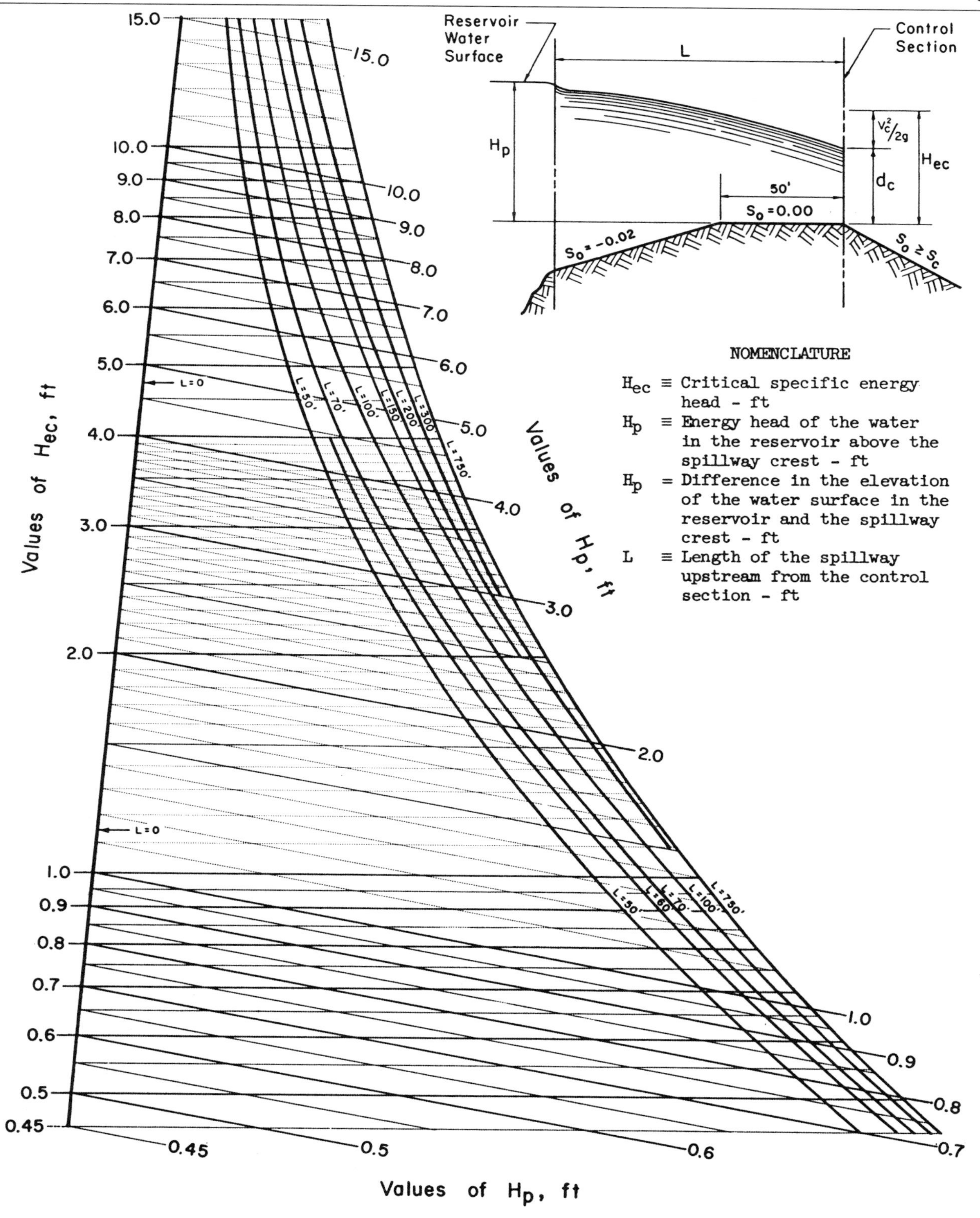

Figure 5D.4 H_{ec} versus H_p for emergency spillways of various lengths L. Case 4, $b = 100$ ft, $z = 2$, $n = 0.04$. (Soil Conservation Service, 1968).

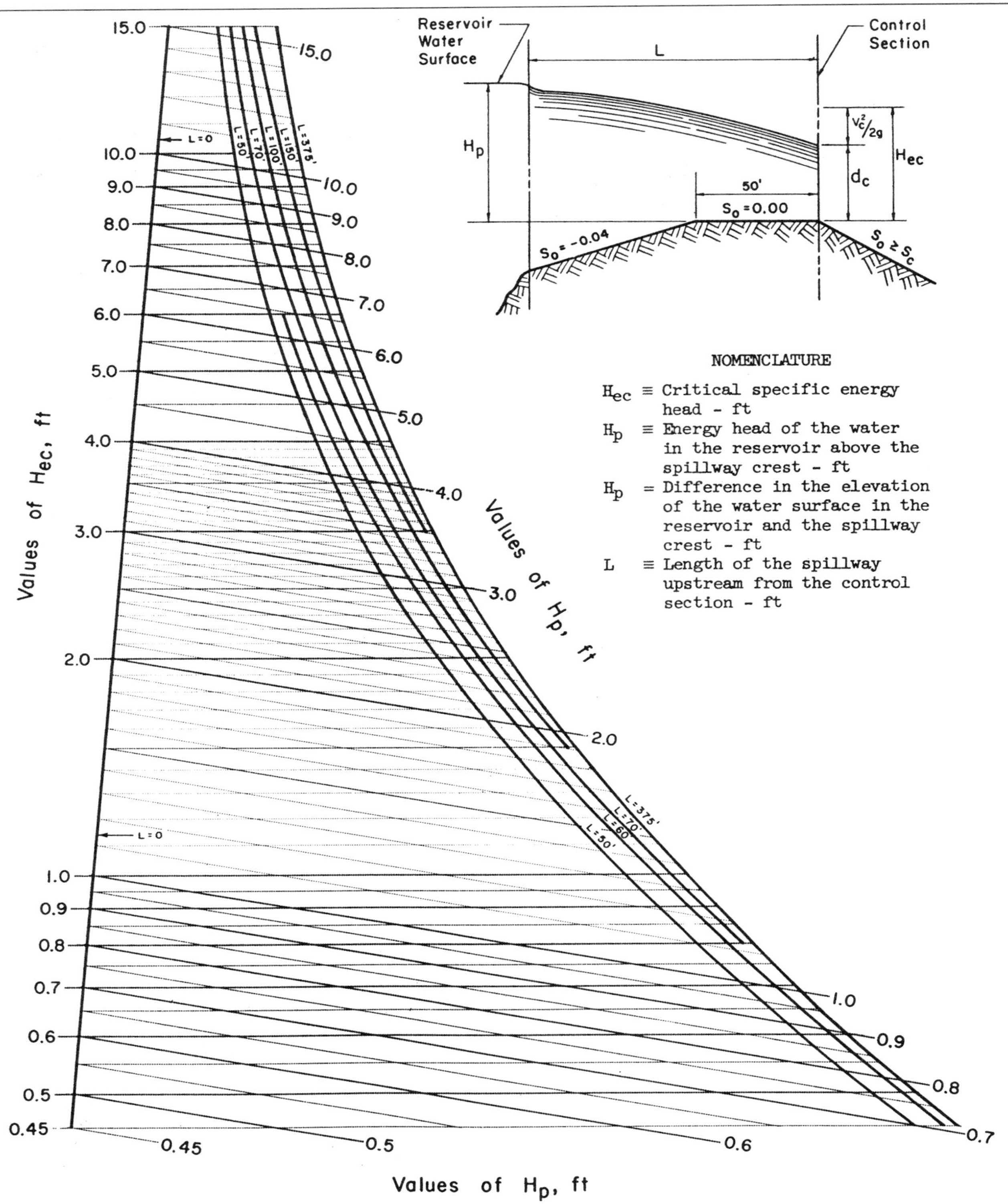

Figure 5D.5 H_{ec} versus H_p for emergency spillways of various lengths L. Case 5, b = 100 ft, z = 2, n = 0.04. (Soil Conservation Service, 1968).

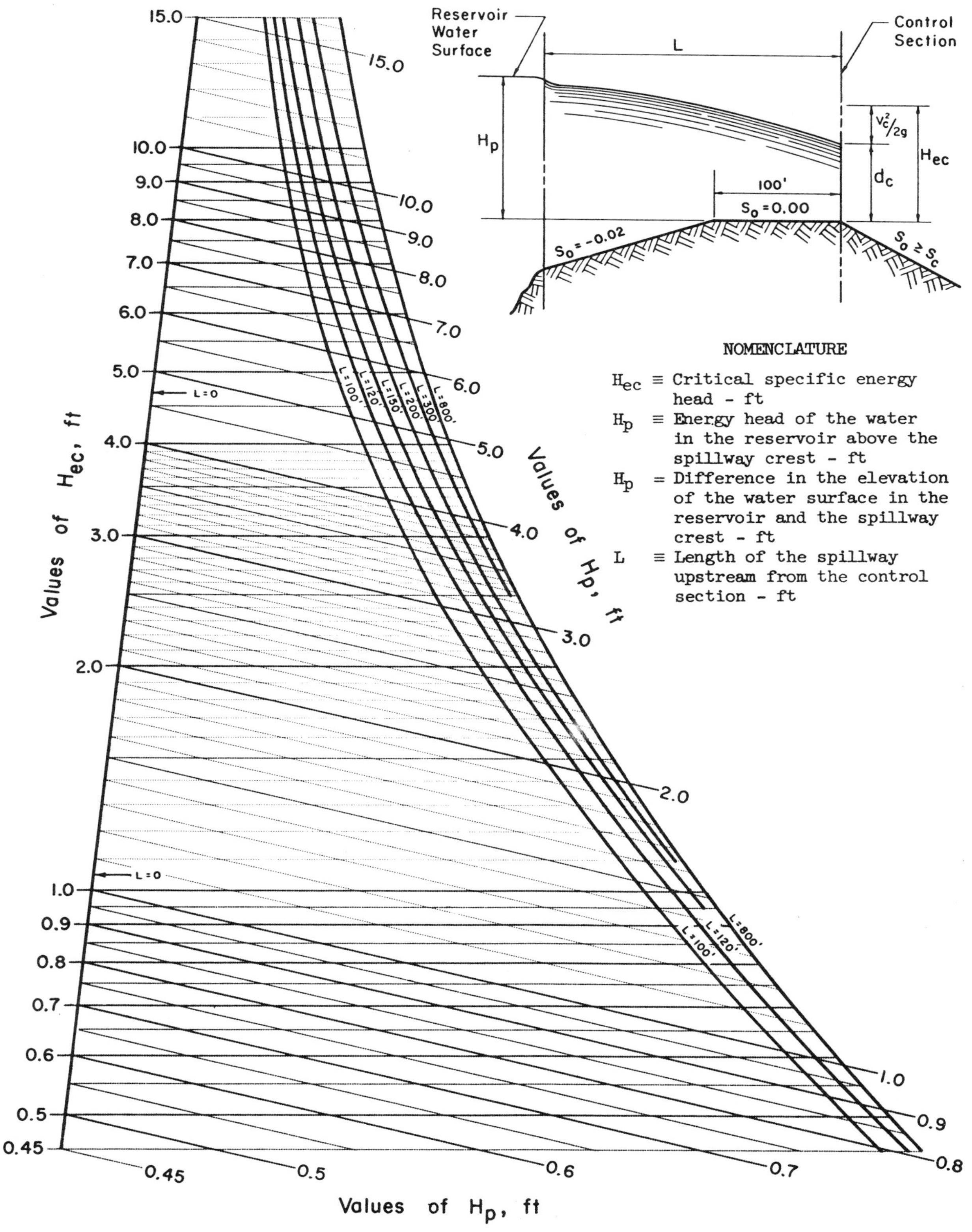

Figure 5D.6 H_{ec} versus H_p for emergency spillways of various lengths L. Case 6, b = 100 ft, z = 2, n = 0.04. (Soil Conservation Service, 1968).

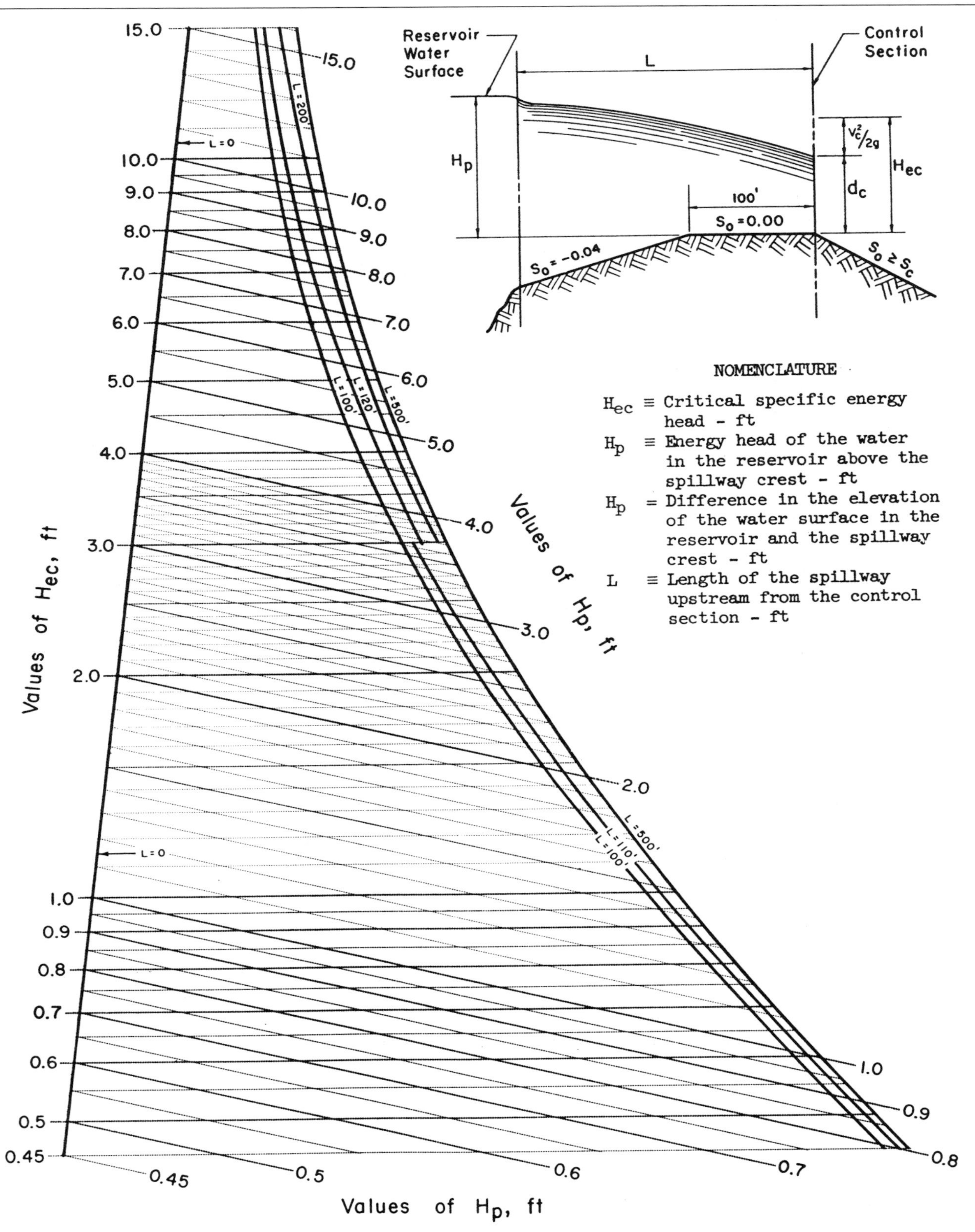

Figure 5D.7 H_{ec} versus H_p for emergency spillways of various lengths L. Case 7, b = 100 ft, z = 2, n = 0.04. (Soil Conservation Service, 1968).

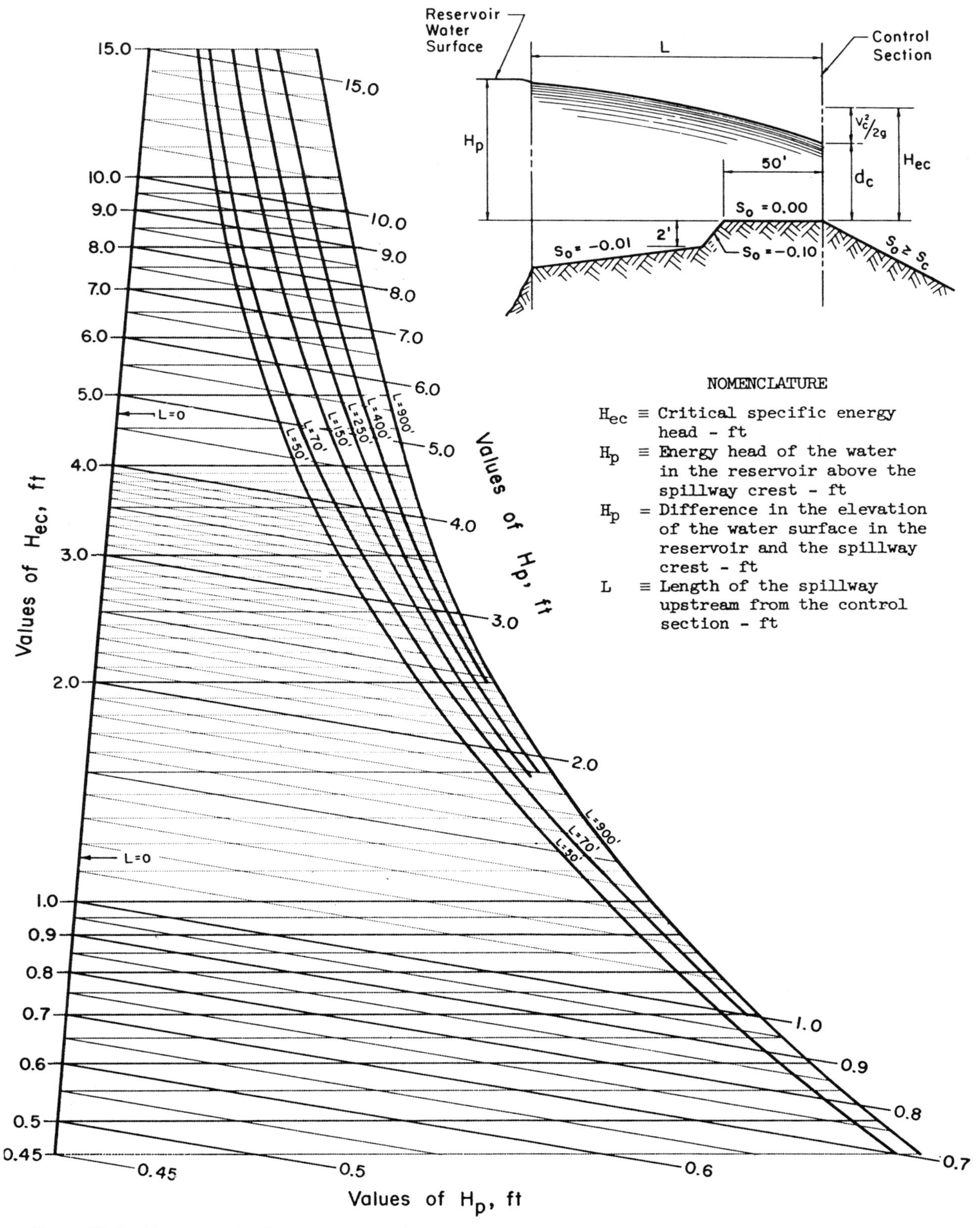

Figure 5D.8 H_{ec} versus H_p for emergency spillways of various lengths L. Case 8, b = 100 ft, z = 2, n = 0.04. (Soil Conservation Service, 1968).

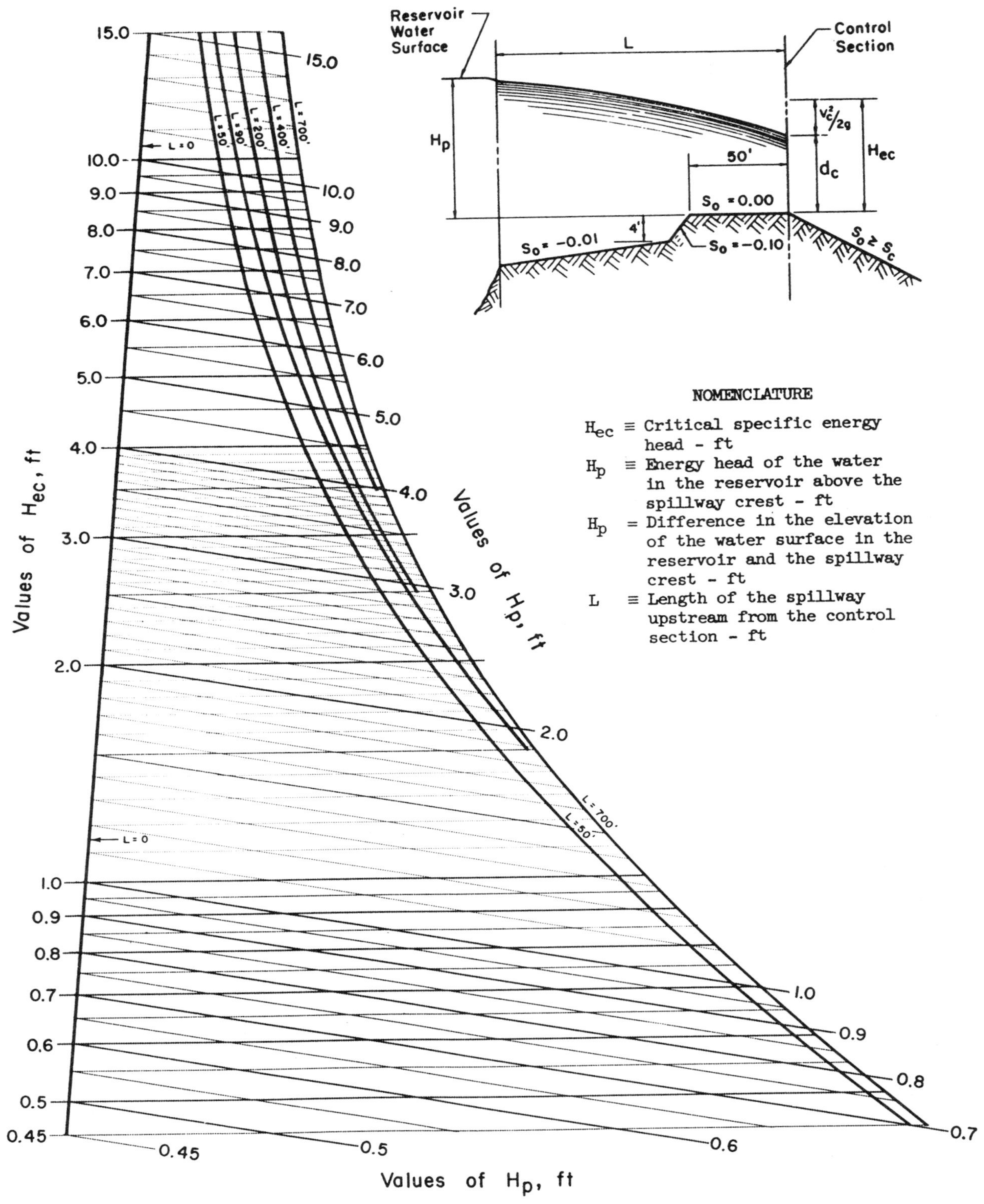

Figure 5D.9 H_{ec} versus H_p for emergency spillways of various lengths L. Case 9, b = 100 ft, z = 2, n = 0.04. (Soil Conservation Service, 1968).

Appendix 5E

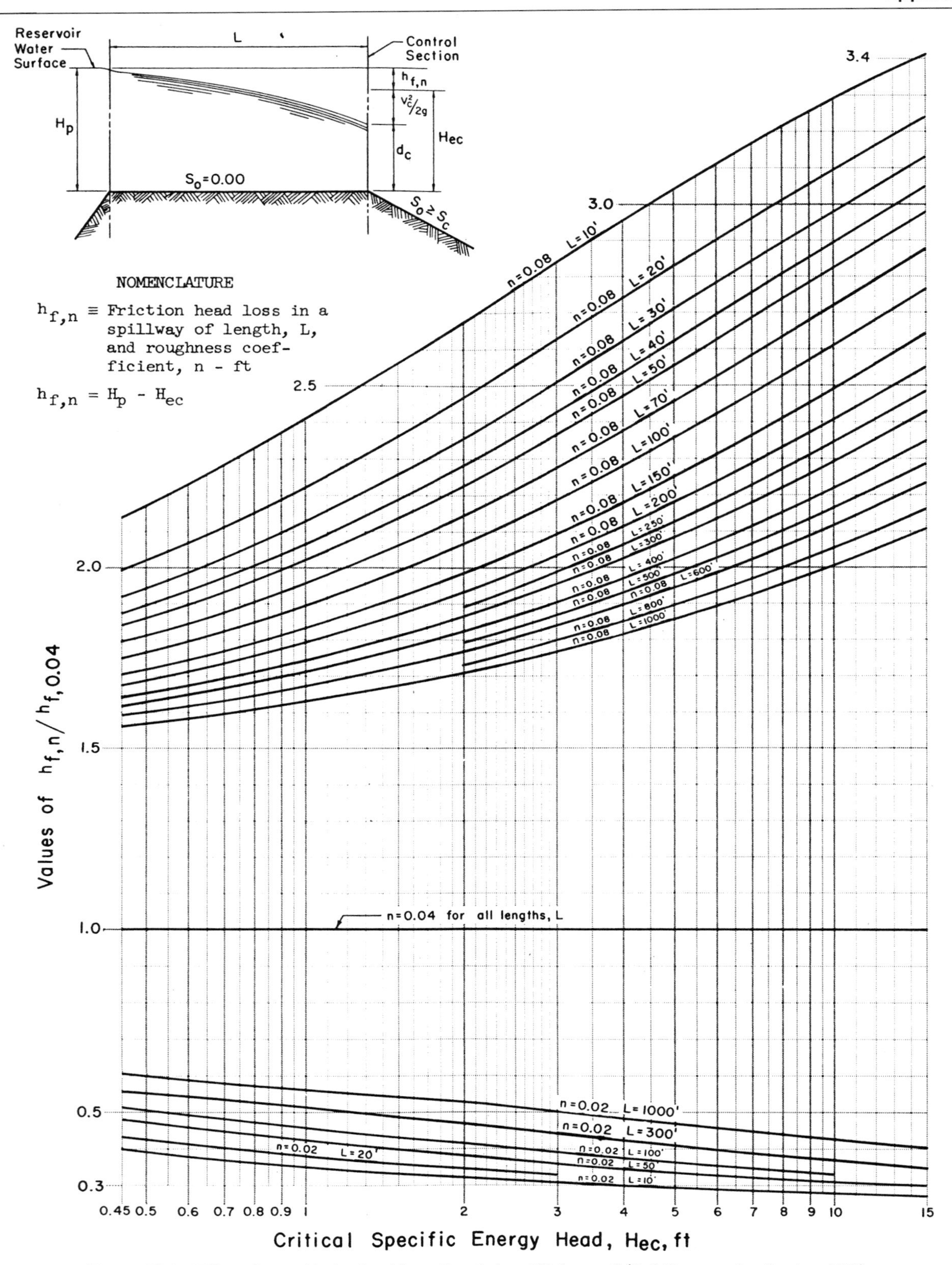

Figure 5E.1 Effect of n on friction head loss. Case 1, b = 100 ft, z = 2 (Soil Conservation Service, 1968).

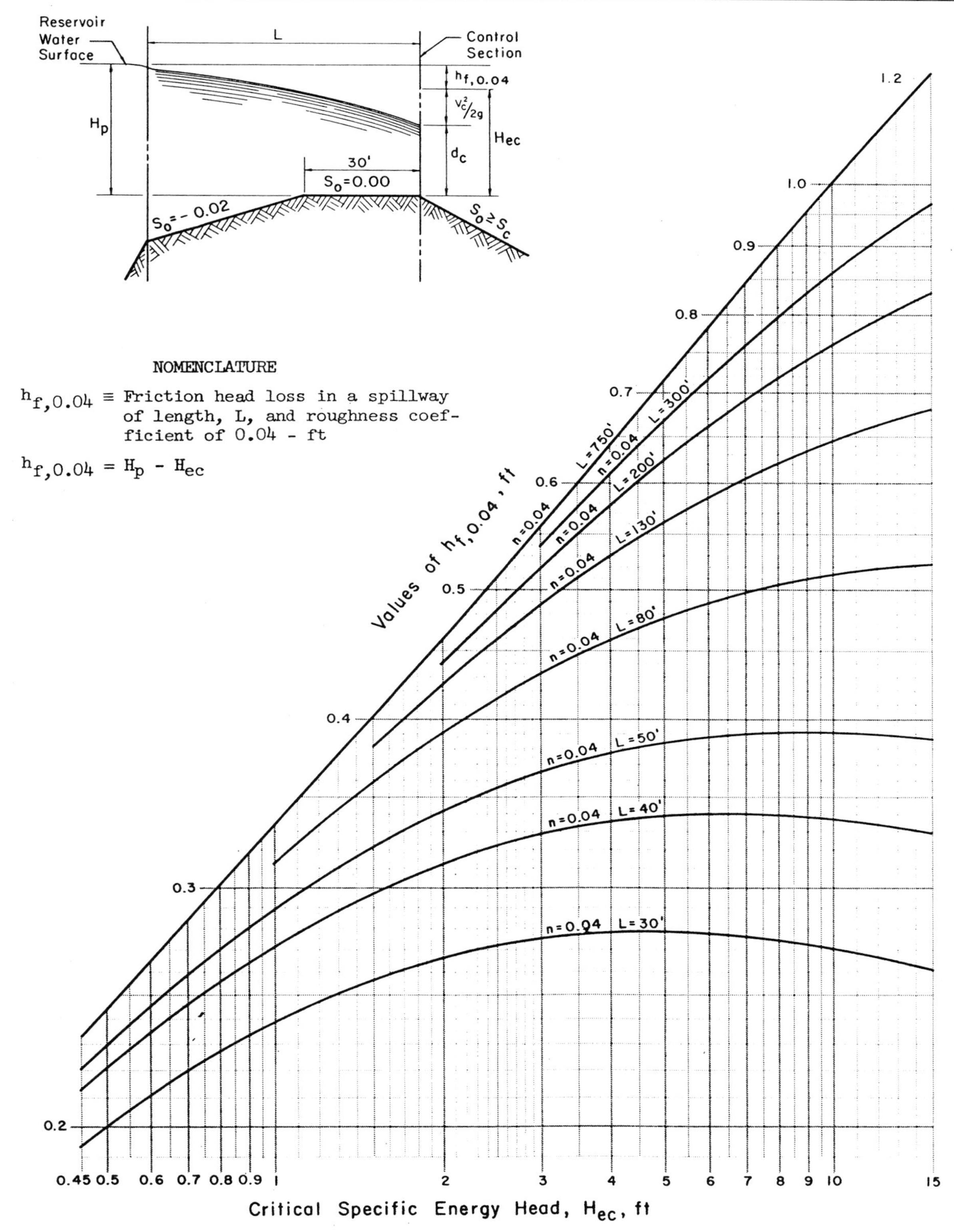

Figure 5E.2 Effect of n on friction head loss for $n = 0.04$. Case 2, $b = 100$ ft, $z = 2$ (Soil Conservation Service, 1968).

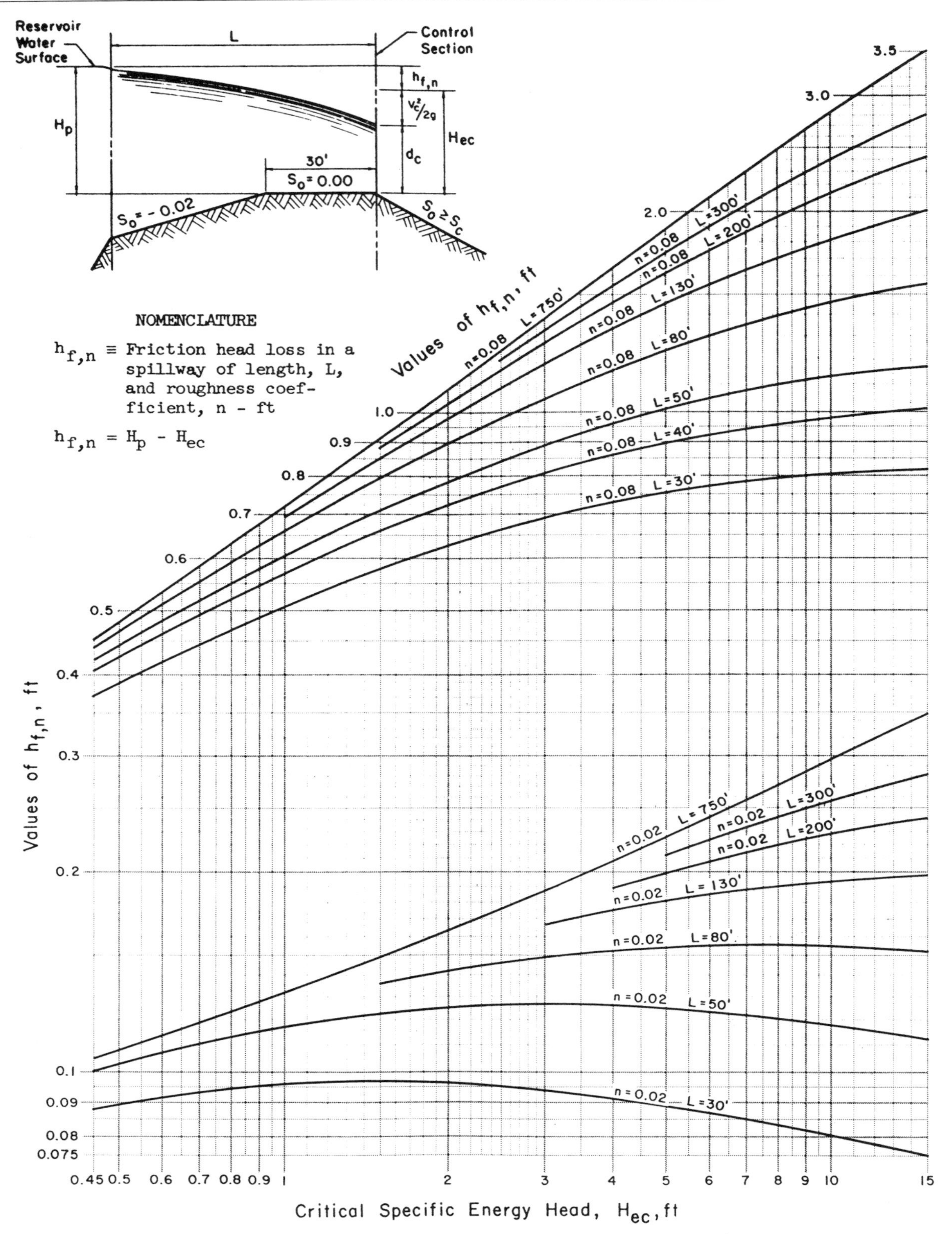

Figure 5E.3 Effect of n on friction head loss for n = 0.02 and 0.08. Case 2, b = 100 ft, z = 2 (Soil Conservation Service, 1968).

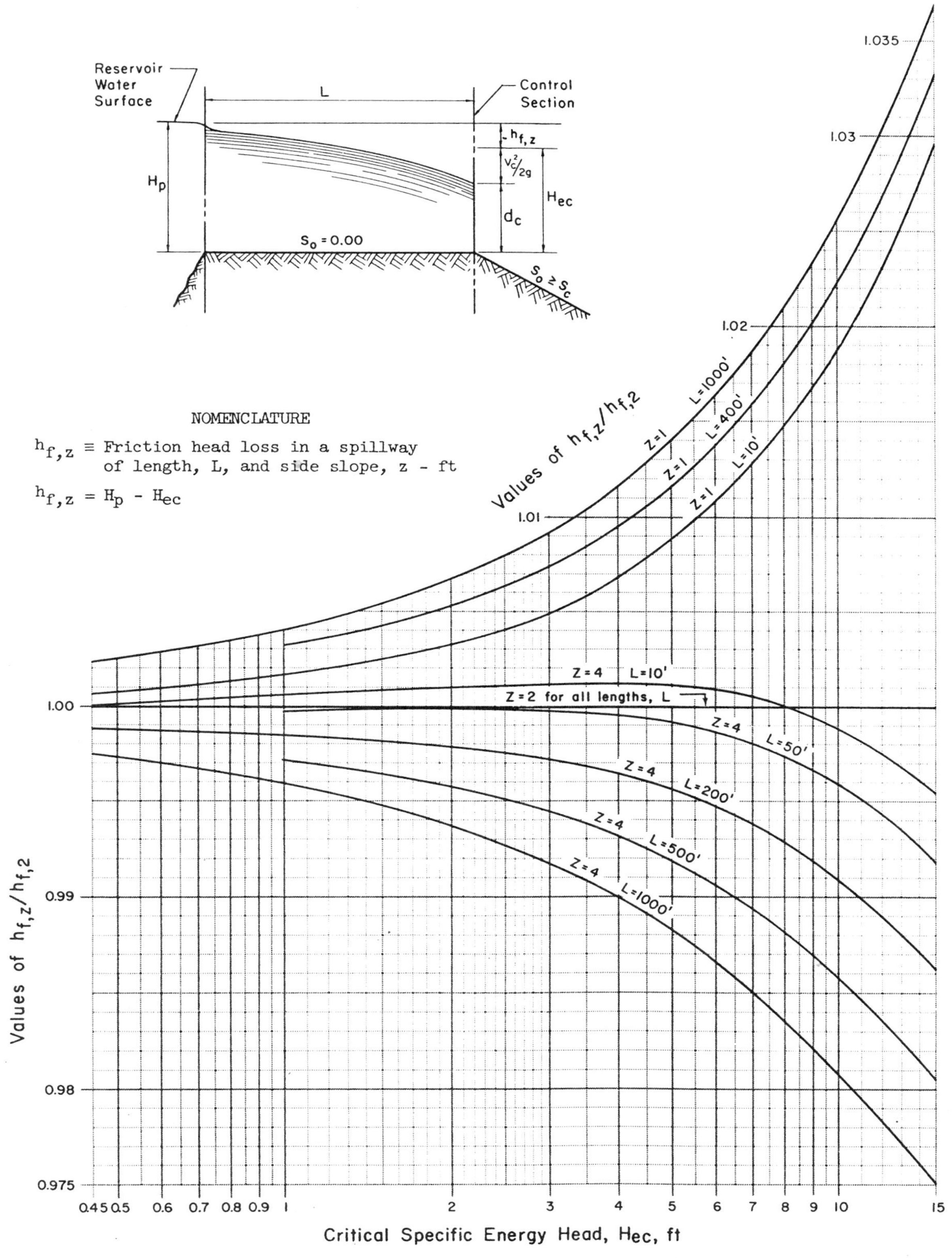

Figure 5E.4 Effect of z on friction head loss. Case 1, b = 100 ft, n = 0.04 (Soil Conservation, 1968).

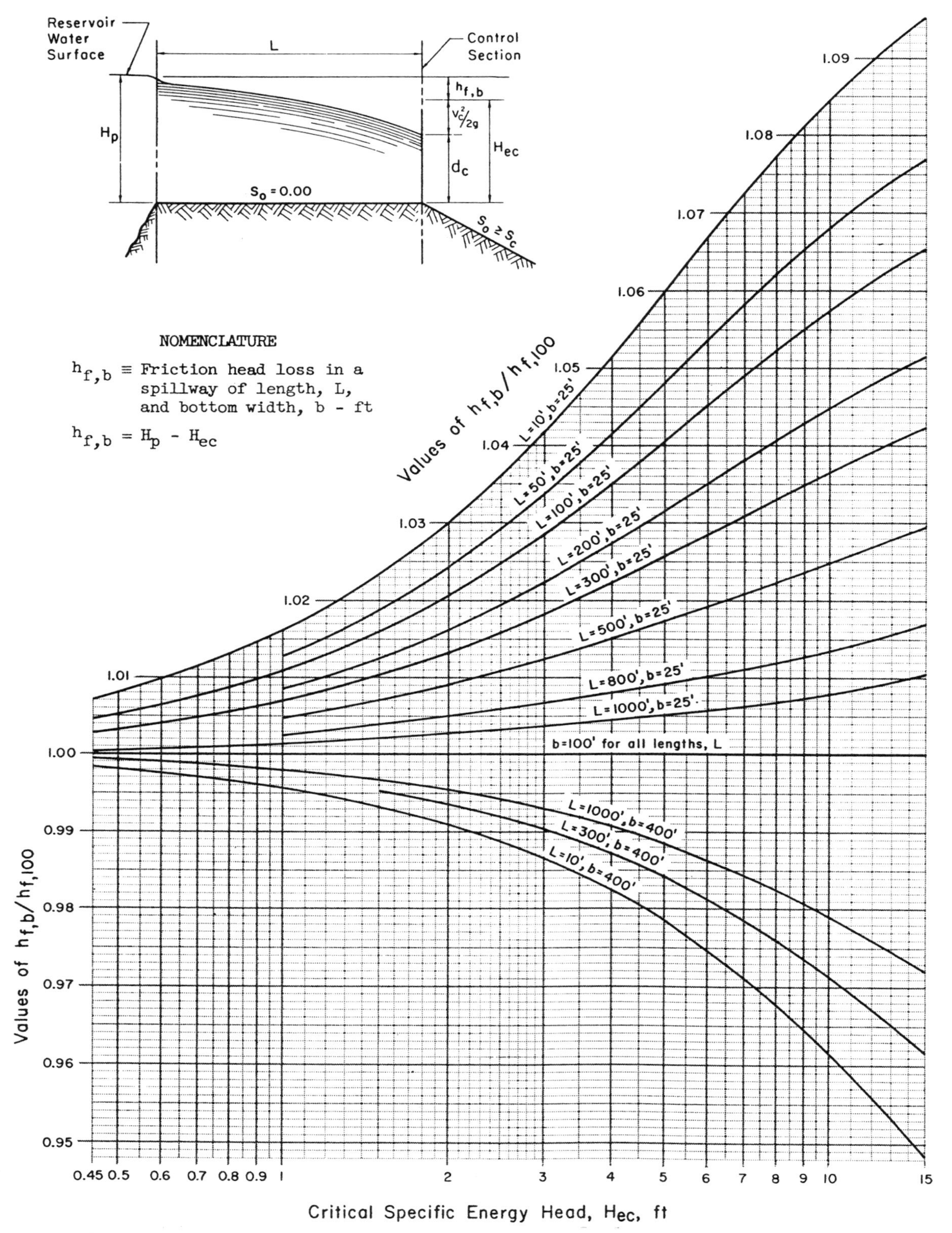

Figure 5E.5 Effect of b on friction head loss. Case 1, $z = 2$, $n = 0.04$ (Soil Conservation Service, 1968).

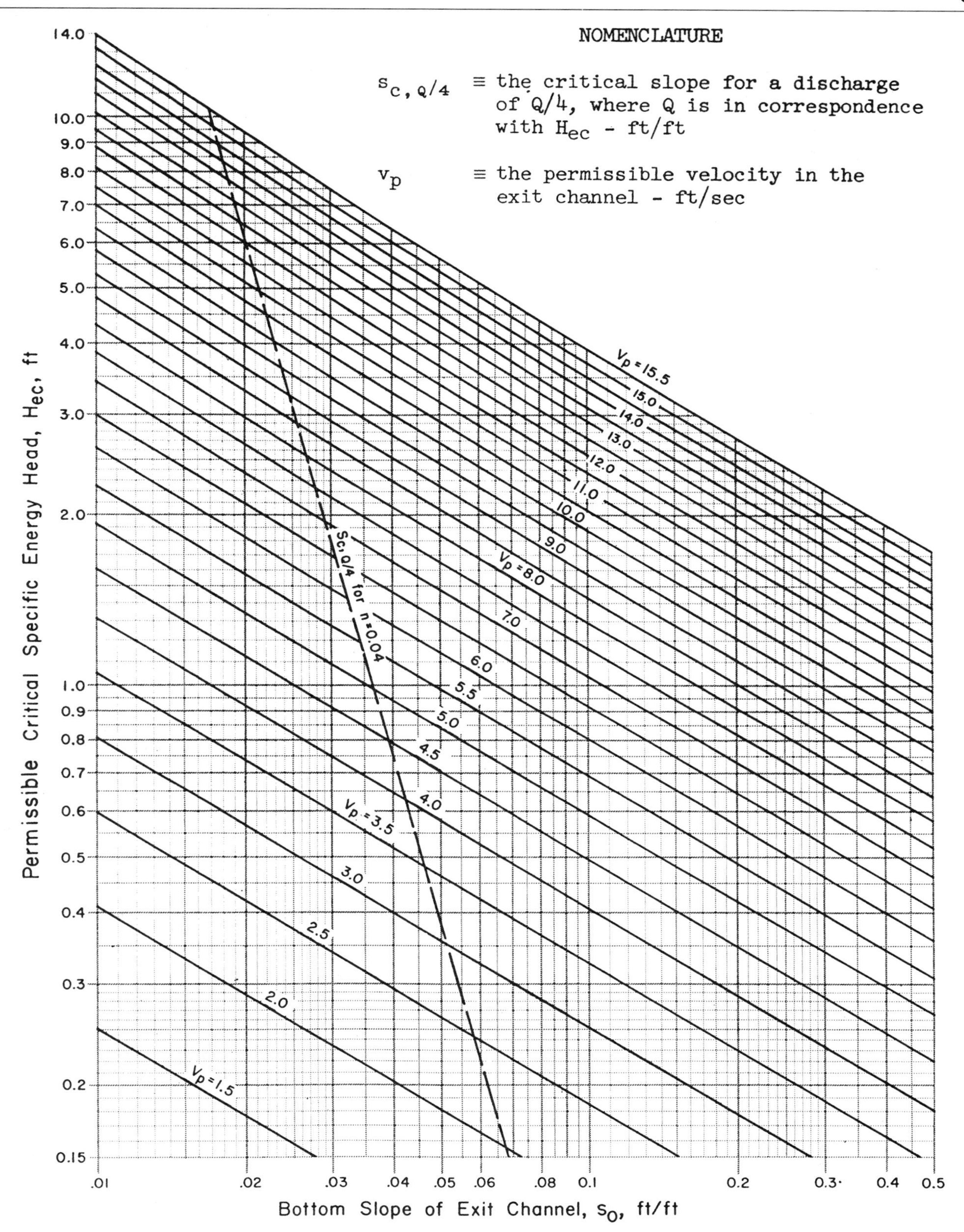

Figure 5E.6 Permissible H_{ec} for various s_o and v_p with b = 100 ft, z = 2, and n = 0.04 (Soil Conservation Service, 1968).

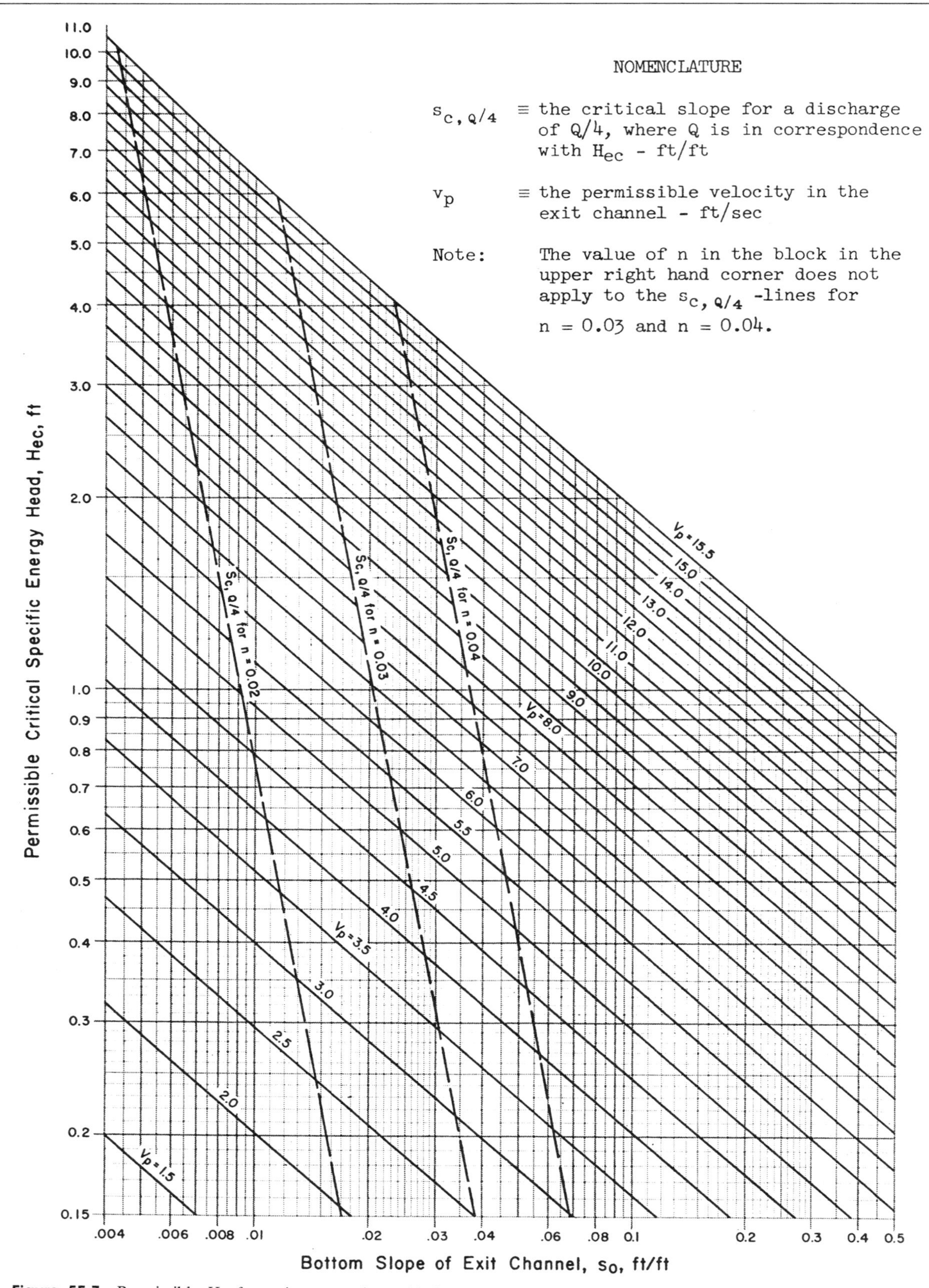

Figure 5E.7 Permissible H_{ec} for various s_o and v_p with b = 100 ft, z = 2, and n = 0.02 (Soil Conservation Service, 1968).

Appendix 8A
R Factor Information

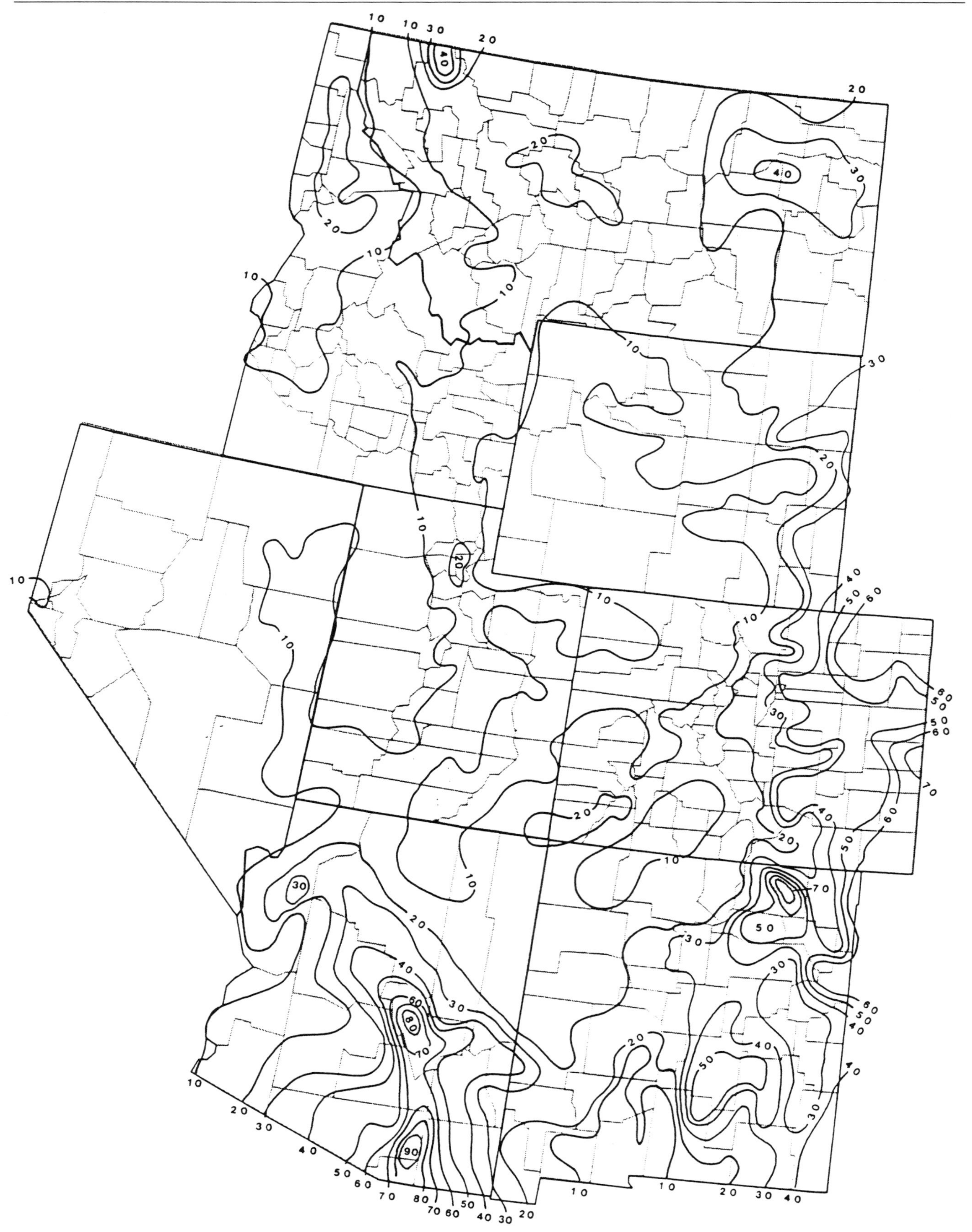

Figure 8A.1 Isolines of *R* factor for Western U.S. (after Renard *et al.*, 1993b). Units on *R* are ft · tonsf · in./acre · hr · year. To convert to metric, MJ · mm/ha · h · year, multiply by 17.02.

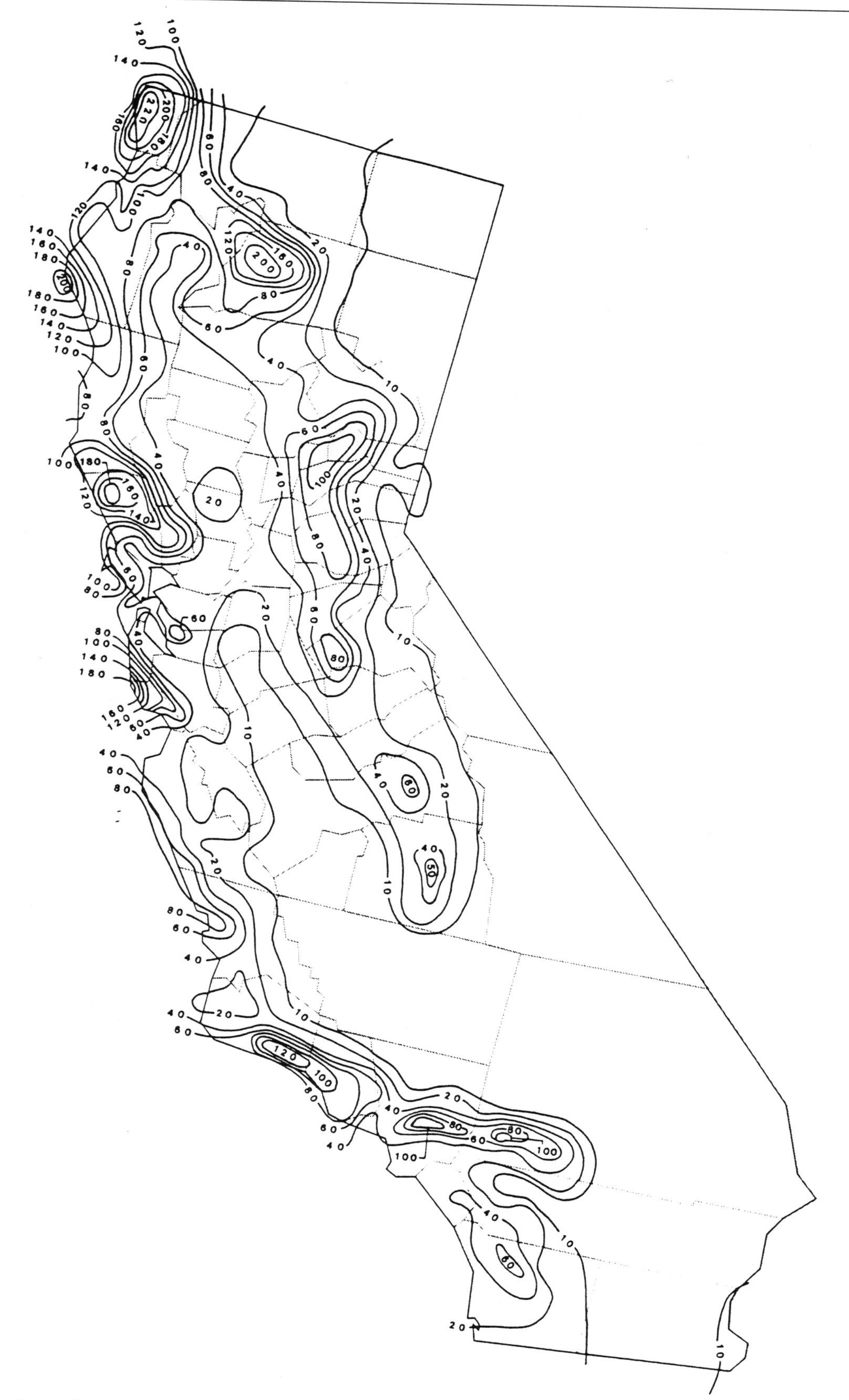

Figure 8A.2 Isolines of *R* factor for California (after Renard, *et al.*, 1993b). Units on *R* are ft · tonsf · in./acre · hr · year. To convert to metric, MJ · mm/ha · h · year, multiply by 17.02.

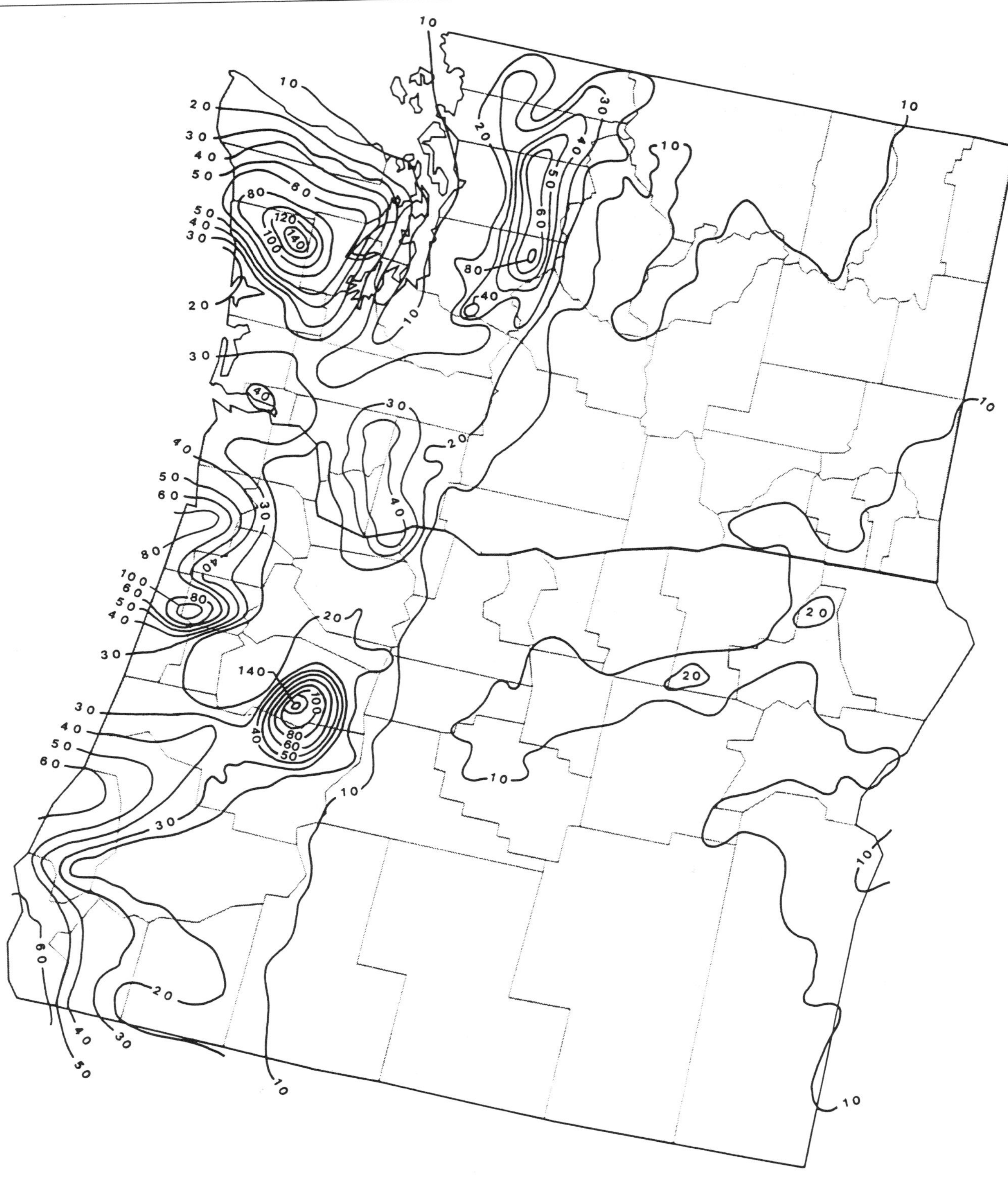

Figure 8A.3 Isolines of R factor for Washington and Oregon (after Renard *et al.*, 1993b). Units on R are ft · tonsf · in./acre · hr · year. To convert to metric, MJ · mm/ha · h · year, multiply by 17.02.

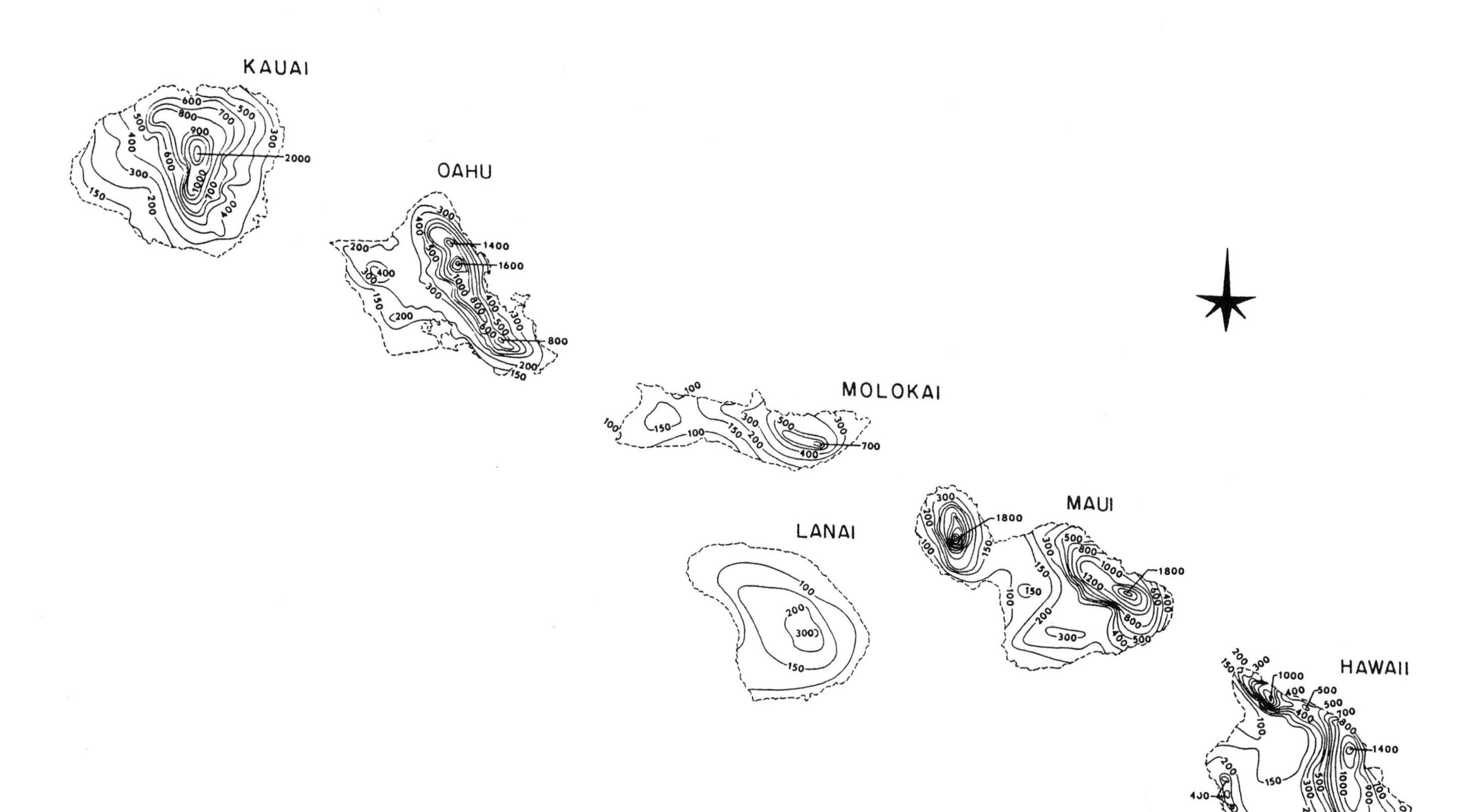

Figure 8A.4 Isolines of R factor for Hawaii (after Renard *et al.*, 1993b). Units on R are ft · tonsf · in./acre · hr · year. To convert to metric, MJ · mm/ha · h · year, multiply by 17.02.

Figure 8A.5 Isolines of 10-year return period single storm R factor for the Eastern U.S. (after Renard *et al.*, 1993b). Units on R are ft · tonsf · in./acre · hr · storm. To convert to metric, MJ · mm/ha · h · storm, multiply by 17.02.

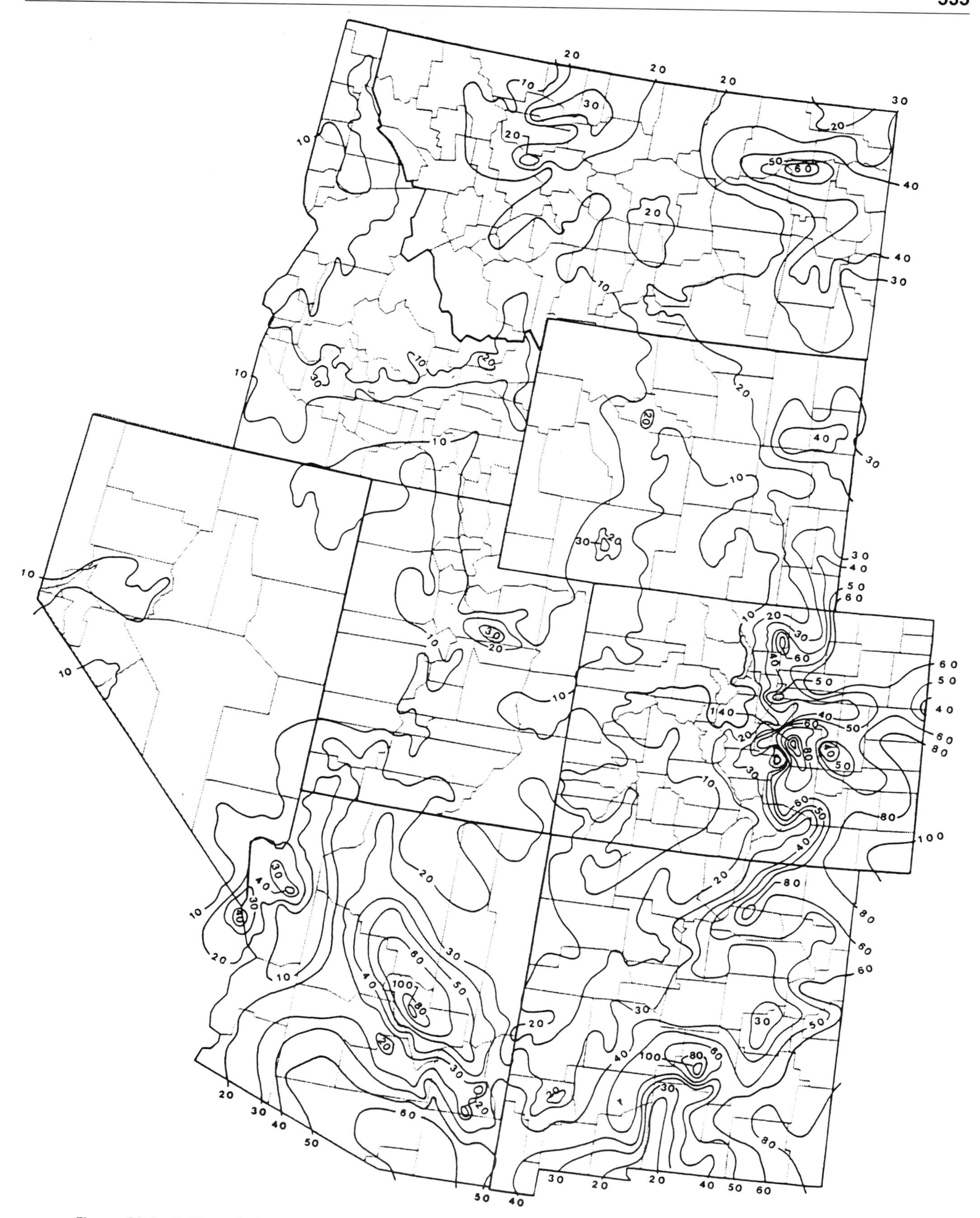

Figure 8A.6 Isolines of 10-year return period single storm R factor for the Western U.S. (after Renard *et al.*, 1993b.) Units on R are ft · tonsf · in./acre · hr · storm. To convert to metric, MJ · mm/ha · h · storm, multiply by 17.02.

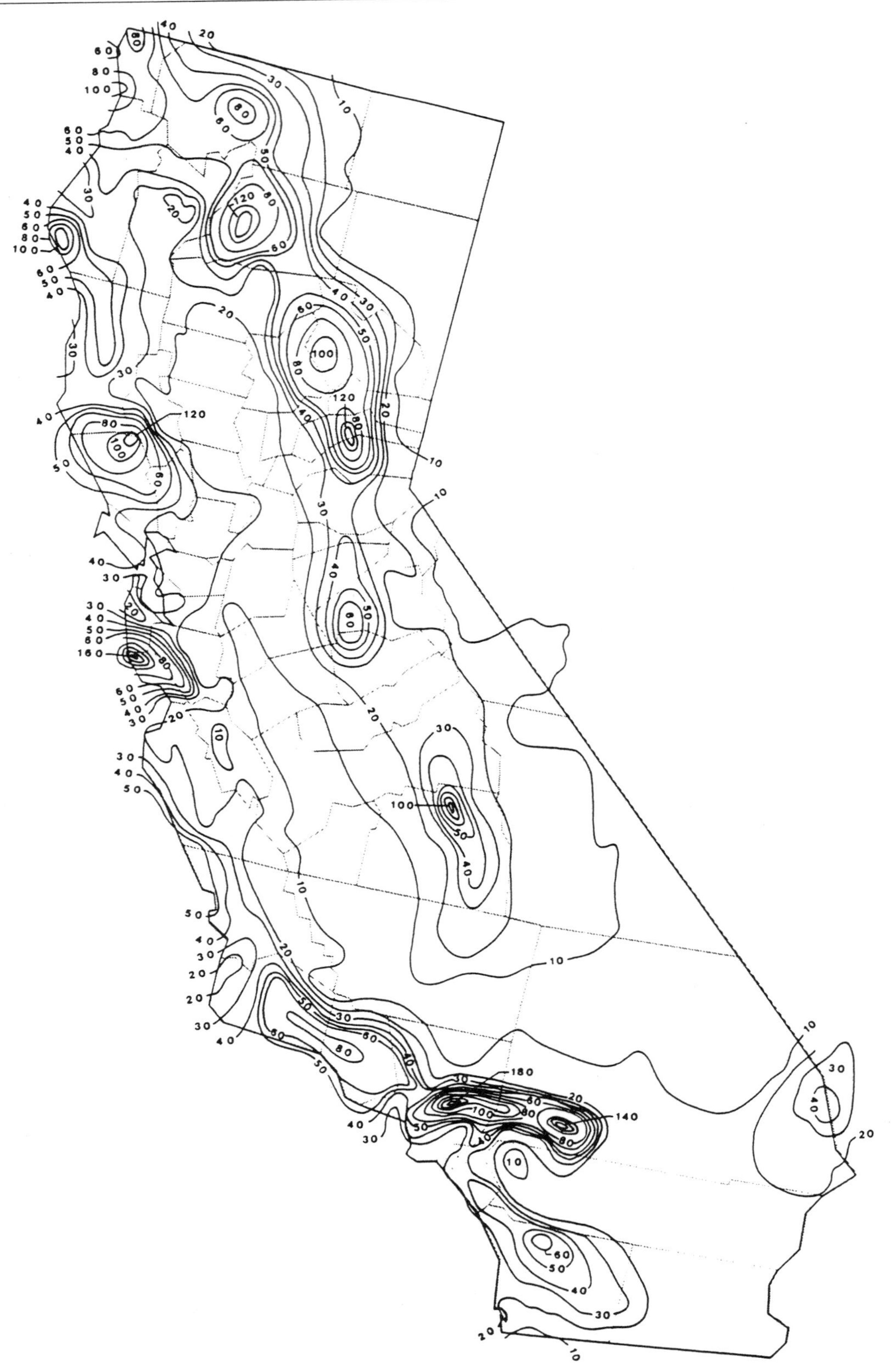

Figure 8A.7 Isolines of 10-year return period single storm R factor for California (after Renard *et al.*, 1993b). Units on R are ft · tonsf · in./acre · hr · storm. To convert to metric, MJ · mm/ha · h · storm multiply by 17.02.

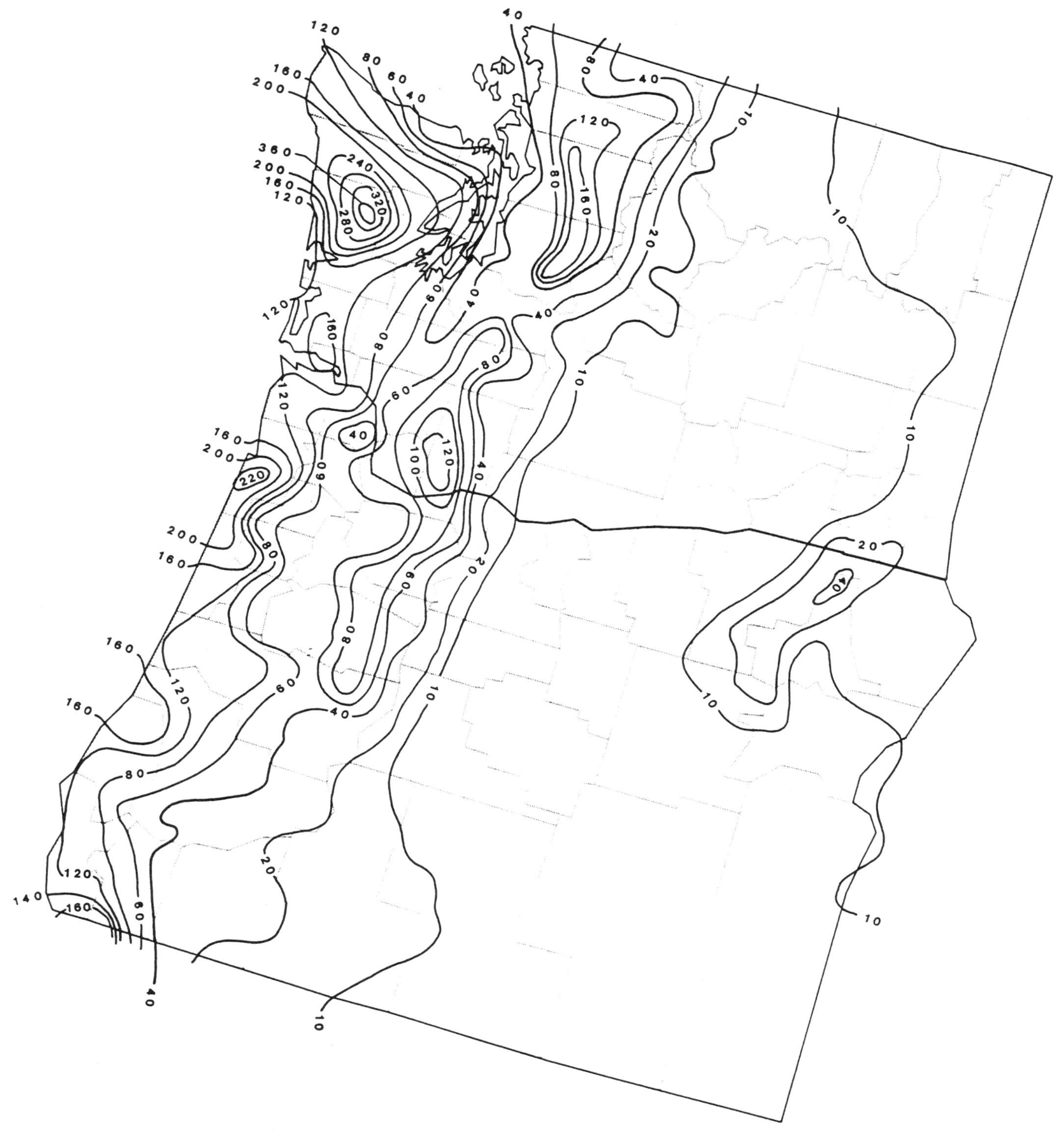

Figure 8A.8 Isolines of 10-year return period single storm R factor for Washington and Oregon (after Renard *et al.*, 1993b). Units on R are ft · tonsf · in./acre · hr · storm. To convert to metric, MJ · mm/ha · h · storm, multiply by 17.02.

Table 8A.1 *R* Factor Distributions for Zones in Fig. 8.9[a]

	Geographic Area									
	Part 1									
Date	1	2	3	4	5	6	7	8	9	10
1/1	0.0	0.0	0.0	0.0	0.0	0.0	0.0	0.0	0.0	0.0
1/15	4.3	4.3	7.4	3.9	2.3	0.0	0.0	0.0	0.8	0.3
2/l	8.3	8.3	13.8	7.9	3.6	0.0	0.0	0.0	3.1	0.5
2/15	12.8	12.8	20.9	12.6	4.7	0.5	0.0	0.0	4.7	0.9
3/1	17.3	17.3	26.5	17.4	6.0	2.0	0.0	0.0	7.4	2.0
3/15	21.6	21.6	31.8	21.6	7.7	4.1	1.2	0.9	11.7	4.3
4/1	25.1	25.1	35.3	25.2	10.7	8.1	4.9	3.6	17.8	9.2
4/15	28.0	28.0	38.5	28.7	13.9	12.6	8.5	7.8	22.5	13.1
5/1	30.9	30.9	40.2	31.9	17.8	17.6	13.9	15.0	27.0	18.0
5/15	34.9	34.9	41.6	35.1	21.2	21.6	19.0	20.2	31.4	22.7
6/1	39.1	39.1	42.5	38.2	24.5	25.5	26.1	27.4	36.0	29.2
6/15	42.6	42.6	43.6	42.0	28.1	29.6	35.4	38.1	41.6	39.5
7/1	45.4	45.4	44.5	44.9	31.1	34.5	43.9	49.8	46.4	46.3
7/15	48.2	48.2	45.1	46.7	33.1	40.0	48.8	57.9	50.1	48.8
8/1	50.8	50.8	45.7	48.2	35.3	45.7	53.9	65.0	53.4	51.1
8/15	53.0	53.0	46.4	50.1	38.2	50.7	64.5	75.6	57.4	57.2
9/1	56.0	56.0	47.7	53.1	43.2	55.6	73.4	82.7	61.7	64.4
9/15	60.8	60.8	49.4	56.6	48.7	60.2	77.5	86.8	64.9	67.7
10/1	66.8	66.8	52.8	62.2	57.3	66.5	80.4	89.4	69.7	71.1
10/15	71.0	71.0	57.0	67.9	67.8	75.5	84.8	93.4	79.0	77.2
11/1	75.7	75.7	64.5	75.2	77.9	85.6	89.9	96.3	89.6	85.1
11/15	82.0	82.0	73.1	83.5	86.0	95.9	96.6	99.1	97.4	92.5
12/1	89.1	89.1	83.3	90.5	91.3	99.5	99.2	100.0	100.0	96.5
12/15	95.2	95.2	92.3	96.0	96.9	99.9	99.7	100.0	100.0	99.0

	Part 2									
Date	11	12	13	14	15	16	17	18	19	20
1/1	0.0	0.0	0.0	0.0	0.0	0.0	0.0	0.0	0.0	0.0
1/15	5.4	3.5	0.0	0.7	0.0	0.0	0.0	0.0	1.0	9.8
2/1	11.3	7.8	0.0	1.8	0.0	0.0	0.0	0.0	2.6	18.5
2/15	18.8	14.0	1.8	3.3	0.5	0.5	0.7	0.6	7.4	25.4
3/1	26.3	21.1	7.2	6.9	2.0	2.0	2.8	2.5	16.4	30.2
3/15	33.2	27.4	11.9	16.5	4.4	5.5	6.1	6.2	23.5	35.6
4/1	37.4	31.5	16.7	26.6	8.7	12.3	10.7	12.4	28.0	38.9
4/15	40.7	35.0	19.7	29.9	12.0	16.2	12.9	16.4	31.0	41.5
5/1	42.5	37.3	24.0	32.0	16.6	20.9	16.1	20.2	33.5	42.9
5/15	44.3	39.8	31.2	35.4	21.4	26.4	21.9	23.9	37.0	44.0
6/1	45.4	41.9	42.4	40.2	29.7	35.2	32.8	29.3	41.7	45.2
6/15	46.5	44.3	55.0	45.1	44.5	48.1	45.9	37.7	48.1	48.2
7/1	47.1	45.6	60.0	51.9	56.0	58.1	55.5	45.6	51.1	50.8
7/15	47.4	46.3	60.8	61.1	60.8	63.1	60.3	49.8	52.0	51.7
8/1	47.8	46.8	61.2	67.5	63.9	66.5	64.0	53.3	52.5	52.5
8/15	48.3	47.9	62.6	70.7	69.1	71.9	71.2	58.4	53.6	54.6
9/1	49.4	50.0	65.3	72.8	74.5	77.0	77.2	64.3	55.7	57.4
9/15	50.7	52.9	67.6	75.4	79.1	81.6	80.3	69.0	57.6	58.5
10/1	53.6	57.9	71.6	78.6	83.1	85.1	83.1	75.0	61.1	60.1
10/15	57.5	62.3	76.1	81.9	87.0	88.4	87.7	86.6	65.5	63.2
11/1	65.5	69.3	83.1	86.4	90.9	91.5	92.6	93.9	74.7	69.6
11/15	76.2	81.3	93.3	93.6	96.6	96.3	97.2	96.6	88.0	76.7
12/1	87.4	91.5	98.2	97.7	99.1	98.7	99.1	98.0	95.8	85.4
12/15	94.8	96.7	99.6	99.3	99.8	99.6	99.8	100.0	98.7	92.4

(continues)

Part 3

Date	21	22	23	24	25	26	27	28	29	30
1/1	0.0	0.0	0.0	0.0	0.0	0.0	0.0	0.0	0.0	0.0
1/15	7.5	1.2	7.9	12.2	9.8	2.0	0.0	0.0	0.6	0.0
2/1	13.6	1.6	15.0	23.6	20.8	5.4	0.0	0.0	0.7	0.0
2/15	18.1	1.6	20.9	33.0	30.2	9.8	1.0	0.0	0.7	0.0
3/1	21.1	1.6	25.7	39.7	37.6	15.6	4.0	0.2	0.7	0.0
3/15	24.4	1.6	31.1	47.1	45.8	21.5	5.9	0.5	1.5	0.2
4/1	27.0	1.6	35.7	51.7	50.6	24.7	8.0	1.5	3.9	0.8
4/15	29.4	2.2	40.2	55.9	54.4	26.6	11.1	3.3	6.0	2.8
5/1	31.7	3.9	43.2	57.7	56.0	27.4	13.0	7.2	10.5	7.9
5/15	34.6	4.6	46.2	58.6	56.8	28.0	14.0	11.9	17.9	14.2
6/1	37.3	6.4	47.7	58.9	57.1	28.7	14.6	17.7	28.8	24.7
6/15	39.6	14.2	48.8	59.1	57.1	29.8	15.3	21.4	36.6	35.6
7/1	41.6	32.8	49.4	59.1	57.2	32.5	17.0	27.0	43.8	45.4
7/15	43.4	47.2	49.9	59.2	57.6	36.6	23.2	37.1	51.5	52.2
8/1	45.4	58.8	50.7	59.2	58.5	44.9	39.1	51.4	59.3	58.7
8/15	48.1	69.1	51.8	59.3	59.8	55.4	60.0	62.3	68.0	68.5
9/1	51.3	76.0	54.1	59.5	62.2	65.7	76.3	70.6	74.8	77.6
9/15	53.3	82.0	57.7	60.0	65.3	72.6	86.1	78.8	80.3	84.5
10/1	56.6	87.1	62.8	61.4	67.5	77.8	89.7	84.6	84.3	88.9
10/15	62.4	96.7	65.9	63.0	68.2	84.4	90.4	90.6	88.8	93.7
11/1	72.4	99.9	70.1	66.5	69.4	89.5	90.9	94.4	92.7	96.2
11/15	81.3	99.9	77.3	71.8	74.8	93.9	93.1	97.9	98.0	97.6
12/1	88.9	99.9	86.8	81.3	86.6	96.5	96.6	99.3	99.8	98.3
12/15	94.7	99.9	93.5	89.6	93.0	98.4	99.1	100.0	99.9	99.6

Part 4

Date	31	32	33	34	35	36	37	38	39	40
1/1	0.0	0.0	0.0	0.0	0.0	0.0	0.0	0.0	0.0	0.0
1/15	0.0	0.1	0.0	0.0	0.0	0.0	0.0	0.0	0.0	0.0
2/1	0.0	0.1	0.0	0.0	0.0	0.0	0.0	0.0	0.0	0.0
2/15	0.0	0.1	0.0	0.0	0.0	0.0	0.0	1.1	0.0	0.0
3/1	0.0	0.1	0.0	0.0	0.0	0.0	0.0	4.3	0.0	0.0
3/15	0.2	0.6	0.6	1.8	2.5	0.9	0.0	7.2	1.6	1.5
4/1	1.0	2.2	2.3	7.3	10.2	3.4	0.0	11.0	6.5	6.2
4/15	3.5	4.3	4.2	10.7	15.9	6.7	1.0	13.9	11.0	10.1
5/1	9.9	9.0	8.8	15.5	22.2	12.7	3.9	17.9	17.8	16.3
5/15	15.7	14.2	16.1	22.0	27.9	18.5	9.1	22.3	24.7	23.3
6/1	26.4	23.3	30.0	29.9	34.7	26.6	19.1	30.3	33.1	32.5
6/15	47.2	34.6	46.9	35.9	43.9	36.3	26.7	43.1	42.8	42.2
7/1	61.4	46.3	57.9	42.0	51.9	46.0	36.3	55.1	50.3	50.1
7/15	65.9	54.2	62.8	48.5	56.9	53.5	47.9	61.3	54.9	55.6
8/1	69.0	61.7	66.2	56.9	61.3	60.2	61.4	65.7	59.7	60.5
8/15	77.2	72.9	72.1	67.0	67.3	68.3	75.1	72.1	68.9	67.5
9/1	86.0	82.5	79.1	76.9	73.9	75.8	84.5	77.9	78.1	74.3
9/15	91.6	89.6	85.9	85.8	80.1	82.6	92.3	82.6	83.6	79.4
10/1	94.8	93.7	91.1	91.2	85.1	88.3	96.0	86.3	87.5	84.1
10/15	98.7	98.2	97.0	95.7	89.6	96.3	99.1	90.3	93.0	91.1
11/1	100.0	99.7	98.9	97.8	93.2	99.3	100.0	93.8	96.5	95.8
11/15	100.0	99.9	98.9	99.6	98.2	99.9	100.0	98.4	99.2	99.1
12/1	100.0	99.9	98.9	100.0	99.8	100.0	100.0	100.0	100.0	100.0
12/15	100.0	99.9	98.9	100.0	99.8	100.0	100.0	100.0	100.0	100.0

(continues)

Part 5

Date	41	42	43	44	45	46	47	48	49	50
1/1	0.0	0.0	0.0	0.0	0.0	0.0	0.0	0.0	0.0	0.0
1/15	0.1	0.0	0.0	1.7	0.2	0.0	0.0	0.0	0.0	0.0
2/1	0.2	0.0	0.0	2.3	0.2	0.0	0.0	0.0	0.0	0.0
2/15	0.2	0.0	0.0	2.4	0.3	0.0	0.0	0.0	0.0	0.0
3/1	0.2	0.0	0.0	2.4	0.3	0.0	0.0	0.0	0.0	0.0
3/15	0.2	0.0	0.0	2.4	0.4	0.0	0.0	0.0	0.0	0.1
4/1	0.2	0.0	0.0	2.4	0.6	0.0	0.0	0.0	0.0	0.4
4/15	0.4	0.2	0.1	2.7	0.8	0.6	0.4	0.0	0.7	2.4
5/1	1.1	0.9	0.4	3.5	1.4	2.6	1.6	0.0	2.7	8.2
5/15	6.8	5.2	2.7	7.6	3.7	7.5	5.8	2.0	8.3	13.7
6/1	22.9	17.3	9.5	18.5	10.2	19.6	17.0	8.1	20.0	23.8
6/15	40.1	33.8	21.9	34.3	22.6	32.9	33.0	15.4	27.5	38.8
7/1	54.9	53.2	42.7	52.5	41.8	48.9	52.5	27.8	35.6	55.1
7/15	63.8	66.5	58.6	64.0	54.0	63.0	66.4	40.7	44.6	66.1
8/1	70.7	75.9	71.1	72.3	64.5	73.5	75.7	52.6	46.0	73.6
8/15	81.5	87.6	84.6	83.3	78.7	83.3	85.5	61.1	70.2	81.8
9/1	89.8	93.7	91.9	90.0	88.4	89.5	91.3	69.3	81.3	87.7
9/15	96.3	97.5	97.1	95.1	96.0	95.6	96.5	82.6	89.2	93.8
10/1	98.7	99.0	99.0	97.3	98.7	98.3	98.8	92.0	93.6	97.0
10/15	99.2	99.7	99.8	98.5	99.4	99.6	100.0	98.0	98.5	99.4
11/1	99.3	100.0	100.0	98.9	99.7	100.0	100.0	100.0	100.0	100.0
11/15	99.4	100.0	100.0	98.9	99.7	100.0	100.0	100.0	100.0	100.0
12/1	99.4	100.0	100.0	98.9	99.8	100.0	100.0	100.0	100.0	100.0
12/15	99.7	100.0	100.0	99.2	99.9	100.0	100.0	100.0	100.0	100.0

Part 6

Date	51	52	53	54	55	56	57	58	59	60
1/1	0.0	0.0	0.0	0.0	0.0	0.0	0.0	0.0	0.0	0.0
1/15	0.0	0.0	0.0	0.0	0.0	0.0	0.0	0.0	0.0	0.0
2/1	0.0	0.0	0.0	0.0	0.0	0.0	0.0	0.0	0.0	0.0
2/15	0.0	0.0	0.0	0.0	0.0	0.0	0.0	0.0	0.0	0.0
3/1	0.0	0.0	0.0	0.0	0.0	0.0	0.0	0.0	0.0	0.0
3/15	0.3	0.0	0.0	0.2	0.0	0.0	0.0	0.2	0.0	0.0
4/1	1.0	0.0	0.0	0.7	0.0	0.0	0.1	0.9	0.0	0.0
4/15	3.1	0.6	0.8	2.4	1.3	1.3	1.0	2.9	2.2	0.4
5/1	8.7	2.5	3.0	7.2	5.4	5.1	3.5	8.0	8.9	1.5
5/15	18.8	6.8	9.5	14.7	13.3	11.4	9.2	13.2	15.6	4.0
6/1	35.8	17.5	24.2	27.2	25.5	22.3	21.5	21.0	24.2	9.5
6/15	49.6	29.8	35.3	37.2	31.6	29.5	31.0	29.1	31.1	13.3
7/1	60.4	46.1	48.0	47.3	38.8	38.5	43.5	38.0	38.3	20.5
7/15	70.2	60.5	63.1	58.8	52.5	51.1	60.4	45.9	46.0	33.6
8/1	77.0	72.7	76.1	67.6	66.8	65.2	75.1	54.5	54.9	52.8
8/15	84.0	86.0	87.7	74.0	75.5	77.8	86.1	65.4	64.2	66.5
9/1	88.8	92.8	93.5	79.2	81.2	85.6	91.6	74.8	73.2	76.7
9/15	93.8	96.8	97.2	86.7	87.9	91.7	96.2	82.1	81.9	88.1
10/1	96.6	98.4	98.6	92.6	92.8	95.0	98.1	87.5	88.5	94.2
10/15	99.1	99.7	99.5	97.9	98.3	98.7	99.4	95.4	95.7	98.6
11/1	100.0	100.0	99.8	99.8	100.0	100.0	99.9	98.8	98.6	100.0
11/15	100.0	100.0	99.9	99.9	100.0	100.0	99.9	99.7	99.4	100.0
12/1	100.0	100.0	100.0	100.0	100.0	100.0	100.0	100.0	99.7	100.0
12/15	100.0	100.0	100.0	100.0	100.0	100.0	100.0	100.0	99.7	100.0

(continues)

Part 7

Date	61	62	63	64	65	66	67	68	69	70
1/1	0.0	0.0	0.0	0.0	0.0	0.0	0.0	0.0	0.0	0.0
1/15	0.0	0.0	0.0	0.0	3.6	0.0	0.0	2.3	2.0	0.5
2/1	0.0	0.0	0.0	0.0	7.0	0.0	0.0	4.5	3.7	0.7
2/15	0.0	0.1	0.0	0.7	9.6	0.0	0.0	7.8	5.7	1.0
3/1	0.0	0.3	0.0	2.8	11.4	0.0	0.0	10.4	7.8	1.3
3/15	0.0	0.8	0.0	7.4	13.0	0.1	0.1	12.0	10.5	1.7
4/1	0.0	2.1	0.0	12.4	14.4	0.5	0.4	13.3	12.4	2.2
4/15	1.3	3.6	0.9	14.4	16.3	1.1	0.9	16.3	13.7	2.8
5/1	5.0	6.5	3.7	15.6	17.7	2.2	1.6	17.7	14.3	3.4
5/15	8.5	9.7	7.8	17.3	18.4	3.6	1.9	18.1	14.7	3.9
6/1	15.5	13.7	13.3	19.4	19.3	6.0	2.4	18.2	15.1	4.7
6/15	29.8	16.5	15.8	21.0	20.5	7.6	5.0	18.3	15.7	5.4
7/1	41.8	20.8	19.9	24.4	23.6	11.1	12.1	18.4	17.1	7.4
7/15	46.0	27.3	29.0	32.3	32.0	19.8	24.8	19.9	22.7	15.7
8/1	49.2	40.1	46.8	48.0	50.0	38.9	48.3	24.5	36.7	36.5
8/15	56.0	56.9	64.7	61.4	66.2	59.7	73.6	35.0	50.4	55.8
9/1	65.1	72.6	78.3	72.1	77.2	74.4	86.5	54.4	63.6	70.3
9/15	71.6	83.4	88.8	81.9	85.4	83.2	92.0	69.4	75.0	80.9
10/1	78.6	89.4	93.9	87.0	88.8	88.1	94.3	78.6	81.8	86.4
10/15	91.1	95.5	98.5	90.1	90.4	94.6	96.6	85.7	87.8	90.9
11/1	97.3	98.1	100.0	92.4	91.3	97.7	97.9	89.2	90.8	93.4
11/15	99.3	99.6	100.0	98.1	92.7	99.4	99.5	91.9	93.2	96.4
12/1	100.0	100.0	100.0	100.0	94.8	100.0	100.0	93.9	94.9	98.1
12/15	100.0	100.0	100.0	100.0	97.0	100.0	100.0	97.0	97.5	99.4

Part 8

Date	71	72	73	74	75	76	77	78	79	80
1/1	0.0	0.0	0.0	0.0	0.0	0.0	0.0	0.0	0.0	0.0
1/15	0.7	0.0	0.0	0.0	0.1	0.0	0.2	0.0	0.0	0.6
2/1	1.2	0.0	0.1	0.0	0.1	0.0	0.3	0.0	0.0	1.2
2/15	1.6	0.0	0.1	0.0	0.1	0.0	0.3	0.0	0.0	1.6
3/1	2.1	0.0	0.2	0.0	0.2	0.0	0.4	0.0	0.0	2.1
3/15	2.8	0.0	0.2	0.1	0.5	0.1	0.8	0.0	0.2	2.5
4/1	3.3	0.1	0.3	0.2	1.3	0.2	1.5	0.2	0.7	3.3
4/15	3.6	0.2	0.6	0.5	1.9	0.6	2.0	0.5	1.3	4.5
5/1	4.0	0.7	1.3	1.2	3.0	1.3	2.8	1.6	2.7	6.9
5/15	4.5	0.8	4.1	2.7	4.1	2.0	3.9	3.8	5.8	10.1
6/1	5.6	1.3	11.5	6.4	6.6	3.5	5.9	8.9	12.7	15.5
6/15	6.5	3.5	18.1	10.2	10.0	4.9	7.2	13.2	18.8	19.7
7/1	9.1	9.9	28.3	18.4	17.6	8.4	10.3	21.8	28.8	26.6
7/15	18.5	24.7	40.2	31.0	28.3	17.4	21.5	35.8	41.6	36.4
8/1	40.6	51.4	54.1	50.7	44.7	37.3	46.5	56.6	58.4	51.7
8/15	59.7	71.5	67.0	68.7	59.4	57.5	66.3	75.4	75.7	67.5
9/1	74.0	83.6	77.2	81.2	71.6	72.9	78.3	86.0	86.5	79.4
9/15	86.3	93.8	87.7	91.6	83.9	83.7	86.5	92.9	94.2	88.8
10/1	91.7	97.7	93.3	96.1	90.3	89.5	90.8	95.9	97.3	93.2
10/15	94.7	99.2	97.5	98.4	94.7	95.8	96.0	98.2	98.9	96.1
11/1	96.0	99.8	99.1	99.2	96.7	98.4	98.2	99.2	99.5	97.3
11/15	96.7	99.9	99.6	99.8	98.8	99.6	99.1	99.8	99.9	98.2
12/1	97.3	99.9	99.8	100.0	99.6	100.0	99.5	100.0	100.0	98.7
12/15	98.8	100.0	100.0	100.0	99.9	100.0	99.8	100.0	100.0	99.3

(continues)

Part 9

Date	81	82	83	84	85	86	87	88	89	90
1/1	0.0	0.0	0.0	0.0	0.0	0.0	0.0	0.0	0.0	0.0
1/15	0.1	0.0	0.0	0.0	0.0	0.0	0.0	0.0	0.0	1.0
2/1	0.1	0.1	0.1	0.1	0.0	0.0	0.0	0.0	1.0	2.0
2/15	0.2	0.1	0.1	0.1	0.0	0.0	0.0	0.0	1.0	3.0
3/1	0.4	0.2	0.1	0.2	0.0	0.0	1.0	1.0	2.0	4.0
3/15	0.5	0.2	0.3	0.3	0.0	0.0	1.0	1.0	3.0	6.0
4/1	0.8	0.5	0.9	0.6	1.0	1.0	2.0	2.0	4.0	8.0
4/15	0.9	1.2	1.6	1.7	2.0	2.0	3.0	3.0	7.0	13.0
5/1	1.5	3.1	3.5	4.9	3.0	3.0	6.0	6.0	12.0	21.0
5/15	3.9	6.7	8.3	9.9	6.0	6.0	10.0	13.0	18.0	29.0
6/1	9.9	14.4	19.4	19.5	11.0	11.0	17.0	23.0	27.0	37.0
6/15	12.8	20.1	30.0	27.2	23.0	23.0	29.0	37.0	38.0	46.0
7/1	18.2	29.8	44.0	38.3	36.0	36.0	43.0	51.0	48.0	54.0
7/15	30.7	44.5	59.2	52.8	49.0	49.0	55.0	61.0	55.0	60.0
8/1	54.1	64.2	72.4	68.8	63.0	63.0	67.0	69.0	62.0	65.0
8/15	77.1	83.1	84.6	83.9	77.0	77.0	77.0	78.0	69.0	69.0
9/1	89.0	92.2	91.2	91.6	90.0	90.0	85.0	85.0	76.0	74.0
9/15	94.9	96.4	96.5	96.4	95.0	95.0	91.0	91.0	83.0	81.0
10/1	97.2	98.1	98.6	98.2	98.0	98.0	96.0	94.0	90.0	87.0
10/15	98.7	99.3	99.5	99.2	99.0	99.0	98.0	96.0	94.0	92.0
11/1	99.3	99.7	99.8	99.6	100.0	100.0	99.0	98.0	97.0	95.0
11/15	99.6	99.8	99.9	99.8	100.0	100.0	100.0	99.0	98.0	97.0
12/1	99.7	99.8	100.0	99.8	100.0	100.0	100.0	99.0	99.0	98.0
12/15	99.9	99.9	100.0	99.9	100.0	100.0	100.0	100.0	100.0	99.0

Part 10

Date	91	92	93	94	95	96	97	98	99	100
1/1	0.0	0.0	0.0	0.0	0.0	0.0	0.0	0.0	0.0	0.0
1/15	0.0	0.0	1.0	1.0	1.0	2.0	1.0	1.0	0.0	0.0
2/1	0.0	0.0	1.0	2.0	3.0	4.0	3.0	2.0	0.0	0.0
2/15	0.0	0.0	2.0	4.0	5.0	6.0	5.0	4.0	1.0	0.0
3/1	1.0	1.0	3.0	6.0	7.0	9.0	7.0	6.0	1.0	1.0
3/15	1.0	1.0	4.0	8.0	9.0	12.0	10.0	8.0	2.0	1.0
4/1	1.0	1.0	6.0	10.0	11.0	17.0	14.0	10.0	3.0	2.0
4/15	2.0	2.0	8.0	15.0	14.0	23.0	20.0	13.0	5.0	3.0
5/1	6.0	6.0	13.0	21.0	18.0	30.0	28.0	19.0	7.0	5.0
5/15	16.0	16.0	25.0	29.0	27.0	37.0	37.0	26.0	12.0	9.0
6/1	29.0	29.0	40.0	38.0	35.0	43.0	48.0	34.0	19.0	15.0
6/15	39.0	39.0	49.0	47.0	41.0	49.0	56.0	42.0	33.0	27.0
7/1	46.0	46.0	56.0	53.0	46.0	54.0	61.0	50.0	48.0	38.0
7/15	53.0	53.0	62.0	57.0	51.0	58.0	64.0	58.0	57.0	50.0
8/1	60.0	60.0	67.0	61.0	57.0	62.0	68.0	63.0	65.0	62.0
8/15	67.0	67.0	72.0	65.0	62.0	66.0	72.0	68.0	72.0	74.0
9/1	74.0	74.0	76.0	70.0	68.0	70.0	77.0	74.0	82.0	84.0
9/15	81.0	81.0	80.0	76.0	73.0	74.0	81.0	79.0	88.0	91.0
10/1	88.0	88.0	85.0	83.0	79.0	78.0	86.0	84.0	93.0	95.0
10/15	95.0	95.0	91.0	88.0	84.0	82.0	89.0	89.0	96.0	97.0
11/1	99.0	99.0	97.0	91.0	89.0	86.0	92.0	93.0	98.0	98.0
11/15	99.0	99.0	98.0	94.0	93.0	90.0	95.0	95.0	99.0	99.0
12/1	100.0	100.0	99.0	96.0	96.0	94.0	98.0	97.0	100.0	99.0
12/15	100.0	100.0	99.0	98.0	98.0	97.0	99.0	99.0	100.0	100.0

(continues)

Part 11

Date	101	102	103	104	105	106	107	108	109	110
1/1	0.0	0.0	0.0	0.0	0.0	0.0	0.0	0.0	0.0	0.0
1/15	0.0	0.0	1.0	2.0	1.0	3.0	3.0	3.0	3.0	1.0
2/1	0.0	1.0	2.0	3.0	3.0	6.0	5.0	6.0	6.0	3.0
2/15	1.0	2.0	3.0	5.0	6.0	9.0	7.0	9.0	10.0	5.0
3/1	2.0	3.0	4.0	7.0	9.0	13.0	10.0	12.0	13.0	7.0
3/15	3.0	4.0	6.0	10.0	12.0	17.0	14.0	16.0	16.0	9.0
4/1	4.0	6.0	8.0	13.0	16.0	21.0	18.0	20.0	19.0	12.0
4/15	6.0	8.0	10.0	16.0	21.0	27.0	23.0	24.0	23.0	15.0
5/1	9.0	11.0	14.0	19.0	26.0	33.0	27.0	28.0	26.0	18.0
5/15	14.0	15.0	18.0	23.0	31.0	38.0	31.0	33.0	29.0	21.0
6/1	20.0	22.0	25.0	27.0	37.0	44.0	35.0	38.0	33.0	25.0
6/15	28.0	31.0	34.0	34.0	43.0	49.0	39.0	43.0	39.0	29.0
7/1	39.0	40.0	45.0	44.0	50.0	55.0	45.0	50.0	47.0	36.0
7/15	52.0	49.0	56.0	54.0	57.0	61.0	53.0	59.0	58.0	45.0
8/1	63.0	59.0	64.0	63.0	64.0	67.0	60.0	69.0	68.0	56.0
8/15	72.0	69.0	72.0	72.0	71.0	71.0	67.0	75.0	75.0	68.0
9/1	80.0	78.0	79.0	80.0	77.0	75.0	74.0	80.0	80.0	77.0
9/15	87.0	85.0	84.0	85.0	81.0	78.0	80.0	84.0	83.0	83.0
10/1	91.0	91.0	89.0	89.0	85.0	81.0	84.0	87.0	86.0	88.0
10/15	94.0	94.0	92.0	91.0	88.0	84.0	86.0	90.0	88.0	91.0
11/1	97.0	96.0	95.0	93.0	91.0	86.0	88.0	92.0	90.0	93.0
11/15	98.0	98.0	97.0	95.0	93.0	90.0	90.0	94.0	92.0	95.0
12/1	99.0	99.0	98.0	96.0	95.0	94.0	93.0	96.0	95.0	97.0
12/15	100.0	100.0	99.0	98.0	97.0	97.0	95.0	98.0	97.0	99.0

Part 12

Date	111	112	113	114	115	116	117	118	119	120	121
1/1	0.0	0.0	0.0	0.0	0.0	0.0	0.0	0.0	0.0	0.0	0.0
1/15	1.0	0.0	1.0	1.0	1.0	1.0	1.0	1.0	2.0	1.0	8.0
2/1	2.0	0.0	2.0	2.0	2.0	3.0	2.0	2.0	4.0	2.0	16.0
2/15	3.0	1.0	3.0	4.0	3.0	5.0	3.0	3.0	6.0	4.0	25.0
3/1	4.0	2.0	4.0	6.0	4.0	7.0	4.0	5.0	8.0	6.0	33.0
3/15	5.0	3.0	5.0	8.0	5.0	9.0	5.0	7.0	12.0	7.0	41.0
4/1	6.0	4.0	6.0	11.0	6.0	12.0	7.0	10.0	16.0	9.0	46.0
4/15	8.0	5.0	8.0	13.0	8.0	15.0	9.0	14.0	20.0	12.0	50.0
5/1	11.0	7.0	10.0	15.0	10.0	18.0	11.0	18.0	25.0	15.0	53.0
5/15	15.0	12.0	13.0	18.0	14.0	21.0	14.0	22.0	30.0	18.0	54.0
6/1	20.0	17.0	17.0	21.0	19.0	25.0	17.0	27.0	35.0	23.0	55.0
6/15	28.0	24.0	22.0	26.0	26.0	29.0	22.0	32.0	41.0	31.0	56.0
7/1	41.0	33.0	31.0	32.0	34.0	36.0	31.0	37.0	47.0	40.0	56.5
7/15	54.0	42.0	42.0	38.0	45.0	45.0	42.0	46.0	56.0	48.0	57.0
8/1	65.0	55.0	52.0	46.0	56.0	56.0	54.0	58.0	67.0	57.0	57.8
8/15	74.0	67.0	60.0	55.0	66.0	68.0	65.0	69.0	75.0	63.0	58.0
9/1	82.0	76.0	68.0	64.0	76.0	77.0	74.0	80.0	81.0	72.0	58.8
9/15	87.0	83.0	75.0	71.0	82.0	83.0	83.0	89.0	85.0	78.0	60.0
10/1	92.0	89.0	80.0	77.0	86.0	88.0	89.0	93.0	87.0	88.0	61.0
10/15	94.0	92.0	85.0	81.0	90.0	91.0	92.0	94.0	89.0	92.0	63.0
11/1	96.0	94.0	89.0	85.0	93.0	93.0	95.0	95.0	91.0	96.0	66.5
11/15	97.0	96.0	92.0	89.0	95.0	95.0	97.0	96.0	93.0	97.0	72.0
12/1	98.0	98.0	96.0	93.0	97.0	97.0	98.0	97.0	95.0	98.0	80.0
12/15	99.0	99.0	98.0	97.0	99.0	99.0	99.0	97.0	97.0	99.0	90.0

(continues)

Part 13

Date	122	123	124	125	126	127	128	129	130	131
1/1	0.0	0.0	0.0	0.0	0.0	0.0	0.0	0.0	0.0	0.0
1/15	7.0	4.0	4.0	7.0	9.0	8.0	8.0	9.0	10.0	8.0
2/1	14.0	8.0	9.0	12.0	16.0	15.0	15.0	16.0	20.0	15.0
2/15	20.0	12.0	15.0	17.0	23.0	22.0	22.0	22.0	28.0	22.0
3/1	25.5	17.0	23.0	24.0	30.0	28.0	29.0	27.0	35.0	28.0
3/15	33.5	23.0	29.0	30.0	37.0	33.0	34.0	32.0	41.0	33.0
4/1	38.0	29.0	34.0	39.0	43.0	38.0	40.0	37.0	46.0	38.0
4/15	43.0	34.0	40.0	45.0	47.0	42.0	45.0	41.0	49.0	41.0
511	46.0	38.0	44.0	50.0	50.0	46.0	48.0	45.0	51.0	44.0
5/15	50.0	44.0	48.0	53.0	52.0	50.0	51.0	48.0	53.0	47.0
6/1	52.5	49.0	50.0	55.0	54.0	52.0	54.0	51.0	55.0	49.0
6/15	54.5	53.0	51.0	56.0	55.0	53.0	57.0	53.0	56.0	51.0
7/1	56.0	56.0	52.0	57.0	56.0	53.0	59.0	55.0	56.0	53.0
7/15	58.0	59.0	53.0	58.0	57.0	53.0	62.0	56.0	57.0	55.0
8/1	59.0	62.0	55.0	59.0	58.0	53.0	63.0	57.0	58.0	56.0
8/15	60.0	65.0	57.0	61.0	59.0	54.0	64.0	57.0	59.0	58.0
9/1	61.5	69.0	60.0	62.0	60.0	55.0	65.0	58.0	60.0	59.0
9/15	63.0	72.0	62.0	63.0	62.0	57.0	66.0	59.0	61.0	60.0
10/1	65.0	75.0	64.0	64.0	64.0	59.0	67.0	61.0	62.0	63.0
10/15	68.0	79.0	67.0	66.0	67.0	63.0	69.0	64.0	65.0	65.0
11/1	72.0	83.0	72.0	70.0	71.0	68.0	72.0	68.0	69.0	69.0
11/15	79.0	88.0	80.0	77.0	77.0	75.0	76.0	73.0	74.0	75.0
12/1	86.0	93.0	88.0	84.0	86.0	83.0	83.0	79.0	81.0	84.0
12/15	93.0	96.0	95.0	92.0	93.0	92.0	91.0	89.0	90.0	92.0

Part 14

Date	132	133	134	135	136	137	138	139	140[b]
1/1	0.0	0.0	0.0	0.0	0.0	0.0	0.0	0.0	0.0
1/15	10.0	8.0	12.0	7.0	11.0	10.0	11.0	8.0	13.0
2/1	18.0	16.0	22.0	15.0	21.0	18.0	22.0	14.0	28.0
2/15	25.0	24.0	31.0	22.0	29.0	25.0	31.0	20.0	43.0
3/1	29.0	32.0	39.0	30.0	37.0	30.0	39.0	25.0	56.0
3/15	33.0	40.0	45.0	37.0	44.0	39.0	46.0	32.0	65.0
4/1	36.0	46.0	49.0	43.0	50.0	46.0	52.0	37.0	69.0
4/15	39.0	51.0	52.0	49.0	55.0	51.0	56.0	42.0	69.4
5/1	41.0	54.0	54.0	53.0	57.0	54.0	58.0	47.0	69.7
5/15	42.0	56.0	55.0	55.0	59.0	57.0	59.0	50.0	70.1
6/1	44.0	57.0	56.0	57.0	60.0	58.0	60.0	53.0	70.4
6/15	45.0	58.0	56.0	58.0	60.0	59.0	61.0	55.0	70.8
7/1	46.0	58.0	56.0	59.0	60.0	59.0	61.0	56.0	71.1
7/15	47.0	59.0	56.0	60.0	60.0	60.0	61.0	58.0	71.5
8/1	48.0	59.0	57.0	61.0	61.0	60.0	61.0	59.0	71.9
8/15	49.0	60.0	57.0	62.0	61.0	60.0	62.0	61.0	72.2
9/1	51.0	60.0	57.0	63.0	61.0	61.0	62.0	63.0	72.6
9/15	53.0	61.0	57.0	65.0	62.0	62.0	62.0	64.0	73.0
10/1	56.0	62.0	58.0	67.0	63.0	63.0	63.0	66.0	73.3
10/15	59.0	64.0	59.0	70.0	64.0	64.0	64.0	68.0	73.6
11/1	64.0	68.0	62.0	74.0	67.0	67.0	66.0	71.0	74.0
11/15	70.0	74.0	68.0	79.0	71.0	72.0	71.0	76.0	76.0
12/1	80.0	83.0	77.0	85.0	78.0	80.0	78.0	85.0	81.0
12/15	90.0	91.0	88.0	92.0	89.0	90.0	89.0	93.0	89.0

[a](Data provided by Renard, USDA–ARS, Tucson, AZ. Units on R are ft•tonsf•in./acre•hr•year. To convert to metric, MJ•mm/ha•hr•year, multiply by 17.02.

[b]Zone 140 is for Pullman, WA and should be used for winter wheat and other dryland grain crops in the Northwest.

Table 8A.2 Return Period Single-Storm *R* Factors for Selected Cities in the U.S.[a]

Location	Index values normally exceeded once in (year)				
	1	2	5	10	20
Alabama					
Birmingham	54	77	110	140	170
Mobil	97	122	151	172	194
Montgomery	62	86	118	145	172
Arkansas					
Fort Smith	43	65	101	132	167
Little Rock	41	69	115	158	211
Mountain Home	33	46	68	87	105
Texarkana	51	73	105	132	163
California					
Red Bluff	13	21	36	49	65
San Luis Obispo	11	15	22	28	34
Colorado					
Akron	22	36	63	87	118
Pueblo	17	31	60	88	127
Springfield	31	51	84	112	152
Connecticut					
Hartford	23	33	50	64	79
New Haven	31	47	73	96	122
District of Columbia	39	57	86	108	136
Florida					
Appalachicola	87	124	180	224	272
Jacksonville	92	123	166	201	236
Miami	93	134	200	253	308
Georgia					
Atlanta	49	67	92	112	134
Augusta	34	50	74	94	118
Columbus	61	81	108	131	152
Macon	53	72	99	122	146
Savannah	82	128	203	272	358
Watkinsville	52	71	98	120	142
Illinois					
Cairo	39	63	101	135	173
Chicago	33	49	77	101	129
Dixon Springs	39	56	82	105	130
Moline	39	50	89	116	145
Rantoul	27	39	56	69	82
Springfield	36	52	75	94	117
Indiana					
Evansville	26	38	56	71	86
Fort Wayne	24	33	45	56	65
Indianapolis	29	41	60	75	90
South Bend	26	41	65	86	111
Terre Haute	42	57	78	96	113
Iowa					
Burlington	37	48	62	72	81
Charles City	33	47	68	85	103
Clarinda	35	48	66	79	94
Des Moines	31	45	67	86	105
Dubuque	43	63	91	114	140
Rockwell City	31	49	76	101	129
Sioux City	40	58	84	105	131
Kansas					
Burlingame	37	51	69	83	100
Coffeyville	47	69	101	128	159
Concordia	33	53	86	116	154
Dodge City	31	47	76	97	124
Goodland	26	37	53	67	80
Hays	35	51	76	97	121
Wichita	41	61	93	121	150
Kentucky					
Lexington	28	46	80	114	151
Louisville	31	43	59	72	85
Middlesboro	28	38	52	63	73
Louisiana					
New Orleans	104	149	214	270	330
Shreveport	55	73	99	121	141
Maine					
Caribou	14	20	28	36	44
Portland	16	27	48	66	88
Skowhegan	18	27	40	51	63
Maryland					
Baltimore	41	59	86	109	133
Massachusetts					
Boston	17	27	43	57	73
Washington	29	35	41	45	50
Michigan					
Alpena	14	21	32	41	50
Detroit	21	31	45	56	68
East Lansing	19	26	36	43	51
Grand Rapids	24	28	34	38	42
Minnesota					
Duluth	21	34	53	72	93
Fosston	17	26	39	51	63
Minneapolis	25	35	51	65	78
Rochester	41	58	85	105	129
Springfield	24	37	60	80	102
Mississippi					
Meridian	69	92	125	151	176
Oxford	48	64	86	103	120
Vicksburg	57	78	111	136	161
Missouri					
Columbia	43	58	77	93	107
Kansas City	30	43	63	78	93
McCredie	35	55	89	117	151
Rolla	43	63	91	115	140

(continues)

Table 8A.2 *(Continued)*

Location	Index values normally exceeded once in (year) 1	2	5	10	20
Springfield	37	51	70	87	102
St. Joseph	45	62	86	106	126
Montana					
Great Falls	4	8	14	20	26
Miles City	7	12	21	29	38
Nebraska					
Antioch	19	26	36	45	52
Lincoln	36	51	74	92	112
Lynch	26	37	54	67	82
North Platte	25	38	59	78	99
Scribner	38	53	76	96	116
Valentine	18	28	45	61	77
New Hampshire					
Concord	18	27	45	62	79
New Jersey					
Atlantic City	39	55	77	97	117
Marlboro	39	57	85	111	136
Trenton	29	48	76	102	131
New Mexico					
Albuquerque	4	6	11	15	21
Roswell	10	21	34	45	53
New York					
Albany	18	26	38	47	56
Binghamton	16	24	36	47	58
Buffalo	15	23	36	49	61
Marcellus	16	24	38	49	62
Rochester	13	22	38	54	75
Salamanca	15	21	32	40	49
Syracuse	15	24	38	51	65
North Carolina					
Asheville	28	40	58	72	87
Charlotte	41	63	100	131	164
Greensboro	37	51	74	92	113
Raleigh	53	77	110	137	168
Wilmington	59	87	129	167	206
North Dakota					
Devils Lake	19	27	39	49	59
Fargo	20	31	54	77	103
Williston	11	16	25	33	41
Ohio					
Cincinnati	27	36	48	59	69
Cleveland	22	35	53	71	86
Columbia	20	26	35	41	48
Columbus	27	40	60	77	94
Coshocton	27	45	77	108	143
Dayton	21	30	44	57	70
Toledo	16	26	42	57	74
Oklahoma					
Ardmore	46	71	107	141	179
Cherokee	44	59	80	97	113
Guthrie	47	70	105	134	163
McAlester	54	82	127	165	209
Tulsa	47	69	100	127	154
Oregon					
Portland	6	9	13	15	18
Pennsylvania					
Franklin	17	24	35	45	54
Harrisburg	19	25	35	43	51
Philadelphia	28	39	55	69	81
Pittsburgh	23	32	45	57	67
Reading	28	39	55	68	81
Scranton	23	32	44	53	63
Puerto Rico					
San Juan	57	87	131	169	216
Rhode Island					
Providence	23	34	52	68	83
South Carolina					
Charleston	74	106	154	196	240
Clemson	51	73	106	133	163
Columbia	41	59	85	106	132
Greenville	44	65	96	124	153
South Dakota					
Aberdeen	23	35	55	73	92
Huron	19	27	40	50	61
Isobel	15	24	38	52	67
Rapid City	12	20	34	48	64
Tennessee					
Chattanooga	34	49	72	93	114
Knoxville	25	41	68	93	122
Memphis	43	55	70	82	91
Nashville	35	49	68	83	99
Texas					
Abilene	31	49	79	103	138
Amarillo	27	47	80	112	150
Austin	51	80	125	169	218
Brownsville	73	113	181	245	312
Corpus Christi	57	79	114	146	171
Dallas	53	82	126	166	213
Del Rio	44	67	108	144	182
El Paso	6	9	15	19	24
Houston	82	127	208	275	359
Lubbock	17	29	53	77	103
Midland	23	35	52	69	85
Nacogdoches	77	103	138	164	194
San Antonio	57	82	122	155	193

(continues)

Table 8A.2 *(Continued)*

Location	Index values normally exceeded once in (year)				
	1	2	5	10	20
Temple	53	78	123	162	206
Victoria	59	83	116	146	178
Wichita Falls	47	63	86	106	123
Vermont					
Burlington	15	22	35	47	58
Virginia					
Blacksburg	23	31	41	48	56
Lynchburg	31	45	66	83	103
Richmond	46	63	86	102	125
Roanoke	23	33	48	61	73
Washington					
Spokane	3	4	7	8	11
West Virginia					
Elkins	23	31	42	51	60
Huntington	18	29	49	69	89
Parkersburg	20	31	46	61	76
Wisconsin					
Green Bay	18	26	38	49	59
LaCrosse	46	67	99	125	154
Madison	29	42	61	77	95
Milwaukee	25	35	50	62	74
Rice Lake	29	45	70	92	119
Wyoming					
Casper	4	7	9	11	14
Cheyenne	9	14	21	27	34

[a](After Wischmeier and Smith, 1978). Units on *R* are ft•tonsf•in./acre•hr•storm. To convert to metric, MJ•mm/ha•h•storm, multiply by 17.02.

Appendix 8B
Universal Soil Loss Equation *C* Factors

Table 8B.1 Typical *C* Factor Values Reported in the Literature for Construction Sites and Disturbed Lands (after Israelson *et al.*, 1980)

Condition	*C* factor
1. Bare soil conditions	
Freshly disked to 6–8 in.	1.00
After one rain	0.89
Loose to 12 in. smooth	0.90
Loose to 12 in. rough	0.80
Compacted root raked	1.20
Compacted bulldozer scraped across slope	1.20
Same except root raked across	0.90
Rough irregular tracked all directions	0.90
Seed and fertilize, fresh, unprepared seedbed	0.64
Same except after 6 months	0.54
Seed, fertilize after 12 months	0.38
Undisturbed except scraped	0.66–1.30
Scarified only	0.76–1.31
Sawdust 2 in. deep, disked in	0.61
2. Asphalt emulsion	
1210 gal/acre	0.01–0.019
605 gal/acre	0.14–0.57
302 gal/acre	0.28–0.60
3 . Dust binder	
605 gal/acre	1.05
1210 gal/acre	0.29–0.78
4. Other chemicals	
Aquatain	0.68
Aerospray 70, 10% cover	0.94
PVA	0.71–0.90
Terra-Tack	0.66
5. Seedings[a]	
Temporary, 0 to 60 days[b]	0.40
Temporary, after 60 days	0.05
Permanent, 2 to 12 months	0.05
6. Brush	0.35

[a]If plantings are used with mulches, use the minimum *C* values.
[b]If dry weather occurs at planting and emergence is a problem, extend the 0–60 days to a period when rainfall normally occurs.

Table 8B.2 *C* Factors for Permanent Pasture, Rangeland, Idle Land, and Grazed Woodlands (after Wischmeier and Smith, 1978)[a]

Vegetal canopy			Cover that contacts the surface Percentage ground cover					
Type and height of raised canopy[b]	Canopy cover (%)	Type[d]	0	20	40	60	80	95–100
No appreciable canopy		G	0.45	0.20	0.10	0.042	0.013	0.003
		W	0.45	0.24	0.15	0.090	0.043	0.011
Canopy of tall weeds or short brush (0.5-m fall height)	25	G	0.36	0.17	0.09	0.038	0.012	0.003
		W	0.36	0.20	0.13	0.082	0.041	0.011
	50	G	0.26	0.13	0.07	0.035	0.012	0.003
		W	0.26	0.16	0.11	0.075	0.039	0.011
	75	G	0.17	0.10	0.06	0.031	0.011	0.003
		W	0.17	0.12	0.09	0.067	0.038	0.011
Appreciable brush or bushes (2-m fall height)	25	G	0.40	0.18	0.09	0.040	0.013	0.003
		W	0.40	0.22	0.14	0.085	0.042	0.011
	50	G	0.34	0.16	0.085	0.038	0.012	0.003
		W	0.34	0.19	0.13	0.081	0.041	0.011
	75	G	0.28	0.14	0.08	0.036	0.012	0.003
		W	0.28	0.17	0.12	0.077	0.040	0.011
Trees, but no appreciable low brush (4-m fall height)	25	G	0.42	0.19	0.10	0.041	0.013	0.003
		W	0.42	0.23	0.14	0.087	0.042	0.011
	50	G	0.39	0.18	0.09	0.040	0.013	0.003
		W	0.39	0.21	0.14	0.085	0.042	0.011
	75	G	0.36	0.17	0.09	0.039	0.012	0.003
		W	0.36	0.20	0.13	0.083	0.041	0.011

[a]All values shown assume: (1) random distribution of mulch or vegetation and (2) mulch of appreciable depth where it exists. Idle land refers to land with undisturbed profiles for at least a period of 3 consecutive years. Also to be used for burned forest land and forest land that has been harvested less than 3 years ago.

[b]Average fall height of waterdrops from canopy to soil surface in meters.

[c]Portion of total surface area that would be hidden from view by canopy in a vertical projection (a bird's-eye view).

[d]G, cover at surface is grass, grasslike plants, decaying compacted duff, or litter at least 2 in. deep. W, cover at surface is mostly broadleaf herbaceous plants (as weeds with little lateral root network near the surface) and/or undecayed residue.

Table 8B.3 *C* Factors for Mechanically Prepared Woodland Sites (after Wischmeier and Smith, 1978)

Percentage of soil covered with residue in contact with soil surface	Soil condition and weed cover[c]							
	Excellent		Good		Fair		Poor	
	NC	WC	NC	WC	NC	WC	NC	WC
None								
A. Disked, raked, or bedded[a]	0.52	0.20	0.72	0.27	0.85	0.32	0.94	0.36
B. Burned[b]	0.25	0.10	0.26	0.10	0.31	0.12	0.45	0.17
C. Drum chopped[b]	0.16	0.07	0.17	0.07	0.20	0.08	0.29	0.11
10% cover								
A. Disked, raked or bedded[a]	0.33	0.15	0.46	0.20	0.54	0.24	0.60	0.26
B. Burned[b]	0.23	0.10	0.24	0.10	0.26	0.11	0.36	0.16
C. Drum chopped[b]	0.15	0.07	0.16	0.07	0.17	0.08	0.23	0.10
20% cover								
A. Disked, baked or bedded[a]	0.24	0.12	0.34	0.17	0.40	0.20	0.44	0.29
B. Burned[b]	0.19	0.10	0.19	0.10	0.21	0.11	0.27	0.14
C. Drum chopped[b]	0.12	0.06	0.12	0.06	0.14	0.07	0.18	0.09
40% cover								
A. Disked, raked or bedded[a]	0.17	0.11	0.23	0.14	0.27	0.17	0.30	0.19
B. Burned[b]	0.14	0.09	0.14	0.09	0.15	0.09	0.17	0.11
C. Drum chopped[b]	0.09	0.06	0.09	0.06	0.10	0.06	0.11	0.07
60% cover								
A. Disked, raked, or bedded[a]	0.11	0.08	0.15	0.11	0.18	0.14	0.20	0.15
B. Burned[b]	0.08	0.06	0.09	0.07	0.10	0.08	0.11	0.08
C. Drum chopped[b]	0.06	0.05	0.06	0.05	0.07	0.05	0.07	0.05
80% cover								
A. Disked, raked, or bedded[a]	0.05	0.04	0.07	0.06	0.09	0.08	0.10	0.09
B. Burned[b]	0.04	0.04	0.05	0.04	0.05	0.04	0.06	0.05
C. Drum chopped[b]	0.03	0.03	0.03	0.03	0.03	0.03	0.04	0.04

[a]Multiply *A* values by following values to account for surface roughness: 0.40, Very rough, major effect on runoff and sediment, storage depressions greater than 6 in.; 0.65, moderate; 0.90, smooth, minor surface sediment storage, depressions less than 2 in. The *C* values for *A* are for the first year following treatment. For *A*-type sites 1 to 4 years old, multiply *C* value by 0.7 to account for aging. For sites 4 to 8 years old, use Table 8B.2. For sites more than 8 years old, use Table 8B.4.

[b]The *C* values for *B* and *C* areas are for the first 3 years following treatment. For sites treated 3 to 8 years ago, use Table 8B.2. For sites treated more than 8 years ago, use Table 8B.4.

[c]NC, no weed cover; WC, weed cover.

Table 8B.4 *C* Factors for Undisturbed Woodlands (after Wischmeier and Smith, 1978)

Effective canopy[a] (% of area)	Forest litter[b] (% of area)	*C* factor[c]
100–75	100–90	0.0001–0.001
75–40	85–70	0.002–0.004
35–20	70–40	0.003–0.009

[a]When effective canopy is less than 20%, the area will be considered as grassland or idle land for estimating soil loss. Where woodlands are being harvested or grazed, use Table 8B.2.

[b]Forest litter is assumed to be at least 2 in. deep over the percentage ground surface area covered.

[c]The range in *C* values is due in part to the range in the percentage area covered. In addition, the percentage of effective canopy and its height has an effect. Low canopy is effective in reducing raindrop impact and in lowering the *C* factor. High canopy, over 13 m, is not effective in reducing raindrop impact and will have no effect on the *C* value.

Table 8B.5 USLE Mulch Factors and Length Limits for Construction Sites (after Wischmeier and Smith, 1978)[a]

Type of mulch	Mulch rate (ton/acre)	Land slope (%)	Factor *C*	Length limit[b] (ft)
Straw or hay, tied	1.0	1–5	0.20	200
down by anchoring	1.0	6–10	0.20	100
and tacking equipment[c]	1.5	1–5	0.12	300
	1.5	6–10	0.12	150
	2.0	1–5	0.06	400
	2.0	6–10	0.06	200
	2.0	11–15	0.07	150
	2.0	16–20	0.11	100
	2.0	21–25	0.14	75
	2.0	26–33	0.17	50
	2.0	34–50	0.20	35
Wood chips	7.0	<16	0.08	75
	7.0	16–20	0.08	50
	12.0	<16	0.05	150
	12.0	16–20	0.05	100
	12.0	21–33	0.05	75
	25.0	<16	0.02	200
	25.0	16–20	0.02	150
	25.0	21–33	0.02	100
	25.0	34–50	0.02	75

[a]Developed by an interagency workshop group on the basis of field experience and limited research data.

[b]Maximum slope length for which the specified mulch rate is considered effective. When this limit is exceeded, either a higher application rate or mechanical shortening of the effective slope length is required.

[c]When the straw or hay mulch is not anchored to the soil, *C* values on moderate or steep slopes having K values greater than 0.30 should be taken at double the values given in this table.

Appendix 8C
Supplemental *C* and *P* Parameters for Revised Universal Soil Loss Equation

Table 8C.1 Typical Crop Parameter Values for Residue Decomposition in Eqs. (8.54) and (8.58) (after Yoder *et al.*, 1993)[a]

Crop	Residue to grain ratio[a]	U/R[b]	C_N[c]	a_w[d] (acre/lb)	Yield[e] per acre	Row spacing (in.)	Plant population (plants/acre)
Alfalfa	0.15	0.0040	30	0.00056	6 tons	drilled	180,000
Brome grass	0.15	0.0040	80	0.00056	4 tons	7 drilled	330,000
Corn	1	0.0017	62	0.00038	130 bu	30	25,000
Cotton	1	0.0022	40	0.00022	900 lb	38	35,000
Oats	2	0.0015	80	0.00106	65 bu	7 drilled	890,000
Peanuts	1.3	0.0022	30	0.00033	2600 lb	36	58,000
Rye	1.5	0.0040	80	0.00056	30 bu	drilled	890,000
Sorghum	1	0.0022	60	0.00034	65 bu	30	41,000
Soybeans	1.5	0.0022	31	0.00058	35 bu	30	110,000
Sunflowers	1.5	0.0012	39	0.00024	1100 lb	30	20,000
Tobacco	1.8	0.0030	80	0.00034	2200 lb	48	6,000
Wheat (spring)	1.7	0.0018	107	0.00060	30 bu	7 drilled	890,000
Wheat (winter)	1.3	0.0018	107	0.00060	45 bu	7 drilled	890,000
Wheat (PNW)	1.3	0.0018	107	0.00060	70 bu	7 drilled	890,000

[a]Larson *et al.* (1978).
[b]Constant for the effect of residue size on decomposition.
[c]Carbon to nitrogen ratio of the residue at harvest.
[d]Ratio of area covered by residue to its weight.
[e]Typical yield value for the crop indicated.

Table 8C.2 Typical Field Operations and Associated RUSLE Parameter Values (after Yoder *et al.*, 1993)

Field operations	R_R random roughness (in.)	Percentage buried residue (%)	Tillage depth (in.)
Moldboard (8 in. deep)	1.9	90	8
Moldboard (5–7 in. deep)	1.3	70	6
Chisel (2-in.shovels)	0.9	25	8
One-way disk (24 to 26 in. disks)	1.2	50	4
One-way disk (18 to 24 in. disks)	1.1	40	4
Large offset disk	2.0	50	6
Blades (36-in. width)	0.5	10	4
Sweeps (24–36 in.)	0.65	15	4
Field cult. (shovels)	0.7	25	4
Field cult. (16–18 in. sweeps)	0.6	20	4
Tandem disk	0.75	50	4
Harrow (tine)	0.4	15	4
Harrow (spike)	0.3	20	4
Fertilizer applicator	0.5	10	4
Rodweeder (plain)	0.4	10	4
Rodweeder (shovel)	0.45	15	4
Broadcast planting	0.4	0	0
Conventional drill	0.4	10	2
Semi-deep furrow drill	0.45	15	2
Deep furrow drill	0.5	20	3
No-till drill	0.4	10	2
Row planter	0.4	15	4
No-till planter	0.4	5	2
Row cultivator	0.6	30	4
Lister	1.0	80	4
Mulch treader	0.6	90	4

Table 8C.3 Growing Season Soil Moisture Depletion and Replenishment Rates for Arid Western Drylands (after Yoder *et al.*, 1993)

A. Depletion rates

Crop	Depletion per 15-day period
Winter Wheat and Other Deep Rooted Crops	1.0
Spring Barley	0.7
Spring Peas and Lentils	0.67
Shallow Rooted Crops	0.50
Fallow	0.0

B. Replenishment rates

$$\text{Replenishment rate per 15-day period} = 0.5 + 0.662\,(P - 1)$$

$$10 \leq P \leq 18$$

P, annual precipitation in inches. Replenishment rate is inches per 15-day period.

Table 8C.4 RUSLE Terrace Factor Values P_{ter} for Conservation Planning (after Foster *et al.*, 1993)[a]

Horizontal terrace interval (ft)	Closed outlets[c]	Open outlet, with percentage grade of[b] 0.1–0.3	0.40–0.7	>0.8
Less than 110	0.5	0.6	0.7	1.0
110–140	0.6	0.7	0.8	1.0
140–180	0.7	0.8	0.9	1.0
180–225	0.8	0.8	0.9	1.0
225–300	0.9	0.9	1.0	1.0
More than 300	1.0	1.0	1.0	1.0

[a]Multiply these values by other P factor values for contouring, strip cropping, or other supporting practices on the interterrace interval to obtain a composite P factor value.

[b]The channel grade is measured on the first 300 ft of terrace or the one-third of total length closest to the outlet, whichever distance is less.

[c]Values for closed outlet terraces also apply to terraces with underground outlets and to level terraces with open outlets.

Appendix 8D
Supplemental *C* Subfactors for Disturbed Forests

Table 8D.1 Effect of bare soil, fine root mat of trees, and soil reconsolidation on C factor (after Dissmeyer and Foster, 1984).

A. Untilled Soils

Percentage bare soil	Percentage of bare soil with dense mat of fine roots in top 3 cm of soil					
	100	80	60	40	20	0
0	0.0000	0.000	0.000	0.000	0.000	0.000
1	0.0004	0.0005	0.0007	0.0010	0.0014	0.0018
5	0.003	0.003	0.005	0.007	0.009	0.012
10	0.005	0.006	0.009	0.013	0.017	0.0123
20	0.011	0.014	0.020	0.028	0.038	0.050
40	0.023	0.027	0.042	0.058	0.079	0.104
60	0.037	0.043	0.067	0.092	0.127	0.167
80	0.055	0.066	0.098	0.141	0.192	0.252
100	0.099	0.117	0.180	0.248	0.342	0.450

B. Tilled soils with good initial fine root mat in topsoil and subsoil has good structure and permeability

Percentage bare soil	Time (months) since tillage			
	0	6	12 and 72+	24+ thru 60
0	0.0000	0.0000	0.0000	0.0000
1	0.0014	0.0017	0.0018	0.0020
5	0.009	0.011	0.012	0.013
10	0.019	0.022	0.023	0.026
20	0.037	0.045	0.049	0.056
40	0.083	0.095	0.104	0.117
60	0.137	0.157	0.172	0.194
80	0.212	0.244	0.267	0.301
100	0.360	0.414	0.450	0.510

C. Tilled soil with poor initial fine root mat in topsoil (subsoil has good structure and permeability)

Percentage bare soil	Time (months) since tillage				
	0	6	12 to 36	48	72+
0	0.0000	0.0000	0.0000	0.0000	0.0000
1	0.0021	0.0023	0.0025	0.0022	0.0018
5	0.014	0.015	0.016	0.015	0.0117
10	0.027	0.031	0.033	0.029	0.023
20	0.058	0.065	0.069	0.060	0.049
40	0.122	0.135	0.144	0.129	0.104
60	0.201	0.224	0.239	0.213	0.171
80	0.313	0.348	0.352	0.330	0.266
100	0.530	0.590	0.630	0.560	0.450

Table 8D.2 Step effect on soil erosion (after Dissmeyer and Foster, 1984).

Percentage slope	Percentage of total slope in steps										
	0	10	20	30	40	50	60	70	80	90	100
5	1.00	0.99	0.99	0.98	0.97	0.96	0.95	0.94	0.94	0.93	0.92
6	1.00	0.97	0.94	0.92	0.89	0.86	0.84	0.81	0.78	0.76	0.73
7	1.00	0.96	0.92	0.88	0.84	0.80	0.75	0.71	0.67	0.63	0.59
8	1.00	0.95	0.90	0.85	0.80	0.75	0.69	0.64	0.59	0.54	0.49
9	1.00	0.94	0.89	0.83	0.77	0.71	0.65	0.60	0.54	0.48	0.42
10	1.00	0.94	0.87	0.81	0.75	0.68	0.62	0.56	0.49	0.43	0.36
12	1.00	0.93	0.85	0.78	0.71	0.63	0.56	0.49	0.42	0.34	0.27
15	1.00	0.92	0.84	0.75	0.67	0.59	0.51	0.43	0.34	0.26	0.18
20	1.00	0.91	0.82	0.74	0.65	0.56	0.47	0.38	0.29	0.20	0.11
30+	1.00	0.91	0.81	0.72	0.63	0.53	0.44	0.35	0.25	0.15	0.06

Table 8D.3 Contour tillage subfactors for forestlands (after Dissmeyer and Foster, 1984).

Percentage slope	On contour 0	Degrees off contour				
		15	30	45	60	90
0–2	0.80	0.88	0.91	0.94	0.96	1.00
3–7	0.70	0.82	0.87	0.91	0.94	1.00
8–12	0.80	0.88	0.91	0.94	0.96	1.00
13–18	0.90	0.94	0.96	0.97	0.98	1.00
19+	1.00	1.00	1.00	1.00	1.00	1.00

Appendix 8E

Roughness Values and Critical Tractive Force Values for CREAMS Equations

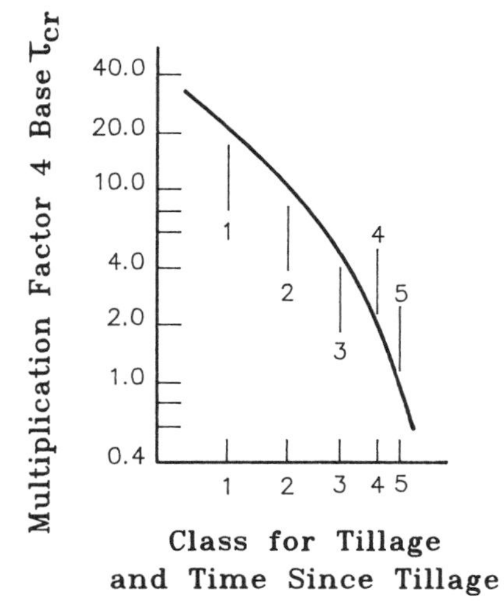

Figure 8E.1 Effect of tillage on critical shear stress. Class 1, long-term without tillage; Class 2, 1 year since seedbed tillage; Class 3, primary tillage in land 1 year since seedbed; Class 4, typical seedbed, Class 5, finely pulverized seedbed (after Foster *et al.*, 1980b).

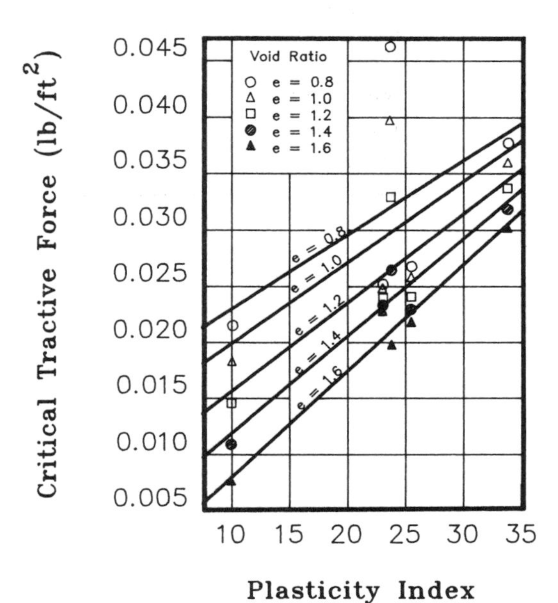

Figure 8E.2 Effect of plasticity index and void ratio on critical shear force (after Lyle and Smerdon, 1965).

Table 8E.1 Overland Flow n Values for CREAMS Equations (after Foster *et al.*, 1980b)

Treatment	Manning's n	Treatment	Manning's n
Cornstalk residue applied to fallow surface		Grass	
1 ton/acre	0.020	Sparse	0.015
2 tons/acre	0.040	Poor	0.023
4 tons/acre	0.070	Fair	0.032
		Good	0.046
Cornstalk residue disk-harrow incorporated		Excellent	0.074
1 ton/acre	0.012		
2 tons/acre	0.020	Dense	0.150
4 tons/acre	0.023	Very dense	0.400
Wheat straw mulch		Rough surface depressions	
0.25 ton/acre	0.015	4 to 5 in. deep	0.046
0.5 ton/acre	0.018	2 to 4 in. deep	0.023
1 ton/acre	0.032	1 to 2 in. deep	0.014
2 tons/acre	0.070	No surface depressions	0.010
4 tons/acre	0.074		
Crushed stone mulch			
15 tons/acre	0.012		
60 tons/acre	0.023		
135 tons/acre	0.046		
240 tons/acre	0.074		
375 tons/acre	0.074		

Small grain (20% to full maturity)	Across slope	Upslope and downslope
Poor stand	0.018	0.012
Moderate stand	0.023	0.015
Good stand	0.032	0.023
Dense	0.046	0.032

Table 8E.2 Concentrated Flow Values for Manning's *n* for Typical Soil Covers[a] for CREAMS Equation (after Foster *et al.*, 1980b)

Cover	Cover density	Manning's *n*
Smooth, bare soil; roughness elements	Less than 1 in. deep	0.030
	1–2 in. deep	0.033
	2–4 in. deep	0.038
	4–6 in. deep	0.045
Corn stalks (assumes residue stays in place and is not washed away)	1 ton/acre	0.050
	2 tons/acre	0.075
	3 tons/acre	0.100
	4 tons/acre	0.130
Wheat straw (assumes residue stays in place and is not washed away	1 ton/acre	0.060
	1.5 tons/acre	0.100
	2 tons/acre	0.150
	4 tons/acre	0.250
Grass (assumes grass is erect and as deep as the flow)	Sparse	0.040
	Poor	0.050
	Fair	0.060
	Good	0.080
	Excellent	0.130
	Dense	0.200
	Very dense	0.300
Small grain (20% to full maturity rows with flow)	Poor, 7-in. rows	0.130
	Poor, 14-in. rows	0.130
	Good, 7-in. rows	0.300
	Good, 14-in. rows	0.200
Rows across floors	Good	0.300
Sorghum and cotton	Poor	0.070
	Good	0.090
Sudangrass	Good	0.200
Lespedeza	Good	0.100
Lovegrass	Good	0.150

[a]Does not include effects of submergence or product of velocity-hydraulic radius.

Table 8E.3 Critical Shear Stress Values as a Function of Tillage and Consolidation for Moderately Erodible Agricultural Soils (after Foster *et al.*, 1980b)[a]

Tillage-consolidation condition	Critical shear stress (lb/ft^2)
Moldboard plowed	0.20
Chisel or disk for primary tillage	0.5
Disking: common corn seedbed or crop cultivation	0.10
Finely pulverized seedbed	0.05
1 month after last tillage of common seedbed	0.20
2 months after last tillage of common seedbed	0.30
3 months after last tillage of common seedbed	0.40
Long term, undisturbed	0.60

[a]To be used as a first estimate when soil information is not available. For soils with dispersion ratio data available, Foster *et al.* recommend that the Smerdon and Beasley (1961) equation be used, or $\tau_c = 0.213/d_r^{0.63}$. These values must be corrected for tillage by factors from Fig. 8E.1.

Appendix 8F

Conveyance Function and Equilibrium Channel Properties for Concentrated Flow Equations

Table 8F.1 Concentrated flow channel parameters (after Lewis, 1990).[a]

Item	X_{*c}/X_{*cf}	W_*	R_*	D_*	$g(X_{*c})$	$g(X_{*cf})$
1	0.00	0.7436	0.1510	0.2523	35.0000	—
2	0.01	0.7287	0.1524	0.2572	26.3603	73.75
3	0.02	0.7138	0.1535	0.2622	13.3985	29.21
4	0.03	0.6990	0.1544	0.2671	9.0856	17.19
5	0.04	0.6841	0.1552	0.2721	6.9351	11.89
6	0.05	0.6692	0.1558	0.2770	5.6497	9.00
7	0.06	0.6543	0.1562	0.2820	4.7971	7.21
8	0.07	0.6395	0.1565	0.2870	4.1918	6.00
9	0.08	0.6246	0.1565	0.2919	3.7413	5.14
10	0.09	0.6097	0.1564	0.2969	3.3941	4.50
11	0.10	0.5949	0.1562	0.3018	3.1194	4.01
12	0.11	0.5800	0.1557	0.3068	2.8974	3.62
13	0.12	0.5651	0.1551	0.3117	2.7152	3.31
14	0.13	0.5502	0.1542	0.3167	2.5637	3.06
15	0.14	0.5354	0.1533	0.3216	2.4365	2.85
16	0.15	0.5205	0.1521	0.3266	2.3288	2.66
17	0.16	0.5056	0.1507	0.3315	2.2371	2.53
18	0.17	0.4908	0.1492	0.3365	2.1587	2.40
19	0.18	0.4759	0.1475	0.3415	2.0915	2.29
20	0.19	0.4610	0.1457	0.3464	2.0340	2.20
21	0.20	0.4461	0.1436	0.3514	1.9849	2.12
22	0.21	0.4313	0.1414	0.3563	1.9430	2.05
23	0.22	0.4164	0.1390	0.3613	1.9076	1.99
24	0.23	0.4015	0.1364	0.3662	1.8781	1.95
25	0.24	0.3867	0.1336	0.3712	1.8539	1.90
26	0.25	0.3718	0.1307	0.3761	1.8345	1.87
27	0.26	0.3569	0.1276	0.3811	1.8197	1.83
28	0.27	0.3420	0.1243	0.3861	1.8092	1.81
29	0.28	0.3272	0.1208	0.3910	1.8029	1.79
30	0.29	0.3123	0.1172	0.3960	1.8005	1.77
31	0.30	0.2974	0.1134	0.4009	1.8021	1.76
32	0.31	0.2826	0.1094	0.4059	1.8076	1.76
33	0.32	0.2677	0.1052	0.4108	1.8172	1.76
34	0.33	0.2528	0.1009	0.4158	1.8309	1.76
35	0.34	0.2379	0.0964	0.4207	1.8490	1.77
36	0.35	0.2231	0.0917	0.4257	1.8718	1.78
37	0.36	0.2082	0.0868	0.4307	1.8998	1.80
38	0.37	0.1933	0.0817	0.4356	1.9334	1.82
39	0.38	0.1785	0.0765	0.4406	1.9736	1.85
40	0.39	0.1636	0.0711	0.4455	2.0212	1.89
41	0.40	0.1487	0.0655	0.4505	2.0778	1.93
42	0.41	0.1338	0.0598	0.4554	2.1454	1.98
43	0.42	0.1190	0.0538	0.4604	2.2267	2.05
44	0.43	0.1041	0.0477	0.4654	2.3260	2.13
45	0.44	0.0892	0.0414	0.4703	2.4497	2.24
46	0.45	0.0743	0.0350	0.4753	2.6083	2.37
47	0.46	0.0595	0.0283	0.4802	2.8209	2.56
48	0.47	0.0446	0.0215	0.4852	3.1272	2.82
49	0.48	0.0296	0.0144	0.4903	3.6289	3.26
50	0.49	0.0145	0.0072	0.4954	4.7181	4.19
51	0.50	0.0000	0.0000	0.5000	5.7322	—

$g(X_*)$ values less than 1.8 indicate insufficient tractive forces to cause scour and the formation of an incised channel. For selected values of $g(X_{*c})$, X_{*c} is a double-valued function. Use of the smaller value will yield a conservatively large erosion estimate.

Appendix 9A
Predicting Turbulence in Sediment Ponds

Characteristics of Interest

Turbulent characteristics are important to prediction and flocculation of sediment. Parameters of interest are *mean square velocities* and *turbulent diffusivities*. Diffusivities are important for prediction of diffusion of sediment. Mean square velocities are important to prediction of flocculation and dispersion.

Models of Turbulence

Using the definition sketch in Fig. 9A.1 the momentum equation, for gradually varied flow in a sediment pond can be written as (Wilson and Barfield, 1986b)

$$\overline{u'v'} = \nu \frac{\partial U}{\partial y} - U_*^2\left(1 - \frac{y}{H}\right), \tag{9A.1}$$

where U_* is the shear velocity and is equal to $(gHS_e)^{0.5}$, $\overline{u'v'}$ is the Reynold's shear stress, ν is kinematic viscosity, U is average velocity at any distance y above the bed, and H is the total depth. An analogy between Reynolds' stresses and turbulent diffusion has been typically assumed for most flows. Using this analogy, the turbulent diffusion coefficient can be estimated as

$$\varepsilon = \frac{\overline{u'v'}}{dU/dy}, \tag{9A.2}$$

where ε is the turbulent diffusion coefficient.

Using a logarithmic velocity profile, $dU/dy = U_*/(ky)$, where k is von Karman's constant (approximately 0.4), the turbulent diffusion coefficient can thus be predicted by

$$\varepsilon = U_* ky\left[1 - \frac{y}{H}\right] - \nu. \tag{9A.3}$$

Conversely, a parabolic velocity would result in (Dobbins, 1944)

$$\frac{dU}{dy} = 2U_s\left[\frac{1 - y/H}{H}\right],$$

where U_S is the velocity at the water surface. The resulting turbulent diffusion coefficient for the parabolic profile is given by

$$\varepsilon = \frac{HU_*^2}{2U_s}. \tag{9A.4}$$

The turbulent diffusion coefficient is predicted by Eq.

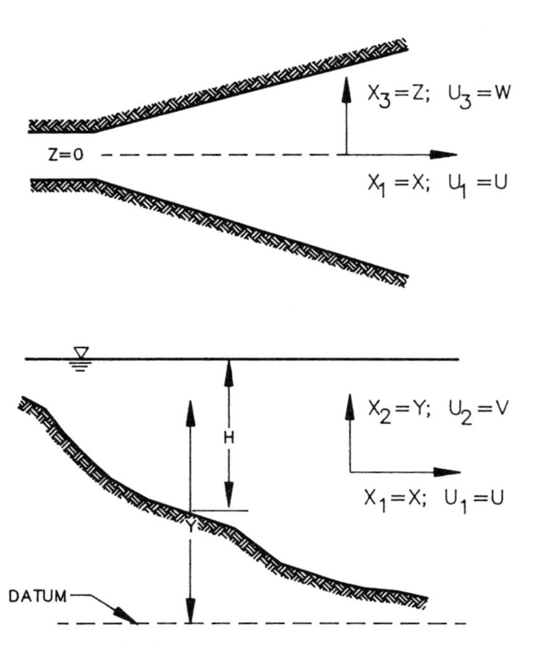

Figure 9A.1 Definition sketch for the momentum equation.

(9A.3) for the log velocity model and by Eq. (9A.4) for the parabolic velocity model.

Model Parameters

The velocity profile models require that the energy slope be calculated. This parameter can be estimated at each point by assuming that (a) the water surface elevation is independent of x and (b) the velocity at a cross section is equal to the inlet volumetric flow rate divided by the corresponding cross-sectional area. Under these conditions, the slope of the energy gradient becomes

$$S_e = \frac{d(U^2/2)}{dx}. \tag{9A.5}$$

The assumption of a horizontal water surface is a standard approximation in reservoir routing. The assumption that the average velocity is Q/A is also reasonable unless major recirculation occurs.

A water surface velocity is required for the parabolic velocity model. Wilson and Barfield (1986a, b) assumed a value for U_S of 1.5 times the average velocity at the cross section.

Model Evaluation

Wilson and Barfield (1986b) evaluated the accuracy of the models on data collected in a model sediment pond. In general, the log model was more accurate than the parabolic model.

Appendix 9B
Rainfall Statistics for EPA Model

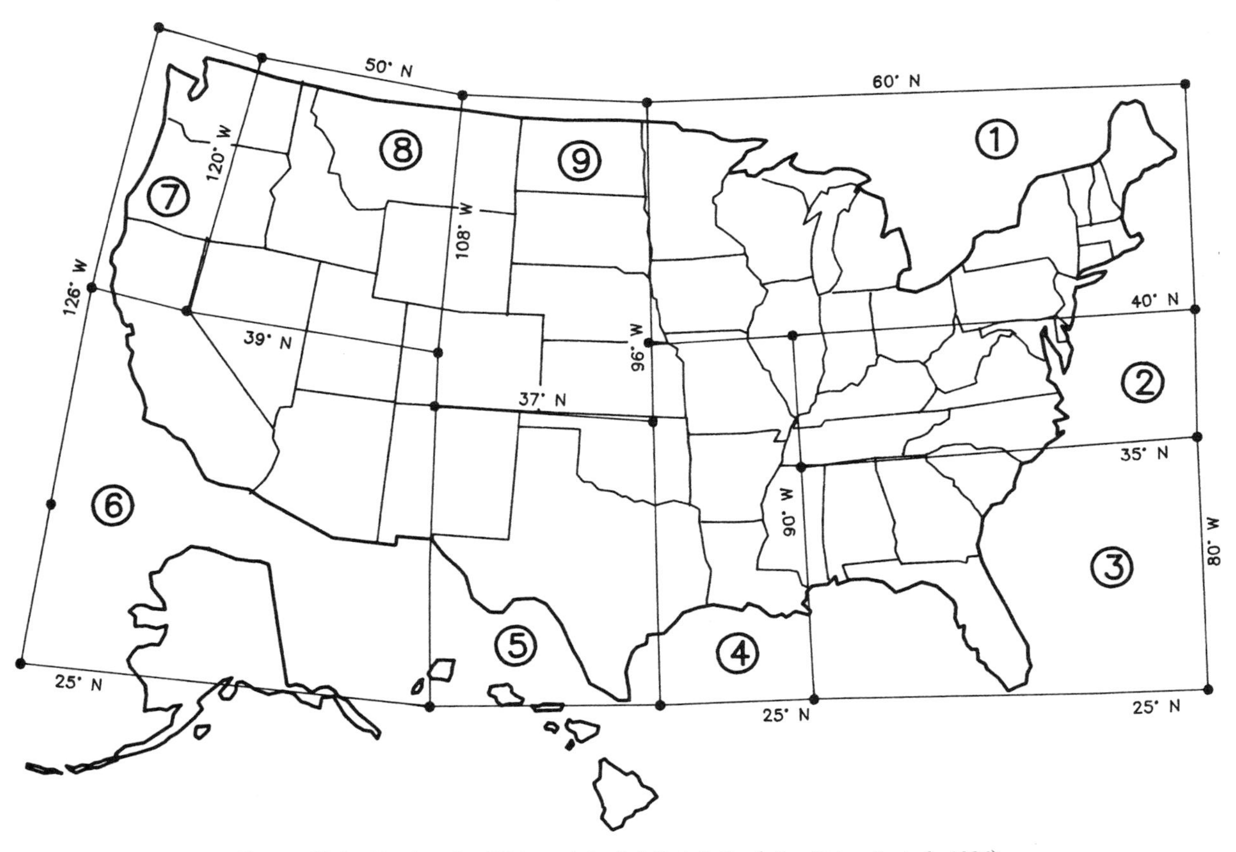

Figure 9B.1 Regions for EPA model rainfall statistics (after Driscoll *et al.*, 1986).

Table 9B.1 Rainfall Statistics for EPA (1986) Model

		Rainfall statistics							
		Q Volume (in.)		I Intensity (iph)		D Duration (hr)		T_{IA} Interval (hr)	
Zone	Period	Mean	C.V.	Mean	C.V.	Mean	C.V.	Mean	C.V.
1	Annual	0.28	1.46	0.051	1.31	5.8	1.05	73	1.07
	Summer	0.32	1.38	0.082	1.29	4.4	1.14	78	1.07
2	Annual	0.36	1.45	0.066	1.32	5.9	1.05	77	1.05
	Summer	0.40	1.47	0.101	1.37	4.2	1.09	77	1.08
3	Annual	0.49	1.47	0.102	1.28	6.2	1.22	89	1.05
	Summer	0.48	1.52	0.133	1.34	4.9	1.33	68	1.01
4	Annual	0.58	1.46	0.097	1.35	7.3	1.17	99	1.00
	Summer	0.52	1.54	0.122	1.35	5.2	1.29	87	1.06
5	Annual	0.33	1.74	0.080	1.37	4.0	1.07	108	1.41
	Summer	0.36	1.71	0.110	1.39	3.2	1.08	112	1.49
6	Annual	0.17	1.51	0.045	1.04	3.6	1.02	277	1.48
	Summer	0.17	1.61	0.080	1.16	2.6	1.01	425	1.26
7	Annual	0.48	1.61	0.024	0.84	20.0	1.23	101	1.21
	Summer	0.26	1.35	0.027	1.11	11.4	1.20	188	1.15
8	Annual	0.14	1.42	0.031	0.91	4.5	0.82	94	1.39
	Summer	0.14	1.51	0.041	1.13	2.8	0.80	125	1.41
9	Annual	0.15	1.77	0.038	1.35	4.4	1.20	84	1.24
	Summer	0.10	1.74	0.058	1.44	3.1	1.14	78	1.13

Appendix 9C

Equations for Predicting the Advance of the Deposition Wedge in Vegetative Filter Strips

The deposition wedge in the grass filter strip is that area where bedload material is deposited, as shown in Figs. 9.40 and 9.41. The following material is a procedure to correct calculated trapping efficiencies for this effect.

Deposition in Zone $B(t)$

Zone $B(t)$ is the area where deposition occurs uniformly along a deposition wedge as shown in Fig. 9C.1. Parameters are estimated at the midpoint of zone $B(t)$. The incoming sediment load to zone $B(t)$ is that coming into the filter, q_{si}, since it is assumed that no sediment is deposited in zone $A(t)$. The sediment load exiting $B(t)$ is equal to the transport capacity for $B(t)$, q_{sd}, as given by Eq. (9.118) or (9.115). Thus, the average sediment load on the deposition wedge, q_{sba}, should be the average of q_{si} and q_{sd}, or

$$q_{sba} = \frac{q_{si} + q_{sd}}{2}, \tag{9C.1}$$

and the fraction of sediment trapped in zone $B(t)$ is given by

$$f = \frac{q_{si} - q_{sd}}{q_{si}}. \tag{9C.2}$$

Other than q_{sba} and f, the major variable of interest is the slope of the deposition wedge, which will be used in subsequent equations to predict the location of the leading edge of the deposition wedge, $X(t)$. The slope of the deposition wedge in zone $B(t)$ is different from channel slope and is defined as the equilibrium slope, S_{et}. Conceptually, it is the slope required for the flow q_{wba} to transport the sediment load q_{sba}. Given the value for q_{sba}, a value for S_{et} can be calculated from a combination of sediment transport rate from Eq. (9.118), velocity from (9.96), and continuity from (9.98). Using these equations and substituting an additional subscript ba for variables that are unique to the deposition wedge,

$$q_{sba} = \frac{K(R_{sba}S_{et})^{3.57}}{d_{pba}^{2.07}}, \tag{9C.3}$$

where K is a constant given by Eq. (9.119). Also

$$V_{mba} = \left(\frac{1.5}{xn}\right) R_{sba}^{2/3} S_{et}^{1/2} \quad \text{(ft/sec)}, \tag{9C.4}$$

where

$$R_{sba} = \frac{S_s d_{fba}}{2d_{fba} + S_s}. \tag{9C.5}$$

From continuity, the flow per unit width, q_{wba}, is given as

$$q_{wba} = V_{mba} d_{fba}. \tag{9C.6}$$

It is assumed that no infiltration occurs in $A(t)$ and $B(t)$; hence

$$q_{wba} = q_{wi}. \tag{9C.7}$$

Given a value of q_{sba}, one simultaneously solves Eqs. (9C.3)–(9C.7) for S_{et} and d_{fba}. The resulting equation for d_{fba} is

$$d_{fba}\left[\frac{d_{fba}S_s}{2d_{fba} + S_s}\right]^{1/6} = C1,$$

where $C1$ is a constant given by

$$C1 = \frac{xnq_{wba}K^{0.14}}{1.5q_{sba}^{0.14}d_{pb}^{0.2898}}. \tag{9C.8}$$

K is defined by Eq. (9.119), q_w is cfs/ft, q_S is lb/sec · ft, and d_{pb} is in mm. Equation (9C.8) must be solved by trial and error.

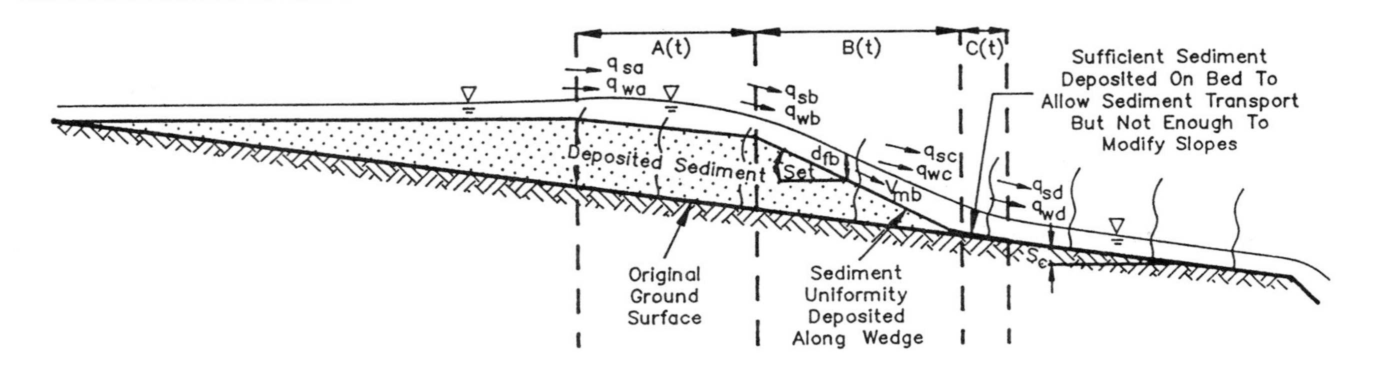

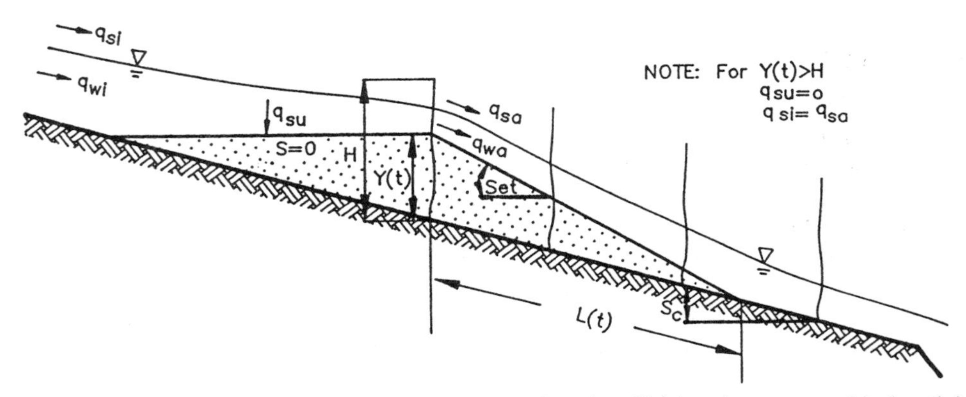

Figure 9C.1 Bedload trapping and deposition in zones $B(t)$, $C(t)$, and upstream delta. Note that $C(t)$ is a short zone and its length is typically ignored.

Location of the Leading Edge of the Deposition Wedge *X*(*t*)

The location of the leading edge of the deposition wedge determines the effective length, $L(t)$, of the suspended load deposition area, $D(t)$, as given in Eq. (9.105). Its location can be determined by sediment mass continuity relationships, as given by Tollner *et al.* (1977). Using the definition given in Fig. 9.41 and the concept of mass continuity, Tollner *et al.* derived the following equations for the depth of deposition and the advance distance $X(t)$, or

$$t^* = \frac{H^2 \gamma_{sb}}{2 f q_{si} S_e} \tag{9C.9}$$

$$Y_f(t_f) = \left[\frac{2}{\gamma_{sb}} \{ f q_s S_e \}_{avg} (t_f - t_i) + Y_i(t_i)^2 \right]^{1/2} \quad \text{for } Y_f(t_f) < H \tag{9C.10}$$

$$X_f(t_f) = \left[\frac{2}{\gamma_{sb}} \frac{f q_{si}}{S_e} (t_f - t_i) + X_i(t_i)^2 \right]^{1/2} \quad \text{for } Y_f(t_f) < H \tag{9C.11}$$

$$Y_f(t_f) = H \quad \text{if } Y_f(t_f) > H \text{ in Eq. (9C.10)} \tag{9C.12}$$

$$X_f(t_f) = X_f(t_f) + (t_f - t_i) \frac{f q_{si}}{H \gamma_{sb}} \quad \text{if } Y_f > H \text{ in Eq. (9C.10)} \tag{9C.13}$$

$$S_e = S_{et} - S_c. \tag{9C.14}$$

In the above terms, t^* is the time required for deposition to reach the height of the media, H, γ_{sb} is the bulk density of the deposited sediment, $Y_f(t_f)$ and $X_f(t_f)$ are the depth of deposition and advance distance, respectively, at time t_f, $Y_i(t_i)$ and $X_i(t_i)$ are the depth of deposition and advance distance, respectively, at an earlier time t_i, and $\{fq_{sba}S_{et}\}_{avg}$ indicates that these values are averaged on the time interval $t_f - t_i$.

To calculate values for $Y_f(t_f)$ and $X_f(t_f)$ at any t_f, one divides the inflow hydrograph and sedimentgraph into discrete time increments; calculates q_{sba}, f, S_{et}, and S_e for each time increment; and then calculates $Y_f(t_f)$ and $X_f(t_f)$ from Eqs. (9C.9)–(9C.14).

Deposition in Zone *A(t)* and Upstream of the VFS

Zone $A(t)$ is the region where sediment has deposited up to the level of the VFS; hence all sediment that enters this zone flows into zone $B(t)$. Backwater effects from the deposition wedge in zone $A(t)$ cause deposition upstream of the VFS; as shown in Fig. 9C.1. Studies by Hayes *et al.* (1984) showed that the slope of the upstream delta was zero. Using this fact and the assumption that the depth of the delta at the interface with the VFS equals the depth of deposition given by Eq. (9C.10), Hayes *et al.* showed that the rate of sediment deposition in the upstream delta was given by

$$q_{su} = \frac{\gamma_{sb}(Y_f^2 - Y_i^2)}{2(t_f - t_i)S_c}, \quad t < t; \tag{9C.15}$$

where γ_{sb} is bulk density of deposited material. The sediment load entering the VFS then is given by

$$q_{sa} = q_{si} - q_{su}, \quad t < t^*, \tag{9C.16}$$

where q_{si} is the sediment load coming from upslope. It should be noted that after the depth of deposition reaches the height of the filter media, q_{su} goes to zero and

$$q_{sa} = q_{si}, \quad t \geq t^*. \tag{9C.17}$$

The fraction of material trapped in the upstream delta can then be calculated as

$$f_u = q_{su}/q_{si}. \tag{9C.18}$$

Again, it is important to point out that the only material assumed to be deposited in the upstream delta and zones $A(t)$–$C(t)$ is coarse particles larger than 0.037 mm. Therefore, the equations for these zones are for coarse particles only, requiring that the inflow sediment load be divided into coarse and fine particles. Procedures for accomplishing this were discussed in the body of Chapter 9 and illustrated in Example Problem 9.15.

Example Problem 9C.1. Rate of advance of the deposition wedge

Estimate the advance distance for the sediment wedge in Problem 9.15 at the end of 5 minutes. Assume γ_{sb} is 1.25 × 62.4 = 78 lb/ft^3. Assume that the VFS is mowed.

Solution:

1. Parameters. From Problem 9.15, $S_S = .0525$ ft, $xn = .056$, $K = 3.68 \times 10^7$, $d_{pd} = .17$ mm, $q_{wi} = q_{wb} = .074$ cfs/ft, $q_{sd}^c = .211$ lb/sec · ft, $q_{si}^c = .32$ lb/sec · ft. From Table 9.8, H = 4 inch = .33 ft

2. Determining values for equilibrium slope S_{et} in zone B(t) (coarse particle only). From Eq. (9C.8), using a superscript c to denote coarse particles,

$$C1 = \frac{xnq_{wba}K^{0.14}}{1.5(q_{sba}^c)^{0.14}d_{pb}^{0.2898}}$$

$$d_{pb} = d_{pd} = 0.17 \text{ mm}.$$

From Eq. (9C.1),

$$q_{sba}^c = \frac{q_{si}^c + q_{sd}^c}{2} = \frac{0.32 + 0.211}{2} = 0.266 \text{ lb/sec} \cdot \text{ft}$$

$$C1 = \frac{(0.056)(0.074)(3.68 \times 10^7)^{0.14}}{1.5(0.266)^{0.14}(0.17)^{0.2898}} = 0.064.$$

From Eq. (9C.8),

$$d_{fba}\left(\frac{d_{fba}s_e}{2d_{fba} + S_s}\right)^{1/6} = C1 = 0.064$$

or

$$d_{fba}\left[\frac{(d_{fba})(0.0525)}{2d_{fba} + 0.0525}\right]^{1/6} - 0.064 = 0.0.$$

By trial and error, d_{fda} is found to be 0.122 ft or 1.464 in. Solving for R_{sb} and S_{et},

$$R_{sb} = \frac{d_{fda}S_s}{2d_{fda} + S_s} = \frac{(0.122)(0.0525)}{(2)(0.122) + 0.0525} = 0.0216.$$

From Eq. (9C.3), S_{et} can be calculated as

$$S_{et} = \frac{q_{sba}^{0.28}d_{pba}^{0.5798}}{R_{sb}K^{0.28}} = \frac{(0.266)^{0.28}(0.17)^{0.5798}}{(0.0216)(3.68 \times 10^7)^{0.28}} = 0.087.$$

3. Ratio between sediment deposition in upstream delta and B(t). Prior to $Y = H$ $(t < t^*)$ in Eq. (9C.9), it is necessary to correct f in Eq. (9C.10) for deposition in the upstream delta. It can be shown from geometry that the ratio between sediment deposited in the upstream delta $\forall_1$, to that in zone $B(t)$, $\forall_2$, is

$$\frac{\forall_1}{\forall_2} \cong \frac{S_{et} - S_c}{S_c} = \alpha,$$

which is the ratio of sediment deposited in $B(t)$ to the total sediment deposited. Also

$$\frac{\forall_1}{\forall_1 + \forall_2} = \frac{\alpha}{1 + \alpha}.$$

The ratio f in Eq. (9C.2) must be modified by this ratio for time prior to t^*. From earlier calculation, $S_{et} = 0.0867$ and S_c is given as 0.08. Hence

$$\alpha = \frac{0.0867 - 0.08}{0.08} = 0.084.$$

If f is the total fraction of coarse particles trapped in the deposition wedge, then the portion trapped in zone $B(t)$ would be

$$f' = f\left(\frac{\alpha}{1+\alpha}\right) = 0.322\left(\frac{0.084}{1+0.084}\right) = 0.025.$$

4. *Time to deposit to top of vegetation.* From Eqs. (9C.14),

$$S_e \cong S_{et} - S_c = 0.0867 - 0.08 = 0.0067.$$

From Eq. (9C.9), using f' in place of f,

$$t^* = \frac{H^2 \gamma_{sb}}{2 f' q_{si}^c S_e}.$$

q_{si}^c was given as 0.32 lb/sec · ft from Example Problem 9.15. Using $\gamma_{sb} = 78$ lb/ft^3 as given,

$$t^* = \frac{(0.33)^2(78)}{(2)(0.025)(0.32)(0.0067)} = 79{,}237 \text{ sec or } 1321 \text{ min}.$$

Obviously, at 5 min, the deposition had not reached the media height.

5. *Advance distance at time 5 min.* $t_i = 0$, $X_i(t) = 0$. At $t_f = 300$ sec, $X_f(t) = ?$ At $t = 5$ min (300 sec), $Y(t) < H$; hence Eq. (9C.11) applies. Using f' instead of f in Eq. (9C.11),

$$X_f(t_f) = \left[\frac{2}{\gamma_{sb}} \frac{f' q_{si}^c}{S_e}(t_f - t_i) + X_i^2(t_i)\right]^{1/2}$$

$$= \left[\left(\frac{2}{78}\right)\frac{(0.025)(0.32)}{0.0067}(300 - 0) + 0\right]^{1/2}$$

$$= 3.03 \text{ ft}.$$

General Appendix
Common Equivalences

Common Equivalencies

Length

Unit	Equivalent					
	Millimeter	Inch	Foot	Meter	Kilometer	Mile
Millimeter	1	0.039 37	0.003 281	0.001 000	1 E–6	0.621 4 E–6
Inch	25.40	1	0.083 3	0.025 40	25.40 E–6	15.78 E–6
Foot	304.8	12	1	0.304 8	304.8 E–6	189.4 E–6
Meter	1,000	39.37	3.281	1	0.001	621.4 E–6
Kilometer	1,000,000	39,370	3,281	1,000	1	0.621 4
Mile	1,609,000	63,360	5,280	1,609	1.609	1

Area

Unit	Equivalent						
	Square inch	Square foot	Square meter	Acre	Hectare	Square kilometer	Square Mile
Square inch	1	0.006 944	645.2 E–6	0.159 4 E–6	64.52 E–9	645.2 E–12	249.1 E–12
Square foot	144	1	0.092 90	22.96 E–6	9.290 E–9	92.90 E–9	35.87 E–9
Square meter	1,550	10.76	1	247.1 E–6	1 E–4	1 E–6	386.1 E–9
Acre	6,273,000	43,560	4,047	1	0.404 7	0.004 047	0.001 563
Hectare	15,500,000	107,600	10,000	2.471	1	0.01	0.003 861
Square kilometer	1.550 E + 9	10,764,000	1,000,000	247.1	100	1	0.386 1
Square mile	4.014 E + 9	27,880,000	2,590,000	640	259	2.590	1

Volume

Unit	Equivalent							
	Cubic inch	Liter	U.S. gallon	Cubic foot	Cubic yard	Cubic meter	Acre-foot	Second-foot-day
Cubic inch	1	0.016 39	0.004 329	578.7 E–6	21.43 E–6	16.39 E–6	13.29 E–9	6.698 E–9
Liter	61.02	1	0.264 2	0.035 31	0.001 308	0.001	810.6 E–9	408.7 E–9
U.S. gallon	231.0	3.785	1	0.133 7	0.004 951	0.003 785	3.068 E–6	1.547 E–6
Cubic foot	1,728	28.32	7.481	1	0.037 04	0.028 32	22.96 E–6	11.57 E–6
Cubic yard	46,660	764.6	202.0	27	1	0.764 6	619.8 E–6	312.5 E–6
Cubic meter	61,020	1,000	264.2	35.31	1.308	1	810.6 E–6	408.7 E–6
Acre-foot	75.27 E + 6	1,233,000	325,900	43,560	1,613	1,233	1	0.504 2
Second-foot-day	149.3 E+6	2,447,000	646,400	86,400	3,200	2,447	1.983	1

Discharge (Flow Rate, Volume/Time)

Unit	Equivalent					
	Gal/min	Liter/sec	Acre-ft/day	Ft^3/sec	Million gal/day	M^3/sec
Gal/min	1	0.06309	0.004419	0.002228	0.001440	63.09 E–6
Liter/sec	15.85	1	0.07005	0.03531	0.02282	0.001
Acre-ft/day	226.3	14.28	1	0.5042	0.3259	0.01428
Ft^3/sec	448.8	28.32	1.983	1	0.6463	0.02832
Million gal/day	694.4	43.81	3.069	1.547	1	0.04381
M^3/sec	15,850	1,000	70.04	35.31	22.82	1

Velocity

Unit	Equivalent				
	Ft/day	Km/hr	Ft/sec	Mile/hr	M/sec
Ft/day	1	12.70 E–6	11.57 E–6	7.891 E–6	3.528 E–6
Km/hr	78,740	1	0.9113	0.6214	0.2778
Ft/sec	86,400	1.097	1	0.6818	0.03048
Mile/hr	126,700	1.609	1.467	1	0.4470
M/sec	283,500	3.600	3.281	2.237	1

Mass

Unit	Equivalent					
	Pound mass	Kilogram	Metric slug	Slug	Metric ton	Long ton
$Pound_{mass}$ (avoird.)	1	0.4536	0.04625	0.03108	453.6 E–6	446.4 E–6
Kilogram	2.205	1	0.1020	0.06852	0.001	984.2 E–6
Metric slug	21.62	9.807	1	0.6721	0.009807	0.009651
Slug	32.17	14.59	1.490	1	0.01459	0.01436
Metric ton	2,205	1,000	102.0	68.52	1	0.9842
Long ton	2,240	1,016	103.7	69.63	1.016	1

Conversion Factors

Temperature

Celcius + 273.15 = Kelvin

Fahrenheit + 459.67 ÷ 1.8 = Kelvin

Rankline ÷ 1.8 = Kelvin

Fahrenheit – 32 ÷ 1.8 = Celcius

Celcius × 1.8 + 32 = Fahrenheit

Viscosity

Centipoise × 0.001 = pascal-seconds

Centistokes × 1.000 E–06 = square meter per second

Square foot per second × 0.09290 = square meter per second

Poise × 0.100 = pascal-seconds

Pound (force) -second per square foot × 47.88 = pascal-seconds

Slug/foot-second × 47.88 = pascal-seconds

Stokes × 1.000 E–04 = square meter per second

Pressure

Atmosphere × 1.013 E 05 = pascal

Bar × 1.000 E 05 = pascal

Feet of water (39.4°F) × 2989 = pascal

Inches mercury × 3386 = pascal

Inches water × 249.1 = pascal

Kilogram (force) per square meter × 9.807 = pascal

Millibar × 100 = pascal

Millimeter mercury × 133.3 = pascal

Pounds per square foot × 47.88 = pascal

Pounds per square inch × 6895 = pascal

Conversion Factors for Universal Soil Loss Equation (USLE) Factors

To convert from:	U.S. customary units	Multiply by:	To obtain:	SI units[a]
Rainfall intensity, *i* or *I*	in./hr	25.4	millimeter / hour	mm / hr
Rainfall energy per unit of rainfall, e	ft•tonf / acre-in.	2.638×10^{-4}	megajoule / hectare•millimeter	MJ / ha•mm
Storm energy, *E*	ft•tonf / acre	0.006701	megajoule / hectare	MJ / ha
Storm erosivity, *EI*	ft•tonf•in. / acre · hr	0.1702	megajoule•millimeter / hectare•hour	MJ•mm / ha•hr
Storm erosivity, *EI*	hundreds of ft•tonf•in.[b] / acre•hr	17.02	megajoule•millimeter / hectare•hour	MJ•mm / ha•hr
Annual erosivity, *R*	hundreds of ft•tonf•in. / acre•hr•year	17.02	megajoule•millimeter / hectare•hour•year	MJ•mm / ha•year
Soil erodibility, *K*[c]	ton•acre•hr / hundreds of acre•ft•tonf•in.	0.1317	metric ton•acre•hour / hectare•megajoule•millimeter	ton•ha•hr / ha•MJ•mm
Soil loss, *A*	ton / acre	2.242	metric ton / hectare	ton / ha
Soil loss, *A*	ton / acre	0.2242	kilogram / square meter	kg/m^2

[a]The prefix mega (M) has a multiplication factor of 1×10^6. To convert ft-tonf to MJ, multiply by 2.712×10^{-3}. To convert to hectare, multiply by 0.4071.

[b]This notation, "hundreds of," means numerical values should be multiplied by 100 to obtain true numerical values in given units. For example, $R = 125$ (hundreds of ft-ton-in./acre-hr) = 12,500 ft-tonf-in./acre-hr. The converse is true for "hundreds of" in the denominator of a fraction. Erosivity, *EI* or *R*, can be converted from a value in U.S. customary units to a value in units of newtons per hour (N/hr) by multiplying by 1.702.

[c]Soil erodibility, *K*, can be converted from a value in U.S. customary units to a value in units of metric ton•hectare per newton•hour (ton•hr/ha•n) by multiplying by 1.317.

Properties of Water

Temperature	Specific weight	Density	Viscosity	Kinematic viscosity
		English units		
(°F)	γ (lb/ft^3)	ρ (slugs/ft^3)	$\mu \times 10^5$ (lb • sec/ft^2)	$\nu \times 10^5$ (ft^2/sec)
32	62.42	1.940	3.746	1.931
40	62.43	1.940	3.229	1.664
50	62.41	1.940	2.735	1.410
60	62.37	1.938	2.359	1.217
70	62.30	1.936	2.050	1.059
80	62.22	1.934	1.799	0.930
90	62.11	1.931	1.595	0.826
100	62.00	1.927	1.424	0.739
110	61.86	1.923	1.284	0.667
120	61.71	1.918	1.168	0.609
130	61.55	1.913	1.069	0.558
140	61.38	1.908	0.981	0.514
150	61.20	1.902	0.905	0.476
160	61.00	1.896	0.838	0.442
170	60.80	1.890	0.780	0.413
180	60.58	1.883	0.726	0.385
190	60.36	1.876	0.678	0.362
200	60.12	1.868	0.637	0.341
212	59.83	1.860	0.593	0.319
		SI units		
(°C)	γ (kN/m^3)	ρ (kg/m^3)	$\mu \times 10^3$ (N•sec/m^2)	$\nu \times 10^6$ (m^2/sec)
0	9.805	999.8	1.781	1.785
5	9.807	1000.0	1.518	1.519
10	9.804	999.7	1.307	1.306
15	9.798	999.1	1.139	1.139
20	9.789	998.2	1.002	1.003
25	9.777	997.0	0.890	0.893
30	9.764	995.7	0.798	0.800
40	9.730	992.2	0.653	0.658
50	9.689	988.0	0.547	0.553
60	9.642	983.2	0.466	0.474
70	9.589	977.8	0.404	0.413
80	9.530	971.8	0.354	0.364
90	9.466	965.3	0.315	0.326
100	9.399	958.4	0.282	0.294

Physical Constants

	English units	SI units
Acceleration of gravity	32.2 ft/sec^2	9.81 m/sec^2
Standard pressure at sea level	14.7 psia	101.32 kN/m^2
	29.92 in. Hg	760 mm Hg
	33.9 ft water	10.3 m water

Index